FLUVIAL SEDIMENTOLOGY VII

Other publications of the International Association of Sedimentologists

SPECIAL PUBLICATIONS

34 Clay Mineral Cements in Sandstones
Edited by R.H. Worden and S. Morad
2003, 512 pages, 246 illustrations

33 Precambrian Sedimentary Environments
A Modern Approach to Ancient Depositional Systems
Edited by W. Altermann and P.L. Corcoran
2002, 464 pages, 194 illustrations

32 Flood and Megaflood Processes and Deposits
Recent and Ancient Examples
Edited by I.P. Martini, V.R. Baker and G. Garzón
2002, 320 pages, 281 illustrations

31 Particulate Gravity Currents
Edited by W.D. McCaffrey, B.C. Kneller and J. Peakall
2001, 320 pages, 222 illustrations

30 Volcaniclastic Sedimentation in Lacustrine Settings
Edited by J.D.L. White and N.R. Riggs
2001, 312 pages, 155 illustrations

29 Quartz Cementation in Sandstones
Edited by R.H. Worden and S. Morad
2000, 352 pages, 231 illustrations

28 Fluvial Sedimentology VI
Edited by N.D. Smith and J. Rogers
1999, 328 pages, 280 illustrations

27 Palaeoweathering, Palaeosurfaces and Related Continental Deposits
Edited by M. Thiry and R. Simon Coinçon
1999, 408 pages, 238 illustrations

26 Carbonate Cementation in Sandstones
Edited by S. Morad
1998, 576 pages, 297 illustrations

25 Reefs and Carbonate Platforms in the Pacific and Indian Oceans
Edited by G.F. Camoin and P.J. Davies
1998, 336 pages, 170 illustrations

24 Tidal Signatures in Modern and Ancient Sediments
Edited by B.W. Flemming and A. Bartholomä
1995, 368 pages, 259 illustrations

23 Carbonate Mud-mounds
Their Origin and Evolution
Edited by C.L.V. Monty, D.W.J. Bosence, P.H. Bridges and B.R. Pratt
1995, 543 pages, 330 illustrations

19 Orbital Forcing and Cyclic Sequences
Edited by P.L. de Boer and D.G. Smith
1994, 571 pages, 320 illustrations

16 Aeolian Sediments
Ancient and Modern
Edited by K. Pye and N. Lancaster
1993, 175 pages, 116 illustrations

3 The Seaward Margin of Belize Barrier and Atoll Reefs
Edited by N.P. James and R.N. Ginsburg
1980, 203 pages, 110 illustrations

1 Pelagic Sediments on Land and Under the Sea
Edited by K.J. Hsu and H.C. Jenkyns
1975, 448 pages, 200 illustrations

REPRINT SERIES

4 Sandstone Diagenesis: Recent and Ancient
Edited by Stuart D. Burley and Richard H. Worden
2003, 648 pages, 223 illustrations

3 Deep-water Turbidite Systems
Edited by D.A.V. Stow
1992, 479 pages, 278 illustrations

2 Calcretes
Edited by V.P. Wright and M.E. Tucker
1991, 360 pages, 190 illustrations

SPECIAL PUBLICATION NUMBER 35 OF THE INTERNATIONAL
ASSOCIATION OF SEDIMENTOLOGISTS

Fluvial Sedimentology VII

EDITED BY

Michael D. Blum, Susan B. Marriott and Suzanne F. Leclair

SERIES EDITOR

Ian Jarvis
School of Earth Sciences and Geography
Centre for Earth and Environmental Science Research
Kingston University
Penrhyn Road
Kingston-upon-Thames KT1 2EE
UK

© 2005 International Association of Sedimentologists
and published for them by
Blackwell Publishing Ltd

BLACKWELL PUBLISHING
350 Main Street, Malden, MA 02148-5020, USA
108 Cowley Road, Oxford OX4 1JF, UK
550 Swanston Street, Carlton, Victoria 3053, Australia

The right of M.D. Blum, S.B. Marriott and S.F. Leclair to be identified as the
Authors of the Editorial Material in this Work has been asserted in
accordance with the UK Copyright, Designs, and Patents Act 1988.

All rights reserved. No part of this publication may be reproduced,
stored in a retrieval system, or transmitted, in any form or by any means,
electronic, mechanical, photocopying, recording or otherwise, except as
permitted by the UK Copyright, Designs, and Patents Act 1988, without
the prior permission of the publisher.

First published 2005 by Blackwell Publishing Ltd

Library of Congress Cataloging-in-Publication Data

Fluvial sedimentology VII / edited by Michael D. Blum, Susan B. Marriott, Suzanne F. Leclair.
    p. cm. — (Special publication number 35 of the International Association
  of Sedimentologists)
  Includes bibliographical references and index.
  ISBN 1-4051-2651-5 (pbk. : alk. paper)
 1. Sedimentology—Congresses. 2. River sediments—Congresses.
I. Title: Fluvial sedimentology 7. II. Blum, Mike D.
III. Marriott, Susan B. IV. Leclair, Suzanne F.
V. Series: Special publication . . . of the International
Association of Sedimentologists; no. 35.

QE471.2.F57 2005
552.5—dc22

2004003106

A catalogue record for this title is available from the British Library.

Set in 9/11½pt Melior
by Graphicraft Limited, Hong Kong
Printed and bound in the United Kingdom
by TJ International Ltd, Padstow, Cornwall.

The publisher's policy is to use permanent paper from mills that operate a
sustainable forestry policy, and which has been manufactured from pulp processed
using acid-free and elementary chlorine-free practices. Furthermore, the publisher
ensures that the text paper and cover board used have met acceptable environmental
accreditation standards.

For further information on
Blackwell Publishing, visit our website:
www.blackwellpublishing.com

Contents

vii   Preface

Fluvial processes and forms

3   Origin of anastomosis in the upper Columbia River, British Columbia, Canada *D. Abbado, R. Slingerland and N.D. Smith*

17   Review of Amazonian depositional systems *A.W. Archer*

41   Kinematics, topology and significance of dune-related macroturbulence: some observations from the laboratory and field *J. Best*

61   Derivation of annual reach-scale sediment transfers in the River Coquet, Northumberland, UK *I.C. Fuller, A.R.G. Large, G.L. Heritage, D.J. Milan and M.E. Charlton*

75   Dune-phase fluvial transport and deposition model of gravelly sand *M.G. Kleinhans*

99   Morphology and fluvio-aeolian interaction of the tropical latitude, ephemeral braided-river dominated Koigab Fan, north-west Namibia *C.B.E. Krapf, I.G. Stanistreet and H. Stollhofen*

121   A qualitative analysis of the distribution of bed-surface elevation and the characteristics of associated deposits for subaqueous dunes *S.F. Leclair and A. Blom*

135   Braided gravel-bed rivers with a limited width: preliminary results of a hydraulic model study *C. Marti and G.R. Bezzola*

145   The morphology and facies of sandy braided rivers: some considerations of scale invariance *G.H. Sambrook Smith, P.J. Ashworth, J.L. Best, J. Woodward and C.J. Simpson*

159   Application of laser diffraction grain-size analysis to reveal depositional processes in tidally influenced systems *P. Siiro, M.E. Räsänen, M.K. Gingras, C.R. Harris, G. Irion, S.G. Pemberton and A. Ranzi*

181   Sedimentology and avulsion patterns of the anabranching Baghmati River in the Himalayan foreland basin, India *R. Sinha, M.R. Gibling, V. Jain and S.K. Tandon*

197   Estimating bedload in sand-bed channels using bottom tracking from an acoustic Doppler profiler *P. Villard, M. Church and R. Kostaschuk*

Experimental and numerical modelling

213   The morphological and stratigraphical effects of base-level change: a review of experimental studies *F.G. Ethridge, D. Germanoski, S.A. Schumm and L.J. Wood*

243   A mass-balance framework for quantifying downstream changes in fluvial architecture *N. Strong, B. Sheets, T. Hickson and C. Paola*

Quaternary fluvial systems

257 The linkage between alluvial and coeval nearshore marine successions: evidence from the Late Quaternary record of the Po River Plain, Italy
A. Amorosi and M.L. Colalongo

277 Depositional processes in latest Pleistocene and Holocene ephemeral streams of the Main Ethiopian Rift (Ethiopia)
M. Benvenuti, S. Carnicelli, G. Ferrari and M. Sagri

295 Fluvio-deltaic floodbasin deposits recording differential subsidence within a coastal prism (central Rhine–Meuse delta, The Netherlands)
K.M. Cohen, M.J.P. Gouw and J.P. Holten

321 Geomorphology and internal architecture of the ancestral Burdekin River across the Great Barrier Reef shelf, north-east Australia
C.R. Fielding, J.D. Trueman, G.R. Dickens and M. Page

349 Quaternary alluvial stratigraphical development in a desert setting: a case study from the Luni River basin, Thar Desert of western India
M. Jain, S.K. Tandon, A.K. Singhvi, S. Mishra and S.C. Bhatt

373 The Middle Valley of the Tiber River, central Italy: Plio-Pleistocene fluvial and coastal sedimentation, extensional tectonics and volcanism
M. Mancini and G.P. Cavinato

Pre-Quaternary fluvial systems

399 Transport modes and grain-size patterns in fluvial basins
P.F. Friend and W.B. Dade

409 Gulf of Mexico Basin depositional record of Cenozoic North American drainage basin evolution
W.E. Galloway

425 Fluvial–estuarine transitions in fluvial-dominated successions: examples from the Lower Pennsylvanian of the Central Appalachian Basin
S.F. Greb and R.L. Martino

453 Palaeogeography and fluvial to estuarine architecture of the Dakota Formation (Cretaceous, Albian), eastern Nebraska, USA
R.M. Joeckel, G.A. Ludvigson, B.J. Witzke, E.P. Kvale, P.L. Phillips, R.L. Brenner, S.G. Thomas and L.M. Howard

481 Improved understanding of fluvial architecture using three-dimensional geological models: a case study of the Westphalian A Silkstone Rock, Pennine Basin, UK
K.J. Keogh, J.H. Rippon, D. Hodgetts, J.A. Howell and S.S. Flint

493 Changing alluvial style in response to changing accommodation rate in a proximal foreland basin setting: Upper Cretaceous Dunvegan Formation, north-east British Columbia, Canada
M.P. Lumsdon-West and A.G. Plint

517 A new evaluation of fining upward sequences in a mud-rock dominated succession of the Lower Old Red Sandstone of South Wales, UK
S.B. Marriott, V.P. Wright and B.P.J. Williams

531 Reservoir scale sequence stratigraphy for hydrocarbon production and development: Tarbat–Ipundu Field, south-west Queensland, Australia
R.S. Root, S.C. Lang and D. Harrison

557 Recognition of a floodplain within braid delta deposits of the Oligocene Minato Formation, north-east Japan: fine deposits correlated with transgression
K. Yagishita and O. Takano

569 Index

Preface

Papers in this IAS Special Publication are derived from oral and poster presentations at the 7th International Conference on Fluvial Sedimentology (ICFS), held in Lincoln, Nebraska (USA), 6–10 August 2001. The ICFS series was initiated in Calgary, Alberta in 1977, has been held every four years, and has become a staple within the international fluvial sedimentology community. The 7th ICFS was attended by 289 professionals and students from 28 countries, who represented universities, government institutions and private enterprise. The next meeting in this series, the 8th ICFS, will be held in 2005 at Delft, The Netherlands, and will be hosted by Delft University of Technology (http://www.8thfluvconf.tudelft.nl/).

The 7th ICFS was hosted by the Department of Geosciences of the University of Nebraska-Lincoln, and, like its predecessors, operated without an umbrella provided by formal affiliation with professional scientific organizations. Indeed, the 7th ICFS could not have taken place without the generous sponsorship and logistical support provided by the University of Nebraska, the hard work contributed by many University of Nebraska faculty and students, the efforts of the many conference field trip leaders, and the financial assistance provided by the International Association of Sedimentologists, Society for Sedimentary Geology, American Association of Petroleum Geologists, Conoco Inc., ExxonMobil Upstream Research Co., Phillips Petroleum Company, Schlumberger Reservoir Technologies and STATOIL.

The 7th ICFS included four days of technical sessions, with 175 oral presentations and 60 posters, plus nine pre-, post- and mid-conference field trips. Conference themes reflected the topical and geographical diversity of exciting research being conducted by fluvial sedimentologists at the beginning of the twenty-first century, and included the following:

1 General topical sessions
  - flow, sediment transport and bedform dynamics
  - fluvial channel systems—modern and ancient
  - fluvial overbank systems—modern and ancient
  - sequence stratigraphy of alluvial successions
  - fluvial systems and economic resources
  - river management
2 Special interest symposia
  - alluvial architecture
  - dryland rivers—process and products
  - deposits in mud-dominated rivers
  - alluvial and tectonic system interactions
  - fluvial system response to climate change through time
  - alluvial responses to accommodation changes
  - response of near-coastal fluvial systems to sea-level change
  - fluvial reservoirs
  - fluvial–estuarine transitions
  - the Late Quaternary Rhine–Meuse system

A similar topical and geographical diversity is reflected in the 29 papers included in this Special Publication. In the first group, one set of papers focuses on flow, sediment transport and bedform dynamics. J. Best presents a series of laboratory and field observations on dune-related macroturbulence. I. Fuller and others quantify reach-scale sediment transfers in the River Coquet, England. M. Kleinhans summarizes results of flume experiments coupled with empirical data from the Rhine to discuss dune-phase bedload transport and the importance of sorting processes. S. Leclair and A. Blom present results of flume experiments designed to decipher controls on the probability distribution of bed-surface elevations, and the structure and texture of the associated deposits, under dune-forming conditions. C. Marti and P. Bezzola present early results of numerical and physical modelling of Alpine streams, which are designed to assist redevelopment of more natural braided patterns in rivers that have been artificially narrowed as a result of river training activities. Last, P. Villard and others discuss the measurement of bedload in sand-bed channels using an acoustic Doppler profiler, and the merits of this technique relative to traditional mechanical samplers.

The second set within the first group focuses on the characteristics of modern fluvial landforms, environments and systems. D. Abbado and others describe the role of high floodplain aggradation rates in promoting anastomosis along a reach of the Columbia River in British Columbia. A. Archer provides a very useful and extensive review of the state of knowledge on Amazonian depositional systems. C. Krapf and others describe the processes, characteristics and importance of fluvial–aeolian interactions for the Koigab Fan in north-west Namibia. G. Sambrook Smith and others discuss the spatial scale invariance of bar shapes and scour depths in modern braided rivers, as well as the difficulties of applying scale invariant concepts to older deposits owing to the importance of temporal evolution and rates of migration of bar forms. P. Siiro and others use laser diffraction analysis to compare the grain-size distributions of Cretaceous and the Miocene epicontinental embayment/seaway systems in South and North America, so as to unravel the formation of sand–mud couplets within tidally influenced inclined heterolithic strata. Finally, R. Sinha and others describe the sedimentological characteristics and avulsion history within an anabranching reach of the Bahgmati River, India, which flows from the Himalayan foothills into the rapidly sudsiding foreland.

A short second group focuses on physical analogue and numerical modelling. F. Ethridge and others provide an overview of a generation of experimental studies at Colorado State University on the morphological and stratigraphical effects of base-level change. This is followed by N. Strong and others who present a new approach for the quantification of downstream changes in alluvial architecture based on mass-balance considerations.

The third group includes papers that address the responses of Quaternary fluvial systems to climate change, active tectonics, and/or sea-level change. A. Amorosi and M. Colalongo discuss alluvial and coeval nearshore marine successions of the Po River Plain, Italy, and consider some implications for sequence-stratigraphy models. M. Benvenuti and others describe depositional processes, facies and the latest Pleistocene to modern evolution of discontinuous ephemeral streams of the main Ethiopian rift. K. Cohen and others discuss how detailed studies of fluvial–deltaic deposits of the Rhine–Meuse delta, The Netherlands, record differential subsidence. C. Fielding and others describe the response of the ancestral Burdekin River, north-east Australia, to sea-level fall, as it cut across the Great Barrier Reef Shelf. M. Jain and others document the Quaternary stratigraphical development of the Luni River system, in the Thar Desert of western India, and the influences of tectonic activity and climate change over a variety of time-scales. Finally, M. Mancini and G. Cavinato summarize the Plio-Pleistocene evolution of the Tiber River system, central Italy, in response to extensional tectonics, volcanism and climate change.

The final group of papers addresses a variety of topics based on studies of pre-Quaternary fluvial systems. The first paper by P. Friend and W.B. Dade presents a model for transport modes and grain-size patterns in fluvial basins. W. Galloway summarizes his conference keynote address, and in doing so provides an overview of Cenozoic North American drainage basin evolution, as recorded in the northern Gulf of Mexico basin, offshore Texas and Louisiana, USA. S. Greb and R. Martino describe fluvial–estuarine transitions in the Lower Pennsylvanian of the Central Appalachian Basin, eastern USA. R.M. Joeckel and others present a detailed summary of fluvial to estuarine architecture and palaeogeography of the Cretaceous Dakota Formation from eastern Nebraska, USA. K. Keogh illustrates the utility of thre-dimensional models of alluvial architecture, using the Westphalian A Silkstone Rock, Pennine Basin, UK as a case study. M. Lumsdon and A.G. Plint discuss how changes in the rate of generation of accommodation affects alluvial style in the Upper Cretaceous Dunvegan Formation, northeast British Columbia, Canada. S. Marriott and others discuss fining upward sequences in a mudrock dominated succession of the Lower Old Red Sandstone of South Wales, UK, and suggest an environment of deposition similar to that of the Channel Country in central Australia, where bedload often consists of sand-sized mud aggregates. R.S. Root and others use a data set from the Tarbat–Ipundu Field in south-west Queensland, Australia, to demonstrate how sequence stratigraphy can be applied at the hydrocarbon reservoir scale. Finally, K. Yagishita and O. Takano describe floodplain strata and their sequence-stratigraphy significance within braid delta deposits of the Oligocene Minato Formation, north-east Japan.

Such a diversity of papers requires a diversity of expertise to provide the critical review necessary to bring them to publication form. Indeed, a large number of individuals, conference attendees and others from the broader community served as reviewers for one or more papers submitted to this IAS Special Publication. These reviewers are J. Abbott, J. Alexander, M. Allison, A. Amorosi, A. Archer, A. Aslan, P. Ashworth, W. Autin, J. Baas, M. Benvenuti, H. Berendsen, S. Bennett, M. Blum, M. Bourke, R. Brenner, J. Bridge, G. Browne, P. Carling, M. Church, I. Cojan, J. Crabaugh, A. Czajka-Kaczka, R. Dalrymple, S. Davies-Vollum, K. Eriksson, K. Farrell, H. Feldman, M. Filgueira-Rivera, J. Friedmann, P. Friend, W. Galloway, P. Ghosh, M. Gibling, Steve Greb, M. Guccione, P. Heller, M. Hicks, J. Holbrook, P. Houben, J. Howell, P. Hudson, J. Isbell, C. James, M. Kleinhans, N. Lancaster, S. Lang, A. Large, J. Laronne, S. Leclair, M. Leeder, G. Lowey, D. Maddy, B. Makaske, S. Marriott, A. Mather, D. May, P. McCarthy, S. McClelland, D. Mohrig, G. Nadon, G. Nanson, C. Paola, F. Pazzaglia, J. Pederson, G. Plint, G. Postma, T. Rittenour, A. Roy, G. Sambrook Smith, G. Saunders, R. Sinha, R. Slingerland, N. Smith, R. Smith, H. Stollhofen, E. Straffin, J. Swenson, S. Tooth, T. Törnqvist, P. Villard, G. Weissmann, A. Wilbers, J. Woodward, C. Wooldridge, P. Wright, and V. Zlotnik. We owe them a debt of gratitude for this valuable service.

MICHAEL BLUM
Baton Rouge, Louisiana, USA

SUSAN MARRIOTT
Bristol, United Kingdom

SUZANNE LECLAIR
New Orleans, Louisiana, USA

Fluvial processes and forms

Origin of anastomosis in the upper Columbia River, British Columbia, Canada

DIMITRI ABBADO*, RUDY SLINGERLAND*,[1]
and NORMAN D. SMITH†

*Department of Geosciences, Pennsylvania State University, 503 Deike Building,
University Park, PA 16802, USA (Email: sling@geosc.psu.edu); and
†Department of Geosciences, University of Nebraska, 214 Bessey Hall, PO Box 880340,
Lincoln NE 68588, USA

ABSTRACT

To understand the origin of anastomosis on the Columbia River between Spillimacheen and Golden, British Columbia, Canada, a geomorphological and sedimentological survey was undertaken during the summer flood of 2000. On the basis of these observations, the study reach can be divided into two sub-reaches: a highly anastomosed section with three to five channels, and a weakly anastomosed section with one to two channels. The highly anastomosed reach occurs immediately downstream from the Spillimacheen tributary and is characterized by a higher channel slope, a higher number of crevasse splays, a larger combined crevasse splay area, a wider valley and a coarser bedload. Higher rates of floodplain aggradation in the highly anastomosed reach are suggested by modern sediment budgets and radiocarbon dates. These geomorphological and sedimentary associations are consistent with the hypothesis that anastomosis of the Columbia River is maintained by a dynamic equilibrium between the rates of channel creation and channel abandonment. Rising base-level, fine bedload and low bed-slope are not necessary immediate conditions for anastomosis of the Columbia River.

INTRODUCTION

Anastomosed rivers consist of two or more interconnected, coexisting channels that typically enclose concave-upwards floodbasins. The channels are usually straight or slightly sinuous, but braided and meandering patterns are also known. Thus, anastomosed rivers are different from braided rivers because the latter contain multiple thalwegs enclosing convex bars within a single channel (Makaske, 2001) whereas anastomosis defines a network of anabranched channels. Although the geomorphological characteristics of anastomosed rivers have been recognized and described (Smith & Putnam, 1980; Smith & Smith, 1980; Rust, 1981; Smith, 1983, 1986; Nanson et al., 1986; Schumann, 1989; Miller, 1991; Knighton & Nanson, 1993; Smith et al., 1997, 1998; Makaske 1998, 2001), the origin of anastomosis is still an unresolved matter (Nanson & Huang, 1999; Makaske 1998, 2001). Indeed, Makaske (1998) argued that understanding the causes of anastomosis 'is one of the major challenges in current fluvial research', and Nanson & Huang (1999) asserted that anabranching rivers (including anastomosed rivers) 'remain the last major category of alluvial systems to be described and explained'.

[1] Corresponding author.

Three hypotheses exist for the origin of anastomosis. In the first, anastomosis is a consequence of frequent avulsions and slow abandonment of earlier channels (see e.g. Makaske (2001) and references therein). According to this point of view, the fluvial system exists in a perpetual transition state consisting of multiple coexisting channels. Anastomosis is thus not a 'graded' state, but rather a by-product of the competition between channel creation and abandonment. Makaske (2001), for example, defined an anastomosed system as the product of a dynamic balance between frequent avulsions that create multiple channels and slow channel abandonment. According to Makaske, the immediate causes of the frequent avulsions are a rise in base-level, subsidence (Smith, 1983), and high rates of aggradation, whether of the channel belt or within the channel. The immediate cause of slow abandonment is conjectured by Makaske (2001) to be low stream power, although few data exist.

In the second hypothesis, anabranching and anastomosed rivers are thought to be an equilibrium form where channels are adjusted in geometry and hydraulic friction to just transmit the imposed water and sediment discharges. In cases where gradient cannot easily be increased to carry a larger sediment load, Nanson & Knighton (1996) and Nanson & Huang (1999) proposed that a shift from single to multiple channels leads to an increase in sediment transport rate per unit water discharge. Thus, like changes in slope and channel form, anastomosis is conjectured to be another mechanism whereby a fluvial system can maintain grade. Makaske (1998) challenged this idea, however, arguing that the multichannel state of the upper Columbia River cannot be taken as a response of the system to maximize water and sediment throughput because, in spite of its anastomosed morphology, the bulk of its water and sediment moves through a single channel.

The third hypothesis was put forward by Galay et al. (1984) from a study of the Columbia River. They postulated that ponding behind alluvial fans led to the formation of large lakes in the upper Columbia Valley. The lakes gradually were filled by river-dominated 'bird's-foot' deltas of which the present anastomosed river system is a final stage. This type is thought to result from contemporaneous filling of shallow lakes and scour of multiple channels in avulsion belts, so the anastomosis should be transitional and short-lived. A subsequent palaeoenvironmental reconstruction of Columbia River deposits (Makaske, 1998) has shown that its anastomosed channels are long-lived and not the result of delta growth into shallow lakes. Therefore this hypothesis will not be considered further here.

The purpose of this paper is to describe the hydraulic and morphological properties of the anastomosed reach of the upper Columbia River in British Columbia, Canada, in order to assess the origin of its anastomosis. The Columbia River near Golden, British Columbia is an appropriate field site, being one of the best-known examples of anastomosis (Locking, 1983; Smith, 1983; Makaske, 1998, 2001; Adams, 1999; Machusick, 2000). Furthermore, hydrological and photographic records are available starting from the first half of the 1900s.

LOCATION AND GEOMORPHOLOGY OF THE STUDY AREA

The study reach is a section of the upper Columbia River near Golden, British Columbia, Canada (Fig. 1). The Columbia River starts at Columbia Lake in southern British Columbia, approximately 80 km south-east of the study reach, and flows north-north-west in a 1–2-km-wide valley for a distance of 160 km along the Rocky Mountain Trench before turning west and south-west. It consists of a single channel between Columbia Lake and the town of Radium, an anastomosed reach between Radium and a kilometre upstream of Golden, and a braided reach at Golden where it flows across the alluvial fan of a tributary, the Kicking Horse River. Anastomosis is particularly evident downstream of Spillimacheen, and this report concentrates on the 55-km reach between Spillimacheen and Golden. Access is provided by Route 95 along the north-east side of the valley, by bridges at Nicholson, Parson and Spillimacheen, and by a railway right-of-way. The area lies within the Cassiar–Columbia Mountain physiographic region and in the Interior Douglas-Fir biogeoclimatic zone (Farley, 1979). Mean annual precipitation varies between 40 and 50 cm yr^{-1}, and mean daily temperature varies from −12 to 15°C in January and July, respectively (Farley, 1979).

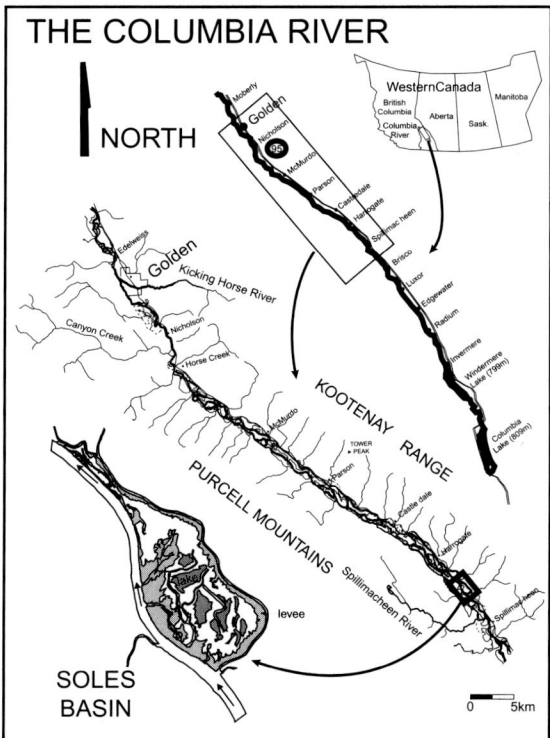

Fig. 1 Location of the study area. The Columbia River flows north-north-west between the Kootenay Range of the Rocky Mountains and the Purcell Mountains. Soles Basin, selected as a typical floodplain of the anastomosing reach, lies immediately downstream of the Spillimacheen River, an important tributary.

Geomorphology and sedimentology of the study reach

In the study reach, the Columbia River consists of multiple, relatively stable channel belts containing low-sinuosity to straight, low-gradient, sand-bed channels. Levees and crevasse splays of the channel belts bound floodbasins containing shallow wetlands and lakes. The channel belts show little lateral migration over their lifetimes, as indicated by the absence of scroll bars on the modern floodplain and by the near vertical accretion of channel facies as seen in cores (see e.g. Makaske, 1998, fig. 3.5). The channels are relatively straight, although smaller channels are slightly more sinuous. Thirty-seven tributaries enter along the reach, forming alluvial fans that narrow the valley and act as local sediment sources. The two largest tributaries are the Spillimacheen River (drainage basin of 1430 km^2) and the Kicking Horse River (drainage basin of 1850 km^2), which respectively define the upstream and downstream limits of the study reach. The Spillimacheen River catchment is a major sediment source for the study reach, contributing silt to fine gravel.

An important observation bearing on the origin of anastomosis is the number and location of channels and their evolution through time. Vibracores show that anastomosed channel deposits in the study reach are characteristically 5–15 m thick, narrow, interconnected stringers of sand (Smith, 1983) that contain sandy crevasse-splay fringes. These facies are stacked vertically, indicating that the channels occupy the same valley location for durations of up to approximately 3000 yr or maybe even longer (Smith, 1983; Makaske, 1998). Vertical aggradation rather than lateral accretion is the dominant sedimentation pattern, a conclusion also supported by the virtual absence of modern oxbow-lake and point-bar deposits. In one cross-valley stratigraphical section (Makaske, 1998), at least nine channels have existed over the past 3000 yr. Of these, six came into existence and three went extinct, indicating the long-term existence of the anastomosed pattern and the episodic nature of channel creation. There is also some indication that the longer lasting channels are wider than 30–50 m (Makaske, 1998), possibly because smaller channels can be occluded by log jams or have their gradient strongly diminished by beaver dams.

The Columbia River sediment load consists of 59 to 82% suspended material (Makaske, 1998), or if wash load is also considered, 89% (Locking, 1983). Locking's sediment budget indicates that at the end of the anastomosed reach near Nicholson (6 km upstream of Golden) the supply of suspended load is much less than the transport capacity of the river. This decline in suspended load is evidence of a significant sediment sink in the anastomosed reach (Locking, 1983). Permanent sequestration of a portion of the bedload also occurs, with channel and crevasse splay storage roughly estimated by Smith to be, respectively, 66% (Smith, 1986) and 10–20% (D.G. Smith, as reported in Makaske, 1998).

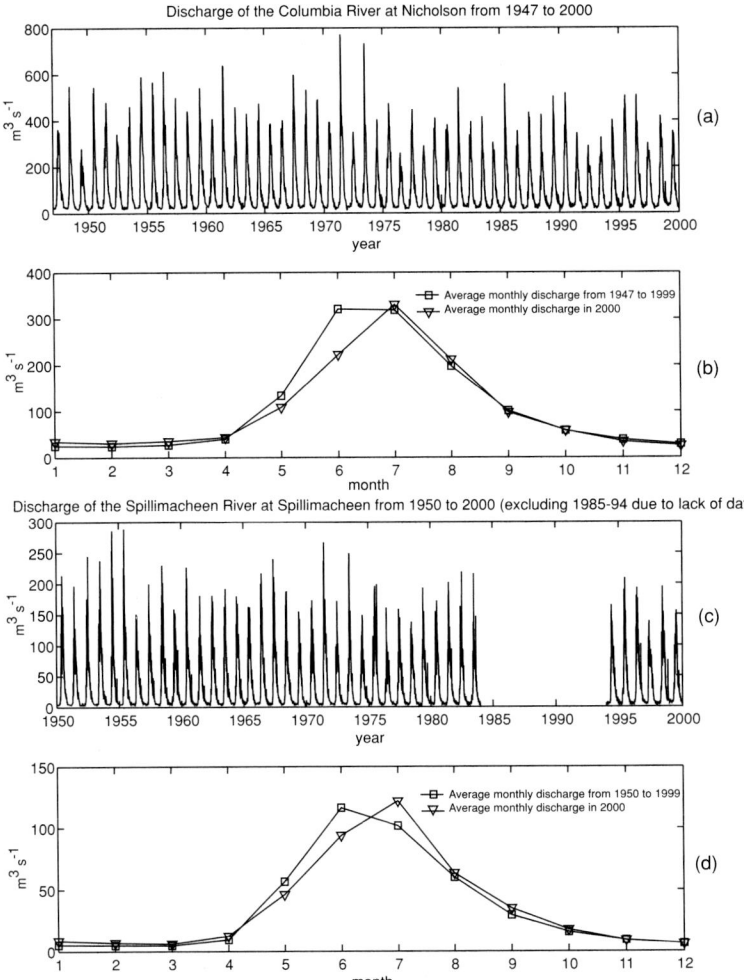

Fig. 2 Discharge data for the rivers studied. (a) Discharge of the Columbia River at Nicholson 1947–2000. (b) Average monthly discharge of the Columbia over the interval 1947–1999 compared with year 2000 average monthly discharge. (c) Discharge of the Spillimacheen River at Spillimacheen 1950–2000. (d) Average monthly discharge of the Spillimacheen over the interval 1950–1999 compared with year 2000 average monthly discharge. Year 2000 flood was average in magnitude but of short duration and delayed owing to cold weather in June (Data from Environment Canada, 2001).

Hydrology

Hydrographs for the Columbia River at Nicholson (1947–present) and the Spillimacheen River near its mouth (1950–present) indicate that discharges for both are highly seasonal (Fig. 2). Minimum discharge for the Columbia occurs in February (average = 24 m^3 s^{-1}), and maximum discharge occurs in June and July (average = 321 m^3 s^{-1}), with overbank discharge of 45 days per year on average occurring almost every year (Locking, 1983). Our field observations were taken during the year 2000 flood, which was average in magnitude but short in duration and somewhat delayed owing to cold weather in June (Fig. 2b & d). The peak flow frequency distribution shows that the maximum peak of 351 m^3 s^{-1} registered at the Nicholson gauging station during the year 2000 occurs on average every 1.2 yr (1-yr flood).

LONGITUDINAL VARIATIONS IN ANASTOMOSIS AND RELATED FEATURES

To better understand the necessary conditions that give rise to anastomosis, the degree of anastomosis of the Columbia River was correlated with channel gradient, crevasse splay distribution, valley width, alluvial fan area, and channel-bed grain size. These parameters were measured during the summer of 2000 or were observed on aerial photographs taken in 1996 at high stage, when the discharge measured at Nicholson was the third highest of the previous 10 yr. Given the hydrological data available from Environment Canada (formerly Water Survey of Canada) from 1947 to present, the 1996 peak discharge (506 $m^3 s^{-1}$) occurs on average every 3.3 yr (3-yr flood).

For our purposes, a channel belt is defined as active, in contrast with non-active, dry, or abandoned, if channels within it contain turbid water on the 1996 aerial photograph, thereby implying at least modest through-flow. Main channels are defined as those wider than 40 m; narrower channels are here termed secondary channels. Figure 3 shows a highly anastomosed section of the study reach where active/non-active channels and crevasse splays are indicated, as well as definition sketches of alluvial fan area, splay area and valley width.

Degree of anastomosis

To quantify the degree of anastomosis, the number of active channels at each of 29 valley cross-sections was counted. The number of channels, used here as a measure of anastomosis, varies from one to five with an average near two (Fig. 4a). On the basis of these differences, the study reach can be divided into an upper highly anastomosed reach (three to five channels), a weakly anastomosed reach (one to three channels), a single channel and a lower braided reach. The braided reach occurs as the Columbia River crosses the alluvial fan of the Kicking Horse River and will not be discussed further here.

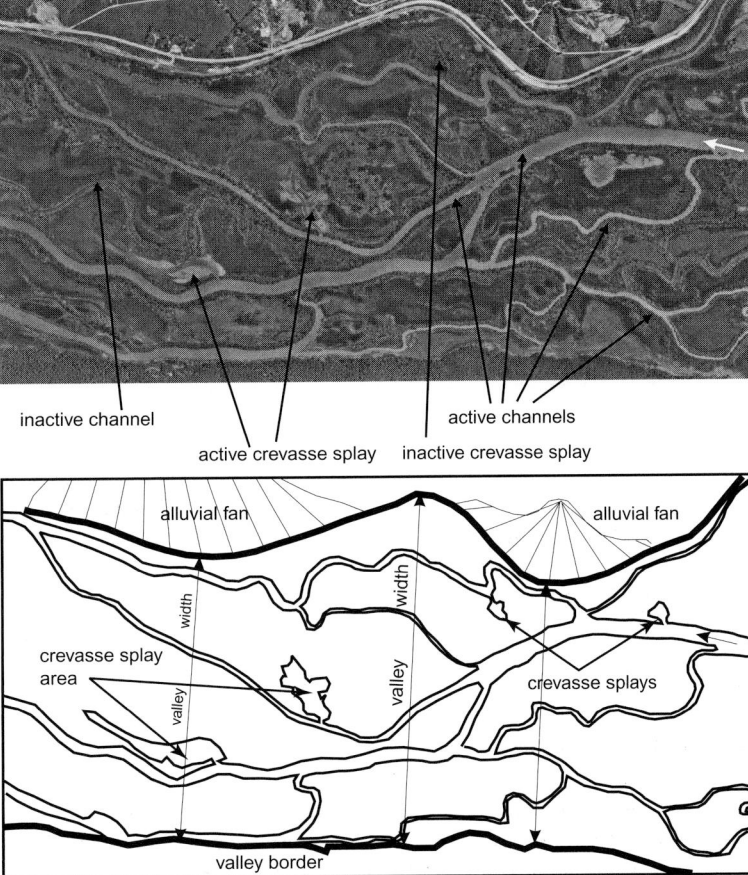

Fig. 3 Morphological elements of the river system. (Top) 1998 aerial photograph of the Columbia River showing active and inactive channels and crevasse splays. (Bottom) Definition sketch showing how crevasse splay area, alluvial fan area, and valley width were computed. See Fig. 6 for location.

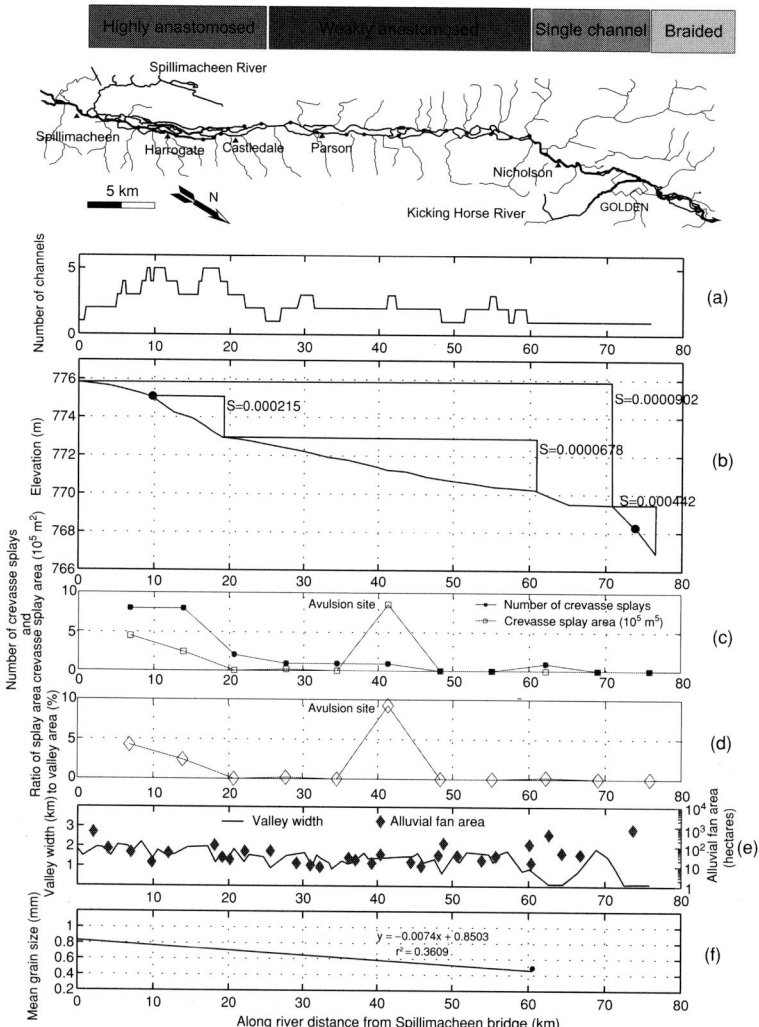

Fig. 4 Comparison of selected morphological parameters of the Columbia River. (a) Number of channels. (b) Elevation. (c) Number and area of splays. (d) Splay/valley-area ratio. (e) Valley width and alluvial fan area. (f) Mean grain size. See text for explanations.

Longitudinal profile

Absolute water-surface elevations were measured at 34 points along the Columbia River using a Leica 500 differential global positioning system (GPS) with a subcentimetre vertical accuracy. The points were measured over the period 13–15 October 2000, between the bridge at Spillimacheen and the Kicking Horse River and corrected for a falling water level of 1 cm day^{-1}. The water elevations (Fig. 4b) are plotted against along-channel distance rather than valley distance to avoid anomalies introduced by variable sinuosity or when the channel flows across the valley. The longitudinal profile is divisible into three sections. A relatively steep section from Spillimacheen to Castledale ($S = 0.000215$) is conjectured to reflect steepening of the Columbia River gradient as a result of sediment input from the Spillimacheen River. A steep section at Golden ($S = 0.000442$) arises as the Columbia River crosses the coarse-grained alluvial fan of the Kicking Horse River. Between Castledale and Golden is a more gentle central portion ($S = 0.000068$) in which the minor fans along the valley show little, if any, effect on channel gradient.

Distribution of active crevasse splays

The study reach was divided into 29 cross-valley swaths, each 2 km wide, in which the numbers of active crevasse splays and their total surface areas were determined. A crevasse splay was considered to be active if turbid water was flowing across its surface in the 1996 aerial photographs. The number of active crevasse splays and total crevasse splay areas are both relatively high in the upper 12 km of the study reach (Fig. 4c). Figure 4d shows the percentage of the valley floor covered by active crevasse splays. There are 12 active crevasse splays in the upper 18 km and only six crevasse splays along the remaining reach. The area covered by active splays decreases monotonically with distance, the exception being an active avulsion site at kilometre 37. At this site, an ongoing avulsion blankets the whole floodbasin with sediment, and small levees have formed since 1960 (Adams, 1999). The study reach therefore can be divided into two sections: an upstream reach with a high number of crevasse splays; and a downstream reach with a low number of crevasse splays.

Valley width

Valley width is potentially an important parameter in determining anastomosis because it defines the maximum available space in which channel belts can form. Variation in valley width is controlled by prograding alluvial fans from side tributaries. Measurements from aerial photographs of valley width and alluvial fan area (Fig. 4e) show little correlation with anastomosis.

Bed material grain size

Bed material was sampled during high stage on 24 June and 6 July 2000 from the mouth of the Spillimacheen River to 5 km upstream of the town of Nicholson. Twenty-five samples were collected along the main thalweg using a bucket sampler (height 15 cm, diameter 10 cm) with three replicates each to capture cross-channel variability. Mean grain size was computed using a self-constructed rapid sediment analyser to obtain a mean fall velocity that was then converted to mean particle diameter using the relationship of Dietrich (1982).

Mean grain size shows considerable scatter (Fig. 4f), probably owing to variations in texture at the crest and troughs of dunes and the occasional introduction of coarse material from tributaries. Nevertheless, the bed material shows a statistically significant fining downstream in the study reach from 1.4–2.2 mm upstream to 0.5–1.1 mm downstream.

Interpretation

The above data indicate that the study reach of the Columbia River (excluding the braided section and single-channel reach) can be divided into two subreaches, a 17-km long, highly anastomosed reach with three to five channels starting immediately below the confluence with the Spillimacheen River, and a 38-km long, weakly anastomosed reach containing one to three channels. The highly anastomosed reach is characterized by a relatively steep channel slope, a higher number of crevasse splays, a higher total crevasse splay area, a higher splay-area/valley-area ratio and coarser bed material (Table 1). These are particularly interesting observations because previous studies have concluded that low gradients and fine grain sizes are necessary conditions for anastomosis (cf. Makaske, 2001). Previous studies also conjectured that rising base-level is a necessary immediate condition for anastomosis (Smith & Smith, 1980), but that is not supported by these data either.

The intensity of crevasse splay activity is interpreted to indicate that alluviation rates are higher in the upstream, highly anastomosed reach. Testing this interpretation with actual measured aggradation rates is difficult, however. The spatially averaged sedimentation rate during the 1982 flood cycle for the entire reach from Spillimacheen to Nicholson was 3.7 mm yr^{-1} (Locking, 1983). This probably is an overestimation of the long-term average because it is based on the 1982 flood, which was well above average. A detailed sediment budget and geomorphological study of a floodbasin in the highly anastomosed reach (Fig. 1, Soles Basin) during the year 2000 flood (Abbado, 2001) shows that it is being actively filled at a rate of 2.2 mm yr^{-1} by a combination of

Table 1 Comparison between the upper and lower anastomosed reaches of the Columbia River in the study area.

Reach	Degree of anastomosis	Number of channels	Slope (cm km^{-1})	Number of crevasse splays*	Area of crevasse splays (m^2 km^{-1})	Splay area/ valley area (%)	Valley width (km)	Grain size (mm)
Upper	High	3–5	21.5	10	c. 60 000	3.3	1.4–2.2	0.5–1.1
Lower	Low	1–3	6.8	1	c. 500†	0.035	0.7–1.8	0.3–0.6

*Average number of crevasse splays per 10 km wide transverse swath.
†Excluding avulsion site.

short-lived crevasse splays, intrafloodbasin channels and settling of grains in temporary lakes. This estimate was obtained by simultaneously measuring the sediment flux into and out of the basin through crevasses and over levee tops during the 2000 flood. This must be considered a minimum because the flood of 2000 was shorter in duration than the average flood (Fig. 2b) and only suspended load was measured. In contrast, 16 km further down the study reach, an average aggradation rate of 1.7 mm yr^{-1} was obtained using a radiocarbon date of 4500 cal. yr BP from *Scirpus lacustris* nuts buried 7.9 m in a floodbasin (Makaske, 1998). Although the data are inconclusive, they are at least consistent with the conjecture that aggradation rates are higher upstream in the more anastomosed reach. Also consistent is the relatively steep slope observable in the longitudinal water profile, which can be interpreted as a wedge of sediments prograding downstream as alluviation occurs. Finally, as Robinson & Slingerland (1998) and Paola (2000) have argued, the downstream-fining itself is suggestive of preferential aggradation in the upstream reach. Although upstream bed-armouring could produce a similar downstream fining trend, it does not adequately explain the present data, because at the time of sampling the pavement appeared to be broken and the bed was in general motion.

SEDIMENT TRANSPORT MODELLING

The observations presented so far do not discriminate between the two hypotheses for the origin of anastomosis in the Columbia because both hypotheses predict that the degree of anastomosis will be correlated with excess sediment supply. Here, the question arises whether the Columbia channels are adjusted to maximize sediment transport rate, as suggested by Nanson & Huang (1999) and Huang & Nanson (2000). In traditional equilibrium channel theory, a river adjusts its slope, geometry and roughness to convey the water supplied and sediment discharge. Nanson & Knighton (1996) and Nanson & Huang (1999) suggested that a river might also change its number of channels to yield the same effect. Based on field observations, they asserted that a reduction in total top-width causes a multichannel network to convey more sediment per unit of total stream power, or, holding slope constant, per unit of discharge, than a single channel. Thus, if an original channel is 100 m wide and, say, 3 m deep, three channels, each 25 m wide and carrying the same discharge at the same slope, will carry more bedload because a reduced width/depth ratio (W/D) is more conducive to water flow and sediment discharge.

The hypothesis to be tested here is that the highly anastomosed reach of the Columbia River is adjusted in channel number and channel width/depth ratios to carry more sediment than a single channel, all other factors such as Manning's n and cross-sectional shape being equal. To test the hypothesis an abstracted Columbia channel network was considered (Fig. 5) in which cumulative top-width, depth and bed-slope are kept constant at 120 m, 3 m and 10^{-4}, respectively, consistent with values observed in the upstream portion of the study reach (Fig. 6 & Table 2). As Table 2 shows, width/depth ratios (defined as top-width divided by the hydraulic depth; see footnote in Table 2) range from 45 to 8. In addition, the distribution of channel widths is bimodal with the minimum occurring between 40 and 50 m. This minimum was used to separate the channels into

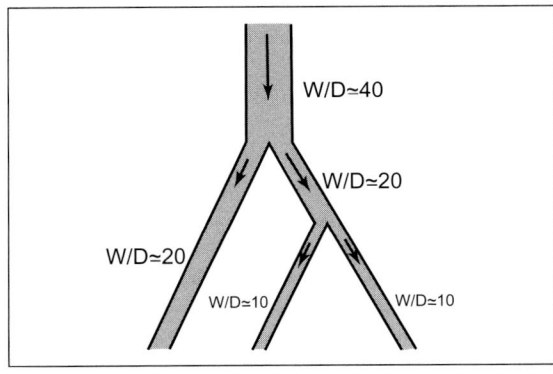

two groups: main channels and secondary channels. Based on these data from the Columbia River, the abstracted model contains a single channel of $W/D \cong 20$, which progressively bifurcates into second-order channels of $W/D \cong 20$ and third-order channels of $W/D \cong 10$. Interestingly, the observed width/depth ratios of main channels decrease

Fig. 5 (*left*) Generic model of an anastomosing river with width/depth (W/D) ratios typical of the Columbia. The W/D ratio progressively decreases with increasing number of channels.

Table 2 Width/depth ratios of cross-sections in the study reach.

Cross-section location	Line number*	Date	Top-width, W (m)	Hydraulic depth, D (m)†	W/D	Number of channels‡	Cumulative top-width (m)§	Source
Main channels¶	1	May 2000	125.7	2.86	44	1	125.7	This study
	4	July 2000	125	2.9	43	2	143	This study
	5	May 2000	88.9	2.87	31	3	120	Filgueira-Rivera**
	6	May 2000	88	2.88	31	3	120	Filgueira-Rivera**
	8	July 2000	90	2.95	31	3	180	This study
	11	May 2000	55	2.82	20	4	188	Filgueira-Rivera**
	12	June 1988	67.6	2.6	26	3	150	Adams, 1999
	17	June 1988	84.5	2.94	29	4	160	Adams, 1999
	20	July 1994	57.5	4.44	13	4	100	Makaske, 1998
	22	May 2000	141	3.12	45	1	141	Filgueira-Rivera**
	23	May 2000	101.6	3.34	30	2	130	Filgueira-Rivera**
Secondary channels	2	May 2000	19	1.96	10	2	143	Filgueira-Rivera**
	3	May 2000	18.7	2.23	8	2	143	This study
	7	May 2000	23.2	2.07	11	3	120	Filgueira-Rivera**
	9	May 2000	13.7	2.25	6	3	180	Filgueira-Rivera**
	10	May 2000	38	2.95	13	4	172	Filgueira-Rivera**
	13	June 1988	30.4	1.58	19	5	Unknown	Adams, 1999
	14	May 2000	38.9	2.9	13	5	Unknown	Adams, 1999
	15	May 2000	31	3.64	9	5	Unknown	Filgueira-Rivera**
	16	May 2000	35	3.49	10	4	Unknown	Filgueira-Rivera**
	18	June 1988	20.3	1.25	16	4	Unknown	Adams, 1999
	19	July 1994	22.5	1.84	12	4	100	Makaske, 1998
	21	July 1994	22.5	2.1	11	4	100	Makaske, 1998
	24	May 2000	21	2.27	9	3	Unknown	Filgueira-Rivera**

*See Fig. 6 for locations.
†D = hydraulic depth, i.e. channel cross-sectional area divided by its top-width.
‡Number of channels equals the sum of main plus secondary channels along a cross-valley transect passing through the particular channel cross-section.
§This is the summed top-width of all channels in a valley-wide transect passing through this location.
¶ Main channels are defined here as having a top width greater than 50 m; all other channels are called secondary channels.
**Personal communication.

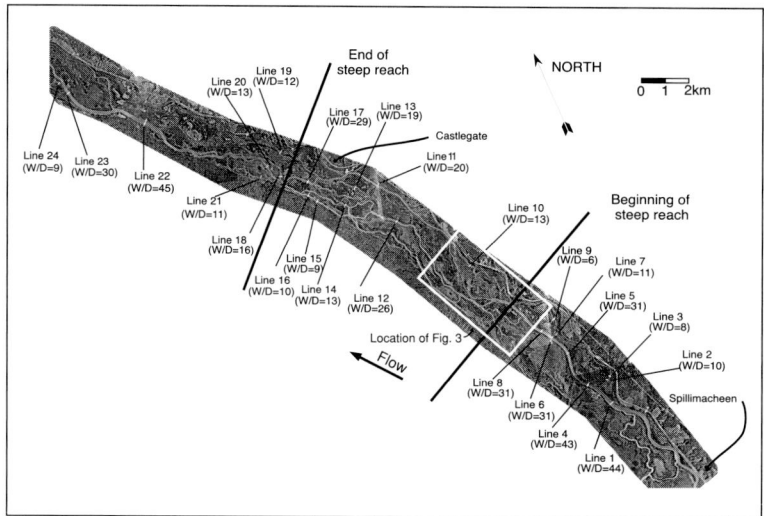

Fig. 6 Aerial photograph collage of the highly anastomosed reach of the Columbia River. Measured cross-sections are indicated by a white line; W/D ratio in parentheses.

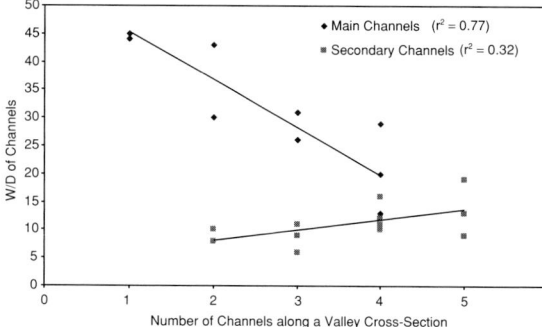

Fig. 7 Width/depth ratios of channels as a function of number of channels passing flow through any valley cross-section. Main channels are defined as those possessing widths greater than 50 m.

with an increasing number of channels along any valley cross-section (Fig. 7), which could be interpreted as consistent with the Nanson–Huang conjecture. In contrast, the ratio for secondary channels increases. An additional important characteristic of secondary channels is that their thalwegs sit at higher elevations compared with the main channels and thus they are active only during high stage.

Sediment transport through this abstracted system is calculated under uniform and steady flow conditions assuming channels of rectangular cross-sectional shape in which:

$$Q = VA \quad (1)$$

where Q (m³ s⁻¹) is water discharge, V (m s⁻¹) is average velocity, and A (m²) is channel cross-sectional area. Velocity, V, is expressed by the Chézy formula:

$$V = C\sqrt{RS} \quad (2)$$

where R (m) is hydraulic radius, S is channel slope and C (m^{1/2}/s) is the Chézy constant. Values for R and C are given by:

$$R = A/(2D + W) \quad (3)$$

$$C = (1/n)R^{1/6} \quad (4)$$

in which D (m) is the water depth, W (m) is the channel width and n is the Manning constant. The system of equations (1)–(4) yields the following fifth-order polynomial in D:

$$k^3 W^5 D^5 - 4D^2 - 4WD - W^2 = 0 \quad (5)$$

where $k = \sqrt{S}/(nQ)$. For fixed Q, n, S and width, equation (5) can be solved for D, thereby yielding a specific width/depth ratio for that combination of values. In the following computations $Q = 270$ m³ s⁻¹, $n = 0.026$, and $S = 0.0001$, consistent with typical values for the Columbia River in the study area.

Once the flow hydraulics are known, bedload and suspended load sediment transport rates are calculated by two methods: (i) the Bagnold (1977) bedload formula coupled with the Rouse (1937)

Table 3 Theoretical sediment transport magnitudes as a function of sediment transport formula and number of channels: $Q \approx 270$ m^3 s^{-1}.

Formula	Number of channels	Width (m)	Depth (m)	W/D	Water velocity (m s^{-1})	Total bedload flux (m^3 s^{-1})	Total suspended load flux (m^3 s^{-1})	Total sediment load (m^3 s^{-1})
Bagnold and Rouse	1	120	2.94	40.8	0.78	0.0061	0.0161	0.0222
	2	60	3.00	20	0.77	0.0058	0.0156	0.0214
	4	30	3.11	9.6	0.74	0.0051	0.0147	0.0198
Van Rijn	1	120	2.94	40.8	0.78	0.0036	0.0336	0.0372
	2	60	3.00	20	0.77	0.0033	0.0310	0.0343
	4	30	3.11	9.6	0.74	0.0028	0.0264	0.0292

suspended sediment formulation, and (ii) the Van Rijn functions for bedload and suspended load (Van Rijn, 1984a,b). These methods were selected because they are appropriate for the grain sizes and slopes observed in the Columbia River, and because they compute both bedload and suspended load.

Solutions of the above system of equations for total sediment transport rate indicate that total sediment load is reduced as the flow is divided into additional channels (Table 3). Total transport rate decreases by approximately 11% and 21% for the Bagnold–Rouse and Van Rijn formulas, respectively. In particular, bedload transport rate, which is more important because it controls in-channel alluviation, decreases by 16% and 22%, respectively. Water velocities decrease by 2% moving from one to two channels, and 5% moving from two to four channels.

In order to generalize these conclusions, it is possible to argue that bed roughness should be greater in the smaller channels because of increased vegetation and because bedform heights there are a greater proportion of the flow depth. This, however, would only further reduce the total sediment load in the multichannel reaches. It could be argued that the idealized model does not capture the greater sinuosity and slightly higher bed elevations of the secondary channels. To address these concerns the steady-state flow field through the actual Columbia River network in the highly anastomosed reach was computed using FESWMS, a two-dimensional, finite element code for non-uniform free-surface flows. Channel geometries were traced from aerial photographs and channel bed elevations were obtained by GPS and cross-section surveys. Water depths and flow velocities at two valley cross-sections, one where the Columbia River consists of a single channel and one where it consists of three channels, were used to recompute total sediment fluxes through the two cross-sections. Predicted sediment fluxes through the reach with three channels were 25 times less than through the single channel, thus supporting the conclusions reached from the idealized model.

DISCUSSION

The multiple channels of the Columbia do not appear to be adjusted in width, depth and number to increase water velocity and sediment transport rates over that of a single channel. These results are interpreted to mean that the Nanson & Huang (1999) hypothesis does not apply to the particular case of the Columbia River. This is not to say that the Nanson–Huang conjecture is everywhere invalidated. In cases where the cumulative top-width of multiple channels is reduced relative to a single channel, sediment transport rates will be increased. In the Columbia, however, the observed width/depth ratios and cumulative top-widths do not effect increases in sediment transport rates as the number of channels is increased.

Rather, it would appear that anastomosis of the Columbia River is a consequence of frequent avulsions (i.e. crevassing) and slow abandonment of earlier channels. High sediment flux from the

Spillimacheen River has overloaded the Columbia, causing high in-channel alluviation rates. These high alluviation rates increase the probability of levee overtopping as well as levee crevassing and crevasse splay formation. Increasing crevassing, in turn, creates numerous new channels through floodbasins. The new channels, flowing generally cross-valley, are usually super-elevated compared with the main channel. For this reason they are mainly active during high stage, and are slowly abandoned because of low flow velocities. Thus, long-lasting channels and complete avulsions of the main channel are tied to gradient advantages. The narrow valley means that cross-valley gradient advantage rarely occurs and the main down-valley channels remain active for thousands of years. In contrast, secondary channels on average are shorter lived. The number of channels active at any time is proportional to the rate of creation of new channels and to their average lifespan, and inversely proportional to their rate of abandonment. If the rate magnitudes are comparable and relatively constant through time, then the number of active channels at any instant is also relatively constant, the exact number being fixed by the channel lifespan. It is in this sense that anastomosis of the Columbia River is a dynamic equilibrium pattern.

It still remains for the reduction of width/depth ratio of main channels as the number of channels in a valley cross-section increases to be explained (Fig. 7). This probably reflects the fact that it is the main channels of the Columbia that transport most of the bedload. The bed elevations of the secondary channels are generally higher than the bed of the main channel, so more water than bedload is siphoned off by secondary channels. The main channel must adjust to carry its bedload with less discharge, and does so by decreasing its width/depth ratio by an amount greater than would arise from the reduction in water discharge alone. In this restricted sense the Columbia main channels are behaving as postulated by Nanson & Huang (1999).

This model of anastomosis is consistent with the correspondence between degree of anastomosis and high slope. As shown by the sediment routing model, anastomosis induces a decrease in sediment transport rates, which is manifested by differential deposition.

CONCLUSIONS

The anastomosed reach of the Columbia River can be divided into highly anastomosed and weakly anastomosed subreaches. The highly anastomosed reach occurs immediately downstream of the confluence with the high-sediment-load Spillimacheen River. The highly anastomosed reach is characterized by a higher channel gradient, a greater number of crevasse splays, a greater crevasse splay area, greater splay-area to valley-area ratio, and coarser channel-bed grain size. Circumstantial evidence indicates that aggradation rates are higher in the highly anastomosed reach as well. A rising base-level downstream does not seem to be a necessary immediate condition for anastomosis.

Calculations using Bagnold, Rouse, and Van Rijn sediment transport formulae show a decrease in sediment flux with increasing number of channels, given typical Columbia channel geometries, bed-slope, and grain size. This is contrary to the predictions of Nanson & Huang (1999), leading us to conclude that anastomosis of the Columbia River is maintained by a dynamic equilibrium between the rates of channel creation and channel abandonment.

ACKNOWLEDGEMENTS

This research was funded by the National Science Foundation under contract EAR 9811860 awarded to Rudy Slingerland and Norman D. Smith. For assistance in the field we thank Ron and Jan van Vugt, Irvine Heintz, Matt Machusick, and especially Manuel Filgueira-Rivera who also shared his cross-section data with us. The manuscript was greatly improved by the insightful comments of reviewers G.C. Nanson and B. Makaske, to whom we are grateful.

REFERENCES

Abbado, D. (2001) The origin of anastomosis in the Upper Columbia River, British Columbia, Canada. Unpublished MSc Thesis, Department of Geosciences, The Pennsylvania State University, 86 pp.

Adams, P.N. (1999) The origin and characteristics of natural levees. Unpublished MSc Thesis, Department of Geosciences, The Pennsylvania State University, 95 pp.

Bagnold, R.A. (1977) Bed load transport by natural rivers. *Water Resour. Res.*, **13**, 303–312.

Dietrich, W.E. (1982) Settling velocity of natural particles. *Water Resour. Res.*, **18**, 1615–1626.

Farley, A.L. (1979) *An Atlas of British Columbia: People, Environment, and Resource Use*. University of British Columbia Press, Vancouver, 135 pp.

Galay, V.J., Tutt, D.B. and Kellerhals, R. (1984) The meandering distributary channels of the upper Columbia River. In: *River Meandering* (Ed. C.M. Elliot), pp. 113–125. Proceedings of the Conference River '83, New Orleans, Louisiana, October 24–26, 1983, American Society of Civil Engineers, New York.

Huang, H.Q. and Nanson, G.C. (2000) Hydraulic geometry and maximum flow efficiency as products of the principle of least action. *Earth Surf. Process. Landf.*, **25**, 1–16.

Knighton, A.D. and Nanson, G.C. (1993) Anastomosis and the continuum of the channel pattern. *Earth Surf. Process. Landf.*, **18**, 613–625.

Locking, T. (1983) Hydrology and sediment transport in an anastomosing reach of the Upper Columbia River, B.C. Unpublished MSc Thesis, Department of Geography, University of Calgary, Canada, 107 pp.

Machusick, M.D. (2000) The effect of floodbasin type on crevasse splay form. Unpublished BSc. Thesis, Department of Geosciences, The Pennsylvania State University, 55 pp.

Makaske, B. (1998) *Anastomosing Rivers: Forms, Processes and Sediments*. Nederlandse Geografische Studies Vol. 249, Koninklijk Nederlands Aardrijkskundig Genootschap/Faculteit Ruimtelijke Wetenschappen, Universiteit Utrecht, Utrecht, 287 pp.

Makaske, B. (2001) Anastomosing rivers: a review of their classification, origin and sedimentary products. *Earth Sci. Rev.*, **53**, 149–196.

Miller, J.R. (1991) Development of anastomosing channels in south-central Indiana. *Geomorphology*, **4**, 221–229.

Nanson, G.C. and Huang, H.Q. (1999) Anabranching rivers: divided efficiency leading to fluvial diversity. In: *Varieties of Fluvial Form* (Eds A. Miller and A. Gupta), pp. 477–494. Wiley, New York.

Nanson, G.C. and Knighton A.D. (1996) Anabranching rivers: their cause, character and classification. *Earth Surf. Process. Landf.*, **21**, 217–239.

Nanson, G.C., Rust, B.R. and Taylor, G. (1986) Coexistent mud braids and anastomosing channels in an arid-zone river: Cooper Creek, Central Australia. *Geology*, **14**, 175–178.

Paola, C. (2000) Quantitative models of sedimentary basin filling: *Sedimentology*, **47** (Suppl. 1), 121–178.

Rouse, H. (1937) Modern conceptions of mechanics of fluid turbulence. *Trans. Am. Soc. Civ. Eng.*, **102**, 436–505.

Robinson, R.A.J. and Slingerland, R.L. (1998) Origin of fluvial grain-size trends in a foreland basin: the Pocono Formation of the Central Appalachian Basin: *J. Sediment. Res.*, **A68**, 473–486.

Rust, B.R. (1981) Sedimentation in an arid-zone anastomosing fluvial system: Cooper's Creek, Central Australia. *J. Sediment. Petrol.*, **51**, 745–755.

Schumann, R.R. (1989) Morphology of Red Creek, Wyoming, an arid-region anastomosing channel system. *Earth Surf. Process. Landf.*, **14**, 277–288.

Smith, D.G. (1983) Anastomosed fluvial deposits: modern examples from Western Canada. In: *Modern and Ancient Fluvial Systems* (Eds J. Collinson and J. Lewin). *Spec. Publs Int. Ass. Sediment.*, **6**, 155–168.

Smith, D.G. (1986) Anastomosing river deposits, sedimentation rates and basin subsidence, Magdalena River, northwestern Colombia, South America. *Sediment. Geol.*, **46**, 177–196.

Smith D.G. and Putnam, P.E. (1980) Anastomosed river deposits: modern and ancient examples in Alberta, Canada. *Can. J. Earth Sci.*, **17**, 1396–1406.

Smith, D.G. and Smith, N.D. (1980) Sedimentation in anastomosing river systems: examples from alluvial valleys near Banff, Alberta. *J. Sediment. Petrol.*, 50, 157–164.

Smith, N.D., McCarty, T.S., Ellery, W.N., Merry, C.L. and Ruther, H. (1997) Avulsion and anastomosis in the panhandle region of the Okavango Fan, Botswana. *Geomorphology*, **20**, 49–65.

Smith, N.D., Slingerland, R.L., Perez-Arlucea, M. and Morozova, G.S. (1998) The 1870s avulsion of the Saskatchewan River. *Can. J. Earth Sci.*, **35**, 453–466.

Van Rijn, L.C. (1984a) Sediment transport, part I. Bed load transport. *J. Hydraul. Eng.*, **110**, 1431–1456.

Van Rijn, L.C. (1984b) Sediment transport, part II. Suspended load transport. *J. Hydraul. Eng.*, **110**, 1613–1641.

Review of Amazonian depositional systems

ALLEN W. ARCHER

Department of Geology, Kansas State University, Manhattan, KS 66506, USA
(Email: aarcher@ksu.edu)

ABSTRACT

Many types of depositional system exist within the Amazon River basin and surrounding areas. These areas provide a number of valuable analogues for large-scale and tropical palaeoriver systems. Only rarely, however, are such important analogues invoked because of a lack of published information. Herein, major components are summarized among the fluvial, estuarine and coastal depositional environments of the Amazon River basin. Of particular importance in the Amazon system are the recurring depositional cyclicities that affect sedimentation. In the upper reaches of the system, yearly water-level fluctuations are related to seasonal variations in rainfall. In some areas these fluctuations exceed 10 m. The resultant flooding of vast areas of rainforest greatly affects sedimentation. In the lower reaches of the system, the dominant water-level fluctuations are related to tides. Tidal ranges are as high as 6 m at the mouth of the Amazon and tidal influences extend more than 800 km up-river. The result is a vast area of tidally influenced, freshwater environments, which are a poorly documented, but important, depositional setting.

There is a variety of river types within the Amazonian system, and the depositional environments along each are greatly affected by their differences in sediment load. Some rivers, which include the tributaries that drain the Andes, are rich in suspended sediment and have alluvial valleys containing features typical of meandering rivers, such as levees, floodplain lakes and nutrient-rich floodplains. Other rivers, with drainages entirely within extensive areas of rainforest, drain deeply weathered terrains with highly leached soils. These rivers lack suspended sediment, but have small amounts of bedload, and do not produce alluvial valleys or levees. From a sequence stratigraphy perspective, the differences among rivers have resulted in considerable variation of response to glacioeustatic-induced sea-level fluctuations and the ensuing inland changes in base level. During low stands, the Amazon deeply incised its valley for thousands of kilometres inland from the present position of the mouth. During transgression, rivers with a high sediment load vertically aggraded and kept pace with base-level rise. At the same time, the valleys of sediment-poor rivers flooded and were transformed into river-mouth lakes. These flooded valleys are not restricted to coast-proximal settings, but can be found at distances more than 1000 km inland from the modern coast.

Sediment supply has also differentially affected coastal settings along the Atlantic north and south of the Amazon mouth. The plume of turbid Amazonian water moves north-westward along the coast because of predominant winds and currents. The longshore drift of this sediment-rich plume has resulted in a low-slope, mud-dominated, prograding coast. South of the Amazon mouth, however, the lack of sediment influx has resulted in a complexly embayed erosional coastline. This southern coast consists of sea cliffs and headlands comprised of Mesozoic and younger rocks, which separate numerous small-scale macrotidal estuaries.

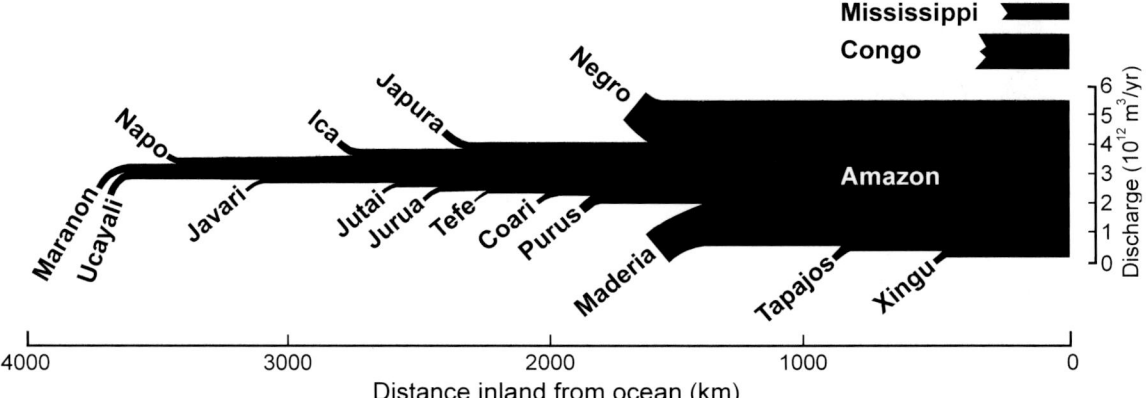

Fig. 1 Comparison of total discharge of the Mississippi, Congo, and Amazon rivers together with relative contribution for the major Amazonian tributaries (adapted from Gibbs, 1967b). Only lower reaches of Mississippi and Congo systems are represented.

INTRODUCTION

Amazon drainage basin

The Amazon River system comprises the largest river system on Earth. As described by Sioli (1984), the large size of the Amazonian drainage basin, which exceeds 7×10^6 km^2, and the equatorial and tropical climate result in tremendous discharge and a high density of tributaries. The Amazon complex comprises the world's largest freshwater system and contains approximately 20% of the global supply of freshwater (Oltman, 1967). The relative discharge of the Amazon (Fig. 1) has been estimated to be five times that of the Congo River and twelve times that of the Mississippi River (Gibbs, 1967a).

The slope of the Amazon drainage basin, east of the Andes Mountains, is exceedingly low. At Iquitos in Peru (Fig. 2), which is a straight-line distance of 2975 km from the Amazon mouth, the elevation of the river during low-water stage is about 114 m above sea level. Following the actual course of the river the distance from the mouth to Iquitos is 3717 km, making this the longest navigable inland waterway in the world. At Manaus in Brazil, which is a straight-line distance of about 1200 km from the mouth, the elevation above sea level is only about 15 m and the average slope to the mouth is only about 2 cm km^{-1}.

The Amazon drainage basin is situated between two ancient crystalline massifs, the Guiana Shield to the north and the Brazilian Shield to the south, with the western margin delineated by the Andes. The basin is elongate and is approximately 3000 km in length in an east–west direction, about 300-km wide in the eastern part and 600- to 800-km wide in the western part. This basin contains a thick sequence of Palaeozoic, Mesozoic, and Cenozoic rocks (Junk & Furch, 1985). It is the largest sedimentary basin in the world and is formed upon a Tertiary plateau consisting of a subdued area of hills and ridges, which are mostly less than 60 m above sea level (Irion, 1984). The elevation of much of the basin lies below 100 m (Fig. 2).

Sequence stratigraphy considerations

From a sequence stratigraphy standpoint, the Amazon depositional system, as defined by the drainage basin and adjacent areas of the coast and continental shelf, is exceedingly complex. During the last glaciation and lowering of sea level, the Amazon deeply incised its valley to a position approximately 2000 km upstream from the current mouth (Tricart, 1977). At that time, sea level was approximately 100 m lower and sediment was

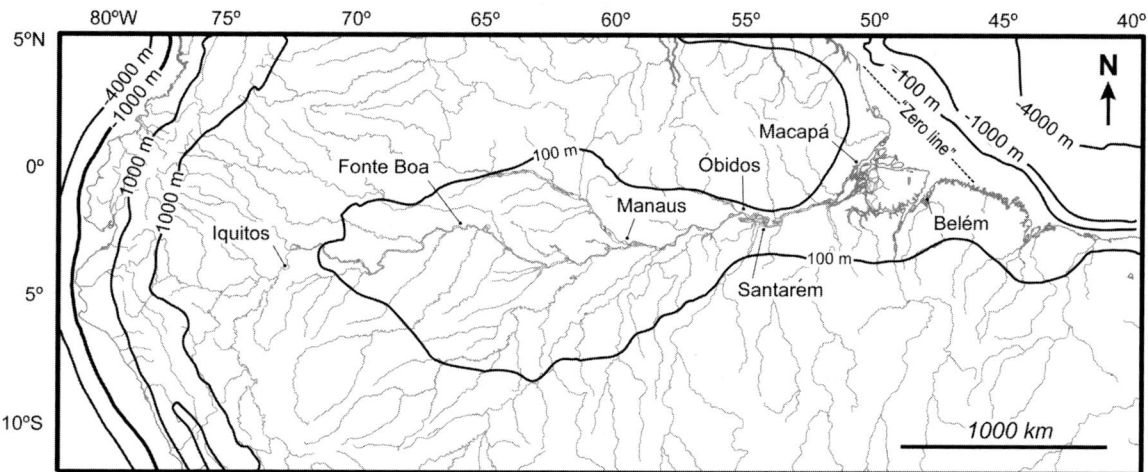

Fig. 2 Relative topography and bathymetry of Amazon Basin and surrounding areas. Much of the basin is less than 100 m above modern sea level. The dashed 'zero line' is used to measure the extent of the inland penetration of tides and comparison of yearly fluvial flux. Locations of major cities, referred to in the text, are shown along the main channel of the Amazonian system. Topography/bathymetry data are from Smith & Sandwell (1997).

carried through what is now a shelfal submarine canyon and deposited in a deep-sea fan termed the *Amazon cone* (Milliman *et al.*, 1975; Damuth, 1977). The glacial-age sediments of this fan are dominated by arkosic sands (25 to 60% feldspar), which suggest the presence of significantly drier climates on the Brazilian and Guiana shields during sea-level low stand (Damuth & Fairbridge, 1970). Milliman *et al.* (1975), however, have questioned such interpretations of dramatic climatic change.

Portions of the modern coast exhibit both submergent and emergent conditions. In general, the coast south of the Amazon mouth is emergent and is characterized by rocky headlines and embayments. In contrast, the coast to the north of the mouth, because of the tremendous amounts of suspended mud, is a depositional coast. Thus there are significant differences between these two coastal areas.

Geological evolution of the Amazon drainage basin

Prior to the Cenozoic Era, neither the Andes Mountains nor the Atlantic Ocean existed and the geological development of these features greatly influenced the subsequent development of the Amazon River system.

The Solimões and Amazon basins are Palaeozoic intracratonic basins bounded by Precambrian shields. The Purús Arch, a basement feature located west of Manaus (Fig. 2), separates these two basins. The Foz do Amazonas Basin, also known as the Marajó Basin, is a Mesozoic feature and is separated from the Amazon Basin by the Gurupá Arch, which is east of the mouth of the Rio Xingu. This arch is a structural high in the basement. The Foz do Amazonas Basin relates to the Late Jurassic to Early Cretaceous reactivation of basement structures, caused by the rifting and subsequent break-up of Gondwanaland.

The upper Amazon, to the west of Manaus, probably drained to the west prior to the uplift of the Andes. During the Miocene, the area to the east was a vast brackish basin, which changed slowly to a freshwater environment (Beurlen, 1970). Miocene deposition of coarse colluvial material within the basin does not agree with present-day depositional conditions related to the dense forest cover. This suggests the former existence of a drier, more seasonal climate and would have occurred when the river profile was graded to a lower sea level.

Types of Amazonian rivers

The rivers and tributaries within the Amazon system exhibit significant differences owing to variations in sediment types and supply, vegetation, and relief and geology of the drainage basins. Wallace (1853) provided a useful and widely accepted river classification, subsequently popularized by Sioli (1950), based upon optical properties. The rivers include *whitewater*, which are light coloured and very turbid owing to high concentrations of suspended sediments, *blackwater*, which lack inorganic suspended sediment but have a significant concentration of humic compounds, and *clearwater*, which lack suspended materials or humic compounds.

Whitewater rivers

This river type includes the Rio Amazonas and its tributaries, primarily the Rios Japurá, Putumayo, Napo, Marañón, Ucayali, and Madeira, which drain the Andes. The Madeira is the most eastern of this series of whitewater rivers. The headwaters consist of readily weathered volcanic and sedimentary rocks, yielding very high sediment loads (Dunne *et al.*, 1998). The light colours in these rivers, which are commonly yellowish or brownish and not actually white, relate to high concentrations of dissolved ions and suspended sediment. Gibbs (1967a,b) estimated that these tributaries comprise only about 12% of the Amazon drainage basin, but supply 86% of the dissolved ions and 82% of the suspended sediment. More recent estimates suggest that 90 to 95% of the suspended sediment is derived from the Andes (Meade *et al.*, 1985; Meade, 1994).

Whitewater rivers are characterized by meandering channel patterns and the bank erosion produces an abundance of floating and partially submerged trees and other types of terrestrial plant debris. The water is relatively rich in nutrients and supports the growth of floating herbaceous vegetation (as both floating meadows and islands). Runoff draining the rainforest does not have typical blackwater colours because organics are adsorbed on to clay particles where they are subsequently decomposed (Jordan, 1985). Erosion and scour along whitewater rivers, such as the Rio Amazonas, can undercut banks and result in long stretches of shoreline catastrophically slumping into the river. These phenomena, termed *terras caidas* (fallen land), can be quite dramatic and the resultant waves have been known to overwhelm small boats (Sioli, 1975).

Floodplains include relict meanders, levees, back swamps and floodplain lakes. The floodplain associated with whitewater rivers is termed *várzea* (Pires & Prance, 1985). Owing to high suspended loads, rapid rates of vertical accretion have kept pace with base-level rise. Bedload of the Rio Amazonas consists primarily of fine-grained sand. In some places there are small amounts of gravel, particularly near bedrock outcrops. Unlike many river systems, however, the average grain size and degree of sorting within the mainstem bed sediments is essentially constant along the 3300 km reach between Iquitos and Belém (Nordin *et al.*, 1980, 1981).

Blackwater rivers

The term blackwater is somewhat misleading because the water within such rivers is actually relatively clear. For the Rio Negro, Sioli (1967) compared the river water to *slightly contaminated distilled water* and stated that the river water has the appearance of very weak tea. The low levels of suspended sediment, lack of dissolved nutrients and acidic conditions prevent the growth of floating vegetation.

Sandy podzolic soils give rise to the blackwater rivers; the colours come from humic acids leached from leaf litter (Leenheer, 1980; Jordan, 1985). The high concentration of organics creates acidic concentrations and the pH of blackwater rivers is commonly about 4. Although blackwater rivers lack suspended sediments, they can have significant amounts of sand-rich bedload. One result of the low pH is that these bedload sands are almost exclusively quartzose because less stable mineral grains are readily dissolved (Leenheer & Santos, 1980).

The combined effects of lack of alluviation and extensive plant stabilization result in the floodplains of blackwater rivers not having levees, backswamps and floodplain lakes. Thus, riverbanks are more stable than in whitewater systems and the influx of plant debris into the river is minimal. Areas of flooded forest are termed *igapó* (Pires &

Prance, 1985). The lack of suspended sediments greatly reduce the potential for vertical accretion, and blackwater rivers have not kept pace with base-level rise. Where these rivers join the mainstem Amazon, the mouths have been flooded to produce river-mouth bays termed *rias*. The largest blackwater ria occurs at the mouth of the Rio Negro near Manaus. The Rio Negro constitutes approximately 10% of the Amazonian drainage basin and contributes about 15% of the discharge of the Rio Amazonas (Goulding *et al.*, 1988). The ria of the Rio Negro attains widths of 15 km and depths as great as 100 m. Bayhead deltas, produced by bedload deposition, commonly occur within the upstream portion of these rias.

Clearwater rivers

Unlike the other river types, clearwater rivers exhibit a wide variation in water chemistry. The largest clearwater rivers, including the Tapajós, Xingu, and Tocantins, are southern tributaries of the Amazon system and drain the Brazilian Shield. Similar to blackwater rivers, these rivers are nearly devoid of suspended sediments, lack alluvial valleys, and are fringed by igapó. The banks are usually stable and there is a lack of floating plant materials in the waters. Various factors influence the flooding cycles within clearwater rivers. In the Rio Tapajós, flooding is controlled by seasonal variations in rainfall. Within the Rio Xingu and Rio Tocantins, periodic flooding relates to tidal effects as well as seasonal variations in rainfall.

Water-level fluctuation

The upper reaches of the Amazon exhibit large-scale variations in river levels related to seasonal rainfall. There is a high-magnitude seasonal river-level fluctuation, which is driven by seasonal variations in rainfall. Conversely, within the estuarine portion of the Amazon, tidal periodicities are the dominant control on river-level changes.

Fluvial discharge fluctuations

The Amazon displays considerable water-level oscillations and, based upon records that extend for most of the twentieth century, discharge has

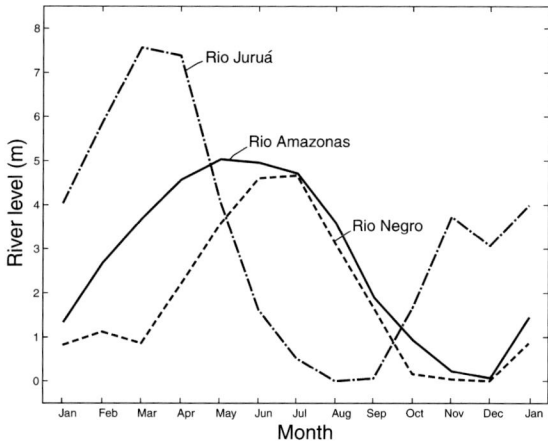

Fig. 3 Seasonal changes in water level on the Rio Juruá at Cruzeiro do Sul for the years 1928–1933, the Rio Amazonas at Óbidos for the years 1928–1933, and the upper Rio Negro at Barcellos (1929–1933). Datum is the lowest-water values (adapted from Pardé, 1936).

not exhibited any significant variations (Richey *et al.*, 1989b). In the main channel, lowest water occurs in late October to early November. River levels rise slowly for about 8 months and the highest levels are attained in late June. The river level then falls for about 4 months (Oltman *et al.*, 1964). The protracted and gradual changes in water level for the Rio Amazonas have been described as the result of a *highly damped hydrograph* (Amarasekera *et al.*, 1997). The smoothness of the annual cycle (Fig. 3) relates to the large drainage basin and the 3-month phase lag between maximum seasonal rainfall in the northern and southern part of the basin (Richey *et al.*, 1989a). In addition, long-term storage of water in rias, floodplains, lakes, and secondary channels serves to greatly dampen short-term discharge fluctuations within the mainstem (Oltman *et al.*, 1964; Richey *et al.*, 1989a).

Goulding *et al.* (1988) discussed the various factors that control the strongly developed seasonality of water levels within the Rio Negro. The Negro drainage basin has rainfall with highly seasonal patterns. The upper parts of the river have the heaviest rains in March to August, the middle parts in March and July, and the lowest reaches have the most rain between December and May. Although this seasonal rainfall affects the water level in the upper parts of the river, water levels

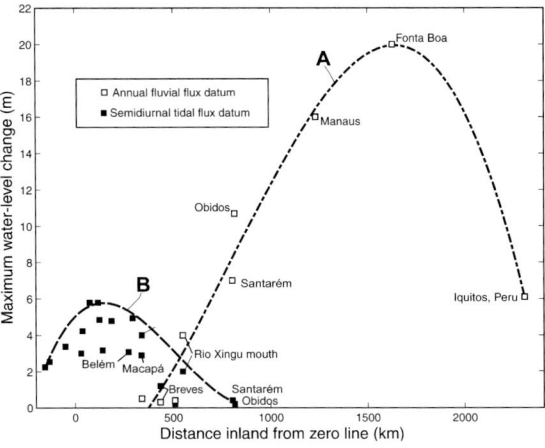

Fig. 4 Relative variation in maximal annual fluvial flux (curve A) and semidiurnal tidal flux (curve B). Inland distances are measured from an arbitrary 'zero line' that is shown on Fig. 2. Curves are second-degree polynomials fitted to the maximal values.

in the lower part are also strongly influenced by the water level of the Rio Solimões. Between the months of November and February, the upper Rio Negro is usually falling when the lower portion is rising because it is effectively impounded by the rising Rio Solimões. This occurs during the dry season, thus flooding in the lower Rio Negro is not related to regional climatological controls. For much of the dry season in Manaus, the Rio Negro is at its highest levels. Waters from the Amazon block the flow of the Rio Negro and the lower bay is essentially transformed into a large, stagnant lake.

Analysis of the annual flood cycle indicates that the highest magnitude of water-level changes occurs within the middle and upper parts of the Amazon basin (Fig. 4). The annual flood cycle averages approximately 9.8 m at Manaus with the highest levels usually reached in June or July. The lower part of the Rio Negro rises for about 8 or 9 months each year and then falls relatively rapidly. Conversely, in the upper parts of the river, which are not influenced by water levels in the Rio Solimões, the rise and fall of the water levels are more symmetrical.

Annual flood cycles affect the distribution of mixing zones that occur at the convergence of the whitewater Rio Amazonas with either blackwater or clearwater tributaries. The most famous mixing zone, known as the *Encontro das Águas* (Mixing of the Waters), occurs near Manaus where the whitewater Rio Solimões joins with the blackwater Rio Negro to form the Rio Amazonas. Per equivalent volume of water, the Rio Negro transports only about 2 to 3% of the suspended solids when compared with the Rio Solimões (Fisher, 1978; Meade et al., 1979). The transparency and darker colour of the Rio Negro waters lead to them being generally warmer than those of the Rio Solimões. In addition, the Rio Solimões is denser because of high suspended-sediment loads. This results in overflow of the whitewater by the blackwater and the development of turbid plumes that ultimately result in complete mixing of the two differing water types. The waters of these two systems maintain their visual individuality for about 20 km, and analyses of water chemistry indicate that complete mixing does not occur for over 100 km downstream (Oltman et al., 1964).

Tidal fluctuations

Changes in water levels of the Amazon related to tidal effects are not restricted to the mouth area, but also occur hundreds of kilometres inland. Within the mouth, tidal ranges can reach nearly 6 m and these extreme tides create ideal conditions for the development of tidal bores throughout the mouth and associated areas (Fig. 5). The presence of bores and the strong tidal effects greatly influence depositional systems within the mouth and hinder development of a subaerial delta. The low slope of the drainage basin and the very large size of the main channels lead to tidal effects being propagated considerable distances inland. It has long been known that tidal influences can be detected at Óbidos (see Fig. 4), which is over 800 km inland from the mouth (Oltman, 1968).

FLUVIAL DEPOSITIONAL SYSTEMS

Meandering channels

The Holocene rise in sea level, in conjunction with very low gradients, elevated base levels throughout much of the Amazon Basin. Sedimentation along the whitewater rivers kept pace with rising base levels and formed extensive forested floodplains, termed *várzea*, which are seasonally

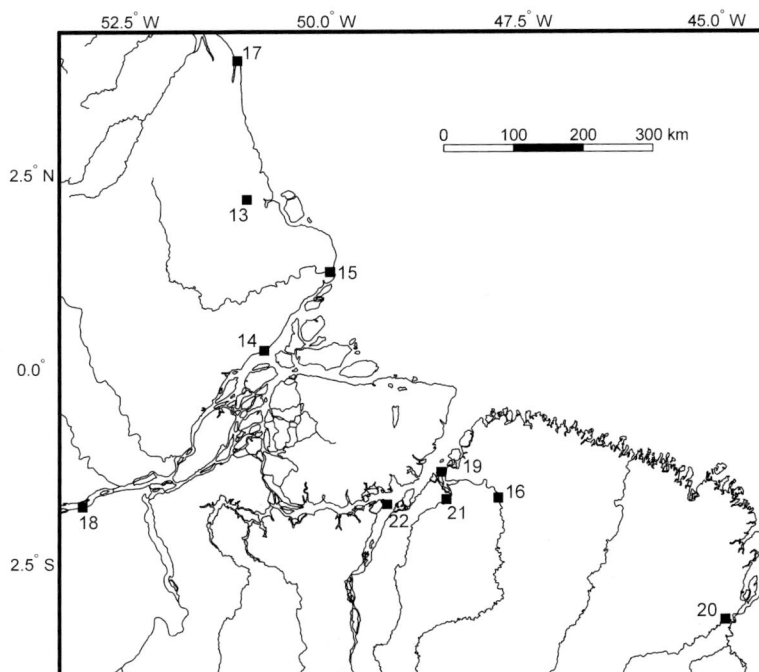

Fig. 5 Location of tidal bores in the Amazon mouth and adjacent areas; data and location numbers from Bartsch-Winkler & Lynch (1988).

inundated. These alluvial floodplains are characterized by extensive meandering, development of scroll-bar topography and formation of oxbow lakes. Even as far west as the pampas of Peru, highly sinuous, meandering whitewater tributaries occur (Fig. 6A). In the central part of the Amazon, tributaries such as the Rio Juruá (Fig. 6B) exhibit high rates of vertical accretion related to seasonal flooding. Within 10 m of the channel of the Rio Juruá, annual flood deposits attain thicknesses of 60 cm (Campbell et al., 1992). The extent of the várzea can be delineated on some radar images owing to differences in vegetation types (Fig. 6c). Várzea forests have lower plant diversity and high percentages of opportunistic taxa because of the fluctuating and unstable environmental conditions.

Along the Rio Amazonas, the várzea extends inland 20 to 100 km from the channel margins (Sioli, 1975). Sedimentation rates and suspended-sediment concentrations on the flooded várzea can be quite high. For a stretch of the Amazon near Manaus, Mertes (1994) estimated that sedimentation rates during flood stages averaged about 1 cm day^{-1}. Above Óbidos, Meade et al. (1979) showed that sediment is stored during rising-water stages and resuspended during falling water. During low discharge, they measured suspended-sediment concentrations of about 50 mg L^{-1} versus about 200 mg L^{-1} at high discharges.

In some areas, meandering appears to be controlled by basement structure. Near Fonte Boa, which is about 700 km west of Manaus, meandering is very well developed in both the mainstem and tributaries (Fig. 7A). This zone of intense meandering may relate to a structural basin bounded to the east and west by N–S trending arches (Dunne et al., 1998). Significant differences in floodplain characteristics occur at the confluence of the Rio Madeira with the Rio Amazonas (Fig. 7B). Upstream of the Rio Madeira mouth, there are relatively few lakes in the floodplain and a general lack of channel migration (Dunne et al., 1998). Downstream of the Madeira, apparently related to the influx of sediment, the river develops more prominent meanders. This area is characterized by a series of large meanders, approximately 20 km in length, with chute-like secondary channels that are termed paranás. With the increased sediment availability, a low and incomplete levee

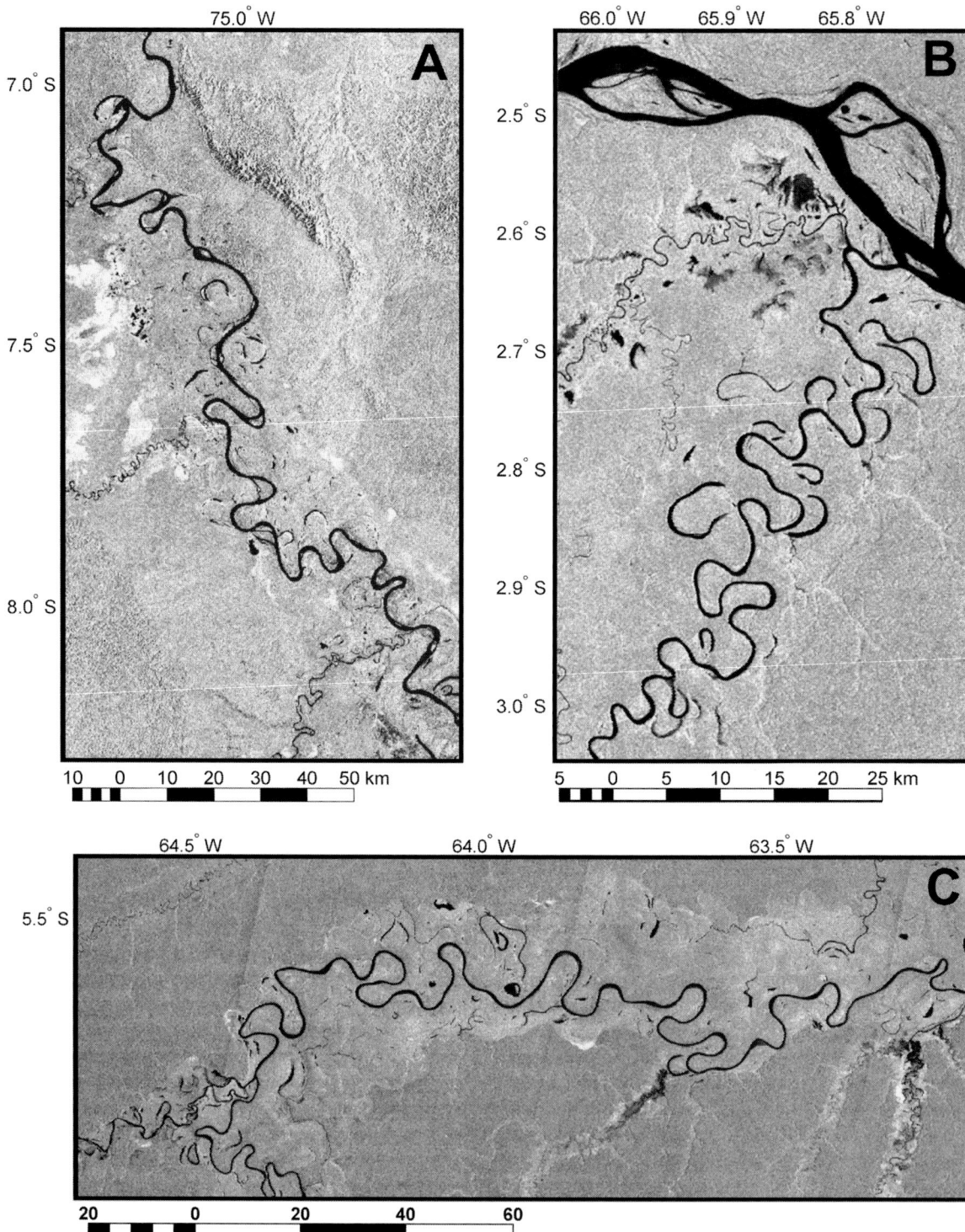

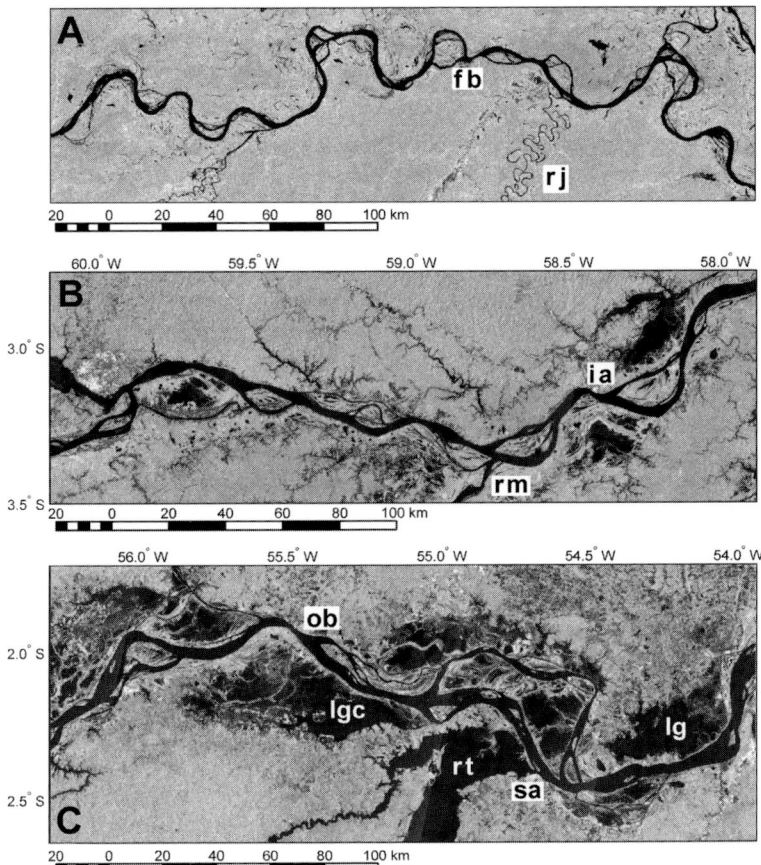

Fig. 7 JERS-1 SAR images of meandering developed within the mainstem of the Amazon River. (A) Very large-scale meanders, some approaching 40 km in length, along the Rio Solimões near the city of Fonte Boa (fb). Chute-like cut-offs, termed *paranás*, remain filled during low-water phases. Oxbow lakes are uncommon in the mainstem, but are common along the smaller-scale Rio Juruá (rj). This area is bounded to the west by the Jutaí Arch and to the east by the Purús Arch (Dunne et al., 1998). (B) Meanders and floodplain in the Rio Amazonas are well developed east of Manaus. At the confluence of the Rio Madeira (rm), the increased sediment flux results in greater development of channel migration and meandering. At the city of Itacoatiara (ia), the mainstream channel is relatively narrow, about 4 km wide, and is about 30 to 40 m deep. The *várzea* ranges from only 25 to 40 km in width in this area. Chute-like secondary channels (paranás) form large islands characterized by extensive scroll-bar topography. (C) Near the cities of Óbidos (ob) and Santarém (sa), the Rio Amazonas exhibits complex meandering. The large ria of the clearwater Rio Tapajós (rt) connects to the Rio Amazonas near Santarém. Low and incomplete levees along the Rio Amazonas result in inundation of large floodplain lakes; the lakes appear related to compactional subsidence (Dunne et al., 1998). Lago Grandé de Curuai (lgc), which is one of the largest floodplain lakes, is south of Óbidos and Lago Grandé (lg) is northeast of Santarém. Immediately east of Lago Grandé, the Amazon develops a straight, non-meandering channel. Meandering and large lakes may be in response to the Monte Alegre intrusions and related uplift, which occur in the subsurface to the east. Image data courtesy of GRFM, © NASDA/MITI.

Fig. 6 (*opposite*) JERS-1 SAR (synthetic aperture radar) images of meandering depositional systems developed within whitewater tributaries of the Amazon drainage basin. (A) The Rio Ucayali, one of the main tributaries of the Rio Marañón, flows northward in the Andean foreland basin. (B) A variety of scale of meandering channels at the confluence of the Rio Solimões and the Rio Juruá, which is well known for its extreme meandering. At its confluence with the mainstem, the Rio Juruá is about 0.8 km wide whereas the Solimões is approximately 2 km wide (Herndon, 1853). (C) The Rio Purús, which joins the mainstem about 200 km upstream from Manaus, has well-developed meandering and numerous floodplain lakes. The lateral extent of the alluvial valley, or *várzea*, is well defined on the image probably because radar reflectivity of the floodplain forests is significantly different from the terra-firma (upland) rainforest. Image data courtesy of GRFM, © NASDA/MITI.

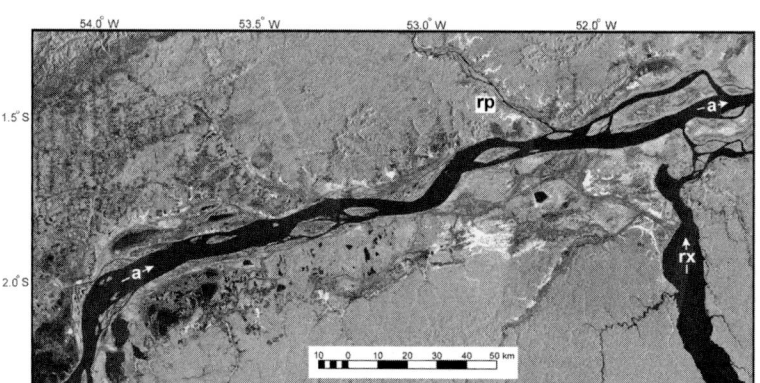

Fig. 8 JERS-1 SAR image of the straight channel of the Rio Amazonas (a) that occurs between Santarém to the west and extends easterly to the confluence of the Rio Xingu (rx), which forms a large ria near its confluence with the Amazon. Flow directions indicated by arrows. Within the straight-channel area, an incipient braided appearance is created by the development of large-scale islands. Other than the Xingu, the only other major tributary in the straight-channel zone is the Rio Paru (rp). Image data courtesy of GRFM, © NASDA/MITI.

system is formed, which is breached by large distributaries (Dunne et al., 1998).

Large-scale meanders, some exceeding 50 km in length, occur near Óbidos (Fig. 7C) and may be related to uplift developed in the subsurface to the east. Low-lying and incomplete levees result in inundation of large floodplain lakes, which are more-or-less equidimensional and thus appear to be related to compactional subsidence (Dunne et al., 1998). At Óbidos, the channel of the Rio Amazonas is relatively narrow (2.5 km wide) and is approximately 60 m deep. There are significant bluffs on each side because the Brazilian and Guiana shields come into their closest proximity in this area. The narrowness of the channel and its position near the inland tidal limit lead to Óbidos being a useful site to measure the discharge of the Amazon (Oltman et al., 1964; Oltman, 1968). Further downstream, tidal fluctuations complicate the estimation of purely fluvial discharges.

Straight channels

East of Santarém, the river develops a straight, non-meandering channel (Fig. 8). This change in channel morphology may relate to uplift related to the Monte Alegre intrusions, which occur in a N–S trend located at about 54°W longitude (Dunne et al., 1998). The large várzea lakes, which are well developed near Santarém, occur to the west of this structural zone. Large islands, with lengths approaching 15 km, occupy the main channel of the Rio Amazonas and create an incipient braided appearance. This change may relate to an increase in load at this point and an inability to erode banks, which is essential for meandering.

RIA DEPOSITION

During the last sea-level low stand, fluvial erosion created an extensive lowstand valley network within the Amazon Basin. During the subsequent sea-level rise, the rates of deposition among the various types of rivers (whitewater, clearwater, blackwater) have exhibited considerable variation. The very small amounts of suspended sediments in clear- and blackwater tributaries have led to them not keeping pace with rising base levels and their lower valleys have been flooded. The largest of these flooded valleys, *rias*, occur at the mouth of the Rios Negro, Tapajós, Xingu and Tocantins. The differences between the post-valley-incision fill of the Rio Negro and Rio Solimões (as the Amazonas is termed upstream of Manaus) is particularly pronounced (Fig. 9). At the upstream limit of these rias, bedload deposition has formed island archipelagos, which are essentially bay-head deltas. Significant differences exist between archipelagos formed primarily under fluvial influences, as compared with the archipelagos formed under tidal influences.

Fluvial archipelagos

Within archipelagos dominated by fluvial deposition, snake-like and sinuous islands have been formed, which are in part related to levee

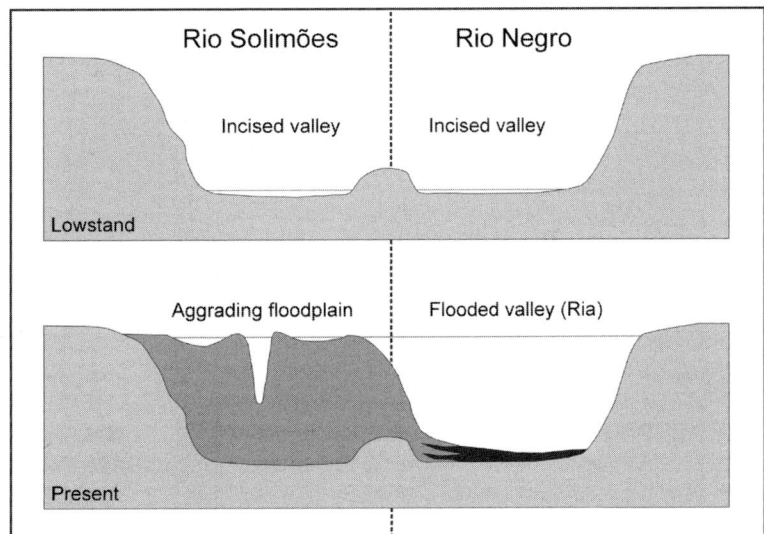

Fig. 9 Hypothetical variations in fill sequences developed between the adjacent valleys of the Rio Solimões and Rio Negro, immediately upstream of Manaus. The Rio Solimões is the major whitewater tributary of the Rio Amazonas and has aggraded during base-level rise. Conversely, the Rio Negro is a major blackwater tributary and, because of a lack of suspended sediments, has not aggraded, resulting in a river-mouth lake, or ria.

deposition during falling-water stages. The larger rias are only partially filled by archipelago deposition. Within some smaller rias, however, the island archipelago can extend to the mouth of the tributary (Fig. 10A).

The largest ria in the Amazon system is formed at the confluence of the Rio Negro, which is the largest tributary, with the mainstem Rio Solimões. Near the city of Manaus, the Rio Negro is about 2.5 km in width, but a few kilometres upstream it widens into a bay-like ria that is over 15 km wide. At the upstream head of the ria, a significant group of islands, termed the *Archipelago de Anavilhanas*, has been formed (Fig. 10B). Here the ria is nearly 30 km wide. The deposition of these islands relates to the influx of suspended sediment from the Rio Branco and other whitewater tributaries (Sioli, 1975). Over 1000 elongate islands, built up of alluvial silt, comprise this area of sedimentation (Goulding *et al.*, 1988). These islands have steep upstream ends and sides and include few sandbars. The low pH and high concentration of humic acids appear to result in flocculation of clay and silt brought in by whitewater Rio Branco (Leenheer & Santos, 1980).

The confluence of the clearwater Rio Tapajós and whitewater Rio Amazonas, near Santarém, provides an excellent example of the different response of these fluvial systems to base-level rise. Because of high sedimentation rates, the Rio Amazonas has vertically aggraded during rising base levels. In contrast, Rio Tapajós has not aggraded and a very large river-mouth bay (ria), with widths as great as 14 km, formed as the river valley flooded during rising base level (Fig. 10C). A longitudinal cross-section through the axis of the Rio Tapajós, through the ria to the confluence with the Rio Amazonas, illustrates the dramatic differences in depositional styles (Fig. 10D). Levees and islands formed by the Rio Amazonas have created dams that impound the waters of the Rio Tapajós within the ria.

At the upstream end of the Tapajós ria is an island archipelago consisting of a series of linear and sinuous islands (Fig. 10E). Although generally similar, this archipelago is not as extensive as the Archipelago de Anavilhanas in the Rio Negro, probably because of reduced bedload. Within the Tapajós the islands achieve lengths of 100 km and are prograding into the open ria, which extends downstream for another 100 km north of the islands (Sioli, 1975).

Floodwaters of the Rio Amazonas periodically enter the ria of the Rio Tapajós. This has produced a small bird's foot delta prograding into the Rio Tapajós mouth bay (Fig. 10F). In general, this delta is prograding in a southerly direction, which is 180° opposed to the actual flow of the Rio Tapajós. Wallace (1853) noted that this rather surprising phenomenon was related to seasonal differences

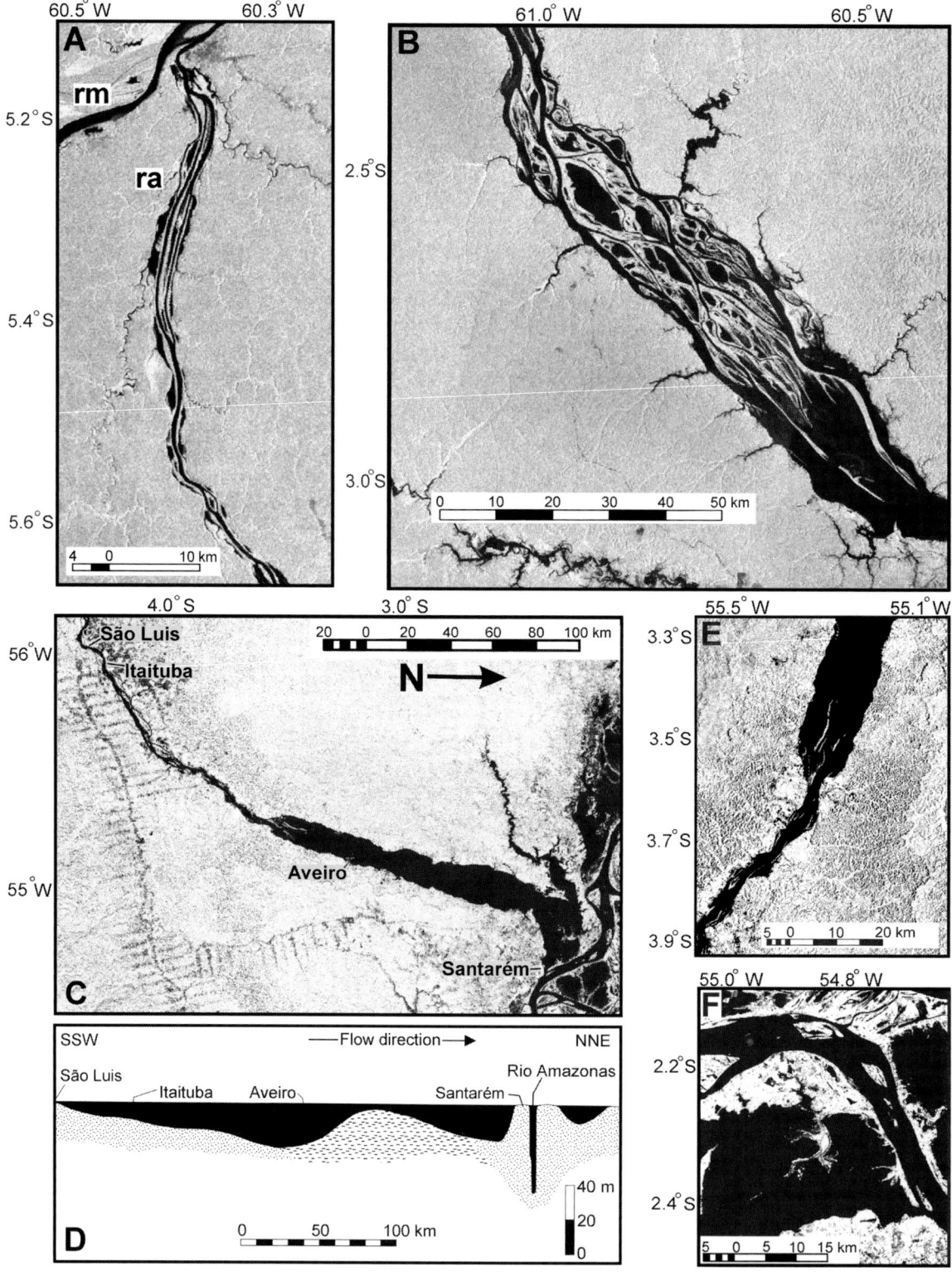

in river levels and to tides. Within the Rio Amazonas at Santarém, the tidal range approaches 50 cm. During the dry season, the water level within the Rio Tapajós mouth bay can be lower than the level of the Rio Amazonas at high tides. During the highest tides, there is flow of sediment-laden water from the Rio Amazonas into the ria of the clearwater Rio Tapajós.

Tidally influenced archipelagos

Downstream of the Rio Tapajós, tidal influences affect deposition within the rias of the Xingu, Tocantins, and other tributaries. The Rio Xingu is a clearwater river and is similar to other rivers with very low sediment loads. Unlike the main-stem Rio Amazonas, the Xingu has not kept pace with base-level rise and the lower valley has been flooded and transformed into an estuary-like ria (Vital, 1996; Vital et al., 1998). The mouth of this ria, at the confluence with the Rio Amazonas, is about 3.5 km wide and attains depths of 30 m. Sediments on the floor of the ria are dominated by silt with minor amounts of clay. Within the mouth, a few large-scale, symmetrical sand waves occur, with heights of up to 14 m and wavelengths of 180 m (Vital et al., 1998). These sand waves are indicative of influence of the Rio Amazonas. At the upstream limit of the ria, an island archipelago has formed (Fig. 11A), which is substantially different to the archipelagos described from the Rio Tapajós and Rio Negro. The tidal influences within the Xingu lead to the islands of its archipelago taking on a more equate, teardrop-shaped form; apparently the tidal effects preclude the development of the typical fluvial levees.

Compared with other depositional settings described herein, the Rio Tocantins has experienced a significant degree of anthropogenic influence. The river has been dammed at a point that is about 260 km in a straight-line distance from the mouth. As this is a clearwater river, however, the direct effects on sedimentation downstream from the dam may not be significant other than to affect the yearly hydrograph. At the upstream limit of the ria, an archipelago of islands has been formed that has a braided appearance (Fig. 11B) similar to those developed in the Rio Xingu.

ESTUARINE DEPOSITIONAL SYSTEMS

Many aspects of the Amazon system are unique and are not easily characterized within existing classification systems. The mouth area of the Amazon system, where it empties into the Atlantic Ocean, is particularly complex and contains several inland deltas and a number of tributaries that fan out at the coast. Defined broadly, the Amazon system actually contains two mouth areas, the Amazon proper to the north that contains the North Channel (An in Fig. 12) and the South Channel (As in Fig. 12), and a secondary mouth area defined by the Rio Pará to the south (B on Fig. 12). The Ilha de Marajó, which is the largest fluvial island in the world, separates these two distinct mouth areas. Erosion and deposition related to tidal currents and tidal bores cause extreme shifting of channels in the Rio Amazonas mouth area; thus commercial shipping within the Amazon system tends to make use of the mouth of the Rio Pará to the south, which includes the

Fig. 10 (opposite) JERS-1 SAR images of rias that contain island archipelagos developed within areas dominated by fluvial deposition. (A) The small-scale ria of the clearwater Rio Aripuaná (ra) is exceptional because the island archipelago, which consists of long, linear islands, extends essentially to its confluence with the whitewater Rio Madeira (rm). (B) The Archipelago de Anavilhanas is formed within the ria of the lower Rio Negro, which is the largest tributary of the Rio Amazonas and the largest blackwater river in the world. The elongate shape of such islands is accentuated by the abundant lakes lying within the island interiors. These lakes attain depths of 20 m and are normally flooded by river water during highwater (Goulding et al., 1988). (C) The flooded river valley, or ria, of the Rio Tapajós is nearly 160 km long and its linear nature suggests a structural control. (D) Cross-section along the longitudinal axis of the Rio Tapajós (adapted from Irion, 1984, p. 208) showing bar of fine-grained sediments near mouth. (E) Incipient fluvial archipelago at the upstream limit of the ria of the Rio Tapajós. (F) Bird's foot delta and crevasse splays developed within the ria of the Rio Tapajós close to its confluence with the Rio Amazonas. Sedimentation occurs during low-water phases of the Rio Tapajós when high tides cause water to flow from the sediment-rich Rio Amazonas. The delta is prograding south and westward, whereas the dominant flow of the Rio Tapajós is to the north and east. Image data courtesy of GRFM, © NASDA/MITI.

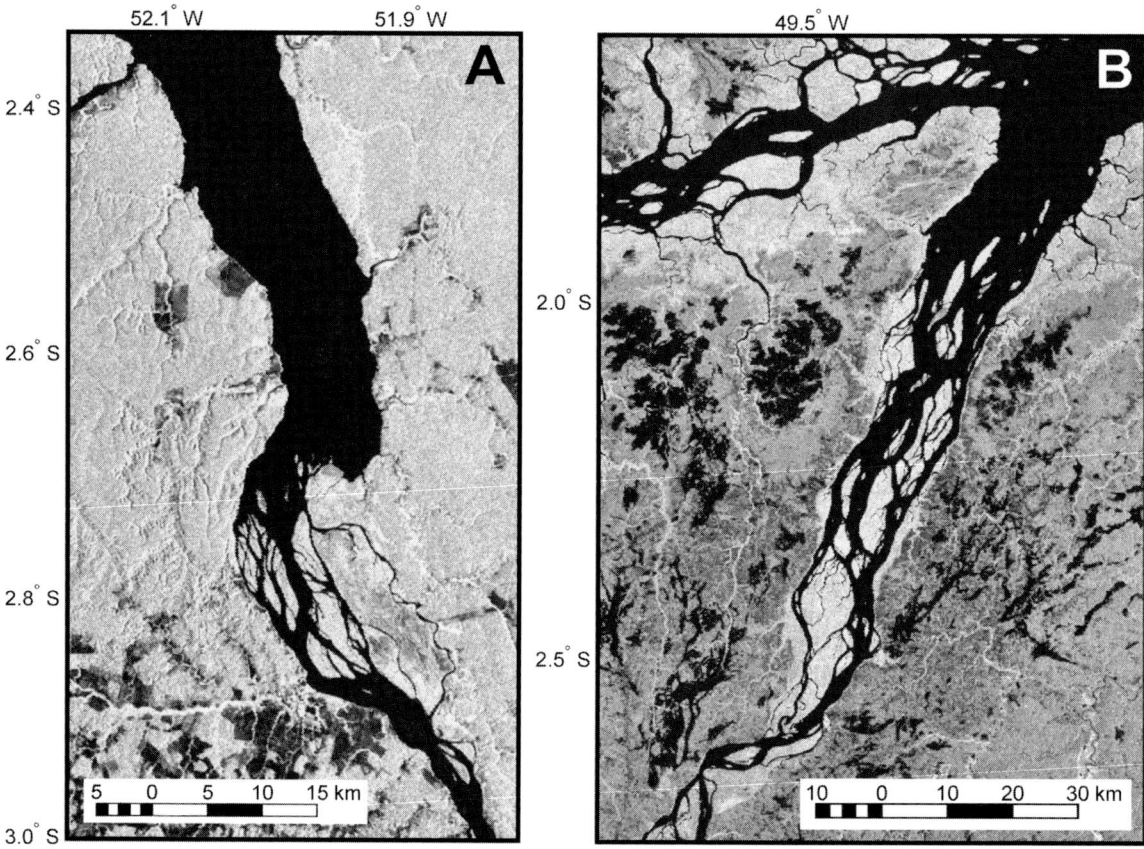

Fig. 11 JERS-1 SAR images of island archipelago formed by tidally influenced fluvial deposition within clearwater rias; (A) braided islands formed at the upstream limit of the ria of the Rio Xingu. The tidal range at the mouth of the Xingu is approximately 2 m (Vital *et al.*, 1998) and a similar meso- to microtidal fluctuation probably affects depositional processes within the archipelago. (B) An archipelago formed by a series of braided islands is formed at the upstream limit of the ria of the Rio Tocantins. Mesotidal ranges predominate in this region. Image data courtesy of GRFM, © NASDA/MITI.

estuary on which the city of Belém (see Fig. 2) is located. This connects to the Amazon proper via a series of channels termed *furos* (C on Fig. 12). These furos are relatively narrow, but are deep and allow passage of ocean-going ships. Water from the Rio Amazonas flows south and east through these furos into the southern mouth area. Thus, the drainage of the Amazon into the Atlantic Ocean includes the mouths of the Rio Amazonas and the Rio Pará and consists of a complex of islands and channels that is hundreds of kilometres wide.

Instead of a subaerial prograding delta, the Amazon has an 'estuarine' mouth. The estuarine coastal embayment that contains the Amazonas and Pará mouths is nearly 300 km in width (Fig. 12). These estuaries, however, are not estuaries *sensu stricto* in terms of saltwater–freshwater mixing (Bowden, 1978). The freshwater outflow of the Amazon is so great that saline water rarely, if ever, intrudes into the northern mouth area and freshwater extends for over 100 km onto the open continental shelf (Gibbs, 1970). Thus, the Amazon system creates problems when applying the definition of an estuary based upon salinity, which is a fundamental criterion used by biologists and other estuary workers. Conversely, a simple geomorphological definition of an estuary, as a drowned river valley, can be readily applied to the

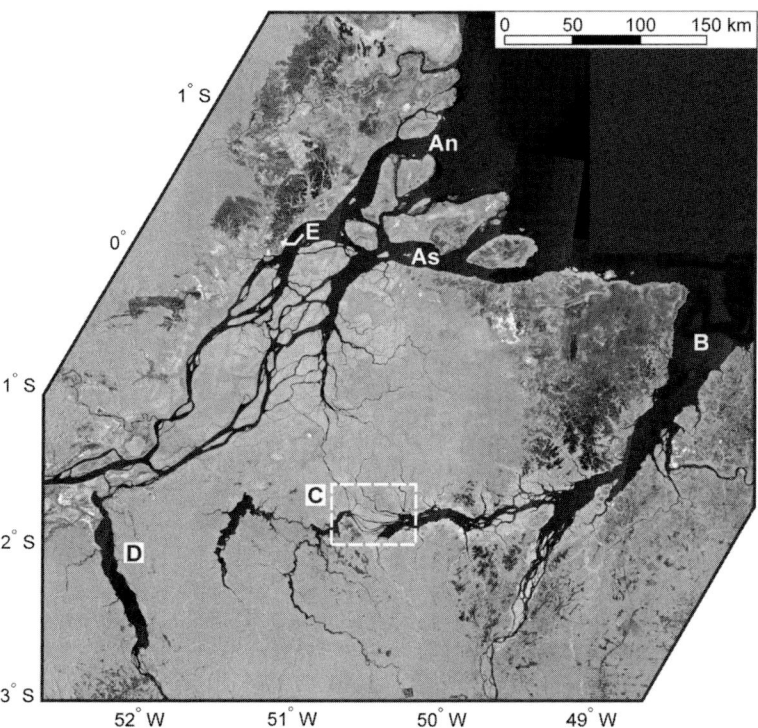

Fig. 12 JERS-1 SAR image of entire Amazon mouth and related areas: An, Amazon North Channel; As, Amazon South Channel; B, Rio Pará embayment; C, Furos area; D, ria at the mouth of the Rio Xingu; E, location of the city of Macapá on the north channel of the Rio Amazonas. Image data courtesy of GRFM, © NASDA/MITI.

lower Amazon and its mouth areas. However, the Amazon system is also definitely not an estuary in the sense that most sequence-stratigraphy literature makes use of the term. Yet another way of defining estuarine conditions is to delineate the inland extent of tidal activity. Tidal influences extend over 800 km upstream to Óbidos (see Fig. 2), where water-level fluctuations of 20 cm have been documented at spring tides during periods of low river flow (Oltman, 1968; Beardsley et al., 1995).

The seaward limit of the estuarine zone extends beyond the mouth to the limits of the freshwater plume. The plume, which extends for considerable distances across the shelf, is the area of estuarine circulation for the Amazon. The estuarine zone includes Ilha de Marajó, with an extent of 40 000 km², which separates the mouths of the Rio Amazonas and Rio Pará. Within the estuarine zone, várzeas are flooded daily by tidal back-up, which is of greater magnitude than seasonal water-level changes (see Fig. 4). The seasonal changes in the lower Amazon are low because the rivers are very wide, there are a great number of outlets to the sea, and Rio Amazonas water can be directed southward within channels along the western margin of the Ilha de Marajó (Huber, 1903; Vital, 1996).

Freshwater systems

Inner Amazon estuary

The confluence of the Rio Xingu (D in Fig. 12) and the Rio Amazonas is commonly considered as the inland limit of the estuarine zone (Vital, 1996; Vital et al., 1998). Here, the Rio Amazonas diverges into a number of separate channels (Fig. 12). Most of the flow is to the north of the Ilha de Marajó, however, some flow is directed around the west end of the island and flows into the Rio Pará system. There are significant tidal ranges throughout this area. At the Rio Xingu mouth, tidal ranges are approximately 2 m (Vital et al., 1998), whereas at Macapá (E in Fig. 12), the tidal range is nearly 3 m. Thus, the entire area is a freshwater fluvial system that is strongly affected by a mesotidal regime. Detailed sedimentology was carried out by Vital

(1996) and discussed in Vital et al. (1998, 1999). Channel sediments are dominated by very fine sands and there is only minor variation in grain size downstream. A lack of downstream grain-size variation has also been noted for the entire length of the Amazon (Nordin et al., 1980, 1981).

Near the mouth of the Rio Xingu, the Amazon splits into a Canal do Norte (North Channel) and Canal do Sul (South Channel) (An and As, respectively, in Fig. 12). As documented by Vital (1996), the channels in this area contain very fine-grained sand whereas channel margins are dominated by silt and clay. In general, the upstream areas have giant, simple asymmetrical sandwaves that are replaced by smaller and compound sandwaves in the downstream reaches. The giant sandwaves attain heights of 10 m and wavelengths of 500 m and are forming in areas where bottom currents reach 2 m s^{-1}. Near Macapá (E in Fig. 12), channel-bottom sand sheets attain thicknesses of 10 m. The upper surface is covered by sinuous- and straight-crested symmetical sandwaves, which reflect fluvial bedload transport influenced by tidal currents. These sandwaves attain 2-m heights and exhibit wavelengths of 20 to 60 m (Vital, 1996). There is a strongly developed seasonal variability of the riverbed. During periods of rising discharge, large scours can be formed that attain depths of 45 m. These scours can be subsequently modified or filled during periods of low discharge. Cross-sections commonly exhibit seasonal variations in river bottom depths of 10 m.

Much of the lower Amazon area is influenced by neotectonic activity. As an example, Vital (1996) noted that fault-like scarps with offsets of as much as 4 m on the Rio Amazonas floor may be related to reactivation of basement structures.

Furos do Breves

The Rio Amazonas is in direct hydrological connection with the Rio do Pará. Water flows from the Amazon through a series of relatively narrow channels, termed *furos* (canals) or *estreitios* (narrows), that pass near the town of Breves (area C in Fig. 12). These channels provide another seaward outlet for excess Rio Amazonas water. The tremendous discharge of the Rio Amazonas results in there being net movement of Amazon water through the furos, which creates turbid conditions in the Rio Pará (Sioli, 1984). Within these furos, reverse flow occurs in the freshwater system in response to tidal forcing. Tidal effects are complex because flood tides originate from both the Rio Amazonas to the north and the Rio Pará from the south.

Detailed investigations of bedforms in this part of the Amazon system were also undertaken by Vital (1996). Sandwaves, which are symmetrical to slightly asymmetrical, occur on the floor of the furos. The lee side is directed towards the Rio Pará, which indicates the dominant transport direction from the Rio Amazonas. Sandwaves exhibit heights ranging from 2 to 5 m and wavelengths of 20 to 60 m. Within the furos, channel-margin sediments are fine-grained and commonly consist of layers of fine sand interlaminated with clays (Vital, 1996). The laminae appear to have been produced by semidiurnal tidal fluctuations. Results of ^{14}C dating suggest that deposition rates of about 6 mm yr^{-1} have occurred over the past 500 to 800 years. Extrapolation would suggest that long-term, fluvial-dominated deposition rates of the order of 6 m kyr^{-1} could potentially occur in this area.

Amazon mouth

The Amazon mouth region is laterally extensive and contains very broad channels (An and As in Fig. 12). The following information has been summarized from Torres (1997). This area is greatly influenced by waves as well as tidal and longshore currents. The two main channels (Canal do Norte and Canal do Sul) attain depths of 80 m and range from 8 to 21 km in width. These main channels are separated by a shallow region (< 5 m depth), which contains an extensive archipelago of fluvio-estuarine islands. The mouth areas of the Amazon are strongly affected by tidal erosion and sedimentation. The rapid rise of the tides produces tidal bores, locally termed *pororoca* ('big roar'), throughout the mainstream and tributaries of the Amazon system (Branner, 1884), the Rio Pará system and the surrounding region (Fig. 5).

Similar to the inner estuarine area described by Vital et al. (1998), the North and South Channels exhibit a depositional bottom consisting of sandwaves and flat topography (Torres, 1997). There are also erosional bottoms floored by palaeomuds, which have been ^{14}C dated to 12 to 13 kyr BP.

Sandwaves consist predominantly of fine-grained sand and attain heights of 11 m and lengths of 300 m. Generally, these large-scale sandwaves are asymmetrical with lee-face inclinations as great as 18° that are orientated in the fluvial-current direction. Sandwaves occur in water depths that are as great as 26 m and tend to overlie flat surfaces consisting of silt and sandy silt.

The Canal Perigoso (Perilous Channel), which is situated between the Canal do Norte and Canal do Sul, contains large-scale sandbanks or sand ridges. These very large features are orientated SE–NW with their steeper, lee face oriented to the north-east. Individual banks attain heights of 70 m, wavelengths as great as 1300 m, and are several kilometres long (Torres, 1997). These banks consist of fine- to medium-grained sand. Similar features, orientated perpendicular to the shore, were described by Nittrouer *et al.* (1986a,b) on the Amazon shelf. The scale and orientation of these banks, as well as their depositional setting, suggest similarity to estuarine tidal sand ridges.

In the more inland reaches of the mouth, the modern sediments are primarily sands. Conversely, on the outer islands of the mouth-area archipelago, the modern sediments are dominantly silt. Within the North and South Channels, suspended-sediment concentrations, as reported by Torres (1997), ranged from 67 mg L^{-1} at low discharge to 232 mg L^{-1} during high discharge. The suspended-sediment concentrations of the North and South Channels suggest that suspended-sediment concentrations are high because of the strong tidal currents that erode sand banks and the shoreline, resulting in an increase in the levels of suspended sediment.

Brackish-water estuaries

Within the Rio Pará near Belém, high tides result in brackish-water intrusion. At the seaward mouth of the Rio Pará and its associated bays, the waters are saline. The rapid rise of the tides, when confined within channels, results in tidal bores, or pororoca. Wallace (1853, p. 89) provided a detailed description of an encounter with a pororoca on the Rio Guama, which is a tributary south of Belém. The tributaries that enter the Rio Pará are clearwater, so this system has not kept pace with rising base-levels and has been converted into a brackish-water estuary. Sedimentation within this area is generally slow, except where there is an influx of sediment brought in by Amazonian water via the Furos de Breves.

Coastal estuarine environments

Mudcapes

North of the Amazon mouth is a mud-dominated coastline that extends nearly 1600 km northwestward to the Orinoco Delta. Owing to longshore currents, 7 to 17% of the suspended sediment of the Amazon is transported along this coast (Kuehl *et al.*, 1986; Nittrouer *et al.*, 1986a; Kineke & Sternberg, 1995). The resulting mud deposition has produced a series of capes, termed *cabos*, which are composed predominantly of very fine-grained sediment. These have been termed *mudcapes* and range from 10 to 100 km in length (Allison *et al.*, 1995a, 2000; Allison & Nittrouer, 1998). A unique aspect of these mudcapes is their predominantly northward orientation, which relates to their formation by longshore transport (Fig. 13).

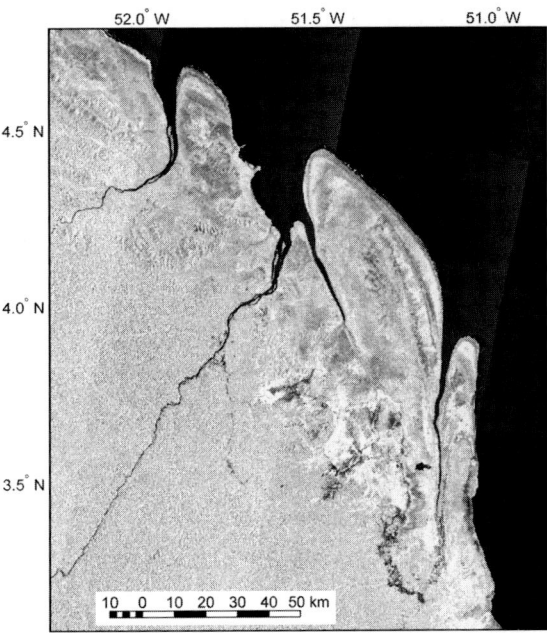

Fig. 13 JERS-1 SAR image of mudcapes developed north of the Amazon mouth. Image data courtesy of GRFM, © NASDA/MITI.

The most southern mudcapes occur near the northern border of Brazil. The Rio Oiapoque, known as the Oyapack in French Guiana, delineates the northern coastal border of Brazil and is related to one of the largest cabos. Funnel-shaped estuarine bays with a prominent northward orientation, such as the Baie d'Oyapack (Fig. 13), are on the inland side of the cabos. Although much of the coastline of eastern South America is retreating owing to sea-level rise, the mudcape-studded coastline is prograding seaward. Over the past 6 kyr, there has been about 10 to 100 km of coastal accretion related to mudcape formation (Allison et al., 1995b, 2000). Large amounts of the sediment from the Amazon plume may be involved in cabo deposition (Kineke & Sternberg, 1995).

Allison et al. (1994, 1995a,b, 2000) performed detailed investigations of the sedimentology of the mudcape Cabo Cassipore, which is near the northern border of Brazil (Fig. 13). Wind-generated waves, ranging from 1 to 2 m in height, are produced by seasonal trade winds. Wave velocities up to 100 cm s^{-1} resuspend individual particles and aggregates of fine-grained sediment. This area is macrotidal, with tidal ranges that locally approach 5 m (Kineke & Sternberg, 1995). Tidal currents reach velocities up to 35 cm s^{-1} and the shore-normal component results in longshore transport of the resuspended sediment. Significant reworking and transport occur during the semidiurnal tidal cycles.

Despite being composed of fine-grained sediment, the mudcape is actively prograding over subaqueous deltaic deposits on an open, non-barred coast. There is a seasonally flooded coastal plain, ranging from 15 to 100 km in width (Allison et al., 1995b). This is an area with high rainfall, which amounts to 3 m annually. Owing to high rainfall and the freshwater content of the Amazon plume, salinity is about 20 psu on the accretionary mudflats.

In the Capo Cassipore area, there is an 85-km-long accretionary mud shoreface that undergoes about 3–4 cm yr^{-1} of vertical accretion. This has a very low gradient and an intertidal zone that ranges from 2 to 5 km in width. The low relief means that only very shallow (< 50 cm deep) and poorly defined drainage channels are produced during low tide. Within the upper intertidal zone, dense stands of mangroves provide stabilization of the mud during high tides. Fluid-rich mud on the mudflats also appears to assist in the dissipation of wave energy (Allison et al., 1995a,b).

There is a strong seasonal variation in mudflat accumulation and deposition, as described by Allison et al. (1995a). During January to June, the mudflats accumulate at rates of as much as 1 cm day^{-1} from fluid-mud suspensions and form a layer of mud that can be 1.5 m thick. Following the period of accumulation, virtually this entire layer of mud is resuspended during July to December and moved via longshore transport to the north-west. Controls on deposition and erosion include seasonal variations in trade winds and sediment supply from the Amazon.

Decadal- to century-scale periods of cabo deposition have alternated with erosional phases, which have produced cheniers on the coastal plain. Vertical sequences, about 4 to 6 m thick, include: basal clinoform mudflat sediments, mangrove-swamp facies, overlain by mud formed in seasonally flooded, supratidal grasslands. Mangroves occur within a narrow and low-diversity littoral belt that is subject to saltwater invasion. Black mangrove (Avicennia) occurs along the shorelines of freshwater inland reaches of the estuaries, whereas the red mangrove (Rhizophora) occurs along estuarine banks (Allison et al., 1995b). Fine-scale laminations, composed of sand and silt, are noted within some of the muds. These laminations appear to record tidal and wave-induced variations in bottom shear stress and sediment supply.

Erosional headlands and estuaries

South-east of the southern Amazon mouth area near Belém, the Atlantic coast of Brazil has a very irregular appearance owing to the presence of rocky headlands that separate a number of relatively small-scale estuaries (Fig. 14). North-west–south-east normal faults, which dip steeply to the north-east, partially control the drainages in this area (Costa et al., 1993). Tidal ranges are macrotidal, with spring tidal ranges that exceed 4.5 m at Salinopolis (see Fig. 14). Lowering of sea level during the Pleistocene resulted in incision of the numerous small rivers. This area, which extends to Maranhão, has been termed a *costa de ria* (coast of flooded river mouths), because of the

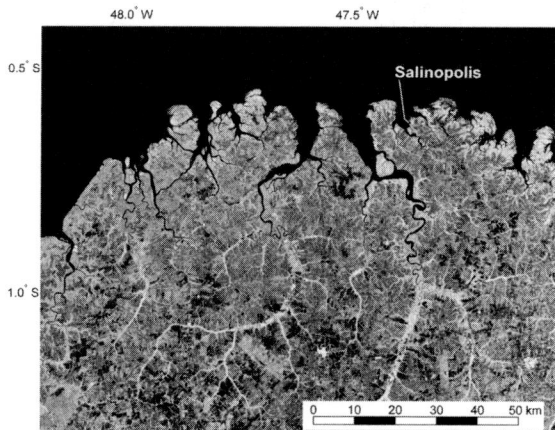

Fig. 14 JERS-1 SAR image of bedrock-walled estuaries developed on coastal areas south of the Rio Pará. Relics of the prior extent of these systems show up as lighter areas on the radar images. Image data courtesy of GRFM, © NASDA/MITI.

numerous funnel-shaped estuaries that were produced during the Holocene sea-level rise. As compared with the Amazon mouth, the southern coastal area has a drier climate and because the freshwater outflow of the Amazon is directed northward, normal marine salinities occur along the coast. Dense stands of mangroves occur within the numerous estuaries. Rising sea level in this area (see Dieter & Neves, 1995) has resulted in the backfilling of relatively small coastal fluvial-drainage systems and development of numerous small-scale estuaries.

COASTAL AND SHELF SETTINGS

Of the sediment discharged by the Amazon, 85–95% is suspended silt and clay-sized particles, dominated by quartz and clay minerals (Kuehl et al., 1988; Kineke & Sternberg, 1995). Discharge from the Amazon forms a plume of turbid, low-salinity water 5 to 10 m thick, termed the *Amazon Plume*, which spreads first offshore and then north-westward to cover the continental shelf (Lentz, 1995). Within the plume, suspended sediments enter into coastal and shelfal environments where a number of processes affects their distribution, redistribution and ultimate sedimentation. These processes include density-driven circulation linked to enormous freshwater discharge, as well as transport related to longshore currents, strong semidiurnal tides, and north-west-orientated trade winds (Nittrouer & DeMaster, 1986; Lentz, 1995).

The world's greatest extent of fluid mud, defined as a high concentration of suspended fine-grained sediments, occurs on the Amazon shelf seaward and down drift of the river's mouth. As discussed by Kineke & Sternberg (1995), the estuarine character of the Amazon shelf, combined with the high level of suspended fine-grained sediment, result in high rates of fluid-mud production. On the Amazon shelf, nearshore surface suspended-sediment concentrations are of the order of 0.5 g L^{-1}. Conversely, within the bottom-hugging fluid mud, concentrations range from 10 to approximately 330 g L^{-1}. The fluid muds can be as much as 7.25 m in thickness, but are generally from 1 to 2 m (Kineke & Sternberg, 1995).

Fluid muds create a transitional boundary between the water column and the sea-floor. These fluid muds are similar to those commonly described in turbidity maximum zones of other estuarine settings, such as the Gironde and Severn estuaries. Within the Amazon shelf, however, the fluid muds are not confined to an enclosed area because estuarine conditions extend for hundreds of kilometres along and seaward of the coast.

Close to the river mouth, the variation in fluid-mud distribution is controlled by fortnightly tides and the location and structure of bottom salinity fronts. Salinity stratification and the best development of salinity fronts occur during neap-tide periods. During this time, the reduced tidal fluctuation and reduction of vertical mixing enhances settling of suspended sediment along the salinity front. Conversely, on the open shelf, variations in fluid-mud distributions appear to be related to seasonal variations in river and sediment discharge and the long-term location of the bottom salinity front (Kineke & Sternberg, 1995).

Subaqueous delta

When compared with typical fluvial deltas, the Amazon has no subaerial feature at the coast (Nittrouer et al., 1986a). High energies within the river mouth, mostly related to tides, result in export of the large amounts of suspended sediments to the shelf. Thus, a substantial subaqueous delta has been produced on the shelf far seaward of the actual coast. Typical fluvial deltas commonly have fine-grained bottomset beds (prodelta). Conversely, within the Amazon the production of large amounts of fluid muds results in mud-rich deposition across much of the subaqueous delta. The subaqueous delta receives about 50% of the sediment transported to the coast by the Rio Amazonas (Kuehl *et al.*, 1986; Nittrouer *et al.*, 1986a). As discussed by Kuehl *et al.* (1988), samples from the top 3 m of the subaqueous delta suggest deposition within the past 100 yr. This yields depositional rates that average about 3 cm yr^{-1}.

Sand percentages are higher immediately offshore of the mouth and then decrease seaward and also decrease to the north. Occurrence of sand is controlled by strong tidal currents, which exceed 1 m s^{-1} close to the sea-bed, and surface waves that have heights of about 1.5 m. Shoals north of the river mouth halt north-western transport of sand. Highest sedimentation rates are about 4 to 10 cm yr^{-1} in the area north-west of the river mouth and near the seaward edge of the topset beds of the subaqueous delta (Nittrouer *et al.*, 1986a). Much of the subaqueous delta consists of millimetre- and centimetre-scale laminations, which have formed in well-oxygenated settings. In general, physical sedimentary structures occur in the topset region where accumulation rates are high (> 4 cm yr^{-1}) or where physical reworking was intense (< 30 m deep). Coarse silt laminae are abundant within the proximal portion of the Amazon subaqueous delta. Laminae are composed of numerous grain types, including quartz, feldspar, heavy minerals, wood fragments and faecal pellets. Macrofauna are generally absent in the topset region. This virtual absence of bioturbating organisms probably relates to intense physical reworking of the sea-floor at these shallow depths. In the distal portions of the delta, laminae are commonly about 0.1 mm thick and consist of quartz silt. On the inner shelf, accumulation rates are low, less than 1 mm yr^{-1}. In the bottomset regions, numerous polychaetes are present. Mottled structures occur seaward of the river mouth and in the bottomset part of the subaqueous delta. In these areas, accumulation rates are low and the degree of physical reworking is minor when compared with the degree of biogenic mixing.

Deep-sea fan

Located seaward of the shelf break, the Amazon deep-sea fan, or *Amazon Cone*, is one of the world's largest. The fan was formed at a time of lowered sea level when the Amazon River flowed to the coast through what is now a submarine canyon, and is thought to have been inactive during the Holocene (Damuth & Fairbridge, 1970). The southern part of this fan also includes contributions from the Rios Pará and Tocantins, which flowed to the shelf break within what is now a secondary submarine canyon (Milliman *et al.*, 1975).

Damuth *et al.* (1988) used side-scan sonar and high-resolution seismic analysis to characterize the details of the fan, which extends from the continental shelf downslope for approximately 700 km, to depths of about 4700 m. The surface of the fan has tributary channels that have intricate meanders with high sinuosities. These include cut-offs and abandoned meander loops, which formed in response to active channel migration. The extensive meander systems suggest large volumes of relatively continuous turbidity flows that were maintained for relatively long periods of time. According to Damuth *et al.* (1988), sporadic turbidity currents, which characterize many active deep-sea fans, would have been too infrequent to create the observed meandering systems.

SUMMARY

Amazon depositional systems are strongly influenced by two dominant periodicities. Within the upper reaches of the system, the yearly cycle of fluvial discharge variation is the dominant influence. Conversely, the lower reaches are influenced by semidiurnal tides, which extend for considerable distances up river. Meandering patterns are best developed within the more inland settings, whereas the tidally influenced portions exhibit generally straight channel systems. Low slopes and large channel sizes lead to tidal influences propagating over 800 km inland.

The main Amazon system, particularly tributaries that drain the Andean highlands, has a very high sediment flux and the river valleys have aggraded with rising base level. River valleys occupied by low-sediment rivers (i.e. blackwater and clearwater systems), however, did not fill with sediment and have been converted into mouth lakes (rias). The upstream end of these lakes may contain archipelagos of elongate, sinuous islands in fluvially dominated rias. Within tidally influenced rias, islands exhibit more equate proportions. Thus, the sequence stratigraphy of incised valley fills within different parts of the Amazon system can be dramatically different because of differences in sediment supply.

The depositional settings of the coastal systems are quite different north and south of the Amazon mouth. Mud-rich, depositional settings and features termed 'mudcapes' characterize the northern coast. To the south, the system is sediment starved and most deposition is occurring within bedrock-confined, brackish-water and small-scale estuarine systems. Overall, the scale of the Amazonian system, as well as its extreme variability, indicate that much more study should be devoted to this unique depositional system.

ACKNOWLEDGEMENTS

This compilation of information on the Amazon Basin was originally initiated in order to develop depositional models for modern analogues of ancient tidal rhythmites. This project was funded by NASA Grant NCC5-234 entitled 'Aspects and Controls of Tidal Sedimentation and the Extraction and Modeling of Tidal Parameters from Modern and Ancient Tidal Rhythmites', which was a collaborative project initiated by Bruce Bills of NASA/GSFC (Goddard Space Flight Center). Subsequent research on the Amazon system has been funded by ACS/PRF (American Chemical Society/Petroleum Research Fund) Grant PRF# 37023-AC8 entitled 'Floodplain Deposition within the Amazon River Mouth, Brazil'.

The altimetry/bathymetry database from Smith & Sandwell (1997) was used for construction of Fig. 2. Radar-image data were originally acquired by JERS-1 SAR (Japanese Earth Resources Satellite 1, Synthetic Aperture Radar), which was sponsored by the Global Rain Forest Mapping (GRFM) project and National Space Development Agency of Japan (NASDA)/MITI. Images Copyright © 1995, 1996 NASDA/MITI with processing by NASA's Alaska SAR Facility and Jet Propulsion Laboratory, and mosaic by NASDA.

Reviews and suggestions made by Meade A. Allison, William E. Galloway, Steven F. Greb, Erik P. Kvale, Admilson M. Torres and Pedro Walfir greatly improved the final version of the manuscript.

REFERENCES

Allison, M.A. and Nittrouer, C.A. (1998) Identifying accretionary mud shorefaces in the geologic record: insights from the modern Amazon dispersal system. In: *Shales and Mudstones*, Vol. I (Eds J. Schieber, W. Zimmerle and P. Sethi), pp. 147–161. E. Schweizerbart'sche Verlagsbuchhandlung, Stuttgart.

Allison, M.A., Nittrouer, C.A., Kineke, G.C. and Sternberg, R.W. (1994) Use of instrumented tripod system to examine sediment dynamics and fine-scale strata formation in muddy surfzone and nearshore environments. *J. Coast. Res.*, **10**, 488–496.

Allison, M.A., Nittrouer, C.A. and Kineke, G.C. (1995a) Seasonal sediment storage on mudflats adjacent to the Amazon River. *Mar. Geol.*, **125**, 303–328.

Allison, M.A., Nittrouer, C.A. and Faria, L.E. (1995b) Rates and mechanisms of shoreface progradation and retreat downdrift of the Amazon River mouth. *Mar. Geol.*, **B125**, 373–392.

Allison, M.A., Lee, M.T., Ogston, A.S. and Aller, R.C. (2000) Origin of Amazon mudbanks along the northeastern coast of South America. *Mar. Geol.*, **163**, 241–256.

Amarasekera, K.H., Lee, R.F., Williams, E.R. and Eltahir, E.A. (1997) ENSO and the natural variability in the flow of tropical rivers. *J. Hydrol.*, **200**, 24–39.

Bartsch-Winkler, S. and Lynch, D.K. (1988) Catalog of worldwide tidal bore occurrences and characteristics. *U.S. Geol. Surv. Circ.*, **1022**, 17 pp.

Beardsley, R.C., Candela, J., Limeburner, R., Geyer, W.R., Lentz, S.J., Castro, B.M., Cacchione, D. and Carneiro, H. (1995) The M_2 tide on the Amazon shelf. *J. Geophys. Res.*, **100**, 2283–2319.

Beurlen, K. (1970) *Geologie von Brasilien*. Gebruder Borntraeger, Berlin, 414 pp.

Bowden, K.F. (1978) Estuary circulation and diffusion problems. In: *Tidal Power and Estuary Management* (Eds R.T. Severn, D.L. Dineley, and L.E. Hawker), pp. 141–150. Scientechnica, Bristol.

Branner, J.C. (1884) The 'pororoca', or bore of the Amazon. *Science*, **4**, 488–490.

Campbell, D.G., Stone, J.L. and Rosas, A., Jr. (1992) A comparison of the phytosociology and dynamics of three floodplain (Várzea) forests of known ages, Rio Juruá, western Brazilian Amazon. *Bot. J. Linn. Soc.*, **108**, 213–237.

Costa, J.B., Borges, M.da S., Bemerguy, R.L., Fernandes, J.M., DaCosta-Junior, P.S. and DaCosta, M.L. (1993) Evolução cenozoica da regiao de Salinopolis, nordeste do Estado do Pará. *Geociencias*, **12**, 373–396.

Damuth, J.E. (1977) Late Quaternary sedimentation in the western equatorial Atlantic. *Geol. Soc. Am. Bull.*, **88**, 695–710.

Damuth, J.E. and Fairbridge, R.W. (1970) Equatorial Atlantic deep-sea arkosic sands and ice-age aridity in tropical South America. *Geol. Soc. Am. Bull.*, **81**, 189–206.

Damuth, J.E., Flood, R.D., Kowsmann, R.O., Belderson, R.H. and Gorini, M.A. (1988) Anatomy and growth pattern of Amazon deep-sea fan as revealed by long-range side-scan sonar (GLORIA) and high-resolution seismic studies. *Am. Assoc. Petrol. Geol. Bull.*, **72**, 885–911.

Dieter, M. and Neves, C.F. (1995) The implications of sea-level rise on the Brazilian coast; a preliminary assessment. *J. Coast. Res.*, **14** (Special Issue), 54–78.

Dunne, T., Mertes, L.A.K., Meade, R.H., Rickey, J.E. and Forsberg, B.R. (1998) Exchanges of sediment between the flood plain and channel of the Amazon River in Brazil. *Geol. Soc. Am. Bull.*, **110**, 450–467.

Fisher, T.R. (1978) Plâncton e produção primária em sistemas aquáticos da bacia da Amazonia Central. *Acta Amazonica*, **8**(4), 43–54.

Gibbs, R.J. (1967a) Amazon River: environmental factors that control dissolved and suspended load. *Science*, **156**, 1734–1737.

Gibbs, R.J. (1967b) The geochemistry of the Amazon River: part 1, the factors that control the salinity and the composition and concentration of the suspended solids. *Geol. Soc. Am. Bull.*, **78**, 1203–1232.

Gibbs, R.J. (1970) Circulation in the Amazon River estuary and adjacent Atlantic Ocean. *J. Mar. Res.*, **28**, 113–123.

Goulding, M., Carvalho, M.L. and Fereira, E.G. (1988) *Rio Negro, Rich Life in Poor Water*. SPB Academic Publishing, The Haugue, 200 pp.

Herndon, W.L. (1853) *Exploration of the Valley of the Amazon, Made under Direction of the Navy Department*. R. Armstrong Press, Washington, DC, 381 pp.

Huber, J. (1903) Dos Furos de Breves, E da parte occidental de Marajó. *Boletim do Museu Paraense de Historia Natural e Ethnographia (Museu Geoldi)*, **3**(1900–1902), 447–498.

Irion, G. (1984) Sedimentation and sediments of Amazonian rivers and evolution of the Amazonian landscape since the Pliocene times. In: *The Amazon, Limnology and Landscape Ecology of a Mighty Tropical River and its Basin* (Ed. H. Sioli), pp. 201–214. W. Junk, Dordrecht.

Jordan, C.F. (1985) Soils of the Amazon rainforest. In: *Key Environments: Amazonia* (Eds G.T. Prance and T.E. Lovejoy), pp. 83–94, Pergamon, Oxford.

Junk, W.J. and Furch, K. (1985) The physical and chemical properties of Amazonian waters and their relationships with the biota. In: *Key Environments: Amazonia* (Eds G.T. Prance and T.E. Lovejoy), pp. 3–17, Pergamon, Oxford.

Kineke, G.C. and Sternberg, R.W. (1995) Distribution of fluid muds on the Amazon continental shelf. *Mar. Geol.*, **125**, 193–233.

Kuehl, S.A., DeMaster, D.J. and Nittrouer, C.A. (1986) Nature of sediment accumulation on the Amazon continental shelf. *Continent. Shelf Res.*, **6**, 209–226.

Kuehl, S.A., Nittrouer, C.A. and DeMaster, D.J. (1988) Microfabric study of fine-grained sediments: observations from the Amazon subaqueous delta. *J. Sediment. Petrol.*, **58**, 12–23.

Leenheer, J.A. (1980) Origin and nature of humic substances in the waters of the Amazon river basin. *Acta Amazonica*, **10**(3), 513–526.

Leenheer, J.A. and Santos, U.M. (1980) Considerações sobre os processos de sedimentação na água preta ácida do rio Negro (Amazônia Central). *Acta Amazonica*, **10**(2), 343–355.

Lentz, S.J. (1995) The Amazon River plume during AMASSEDS: subtidal current variability and the importance of wind forcing. *J. Geophys. Res.*, **199**, 2377–2390.

Meade, R.H. (1994) Suspended sediments of the modern Amazon and Orinoco Rivers. *Quat. Int.*, **21**, 29–39.

Meade, R.H., Nordin, C.F., Curtis, W.F., Costa-Rodrigues, F.M. and Edmond, J.M. (1979) Transporte de sedimentos no rio Amazonas. *Acta Amazonica*, **9**(3), 529–547.

Meade, R.H., Dunne, T., Richey, J.E., Santon, U.M. and Salati, E. (1985) Storage and remobilization of suspended sediment in the lower Amazon River of Brazil. *Science*, **228**, 488–490.

Mertes, L.A. (1994) Rates of flood-plain sedimentation on the central Amazon River. *Geology*, **22**, 171–174.

Milliman, J.D., Summerhayes, C.P. and Barretto, H.T. (1975) Quaternary sedimentation on the Amazon continental margin: a model. *Geol. Soc. Am. Bull.*, **86**, 61–614.

Nittrouer, C.A. and DeMaster, D.J. (1986) Sedimentary processes on the Amazon continental shelf: past, present and future research. *Continent. Shelf Res.*, **6**, 5–30.

Nittrouer, C.A., Kuehl, S.A., DeMaster, D.J. and Kowsmann, R.O. (1986a) The deltaic nature of the Amazon shelf sedimentation. *Geol. Soc. Am. Bull.*, **97**, 444–458.

Nittrouer, C.A., Curtin, T.B. and DeMaster, D.J. (1986b) Concentration and flux of suspended sediment on the Amazon continental shelf. *Continent. Shelf Res.*, **6**, 151–174.

Nordin, C.F., Meade, R.H., Curtis, W.F., Bósio, N.J. and Landim, P.M. (1980) Size distribution of Amazon River bed sediment. *Nature*, **286**, 51–52.

Nordin, C.F., Meade, R.H., Curtis, W.F., Bósio, N.J. and Landim, P.M. (1981) Distribuição do sedimento do leito do rio Amazonas—nenhuma mudança apreciável rio abaixo. *Acta Amazonica*, **11**, 769–772.

Oltman, R.E. (1967) Reconnaissance investigations of the discharge water quality of the Amazon. *Atas simpósio sôbre biota Amazônica, Rio de Janeiro*, **3** (Limnologia), 163–185.

Oltman, R.E. (1968) Reconnaissance investigations of the discharge and water quality of the Amazon River. *U. S. Geol. Surv. Circ.*, **552**, 16 pp.

Oltman, R.E., Sternberg, H.O'R., Ames, F.C. and Davis, L.C., Jr. (1964) Amazon River investigation, Reconnaissance measurements of July, 1963. *U. S. Geol. Surv. Circ.*, **486**, 15 pp.

Pardé, M. (1936) Les variations saisonnières de l'Amazone. *Annal. Géogr.*, **45**, 502–511.

Pires, J.M. and Prance, G.T. (1985) The vegetation types of the Brazilian Amazon. In: *Key Environments: Amazonia* (Eds G.T. Prance and T.E. Lovejoy), pp. 109–145, Pergamon, Oxford.

Richey, J.E, Mertes, L.A., Dunne, T., Victoria, R.L., Forsberg, B.R., Tancredi, A.C. and Oliveira, E. (1989a) Sources and routing of the Amazon River flood wave. *Global Biogeochem. Cycl.*, **3**, 191–204.

Richey, J.E., Nobre, C. and Deser, C. (1989b) Amazon River discharge and climate variability: 1903 to 1985. *Science*, **246**, 101–103.

Sioli, H. (1950) Das Wasser im Amazonasgebiet. *Forsch. Fortschr.*, **26**, 274–280.

Sioli, H. (1967) Studies in Amazonian waters. *Atlas do Simpósis sobre a Biota Amazônica*, **3** (Limnologia), 39–50.

Sioli, H. (1975) Amazon tributaries and drainage basins. In: *Ecological Studies—Analysis and Synthesis* (Eds J. Jacobs, O.L. Lange, J.S. Olson and W. Wieser), Vol. 10, pp. 199–213, Springer-Verlag, New York.

Sioli, H. (1984) The Amazon and its main affluents: hydrography, morphology of the river courses, and river types. In: *The Amazon, Limnology and Landscape Ecology of a Mighty Tropical River and its Basin* (Ed. H. Sioli), pp. 127–165, W. Junk, Dordrecht.

Smith, W.H.F. and Sandwell, D.T. (1997) Global seafloor topography from satellite altimetry and ship depth soundings. *Science*, **277**, 1956–1962.

Torres, A.M. (1997) *Sedimentology of the Amazon mouth: North and South Channels, Brazil*. Report No. 82, Geologisch-Paläontologisches Institut, Christian-Albrechts-Universität, Kiel, 145 pp.

Tricart, J. (1977) Types des lits fluvieux en Amazonie brésiliene. *Annal. Geogr.*, **473**, 1–53.

Vital, H. (1996) Sedimentology of the lowermost Amazon (Rio Xingu—Macapa) and the 'Estreitos de Breves'—Brazil. Unpub. PhD thesis, Christian Albrechts University, 189 pp.

Vital, H. Stattegger, K., Posewang, J. and Theilen, F. (1998) Lower Amazon River: morphology and shallow seismic characteristics. *Mar. Geol.*, **152**, 277–294.

Vital, H., Stattegger, K. and Garbe-Schöenberg, C.D. (1999) Composition and trace-element geochemistry of detrital clay and heavy-mineral suites of the lowermost Amazon River: a provenance study. *J. Sediment. Res.*, **69**, 563–575.

Wallace, A.R. (1853) *Narrative of Travels on the Amazon and Rio Negro*. Reeve and Company, London, 363 pp.

Kinematics, topology and significance of dune-related macroturbulence: some observations from the laboratory and field

JIM BEST

Earth and Biosphere Institute, School of Earth and Environment, University of Leeds, Leeds, West Yorkshire LS2 9JT, UK (Email: j.best@earth.leeds.ac.uk)

ABSTRACT

Macroturbulence, which may advect through the entire water depth, dominates the flow field associated with alluvial sand dunes and has long been regarded as the principal mechanism for suspending bedload sediment over dunes. The origin of this macroturbulence has been linked to shear layer development in the dune lee, often associated with flow separation, and the form of these coherent flow structures has been noted as 'boils' that erupt onto the water surface. Although past work has quantified the mean and turbulent flow characteristics of flow over dunes using at-a-point measurements, these studies have not been able to trace the *evolution* of such macroturbulent events over a dune-covered bed. Additionally, the topology of the dune-related macroturbulence has not been explained in relation to the structure of the developing surface boils. This paper tackles both of these issues using a twofold approach: (i) use of whole flow field quantification using particle imaging velocimetry (PIV) over a series of fixed laboratory dunes; (ii) observations of the water surface over large sand dunes in the Jamuna River, Bangladesh.

Particle imaging velocimetry results are in good agreement with past work detailing the mean flow field over sand dunes. The PIV images over the lee and stoss sides of an experimental dune, and observation of the water surface above natural dunes, reveal four key dynamic attributes to flow.

1 The shear layer and separation zone associated with dunes are spatially and temporally dynamic, and 'flapping' of the shear layer may be modulated by turbulent coherent flow structures generated upstream.
2 Reynolds stresses in the lee side are dominated by the free shear layer associated with the separation zone.
3 Ejections of low downstream momentum fluid away from the bed dominate the instantaneous flow field over the crestal regions of the dune. These ejections, in turn, however, create return flows towards the bed both in front of, and behind, the ejection. The highest instantaneous Reynolds stresses are associated with these ejections and inrushes.
4 'Boils' on the water surface over a natural dune field often consist, firstly, of a central upwelling that has a spanwise axis of rotation and, secondly, later secondary vortices that possess a vertical axis of rotation. This pattern of flow can be explained by the interaction of a vortex loop with the free surface.

These results provide a mechanism that links the flow fields of adjacent dunes and highlight how dune-related macroturbulence may dominate the entrainment of sediment into both suspended and bedload transport. Additionally, the mechanism identified here may also be applied to explain the sequence of turbulent events present over other large grain and form roughness in depth-limited alluvial channels.

INTRODUCTION AND AIMS

Dunes are one of the most common alluvial bedforms and have received considerable attention over the past three decades in relation to their occurrence and morphology (e.g. Dinehart, 1989; Gabel, 1993; Dalrymple & Rhodes, 1995; Roden, 1998; Carling, 1999; ten Brinke et al., 1999; Wewetzer & Duck, 1999; Carling et al., 2000a,b), flow dynamics (Jackson, 1976; Müller & Gyr, 1983, 1986, 1996; Yalin, 1992; Kostaschuk & Church, 1993; Lyn, 1993; Nelson et al., 1993; McLean et al., 1994; Bennett & Best, 1995; Kadota & Nezu, 1999; Schmeeckle et al., 1999; Kostaschuk, 2000; Best et al., 2001; Best & Kostaschuk, 2002; Kleinhans, 2002), relationship to sediment transport (Engel & Lam Lau, 1980, 1981; Itakura & Kishi, 1980; Soulsby et al., 1991; Kostaschuk & Ilersich, 1995; Mohrig & Smith, 1996; Bennett & Vendetti, 1997; Villard & Kostaschuk, 1998; Vionnet et al., 1998; Kostaschuk & Villard, 1999; McLean et al., 1999; Shimizu et al., 1999; Vendetti & Bennett, 2000) and effect upon flow resistance (Klaassen, 1979; Wijbenga, 1990; Julien & Klaassen, 1995). Past research has illustrated the nature of the mean and turbulent flow field over sand dunes and shown that these bedforms, which scale with flow depth, are associated with large-scale turbulence or 'macroturbulence' that may be generated over the dune and rise to erupt at the water surface; these eruptions are termed 'boils' by Matthes (1947) and have been the subject of much subsequent work (e.g. Coleman, 1969; Jackson, 1976; Rood & Hickin, 1989; Babakaiff & Hickin, 1996). Both Coleman (1969) and Babakaiff & Hickin (1996) documented the form of these boils as they erupted on the water surface and began to provide indications as to the nature of fluid movement within these macroturbulent coherent structures. Babakaiff & Hickin (1996) related the intensity of eruption of the boil to the relative roughness of the dune, finding that larger dunes produced more intense macroturbulence.

Although macroturbulence has often been associated with flow separation in the lee of steep, angle-of-repose slipface dunes, it has become apparent from studies of low-angle dunes in the field and flume (Smith & McLean, 1977; McLean & Smith, 1979; Kostaschuk & Villard, 1996, 1999; Roden, 1998; Best et al., 2001; Best & Kostaschuk, 2002) that macroturbulence can also be generated by dunes that have much lower angle leeside slopes and possible intermittent flow separation. Additionally, the importance of macroturbulence has been stressed by many authors in relation to the occurrence of dunes and their distinction from other bedforms (e.g. Jackson, 1976; Bennett & Best, 1995, 1996; Best, 1993, 1996; Robert & Ulhman, 2001), as well as to their influence on sediment transport (e.g. Schmeeckle et al., 1999; Vendetti & Bennett, 2000) and the growth of other larger channel topography, such as mid-channel bars (Ashworth et al., 2000). Bennett & Best (1996) hypothesized that the evolution of dunes from a rippled bedstate was associated with the increasing importance of macroturbulence that could penetrate through the entire water depth. They argued that once this occurred, then the subsequent bed-directed inrush of fluid, required to meet continuity to replace the upwelling 'boil', would be sourced from higher in the flow than over ripples and would thus have a higher velocity. They further reasoned that this would subsequently increase the downstream bedload sediment transport rate at the next bedform crest as compared with a rippled bedstate. Recently, Robert & Uhlman (2001) have documented the large increase in turbulence intensity and momentum exchange across the ripple–dune transition and also noted the increased spatial variability in the turbulent flow characteristics as the transition takes place.

Macroturbulence is thus fundamental to dune occurrence and is a common feature visible on the surface of rivers with dune-covered beds. The objective of the current study was thus to examine the *temporal* characteristics of flow over dunes and examine the relationship between macroturbulence generation in the dune leeside and fluid motion at the next downstream crest, to enable testing of the hypothesis proposed by Bennett & Best (1996). The present study provides qualitative and quantitative evidence of the kinematics and topology of dune-related macroturbulence from two sources:

1 quantitative laboratory results that have been obtained using particle imaging velocimetry (PIV) of whole flow field fluid motions associated with dunes;
2 field observations of the patterns of fluid motion on the river surface above dune-covered beds.

These results provide support for past hypotheses concerning the interactions of dune-related macroturbulence with the outer flow, and allow proposition of a qualitative model of macroturbulence interactions with the water surface that is used to interpret the topology of the macroturbulent structures. This model is of more widespread application in explaining the origin and topology of macroturbulence generated over large-scale bed roughness in depth-limited alluvial channels.

METHODS

Laboratory

Laboratory experiments were conducted in a recirculating hydraulic flume that was 10 m long by 0.30 m deep and 0.30 m wide. The dunes studied were cast in smooth fibreglass and had an identical shape to the angle-of-repose dunes that are fully reported in Bennett & Best (1995). The entire length of the flume was filled with identical fibreglass dunes that were 0.04 m high and 0.63 m wavelength (see schematic layout, Fig. 1A). A steady, uniform flow was then established over these dunes, with the mean velocity over the dune crest being 0.44 m s^{-1} (see Table 1 for mean flow

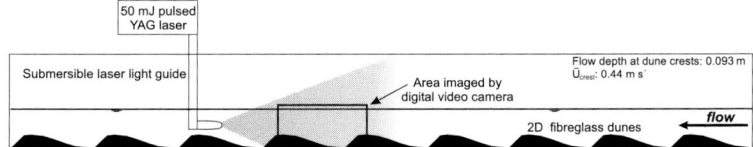

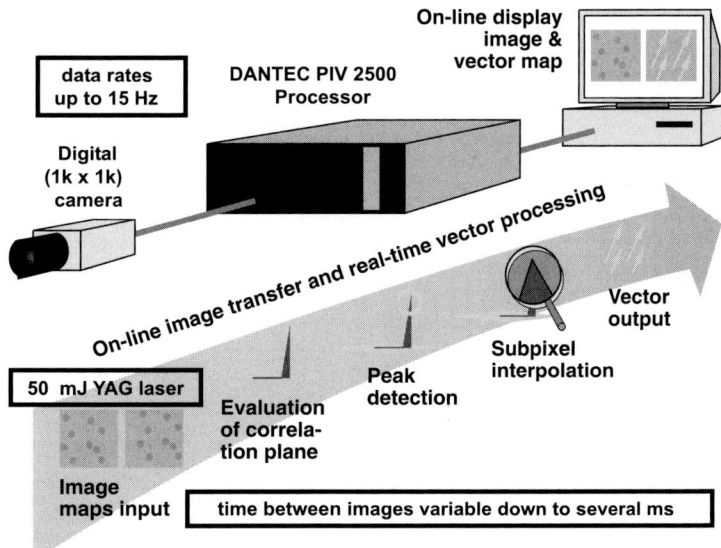

Fig. 1 (A) Schematic diagram of the experimental set-up with the bed of the flume covered with identical fibreglass dunes with a morphology identical to those used in Bennett & Best (1995). (B) Schematic flow diagram of the DANTEC particle imaging velocimetry (PIV) system (modified from DANTEC, 2000).

Table 1 Summary of hydraulic conditions and bedform morphology used in this study and that of Bennett & Best (1995).

	This study	Bennett & Best (1995)
Flow depth at crest, Y_{crest} (m)	0.093	0.10
Mean downstream velocity at crest, $\bar{U}_{crest}$ (m s^{-1})	0.44	0.57
Froude number, Fr	0.46 (crest)	0.58 (crest)
Dune wavelength, λ (m)	0.63	0.63
Dune height, h (m)	0.04	0.04

conditions), this being slower than the flow studied by Bennett & Best (1995; Table 1).

Quantification of flow in these experiments was achieved using PIV, which uses the displacement of neutrally buoyant particles over a known time period to track their movement and velocity (see Adrian (1991, 1996) for reviews, and Schmeeckle et al. (1999) and Tait et al. (1996) for examples of PIV results over dunes and a gravel bed, respectively). Roy et al. (1999) and Roy & Buffin-Bélanger (2001) have highlighted the great benefits of utilizing flow visualization combined with multipoint measurements. Their field studies, however, where visualization is far more problematic than in controlled laboratory conditions, used dye visualization and three at-a-point measurements, which, although allowing fuller investigation of the link between velocity time-series and coherent flow structure, were still necessarily limited in their true quantitative scope. Indeed, the use of limited sampling and also choice of threshold criteria for distinguishing turbulent events may encourage the interpretation of two-dimensional structure from what are truly three-dimensional vortices (Smart, 2001). The use of whole flow field PIV overcomes these significant problems and allows quantitative visualization of the flow by monitoring several thousand individual points *simultaneously*. This permits far fuller appreciation of the kinematics of the flow field and reconstruction of the advection and interaction of vortices. The laboratory work reported here represents one of the first applications of this technique to flow over dunes and has many applications to flow over appreciable bed roughness in depth-limited flows.

A DANTEC 2500 PIV system was used in the present study that is capable of providing whole flow field quantification at data rates of up to 15 Hz. Flow seeding was provided by natural impurities in the water and 10 μm titanium-coated mica particles. Two successive digital images are taken of a flow field area, with the images being separated by a short, user-defined interval (here 750 ns). The DANTEC system then calculates the maximum cross-correlation between particle positions in a given interrogation area (e.g. 16 × 16 pixels) and then uses this maximum correlation peak to produce a mean particle displacement, and hence velocity, for each interrogation area (Fig. 1B). A minimum of six particle pairs is required in each interrogation area in order for the data to be validated. The PIV system can thus produce quantitative whole flow field maps: in this study a two-dimensional plane (x–y or horizontal–vertical plane) was examined using one camera; however, use of two cameras can permit quantification of three-dimensional velocities in a two-dimensional plane (DANTEC, 2000). The raw displacement vectors were then filtered to remove large outliers in the data set caused by spurious correlation peaks. Tests showed that a threshold of > 0.9 m s^{-1} was suitable to represent twice the velocity over the crest and this removed less than 2% of the raw vectors, largely near the flow surface where image quality, owing to reflection, was poorer. A moving average of three adjacent pixels was used to yield a data grid with approximately 6500 points in each image. A 50 mJ YAG laser provided a pulsed laser light sheet into the test section through a streamlined submersed light guide (Fig. 1A) that was located 1 m downstream of the test section; the time between laser pulses for these flow conditions was optimized to produce the best validation of velocities (i.e. the time gap should be large enough to allow detectable differences in particle position between images but not so large that the correlation of particles begins to deteriorate) and a time gap of 750 ns between laser pulses was used. In this paper, the flow field in the immediate dune leeside and dune crest were examined (Fig. 2) and data rates of between 1 and 15 Hz were used to examine the mean flow field and the instantaneous flow structure. Each flow field was interrogated using either a 16 × 16 or 32 × 32 pixel grid, with between 25 and 50% overlap between

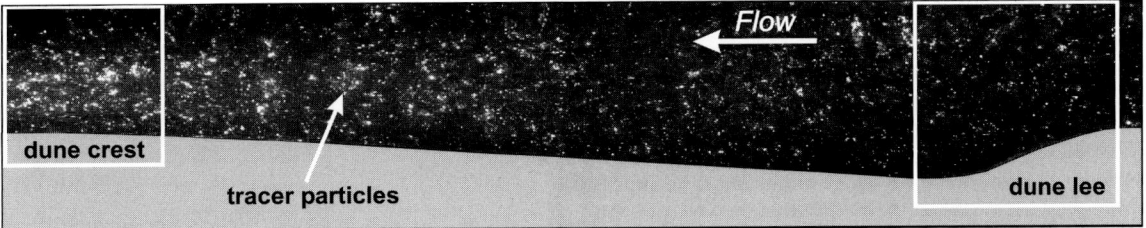

Fig. 2 Particle imaging velocimetry image of the whole experimental dune, with the two areas of study in dune lee and dune crest highlighted. The fields of view are 0.204 m long by 0.205 m deep for the leeside, and 0.148 m wide by 0.149 m deep for the crestal image.

areas and subpixel interpolation to increase the data density (DANTEC, 2000). The mean velocity was derived for each interpolation area and this value was used to calculate the velocity fluctuations and turbulence parameters.

Particle imaging velocimetry can provide an unrivalled method for quantitative visualization of flow, but several factors require care and consideration in the experimental set-up.

1 The system used herein has a maximum sampling rate of 15 Hz, although recent progress has enabled development of PIV that can sample up to 16 kHz for short time periods. The operator can select the sampling frequency and this must be considered at the outset: greater sampling rates require more data storage and thus the length of sample possible is set by the storage available in the PIV processor and the sampling frequency. As each image pair captured is approximately 2 Mb in size, 1 min of data at 15 Hz requires $c.$ 1.8 Gb of memory on the processor. Data storage is thus a major consideration in experimental design.

2 Great care is required in illumination of the test section, and reflections from objects or the water surface can cause problems in particle detection.

3 The spatial resolution of measurements is a function of the flow field area imaged and the pixel resolution of the camera, i.e. greater resolution will be obtained by imaging a smaller physical area using the same pixel image size. The operator may thus face a choice depending on the size of the flow field structures of interest: in this study detailed images were taken of the lee and stoss side of the dune that were of a higher spatial resolution than images taken across the entire dune wavelength.

4 Imaging through a moving, and variable height, water surface is difficult owing to refraction of light and changing water depth: this can be solved if a transparent plate can be placed on the surface but this is problematic in experiments such as those reported herein where water surface interactions are important: this problem thus only allowed visualization from the side of the flume in a two-dimensional plane in the experiments reported herein.

Field

Qualitative observations and photographs of turbulent structures associated with flow over natural dune-covered sand-bed rivers were obtained from the surface of the Jamuna River, Bangladesh. Photographs and video records were obtained during high flow stage in August 1994 of flow over 2.5–4-m-high sand dunes in the main channel of the Jamuna River near Bahadurabad, Bangladesh. The flow depth here was $c.$ 12 m and the mean flow velocity was $c.$ 1.5 m s^{-1}. Further details of this study reach are given in McLelland et al. (1999) and Ashworth et al. (2000), and details of the dune kinematics and flow fields are given in Roden (1998). The large size of these sand dunes resulted in very large-scale macroturbulence (see the earlier work of Coleman, 1969), with surface boils above large dunes being regular in periodicity (Roden, 1998) and ranging between 5 and 50 m in diameter. As the evolution of the surface eruption of these boils occurred over a period of 5–25 s, detailed observations of video tapes and 35 mm photographs allowed the evolution of these macroturbulent patches and their internal fluid motions to be discerned. The higher sediment concentrations in the surface boils also aided observation of their internal structure.

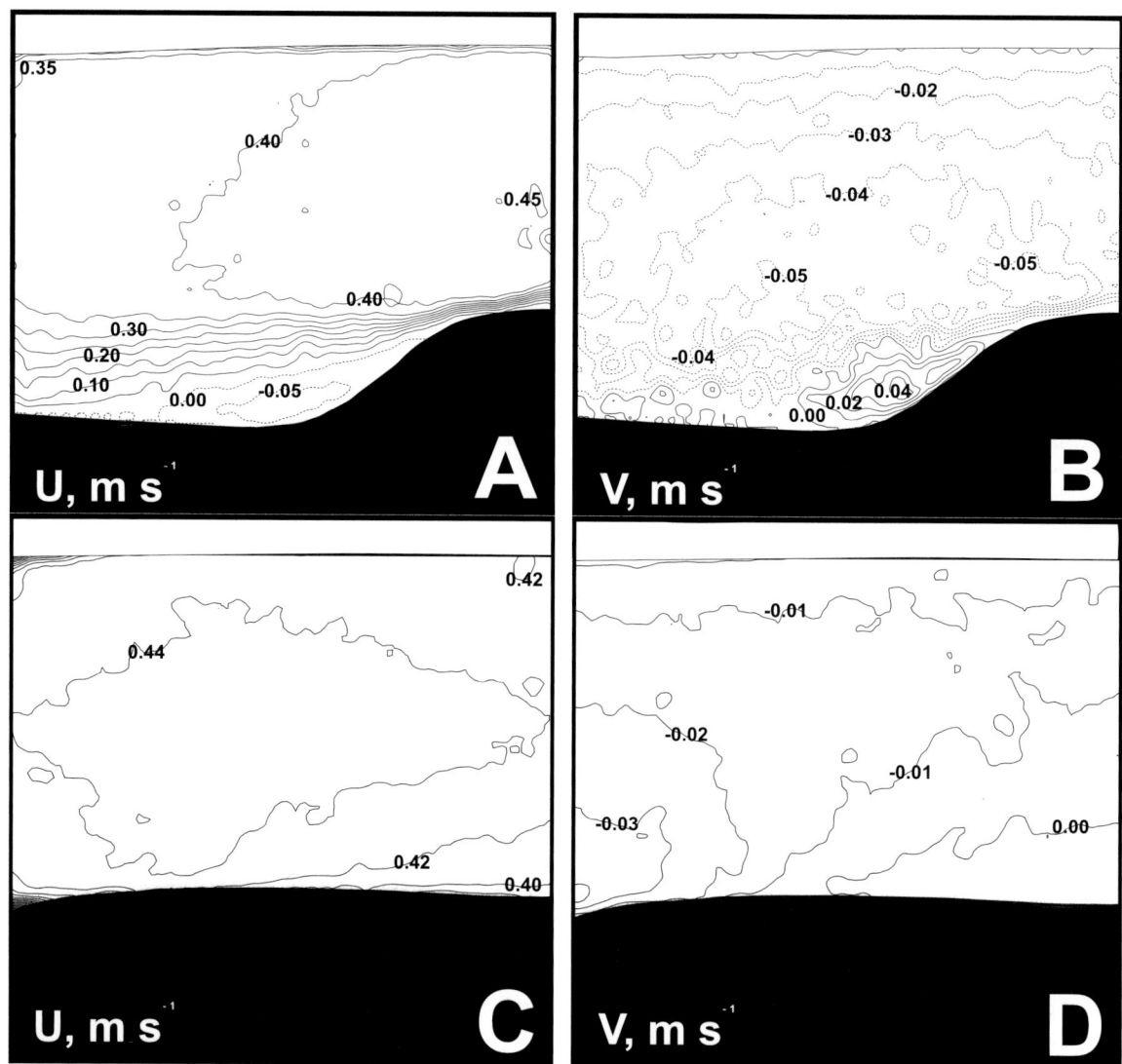

Fig. 3 Mean flow maps for the leeside and crestal regions (see Fig. 2 for location and dimensions of fields of view). (A & B) Mean downstream, U, and vertical, V, components of flow in the dune leeside. (C & D) Mean downstream, U, and vertical, V, components of flow over the dune crest. Negative values in (A) and (B) are indicated by dashed contours, and indicate upstream flow and flow towards the bed for the U and V components respectively.

RESULTS

Laboratory studies of dune macroturbulence

Maps of the mean downstream and vertical velocities in the dune lee (Fig. 3) show the pattern of flow that is normally associated with negative steps and bedforms with steep leesides. Flow deceleration in the leeside (Fig. 3A) is associated with flow separation that produces a recirculating region of fluid that has reverse velocities of up to 0.05 m s^{-1}, with the maximum downstream velocity over the dune crest being $c.\ 0.45$ m s^{-1}. Vertical velocities (Fig. 3B) show positive values (i.e. fluid moving away from the bed) in the lower part of the leeside flow separation zone (maximum vertical velocity of 0.05 m s^{-1}) but flow directed towards the bed (i.e. negative v) over the rest of the flow depth. Flow over the downstream dune crest (Fig. 3C & D) is far more uniform and the mean flow has largely recovered from the immediate effects of the upstream dune wake. The mean downstream velocity is fairly uniform (Fig. 3C), with the vertical velocity showing the effects of topographic acceleration over the upstream stoss side of the dune and the influence of the next downstream leeside in beginning to direct flow towards the bed over the crest, producing more negative vertical velocities over the crestal region (Fig. 3D).

The pattern of mean flow described above corresponds well with past studies over identical dunes (Bennett & Best, 1995) and also other dune morphologies in both laboratory (Nelson et al., 1993; McLean et al., 1994) and field (Roden, 1998; Kostaschuk, 2000; Best et al., 2001). Figure 4 shows a comparison of three downstream velocity profiles with the data of Bennett & Best (1995) for positions (A) in the dune lee, (B) just downstream of reattachment and (C) on the dune crest, these corresponding to profiles 21, 31 and 68 given by Bennett & Best (1995, fig. 3). The profiles confirm the patterns described above and show a good agreement, especially in the dune lee (Fig. 4A), between the two studies. The PIV system is able to resolve flow closer to the bed than the laser Doppler anemometer (LDA) data reported by Bennett & Best (1995), with the present experiment revealing slightly more near-bed retardation of flow near reattachment, but greater near-bed flow acceleration near the crest (Fig. 3). These small differences may be attributed to both the different mean flow conditions between studies (see Table 1) and any slight non-coincidence of the profiles compared. Nevertheless, the close agreement between these profiles and the mean flow fields demonstrates that similar flow conditions have been investigated herein compared with the comprehensive LDA study of Bennett & Best (1995). The instantaneous

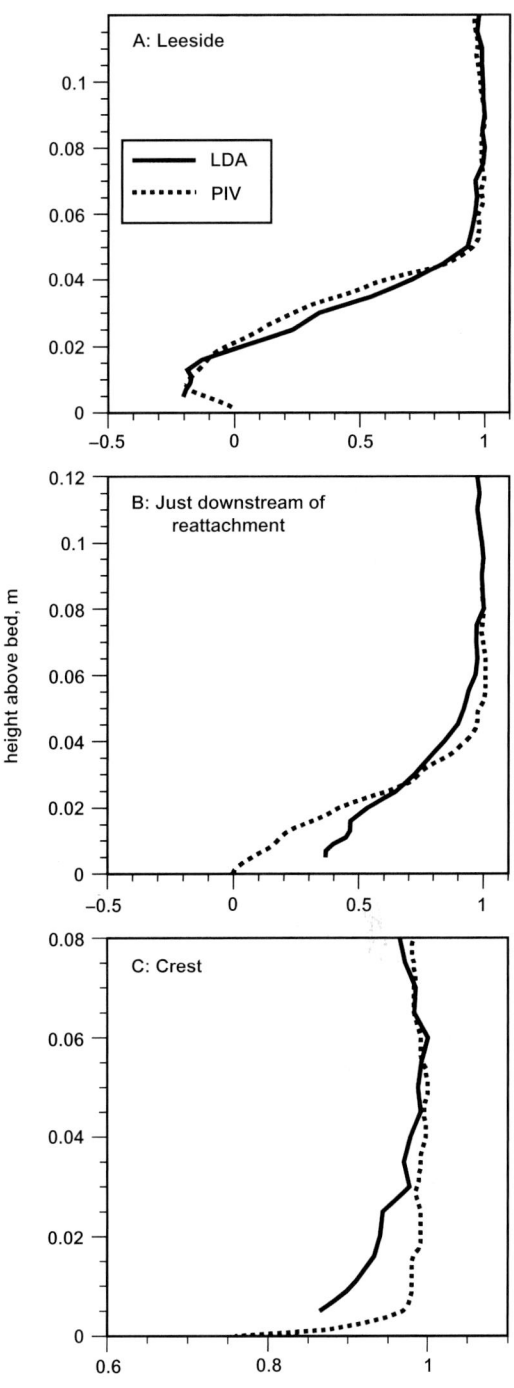

Fig. 4 (*right*) Comparison of downstream velocity profiles at three locations over the dune between this study and the LDA study of Bennett & Best (1995). Profiles are shown for: (A) leeside, (B) just downstream of reattachment and (C) the crestal region, and correspond to profiles 21, 31 and 68 of Bennett & Best (1995). Mean downstream velocity at-a-point, U, is made dimensionless through division by the maximum downstream velocity in the profile, U_{max}.

flow fields quantified in the present study can thus be compared directly to the maps of mean flow and turbulence parameters presented by Bennett & Best (1995).

The broad pattern of flow thus corresponds to expanding, separated flow in the dune lee and this pattern of flow has been shown in many past studies to dominate the production of macroturbulence associated with dunes. These pictures of flow, however, present the *mean* conditions and do not reveal the vital *instantaneous* patterns of flow. Although detailed at-a-point time series can provide abundant information concerning the nature of turbulence (Nelson *et al.*, 1993; McLean *et al.*, 1994; Bennett & Best, 1995; Buffin-Bélanger & Roy, 1998), importantly they cannot provide images of the entire flow field. It is thus often difficult, if not impossible, to relate at-a-point time series to whole flow field motions and the nature of individual turbulent events, their origin and evolution. A unique capability of PIV, however, is the provision of this holistic flow field mapping and the quantitative visualization of evolving coherent structures. In order to display some of the important coherent flow structures associated with dunes, Plates 1 & 2 present images of the motion of coherent flow structures, as shown by regions of high and low velocity flow, in the dune leeside and above the dune crest. Each figure shows a series of whole flow field images at different times over the duration of one coherent flow structure and illustrates maps of downstream velocity, vertical velocity and Reynolds stress. These maps are the first images of their kind ever produced over dune bedforms and show the close interplay between coherent events in the dune boundary layer. A fuller appreciation of these movements, however, is also obtained from animations of the dune flow field, which can be accessed at http://earth.leeds.ac.uk/research/seddies/best/7icfsdunes.htm. The motions associated with each zone are described below.

Dune lee

Fluid motion in the dune lee (Plate 1) is dominated by flow separation that creates a zone of slowly recirculating fluid, which is separated from the freestream flow by a shear layer with high velocity gradient. It is clear, however, that the size of the separation zone varies greatly in time and expands and contracts, this motion depicting the 'flapping' of the shear layer associated with zones of flow separation (e.g. Simpson, 1989; Best & Kostaschuk, 2002). Similar observations have been made by Schmeeckle *et al.* (1999), who noted the difficulty in defining a reattachment point at any one instant. Plate 1 shows one such contraction event where the separation zone, defined by the zero contour, is at its largest at 0 s but then contracts to a minimum at 0.60–0.87 s, before expanding once again. These figures show that the contraction leads to a change in position of the reattachment point (see zero u contour, Plate 1; 0.27–0.60 s), with the zone of downward flow at the bed moving closer to the leeside (e.g. 1.00 s). The contraction of the separation zone appears to coincide with an inrush of fluid towards the upstream crest (see larger patch of bed-directed fluid at crest at 0.27–0.60 s), which subsequently causes contraction of the separation zone, leading to a smaller zone of recirculating flow at 0.87–1.00 s. The nature of this region is thus controlled, in part, by the upstream flow behaviour, which in turn is related to separation zone effects upstream. Hence the nature of the separation zone turbulence and periodicity at one dune is significantly controlled by the temporal turbulent pattern imparted from the upstream bedform.

The modulation of turbulence and destabilization effects of coherent flow structures on separation zone stability have been investigated over negative steps by Mullin *et al.* (1980), and this points to the probable significant interaction between flow fields of dune bedforms. Hence, larger macroturbulence across the ripple–dune transition (Bennett & Best, 1995; Robert & Uhlman, 2001) may lead to the greater interaction between the flow fields of adjacent bedforms and cause greater temporal and spatial variability in the separation zone behaviour. All the maps of Reynolds stress (Plate 1) show that the highest stresses are located within the shear layer, with positive values being associated with the inrush of high velocity fluid towards the bed (quadrant-4 events) (quadrant events are defined by their instantaneous velocities u' and v' (see Best, 1996): quadrant 1, $u' > 0$ and $v' > 0$; quadrant 2, $u' < 0$ and $v' > 0$; quadrant 3, $u' < 0$ and $v' < 0$; quadrant 4, $u' > 0$ and $v' < 0$) and also ejection of lower momentum fluid into the outer flow (quadrant-2 events). Negative stresses are also high in parts of the leeside shear layer and associated with the ejection of high momentum fluid (quadrant-1 events or outward interactions). The highest stresses in the shear layer, however, are at least up to 40 times greater than in the surrounding fluid and would thus dominate the instantaneous transport of sediment. Temporal variations in the position of the shear layer would thus greatly affect the location of sediment erosion and transport. It is also interesting that the maps of flow within the separation zone (Plate 1) reveal a complex pattern, with the presence of several regions of high and low velocity and Reynolds stress. This supports the previous observations of Schmeeckle et al. (1999), who suggested that, at any one given time, the separation zone consists of several large vortices that extend from the shear layer to the bed.

Dune crest

The instantaneous flow fields at the dune crest (Plate 2) are dominated by ejection of low downstream-velocity fluid into the outer flow, which has originated from the upstream dune, and the subsequent reaction of the flow field to this ejection. One such ejection, which occurs over a period of 0.80 s, is shown in Plate 2 and consists of the passage of a zone of low downstream velocity fluid away from the bed (0.13–0.40 s); the region of positive v can be seen to extend up to near the flow surface (0.40 s) with the low u-velocity area becoming more detached from the bed at 0.40 s. This ejection of fluid away from the bed is also associated with significant changes in the temporal signal of the vertical velocity. A distinct area of $-v$ is associated with flow in front of, and behind, the ejection (e.g. 0.27–0.53 s) and reflects the likely three-dimensional topology of flow associated with the ejection. A plot of the ***uv*** vectors at 0.27 s (Fig. 5A) shows the strong bed-directed flow downstream of the region of $+v$, whereas a vector plot of flow at 0.40 s (Fig. 5B) illustrates the strong bed-directed flow upstream of the ejection. This pattern can be explained by considering the reaction of the outer flow to the passage of an ejection. As low-momentum fluid is ejected away from the bed, fluid from higher in the flow is induced to move towards the bed in order

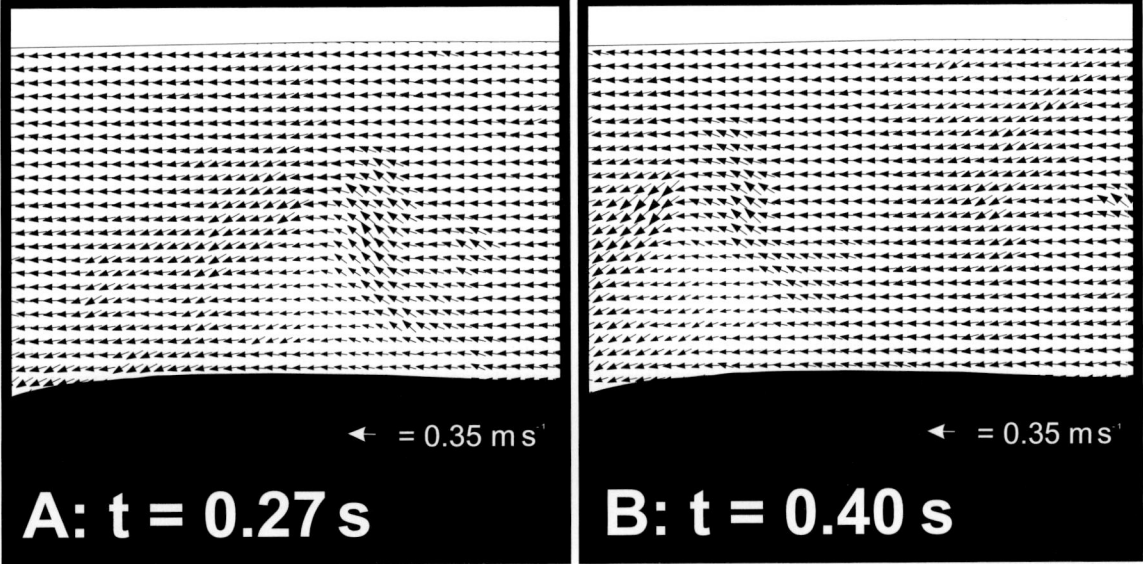

Fig. 5 Plots of ***uv*** vectors for the crestal area at time frames 0.27 and 0.40 s (see Plate 2). Note that the vertical velocities in these vector plots have been multiplied by a factor of five to enable these motions to be visualized. These vector maps illustrate the upward flow associated with the ejection and the flow towards the bed both in front of, and behind, this ejection event.

to satisfy continuity and accounts for the inrush of fluid behind the ejection. These plots provide graphic confirmation of the hypothesis of Bennett & Best (1996) that proposed a link between ejection events and subsequent inrushes sourced from higher in the flow. Additionally, the inrush of flow towards the bed in front (i.e. downstream) of the ejection (Plate 2 & Fig. 5; 0.27–0.53 s) may be explained by considering the likely three-dimensional shape of the ejection (see field observations below and also Schmeeckle et al., 1999, p. 267) that would also cause fluid to be entrained from around the *sides* of the ejection and create a return flow towards the bed as the ejection ascended in the flow. The maps of Reynolds stress (τ_R: Plate 2) also clearly illustrate the dominance of the ejection event through the flow field and that the bed-directed flow in front of the ejection is associated with higher Reynolds stresses (i.e. see higher near-bed τ_R near dune crest at 0.40–0.53 s). The ejection depicted here (Plate 2; 0.27–0.40 s) occupies at least half the flow depth, in good agreement with the observations of Schmeeckle et al. (1999).

In order to illustrate further the temporal nature of flow at several points in this flow field, the at-a-point time series in u, v and τ_R for four points are given in Fig. 6. These points are located (see inset Fig. 6) in mid-flow (1 and 2), near the surface (point 3) and near the bed in close proximity to the dune crest (point 4), and reflect the passage of the ejection shown in Plate 2. Point 1 shows the passage of the ejection that is shown by a marked decrease in u, a positive v and large τ_R at 0.27 s (Fig. 6; arrowed 'a'). Point 2, located 41 mm downstream of point 1, shows the same response as the ejection advects past this region at 0.33 s (Fig. 6; arrowed 'b'), but additionally shows the strong downflow of fluid towards the bed in front of the advancing ejection at 0.27 s (Fig. 6; arrowed 'c'). Flow near the surface (point 3) also shows the presence of the ejection at a slightly later time (c. 0.40 s; Fig. 6; arrowed 'd'), although it is noticeable here that the decrease in u is less, the peak in τ_R is much smaller in magnitude than lower in the flow and also there is a lag in the peaks in v and τ_R (see label 'd' arrows, Fig. 6). This indicates the significant mixing of the fluid as it rises into this region of higher mean velocity flow and that the peaks in τ_R may be associated with a structure that is larger here than that present lower in the flow. Point 4, located near the bed, shows a peak in τ_R at c. 0.47 s (Fig. 6; arrowed 'e') that is associated with the inrush of flow towards the bed in front (downstream) of the propagating ejection (see Plate 2). Schmeeckle et al. (1999) also reported that large-scale vortices frequently break up into smaller scale vortical motions in the mid-stoss region, with only some of the largest vortices advecting through the entire flow depth. Schmeeckle et al. (1999) and Shimizu et al. (1999) found that this break-up occurs in the region of maximum topographically induced flow acceleration, in a similar position to that documented here. This break-up process, however, will clearly be aided by the inrush events towards the bed documented in the present experiments that are associated with large-scale ejections.

Field observations of water surface structure

Observations, video recordings and series of still photographs of the river surface above dune-covered beds in the Jamuna River allow a summary of the common series of events as large-scale dune-related macroturbulence erupts at the flow surface. These observations complement and extend those previously made by Coleman (1969) in the Jamuna, Jackson (1976) in the Wabash River and Babakaiff & Hickin (1996) in the Squamish River. Photographs from several upwellings on the surface of the Jamuna are shown in Fig. 7, and a summary schematic of the common fluid motions is illustrated in Fig. 8.

The initial eruption of the ejection on the water surface (Fig. 7A; $t = c.$ 0–3 s, Fig. 8) often appears as a patch of fluid that is upwelling at its rear edge and then downwelling at its downstream front edge. The movement of this fluid thus has a clear axis of rotation in the downstream-spanwise (x–z) plane and mirrors the observations of Babakaiff & Hickin (1996), who termed this a 'roller structure' (see Fig. 9A). Continued upwelling of the ejection produces an expanding area of upwelling fluid that begins to show a more radial pattern of flow, with the major upwellings being towards the centre of the ejection (Fig. 7B; $t = c.$ 3–5 s, Fig. 8), but clear downwelling at the front edge. Shear at the edges of this upwelling may begin to generate some rotation in the vertical-spanwise (y–z)

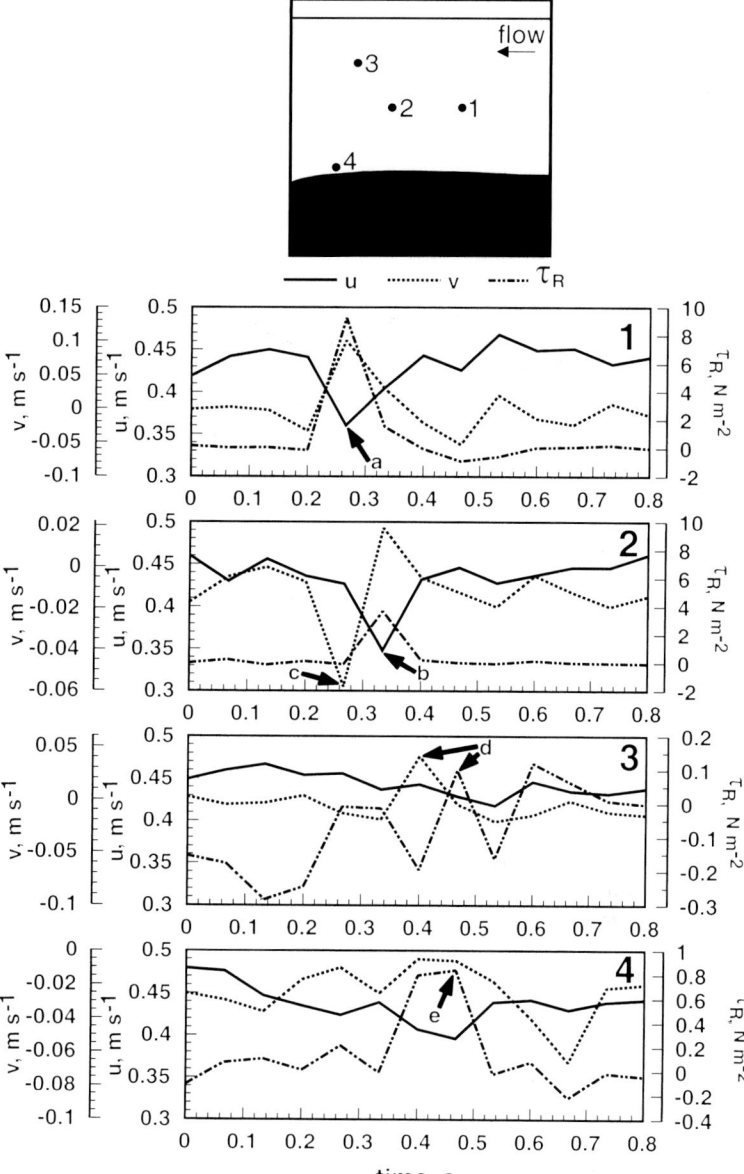

Fig. 6 Time series of instantaneous downstream velocity (u), vertical velocity (v) and Reynolds Stress (τ_R) for four positions above the dune crest (see inset diagram) for the time periods illustrated on the flow field maps in Plate 2. The arrows on each graph and labels a–e refer to the ejection–inrush event at each point in the flow discussed within the text. Note changing ordinate scales for v and τ_R between graphs.

plane ($t = c.$ 5 s; Fig. 8). In the photographs of the Jamuna shown here, an opposing slight wind encouraged development of distinct waves, up to $c.$ 0.3 m high, at the downstream edge of the upwelling (Fig. 7C & D; $t = c.$ 1–5 s, Fig. 8). At this stage, a vertical vorticity also begins to be well-developed both at the edges of the upwelling, owing to shear with the surrounding flow (Fig. 7E; $t = c.$ 5–10 s, Fig. 8), and also at the trailing edge of the upwelling where distinct, separate vortices with a vertical axis of rotation start to form ($t = c.$ 10 s; Fig. 8). The rotation direction of these vertical vortices on each side of the upwelling often tends to be in towards the upwelling ($t = c.$ 10 s; Fig. 8), this again mirroring the observation of Babakaiff & Hickin (1996; Fig. 9A) who observed 'horns' associated with surface macroturbulence that also rotated in towards the upwelling ejection event.

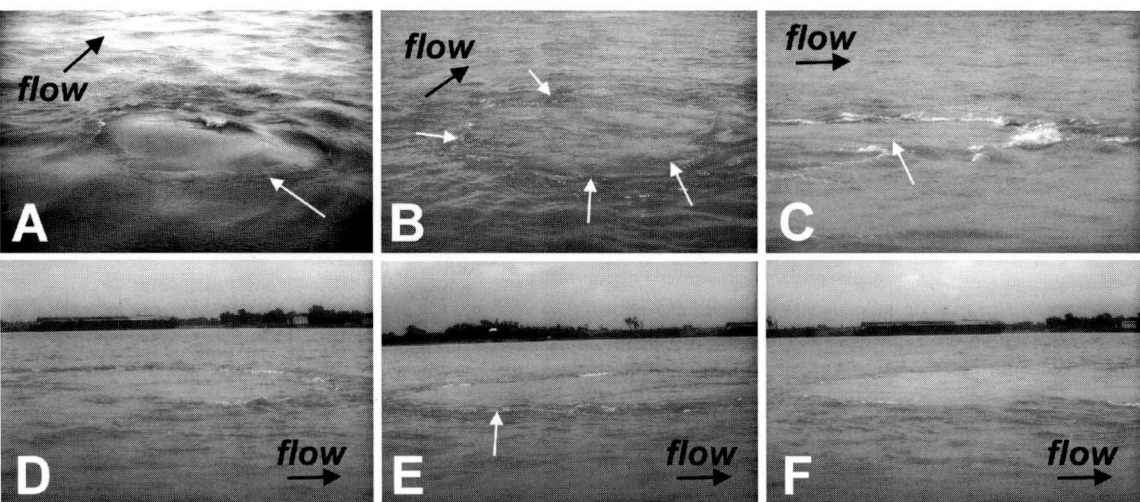

Fig. 7 Photographs of vortex–free-surface interactions in the Jamuna River, Bangladesh. (A–D) are images of separate events whereas (E) and (F) show the same boil. (A) Initial upwelling showing water surface waves on downflow side of boil. Width of image $c.$ 4 m. (B) Complex upwelling with multiple points of upwelling (arrowed). Width of image $c.$ 8 m. (C) Violent upwelling (arrowed) and development of strong water surface waves at downflow side of boil. Width of image $c.$ 3 m. (D) Water surface waves around large boil that is $c.$ 30 m wide. (E & F) Two images taken approximately 5 s apart showing final stage of development of a large-scale eruption which at its maximum reaches $c.$ 30–35 m across. (E) Large boil on the water surface with developing vertical vortices (arrowed) and diminishing water surface waves. (F) Boil eruption wanes further and the event begins to lose its distinct water surface expression.

Continued development of the ejection (Fig. 7F; $t = c.$ 15 s, Fig. 8) sees further radial expansion of the patch of upwelling fluid, although the intensity of upwelling decreases and shear at its edges produces a series of smaller vortices with a vertical axis of rotation. The larger vertical vortices generated to the back and rear of the upwelling also at first increase, but then decrease, in their size and rotational velocity as the upwelling continues ($t = c.$ 15 s; Fig. 8). Dissipation of the upwelling causes these features to become more indistinct until the eruption is finished. The width on the water surface of these upwellings associated with dune-related macroturbulence ranged from 5 to approximately 50 m in the Jamuna River.

DISCUSSION: THE TOPOLOGY OF DUNE-RELATED MACROTURBULENCE

These results of PIV quantification of the whole flow field dynamics associated with sand dunes and synthesis of observations of macroturbulence– water-surface interactions above natural dune-covered sand beds—reveal four key features.

1 The shear layer and separation zone associated with dunes are spatially and temporally dynamic, and 'flapping' of the shear layer may be caused and modulated by turbulent coherent flow structures generated upstream.

2 Reynolds stresses in the leeside are dominated by the free shear layer associated with the separation zone in dunes with angle-of-repose leesides.

3 Ejections of low downstream-momentum fluid away from the bed dominate the instantaneous flow field over the crestal regions of the dune. These ejections create return flows towards the bed both in front of, and behind, the ejection. The highest instantaneous Reynolds stresses are associated with these ejections and inrushes (quadrant 2 and 4 events respectively).

4 Although the water surface over a dune field does show a great variability in the patterns of upwellings or 'boils' associated with dunes, a common pattern can be discerned that consists of a central upwelling that has a spanwise axis

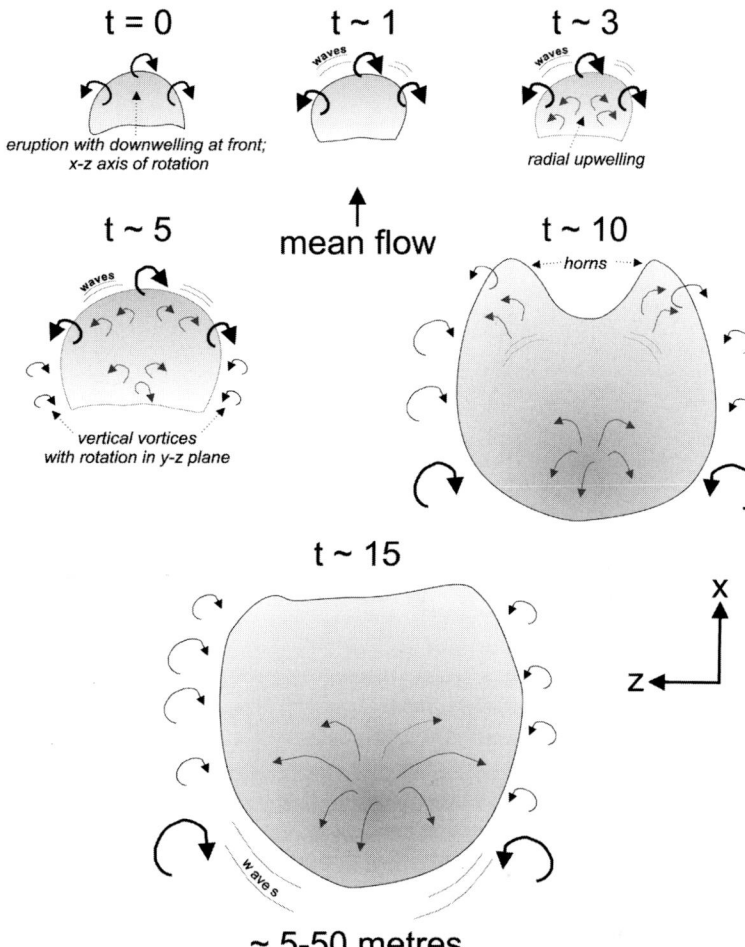

Fig. 8 Sketches of the patterns of flow associated with eruption of ejections onto the water surface, Jamuna River, Bangladesh. These sketches represent a summary of still and video photography over a number of large sand dunes and depict the most commonly observed sequence of events: t, time from start of boil eruption (s). These surface boils can reach up to 50 m in diameter over the largest sand dunes. x, y and z refer to downstream, vertical and transverse axes respectively.

of rotation and is later associated with secondary vortices that possess a vertical axis of rotation.

Previous work examining the coherent flow structures associated with dunes (Müller & Gyr, 1986; Nezu & Nakagawa, 1993; Kadota & Nezu, 1999) has suggested that in three dimensions these structures have a loop or horseshoe morphology (Fig. 9B & C), although these studies show that such vorticity can arise either along the shear layer or from the reattachment region (Fig. 9B & C), a feature also observed in the present experiments. Acarlar & Smith (1987) also demonstrated a similar morphology for vortices generated at low flow Reynolds number downstream of hemispherical obstacles. Such three-dimensional vortices may be more easily attained over bedforms that have a sinuous crestline that encourage production of vortices with a limited spanwise extent. Experimental work has shown how leeside flow associated with cylinders (Délery, 2001) and rectangular roughness (Fig. 9D; Martinuzzi & Tropea, 1993) may also lead to production of a flow structure associated with the separation zone that has a distinct 'horseshoe' morphology and thus could be expected to shed structures with a 'loop' shape.

Additionally, work on the interactions between coherent flow structures with a free surface (e.g. Rashidi & Banerjee, 1988; Rood, 1995; Sarpkaya, 1996; Kumar et al., 1998) has included focus on the interaction of a vortex loop with a free surface (see Sarpkaya, 1996). This work has shown the initial contact of the loop head with the free

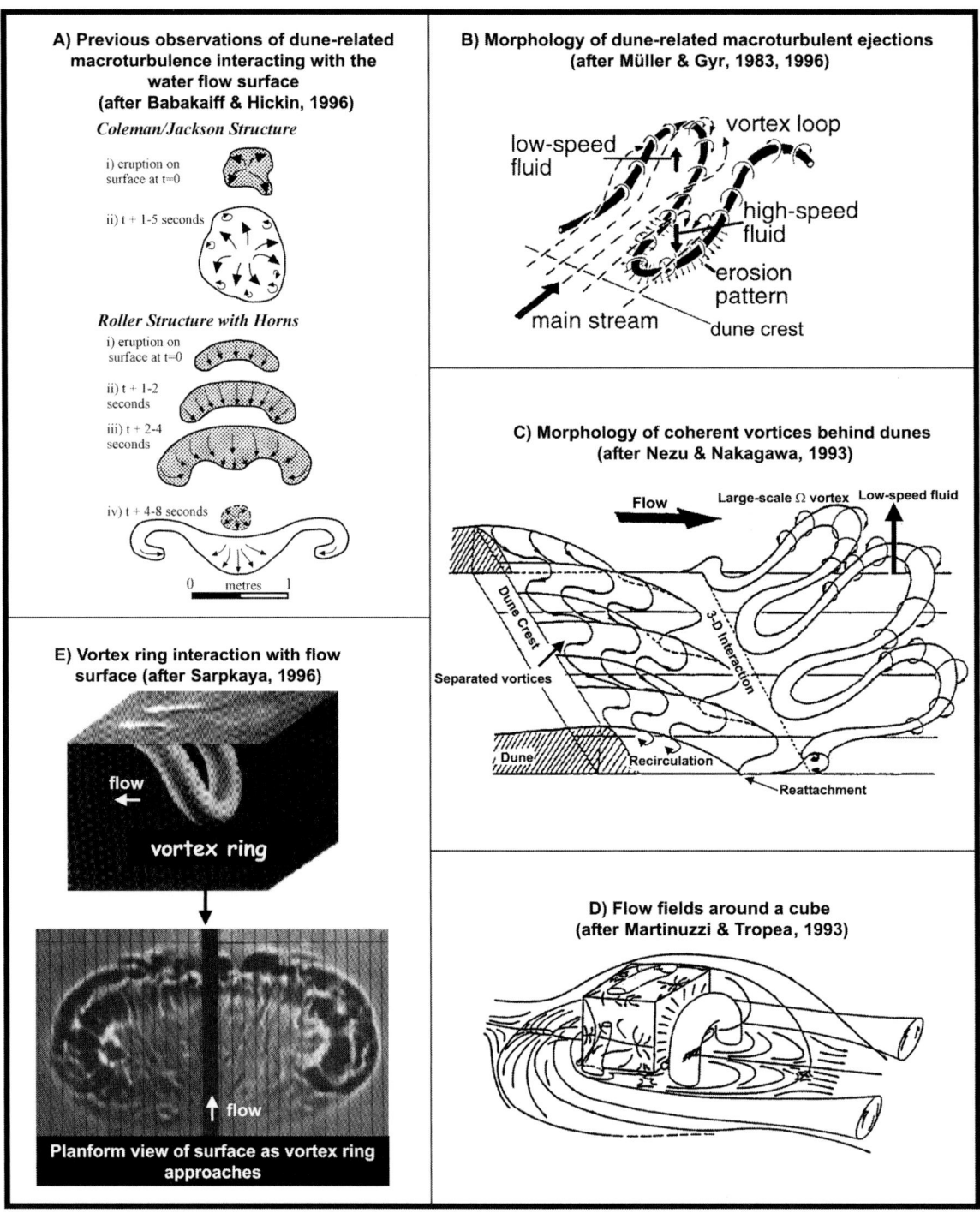

Fig. 9 Vorticity associated with dunes and surface obstructions, and vortex interactions with a free surface. (A) Evolution of boils on the water surface (from Babakaiff & Hickin (1996) and after Coleman (1969) and Jackson (1976)). (B) Three-dimensional flow around a boil and the structure of dune-related macroturbulence proposed by Müller & Gyr (1983, 1986, 1996). (C) The morphology of coherent vortices behind dunes (after Nezu & Nakagawa, 1993). (D) Vortex systems associated with flow around a surface-mounted cube (after Martinuzzi & Tropea, 1993). (E) Interactions of a vortex ring with the free surface (after Sarpkaya, 1996): vortex legs interacting with the surface (upper) and plan view of loop vortex approaching surface (lower).

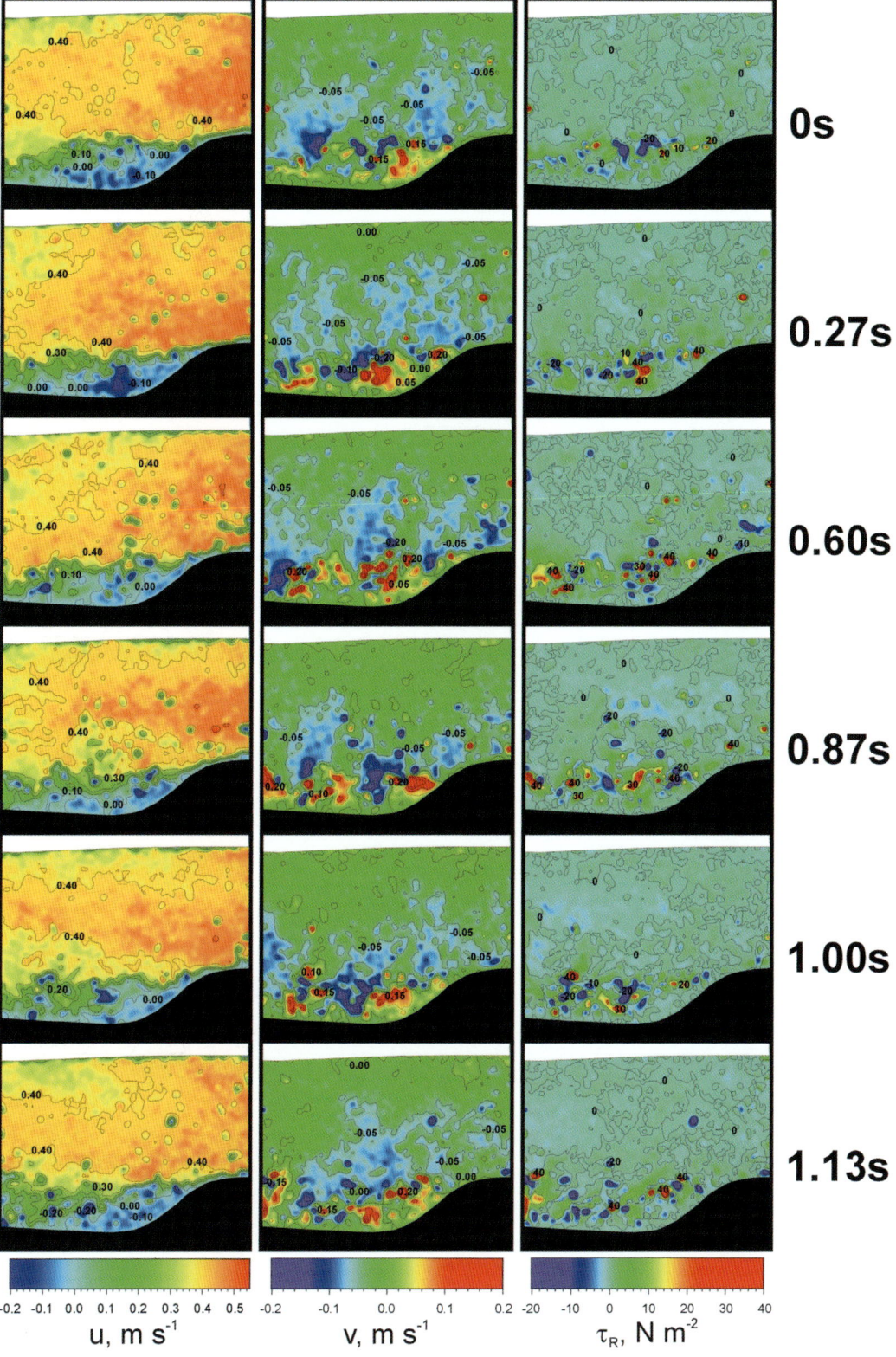

Plate 1 Whole flow field maps in the dune leeside of instantaneous downstream velocity (u; left), vertical velocity (v; middle) and Reynolds stress (τ_R; right) over a period of 1.13 s. Images were obtained at 15 Hz and the field of view for each image is 0.204 m long by 0.205 m deep. The maps were produced using a kriging method of interpolation onto a 1.25 mm² grid with a spherical distribution and 50 samples per kriging pass.

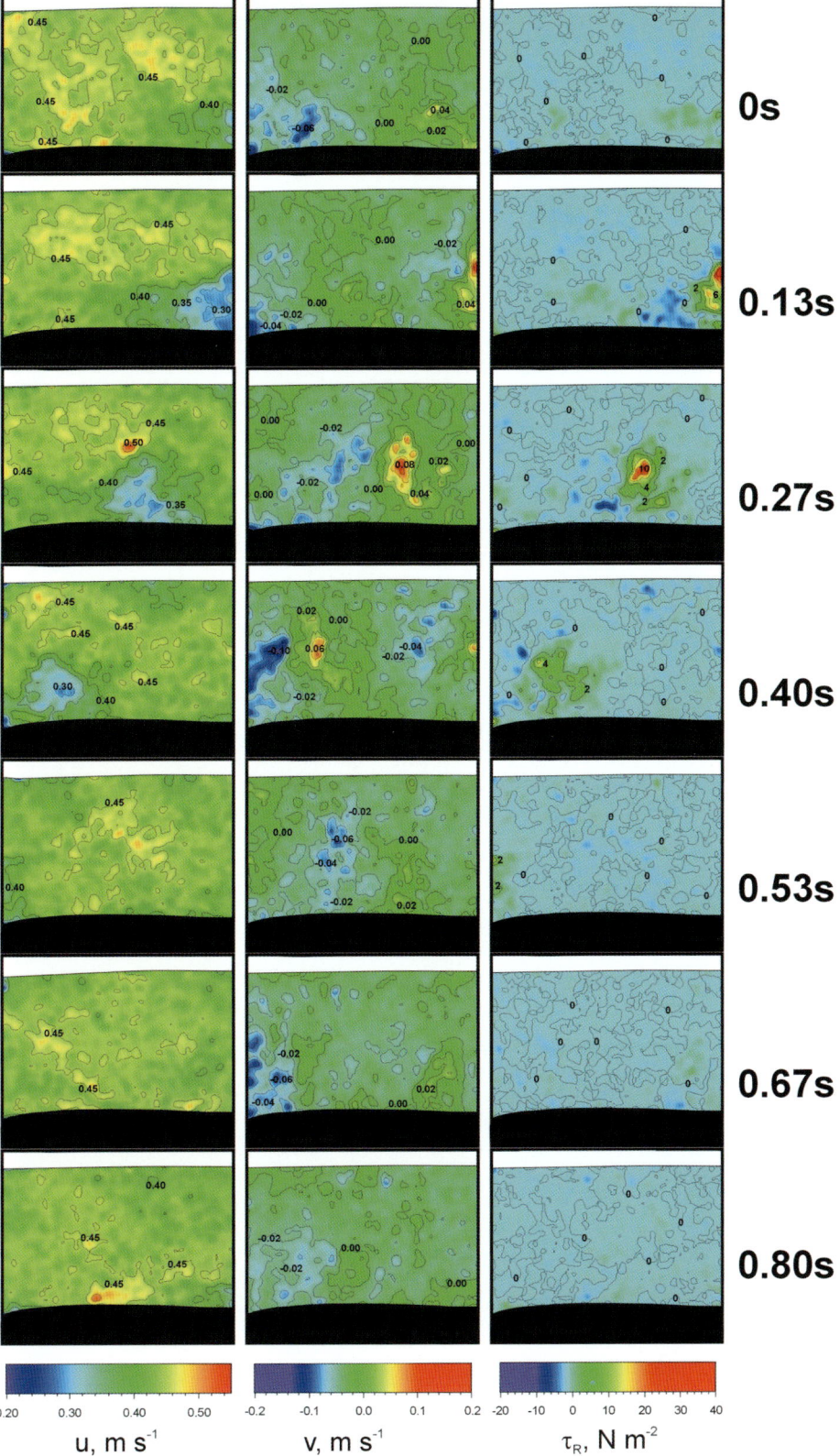

Plate 2 Whole flow field maps over the dune crest of instantaneous downstream velocity (u; left), vertical velocity (v; middle) and Reynolds stress (τ_R; right) over a period of 0.80 s. Images were obtained at 15 Hz and the field of view for each image is 0.148 m wide by 0.149 m deep. The maps were produced using a kriging method of interpolation onto a 1.25 mm² grid with a spherical distribution and 50 samples per kriging pass.

surface (Fig. 9E) and that as the loop continues to rise then the legs of the vortex loop become connected to the free surface (Fig. 9E). This pattern of large-scale connection of the vortex loop head with an x–z axis of rotation, which is followed by emergence of vertical vortices with a y–z axis of rotation, bears striking similarity to the observations made on the surface of rivers with dune-covered beds. These observations suggest that the common topology of macroturbulence associated with dunes with distinct flow separation zones may be a vortex loop that, when interacting with the water surface, provides an interaction similar to that of classical vortex loops. This similarity of interaction allows proposition of a conceptual model to explain the patterns of upwelling observed on the water surface above dune beds (Fig. 10).

Once the vortex has been generated and shed into the outer flow, possibly assisted by flapping of the shear layer, which will aid expulsion of the ejection, the vortex approaches the surface (Fig. 10A). The first interaction of the ejection with the surface occurs as the head of the vortex loop encounters the surface, this producing an upwelling on the flow surface on the upflow side of the vortex and a downwelling on its downflow margin (Fig. 10B). As the eruption continues (Fig. 10C) the legs of the loop vortex interact with, and become connected to, the flow surface (see Sarpkaya, 1996; Kumar et al., 1998), producing vortices with a vertical axis of rotation. Continued eruption at the centre of the vortex produces an upwelling in its centre and shear with the surrounding flow also produces secondary vortices,

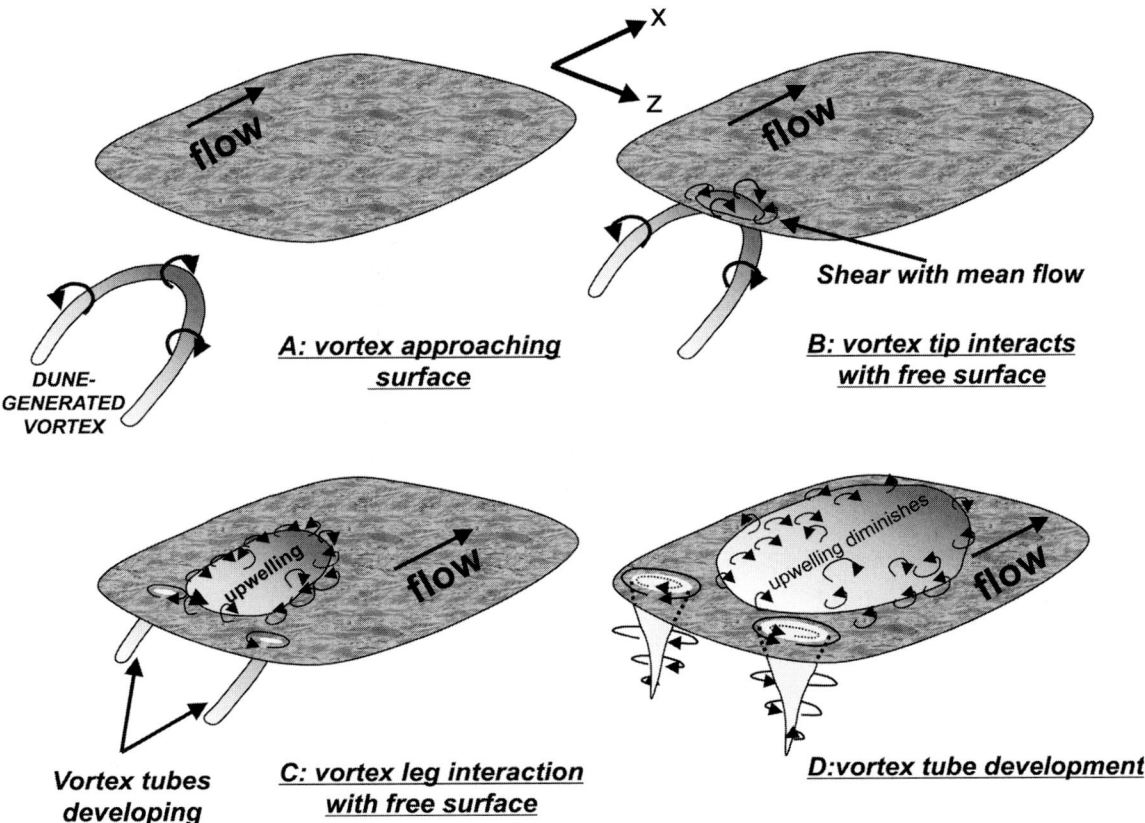

Fig. 10 A schematic model of flow to illustrate the interaction of dune-related macroturbulence with the water surface. This three-dimensional sketch illustrates the various stages of interaction of the boil with the flow surface: however, this interaction will also be associated with the inrush of fluid back towards the bed as quantified in the laboratory PIV experiments.

with vertical axes, at the edges of the main boil (Fig. 10C & D). As the upwelling continues, the velocity of the ejection decreases and the boil begins to diminish in intensity, with the vertical vortices of the legs of the vortex loop also enlarging in size but decreasing in rotational velocity (Fig. 10D).

This general pattern of flow can be observed on the water surface over most dune-covered beds, and extends previous visual observations of boil morphology (e.g. Jackson, 1976; Babakaiff & Hickin, 1996). These macroturbulent events are well-known to be a dominant mechanism for the suspension of sediment over dune beds (Jackson, 1976; Lapointe, 1992; Shimizu et al., 1999; Schmeeckle et al., 1999; Vendetti & Bennett, 2000) and in the present study were visually associated with higher quantities of suspended sediment that can quantitatively be linked to higher sediment concentrations in the Jamuna (Roden, 1998). Another significant corollary of the macroturbulent ejections demonstrated here, however, is their significant influence on inrushes of fluid towards the bed that are required to satisfy continuity (Plate 2). This mechanism of surface interaction has also been proposed in other studies to possibly induce subsequent inrushes (Rashidi & Banerjee, 1988), and these may dominate the instantaneous Reynolds stresses and bed sediment transport. The role of the ejection–inrush sequence may thus be vital in increasing sediment transport rates over the dune and in causing the ripple–dune transition (Bennett & Best, 1996). Additionally, it is also clear that the spatial variability in flow is greater over a dune bed (Robert & Uhlman, 2001) and this may aid the ripple–dune transition.

The present results also lend support to past work that has highlighted the importance of quadrant-4 events in dominating bedload sediment transport (Nelson et al., 1995). This ejection–inrush process, and the series of events outlined here, may also be expected to be important over other types of large-scale bed roughness in alluvial channels. For instance, in gravel-bed rivers this process may provide a mechanism to explain the high and low speed 'wedges' of flow observed by Buffin-Bélanger & Roy (1998), Buffin-Bélanger et al. (2000) and Roy & Buffin-Bélanger (2001). These authors have described observations of eddy shedding associated with pebble clusters and proposed that the flow is dominated by a series of low and high speed 'wedges', of low frequency (c. 0.15 Hz), with the high-speed wedges being dominant in producing large bed shear stresses, which were reasoned to be the key element in sediment transport (Buffin-Bélanger et al., 2000). However, these representations of the flow field were based on only three simultaneous measurements that each had a sampling volume of approximately 0.1 Y, which makes interpretation of the precise flow structure and the behaviour of *individual* turbulent events problematic. The present study has yielded a holistic picture of flow that provides a mechanism to explain these previous observations. Wedges of low-speed flow, produced by ejection/flapping of the shear layer, are followed by the inrush of high-speed wedges from the outer flow, which, because they move towards the bed from the outer flow, arrive later at the bed than higher in the boundary layer.

The visualizations shown here (Plate 2) may be used to provide a mechanism to explain the pattern of u and v signals presented by Ferguson et al. (1996) and flow patterns conjectured by Yalin (1992). Buffin-Bélanger et al. (2000) and Roy & Buffin-Bélanger (2001) highlight, however, that there is often no discernible pattern to the quadrant structure in their time series. The present study suggests how this may reflect the complex three-dimensional topology of the coherent flow structures that will produce varying quadrant signatures depending on how they translate through any measurement volume. The response of the flow to ejection and shear layer flapping will involve a three-dimensional flow around the ejection that will induce inrushes both in front of, and behind, the advecting ejection (Plate 2). Future PIV work will focus on the two- and three-dimensional flow field of dunes and grain roughness to fully quantify the characteristics of this macroturbulence, which may often dominate the entire boundary layer in depth-limited flows.

ACKNOWLEDGEMENTS

The laboratory work and ideas expressed in this paper have been made possible through research enabled and funded by the UK Natural Environment Research Council for which I am extremely grateful, including a JREI grant to establish the PIV facility (GR3/JE140), grants to develop and apply phase and acoustic Doppler anemometry (GR3/8235; GR3/10015) and a fieldwork grant to JLB and Phil Ashworth for research on the Jamuna River (GR9/2034). The support and funding of Flood Action Plan 24 in Bangladesh was also central to research on the Jamuna River and I am indebted to all involved in the organization and execution of this huge river survey project. A Royal Society travel support grant enabled presentation of this paper at the Seventh International Conference on Fluvial Sedimentology. I would like to thank Mark Franklin for his support in the laboratory and his central part in the PIV experiments. Additionally, I am extremely grateful for the advice and expertise of Robert Jaryczewski and Graham Hassall of DANTEC, UK, in setting up the Leeds PIV system. I would also like to acknowledge many discussions on different aspects of macroturbulence in alluvial channels over a number of years with Ray Kostaschuk, Sean Bennett, Phil Ashworth, John Bridge, Paul Villard, Julie Roden, Mark Lawless, Stuart Lane and Alistair Kirkbride, which have helped greatly to shape my views. I am very grateful to reviewers André Roy and Maarten Kleinhans for their detailed and thought-provoking reviews that helped considerably to sharpen the paper. Finally, I am thankful to my colleagues in the field for their patience in continuing to put up with me taking yet more photographs of river surfaces and boils!

NOMENCLATURE

Fr Froude number ($=\bar{U}_{crest}/\sqrt{g Y_{crest}^{0.5}}$)
g gravitational acceleration (m s^{-1} s^{-1})
h dune height (m)
t time (s)
u instantaneous downstream velocity at-a-point (m s^{-1})
u' deviation of instantaneous downstream velocity from mean ($= u - U$; m s^{-1})
U mean downstream velocity at-a-point (m s^{-1})
$\bar{U}_{crest}$ mean downstream velocity at dune crest (m s^{-1})
U_{max} maximum downstream velocity in a vertical profile (m s^{-1})
v instantaneous vertical velocity at-a-point (m s^{-1})
v' deviation of instantaneous vertical velocity from mean ($= v - V$; m s^{-1})
V mean vertical velocity at-a-point (m s^{-1})
Y flow depth (m)
Y_{crest} flow depth at dune crest (m)
λ dune wavelength (m)
ρ fluid density (kg m^{-3})
τ_R Reynolds stress ($\tau_R = -\rho \overline{u'v'}$; N m^{-2})

x, y and z refer to the downstream, vertical and transverse axes respectively.

REFERENCES

Acarlar, M.S. and Smith, C.R. (1987) A study of hairpin vortices in a laminar boundary layer. Part 1. Hairpin vortices generated by a hemispherical protuberance. *J. Fluid. Mech.*, **175**, 1–41.

Adrian, R.J. (1991) Particle-imaging techniques for experimental fluid dynamics. *Ann. Rev. Fluid Mech.*, **23**, 261–304.

Adrian, R.J. (1996) Laser velocimetry. In: *Fluid Mechanics Measurements* (Ed. R.J. Goldstein), pp. 175–299, Taylor and Francis, London.

Ashworth, P.J., Best, J.L., Roden, J., Bristow, C.S. and Klaassen, G.J. (2000) Morphological evolution and dynamics of a large, sand braid-bar, Jamuna River, Bangladesh. *Sedimentology*, **47**, 533–555.

Babakaiff, C.S. and Hickin, E.J. (1996) Coherent flow structures in Squamish River Estuary, British Columbia, Canada. In: *Coherent Flow Structures in Open Channels* (Eds P.J. Ashworth, S.J. Bennett, J.L. Best and S.J. McLelland), pp. 321–342. Wiley, Chichester.

Bennett, S.J. and Best, J.L. (1995) Mean flow and turbulence structure over fixed, two-dimensional dunes: implications for sediment transport and dune stability. *Sedimentology*, **42**, 491–513.

Bennett, S.J. and Best, J.L. (1996) Mean flow and turbulence structure over fixed ripples and the ripple-dune transition. In: *Coherent Flow Structures in Open Channels* (Eds P.J. Ashworth, S.J. Bennett, J.L. Best and S.J. McLelland), pp. 281–304. Wiley, Chichester.

Bennett, S.J. and Vendetti, J. (1997) Turbulent flow and suspended sediment transport over fixed dunes. In: *Proceedings, Conference on Management of Landscapes Disturbed by Channel Incision* (Eds S.S.Y. Wang, E.J. Langendoen and F.D. Shields, Jr), pp. 949–954. Center for Computational Hydroscience and Engineering, University of Mississippi, Oxford, USA.

Best, J.L. (1993) On the interactions between turbulent flow structure, sediment transport and bedform development: some considerations from recent experimental research. In: *Turbulence: Perspectives on Flow and*

Sediment Transport (Eds: N.J. Clifford, J.R. French and J. Hardisty), pp. 61–92. Wiley, Chichester.

Best, J.L. (1996) The fluid dynamics of small-scale alluvial bedforms. In: *Advances in Fluvial Dynamics and Stratigraphy* (Eds P.A. Carling and M.R. Dawson), pp. 67–125. Wiley, Chichester.

Best, J.L. and Kostaschuk, R.A. (2002) An experimental study of turbulent flow over a low-angle dune. *J. Geophys. Res.*, **107**(C9), 3135–3154.

Best, J.L., Kostaschuk, R.A. and Villard, P.V. (2001) Quantitative visualization of flow fields associated with alluvial sand dunes: results from the laboratory and field using ultrasonic and acoustic Doppler anemometry. *J. Vizualization*, **4**, 373–381.

Buffin-Bélanger, T. and Roy, A.G. (1998) Effects of a pebble cluster on the turbulent structure of a depth-limited flow in a gravel-bed river. *Geomorphology*, **25**, 249–267.

Buffin-Bélanger, T., Roy, A.G. and Kirkbride, A.D. (2000) On large-scale flow structures in a gravel-bed river. *Geomorphology*, **32**, 417–435.

Carling, P.A. (1999) Subaqueous gravel dunes. *J. Sediment. Res.*, **69**, 534–545.

Carling, P.A., Gölz, E., Orr, H.G. and Radecki-Pawlik, A. (2000a) The morphodynamics of fluvial sand dunes in the River Rhine near Mainz, Germany, Part I: sedimentology and morphology. *Sedimentology*, **47**, 227–252.

Carling, P.A., Williams, J.J., Gölz, E. and Kelsey, A.D. (2000b) The morphodynamics of fluvial sand dunes in the River Rhine near Mainz, Germany, Part II: hydrodynamics and sediment transport. *Sedimentology*, **47**, 253–278.

Coleman, J.M. (1969) Brahmaputra River: channel processes and sedimentation. *Sediment. Geol.*, **3**, 129–239.

Dalrymple, R.W. and Rhodes, R.N. (1995) Estuarine dunes and bars. In: *Geomorphology and Sedimentology of Estuaries* (Ed. G.M.E. Perillo), pp. 359–422. Elsevier, Amsterdam.

DANTEC (2000) *FlowMap Particle Image Velocimetry Instrumentation, Installation and Users Guide.* DANTEC, Lyngby, Denmark.

Délery, J.M. (2001) Robert Legendre and Henri Werlé: toward the elucidation of three-dimensional separation. *Annu. Rev. Fluid Mech.*, **33**, 129–154.

Dinehart, R.L. (1989) Dune migration in a steep, coarse-bedded stream. *Water Resour. Res.*, **25**, 911–923.

Engel, P. and Lam Lau, Y. (1980) Computation of bed load using bathymetric data, *J. Hydraul. Div., ASCE*, **106**, 369–380.

Engel, P. and Lam Lau, Y. (1981) Bed load discharge coefficient, *J. Hydraul. Div., ASCE*, **106**, 1445–1454.

Ferguson, R.I., Kirkbride, A.D. and Roy, A.G. (1996) Markov analysis of velocity fluctuations in gravel-bed rivers. In: *Coherent Flow Structures in Open Channels* (Eds P.J. Ashworth, S.J. Bennett, J.L. Best and S.J. McLelland), pp. 165–183. Wiley, Chichester.

Gabel, S.L. (1993) Geometry and kinematics of dunes during steady and unsteady flows in the Calamus River, Nebraska, USA. *Sedimentology*, **40**, 237–269.

Itakura, T. and Kishi, T. (1980) Open channel flow with suspended sediment on sand waves. In: *Proceedings of the Third International Symposium on Stochastic Hydraulics* (Eds H. Kikkawa and Y. Iwasa), pp. 599–609. Tokyo

Jackson, R.G. (1976) Sedimentological and fluid-dynamic implications of the turbulence bursting phenomenon in geophysical flows, *J. Fluid Mech.*, **77**, 531–560.

Julien, P.Y. and Klaassen, G.J. (1995) Sand–dune geometry of large rivers during flood. *J. Hydraul. Eng.*, **121**, 657–663.

Kadota, A. and Nezu, I. (1999) Three-dimensional structure of space–time correlation on coherent vortices generated behind dune crest, *J. Hydraul. Res.*, **37**, 59–80.

Klaassen, G.J. (1979) *Sediment Transport and Hydraulic Roughness in Relation to Bed Forms.* Publication 213, Delft Hydraulics, Delft, 33 pp.

Kleinhans, M.G. (2002) Sorting out sand and gravel: sediment transport and deposition in sand–gravel bed rivers. PhD Thesis, Royal Dutch Geographical Society/Faculty of Geographical Sciences, Utrecht University. *Neth. Geogr. Stud.*, **293**, 317 pp.

Kostaschuk, R.A. (2000) A field study of turbulence and sediment dynamics over subaqueous dunes with flow separation. *Sedimentology*, **47**, 519–531.

Kostaschuk, R.A. and Church, M.A. (1993) Macroturbulence generated by dunes: Fraser River, Canada. *Sediment. Geol.*, **85**, 25–37.

Kostaschuk, R.A. and Ilersich, S.A. (1995) Dune geometry and sediment transport: Fraser River, British Columbia. In: *River Geomorphology* (Ed. E.J. Hickin), pp. 19–36. Wiley, Chichester.

Kostaschuk, R.A. and Villard, P.V. (1996) Flow and sediment transport over large subaqueous dunes: Fraser River, Canada. *Sedimentology*, **43**, 849–863.

Kostaschuk, R.A. and Villard, P.V. (1999) Turbulent sand suspension over dunes. In: *Fluvial Sedimentology VI* (Eds N.D. Smith and J. Rogers). *Spec. Publs Int. Assoc. Sedimentol.*, **28**, 3–14.

Kumar, S., Gupta, R. and Banerjee, S. (1998) An experimental investigation of the characteristics of free-surface turbulence in channel-flow. *Phys. Fluids*, **10**, 437–456.

Lapointe, M. (1992) Burst-like sediment suspension events in a sand bed river. *Earth Surf. Process. Landf.*, **17**, 253–270.

Lyn, D.A. (1993) Turbulence measurements in open-channel flows over artificial bedforms. *J. Hydraul. Eng.*, **119**, 306–326.

Martinuzzi, R. and Tropea, C. (1993) The flow around surface-mounted, prismatic obstacles placed in a fully developed channel flow. *J. Fluids Eng.*, **115**, 85–92.

Matthes, G.H. (1947) Macroturbulence in natural stream flow. *Trans. Am. Geophys. Union*, **28**, 255–262.

McLean, S.R. and Smith, J.D. (1979) Turbulence measurements in the boundary layer over a sand wave field. *J. Geophys. Res.*, **84**, 7791–7808.

McLean, S.R., Nelson, J.M. and Wolfe, S.R. (1994) Turbulence structure over two-dimensional bedforms: implications for sediment transport. *J. Geophys. Res.*, **99**, 12729–12747.

McLean, S.R., Wolfe, S.R. and Nelson, J.M. (1999) Predicting boundary shear stress and sediment transport over bedforms. *J. Hydraul. Eng.*, **125**, 725–736.

McLelland, S.J., Ashworth, P.J., Best, J.L., Roden, J. & Klaassen, G.J. (1999) Flow structure and spatial distribution of suspended sediment around an evolving braid bar, Jamuna River, Bangladesh. In: *Fluvial Sedimentology VI* (Eds N.D. Smith and J. Rogers). *Spec. Publs Int. Assoc. Sedimentol.*, **28**, 43–57.

Mohrig, D. and Smith, J.D. (1996) Predicting the migration rates of subaqueous dunes. *Water Resour. Res.*, **10**, 3207–3217.

Müller, A. and Gyr, A. (1983) Visualisation of the mixing layer behind dunes. In: *Mechanics of Sediment Transport* (Eds B.M. Sumer and A. Müller), pp. 41–45. A.A. Balkema, Rotterdam.

Müller, A. and Gyr, A. (1986) On the vortex formation in the mixing layer behind dunes. *J. Hydraul. Res.*, **24**, 359–375.

Müller, A. and Gyr, A. (1996) Geometrical analysis of the feedback between flow, bedforms and sediment transport. In: *Coherent Flow Structures in Open Channels* (Eds P.J. Ashworth, S.J. Bennett, J.L. Best and S.J. McLelland), pp. 237–247. Wiley, Chichester.

Mullin, T., Greated, C.A. and Grant, I. (1980) Pulsating flow over a step. *Phys. Fluids*, **23**, 669–674.

Nelson, J.M., McLean, S.R. and Wolfe, S.R. (1993) Mean flow and turbulence fields over two-dimensional bedforms. *Water Resour. Res.*, **29**, 3935–3953.

Nelson, J.M., Shreve, R.L., McLean, S.R. and Drake, T.G. (1995) Role of near-bed turbulence structure in bed load transport and bed form mechanics. *Water Resour. Res.*, **31**, 2071–2086.

Nezu, I. and Nakagawa, H. (1993) Three-dimensional structures of coherent vortices generated behind dunes in turbulent free-surface flows. In: *Proceedings, 5th International Symposium on Refined Flows Modelling and Turbulence Measurements*, pp. 603–612.

Rashidi, M. and Banerjee, S. (1988) Turbulence structure in free-surface channel flows. *Phys. Fluids*, **31**, 2492–2503.

Robert, A. and Uhlman, W. (2001) An experimental study on the ripple–dune transition. *Earth. Surf. Proc. Landf.*, **26**, 615–629.

Roden, J.E. (1998) The sedimentology and dynamics of mega-dunes, Jamuna River, Bangladesh. Unpublished PhD thesis, Department of Earth Sciences and School of Geography, University of Leeds, 310 pp.

Rood, E.P. (1995) Free-surface vorticity. In: *Fluid Vortices* (Ed. S. Green), pp. 687–730. Kluwer, Norwell, MA.

Rood, K.M. and Hickin, E.J. (1989) Suspended sediment concentration in relation to surface-flow structure in Squamish River estuary, southwestern British Columbia. *Can. J. Earth Sci.*, **26**, 2172–2176.

Roy, A.G. and Buffin-Bélanger, T. (2001) Advances in the study of turbulent flow structures in gravel-bed rivers. In: *Gravel-bed Rivers V* (Ed. M.P. Mosley), pp. 375–397. New Zealand Hydrological Society, Wellington.

Roy, A.G., Biron, P., Buffin-Bélanger, T. and Levasseur, S. (1999) Combined visual and quantitative techniques in the study of natural flows. *Water Resour. Res.*, **35**, 871–877.

Sarpkaya, T. (1996) Vorticity, free surface, and surfactants. *Ann. Rev. Fluid Mech*, **28**, 83–128.

Schmeeckle, M.W., Shimizu, Y., Hoshi, K., Baba, H. and Ikezaki, S. (1999) Turbulent structures and suspended sediment over two-dimensional dunes. In: *River, Coastal and Estuarine Morphodynamics, Proceedings International Association for Hydraulic Research Symposium*, pp. 261–270. Genova, September.

Shimizu, Y., Schmeeckle, M.W., Hoshi, K. and Tateya, K. (1999) Numerical simulation of turbulence over two-dimensional dunes. In: *River, Coastal and Estuarine Morphodynamics, Proceedings International Association for Hydraulic Research Symposium*, pp. 251–260. Genova, September.

Simpson, R.L. (1989) Turbulent boundary-layer separation. *Ann. Rev. Fluid Mech.*, **21**, 205–234.

Smart, G.M. (2001) Discussion of 'Advances in the study of turbulent flow structures in gravel-bed rivers' by Roy, A.G. and Buffin-Bélanger, T. In: *Gravel-bed Rivers V* (Ed. M.P. Mosley), pp. 399–400. New Zealand Hydrological Society, Wellington.

Smith, J.D. and McLean, S.R. (1977) Spatially-averaged flow over a wavy surface. *J. Geophys. Res.*, **82**, 1735–1746.

Soulsby, R.L., Atkins, R., Walters, C.B. and Oliver, N. (1991) Field measurements of suspended sediment over sandwaves. In: *Euromech 262—Sand Transport in Rivers, Estuaries and the Sea* (Eds R.L. Soulsby and R.E. Bettess), pp. 155–162. Balkema, Rotterdam.

Tait, S.J., Willetts, B.B. and Gallagher, M.W. (1996) The application of particle image velocimetry to the study of coherent flow structures over a stabilizing sediment bed. In: *Coherent Flow Structures in Open Channels* (Eds P.J. Ashworth, S.J. Bennett, J.L. Best and S.J. McLelland), pp. 185–201. Wiley, Chichester.

Ten Brinke, W.B.M., Wilbers, A.W.E. and Wesseling, C. (1999) Dune growth, decay and migration rates during a large-magnitude flood at a sand and mixed sand-gravel bed in the Dutch Rhine river system. In: *Fluvial Sedimentology VI* (Eds N.D. Smith and J. Rogers). *Spec. Publs Int. Ass. Sedimentol.*, **28**, 15–32.

Vendetti, J.G. and Bennett, S.J. (2000) Spectral analysis of turbulent flow and suspended sediment transport over dunes. *J. Geophys. Res.*, **105**, 22 035–22 047.

Vionnet, C., Marti, C., Amsler, M. and Rodriguez, L. (1998) The use of relative celerities of bed forms to compute sediment transport in the Paraná River. In: *Modelling Soil Erosion, Sediment Transport and Closely Related Hydrological Processes. Int. Assoc. Hydraul. Res. Spec. Publ.*, **249**, 399–406.

Villard, P.V. and Kostaschuk, R.A. (1998) The relation between shear velocity and suspended sediment concentration over dunes: Fraser Estuary, Canada. *Mar. Geol.*, **148**, 71–81.

Wewetzer, S.F.K. and Duck, R. (1999) Bedforms of the middle reaches of the Tay Estuary, Scotland. In: *Fluvial Sedimentology VI* (Eds N.D. Smith and J. Rogers). *Spec. Publs Int. Assoc. Sedimentol.*, **28**, 33–41.

Wijbenga, J.H.A. (1990) *Flow Resistance and Bed-form Dimensions for Varying Flow Conditions*. Report Q785, Delft Hydralics, Delft, 73 pp. and annexes.

Yalin, M.S. (1992) *River Mechanics*. Pergamon Press, Oxford, 219 pp.

Derivation of annual reach-scale sediment transfers in the River Coquet, Northumberland, UK

IAN C. FULLER[*,1], ANDREW R.G. LARGE[†], GEORGE L. HERITAGE[‡], DAVID J. MILAN[§] and MARTIN E. CHARLTON[†]

*School of People, Environment and Planning, Massey University, Palmerston North, New Zealand (Email: I.C.Fuller@massey.ac.nz);
†Department of Geography, University of Newcastle, Newcastle upon Tyne, NE1 7RU, UK;
‡Division of Geography, School of Environment and Life Sciences, University of Salford, Manchester, M5 4WT, UK; and
§Geography and Environmental Management Research Unit, University of Gloucestershire, Cheltenham, GL50 4AZ, UK

ABSTRACT

Measurement of three-dimensional morphological change in a river channel provides a useful means of assessing reach-scale rates and patterns of fluvial sediment erosion, transfer and deposition. The morphological sediment budgeting techniques used to generate these estimates of sediment flux may be particularly valuable in unstable gravel bed rivers, owing to the inherent difficulties associated with direct measurement of bedload transport. This paper compares two approaches used to derive a measure of annual sediment transfers within a 1 km long piedmont reach of the gravel-bed River Coquet in Northumberland, northern England, which has a locally braided channel planform and has experienced lateral instability over the past 150 yr. The first technique utilized channel planform and cross-profile surveys based on theodolite-EDM survey of (i) 21 monumented channel cross-profiles and (ii) channel and gravel bar margins. The second method used theodolite–EDM surveys to generate a series of xyz coordinates for the channel and bars within the reach, from which digital elevation models (DEMs) can be constructed. Calculating the difference between two DEM surfaces provides a measure of volumetric change between surveys. The compatibility between the two exercises, carried out during the springs of 1999 and 2000, was assessed by an error analysis comparing the surveyed cross-profiles with sections abstracted from the DEMs. This indicates a mean gross error between surveyed and DEM profiles of approximately twice the value of the D_{50} of the surface sediment in the reach. The accuracy of the DEMs as a representation of the terrain surface is quantified using residual analysis, which indicates that more than 96.3% of the interpolated surface is accurate to ±5 cm (equivalent to the surface sediment D_{50}) for both the 1999 and 2000 DEM surfaces. Comparison of sediment volumes derived from the two approaches suggests that, relative to the higher resolution DEM survey, estimation of sediment transfers using monumented profiles and planform underestimates the magnitude of volumetric changes that occur within the reach. The degree of underestimation is reach and time dependent. The DEM method using a carefully designed sampling strategy based on morphological units provides a rigorous identification of spatial patterns of erosion and deposition, which cross-section-based approaches may fail to include.

INTRODUCTION

In gravel-bed rivers, changes in channel morphology reflect the movement of bed material (Leopold, 1992; Martin & Church, 1995; Ashmore & Church, 1998), which, once entrained, moves short distances by traction, saltation or suspension (Martin & Church, 1995). Flux of bedload, however, is notoriously difficult to measure (Reid et al., 1999). Bedload transport is random and discontinuous (Einstein, 1937), and often occurs in a series of pulses, reflecting the downstream passage of sediment waves (Gomez et al., 1989; Hoey, 1992; Nicholas et al., 1995; Paige & Hickin, 2000), and variable bed patchiness during initiation of motion (Garcia et al., 2000). Bedload formulae may fail to predict transport rates and thresholds of entrainment accurately, particularly in armoured beds (Gomez & Church, 1989). This is a product of the wide range of sediment properties that condition sediment mobilization, together with the inherent natural variability of bed-sediment and flow structure (Milan et al., 1999; Buffin-Bélanger et al., 2000). Although transport rates have been accurately predicted for unarmoured beds (e.g. Reid et al., 1996), to date only one accurate prediction has been obtained for an armoured bed (Habersack & Laronne, 2002).

Attempts at estimating bed material transport rates have also been made by combining tracer data with hypothesized depths of bed activity (Mosley, 1978; Kondolf & Matthews, 1986) or measured scour-and-fill (Laronne et al., 1992). However, in an unstable braided channel, tracer recovery rate may be as little as 5% by volume, and channel instability precludes precise linkage between sediment transfers and bedforms in such systems (Mosley, 1978). Channel morphology nevertheless influences tracer distribution, and the role of in-channel storage is reason for caution in using short-term tracer studies to compute rates of sediment flux (Kondolf & Matthews, 1986). Furthermore, direct measurement of scour-and-fill has proved to be a very time consuming process, and is realistically applicable only to small ephemeral or shallow (< 0.2 m) perennial rivers (Laronne et al., 1992).

Quantification of changes in channel morphology, which take place in response to sediment erosion, transfer and deposition (Ashmore & Church, 1998), provides a further means by which bedload flux can be monitored (Davies, 1987). This (i) avoids some of the complications caused by spatial and temporal variability of bedload transport (Hubbell, 1987), (ii) overcomes difficulties associated with use of tracers and direct measurement of scour-and-fill, and (iii) provides an approach for estimating sediment transport in the context of storage and displacement of bed material and channel stability (Ashmore & Church, 1998). Investigation of three-dimensional morphological change in gravel-bed rivers requires repeat survey of bed elevation and planform adjustments. The majority of research in this field has focused on the bar-scale (Neill, 1987; Ferguson & Ashworth, 1992; Ferguson et al., 1992; Lane et al., 1994).

In relation to larger scale investigations, Goff & Ashmore (1994) sought to determine bedload transport rates between specific areas of the bed at a scale (around 60 m reach length) extending beyond the individual bar complex in a proglacial braided river using field-based sediment budgeting. Channel planform in their study was derived from photography, rather than high-resolution topographic survey, and budgets were derived exclusively from repeat cross-profile data. In addition, Goff & Ashmore's (1994) research was limited to a 2-month (summer melt) period, albeit providing a high-resolution dataset for a short period. By comparison, a much lower resolution, but more spatially and temporally extensive, study was conducted by Martin & Church (1995). This used a series of channel surveys based on cross-profiles and planform maps to estimate bedload transport on the Vedder River, British

Columbia, in an 8 km reach over 10 yr. Their measurements examined sediment transfer through 800 m subreaches, which exceeded average bar spacing in the Vedder River, suggesting that the identified patterns of sediment transfer were of a coarse resolution. Furthermore, although survey data were collected over a 10-yr period, some surveys were separated by as much as 3 yr, making an accurate assessment of annual fluxes problematic. Similar approaches estimating sediment flux over the longer-term, using changes in channel morphology over tens of years, and at a relatively coarse scale, have been used by Ham & Church (2000) and McLean & Church (1999).

Brewer & Passmore (2002) presented budgeting techniques designed to evaluate sediment transfers at intermediate reach scales (1–3 km), which are more typical of major instability zones in UK piedmont rivers. This approach also discriminates between morphological units (channel and bar) within a reach. In assessing sediment transfers at this higher, reach-scale resolution, it is possible to identify more accurately sediment flux and storage in instability zones, which may be overlooked by larger scales of study. Wathen et al. (1997) noted that reach-scale sediment storage is rarely quantified in sediment budget studies, but recognized that it has a significant effect on the accuracy of morphological methods of bedload estimation at the reach scale. An application of Brewer & Passmore's (2002) methods has quantified sediment transfers in the same 1 km reach of the Coquet as this study for 1997–98 and 1998–99 (Fuller et al., 2002). Estimation of changes observed over the period 1999–2000, and presented here, represents a natural extension to this work.

To date, application of the 'morphological approach' (sensu Ham & Church, 2000) to reach-scale sediment budgeting in its various forms (e.g. Goff & Ashmore, 1994; Martin & Church, 1995; Ham & Church, 2000; Paige & Hickin, 2000; Brewer & Passmore, 2002; Fuller et al., 2002) has largely relied upon channel cross-sections to generate the elevation changes required to calculate volumetric sediment transfers. Lane et al. (1994) suggested that this emphasis on the channel cross-section is a legacy of hydraulic geometry, and may represent a weakness in the morphological approach. Extrapolating changes measured along a cross-profile to areas of channel and bars, which may be tens of metres from a section, may be problematic, as this assumes the profile is an adequate representation of the units it bisects (Fuller et al., 2002). Furthermore, the cross-section may not adequately reflect downstream change in channel morphology (Lane et al., 1994), and may fail to accommodate the potential effects of downstream sedimentological structures on channel processes (Naden & Brayshaw, 1987). This has led Brasington et al. (2000) to note that, 'the implicitly cross-stream emphasis gives rise to a high degree of uncertainty in reach-scale sediment budgets derived through the interpolation of cross-section data'. Although some morphological approaches (e.g. Lane et al., 1994; Eaton & Lapointe, 2001) have moved towards a 'distributed terrain-sensitive survey' (Ashmore & Church, 1998) to generate reach digital elevation models (DEMs), no comparison between this and the cross-section approach has yet been made. This paper, therefore, seeks to assess the accuracy of a cross-profile-dependent morphological approach to quantifying sediment transfers, using DEMs constructed for a reach of the River Coquet that has been the subject of budgeting using channel cross-profiles since 1997.

One of the reasons for the traditional reliance on cross-stream approaches to reach-scale sediment budgeting lies in the practical difficulties of topographic data acquisition, storage and analysis at resolutions which adequately reflect channel morphology (Brasington et al., 2000). Cross-section-derived budgets, which can be acquired and processed rapidly (Brewer & Passmore, 2002), have thus formed the traditional 'backbone' of this research to date. Development of increasingly sophisticated field survey equipment (Total Station, GPS, photogrammetry) has facilitated collection of highly detailed field datasets. In turn these have led to the production of high-resolution DEMs of fluvial environments (e.g. Lane et al., 1994; Milne & Sear, 1997; Heritage et al., 1998; Brasington et al., 2000). Acquisition of high-resolution topographic information is central to effective construction of such DEMs (Westaway et al., 2000). In this study, such data were acquired using a Total Station.

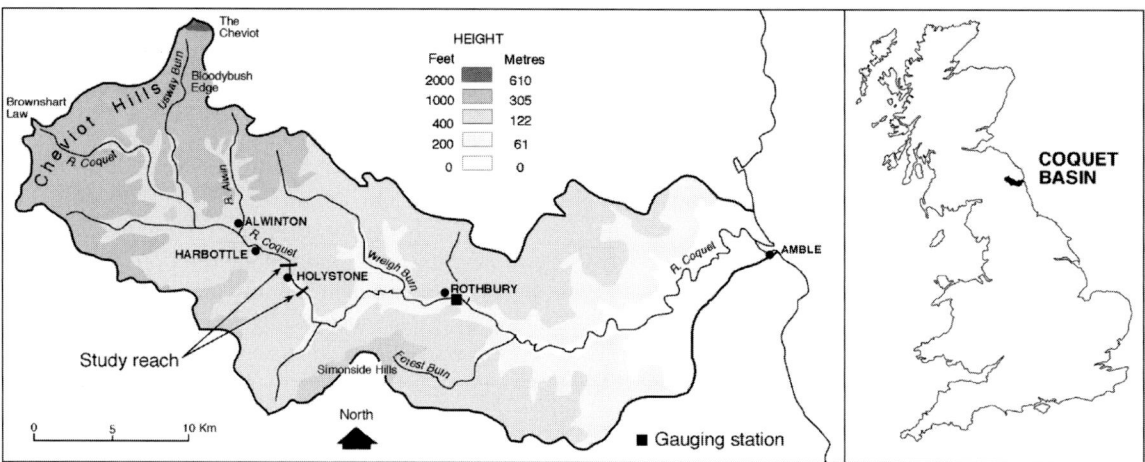

Fig. 1 River Coquet catchment, showing location of the study reach at Holystone.

STUDY SITE

The River Coquet rises in the Cheviot Hills (776 m a.s.l.) in northern England (Fig. 1). The focus of this research has been an instability zone located at Holystone (UK National Grid Reference NY 958027), where the river has a slope of 0.0093 in a piedmont setting at the upland fringe in the catchment, draining an area of approximately 255 km² (Fig. 1). The Coquet at Holystone has been characterized by a high degree of lateral instability and channel avulsion over the past 150 yr (Fuller et al., 2002). The contemporary active channel at Holystone comprises features common to both braided and meandering rivers, being flanked and locally divided by expanses of bare gravel (of variable width, up to 80 m), but having well-defined pool–riffle units (Fuller et al., 2002); as such it may be classed as a wandering river (Ferguson & Werritty, 1983). The 400–500-m-wide valley floor at Holystone is classified as a medium-energy non-cohesive, wandering gravel-bed river floodplain (Nanson & Croke, 1992). The valley as a whole is similar to other gravel-bed rivers in northern England, in that it displays a characteristic 'hourglass' valley morphology, with alternating confined and unconfined sections (Macklin, 1999). The D_{50} of the surface sediment in the reach is 51 mm (Fuller et al., 2002). During the study period (1999–2000) discharges equivalent to bankfull occurred on six separate occasions.

METHODS

Cross-profiles and planform

Theodolite mapping using a Sokkia Set 5F Total Station (with a precision of ±5 mm) was used to survey channel planform (channel boundaries, barforms and major chute channels) and monumented cross-profiles in March 1999 and 2000. The spacing of the cross-profiles along the study-reach was designed to ensure that channel-bed elevation was measured at regular intervals through the reach and that each bar in the reach was crossed by at least one profile. This provided the means of assessing the vertical changes, produced by sediment gain or loss, that occurred across the morphological units in the reach. Survey of the cross-profiles generated mean point densities of 0.41 (SD 0.09) and 0.44 (SD 0.09) points per metre length of profile in 1999 and 2000 respectively. The locations of the cross-profiles dividing the site into subreaches are shown overlaying the 1999 channel planform (Fig. 2). Cross-profile data for each survey were incorporated into an Arc/Info™ GIS.

Sediment budgets were calculated using the morphological budget described in detail by

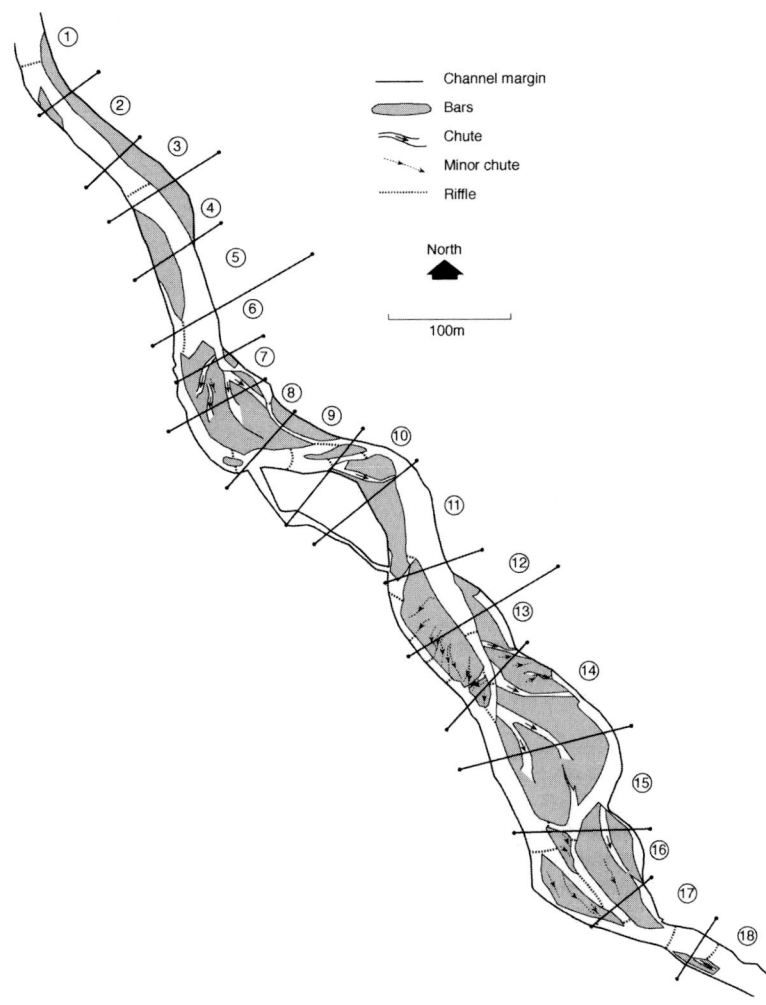

Fig. 2 Channel planform at Holystone in 1999, showing the division of the reach into subreaches by cross-sections. Arrows indicate direction of flow.

Brewer & Passmore (2002) and Fuller *et al.* (2002; Fig. 3). This approach integrates both planform and cross-profile data at the resolution of the discrete morphological unit within each subreach. Within Arc/Info™ GIS, vertical changes in area along each cross-profile were calculated and standardized to net gain/loss values per square metre. These values were then multiplied by the corresponding planform area values to provide a net gain/loss value (cubic metres) for each morphological unit within the subreach. Where morphological units are bounded by two cross-profiles (e.g. large bar and main channel units), the planform area was halved and multiplied by the upstream and downstream cross-profile data respectively (Fig. 3).

DEM generation

Coincident with the planform mapping exercise described above, additional points were surveyed from a common local datum across the channel and bar units using the same survey equipment. Sampling of the topographic surface using these points was conditioned by breaks in slope within and between morphological units throughout the reach. The data points used in DEM generation were not therefore tied to grids, but followed

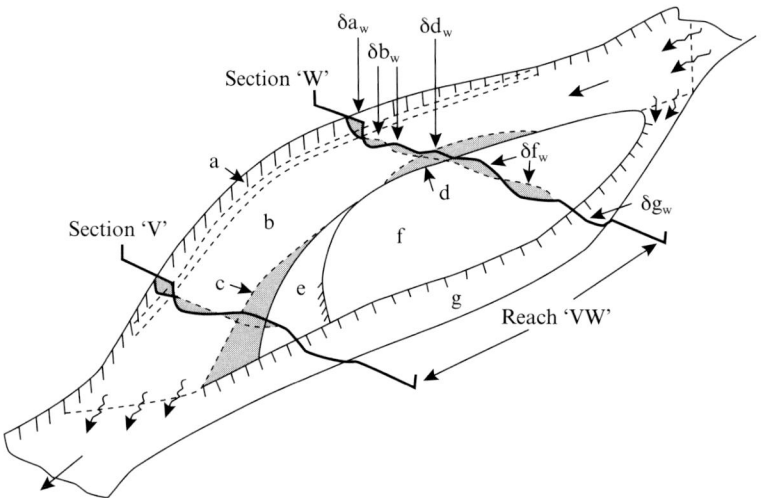

Stage 1: Calculation of volumetric change for each morphological unit

for morphological units traversed by:

1 channel profile
(eg unit 'd' on profile 'W')

$$\delta V_d = \left[\frac{\delta A_{dw}}{L_{dw}} \times A_d \right]$$

2 channel profiles
(eg unit 'a' on profile 'W' and 'V')

$$\delta V_a = \left[\frac{\delta A_{aw}}{L_{aw}} + \frac{\delta A_{av}}{L_{av}} \right] \times \frac{A_a}{2}$$

Where
δV_d = volumetric change for morphological unit 'd' (m^3)
δA_{dw} = change in cross-sectional area of morphological unit 'd' (m^2)
L_{dw} = length of morphological unit 'd' across profile 'W' (m)
A_d = planform area of morphological unit 'd' at end of the budget period (m^2)
δV_a = volumetric change for morphological unit 'a' (m^3)
δA_{aw} = change in cross-sectional area of morphological unit 'a' across profile 'W' (m^2)
δA_{av} = change in cross-sectional area of morphological unit 'a' across profile 'V' (m^2)
L_{aw} = length of morphological unit 'a' across profile 'W' (m)
L_{av} = length of morphological unit 'a' across profile 'V' (m)
A_a = planform area of morphological unit 'a' at end of the budget period (m^2)

Stage 2: Calculation of volumetric change for each sub-reach

eg for sub-reach 'VW': $\delta V_{vw} = \Sigma (\delta V_a + \delta V_b \ldots \ldots + \delta V_g)$

Fig. 3 Derivation of sediment budgets using channel cross-sections and outline planform (after Brewer & Passmore, 2002).

natural surfaces and features within the channel and bars. In 1999, 2661 points were surveyed within the active channel area (50 509 m^2), producing an overall point density in the reach of 0.05 points m^{-2}. In 2000, 2985 points were surveyed within the same area, producing a point density of 0.06 points m^{-2}. The construction of a DEM for each survey was undertaken within SURFER™ GIS (Fig. 4). Data interpolation was based on kriging at a grid interval of 0.25 m. Although this interpolation interval is not consistent with the reach-averaged sampling densities, it accommodates those areas of the reach where topography was sampled at an

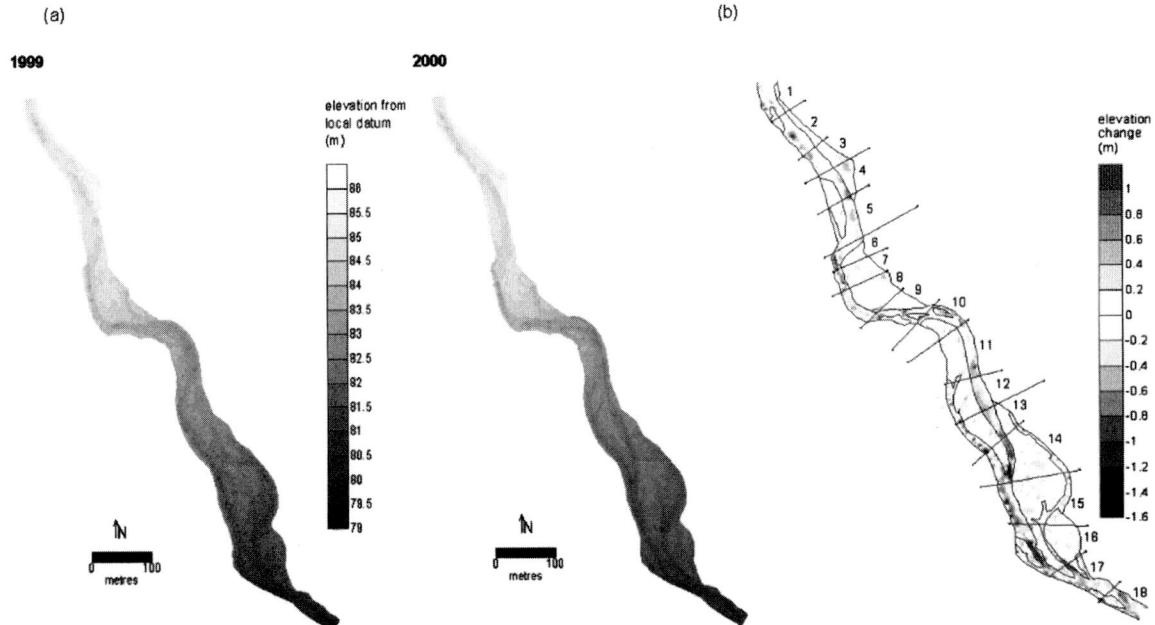

Fig. 4 (a) Digital elevation models of the Holystone reach in 1999 and 2000. (b) Digital elevation model of difference between 1999 and 2000, showing subreaches and bar outlines in 2000.

approximately equivalent resolution to this grid interval (e.g. breaks in slope). This grid interval therefore ensured the construction of a DEM that is sensitive to more subtle changes in morphology, as coarser grid intervals may lead to the loss of potentially important breaks in slope within the channel environment (Brasington et al., 2000).

Although finer grid intervals (requiring more computational time for calculation and memory for storage) could have been used, Brasington et al. (2000) found very little difference between the surface representation of the 0.1 m and 0.25 m grid for DEMs constructed for a reach of the gravel-bedded River Feshie (D_{50} 65 mm). Furthermore, Brasington & Richards (1998) suggested that the use of very fine grids may incorporate 'spurious artefacts' within the DEM, which are not representative of the real elevation surface. Kriging represents a geostatistical gridding method that uses irregularly spaced data (Dixon et al., 1998), thus it is ideally suited to the morphologically driven datasets collected in the Coquet, which are irregularly spaced. This contrasts with a grid sampling approach, which is better served using triangulation with linear interpolation (TIN) (Dixon et al., 1998; Brasington et al., 2000). The accuracy of the resulting DEMs derived from these data points is discussed below. Subtraction of the 2000 DEM surface from the 1999 surface produced a DEM of difference between the surfaces, allowing clear identification of the most important changes in channel morphology within the reach for the period 1999–2000 (Fig. 4b).

DEM ACCURACY

Before a comparison of budgeting methods can be made, the accuracy of the DEMs (as representation of topographic surface within the reach at Holystone) must be determined. In order to independently assess the quality of DEMs, which are controlled by the precision, reliability and accuracy of individual points and their spatial coverage (Lane, 1998), a comparison of DEM-extracted information with independently acquired ground measurements is required (Westaway et al., 2000). In the first instance, the independently acquired information here is provided by the channel

Table 1 DEM error analysis.

Cross-profile	1999		2000	
	Gross difference (m^2)	Difference per unit length (m m^{-1})	Gross difference (m^2)	Difference per unit length (m m^{-1})
1	3.240	0.109	1.314	0.044
2	1.880	0.037	3.352	0.066
3	7.753	0.175	3.532	0.080
4	3.914	0.081	1.634	0.034
5	5.787	0.114	3.160	0.075
7	6.453	0.100	7.223	0.112
8	3.842	0.091	4.048	0.096
9	3.622	0.101	3.574	0.099
10	6.827	0.156	5.789	0.132
11	6.005	0.121	0.813	0.016
11a	7.990	0.126	11.201	0.176
12	6.863	0.111	4.734	0.077
13	6.911	0.081	15.496	0.182
14	19.291	0.193	19.812	0.198
15	11.860	0.135	6.786	0.077
15a	9.277	0.104	4.275	0.048
16	4.873	0.093	5.209	0.099
17	3.454	0.131	0.767	0.029
18	1.961	0.092	2.276	0.107
Mean	Not determined	0.113	Not determined	0.092
SD	Not determined	0.030	Not determined	0.050

cross-profile surveys, which were surveyed separately (but within a week) from the higher resolution topographic survey used to provide co-ordinates for the DEMs. No geomorphologically significant flows occurred during this brief period. The precise location of the channel cross-profiles within the high-resolution survey is known, as the pegs used to monument each cross-profile were surveyed using the same equipment and local datum. The DEM cross-profiles were abstracted at the 0.25 m grid interval used in DEM construction, as a series of x, y, z points in a straight line between the co-ordinates of each monumented peg.

Both the DEM and surveyed cross-profiles were incorporated into an Arc/Info™ GIS in order to generate overlays of each cross-section for the two techniques in both the 1999 and 2000 surveys. The files generated in Arc/Info™ could then be viewed in Arc/View™, which enabled visualization of the areas of difference between the DEM and survey cross-profiles using a series of difference polygons. The gross areas of difference between these profiles (where DEM abstraction lies above or below the surveyed profile) were summed and then standardized to a measure of gross area difference per unit length along the cross-profile across the active channel (Table 1). This served to assess the compatibility between DEM and surveyed cross-profiles; a large difference would negate comparison of sediment budgets derived from the two approaches.

The error analysis in Table 1 indicates that the mean difference per unit length between the surveyed cross-profiles and those abstracted from the DEM is slightly larger than the D_{84} of the surface sediment, measured at 83 mm (Fuller et al., 2002). Allowing for a 12% error associated with the sediment size statistic (Fuller et al., 2002), the 2000 mean difference corresponds with the D_{84}. However, a close correspondence between the surveyed and DEM-derived profiles is not necessarily expected. Both surveys (profile and x, y, z DEM points) will be subject to errors associated with surface roughness (e.g. placement of the detail pole used in survey will vary between the top and bottom of clasts). Taking the D_{50} (51 mm) into account, a comparison between these surveys will, in effect, compound the error induced by the irregularity of the sediment surface. This suggests that the accuracy of either one survey will be ±5 cm, producing a compound error of ±10 cm when comparing the two. Furthermore, this figure is reach-averaged and does not take into account local variability in surface roughness, which is a significant source of error (Brasington et al., 2000), and may explain the variation in the differences per unit length in Table 1. This analysis, therefore, suggests as good as possible a correspondence between the methods of survey, within the limitations imposed by surface roughness.

Residual analysis

A quantitative measure of how well the interpolated surface agrees with the original survey data is obtained using a surface of residual points. The vertical difference between the z value of the surveyed data points and the interpolated z value on the DEM surfaces at the same xy location on the interpolated surface was computed for the 1999 and 2000 DEMs using equation (1)

$$z_{res} = z_{surv} - z_{interp} \qquad (1)$$

where z_{res} is the residual value, z_{surv} is the surveyed z value and z_{interp} is the z value of the interpolated surface at the xy co-ordinate. A residual is defined as the difference between the z value of a surveyed data point and the interpolated z value. This residual analysis indicates that 96.3% of the 2000 DEM surface and 96.8% of the 1999 DEM is accurate to ±5 cm (equivalent to the surface sediment D_{50}). There is no systematic pattern to the residual values through the reach.

DEM differencing

A further means of quantifying the reliability of the DEMs is to calculate the net volume between the upper and lower surface using three methods that are provided in SURFER™ GIS: (i) the Trapezoidal rule, (ii) Simpson's rule and (iii) Simpson's 3/8 rule. Close agreement between the volumes computed using the three methods

Table 2 Net volumes between 1999 and 2000 DEMs calculated using three numerical integration algorithms (SURFER™ 1999).

Volumetric method	Net volume (m³)
Trapezoidal rule	−7518.11
Simpson's rule	−7518.80
Simpson's 3/8 rule	−7518.51

(Table 2) indicates an appropriate grid resolution of the DEMs (SURFER™, 1999).

Volumetric calculations using SURFER™ GIS

Mathematically, the volume under a function is defined by a double integral (equation 2)

$$\text{volume} = \int_{x_{max}}^{x_{max}} \int_{y_{max}}^{y_{max}} f(x,y) \, dx \, dy \qquad (2)$$

In SURFER™, this is computed by first integrating over X (the columns) to obtain the areas under the individual rows, and then integrating over Y (the rows) to obtain the final volume. This approach is explained in detail in Press et al. (1988). SURFER™ approximates the necessary one-dimensional integrals using three classic numerical integration algorithms (SURFER™, 1999).

The reliability of a DEM is dependent upon the means of data acquisition and processing (Lane, 1998). The data used to generate the DEMs presented here are acquired using precise instrumentation to survey irregular, but carefully defined points to best represent channel and bar morphology. Processing is based on an appropriate (Brasington et al., 2000) grid using effective (Dixon et al., 1998) interpolation. Although the data point density used here is relatively low (compare values of 0.06 points m^{-2} with 1.10 points m^{-2} as used by Brasington et al., 2000), the reliability of the DEMs, as measured using independent comparison, residual analysis and volume differencing appears to be acceptable within the errors imposed by surface grain roughness. This may have implications concerning the sampling of the topographic surface, questioning

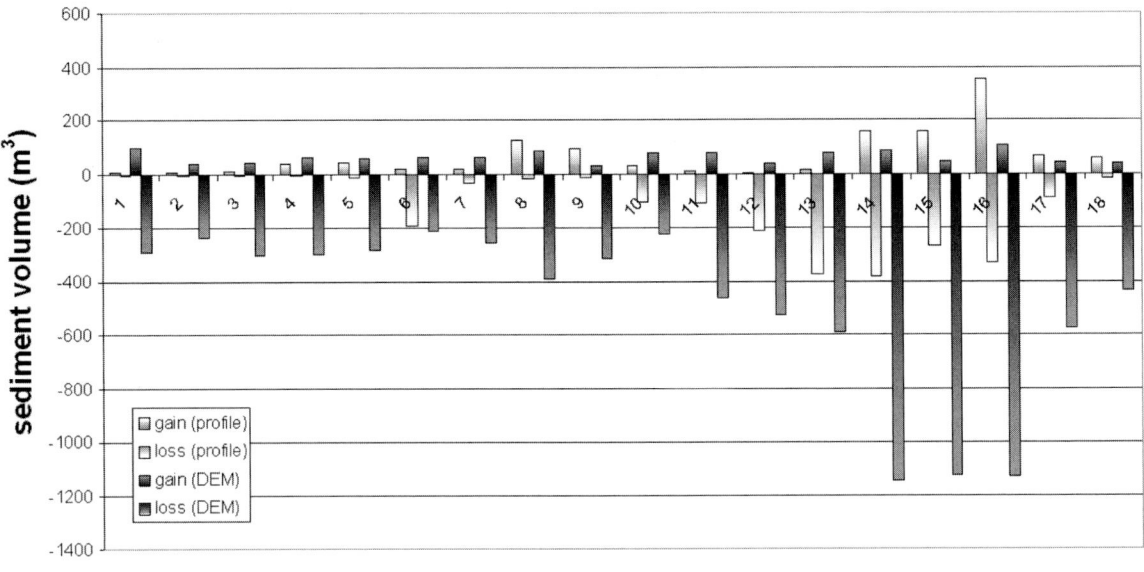

Fig. 5 Subreach sediment transfers calculated using channel cross-section and outline planform budgeting (method outlined in Fig. 3) and DEM subtraction.

the often cited belief that the more data points, the more reliable the surface. However, the assessment of this argument is not within the scope of the paper (see Chappell *et al.*, 2003). The accepted reliability of the DEM suggests that it is a valid exercise to compare volumetric estimates of sediment transfer using DEM subtraction and channel cross-profile budgeting. This is further supported by a more rigorous technical assessment of DEM quality in this reach, which is reported elsewhere (Fuller *et al.*, 2003).

SEDIMENT TRANSFERS

DEM and profile budgeting compared

To provide an effective comparison between the profile and DEM budgeting approaches requires division of the reach DEM into subreaches exactly equivalent to the cross-profile driven subreaches, followed by a DEM subtraction (2000–1999) for each discrete subreach. Volumetric changes were, therefore, calculated for identical subreaches to those defined by the cross-profiles (cf. Fig. 2).

Sediment transfers calculated using the cross-profile dependent approach are shown alongside those from the DEM subtraction approach for each subreach at Holystone (Fig. 5). This shows considerable (and consistent) underestimation of the magnitude of sediment losses using profiles, but a tendency for sediment gains to be overestimated by the profile approach in some subreaches. Given the compatibility of the two methods of topographic representation (surveyed profile and DEM), this must result from significant morphological changes taking place between the cross-profiles, which cannot be accounted for by the cross-profiling technique. Where it occurs, localized deposition measured on a cross-profile is being extrapolated across too large an area in those reaches. This results in an overestimation of sediment gain in these locations (notably sub-reaches 14 to 17). Discrepancy in sediment losses between the DEM and profile budgets (Fig. 5), however, is the result of localized channel scour occurring *between* channel profiles. Figure 4b highlights this failure of cross-profiles to adequately represent the extent of scour in most subreaches, with intensive scour often taking place between profiles. The residuals analysis and reach-volume differencing verify the reliability of the DEM.

Lane (1998) noted that the use of a DEM-based approach to monitor the three-dimensional nature of river channel form has a major advantage over the cross-profile approach in that data collection is now topographically defined, resulting in better surface representation. Furthermore, change in location of scour/deposition from the originally identified 'sensitive' areas, which directed cross-profile location, may account for some errors; as cross-profiles are carefully selected at the outset of the investigation (here 1997), these positions remain fixed, irrespective of later channel development and shifts in areas of activity. The consequence of this dynamism was such that by 2000 some profiles were no longer effectively identifying geomorphological activity in the Coquet channel. This does not suggest that profile-derived budgets as a whole have a tendency to overestimate gains and underestimate losses, but reflects local processes operating within the Holystone reach during the survey period. Nevertheless, Fig. 4b demonstrates the localized nature of both scour and fill, both of which are highly discontinuous at this site. In a system with such a complex pattern of sediment losses and gains, it is hardly surprising that fixed channel cross-profiles fail to provide an adequate measure of sediment transfer volumes. No single cross-profile within the reach can be said to be truly representative of the sub-reach it represents.

Discussion

The DEM of difference (Fig. 4b) permits identification of the locations of net erosion and deposition within the reach between 1999 and 2000. Immediately apparent (and consistent with Fig. 5) is the predominance of sediment losses (where the 2000 surface is lower than 1999). In particular, losses are concentrated within the channel, and there appears to be little loss of material from bar surfaces in the reach. Total channel losses equate to 5668 m^3, whereas 1850 m^3 has been lost from bars within the reach. This volume loss associated with the channel includes losses generated by bank erosion, as the retreat of banks 1–2 m high produced a surface 1–2 m lower at the equivalent x,y point in the subsequent (2000) DEM. Losses resulting from bank erosion are consistently greatest on the outside of the bends throughout the reach. Mid-channel bars within the reach (within subreaches 16 and 9/10; cf. Fig. 4) have been subject to notable change, the downstream feature having been removed completely, revealing a considerable loss in the DEM of difference, equivalent to 620 m^3 (Fig. 4b).

Sediment gains appear to be highly localized, and may be identified both within the channel and at the margin of bars. Locally significant in-channel accumulation (59 m^3) has taken place within subreach 6 at the head of a major left-bank lateral bar complex and associated riffle crest (Fig. 4b). In addition, localized accumulation is linked to the small-scale progradation of the left-bank lateral bar in subreach 8/9 (56 m^3) and the development of a mid-channel feature in subreach 10 (also 56 m^3). Furthermore, although destruction of the mid-channel bar in subreach 16 resulted in considerable erosion, the deposition of 108 m^3 of sediment in this reach created both a new lateral feature (right-bank) and extended a lobe of sediment downstream (Fig. 4b).

Comparison of DEMs in Fig. 4a demonstrates that the 2000 DEM discerns the principal (wetted) channel in subreaches 10–14 much more clearly than 1999. This comparison indicates that there has been channel deepening by incision, representing continued adjustment to an avulsion which straightened the planform in early 1999. This avulsion cut a new channel at the right-bank margin of the active channel zone within reaches 14 and 15, and was associated with extensive deposition at the head of the abandoned channel in subreach 14 (Fuller et al., 2002). The channel in this part of the Holystone reach has continued to adjust to these morphological changes, notably by incising the deposits at the head of subreach 14 (eroding 482 m^3), a process facilitated by a steeper slope associated with lower local base-level provided by the shortened right-bank channel. This incision also appears to have propagated upstream as far as subreach 10, deepening the channel as it rationalizes post-avulsion and eroding 752 m^3 of sediment. The new (right-bank) channel has also been subject to consistent and extensive scour along its length (loss of 1108 m^3) as this feature has continued to develop. A further product of the steeper slope in subreaches 14 and 15 is the continued high degree of activity in the channel downstream (subreaches 16 and 17), where both bar destruction and formation have taken place. Again, this may be attributed to some degree of post-avulsion channel rationalization (both lateral and vertical). Although bar destruction and reformation have taken place in this part of the reach, the overall tendency for channel incision means that parts of the new bar surfaces are lower than the channel surface of 1999 (cf. Fig. 4b). This serves to further illustrate the intensity of channel incision that has taken place in this sensitive zone. Use of DEMs clarifies such details of channel development, as channel cross-profiles and planform outlines (at the scale used in this study) would not be sufficient to fully discern these changes. However, as surveys are separated by a year, it is not possible to provide further information on the timing or number of episodes of scour and fill responsible for these changes.

Farther upstream (e.g. subreaches 6–10), sediment transfer patterns are principally associated with quasi-equilibrium bend development. Here bank erosion and associated bed scour on bend apices (e.g. eroding 275 m^3 in subreach 6 and 133 m^3 in subreach 10) has extended the bends laterally (Fig. 4). This erosion has been accompanied by localized accumulation at the margins of a lateral bar, consistent with models of bend development (e.g. Hooke, 1995). The localized nature of sediment accumulation in these zones may reflect deposition of gravel in lobate or sheet form, as this material is known to travel as a series of pulses/waves within the reach (Fuller et al., 2002). Such pulsing has been observed elsewhere (e.g. Ferguson & Werritty, 1983; Wathen et al., 1997). As such, the processes conditioning behaviour of the Coquet at Holystone can be construed as being typical of a wandering gravel-bed river (Ferguson & Werritty, 1983).

CONCLUSIONS

Digital elevation models in a 1 km reach of the River Coquet at Holystone provide an accurate representation of channel morphology. This paper demonstrates that DEMs derived from relatively low point density sampling of channel topography are a reliable representation of morphological units (within the limitations imposed by surface sediment characteristics), and can be used effectively to quantify meaningful volumes of sediment transfer at the 1 km reach scale. This addresses the need identified by Lane et al. (1994) that much more detailed information on river bed topography is needed to accommodate the increasing recognition of spatially distributed form–process feedback in fluvial environments, and provide a more rigorous linkage between channel topography and sediment transport processes.

The DEM subtraction method has permitted discernment of unit-by-unit behaviour contributing to within-reach sediment transfers. This shows the greatest changes and transfers taking place within the channel, as opposed to bars, during the survey period. Whilst this paper has assessed the validity of a cross-profile budgeting method at the site, it therefore also represents the next logical step in this research towards providing that more rigorous linkage between channel topography and sediment transport processes.

The DEM method presented, which uses a carefully designed sampling strategy (based on morphological units), provides a rigorous characterization of spatial patterns of erosion and deposition, which cross-section-based approaches at the site have failed to identify. In particular, the use of DEMs has permitted the identification of long-stream three-dimensional channel (and bar) development for this reach of the River Coquet, as the channel morphology adjusts to post-avulsion quasi-equilibrium conditions. The DEM method therefore should be adopted for the most effective identification of sediment transfers at the relevant reach scales in dynamic gravel-bed rivers.

ACKNOWLEDGEMENTS

We thank Mr. Guy Renwick (Holystone Grange) and Brian Little (Sharperton) for granting access to the study reach. We also thank Ann Rooke for drafting Figs 1–3. Helpful comments on the submitted draft were received from referees J. Laronne and P.V. Villard and the editors. ICF acknowledges financial support for fieldwork from 1996 RAE monies to the Division of Geography & Environmental Management, University of Northumbria.

REFERENCES

Ashmore, P.E. and Church, M.A. (1998) Sediment transport and river morphology: a paradigm for study. In: *Gravel-bed Rivers in the Environment* (Eds P.C. Klingemann, R.L. Beschta, P.D. Komar and J.B. Bradley), pp. 115–148. Water Resources Publications, Highlands Ranch, CO.

Brasington, J. and Richards, K.S. (1998) Interactions between model predictions, parameters and DTM scales for TOPMODEL. *Comput. Geosci.*, **24**, 299–314.

Brasington, J., Rumsby, B.T. and McVey, R.A. (2000) Monitoring and modelling morphological change in a braided gravel-bed river using high resolution GPS-based survey. *Earth Surf. Process. Landf.*, **25**, 973–990.

Brewer, P.A. and Passmore, D.G. (2002) Sediment budgeting techniques in gravel bed rivers. In: *Sediment Flux to Basins: Causes, Controls and Consequences* (Eds S. Jones and L.E. Frostick). *Geol. Soc. London Spec. Publ.*, **191**, 97–113.

Buffin-Bélanger, T., Roy, A.G. and Kirkbride, A.D. (2000) On large-scale flow structures in a gravel-bed river. *Geomorphology*, **32**, 417–435.

Chappell, A., Heritage, G.L., Fuller, I.C., Large, A.R.G. and Milan, D.J. (2003) Geostatistical analysis of ground-survey elevation data to elucidate spatial and temporal river channel change. *Earth Surf. Process. Landf.*, **28**, 349–370.

Davies, T.H.R. (1987) Problems of bedload transport in braided gravel-bed rivers. In: *Sediment Transport in Gravel-bed Rivers* (Eds C.R. Thorne, J.C. Bathurst and R.D. Hey), pp. 793–828. Wiley, Chichester.

Dixon, L.F.J., Barker, R., Bray, M., Farres, P., Hooke, J., Inkpen, R., Merel, A., Payne, D. and Shelford, A. (1998) Analytical photogrammetry for geomorphological research. In: *Landform Monitoring, Modelling and Analysis* (Eds S.N. Lane, K.S. Richards and J.H. Chandler), pp. 63–94. Wiley, Chichester.

Eaton, B.C. and Lapointe, M.F. (2001) Effects of large floods on sediment transport and reach morphology in the cobble-bed Sainte Marguerite River. *Geomorphology*, **40**, 291–309.

Einstein, H.A. (1937) Bedload transport as a probability problem. PhD thesis, Zurich. (English translation by Sayre, W.W. (1972) in: *Sedimentation* (Ed. H.W. Shen) Fort Collins, Colorado.)

Ferguson, R.I. and Ashworth, P.J. (1992) Spatial patterns of bedload transport and channel change in braided and near braided rivers. In: *Dynamics of Gravel-bed Rivers* (Eds P. Billi, R.D. Hey, C.R. Thorne and P. Tacconi), pp. 477–496. Wiley, Chichester.

Ferguson, R.I. and Werritty, A. (1983) Bar development and channel changes in the gravelly River Feshie. In: *Modern and Ancient Fluvial Systems* (Eds J. Collinson and J. Lewin). *Spec. Publs Int. Ass. Sediment.*, **6**, 133–143.

Ferguson, R.I., Ashmore, P.E., Ashworth, P.J., Paola, C. and Prestegaard, K.L. (1992) Measurements in a braided river chute and lobe, I, flow pattern, sediment transport and channel change. *Water Resour. Res.*, **28**, 1877–1886.

Fuller, I.C., Passmore, D.G., Heritage, G.L., Large, A.R.G., Milan, D.J. and Brewer, P.A. (2002) Annual sediment budgets in an unstable gravel bed river: the River Coquet, northern England. In: *Sediment Flux to Basins: Causes, Controls and Consequences* (Eds S. Jones and L.E. Frostick). *Geol. Soc. London Spec. Publ.*, **191**, 115–131.

Fuller, I.C., Large, A.R.G., Charlton, M.E., Heritage, G.L. and Milan, D.J. (2003) Reach-scale sediment transfers: an evaluation of two morphological budgeting approaches. *Earth Surf. Process. Landf.*, **28**, 889–903.

Garcia, C., Laronne, J.B. and Sala, M. (2000) Continuous monitoring of bedload flux in a mountain gravel-bed river. *Geomorphology*, **34**, 23–31.

Goff, J.R. and Ashmore, P.E. (1994) Gravel transport and morphological change in braided Sunwapta river, Alberta, Canada. *Earth Surf. Process. Landf.*, **19**, 195–213.

Gomez, B. and Church, M. (1989) An assessment of bed load sediment transport formulae for gravel bed rivers. *Water Resour. Res.*, **25**, 1161–1186.

Gomez, B., Naff, R.L. and Hubbell, D.W. (1989) Temporal variations in bedload transport rates associated with migration of bedforms. *Earth Surf. Process. Landf.*, **14**, 135–156.

Ham, D.G. and Church, M. (2000) Bed-material transport estimated from channel morphodynamics: Chilliwack River, British Columbia. *Earth Surf. Process. Landf.*, **25**, 1123–1142.

Habersack, H.M. and Laronne, J.B. (2002) Evaluation and improvement of bedload discharge formulas based on Helley-Smith sampling in an alpine gravel bed river. *J. Hydraul. Eng.*, **128**, 484–499.

Heritage, G.L., Fuller, I.C., Charlton, M.E., Brewer, P.A. and Passmore, D.G. (1998) CDW photogrammetry of low relief fluvial features: accuracy and implications for reach-scale sediment budgeting. *Earth Surf. Process. Landf.*, **23**, 1219–1233.

Hoey, T.B. (1992) Temporal variations in bedload transport rates and sediment storage in gravel-bed rivers. *Prog. Phys. Geogr.*, **16**, 319–338.

Hooke, J.M. (1995) Processes of channel planform change on meandering channels in the UK. In: *Changing River Channels* (Eds A. Gurnell and G.E. Petts), pp. 87–115. Wiley, Chichester.

Hubbell, D. (1987) Bed load sampling and analysis. In: *Sediment Transport in Gravel-bed Rivers* (Eds C.R. Thorne, J.C. Bathurst and R.D. Hey), pp. 89–118. Wiley, Chichester.

Kondolf, G.M. and Matthews, W.V.G. (1986) Transport of tracer gravels on a coastal California river. *J. Hydrol.*, **85**, 265–280.

Lane, S.N. (1998) The use of digital terrain modelling in the understanding of dynamic river channel systems. In: *Landform Monitoring, Modelling and Analysis* (Eds S.N. Lane, K.S. Richards and J.H. Chandler), pp. 311–342. Wiley, Chichester.

Lane, S.N., Chandler, J.H. and Richards, K.S. (1994) Developments in monitoring and terrain modelling small-scale river-bed topography. *Earth Surf. Process. Landf.*, **19**, 349–368.

Laronne, J.B., Outhet, D.N. and Duckham, J.L. (1992) Determining event bedload volumes for evaluation of potential degradation sites due to gravel extraction, N.S.W., Australia. *Proceedings, Symposium on Erosion and Sediment Transport Monitoring Programmes in River Basins*, IAHS Publication No. **210**, pp. 87–94. Oslo, August.

Leopold, L.B. (1992) Sediment size that determines channel morphology. In: *Dynamics of Gravel-bed Rivers* (Eds P. Billi, R.D. Hey, C.R. Thorne and P. Tacconi), pp. 297–311.Wiley, Chichester.

Macklin, M.G. (1999) Holocene River Environments in Prehistoric Britain: Human interaction and impact. *Quaternary Proceedings*, **7**, 521–530.

Martin, Y. and Church, M. (1995) Bed-material transport estimated from channel surveys: Vedder River, British Columbia. *Earth Surf. Process. Landf.*, **20**, 347–361.

Milne, J.A. and Sear, D.A. (1997) Modelling river channel topography using GIS. *Int. J. Geographical Inform. Sci.*, **11**, 499–519.

McLean, D.G. and Church, M. (1999) Sediment transport along lower Fraser River 2. Estimates based on the long-term gravel budget. *Water Resour. Res.*, **35**, 2549–2559.

Milan, D.J., Heritage, G.L., Large, A.R.G. and Brunsdon, C.F. (1999) Influence of particle shape and sorting upon sample size estimates for a coarse-grained upland stream. *Sediment. Geol.*, **129**, 85–100.

Mosley, M.P. (1978) Bed material transport in the Tamaki River near Dannevirke, North Island, New Zealand. *N. Z. J. Sci.*, **21**, 619–626.

Naden, P. and Brayshaw, A.C. (1987) Bedforms in gravel-bed rivers. In: *River Channels: Environment and Process* (Ed. K.S. Richards), IBG Special Publication **18**, pp. 249–271. Blackwell, Oxford.

Nanson, G.C. and Croke, J.C. (1992) A genetic classification of floodplains. *Geomorphology*, **4**, 459–486.

Neill, C.R. (1987) Sediment balance considerations linking long-term transport and channel processes. In: *Sediment Transport in Gravel-bed Rivers* (Eds C.R. Thorne, J.C. Bathurst and R.D. Hey), pp. 225–240. Wiley, Chichester.

Nicholas, A.P., Ashworth, P.J., Kirkby, M.J., Macklin, M.G. and Murray, T. (1995) Sediment slugs: large scale fluctuations in fluvial sediment transport rates and storage volumes. *Prog. Phys. Geogr.*, **19**, 500–519.

Paige, A.D. and Hickin, E.J. (2000) Annual bed-elevation regime in the alluvial channel of Squamish River, southwestern British Columbia, Canada. *Earth Surf. Process. Landf.*, **25**, 991–1009.

Press, W.H., Flannery, B.P., Teukolsky, S.A. and Vetterling, W.T. (1988) *Numerical Recipes in C.* Cambridge University Press, Cambridge.

Reid, I., Powell, D.M. and Laronne, J.B. (1996) Prediction of bedload transport by desert flash-floods. *J. Hydraul. Eng.*, **122**, 170–173.

Reid, I., Laronne, J.B. and Powell, D.M. (1999) Impact of major climate change on coarse-grained river sedimentation: a speculative assessment based on measured flux. In: *Fluvial Processes and Environmental Change* (Eds A.G. Brown and T.A. Quine), pp. 105–115. Wiley, Chichester,

SURFER™ (1999) *Manual*. Golden Software, Golden, CO.

Wathen, S.J., Hoey, T.B. and Werritty, A. (1997) Quantitative determination of the activity of within-reach sediment storage in a small gravel-bed river using transit time and response time. *Geomorphology*, **20**, 113–134.

Westaway, R.M., Lane, S.N. and Hicks, D.M. (2000) The development of an automated correction procedure for digital photogrammetry for the study of wide, shallow, gravel-bed rivers. *Earth Surf. Process. Landf.*, **25**, 209–226.

Dune-phase fluvial transport and deposition model of gravelly sand

MAARTEN G. KLEINHANS

Utrecht University, Faculty of Geography, Department of Physical Geography, P.O. Box 80115,
3508 TC Utrecht, The Netherlands (Email: m.kleinhans@geog.uu.nl)

ABSTRACT

The importance and role of sorting processes in dune-phase bedload transport and deposition is demonstrated with flume experiments, vibracores from the River Rhine (The Netherlands) and with sediment transport and dune data from the River Rhine. The entrainment and deposition depth of the sediment depend on dune trough levels below the average bed level and therefore on the dune height. As bedload sediment transport depends partly on grain size, it will be dependent upon the relict vertical sorting left in the bed by former discharge waves. The vertical sorting is created by sorting in grain flows on the lee side of dunes and by gravel lag formation in the dune troughs. Based on these principles, a simple reach-representative process model is developed for the prediction of bed sediment reworking, vertical sorting and deposition by dunes. The model is applied to two successive discharge waves of different magnitude, and predicts qualitatively the same vertical sorting characteristics as observed in the vibracores from the River Rhine (The Netherlands) after two successive discharge waves. The effect of the sorting on the sediment transport, and how to include this feedback in future models, is discussed.

INTRODUCTION

Relevance of sediment sorting and scope

It has been established that sediment transport and deposition in channel beds with sand and gravel entails consideration of the different sediment sizes and their sorting in the river bed (e.g. Klaassen, 1987, 1991; Klaassen et al., 1987; Bridge & Bennett, 1992; Wilcock, 1993, 2001; Wathen et al., 1995). In sand–gravel-bed rivers with both large dunes and significant grain-size variation, the sediment is sorted during the development, migration and decay of dunes during discharge waves. As sediment transport depends on grain size, the sorted sediment in turn will affect sediment transport rates in the course of a discharge wave. Sorting during deposition and successive entrainment therefore must be linked in morphodynamic models (Kleinhans, 2001). In the past, mathematical, so-called 'active layer' models have been developed to link dune migration, sediment transport and vertical sediment sorting (e.g. Ribberink, 1987). Usually the sediment involved in dune migration is modelled as a single or at best a few active layers, which has the disadvantage that vertical sediment sorting trends related to the dunes cannot be well represented. Recently Parker et al. (2000) developed a mathematical morphological model concept with a continuous (instead of discrete) description of sediment sorting in the dunes and the bed, but could not implement it because a general predictor for the vertical sediment sorting was not available. A generic understanding of vertical sorting mechanisms is important for the further development of morphological models for sand–gravel-bed river channel behaviour.

This paper describes necessary elements for a predictor of stratigraphy and sorting in channel bed deposits, and the role of dune height variation in this sorting. Furthermore, the effect of this vertical sorting in the channel bed on bedload transport is discussed. Flume experiments and field data from the River Rhine are used to verify concepts of sorting and their relations with transport. A process model is developed for the sorting in the bed, and the necessary elements are identified for a future mathematical model by which both the sorting patterns and the relation between sorting and the bedload transport can be predicted. The model developed is an extension of Klaassen (1987) and Klaassen *et al.* (1987), which may be applicable in a model like the one described by Parker *et al.* (2000) for the case of rivers with dunes.

This study is focused on reach-representative conditions and daily averaged sediment transport for three practical reasons: data collection of sediment sorting and sediment transport in the River Rhine were done at these scales, dune population characteristics must be studied over a number of dunes, and morphodynamic models necessary for large-river management often are schematized for these time- and spatial scales.

Cross-bedded deposits and lag deposits

Two types of deposits are distinguished genetically, according to Boersma *et al.* (1968), Allen (1970), Ribberink (1987) and Kleinhans (2001): the cross-bedded deposit with an upward fining, and the lag deposit. Cross-bedded deposits are formed by the propagation of dunes by discontinuous grain flows of bedload sediment at the lee side of a dune. The sediment is sorted vertically in the grain flow process (Fig. 1): the gravel is deposited mainly on the lower half of the lee slope, whereas the finer grades are predominantly deposited in the upper half. The result is an upward fining deposit. Allen (1963, 1970), Boersma *et al.* (1968) and many others found a distinct upward fining within subaqueous dunes in the laboratory and in the field, indicating that sorting in the grain flow plays an important role. Although this sorting principle related to grain flows is well known, a mathematical description of the process is not available. Hunter (1985) studied grain flow

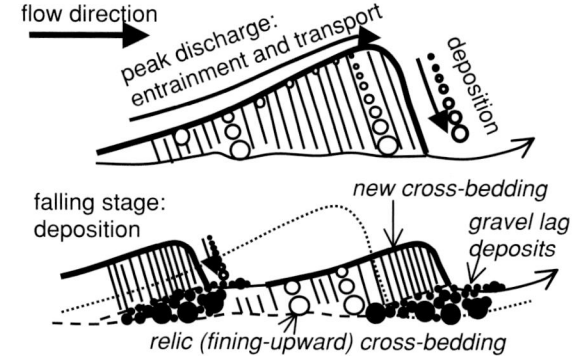

Fig. 1 Hypothesized channel-bed deposits. During peak discharge the transported sediment is sorted in the grain flows on the lee side of a dune. During waning discharge, the dunes diminish and leave cross-bedding relics. Owing to selective deposition, an upward fining accumulation of lag deposits is created in addition.

dynamics but referred only to well-sorted sand. Jopling (1965) developed a model for the composition of foresets and toesets as a result of settling from suspension at the lee side of a laboratory delta, but focused mainly on conditions with significant suspension, which probably is not relevant for the gravelly sand under study here. When the deposition on slipfaces is dominated by suspension, vertical sorting may disappear or even become coarsening upward (Hunter, 1985).

This study is limited to grain-flow-dominated conditions, which reasonably can be assumed to play a key role in rivers with coarse sediment. Makse *et al.* (1997) and Koeppe *et al.* (1998) created stratification and vertical sorting in a bimodal mixture of sand and sugar (in air), and were able to model the sorting patterns. The fining upward pattern was caused by kinematic sorting on the slipface, in which the small grains move downward through the pores of the dilated sediment mixture on the top of the slipface, but the large grains cannot move down. The large grains then are on top of the slipface and have more opportunity to roll down the slope over the small grains. In their models and experiments, however, the cross-stratification and fining upward sorting seem to exclude each other, which raises the question whether their models could be applied to natural sediments and conditions (see Kleinhans (2002) for an extensive review). Kleinhans (2002) determined fining upward sorting in a cross-stratified subaqueous delta deposit formed in natural sediments in a narrow flume. It was observed that kinematic sorting indeed caused the large grains to emerge on the top of grain flow on the slip face, and this process dominated the sorting even in significant suspension of the finer grades. Contrary to the experiments of Makse and Koeppe, the net sorting in the delta was distinctly fining upward.

Lag deposits are formed as a result of size-selective entrainment and deposition during the different stages of a discharge wave. The winnowing of fine and intermediate grades from the bed and their reworking into dunes leads to a lag layer consisting of the coarsest grades that were not in motion (Carling et al., 2000). In addition, the largest grains that are in motion during the peak discharge will be deposited in lowering discharge, whereas smaller grains remain in transport. This results in an upward fining accumulation of lag deposits without cross-bedding. Obviously the process of sorting in grain flows on the lee side of dunes may help to transfer the coarser grades down into the active layer (Fig. 1). However, if cross-stratification cannot be observed in deposits because of disturbance in the sampling procedure or because the sediment is too coarse for cross-stratification, then it is impossible to distinguish between the gravelly layers formed by grain flows and lag deposits. The importance of these two sorting mechanisms and resulting deposits will be demonstrated in the next sections.

Sediment transport and sorting during and after successive discharge waves

These sorting patterns are erased and recreated by the dunes of a next discharge event, down to the depth of the deepest troughs of the new dunes. For discharge waves in order of decreasing magnitude, the lower part of their deposits may therefore be preserved. Suppose that two successive discharge waves rework the river bed (no measurable net aggradation or degradation), and the first discharge wave is a large one (e.g. recurrence interval of 10 yr or so, panel 2 in Fig. 2) and the second is smaller (panel 4 in Fig. 2). The resulting deposit (panel 5) may consist of two lag layers, one at the trough depth of dunes which occurred in the first discharge wave (panel 3), and one at the trough depth of the second. In between and above the lag

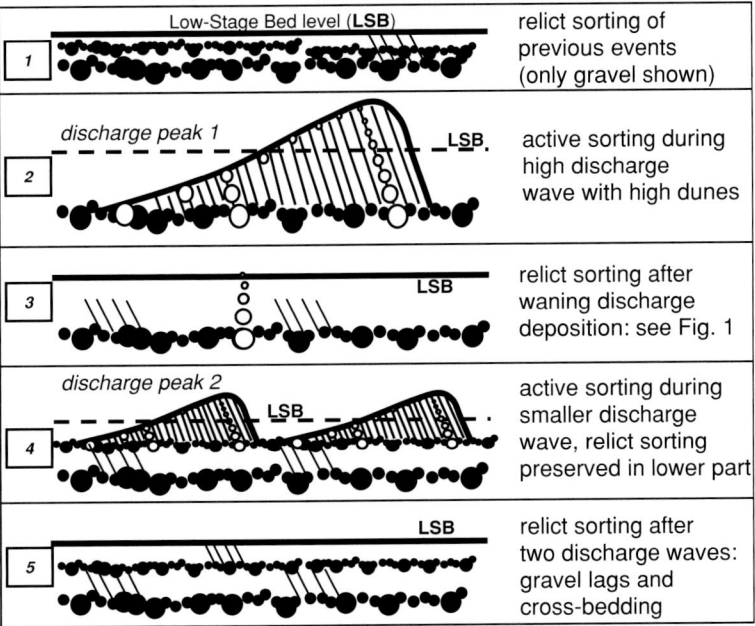

Fig. 2 Conceptual model of sediment erosion and deposition with emphasis on the role of dunes in two successive discharge waves in order of decreasing magnitude. The vertical sorting (here only the gravel lag layers are shown) is as described in Fig. 1. The smaller dunes of the second discharge wave erase part of the sorting resulting from the first wave, and leave a comparable but less deep vertical sorting.

layers, the channel bed may consist of both cross-bedded and lag deposits. In Fig. 2, only the lag layers are shown, which are the hypothesized result of two successive discharge waves of which the second one was smaller than the first one. Note that consideration of an inverse order of discharge waves (first small ones then a large one) is not useful for studying the deposits, because then the deposits of the small ones are erased and reworked by the large one.

Dunes are fluctuations of the local bed level around the average bed level. Assuming no net deposition or erosion, the latter is equal to the low-flow plane-bed level. The deepest dune troughs locally determine the depth or lowest level of gravel deposition or entrainment in the bed. The variation in dune height and dune-trough-scour depth may be large (Paola & Borgman, 1991; Leclair, 2000; Leclair & Bridge, 2001), thus there may only be a few deep troughs during a certain period of time. Expressed alternatively, it takes a long time and many dune passages before the gravel has uniformly been worked down to (or entrained from) the depth of the deepest troughs (Ribberink, 1987). Thus, on a reach-representative length scale (c. 1 km), with rather regular dunes that move at best a few dune lengths during a discharge wave, and during the non-equilibrium conditions of a discharge wave, the deepest troughs are not likely to dominate the vertical sorting pattern but occur only locally. In that non-equilibrium case, the average or some measure between the average and deepest dune trough depths is more representative of the depth of scour and gravel deposition than in the cases studied, for example, by Leclair (2000) and Leclair & Bridge (2001). As the sediment transport and deposition on a reach-representative scale are of interest here, the average dune height with concurrent average scour depth is assumed to be representative. An appropriate active layer is then defined here as the layer of moving dunes (represented by their average dimension) in which the sediment is sorted during the time of dune passage (e.g. Ribberink, 1987; Klaassen, 1991). This concept is only a slight simplification of that of Paola & Borgman (1991) Leclair & Bridge (2001) and Parker et al. (2000) because in the non-equilibrium conditions considered here the impact of the deepest troughs is extremely limited. Had the discharge waves in the Rhine been longer, then the concept here would be oversimplified. Further consequences of using the average dune dimensions are discussed later.

Hypothesizing, the bedload transport during a discharge wave will be affected by gravelly layers deposited in a previous event as follows:
1 in rising discharge, the dunes attempt to lower their troughs through a relict gravelly layer, which may reduce the natural variation in trough scour depth;
2 as a consequence, this gravel is being incorporated in the active layer;
3 thus the sediment available for transport (i.e. active layer) is relatively coarse, and the bedload transport rate consequently is relatively small;
4 in lowering discharge, the gravel is worked down deeper into the active layer by the grain flow process and in the form of lag deposits;
5 as the dunes become lower and rework an increasingly shallower active layer, the gravel layers are abandoned and the active layer becomes finer;
6 as a result of the fining of the active layer, the bedload rate in falling stages is relatively larger than in rising stages, leading to a hysteresis of bedload rate as a function of discharge.

The bedload transport (capacity) in turn affects the sorting of the bed. If the transport capacity of gravel is high enough to transport all available gravel in the dune layer, then a gravel layer in the dune troughs will not be very thick. Thus bedload transport capacity, availability of sediment in the dune layer and vertical sorting in the deposits are intimately linked.

FLUME EXPERIMENTS

The concepts introduced in the previous section are in part demonstrated with controlled laboratory experiments. Three experiments have been carried out recently (T5, T7 and T9; in Blom & Kleinhans, 1999; Kleinhans, 2000, Kleinhans & Van Rijn, 2002) with a slightly bimodal sand–gravel mixture (D_5 = 0.1 mm, D_{50} = 1.8 mm, D_{100} = 16 mm), which was dredged from the field site in the River Rhine that is described below, and cut-off at 16 mm to prevent choking of the sediment recirculation system. The flume was 50 m long

Table 1 Basic data from flume experiments presented by Kleinhans & Van Rijn (2002) and Blom & Kleinhans (1999).

Number	Condition*	Hydraulic radius† (m)	Discharge ($m^3 s^{-1}$)	Flow velocity ($m s^{-1}$) (Q/A)	Water surface slope (10^{-3})
T5	Selective transport	0.19–0.23	0.22–0.26	0.69	−1.472
T7	All grain sizes in motion	0.30–0.32	0.41–0.43	0.79	−1.520
T9	Selective transport	0.22–0.25	0.26–0.28	0.70	−1.694

Number	Grain shear stress (Pa)	Total bedload transport‡ ($g s^{-1} m^{-1}$)	Gravel§ fraction in bedload (%)	Dune height‡ (10^{-3} m)	Final bed state
T5	2.0	42.0 ± 0.4	21	28 ± 11	Small barchans over armour layer
T7	2.3	66.0 ± 0.8	25	57 ± 21	Large dunes over armour layer
T9	2.0	50.8 ± 0.6	18	49 ± 21	Large dunes over buried armour layer

*Initial condition: bed was mixed and bedslope installed at -1.400×10^{-3}.
†Corrected for side-wall roughness with the Vanoni-Brooks method.
‡Given with standard deviations (67th percentiles).
§Gravel is defined as $D > 2$ mm. In the sediment put into the flume the gravel fraction is 40%. Values have been determined from average sieve curve of bed and recirculated sediments and have a standard error of ±0.2%.

and 1.5 m wide and had electromagnetic bed and water surface profilers that were automatically moved along the flume during the experiments. Basic data are given in Table 1. Uniform flow conditions were maintained to avoid large-scale erosion and sedimentation. The transported sediment was recirculated and the submerged weight in transport measured every few minutes. The standard errors in measuring sediment transport rates are less than 3%. The tests were continued until a morphological equilibrium was reached (change became smaller than noise), which took a few days. The flow, sediment transport and dune parameters were determined by averaging all measurements (sediment transport in the recirculation, fixed flow discharge, water surface and bed profiles over the downstream half of the flume) during the equilibrium phase of the experiments.

The main difference between the experiments is their history. Experiments T5 and T9 had approximately the same flow conditions (0.7 m s^{-1}), whereas T7 had a larger flow velocity of 0.8 m s^{-1}. The bed shear stresses were just above the initial motion threshold of the coarsest sediment sizes for T7 (but not for T5 and T9), whereas suspension of sand was negligible. Whereas T5 starts from a fully mixed bed, T9 initially has the history of sorting from T5 and T7. This leads to a different final sorting for T9 although the flow velocity was the same as in T5. In T5, the preferential entrainment of sediment finer than the bulk bed sediment led to the formation of a continuous armour layer at the level of dune troughs below the propagating dunes (Fig. 3) (Kleinhans, 2000). Furthermore the sediment was vertically sorted in the dunes. T7 had higher dunes, while the armour at the level of the troughs was lowered due to the winnowing of the extra sediment used to increase the dune volume. Due to the higher shear stress of T7, the entrained sediment also was coarser than in T5. Comparable findings are presented by Klaassen et al. (1987) and Klaassen (1991). In T9 the largest particles from the dunes of T7 were only partially or no longer mobile. This led to a decrease of the dune height and deposition in the dune troughs. Thus the armour layer of the previous experiment was buried below a growing lag deposit of fine gravel (fining upward), seen in Fig. 3 as a fining upward trend below the dune trough level of T9.

The transport rate and composition are affected by the sorting in the bed, as revealed by sediment transport measurements. The sediment transport and the fraction of gravel in transport were highest in T7, and were considerably different in T5 and T9 despite their equal flow velocities (Kleinhans & Van Rijn, 2002). The sediment transport in T9 was 15% higher than in T5, whereas the fraction of gravel in T9 was about the same factor lower

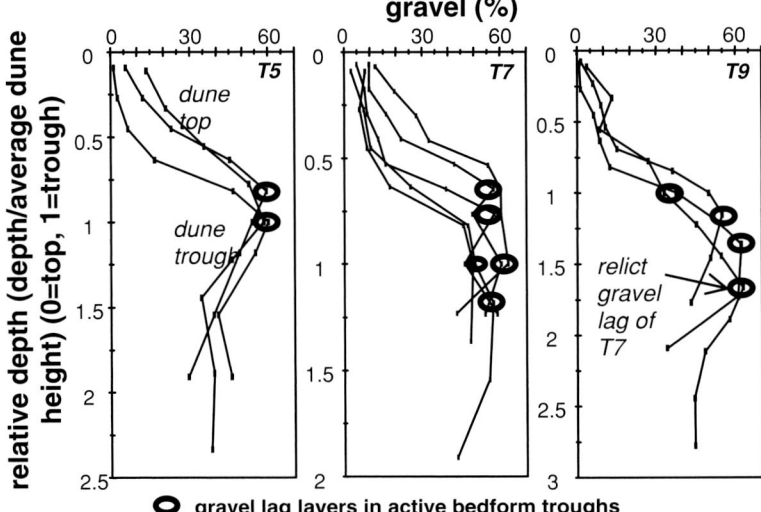

Fig. 3 Vertical sorting measured in the dunes in the flume experiments (Kleinhans, 2000), expressed as the gravel fraction in the sediment ($D_{gravel} > 2$ mm, remainder is sand). The depths of the dune troughs (and armour layers in T5 and T7) are given as circles. The gravel fraction clearly decreases to the dune top (upward fining). For T9 there is also a relict upward fining deposit below the dune trough level, which consists of relict cross-bedding and the accumulation of lag deposits (Fig. 1). (Data from Blom & Kleinhans, 1999.)

(Table 1). The entrained gravel of T5 and T7 was deposited in the troughs of the dunes in T9, and of course the finer sediment of T9 was more easily transported. This was also found by Klaassen et al. (1987) and Klaassen (1991). The experiments showed that a coarse layer is left in the bed that reflects the level of the troughs of the highest dunes in the recent past. The duration of the experiments and the two-dimensionality of the dunes allowed the vertical sorting to attain approximate equilibrium, thus the observed gravel lag layer was plane and continuous at its base, with some cross-stratification of T7 preserved above it.

DATA FROM THE RIVER RHINE

Field site description

The effect of successive discharge waves on reach-representative vertical sorting and sediment transport is studied with data collected in the River Rhine, using vibracores of the channel bed (Gruijters et al., 2001), statistics on dune dimensions and migration rates in 1995 and 1998, as well as suspended and bedload sediment transport during the discharge wave in 1998 (Wilbers, 1998, 1999; Kleinhans, 1999, 2000; Kleinhans & Ten Brinke, 2001). These datasets were all collected near the bifurcation Pannerdensche Kop, where the Bovenrijn splits into a large branch, the River Waal (two-thirds of the discharge), and a small branch, the Pannerden Channel (one-third of the discharge, Fig. 4). The Rhine upstream of the bifurcation point is called Bovenrijn. The sediment at that point consists of sand and gravel with $D_{50} = 2$ mm and $D_{90} = 14$ mm. The basic data for 1998 are given in Table 2. In 1995 the largest discharge wave of the past two decades occurred, in 1997 a small discharge wave occurred, and in 1998 a larger one occurred, and between 1998 and the moment of vibracoring, no significant discharge wave occurred (the discharge was determined daily from water levels and calibrated relations between discharge and water level throughout the years).

Dune height and sediment transport measurements

The dune dimensions were obtained from three-dimensional echo soundings. The echo soundings of 1998 were also used for dune tracking to estimate the bedload transport rate from the propagation and dimensions of the dunes with the method of Ten Brinke et al. (1999). Both the bedload sampling method with the Helley-Smith-type sampler and the dune tracking method in 1998 show small bedload transport rates (of sand plus gravel) at rising stage and large transport at falling stage (Fig. 5), i.e. a considerable hysteresis. At falling stage the transport measured with the Helley-

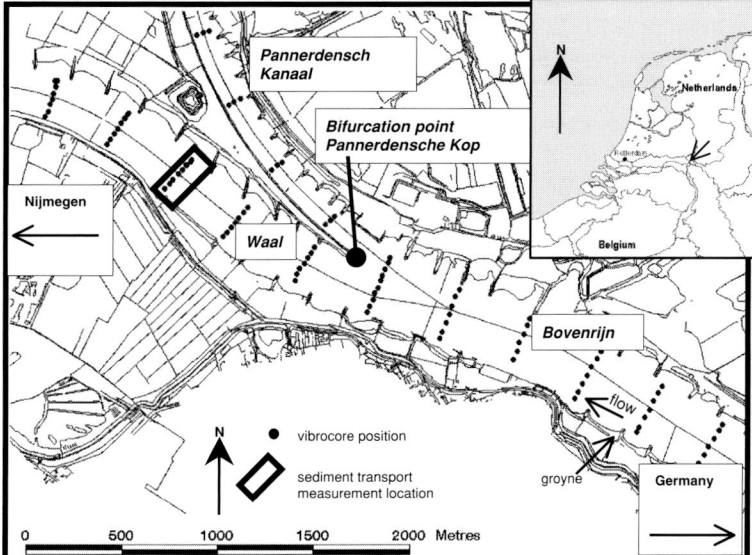

Fig. 4 Map of the measurement positions of direct sampling of the sediment transport and flow in the River Waal and of dune tracking in the Waal and Bovenrijn, as well as the vibracore positions. The whole area shown in detail was mapped with multibeam echo-sounders.

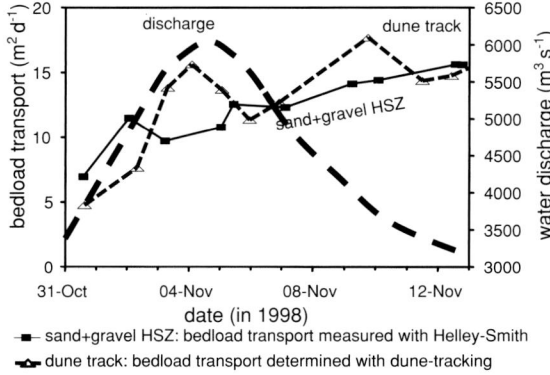

Fig. 5 Measured sediment transport in the Waal during the discharge wave of 1998, given as total bedload divided by channel width. The measurements with the Helley-Smith-type bedload sampler (HSZ) have also been split in the gravel and sand bedload transport.

Smith and determined with the dune tracking method remained more or less constant and large for days while the flow velocity decreased. The measurements with the Helley-Smith have a high accuracy: the uncertainty of cross-channel averaged sediment transport related to natural variability and as a result of the sediment transport measurement methods is 15–20% (Kleinhans & Ten Brinke, 2001). No significant degradation or aggradation of the measurement location could be detected from the echo-soundings.

The dunes were observed to become inactive at falling stage, and small secondary dunes emerged on top of the large primary dunes while reworking and smoothing the tops of the primary dunes. This phenomenon has been observed by others in large rivers and has been attributed to the tardy reaction of large dunes to fast changes in discharge, although a more physical explanation is still wanting (e.g. Allen & Collinson, 1974). Interestingly, the phenomenon is much more prominent in the Waal than in the upstream Rhine. This may be related to the mobility of the sediment, of which the diameter is only slightly larger in the upstream Rhine whereas the flow discharge is 1.5 times as large. Given the rather slow migration celerity of the large dunes relative to their own length and their becoming inactive, much of the sediment in these dunes was not reworked in the remainder of the discharge wave. Thus the opportunity for preservation of upward fining cross-stratified deposits is considerable (Fig. 1) in the whole area. In general, the bedforms are much smaller than expected from empirical relations between water depth and dune height, which may be due to the non-equilibrium conditions and the presence of groynes.

Table 2 (a) Depth- and width-averaged flow parameters and sediment transport in the River Waal, 1998. (Data from Kleinhans, 2002.)

Date	Time of day (hours)	Discharge ($m^{-3}\ s^{-1}$)	Depth-averaged velocity† ($m\ s^{-1}$)	Chezy† ($m^{0.5}\ s^{-1}$)	Shear velocity u_*†‡ ($m\ s^{-1}$)	Roughness length k_s† (m)	Flow depth (m)	Suspended load† ($g\ s^{-1}\ m^{-1}$)	Bedload transport‡ ($g\ s^{-1}\ m^{-1}$)	Bedload sand ($g\ s^{-1}\ m^{-1}$)	Bedload gravel ($g\ s^{-1}\ m^{-1}$)
31 October	13:55:43	3993	1.61	50.6	0.098	0.17	9.2	476	130	80	50
02 November	01:27:51	5032	1.70	40.5	0.126	0.65	9.7	650	214	141	73
03 November	05:21:26	5741	1.80	39.7	0.137	0.76	10.2	737	181	122	59
05 November	01:04:17	6097	1.86	38.9	0.129	0.88	10.7	715	201	144	57
05 November	10:38:34	5891	1.79	39.3	0.124	0.83	10.5	606	234	171	63
07 November	02:45:00	5114	1.62	40.9	0.115	0.65	10.1	351	230	148	82
09 November	05:53:34	4198	1.51	39.1	0.117	0.76	9.4	199	264	181	83
10 November	02:12:51	3899	1.48	39.4	0.112	0.72	9.2	143	269	197	71
12 November	13:40:43	3407	1.42	44.4	0.092	0.35	8.7	83	291	222	69
12 November	18:34:17	3372	1.42	44.5	0.092	0.35	8.6	83	291	222	69

†Flow parameters (except discharge) and suspended sediment transport were measured with the Acoustic Sand Transport instrument ASTM (Kleinhans & Ten Brinke, 2001).
‡Calibration factor 2.74 of Helley-Smith (so all sample volumes were multiplied with this factor; Kleinhans, 2002).

Table 2 (b) Dune parameters and bedload transport from dune tracking in the Waal and Bovenrijn, 1998. (Data from Wilbers, 1999; also see Kleinhans, 2002.)

Date	Discharge (m^{-3} s^{-1})	Dune height (m)	Dune height variation (95% interval) (m)	Dune length (m)	Dune celerity (m day^{-1})	Dune track (m^2 day^{-1})	Bedload transport* (g s^{-1} m^{-1})
	Waal						
30 October	3166	0.12	0.11–0.14	3.29	250	3	59
31 October	4056	0.12	0.11–0.13	3.93	210	3	55
2 November	5363	0.22	0.20–0.25	6.76	70	5	76
3 November	5946	0.34	0.32–0.37	8.39	93	14	233
4 November	6168	0.47	0.43–0.50	10.87	73	15	261
5 November	6002	0.53	0.47–0.59	13.09	59	14	233
6 November	5536	0.53	0.47–0.58	17.75	45	12	197
7 November†	4863	0.49	0.43–0.54	8.65	80	13	217
10 November	3770	0.30	0.27–0.32	6.57	130	17	290
12 November	3436	0.29	0.27–0.32	6.31	105	14	240
13 November	3267	0.28	0.26–0.30	6.07	105	14	243
16 November	3045	0.31	0.29–0.34	5.98	119	17	287
19 November	3042	0.20	0.19–0.22	5.58	110	7	126

Date	Discharge (m^3 s^{-1})	Dune height (m)	Dune height variation (95% interval) (m)	Dune length (m)	Dune celerity (m day^{-1})	Dune track (m^2 day^{-1})	Bedload transport* (g s^{-1} m^{-1})
	Bovenrijn						
30 October	4783	0.34	0.32–0.36	7.84	120	19	313
31 October	6180	0.48	0.44–0.51	10.97	61	18	311
2 November	8119	0.72	0.66–0.77	15.99	57	20	341
3 November	9045	0.90	0.84–0.97	19.99	55	22	379
4 November	9464	0.98	0.90–1.06	21.92	51	22	378
5 November	9149	1.07	0.98–1.17	24.30	41	20	337
6 November	8267	1.13	1.03–1.23	26.03	31	16	264
7 November	7273	1.19	1.09–1.29	29.16	25	13	223
10 November	5640	0.92	0.83–1.02	32.30	18	8	128
12 November†	5122	0.27	0.22–0.32	6.80	108	12	205
13 November	4851	0.26	0.22–0.30	6.68	110	12	204
16 November	4522	0.29	0.26–0.32	6.60	111	14	239
19 November	4527	0.23	0.21–0.25	7.52	96	11	181

*Unit of transport (g s^{-1} m^{-1}): grammes per second per metre width. Density of sediment including pores is 1460 kg m^{-3}.
†Small dunes become active (and large inactive) on 7 November in the Waal and on 12 November in the Bovenrijn. From this date on, the parameters of the small dunes are given.

The bedload sediment was slightly bimodal (like the bed material), with mode diameters of 0.5 mm for sand and 10 mm for gravel, allowing a convenient division between sandy and gravelly at 2 mm. The bedload at rising stage consisted of 60% sand and 40% gravel, whereas it became sandier during falling stages, with the sand content rising to 75% near the end of the discharge wave (Kleinhans, 1999, 2000). Thus, the bedload transport was larger after the discharge peak than before. This relative rise was solely due to the absolute rise in rate of sand transport, since the rate of gravel transport decreased after the discharge peak.

Explanations for hysteresis of transport rate

The hysteresis observed in the River Rhine is much larger than that observed in the flume experiments, and larger than can reasonably be expected from differences in grain sizes (bedload predictors are not overly sensitive to grain size). It is likely that other factors amplify the hysteresis. The possible causes of this hysteresis are discussed in detail in Kleinhans (2002) and are summarized here. For lack of more definite data or modelling, the explanations remain hypothetical to some extent. However, a key point to be kept in mind is that in the Rhine upstream of the bifurcation point, the hysteresis in bedload (measured with the dune-track method) was opposite to that in the lower Rhine branch, which considerably limits the possibilities. The causes of hysteresis are probably as follows.

1 Hysteresis of hydraulic roughness owing to dune development (cf. Allen & Collinson, 1974) may contribute to the transport hysteresis as follows. The tardy reaction of large dunes to changing flow causes a time lag (i.e. hysteresis) between dune height and flow discharge. Thus the dune height is temporarily larger after the discharge peak than before. The energy that is available for the bedload transport is the difference between the total energy (total bed shear stress) and the energy dissipated by dunes, whereas higher dunes generally dissipate more energy. At some point after the discharge peak, secondary dunes emerge superimposed on the primary dunes (Kleinhans, 2002), and are no longer destroyed as they arrive at the top and lee side of the primary dunes. This means that there is probably no flow separation related to the primary dunes and that their hydraulic roughness is negligible. Thus the hydraulic roughness is lower than it was before the discharge peak, and there is more energy available for the transportation of bedload after the discharge peak, leading to counter-clockwise hysteresis of the bedload transport rate. The suspended load transport on the other hand is the largest before the discharge peak (clockwise hysteresis) because the suspension depends on the turbulence generation by dunes.

2 Vertical sorting of bedload sediment in the dunes combined with the dune height development resis causes hysteresis as follows. Because the dune height is lagging behind the discharge, the bed shear stress in falling stages is lower than it was before the discharge peak at the same dune height. Owing to the lower bed shear stress, the gravel is worked down to form a lag deposit. As the dunes further diminish in height, this gravel layer is abandoned. As a result, the sediment in the active layer is finer than it was before the discharge peak. This leads to higher sediment transport rates after the peak than before, i.e. counter-clockwise hysteresis.

3 The erosion of fine sediment from sand deposits below the active river bed upstream of the measurement section may lead to hysteresis as follows. This fine sediment moves downstream as a sand wave, and is more easily transported than the sand–gravel mixture in the active river bed. Depending on the arrival time of the sand wave (and thus the measurement location), clockwise or counter-clockwise hysteresis of bedload results. The origin of the sand wave could be from low-flow sand deposits in the upstream meander pool or sediment derived from an (unknown) bank collapse or gravel mining activity upstream, but this was unfortunately not observed. It is unlikely that the sand was mobilized from between the groynes, as echo-sounding and flow and concentration measurements show that sand enters these areas during high discharge and leaves it during low discharge (Schans, 1998). It is also unlikely that the sand originates from suspension fall-out after the discharge peak, because the concentrations are far too low to contribute a significant volume of sand (Kleinhans, 2002).

Presumably all these hypotheses together are responsible for some part of the hysteresis. The first hypothesis should be tested with a mathematical turbulence model for flow over dunes, which is outside the scope of this paper. The second and third potentially explain the observed fining of bedload sediment. Only the last hypothesis, however, is able to explain why the hysteresis in the Rhine branch (Bovenrijn) upstream of the bifurcation point is opposite to that in the lower branch (Waal). Unfortunately, there are no measurements that could indicate the plausibility of a migrating sand wave, therefore this hypothesis remains unfounded for the time being. As neither of the processes responsible for the hysteresis are incorporated in the model presented in this paper, it is not expected that the hysteresis of bedload transport, especially the sandy part of it, is correctly hindcast by the model.

Gravelly layers in the Rhine

In the Waal, 42 vibracores were collected and in the Bovenrijn 35 (Fig. 4). The total depth of the vibracores commonly exceeded 3 m depth below the low stage plane bed. From the measured dune height, the likely depth of gravel lag deposition below the low stage bed level (LSB) can be estimated. This estimate will be compared here with the observed depth of gravelly layers in the cores. When the vibracores were taken, the relict dune tops of previous discharge waves had been smeared out completely and the river bed was plane. Thus, the level of the troughs and expected gravelly layers is 0.5 times or slightly more the dune height below the LSB. With the echo-soundings, no measurable net erosion or aggradation was found. The coarse layers from the trough levels of the largest dunes of 1995 and 1998 are thus predicted at 0.5 to 0.6 times the dune height below the top of the vibracores (Table 3).

The depths of the bottom of the gravelly layers below the top of the cores were measured, rounded to 0.1 m precision, and are given as frequency distributions for the Waal and the Bovenrijn (Fig. 6). Two maxima are found in both the Bovenrijn and the Waal roughly at the expected depth of the dune troughs (Fig. 6 & Table 3). It is deemed unlikely that the dunes of only the 1998 event were the cause of this bimodal pattern of gravelly layers, because then the bimodality is unexpected for this large number of vibracores (also see Fig. 7, where the scour depth distribution is clearly unimodal). It is therefore concluded that these layers were deposited by the dunes of 1995 and 1998. There is considerable variation in gravel layer depth around each maximum, which is probably due to trough scour depth variation.

The bimodal distribution of gravel layers was reproduced in experiments with dunes in a sediment mixture in fast changing discharges carried out at St Anthony Falls Laboratory in Minneapolis (USA) (Kleinhans, 2002). In that case, there was no overlap at all between the gravel layer distributions in depth of the two subsequent discharge waves. It is acknowledged that planar gravel layers could also be the lowest part of cross-stratified deposits truncated at a few grain diameters thickness, but the grain size of the planar gravel layers was considerably larger than in the overlying

Table 3 Predicted and observed reach-averaged depths (in m ±0.1 m) below the low-stage bed level of gravel layers in the Rhine branches for two successive discharge waves.

Year of discharge wave	Bovenrijn			Waal		
	Dune height	Gravel-layer depth		Dune height	Gravel-layer depth	
		Predicted	Observed		Predicted	Observed
1995	1.5	0.8	0.7	1.4*	0.7*	0.7
1998	1.2	0.5	0.3	0.6	0.3	0.3

*In 1995 no dune height measurements were done in the Waal. The dune height used here is estimated from the measured dune heights in 1993, during a comparable discharge wave (Wilbers, 1998). The maximum reach-representative dune height in the Waal was 0.6 m in 1998. The maximum reach-representative dune height in the Bovenrijn was 1.0 m in 1998 and 1.5 m in 1995. In 1995 no echo-soundings were done at the measurement location but from measurements carried out in 1993 at more or less the same discharge peak both upstream and downstream of the measurement location, a probable reach-representative daily maximum dune height of 1.4 m is obtained.

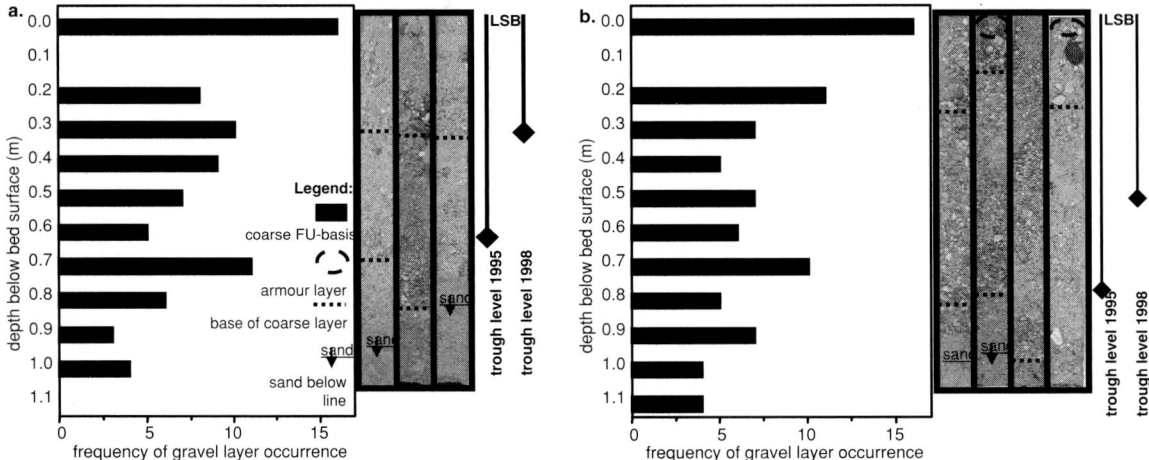

Fig. 6 Observations of the deposits in the River Waal (a) and Bovenrijn (b). The depths of the gravelly layers were measured and used for the histograms (given as the frequency of gravel layer occurrence in the total number of borings used: 35 in Bovenrijn and 42 in Waal). A few examples of vibracores are given, which show the gravel layers. The gravel layers at the bed surface are armour layers, whereas the lower gravel layers are interpreted as the bases of relict cross-bedding and lag deposits (Fig. 1). The lower gravel layers are dominantly found at the depth of dune troughs observed during the discharge waves of 1995 and 1998 (Fig. 2).

sediment (sudden change of grain size in depth), indicating a lag origin.

The observations of both planar gravel layers of a few grains thick and of fining-upward cross-stratification corroborate that the vertical sorting was generated by both sorting mechanisms as described in the introduction. This indicates the importance of the dune height adaptation time and dune celerity. On the one hand, there

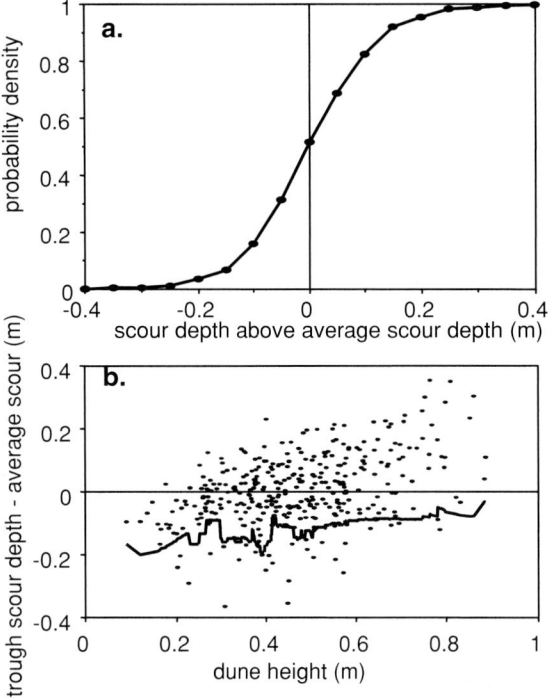

Fig. 7 (*left*) Statistics of dune scour depth of 417 dunes at positions varying from −100 m to +100 m along the river axis over a 0.5 km stretch of the River Waal on 4 November (peak discharge). The average dune height was about 0.5 m. The deviation of scour depth of individual dunes above or below the average scour depth is determined by subtracting the moving average over 10 dunes (to remove the large-scale morphology) from the individual scour depth. (a) Probability distribution of the deviation of scour depth. (b) Deviation of scour depth as a function of dune height (dots). The line denotes the 90th percentile of deepest trough scours calculated from a moving bin of 20 dunes on the list of all dunes ordered from small to large dune height.

are dunes that adapt slowly to changing flow, and become inactive in lowering discharge (Allen & Collinson, 1974; Wilbers, 1998, 1999). In this case, truncated primary cross-bedding of the relict dunes and secondary cross-bedding of the trough-infills by secondary dunes are both preserved, as well as lag deposits that vary in thickness in the flow direction (thin below relict dunes and thick below trough-infills). On the other hand, for dunes that adapt very quickly to changing flow (no significant hysteresis), for instance in flume experiments, the secondary cross-bedding will not occur, but cross-bedding and fining-upward lag deposits will be preserved.

The experiments with dunes in a sediment mixture in fast changing discharges carried out at St Anthony Falls Laboratory in Minneapolis (USA) (Kleinhans, 2002) show that this effect indeed occurs, but that the lag deposits and primary foreset deposits for the experimental conditions occur much more frequently in the bed than the fine secondary cross-bedding. This suggests that the effect of the secondary cross-bedding is not of first-order importance. The vibracores of the River Rhine show a considerable variation in grain size and sorting, but the core-spacing is much larger than the dune length and it is not clear whether the variation is the result of the primary and secondary cross-bedding or has other causes such as sorting of sediment in the meander bends. The question of importance of the secondary cross-bedding in the Rhine therefore remains unresolved.

Effect of variability of dune height and scour depth on gravel-lag-layer depth

The statistics of the trough level (comparable to the term scour depth in Leclair & Bridge, 2001) have been determined from the longitudinal profiles of dunes on 4 November 1998 (peak discharge) (Fig. 7). Three-dimensional dunes are more irregular than two-dimensional dunes and therefore have deeper troughs every now and then. According to the bedform stability diagrams of Southard & Boguchwal (1990), dunes become more three-dimensional with increasing flow velocity. Using the data from the peak discharge in the Waal ensures that the dunes are as near the condition for three-dimensional dunes as they will get in the River Waal. The coordinates of the dune troughs were determined with the software described earlier. A moving average of 10 trough scour depths was subtracted from each individual (mid-point) scour depth to obtain normalized trough scours. The normalized trough scours were used to compute the trough scour probability distribution (Fig. 7 first panel). In addition, the individual normalized trough scour depths are shown as a function of the concurrent dune height (Fig. 7 second panel).

Only 10% of all dunes scour more than 0.13 m below the average scour depth of the surrounding 10 dunes. This variation encompasses at best two classes in the histogram of gravel layer depth (Fig. 6). In addition, the reach-representative statistics are of interest here, although individual deep scours do not propagate fast enough to cover major parts of the river reach. As pointed out earlier, this means that the dune height and scour depth variation is not very relevant for the modelled depth of gravel deposition in this case, where the depths of the gravel layers from the two events are sufficiently far apart.

According to Leclair & Bridge (2001), the average cross-set thickness is approximately one-third (1 over 2.9 [±0.7]) times the dune height. For a dune height of 0.5 m this yields cross-sets of $[1/(2.9 - 0.7) = 0.45 \times 0.5$ m$]$ 0.23 m thickness for scour depths that are one standard deviation (67th percentile) larger than the average scour depth. The 10th percentile of scour depth of the dunes in the Waal, however, is less than 0.2 m below the average scour depth. It therefore must be concluded that the scour-depth variation, and therefore cross-set thickness, in the Waal is considerably less than predicted with the Leclair & Bridge (2001) parameter. Indeed double gravel lag layers with a distance of about 0.1 m or more were only observed in about 10% of the cores in the Waal; in most cases only one gravel lag layer was observed at the level created by a single discharge event.

Figure 7 shows that the deepest scour depths are not necessarily related to the largest dunes. The 10% deepest scours are only 0.1 m extra below the average scour depth for the largest dunes, whereas this is 0.2 m for the average and smaller dunes. Thus the slightly smaller dunes in the dune train may more often have deeper scour holes than the largest dunes.

PROCESS MODEL OF VERTICAL SORTING

Model description

A process model for the prediction of vertical sorting in the river bed is described below. It predicts the potential for the formation of gravel lag layers and the fining upward pattern in channel deposits as a result of two discharge waves. The cross-bedding is assumed to be preserved over the full depth of the active layer in the bed (details explained below), except when a certain portion of the active layer is filled with lag deposits (from base upwards). A highly simplified method for the prediction of lag deposits is followed here. From flow parameters, the bedload transport capacity of that flow is predicted with the method of Kleinhans & Van Rijn (2002). The bedload transport sediment composition is compared with that of the active layer at the same moment. If there is more gravel present in the active layer than can be transported by the flow, then the difference is potentially available for the formation of a gravel lag layer. The objective for modelling is to hindcast the sorting in the River Rhine for the discharge waves of 1995 and 1998, and determine the relative importance of cross-bedding and gravel lag layer deposits.

This model is only intended to predict the sediment sorting in the channel bed and serves to illustrate the potential importance of gravel lag layers in the channel deposits, but cannot be used for accurate quantitative modelling of the sediment transport. The reason is that bedload predictors always predict that bedload sediment is finer than bed sediment (except for extremely high discharges). If this process were allowed to continue for a long time, the model concept presented here would eventually work down all the gravel into a lag layer, which is not realistic. Therefore the model is not allowed to do this, and the potential gravel lag layer is predicted from the composition of the whole active layer (including the mobile and immobile gravel of the previous time step) in every time-step.

Furthermore, in this model it is assumed that the gravel lag layer forms instantaneously, and is (for increasing dune height) instantaneously entrained into transport in the next time-step. Obviously the deposition and entrainment processes take time in reality. In point of fact, the entrainment of a gravel lag layer by growing dunes may be so difficult that it causes the previously discussed hysteresis of bedload transport. The bedload transport computation in the model will not give hysteresis, because this effect is not incorporated in this simplified process model. Thus, the 'potential' nature of the gravel lag layer is emphasized here.

The model (Fig. 8) (i) computes flow parameters, (ii) predicts the dune height, (iii) derives the active layer from the dune height, (iv) sorts the sediment vertically in the active layer, (v) predicts the bedload transport from the grain-related shear stress and the sediment in the active layer, and (vi) deposits the gravel that cannot be transported in a lag deposit at the base of the active layer.

The active layer thickness is calculated from the (average) dune height (H) as $0.5 H$. The choice of the average dune height (rather than the 90th percentile of the dune height distribution as representing the deepest scour depth) is evaluated later. The dune height is usually observed to lag behind the discharge wave: the dune height before the discharge peak is smaller than after the discharge peak, leading to a hysteresis. The dune height is therefore predicted including this hysteresis of dune height as follows. During rising discharge, the reach-averaged dune height is computed with an empirical predictor derived from measurements of 1995, 1997 and 1998 discharge waves at this location:

$$H_{\text{rise}} = h\, 0.025(D_{50}/h)^{0.3} T^{1.7} \quad \text{(in m)} \qquad (1)$$

For falling discharge the dune height predictor of Van Rijn (1993) appears to do well and is used:

$$H_{\text{fall}} = h\, 0.11(D_{50}/h)^{0.3}(1 - e^{-0.5T})(25 - T) \qquad (2)$$

(in m) in which D_{50} is the median grain size of the average bed sediment. The transport parameter T (Van Rijn, 1984) is given as $T = (\tau' - \tau_{\text{cr}})/\tau_{\text{cr}}$ (dimensionless), with τ_{cr} based on the Shields criterion of the D_{50} of the average bed sediment. Although the trend of the dune height during lowering discharge is thus predicted well, the dune height itself is systematically overpredicted in the Waal. The reason is the time needed for adaptation of the dune dimensions; the actual

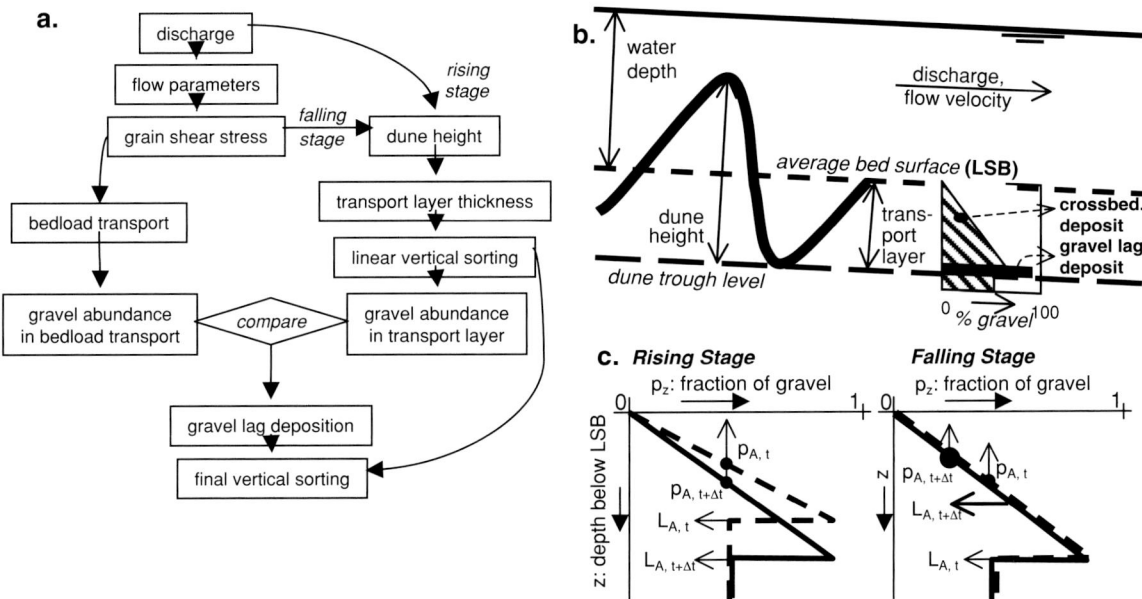

Fig. 8 (a) The structure of the process model for bedload transport and vertical sorting. (b) A sketch of the principal parameters. (c) The different response of cross-bedded sorting in the model in the rising and the falling stages. In the rising stage, the sorting is worked down deeper in the bed, whereas in the falling stage, only the active part of the antecedent sorting is reworked, leading to the same sorting curve, but a lower gravel abundance in the active layer ($p_{A,t+\Delta t}$). Other notations are explained in the text.

dune height attained at the discharge peak is lower than the equilibrium dune height at the same discharge. Therefore the second expression for dune height is matched to the first to give the same dune height at peak discharge by adjusting the constant factor (0.11).

The grain shear stress τ' is based on the grain roughness computed from the White–Colebrook equation in which the Nikuradse grain roughness length is given as $k'_s = D_{90} = 0.012$ m with the D_{90} of the average bed sediment. This method, which is also used in the transport prediction, has been shown to work well on the Rhine sediment (Kleinhans & Van Rijn, 2002). Alternative methods (e.g. Engelund & Hansen, 1967 (in Van Rijn, 1993); McLean et al., 1999) would predict the dune-related shear stress and subtract this from the total shear stress (from the depth–slope product), but in dune-dominated conditions such as those of the Rhine, the small residual grain shear stress would have an unacceptably large uncertainty owing to errors in the dune-related roughness predictors and water surface slope measurements.

Sorting modelling

Two depositional units are modelled: the cross-bedded deposit with a linear upward fining sorting (linear decrease of gravel content in upward direction), and the lag deposit (deposit of gravel only). In the model, the average bed level (equal to the initial plane bed level) is at vertical coordinate $z = 0$ (Fig. 8), and z is positive in the downward direction. For simplicity, the sediment is represented in only two grain-size fractions (i): sand and gravel, where D_i refers either to the sand or the gravel fraction. The sand abundance (p_{sand}, p = fraction, 100 p = percentage) is by definition $p_{sand} = 1 - p_{gravel}$. In the model, only the gravel fraction is actually computed.

The grain-related shear stress and the D_{50} and gravel abundance in the active layer are the input parameters for the prediction of bedload transport for each size fraction with a Meyer-Peter & Mueller (1948) type predictor, adapted for grain-size fractions and hiding–exposure effects, and accounting for near-bed turbulence according to

Kleinhans & Van Rijn (2002). The chosen hiding–exposure function (Egiazaroff, 1965) was shown to represent the effects well in the flume experiments with Rhine sediment described earlier, and therefore is assumed to be representative for the measurement location as well.

Initially, a completely uniform distribution of gravel in the vertical direction in the bed is assumed:

$$dp_z/dz = 0 \qquad (3)$$

in which p is the fraction of gravel. The active layer thickness (L_A) at time t is computed from:

$$L_{A,t} = 0.5H_t \qquad (4)$$

Now the available gravel is linearly distributed over the depth of the active layer (L_A below $z = 0$) with:

$$dp_z/dz = p_{A,t}/(0.5L_{A,t}) \qquad (5)$$

in which $p_{A,t}$ is the depth-averaged gravel fraction in the active layer at time t. The factor 0.5 determines that the gravel fraction at $z = 0.5L_{A,t}$ is equal to $p_{A,t}$ because the depth-averaged gravel fraction in a linear distribution is by definition equal to p_z at $z = 0.5L_{A,t}$. Obviously, this only works for gravel fractions smaller than 50%, which is the case in the application of this model in this paper. Otherwise, the additional constraint is that $0 < p_z < 1$ for all z. The use of a linear sorting function is justified by flume experiments devised to measure vertical sorting resulting from grain flows in natural sediments (Kleinhans, 2002).

In time, the dune height and active layer thickness change. In the active layer at $t + \Delta t$, only the sediment in the new active layer is assumed to be entrained from the sorted bed from the previous time-step. To calculate the depth-averaged gravel fraction (p_A) of the new active layer, the gravel abundance ($p_{z,t}$) is therefore averaged over the depth of the new active layer at $t + \Delta t$:

$$p_{A,t+\Delta t} = 1/L_{A,t+\Delta t} \int_{z=0}^{L_{A,t+\Delta t}} p_{z,t} \qquad (6)$$

Now the vertical sorting (p_z) in the channel bed at $t + \Delta t$ is computed as:

$$p_{z,t+\Delta t} = p_{z,t} \quad \text{for } z > L_{A,t+\Delta t} \qquad (7)$$

$$p_{z,t+\Delta t} = z\, dp_{z,t+\Delta t}/dz = zp_{A,t+\Delta t}/(0.5L_{A,t+\Delta t})$$
$$\text{for } z > L_{A,t+\Delta t} \qquad (8)$$

in which equation (7) accounts for antecedent gravel abundance below the active layer, and equation (8) accounts for the linearly distributed gravel that is sorted anew for time $t + \Delta t$.

Now the gravel lag layer can be modelled. First, the bedload sediment transport rates of sand, gravel and total (sum of sand and gravel) are predicted, based on the depth-averaged fractions, using $p_{A,t+\Delta t}$ for gravel and $1 - (p_{A,t+\Delta t})$ for sand. The gravel abundance $f_{A,g}$ that potentially can be transported by the flow (and depth-averaged in the active layer) is computed with:

$$f_{A,g} = q_{b,gravel}/q_{b,total} \qquad (9)$$

in which the bedload transport q_b for sand and gravel is computed with the bedload predictor described in the previous section for each time step t. The gravel that is potentially transported by the flow can now be compared with the gravel actually present in the active layer, to find the gravel fraction p_L that will potentially form a gravel lag layer

$$p_{L,t} = p_{A,t} - f_{g,A,t} \qquad (10)$$

This value is used to compute the potential gravel lag layer thickness per unit width and length in the channel bed deposit. Assuming that a gravel lag layer consists of gravel only ($p = 1$), the thickness L_G of the potential gravel lag layer can be computed as:

$$L_{G,t} = p_{L,t}L_{A,t} \qquad (11)$$

This gravel layer is then deposited in the model at depth $(L_{A,t} - L_{G,t}) < z \leq L_{A,t}$, so the final sorting in the model becomes then:

$$p_{final,z,t+\Delta t} = p_{z,t} \quad \text{for } z > L_{A,t+\Delta t} \qquad (12)$$

$$p_{final,z,t+\Delta t} = z\, dp_{z,t+\Delta t}/dz$$
$$= zp_{A,t+\Delta t}/(0.5L_{A,t+\Delta t})$$
$$\text{for } z < (L_{A,t+\Delta t} - L_{G,t+\Delta t}) \qquad (13)$$

$$p_{final,z,t+\Delta t} = 1$$
$$\text{for } (L_{A,t+\Delta t} - L_{G,t+\Delta t}) \leq z \leq L_{A,t+\Delta t} \qquad (14)$$

in which equation (12) accounts for antecedent gravel abundance below the active layer, equation

(13) the linearly distributed gravel that is sorted anew for time $t + \Delta t$, and equation (14) the gravel lag deposition ($p_{gravel} = 1$).

Obviously, the resulting distribution of gravel in the model does not conserve the mass of the gravel completely (errors of a few per cent), because it adds gravel to the sediment in the gravel lag layer but does not subtract this amount from the gravel abundance in the active layer. The reason is that no transfer function is known and has been specified for working down the gravel from the foresets to the lag deposits, and therefore it is not known how to subtract the lag gravel from the cross-bedded gravel units. This error does not propagate in the model, however, because the composition of the active layer p_A at $t + \Delta t$ is not computed from $p_{final,z,t}$ but from $p_{z,t}$, in which this gravel has not yet been added. For future purposes of bedload transport and morphological modelling, mass must obviously be conserved.

APPLICATION OF THE MODEL TO THE RIVER RHINE

Boundary conditions

The sediment transport and vertical sorting of two recent discharge waves in the Waal branch of the River Rhine are roughly hindcast with the model. The exact form of the discharge wave is of lesser interest and was therefore generalized and smoothed: mainly the maximum discharge and the maximum attained dune height are of importance here. The first modelled discharge wave must be the largest previous discharge wave of the period of interest. This is because the sorting in the bed is reset at that time and the deposit is predictable as a singular upward fining cross-bedded set. For the Waal, the magnitude of the discharge wave of January 1995 (maximum discharge about 10 000 m³ s⁻¹) is chosen as a boundary condition. The next discharge wave was in November 1998 (maximum discharge about 6000 m³ s⁻¹). During this event, extensive field measurements were carried out as described above. These measurements will be used to evaluate the modelling results. The initial average gravel abundance in the bed sediment in 1995 is 45%, from the grain size analysis of the vibracores.

The model was implemented in a spreadsheet. The vertical grid spacing and time-step were chosen such that the change in lag deposit thickness is smaller than the grid spacing for all time steps to prevent unrealistic gaps in the lag deposits. In this case, a vertical grid spacing of 13 mm was used, which is about equal to the D_{90} of the sediment mixture. Gravel lag layers thinner than 13 mm have been ignored. The time-step was 1 day.

There are no field data available of water depths and flow velocities for the whole range of modelled discharges. Therefore a readily available one-dimensional flow model of the Rhine branches was used to compute width- and depth-averaged water depth (h) and flow velocity (u) for a large range of discharges (Q) in the Waal (A. Wolters, SOBEK-model, Rijkswaterstaat-RIZA, personal communication). This flow model has been calibrated on water levels at several stations along the river as well as discharge measurements over a large range of low and high discharges.

Modelled flow parameters, dune height and bedload transport

The flow velocity and grain shear stress follow the pattern of the discharge (Fig. 9a) as expected. The water depth is attenuated for a discharge above 8000 m³ s⁻¹ owing to the flooding of the embanked floodplains, which leads to a sudden increase of the width of the river. In Fig. 9a the measured flow velocities of the second discharge wave are given, and are well reproduced by the model.

The simulated maximum dune height (Figs 9b & 10a) in the second wave is somewhat smaller than the observed height. Note that, in the model, hysteresis is due to changing the algorithm for dune height, not due to a modelled dynamic effect. The predicted sand and gravel transport are in the same order of magnitude as the observed transport (Figs 9c & 10b), but the trend is obviously totally wrong, and cannot be predicted with this model for reasons mentioned earlier.

Modelled sediment sorting and lag deposits

The modelled sorting in the bed is shown in Fig. 11(a & b). During rising stages and at peak flow, a thin gravel lag layer formed in the dune troughs and moved downwards in the bed as the

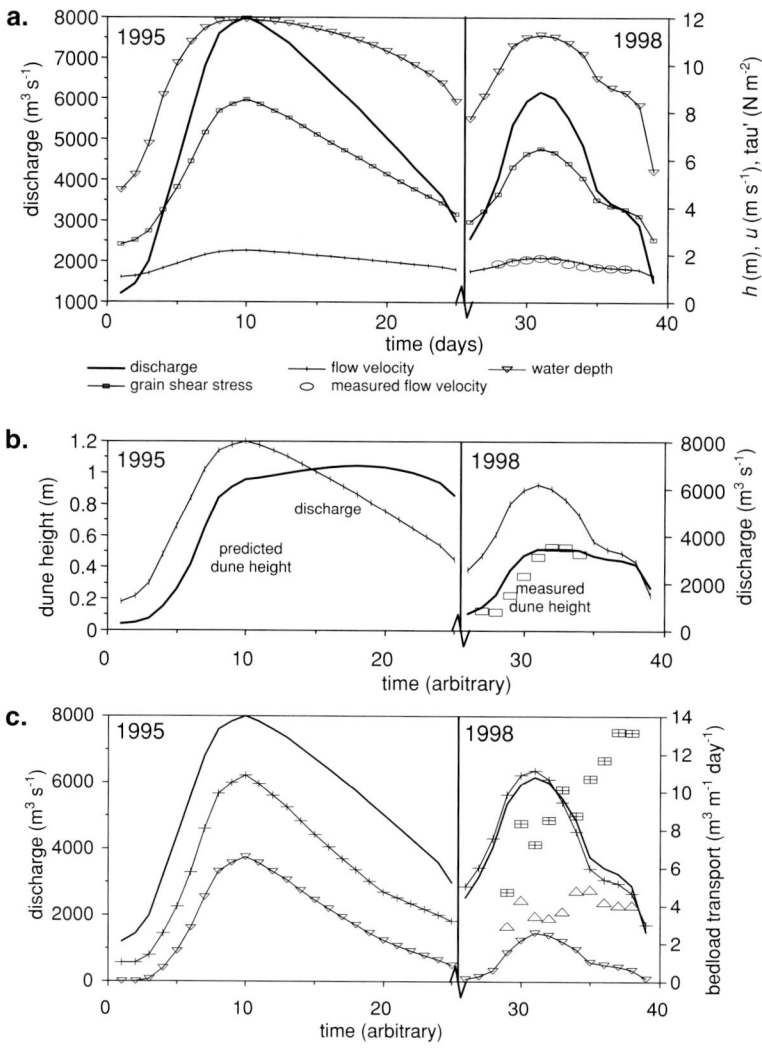

Fig. 9 (a) Discharge, water depth, flow velocity and grain shear stress, as well as the measured velocities. (b) Dune height, as well as the measured dune heights. (c) Bedload transport of sand and gravel, as well as the measured transport (using Helley-Smith-type bedload sampler).

dunes grew until peak flow. These thin layers result from the intrinsic difference between active layer sediment and predicted bedload sediment with the bedload predictor, as explained earlier. In falling stages, the modelled gravel lag became thicker at 0.4–0.5 m below the bed surface. During the second discharge wave, the same pattern was found with a gravel lag layer at 0.2–0.25 m below the bed surface, except that it was thinner. An important observation is that the potential for gravel lag formation is largest during falling stage. The reason is that during falling stage the bed shear stress is decreasing, which leads to increasing demobilization and deposition of the larger grain sizes, while the dune height is still increasing.

The predicted, reach-representative sorting agrees with the results from vibracores from the Waal, in which also two layers could be found at depths of about 0.3 and 0.7 m that were the result of the discharge waves of 1995 and 1998 (Fig. 6). The difference in depth of the gravel layers is related to small errors in the dune height predictions. These could be adjusted arbitrarily to improve the predictions, but the essential feature of the vibracores reproduced here is not the exact

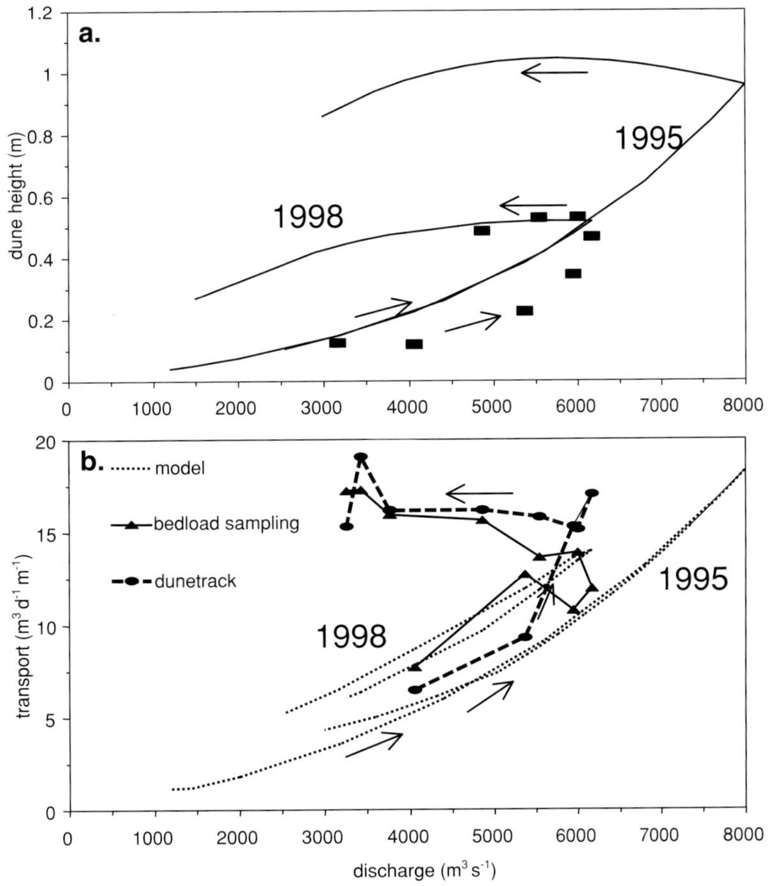

Fig. 10 (a) Dune height and (b) sediment transport as a function of discharge. Both the predictions (lines) and the measurements (symbols) show a strong hysteresis. The bedload transport as measured with the sampler increases even further in the falling stage, while the bedload transport determined with the dune-tracking method decreases. The predicted bedload transport curve for 1998 lies higher than for 1995, which is caused by the antecedent sorting from 1995 (which itself had not had antecedent sorting included in the model).

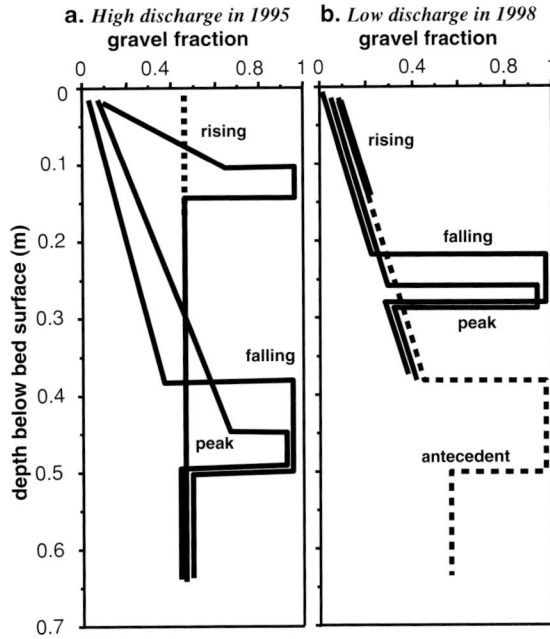

Fig. 11 (*right*) Model output: the final sorting of the river bed after the first two discharge waves (a & b). A gravel lag in the dune trough sinks into the bed as the dunes grow. In the falling stage, a much thicker gravel lag is formed. In the second discharge wave the same happens, but at a shallower depth. The lines have been given small offsets to be able to distinguish between different steps. Compare with Fig. 6.

depth but the bimodality of the gravel distribution in depth.

Sensitivity analysis

For two parameters the sensitivity was roughly tested: the maximum dune height and the vertical sorting function.

1 The sensitivity to the maximum dune height for the results is marginal. When this maximum dune height is decreased or increased, then the active layer thickness and depth of the gravel layers decreases or increases with the same factor. The sensitivity to the dune height hysteresis was not tested, but a decreased hysteresis will have the effect that the potential gravel layers become less thick.

2 The vertical sorting function as given here has the largest slope possible, which means that a perfect sorting efficiency is assumed. When this slope is decreased by 5%, then the gravel layers become thinner. Between 5% and 6% decrease, the gravel layer of the second discharge wave disappears altogether. Thus the results are rather sensitive to the vertical sorting function.

Concluding, the model is able to reproduce the sorting in the river bed, and this sorting is genetically traceable to discharge events, but the sensitivity to the vertical sorting function is large.

DISCUSSION

Effect of sorting function for cross-bedded (foreset) deposits

The vertical sorting at the lee side of the dunes is assumed to be linear, therefore the transition from cross-stratified deposit to lag deposit seems rather abrupt in the model outcome, which is only partly supported by the field and flume data. In the present model, the sorting is maximal in the sense that it yields the least possible gravel abundance in the top layer, and the maximum possible gravel abundance in the bottom layer. It could well be that the sorting within grain flows in nature is less effective than assumed here. As shown in the sensitivity analysis, this would lead to a less clear sorting in the cross-bedded deposits. At the same time, if the gravel accommodation in the top part of the active layer was higher in falling stage, then the potential gravel lag deposits would be thinner or vanish altogether (see section on sensitivity analysis). Concluding, the sorting in the active layer depends greatly on the vertical sorting function. Yet, the most straightforward parameter choices and assumptions give reasonable results, indicating that the conclusions herein are at least in the direction of what may be found with further research. To be able to predict the sorting for more than two grain-size fractions, a quantitative predictor must be found for the vertical sorting in grain flows and due to selective deposition. The results of experiments by Kleinhans (2000; see also Blom & Kleinhans, 1999), as given in Fig. 3, indicate that the sorting function deviates only slightly from linear and indeed comes close to zero gravel in the top of the dunes. Experiments to study vertical sorting (carried out at St Anthony Falls Laboratory in Minneapolis, USA) by Kleinhans (2002) also show that the optimal sorting does not deviate very much from a linear profile.

Future model development

The problems notwithstanding, the main asset of this model is that it demonstrates a possibility for the prediction of vertical sediment sorting by two processes. The model demonstrates that essential elements for the prediction of sediment transport and vertical sorting in rivers with dunes and sand–gravel mixtures are:

1 a vertical sorting function for equilibrium sorting in the foreset (cross-bedded) deposits;
2 a time-dependent approach to this equilibrium to account for the time necessary to deposit or entrain the gravel at the base of dunes;
3 a dune height predictor including a time-dependent approach to the equilibrium dune height.

As a result, the model should be able to reproduce:

1 sediment transport hysteresis during a single discharge wave;
2 relative sediment transport differences between successive discharge events (for the same discharge);
3 secondary, sandy cross-stratification (infilling of dune troughs with sediment from the tops of

these dunes) in the case of strong dune height hysteresis, in addition to primary cross-stratification and lag deposits, and their relative abundance in the channel bed for different dune height adaptation time-scales.

CONCLUSIONS

River channel bed deposition of sand–gravel mixtures is strongly coupled to dune development and the sorting of sediment in the grain flow at the lee side of dunes, as well as grain-size-selective deposition in the dune troughs (gravel lags). Data from the River Rhine clearly indicate that vertical sorting is generated in the river bed down to a depth depending on the maximum (reach-averaged) dune height attained in a discharge wave.

Consequently, strong feedbacks between bedload sediment transport rate and composition versus deposition in the bed are expected. The bedload sediment transport rate and composition are dependent on the sediment in the active layer (down to the dune troughs), which is determined by the sorting in previous events. The deposits on the other hand are governed by the selective transport and by sorting in dunes, as well as the dune height development during a discharge event. Consequently, the vertical sorting of previous discharge waves may cause hysteresis of the sediment transport, which can only be hindcast by incorporating this sorting.

The process model developed in this paper is able to reproduce the main characteristics of vertical sorting in the bed of the River Rhine resulting from two recent discharge waves. This approach should be seen as a first exploratory attempt to model vertical sorting. The model shows that both vertical sorting resulting from grain flows on the lee side of dunes and from selective deposition play an important role in the sediment transport and deposition of sand–gravel sediment. The dune height adaptation to changing flow and dune celerity determine the ratios of lag deposits and cross-stratification. The deposition of gravel lag layers in the bed occurs mainly during falling stage, because the grain shear stress is then decreasing while the dune height is still increasing. Sediment transport is not correctly predicted with this model, however, because the phenomena that cause strong transport hysteresis were not included in the model. The order and magnitude of discharge waves are important, as the largest wipes out all antecedent sorting, whereas the processes in smaller ones depend on the depositions of previous waves.

ACKNOWLEDGEMENTS

The present research is part of a PhD research programme at Utrecht University on sediment transport and dune behaviour in sand–gravel-bed rivers during high discharges. The investigations were in part supported by The Netherlands Earth and Life Sciences Foundation (ALW) with financial aid from The Netherlands Organization for Scientific Research (NWO). The National Institute for Inland Water Management and Waste Water Treatment (RIZA) and the Directorate Eastern Netherlands of Rijkswaterstaat in The Netherlands financed and carried out the measurements in the rivers Waal and Bovenrijn, as well as the vibracores (in cooperation with the Dutch Institute for Applied Geology (TNO-NITG). RIZA is gratefully acknowledged for the permission to use the vibracores. Antoine Wilbers (Utrecht University) is thanked for providing the dune statistics from the multibeam echo soundings. The flume experiments were financed by (i) the Transport and Mobility of Researchers Programme of the European Commission and (ii) the consortium of Twente University, The Institute for Inland Water Management and Waste Water Management (RIZA) and WL Delft Hydraulics. Ard Wolters is thanked for providing the SOBEK model output. Thoughtful comments by Janrik van den Berg, Leo van Rijn, Suzanne Leclair, Gary Parker and the manuscript reviewers John Bridge and Mike Church were much appreciated. Some views in this paper, however, do not agree with those of the manuscript reviewers, notably the role of the deepest dune trough scour depth and the dune migration velocity in the vertical sorting. More complete analyses of the present dataset in non-uniform sediment and also of existing datasets in more uniform sediment (currently underway) are deemed necessary and might change some of the presented interpretations in the future.

NOMENCLATURE

C Chezy coefficient ($m^{0.5}\ s^{-1}$)
D grain size (m)
f_i fraction of grain size i in bedload sediment
g gravitational acceleration (9.81 m s^{-2})
h water depth (m)
H dune height (m)
k_S hydraulic roughness (m)
L_A thickness of active layer
L_G thickness of gravel lag layer
LSB   low-stage bed level (plane bed)
p probability/abundance of grain size fraction (100 times p yields p in %) (-)
q_b bedload transport ($m^3\ m^{-1}\ s^{-1}$)
Q flow discharge ($m^3\ s^{-1}$)
T relative excess shear stress parameter: $(\tau' - \tau_{cr})/\tau_{cr}$ (-)
t time coordinate
u flow velocity (m s^{-1})
z depth coordinate, z = 0 at LSB
θ Shields parameter (subscript 'cr' refers to critical Shields parameter) (-)
ξ hiding-exposure coefficient (-)
ρ density (subscript 's' refers to sediment) (kg m^{-3})
τ bed shear stress (N m^{-2})

Prime (') indicates 'related to grains'.
Subscripts i, 50, 90, etc. refer to grain-size fractions i or grain-size distribution percentiles.

REFERENCES

Allen, J.R.L. (1963) Sedimentation to the lee of small underwater sand waves: an experimental study. *J. Geol.*, **73**, 95–116.

Allen, J.R.L. (1970) The avalanching of granular solids on dune and similar slopes. *J. Geol.*, **78**, 326–351.

Allen, J.R.L. and Collinson, J.D. (1974) The superimposition and classification of dunes formed by unidirectional aqueous flows. *Sediment. Geol.*, **12**, 169–178.

Blom, A. and Kleinhans, M.G. (1999) *Non-uniform Sediment in Morphological Equilibrium Situations. Data Report Sand Flume Experiments 97/98.* University of Twente, Rijkswaterstaat RIZA, WL/Delft Hydraulics, University of Twente, Civil Engineering and Management, Twente, 50 pp.

Boersma, J.R., Van de Meene, E.A. and Tjalsma, R.C. (1968) Intricate cross-stratification due to interaction of a mega ripple with its lee-side system of backflow ripples (upper-pointbar deposits, Lower Rhine). *Sedimentology*, **11**, 147–162.

Bridge, J.S. and Bennett, S.J. (1992) A Model for the Entrainment and Transport of Sediment Grains of Mixed Sizes, Shapes and Densities. *Water Resour. Res.*, **28**, 337–363.

Carling, P.A., Goelz, E., Orr, H.G. and Radecki-Pawlik, A. (2000) The morphodynamics of fluvial sand dunes in the River Rhine, near Mainz, Germany. I. Sedimentology and morphology. *Sedimentology*, **47**, 227–252.

Egiazaroff, I.V. (1965) Calculation of nonuniform sediment concentrations. *J. Hydraul. Div. ASCE*, **91**, 225–248.

Gruijters, S.H.L.L., Veldkamp, J.G., Gunnik, J. and Bosch, J.H.A. (2001) *De lithologische en sedimentologische opbouw van de ondergrond van de Panner-densche Kop.* TNO-report NITG-01-166-B, Utrecht, 41 pp. (In Dutch.)

Hunter, R.E. (1985) A kinematic model for the structure of lee-side deposits. *Sedimentology*, **32**, 409–422.

Jopling, A.V. (1965) Laboratory study of the distribution of grain sizes in cross-bedded deposits. In: *Primary Sedimentary Structures and their Hydrodynamic Interpretation* (Ed. G.V. Middleton). *SEPM Spec. Publ.*, **12**, 53–65.

Klaassen, G.J. (1987) Experiments on the effect of gradation on sediment transport. *Euromech 215 Colloquiem*, pp. 127–146. Genova, 15–19 September.

Klaassen, G.J. (1991) Experiments on the effect of gradation and vertical sorting on sediment transport phenomena in the dune phase. *Grain Sorting Seminar*, 21–25 October, Ascona, Switzerland.

Klaassen, G.J., Ribberink, J.S. and De Ruiter, J.C.C. (1987) On the transport of mixtures in the dune phase. *Euromech 215 Colloquiem*, Genova, September 15–19.

Kleinhans, M.G. (1999) *Sediment Transport in the River Waal: High Discharge Wave, November, 1998.* ICG 99/6, Netherlands Centre for Geo-ecological Research/Utrecht University Physical Geography, Utrecht, 101 pp. (In Dutch.)

Kleinhans, M.G. (2000) The relation between bedform type, vertical sorting in bedforms and bedload transport during successive discharge waves in large sand–gravel bed rivers with fixed banks. *Proceedings, Gravel Bed Rivers Conference 2000*, 28 August–3 September, Christchurch, New Zealand (Eds T. Nolan and C. Thorne), Special Publication CD-rom of the New Zealand Hydrological Society.

Kleinhans, M.G. (2001) The key role of fluvial dunes in transport and deposition of sand–gravel mixtures, a preliminary note. *Sediment. Geol.*, **143**, 7–13.

Kleinhans, M.G. (2002) *Sorting out Sand and Gravel; Sediment Transport and Deposition in Sand–gravel Bed Rivers.* Royal Dutch Geographical Society, Utrecht, *Neth. Geogr. Stud.*, **293**, 317 pp.

Kleinhans, M.G. and Ten Brinke, W.B.M. (2001) Accuracy of cross-channel sampled sediment transport in large sand–gravel-bed rivers. *J. Hydr. Eng.*, **127**, 258–269.

Kleinhans, M.G. and Van Rijn, L.C. (2002) Stochastic prediction of sediment transport in sand–gravel bed rivers. *J. Hydraul. Eng.*, **128**, 412–425.

Koeppe, J.P., Enz, M. and Kakalios, J. (1998) Phase diagram for avalanche stratification of granular media. *Phys. Rev. E.*, **58**.

Leclair, S.F. (2000) Preservation of cross-strata due to migration of subaqueous dunes. Unpublished PhD Dissertation, Binghamton University, New York.

Leclair, S.F. and Bridge, J.S. (2001) Quantitative interpretation of sedimentary structures formed by river dunes. *J. Sediment. Res.*, **71**.

Makse, H.A., Havlin, S., King, P.R. and Stanley, H.E. (1997) Spontaneous stratification in granular mixtures. *Nature*, **386**, 379–381.

McLean, S.R., Wolfe, S.R. and Nelson, J.M. (1999) Predicting boundary shear stress and sediment transport over bedforms. *J. Hydraul. Eng.*, **125**, 725–736.

Meyer-Peter, E. and Mueller, R. (1948) Formulas for bed-load transport. *Second Conference, International Association of Hydraulic Research*, pp. 39–64. Stockholm.

Paola, C. and Borgman, L. (1991) Reconstructing random topography from preserved stratification. *Sedimentology*, **38**, 553–565.

Parker, G., Paola, C. and Leclair, S. (2000) Probabilistic Exner sediment continuity equation for mixtures with no active layer. *J. Hydraul. Eng.*, **126**, 818–826.

Ribberink, J. (1987) Mathematical modelling of one-dimensional morphological changes in rivers with non-uniform sediment. PhD thesis, Delft University, Delft.

Schans, H. (1998) *Representativity of Measurements in between Groynes in 1996 and 1997 for the Whole River Waal.* ICG 98/15, Netherlands Centre for Geo-ecological Research/Utrecht University Physical Geography, Utrecht, 145 pp. (In Dutch.)

Southard, J.B. and Boguchwal, A. (1990) Bed configurations in steady unidirectional water flows. Part 2. Synthesis of flume data. *J. Sediment. Petrol.*, **60**, 658–679.

Ten Brinke, W.B.M., Wilbers, A.W.E. and Wesseling, C. (1999) Dune growth, decay and migration rates during a large-magnitude flood at sand and mixed sand–gravel bed in the Dutch Rhine river system. In: *Fluvial Sedimentology VI* (Eds N.D. Smith and J. Rogers). *Spec. Publs Int. Assoc. Sedimentol.*, **28**, 15–32.

Van Rijn, L.C. (1984) Sediment transport, part I: bed load transport. *J. Hydraul. Eng.*, **110**, 1431–1456.

Van Rijn, L.C. (1993) *Principles of Sediment Transport in Rivers, Estuaries and Coastal Seas.* Aqua Publications, Oldemarkt, The Netherlands, 335 pp.

Wathen, S.J., Ferguson, R.I., Hoey, T.B. and Werritty, A. (1995) Unequal mobility of gravel and sand in weakly bimodal river sediments. *Water Resour. Res.*, **31**, 2087–2096.

Wilbers, A.W.E. (1998) *Bedload Transport and Dune Development during Discharge Waves in the Bovenrijn and the Waal.* ICG 98/12, Netherlands Centre for Geo-ecological Research/Utrecht University Physical Geography, Utrecht, 60 pp. (In Dutch)

Wilbers, A.W.E. (1999) *Bedload Transport and Dune Development during Discharge Waves in the Rhine Branches, Echo Soundings of the Flood in November 1998.* ICG 99/10, Netherlands Centre for Geo-ecological Research/Utrecht University Physical Geography, Utrecht, 60 pp. (In Dutch.)

Wilcock, P.R. (1993) Critical shear stress of natural sediments. *J. Hydraul. Eng.*, **119**, 491–505.

Wilcock, P.R. (2001) The flow, the bed and the transport: interaction in the flume and field. In: *Gravel-bed Rivers V* (Ed. M.P. Mosley), pp. 183–220. New Zealand Hydrological Society, Wellington.

# Morphology and fluvio-aeolian interaction of the tropical latitude, ephemeral braided-river dominated Koigab Fan, north-west Namibia

CARMEN B.E. KRAPF*,[1], IAN G. STANISTREET† and HARALD STOLLHOFEN*

*Geologisches Institut der RWTH Aachen, Wüllnerstrasse 2, 52056 Aachen, Germany; and
†Department of Earth Sciences, University of Liverpool, Brownlow Street, PO Box 147, Liverpool L69 3BX, UK

ABSTRACT

The Koigab Fan is the largest of the active fan systems formed by some of the west-south-west flowing ephemeral river systems of the Skeleton Coast area, north-west Namibia. Issuing from the volcanic Etendeka Plateau, the Koigab River flows towards the Atlantic Ocean across a considerable climatic gradient from semi-arid summer rainfall in the mountainous catchment, to hyperarid in the coastal depositional setting.

The morphology of channels can be discerned over the whole fan surface (gradient 1.011), the majority of which appears as a vast deflation surface on which lithic and heavy mineral grains are concentrated by aeolian removal of fines. The Koigab catchment restricts source-rock lithologies to flood basalts and interleaved quartz latites of the Etendeka Plateau, so components that unequivocally relate to a volcanic source (e.g. volcanic lithics, Ti-magnetite, pyroxenes) indicate fluvial transport, whereas grains reflecting a metamorphic basement source (e.g. garnet, muscovite, staurolite) must be aeolian derived. Both heavy mineral and grain-size data were used to estimate the amount of fluvio-aeolian interaction at the Koigab Fan surface. This aspect is significant because it comprises not only winnowing of the fan surface and of 'fresh' sandy channel deposits but also fluvial recycling of aeolian material. The contribution of aeolian-derived grains to river deposits increases from 5% in the fan apex area to as much as 50% in the distal fan reaches.

In the spectrum of fan types, the Koigab Fan takes an intermediate position both in size and in terms of the braided river style between debris flow and low sinuosity meandering fan systems. Within the braided fluvially dominated fan class itself the Koigab Fan is also intermediate in size, but its ephemeral channels contrast sharply with those of perennial glacial outwash fans previously described from the sub-Arctic. Within low-latitude fan systems, the Koigab also contrasts with other highly vegetated fans in the tropics, for example the sub-aerial portion of the Yallahs Fan-delta, Jamaica. Thus, the Koigab Fan is important as a potential analogue for Precambrian and early Palaeozoic low-latitude fan systems that lacked surface vegetation prior to the evolution of land plants.

Present address: ASP—Australian School of Petroleum, Santos Petroleum Engineering Building, The University of Adelaide, SA 5005, Australia (Email: ckrapf@asp.adelaide.edu.au).

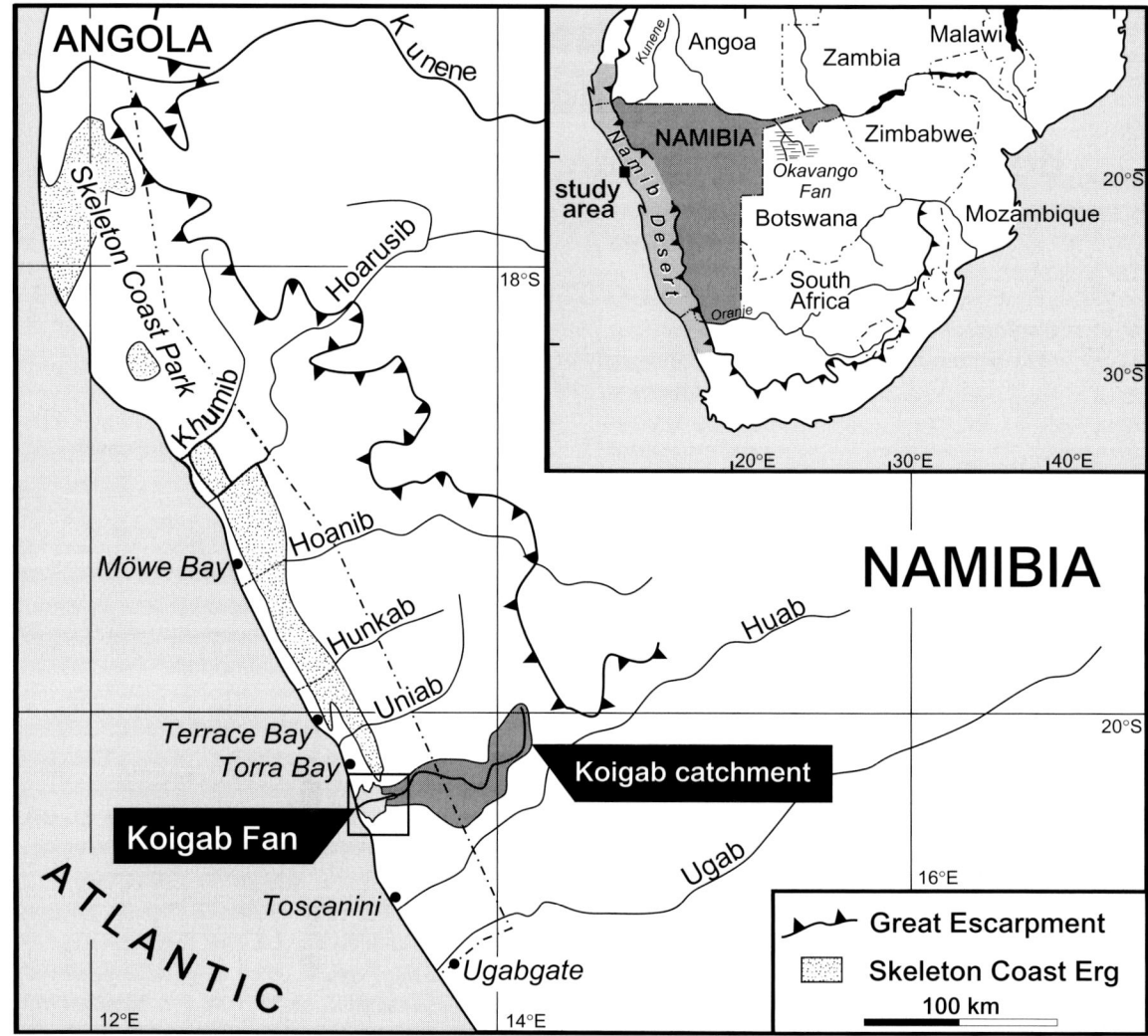

Fig. 1 Location of the Koigab Fan and the Koigab catchment area in north-west Namibia.

INTRODUCTION

The Koigab Fan is located in the Namib Desert in north-western Namibia (Fig. 1). This paper describes processes and environments of the Kogab Fan and, in doing so, aims to extend the knowledge of the variability of the braided, fluvially dominated fans, particularly fans that occur in arid climatic settings, are relatively unvegetated and are dominated by ephemeral braided rivers. Furthermore, the setting of the Koigab Fan provides an opportunity to study fluvial interactions with aeolian sand flux into and away from such a body.

THE NAMIB DESERT ENVIRONMENT

The Koigab Fan is located within southern African tropical latitudes (20°S) in a western continental, coast-parallel arid zone. The fan (Fig. 2) was first referred to by Van Zyl & Scheepers (1993), and is intermediate in size for braid-

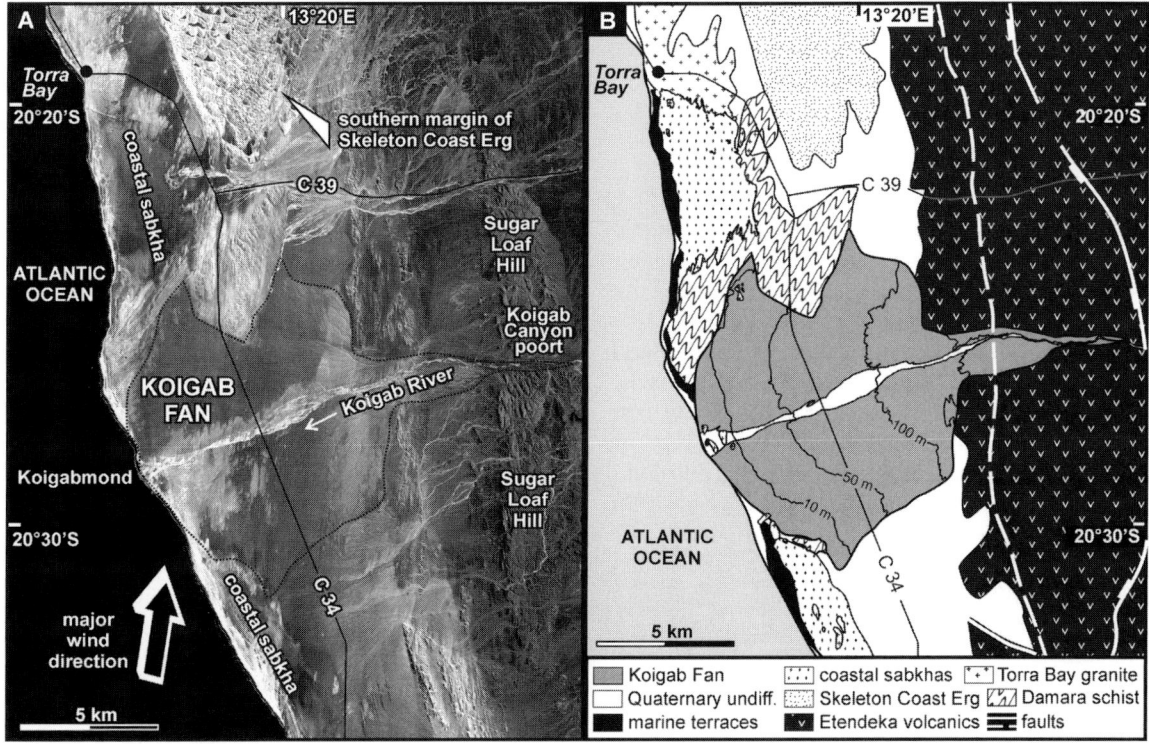

Fig. 2 The appearance of the Koigab Fan and parts of its Etendeka Plateau catchment in (A) a Landsat TM-5 image 171-074 and (B) a simplified geological map.

dominated alluvial fans (cf. Stanistreet & McCarthy, 1993), measuring 15 km from apex to toe and 23 km across its maximum lateral extent. Its depositional area has been relatively little disturbed by human activity owing to its position within the Skeleton Coast National Park, an area characterized by the hyperarid but occasionally foggy coastal Namib desert climate (Lancaster, 1982; Lancaster et al., 1984).

Regional climate is controlled by the intensity of the cool, northward flowing Benguela Current offshore Namibia, the subtropical South Atlantic anticyclone and monsoonal influences from the north-east, which are associated with disturbances of the intertropical convergence zone (Tyson, 1986; Jury, 1996; Ward & Swart, 1997; McCarthy et al., 2000; Stanistreet & Stollhofen, 2002). Climatic effects resulting from the variable strength of the Benguela Current are analogous to effects of the Humboldt Current on the Atacama Desert arid zone of Chile (cf. Messerli et al., 1993).

With the exception of the Kunene River, all of the Skeleton Coast rivers are ephemeral (Jacobson et al., 1995). Experiencing flood pulses of short duration, typically a few days only, the rivers flow to the South Atlantic coast across a considerable climatic gradient from summer-rainfall-dominated mountainous source regions (300–600 mm yr^{-1}) to the arid (< 50 mm yr^{-1}) desert over a distance of merely 150 km (Jacobson et al., 1995). No discharge records are available because of the remote setting and flashy nature of flooding. Minor river flows normally occur during November to April, during the annual rainy season of the Southern Hemisphere summer, whereas high-magnitude floods have been observed at a 9 yr average during the past 63 yr (cf. Shannon et al., 1986; Jacobson et al., 1995; Stanistreet & Stollhofen, 2002), preferentially between February and April. Considerable precipitation can then supply large volumes of water and sediment discharge from the mountainous hinterland catchment (Fig. 1). Thus the

Koigab River may rarely undergo major flooding, but frequently experiences several years of drought.

Wind as a geological factor

During the whole year, strong onshore south to south-south-westerly winds dominate the Skeleton Coast area (Lancaster, 1982; Barnard, 1989). Average wind speed measured at Möwe Bay weather station (Fig. 1) during the periods October 1994 to March 1995 and September 2000 to March 2001 is 4.34 m s^{-1}, with a maximum of 31 m s^{-1} (Namibian Weather Bureau, Windhoek). In contrast, the 'Bergwinds' from easterly directions blow only during several days between April and July (Tyson, 1969) and achieve wind speeds of up to 17 m s^{-1}. Such winds play an important role in the transport of aeolian dust (Whitaker, 1984) and resulting 'dust plumes' extend over 100 km offshore. However, because of the dominance and persistence of the southerly winds, the effects of bergwinds are not as obvious in the study area as they are in the Central Namib Desert (Lancaster et al., 1984) further to the south along the Namibian coastline. During the 1973–1977 period, 98% of the annual sand flow was recorded from south-south-easterly to south-south-westerly directions, with winds able to move sand for almost 50% of the time and maximum sand flow occurring during August and October (Lancaster, 1982).

Geodynamic framework, tectonics and drainage evolution

The river catchment areas of north-west Namibia are developed largely outboard of the Great Escarpment (Fig. 1), a pronounced rise in topography, which in north-west Namibia separates a coastal strip, the Skeleton Coast, varying in height between sea-level and 400 m, from a mountainous hinterland, typically varying between 900 m and 1300 m. The Great Escarpment fronts the highveld of the continental interior, maintaining heights between 1300 m and 1700 m, and encircles the entire southern African subcontinent (King, 1951; Partridge & Maud, 2000). The initiation of the Great Escarpment followed the 'unzipping' of the oceans that presently surround southern Africa: the Indian Ocean to the east in the Mid- to Late Jurassic at 157 Ma, and the Atlantic Ocean to the west during the Early Cretaceous (Dingle, 1992/1993).

Subsequent uplift has affected the Namibian coastal tract, occurring particularly during the Late Cretaceous (Ward, 1987; Brown et al., 2000; Cockburn et al., 2000; Raab, 2001), followed by less well-known phases of uplift during the Tertiary (Ward, 1987) and Quaternary (Klein, 1980; K. Weber, Geoscience Centre, University of Goettingen, personal communication, 2002). Such phases of uplift have been related to the headward incision of deep canyons by westward draining river systems such as the Kuiseb in southern Namibia (Korn & Martin, 1957; Ward, 1987) and the Omaruru (Klein, 1980) just to the south of the Skeleton Coast Park. Subsequent aggradational and degradational cycles have been interpreted to reflect an overall but fluctuating decrease in hydrological competency from the Early–Middle Pleistocene (Oswater Conglomerate) to the present day (Ward, 1987). Of the rivers running off the western margin of southern Africa, only the Orange and Kunene River catchments progressed substantially into the continental interior to the east, their headwaters capturing river systems throughout the subcontinental interior (de Wit et al., 2000; Jacob, 2001).

Unlike some other rivers of the northern Namibian coastal strip (e.g. Hoanib, Hoarusib), the Koigab River (Fig. 1) does not make a major breach back through the Great Escarpment region (Van Zyl & Scheepers, 1993; Jacobson et al., 1995). In particular the system is aggradational in character in its lower course, despite widespread uplift of the coastal region during the Quaternary (Klein, 1980). The presently active Koigab channel extends headwards into the Etendeka Plateau to generate a pronounced gorge incision of about 24 m at Koigab poort, 20 km east of the coast. The area through which the poort is cut (Fig. 2) comprises a series of down-to-the-west imbricate fault blocks which relate to the major north–south trending Ambrosiusberg fault zone (Milner, 1986). Cretaceous volcanics exposed in the fault blocks dip between 5° and 30° towards the east (Milner, 1986; Schlicker, 1999) whereas capping terraces of Lower to Middle Pleistocene (Ward, 1987) conglomerates (van Zyl & Scheepers, 1993) have not been tilted.

SIGNIFICANCE AND SETTING OF THE KOIGAB FAN SYSTEM

The Koigab Fan is the largest of the fan systems formed by some of the west-south-west flowing ephemeral braided rivers (Fig. 2). The fan bears little surface vegetation, which is restricted to shrubby vegetation near the fan toe, associated mainly with small-scale aeolian bedforms such as coppice dunes. This scarce peripheral vegetation relies upon the mists of the Skeleton Coast that drift in daily, especially during winter months (Lancaster, 1982), maintained by the cold Benguela Current offshore from the Atlantic coastline of Namibia.

This fan system was chosen for study because it differs from many described systems of equivalent dimension in that it: (i) is located at a low tropical latitude within an arid climatic setting; (ii) is dominated by mixed gravel–sand ephemeral braided river channels; (iii) lacks significant surface vegetation; (iv) has been relatively unaffected by human activity; (v) is a perfect study site for recording various types of fluvio-aeolian interaction; and (vi) thereby acts additionally as a model for certain Precambrian and Early Palaeozoic fan depositional systems deposited prior to the evolution of land plants.

Koigab catchment characteristics

With a catchment area of about 2400 km^2, the Koigab River is one of the smallest of the ephemeral rivers of the Skeleton Coast (Jacobson et al., 1995). The highest elevation within the hinterland catchment is 1571 m a.s.l., related to one of the famous table mountains of the Lower Cretaceous Etendeka Plateau in the Khorixas District. There, tholeiitic flood basalts and interbedded quartzlatitic rheoignimbrites (Marsh et al., 2001; Milner et al., 1992), which erupted between 131 and 133 Ma, immediately preceding the opening of the South Atlantic Oceanic (Renne et al., 1996), are the only source rocks in the catchment. Weathering products of the Etendeka volcanics comprise dominantly volcaniclastic lithics, plagioclase and clinopyroxene plus minor amounts of volcanic magnetites, orthopyroxenes and chlorite. The few quartz grains are derived from vesicle fills of latites, such as agate.

Only in the Koigabmond area do Upper Proterozoic Damara metamorphic basement rocks crop out (Ahrendt et al., 1983) within the active Koigab channel and the adjacent fan areas (Fig. 2B). Essentially these are schists and migmatite gneisses, composed mainly of quartz, biotite and garnet. The schists and migmatites are intruded by the Torra Bay Granite and pegmatites, both composed of quartz, feldspar and garnets. These contrasting source rock types are important for the understanding of fluvio-aeolian processes on the Koigab Fan.

FAN MORPHOLOGY

The principal anatomy

On its 130 km route to the Atlantic Ocean the Koigab River traverses a height difference of about 1200 m, which results in an overall river gradient of about 0.009. The regional gradient, for comparison, is about 0.017. The Koigab River, leaving the area of the Etendeka Plateau via the Koigab Canyon poort, enters on to the fan surface through an entrenched fan apex (Fig. 3) at a height of 200 m a.s.l. and then flows down to sea level over a distance of 15 km (Fig. 4).

The distal fan base has a width of 15 km, forming an arc of 110° (Fig. 2). Only a central sector of 15° currently debouches directly into the sea, through an area of active beach berm, backshore and foreshore (Fig. 2A). Here, river discharge passes into the ocean, to feed an area of offshore shoal 5 km by 4 km. Figure 4 shows a resulting small sediment prism on the shelf area immediately in front of the mouth of the active river channel. To either side of the river mouth the foreshore is sandy and scattered with articulated shells and shell debris, including inarticulate brachiopods.

The distal fan areas to either side of the central sector pass into different settings. To the north the fan rests on a basement area of Damaran schists, which is backed by a large coastal sabkha that occasionally is flooded by marine as well as by fluvial processes. To the south the distal fan debouches into coastal sabkhas (Figs 2 & 4), where fine marine-flood sediments are heavily indurated by subsurface precipitation of salts, particularly gypsum. The coastal sabkhas are separated from

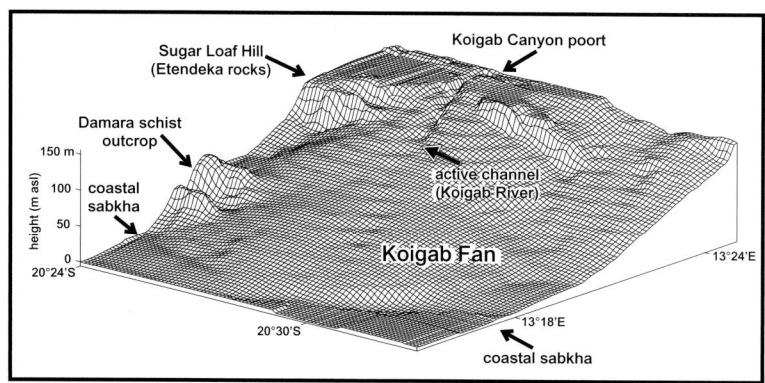

Fig. 3 Three-dimensional digital elevation model shows upstream view from the Koigabmond towards the north-east, illustrating geomorphological elements of the Koigab Fan. Data derived from topographic sheet 2013 AC and AD TORRABAAI and sheet 2013 CB, 1 : 50 000; Surveyor General, Windhoek, Namibia.

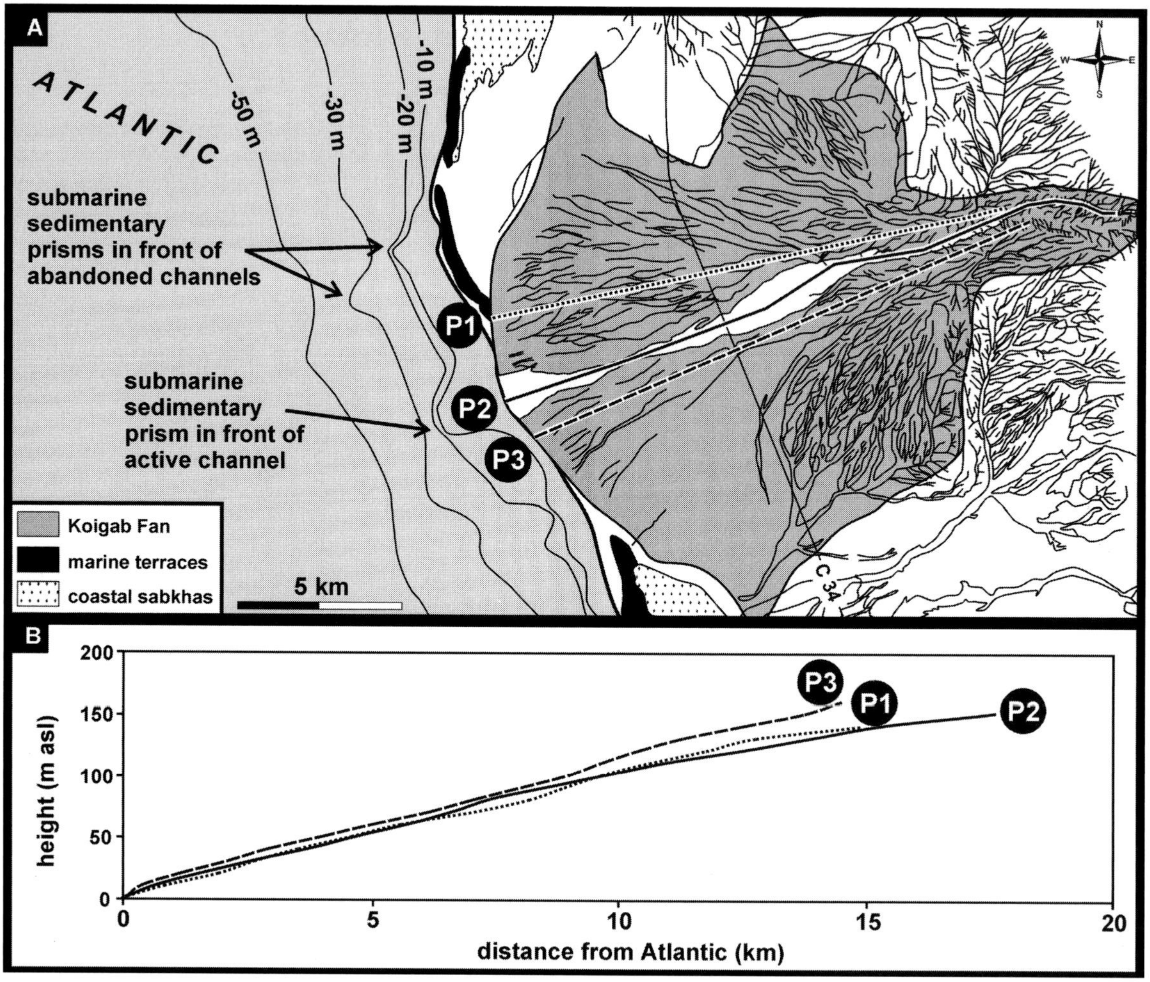

Fig. 4 (A) Drainage network of the Koigab Fan area with the location of radial fan profiles and bathymetric offshore data. (B) Contrast between downfan profiles of the active Koigab River (P2) and the adjacent fan surface (P1 and P3). Data derived from topographic sheet 2013 AC and AD TORRABAAI and sheet 2013 CB, 1 : 50 000; Surveyor General, Windhoek, Namibia. Bathymetric data derived from SAN 103, 1 : 150 000, Hydrographic Office, South African Navy.

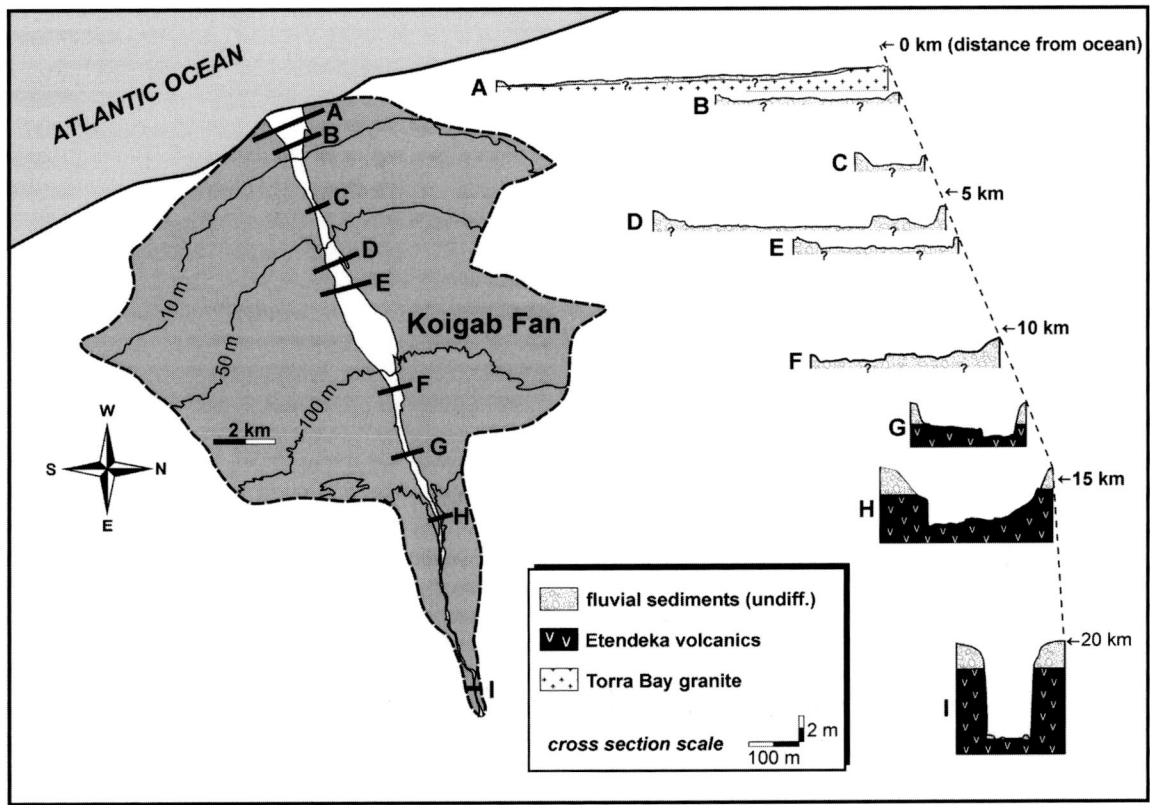

Fig. 5 Map showing locations of surveyed channel cross-sectional profiles by theodolite level and tape. Sections I to A to the right of the fan outlines were measured at various positions down the Koigab River from the fan apex to the river mouth, with letters to the right of the sections indicating their distance from the Atlantic Ocean.

the ocean over large areas by shore-parallel beach ridges, which have been uplifted beyond the reach and effects of modern sea-level fluctuations. Their heights correspond to the Late Pleistocene and early Holocene raised shorelines reported by Pickford & Senut (1999). The beach ridges front on to the presently active beach berm, backshore and foreshore systems similar to those close to the actual river mouth.

Koigab channel characteristics and the fan surface

The fan-head is heavily entrenched by the active river channel, which presently emerges on to the fan surface only near the outermost fan perimeter (Fig. 2). Midway up the fan the Koigab River channel is incised 3–6 m into the fan surface, whereas at the apex the channel is incised to a depth of 10–15 m (Fig. 3). Upstream from the apex, downcutting into bedrock has developed the Koigab Canyon, a gorge cut back as a major nick-point into the faulted Etendeka Plateau to a maximum depth of 24 m (Fig. 5, section I).

Active channels

The contours of channels visible on the satellite image (Figs 2A & 6A), and verified by ground inspection, reveal that the Koigab Fan surface includes both long-term abandoned inactive (Fig. 6B) and presently active (Fig. 6C & D) channels. Sections across the active Koigab River channel were surveyed at various positions along

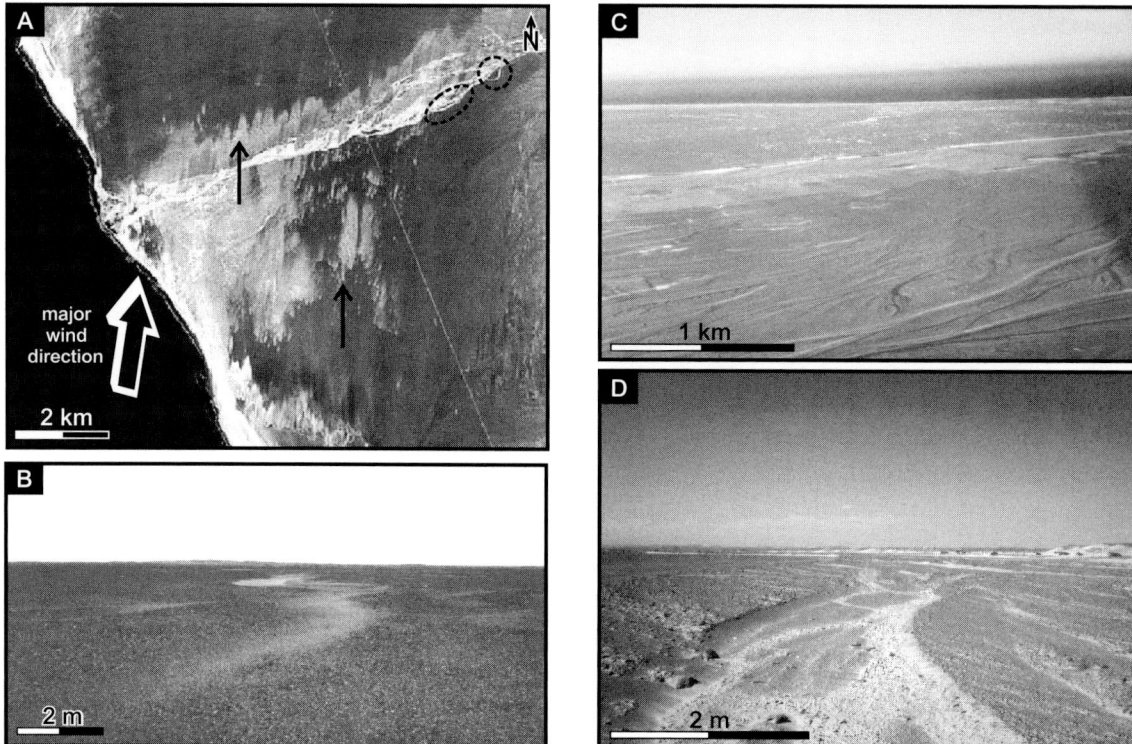

Fig. 6 Photographs of the Koigab Fan surface showing (A) the major part of the Koigab Fan on a Landsat TM-5 image 1 month after the major flood event of February 1995. The flood followed the main channel, overtopping the banks locally (e.g. circles). After the flood event the channel was subjected to aeolian winnowing (arrows) and sediment was blown on to the northern part of the fan by wind from dominantly southern directions (data derived from Landsat TM-5 scene 181-074, 30.03.1995). (B) Shallow, slightly incised channel on the central southern part of the fan. Fine-grained material is blown out by deflation, and sand accumulates only in the lee of small cut-walls. (C) Oblique view towards west-south-west illustrating the drainage network of the fan surface. The fan surface in the foreground displays less active and/or abandoned channels, and the centre ground shows the main channel (Koigab River) 2 weeks after a smaller flood event in April 2000. The southern part of the fan surface is characterized by shrub coppice fields (white spots). (D) The typical field appearance of the Koigab River: small, slightly incised channels with longitudinal gravel bars and the thalweg is partly covered by thin waning flood stage sand and mud layers.

the length of the active channel (Fig. 5) as it traverses the fan surface, using a survey level, staff and measuring tape. These surveys were undertaken in order to collect information concerning channel geometry, bar and depositional forms within the channel, clast textural data, and to locate samples of sand-grade material deposited during waning floods. Figure 5 illustrates changes in the shape and style of the channel cross-sections at locations A to I from the river mouth to the fan apex.

The Koigab River itself is a low-relief, braided river with longitudinal bars that separate medium to low flow channels (Fig. 6D). The channels are filled with medium sand to boulders (usually < 50–60 cm), and the uppermost parts of the channels are modified during long-lasting interflood periods by wind deflation of finer grains. The most common bar form in the active channel is that of vegetation-free longitudinal gravel bars (averaging 100–300 m long and 20–50 m wide) composed of subangular to subrounded quartz latite and basalt clasts. They are poorly sorted to unsorted, comprising a large variety of grain sizes from fine sand to boulders. Slight imbrication, caused by the westerly flow direction, is evident in some places. Winnowing of finer material by wind causes the development of a deflation surface on the top of the bars. Bar height decreases from about 70 cm at the fan apex to less than 10 cm near the river mouth. After flood events, small aeolian sand ramps form in the lee of the bar flanks. Fluvially generated sandy bar forms tend to be eroded and reworked by the strong southerly winds, a factor contributory to the relative lack of sandy fluvial bars documented in the cross-channel sections and also in older sequences. This has also been described by Stanistreet & Stollhofen (2002) from the modern Hoanib River.

A comparison of the topographic profiles downfan and along the active river channel (Fig. 4) shows that they have approximately the same gradient; only in the lowermost part do the fan profiles become shallower. The main channel widens towards the middle part of the fan and becomes narrower and more deeply incised 3.5 km away from the river mouth. At this point a prominent Damaran basement ridge impinges into the fan from the north (Figs 1–3). It is speculated that this is a surface expression of a basement high that causes this constriction. Downfan from this area the river bed becomes wider and less incised again.

The restricted source terrain is of considerable advantage in assessing down-channel variation in clasts entrained by Koigab River floods. Only two types of clast are apparent, those composed of massive to vesicular but intensively weathered basalt and those composed of varieties of quartz latite. Data derived from different clast types/lithologies were collected separately, so that comparative assessments could be made from one cross-sectional profile to the next. Figure 7 shows graphs of various gravel and sand sample parameters measured at specific cross-sectional profiles across the typical convex-up fan surface (Fig. 7A) and down the Koigab channel (Figs 7B–D). The analysis of the maximum clast size indicates that there is little significant reduction in clast size downstream (Fig. 7B). If the grain-size parameters from channel samples are compared, neither changes in sorting nor in mean value are significant (Fig. 7C & D). This reflects the ephemeral flash-flood style of the active river (cf. Williams & Rust, 1969).

Abandoned Koigab channels

The morphology of abandoned channels is variably discernible on the fan surface. The Koigab River bisects the fan surface along its central axis (Fig. 3). To the north, channels are least modified by subsequent flooding and aeolian fill. The most recently abandoned channel is immediately north of the active channel and channels become progressively less well-defined (Figs 2 & 6A) the further they are situated northward away from the active channel. South of the present Koigab River channel, the abandoned channels are far less well-preserved, defined as shallow troughs in the surface. We therefore deduce that the active channel is in the process of migrating southwards through successive lateral avulsions, slowly reworking an earlier fluvially generated surface, in which only rather ancient channel remnants are preserved. This reasoning also explains why aeolian bedforms (shrub coppices, sand ribbons, sand ramps and barchanoid forms) are also best developed (Fig. 8A) within the southern half of the fan. In the northern area such forms would have been reworked by the laterally migrating active channel. There is clearly a relationship between the setting of the Koigab Fan and one of the most important input areas of sand into the southern end of the northern Namib Sand Sea (Lancaster, 1982; Krapf et al., 2000), or Skeleton Coast Erg. The fluvial disturbance and windblown reworking of large volumes of sand from the northern half of the fan appears to represent an important source, in addition to that of the shoreline, for sand provided to that part of the dune field.

Ancient fluvial sequences of the Koigab Fan

It is difficult to record the depositional style of abandoned channel systems because of the deflationary nature of the fan surface (Fig. 8B). Fan deposits are exposed only in cut-walls of the main active channel (Fig. 9) and in terraces in the mid-fan, but most are covered by sand which is blown from the beach over the fan surface. These subrecent fan deposits consist mostly of cobble

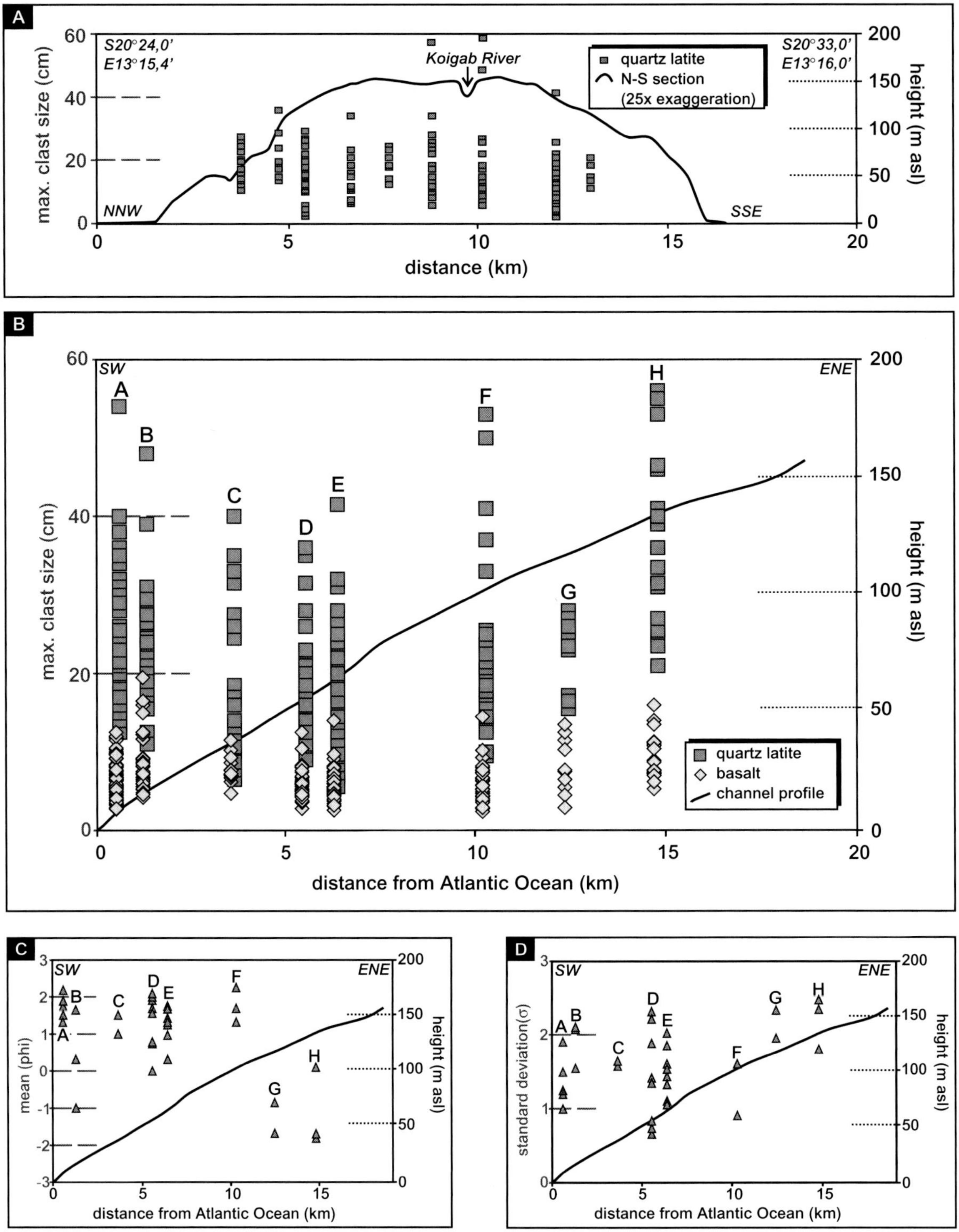

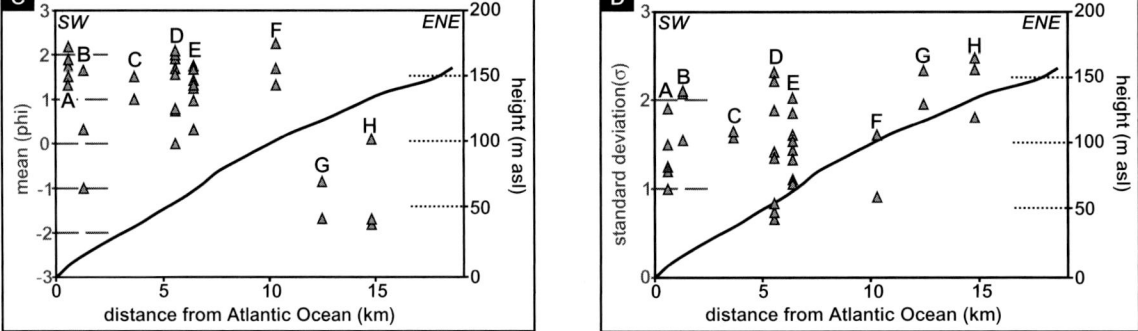

Fig. 7 (A) Maximum clast-size variation along a NNW–SSE section across the central part of the fan parallel to the main road C34 (see Fig. 4A). Along-channel variations from fan apex to river mouth of (B) maximum clast size, (C) mean and (D) sorting of samples from channel sands. Sample locations of A–D are shown in Fig. 5.

SEDIMENTOLOGICAL PROCESSES SHAPING THE KOIGAB FAN

Fluvio-aeolian interaction affecting the Koigab Fan surface

Long-lasting interflood periods, the hyperarid setting and permanently strong south-south-westerly onshore winds provide a setting where aeolian processes operate throughout the year and cause deflation, particularly of sand-sized material at the fan surface. At an advanced stage, this finally leaves a serir-like lag deposit of coarse pebbles behind (Fig. 8B). The provision of abundant sand during ephemeral river flooding is usually followed by enhanced aeolian winnowing of sand out of dry channelways at the fan surface (Krapf & Stollhofen, 2000). These sands either accumulate as sandramps within another downwind channel, preferentially at its leeward side, or they form shrub coppices (Fig. 8A), sand ribbons and barchan trains which add material continuously to the Skeleton Coast Erg, developed 6.3 km farther downwind of the active Koigab channel. Such winnowing out of a fluvial channel is well illustrated in Fig. 6A & B where the bright greyish shading shows the widespread development of shrub coppices. An additional, more constant source of aeolian sand is provided in the coastal areas by a sandy backshore strip of several tens to hundreds of metres width.

River flooding not only transports 'juvenile' material from inland volcanic source rocks but also reworks and mixes with aeolian sands that are accumulated within the river channels. This may result finally in multiple cycles of fluvial and aeolian recycling of sand in the fan area, which is also recorded by ancient fluvial sequences of the Koigab Fan (Fig. 9). Heavy mineral assemblages within fluvial deposits at sampling localities I to A (cf. Fig. 5) downstream along the active Koigab channel have been tested as a tool to demonstrate the amount of aeolian material input into the river system (Schlicker, 2000). Components that unequivocally relate to a volcanic source (e.g. volcanic lithics, Ti-magnetite, pyroxenes and yellowish epidote) can be distinguished from grains that require a metamorphic basement source (e.g. garnet, biotite, muscovite, staurolite, chloritoid, tourmaline, titanite). As metamorphic rocks crop

Fig. 8 Photographs of (A) a typical shrub coppice field on the southern Koigab Fan surface (oblique windward view) developed on (B) the deflated fan surface.

to boulder gravel with well-developed low-angle planar cross-bedding (Fig. 10A) and plane-bedded stratification (Fig. 10B, lower part) related to longitudinal gravel bar migration. The gravel is dominated by quartz latite clasts, with minor basalt clasts. The gravels are interbedded with medium-grained sand layers (Figs 9 & 10B upper part) representing waning flood stage deposits of eroded aeolian sands and sandramps. These fine- to coarse-grained sand layers (< 1.2 m) show transitional flow plane bedding to low-angle cross-bedding in the lower parts and climbing ripple cross-lamination with convolute deformation structures in the uppermost part (Fig. 10C). Thus braided fluvial processes deposited gravels and sands to cause the overall aggradation of the subrecent Koigab Fan in a manner similar to those of the modern Koigab River at present.

Fig. 9 Photograph and sedimentary log of 'older' Koigab river deposits, exposed in a sidewall of the active channel close to cross-sectional profile E (Fig. 5).

Fig. 10 Photographs of terrace and cut-walls showing ancient sequences of fan sedimentation. (A) Cross-bedded gravel fill of a river channel. (B) Plane laminated sand channel fill above a channel gravel bar deposit. (C) Sand flood depositional unit deposited in the lower fan area. Transitional flow regime plane lamination passes into pronounced climbing ripple cross-lamination to be topped by antidune bedding. Hammer shaft in (A) and (B) is 30 cm; ranging pole in (C) is 1 m.

out in the Koigabmond area, an increase in these components would be expected in the distal part of the fan. There are no outcrops of metamorphics in the upstream area, therefore the presence of heavy minerals derived from metamorphic rocks in the fan apex is limited and would reflect aeolian processes. Thus, the suite of heavy minerals from metamorphic rocks reflects the minimum amount of aeolian input, which increases from 5% in the fan apex area (Fig. 11, section I) to as much as 50% in the distal fan reaches (Fig. 11, section C). In fact, the true amount of material added to the river system by aeolian processes may be even higher if the amount of aeolian recycling of fluvial sands is considered. However, this process could not be accounted for on the basis of petrographic data.

Figure 11 distinguishes fluvial sand samples taken at the downwind (northern) flank (right column of each pair) from those derived from the upwind (southern) side of the channel (left column of each pair), with most of the latter recording a higher degree of 'aeolian' material input. This may illustrate that fluvial deposits at the southern channel margin represent erosion and incorporation of aeolian sands that originally accumulated as sandramps in the lee of channel-cut walls during interflood episodes.

There is also good evidence from grain-size data that suprafan aeolian deposits frequently involve grains winnowed out of fluvial deposits. Figure 12 demonstrates that the mean grain size of aeolian sediments (samples of shrub coppice and sand-sheets) varies considerably along a N–S orientated section across the Koigab Fan, but this variation fits a general trend. At the southern margin of the fan, the majority of aeolian deposits are composed of medium- to coarse-grained sand (mean: 1.45 ϕ), which fines over 15 km distance downwind to fine-grained sand (mean: 2.2 ϕ). This development is disturbed by the provision of coarser, 'juvenile' material through the active Koigab channel, which records a downstream fining trend running oblique to the aeolian trend. Aeolian sands sampled immediately downwind of the Koigab River channel are markedly coarser-grained again (mean 1.6–1.9 ϕ) but show another downwind fining trend to reach a mean of 2.2 ϕ.

The Koigab Fan thus compares well with studies of Bullard & Livingstone (2002), where moderate

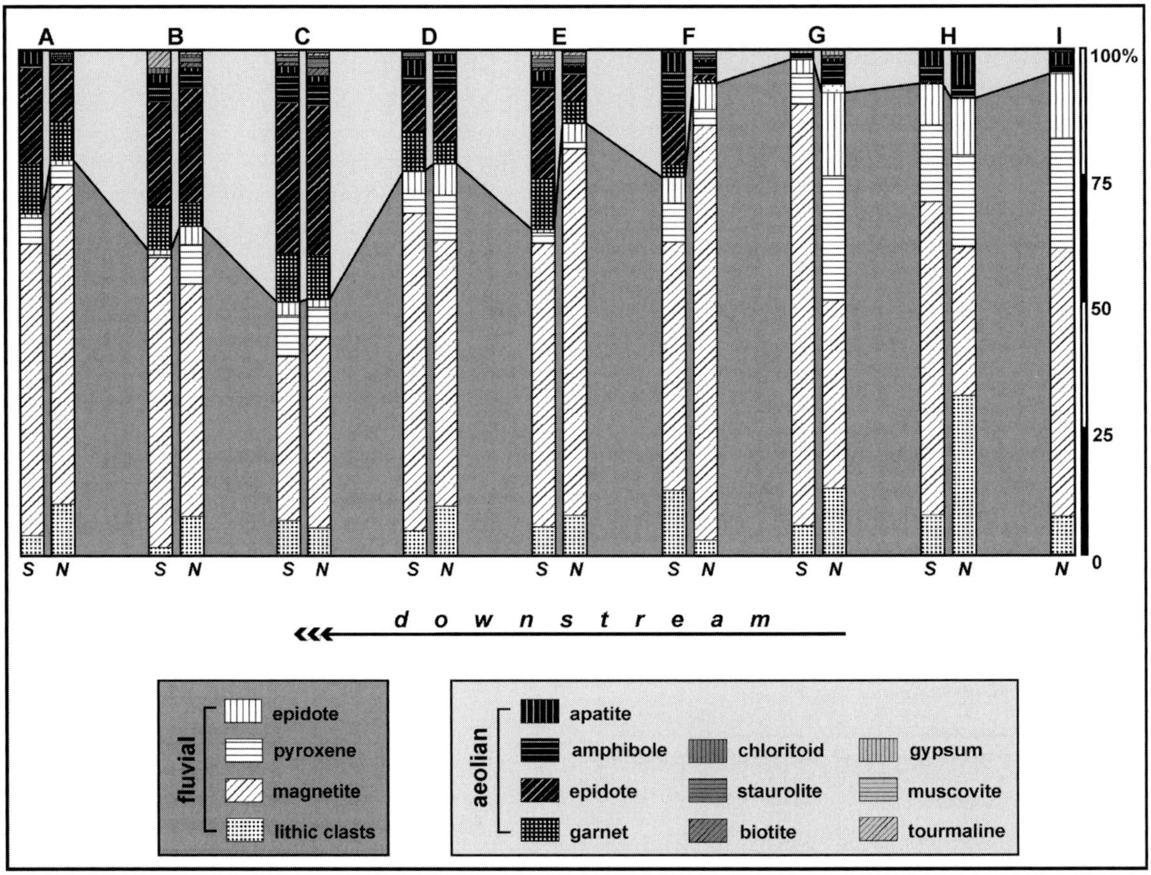

Fig. 11 Downstream variation of fluvial versus aeolian derived heavy mineral spectra of Koigab river deposits (63–125 μm fraction). Sampling locations A–I are shown in Fig. 5. Left-hand columns represent heavy mineral assemblages from upwind southern channel flanks, and right-hand columns from downwind northern channel flanks. See text for further explanation.

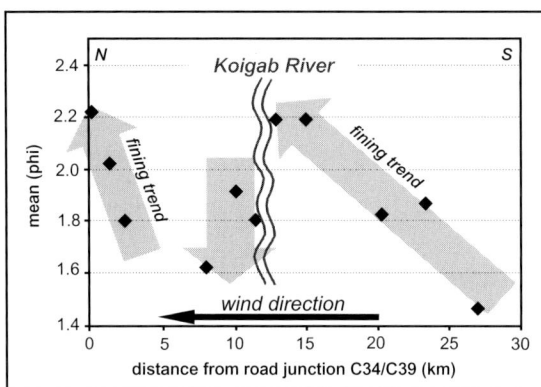

Fig. 12 (*left*) Mean grain-size trends (based on Folk & Ward parameters) showing downwind fining in aeolian sand accumulations (shrub coppices and sand ribbons) along a N–S section across the Koigab Fan surface parallel to the main road C34. Downwind fining trends are interrupted by fluvial input of 'juvenile', less mature grains.

to high intensity fluvial events occur at low frequency but aeolian processes operate at high frequency and with moderate intensity.

Interaction of the fan with the marine environment

Active channel and oceanic interaction

Only a small sector of the fan 2 km to either side of the presently active Koigab channel interacts directly with the South Atlantic Ocean. Ephemeral floods along the channel have deposited a small sedimentary prism on to the adjacent shelf area, directly in front of the Koigab river mouth (Fig. 4). Also clear on Fig. 4 are the positions of previous prisms, although not as well defined as the most recent one, at the mouths of previously abandoned channels. Thus reworking of these prisms is a relatively slow process. This is surprising given the high wave energy nature of the Namibian coastline, the narrow width of the shelf and the storm-prone nature of the region. It is tempting to speculate that boulders of Etendeka volcanics are transported far offshore to form the framework of the offshore flood prism, in order to explain its longevity as a topographic feature. The distribution and form of the abandoned flood prisms also support the notion that the Koigab River is migrating southwards. First, the topographic form of the prisms becomes less pristine moving northwards away from the active channel. Second, the shelf to the south of the active channel shows no prisms. This is interpreted to indicate that this area has had the longest opportunity to be resculpted by wave and storm activity.

Marine beach and washover sediments occur at the present river mouth. The low angle cross-bedded foreshore sediments are characterized by intensive heavy mineral laminae consisting mainly of concentrated garnet and magnetite grains (Fig. 13). Often driftwood of sizes up to 1 m and 20 cm thickness and single beach cobbles of quartz latites with sizes up to 30 cm are embedded in the otherwise very well-sorted fine to coarse sand. These sediments are draped by a thin mud layer of less than 3 cm thickness, documenting the very last stage of a fluvial flood event. Single layers of shell beds up to 10 cm thick, comprising mainly inarticulate brachiopod shells, are good indicators of marine washover at the river mouth (Fig. 13).

Fig. 13 Photograph of marine beach and washover sediments at the present-day river mouth. The low angle cross-bedded, well-sorted, fine- to coarse-grained foreshore sands are characterized by intensive heavy mineral laminae consisting mainly of concentrated garnet and magnetite grains. The sediments are draped by a thin, < 3 cm thick, mud layer, documenting the very last stage of the fluvial flood event of April 2000. The hammer (circle; shaft = 30 cm) shows the position of a shell-bed < 10 cm thick, mainly comprising inarticulate brachiopod shells, indicating marine washover.

Flanking shoreline barriers and lagoons

Shoreline barrier systems and large lagoons front large areas of the Koigab Fan. The barrier systems show evidence of recent uplift, as progressively older barrier crests are successively higher (Fig. 14). The highest barrier crest ranges between 4 and 6 m and could be correlated with heights of Pleistocene raised marine complexes from southern Namibia (Pickford & Senut, 1999). The most recent barrier sits uplifted on a wave-cut rock pediment about 2–3 m above sea level, which could be correlated with the latest Pleistocene to Holocene 2 m raised beaches in southern Namibia (Pickford & Senut, 1999). Only the most recent barrier in a position furthest away from the fan has washover channels incised through them, feeding subaerial washover fans that have aggraded to present day sea-level in that area. In contrast, the multibarrier system developed closer to Kuiseb schist outcrop and the northern fan base acts as a permanent barrier, and the lagoonal salt marsh area there, representing an earlier sabkha sediment–water interface, has been surveyed at a level 1 m below present sea level (Fig. 14, section 2).

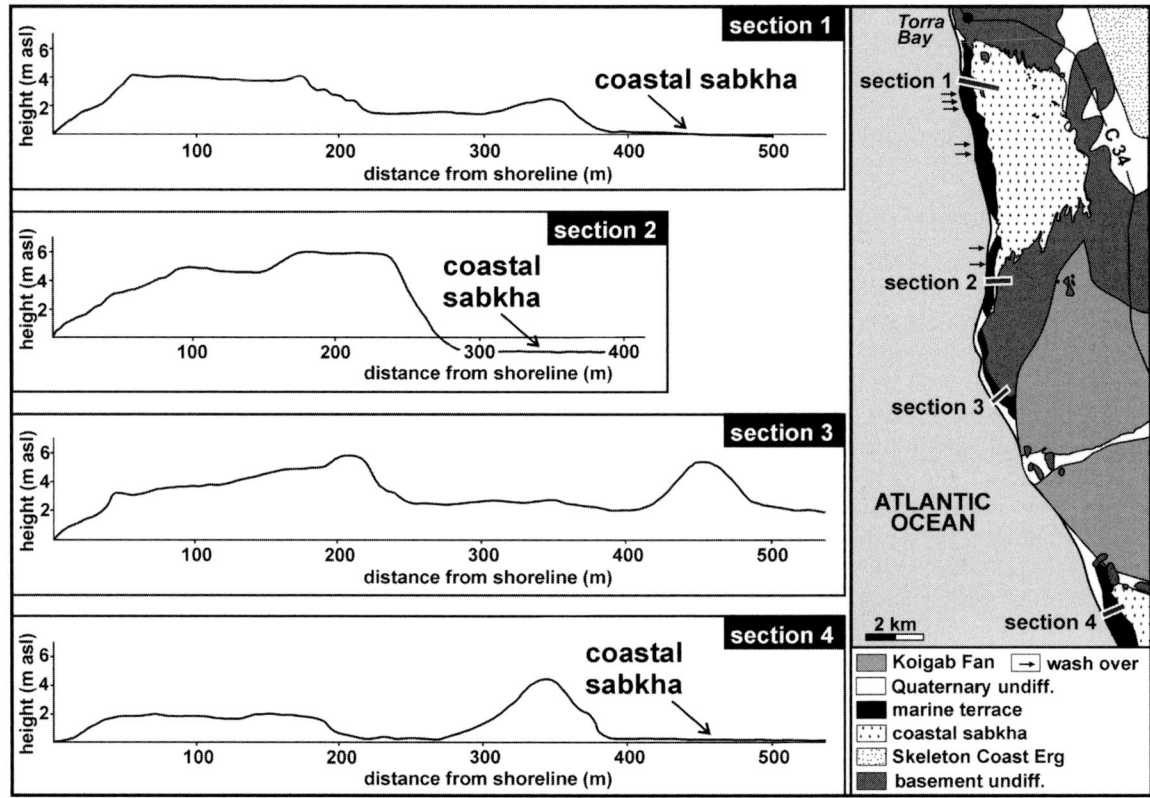

Fig. 14 Topographic profiles 1 to 4 measured by survey level and measuring tape through barrier systems and continuing on to the (1 and 2 northern, 4 southern) lagoonal sediment surface. In 1 and 4 the lagoon surface relates to modern sea-level as a result of recent washover effects and subaerial fans. In 2 the barriers partially protect the lagoon from marine inundation.

This height differential perhaps reflects the isolation of this part of the lagoon from marine flooding during the present interglacial sea-level rise.

THE KOIGAB FAN COMPARED WITH OTHER FANS AND RIVERS ALONG THE NAMIBIAN COAST

Smaller fan systems of similar type adjacent to the Koigab Fan

To the south of the Koigab Fan four similar fan systems have been recognized, issuing from the Etendeka Plateau and neighbouring areas, and subsidiary to the Koigab catchment. These fans define the size and gradient range that can be achieved by this type of braided river dominated alluvial fan system.

Two smaller fan systems, the Sout and the Salt Fans, can be identified directly south of the Koigab Fan. South of the Skeleton Coast Park, two other fan systems have been recognized: the Horingbaai and the Messum Fans. These have similar dimensions to the Koigab Fan, but they differ in river length and size of the catchment area (Fig. 15). All the above fans are characterized by gradients between 0.016 and 0.009, are fed by braided ephemeral rivers and the deposits consist predominantly of gravels. The catchment areas of the Horingbaai and Messum Fans are not restricted to the Etendeka Plateau, rather parts of the catchments are situated on Damaran basement rocks. In contrast with the Koigab, Messum and

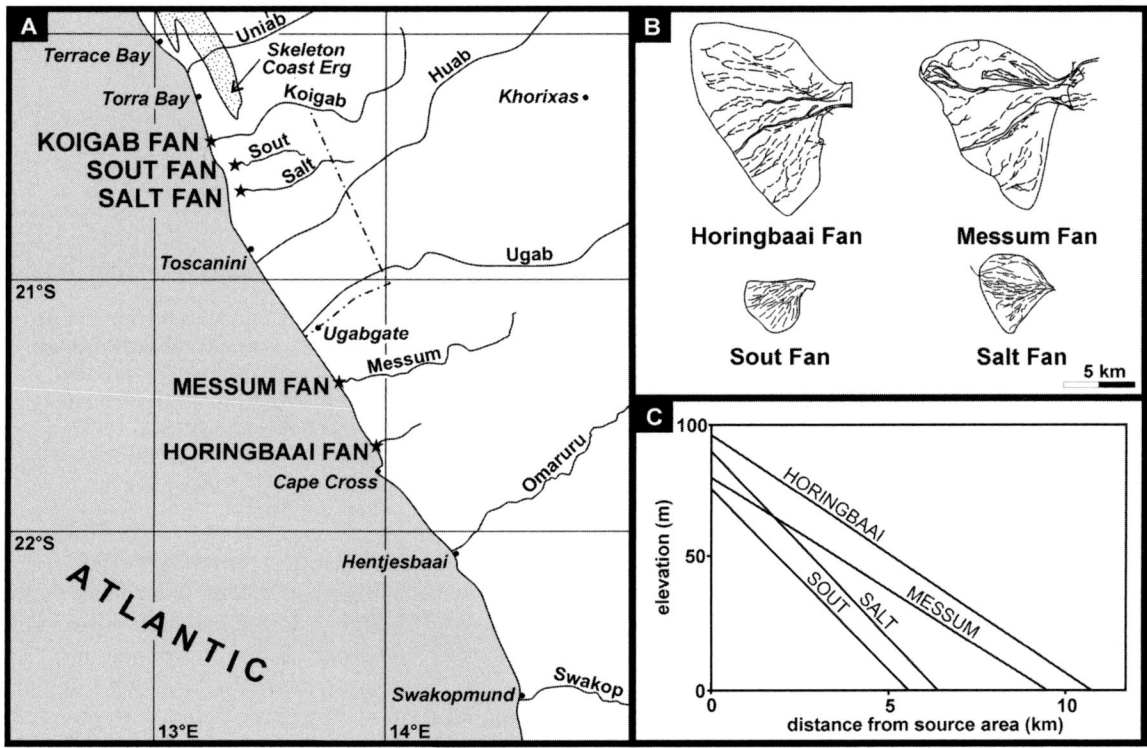

Fig. 15 (A) Map locating the Horingbaai, Messum, Sout and Salt Fan systems south of the Koigab Fan. These fan systems are similar in order of magnitude to the Koigab Fan in terms of (B) their size and (C) their downfan gradients.

Horingbaai Fans, the Sout and Salt Fans do not debouch into the ocean but into the lagoonal system 1.5 to 2.5 km east of the Atlantic Ocean.

CLASSIFICATION OF THE KOIGAB FAN AND COMPARISION WITH WORLD-WIDE FANS AND FAN SYSTEMS

Fan, terminal fan or fan-delta?

The Koigab Fan has been described as a coastal terminal fan by Van Zyl & Rust (1993), but as ephemeral flows down the Koigab channel frequently reach the fan perimeter, and through it debouch into the Atlantic, it is not a typical terminal fan (cf. Kelly & Olson, 1993). Nevertheless, the uplifted barrier–lagoon system and basement outcrop divorce most of the outer fan perimeter from marine processes. The effects of fan sedimentation on the adjacent shelf are minimal and ultimately ephemeral. Flood prisms are of the same dimensions as the width of the active channel and do not survive under the high wave-energy conditions much beyond one succeeding channel abandonment. On the other hand, the area of the Koigab river mouth does provide a small prism of sediment to the neighbouring shelf. Stanistreet & McCarthy (1993) have suggested that for every fan type there is a corresponding fan-delta. On balance, however, the Koigab Fan system exhibits the characteristics of an alluvial fan rather than those of a fan-delta.

Position and contrasts within alluvial fan classifications

In terms of subaerial fan types, the Koigab Fan classifies as a braided alluvial fan, showing no

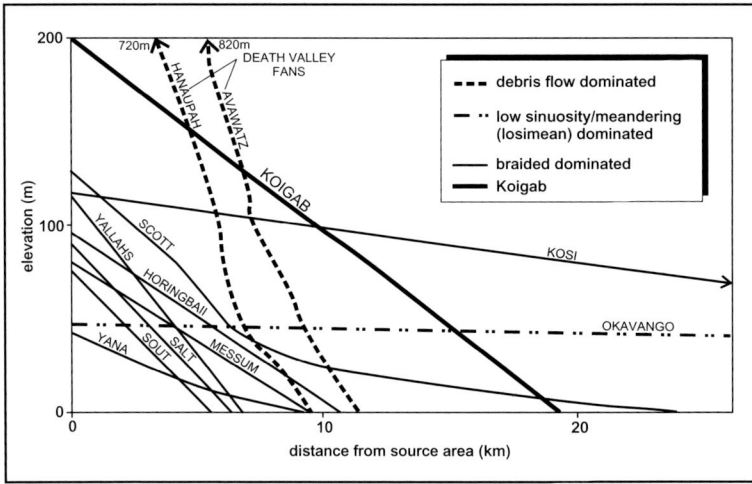

Fig. 16 Comparative gradients of various fans and subaerial fan-delta surfaces world-wide (compiled from own data and the diagrams of Boothroyd & Ashley, 1975; Wescott, 1990; Stanistreet & McCarthy, 1993).

evidence of debris flows or sheetfloods (Blair & McPherson, 1994a,b) nor of meandering or low sinuosity river channels typical of another type (Stanistreet et al., 1993). It is therefore an end member in the classificatory triangle proposed by Stanistreet & McCarthy (1993). In terms of size within the braided alluvial fan class, the Koigab Fan is intermediate between small-scale braided river dominated systems (Figs 16 & 17), such as the Yana Fan, Alaska (Boothroyd & Ashley, 1975), and braided river dominated megafans such as the Kosi Fan, developed south of the Himalayas (Gohain & Parkash, 1990; Wells & Dorr, 1987a,b). The four other fan systems described from the Skeleton Coast (Fig. 15) are of a similar order of magnitude to the larger Koigab Fan system, but show variations from 5 to 15 km from apex to periphery and 4.25 to 23 km across their broadest dimension.

In terms of gradient the Koigab Fan is rather steep (0.011) in contrast (Fig. 16) to other braided river dominated fans, such as the Kosi (0.0002 Gohain & Parkash, 1990) and Alaskan fan systems (typically 0.0066 (proximal) to 0.0033 (distal), Boothroyd & Nummedal, 1978), and shows profiles closer to debris flow dominated fans (Blair & McPherson, 1994a). The four other Skeleton Coast fans also show steeper slope gradients, from 0.016 to 0.0093 (Figs 15 & 16).

In terms solely of geometry, size and depositional setting the Koigab Fan is close to those fans and fan-delta systems described from the Arctic and sub-Arctic (Boothroyd & Ashley, 1975; Boothroyd & Nummedal, 1978). In terms of depositional style, slope (Fig. 16), climatic setting and fan surface, however, there could not be more contrast. The Arctic fans are proglacial, fed by perennial river systems associated with ice melt, whereas the Koigab Fan is fed by an ephemeral river providing both sediment and water discharge in short energetic bursts. This may to some extent explain differences in fan gradient between the two types of fans.

In contrast with other fans described in the literature, the Koigab Fan and the four other Skeleton Coast fans show no evidence for subdivision into upper, middle and lower fan segments. This is also confirmed by fluvial bedforms, which appear not to change considerably in style between apex and fan base.

The Koigab Fan and the four other Skeleton Coast fans (Fig. 16) thus bridge a gap in both size and gradient between high gradient small fans and lower gradient fans and megafans.

Koigab Fan climate and vegetation and significance for Precambrian fans

In terms of climate, unlike other braided river dominated alluvial fan examples, the Koigab Fan is deposited in a hyperarid setting, and it is only precipitation in the mountainous catchment of the Etendeka Plateau that allows its present style of development. This negates concepts of only

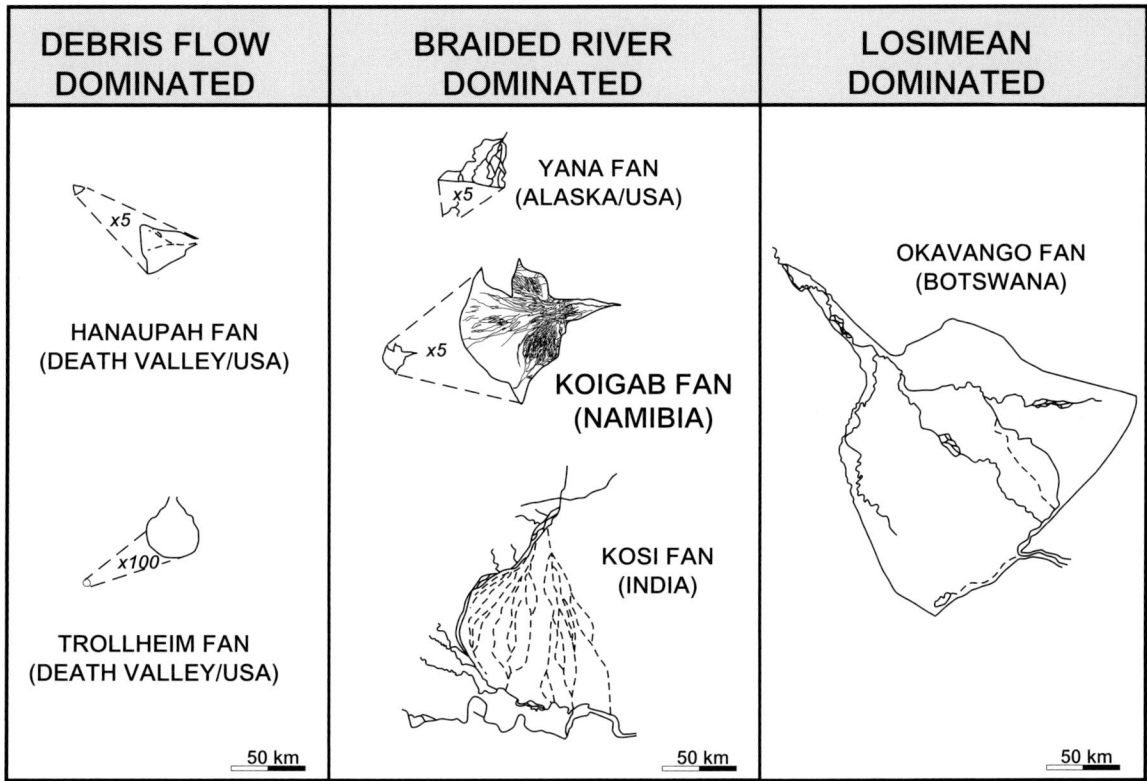

Fig. 17 Classification of the Koigab Fan within the subaerial fan classes of the fan classification scheme of Stanistreet & McCarthy (1993).

debris-flow and sheetflood dominated fans developing in arid to semi-arid climates. Like the Okavango Fan, the Koigab Fan provides an important caveat in utilizing fan types as climatic indicators in ancient sequences, because fan types may depend as much on catchment conditions to define their depositional characteristics (Moscariello et al., 2002) as on the climatic conditions in their depositional area. Other subaerial braided fan systems have developed in low-latitude tropical locations, for example, the subaerial portion of the Yallahs Fan-delta of Jamaica (Burke, 1967; Wescott & Ethridge, 1980; Wescott, 1990) are heavily vegetated, even to the extent of tropical rain forest. In contrast, the arid tropical-latitude setting of the Koigab Fan ensures that vegetation is restricted to aeolian bedforms, such as shrub coppices. The lower delta plain of the Yallahs Fan, which is covered by marshes, swamps and open ponds, is characterized by muddy sands and organic-rich mud deposits. The Koigab Fan, by contrast, bears little vegetation and low rates of chemical weathering ensure a much lower supply of clay-grade material.

Difficulties are encountered in finding examples of modern braided river dominated fans that can serve as suitable analogues for ancient alluvial fan sequences. This problem is exacerbated when dealing with fluvial sequences and fan systems (Rust, 1978) that are older than the Silurian/Devonian, when plants initially colonized the land surface. On the one hand, proglacial alluvial fans represent a rather specialized set of circumstances that, although providing a picture of how perennial rivers may operate on a fan surface, were uncommonly encountered during Earth history except during Permo-Carboniferous, Late Ordovician, Neoproterozoic and possibly Palaeoproterozoic ice ages. By contrast, modern non-glacial braided river dominated fans are commonly densely vegetated (e.g. subaerial portions of the Yallahs Fan-delta; Westcott & Ethridge, 1980), and are often heavily affected by human occupation (e.g. Kosi Fan) and agricultural activity (e.g. Colorado Fan, Texas; McGowen & Garner, 1970). The Koigab Fan therefore offers an unusual opportunity to view a fan surface and system that is similar in some ways to how Precambrian fans might have appeared and acted.

CONCLUSION

The unique climatic and hydrological setting of the Koigab Fan, with its hyperaridity, permanently strong south-westerly onshore winds, long-lasting interflood periods and flood events operating at high intensity but low frequency, favours fluvio-aeolian interaction on its surface. Aeolian processes operate throughout the year and cause deflation and the development of lag deposits. Fluvial processes provide abundant sand during and after flood events, which are followed by enhanced aeolian winnowing of sand out of dry channelways on to and over the fan surface. Heavy mineral assemblages within fluvial deposits and grain-size analysis from aeolian sand samples demonstrate the amount of aeolian material input into the river system and the amount of fluvial material output from the river system.

The Koigab Fan extends knowledge of the variability of the braided fluvially dominated fan class. In terms of form, magnitude and characteristics it represents an important intermediate type both within that class and within the range of world-wide subaerial fans as a whole. In terms of the latter, it and nearby fans in Namibia provide an important bridge between ultrahigh- and ultralow-gradient fan systems.

As the Koigab Fan lacks significant surface vegetation it can act additionally as a model for certain Precambrian and early Palaeozoic fan depositional systems deposited prior to the evolution of land plants.

ACKNOWLEDGEMENTS

Funding was provided by the German Research Foundation (DFG) through the Postgraduate Research Programme 'Interdisciplinary Geoscience Research in Africa'. The Geological Survey of Namibia, Namibian Nature Conservation and the Ministry of Environment and Tourism provided logistic support for our Skeleton Coast research. Special thanks to the Skeleton Coast Park conservation staff John Paterson, Alvin Engelbrecht, Mareli Grobler and the staff at Ugab-Gate for accommodation and allowing us essential access to remote areas of the study area. Special thanks to Bernd Schlicker and Mario Werner for their supportive field and laboratory work. We thank Mary Bourke and Nick Lancaster for their critical reviews of the manuscript and Mike Blum for his editorial support.

REFERENCES

Ahrendt, H., Behr, H.J., Clauer, N., Hunziker, J.C., Porada, H. and Weber, K. (1983) The Northern Branch. Depositional development and timing of the structural and metamorphic evolution within the framework of the Damara Orogen. In: *Intracontinental Fold Belts; Case Studies in the Variscan Belt of Europe and the Damara Belt in Namibia* (Eds H. Martin and F.W. Eder), pp. 723–743. Springer-Verlag, Berlin.

Barnard, W.S. (1989) Die sandseë van die Namib en hul verbreiding. *S. Afr. Geogr.*, **16**(1/2), 14–38.

Blair, T.C. and McPherson, J.G. (1994a) Alluvial fans and their natural distinction from rivers based on morphology, hydraulic processes, sedimentary processes, and facies assemblages. *J. Sediment. Res.*, **A64**(3), 450–489.

Blair, T.C. and McPherson, J.G. (1994b) Alluvial fan processes and forms. In: *Geomorphology of Desert Environments* (Eds A.D. Abrahams and A.J. Parson), pp. 354–402. Chapman and Hall, London.

Boothroyd, J.C. and Ashley, G.M. (1975) Process, bar morphology and sedimentary structures on braided outwash fans, North-eastern Gulf of Alaska. In: *Glaciofluvial and Glaciolacustrine Sedimentation* (Eds A.V. Jopling and B.C. McDonald). *Soc. Econ. Paleontol. Mineral. Spec. Publ.*, **23**, 193–222.

Boothroyd, J.C. and Nummedal, D. (1978) Proglacial braided outwash: a model for humid alluvial fan deposits. In: *Fluvial sedimenology* (Ed. A.D. Miall), *Mem. Can. Soc. Petrol. Geol.*, **5**, 641–668.

Brown, R.W., Gallagher, K., Gleadow, A.J.W. and Summerfield, M.A. (2000) Morphotectonic evolution of the South Atlantic margins of Africa and South America. In: *Geomorphology and Global Tectonics* (Ed. M.A. Summerfield), pp. 255–281. Wiley, Chichester.

Bullard, J.E. and Livingstone, I. (2002) Interactions between aeolian and fluvial systems in dryland environments. *Area*, **34**(1), 8–16.

Burke, K. (1967) The Yallahs Basin: a sedimentary basin southeast of Kingston Jamaica. *Mar. Geol.*, **5**, 45–60.

Cockburn, H.A.P., Brown, R.W., Summerfield, M.A. and Seidl, M.A. (2000) Quantifying passive margin denudation and landscape development using a combined fission-track thermochronology and cosmogenic isotope analysis approach. *Earth Planet. Sci. Lett.*, **179**, 429–435.

De Wit, M.C.J., Marshall, T.R. and Partridge, T.C. (2000) Fluvial deposits and drainage evolution. In: *The Cenozoic of Southern Africa* (Eds T.C. Partridge and R.R. Maud), pp. 55–72. Oxford University Press, Oxford.

Dingle, R.V. (1992/93) Structural and sedimentary development of the continental margin off southwestern Africa. *Comm. Geol. Surv. Namibia*, **8**, 35–43.

Gohain, K. and Parkash, B. (1990) Morphology of the Kosi Megafan. In: *Alluvial Fans—A Field Approach* (Eds A.H. Rachocki and M. Church), pp. 151–178. Wiley, Chichester.

Jacob, J.R. (2001) Incision and aggradation in the Orange River valley, southwestern Africa. In: *Program and Abstracts, 7th International Conference on Fluvial Sedimentology* (Eds J.A. Mason, R.F. Diffendal Jr. and R.M. Joeckel), p. 138. Lincoln, Nebraska.

Jacobson, P.J., Jacobson, K.M. and Seely, M.K. (1995) *Ephemeral Rivers and their Catchments: Sustaining People and Development in Western Namibia.* Desert Research Foundation of Namibia, Windhoek, 160 pp.

Jury, M.R. (1996) Regional teleconnection pattern associated with summer rainfall over South Africa, Namibia and Zimbabwe. *Int. J. Climatol.*, **16**(2), 135–153.

Kelly, S.B. and Olsen, H. (1993) Terminal fans—a review with reference to Devonian examples. *Sedimentary Geolgy*, **85**, 339–374.

King, L.C. (1951) *South African Scenery; a Textbook of Geomorphology*, 2nd edn. Oliver and Boyd, Edinburgh, 379 pp.

Klein, J.A. (1980) Pleistocene to recent faulting in the area west of Omaruru SWA/Namibia). *Geol. Surv. Namibia, Regional Geol. Ser., Open File Rep.*, **RG 4**, 1–26.

Korn, H. and Martin, H. (1957) The Pleistocene in southwest-Africa. *Proceedings of the Third Pan-African Congress on Prehistory*, Livingstone, 1955, 14–22.

Krapf, C.B.E. and Stollhofen, H. (2000) Fluvio-aeolian interaction in the Koigab Fan area—Skeleton Coast, Namibia. *Mitt. Ges. Geol. Bergbaustud. Österr.*, **43**, 76–77.

Krapf, C.B.E., Stollhofen, H. and Stanistreet, I. (2000) Contrasting styles of fluvio-aeolian interaction between ephemeral river systems and dunes of the Skeleton Coast Erg, Namibia. In: *Abstract IGCP 413 Meeting: Linkages between Fluvial, Lacustrine and Aeolian Systems*, Desert Study Center Zzyzx, California (Ed. N. Lancaster), pp. 11–12.

Lancaster, N. (1982) Dunes on the Skeleton Coast, Namibia (South West Africa): geomorphology and grain size relationships. *Earth Surf. Process. Landf.*, **7**, 575–587.

Lancaster, J., Lancaster, N. and Seeley, M.K. (1984) Climate of the central Namib Desert. *Madoqua*, **14**(1), 5–61.

Marsh, J.S., Ewart, A., Milner, S.C., Duncan, A.R. and Miller, R. McG. (2001) The Etendeka Igneous Province: magma types and their stratigraphic distribution with implications for the evolution of the Paraná-Etendeka flood basalt province. *Bull. Volcanol.*, **62**, 464–486.

McCarthy, T.S., Cooper, G.R.J., Tyson, P.D. and Ellery, W.N. (2000) Seasonal flooding in the Okavango Delta, Botswana; recent history and future prospects. *S. Afr. J. Sci.*, **96**, 25–33.

McGowen, J.H. and Garner, L.E. (1970) Physiographic features and stratification types of coarse-grained point bars: modern and ancient examples. *Sedimentology*, **17**, 77–111.

Messerli, B., Grosjean, M., Bonani, G., Buergi, A., Geyh, M.A., Graf, K., Ramseyer, K., Romero, H., Schotterer, U., Schreier, H. and Vuille, M. (1993) Climate change and natural resource dynamics of the Atacama Altiplano during the last 18 000 years; a preliminary synthesis. In: *Mountain Geoecology and the Andes; Resources Management and Sustainable Development* (Eds J.D. Ives and P. Ives). *Mountain Res. Devel.*, **13**(2), 117–127.

Milner, S.C. (1986) The geological and volcanological features of the quartz latites of the Etendeka Formation. *Comm. Geol. Surv. S.W.Afr./Namibia*, **2**, 109–116.

Milner, S.C., Duncan, A.R. and Ewart, A. (1992) Quartz latite rheoignimbrite flows of the Etendeka Formation, north-western Namibia. *Bull. Volcanol.*, **54**, 200–219.

Moscariello, A., Marchi, L., Maraga, F. and Mortara, G. (2002) Alluvial fan activity in the Italian Alps. Sedimentary facies, processes and related hazards. In: *Flood and Megaflood Deposits: Recent and Ancient* (Eds I.P. Martini, V.R. Baker and G. Garzon), *Spec. Publs Int. Assoc. Sediment.*, **32**, 141–166.

Partridge, T.C. and Maud, R.R. (2000) Macro-scale geomorphic evolution of Southern Africa. In: *The Cenozoic of Southern Africa* (Eds T.C. Partridge and R.R. Maud), pp. 3–18. Oxford University Press, Oxford.

Pickford, M. and Senut B. (1999) Geology and palaeobiology of the Namib Desert Southwestern Africa. Volume 1: geology and history of study. *Mem. Geol. Surv. Namibia*, **18**, 155 pp.

Raab, M. (2001) The geomorphic response of the passive continental margin of northern Namibia to Gondwana break-up and global scale tectonics. Unpubl. PhD thesis, University of Göttingen, 253 pp. (online publication: http://www.sub.uni-goettingen.de/)

Renne, P.R., Glen, J.M., Milner, S.C. and Duncan, A.R. (1996) Age of Etendeka flood volcanism and associated intrusions in southwestern Africa. *Geology*, **24**, 659–662.

Rust, B.R. (1978) Depositional model for braided alluvium. In: *Fluvial Sedimentology* (Ed. A.D. Miall). *Mem. Can. Soc. Petrol. Geol.*, **5**, 605–625.

Schlicker, B. (1999) *Erläuterungen zur Geologischen Karte 1.25.000 des Koigab-Canyon und seiner Umgebung, Skelettküste, NW Namibia.* Unpublished mapping project, University of Würzburg, 64 pp.

Schlicker, B. (2000) Sedimentfracht und Teilablagerungsräume des Koigab-Reviers—Skelettküste, NW Namibia. Unpublished diploma-thesis, University of Würzburg, 109 pp.

Shannon, L.V., Boyd, A.J., Brundrit, G.B. and Taunton-Clark, J. (1986) On the existence of an El Niño-type phenomenon in the Benguela System. *J. Mar. Res.*, **44**, 495–520.

Stanistreet, I.G. and McCarthy, T.S. (1993a) The Okavango Fan and the classification of subaerial fan systems. *Sediment. Geol.*, **85**, 115–133.

Stanistreet, I.G. and Stollhofen, H. (2002) Hoanib River flood deposits of Namib Desert interdunes as analogues for thin permeability barrier mudstone layers in aeolianite reservoirs. *Sedimentology*, **49**, 719–736.

Stanistreet, I.G., Cairncross, B. and McCarthy, T.S. (1993b) Low sinuosity and meandering bedload rivers of the Okavango Fan; channel confinement by vegetated levees without fine sediment. *Sediment. Geol.*, **85/1–4**, 135–156.

Tyson, P.D. (1969) Athmospheric circulation and precipitation over Africa. *Environ. Stud. Occas. Pap.*, **2**, 1–22.

Tyson, P.D. (1986) *Climatic Change and Variability in Southern Africa.* Oxford University Press, Cape Town, 220 pp.

Van-Zyl, J.-A. and Scheepers, A.C.T. (1993) The geomorphic history and landforms of the Lower Koigab River, Namibia. *S. Afr. Geogr.*, **20**(1/2), 12–22.

Ward, J.D. (1987) The Cenozoic succession in the Kuiseb Valley, Central Namib Desert. *Mem. Geol. Surv. S.W.Afr./Namibia*, **9**, 45 pp.

Ward, J.D. and Swart, R. (1997) Flash-flood fluvial systems of the Central Namib Desert. *Field Guide, 6th International Conference Fluvial Sedimentology*, Cape Town, SA, 37 pp.

Wells, N.A. and Dorr, J.A., Jr. (1987a) Shifting of the Kosi River, northern India. *Geology*, **15**, 204–207.

Wells, N.A. and Dorr, J.A., Jr. (1987b) A reconnaissance of sedimentation on the Kosi alluvial fan of India. In: *Recent Developments in Fluvial Sedimentology* (Eds F.G. Ethridge, R.M. Flores and M. Harvey), *Soc. Econ. Paleontol. Mineral. Spec. Publ.*, **39**, 51–62.

Wescott, A.W. (1990) The Yallahs Fan Delta: a coastal fan in a humid tropical climate. In: *Alluvial Fans—A Field Approach* (Eds A.H. Rachocki and M. Church), pp. 213–225. Wiley. Chichester.

Wescott, W.A. and Ethridge, F.G. (1980) Fan-delta sedimentology and tectonic setting—Yallahs fan-delta, southeast Jamaica. *Bull. Am. Assoc. Petrol. Geol.*, **64**, 374–399.

Whitaker, A. (1984) Dust transport by Bergwinds of the coast of south west Africa. Unpubl. B.Sc. (Hons.) Thesis, University of Cape Town, 31 pp.

Williams, P.F. and Rust, B.R. (1969) The sedimentology of a braided river. *J. Sediment. Petrol.*, **39**(2), 649–679.

A qualitative analysis of the distribution of bed-surface elevation and the characteristics of associated deposits for subaqueous dunes

SUZANNE F. LECLAIR* and ASTRID BLOM†,[1]

*Department of Earth and Environmental Sciences, Tulane University, Dimwiddie Hall, New Orleans, LA 70118, USA (Email: sedimentologist@hotmail.com/leclair@tulane.edu); and †WL/Delft Hydraulics, P.O. Box 177, 2600 MH Delft, The Netherlands

ABSTRACT

This paper analyses controls on the probability distribution of bed-surface elevations, P_s, and the structure and texture of the associated deposits, in dune-forming conditions. It is important to understand these controls in order to develop predictors required to implement a new theory that is based on a probabilistic approach of the Exner equation for sediment continuity, and to improve the interpretation of sedimentological records of fluvial origin. Experiments were conducted in three flumes of different sizes, with flow depth ranging from 0.15 to 0.87 m, flow velocity ranging from 0.5 to 0.84 m s^{-1}, and with sand- to gravel-dominated mixtures. Distributions of bed-surface elevations were measured from time-series and/or successive bed profiles. Vertical profiles of bed composition (i.e. vertical sorting) and/or structure (i.e. cross-sets) of the deposits were analysed.

Results show that the dimensionless bed shear stress and vertical sorting are the major controls on the range and shape of the non-dimensional P_s curve. Here, the dimensionless bed shear stress is used as a component of the sediment transport stage. Low values of sediment transport stage produce P_s curves that are short-ranged and quite evenly spread around the average bed level. The presence of a coarse bed layer, which affects the composition of the transported sediment, has a similar effect on the P_s curve. Otherwise, the P_s curves tend to be skewed toward the lowest values of bed surface elevation. Interactions between sediment transport stage, vertical sorting, dune height, and the P_s curves mean that the last of them can be partially reconstructed from the analysis of the geometry and texture of cross-sets. The distribution of elevation of cross-set lower boundaries mimics the P_s curve toward its lower limit. The ratio of mean dune height over flow depth indicates approximately the upper limit of the P_s curve, and these variables can be estimated via cross-set thickness distribution. It is hoped that this study will help develop a comprehensive theory that could predict sediment transport from the characteristics of dunes and their deposits, and vice-versa.

---

[1]Present address: Department of Civil Engineering, University of Twente, P.O. Box 217, 7500AE Enschede, The Netherlands.

INTRODUCTION

Dunes are common bedforms in rivers, and changes in their geometry as they migrate under various flow conditions affect the bed morphology and the characteristics of their preserved deposits (e.g. Allen, 1982). The nature of dunes is of interest to scientists working with different perspectives, e.g. engineers concerned about river-bed erosion or sedimentologists eager to interpret the fluvial record. For that reason, the authors have independently developed similar databases on dune characteristics, although pursuing different objectives, such as dune migration and formation of cross-sets (Leclair et al., 1997; Leclair, 2000, 2002; Leclair & Bridge, 2001) or vertical sorting to improve sediment continuity concepts for non-uniform sediment (Blom, 2000; Blom et al., 2001, 2003). The present study stems from a shared curiosity about dune-migration processes and their effects on bed topography and the characteristics of associated deposits.

The two studies discussed here were conducted separately, so many differences as well as similarities exist between them. Differences occurred in the:

1 choice of flow and sediment conditions for the experiments, resulting in large ranges in flow depth, velocity and sediment transport stage;
2 sampling scheme for dune deposits, which favoured structural (Leclair) or textural (Blom) analysis;
3 overall analytical procedures, which highlighted different aspects of the relationships between sediment transport and the characteristics of subaqueous dunes and their deposits (more below).

The major similarities arise from a common particular attention to the occurrence of the deepest dune troughs, as these represent the possible lower boundaries of cross-sets, as well as the location of the coarser particles. Hence, extensive, high-resolution sampling of variation in bed-surface elevation was produced, as well as detailed measurements of dune geometry, including matching variables rarely available in other data sets, such as the height of individual dunes and the depth of their associated trough scour relative to mean bed level. Moreover, the differences between experimental settings provide an interesting range of conditions that might have never been planned otherwise. This paper integrates these two studies by applying some of the analytical methods developed with one data set to the other set, and presents new insights resulting from this integration effort.

Cross-set formation model

Figure 1 illustrates the concept of cross-set formation at a given point. The variability of the height and trough-scour depth of successive bedforms can produce a stack of cross-sets, even without net deposition (Paola & Borgman, 1991; Leclair, 1997, 2002). In the Paola–Borgman model (1991) for the probability density function (PDF) of cross-set thickness, s,

$$p(s) = ae^{-as}(e^{-as} + as - 1)/(1 - e^{-as})^2 \quad (1)$$

the parameter a represents the mean and standard deviation of trough-scour depth, ts, because by definition $a = 1/\beta$ and $\beta = ts_{sd}^2/ts_m$. Leclair & Bridge (2001) proposed a modified Paola-Borgman model

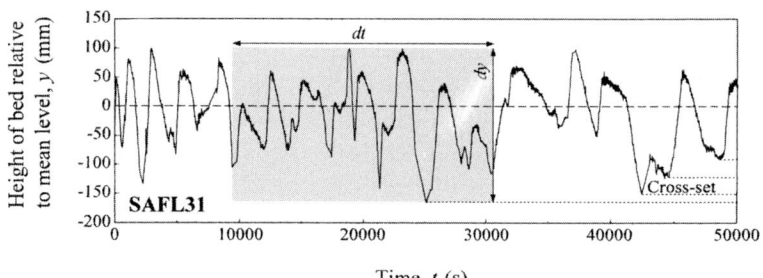

Fig. 1 Time-series of variation in bed-surface elevation from an experiment of Leclair (2000). Shaded block illustrates the concept behind the probabilistic Exner sediment continuity equation of Parker et al. (2000), based on the probabilistic nature of variations of bed-surface elevations owing to the migration of bedforms. These variations could be either a function of time or distance, as long as the flow conditions are steady and the bed-elevation series is stationary.

where the predicted mean cross-set thickness, s_m^{pred}, is defined by

$$s_m^{pred} = lr/c + 1.64493/(h_{sd}/ts_{sd})a \qquad (2)$$

where l is mean dune length, r is mean aggradation rate, c is mean dune migration rate, h_{sd} and ts_{sd} are the standard deviation of dune height and trough-scour depth below mean bed level, respectively, and $a = 1/\beta$, estimated here from the gamma function describing dune-height probability density distribution. Therefore, the distribution of observed cross-set thicknesses reflects the distributions of dune height and trough-scour depth, and both of these contribute to the variation in bed topography that will be considered in the following section.

Sediment continuity models

Fundamental to predicting morphological changes is the Exner equation of sediment continuity, which can be expressed in the form

$$(1 - \lambda)(\partial \eta/\partial t) = D - E \qquad (3)$$

where λ is the bed porosity, η is the average bed level, t is time, D is the volume of sediment deposited on to the bed, per unit area and time, and E is the volume of sediment entrained from the bed (Parker et al., 2000, equation 11). Parker et al. (2000) derived a probabilistic form of the Exner equation for sediment continuity, based on the probabilistic nature of variations of bed-surface elevations due to the migration of bedforms (Fig. 1). For uniform sediment, they derived

$$(1 - \lambda)(\partial P_s/\partial t) = D_e - E_e \qquad (4)$$

where P_s is the probability distribution of bed-surface elevations (e.g. Parker et al., 2000, fig. 7), and D_e and E_e are elevation-specific values of deposition and entrainment densities, respectively, defined such that $D_e dxdz$ and $E_e dxdz$ are the volumes of sediment deposited and entrained from a bed element with sides dx and dz (Parker et al., 2000, equation 21). This new conceptual framework brought up the need for: (i) formulations for the elevation-specific deposition and entrainment densities for uniform and non-uniform sediment; and (ii) an understanding of the controls on the shape of the P_s curve in the case of dunes.

Recent developments

A first step towards new formulations for the elevation-specific deposition and entrainment densities in dune conditions and, as such, toward the development of a new sediment continuity concept for non-uniform sediment was accomplished by Blom et al. (2001). They considered the evolution of the sorting profile of tracer particles in uniform sediment. Then, they considered the possible controls on the shape of the P_s curve when the bed is covered with dunes. The active bed was defined as the range of bed-surface elevations that are exposed to the flow, and this definition is used in the present paper. Their experiments showed that for similar discharge the P_s curve covered different ranges of bed-surface elevations, depending on the initial vertical sorting (more below). They also related the P_s curve to the vertical sorting in dune deposits and their results indicate that the coarser grains were usually associated with the dune-trough levels or the lower elevations of the active bed, that is, the initial vertical sorting has been modified by the dune-migration processes.

In the experiments of Blom et al. (2001), it seems that the difference in total volume of each size of sediment in the active bed affected the shape of the probability distribution of bed heights for a given discharge: the range of values for bed-surface elevation was larger in run B2, where an initial coarse bed laid over sediment composed only of the finest fraction, than in run A2, where a coarse bed laid over the tri-modal mixture (see details on grain-size distribution in Method section). Such results stress the importance of investigating the control of characteristics of the sediment mixture on the shape of the P_s curve for similar flow conditions. In addition, the P_s curve typically increased in its range of bed-surface elevations with increasing discharge (Blom et al., 2003). The predictive possibilities of equation (4) require a better understanding of how the P_s curve changes with temporal variation in flow conditions (e.g. flow depth, velocity, energy slope; see Parker et al., 2000, equation 27).

The goal of this paper is to take advantage of the existence of two new, parallel data sets on dune migration (Blom et al., 2003; Leclair, 2000) to:

1 investigate the controls on the P_s curve by comparing results from experiments conducted under different flow and sediment conditions;
2 identify which parameters of the P_s curve could be estimated from the geometry and grain sorting of cross-sets;
3 attempt to set bases for the development of a comprehensive theory of dune bed-surface elevation PDF linked to cross-set thickness PDF.

METHODS

Experimental methods

The authors independently conducted experiments under dune conditions in sediment-recirculating flumes:
1 BU runs at Binghamton University, NY (length 7.6 m, width 0.6 m);
2 SAFL runs at Saint-Anthony Falls Laboratory, University of Minnesota (length 76 m, width 2.7 m);
3 Runs A1, A2, B1, B2, and T10 at WL/Delft Hydraulics (WL/DH), in The Netherlands (length 50 m, width 1 or 1.5 m).

Extensive descriptions of the experimental designs are given by Blom & Kleinhans (1999), Blom (2000), Leclair (2000, 2002) and Blom et al. (2003). The present paper considers a selected set of experiments from these studies. Flow conditions of the selected experiments are summarized in Table 1. A trimodal mixture of sediment was used for the A and B experiments at WL/DH, a sediment mixture from the River Rhine was used for run T10, and moderately well-sorted and poorly sorted sediment were used for the BU and SAFL experiments, respectively (Fig. 2).

The BU bed-elevation data, as well as all SAFL data, come from at-a-point time-series (Fig. 1; Leclair, 2000, 2002). The BU data on dune height and trough-scour depth below mean bed level were determined from successive bed-elevation profiles (Leclair, 2000, 2002), as were all WL/DH data (Blom, 2000; Plate 1). In Table 1, the symbols h_m and ts_m are the mean values, for a given run, of h_i and ts_i, which are the height and trough-scour depth of an individual dune at a given location on the bed profile and/or at a given time. Similarly, the ratio ts_m/h_m is the mean value of ts_i/h_i that indicates at which bed level an individual dune is

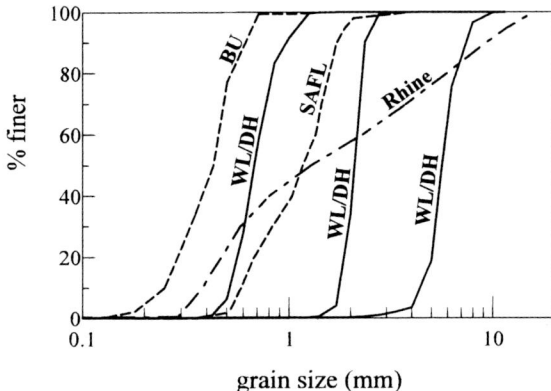

Fig. 2 Grain-size distribution in the experiments at Binghamton University (BU), Saint-Anthony Falls Laboratory (SAFL), and in runs T10 (Rhine) and A1-A2-B1-B2 at WL Delft Hydraulics (WL/DH).

migrating (e.g. Leclair, 2002). The ratio ts_m/h_m is equivalent to the mean dune-shape factor (compare values in Table 1 for WL/DH runs with data in Blom et al., 2003, table 2).

Cross-set thicknesses on sediment peels were measured for BU and SAFL experiments and mean cross-set thickness, s_m^{obs}, was computed from these measurements (Table 1; Leclair, 2001, 2002), whereas vertical profiles of the bed composition were described for WL/DH runs (Blom et al., 2003). No sediment peels were made at WL/DH and it was not possible to perform a reliable quantitative description (e.g. using image-analysis software) of the vertical sorting in deposits from BU and SAFL peels. Finally, sediment transport rate was not measured at BU and SAFL.

Computations

The Froude number was computed as

$$Fr = U/(gd)^{1/2} \qquad (5)$$

where U is mean flow velocity, g is acceleration due to gravity and d is mean flow depth. The spatially averaged bed shear stress was computed as

$$\tau_o = \rho g d S \qquad (6)$$

where ρ is fluid density, and dimensionless spatially-averaged bed shear stress was computed as

Table 1 Experimental conditions and results for selected runs from Blom *et al.* (2003) and Leclair (2000).

Run	d (m)	U (m s^{-1})	D_{50} (mm)	Fr	S (10^{-3})	Ω	h_m/d (mm)	h_m (mm)	h_{sd} (mm)	h_{sd}/h_m	ts_m (mm)	ts_{sd}	ts_m/h_m $(x-1)$	h_{sd}/ts_{sd}	a (mm^{-1})	s_m^{pred} (mm)	s_m^{obs} (mm)
A1	0.154	0.64	TriM*	0.52	2.0	0.33	0.11	17	5.4	0.30	−8	2.6	0.47	2.0	0.58	1.4	ND
A2	0.320	0.83	TriM	0.47	1.8	0.60	0.15	49	15.7	0.32	−29	14.2	0.59	1.1	0.198	7.5	ND
B1	0.155	0.63	TriM	0.51	1.9	0.29	0.12	18	4.7	0.26	−9	2.4	0.50	1.95	0.81	1.1	ND
B2	0.389	0.69	TriM	0.35	2.2	0.89	0.31	122	32.9	0.26	−73	29.2	0.59	1.13	0.11	13.2	ND
BU9	0.15	0.5	0.43	0.40	3.7	0.94	0.29	43.2	16.9	0.39	−21.7	21.6	0.50	0.78	0.13	16.2	17.9
BU14	0.15	0.6	0.43	0.50	4.5	0.95	0.37	54.9	24.6	0.44	−38.3	24.6	0.69	1.0	0.077	21.4	15.1
BU21	0.15	0.75	0.43	0.60	4.1	0.95	0.32	48.2	21.0	0.43	−31.0	23.2	0.64	0.91	0.10	18.1	23.4
SAFL27	0.19	0.84	0.81	0.60	2.6	0.89	0.35	67	34	0.50	−50	37	0.74	0.9	0.07	26.1	26.1
SAFL29	0.21	0.5	0.81	0.34	3.0	0.90	0.36	76	43	0.56	−35	46	0.46	0.95	0.064	27.1	27.1
SAFL31	0.53	0.6	0.81	0.26	2.0	0.94	0.26	137	53	0.38	−76	75	0.55	0.71	0.13	17.8	32
SAFL32	0.54	0.8	0.81	0.35	2.0	0.94	0.21	115	45	0.39	−62	44	0.54	1.03	0.06	26.6	20
SAFL33	0.87	0.8	0.81	0.27	2.1	0.97	0.15	128	48	0.37	−75	58	0.59	0.83	0.04	49.5	32
T10	0.193	0.59	Rhine*	0.43	1.2	0.38	0.09	17	ND	ND	ND7	ND	0.41	ND	ND	ND	ND

*See Fig. 2; TriM = trimodal mixture.

$$\theta = \tau_0/(\sigma - \rho)gD_{50} \qquad (7)$$

where σ is sediment density and D_{50} is median sediment grain size. Then, with θ_c being the value of θ at the threshold of sediment motion, a sediment-transport stage, Ω, was calculated as

$$\Omega = 1 - (\theta_c/\theta) \qquad (8)$$

This probably appears to be an unusual form of excess shear stress, but it is intended to be to the format used in Gill (1971), for analysis purposes. Predicted mean cross-set thickness, s_m^{pred}, was computed from equation (2). Data on experimental conditions and geometrical characteristics of dunes and cross-sets are in Table 1.

RESULTS

Bed-elevation probability distribution and dune geometry

Effect of initial vertical sorting

The present analysis provides a new illustration of the results from Blom et al. (2003) where different P_s curves were produced from runs with similar discharge but different initial vertical sorting (see e.g. runs A2 and B2 in Fig. 3). The wide range of bed-surface elevations in run B2 was due to mean dune height and trough-scour depth being more than twice as large as in run A2 (Table 1; Fig. 4). Figure 4 shows that the distribution of ts_i/h_i is more even relative to mean bed level (i.e. where $ts_i/h_i = -0.5$) in run A2 than in run B2. Moreover, Fig. 4 shows that many individual dunes with $h_i/d < 0.3$ in run B2 (i.e. smaller than the largest dunes of run A2) have trough scours attaining lower bed-surface elevations than they did in run A2 (compare dots for $ts_i/h_I < -0.5$ between the two graphs in Fig. 4). The reason for this variability in dune height and trough-scour depth between runs is because all size fractions in run B2 are being fully transported, whereas in run A2 partial transport prevailed and a coarse bed layer was present below the dunes. This new analysis highlights the interdependency between vertical sorting, grain-size-selective sediment transport, dune height distribution and P_s curve.

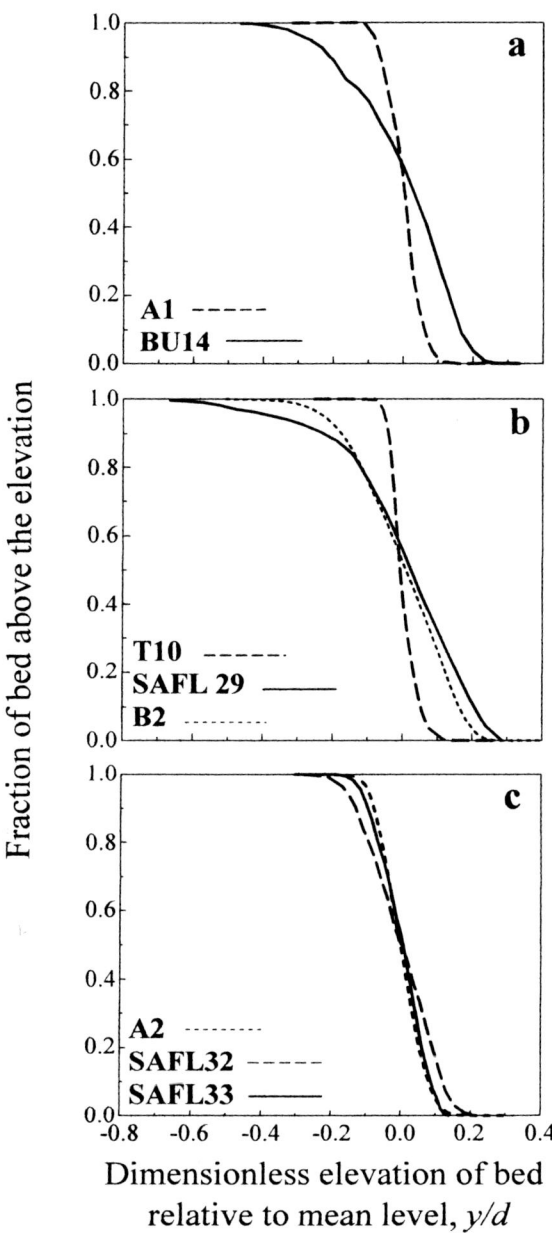

Fig. 3 Function $P_s(y/d)$ denoting fraction of bed recorded at a point above bed surface elevation, y; where y is measured relative to mean bed level, which = 0, and non-dimensionalized using mean flow depth, d.

Effect of sediment mixture

Even without initial sorting, matching runs with different sediment mixtures but quite similar flow depths and velocities also produce different

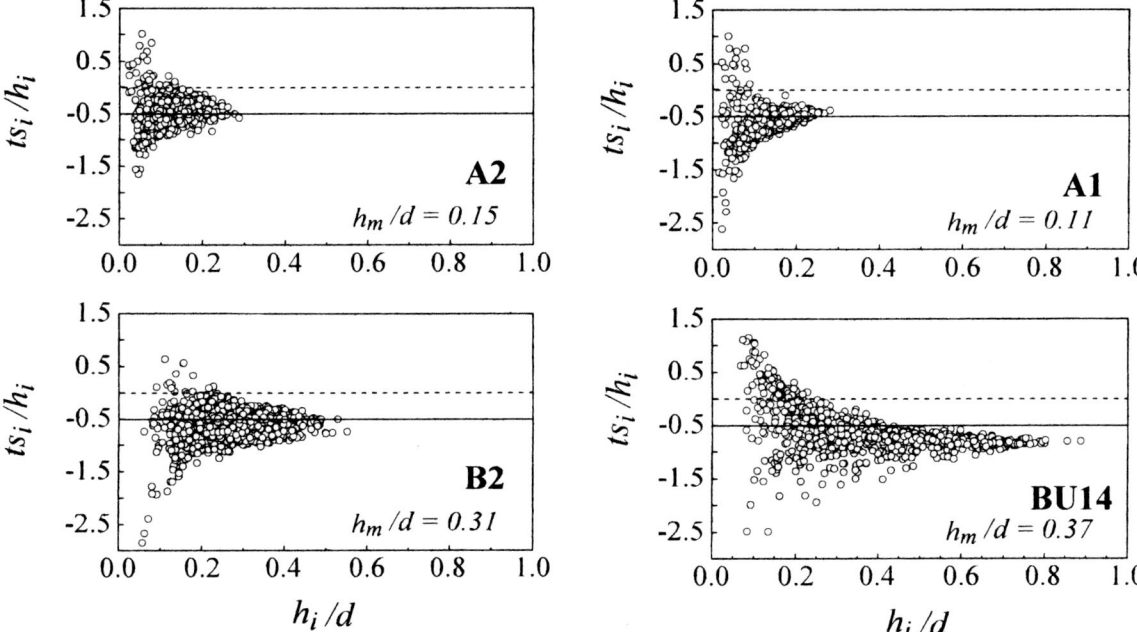

Fig. 4 Non-dimensionalized height of individual dunes, h_i/d versus ratio of trough-scour depth/dune height, ts_i/h_i, which indicates the relative elevation of the dune for experiments A2 and B2 with similar discharge but different initial vertical sorting (see text for details, and Blom et al., 2003). The solid line at −0.5 indicates that dune-scour depth is half the dune height. Dunes migrating completely above mean bed-level plot above the dashed line.

Fig. 5 Non-dimensionalized height of individual dunes, h_i/d, versus ratio of trough-scour depth/dune height, ts_i/h_i, which indicates the relative elevation of the dune for experiments A1 and BU14 with similar flow depth and velocity, but different sediment grain size. The solid line at −0.5 indicates that dune-scour depth is half the dune height. Dunes migrating completely above mean bed-level plot above the dashed line.

probability distributions of bed-surface elevation, P_s (runs A1 and BU14, Fig. 3a; runs T10 and SAFL29, Fig. 3b; Table 1). The range of values of dimensionless bed-surface elevation is smaller in the case of run A1 and run T10, which are experiments with coarser sediment mixtures, smaller values of energy slope, and hence lower sediment transport stages, than their respectively associated runs B14 and SAFL 29. Furthermore, the shape of the P_s curves for runs BU14 and SAFL29 is not spread evenly around mean bed-surface elevation, but is more skewed toward the low elevations than for their matching runs with a coarser sediment mixture. This is because significant differences in dune height and trough-scour depth are associated with the change in sediment transport stage: e.g. in run A1 (Fig. 5), the largest dunes of the distribution plot about evenly above and below the −0.5 line, whereas in run BU14 (Fig. 5)

the large dunes ($h_i/d > \approx 0.4$) all have a ratio of trough-scour depth over dune height $ts_i/h_I < -0.5$. The fact that dunes may or may not erode the bed at their troughs toward deeper levels depends on the vertical sorting profile and leads to different distributions of elevations relative to the mean bed-surface elevation, thus affecting the shape of the P_s curve.

Effect of mean dune height/flow depth

In the present experiments, the ratio of mean dune height over flow depth (h_m/d) ranges from 0.09 to 0.37 (Table 1) and these values are comparable to those observed in other flumes and rivers (e.g. Bridge & Jarvis, 1982; Iseya, 1984; Gabel, 1993; Mohrig & Smith, 1996). The interdependency between sediment transport stage, the P_s curve and dune height is illustrated by the variation of the ratio of mean dune height over flow depth (Fig. 3 and Table 1). Runs with similar sediment transport stage produce similar P_s curves: the curves of runs A1 and T10 reflect very well a low sediment transport stage, and runs with sediment transport stages > 0.90 (all BU runs and SAFL29) show only moderate variation in the range and shape of their P_s curves. Even runs such as B2 and SAFL29, which owe their similar values of sediment transport stage to different combinations of d, S and D_{50} (see equations 6 and 7), produce very similar P_s curves. The P_s curves for the experiments with the largest flow depths, however, do not reflect their high sediment transport stage. Indeed, the P_s curve for run SAFL33 is similar to those of runs A2, A1 and T10, despite its sediment transport stage of 0.97. Runs SAFL 32 and 33 have a relatively low ratio of mean dune height over flow depth, slightly above those for runs A1 and T10. Otherwise, it is worthy of note that the difference between the P_s curves of runs B2 and SAFL29 exists essentially in the extent of the curve toward the low-elevation values, which seems to be related to the difference in the variability of dune height, with $h_{sd}/h_m = 0.26$ and 0.56, respectively. No strong relationship (i.e. $r^2 = 0.49$) between h_{sd}/h_m and h_m/d was found in our combined set of data, however, suggesting that the real control might be the vertical sorting.

Dune deposits as clues for reconstructing bed-elevation probability distribution

Potential for preservation of cross-sets

For the WL/DH experiments, the model predicts a mean cross-set thickness that is about equal to, or much smaller than, the largest grains in the sediment mixture (i.e. c. 10 mm). Therefore, no or very few real cross-sets could be preserved in these runs. This limitation on the preservation of cross-sets is due to the low variability of dune height, i.e. to the value of the coefficient of variation of dune height (h_{sd}/h_m), which averages 0.28 for the WL/DH experiments, compared with 0.43 at BU and SAFL (Table 1). This situation for the WL/DH experiments is likely to be due both to the low conditions of sediment transport stage and to the formation of a coarse bed layer at the base of dunes. A similar case occurred at the North Fork Toutle River, however, where gravel dunes with $h_{sd}/h_m \approx 0.23–0.3$ left no cross-set, under a sediment transport stage of about 0.62–0.79 (estimated from Dinehart, 1992). In another published case, deposits from the Calamus River (Leclair & Bridge, 2001) were formed by dunes with $h_{sd}/h_m \approx 0.40–0.53$ and sediment transport stage of about 0.90–0.94 (estimated from Gabel, 1993). These values are in the same range as those from BU and SAFL experiments with preserved cross-sets. The present results therefore would seem to indicate that most preserved dune (and only dune) deposits record events with high sediment transport stages.

Cross-sets and vertical sorting

The interface between the dune deposits and the underlying sediment may indicate the nature of bedload transport that prevailed during their migration (assuming no change in sediment mixture occurred in the system). In the WL/DH experiments, the bedload material was always finer than the original material because conditions of selective transport prevailed (Blom et al., 2003). This was clearly not the case in the BU and SAFL experiments. On sediment peels of deposits left by dunes in BU runs (moderately well-sorted sediment; Fig. 6), the cross-sets have more relief, indicating that they are composed of coarser grained sediment than the underlying non-transported sand (i.e. the dune deposit has larger pores in which the epoxy-resin flows). Note that the base of the deposit remains at about the same depth, although it is defined by different dunes (Fig. 6). This is because bed-height variation in time (e.g. Fig. 1), and hence the elevation of the deepest scours, is very similar at any given point along the profile (Leclair, 2000). Plate 2 shows that the dune deposit formed during run SAFL33 (poorly sorted sediment) is, like that of the BU runs (Fig. 6), coarser than the underlying sediment. As in the WL/DH experiments, however, coarser grains are found mostly at the base of the deposit (Plate 2). This is because coarse grains tend to settle at the base of individual dunes (Blom et al., in press; Kleinhans, 2001), and because the very deepest dunes in a series are usually partially preserved (Leclair, 2000). Otherwise, in the upper part of the deposit, the vertical sorting varies markedly

Fig. 6 Epoxy-resin sediment peel from deposit at the end of non-aggradational run BU8 (a replicate of run BU9; no photographs of peels from run BU9 are available). Flow is from right to left. Peel is 1 m long. Non-transported sediment (bottom) and filling (top) have been painted (dark grey) in order to enhance the cross-sets (light grey) formed by the migration of dunes with various height and trough-scour depth.

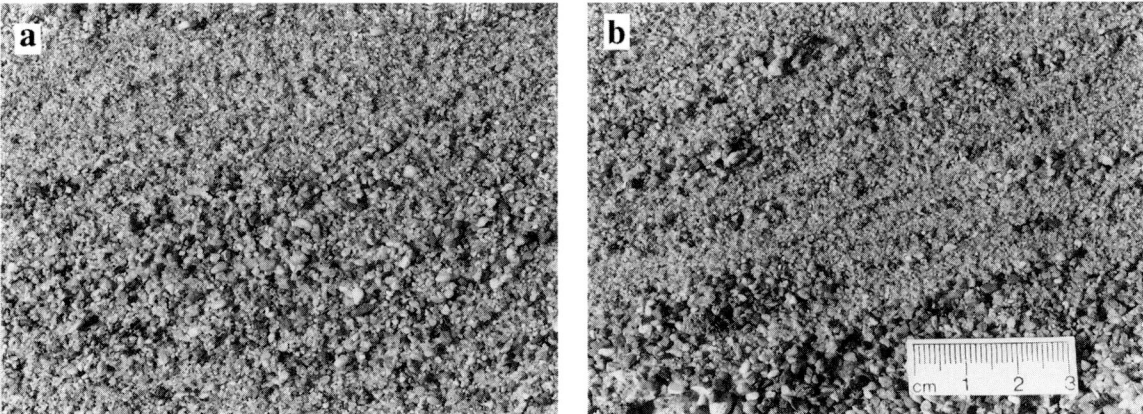

Fig. 7 Details of well-preserved cross-sets from (a) run SAFL 29 (low flow velocity) and (b) run SAFL 27 (high flow velocity). The upper boundaries of the cross-sets are seen as delicate lines at the level of the labels. Flow depth was approximately the same in both experiments. Flow direction is from right to left. In both cases, coarser grains gather at the base of the cross-set, but in the higher velocity run (b), coarser grains can be seen at top of cross-set and anywhere along the cross-strata, whereas at lower flow velocity (a), only finer grains compose the top part of the cross-set.

downstream, as does the thickness of individual cross-sets (Plate 2).

When dunes are well preserved, as they were in the experiment of Leclair (2000), vertical sorting can be observed within cross-sets. In matching flow-depth runs at low flow velocity (SAFL29, Fig. 7a) and high flow velocity (SAFL27, Fig. 7b), coarse grains can be observed at the base of cross-sets. At higher flow velocity (SAFL27), however, coarser grains are also seen along many cross-strata, and even at the very top of the cross-set (Fig. 7b). The presence of coarse grains at different depths seems to be due to the lateral variability in sediment transport in the case of more three-dimensional dunes at higher flow velocity (M. Kleinhans, personal communication, 2001). At lower flow velocity (SAFL29), coarse grains typically gather only at the bases of cross-sets (Fig. 7a). These findings may permit a qualitative interpretation of flow velocity from the internal textural features of the cross-sets. This information could be added to, or used instead of (depending on availability of data), similar interpretations from the ratio of mean cross-set thickness/mean cross-set length that decreases as the Froude number increases (Leclair, 2002).

From cross-sets to P_s

In the present analysis of bed-elevation probability distribution, the effect of flow depth on dune height (i.e. dune height tends to increase with flow depth; see Allen, 1982), and hence on the range of elevations in P_s curves, is removed by the use of dimensionless elevation, y/d. Therefore, a value for flow depth should be found in order to reconstruct the P_s curves from the characteristics of dune deposits. The estimation of flow depth from mean dune height is often based on commonly observed values, although it is well known that Yalin (1964) proposed a theoretical value of

$$h_m = 0.167d \qquad (9)$$

Gill (1971) demonstrated that equation (9) can be derived from the Exner equation for sediment continuity, and is valid only for certain rates of sediment transport. The derivation proposed by Gill (1971, equation 16) is

$$h_m = (d/2n\alpha)[1 - (\tau_c/\tau_o)] \qquad (10)$$

which includes controlling variables such as dune-shape factor α (our ts_m/h_m), sediment-transport stage, $1 - (\tau_c/\tau_o)$, plus a constant and parameter n from a formula for bedload-transport rate. Following this approach, the relationship of dune height with flow depth, dune-shape factor and the present parameter for sediment-transport stage were investigated (Fig. 8), combined here in a variable G, defined as

$$G = (d/\alpha)\Omega \qquad (11)$$

In these experiments, the shape factor ranges from 0.41 to 0.74 and shows no clear relationship with the sediment-transport stage (Table 1). Roden et al. (1997) reported an increase in the shape factor, however, from 0.45 to 0.62, through a flood hydrograph in the Jamuna River, Bangladesh. It was not possible to include data from other studies in this functional analysis because no complete set of correspondent values of flow depth, dune height and shape factor, median grain size, shear stress and energy slope were found. The result from the reduced major axis regression on our combined experimental data set is

$$h_m = 0.15G^{0.9} \quad r^2 = 0.89 \qquad (12)$$

or

$$G = 4.1h_m^{1.1} \quad r^2 = 0.91 \qquad (13)$$

Equation (12) is close to Yalin's theory (equation 9) if $G \approx d$, i.e. if the value of the sediment-transport rate is about that of the dune-shape factor (see equation 11). Given that the dune-shape factor roughly ranges 0.5–0.7, and that $\theta_c \approx 0.045$ (Bridge & Bennett, 1992), the Shields parameter, θ, would vary from 0.09 to 0.15 in the case of $G \approx d$, and this is a limited range of θ for bedload-transport conditions. Such conditions may not lead to significant preservation of dune deposits (see above), hence high values of sediment transport stage (e.g. $\Omega \approx 0.9$) could be assumed if deposits are preserved at all. In this case, estimation of flow depth from cross-set thickness and dune height (using equation 13 then equation 11) could be constrained to the variability of the dune-shape factor ($\alpha \approx 0.5$–0.7). In addition, although the coefficient of determination, r^2, is high in the above equations, data with $G \leq 0.2$ m (runs A1, B1 and T10) depart markedly from the trend, indicating that there may be a cut-off value of G below which value dunes may not fully develop.

The lower and upper limits of the P_s curve also can be estimated from analysis of the distribution of the relative elevation of cross-set lower boundaries and cross-set thicknesses, respectively. Here, an example of how this could be achieved is presented (Fig. 9), using only one set of data on trough-scour depths of formative dunes from Leclair (2000). The distribution of the elevation

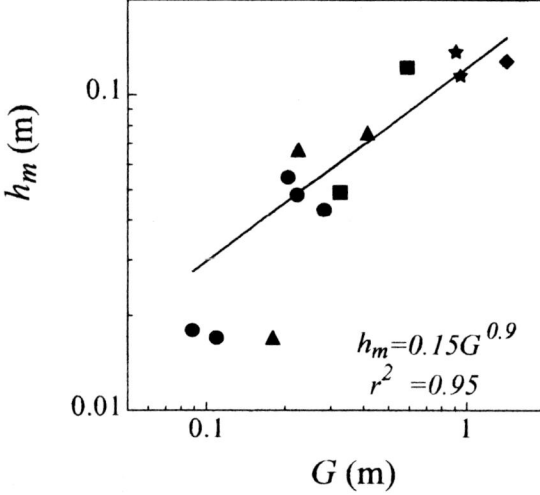

Fig. 8 Variation of mean dune height, h_m, with parameter $G = (d/\alpha)[1 - (\theta_c/\theta)]$, with reduced major axis regression line. Similar symbols indicate experiments with similar flow depth, and show that mean dune height does not correlate well with flow depth, d, alone.

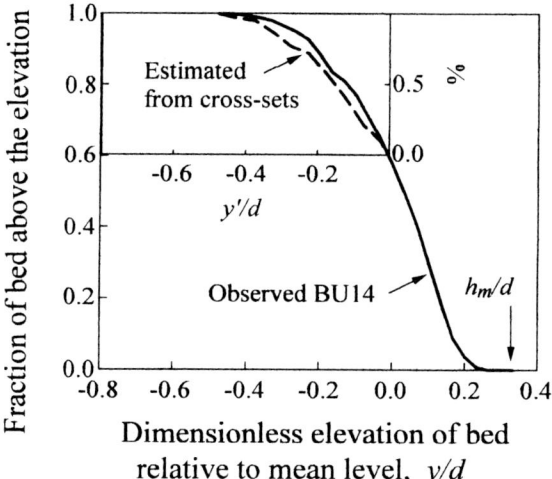

Fig. 9 Function $P_s(y/d)$ for run BU14 shown along with function $P_s(y'/d)$, where y' is the elevation of the lower boundary of cross-sets; y' is measured relative to the shallowest cross-set boundary (proxy for mean bed level) over the 3-m-long test section.

of dune trough-scour paths (e.g. as seen on Fig. 6) was measured relative to a surrogate of mean bed level, i.e. the elevation of the highest cross-set lower boundary (based on Fig. 1, it is very unlikely that the shallower cross-set boundary would be higher than mean bed level). Therefore, the probability distribution of these elevations (also made dimensionless) is a subsample of the P_s curve below $y/d = 0$. This subsample, y'/d, reproduces very well the observed lower limit, and the shape of the P_s curve on the low-elevation side (Fig. 9). The observed P_s curve has a higher fraction of bed above a given elevation because dune-trough scours are not the only bed-surface elevations lower than the mean. The upper limit can now be approximated from the estimation of mean dune height from cross-set thickness distribution (Leclair & Bridge, 2001). It seems that dunes with heights equal to, or larger than, the mean dune height ($h_i \geq h_m$) often cannot migrate higher than mean bed level, i.e. their trough could not be above mean bed level (i.e. for such dunes, $ts_i/h_i < 0$; see Fig. 5), owing to limitation by flow depth. The ratio of mean dune height over flow depth could then be an indicator of the upper limit of the P_s curve, and Fig. 3 shows clearly that, even for runs without any preserved deposit, this upper limit approximates the observed h_m/d (Table 1). At this time, insufficient results exist to identify clear controls for the rest of the P_s curve ($0 < y/d < h_m/d$).

DISCUSSION

Sediment transport stage and vertical sorting appear to be major controls on dune-bed topography, and hence on the shape of P_s. This is somehow expected, as the effects of the dimensionless bed shear stress on dune geometry, especially dune steepness, have been well known for decades (e.g. Yalin & Karahan, 1979). However, this large body of data is mostly about relationships between mean values (e.g. mean dune height and length), not about variability within a distribution, and hence is less relevant for developing functional relationships for P_s, or for its interpretation. In the present experiments, it is shown that low variability of dune height and short-range P_s curves tend to be associated with low sediment transport stage (Table 1). Although it is clear that a higher shear stress means higher dunes, and thus a wider range of bed elevations exposed to the flow, the controls on dune height variability for comparable sediment transport stages remain an enigma: this is critical for predicting the left-hand side of the P_s curve with refinement. The results presented also extend the range of h_{sd}/ts_{sd} previously observed (i.e. 0.7–1.1; Leclair & Bridge, 2001) and used for predicting dune height from cross-set thickness. Table 1 shows that runs with $h_{sd}/ts_{sd} \approx 2.0$, which is a very commonly assumed value, actually produced no preserved cross-sets. This indicates that the reliability of the interpretative method proposed by Leclair & Bridge (2001), and hence the reconstruction of the P_s curve, should not be affected by these new results.

It is expected that in nature the effect of sediment transport stage on the P_s curve would vary mostly with grain size (given y/d), as an increase in slope between different rivers would be counterbalanced by a decrease in grain size (Schumm, 1960; Williams, 1978). Moreover, for any given river, the changes in sediment mixture over time (for time scales relevant to dune migration) would be more important than changes in slope (Leopold & Maddock, 1953). However, the variability in vertical sorting is as important as mean grain size (which is used for computing sediment transport stage), as suggested by the differences between runs B2 and SAFL29 (Fig. 3, Table 1). The effect of energy slope, s, has not been elaborated because values for all selected runs varied within only one order of magnitude, whereas D_{50} varied over two orders of magnitude (Table 1).

Non-dimensional PDFs directly facilitate comparison among experiments, but their use would also allow the prediction of several possible P_s curves for a range of h_m/d, e.g. depending on the range of dune shape factor and sediment transport stage used in equation (1), or on the estimated equilibrium status of dune height with flow conditions. High values of h_m/d (i.e. up to 0.37) are not only an effect of experimental conditions: they also have been observed in rivers with flow depths of 0.34–0.61 m (Gabel, 1993) and 7.8–9.1 m (Kostaschuk, 2000). In these two cases, the channels were rather straight (> 2 km long in Kostaschuk, 2000) and some of the lowest values (as small as 0.05) occurred in channel bends (Bridge & Jarvis, 1982). Otherwise, Mohrig & Smith (1996) observed $h_m/d \approx 0.25$ for dunes migrating over a bar. Apart from the obvious flow unsteadiness and non-uniformity in nature, it may well be that the 'fetch' in the alongstream direction, which is required for dunes to reach their equilibrium geometry, is often insufficient. It is likely that dunes from run SAFL33 would have continued to grow downstream if the flume had been longer. Therefore, when developing P_s curves, it may be necessary to adjust estimates of flow depth, depending on the estimated channel morphology and dune equilibrium with the flow in the problem of interest.

It is hoped that results from this research will help implement formulations for river bed variation in dune-forming conditions, such as the probabilistic Exner sediment continuity equation proposed by Parker et al. (2000). The present study provides a step towards developing a more robust theory that relates the dynamics of erosion and deposition during dune migration to bed topography, the geometry and vertical grain sorting of the deposits, and preserved cross-sets. For now, there is a particular need for more integrated data sets on grain-size selective transport rates, dune geometry, PDF of dune-trough elevation and of bed-surface elevation, vertical sorting profiles and the potential for preservation of the deposits.

CONCLUSION

Analysis of a combined set of data from two separate dune studies has identified certain controls on the bed-surface-elevation probability distribution (P_s), as this parameter relates to a new formulation of the Exner equation for sediment continuity. Moreover, this analysis also has revealed some indicators for reconstructing this distribution from preserved deposits.

Results show that the dimensionless bed shear stress, used here as a component of the sediment transport stage, and vertical sorting are the major controls on the range and shape of the non-dimensional Ps curve. Low values of sediment transport stage produce P_s curves that are short-ranged and quite evenly spread around the average bed level. The presence of a coarse bed layer, which affects the composition of the transported sediment, has a similar effect on the P_s curve. Otherwise, P_s curves tend to be skewed toward the lowest values of bed-surface elevation. The entire distribution of grain size is important and seems to control the P_s curve near its lower limit, such that for similar values of sediment transport the variability in dune height and trough-scour depth would depend on sorting (and initial vertical sorting). The P_s curves from dune-covered beds that are not in equilibrium with flow conditions do not reflect the actual value of sediment transport. Other controls on the overall shape of the P_s curve could not be clearly assessed.

The interpretation of the geometry of the preserved cross-sets can be used to partially reconstruct P_s curves. The low-elevation side of the curve can be approximated by analysis of the probability distribution of the elevation of the lower boundaries of cross-sets. The upper limit of P_s roughly corresponds to the ratio of mean dune height over flow depth. As a method for estimating mean dune height from the distribution of cross-set thicknesses already exists, here an empirical relationship for estimating flow depth from dune height and the dune shape factor, and sediment transport stage, is proposed. Moreover, grain sorting, among and within cross-sets, can allow a qualitative interpretation of flow velocity and sediment transport stage. It is hoped that this study will assist the development of a more comprehensive theory that ultimately can predict the characteristics of dunes and their deposits from data on sediment transport or, conversely, reconstruct sediment transport from the characteristics of preserved dune deposits.

ACKNOWLEDGEMENTS

Leclair's PhD research was funded by a grant to John S. Bridge from the Petroleum Research Fund— American Chemical Society, by a grant to S.L. from Le Fonds pour la Formation de Chercheurs et l'Aide à la Recherche (FCAR, Québec, Canada) by Saint-Anthony Falls Laboratory (SAFL), and by the Geological Society of America (GSA) Robert K. Fahnestock Memorial Research Award. Blom's PhD research was funded by RIZA-WL-UT (the Institute for Inland Water Management and Waste Water Treatment of the Ministry of Transport, Public Works and Water management in The Netherlands, WL Delft Hydraulics, and the University of Twente). Thanks to Chris Paola and Gary Parker for inviting S. Leclair to run dunes at SAFL. Thanks to Maarten Kleinhans for his participation in experiment T10. Official comments by Chris Paola and an anonymous reviewer were more than appreciated. Finally, the authors are very grateful to Gary Parker for introducing them to each other.

NOMENCLATURE

a	parameter = $1/\beta$ from gamma PDF of dune height distribution [L^{-1}]
d	mean flow depth [L]
D_{50}	median sediment grain size [L]
Fr	Froude number
g	gravitational acceleration [L T^{-2}]
G	compound variable controlling dune height [L]
h_i	height of an individual dune [L]
h_m	mean dune height [L]
h_{sd}	standard deviation of dune height [L]
S	mean water surface slope
s_m^{obs}, s_m^{pred}	(observed, predicted) mean cross-set thickness [L]
t	time [T]
ts_i	dune trough-scour depth below mean bed level for an individual dune [L]; $ts_i < 0$ and > 0 when below and above mean bed level, respectively
ts_m	mean dune trough-scour depth below mean bed level [L]
ts_{sd}	standard deviation of dune trough-scour depth [L]
U	mean flow velocity [L T^{-1}]
y	bed-surface elevation relative to mean bed level [L]
y'	elevation of cross-set lower boundary relative to shallowest boundary [L]
λ	bed porosity
θ, θ_c	spatially averaged dimensionless bed shear stress, critical value at threshold of entrainment
ρ	fluid density [M L^{-3}]
σ	sediment density [M L^{-3}]
τ_o, τ_c	spatially averaged bed shear stress, critical value at threshold of entrainment [M L^{-1} T^{-2}]
Ω	sediment transport stage

REFERENCES

Allen, J.R.L. (1982) *Sedimentary Structures; their Character and Physical Basis*, Vol.1. Amsterdam, Elsevier, 663 pp.

Blom, A. (2000) *Flume Experiments with a Trimodal Mixture*. Data Report Sand Flume Experiment 1999/2000, Research Report CiT: 2000R-004/MICS-013, University of Twente.

Blom, A. and Kleinhans, M.G. (1999) *Non-uniform sediment in morphological equilibrium situations*. Data Report Sand Flume Experiments 97/98, University of Twente, Rijkswaterstaat RIZA, WL/Delft Hydraulics, University of Twente, Civil Engineering and Management.

Blom, A., Parker, G. and Ribberink, J.S. (2001) Vertical exchange of tracers and non-uniform sediment in dune situation. *Proceedings, River, Coastal and Estuarine Morphodynamics*, International Association for Hydrological Research, Obihiro, Japan.

Blom, A., Ribberink, J.S. and Vriend, H.J. (2003) Vertical sorting in bed forms: Flume experiments with a natural and a tri-modal sediment mixture. *Water Resour. Res.*, **39**(3), 1025–1038.

Bridge, J.S. and Bennett, S.J. (1992) A model for the entrainment and transport of sediment grains of mixed sizes, shapes, and densities. *Water Resour. Res.*, **28**, 337–363.

Bridge, J.S. and Jarvis, J. (1982) The dynamics of a river bend: a study in flow and sedimentary processes. *Sedimentology*, **29**, 499–541.

Dinehart, R.L. (1992) Evolution of coarse gravel bed forms: Field measurements at flood stage. *Water Resour. Res.*, **28**, 2667–2689.

Gabel, S.L. (1993) Geometry and kinematics of dunes during steady and unsteady flows in the Calamus River, Nebraska, USA. *Sedimentology*, **40**, 237–269.

Gill, M.A. (1971) Height of sand dunes in open channel flows. *Proc. Am. Soc. Civ. Eng., J. Hydraul. Div.*, **97**(HY12), 2067–2073.

Iseya, F. (1984) *An Experimental Study of Dune Development and its Effect on Sediment Suspension*. Environmental Research Center Paper No. 5, The University of Tsukuba, Japan, 56 pp.

Kleinhans, M.G. (2001) The key role of fluvial dunes in transport and deposition of sand-gravel mixtures, a preliminary note. *Sediment. Geol.*, **243**, 7–13.

Kostaschuk, R. (2000) A field study of turbulence and sediment dynamics over subaqueous dunes with flow separation. *Sedimentology*, **47**, 519–531.

Leclair, S.F. (2000) Preservation of cross-strata due to migration of subaqueous dunes. Unpublished PhD thesis, Binghamton University, New York, 422 pp.

Leclair, S.F. (2002) Preservation of cross-strata due to migration of subaqueous dunes: an experimental investigation, *Sedimentology*, **49**, 1157–1180.

Leclair, S.F. and Bridge, J.S. (2001) Quantitive interpretation of sedimentary structures formed by river dunes. *J. Sediment. Res.*, **71**, 713–716.

Leclair, S.F., Bridge, J.S. and Wang, F. (1997) Preservation of cross-strata due to migration of subaqueous dunes over aggrading and non-aggrading beds: Comparison of experimental data with theory. *Geosci. Can.*, **24**, 55–66.

Leopold, L.B. and Maddock, T. (1953) The hydraulic geometry of stream channels and some physiographic implications. *Prof. Pap. U.S. Geol. Surv.*, **252**, 57 pp.

Mohrig, D. and Smith, J.D. (1996) Predicting the migration rates of subaqueous dunes, *Water Resour. Res.*, **32**, 3207–3217.

Paola, C. and Borgman, L. (1991) Reconstructing topography from preserved sratification. *Sedimentology*, **38**, 553–565

Parker, G., Paola, C. and Leclair, S. (2000) Probabilistic Exner sediment continuity equation for mixture with no active layer. *J. Hydraul. Eng., Am. Soc. Civ. Eng.*, **126**, 818–826.

Roden, J., Best, J. and Ashworth, P. (1997) Morphology, kinematics and flow dymamics of mega-dunes, Jamuna River, Bangladesh. *Sixth International Conference on Fluvial Sedimentology Abstract Volume* (Ed. J. Rogers), University of Cape Town, p. 175.

Schumm, S.A. (1960) The shape of alluvial channels in relation to sediment type. *Prof. Pap. U.S. Geol. Surv.*, **352-B**, 30 pp.

Williams, G.P. (1978) Hydraulic geometry of river cross-sections-theory of minimum variance. *Prof. Pap. U.S. Geol. Surv.*, **1029**, 47 pp.

Yalin, M.S. (1964) Geometrical properties of sand waves. *J. Hydraul. Eng.*, **90**, 105–119.

Yalin, M.S. and Karahan, E. (1979) Steepness of sedimentary dunes. *J. Hydraul. Eng.*, **105**, 381–392.

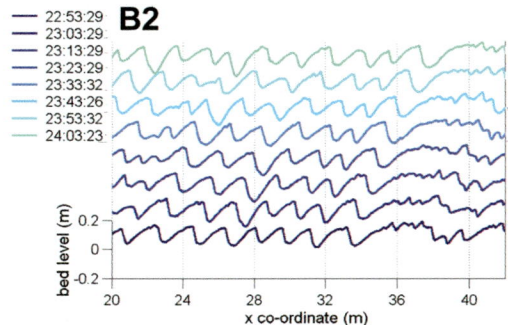

Plate I Selected successive bed profiles from run B2 of Blom (2000). Time of each profile is in hours: minutes: seconds from the beginning of the experiments.

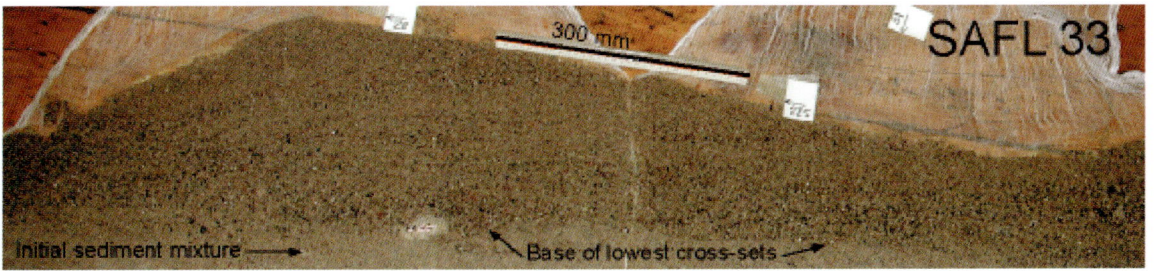

Plate 2 Latex sediment peels from deposit at the end of run SAFL33. Flow is from right to left.

Braided gravel-bed rivers with a limited width: preliminary results of a hydraulic model study

CHRISTIAN MARTI and GIAN RETO BEZZOLA

Laboratory of Hydraulics, Hydrology and Glaciology (VAW), Swiss Federal Institute of Technology (ETH) Zurich, Gloriastr. 37-39, ETH Zentrum, CH-8092 Zurich, Switzerland (Email: marti@vaw.baug.ethz.ch)

ABSTRACT

For ecological and flood protection reasons, the current policy of the Swiss Federal Office for Water and Geology is to give rivers more space. In contrast to past river training measures, the spatial needs of a river are no longer defined as being a narrow straight channel. As a result, in some areas, rewidening projects are planned or have already been realized. To understand the morphologically dynamic processes of such rewidened sections, which in most cases lead to braided rivers with a limited width, a research project has been started at the Laboratory of Hydraulics, Hydrology and Glaciology (VAW) of the Swiss Federal Institute of Technology (ETH) in Zurich. This project aims to describe and quantify aggradation, degradation and channel rearrangement during floods by means of numerical and physical modelling of braided rivers. A brief overview of existing work, useful in the design of rewidened channels, is presented, and the experimental set-up at VAW as well as the planned experimental concept for the new study are introduced. The first experiments, accomplished in a laboratory flume with constant discharge, are described. Some preliminary results allow the assumption that a sediment transport formula, developed at VAW, is also applicable for widened areas in steep rivers. It is concluded that a low braiding index in the widened river section implies a high transport capacity and vice versa. The highest transport rate was always observed when a dominant single channel was moving laterally.

INTRODUCTION

Since the late eighteenth century most rivers in the Alpine region have been constricted between embankments. This river training brought many advantages, such as flood control, increased land availability and malaria control, but in the past few decades problems have appeared. The transport capacity in these constricted stretches increased, and in some areas the natural bedload input decreased owing to human activities such as dredging and installation of hydro-electric power plants. Both impacts are leading to continuous degradation of the river bed and consequently to local levee and bridge pier scouring. A further consequence is lowered groundwater tables. Additionally, conventional river training often results in monotonous river landscapes with poor variation in aquatic habitat.

Experience gathered from recent flood events has shown the limits of conventional river training. Floods that exceed the capacity of the channel have resulted in the failure of levees, often causing severe damage. To avoid these disadvantages, more differentiated river training concepts are a necessity. Modern river training should be nature-orientated. It has to guarantee appropriate protection against floods and at the same time minimize

Fig. 1 First realized rewidening project at the River Emme near Utzensdorf, Berne, Switzerland. Photograph (a) taken in 1991 before and photograph (b) in 1992 after the rewidening works. Photograph (c) after a flood event in 1995. (Photographs (a) and (b) H. Kobi, and (c) R. Huber, © Documenta Natura, Berne, Switzerland.)

negative impacts on ecosystems. This means that present-day rivers may need to be modified to allow at least part of their original dynamics to operate, thereby resulting in an increased morphological diversity. These demands can be satisfied to a large extent by increasing the width of constricted rivers again (Fig. 1). In western Europe the space for rewidening will always be limited. Therefore, even if the development of a braided pattern is allowed again, the rewidened river section will have a limited width.

At the Laboratory of Hydraulics, Hydrology and Glaciology (VAW) two dissertations have dealt with river widenings and wide rivers themselves (Zarn, 1997; Hunzinger, 1998). Based on flume experiments under steady-state conditions, conducted with slopes between 0.2 and 1.5% and a model mixture with a narrow range of grain sizes, the results of these studies are primarily applicable to midland rivers with moderate hydrographs during a flood. In the present work the effects of bedload transport and scouring will be examined for steeper slopes and bed material with a wider range of grain size. Additionally, the focus will be widened by studying the impact of transient conditions, first on the morphological processes in general, and second on the applicability of the approaches of Zarn (1997) and Hunzinger (1998). Special attention will be paid to the influence of long periods with little bedload input, which may transform a braided pattern into a single incised channel. To address these objectives, experiments in a new laboratory flume were carried out. Furthermore, it is planned to adapt a two-dimensional finite-volume program with a mobile bed routine to model the above-mentioned processes.

Before explaining the experimental set-up and procedure, a short overview of the studies by Zarn (1997) and Hunzinger (1998) will be presented, and some other problematic aspects of rewidening, as proposed in literature, will be given.

KNOWN APPROACHES FOR SOME PROBLEMATIC ASPECTS OF REWIDENING

Channel morphology in long, wide river reaches

The conditions under which a river will restore a braided pattern can be assessed by using one of the common meandering–braiding transition criteria (Bridge, 1993). Alternatively river-bed morphology can be estimated using the method developed by da Silva (1991) and also described by Yalin (1992). They introduced relative bed width $Y = w_{Bf}/h$ and relative flow depth $Z = h/D$ as being the relevant parameters that control river morphology (Fig. 2), where w_{Bf} is the bankfull width, h the averaged flow depth during bankfull stage and D the characteristic grain diameter of the river bed. Similar diagrams were presented by Anderson et al. (1975) and Jaeggi (1983).

Channel morphology in wide river reaches of finite length

If the widened reach chosen is too short, the flow cannot expand to the total width and no braiding will occur. Therefore, determination of the minimum length of a widened reach in order to obtain a braided morphology is of major importance. In addition, it has to be kept in mind that aggradation will take place within the widened area. Both phenomena were investigated by Hunzinger (1998), as well as the maximum scour depth under such circumstances. A compilation of the most important results is presented in Hunzinger & Zarn (1997) and Hunzinger (1999). Applying these results, local rewidening has been realized successfully in Switzerland and in Austria (e.g. Zarn, 1993; Zarn & Hunzinger, 1997; Habersack et al., 2000).

Bedload capacity in braided rivers with limited width

Common bedload transport formulae are usually based on either excess shear stress (e.g. Meyer-Peter & Müller, 1948; Smart, 1984; van Rijn, 1984), excess stream power (Bagnold, 1980) or excess specific discharges (Schoklitsch, 1962). Such formulae are derived from model or prototype data gathered in mostly straight single channels and are therefore not directly applicable to braided rivers. Other approaches stemming from the regime theory are primarily developed for rivers with alternating bars and braided rivers (e.g. Parker, 1979; Ramette, 1990). Applied to model data gathered by Zarn (1997) and Ashmore (1988), and to prototype data from the Ohau River, presented by Mosley (1982) and Thompson (1985), they either underestimate (Parker, 1979) or overestimate (Ramette, 1990) the effective transport.

For a better estimation of the bedload transport, Jaeggi (1992) suggested the introduction of efficiency coefficients depending on both the river morphology and the river width. From his model tests, Zarn (1997) derived two empirical

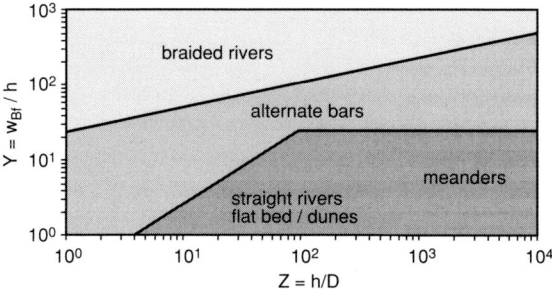

Fig. 2 Channel morphology as a function of $Y = w_{Bf}/h$ and $Z = h/D$, after da Silva (1991).

approaches for braided river sections. The first approach was based on a surrogate channel with a flat bed and the width of the averaged wetted width (w_w) of the braided section. The second approach was based on the optimum width (w_{Opt}) which was defined as the width at which the maximum bedload transport (Q_{bmax}) is calculated. Given a discharge, the slope and the bed material, he suggested calculating the optimum width based on a slightly modified form of the Meyer-Peter & Mueller formula. (The modified dimensionless form is $\Phi = 5(\theta' - \theta_{cr})^{1.5}$, where $\Phi = q_b/\rho_s[(\rho_s/\rho_w - 1)gD^3]^{-0.5}$ and $\theta' = v_*^2/[(\rho_s/\rho_w - 1)gD]$, in contrast to the original form of $\Phi = 8(\theta' - \theta_{cr})^{1.5}$.) Using w_{Opt} and Q_{bmax}, the bedload transport for a braided river with a width of w_{Bf} can be calculated as follows

$$Q_b = Q_{bmax}(3.65e^{-0.86U} - 4e^{1.5U} + 0.35) \quad (1)$$

where $U = w_{Bf}/w_{Opt}$.

This equation was derived from experimental data by Zarn (1997), but also compares favourably with the data of Ashmore (1988) and the data gathered from the first experiment of the present study (e.g. Fig. 3). Nevertheless, in practice, difficulties might occur when calculating the bedload over different sequences of hydrographs (Zarn, 2003).

To improve these calculations for rivers with alternating bars or braided morphology, Carson & Griffiths (1987) suggested using local hydraulic parameters instead of spatially averaged parameters. Nonetheless, without powerful computer programs that are capable of considering the continuous changing of these local parameters, it remains impossible to estimate proper transport rates. To bridge this gap a two-dimensional finite-volume computer program developed at VAW (Beffa & Faeh, 1994), extended by a mobile bed routine (Faeh, 1996), will be tested. The ability of the program to simulate the formation of a braided river pattern starting from an initially planar bed has been successfully assessed by McArdell & Faeh (2001).

EXPERIMENTAL SET-UP

The use of hydraulic modelling is a well-known technique in civil engineering studies, and some geomorphological studies rely on this method (e.g. Ashmore, 1982; Young & Davies, 1991; Ashworth et al., 1992). The basic principles underlying such an investigation are summarized in Young & Warburton (1996).

The experiments presented here were conducted in a concrete flume 28.5 m long and 3.2 m wide. The channel width could be varied by laterally moveable side-elements. The surface roughness of these side-elements matched approximately the grain roughness of the model bed material, D_{min} being 0.2 mm and D_{max} being 16.5 mm. The model mixture used had a rather wide range of grain size. The median grain size D_{50} was 2.9 mm and D_{90} = 9.2 mm. The standard deviation of the mixture, as characterized by $\sigma = (D_{84}/D_{16})^{0.5}$, was slightly above 3.5. The use of this mixture resulted in high grain Reynolds numbers ($Re^* > 100$) and therefore fully turbulent conditions. This fact simplified the conversion from the model to prototype data.

The slope of the flume was 2.1%. Vertically moveable sills at the inflow and the outflow allowed for bed slopes between 1.4% and 2.8%. The maximum discharge in the flume was more than 70 L s^{-1}. A computer-controlled valve allowed simulation of any desired hydrograph. The discharge is measured by a magnetic inductive discharge (MID) meter. The sediment feeder at the upstream end of the flume was computer-controlled and allowed feeding rates between 0 g s^{-1} and 1200 g s^{-1} (oven-dry mass). At the outlet the bedload was caught in a filtering basket and continuously weighed.

To survey the bed topography, laser distance meters were installed on an xyz-positioning

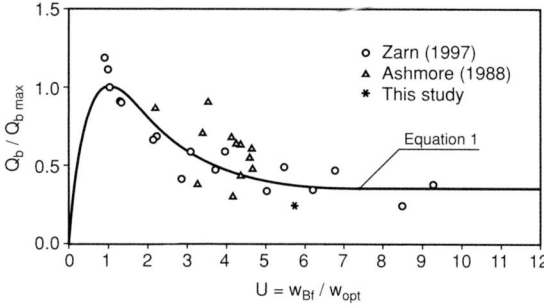

Fig. 3 Q_b/Q_{bmax} as a function of w_{Bf}/w_{Opt} according to equation (1) derived from the experiments of Zarn (1997). The data of Ashmore (1988) and of this study fit the function well.

system. Over a length of 26.5 m, every point of the flume could be reached by this system. Additionally, local water levels could be recorded by ultrasonic distance meters. As the laser sensors were able to detect the river bed through water of a maximum thickness of about 10 cm, the local bed and water level could be measured simultaneously. For the laser measurements, in this case, a compensation for the differing refractive index had to be applied. An 11-m-long section of the lower flume was documented using a digital camera over the full experimental duration (1 shot per min). An overview of the experimental set-up is given in Fig. 4.

EXPERIMENTAL PROCEDURE

Several series of experiments were conducted, each starting from a different initial slope of the river bed. The initial slopes represented reaches of four specific alpine rivers with a slightly braided morphology (Rhône 1.45%, Melezza 1.85%, Gérine 2.25% and Brenno 2.65%). As the grain-size distributions of these prototype rivers, normalized by D_{90}, coincide sufficiently well for all experimental series, the same sediment mixture could be used. With the mixture chosen, the scaling factor varies, depending on the river, between 21 (Gérine) and 50 (Rhône).

Each of the four series was subdivided into three phases (Fig. 5). In the first phase, the development of the braided pattern from the formerly initial planar surface was initiated by steady-state conditions. The discharge during this phase was approximately 0.4 times the mean annual flood of the river modelled (Fig. 5a). The aim of this phase

Fig. 4 Overview of the laboratory flume: 1, laterally moveable side-elements; 2, water inlet; 3, sediment feeder; 4, *xyz*-positioning system; 5, laser and ultrasonic sensors; 6, filtering basket with automatic scales; 7, control unit.

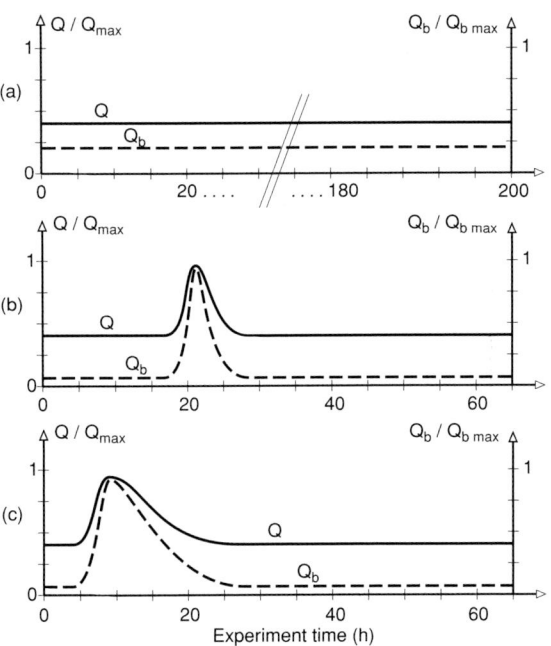

Fig. 5 Applied hydrographs during one experimental series. (a) Phase 1: steady-state conditions. (b) Phase 2: flood event of 5 h duration. (c) Phase 3: flood event of 15 h duration. Before and after the flood event in phases 2 and 3, the sediment input was five times smaller than in phase 1.

was to reach an equilibrium state. (In this case equilibrium state means that the sediment output equals the input and the spatially averaged bed slope remains more or less the same during the time span observed.) This enabled assessment of the applicability of available bedload transport formulae for braided rivers (e.g. Pickup & Higgins, 1979; Zarn, 1997). The expected duration for phase 1 was around 200 h.

Given the now braided bed, the second phase started with the same discharge as used in phase 1, but with a sediment feed rate reduced by five times, to simulate a period of low sediment input. Subsequently, the discharge and the sediment feed rate were raised to values representing a mean annual flood. After the flood peak had passed, the same conditions as at the beginning of phase 2 were held again over a longer period (Fig. 5b). Phase 3 was similar to phase 2 except that the duration of the flood hydrograph was three times longer (Fig. 5c).

Every 15 h an experiment was interrupted to empty the sediment basket, to refill the sediment feeder and to record the bed topography in a dry state. The grid spacing for the topography recording, which was done by laser, was 20 mm (lateral) by 100 mm (longitudinal). During the experimental runs, additional data were gathered, including ultrasonic readings of the water level. As the river bed was in motion and a certain time elapsed between the measurements of each cross-section, the data do not represent a single moment in time. The assumption of steady conditions is valid only for a single cross-section.

PRELIMINARY RESULTS

As the experimental investigation is in its initial stages, only some preliminary results can be presented at this time, all of them derived from two runs with constant discharge. The discharge during the first run was 26.5 L s^{-1} and during the second run 21.4 L s^{-1}, whereas the other parameters, such as the initial slope (1.85%) and the feeding rate (17.7 g s^{-1}), were equal for both. The initial slope was determined for the river represented, the feeding rate by the volumetric capacity of the sediment feeder and the experiment duration. The discharges were calculated according

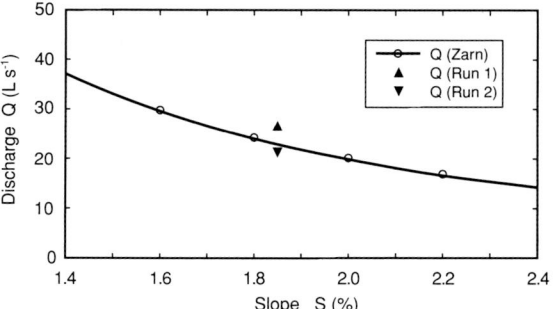

Fig. 6 Necessary water discharge as a function of the slope for a given feeding rate of 17.7 g s^{-1}, a characteristic grain diameter of 4.1 mm and a given bankfull river width of 3.0 m according to the surrogate channel approach of Zarn (1997). The triangles show the chosen discharges for runs 1 and 2.

to the surrogate channel approach by Zarn (1997). With this approach it is possible to calculate the necessary water discharge as a function of slope for a given feeding rate and a given bankfull river width (Fig. 6). Based on preliminary tests, the calculated discharge, however, was increased for run 1, as—compared with the sediment feeding rate —insufficient bedload transport was expected. As shown below, this assumption turned out to be wrong.

The cumulative sediment output curves of runs 1 and 2 are shown in Fig. 7. In the first run, the

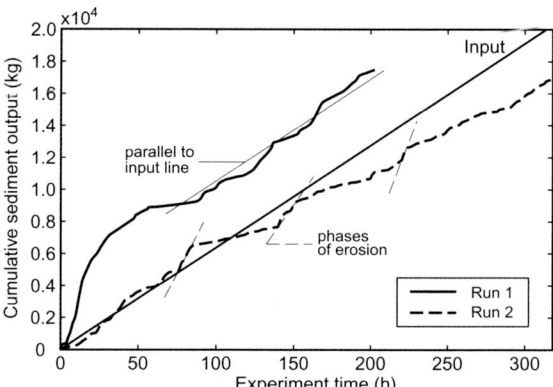

Fig. 7 Cumulative sediment output of runs 1 and 2 compared with sediment input. The thin line parallel to the input line marks a phase of equilibrium during run 1. The steep, thin dashed lines mark phases of heavy erosion during run 2.

sediment output was greater than the input during the entire experiment. Hence, the channel was in a state of erosion. Focusing on the time span between the 70 h and the end of the run, the output curve oscillates around a line parallel to the input line. The averaged sediment output of 17.9 g s^{-1} within this span was only slightly above the input rate of 17.7 g s^{-1}. In addition, the spatially averaged bed slope, measured every 15 h, remained almost stable, oscillating around a temporal mean of 1.55% with a maximum of 1.59% and a minimum of 1.50%. Therefore, an equilibrium state can be assumed. In terms of bedload, these data match quite well the results of Ashmore (1988) and Zarn (1997), which were derived under similar equilibrium state conditions, but for lower slopes and finer bed material (Fig. 3).

Interesting observations were made concerning the morphological changes of the river bed during the second run ($Q = 21.4$ L s^{-1}). Shortly after the start with the initially planar bed, a slightly braided bed pattern developed, followed by a phase of heavy erosion between hour 63 and 83, ending in a dominant channel. The spatially averaged slope decreased in the phase of erosion from 1.85% to 1.72%. After that, the mean bedload output was lower than the input until hour 139. Consequently, aggradation took place and a braided pattern redeveloped. Following another phase of erosion ($t = 139$–152 h) the cycle restarted. During the whole experiment it was possible to observe the same process once again ($t = 152$–227 h) and then another period when the phase of erosion was not that distinct. Until the end of the experiment the spatially averaged bed slope rose again to 1.9%. Over the total duration of the experiment, therefore, the channel was in a state of aggradation.

The change between erosion and aggradation in the second run is recognizable in Fig. 7 by the stepped cumulative output line, where the steep zones mark phases of erosion. This development cycle also can be seen in Fig. 8 where a time series of the bedload transport rate is plotted and the phases of erosion fall together with the bedload transport peaks. Similar oscillating patterns have been observed in model tests (e.g. Hoey & Sutherland, 1991; Warburton & Davies, 1994) and field measurements (Hoey *et al.*, 2001). The transport peaks in Fig. 8 mark periods with a dominant

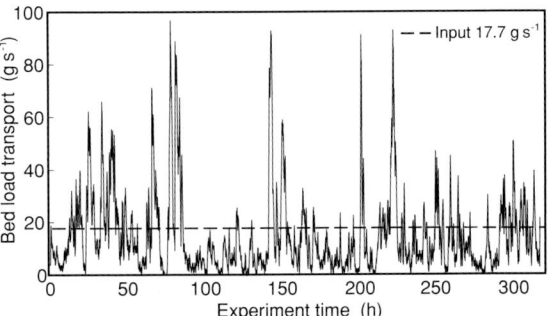

Fig. 8 Time series of bedload transport during run 2. Plotted data are a moving mean with a time window of 30 min.

single channel, whereas the minima occur when a braided pattern is developing. Figure 9a–d shows this development: (a) and (c) were taken at $t = 220$ h shortly after a transport peak; (b) and (d) were taken at $t = 275$ h during a period of low bedload transport.

CONCLUSIONS AND OUTLOOK

According to the transport calculation of Zarn (1997), in the first run erosion had to be expected, as the discharge was rather high for the given initial slope and sediment feed rate. The chosen discharge can be understood as an upper limit. At the same time, the tendency for aggradation in the second run is explainable, as the discharge of 21.4 L s^{-1} represents a lower limit in the same context (Fig. 6). So, as predicted by the calculations, the discharge at which the equilibrium state is reached quickly, given the slope of 1.85% and the feeding rate of 17.7 g s^{-1}, lies between 21.4 L s^{-1} and 26.5 L s^{-1}. Obviously the equation of Zarn (1997) may be used for slopes steeper and grain-size distributions coarser than the range used to derive it. The investigations presented here, however, have been carried out under steady-state conditions. Whether the formula also can be applied to cases with a variable discharge will have to be assessed by analysing the data from the experiments based on a hydrograph and further planned runs.

The observation in the second run ($Q = 21.4$ L s^{-1}) concerning the bedload peaks during periods of a single dominant channel leads to the assumption

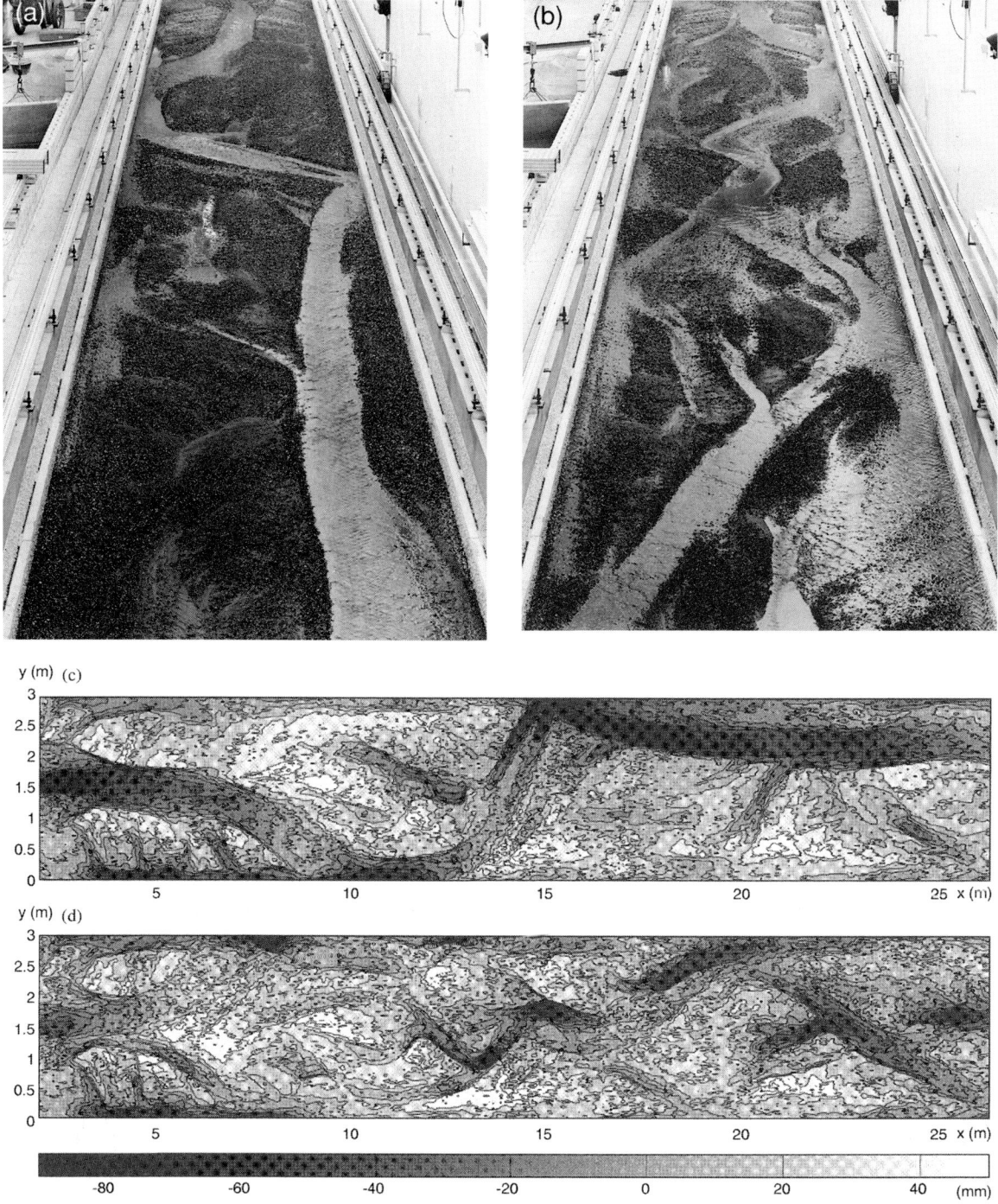

Fig. 9 Comparison of the bed morphology during high bedload transport (a & c) and low bedload transport (b & d). The photographs (a & b) are taken at a discharge 10 times lower than the actual experimental discharge. Figures (c) and (d) are plots of the differential elevation of the river bed, compared with the initial planar bed, measured at the same time as the photographs were taken. Flow direction for photographs (a & b) is from top to bottom, and for greyscale plots (c & d) is from left to right.

that a lower braiding index implies a higher transport rate and vice versa. This was also concluded by Warburton (1996), in contrast to the earlier work of Warburton & Davies (1994). Nevertheless, the presence of a dominant single channel alone does not seem to be sufficient to explain the peaks in the bedload output. Analysis of the time-lapse pictures suggests that such peaks may occur only when the single channel is moving laterally.

Detailed evaluation of existing and future experiments will have to assess possible correlations between statistical values describing the bed topography and morphology (e.g. mean number of channels per cross-section, braiding index, total sinuosity) and the bedload transport rate. Furthermore, the capability of the two-dimensional numerical model to predict the bedload transport rate will be assessed by trying to numerically simulate the physical experiments. It will also be of major interest whether these simulations lead to a bed morphology comparable to those observed in the physical experiments.

ACKNOWLEDGEMENTS

This study is funded by the ETH-Zurich and the experimental work is supported by the workshop staff of VAW. Much translation assistance came from Urs Keller.

REFERENCES

Anderson, A.G., Parker, G. and Wood, A. (1975) *The Flow and Stability Characteristics of Alluvial River Channels*. Project Report No. 161, St. Anthony Falls Hydraulic Laboratory, Minneapolis, MN, 115 pp.

Ashmore, P.E. (1982) Laboratory modelling of gravel braided stream morphology. *Earth Surf. Process. Landf.*, **7**, 201–225.

Ashmore, P.E. (1988) Bed load transport in braided gravel-bed stream models. *Earth Surf. Process. Landf.*, **13**, 677–695.

Ashworth, P.J., Ferguson, R.I. and Powell, M.D. (1992) Bedload transport and sorting in braided channels. In: *Dynamics of Gravel-bed Rivers* (Eds P. Billi, R.D. Hey, C.R. Thorne and P. Tacconi), pp. 498–513. Wiley, Chichester.

Bagnold, R.A. (1980) An empirical correlation of bed load rates in natural rivers. *Proc. R. Soc. Lond., Ser. A*, **372**, 453–473.

Beffa, C. and Faeh, R. (1994) Flood propagation on mobile beds under mountainous flow conditions. In: *Speciality Conference 'Modelling of Flood Propagation over Initially Dry Areas'*, pp. 327–341. American Society of Civil Engineers, Milan.

Bridge, J.S. (1993) The interaction between channel geometry, water flow, sediment transport and deposition in braided rivers. In: *Braided Rivers* (Eds J.L. Best and C.S. Bristow). *Geol. Soc. Lond. Spec. Publ.*, **75**, 13–72.

Carson, M.A. and Griffiths, G.A. (1987) Influence of channel width on bed load transport capacity. *J. Hydraul. Eng.*, **113**(12), 1489–1509.

Da Silva, A.M.A.F. (1991) Alternate bars and related alluvial processes. MSc thesis, Queen's University, Kingston, Ontario, Canada.

Faeh, R. (1996) Erosion-based dambreak simulation. In: *2nd International Conference on Hydroinformatics*, Vol. 2, pp. 681–688, Zurich.

Habersack, H.M., Koch, M. and Nachtnebel, H.-P. (2000) Flussaufweitungen in Oesterreich—Entwicklung, Stand und Ausblick. *Oesterr. Wasserwirtsch.*, **52**(7/8), 143–153.

Hoey, T. and Sutherland, A.J. (1991) Channel morphology and bed load pulses in braided rivers: a laboratory study. *Earth Surf. Process. Landf.*, **16**, 447–462.

Hoey, T., Cudden, J. and Shvidchenko, A. (2001) The consequences of unsteady sediment transport in braided rivers. In: *Gravel-bed Rivers V* (Ed. M.P. Mosley), pp. 121–140. New Zealand Hydrological Society, Wellington, New Zealand.

Hunzinger, L.M. (1998) *Flussaufweitungen—Morphologie, Geschiebehaushalt und Grundsätze zur Bemessung*. Mitteilung No. 159, Versuchsanstalt für Wasserbau, Hydrologie und Glaziologie, ETH-Zürich, Zürich, 206 pp.

Hunzinger, L.M. (1999) Morphology in river widenings of limited length. In: *28th International Association for Hydraulic Research Congress*, Graz, Austria. (Paper on CD-Rom.)

Hunzinger, L.M. and Zarn, B. (1997) Morphological changes at enlargements and constrictions of gravel bed rivers. In: *3rd International Conference on River Flood Hydraulics*, pp. 227–236, HR Wallingford, Stellenbosch, South Africa.

Jaeggi, M. (1983) *Alternierende Kiesbänke*. Mitteilung No. 62, Versuchsanstalt für Wasserbau, Hydrologie und Glaziologie, ETH-Zürich, Zürich, 286 pp.

Jaeggi, M. (1992) *Sediment Haushalt und Stabilitaet von Flussbauten*. Mitteilung No. 119, Versuchsanstalt für Wasserbau, Hydrologie und Glaziologie, Zürich, pp. 105.

Mcardell, B.W. and Faeh, R. (2001) A computational investigation of river braiding. In: *Gravel-bed Rivers V* (Ed. M.P. Mosley), pp. 73–86. New Zealand Hydrological Society, Wellington, New Zealand.

Meyer-Peter, E. and Müller, R. (1948) Formulas for bed load transport. In: *2nd Conference of the International Association for Hydraulic Research*, pp. 39–64, Stockholm, Sweden.

Mosley, M.P. (1982) Analysis of the effect of changing discharge on channel morphology and instream uses in a braided river, Ohau river, New Zealand. *Water Resour. Res.*, **18**(4), 800–812.

Parker, G. (1979) Hydraulic geometry of active gravel rivers. *J. Hydraul. Div. Am. Soc. Civ. Eng.*, **105**(HY9), 1185–1201.

Pickup, G. and Higgins, R.J. (1979). Estimating sediment transport in a braided gravel channel—the Kawerong river, Bougainville, Papua New Guinea. *J. Hydrol.*, **40**, 283–297.

Ramette, M. (1990) Essai d'explication des morphologies fluviales à partir de la théorie du régime. *La Houille Blanche*, **45**(1), 43–60.

Schoklitsch, A. (1962) *Handbuch des Wasserbaus*. Springer-Verlag, Vienna, Austria.

Smart, G.M. (1984) Sediment transport formula for steep channels. *J. Hydraul. Eng.*, **110**(3), 267–276.

Thompson, S.M. (1985) Transport of gravel by flows up to 500 m^3/s, Ohau river, Otago, New Zealand. *J. Hydraul. Res.*, **23**(3), 285–303.

Van Rjin, L.C. (1984) Sediment transport, part I: bed load transport. *J. Hydraul. Eng.*, **110**(10), 1431–1456.

Warburton, J. (1996) Active braidplain width, bed load transport and channel morphology in a model braided river. *J. Hydrol. (NZ)*, **35**(2), 259–285.

Warburton, J. and Davies, T.R.H. (1994) Variability of bed load transport and channel morphology in a braided river hydraulic model. *Earth Surf. Process. Landf.*, **19**, 403–421.

Yalin, M.S. (1992) *River Mechanics*. Pergamon Press, Oxford, 219 pp.

Young, W.J. and Davies, T.R.H. (1991) Bed load transport processes in a braided gravel-bed river model. *Earth Surf. Process. Landf.*, **16**, 499–511.

Young, W.J. and Warburton, J. (1996) Principles and practice of hydraulic modelling of braided gravel-bed rivers. *J. Hydrol. (NZ)*, **35**(2), 175–198.

Zarn, B. (1993) Stabilising a riverbed by local widening—a case study of the river Emme, Switzerland. In: *Sediment Transport Mechanisms in Coastal Environments and Rivers, Euromech 310*, pp. 388–396. World Scientific Publication, 1994, Le Havre, France.

Zarn, B. (1997) *Einfluss der Flussbettbreite auf die Wechselwirkung zwischen Abfluss, Morphologie und Geschiebetransportkapazitaet*. Mitteilung No. 154, ersuchsanstalt für Wasserbau, Hydrologie und Glaziologie, ETH-Zürich, Zürich, pp. 240.

Zarn, B. (2003) Alpenrhein. In: *Feststofftransportmodelle für Fliessgewässer*, pp. 195–212. Deutsche Vereinigung für Wasserwirtschaft, Abwasser und Abfall e. V., Arbeitsbericht WW-2.4, Hennef, Germany.

Zarn, B. and Hunzinger, L.M. (1997) Erfahrungen mit Flussaufweitungen—das Beispiel Birne Emme. In: *Massnahmen zur naturnahen Gewaesserstabilisierung*, pp. 278–294. Deutscher Verband für Wasserwirtschaft und Kulturbau, Schriften 118.

The morphology and facies of sandy braided rivers: some considerations of scale invariance

GREGORY H. SAMBROOK SMITH*, PHILIP J. ASHWORTH†,
JAMES L. BEST‡, JOHN WOODWARD‡,[1]
and CHRISTOPHER J. SIMPSON§

*School of Geography, Earth and Environmental Sciences, University of Birmingham,
Birmingham, B15 2TT, UK (Email: g.smith.4@bham.ac.uk);
†Division of Geography, School of the Environment, University of Brighton,
Lewes Road, Sussex, BN2 4GJ, UK;
‡Earth and Biosphere Institute, School of Earth and Environment, University of Leeds, Leeds,
West Yorkshire, LS2 9JT, UK; and
§Department of Geography, Simon Fraser University, 8888 University Drive, Burnaby,
British Columbia, V5A 1S6, Canada

ABSTRACT

A fundamental and unresolved question in fluvial sedimentology concerns the nature of scale invariance and whether it is appropriate to apply data from a single river or outcrop of alluvial sediments to others of a different size. This issue is addressed herein by: (i) examining the similarity in aspects of the morphology of modern braided rivers: (ii) comparing the subsurface facies of three sandy braided rivers of differing scale (30–2000 m channel width), as revealed by ground-penetrating radar (GPR). Measurement of braid-bar shape in 15 rivers, covering four orders of magnitude in spatial scale, demonstrates that a simple index of bar planform shape, the width : length ratio, is scale invariant. Additionally, scour depths at channel confluences are similar in their relative scale across channels of greatly differing size. Comparison of the subsurface sedimentary facies of three sandy braided rivers using GPR demonstrates that sandy braided rivers exhibit a degree of scale invariance, with the ubiquitous occurrence of trough cross-stratification associated with migrating dunes. Significant differences exist in the occurrence of other facies, however, both between rivers and between bars within the same river, most notably in the predominance of either high-angle planar cross-stratification or low-angle stratification. These differences are controlled by a wide range of factors, which may include the discharge regime, local bar and channel topography, anabranch width : depth ratio and the abundance of vegetation. Hence, although rivers and individual bars within the same river may have similar surface planform shapes, their subsurface facies may be very different. A single, universal facies model for sandy braided rivers is thus probably inappropriate and will remain elusive.

---

[1]Present address: Division of Geography, Northumbria University, Lipman Building, Newcastle upon Tyne NE1 8ST, UK

INTRODUCTION

Some issues concerning scaling have been addressed by fluvial sedimentologists, such as the relationship between dune height, dune length and flow depth (e.g. Jackson, 1976; Yalin, 1992), bar length, bar width and bankfull channel-width (e.g. Bridge, 1985; Ashley, 1990), the hierarchy of bars and channels within rivers (e.g. Jackson, 1975; Bristow, 1987; Bridge, 1993a) and the different hierarchies of strata and their bounding surfaces (e.g. Miall, 1988a,b; Bridge, 1993b; Holbrook, 2001). There has been little consideration, however, of the upscaling (the application of results and models from a system to one of greater size) of the form and process relationships operating in alluvial environments and their link to the development of facies models (Bristow & Best, 1993). Such an understanding of scaling is needed to enhance the interpretation of ancient fluvial sediments, especially when applying one facies model to different localities or outcrops. This issue attains especial significance in quantitative reservoir modelling, where it is essential to select a matching size, and form, of analogue for the interpreted depositional environment (Bridge & Tye, 2000; Moreton et al., 2002).

One aspect of scaling that has been investigated is the application of fractal analysis to the geometry of meandering (Stölum, 1996) and braided (Sapozhnikov & Foufoula-Georgiou, 1996) river planforms. For example, Sapozhnikov & Foufoula-Georgiou (1997, p. 1983) concluded that 'if a small part of a braided river is stretched in a certain way along the mainstream direction and a certain way along the perpendicular direction, then this stretched part looks statistically the same as a bigger part of the river'. There has been no work, however, to determine whether these two-dimensional fractal relationships of planform apply to the complexity of preserved fluvial deposits. This is an important gap in knowledge, because to interpret many alluvial deposits, facies models are commonly derived from contemporary environments, and then applied to ancient deposits and other modern rivers (e.g. Miall, 1978, 1996; Bridge, 1993a; Bridge & Tye, 2000). This approach assumes that a facies model derived for a small part of a river can be applied to the whole river as well as to other rivers, either smaller or larger in size. For example, the influential sedimentary vertical profiles proposed by Miall (1978) were named after the specific rivers from which they were derived (i.e. Trollheim, Scott, Donjek, South Saskatchewan, Platte and Bijou Creek) and each is presented as a 'type site' that can be applied to any river with broadly similar characteristics, regardless of scale. Although these schemes have since been refined (e.g. Miall, 1996), they are often still not considered in relation to any scale-dependence of the sedimentary sequences present.

Bridge (1993a), in a comprehensive review of braided rivers, provided three figures that highlight the key aspects of braid-bar sedimentology (Bridge, 1993a, figs 21–23, pp. 56–57). The figures are presented without a scale and a caption that states '. . . channel belt widths in nature will range from tens to thousands of metres', implying that this model and hence braided river deposits are scale invariant. Some authors, however, have expressed concern regarding the universal application of a facies model from one river to others of different scale. For example, Cant (1978a, p. 638) stated that the facies model developed from the modern South Saskatchewan River and Devonian Battery Point Formation '. . . does not extend to very large rivers . . . differences between the Brahmaputra and the smaller rivers may be reflected not only in the scales of the deposits, but also in their facies types'. This caveat, however, has often been ignored in subsequent widespread applications of the model.

The present paper seeks to highlight and discuss some issues of scale invariance in fluvial sedimentology. The discussion is limited to the case of braided rivers and, in the majority of cases, sand-bed rivers. The aims of the paper are to:

1 analyse data on two key aspects of braided river morphology, the aspect ratio of bars and confluence scour depth, to establish if these properties are scale invariant;

2 compare GPR profiles from three classic braided river 'type sites' (Calamus, South Saskatchewan and Jamuna) in order to examine whether braided river deposits are scale invariant.

This analysis will then be used to discuss issues of scaling in facies models within fluvial sedimentology.

SCALE INVARIANCE AND THE SURFACE MORPHOLOGY OF BRAIDED RIVERS

To assess the nature of scale invariance, two aspects of braided rivers were examined: the shape of braid bars and the depth of scours at channel confluences. These two features are important because they represent the surface geometry of the largest scale of macroforms and the maximum depth of erosion above which the sediments of a braided channel are deposited.

Bar shape

The aspect ratio of mid-channel bars was measured from a series of images from laboratory experiments, aerial photographs and satellite images (Table 1). Aspect ratio is defined here as B/A, where B is the widest part of the bar on a line perpendicular to the interpreted long axis (A). Figure 1 and Table 1 show a compilation of 15 data sets from rivers and experimental braided channels that span four orders of magnitude in size. The slope of a least-squares regression line through this aggregated data set ($n = 1090$ points) is 0.97 and not significantly different from 1 ($P = 0.05$). Thus, bar shape ($B : A$ ratio) is independent of river size. A similar conclusion was put forward by Komar (1984) for 38 streamlined bars of the meandering Mississippi, Missouri and Columbia Rivers. Although the data plotted in Fig. 1 contain measurements of bar shape taken at a range of flow stages, the consistency in average bar shape between rivers suggests the bar $B : A$ ratio may be independent of flow stage or discharge.

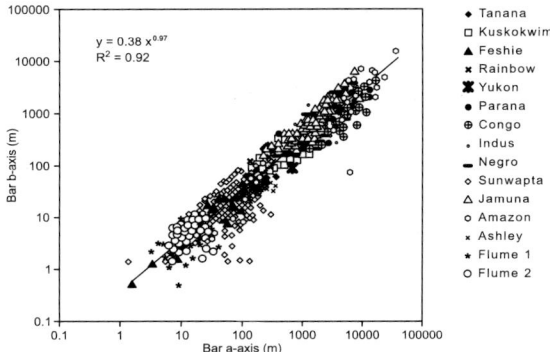

Fig. 1 Bar shape, as defined by A and B axis for 15 contrasting braided rivers covering four orders of magnitude of size. Data from flume experiments are described in Ashworth *et al.* (1994). Of the rivers, seven were measured from aerial photographs from scales of 1 : 5000 to 1 : 40 000 and six from satellite photographs at 1 : 250 000.

Table 1 Characteristics of rivers and summary of bar shape data plotted in Fig. 1.

River	Country	Latitude (°)	Longitude (°)	Catchment (km²)*	Mean A (m)	Mean B (m)	Mean B/A	n
Flume 1					14	5	0.36	64
Flume 2					15	6	0.41	72
Feshie	UK	3.9W	57.1N	90	73	22	0.35	41
Kuskokwim	USA	153.8W	62.7N	252	516	190	0.41	75
Sunwapta	Canada	117.7W	52.4N	329	69	20	0.37	144
Ashley	NZ	172.4E	43.4S	666	130	32	0.27	83
Rainbow	USA	154.8W	58.5N	735	769	233	0.35	44
Tanana	USA	147.4W	66.3N	53 700	344	145	0.45	74
Yukon	USA	143.0W	65.5N	254 000	746	293	0.42	34
Indus	India	70.0E	28.0N	475 000	2459	1807	0.40	119
Negro	Brazil	61.5W	1.0S	563 000	2136	723	0.35	47
Congo	Congo	22.0W	2.0N	911 000	2839	587	0.25	115
Ganges	Bangladesh	88.5E	24.5N	1 070 000	1986	983	0.50	82
Parana	Argentina	22.0E	31.0S	2 090 000	2888	888	0.35	59
Amazon	Brazil	59.5W	3.0S	4 070 000	6258	2001	0.30	37

*Catchment area to location of river image.

Scour depth

Another key component of braided rivers is the depth of significant scour generated within channels, including sites at the outside of bends, bank protrusions, channel constrictions and channel confluences (Ashmore & Parker, 1983; Klaassen et al., 1988; Best & Ashworth, 1997). In order to examine the depth of channel confluence scour, Fig. 2 shows examples of the relationship between relative confluence scour depth, d_r (expressed as a ratio with the mean upstream flow depths), and junction angle, α, taken from a number of studies. These studies encompass experimental and field studies of channels that are between 0.01 m and 13.20 m in depth. The plot shows the similarity in the range of d_r across all channel depths. Some scatter in this plot is due to the influence of other key controlling variables at channel confluences, such as the discharge ratio, depth differential between the incoming tributaries and the exact confluence planform. The scaling of scour depth, however, as well as the similarity in aspects of the planform morphology (Paola, 1997), suggest that as a first approximation maximum confluence scour-depth can be estimated from this type of plot. Furthermore, this relationship holds from the scale of experimental junctions (scour depth of c. 0.1 m) to some of the world's largest natural confluences (scour depths c. 35 m). The maximum depth of erosion, or 'combing' depth (Paola & Borgman, 1991) of the braided river may thus be set by this scour and hence the scaling of the thickness of individual alluvial channel storeys may scale to the maximum channel depth, this being vital in interpretation of erosive boundaries in fluvial successions (Salter, 1993; Best & Ashworth, 1997).

The data presented in Figs 1 & 2 and Table 1 support the work on planform morphology by Stölum (1998), Foufoula-Georgiou & Sapozhnikov (1998, 2001), Sapozhnikov & Foufoula-Georgiou (1997, 1999), Sapozhnikov et al. (1998) and Nykanen et al. (1998) in that several aspects of the *surface* morphology of braided rivers can be considered scale invariant. A key question that follows from this, however, is whether this scale invariance is reflected in the subsurface sedimentology and hence if similar facies models can be applied between different scales of river. As a first step towards addressing this issue, ground-penetrating radar (GPR) data sets are compared from the Calamus, South Saskatchewan and Jamuna rivers, where channel size covers three orders of magnitude.

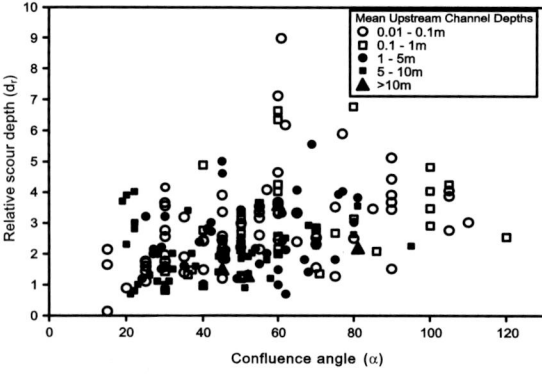

Fig. 2 Plot of relative scour depth at channel confluences, d_r, against the confluence angle, α, for channel depths ranging over three orders of magnitude (0.01 m to 13.2 m; n = 198). Relative scour depth, d_r, is defined as the depth of the bed scour in relation to the average flow depth in the upstream tributaries (i.e. d_r = 0 when there is no scour). Data sources are from experimental studies of channel junctions and a range of field studies. Data from Mosley (1975, 1976, 1982), Ashmore & Parker (1983), Best (1985, 1988), Klaassen & Vermeer (1988), Roy & De Serres (1989), Orfeo (1995), Best & Ashworth (1997), Roy et al. (1988), McLelland et al. (1996), Rhoads & Sukhodolov (2001) and from the authors' surveys in Bangladesh (see Sarker, 1996; Delft Hydraulics & DHI, 1996).

SCALE INVARIANCE IN SUBSURFACE ALLUVIAL ARCHITECTURE

In order to set the discussion of subsurface sedimentology in context, brief details of the morphological characteristics of the three rivers (Fig. 3) are presented below.

Calamus River, Nebraska, USA

Full details of the Calamus River (c. 30 m wide) can be found in Bridge et al. (1986, 1998), Bridge & Gabel (1992) and Gabel (1993). Bridge et al. (1998, p. 978) describe the Calamus channel pattern as 'braided, low-sinuosity' although an earlier

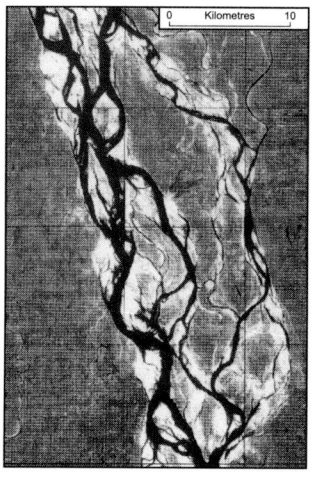
Jamuna River

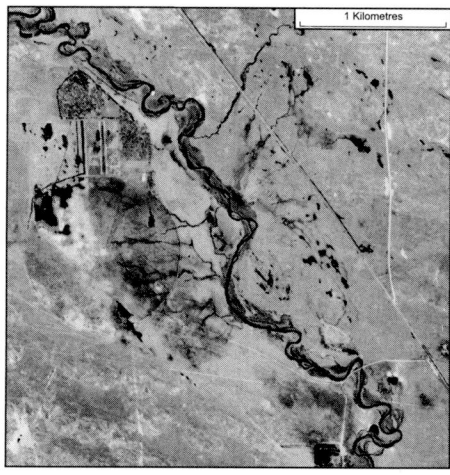
Calamus River

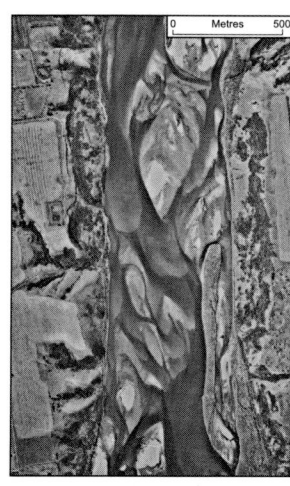
South Saskatchewan River

Fig. 3 Aerial photographs of the Calamus (1999, supplied by US Geological Survey Water Resources Division), South Saskatchewan (2000, commissioned from Information Services Corporation of Saskatchewan), and Jamuna (1996, supplied by Center for Environmental and Geographic Information Services, Bangladesh). Flow direction in all three rivers is from top to bottom.

description of it as 'transitional between classical "meandering" and "braided" patterns' (Bridge et al., 1986, p. 852) more closely reflects the predominantly meandering planform with occasional, often well vegetated, mid-channel bars (see Fig. 3). Typical dimensions of the rhomboidal shaped mid-channel bars are 45 m × 10 m × 1 m, with average downstream migration rates of 1–2 m yr^{-1}. Bars evolve from the emergent parts of alternate bars and point bars, and sand dunes are ubiquitous and reach up to 0.5 m in height. Bed material is predominantly medium-grained sand and bank erosion is of the order of decimetres per year, with low sinuosity reaches eroding at less than this rate.

South Saskatchewan River, Canada

The work of Cant (1978b) and Cant & Walker (1978) provides background information on the characteristics of the braided South Saskatchewan River. The braidplain is incised into Cretaceous and more recent Quaternary surface deposits that restrict the lateral movement of channels. The river is c. 500 m wide (Fig. 3) with four main types of mid-channel bar:

1 lobate-fronted, often submerged bars;
2 emergent bars with a characteristic horseshoe planform, i.e. downstream-elongate bartail limbs;
3 emergent bars with variable planform and cross-bar channels;
4 stable vegetated bars.

Typical dimensions of the second bar type are 175 m × 140 m × 2 m, with downstream migration rates of up to 100 m yr^{-1}. These bars may evolve from submerged lobate bars that are often organized in an alternate bar pattern as shown in the morphological model of Bridge (1993a). Dunes migrate through the channels of the South Saskatchewan during most flow stages and can reach up to 1.5 m high (Cant & Walker, 1978). Bed grain size is predominantly medium sand with an average D_{50} of 0.30 mm.

Jamuna River, Bangladesh

Descriptions of the Jamuna River (also known as the Brahmaputra in certain reaches) can be found in Coleman (1969), Klaassen et al. (1988), Bristow (1993) and Ashworth et al. (2000). The Jamuna is often cited as an analogue for ancient large sandy braided-river deposits (e.g. Conaghan & Jones, 1975;

Rust & Jones, 1987; Gardiner et al., 1990). Channels can be up to 2000 m wide (Fig. 3) with a typical, actively migrating mid-channel bar having dimensions of 1500 m × 500 m × 12 m (Ashworth et al., 2000). Such bars often have a horseshoe planform similar to those in the South Saskatchewan, migrate downstream and laterally by up to 500 m yr^{-1} and can cause significant adjacent bank erosion up to 1000 m yr^{-1} (Ashworth et al., 2000). Submerged lobate alternate bars that are present in the Calamus and South Saskatchewan rivers are rare in the Jamuna. Dunes are ubiquitous throughout the channels of the Jamuna, however, and may reach 6 m in height (Roden, 1998). Grain size for the reach reported in this paper is predominantly fine sand with a D_{50} of 0.14 mm.

Sedimentological comparison

Comparison of the subsurface facies is achieved by analysis of the three GPR data sets (Fig. 4) from each of these rivers. The Calamus GPR data are reproduced from Bridge et al. (1998), the Jamuna from Best et al. (2003) and the South Saskatchewan from a new dataset. Table 2 details the GPR survey methodology, post-processing and ground-truthing for the three studies to highlight the variation in data collection techniques. Woodward et al. (2003) provide a review of the implications of using different GPR data collection techniques on the data quality and subsequent interpretation. The Calamus survey was undertaken in continuous (i.e. not 'stop-and-collect') mode that degrades data quality, and together with the low contrast in dielectric permittivity, this probably explains the slight loss in clarity of the final profiles compared with the South Saskatchewan and Jamuna. However, the Calamus study had by far the most comprehensive ground-truth programme achieved through vibracore analysis. Comparison of the data sets reveals the following four broad groups of facies.

Trough cross-stratification

Trough cross-stratification associated with migrating dune bedforms is the dominant facies in all of the rivers. The size of dune cross-stratification scales with flow depth, with dune height diminishing from thalweg to bartop. Figure 4 demonstrates the similarity of trough cross-stratification between the South Saskatchewan and Jamuna. Although not as obvious in the Calamus River GPR profiles, this type of stratification is discernible, and extensive vibracoring (Bridge et al., 1998) has verified that it is frequently found within the Calamus River deposits. The thickness of dune cross-sets scales with dune height (LeClair & Bridge, 2001) and dune height scales approximately with flow depth (Ashley, 1990). Thus, on average, the size of the sets of dune cross-stratification preserved in the subsurface decreases from the Jamuna to South Saskatchewan to Calamus.

High-angle planar cross-stratification

A striking characteristic of both the Jamuna and South Saskatchewan rivers is the presence of high-angle, up to c. 35°, planar cross-stratification that is not seen in the Calamus River. This stratification develops either by the accretion of bar-margin slipfaces or downstream progradation of bartail limbs. The largest cross-sets have formed by bar-margin slipface accretion (Fig. 4) as bars migrated in an oblique direction (i.e. both downstream and laterally) and into an adjacent deep thalweg (Ashworth et al., 2000; Best et al., 2003). This process of deposition can produce cross-sets that extend from the thalweg to bartop, c. 8 m in the Jamuna and c. 2 m in the South Saskatchewan. These sets extend for c. 100 m in the Jamuna and c. 20 m in the South Saskatchewan. Where these high-angle sets result from downstream progradation of bartail limbs (Fig. 4), they have reduced thicknesses, c. 3 m in the Jamuna and c. 1 m in the South Saskatchewan, and can extend laterally c. 100 m in both rivers. Additionally, this high-angle planar cross-stratification may be bounded by an erosion surface at the base and/or top of the sequence. In the South Saskatchewan and Jamuna, high-angle planar cross-stratification can comprise a significant part of the internal structure of braid bars. Although high-angle planar cross-stratification can also form by dune migration over and around bars, the examples in Fig. 4 show that the sets created by bar-margin deposition are much thicker than would be deposited by dunes. The dimensions of the sets of high-angle planar cross-stratification, in relation to channel depth, therefore can be used to discriminate

Table 2 Comparison of GPR methodology for the three river studies. Data from Bridge et al. (1998) for the Calamus, Woodward et al. (2003) for the South Saskatchewan and Best et al. (2003) for the Jamuna.

Data collection	Calamus	South Saskatchewan	Jamuna
GPR Unit	GSSI SIR-SYSTEM 10	PulseEkko 100	PulseEkko 100
Survey	Common-offset	Common-offset	Common-offset
Antennae frequency	500 MHz	200 MHz	100 MHz
Antennae separation	Not known	0.75 m	1.00 m
Antennae orientation	Both perpendicular to survey line	Both perpendicular to survey line	Both perpendicular to survey line
Station spacing	c. 0.06 m	0.10 m	0.50 m
Mode of data collection	Continuous	Stop-and-collect	Stop-and-collect
Stacks	8	64	64
Processing			
Software	Radan Win3	Gradix 1.10	PulseEkko
Dewow	No	8.71 MHz	Yes
Drift removal	No	Yes	Yes
Set time-zero	Yes	Yes	Yes
Bandpass filter	Low and highpass at 80 and 20 cycles scan^{-1}	Trapezoidal filter with 50, 100, 220 and 440 MHz gates	Lowpass at 40% Nyquist frequency
Background removal	32 traces	200 traces	No
Depth conversion velocity	0.055–0.06 m ns^{-1}	0.05 m ns^{-1}	0.12–0.17 m ns^{-1}
Elevation statics	Yes	Yes	Yes
Gain	Autogain	Automatic gain control 25 ns	Automatic gain control 100 ns
Ground-truth control	Comparison with cores	Cut-face experiment	Comparison with cores/trench

between the two different formative processes (Bridge & Tye, 2000).

Low-angle stratification

Low-angle stratification (< 10°) is present in all three rivers (Fig. 4) and is associated with upstream, downstream, vertical and lateral accretion. Although present in the Jamuna and South Saskatchewan, this type of stratification is the dominant large-scale structure found within the Calamus, where it is present throughout most of the subsurface. The average downstream dip of these low-angle sets in the Calamus is c. 2°, with cross-stream dips up to 6°. Bridge et al. (1998) interpreted these sets as products of lateral and downstream growth of bars coupled with vertical accretion.

Minor facies

Channel-fill features (Fig. 4) can be found in all three rivers and possess a concave basal erosion surface, with a variety of angles of stratification forming the subsequent fill. In the South Saskatchewan, 'bartop scours' (Fig. 4) also produce a similar pattern of reflectors, but form in bartail areas where oblique accretion of the downstream limb isolates a depression that existed between, or in the lee of, the bartail limbs. Also present in the GPR traces from the Jamuna and South Saskatchewan are very strong undulating reflectors (Fig. 4) associated with the deposition of clay drapes, often in the bar-lee, at low stage. A comparable feature in the Calamus is associated with the lack of reflectors at the top of the profile because of fine-grained deposition and subsequent soil development. Convex-up reflectors are occasionally present in the Calamus and South Saskatchewan and are related to the accretion of scroll bars onto the main bar nucleus.

IMPLICATIONS FOR THE SCALE INVARIANCE OF SEDIMENT DEPOSITION AND FACIES

From the preceding results, a number of similarities and differences between these three sand-bed braided rivers is apparent. Trough cross-stratification generated by dune migration is ubiquitous in all three rivers, and shows a scale dependence with flow depth. However, significant differences in the larger-scale stratification are present, with the Jamuna and South Saskatchewan possessing high-angle, planar cross-stratification that is largely absent in the Calamus.

Analysis of sequential aerial photographs, coupled with the GPR results discussed above, demonstrates that the high-angle planar cross-stratification common in the Jamuna and South Saskatchewan is caused by the downstream and/or lateral migration of a bar margin and subsequent deposition at the margin avalanche face. The thickest and most extensive sets are found where the bar migrates into an adjacent deep thalweg. This form of deposition has been reported in other rivers, for example, the Platte River (Smith, 1972; Crowley, 1983; Blodgett & Stanley, 1980), and may be a common feature of sandy braid-bar deposition. For example, quantification of the 1.13 km of GPR lines collected from a bar in the South Saskatchewan (Fig. 5a) reveals that high-angle planar cross-stratification comprises 39% of the bar margin facies. Aerial photographic records of this bar (Fig. 5a) demonstrate that it had migrated downstream and laterally towards the left channel thalweg between April and October 2000 (GPR surveys were undertaken in June 2000), and Total Station surveys showed that the left thalweg around the bar was $c.$ 3 m below the bar surface.

Low-angle stratification develops in the South Saskatchewan as thin sheets of sand, or fields of low-relief dunes, that migrate up onto the bar from the adjacent channel. This depositional style is commonly associated with a bar that is relatively stable and aggrades principally through lateral and vertical accretion (see example in Fig. 5b). High-angle planar cross-stratification represented only 1% of the facies present in the 0.24 km of GPR lines taken over the stable bar shown in Fig. 5b, compared with 46% for low-angle stratification.

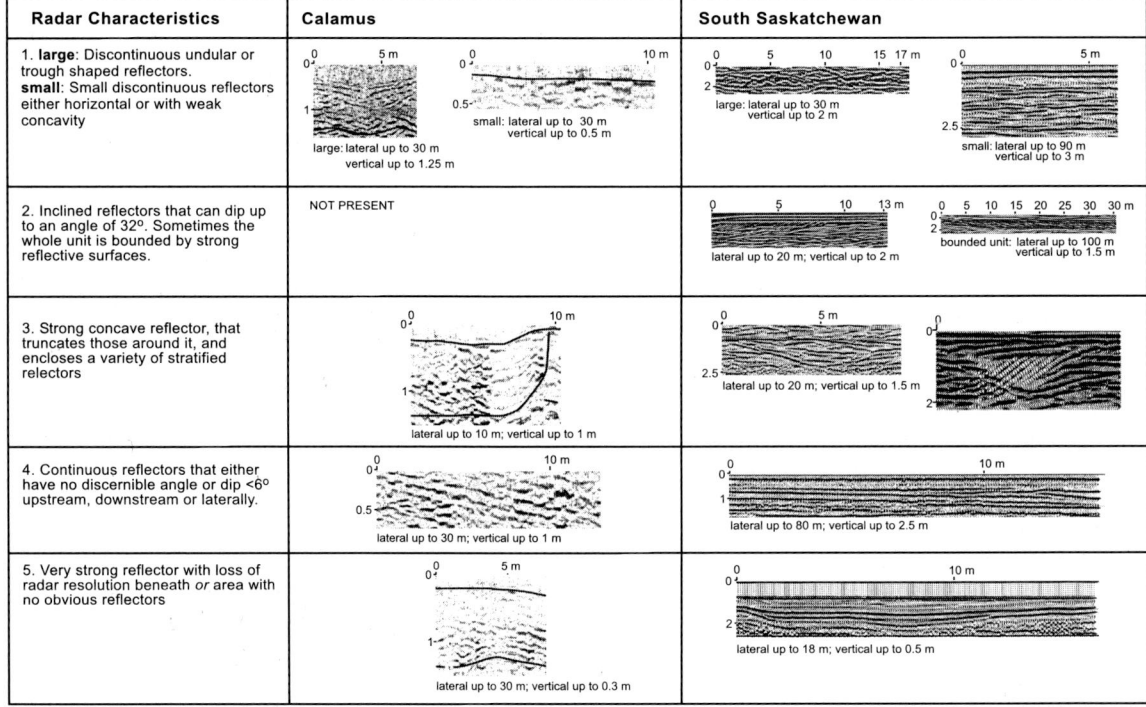

Fig. 4 Comparative GPR facies and interpretations for the Calamus (data from Bridge et al., 1998), South Saskatchewan (data collected by the authors in June 2000) and Jamuna (data from Best et al., 2003).

The explanation for this contrast in facies, both within and between the three rivers, appears to lie in the nature of, and controls on, bar evolution. A range of factors may be important in these controls.

1 The discharge regime will have a strong influence on whether large-scale high-angle planar cross-stratification will develop extensively. As stated above, in both the Jamuna and South Saskatchewan this facies occurred most commonly when bars were overtopped and sediment was transported *over* the bar surface and then avalanched down the bar margin, thus forming a high-angle slipface. Rivers in which large variations in discharge are present, and in which bars are overtopped for relatively long periods, may therefore have a high probability of producing high-angle planar cross-stratification that will be preserved in the subsurface. The presence of cross-bar, bar-top flow thus appears critical in this respect. For example, the Jamuna is subject to a pronounced monsoonal wet season and large fluctuations in discharge, whereas in contrast the Calamus has a more uniform discharge regime. Studies of flow structure associated with a braid bar in the Jamuna (McLelland *et al.*, 1999) have clearly shown the presence of cross-bar flow at high stage (McLelland *et al.*, 1999, plate 1c) that is associated with a steep, bar-margin slipface. Velocity vectors documented around a bar in the Calamus at high flow stage (Bridge & Gabel, 1992; Bridge, 1993a), however, show that flow is constrained to the channels, with flow being similar to that in individual curved channels and with no evidence of cross-bar, bar-top flow.

2 The frequency with which bars are overtopped, and hence the likelihood of high-angle planar cross-stratification forming, will also be a function of the bar and channel cross-section morphology. If the bar elevation is relatively low with respect to channel depth, then small increases in stage are more likely to lead to bars being overtopped and subsequent deposition of an angle-of-repose slipface. For example, bar surfaces in the South

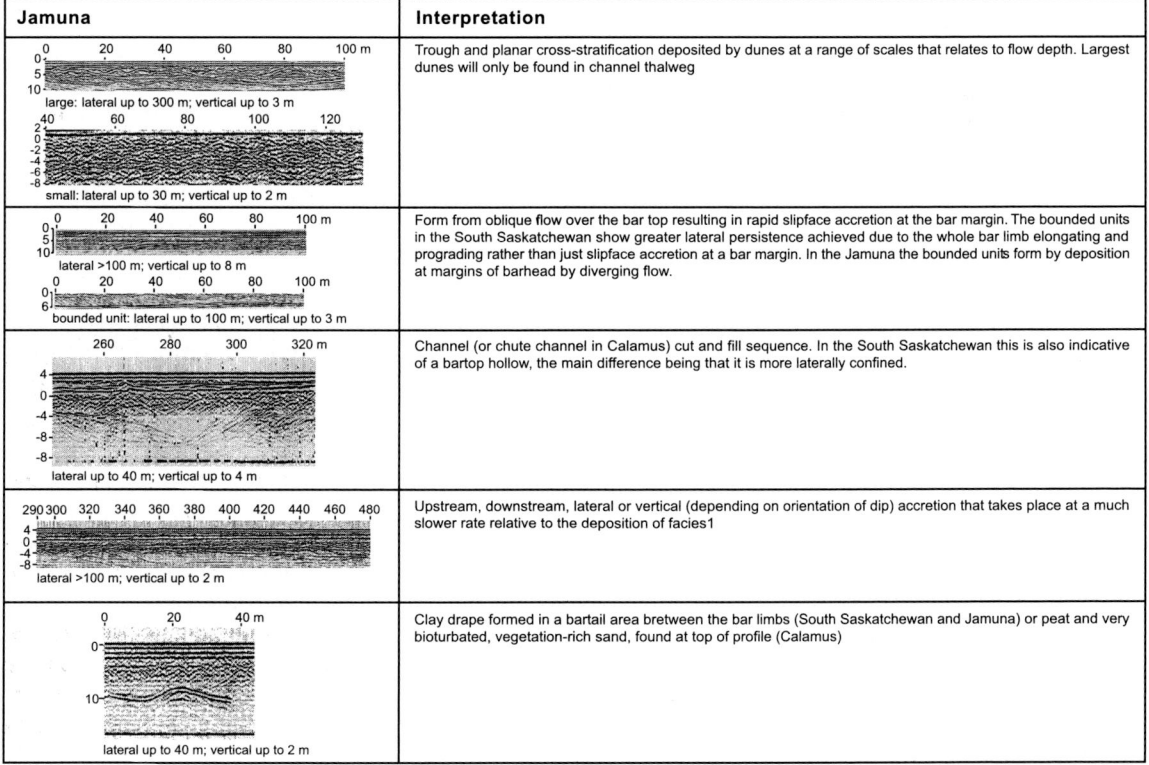

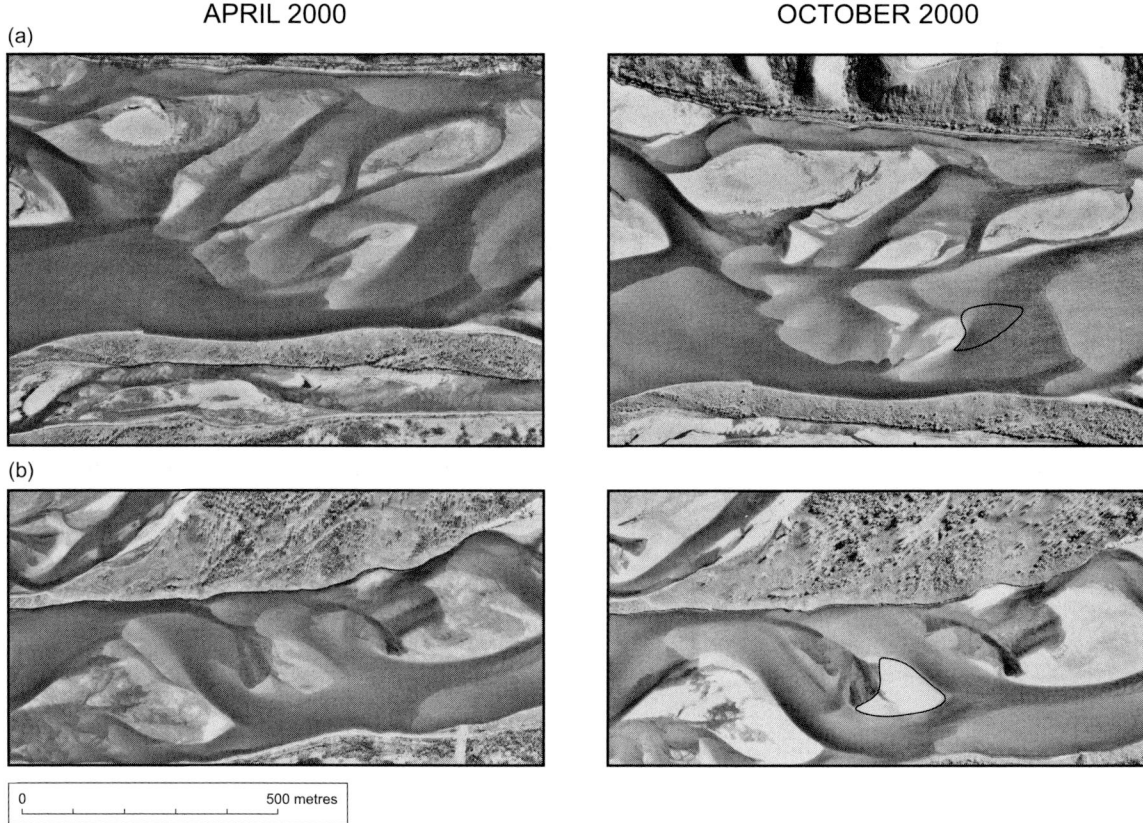

Fig. 5 Aerial photographs of two study bars in the South Saskatchewan River taken in April 2000 and October 2000. The black solid line on the October image indicates the position of the bar in April. Note that the bar in (a) has migrated a distance equivalent to its length, whereas the bar in (b) shows negligible movement. Flow is right to left in all images. Aerial photographs were commissioned from Information Services Corporation of Saskatchewan.

Saskatchewan can be overtopped by even modest flows during the summer when discharge is low.

3 Crowley (1983) observed that bar slipfaces are poorly defined near bank edges, and hence high-angle planar cross-stratification was less prevalent. Thus, where width : depth ratios are relatively low then the influence of the channel banks may be more likely. For example, the Calamus has a lower width : depth ratio (of the order 10^0) than either the Jamuna or South Saskatchewan (of the order 10^1).

4 The stabilizing influence of vegetation on a bar may inhibit the generation of steep bar-margin slipfaces and favour deposition on the bar-top and bar margin that is preserved as low-angle accretion surfaces. For example, the bars studied by Bridge *et al.* (1998) on the Calamus had a significant surface vegetation cover compared with those studied in the South Saskatchewan and Jamuna, and the Calamus possesses a large component of low-angle accretion rather than steep bar margins. Additionally, bar-top vegetation may also impart greater stability to the bar and lead to a lower rate of lateral bar migration, thus increasing the relative importance of both low-angle lateral accretion and vertical accretion in the preserved subsurface deposits.

5 Some of the differences in facies between the rivers may be due largely to variability in the degree of braiding. The Calamus River has a predominantly meandering channel pattern with frequent, highly sinuous bends (Fig. 3) and only the occasional midstream island. Thus, although Bridge & Tye (2000, fig. 2) used the planform and GPR interpretations for the Calamus as presented in Bridge et al. (1998) to propose a depositional model for braided rivers (see wording in caption to fig. 2 in Bridge & Tye, 2000), the Calamus is probably an atypical example of a sandy, braided river. Hence, the abundance of low-angle cross-stratification in the Calamus may simply be associated with the gentle channelward-sloping surfaces of rather sinuous, stable reaches rather than being a product of a slowly migrating 'braid' bar.

Comparison of the surface planform morphology and subsurface facies thus reveals a conflicting picture of scale invariance. Although features of the surface morphology, such as bar shape (Fig. 1) and maximum scour depth (Fig. 2), show remarkable consistency regardless of the size or type of braided river, this is not always displayed in the subsurface facies. For example, inspection of the South Saskatchewan River (Fig. 3) indicates a dominance of submerged lobate-fronted alternate bars, which then develop into, and form the basis of, emergent bars in a similar manner to the model for braid-bar formation proposed by Bridge (1993a, fig. 1). Bar growth also proceeds from alternate bars in a similar manner to the Calamus River (Bridge et al., 1986). In contrast, submerged alternate bars are infrequent in the larger Jamuna River, although it has been suggested that mid-channel braid bars may develop into lobate alternate bars over time (Ashworth et al., 2000, fig. 11). Despite the fact that braid bars evolve by a similar mechanism of alternate bar growth in the Calamus and South Saskatchewan, this is not expressed in their subsurface stratigraphy, where the South Saskatchewan and Jamuna appear more similar. The mechanism by which a braid bar initially evolves may thus not have any direct relevance to the final preserved architecture. Hence, although alluvial facies models are often applied from one river to another on the basis of a similarity in surface morphology, the subsurface sedimentary facies are an expression of a much more complex range of factors, as discussed above. As only some features of both the surface and subsurface deposition are scale independent, caution therefore should be taken in applying facies models between rivers. The variation in river surface and subsurface depositional form within and between rivers may be so great that a universal facies model for sandy braided alluvium is both inappropriate and unachievable.

CONCLUSIONS

Data across a wide range of scales (decimetre to kilometre) of braided rivers show:

1 The surface planform morphology of braid bars and the maximum relative depth of confluence scour display a scale invariance over many different types and sizes of braided river.

2 Sandy braided rivers display a degree of invariance in the occurrence of their subsurface deposits in that trough-cross stratification associated with dunes is the dominant feature of all three rivers detailed herein. Since dune height, and thus the thickness of dune cross-sets, scales with flow depth, bigger rivers have larger sets of dune cross-stratification.

3 The occurrence of high-angle planar cross-stratification and low-angle stratification is variable both between rivers and between bars within the same river. The relative presence of these two facies within the stratigraphy is related to a wide range of factors that may include the discharge regime, local bar and channel topography, anabranch width : depth ratio and the abundance of vegetation.

4 The complexity of bar evolution over a range of braided rivers negates derivation of a single, universal facies model for sandy braided alluvium. More data are required to test whether different scales of braided river, with broadly the same range and frequency of bar types, produce the same distribution of facies in the subsurface.

The traditional classification of alluvial channels and their deposits by channel pattern may be inappropriate and classification may be better served by a fuller consideration of bar dynamics, and likely preservation, both within and between different rivers. One fruitful approach may lie in the application of modelling using bed-height distributions (e.g. Paola & Borgman, 1991; Bridge & Best, 1997; LeClair & Bridge, 2001) if the full three-dimensional topography of braided rivers can be quantified repeatedly using remote sensing techniques. This will allow both direct comparison to three-dimensional GPR datasets and allow testing of predictive modelling of subsurface fluvial architecture. Such future analysis holds many important implications for the interpretation of ancient outcrops, subsurface well analysis and quantitative reservoir modelling.

ACKNOWLEDGEMENTS

PJA and JLB gratefully acknowledge the support of BP Exploration for provision of summer and PhD studentships to investigate the scaling of braid bar geometry and Delft Hydraulics and the Danish Hydraulics Institute as main contractors of the Flood Action Plan 24 (Bangladesh) for logistical and financial support for the Jamuna River GPR surveys. The Jamuna River GPR study was also supported by NERC grant GR9/02034 to JLB, PJA and Charlie Bristow. The South Saskatchewan GPR surveys were supported by NERC grant GR9/04273 to PJA, JLB and GSS. Many people have helped produce some of the data presented here including Dave Ashley, John Couperthwaite, Lucy Ellis, Angus Jackson, Gerrit Klaassen, Tavi Murray, Jeff Peakall, Julie Roden, Derald Smith and Ruth Stephens. We are also grateful to Oscar Orfeo for permission to use his data concerning the confluence of the Parana and Paraguay rivers. We thank John Bridge and an anonymous referee for their challenging and thought-provoking reviews.

REFERENCES

Ashley, G.M. (1990) Classification of large-scale subaqueous bedforms: a new look at an old problem. *J. Sediment. Petrol.*, **60**, 160–172.

Ashmore, P.E. and Parker, G. (1983) Confluence scour in coarse braided streams. *Water Resour. Res.*, **19**, 392–402.

Ashworth, P.J., Best, J.L., Leddy, J.O. and Geehan, G. (1994) The physical modelling of braided rivers and deposition of fine-grained sediment In: *Process Models and Theoretical Geomorphology* (Ed. M.J. Kirkby), pp. 115–139. Wiley, Chichester.

Ashworth, P.J., Best, J.L., Roden, J., Bristow, C.S. and Klaassen, G.J. (2000) Morphological evolution and dynamics of a large, sand braid-bar, Jamuna River, Bangladesh. *Sedimentology*, **47**, 533–555.

Best, J.L. (1985) Flow and sediment transport at river channel confluences. Unpublished PhD thesis, University of London.

Best, J.L. (1988) Sediment transport and bed morphology at river channel confluences. *Sedimentology*, **35**, 481–498.

Best, J.L. and Ashworth, P.J. (1997) Scour in large braided rivers and the recognition of sequence stratigraphic boundaries. *Nature*, **387**, 275–277.

Best, J.L., Ashworth, P.J., Bristow, C.S. and Roden, J. (2003) Three-dimensional sedimentary architecture of a large, mid-channel sand braid bar, Jamuna River, Bangladesh. *J. Sediment. Res.*, **73**, 516–530.

Blodgett, R.H. and Stanley, K.O. (1980) Stratification, bedforms and discharge relations of the Platte braided river system, Nebraska. *J. Sediment. Petrol.*, **50**, 139–148.

Bridge, J.S. (1985) Paleochannels inferred from alluvial deposits: a critical evaluation, *J. Sediment. Petrol.*, **55**, 579–589.

Bridge, J.S. (1993a) The interaction between channel geometry, water flow, sediment transport and deposition in braided rivers. In: *Braided Rivers* (Eds J.L. Best and C.S. Bristow). *Geol. Soc. Lond. Spec. Publ.*, **75**, 13–71.

Bridge, J.S. (1993b) Description and interpretation of fluvial deposits: a critical perspective. *Sedimentology*, **40**, 801–810.

Bridge, J.S. and Best, J.L. (1997) Preservation of planar laminae due to migration of low relief bed waves over aggrading upper stage plane beds: comparison of experimental data with theory. *Sedimentology*, **44**, 253–262.

Bridge, J.S. and Gabel, S.L. (1992) Flow and sediment dynamics in a low sinuosity, braided river: Calamus River, Nebraska Sandhills. *Sedimentology*, **39**, 125–142.

Bridge, J.S. and Tye, R.S. (2000) Interpreting the dimensions of ancient fluvial channel bars, channels, and channel belts from wireline-logs and core. *Bull. Am. Assoc. Petrol. Geol.*, **84**, 1205–1228.

Bridge, J.S., Smith, N.D., Trent, F., Gabel, S.L. and Bernstein, P. (1986) Sedimentology and morphology of a low sinuosity river: Calamus River, Nebraska Sand Hills. *Sedimentology*, **33**, 851–870.

Bridge, J.S., Collier, R. and Alexander, J. (1998) Large-scale structure of Calamus deposits (Nebraska, U.S.A.) revealed using ground-penetrating radar. *Sedimentology*, **45**, 977–986.

Bristow, C.S. (1987) Brahmaputra River: channel migration and deposition. In: *Recent Developments in Sedimentology* (Eds F.G. Ethridge, R.M. Flores and M.D. Harvey). *Soc. Econ. Paleontol. Mineral. Spec. Publ.*, **39**, 63–74.

Bristow, C.S. (1993) Sedimentary structures in bar tops in the Brahmaputra River, Bangladesh. In: *Braided Rivers* (Eds J.L. Best and C.S. Bristow). *Geol. Soc. Lond. Spec. Publ.*, **75**, 277–289.

Bristow, C.S. and Best, J.L. (1993) Braided rivers: perspectives and problems. In: *Braided Rivers* (Eds

J.L. Best and C.S. Bristow). *Geol. Soc. Lond. Spec. Publ.*, **75**, 1–11.

Cant, D.J. (1978a) Development of a facies model for sandy braided river sedimentation: comparison of the South Saskatchewan River and the Battery Point Formation. In: *Fluvial Sedimentology* (Ed. A.D. Miall). *Can. Soc. Petrol. Geol. Mem.*, **5**, 627–639.

Cant, D.J. (1978b) Bedforms and bar types in the South Saskatchewan River. *J. Sediment. Petrol.*, **48**, 1321–1330.

Cant, D.J. and Walker, R.G. (1978) Fluvial processes and facies sequences in the sandy braided South Saskatchewan River, Canada. *Sedimentology*, **25**, 625–648.

Coleman, J.M. (1969) Brahmaputra River: channel processes and sedimentation. *Sediment. Geol.*, **3**, 129–239.

Conaghan, P.J. and Jones, J.G. (1975) The Hawkesbury Sandstone and the Brahmaputra: a depositional model for continental sheet sandstones. *J. Geol. Soc. Austral.*, **22**, 275–283.

Crowley, K.D. (1983) Large-scale bed configurations (macroforms), Platte River Basin, Colorado and Nebraska: primary structures and formative processes. *Geol. Soc. Am. Bull.*, **94**, 117–133.

Delft Hydraulics and DHI (1996) *FAP24 River Survey Project, Final Report, Main Volume.* Prepared for Flood Plan Coordination Organisation, Dhaka, Bangladesh, 280 pp.

Foufoula-Georgiou, E. and Sapozhnikov, V.B. (1998) Anisotropic scaling in braided rivers: an integrated theoretical framework and results from application to an experimental river. *Water Resour. Res.*, **34**, 863–867.

Foufoula-Georgiou, E. and Sapozhnikov, V. (2001) Scale invariances in the morphology and evolution of braided rivers. *Math. Geol.*, **33**, 273–291.

Gabel, S.L. (1993). Geometry and kinematics of dunes during steady and unsteady flows in the Calamus River, Nebraska, USA. *Sedimentology*, **40**, 237–269.

Gardiner, S., Thomas, D.V., Bowering, E.D. and McMinn, L.S. (1990) A braided fluvial reservoir, Peco field, Alberta, Canada. In: *Sandstone Petroleum Reservoirs* (Eds J.H. Barwis, J.G. McPherson and J.R.J. Studlick), pp. 7–29. Springer-Verlag, New York.

Holbrook, J. (2001) Origin, genetic interrelationships, and stratigraphy over the continuum of fluvial channel form bounding surfaces: an illustration from middle Cretaceous strata, southeastern Colorado. *Sediment. Geol.*, **144**, 179–222.

Jackson, R.G. (1975) Hierarchical attributes and a unifying model of bed forms composed of cohesionless material and produced by shearing flow. *Bull. Geol. Soc. Am*, **86**, 1523–1533.

Jackson, R.G. (1976) Sedimentological and fluid-dynamic implications of the turbulence bursting phenomenon in geophysical flows. *J. Fluid Mech.*, **77**, 531–560.

Komar, P.D. (1984) The Lemniscate Loop—comparisons with the shapes of streamlined landforms. *J. Geol.*, **92**, 133–145.

Klaassen, G.J. and Vermeer, K. (1988) Confluence scour in a large braided river with fine bed material. In: *Proceedings International Conference on Fluvial Hydraulics*, Budapest, Hungary, 395–408.

Klaassen, G.J., Vermeer, K. and Uddin, N. (1988) Sedimentological processes in the Jamuna (Lower Brahmaputra) river, Bangladesh. *Proceedings of the International Conference on Fluvial Hydraulics*, Budapest, 381–394.

LeClair, S.F. and Bridge, J.S. (2001) Quantitative interpretation of sedimentary structures formed by river dunes. *J. Sediment. Res.*, **71**, 713–716.

McLelland, S.J., Ashworth, P.J. and Best, J.L. (1996) The origin and downstream development of coherent flow structures at channel junctions. In: *Coherent Flow Structures in Open Channels* (Eds P.J. Ashworth, S.J. Bennett, J.L. Best and S.J. McLelland), pp. 459–490. Wiley, Chichester.

McLelland, S.J., Ashworth, P.J., Best, J.L., Roden, J. and Klaassen, G.J. (1999) Flow structure and spatial distribution of suspended sediment around an evolving braid bar, Jamuna River, Bangladesh. In: *Fluvial Sedimentology VI* (Eds N.D. Smith and J. Rogers). *Spec. Publs Int. Assoc. Sedimentol.*, **28**, 43–57.

Miall, A.D. (1978) Lithofacies types and vertical profile models in braided river deposits: a summary. In: *Fluvial Sedimentology* (Ed. A.D. Miall). *Can. Soc. Petrol. Geol. Mem.*, **5**, 597–604.

Miall, A.D. (1988a) Facies architecture in clastic sedimentary basins. In: *New Perspectives in Basin Analysis* (Eds K.L. Kleinspehn and C. Paola), pp. 67–81. Springer-Verlag, Berlin.

Miall, A.D. (1988b) Architectural elements and bounding surfaces in fluvial deposits: anatomy of the Kayenta Formation (Lower Jurassic), southwest Colorado. *Sediment. Geol.*, **55**, 233–262.

Miall, A.D. (1996) *The Geology of Fluvial Deposits.* Springer-Verlag, New York, 582 pp.

Moreton, D.M., Ashworth, P.J. and Best, J.L. (2002) The physical scale modelling of braided alluvial architecture and estimation of subsurface permeability. *Basin Res.*, **14**, 1–21.

Mosley, M.P. (1975) An experimental study of channel confluences. Unpublished PhD thesis, Colorado State University.

Mosley, M.P. (1976) An experimental study of channel confluences. *J. Geol.*, **84**, 535–562.

Mosley, M.P. (1982) Scour depths in branch channel confluences, Ohau River, Otago, New Zealand. *Trans. NZ Inst. Prof. Eng.*, **9**, 17–24.

Nykanen, D.K., Foufoula-Georgiou, E. and Sapozhnikov, V.B. (1998) Study of spatial scaling in braided river patterns using synthetic aperture radar imagery. *Water Resour. Res.*, **34**, 1795–1807.

Orfeo, O. (1995) Sedimentología del río Paraná en el área de confluencia con el río Paraguay. Unpublished PhD thesis, Facultad de Ciencias Naturales y Museo, Universidad Nacional de La Plata, 290 pp.

Paola, C. (1997) Geomorphology—when streams collide. *Nature*, **387**, 232–233.

Paola, C. and Borgman, L. (1991) Reconstructing random topography from preserved stratification. *Sedimentology*, **38**, 553–565.

Rhoads, B.L. and Sukhodolov, A.N. (2001) Field investigation of three-dimensional flow structure at stream confluences: 1. Thermal mixing and time-averaged velocities. *Water Resour. Res.*, **37**, 2393–2410.

Roden, J.E. (1998) *The sedimentology and dynamics of mega-dunes, Jamuna River, Bangladesh*. Unpublished PhD thesis, Department of Earth Sciences and School of Geography, University of Leeds, 310 pp.

Roy, A.G. and De Serres, B. (1989) Morphologie du lit et dynamique des confluents de cours d'eau. *Bull. Soc. Geogr. Liege*, **25**, 113–127.

Roy, A.G., Roy, R. and Bergeron, N. (1988) Hydraulic geometry and changes in flow velocity at a river confluence with coarse bed material. *Earth Surf. Process. Landf.*, **13**, 583–598.

Rust, B.R. and Jones, B.G. (1987) The Hawkesbury Sandstone south of Sydney, Australia: Triassic analogue for the deposit of a large, braided river. *J. Sediment. Petrol.*, **57**, 222–233.

Salter, T. (1993) Fluvial scour and incision: models for their influence on the development of realistic reservoir geometries. In: *Characterisation of Fluvial and Aeolian Reservoirs* (Eds C.P. North and D.J. Prosser). *Geol. Soc. Lond. Spec. Publ.*, **73**, 33–51.

Sapozhnikov, V.B. and Foufoula-Georgiou, E. (1996) Self-affinity in braided rivers. *Water Resour. Res.*, **33**, 1983–1991.

Sapozhnikov, V.B. and Foufoula-Georgiou, E. (1997) Experimental evidence of dynamic scaling and indications of self-organized criticality in braided rivers. *Water Resour. Res.*, **32**, 1429–1439.

Sapozhnikov, V.B. and Foufoula-Georgiou, E. (1999) Horizontal and vertical self-organization of braided rivers toward a critical state. *Water Resour. Res.*, **35**, 843–851.

Sapozhnikov, V.B., Murray, A.B., Paola, C. and Foufoula-Georgiou, E. (1998) Validation of braided-stream models: spatial state–space plots, self-affine scaling, and island shapes. *Water Resour. Res.*, **34**, 2353–2364.

Sarker, M.H. (1996) Morphological processes in the Jamuna River. Unpublished MSc thesis, International Institute for Hydraulic and Environmental Engineering, Delft, 175 pp.

Smith, N.D. (1972) Some sedimentological aspects of planar cross-stratification in a sandy braided river. *J. Sediment. Petrol.*, **42**, 624–634.

Stölum, H-H. (1996) River meandering as a self-organization process. *Science*, **271**, 1710–1713.

Stölum, H-H. (1998) Planform geometry and dynamics of meandering rivers. *Geol. Soc. Am. Bull.*, **110**, 1485–1498.

Woodward, J., Ashworth, P.J., Best, J.L., Sambrook Smith, G.H. and Simpson, C.J. (2003). The use and application of GPR in sandy fluvial environments: methodological considerations. In: *Ground Penetrating Radar in Sediments* (Eds C.S. Bristow and H.M. Jol). *Geol. Soc. Lond. Spec. Publ.*, **211**, 127–142.

Yalin, M.S. (1992) *River Mechanics*. Pergamon Press, Oxford, 219 pp.

Application of laser diffraction grain-size analysis to reveal depositional processes in tidally influenced systems

PETRI SIIRO*,[1], MATTI E. RÄSÄNEN*, MURRAY K. GINGRAS†, CHAD R. HARRIS‡,[2], GEORGE IRION§, S. GEORGE PEMBERTON‡ and ALCEU RANZI¶

*Department of Geology, University of Turku, 20014 Turku, Finland;
†Department of Geology, University of New Brunswick, Fredericton, NB, E3B 5A3, Canada;
‡Department of Earth and Atmospheric Sciences, University of Alberta, Edmonton, T6G 2E3, Canada;
§Forschungsinstitut Senckenberg, Schleusenstrasse 39a, D-263B2, Wilhelmshaven, Germany; and
¶Laboratório de Pesquisas Paleontológicas, Departamento de Ciências da Natureza, Universidade Federal do Acre, Rio Branco, AC, CEP 69915-900, Brasil

ABSTRACT

The differentiation of depositional environments using the grain-size distribution patterns of estuarine channel sediments is investigated in two parallel depositional environments of different ages: the Cretaceous McMurray Formation in the Western Canadian Sedimentary Basin; and the Miocene Pebas Formation in the Amazonia foreland basin. Channel sediments from fluvial-dominated coastal plain/inner estuary settings are compared with marine-influenced middle estuarine channel deposits. The upward decreasing grain-size trend and simultaneous change to more poorly sorted grain-size distributions of the coastal plain estuary depositional sequence indicate the transition from a fluvial-dominated regime to a mixed fluvial and tidal regime. The fluvial-dominated coastal plain/inner estuary sands have clear bimodal grain-size distributions with a minor mode in coarse sand. They are coarser grained, better sorted and more negative skewed than the inclined heterolithic stratified channel sands of the mixed fluvial and tidal regime of the middle estuary. X-ray diffraction results show that decreases in the proportion of kaolinite from fluvial to estuarine sediments and then a further decrease to shoreface sediments combined with a simultaneous increase in the proportions of smectite, illite and chlorite are indications of the change of the depositional environments from riverine to more marine during the ongoing transgression.

INTRODUCTION

The grain-size distribution of sediments is potentially a useful tool for the differentiation of ancient depositional environments, because sediments that settle in different depositional environments may possess distinctive grain-size distributions as a result of changes in erosion, transportation and deposition processes. In the past, grain-size distributions were used extensively as environmental

[1]Present address: Kankilantie 1 B, 36600 Pälkäne, Finland (Email: petri.siiro@ymparisto.fi).

[2]Present address: Canadian Natural Resources Limited, 855 – 2 Street S.W., Calgary, AB, T2P 4J8, Canada.

indicators (e.g. Mason & Folk, 1958; Friedman, 1961, 1979; Visher, 1969; Buller & McManus, 1972; Coldberg, 1980), but the reliability of grain-size data for the description and identification of depositional environments and processes was contested in the 1980s and 1990s (e.g. Sedimentation Seminar, 1981; Gale & Hoare, 1991). The use of grain-size as an environmental discriminator is re-examined here. Modern automatic grain-size analysers give more accurate and repeatable information on the grain-size properties of sediments than traditional mechanical sieving methods, offering potential as a robust analytical tool.

The objective of this study was to characterize and compare the grain-size distribution patterns of sediments from two parallel depositional environments of different ages: Cretaceous and Miocene epicontinental embayment–seaway systems in North and South America, respectively. In the Cretaceous McMurray Formation (Alberta, Canada) the transition from fluvial-dominated coastal plain–inner estuary to marine-dominated outer estuary are scrutinized in vertical sections. This was accomplished using cores and outcrops from the same geographical region. In the Miocene Pebas Formation (Peru and Brazil) the same facies transition was studied in outcrops. Biostratigraphical data were used to identify strata that were deposited at the same time in the different parts of the embayment–estuary system (Hoorn, 1993, 1994a,b; Räsänen et al., 1995).

The textural characteristics of the sediments from the above ancient depositional environments were also compared with modern sediments from the lowermost Amazon River and Amazon mouth. These were deposited in a fluvial-dominated environment with minor tidal influence. This comparison was carried out to ascertain whether the ancient and modern environments show similar patterns in sedimentation, or if they present contrasting data.

GEOLOGICAL SETTING

McMurray Formation

The Cretaceous McMurray Formation in Alberta, Canada was deposited in a large north trending valley. The valley was cut into Devonian carbonates during a low stand of sea level (lowstand system tract, LST) and was filled with McMurray Formation sediments during the subsequent sea-level rise accompanying the late Aptian to early Albian transgression of the Boreal Sea from the north (Flach, 1984; Wightman et al., 1989, 1991). The interplay between rising sea level and northward flowing rivers from the Canadian Shield provided an overall transitional environment from fluvial to estuarine and marine shoreline (Wightman & Pemberton, 1997). The McMurray Formation represents a classic example of estuarine and marginal marine sedimentation in a non-barred coastal environment during the first major transgression of the Boreal Sea into the western Canadian sedimentary basin. Parasequences within the McMurray Formation almost certainly represent eustatic sea-level fluctuations, as the active orogenic belt of western North America and the influence of the major tectonic subsidence of the foreland basin were too far to the west to explain these depositional sequences (Ranger & Pemberton, 1992).

Early studies proposed that the McMurray Formation was deltaic and had a freshwater origin (Carrigy, 1971; Nelson & Glaister, 1978). Carrigy (1959) had previously established an informal three-fold stratigraphy of the McMurray Formation consisting of lower, middle and upper units. Steward & MacCallum (1978) suggested that the McMurray Formation accumulated under estuarine conditions, with a lower fluvial unit, a thick middle estuarine unit, and an upper marginal marine unit. Mattison et al. (1989) confirmed that fluvial channel facies dominate the lower unit, a complex of estuarine channel and tidal flat sediments make up the middle unit, and bioturbated shoreface sands deposited seaward of the main estuary channel complexes represent the upper unit of the formation.

The lower McMurray is preserved only in the deepest valleys above the basal unconformity. Inclined heterolithic stratification (IHS) of tidal point-bars in estuarine channels and a brackish suite of ichnofossils (Pemberton et al., 1982; Ranger & Pemberton, 1988) demonstrate that marine processes strongly influenced the middle and upper McMurray Formation sediments. At the end of McMurray sedimentation, as the Boreal Sea transgressed southward, glauconitic shallow-marine sands of the Wabiskaw Formation were deposited on top of the McMurray Formation. The deeper offshore mudstones of the Clearwater Formation overlie the Wabiskaw Formation (Jeletzky, 1971; Jardine, 1974).

Pebas Formation

The Pebas Formation (Solimões Formation/Group in Brazil) is a kilometre to few hundreds of metres thick, informal lithostratigraphical formation composed of laterally extensive blue smectitic clay beds intercalated with sand and lignite beds. The formation covers about 1×10^6 km^2 of the Amazonian foreland and adjacent intracratonic basins in Peruvian, Brazilian, Colombian and Bolivian Amazonia (Räsänen et al., 1998; Gingras et al., 2002). The Pebas Formation was deposited in the actively subsiding foreland basin. Most of the subsidence was related to the flexure of the South American craton under the stress of the developing Andean thrust belt. The formation has been palynologically dated to late Early Miocene to Late Miocene (17–10 Ma) by Hoorn (1993, 1994a,b). This time interval partly coincides with the Serravallian eustatic sea-level highstand around 14 Ma (Haq et al., 1988). Thus, the formation of the epicontinental embayment seems to be a result of both tectonic and eustatic processes.

The sequence stratigraphy of the Pebas Formation shows repeated prograding shoreface parasequences in a marginal marine environment (Gingras et al., 2002). During the Miocene, western Amazonia was influenced by several marine incursions. These parasequences have lithofacies similar to the parasequences described from the McMurray Formation. However, they are generally thinner and shorefaces are less commonly truncated by channels (Räsänen et al., 1995, 1998). Although alternative interpretations for the origin of the Pebas/Solimões Formation have been published (Hoorn, 1994a; Wesselingh et al., 2002), new sedimentological, stratigraphical and palaeoichnological data (Räsänen et al., 1995, 1998; Gingras et al., 2000, 2002) support a marginal marine interpretation for the Pebas Formation in Peru.

The marginal marine conditions during the Miocene in the Amazonian foreland basin were probably caused by marine incursions from the Caribbean Sea to the north (Hoorn, 1993; Hoorn et al., 1995). Other possible marine connections include: (i) the west Pacific gateway; (ii) the Río de la Plata area; and (iii) from the east through the lower Amazon valley (von Ihering, 1927; Nuttal, 1990; Bolthovskoy, 1991). The sediments infilling the subsiding foreland basin were mainly supplied from the Guyana shield during the Early Miocene. Between the Early and Middle Miocene, provenance changed, with sediment becoming primarily of Andean origin (Hoorn, 1993).

Lowermost Amazon River and Amazon mouth

The modern Amazon River mouth is an extensive plain between the Amazon River and Amazon continental shelf. High river discharge and oceanographic processes in the Atlantic Ocean influence sedimentation in the Amazon mouth. The discharge of the Amazon River is so large that seawater does not enter the river mouth, despite the tidal effect, and most of the estuarine circulation occurs on the continental shelf (Gibbs, 1970; Geyer et al., 1991). About 85–95% of the suspended sediment discharged from the Amazon River comprises silt- and clay-size particles with a mean diameter of approximately 4 µm (Gibbs, 1967). Sedimentological variation in Amazon River mouth sediments is caused by the interaction of estuarine circulation, high suspended-sediment concentrations and tidal energy.

METHODS

Field studies

McMurray Formation deposits were studied in outcrop along the Ells River in Alberta, Canada (Figs 1 & 2). The outcrop represents typical McMurray Formation sediments and is composed of channel and IHS channel-fill deposits. In addition, two McMurray Formation cores were studied at the Alberta Energy and Utilities Board (EUB) Core Laboratory. One core (McMurray Core 1: 11-3-96-11w4) consisted mainly of estuarine channel deposits; in the other core (McMurray Core 3: 7-26-76-12w4) shoreface deposits overlie estuarine channel deposits. The outcrop and the cores were logged in detail to define lithological units, and record sedimentary structures and amounts of

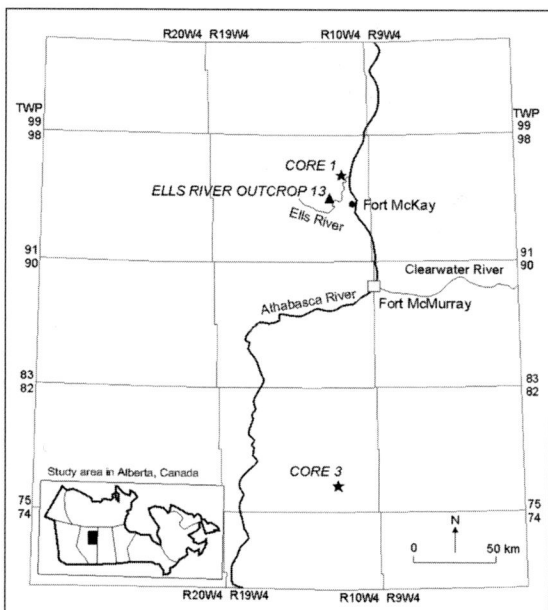

Fig. 1 Location map of the McMurray Formation cores and Ells River outcrop. (Core 1 = 11-3-96-11w4; Core 3 = 7-26-76-12w4).

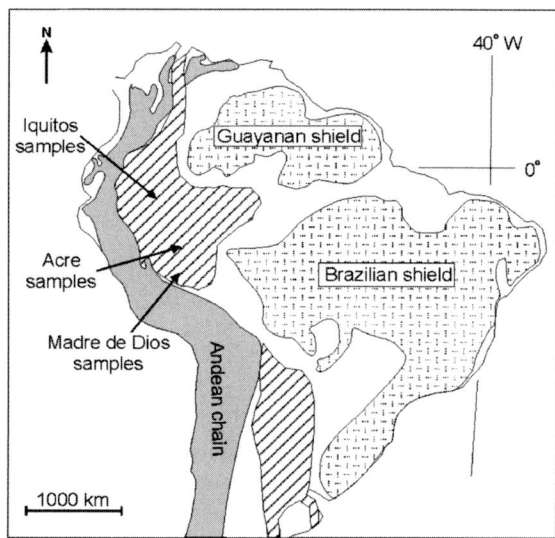

Fig. 3 Map of South America showing locations where Miocene IHS channel samples were collected in western Amazonia. The oblique shading shows the approximate maximum area of the Late Miocene Pebasian marine incursion, and the documented limits of the roughly contemporaneous Paranan Sea in the south. Modified from Räsänen et al. (1995) and Marshall & Lundberg (1996).

Fig. 2 Ells River outcrop 13 on the Ells River. The total height of exposure shown in the photograph is approximately 70 m.

bioturbation. Both the outcrop and cores were sampled for grain-size analysis.

Pebas Formation sediment samples were collected from outcrops along the Amazon River and its tributaries in Peru and Brazil (Fig. 3). The IHS sand–mud samples were collected from the Iquitos and Acre regions. In addition, coastal plain fluvial or inner estuary fluvial-dominated channel sands of the Ipururo Formation were collected from Madre de Dios region. The Ipururo Formation is considered to be laterally equivalent to the Pebas Formation (Kummel, 1948).

To relate grain-size distribution data to recent current measurements in a modern environment, a small set of channel sands and muds from the lowermost Amazon and Amazon mouth were analysed (Fig. 4). These samples were collected by Professor Stattegger and Dr Vital of the Geologisch-Paläontologisches Institut of Kiel University (Vital, 1996).

Laboratory methods

Sand samples from the Ells River outcrop and from the two McMurray cores were bitumen stained. The bitumen was removed before carrying out the grain-size analysis by extracting the samples with cold toluene. The samples were first

Grain-size analysis and depositional processes

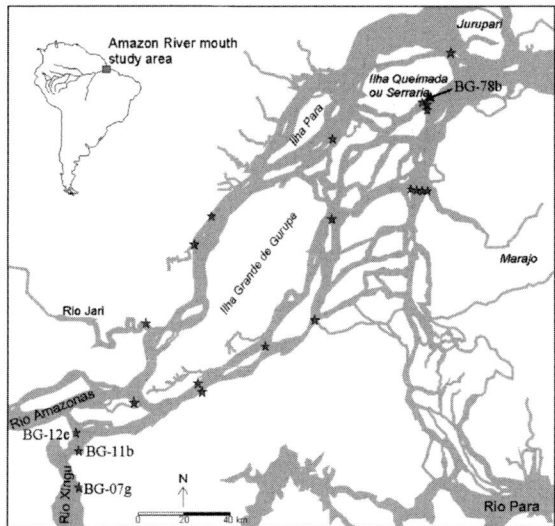

Fig. 4 Map of the lowermost Amazon and Amazon mouth with sample locations (stars).

extracted with toluene and centrifuged (30 min at 2500 r.p.m.). The oily toluene was then removed using a pipette. The procedure was repeated 5–10 times depending on the amount of the bitumen in the samples. Prior to grain-size analysis, all samples were first treated with 10% hydrogen peroxide to remove organic matter, and with 1 M HCl to dissolve carbonates. Samples were then disaggregated using 0.05 M sodium pyrophosphate with ultrasonification prior to analysis. Grain-size determinations were performed with a laser diffraction grain-size analyser (Coulter LS-200). Sediments were classified and named according to the particle-size classification of Folk (1954). Statistical variables—mean grain-size, inclusive graphic standard deviation (sorting) and skewness—were calculated from grain-size distributions using the formulations of Folk & Ward (1957). Median grain-size and mode were measured from grain-size distribution curves.

Clay mineralogical analyses were performed at the Senckenberg Institut in Wilhelmshaven, Germany, following the method of Räsänen et al. (1998). The < 2 μm fractions were separated using settling tubes (Attenberg Cylinder). The mineral compositions of the Mg^{2+}, K^+ and ethylene glycol treated clays were determined by X-ray diffractometry (XRD).

RESULTS AND INTERPRETATION

McMurray Formation

Sedimentary facies, environments and grain-size distributions of Core 3

An interval of 46.3 m of Core 3 (7-26-76-12w4) was logged and sampled (Fig. 5). The lower 23.7 m is composed of four upward-fining units of IHS

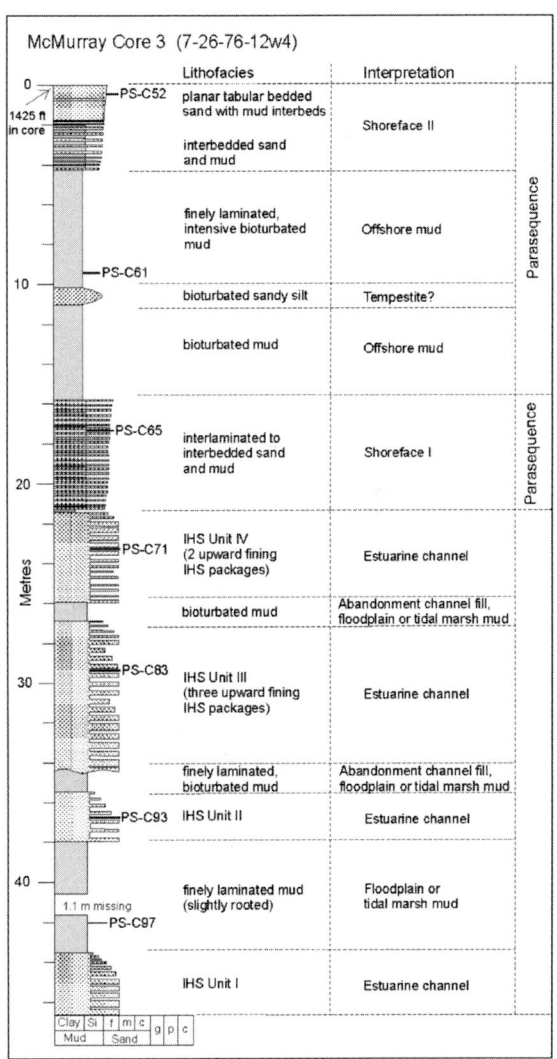

Fig. 5 Measured section of McMurray Formation Core 3 (7-26-76-12w4). Samples for clay mineral analysis are indicated (PS-C52 to PS-C97).

Table 1 Average grain-size parameters for sediment samples of the McMurray and Pebas/Ipururo formations.

Core/outcrop	Lithology	Mean (φ)	Mode (φ)	Sorting SD (φ)
Core 3	Shoreface II sands	5.0	3.2	2.2
	Shoreface I sands	4.0	2.8	2.0
	IHS unit IV sands	4.1	3.1	1.5
	IHS unit III sands	3.8	3.0	1.5
	IHS unit II sands	3.4	3.1	1.1
	IHS unit I sands	3.3	2.8	1.3
	Shoreface II muds	7.3	6.7	2.0
	Offshore muds	7.9	7.1	1.7
	Shoreface I muds	7.2	7.0	2.3
	IHS unit IV muds	6.7	5.5	1.8
	IHS unit III muds	6.1	3.0	2.5
Core 1	IHS unit II sands	3.2	2.8	1.0
	Unit I sands	2.2	2.7	0.7
	IHS unit II muds	6.4	6.7	2.2
	Unit I muds	6.9	7.0	2.4
Ells River outcrop 13	IHS units IV–VI sands	3.4	2.7	1.3
	IHS unit III sands	3.7	3.0	1.4
	IHS unit II sands	3.0	2.7	0.7
	IHS unit I sands	2.7	2.6	0.4
Pebas/Ipururo Formation	Iquitos IHS sands	4.2	3.3	1.8
	Acre IHS sands	4.2	3.1	1.9
	Madre de Dios sands	1.8	1.7	1.4
	Late Miocene IHS sands	4.4	3.1	2.0
	Middle Miocene IHS sands	3.1	3.0	0.8

SD, standard deviation; IHS, inclined heterolithic sand.

sands and muds interpreted to represent deposits of individual estuarine channels. These IHS successions are capped by bioturbated and laminated muds interpreted to be abandonment channel-fill, floodplain or tidal-marsh muds. The upper 21.6 m of Core 3 consists of two separate units of horizontally interlaminated to interbedded upward-coarsening sands and muds interpreted to represent prograding shoreface parasequences. These two shoreface parasequences are separated by > 10 m of thick, intensively bioturbated mud, interpreted to represent a maximum flooding offshore deposit.

The average grain-size distributions of the IHS sands in Core 3 grade upwards from unimodal into slightly bimodal from the lower to the upper IHS channel unit (I–IV, Figs 5 & 6). Through the same interval, the average mode and mean grain-size trends are upward fining (Table 1). Within each IHS unit and each intrachannel package the overall grain-size trends are also upward fining (Table 2). The average grain-sizes of the shoreface sands grade upwards into more bimodal and finer-grained distributions from unit I to unit II (Fig. 6 & Table 1). Within both shoreface sand units the grain-size trends coarsen upwards (Table 2).

The average grain-size distributions of the IHS muds in Core 3 grade upward from polymodal to unimodal from unit III to unit IV (Fig. 6), and the average mode and mean grain-size trends are upward-fining (Table 1). The average grain-size distributions of the overlying shoreface muds are polymodal (Fig. 6), and the average mode coarsens upwards from unit I to unit II, although the average mean grain-size trend is slightly upward-fining (Table 1). Between the shoreface units is an offshore mud layer, which in contrast to the shoreface muds (Fig. 6) is bimodal, better sorted and finer grained (Table 1).

Table 2 Grain-size parameters of McMurray Formation Cores 3 and 1 sand samples.

Core	Unit	Sample	Material	Mean (φ)	Median (φ)	Mode (φ)	Sorting SD (φ)	Skew (φ)	Sand/silt/clay (%)
Core 3	Shore-face unit II	PS-C52	Silty sand	4.63	3.61	3.24	2.09	0.73	60.7/27.2/12.1
		PS-C53	Sandy silt	5.18	4.21	3.37	2.25	0.61	45.8/37.2/17.0
		PS-C55	Sandy silt	5.10	4.35	3.24	2.14	0.53	41.3/44.4/14.3
		PS-C57	Sandy silt	5.16	4.44	3.37	2.14	0.51	40.5/44.9/14.6
	Shore-face unit I	PS-C64	Muddy sand	4.51	3.35	3.10	2.15	0.79	66.9/20.9/12.2
		PS-C67	Sand	2.68	2.64	2.57	0.94	0.47	92.3/3.9/3.8
		PS-C68	Muddy sand	4.01	2.95	2.70	2.02	0.82	77.3/12.4/10.3
		PS-C69	Muddy sand	4.37	3.07	2.84	2.30	0.81	71.4/15.3/13.3
	IHS unit IV	PS-C72 (PII)	Sand	3.26	3.22	3.10	0.53	0.32	90.5/7.5/2.0
		PS-C74 (PII)	Silty sand	3.22	3.17	3.10	0.62	0.40	88.9/8.9/2.2
		PS-C75 (PI)	Silt	5.49	5.02	4.58	1.61	0.51	8.8/79.7/11.5
		PS-C77 (PI)	Sandy silt	4.69	4.39	4.05	1.39	0.51	29.6/63.0/7.4
	IHS unit III	PS-C79 (PIII)	Silt	5.99	5.81	5.53	1.28	0.37	1.5/87.9/10.6
		PS-C80 (PIII)	Sand	3.09	3.05	2.97	0.56	0.33	92.3/5.7/2.0
		PS-C81 (PII)	Silty sand	3.95	3.75	3.51	1.28	0.46	61.2/33.9/4.9
		PS-C82 (PII)	Silty sand	2.87	2.67	2.43	1.12	0.61	85.3/11.2/3.5
		PS-C84 (PII)	Silty sand	3.20	3.13	2.97	0.87	0.45	87.1/10.1/2.9
		PS-C85 (PI)	Sandy silt	4.59	4.13	3.78	1.55	0.58	43.1/49.0/7.9
		PS-C87 (PI)	Sand	2.76	2.74	2.70	0.43	0.15	97.1/2.1/0.8
		PS-C88 (PI)	Silty sand	4.03	3.35	2.97	1.70	0.69	69.3/23.7/7.0
		PS-C90 (PI)	Sand	3.11	3.07	2.97	0.53	0.30	92.5/5.7/1.8
	IHS unit II	PS-C92	Silty sand	3.26	3.21	3.10	0.66	0.40	89.6/7.9/2.6
		PS-C94	Silty sand	3.38	3.23	3.10	1.06	0.53	82.0/14.1/3.9
		PS-C95	Silty sand	3.80	3.26	2.97	1.53	0.67	72.1/22.3/5.6
	IHS unit I	PS-C99	Sandy silt	5.46	4.69	3.91	2.10	0.54	29.0/54.3/16.7
		PS-C100	Sand	3.01	2.98	2.84	0.49	0.26	93.8/4.4/1.8
		PS-C101	Sand	2.55	2.51	2.43	0.44	0.22	97.4/2.0/0.6
		PS-C102	Sand	3.05	2.99	2.84	0.59	0.33	91.5/6.6/2.0
		PS-C103	Sand	3.14	3.09	2.97	0.54	0.30	91.8/6.3/1.9
Core 1	IHS unit II	PS-C27	Silty sand	3.19	3.09	2.97	0.90	0.50	87.3/9.6/3.1
		PS-C25	Sand	3.19	3.14	3.10	0.50	0.30	91.7/6.3/2.0
		PS-C23	Sandy silt	3.17	2.83	2.70	1.26	0.70	83.1/12.3/4.6
		PS-C22	Silty sand	3.35	3.22	2.97	0.93	0.48	81.8/15.2/3.0
	Planar tabular inter-bedded sand and mud unit I	PS-C15	Sand	2.41	2.33	2.16	0.91	0.41	92.7/4.6/2.7
		PS-C13	Sand	2.39	2.33	2.16	0.70	0.22	94.6/3.7/1.7
		PS-C11	Sand	2.49	2.42	2.30	0.67	0.38	93.7/4.3/2.0
		PS-C9	Sand	2.12	2.11	2.03	0.51	0.03	98.3/1.2/0.5
		PS-C7	Sand	2.00	2.00	2.03	0.56	−0.03	98.6/1.0/0.4
		PS-C5	Sand	1.95	1.97	2.03	0.63	−0.12	98.0/1.2/0.8

SD, standard deviation; IHS, inclined heterolithic sand; PI–PIII, depositional packages.

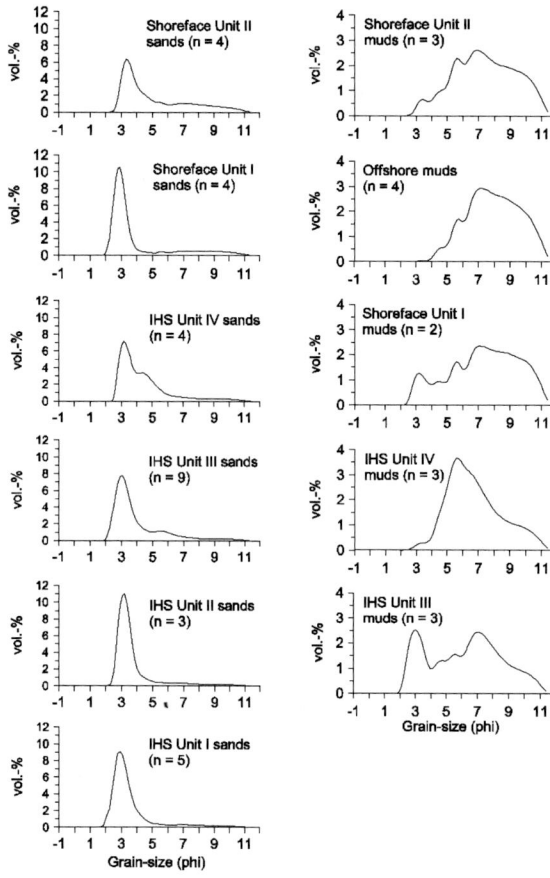

Fig. 6 Average grain-size distribution curves of McMurray Formation Core 3 sediments: IHS channel sand (units I–IV), IHS muds (units III and IV), and the overlying shoreface sands and muds (units I and II), and offshore muds are shown. See Fig. 5 for unit divisions. n = number of samples analysed.

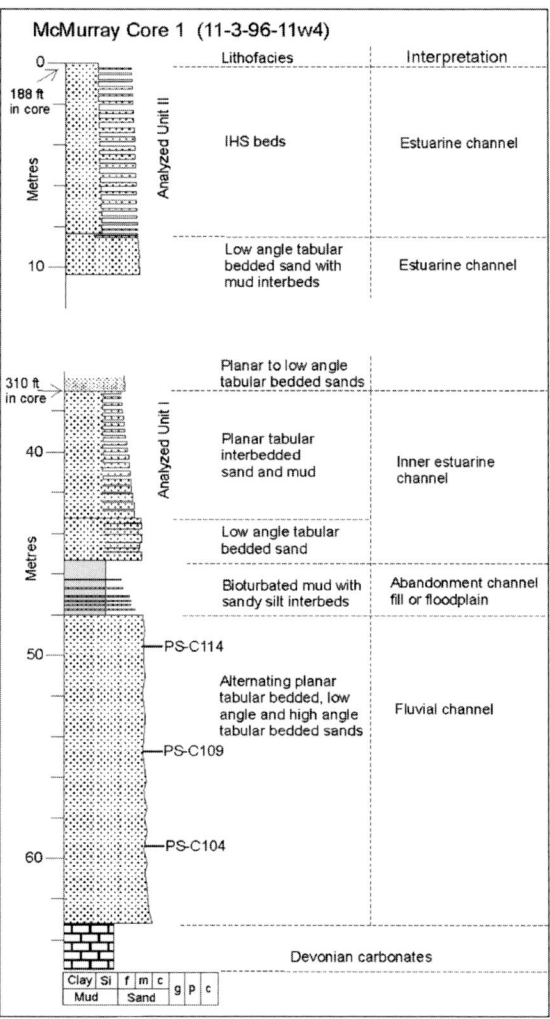

Fig. 7 Measured sections of McMurray Formation Core 1 (11-3-96-11w4). Samples for clay mineral analysis are indicated (PS-C104 to PS-C114).

Sedimentary facies, environments and grain-size distributions of Core 1

Three intervals of Core 1 (11-3-96-11w4) were sampled (Fig. 7). The base of the core comprises Devonian carbonates that are unconformably overlain by over 15 m of alternating planar, low-angle and high-angle tabular bedded sands. The sands are interpreted to represent the fluvial unit of the lower McMurray Formation. Samples from the lower core included grains over $-1.0\,\phi$ (2000 μm) diameter; these were too coarse for analysis with the Coulter LS-200 and the collected samples were too small for sieving. The sands are overlain by a bioturbated mud bed with sandy silt interbeds that are ultimately covered with low-angle tabular bedded sands. The lowest interval analysed (unit I) consists of interbedded sands and muds. The mud beds of this unit thicken upwards and the sands fine upwards. Unit I is interpreted to have been deposited in an inner estuary channel. Unit II is interpreted to represent an upward-fining IHS succession, and represents lateral accretion deposits on point bars in estuarine channels.

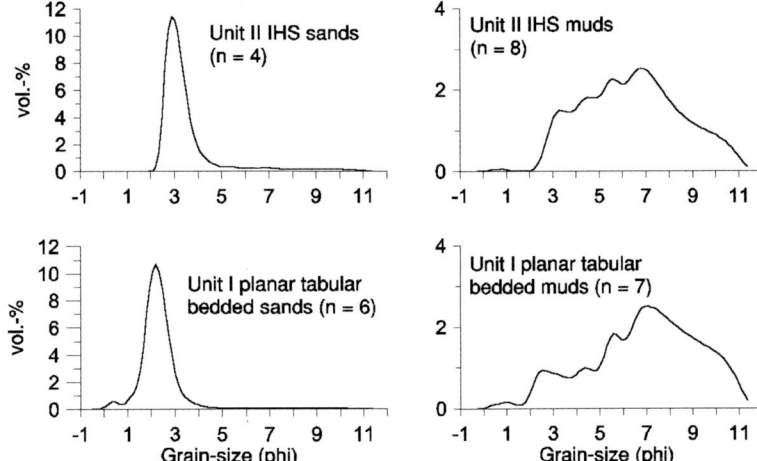

Fig. 8 Average grain-size distribution curves of units I and II sands and muds from McMurray Formation Core 1. See Fig. 7 for unit divisions. n = number of samples analysed.

The average grain-size distributions of the sands in Core 1 grade upward from bimodal into unimodal from planar tabular interbedded Unit I to IHS Unit II (Fig. 8). Unit I has a minor peak around 0.3 ϕ. The average mode and mean grain-size of unit I sands is clearly coarser grained than the average mode and mean grain-size of the IHS sands in unit II (Table 1). The sands of Unit I are negative coarse-skewed to positive fine-skewed (average skewness 0.15 ϕ; Table 2). This is distinctively different from the positive fine-skewed sands in unit II at the same location (average skewness 0.50 ϕ). The average grain-size distributions of the muds in Core 1 are polymodal in both units (Fig. 8). The average mode and mean grain-size coarsens upwards from unit I to unit II (Table 1).

Sedimentary facies, environments and grain-size distributions of the Ells River outcrop

Six IHS units were sampled from a 57 m section of Ells River outcrop 13 (Fig. 9, units I to VI). The IHS beds are interpreted to have accumulated as lateral accretion deposits on point bars in estuarine channels. Tidal marsh muds are normally deposited on top of IHS deposits on point bars of migrating channels in tidally-influenced estuaries (Smith, 1988a). However, at the Ells River, tidal marsh muds are generally absent and they only separate IHS units in the uppermost part of the section.

The average grain-size distributions of the IHS sands in the Ells River outcrop are all unimodal (Fig. 10). The average mode and mean grain-size trends first fine upwards from Unit I to Unit III, but then coarsen slightly in units IV–VI (Table 1). Within the individual IHS units the overall grain-size trend is upward fining (Table 3). In Unit I the mode varies without clear upward fining or upward coarsening.

Pebas/Ipururo formations

Sedimentary facies, environments and grain-size distributions

The Pebas Formation sand samples were collected from IHS channel-fills and from coastal-plain fluvial channel or fluvial-dominated inner-estuary deposits. They come from several separate channels in three regions of Peruvian and Brazilian Amazonia (Fig. 11):

1 coastal plain fluvial channel sands were sampled at three locations near Madre de Dios and represent the most proximal coastal channels;
2 Acre IHS sands originated from three localities 300–400 km north and basinward of Madre de Dios;
3 the Iquitos IHS sands were collected at four sites and represent the most distal deposits in the Pebasian system.
The localities sampled from Madre de Dios, Acre and the south-western part of the Iquitos region

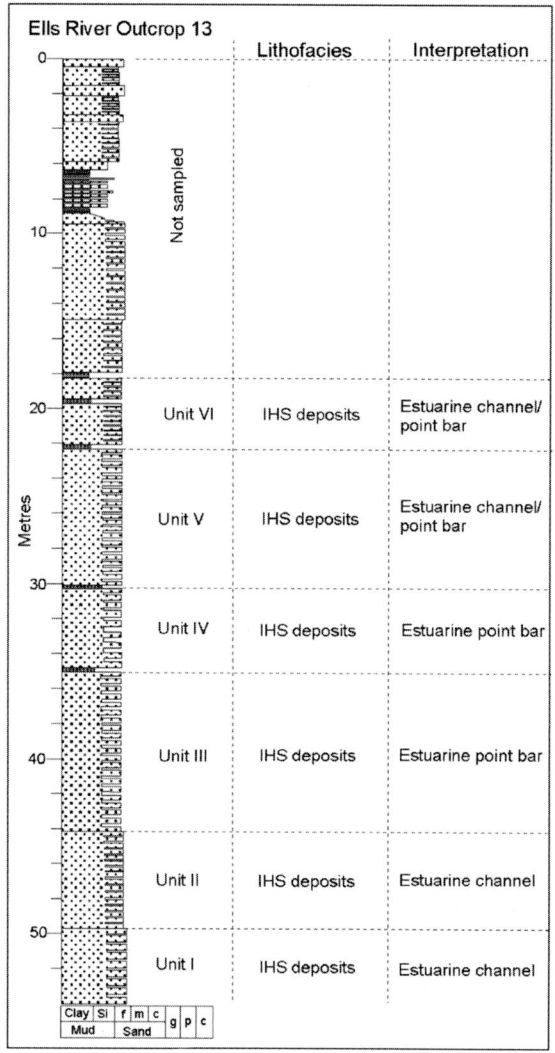

Fig. 9 Measured section of Ells River outcrop 13.

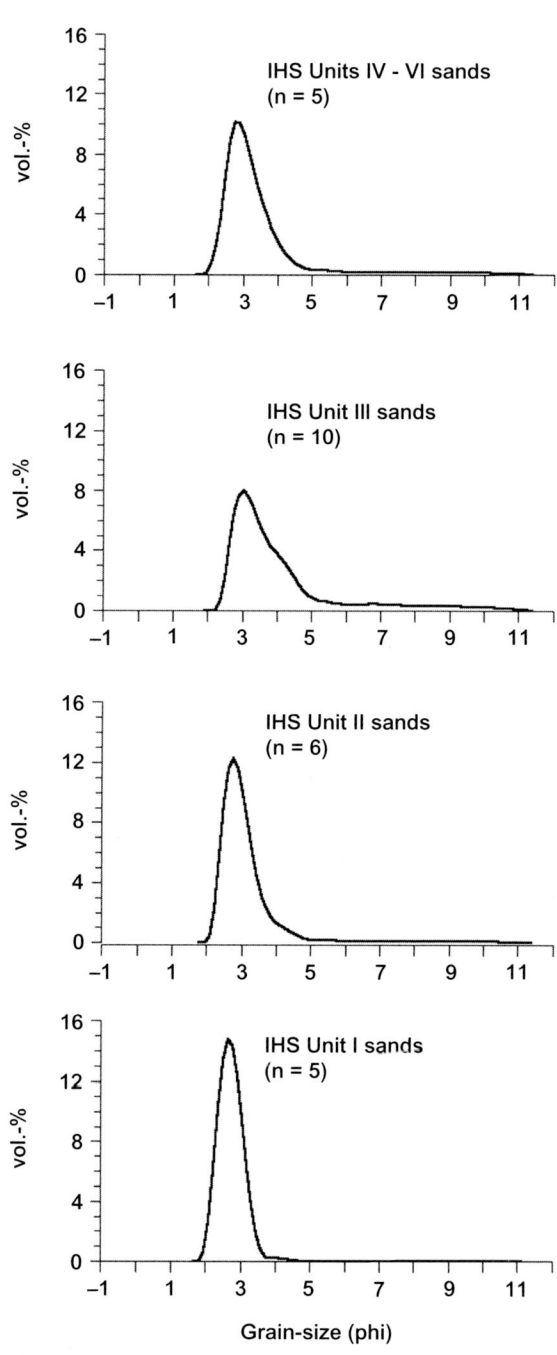

Fig. 10 Average grain-size distribution curves of IHS sands (units I to III and IV–VI) from the Ells River outcrop 13. See Fig. 9 for unit divisions. n = number of samples analysed.

are late Miocene in age and therefore can be thought to roughly represent a palaeogeographical proximal to distal transect in the late Miocene Pebasian embayment. The samples from the north-eastern part of the Iquitos region represent the middle Miocene. The estimated dates are imprecise, however, but the calculated average values of grain-size data are thought to smooth the possible temporal heterogeneity.

The average grain-size distributions of the Madre de Dios coastal plain fluvial channel sands and the IHS channel sands of the Acre and Iquitos

Table 3 Grain-size parameters of Ells River outcrop 13 sand samples.

Unit	Sample	Material	Mean (φ)	Median (φ)	Mode (φ)	Sorting SD (φ)	Skew (φ)	Sand/silt/clay (%)
IHS IV–VI	PS-00-50	Silty sand	3.77	3.61	3.51	1.17	0.48	70.6/24.7/4.7
	PS-00-51	Sand	2.95	2.93	2.84	0.40	0.19	95.9/2.8/1.3
	PS-00-52	Silty sand	3.19	3.07	2.84	0.80	0.41	85.3/12.4/2.3
	PS-00-54	Sand	2.58	2.56	2.57	0.35	0.15	97.5/1.6/0.9
	PS-00-56	Silty sand	3.22	3.07	2.84	0.99	0.54	84.9/11.8/3.3
IHS III	PS-00-59	Silty sand	3.84	3.59	3.24	1.35	0.51	66.4/28.1/5.5
	PS-00-61	Silty sand	3.68	3.36	2.97	1.37	0.58	71.5/23.3/5.2
	PS-00-63	Sandy silt	4.57	4.04	3.91	1.66	0.59	47.3/44.3/8.4
	PS-00-65	Sandy silt	5.00	4.38	4.18	1.67	0.62	28.2/61.9/9.9
	PS-00-66	Silty sand	3.52	3.37	2.97	1.10	0.49	76.1/20.0/3.9
	PS-00-67	Sand	3.01	2.98	2.84	0.42	0.24	94.9/3.6/1.4
	PS-00-68	Sand	2.90	2.87	2.84	0.41	0.21	95.9/2.9/1.2
	PS-00-69	Silty sand	3.49	3.33	2.97	1.07	0.49	77.1/19.3/3.6
	PS-00-70	Silty sand	4.47	3.85	2.97	1.91	0.56	53.9/37.1/9.0
	PS-00-71	Silty sand	3.45	3.27	2.97	1.10	0.53	78.5/17.6/3.9
IHS II	PS-00-72	Sand	2.93	2.91	2.84	0.39	0.18	96.6/2.4/1.0
	PS-00-74	Sand	2.95	2.90	2.84	0.50	0.31	93.4/5.0/1.6
	PS-00-76	Sand	2.84	2.80	2.70	0.45	0.25	95.6/3.2/1.2
	PS-00-77	Sand	2.96	2.87	2.70	0.62	0.35	91.4/6.8/1.8
	PS-00-78	Silty sand	3.10	2.80	2.43	1.16	0.62	81.3/15.4/3.3
	PS-00-79	Silty sand	3.11	2.87	2.70	1.09	0.62	84.1/12.4/3.5
IHS I	PS-00-80	Sand	2.61	2.60	2.57	0.36	0.12	97.7/1.6/0.6
	PS-00-81	Sand	2.79	2.77	2.70	0.35	0.12	97.5/1.8/0.8
	PS-00-82	Sand	2.52	2.51	2.43	0.34	0.08	98.5/1.1/0.4
	PS-00-83	Sand	2.76	2.74	2.70	0.34	0.12	97.8/1.6/0.7
	PS-00-84	Sand	2.63	2.62	2.57	0.36	0.11	98.2/1.3/0.5

SD, standard deviation; IHS, inclined heterolithic sand.

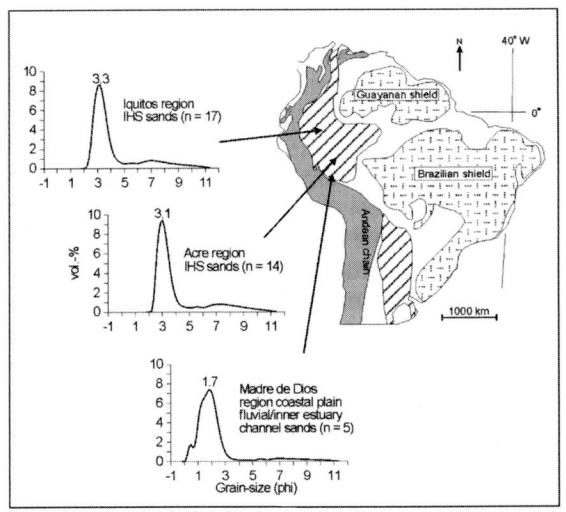

Fig. 11 (*left*) Average grain-size distribution curves of Pebas coastal plain fluvial–inner estuary channel sands from the Madre de Dios region, and IHS channel sands from the Iquitos and Acre regions. n = number of samples analysed.

regions are slightly bimodal. The average mode and mean grain-size fine basinward from Madre de Dios to Acre and Iquitos (Table 1). The Madre de Dios sands have a minor peak at 0.4 φ and the Acre and Iquitos sands have a small peak at the silt–clay boundary; the Madre de Dios sands are also better sorted than IHS sands from Acre and Iquitos (Table 4).

Temporally, the average grain-size distributions of the middle Miocene estuarine IHS channel sands to the late Miocene estuarine IHS channel

Table 4 Grain-size parameters of western Amazonia outcrop sediment samples.

Unit	Sample	Material	Mean (φ)	Median (φ)	Mode (φ)	Sorting SD (φ)	Skew (φ)	Sand/silt/clay (%)
Iquitos IHS sands	PS-99-3	Silty sand	4.43	3.69	3.51	1.69	0.72	65.7/25.9/8.4
	PS-99-5	Silty sand	4.40	3.48	3.24	1.85	0.73	71.1/20.6/8.3
	PS-99-7	Silty sand	4.38	3.45	3.24	1.88	0.73	70.9/20.8/8.3
	PS-99-9	Sand	3.18	3.14	3.10	0.79	0.43	90.9/6.6/2.5
	PS-99-11	Silty sand	3.11	3.05	2.97	0.91	0.49	89.4/7.1/3.5
	PS-99-13	Sand	2.88	2.84	2.70	0.71	0.44	93.1/4.4/2.6
	PS-99-20	Sand	3.19	3.15	3.10	0.70	0.39	91.1/5.9/3.1
	PS-99-22	Sand	2.81	2.74	2.70	0.65	0.45	92.2/5.6/2.2
	PS-99-38	Sand	3.12	3.07	2.97	0.93	0.45	90.0/6.8/3.2
	PS-99-40	Silty sand	3.32	3.21	3.10	1.08	0.50	84.0/11.8/4.2
	PS-99-42	Silty sand	3.79	3.31	3.10	1.51	0.65	75.7/18.2/6.1
	PS-99-50	Silty sand	4.11	3.60	3.24	1.51	0.65	66.2/27.3/6.5
	PS-99-51	Silty sand	4.23	3.23	2.97	1.94	0.79	72.5/18.4/9.1
	PS-99-52	Silty sand	4.67	3.73	3.37	1.89	0.74	61.5/28.2/10.3
	PS-99-53	Silty sand	4.90	3.97	3.24	2.07	0.64	50.3/37.1/12.6
	PS-99-54	Sandy silt	6.30	6.46	6.74	2.12	−0.04	16.4/61.5/22.1
	PS-99-55	Sandy silt	5.82	5.93	3.37	2.17	0.02	28.5/53.8/17.7
	PS-99-77	Silty sand	3.33	3.28	3.24	0.85	0.45	87.9/8.8/3.3
	PS-99-79	Silty sand	3.22	3.17	3.10	0.85	0.46	89.7/7.1/3.3
	PS-99-81	Sand	3.10	3.06	2.97	0.51	0.32	93.1/4.5/2.4
	PS-99-83	Sand	3.11	3.07	2.97	0.50	0.31	93.1/4.7/2.3
	PS-99-85	Sand	3.01	2.98	2.84	0.50	0.33	93.7/4.2/2.1
	PS-99-87	Muddy sand	3.17	3.09	2.97	0.97	0.51	88.3/7.7/4.0
	PS-99-89	Sand	2.98	2.94	2.84	0.67	0.41	92.6/5.0/2.4
Acre IHS sands	Y-93-1f	Muddy sand	4.84	3.69	3.24	2.23	0.72	59.0/26.1/14.9
	Y-93-1g	Silty sand	4.89	3.78	3.37	2.12	0.72	57.1/28.9/14.0
	Y-93-2a	Silty sand	4.39	3.42	3.10	1.88	0.75	69.5/22.2/8.3
	Y-93-2c	Muddy sand	5.05	3.91	3.24	2.29	0.68	51.4/30.8/17.8
	Y-93-2e	Sandy silt	5.30	4.44	3.37	2.21	0.55	43.9/38.0/18.1
	Y-93-4a1	Muddy sand	3.41	3.21	3.10	1.24	0.62	82.7/10.0/7.3
	Y-93-4b1	Silty sand	4.46	3.39	3.10	1.95	0.76	70.0/21.2/8.8
	Y-93-4c1	Silty sand	4.46	3.48	3.24	1.92	0.74	70.1/20.1/9.8
	Y-93-6a	Muddy sand	4.39	3.09	2.84	2.24	0.79	67.0/21.0/12.0
	Y-93-6b	Silty sand	3.54	3.08	2.84	1.40	0.68	80.4/14.8/4.8
	Y-93-11b	Muddy sand	3.79	2.99	2.84	1.77	0.78	79.6/12.5/8.0
	Y-93-11c	Silty sand	4.17	3.37	3.10	1.78	0.72	70.1/22.4/7.5
	Y-93-11d	Silty sand	4.14	3.27	2.97	1.84	0.74	69.2/22.9/7.9
	Y-93-11e	Silty sand	3.90	3.12	2.84	1.67	0.75	75.7/18.8/5.5
Madre de Dios coastal plain fluvial sands	TL-91-8a	Muddy sand	1.63	1.33	1.22	1.60	0.63	86.5/8.5/5.0
	TL-91-8b	Sand	1.50	1.37	1.22	1.41	0.49	90.1/6.3/3.7
	MD-98-3	Silty sand	1.89	1.80	1.76	1.41	0.40	88.5/7.8/3.7
	MD-98-19	Sand	1.86	1.85	1.89	0.78	0.20	94.8/3.4/1.8
	00-9	Silty sand	2.28	2.25	2.16	0.88	0.39	53.6/43.7/2.7

SD, standard deviation; IHS, inclined heterolithic sand.

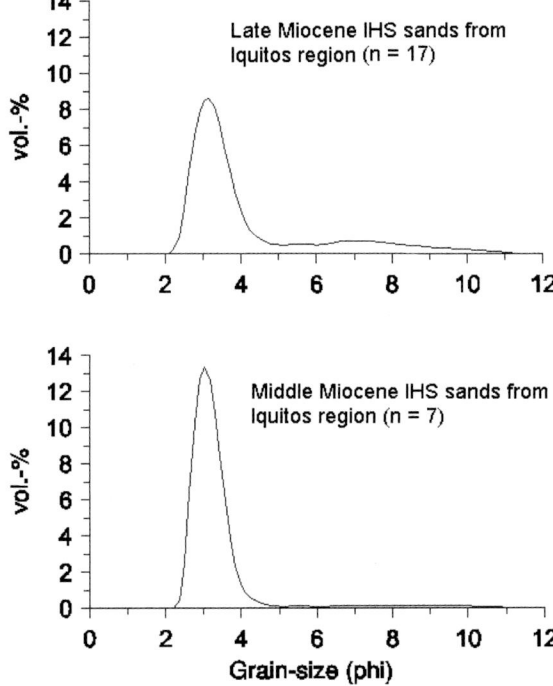

Fig. 12 Average grain-size distribution curves of middle Miocene estuarine IHS channel sands, and late Miocene estuarine IHS sands from Iquitos region. n = number of samples analysed.

sands, change from unimodal to slightly bimodal in the Iquitos region (Fig. 12). The average mode and mean grain-size decrease simultaneously and the sorting changes from moderately-sorted to poorly-sorted (Table 1).

Lowermost Amazon River and Amazon mouth

No systematic downstream variation in the grain-size parameters of surface sediments within the lowermost Amazon and Amazon mouth is seen. The mean grain-size of the sands varies between 1.4 and 3.7 ϕ and the mean grain-size of the muds varies between 5.0 and 6.3 ϕ (Table 5). The grain-size distributions of the sands are mainly unimodal and well- to moderately-sorted with modes varying between 1.0 and 3.5 ϕ. Four sand samples are bi- or polymodal (Fig. 13) with modes varying between 1.0 and 2.4 ϕ. The grain-size distributions of the muds are uni- to polymodal and poorly to very poorly sorted with modes varying between 3.2 and 5.5 ϕ.

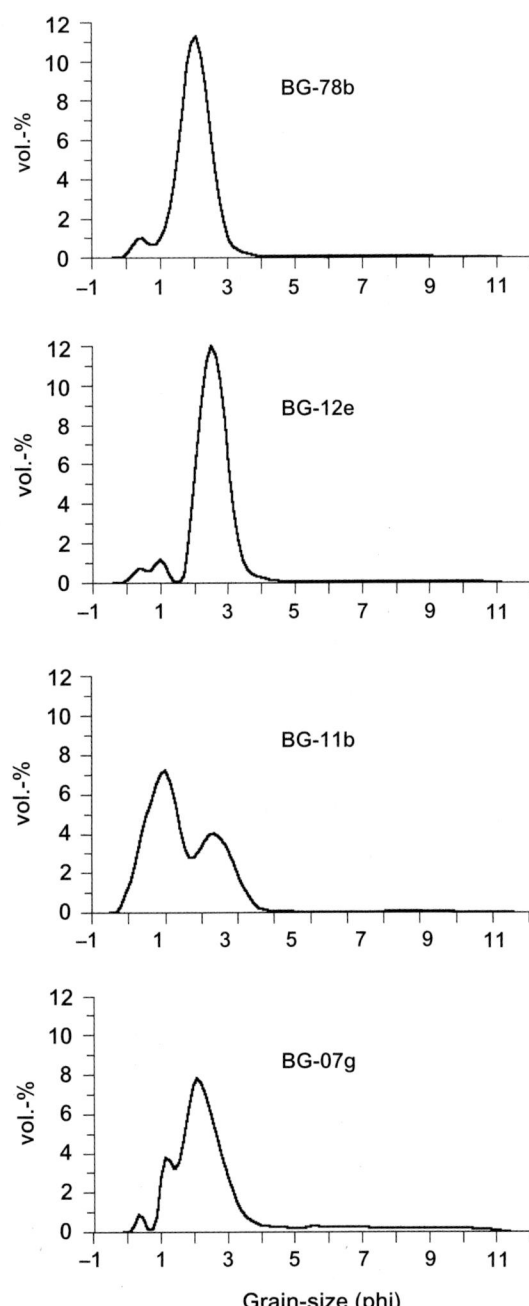

Fig. 13 Grain-size distribution curves of the bi- and polymodal sand samples from lowermost Amazon and Amazon mouth. See Fig. 4 for sample locations.

Table 5 Grain-size parameters of lowermost Amazon and Amazon mouth modern sediment samples.

Unit	Sample	Material	Mean (ϕ)	Median (ϕ)	Mode (ϕ)	Sorting SD (ϕ)	Skew (ϕ)	Sand/silt/clay (%)
BG-03	Channel?	Sand	2.38	2.35	2.30	1.21	0.33	91.1/5.7/3.2
BG-04c	Channel, bar	Sand	2.89	2.86	2.84	0.41	0.26	95.2/3.1/1.7
BG-07g	Beach?	Muddy sand	2.24	2.16	2.03	1.36	0.37	89.3/7.0/3.7
BG-08a	Channel	Sand	2.05	2.00	1.89	0.65	0.37	94.8/3.2/2.0
BG-08c	Channel	Silty sand	3.47	3.33	3.24	1.14	0.53	80.6/14.5/4.9
BG-11b	Beach	Sand	1.41	1.22	0.95	0.92	0.28	98.5/1.0/0.5
BG-12a	Margin	Sandy silt	6.31	6.17	5.39	2.11	0.12	13.0/64.6/22.4
BG-12b	Channel	Sand	2.89	2.84	2.70	0.61	0.34	93.2/4.6/2.2
BG-12c	Channel	Sand	2.67	2.60	2.43	0.68	0.42	93.2/4.7/2.1
BG-12e	Channel, bar	Sand	2.48	2.47	2.43	0.60	−0.12	97.1/1.9/1.0
BG-14c	Channel	Silty sand	3.71	3.59	3.51	1.12	0.48	73.4/21.5/5.1
BG-16c	Channel	Sandy silt	6.24	5.95	5.39	2.05	0.23	11.0/68.2/20.8
BG-19	Channel, bar?	Silty sand	3.31	3.14	2.84	1.15	0.52	81.9/14.0/4.1
BG-20	Channel, bar?	Sandy silt	5.23	4.69	4.05	1.96	0.45	29.9/57.4/12.7
BG-26c	Channel	Sand	2.93	2.86	2.70	0.88	0.49	90.6/6.4/3.0
BG-28b	Channel	Sand	2.67	2.65	2.57	0.37	0.15	97.5/1.7/0.8
BG-60	Channel	Sand	2.82	2.79	2.70	0.42	0.22	96.0/2.7/1.3
BG-78b	Channel, bar	Sand	1.97	1.98	2.03	0.57	−0.05	97.6/1.7/0.7
BG-78c	Channel	Sandy silt	5.02	4.39	3.24	2.06	0.49	40.4/47.1/12.5
BG-78e	Channel?	Sand	2.85	2.80	2.70	0.69	0.41	92.6/5.1/2.3
BG-79a	Channel, bar	Sandy silt	6.11	5.99	5.53	2.04	0.12	14.6/66.8/18.6
BG-79b	Channel, bar	Sand	2.62	2.59	2.57	0.43	0.22	96.0/2.7/1.3
BG-79c	Channel, bar	Silty sand	2.77	2.66	2.57	0.91	0.55	89.2/8.2/2.6
BG-79d	Channel, bar	Sandy silt	6.09	5.81	5.26	2.06	0.23	14.2/65.7/20.1

SD, standard deviation.

Clay mineralogy

Three samples from the coarse-grained (fluvial) sand of the McMurray Core 1 were analysed (PS-C104, PS-C109, PS-C114; Fig. 7). Four IHS mud samples (PS-C97, PS-C93, PS-C83, PS-C71), one shoreface mud sample (PS-C65), one offshore mud sample (PS-C61) and one shoreface sand sample (PS-C52) from the McMurray Core 3 were analysed (Fig. 5). One Ipururo Formation sample (MD-98-02) from Madre de Dios was also analysed to compare its clay mineralogy with an equivalent McMurray Formation sample (PS-C109). Sample MD-98-02 was taken from the same location and from the same lithological unit as sample MD-98-03, on which a grain-size analysis was performed.

X-ray diffraction analyses show that kaolinite is the dominant clay mineral in both the fluvial sand and IHS mud samples (Fig. 14). Illite is as abundant as kaolinite in the Ipururo Formation coastal plain fluvial or inner estuary fluvial-dominated channel sands, whereas smectite is the dominant clay mineral in the shoreface sediments of the McMurray and Pebas formations (Räsänen et al., 1998). In the McMurray Formation the kaolinite : smectite ratio decreases upward, from fluvial sands to estuarine IHS muds, and then decreases further in the shoreface sediments. Illite is more abundant in the IHS and shoreface muds than in the fluvial sediments.

DISCUSSION

Fluvial deposits

The fining upward unit I in McMurray Core 1 consists of planar tabular interbedded sands and

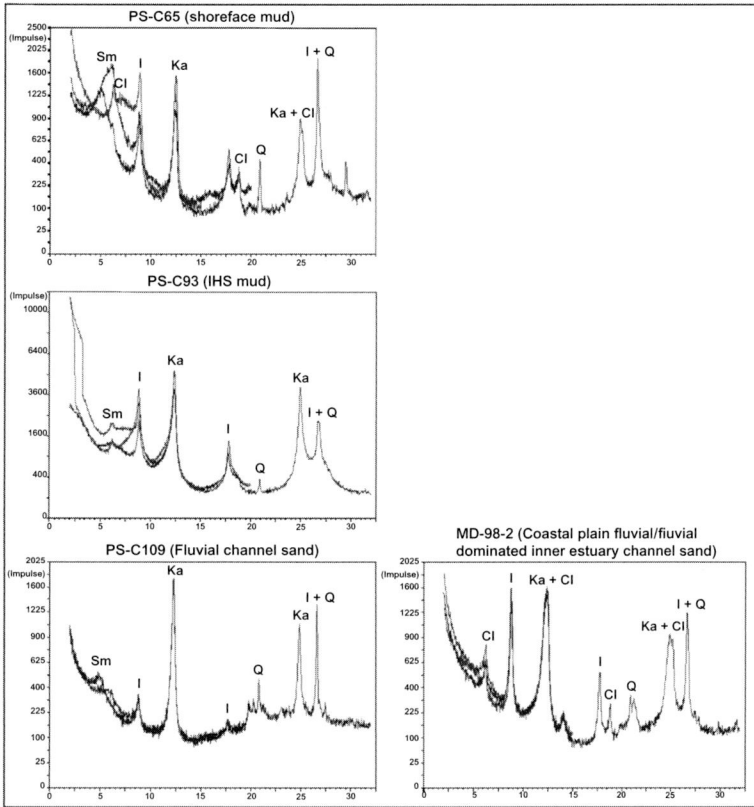

Fig. 14 Typical X-ray diffraction patterns of clay-mineral assemblages of the McMurray Formation (PS-C65, PS-C93, PS-C109) and Ipururo Formation (MD-98-2) samples. Sm = smectite, I = illite, Ka = kaolinite, Q = quartz, Ca = calcite, Cl = chlorite. Note the different scales on the y axes of the diagrams.

muds. It is interpreted to represent a fluvial-dominated inner estuary depositional environment of the Dalrymple et al. (1992) estuary classification. The fining and thinning upward nature of the sands, and the preservation of interbedding with annual cyclicities, suggest gradual change from a high-energy to relatively low-energy environments within the estuary (Ehlers & Chan, 1999). The Ipururo Formation sands from the Madre de Dios region are interpreted to have been deposited in a similar environment, in a fluvial-dominated inner estuary channel or in a more landward position in a coastal plain fluvial channel. The mean grain-size of these channel sands varied between 2.0 and 2.5 ϕ in the McMurray Formation, and between 1.5 and 2.3 ϕ in the Ipururo Formation (Table 6).

IHS deposits

The mean grain-size of the IHS sands varies between 2.5 and 6.0 ϕ in the McMurray Formation and between 2.8 and 6.3 ϕ in the Pebas Formation (Table 6). Within the IHS units and their internal packages, the grain-size profiles of the sand samples decrease upwards. This results from the lateral migration of the channel margin and reflects deposition in the progressively marginal portions of the channel with less current energy. In the latest phase of channel deposition these IHS units are capped by abandoned/intertidal channel-fill, floodplain or tidal marsh muds. This characteristic fining upward IHS succession has been widely reported from other McMurray Formation outcrops and cores (e.g. Strobl et al., 1997; Wightman & Pemberton, 1997) and from the Glauconitic Member near Drayton Valley, central Alberta (Brownridge & Moslow, 1991). It is also reported from modern environments including Willapa River, USA and the Athabasca River delta, Canada (Smith, 1987, 1988a,b). The overall fining upward trend in cores and outcrops through the whole stacked IHS channel succession indicates that the zone of deposition within the estuary has

Table 6 Measured grain-size parameters for the different lithological units, and processes governing their formation, of the Pebas/Ipururo formations and the McMurray Formation. Averages are shown in parentheses.

Deposits	Environment	Lithology	Mean (φ)	Mode (φ)	Sorting SD (φ)	Skewness (φ)	Clay (%)	Major processes
McMurray Formation	Fluvial	Sands	2.0–2.5 (2.2)	2.0–2.3 (2.1)	0.51–0.91 (0.66)	−0.12–0.41 (0.15)	0.4–2.7 (1.4)	Fluvial
		Muds	5.4–7.8 (6.9)	2.3–7.1 (5.7)	1.69–2.96 (2.26)	−0.16–0.09 (−0.01)	22.4–44.3 (34.3)	
	IHS	Sands	2.5–6.0 (3.4)	2.4–5.6 (3.1)	0.34–2.10 (0.91)	0.08–0.70 (0.40)	0.4–16.7 (3.8)	Fluvial (seasonal/fluctuating discharge) Mesotidal Biogenic reworking
		Muds	3.4–7.3 (5.8)	2.6–7.0 (4.7)	0.96–2.61 (1.91)	−0.01–0.59 (0.23)	3.3–31.7 (18.9)	
	Shoreface	Sands	2.7–5.2 (4.5)	2.6–3.4 (3.1)	0.94–2.30 (2.00)	0.47–0.82 (0.66)	3.8–17.0 (12.2)	Waves (oscillatory and shoaling) Storms Longshore and rip currents Biogenic reworking
		Muds	6.7–7.7 (7.3)	6.7–7.3 (7.0)	1.82–2.35 (2.12)	−0.13–0.11 (−0.09)	31.1–44.6 (38.8)	
	Offshore	Muds	7.6–8.2 (7.9)	6.7–8.2 (7.4)	1.62–1.77 (1.71)	−0.05–0.21 (0.08)	37.5–55.4 (47.4)	Biogenic reworking Storms
Pebas/Ipururo formations	Fluvial	Sands	1.5–2.3 (1.8)	1.2–2.2 (1.7)	0.78–1.60 (1.22)	0.20–0.63 (0.42)	1.8–5.0 (3.4)	Fluvial
	IHS	Sands	2.8–6.3 (4.0)	2.7–6.7 (3.1)	0.50–2.29 (1.47)	−0.04–0.79 (0.57)	2.1–22.1 (7.8)	Fluvial (seasonal/fluctuating discharge) Microtidal Biogenic reworking
Modern Amazon River mouth	Fluvial	Sands	1.4–3.7 (2.7)	1.0–3.5 (2.5)	0.37–1.36 (0.79)	−0.12–0.55 (0.32)	0.5–5.1 (2.4)	Fluvial (weak seasonality) Weak tidal influence
		Muds	5.0–6.3 (5.8)	3.2–5.5 (4.8)	1.96–2.11 (2.05)	0.12–0.49 (0.27)	12.5–22.4 (17.9)	

SD, standard deviation; IHS, inclined heterolithic stratified.

changed from inner to middle estuary during the ongoing transgression.

Depositional processes forming the IHS are interpreted to be fluvial with a tidal imprint. Sand beds in the IHS can be ascribed to major seasonal river-flooding periods (rainy season); the mud beds are interpreted to have been deposited during low-water periods (dry season) when there was reduced flow in the river (e.g. Smith, 1985, 1987; Ranger, 1994). Clifton & Phillips (1973) documented IHS being deposited over long time periods as a result of seasonal fluctuations in the turbidity maximum.

In the Pebas Formation IHS sands, the grain-size trend from the middle Miocene IHS deposits to the late Miocene IHS deposits shows fining and a shift to more poorly sorted distributions (Fig. 13). This indicates that the depositional environment had shifted to the middle portions of the embayment or estuary during the transgression, at the same time as tidal influence increased.

The clay mineral distribution in the McMurray Formation clearly reveals differences between shoreface and estuarine fine-grained sediments. As is the case in many estuaries (e.g. Elbe estuary, Irion et al., 1987), fluvial sediments are either trapped in the estuary or they partly mix with marine sediments. During rising sea level the lower sections of rivers become estuaries, with the increasing depositional area encouraging aggradation of large volumes of sediments. This is the main reason why sediments off the estuaries may not originate from the inflowing rivers and why the clay mineralogy of the shoreface muds of the McMurray and Pebas formations differs significantly from that of the upper estuaries.

The clay mineral assemblage of the Ipururo Formation coastal plain fluvial or inner estuary fluvial-dominated channel sands corresponds well to the clay mineralogy of the McMurray Formation fluvial sands. The more abundant illite and chlorite contents in the Ipururo Formation sand are probably the result of a more seaward depositional environment within the estuary. The seaward increasing abundance of illite, chlorite and smectite with a concurrent decrease in the abundance of kaolinite is a typical pattern for estuarine depositional environments (Feuillet & Fleischer, 1980; Chamley, 1989).

Shoreface deposits

Shoreface deposits were present only in McMurray Core 3, where they overlie estuarine IHS channel deposits. The upward-coarsening trend in the shoreface deposits is typical of shoreface successions, which typically show progradation of the lower and upper shoreface (e.g. Van Wagoner et al., 1992). Thus shoreface grain-size trends are easily separated from upward-fining IHS channel successions. The rhythmic deposition of sand and mud on the shoreface might also have been caused by seasonal variations in wind activity, current velocity and sediment input.

The modes of the shoreface sands in McMurray Core 3 are similar to the modes of the IHS sands in the same core, but the mean grain-size of the shoreface sands is finer and they are more poorly sorted (Table 1). The amount of clay is also greater in the shoreface sands than in the underlying IHS sands (Table 6). All the IHS and shoreface sands are positive skewed, but the shoreface sands reveal a more positive fine-skewed grain-size population, which is distinctively different from the IHS channel sands at the same site.

The poorly sorted and positively skewed grain-size distributions of the shoreface sands indicate that they were deposited below wave base on the lower shoreface, in a transition zone where mud is available together with sand (Reineck & Singh, 1975). Deposition might have also occurred in sheltered lagoonal environments free of significant wave and current action. The shoreface muds are clearly finer than the IHS muds, and the amount of clay is also greater in the shoreface muds than in the underlying IHS muds (Table 6). Both the shoreface and the IHS muds are poorly to very poorly sorted.

Modern channel sediments from the Amazon River

No systematic downstream grain-size variation was observed within the lowermost Amazon and Amazon mouth sediments because the sediments are influenced by similar processes (high fluvial discharge with weak seasonality and weak tidal effect) across the entire study area. Vital (1996) and Torres (1997) also found this to be the case. Vital (1996) concluded, in her study of surficial sediments from the lowermost Amazon, that sand is being deposited in the central part of the channel (as a corridor of sands), and silty sediments are being deposited at the channel margins. The bedforms on the sand surface also reflected strong bottom currents and sediment transport, as well as tidal influence (Vital, 1996).

Coastal plain/inner estuary and middle/outer estuary sediments

The transition from a coastal plain–inner estuary (fluvial-dominated) to the mixed fluvial and tidal regime of a middle–outer estuary can be inferred from the variation in the grain-size distribution curves and grain-size parameters of both the Miocene Pebas/Ipururo formations sands (Fig. 11 & Table 6) and the Cretaceous McMurray Formation sands (Figs 6, 8 & 10, & Table 6).

The distinct change upward in grain-size distributions in McMurray Formation Core 1 from bimodal to unimodal, and the fining upward trend from the lower unit I sands to the upper unit II sands, are indicative of a clear change in the depositional regimes of these units. The decreasing grain-size trend of the sands through units I–II indicates that the strength of the current during the deposition of the sands was decreasing. This is attributed to a middle estuarine position for the unit II channel during the later phases of the transgression. The unit I channel represents an inner estuary channel with strong fluvial influences, whereas unit II represents a more tidally influenced middle estuary channel. The unit I sands are negative coarse-skewed to positive fine-skewed, and unit II sands are more positive fine-skewed. According to Sahu (1964) negatively skewed sediments are affected by a higher energy depositing agent or were subjected to transportation for a greater length of time. The coarsening upward trend in the muds from unit I to unit II may indicate increased tidal currents affecting mud deposition.

The notable change in grain-size distributions and parameters from the coastal plain fluvial or fluvial-dominated inner estuary sands of Madre de Dios, to Acre and Iquitos IHS sands (Fig. 11, Table 6), together with the sedimentological evidence (Räsänen et al., 1995, 1998), suggest that these sediments were deposited in different environments within the epicontinental embayment or large estuary system. The Madre de Dios sediments are interpreted to have been deposited in coastal plain or inner estuary environments closer to the uplifting Andean mountains with more fluvial influence. The Acre and Iquitos IHS sediments are interpreted to have been deposited as point bars of estuary channels in the middle portions of the Pebasian basin, where both tidal and fluvial processes influenced deposition.

The fluvial coastal plain sands or fluvial-dominated inner estuary sands are coarser grained, better sorted, more clearly bimodal (with a minor mode in the coarse sand around 0.3 ϕ) and more negative coarse-skewed than the IHS sands from the mixed fluvial–tidal regime of the middle estuary (Table 6). The textural differences between the coastal plain fluvial or fluvial-dominated inner estuary sands and the IHS sands of the middle estuary (mixed fluvial and tidal regime) are also seen in Fig. 15 where sands from these environments are plotted as separate groups.

By contrast, in the McMurray Formation, the fluvial coastal plain muds or fluvial-dominated inner estuary muds are finer grained, more poorly sorted and more negative coarse-skewed than the IHS muds from the mixed fluvial tidal regime of the middle–outer estuary (Table 6). The IHS sands from the Pebas Formation are finer grained and more poorly sorted than the McMurray Formation IHS sands (Table 6). This may indicate less fluvial or/and tidal currents in the larger western Amazonian basin, which also had gentler gradients than the western Canada basin. Tidal transport was evidently less in the Pebas Formation deposits so fluvial transport dominated.

Comparison of textural characteristics: Miocene channel sediments and modern channel sediments from the lowermost Amazon

Modern channel sands from the lowermost Amazon and Amazon mouth are finer grained and better sorted than the (fluvial-dominated) coastal plain–inner estuary channel sands of the Pebas Formation (Table 6). In contrast, the modern channel sands are coarser grained and also better sorted than the IHS channel sands of the Pebas Formation. Some modern channel sands from the lowermost Amazon and Amazon mouth show

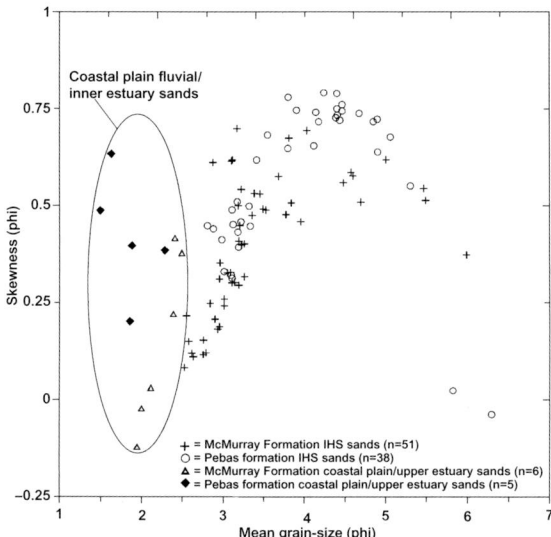

Fig. 15 Plot of mean grain-size versus skewness for IHS channel sands and fluvially dominated coastal plain–inner estuary channel sands from the McMurray and Pebas formations.

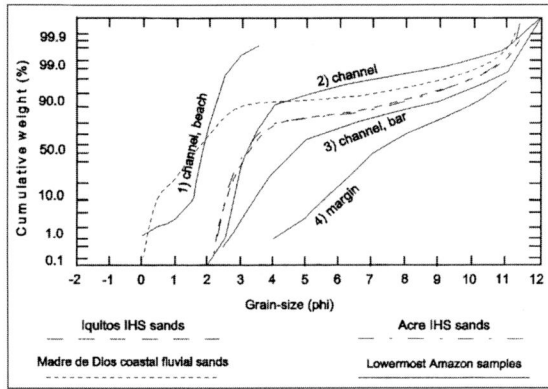

Fig. 16 Cumulative average grain-size distribution curves of the Miocene Pebas sands from the Acre, Iquitos and Madre de Dios regions, and cumulative curves of modern lowermost Amazon surficial sediment samples. Modified from Vital & Stattegger (2000).

similar bimodal grain-size distribution patterns to the (fluvial-dominated) coastal plain–inner estuary sands of the Pebas and McMurray formations. This indicates that the lowermost Amazon and Amazon mouth reflect the dominant discharge of the Amazon River and therefore fluvial processes are dominating on the lowermost Amazon system.

To compare the Pebas Formation channel sediments from western Amazonia with modern channel sediments from the lowermost Amazon River, the cumulative grain-size curves plotted on a probability scale were combined on one graph (Fig. 16). The similarity of the Miocene IHS sands and modern channel and bar sediment grain-size distributions (curves 2 and 3) is obvious. The modern channel sediments in the lowermost Amazon are by these characteristics similar to the studied Miocene IHS sediments from western Amazonia. The Madre de Dios sands are coarser and more poorly sorted than the modern channel and beach sands from the lowermost Amazon (Table 6), and are easily differentiated.

The lowermost Amazon and Amazon mouth sediments do not show as distinct seasonal variations in the deposition of sand and mud beds as interpreted here for the IHS sediments of the McMurray and Pebas formations. The reason for this could be the small difference between the seasonal maximum and minimum water discharge in the Amazon River (a factor of only about two to four), and because the large drainage basin extends both north and south of the Equator and has segments that flood at different times of the year.

CONCLUSIONS

Detailed grain-size analyses, performed by laser diffractometer, reveal information about the textural properties of sediments that may have gone unrecognized using older grain-size analysis techniques. Grain-size analysis continues to be a useful tool when used together with sedimentological observations. In a case study of Cretaceous, Miocene and modern sediments, grain-size analysis and sedimentological data demonstrate:

1 An upward-fining grain-size trend and a simultaneous change to more poorly sorted grain-size distributions in a coastal plain estuary depositional sequence is indicative of the transition from a fluvial-dominated to a mixed fluvial and tidal regime.

2 The transition can be observed stratigraphically in the Cretaceous McMurray Formation associated with the late Aptian to early Albian transgression,

and is seen regionally in the Miocene Pebas Formation outcrops across the depositional basin.

3 Fluvial-dominated coastal plain/inner estuary sands have clear bimodal grain-size distributions with a minor mode in the coarse sand fraction around 0.3–0.4 ϕ. They are coarser grained, better sorted and more negative coarse-skewed than inclined heterolithic stratified (IHS) channel sands in the mixed fluvial and tidal regime of the middle estuary. These features are detectable in both of the studied ancient epicontinental embayments or estuarine systems.

4 The IHS channel sands from the Miocene Pebas Formation are finer grained and more poorly sorted than comparable Cretaceous McMurray Formation IHS channel sands. This indicates less fluvial flow and/or reduced tidal influences in the larger Miocene western Amazonian Basin, which had gentler gradients and was positioned farther from the ocean than the Cretaceous Western Canadian Basin.

5 Grain-size distribution patterns of the modern channel sands from the lowermost Amazon and Amazon mouth show similarities and differences to Late Miocene Pebas Formation fluvial and IHS channel sands. It is concluded, however, that the lowermost 250 km of the Amazon River studied here do not fully represent the transition from fluvial to estuarine environments because the entire system is so large and because estuarine circulation occurs further out on the shelf. It is not considered to be a realistic recent analogue for the IHS deposits of the Pebas and McMurray formations studied here.

6 X-ray diffraction determinations of clay mineral assemblages in the McMurray Formation show that the proportion of kaolinite decreases upward in the succession from fluvial to IHS estuarine sediments to shoreface sediments and the proportions of smectite, illite and chlorite increase. This is attributed to a change of the depositional environments from riverine to more marine during the late Aptian to early Albian transgression.

ACKNOWLEDGEMENTS

This research was funded by the Neste Inc. Foundation and Ingemmet (Instituto Geológico Minero y Metalúrgico, Peru). We are grateful to all Amazon Research Team members (University of Turku) who provided help in this research. We also thank Professor K. Stattegger and Dr Vital from the Geologisch-Paläontologisches Institut of Kiel University for letting us use their sediment samples collected from the Amazon River mouth.

REFERENCES

Bolthovskoy, E. (1991) Ihering's hypothesis in the light of foraminiferological data. *Lethaia*, **24**, 190–197.

Brownridge, S. and Moslow, T.F. (1991) Tidal estuary and marine facies of the Glauconitic Member, Drayton Valley, central Alberta. In: *Clastic Tidal Sedimentology* (Eds D.G. Smith, G.E. Reinson, B.A. Zaitlin and R.A. Rahmani). *Can. Soc. Petrol. Geol. Mem.*, **16**, 107–122.

Buller, A.T. and McManus, J. (1972) Simple metric sedimentary statistics used to recognize different environments. *Sedimentology*, **18**, 1–21.

Carrigy, M.A. (1959) Geology of the McMurray Formation, Part III, general geology of the McMurray area. *Alberta Res. Council Mem.*, **I**, 130 pp.

Carrigy, M.A. (1971) Deltaic sedimentation in Athabasca Tar Sands. *Am. Assoc. Petrol. Geol. Bull.*, **55**, 1155–1169.

Chamley, H. (1989) *Clay Sedimentology*. Springer-Verlag, Berlin–Heidelberg, 623 pp.

Clifton, E.H. and Phillips, R.L. (1973) Lateral trends and vertical sequences in estuarine sediments, Willapa Bay, Washington. *U.S. Geol. Surv. Prof. Pap.*, **850**, 124 pp.

Coldberg, R. (1980) Use of grain-size frequency data to interpret the depositional environment of the Pliocene Pleshet Formation, Beer Sheva, Israel. *J. Sediment. Petrol.*, **50**, 843–856.

Dalrymple, R.W., Zaitlin, B.A. and Boyd, R. (1992) Estuarine facies models: conceptual basis and stratigraphical implications. *J. Sediment. Petrol.*, **62**, 1130–1146.

Ehlers, T.A. and Chan, M.A. (1999) Tidal sedimentology and estuarine deposition of the Proterozoic Big Cottonwood Formation, Utah. *J. Sediment. Res.*, **69**, 1169–1180.

Feuillet, J.-P. and Fleischer, P. (1980) Estuarine circulation: controlling factor of clay mineral distribution in James River estuary, Virginia. *J. Sediment. Petrol.*, **50**, 267–279.

Flach, P.D. (1984) Oil sand geology—Athabasca deposit north. *Alberta Res. Council Bull.*, **46**, 31 pp.

Folk, R.L. (1954) The distinction between grain size and mineral composition in sedimentary-rock nomenclature. *J. Geol.*, **62**, 344–359.

Folk, R.L. and Ward, W.C. (1957) Brazos River bar: a study in the significance of grain size parameters. *J. Sediment. Petrol.*, **27**, 3–26.

Friedman, G.M. (1961) Distinction between dune, beach, and river sands from their textural characteristics. *J. Sediment. Petrol.*, **31**, 514–529.

Friedman, G.M. (1979) Address of the retiring president of the International Association of Sedimentologists: differences in size distributions of populations of particles of sand of various origins. *Sedimentology*, **26**, 3–32.

Gale, S.J. and Hoare, P.G. (1991) *Quaternary Sediments*. Wiley, New York, 323 pp.

Geyer, W.R., Beardsley, R.C., Candela, J., Castro, B.M., Legeckis, R.V., Lentz, S.J., Limeburner, R., Miranda, L.B. and Trowbridge, J.H. (1991) The physical oceanography of the Amazon outflow. *Oceanography*, **4**, 8–14.

Gibbs, R.J. (1967) The geochemistry of the Amazon River system: part I. The factors that control the salinity and the composition and concentration of the suspended solids. *Geol. Soc. Am. Bull.*, **78**, 1203–1232.

Gibbs, R.J. (1970) Circulation in the Amazon River estuary and adjacent Atlantic Ocean. *J. Mar. Res.*, **28**, 113–123.

Gingras, M.K., Räsänen, M. and Pemberton, S.G. (2000) Ichnological evidence for marine and marginal marine deposition in the Miocene Pebas Formation, Perú. *II Congreso Latinoamericano de Sedimentología. VIII Reunión Argentina de Sedimentología*, Mar del Plata, Argentina, 83–84.

Gingras, M.K., Räsänen, M.E., Pemberton, S.G. and Romero, L.P. (2002) Ichnology and sedimentology reveal depositional characteristics of shoreface parasequences in the Miocene Amazonian Foreland Basin. *J. Sediment. Res.*, **76**, 871–883.

Haq, B.U., Hardenbol, J. and Vail, P.R. (1988) Mesozoic and Cenozoic chronostratigraphy and cycles of sea level change. In: *Sea-Level Changes—an Integrated Approach* (Eds C.K. Wilgus, B.S. Hastings, C.G. St C. Kendall, H.W. Posamentier, C.A. Ross and J.C. Van Wagoner). *Soc. Econ. Paleontol. Mineral. Spec. Publ.*, **42**, 71–108.

Hoorn, C. (1993) Marine incursions and the influence of Andean tectonics on the Miocene depositional history of northwestern Amazonia: results of a palynostratigraphic study. *Palaeogeogr., Palaeoclimatol., Palaeoecol.*, **105**, 267–309.

Hoorn, C. (1994a) Fluvial palaeoenvironments in the intracratonic Amazonas Basin (Early Miocene–early Middle Miocene, Colombia). *Palaeogeogr., Palaeoclimatol., Palaeoecol.*, **109**, 1–54.

Hoorn, C. (1994b) An environmental reconstruction of the palaeo-Amazon River system (Middle–Late Miocene, NW Amazonia). *Palaeogeogr., Palaeoclimatol., Palaeoecol.*, **112**, 187–238.

Hoorn, C., Guerrero, J., Sarmiento, G.A. and Lorente, M.A. (1995) Andean tectonics as a cause for changing drainage patterns in Miocene northern South America. *Geology*, **23**, 237–240.

Irion, G., Wunderlich, F. and Schwedhelm, E. (1987) Transport of clay minerals and anthropogenic compounds into the German Bight and the provenance of the fine-grained sediments SE of Helgoland. *J. Geol. Soc. Lond.*, **144**, 153–160.

Jardine, D. (1974) Cretaceous oil sands of Western Canada. In: *Oil Sands: Fuel of the Future* (Ed. L.V. Hills). *Can. Soc. Petrol. Geol. Mem.*, **3**, 50–67.

Jeletzky, J.A. (1971) Marine Cretaceous biotic provinces and paleogeography of western and arctic Canada. *Geol. Surv. Can. Pap.*, **70-22**, 92 pp.

Kummel, B. (1948) Geological reconnaissance of the Contamana Region, Peru. *Geol. Soc. Am. Bull.*, **59**, 1217–1266.

Marshall, L.G. and Lundberg, J.G. (1996) Miocene Deposits in the Amazonian Foreland Basin. *Science*, **273**, 123–124.

Mason, C.C. and Folk, R.L. (1958) Differentiation of beach, dune and aeolian flat environments by size analysis, Mustang Island, Texas. *J. Sediment. Petrol.*, **28**, 211–226.

Mattison, B.W., Fox, A.J. and Pemberton, S.G. (1989) Sedimentologic, paleontologic and ichnologic criteria for the recognition of ancient estuarine deposits: an example from the Lower Cretaceous McMurray Formation in the Athabasca Oils Sands area of northeastern Alberta. In: *Modern and Ancient Examples of Clastic Tidal Deposits—a Core and Peel Workshop* (Ed. G.E. Reinson), pp. 66–79. Calgary, Alberta.

Nelson, H.W. and Glaister, R.P. (1978) Subsurface environmental facies and reservoir relationships of the McMurray oil sands, northeastern Alberta. *Bull. Can. Petrol. Geol.*, **26**, 177–207.

Nuttal, C.P. (1990) A review of the Tertiary non marine molluscan faunas of the Pebasian and other inland basins of north-western South America. *Bull. Br. Mus. Nat. Hist. Geol.*, **45**(2), 165–371.

Pemberton, S.G., Flach, P.D. and Mossop, G.D. (1982) Trace fossils from the Athabasca Oil Sands, Alberta, Canada. *Science*, **217**, 825–827.

Ranger, M.J. (1994) A basin study of the southern Athabasca Oil Sands deposit. PhD thesis, University of Alberta, Edmonton, Alberta, 290 pp.

Ranger, M.J. and Pemberton, S.G. (1988) Marine influence on the McMurray Formation in the Primrose area, Alberta. In: *Sequences, Stratigraphy, Sedimentology: Surface and Subsurface* (Eds D.P. James and D.A. Leckie). *Can. Soc. Petrol. Geol. Mem.*, **15**, 439–450.

Ranger, M.J. and Pemberton, S.G. (1992) The sedimentology and ichnology of estuarine point bars in the McMurray Formation of the Athabasca Oil Sands deposits, northeastern Alberta, Canada. In: *Applications of Ichnology to Petroleum Exploration* (Ed. S.G. Pemberton). *Soc. Econ. Paleontol. Mineral. Core Workshop*, **17**, 401–421.

Räsänen, M.E., Linna, A.M., Santos, J.C.R. and Negri, F.R. (1995) Late Miocene tidal deposits in the Amazonian Foreland Basin. *Science*, **269**, 386–390.

Räsänen, M., Linna, A., Irion, G., Rebata Hernani, L., Vargas Huaman, R. and Wesselingh, F. (1998) Geología y Geoformas de la Zona de Iquitos. In: *Geoecología y desarrollo Amazónico: estudio integrado en la zona de Iquitos, Perú* (Eds R. Kalliola, and S. Flores Paitán). *Ann. Univ. Turkuensis Ser. A II*, **114**, 59–137.

Reineck, H.-E. and Singh, I.B. (1975) *Depositional Sedimentary Environments*. Springer-Verlag, Berlin–Heidelberg, 439 pp.

Sahu, B.K. (1964) Depositional mechanism from the size analysis of clastic sediments. *J. Sediment. Petrol.*, **34**, 73–83.

Sedimentation seminar (1981) Comparison of methods of size analysis for sands of the Amazon Solimões Rivers, Brazil and Peru. *Sedimentology*, **28**, 123–128.

Smith, D.G. (1985) *Modern Analogues of the Mcmurray Formation Channel Deposits, Sedimentology of Mesotidal-influenced Meandering River Point Bars with Inclined Beds of Alternating Mud and Sand*. Final Report for Research Project No. 391, Alberta Oil Sands Technology and Research Authority, Calgary, Alberta, 78 pp.

Smith, D.G. (1987) Meandering river point bar lithofacies models: modern and ancient examples compared. In: *Recent Developments in Fluvial Sedimentology* (Eds F.G. Ethridge, R.M. Flores and M.D. Harvey). *Soc. Econ. Paleontol. Mineral. Spec. Publ.*, **39**, 83–91.

Smith, D.G. (1988a) Mesotidal estuary point bar deposits: a comparative sedimentology of modern and ancient examples in peels and core (abs.). In: *Sequences, Stratigraphy, Sedimentology: Surface and Subsurface* (Eds D.P. James and D.A. Leckie). *Can. Soc. Petrol. Geol. Mem.*, **15**, 584.

Smith, D.G. (1988b) Modern point bar deposits analogous to the Athabasca oil sands, Alberta, Canada. In: *Tide-influenced Sedimentary Environments and Facies* (Eds P.L. de Boer, A. van Gelder and S.D. Nio), pp. 417–432. D. Reidel, Amsterdam.

Steward, G.A. and MacCallum, G.T. (1978) Athabasca Oil Sands guide book. *Canadian Society of Petroleum Geologists' International Conference on Facts and Principles of World Oil Occurrence*, Calgary, Alberta, 33 pp.

Strobl, R.S., Muwais, W.K., Wightman, D.M., Cotterill, D.K. and Yuan, L. (1997) Geological modelling of McMurray Formation reservoirs based on outcrop and subsurface analogues. In: *Petroleum Geology of the Cretaceous Mannville Group, Western Canada* (Eds S.G. Pemberton and D.P. James). *Can. Soc. Petrol. Geol. Mem.*, **18**, 292–311.

Torres, A.M. (1997) *Sedimentology of the Amazon Mouth: North and South Channels, Brazil*. Berichte-Report 82, Geologishe–Paläontologisches Institüt, University of Kiel, 159 pp.

Van Wagoner, J.C., Mitchum, R.M., Campion, K.M. and Rahmanian, V.D. (1992) Siliciclastic sequence stratigraphy in well logs, cores, and outcrops. *Am. Assoc. Petrol. Geol. Methods Explor. Ser.*, **7**, 55 pp.

Visher, G.S. (1969) Grain size distributions and depositional processes. *J. Sediment. Petrol.*, **39**, 1074–1106.

Vital, H. (1996) Sedimentology of the lowermost Amazon (Rio Xingu–Macapa) and the 'Estreitos de Breves'–Brazil. PhD thesis, University of Kiel, 189 pp.

Vital, H. and Stattegger, K. (2000) Sediment dynamics in the lowermost Amazon. *J. Coastal Res.*, **16**, 316–328.

Von Ihering, H. (1927) *Die Geschichte des Atlantischen Ozeans*. Fisher, Jena, 237 pp.

Wesselingh, F.P., Räsänen, M.E., Irion, G., Vonhof, H.B., Kaandorp, R., Renema, W., Romero Pittman, L. and Gingras, M. (2002) Lake Pebas: a palaeoecological reconstruction of a Miocene long-lived lake complex in Western Amazonia. *Cainozoic Res.*, **1**, 35–81.

Wightman, D.M. and Pemberton, S.G. (1997) The Lower Cretaceous (Aptian) McMurray Formation: an overview of the Fort McMurray area, northeastern, Alberta. In: *Petroleum Geology of the Cretaceous Mannville Group, Western Canada* (Eds S.G. Pemberton and D.P. James). *Can. Soc. Petrol. Geol. Mem.*, **18**, 312–344.

Wightman, D.M., Rottenfusser, B., Kramers, J. and Harrison, R. (1989) Geology of the Alberta oil sands deposits. In: *AOSTRA Technical Handbook on Oil Sands, Bitumen and Heavy Oils* (Eds L.G. Hepler and C. Hsi). *Alberta Oil Sands Technol. Res. Auth. Tech. Publ. Ser.*, **6**, 1–9.

Wightman, D.M., MacGillivray, J.R., McPhee, D., Berhane, H. and Berezniuk, T. (1991) McMurray Formation and Wabiskaw Member (Clearwater Formation): regional perspectives derived from the North Primrose Area, Alberta, Canada. In: *5th UNITAR International Conference on Heavy Crude and Tar Sands* (Ed. R.F. Meyer), **1**, pp. 285–320. UNITAR Centre for Heavy Crude and Tar Sands, Tulsa, TX.

Sedimentology and avulsion patterns of the anabranching Baghmati River in the Himalayan foreland basin, India

R. SINHA*, M.R. GIBLING†, V. JAIN* and S.K. TANDON‡

*Engineering Geology Group, Indian Institute of Technology Kanpur 208016, India (Email: rsinha@iitk.ac.in);
†Department of Earth Sciences, Dalhousie University, Halifax, Nova Scotia, B3H 3J5, Canada; and
‡Department of Geology, University of Delhi, Delhi 110007, India

ABSTRACT

The Baghmati River, a foothills-fed system in the Himalayan foreland basin of north Bihar, has an anabranching mid-stream reach and floodplains that aggraded rapidly during the late Holocene. The river is characterized by variable discharge, frequent and widespread overbank flooding, and high sediment load. Changes in river course on a decadal time-scale have resulted in temporarily abandoned reaches that are periodically reoccupied. Chute and neck cutoffs, and crevasse splays are also prominent. Borehole logs show that the anabranching reach is underlain by sandy channel bodies up to 25 m thick, separated by mudstone units up to 30 m thick. Extrapolation of floodplain accumulation rates to the mudstones suggests that channels were stably positioned for thousands to tens of thousands of years, allowing thick muds to accumulate. Repeated reoccupation of pre-existing drainage lines may have promoted the creation of thick, narrow channel bodies. Stacked overbank deposits probably form the bulk of the floodplain sediments, but channels that avulse into floodplain lakes (tals) may generate associated avulsion deposits. The Baghmati River sediments are a modern analogue for the deposits of rapidly subsiding extensional and foreland basins in the ancient record.

INTRODUCTION

The Baghmati River of the Himalayan foreland basin rises in the foothills of the Himalaya and drains the plains of north Bihar, eastern India, ultimately feeding into the Ganga River system (Fig. 1). The Baghmati River is located in an interfan position between the large Kosi and the Gandak fan systems but is a major river system in its own right. The river system displays an anabranching pattern in its mid-stream reaches and the system is aggrading rapidly in a region of rapid subsidence. Although fluvial deposits attributed to anabranching rivers are widespread in the geological record of foreland basins (Nadon, 1994), few modern examples of such rivers have been described.

This paper builds on earlier geomorphological and hydrological studies of the Baghmati River (Sinha & Friend, 1994; Sinha, 1996) to describe the near-surface sediments of the anabranching reach. Available deep borehole records have been examined to decipher the long-term sedimentation history. Of particular interest are the patterns of avulsion shown by the river (Jain, 2000; Jain & Sinha, 2003, 2004) and their implications for the

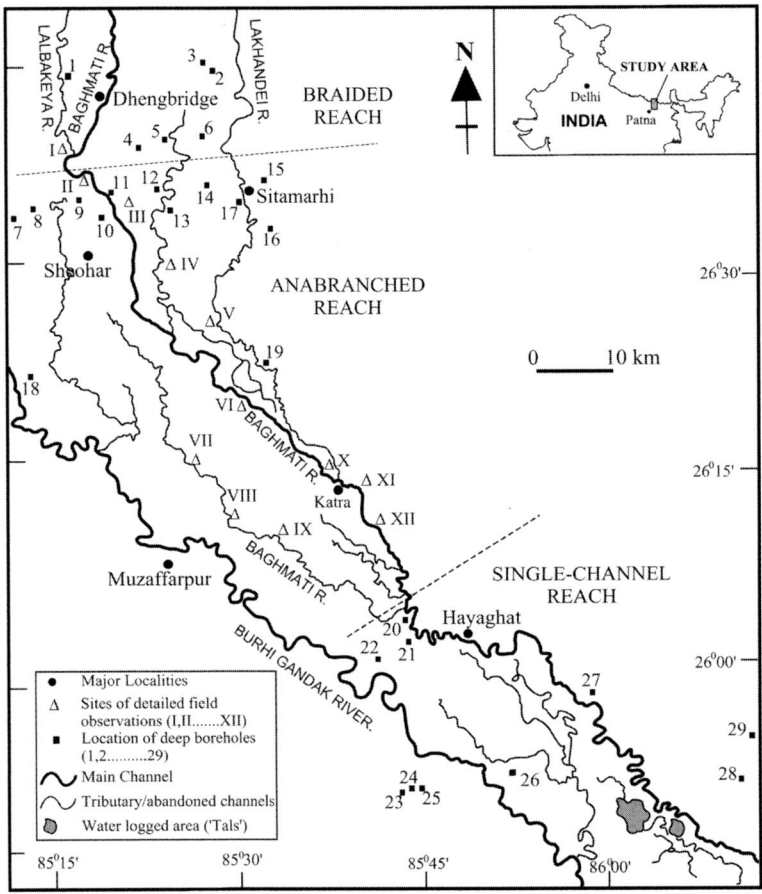

Fig. 1 Location map of the Baghmati River in the Himalayan foreland basin of north Bihar. Sites of field observation (I–XII) and boreholes (1–29) are marked; I, Dhengbridge; II, Belwa; III, Kola river bridge; IV and V, Sheohar road; VI, Janar; VII, Maksudpur; VIII–IX, Brick-pit sections; X, Katra; XI, Nawada; XII, Benibad.

stratigraphical record. Despite previous research on this theme (Smith *et al.*, 1989; Kraus, 1996; Jones & Schumm, 1999), more studies are needed on avulsions and their relationship with floodplain sedimentation, particularly from tectonically active and rapidly subsiding basins. The Baghmati River of the north Bihar plains offers such a setting and provides an opportunity to study the avulsion process and its implications for the development of sedimentary records on a time-scale of 10^1–10^3 yr from near-surface sedimentology, and $>10^4$ yr from borehole records. The study has important implications for the interpretation of ancient fluvial sequences, as the river shows a distinctive proximal–distal trend of braided–anabranching–meandering transitions that has not been recognized to date in the stratigraphical record.

THE BAGHMATI RIVER

The Baghmati River rises in the Sheopuri range in Nepal, and has a basin area of about 8848 km² (Sinha & Friend, 1994). The river is braided in its upstream reaches, south of the Himalayan Front, and is typically single-channel and meandering in its downstream reaches, with the sinuosity of local reaches ranging from 1.0 to 2.41, before its junction with the Kosi River. The Baghmati River has numerous anabranches in a middle-stream reach that is 75 km long (Fig. 1). Channel shape changes from wide and shallow (average w/d ratio 109) at an upstream station (Dhengbridge) to narrow and deep (average w/d ratio 16) at a downstream station (Hayaghat). In

Plate I Main anabranch of the Baghmati River at Benibad (site V, Fig. 1). The river displays a sinuous channel about 8 m deep and 100 m wide with a prominent cut-bank on the concave side (foreground). Note the chute channel on the inner side of the point bar.

Plate 2 Proximal floodplain (probably levee) strata. Close-up of a cut-bank section at Benibad (site XII, Fig. 1). Note sand and mud layers, ripple formsets with mud drapes, and scour fills. Coin is 20 mm in diameter.

the Dhengbridge–Hayaghat reach, channel slope decreases downstream from 3.07 m km^{-1} at Dhengbridge to 0.11 m km^{-1} at Hayaghat close to the low-gradient Ganga plains. Localized tectonic influence in the river basin is evidenced from major earthquakes in 1934 and 1988 and from subsurface faults mapped by earlier workers (Arita et al., 1973; Valdiya, 1976; Dasgupta et al., 1987; Dasgupta, 1993). Many reaches display features that are generally considered to indicate neotectonic influence, including straight channel reaches, compressed meanders, angular drainage patterns and sudden changes in flow direction (Jain, 2000; Jain & Sinha, 2001).

The channels of the Baghmati River are anastomosing in the sense of Knighton & Nanson (1993) and Makaske (2001), with two or more interconnected channels separated by floodplain segments. Anabranching is initiated downstream of the confluence of Lalbakeya and Baghmati rivers. The right anabranch flows southward and the left anabranch initially flows south-eastward for 15 km, before turning south to close the anabranching reach. Sequential reconstruction of channel configurations since 1924, based on maps and satellite images, shows that the right anabranch has been unstable, with several avulsions recorded. At present, the right anabranch joins the Burhi Gandak River to the west (Fig. 1) instead of rejoining the anabranching system. In addition to the major anabranches, there are numerous smaller anabranches that are connected to the major drainage network for short periods. Major changes in river course have been documented (Sinha, 1996; Jain, 2000), resulting in temporarily abandoned channel reaches, some many kilometres long. Most abandoned channels are partially filled from rainfall during the monsoon season. Crevasse splays, levees, large point bars, and floodplain lakes (tals) can be identified in the field and are visible on satellite images.

Hydrological data for the Baghmati River are available for two stations, namely Dhengbridge (upstream) and Hayaghat (downstream) (see Fig. 1 for locations), collected by the Central Water Commission (CWC), Government of India. The Baghmati River is characterized by variable discharge (average annual discharge 156–189 m^3 s^{-1}) and frequent, widespread overbank floods (mean annual flood 1100–1450 m^3 s^{-1}). The downstream station invariably shows a higher discharge, particularly in monsoon months, and this essentially reflects tributary influence and base-flow contribution. The peak discharge is extremely variable from year to year, with no apparent relationship in pattern or range of variation between the upstream and downstream stations. The total sediment load is higher at the upstream station (Dhengbridge, 10.41 Mt yr^{-1}) than at the downstream station (Hayaghat, 7.21 Mt yr^{-1}). There is a marked increase in sediment concentration during monsoon months related to intense rainfall and erosion in the Himalayan foothills. At the downstream station, much of the sediment is transported as wash load (generally 60–70%, but as high as 95% during some seasons), contributed in part from the erosion of muddy banks and reworking of floodplain sediments within the basin (Sinha & Friend, 1994).

Channel migration through avulsion is the most important fluvial process operating in the basin (Sinha, 1996). Based on comparison of maps and satellite images, Jain & Sinha (2003, 2004) documented avulsions on a decadal to century scale over the past century. Eight major and several minor avulsions have occurred in the past 250 yr, indicating a high degree of channel instability. The period of channel activity (time for which a particular channel is active) and lengths of channel abandoned are variable. Some of these avulsions, such as the Belwa avulsion (Fig. 1), have occurred around a nodal point, but others are essentially of abandonment and reoccupation type, triggered by major flood events or local sedimentological readjustments. A few examples of avulsions linked to neotectonic fault positions and tilting have been reported (Jain & Sinha, 2001, 2003), causing far-field shifts of relatively more stable channels (those with residence periods of 100 yr or more). In addition, neck and chute cut-offs are frequently observed in the field and on satellite images. Many cut-offs still receive water through the adjoining channel during high flows.

Overbank flooding is very frequent in the Baghmati plains, a consequence of the variable and periodically high peak discharge, coupled with a mean annual flood that exceeds the bank-full discharge at both upstream and downstream stations. Many floodplain areas are flooded every year and the average period of inundation is as high as 30 days. Thus, most of the depositional part of the river basin receives a frequent and abundant supply of overbank mud. Large floodplain lakes (tals) are present in the anabranching and single-channel reaches, with most of the lake water and sediment provided by small, plains-fed rivers with a high suspended load.

NEAR-SURFACE SEDIMENTS OF THE BAGHMATI PLAINS

Exposures in channel-cut banks, brick and sand pits, and small trenches were examined during investigations in the Baghmati plains at low stage in February 2000 (see Fig. 1 for sites of field investigations). These exposures provide information about most geomorphological elements of the river system, and are especially useful for comparison with lithological information from borehole logs. Sinha (1996), Sinha et al. (1996) and Sinha & Friend (1999) briefly described Baghmati channel and floodplain deposits. Descriptions of fluvial facies elsewhere in the Ganga plains have been presented by Kumar & Singh (1978), Wells & Dorr (1987), Singh & Bhardwaj (1991) and Singh et al. (1993), among others, and are broadly similar to those described here. The channels are not deeply incised in the study area, and exposed bank sediments are those of the modern, aggrading floodplain system, rather than earlier Holocene or late Pleistocene sediments.

Channel deposits

The upper braided reaches were studied only briefly, for example, at Belwa village (site II, Fig. 1). Here, the river is of shallow sand-bed type, with active channels about 200 m wide, and bank-attached and in-channel (braid) bars. Exposed sandy braid bars have ripples and two-dimensional dunes and mud veneers, with algal growth in adjacent sloughs. In a 1 m cut-bank section, shallow troughs, 100 mm deep, are filled with fine-grained sand that shows planar laminations below and ripple-drift cross-lamination above. The troughs are cut into large-scale inclined strata that dip towards the floodplain. These sediments probably represent a linear swale that was filled by lateral accretion of a bar margin. Towards the floodplain, channel sands pass into proximal floodplain (levee?) deposits of flat-lying sand and mud with in-phase climbing ripples and soft-sediment deformation. These deposits provide evidence for rapid bar and bank accretion with periods of upper-regime flow during near-bankfull stages. In the anabranching reach, the largest channel observed is about 8 m deep and 100 m wide at bankfull level. Most channels are smaller, typically 3–4 m deep and 50–100 m wide. The smallest channel is 2 m deep and 15 m wide. All the channels observed are sinuous, with point bars, chute channels on the point bars and low levees. Within the largest channel of the Baghmati system at Benibad (site XII), east of Muzaffarpur, the lowermost part of a prominent point bar is covered with two-dimensional dunes with superimposed ripples (Plate 1). A prominent chute channel separates lower and higher point-bar levels. The floodplain is under erosion at concave bends, with prominent cut-banks and slump blocks, and one small concave-bank bench was observed.

Proximal floodplain deposits

These deposits were observed in several channel-cut banks on concave bends. The banks are typically higher than the adjoining floodplain, suggesting that the sections represent levees, linked to the present channel course, that have been transected during channel migration. The geomorphological setting during the deposition of these sediments is uncertain, and we use the general term 'proximal floodplain deposits' for descriptive purposes. In the main Baghmati channel at Benibad (site XII), a 2 m section consists of intercalated sheets of very fine-grained sand and brown silt and clay, 10–30 cm thick (Plate 2). The section coarsens upwards, with progressively thicker and more sandy beds above. The sand sheets comprise stacked sets of ripple-drift cross-lamination and form sets of isolated ripples with mud drapes, and with mud- and sand-filled scours a few decimetres thick. Flow directions are away from the modern channel, consistent with an origin from channel spillover. Deposition took place rapidly under lower-regime conditions from flows with a large suspended-sediment load. Sand/mud couplets may represent individual overbank events or a series of events

Janar sand pit

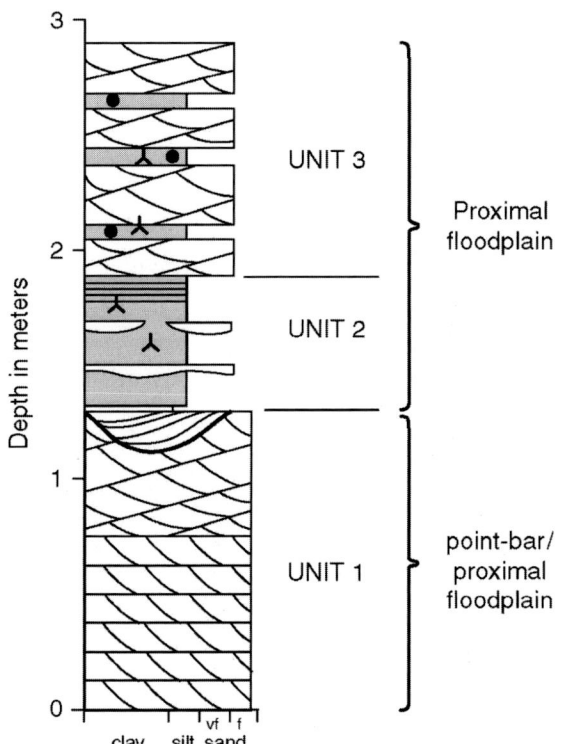

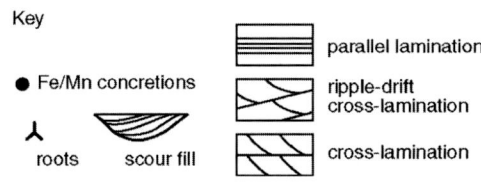

Fig. 2 Log of section in sand pit at Janar (site X, Fig. 1) interpreted as proximal floodplain deposits.

Muzaffarpur brick pit

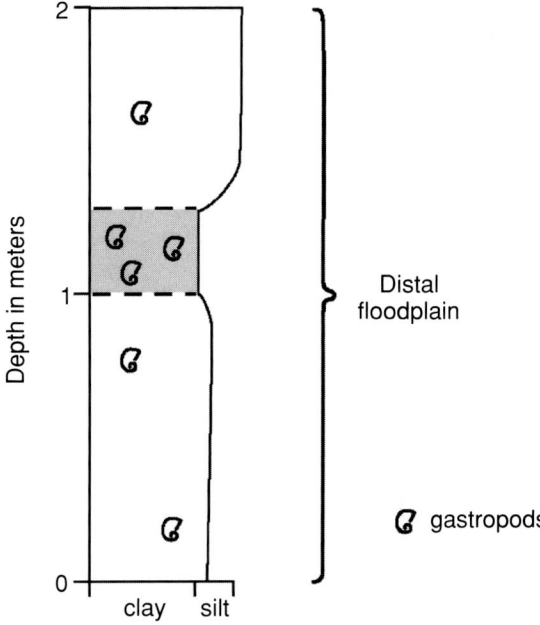

Fig. 3 Representative log of distal floodplain deposits in brick pits near Muzaffarpur (sites VIII and IX, Fig. 1).

during a single flood. Local lows in the modern levee mark possible future crevasse sites, and farmers have blocked some with clay walls.

At Janar village (site X), a small sand pit on the south bank of the Baghmati main channel was examined (Fig. 2). The lowermost unit consists of fine to very fine-grained, ripple cross-laminated sand with a 0.5 mm thick ripple-drift coset and a scoured top, and is interpreted as a point-bar top or proximal floodplain deposit. The middle unit consists of brown clay and silt, rooted and weakly stratified, with 50–100 mm lenses of very fine-grained sand. The topmost unit has 0.4 m-thick layers of fine- to very fine-grained sand with ripple-drift cross-lamination, interbedded with silty clays that contain roots and ferruginous or manganiferous concretions. Both middle and upper units are interpreted as proximal floodplain deposits.

Distal floodplain deposits

These were examined in pits east of Muzaffarpur (sites VIII and IX), excavated preferentially in clay-rich deposits suitable for brick manufacture. In a 2 m section (Fig. 3), which appears representative of the many pits in the area, the basal 1 m is grey silt-rich mud with sparse shells of high-spired gastropods, overlain by 0.3 m of dark brown, clay-rich mud with abundant scattered gastropods. The topmost 0.7 m is grey-yellow silt. Stratification is weak throughout, with gradational unit contacts. One pit about 200 m from a modern channel includes a 150 mm layer of very fine-grained sand within brown, shelly clay. The layer has a sharp, erosive base, a gradational top, and appears unstratified; it thickens into a shallow depression within the clay host. These muds have been interpreted as cumulative floodplain soils based on soil micromorphological features (Sinha & Friend, 1999) with accumulation rates of 0.7–1.5 mm yr^{-1} measured over periods of c. 1 kyr (Sinha et al., 1996). These areas receive frequent additions from floods and rarely dry out completely, resulting in only a modest degree of pedogenic alteration. The prominent sand bed is interpreted as a thin crevasse-splay sheet. The alternation of silt- and clay-rich intervals, which do not have the appearance of discrete splay events, could either represent alternate phases of reactivation and abandonment of nearby channels or variable flood magnitude.

Crevasse splay deposits

Thick sandy splay deposits were observed at a site north-east of Muzaffarpur, where they represent an incomplete stage of levee crevassing and splay development adjacent to large channels. At Nawada village (site XI), a crevasse and splay had developed during the previous flood season, less than 4 months before our visit. The channel at this site is > 6 m deep and c. 100 m wide. Previous erosion at the concave bank had exhumed a tough, clay-rich palaeosol, with roots, shells and charcoal, that forms a bench within the channel. The palaeosol is overlain in the cut bank by 2.5 m of brown silty clay with gastropods, articulated bivalves, woody fragments and roots. The topmost 0.6 m of bank sediments consists of ripple cross-laminated, very fine-grained sand with 50–100 mm clay layers. The crevasse had cut through these soft topmost sands to the more cohesive mud below.

The splay, which had invaded an orchard, was elongate, about 200 m long and 100 m wide, and excavation by villagers had revealed much of the deposit. At one site (Fig. 4), 1 m of sediment accumulated during the previous flood season, according to local information. The crevasse sediment rests on tough, rooted floodplain clay similar to the cut-bank mud, and contains five sand layers 0.05–0.4 m thick, with sharp or gradational bases and sharp tops. The sands contain

Fig. 4 Crevasse-splay section 1 m thick at Nawada (site XI, Fig. 1) deposited in the previous flood season. Note multiple sand layers with ripple-drift cross-lamination, mud sheets and drapes. Dark basal layer is underlying floodplain clay.

sets of ripple-drift cross-lamination 50 mm thick, with sigmoidal foresets and angle of climb decreasing upwards within the layers, as well as isolated ripple formsets with thin clay drapes. The sands are interbedded with brown mud layers < 100 mm thick, implying multiple flow pulses during the seasonal flood, with rapid deposition of abundant suspended sediment. Flow directions determined by excavation of ripple cross-sets varied considerably across the area, and flow in the upper part of the splay near the main channel indicated that water drained back into the channel at a late stage. At another site (Janar, site VI), a modern crevasse channel was observed, as well as older crevasse splay deposits exposed in pits similar to those in Nawada splay. A satellite image of this part of the

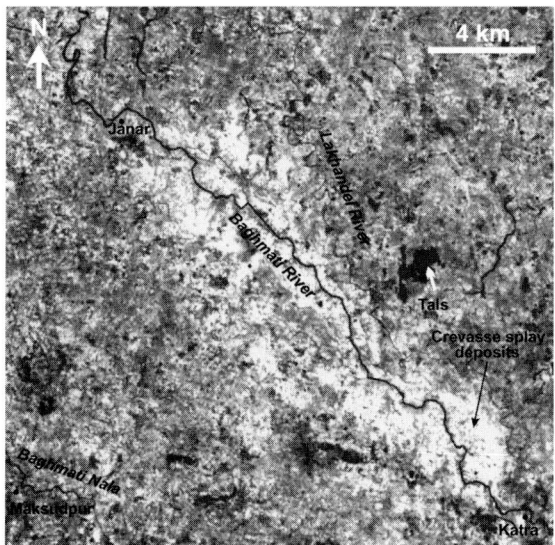

Fig. 5 Satellite image of 1989, showing the abundance of crevasse-splay deposits (pale areas) along the Baghmati River. Dark areas are 'tals'.

Baghmati channel system (Fig. 5) shows many white sand areas along the banks, suggesting that crevassing and splay emplacement are important processes.

Abandoned-channel fills

Temporarily abandoned channel reaches are common, but little sedimentological information was obtained because most fills are vegetated and are not exposed. A prominent abandoned channel was observed at the Kola River bridge (site III), built in 1906 over an active channel that was abandoned about 15 yr ago. Much of the abandoned channel is filled with sediment to the level of the bridge platform, and the remaining channel is 20–30 m wide and 5 m deep at concave bends. Pools and small vegetated point bars are present in some bends, and the surface channel fill consists of at least 0.5 m of brown, slightly silty clay with a thick vegetative cover. The width of the bridge (designed for an active channel) is almost double the width of the present, inactive channel, suggesting that much of the original channel has been occluded since abandonment. Brick pits on the floodplain close to a convex bend downflow from the bridge show 1.5 m of clay-rich mud, with a few sand layers that pinch out towards the floodplain.

Lake (tal) deposits

Lakes appear as dark patches on satellite images owing to their extremely low reflectivity (Fig. 5). These lakes are present mainly in the single-channel (downstream) reaches of the Baghmati River, particularly near its confluence with the Kosi River and in the vicinity of the main axial river system, the Ganga. The water depth of these lakes varies from 1 to 2 m for smaller lakes (< 5 km^2) and 4 to 5 m for larger lakes (> 25 km^2) during high-flow season. During the low-flow season, most of these lakes are swamps or marshy land. Lake sediments were observed at Baraila Tal, a clay-rich playa lake, fed only by small, plains-fed rivers, between the main Baghmati and Burhi Gandak systems. The study area was a zone 100 m wide along the lake margin, between standing water and crops, that had been inundated during the previous monsoon season and had a sparse cover of creeping plants. Excavation to 0.5 m depth revealed brown to grey clay, slightly silty and fairly cohesive, with yellow concretions up to 5 mm in diameter, roots and some preserved stems. Desiccation cracks up to 10 mm wide had generated polygons 0.3–0.5 m in diameter. The clay contains gastropods, shrimps, fish and insect remains, and reed-like plants grow in nearby standing water. Local information suggests that several metres of near-surface clay are underlain by sand and silt. The chimneys of crustacean burrows were abundant in this zone, composed of mamillated grey-yellow silty clay, the colour and silt content of which probably reflects burrow penetration through the underlying coarser sediment.

SUBSURFACE RECORDS

Twenty-nine lithological logs of deep boreholes, drilled in the Baghmati floodplains by water resource agencies between 1952 and 1977, were available for the study area (Fig. 1). Interpretation of the logs was greatly assisted by observations in the modern river system, outlined above. Sand and clay units (aquifers and aquicludes) were clearly distinguished in all logs, but the minimum layer thickness recorded is 0.5 m and the data quality is highly variable. Most boreholes were drilled at village sites in the floodplain some distance from major channels, and thus they tend to be mud-dominated at shallow depth. Sands are commonly recorded as very fine, fine, medium and coarse, and are mainly grey with a few brown intervals. They range in thickness from a few metres to > 40 m, and most are interpreted as channel bodies.

Clays are variably recorded as yellow, brown, light or dark grey and black, locally sandy or silty. Colour designations are likely to be highly subjective, but dark grey and black clays, which may represent distal floodplain and tal deposits, are indicated separately on the logs of Fig. 6 (although many brickpit clays appear dark brown at outcrop). Silty and sandy clays probably represent proximal floodplain and crevasse-splay deposits. Kankar, or nodular carbonate material, is recorded in both clays and sands. Kankar is probably present in the clays as isolated, *in situ* nodules, such as those described by Sinha & Friend (1999), and the logs contain no indication of solid kankar layers. Kankar is probably present mainly as reworked soil fragments in sands. Pebbles and 'stones' are recorded in some sands, but it is not clear if these are of extra- or intrabasinal origin.

Borehole locations are often imprecise (e.g. village names are the most common locator), no samples or age estimates are available, and the logs are too widely spaced for detailed correlation and determination of width/thickness ratios of channel bodies. Nevertheless, the log suite provides important information about major sandy and muddy units to about 100 m depth.

Lithologs from the upper braided reach (Fig. 6a) show interbedded coarse sand and clay down to 30–50 m depth, with sand bodies < 15 m thick and some thick units of dark clay. The logs were obtained at floodplain sites distant from the main Baghmati channel (Fig. 1), and the channel bodies probably represent small plains-fed rivers. At greater depth, most logs record thick sand bodies (locally > 40 m thick). These thick sands appear to be correlative between closely spaced logs (e.g. logs 4–6 in a 10 km cross-river transect) and probably constitute a widespread sand sheet generated by the Baghmati River.

Lithologs from the northern part of the anabranched reach (Fig. 6b) show a similar pattern of sand–clay intercalation in the top 50 m, with about equal proportions of the two sediment types. Sand units are predominantly in the fine and medium categories and are a few metres to 40 m thick, with a considerable number in the 10–25 m range. Clays are typically < 10 m but locally up to 30 m thick, and include thin dark units. Thin sand units within thick clays are probably crevasse splays such as those recognized in brick pit sections (sites VIII and IX). Where a reasonably spaced transect is present, as in logs 9–15, sand bodies in the upper part of the logs appear laterally discontinuous with intervening thick clays, suggesting that channel-body widths are a few kilometres or less. Below about 50 m depth, thick medium to coarse sand bodies are present, and probably represent a downstream continuation of the sand sheet beneath the braided reach.

Lithologs from the downstream single-channel reach (Fig. 6c) show a high proportion of clay throughout, with clay units (some of them dark) up to 50 m thick, especially in areas distant from the modern channel (logs 23–25); some dark clays may represent tals, which are common in this region. Thin sands intercalated with clays may represent crevasse splays or the occasional influx of sand-grade material into tals. Sands in the top 50 m range from a few metres to 25 m thick, with several in the 12–25 m range. Where lithologs are closely spaced (e.g. logs 20–22 in a 7 km cross-river transect), sand bodies are present at comparable levels and probably continuous, suggesting that some channel bodies are wider relative to their thickness than in the anabranched reach. There is little indication of extensive sand sheets at depth, although sand bodies at deeper levels are > 30 m thick locally.

Subsurface data from the Kosi Fan (Singh *et al.*, 1993) show a sheet of gravel and sand > 60 m thick capped by a surficial unit of sand and mud, typically up to 10 m thick but locally up to 40 m. The authors interpreted the lower unit as a braided-river deposit and the upper unit as a 'megafan sweep' succession, generated by migration of the active zone of smaller channels across the fan. These units broadly resemble the thick sand units and sand–mud units below the Baghmati braided and anabranched reaches. This similarity suggests

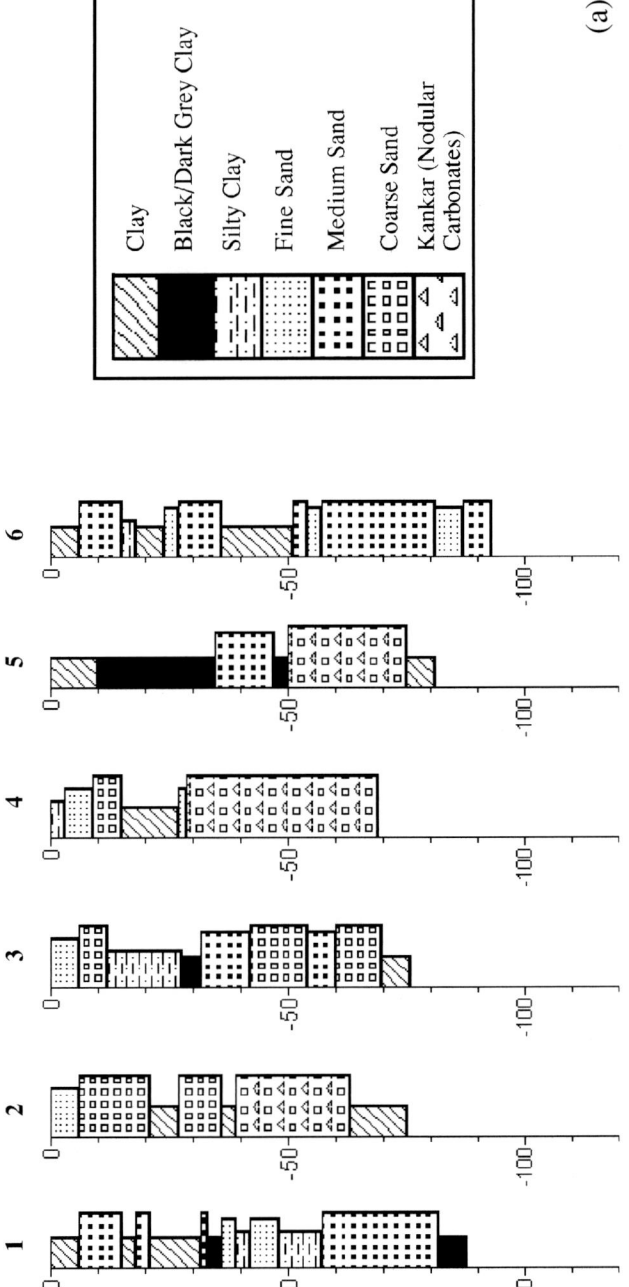

Fig. 6 Borehole logs (depth in metres) through the Baghmati River plains. Numbers at top indicate borehole locations, shown in Fig. 1: (a) upper braided reach.

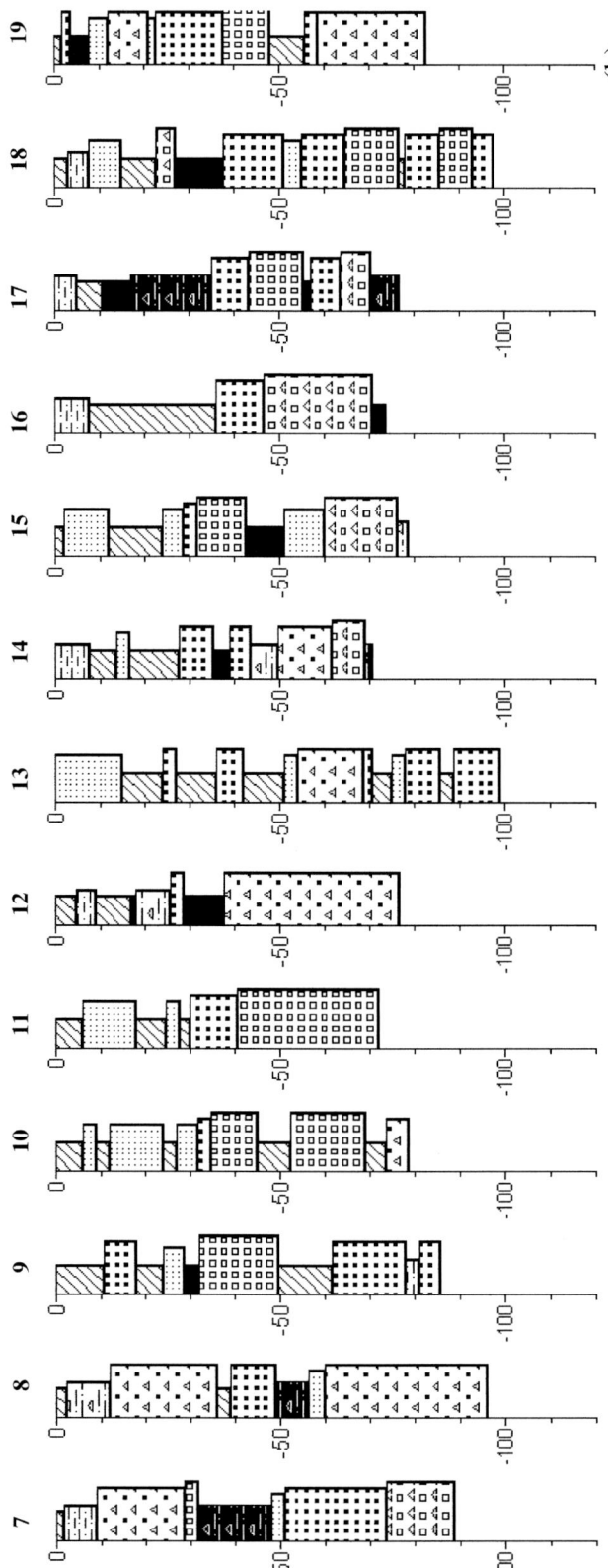

Fig. 6 (b) Middle anabranched reach.

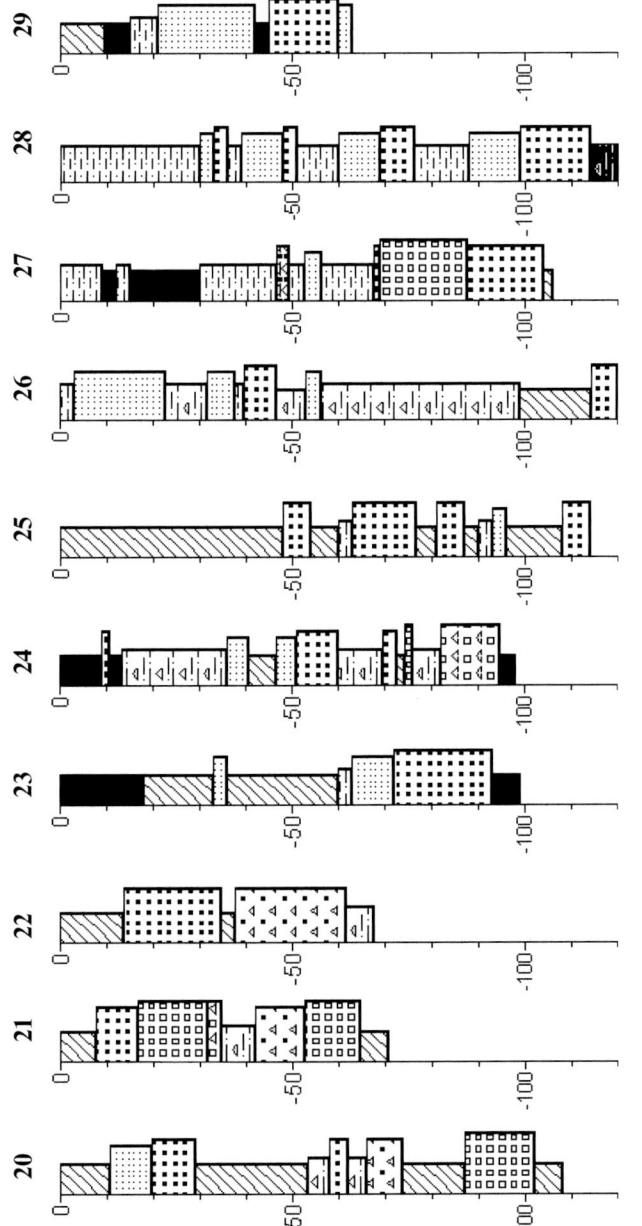

Fig. 6 (c) Lower single-channel reach.

that gravel–sand sheets and sand–mud units are widespread components of the northern Gangetic plains. The sand sheet at depth below the Baghmati braided and anabranching reaches probably represents braided facies of the Baghmati system, but might represent in part contributions from the adjoining Gandak and Kosi systems.

AVULSION DYNAMICS OF THE BAGHMATI RIVER

Maps and satellite images indicate that the anabranching Baghmati reach exhibits hyperavulsive behaviour, with prominent avulsion events recorded on a decadal to century scale (Jain, 2000; Jain & Sinha, 2003). Crevassing is widespread along most channels and commonly results in breakouts within a few flood seasons (as perhaps will happen at Nawada), assisted by frequent, large-flow events that result in prolonged overbank periods. Bank sediments are not strongly cohesive, and probably provide only limited resistance to erosion. The presence of point bars in most channels suggests that lateral migration and accretion are also active.

Jones & Schumm (1999) distinguished local and regional avulsions, in which the new course rejoins or fails to rejoin the previous channel, respectively. Although Baghmati anabranches occasionally avulse to locations beyond the present course of the river, for example, into the Burhi Gandak system to the west (see Fig. 1), most avulsive events are local. New, short channel segments are frequently cut, but most avulsive events over the past century have resulted in reoccupation of channels that were abandoned years to decades earlier (Jain, 2000).

Channel reoccupation during avulsion has been documented in the Texas Gulf Coastal Plain (Aslan & Blum, 1999), where a low sediment supply and slow floodplain aggradation promote the reoccupation of prominent Pleistocene palaeochannels. In the Cumberland Marshes of Saskatchewan, avulsion reoccupies older channels under conditions of stable lake level, but tends to create new courses where channels avulse into lakes with rising levels (Morozova & Smith, 2000). Reoccupation is also typical of the avulsions in the Rhine–Meuse system (Stouthamer & Berendsen, 2000). Neither scenario fits the Baghmati setting, where numerous courses coexist (and are periodically occupied and abandoned) on a rapidly aggrading floodplain with only small wetland areas. Avulsion frequencies recorded in river systems such as the Mississippi, Rhine–Meuse and Saskatchewan are typically much longer than those documented in the Baghmati—about 1600–2300 yr (Tornqvist, 1994; Morozova & Smith, 1999, 2000; Stouthamer & Berendsen, 2000). Avulsions on a decadal scale were documented by Van Gelder et al. (1994) on the rapidly aggrading Yellow River delta and by Jones & Harper (1998) in the Rio Grande River, which, during the past century, avulsed frequently over distances of less than 1 km.

In the Cumberland Marshes, crevasse splays are common only where avulsion reoccupies smaller pre-existing channels that were unable to accommodate the increased discharge (Morozova & Smith, 2000). Although progressive filling of abandoned channels in the Baghmati system reduces their dimensions, as at the Kola Bridge site, the abundance of crevasse splays along many channels suggests that crevassing is a routine process linked to poorly cohesive banks and a regime of widespread overbank floods.

Tectonic activity may cause avulsion, either instantaneously from earthquakes or over longer periods through changes in river and floodplain gradients (see discussions in Jones & Schumm, 1999, and Stouthamer & Berendsen, 2000). In the tectonically active Baghmati system, both instantaneous and longer term effects may cause avulsion. The positions of faults below this rapidly aggrading area are not known with great precision, however, and it is difficult to link avulsions to individual faults or to determine whether avulsion locations are clustered at nodal points on fault systems. Crevasse scenarios, such as that at Nawada, suggest that major floods are the most common avulsion triggers, as noted elsewhere by Jones & Harper (1998) and Makaske (2001).

Recent research has recognized the importance of avulsion deposits in modern river basins and in the ancient record. In some systems, avulsion and splay progradation into wetlands causes rapid, short-term sediment accumulation while the new reach is established (Smith et al., 1989; Kraus, 1996; Morozova & Smith, 2000). In consequence, some floodplain deposits are dominated by sediments laid down during short avulsive periods, rather than by sediments laid down over long peri-

ods by repeated overbank floods. In the Baghmati River, avulsion deposits may be generated where channels avulse into tals, perhaps indicated by the intercalation of channel bodies with exceptionally thick mud bodies. Several lines of evidence, however, suggest that many Baghmati floodplain accumulations consist of sediments that we term *repeated flood deposits*:
1 the fluvial regime involves near-annual, extensive, and sediment-charged floods;
2 most crevassing feeds sediment through levees onto floodplains that are dry outside the monsoon season, and no progradational splay complexes into lakes and wetlands (such as those of the Cumberland Marshes) have been identified;
3 geochronological evidence (Sinha et al., 1996) shows that extensive near-surface areas of the plains have experienced rapid accumulation from relatively stable channels;
4 'channel reoccupation', rather than 'new course construction', appears to be the main avulsive process, although both modes of floodplain accumulation are probably operative locally.

LONG-TERM HISTORY OF THE BAGHMATI ANABRANCHING REACH

The modern anabranching reach of the Baghmati is underlain by thick sand units (typically 10–25 m) and thick mud units (up to 25 m), with widths of some channel bodies constrained to less than a few kilometres, probably much less. Although the channel-body architecture is incompletely known, a geometric pattern of numerous bodies with a relatively low width : thickness ratio and encased in floodplain muds accords well with models for fixed-channel systems that commonly have an anabranching planform (Friend, 1983; Gibling et al., 1998).

Floodplain accumulation estimates of 0.7–1.5 mm yr^{-1} over the past c. 2400 yr (Sinha et al., 1996) imply rapid aggradation during the late Holocene. It is not clear whether sediment accumulation was continuous at the study sites, nor whether rates have been enhanced through time in this intensely agricultural area as a result of deforestation and an increased sediment yield. These rates are much higher than those documented for other near-surface parts of the Gangetic plains. Joshi & Bhartiya (1991) estimated a rate of 0.2 mm yr^{-1} over a period of 10 kyr based on ^{14}C dates from eastern Uttar Pradesh, and Chandra (1993) estimated 0.2–0.05 mm yr^{-1} over 10 kyr based on luminescence dates from the middle Gangetic plains. The eastern Uttar Pradesh plains also show mature soils, 3–4 m thick, with well-developed carbonate horizons, estimated to be as old as 13 500 yr BP (Srivastava et al., 1994; Srivastava, 2001).

In view of the apparent similarity of near-surface facies to lithological units in boreholes, it is reasonable to extrapolate this accumulation rate to longer periods. Large floods may cause rapid short-term accretion, for example, a rate of 93 mm yr^{-1} was estimated by Sinha et al. (1996) at a Baghmati bank section near Minapur. Given such uncertainties and the probability of human influence, the time frame of these deposits can be estimated only in general terms. The maximum thickness of continuous muddy floodplain deposits that underlie the middle anabranching reach is 30 m, excluding the sandy units. Assuming a maximum accretion rate of 1.5 mm yr^{-1} and disregarding compaction of these relatively cohesive units, the muddy units would have accumulated in about 20 kyr. The period represented by these deposits might be considerably less if exceptional floods contributed periodic pulses of fine sediment. Nevertheless, this estimate suggests that floodplain fines accumulated locally for periods of thousands to tens of thousands of years without a channel avulsing into the area, and also implies that channels within the anabranching reach were relatively stably positioned.

The thickness of many channel-sand bodies in the upper parts of the logs (commonly 10–25 m) is much greater than the depth of the modern Baghmati channels (maximum observed c. 8 m). Comparisons of width are more difficult to make in view of the large borehole spacing. Several explanations for this disparity are possible.
1 Channels with similar dimensions to the modern anabranching network persistently reoccupied older drainage lines while sediments accumulated on the adjacent plains, generating thick channel bodies with superimposed storeys. This explanation is in accord with the recent history of the Baghmati system.
2 Levee buildup allowed prolonged aggradation of sediments within the channels. Although likely to be a contributing factor, such an explanation

implies a long-term balance between levee and channel sedimentation that may be difficult to achieve (see discussion in Makaske, 2001, p. 185).

3 Channel dimensions changed through time, perhaps owing to secular variation in discharge, sediment load, or subsidence rate. This cannot be ruled out, although no firm evidence currently supports this possibility.

4 The natural riparian vegetation may have stabilized banks more effectively than is currently the case in this agricultural region, perhaps promoting aggradation within the channels. Although vegetation change must have affected the Baghmati plains, the effect of such change on channel dimensions is difficult to quantify.

APPLICATION TO THE ANCIENT RECORD

Nadon (1994) noted that ribbon-channel bodies, many of which may represent anabranching rivers, are especially common in the fills of rapidly subsiding ancient basins, especially extensional and foreland basins. To date, few modern anabranching examples have been described from such settings, exceptions being the Magdalena River of the Andean Foreland Basin (Smith, 1986) and the Okavango River in the African Rift System (McCarthy et al., 1992). Many of the best studied anabranching rivers are from cratonic and other low-subsidence settings.

Makaske (2001), in a review of anastomosing rivers, identified a group of 'long-lived, highly dynamic anastomosing systems, constantly rejuvenated by frequent avulsions', and noted their great importance for the geological record because such systems aggrade rapidly and produce thick fluvial successions. The Baghmati River, with a foreland-basin setting, high sediment load, and relatively narrow channel bodies, may be a particularly suitable modern analogue for narrow channel bodies in ancient high-subsidence settings. Figure 7 shows a provisional model for the Baghmati River system, with particular reference to major processes and architectural features of the anabranching reach. Anabranching channel patterns are also present on parts of the Kosi Fan (Gohain & Parkash, 1990), and many parts of the Himalayan Foreland Basin may contain deposits of this type.

CONCLUSIONS

The Baghmati River of the Himalayan Foreland Basin in north Bihar, India, has a high sediment load and peak discharges that frequently inundate the floodplain for long periods. An upstream braided reach and downstream single-channel reach are separated by a 75-km-long anabranching reach with sinuous channels and a floodplain that has aggraded rapidly during the late Holocene. This reach has experienced avulsions on a decadal scale over the past century, with frequent crevassing through levees and poorly stabilized banks. Avulsion periodically abandons and reoccupies long channel reaches.

Borehole logs down to about 100 m show that the anabranching reach is underlain by sandy units, 10–25 m thick, that are probably relatively narrow and separated by mudstone units up to 30 m thick. Extrapolation of near-surface floodplain accumulation rates to these mudstones suggests that channels were stably positioned for tens of thousands of years, allowing thick fine-grained units to build up. Repeated reoccupation of drainage lines with depths comparable to those of the modern channels may have promoted the creation of thick channel bodies. The floodplain deposits probably include both repeated (seasonal) flood deposits, from floods such as those that inundate the Baghmati plains in most years, and avulsion deposits into floodplain lakes (tals). The Baghmati River sediments constitute a good modern analogue for many extensional and foreland basin deposits in the ancient record.

ACKNOWLEDGEMENTS

Funding from the All India Council of Technical Education in the form of a Career Award to R. Sinha is thankfully acknowledged. Funding from a grant to M. Gibling from the Natural Sciences and Engineering Research Council of Canada is gratefully acknowledged. The manuscript was written while R. Sinha was in receipt of the Alexander von Humboldt Fellowship at the Institute of Mineralogy and Geochemistry, Karlsruhe University, Germany, and while M. Gibling was on leave at the Department of Geography, Exeter University, UK; we thank both these institutions for their generous assistance. We thank M. Jain,

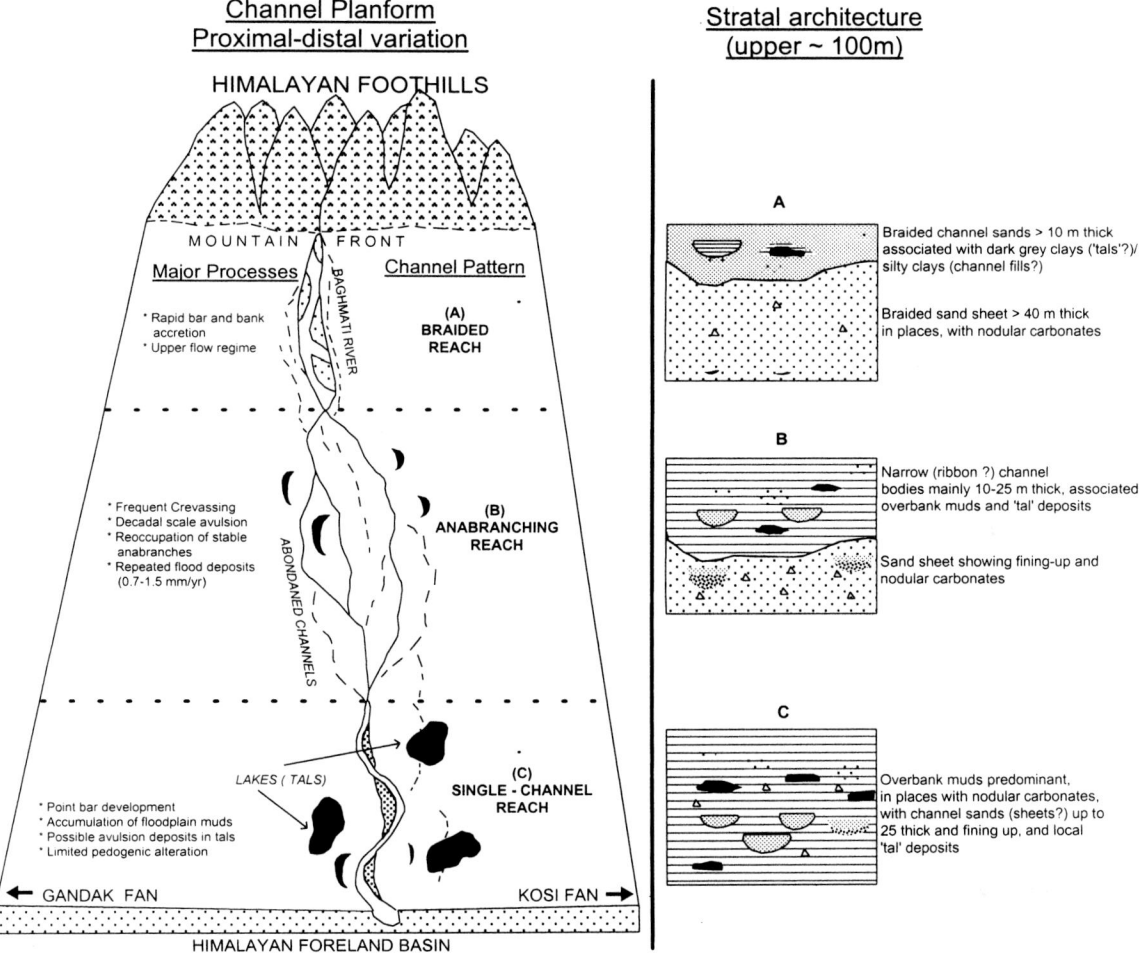

Fig. 7 Sedimentation model for the hyperavulsive Baghmati River in the Himalayan Foreland Basin.

University of Wales, Aberystwyth, for fruitful discussions in the field.

REFERENCES

Arita, K., Ohta, Y., Akiba, C. and Maruo, Y. (1973) Kathmandu Region. In: *Geology of the Nepal Himalayas* (Ed. Y. Ohta and C. Akiba), pp. 99–146, Saikon Publishing Company, Tokyo.

Aslan, A. and Blum, M.D. (1999) Contrasting styles of Holocene avulsion, Texas Gulf coastal plain, USA. In: *Fluvial Sedimentology VI* (Eds N.D. Smith and J. Rogers). Int. Assoc. Sedimentol. Spec. Publ., **28**, 193–209.

Chandra, S. (1993) Fluvial landforms and sediments in the north-central Gangetic Plain, India. Unpublished PhD Thesis, Darwin College, University of Cambridge, Cambridge.

Dasgupta, S. (1993) Tectono-geologic framework of the eastern Gangetic foredeep. *Geol. Surv. India Spec. Publ.*, **31**, 61–69.

Dasgupta, S., Mukhopadhyay, M. and Nandy, D.R. (1987) Active transverse features in the central portions of the Himalaya. *Tectonophysics*, **136**, 255–264.

Friend, P.F. (1983) Towards the field classification of alluvial architecture or sequence. In: *Modern and Ancient Fluvial Systems* (Eds J. Collinson and J. Lewin). Spec. Publs Int. Assoc. Sediment., **6**, 345–354.

Gibling, M.R., Nanson, G.C. and Maroulis, J. (1998) Anastomosing river sedimentation in the channel country of central Australia. *Sedimentology*, **45**, 595–619.

Gohain, K. and Parkash, B. (1990) Morphology of Kosi megafan. In: *Alluvial Fans: a Field Approach* (Eds A.H. Rachocki and M. Church), pp. 151–178. Wiley, Chichester.

Jain, V. (2000) Fluvio-geomorphological analysis of Baghmati river system, North Bihar with special reference to fluvial hazards. Unpublished PhD thesis, Indian Institute of Technology, Kanpur.

Jain, V. and Sinha, R. (2001) Response of neotectonics on the anabranching Baghmati river system, Himalayan foreland basin, eastern India. *Proceedings of the International Conference on Fluvial Sedimentology* (Eds J.A. Mason, R.F. Diffendal Jr and R.M. Joeckel), Nebraska, USA, August 2001, 141.

Jain, V. and Sinha, R. (2003) Hyperavulsive-anabranching Baghmati river system, north Bihar plains, eastern India. *Z. Geomorph.*, **47**(1), 101–116.

Jain, V. and Sinha, R. (2004) Fluvial dynamics of an anabranching river system in Himalayan foreland basin, Baghmati river, north Bihar plains, India. *Geomorphology*, **60**, 147–170.

Jones, L.S. and Harper, J. (1998) Channel avulsions and related processes, and large-scale sedimentation patterns since 1875, Rio Grande, San Luis Valley, Colorado. *Geol. Soc. Am. Bull.*, **110**, 411–421.

Jones, L.S. and Schumm, S.A. (1999) Causes of avulsion: An overview. In: *Fluvial Sedimentology VI* (Eds N.D. Smith and J. Rogers). *Spec. Publ. Int. Assoc. Sedimentol.*, **28**, 171–178.

Joshi, D.D. and Bhartiya, S.P. (1991). Geomorphic history and lithostratigraphy of a part of the eastern Gangetic plain, Uttar Pradesh. *J. Geol. Soc. India*, **37**, 569–576.

Knighton, A.D. and Nanson, G.C. (1993) Anastomosis and the continuum of channel pattern. *Earth Surf. Process. Landf.*, **18**, 613–625.

Kraus, M.J. (1996). Avulsion deposits in lower Eocene alluvial rocks, Bighorn basin, Wyoming. *J. Sediment Res.*, **66B**, 354–363.

Kumar, S. and Singh, I.B. (1978) Sedimentological study of Gomti river sediments, Uttar Pradesh, India: example of a river in alluvial plain. *Senckenberg. Marit.*, **10**, 145–211.

Makaske, B. (2001) Anastomosing rivers: a review of their classification, origin and sedimentary products. *Earth Sci. Rev.*, **53**, 149–196.

McCarthy, T.S., Ellery, W.N. and Stanistreet, I.G. (1992) Avulsion mechanisms on the Okavango fan, Botswana: the control of a fluvial system by vegetation. *Sedimentology*, **39**, 779–795.

Morozova, G.S. and Smith, N.D. (1999) Holocene avulsion history of the lower Saskatchewan fluvial system, Cumberland Marshes, Saskatchewan–Manitoba, Canada. In: *Fluvial Sedimentology VI* (Eds N.D. Smith and J. Rogers). *Spec. Publ. Int. Assoc. Sedimentol.*, **28**, 231–249.

Morozova, G.S. and Smith, N.D. (2000) Holocene avulsion styles and sedimentation patterns of the Saskatchewan river, Cumberland marshes, Canada. *Sediment. Geol.*, **130**, 81–105.

Nadon, G.C. (1994) The genesis and recognition of anastomosed fluvial deposits: data from the St. Marr River Formation, southwestern Alberta, Canada. *J. Sediment. Res.*, **B64**, 451–463.

Singh, A. and Bhardwaj, B.D. (1991) Fluvial facies model of the Ganga river sediments, India. *Sediment. Geol.*, **72**, 135–146.

Singh, H., Parkash, B. and Gohain, K. (1993) Facies analysis of the Kosi megafan deposits. *Sediment. Geol.*, **85**, 87–113.

Sinha, R. (1996) Channel avulsion and floodplain structure in the Gandak–Kosi interfan, north Bihar plains, India. *Z. Geomorph.*, **103**, 249–268.

Sinha, R. and Friend, P.F. (1994) River systems and their sediment flux, Indo-Gangetic plains, northern Bihar, India. *Sedimentology*, **41**, 825–845.

Sinha, R. and Friend, P.F. (1999) Pedogenic Alteration in the Overbank Sediments, North Bihar Plains, India. *J. Geol. Soc. India*, **53**, 163–171.

Sinha, R., Friend, P.F. and Switsur, V.R. (1996) Radiocarbon dating and sedimentation rates of Holocene alluvial sediments of northern Bihar plains, India. *Geol. Mag.*, **133**(1), 85–90.

Smith, D.G. (1986) Anastomosing river deposits, sedimentation rates and basin subsidence, Magdalena river, northwestern Colombia, South America. *Sediment. Geol.*, **46**, 177–196.

Smith, N.D., Cross, T.A., Dufficy, J.P. and Clough, S.R. (1989) Anatomy of an avulsion. *Sedimentology*, **36**, 1–23.

Srivastava, P. (2001) Palaeoclimatic implications of pedogenic carbonates in Holocene soils of the Gangetic Plains, India. *Palaeogeogr., Palaeoclimatol., Palaeoecol.*, **72**(3–4), 207–222.

Srivastava, P., Parkash, B., Sehgal, J.L. and Kumar, S. (1994) Role of neotectonics and climate in development of the Holocene geomorphology and soils of the Gangetic Plains between the Ramganga and Rapti rivers. *Sediment. Geol.*, **94**, 129–151.

Stouthamer, E. and Berendsen, H.A.J. (2000) Factors controlling the avulsion history of the Rhine–Meuse Delta (The Netherlands). *J. Sediment. Res.*, **70**(5), 1051–1064.

Tornqvist, T.E. (1994) Middle and late Holocene avulsion history of the River Rhine (Rhine–Meuse Delta, Netherlands). *Geology*, **22**, 711–714.

Valdiya, K.S. (1976) Himalayan transverse faults and folds and their parallelism with subsurface structures of north Indian Plains. *Tectonophysics*, **32**, 353–386.

Van Gelder, A., Van Den Berg, J.H., Cheng, G. and Xue, C. (1994) Overbank and channel fill deposits of the modern Yellow River Delta. *Sediment. Geol.*, **90**, 293–305.

Wells, N.A. and Dorr, J.A. (1987) Shifting of Kosi River, northern India. *Geology*, **15**, 204–207.

Estimating bedload in sand-bed channels using bottom tracking from an acoustic Doppler profiler

PAUL VILLARD*,[1], MICHAEL CHURCH† and
RAY KOSTASCHUK‡

*Department of Geography, The University of British Columbia, Vancouver,
British Columbia, V6T 1Z2, Canada;
†Department of Geography and Peter Wall Institute for Advanced Studies, The University of
British Columbia, Vancouver, British Columbia, V6T 1Z2, Canada; and
‡Department of Geography, University of Guelph, Guelph, Ontario, N1G 2W1, Canada

ABSTRACT

A method to estimate bedload in sand-bed rivers using bottom tracking from an acoustic Doppler profiler (ADP) is outlined and tested. The velocity of a mobile sand bed is related to the difference between the 'apparent' boat velocity with respect to the bed measured with ADP bottom tracking and the 'actual' boat velocity measured with a differential global positioning system (DGPS). Under simple assumptions the velocity of a mobile sand bed can be converted to an estimate of bedload. Estimates of bedload using this method were acquired from a launch anchored over the upper stoss face of large sand dunes in the Fraser River, British Columbia. These estimates are compared with measurements from a Helley–Smith sampler and predictions from the Van Rijn bedload formula. All three methods produce consistent and comparable values. Near concordance was observed between the mechanical and acoustic estimates of bedload transport, although with a large degree of scatter, indicating that both instruments are measuring similar fractions of near-bed transport. In comparison, the correlation with computed transport, although stronger, was more strongly biased. Part of the difficulty in reconciling the measurements lies in the arbitrary nature of the division between 'bedload' and near-bed suspension in transport over a sand bed. Within this constraint, ADP technology shows potential to yield remote measurements of bedload in sand-bed channels, escaping the limitations of mechanical samplers.

INTRODUCTION

Establishing sediment transport in sand-bed channels is of vital importance to both sedimentological and engineering studies. Sand transport is commonly divided into suspended load and bedload, owing both to the physics of the transport process and the workings of traditional mechanical samplers. By definition, suspended load is material that is buoyed by the water motion

[1]Present address: PARISH Geomorphic Ltd, Georgetown, Ontario, L7G 4J9 Canada. Email: pvillard@parishgeomorphic.com

and bedload is the material that moves near the bed by traction and saltation. In practice, owing to limitations of mechanical samplers, a portion of suspended load near the bed (< 0.10 m) is also included in bedload measurements (e.g. Emmett, 1980). Although methods for measuring both components are available, development of reliable methods for measuring bedload has proven difficult. Several mechanical methods have been proposed for measuring bedload transport (e.g. Helley & Smith, 1971; Langley, 1971; Klingeman & Milhous, 1971). In sand-bed channels, one of the more commonly used mechanical samplers is the pressure difference Helley–Smith (HS) sampler (Ryan & Porth, 1999) but, despite modifications to the original design, it has—like all mechanical samplers—several significant limitations. Mechanical samplers disturb the flow and bed, offer only a point measurement, and are difficult to deploy in deep channels. The resulting measurements have large associated errors and are prone to overestimations (e.g. Emmett, 1980; Gomez et al., 1989, 1990; Glysson, 1993; Ryan & Porth, 1999).

Mechanical methods for measuring suspended sediment concentration and flux in open channels have, to a large extent, been superseded by optical and acoustic instruments, which allow rapid, high-resolution, remote sampling (e.g. Downing, 1983; Hay, 1983; Thorne et al., 1991; Zedel & Hay, 1999; Agrawal & Pottsmith, 2000). There have also been attempts to use acoustics to measure bedload transport (e.g. Lowe et al., 1991; Sutton & Jaffe, 1992; Rennie et al., 2002).

Rennie et al. showed that the Sontek® Acoustic Doppler Profiler (ADP), an off-the-shelf instrument, has the potential for sensing bedload transport remotely in cobble-bed channels. The method estimates bedload velocity by utilizing the bottom-tracking technology available in the ADP. Unfortunately the heterogeneous nature of transport in cobble-bed channels, coupled with the firmware associated with the bottom-tracking, results in very large error estimates and necessitates long sampling periods. The proprietary pulse-to-pulse incoherent 'narrow band' technique used in the bottom tracking is optimized for reflection off a hard, immobile substrate, where a narrow peak in return frequencies is expected (Rennie et al., 2002). This complicates the application of the technique in partial transport environments, such as cobble-bed channels, where the range of particle speeds and, therefore, return frequencies is broad. The method might be more appropriately applied in transport environments where a narrower spectrum of signal return frequencies can be expected, such as in sand-bed channels where sediment transport is likely to be more continuous locally.

The objectives of this paper are:
1 to examine the utility of using bottom tracking by a commercially available ADP to estimate the velocity of mobile sand over the bed;
2 to outline a simple method for converting the speed of the near-bed mobile sand layer to an estimate of bedload transport;
3 to compare these estimates with measurements from a mechanical Helley–Smith bedload sampler and estimates from the Van Rijn (1984) bedload transport formula.

It is intended, thereby, to demonstrate the feasibility to obtain estimates of bedload transport in sand-bed rivers using remote acoustic technology.

STUDY AREA

Measurements were conducted in Canoe Pass, a small sand-bedded distributary that branches from the South Arm of the Fraser River and discharges onto Roberts Bank and into the Strait of Georgia, a meso-tidal marine basin on the southwest coast of Canada. The Fraser River has a mean annual discharge of 3400 m^3 s^{-1}, with maximum freshet-driven flows over 11 000 m^3 s^{-1}. The portion of the reach where measurements were conducted has been the site of several hydraulic and sand transport studies (e.g. Villard, 1995; Kostaschuk, 2000). The channel is approximately 350 m wide and the deepest sections are < 10 m deep at high tide. Although the flows are tidally influenced, there is little evidence of salt-wedge intrusion during the freshet, and flows are consistently seaward (Villard, 1995). Large, steep asymmetric dunes reach heights of 3 m and lengths of 50 m (Kostaschuk, 2000).

METHODS

Observations were collected on 8, 13 and 16 June 2000. Over the study period, river discharge was rising towards the annual snow melt freshet peak. Daily average discharge measured at Hope (140 km up-river) was 6320 m^3 s^{-1} on 8 June, 7560 m^3 s^{-1} on 13 June and 7180 m^3 s^{-1} on 16 June. Discharge peaked at 8050 m^3 s^{-1} on 6 July.

Measurements were collected from a launch on the rising and falling tide, giving a range of velocities, depths and transport rates. Water depth and bedform geometry were measured with a 200 kHz Bathy® 1500 digital echo-sounder, bedload transport measurements were made with a HS sampler, surficial sediment samples were collected with a pipe-dredge, and a Sontek® 1500 kHz, three-beam acoustic Doppler profiler (ADP) was used to measure the velocity of the mobile bed and water column. Both the echo-sounder and ADP were tied to a Trimble® AgGPS Model 122 DGPS. The echo-sounder collected a measurement every second and has a stated vertical accuracy of ±0.01 m.

The HS sampler has a 0.076 m by 0.076 m orifice and a 0.208-mm mesh bag. Repeated sampling with the HS sampler in the South Arm of the Fraser River indicates that estimates can vary around the mean by ±35% (99% confidence limits; ±23% for 95% confidence) at constant flow conditions (Kostaschuk & Ilersich, 1995). A part of the scatter may represent real variations in transport related to the position of the sampler on the dune.

Initially, default settings on the bottom tracking were used to measure the velocity of the mobile bed, giving an acoustic pulse length of 3 m. With this pulse length the ADP had difficulty resolving the bed position in depths less than 7 m. The pulse length was reduced to 0.20 m, which removed this problem. Owing to the problems with the bottom tracking before the settings were modified, measurements from six samples collected on 8 June could not be used.

Downstream transects with the echo-sounder were run before commencing sampling. From the transect a large dune was selected for closer examination. The dune was surveyed and the launch was anchored over the dune crest. Over the next several hours, 5-min mechanical bedload measurements along with simultaneous ADP measurements (5-min bursts of 5-s averages) of bed speed and water velocities were collected. The long sampling period was to allow collection of a significant number of bottom-tracking estimates. The HS sampler was deployed near the edge of the sampling circle defined by the ADP beams. The constant displacement (of about 3 m) between the centroid of the bottom-track measurements and the bedload sampling was dictated by the arrangements on the launch for instrument deployment and may have introduced an unknown measure of variability between the two sets of measurements. The bedload samples never filled more than one-third of a Helley–Smith mesh bag, so the hydraulic performance of the sampler is considered not to have been compromised. Through each 5-min run the bed position with respect to the launch was monitored with the echo sounder. After several stationary measurements the position of the launch with respect to the dune crest was checked and the boat was repositioned if necessary. After anchored measurements were completed a surficial bed sample was collected with the pipe-dredge and the dunes were resurveyed.

Proximity to the upper side of the dune was maintained to allow the collection of velocity profiles that best relate to the sediment transport (Villard & Kostaschuk, 1998; McLean et al., 1999a,b). Bedload and near-bed suspension dominate in this region (Kostaschuk & Villard, 1996, 1999; Carling et al., 2000). Furthermore, bedload transport is expected to be relatively homogeneous (spatially) across the upper stoss towards the dune crest (Kostaschuk & Villard, 1996), a consideration in respect of the large sampling area and requirement of consistency between transducers to resolve the bottom tracking, and because of the displacement of the mechanical sampler.

ESTIMATING BEDLOAD TRANSPORT WITH AN ACOUSTIC DOPPLER PROFILER

The ADP uses an internal compass (resolution ±2°) to define flow directions and a tilt sensor to correct for ship pitch and roll (resolution ±1°). The three ADP transducers are set at 25° from the vertical axis and are equally spaced radially (120°), producing profiles of velocity components in the along-beam direction measured from scatterers in the ensonified field (Fig. 1). As the relative orientation of the three transducers is known, the three along-beam velocities can be

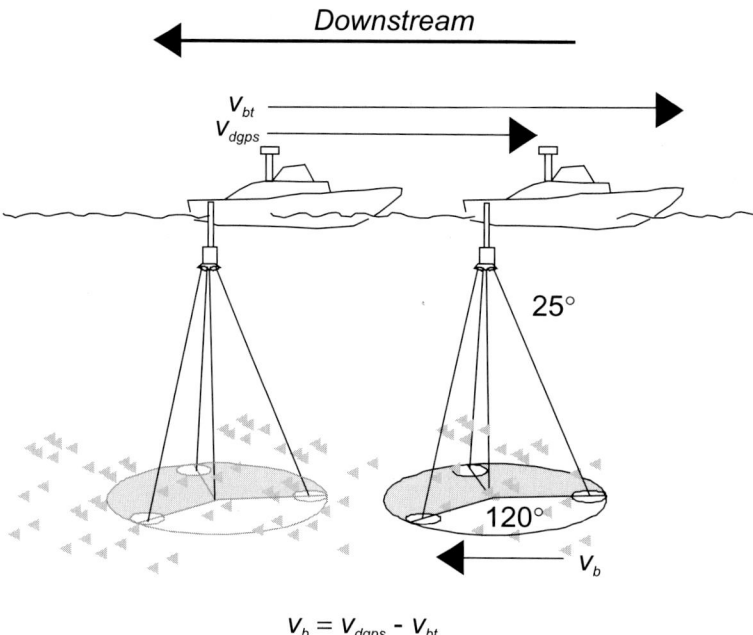

Fig. 1 An idealized sketch (not to scale) of bottom tracking and DGPS used in coordination to estimate bedload. Left vessel represents the initial position; right vessel represents true final position. Top arrows indicate apparent distance measured by the two methods.

transformed to Cartesian velocities referenced to north or to the instrument orientation.

When mounted on a moving vessel, correction for the vessel's velocity is necessary to obtain absolute water velocities. Two methods of correction are available. One is to determine the vessel's velocity directly from estimates of the vessel position at successive time-steps using a differentially corrected global positioning system (DGPS). The second method uses bottom tracking, which involves measurement of the Doppler shift of an independent acoustic echo from the substrate. Although bottom tracking is considered more accurate than the DGPS, it produces an absolute correction only over hard, immobile substrates. If the bed is mobile, the Doppler shift is a function of the velocity of the vessel and of the mobile bed. The resulting bias in the estimate of vessel velocity is used to estimate the velocity of the mobile bed. Rennie *et al.* (2002) showed that the velocity of a mobile bed, v_b, is given by the difference between the 'apparent' launch velocity with respect to the bed, v_{bt}, measured with ADP bottom tracking, and the 'actual' launch velocity, v_{dgps}, measured with the DGPS (Fig. 1).

Potential instrument-related sources of error include acoustic noise, the proprietary internal broad-band filter, and compass and DGPS positioning (Rennie *et al.*, 2001). A measure of standard deviation can be calculated for each velocity profile that is a combination of instrument noise and true variation in current velocity. Sontek (1998) gives the instrument noise as a velocity standard deviation, σ_n, defined as

$$\sigma_n = 140 c / F \Delta z N^{0.5} \qquad (1)$$

where c is speed of sound in water, F is acoustic frequency, Δz is bin size and N is the number of measurements over the sampling interval. In our study: c is assumed to be 1500 m s^{-1}; F is the ADP frequency of 1.5 MHz; Δz was set at the minimum bin size of 0.25 m; and N, the number of pulses in the sample, is the product of the sampling interval and the pulse rate, which was set at the maximum of 9 pulses s^{-1}. These parameters yield $\sigma_n = 0.56/N^{0.5}$. From laboratory observation over an artificial cobble bed (the only critical test available at present), Rennie *et al.* (2001) defined an empirically fit standard deviation estimate for the bottom tracking

$$\sigma_{bn} = 1.3 v_{bt}/N^{0.5} \qquad (2)$$

which, for $v_{bt} < 0.43$ m s^{-1}, yields lower estimates of error than equation (1). Another potential source of error is the accuracy of the compass, as the raw ADP velocity data are given in easting/northing and up/down. The compass accuracy is reported to be ±2° (Sontek, 1998). Rennie et al. (2001) calculated the error related to the compass uncertainty to be 3.5% of the magnitude of the velocity vector from the mobile bed.

The final source of instrument error arises from the boat motion estimate from the Trimble® DGPS. The DGPS utilizes differentially corrected signals from a navigation beacon located approximately 20 km from the study area, allowing submetre accuracy in position. During earlier measurements in the same reach, maximum *potential* error in vessel speed from the DGPS from a 17-min anchored test with 2-s averages was ±0.02 m s^{-1} with no systematic bias.

A sediment-transport-related consideration is the depth of penetration of the acoustic pulse. The depth of penetration is a function of the frequency of the transducer and the bed finding algorithm applied. A complicating factor is the sediment concentration gradient within a mobile sand layer, which makes differentiation between near-bed suspension, bedload and the immobile sand bed difficult. It is assumed that the echo returns from some depth within the mobile sand layer, although further testing of this assumption is warranted.

Another consideration is the homogeneity of the bed velocities measured among the three transducers. Each transducer has its individual acoustic footprint (Fig. 1). The transducers have diameters of 0.06 m. Owing to the angle of the beam with respect to the bed, the sampling area is quasi-elliptical with major axis $0.032d + 0.06$ and minor axis $0.029d + 0.06$ (Rennie et al., 2002), d being water column depth. The static sampling area of the bottom tracking is related to the diameter of spread amongst the three beams, which, with this transducer configuration, increases with depth to a maximum of approximately $0.93d$, at the bed.

The bed velocities measured in the three beams should be similar, as the ADP resolves the bottom-tracking velocity from components of the along-beam velocity measured by each transducer. Therefore, any variation in bedload between the three beams produces measurement error. Also, the comparison between mechanical and acoustic methods assumes homogeneous transport within the sampling areas of the two instruments. Transport heterogeneity within the ADP beam spread contributes an irreducible element of variability associated with the method. The two instruments were located as close together as boat arrangements permitted.

VAN RIJN BEDLOAD TRANSPORT FORMULA

The bedload component of the semi-empirical total sediment transport formula proposed by Van Rijn (1984) was used for a comparison with the estimates from the two measurement techniques. This formula was chosen as it has been found to produce reasonable approximations of bedload for a range of river and estuary conditions (Soulsby, 1997). Moreover, the Van Rijn (1984) formula defines sediment transport as a function of mean flow velocity, whereas in many of the other formulae, sediment transport is a function of water slope, which is much more difficult to measure in estuaries.

Bedload transport q_{vr} (kg m^{-1} s^{-1}) is expressed as

$$q_{vr} = 0.053 \rho_s T^{2.1}/D_*^{0.3} [(s-1)g]^{0.5} D_{50}^{1.5} \qquad (3)$$

where ρ_s is the sediment density, s is the sediment density ratio, g is acceleration due to gravity, D_{50} is the median grain size of the surficial bed sediment, T is the transport stage parameter and D_* is a scaled particle parameter

$$D_* = D_{50}[(s-1)g/\upsilon^2]^{1/3} \qquad (4)$$

where υ is the kinematic viscosity. Transport stage parameter T is defined as

$$T = [(u'_*)^2 - (u_{*cr})^2]/(u_{*cr})^2 \qquad (5)$$

where u'_* is the bed shear velocity related to the grains and u_{*cr} is the critical bed shear velocity. The bed shear velocity scaled to the grains, u'_*, is expressed as

$$u'_* = u_{avg}/5.75\log(12d/3D_{90}) \qquad (6)$$

where u_{avg} is the depth-averaged velocity, d is depth and D_{90} is the 90th percentile of the grain size of the surficial bed sediment. The term in parentheses is found to be appropriate for sand grain roughness. In equation (6) d is used, following Soulsby (1997), as a surrogate for hydraulic radius, which was used in the original formula to define a cross-section average transport instead of a point estimate. The critical bed shear velocity, u_{*cr}, is defined from an empirical fit of the Shields curve (Van Rijn, 1984), which is represented for the range in particle sizes in this study as

$$u_{*cr} = [0.14 D_*^{-0.64}(s-1)gD_{50}]^{0.5} \qquad (7)$$

RESULTS

Bed velocity

Surficial sediments were well-sorted fine to medium sands with median grain sizes of 0.20 mm, 0.21 mm and 0.30 mm from samples collected on 8, 13 and 16 June. Bedform size also varied on the three days of measurements. Figure 2 shows bed profiles and launch positions on each day. In each case the position was on the stoss-side near the crest of the local dune form. Owing to variation of the launch position in the channel and the antecedent flow, observed bedforms varied in size on the three days. Heights and lengths of the dunes directly below the launch on the 8, 13 and 16 June were 1.21 m and 35.43 m, 0.45 m and 12.50 m, and 1.23 m and 37.31 m, respectively.

The average depth, d, and depth-averaged horizontal velocity, u_{avg}, measured over the bedform during the collection of each bedload sample are plotted in Fig. 3. Water depths ranged from 5.22 m to 7.64 m and u_{avg} ranged from 0.36 m s^{-1} to 0.97 m s^{-1}. The plot illustrates the influence of the tidal cycle on channel velocity. Shear velocity related to the grains, u'_*, ranged from 0.013 m s^{-1} to 0.035 m s^{-1}, calculated using equation (6).

The diameters of the longest axis of the acoustic footprints were calculated from the depth measurements and found to range from 0.23 m to 0.30 m. The diameter of the sampling area circle surrounding the three acoustic footprints ranged from 4.85 m to 7.11 m. Using the maximum sampling diameter from each day the proportion of the bedform length sampled on the 8, 13 and 16 June was 0.17, 0.46 and 0.18, respectively. The chief source of variation in the ratio of sampling diameter to bedform length is variation in dune length.

Northing and easting components of vessel velocity measured with the DGPS from a single sample run collected on the 16 June are plotted in Fig. 4a. The scatter in the vessel speeds is an order of magnitude smaller and more systematic than that measured by the bottom tracking (Fig. 4b). The direction of scatter is perpendicular to the mean flow and is probably a product of the vessel swaying on the anchor. Mean northing and easting components of the vessel speeds measured with the DGPS, v_{dgps}, were 0.0006 m s^{-1} and −0.0008 m s^{-1} with standard deviations of 0.0040 m s^{-1} and 0.0032 m s^{-1}. These values indicate no resolvable net movement of the anchored launch during the run. Mean northing and easting components of the mobile bed velocity, v_b (Fig. 5b), were −0.023 m s^{-1} and −0.046 m s^{-1}, with standard deviations about the mean of 0.122 m s^{-1} and 0.144 m s^{-1}. The

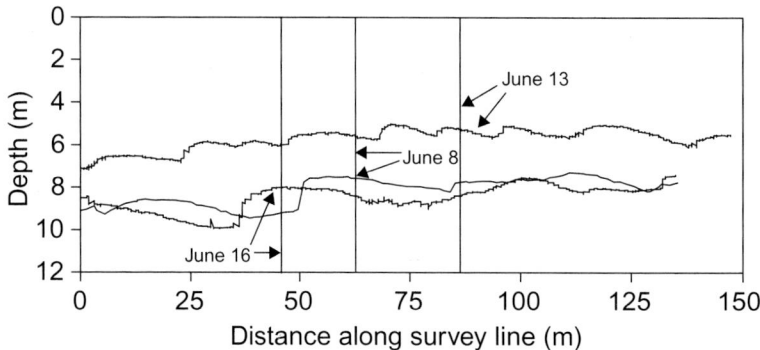

Fig. 2 Bed profiles conducted at the end of measurements on the three survey days. Vertical lines mark the approximate location of the launch with respect to the bed.

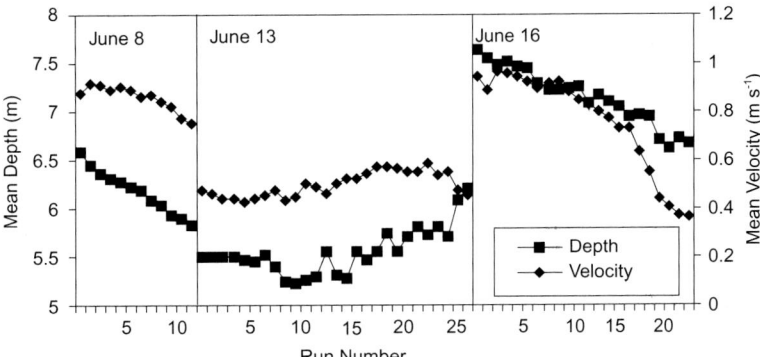

Fig. 3 Average depths and depth-averaged horizontal velocities from each 5-min sample on the three days measurements were conducted.

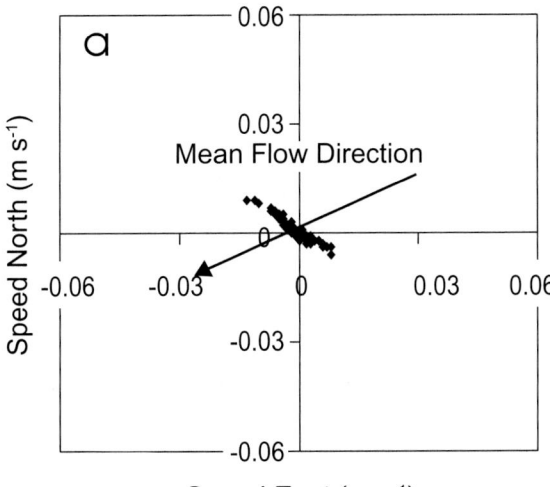

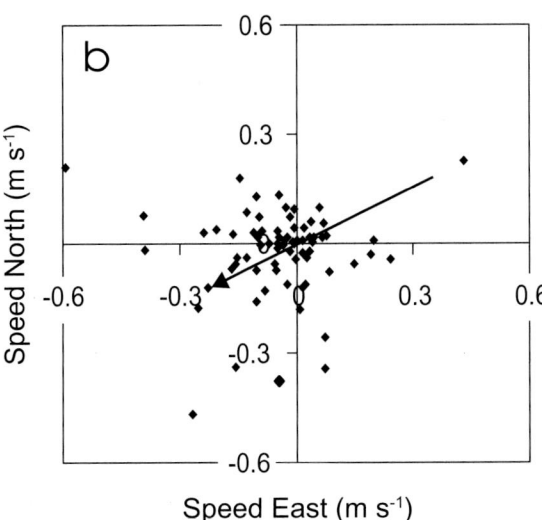

Fig. 4 (*left*) Five-second averaged easting and northing components of (a) vessel velocity measured by the DGPS, v_{dgps}, and (b) the velocity of the mobile bed, v_b, during a 5-min sample collected on 16 June. The arrow on both plots is the direction of depth-averaged flow during the run.

mean velocity is in the direction of the depth-averaged flow (Fig. 5a), both plots showing a trend towards the south-west.

A comparison was made between the standard deviation of the velocity of the mobile bed measured during each run and the potential instrument-related standard deviations. The maximum, average and minimum standard deviations measured from all runs are 0.148 m s^{-1}, 0.060 m s^{-1} and 0.020 m s^{-1}. Assuming that the average standard deviation of the vessel velocities measured by the DGPS (0.0053 m s^{-1}) is representative of the maximum *potential* error associated with the DGPS, on average only 9% of the *potential* error is associated with the DGPS. Therefore, most of the variation must be associated with the bottom tracking.

Estimates of standard deviation using equation (1) are smaller than observed from the velocity of the mobile bed, with maximum, average and minimum values of 0.014 m s^{-1}, 0.011 m s^{-1} and 0.010 m s^{-1}, respectively. The average estimated value is approximately one-fifth of the average measured value. The largest measured standard deviations are related to the most energetic transport periods on 16 June. As $v_{bt} < 0.4$ m s^{-1}, the function defined by Rennie *et al.* (2001) (equation 2) produces much lower estimates of standard

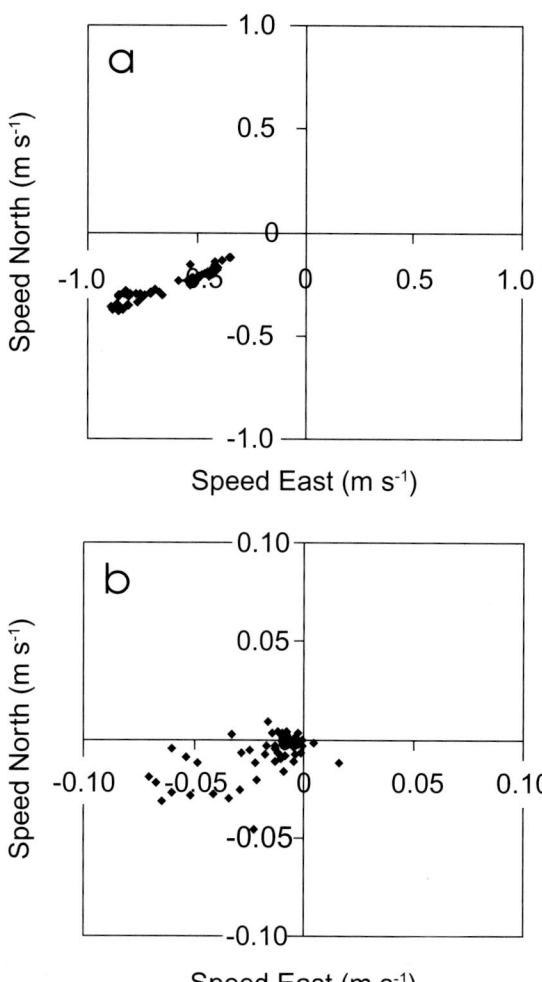

Fig. 5 Run-averaged easting and northing components of (a) depth-averaged water velocity and (b) bedload velocity, for all sample runs.

deviation than either the measurements or estimates from equation (1).

A conservative estimate of the instrument-related component of the standard deviations observed in the velocity of the mobile bed can be approximated by pooling the predicted standard deviation from equation (1) and the observed standard deviation from the DGPS. Comparison between this and the observed value indicates that approximately 30% of the variation can be related to the instrumentation. It is likely that the remainder of the variation is due to the heterogeneous nature of near-bed sediment transport.

Bedload

The bedload velocity u_b can be applied to estimate bedload transport q_{bt} through

$$q_{bt} = u_b \delta_b c_b \qquad (8)$$

where δ_b is the thickness of the bedload layer and c_b is bedload concentration, which here is $c_b = r\rho_s$ where r is the volumetric concentration.

Although neither variable is known, they can be approximated within reasonable bounds. Two methods were used to characterize them. First, constant values of δ_b and c_b were chosen. The bedload thickness, δ_b, was set using the classic definition of bedload being within two grain heights of the bed, $2D_{50}$ (Einstein, 1950; Bagnold, 1973). This definition gives δ_b values of 0.0004 to 0.0006 m. The bedload concentration, $c_b = r\rho_s$, was set at 0.40×2650 kg m^{-3}, approximately 60% of the maximum volumetric bedload concentration possible (0.65; Van Rijn, 1984). Second, δ_b and c_b were calculated as a function of the particle parameter, D_*, and the transport stage parameter, T, following equations developed by Van Rijn (1984)

$$\delta_b = 0.3 D_*^{0.7} T^{0.5} D_{50} \qquad (9)$$

$$c_b = 0.18 \rho_s T/D_* 0.65 \qquad (10)$$

For all the runs maximum, average and minimum values of δ_b and c_b were, respectively: 0.0005 m, 0.0003 m and 0 m; and 0.128×2650 kg m^{-3}, 0.049×2650 kg m^{-3} and 0 kg m^{-3}. Modelled values predict a thinner bedload layer with lower concentration than the constant values that follow a classic definition of bedload.

The relations among the different methods of estimating bedload are calibrations. The appropriate means of determining the parameters of a calibration is functional analysis (Kendall & Stuart, 1965). Least-squares regression is commonly used to define relations between variables. However, when correlation values are not close to one and there is error in the estimate of the independent variable, regression creates a systematic bias in the calibration (Mark & Church, 1977). If there is knowledge of the errors in the estimates of the variables, functional analysis produces an unbiased estimate of the parameters. Bedload estimates from bottom tracking with constant δ_b and c_b, q_{bts}, and bottom tracking with modelled values of δ_b

and c_b, q_{btm}, varied around the mean by ±31% and ±33% (95% confidence limits) respectively. These values are of the same order as the variance of ±23% (95% confidence limits) reported for HS samples (Kostaschuk & Ilersich, 1995). Adopting these variance measures as indicative of measurement error, it is found that the errors around the independent variable, e_x, and dependent variable, e_y, are of similar proportion, so the error ratio $e_y/e_x \approx 1$ and the linear slope of the functional relation, m_f, is given by (Mark & Church, 1977)

$$m_f = \{(m_r^2/R^2 - 1) + [(m_r^2/R^2 - 1)^2 + 4m_r^2]^{0.5}\}/2m_r \quad (11)$$

where m_r is the slope of the least-squares regression and R^2 is the coefficient of determination. Confidence limits around the slope and intercept assuming $e_y/e_x \approx 1$ are given by

$$\tan\{\tan^{-1}(b_f) \pm 0.5\sin^{-1}(2t[(1 - R^2)/((n-2)(m_r^2/R^2 + R^2/m_r^2 + 4R^2 - 2))]^{0.5})\} \quad (12)$$

(cf. Davies & Goldsmith, 1972, p. 209), where t is the one-tailed value of Student's t for $n - 2$ degrees of freedom. Table 1 contains results of linear functional analysis among bedload estimates from bottom tracking, q_{bts}, with constant δ_b and c_b, bottom tracking, q_{btm}, with modelled values of δ_b and c_b, HS sampler, q_{hs}, and Van Rijn bedload formula, q_{vr}, and between u_b and Van Rijn bedload formula, q_{vr}. A significant relation is found between the bedload estimates from the HS sampler and Van Rijn formula, although the formula overpredicts the mechanical measurements of bedload by approximately 2.5 times (Table 1).

Figure 6a is bedload estimated from the ADP using constant δ_b and c_b as a function of bedload measured with the HS sampler (potential error bars for the Helley–Smith measurements are based on those reported by Kostaschuk & Ilersich (1995)). The ADP estimates using constant δ_b and c_b are 1.5 times those of the HS sampler and there is also a large amount of scatter, yielding the modest value of R^2 (Table 1). Bias is presumably related to the choice of values for δ_b and c_b; if c_b is reduced by 33% (to approximately $0.25\rho_s$) or if δ_b is adjusted by a similar quantity, the bias with respect to q_{hs} disappears. These essential unknowns easily cover the range of the bias.

Figure 6b is a plot of bedload estimates from the ADP with constant values of δ_b and c_b as a function of the Van Rijn bedload estimates. The linear relation between the ADP and formula estimates is stronger, as illustrated by comparison between the R^2 values, than that estimated between the mechanical sampler and modelled estimates (Table 1). This improvement may be due to the co-location of bed and water column velocity measurements. However, there is now a strong bias in the opposite sense, the ADP estimates being only about 0.5 times those derived from the formula.

There is no improvement in either relations, as indicated by the R^2 values (Table 1), when modelled values of δ_b and c_b are used. Moreover, the estimates are significantly smaller than the

Table 1 Results of linear functional analysis among bedload estimates from the ADP, Helley–Smith sampler and the Van Rijn formula.

Methods*	Functional analysis equation	R^2	95% confidence range on the slope	95% confidence range on the intercept	n
q_{bts} versus q_{hs}	$y = 1.555x + 0.002$	0.422	1.194 to 2.085	0 to 0.004	53
q_{hs} versus q_{vr}	$y = 0.430x - 0.001$	0.606	0.362 to 0.499	0 to −0.002	54
q_{bts} versus q_{vr}	$y = 0.538x + 0.002$	0.749	0.468 to 0.611	0.001 to 0.003	54
q_{btm} versus q_{hs}	$y = 0.195x + 0.0005$	0.408	0.140 to 0.249	0 to 0.001	53
q_{btm} versus q_{vr}	$y = 0.1079x - 0.00013$	0.738	0.093 to 0.123	−0.002 to 0	54
u_b versus q_{vr}	$y = 0.954x - 0.003$	0.662	0.805 to 1.121	−0.005 to −0.001	53

*Bedload transport estimates from bottom tracking with constants, q_{bts}, bottom tracking with model component, q_{btm}, Helley–Smith sampler, q_{hs}, Van Rijn bedload formula, q_{vr}, and ADP estimate of bedload velocity, u_b. $P < 0.001$ in all cases.

206 P. Villard, M. Church and R. Kostaschuk

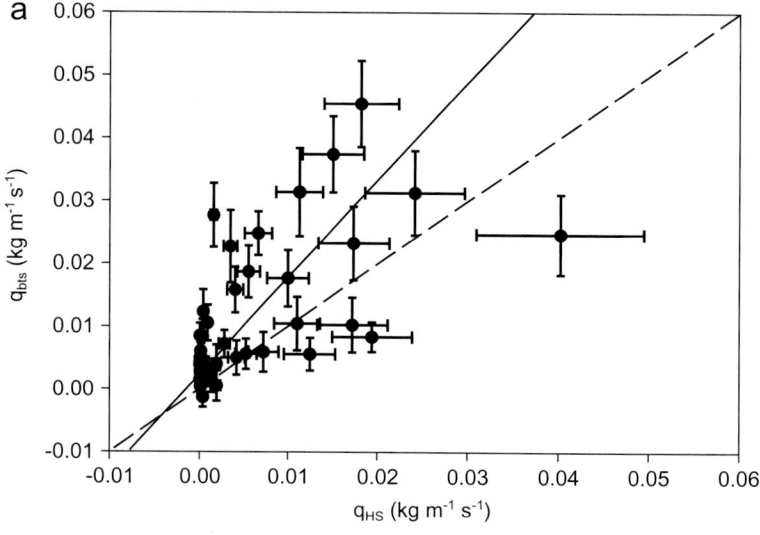

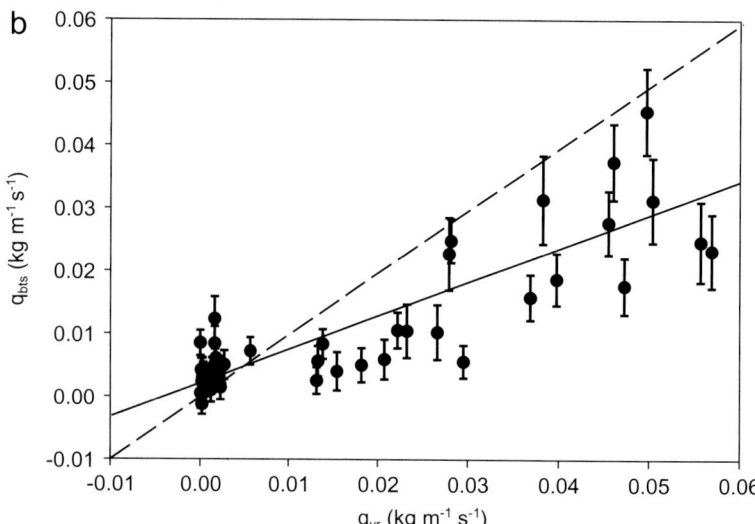

Fig. 6 (a) Acoustic Doppler profiler bedload estimates using constant bedload layer thickness and concentration versus HS bedload estimates, and (b) ADP bedload estimates using constant bedload layer thickness and concentration versus Van Rijn bedload formula estimates. Bars are 95% confidence intervals. The solid line is the functional analysis fit and the dashed line is the one-to-one line.

mechanical measurements, ADP estimates with constant values of δ_b and c_b and the Van Rijn estimates. Furthermore, as T and D_* appear in both the acoustic and the modelled estimates, a component of the correlation is spurious. The proportion of the correlation that is spurious can be evaluated by obtaining the functional relation between u_b and q_{vr}. This relation has $R^2 = 0.662$, slightly better than the relation between q_{hs} and q_{vr}. The best linear fit, as measured by the R^2 values, among the different methods is between ADP estimates with constant values of δ_b and c_b and the Van Rijn estimates.

DISCUSSION

It can be expected that the acoustic approach will lend itself to the measurement of bedload or, at least, near-bed transport because the intensity of the acoustic return from particulate scatterers within an ensonified volume is directly proportional to the concentration and the volume of the individual particles (Lynch & Agrawal, 1991). The acoustic signal is biased toward the larger sediment fractions, which generally also compose the bedload component, because of the volume sensitivity. In the experiments, the acoustic pulse length was 0.20 m, so all the scatterers that bias the bottom tracking are within 0.20 m of the bed. This thickness is about 2.5 times the height of the Helley–Smith sampler (0.078 m). Still, exactly which components of the near-bed sediments are biasing the bottom tracking remains unknown. The linear calibrations between the different methods (Table 1) indicate that consistent signals are being returned and permit some inference about the materials 'sampled' by the ADP (Table 1). The ADP-estimated bedload rates using modelled values of δ_b and c_b did not provide improvements in fits with estimates from the other methods and produced significant underpredictions compared with the other methods. This is due to the lower than expected estimates of δ_b and c_b. Therefore, the remaining discussion will concentrate on the simpler acoustic method. The ADP-estimated bedload rates with constant values of δ_b and c_b are systematically higher than those from the HS sampler by about 1.5 times. Even with this mean bias, paired estimates from HS sampling and the ADP fall around the one-to-one line (Fig. 6a) and the intercept from functional analysis is not significantly different than zero. This suggests that the HS sampler and ADP were systematically 'sampling' similar materials. It has been shown that the HS sampler overestimates bedload in sands due to the capture of suspended materials (Emmett, 1980). In the grain-size range examined here it is expected that this overestimation is in the range of 1.75 times (Emmett, 1980). Therefore, it is likely that the ADP is also biased by suspended material—conceivably by as much as 2.5 times. The strong linear relation between HS and ADP estimates with constant values of δ_b and c_b suggests that the suspension bias is systematic.

The Van Rijn bedload formula was found to predict bedload rates approximately 1.9 times and 2.3 times higher than estimated from the ADP and the HS sampler respectively. At face value, this consistent overestimation suggests that the coefficients in the formula need to be adjusted for the conditions observed in Canoe Pass.

A radically different construction may be placed upon the foregoing results. It might be supposed that each of the methods is performing reasonably, but that it is sampling a different fraction of the near-bed sediment transport. In the case of the HS sampler, the sample population is all material moving within 76 mm of the solid surface, by whatever mode. The van Rijn formula, in this view, is estimating the transport within a region defined by δ_b. In fact, the two estimates remain inconsistent in our measurements, as $\delta_b \ll 76$ mm but $q_{vr} > q_{hs}$. The interpretation of the ADP results remains less clear, but the relation between q_{bts} and q_{hs} is not inconsistent with the notion that the ADP is sampling a zone that may extend up to 200 mm from the bed. These observations, in turn, suggest that the essentially arbitrary division between 'bedload' and near-bed suspension in established transport over a sand bed must be ignored in any practical measure of sand transport.

In the present trials, the strongest correlations were found between the ADP estimates of u_b and q_{bts}, and q_{vr} (Table 1). The reduced correlation with q_{hs} may be a function of the inherent variability within mechanical sampling methods, or of the displacement between the ADP sampling volume and the mechanical trap. The best results were obtained between the Helley–Smith sampler and ADP by constant values of δ_b and c_b. These results indicate that the outlined acoustic method shows potential for estimating near-bed sand transport quickly and accurately, although at this point it appears little may be gained from modelling δ_b and c_b. The technique needs to be tested over a wider range of bed material and flow conditions, a more thorough examination of the suspension bias is needed, and instrument errors need to be properly understood.

CONCLUSIONS

The acoustic method estimates sand transport velocities within the near-bed mobile sand layer. There is also near concordance between mechanical and the sample, acoustic estimates of bedload transport, but with large scatter, indicating that both instruments apparently are measuring similar fractions of near-bed transport. Additional work is needed to understand the remaining bias between the acoustic measurements and the mechanical sampler. Problems may lie in the instrument signal processing, in the variable performance of the mechanical sampler, or in the likely circumstance that the two instruments are systematically sampling different integrations of the near-bed transport. Our ADP was an off-the-shelf velocity profile instrument not optimized for near-bed sediment transport observations. The purpose of our study was to demonstrate the feasibility of acoustic measurements using such equipment. The results indicate that the acoustic method shows potential for allowing rapid collection of remotely sensed estimates of bedload in sand-bed channels without many of the limitations associated with mechanical samplers.

ACKNOWLEDGEMENTS

We are grateful to Arjoon Ramnarine for his expert command of the UBC Oceanography Launch in Canoe Pass. We also thank Jason Blair and Colin Rennie for help during the field deployment. Financial support was through Church's and Kostaschuk's NSERC research grants and an NSERC Post-doctoral fellowship provided support for Villard. The paper was improved substantially as the result of careful reviews by Manuel Figueira-Rivera and Suzanne Leclair.

NOMENCLATURE

c	speed of sound in water
c_b	bedload concentration
C'	Chezy coefficient related to the sand grain
d	depth
D_{50}	median grain size of the surficial sediment
D_{90}	90th percentile of the grain size of the surficial sediment
D_*	particle parameter
e_y, e_x	error around the independent and dependent variables
F	acoustic frequency
g	acceleration due to gravity
m_f	slope from functional analysis
m_r	slope from least-squares regression
n	number of samples
N	number of measurements over the sampling interval
q_{bt}	bedload transport measured by the ADP
q_{bts}	bottom tracking with constants
q_{hs}	bedload transport estimates from bottom tracking with model component
q_{hs}	bedload transport estimates from Helley–Smith sampler
q_{vr}	bedload transport estimates from Van Rijn formula
r	volumetric bedload concentration
s	sediment density ratio
t	one-tailed value of Student's t
T	transport stage parameter
u_{avg}	horizontal depth-averaged velocity
u_b	horizontal bedload velocity referenced to u_{avg}
u'_*	bed shear velocity related to the grains
u_{*cr}	critical bed shear velocity
v_b	velocity of a mobile sand bed
v_{bt}	'apparent' boat velocity with respect to the bed measured with ADP bottom tracking
v_{dgps}	'actual' boat velocity measured with DGPS
Δz	bin size
ρ_s	sediment density
υ	kinematic viscosity
δ_b	depth of bedload layer

REFERENCES

Agrawal, Y.C. and Pottsmith, H.C. (2000) Instruments for particle size and settling velocity observations in sediment transport. *Mar. Geol.*, **168**, 89–114.

Bagnold, R.A. (1973) The nature of saltation and bed-load transport in water. *Proc. Roy. Soc. Lond.*, Ser. A, **332**, 473–504.

Carling, P.A., Williams, J.J., Gölz, E. and Kelsey, A.D. (2000) The morphodynamics of fluvial sand dunes in the Rhine River near Mainz, Germany. II. Hydrodynamics and sediment transport. *Sedimentology*, **47**, 253–278.

Davies, O.L. and Goldsmith, P.L. (1972) *Statistical Methods in Research and Production*. Oliver and Boyd, Edinburgh.

Downing, J.P. (1983) An optical instrument for monitoring suspended particulates in ocean and laboratory. In: *Proceedings of Oceans '83*, pp. 199–202. American Society of Civil Engineers, San Francisco, CA.

Einstein, H.A. (1950) *Bed-load Functions for Sediment Transport in Open Channel Flows*. Technical Bulletin No. 1026, U.S. Department of Agriculture, Washington, DC.

Emmett, W.W. (1980) A field calibration of the sediment-trapping characteristics of the Helley–Smith bed load sampler. *U. S. Geol. Surv. Prof. Pap.*, **1139**, 44 pp.

Glysson, G.D. (1993). U.S. Geological Survey bedload sampling policy. In: *Hydraulic Engineering '93*, Part 1 (Eds Shen, H.W., Su, S.T. and Wen, F.), pp. 701–706.

American Society of Civil Engineers, San Francisco, CA.

Gomez, B., Naff, R.L. and Hubbell, D.W. (1989) Temporal variations in bedload transport rates associated with the migration of bedforms. *Earth Surf. Process. Landf.*, **14**, 135–156.

Gomez B., Hubbell, D.W. and Stevens, H.H. (1990) At-a-point bed load sampling in the presence of dunes. *Water Resour. Res.*, **26**, 2717–2731.

Hay, A.E. (1983) On the remote acoustic detection of suspended sediment at long wavelengths. *J. Geophys. Res.*, **88**(C12), 7525–7542.

Helley, E.J. and Smith, W. (1971) Development and calibration of a pressure difference bedload sampler. *U.S. Geol. Surv. Water Resour. Div. Open-file Rep.*, 18 pp.

Kendall, M.G. and Stuart, A. (1965) *The Advanced Theory of Statistics*, Vol. 2. Hafner, New York.

Klingeman, P.C. and Milhous, R.T. (1971). Oak Creek vortex bed-load sampler. *EOS, Trans. Am. Geophys. Union*, **52**(5), 434.

Kostaschuk, R.A. (2000). A field study of turbulence and sediment dynamics over subaqueous dunes with flow separation. *Sedimentology*, **47**, 519–531.

Kostaschuk, R.A. and Ilersich, S.A. (1995) Dune geometry and sediment transport: Fraser River, British Columbia. In: *River Geomorphology* (Ed. E.J. Hickin), pp. 19–36. Wiley, Chichester.

Kostaschuk, R.A. and Villard, P.V. (1996). Flow and sediment transport over large subaqueous dunes: Fraser River, Canada. *Sedimentology*, **43**, 849–863.

Kostaschuk, R.A. and Villard, P.V. (1999). Turbulent sand suspension over dunes In: *Fluvial Sedimentology VI* (Eds N.D. Smith and J. Rogers). *Spec. Publ. Int. Assoc. Sedimentol.*, **28**, 3–14.

Langley, H.I. (1971) *An Experiment in Bedload Sampling by Pumping*. Department of Public Works, Vancouver, BC.

Lowe, R.L., Inman, D.L. and Drake, T.G. (1991). A bedload sensor for wave and current regimes. *EOS, Trans. Am. Geophys. Union*, **72**(44), 230.

Lynch, J.F. and Agrawal, Y.C. (1991) A model-dependent method for imverting vertical profiles of scattering to obtain particle size spectra in boundary layers. *Mar. Geol.*, **99**, 387–401.

Mark, D.M. and Church, M. (1977) On the misuse of regression in earth science. *Math. Geol.*, **9**, 63–75.

McLean, S.R., Wolfe, S.R. and Nelson, J.M. (1999a) Predicting boundary shear stress and sediment transport over bedforms. *J. Hydraul. Eng.*, **125**, 725–736.

McLean, S.R., Wolfe, S.R. and Nelson, J.M. (1999b) Spatially averaged flow over a wavy boundary revisited. *J. Geophys. Res.*, **104**, 15743–15753.

Novak, P. (1957) Bed load meters—development of a new type and determination of their efficiency with the aid of scale models. *Transactions, Meeting of the Association of Hydraulics Research, 7th General Meeting*, pp. A9–A11, Lisbon.

Rennie, C.D., Millar, R.G. and Villard, P.V. (2001) Laboratory measurements of bedload transport velocity using an acoustic Doppler current profiler. In: *Proceedings, 15th Canadian Hydrotechnical Conference* (Ed. R. Balachandar), Canadian Society for Civil Engineering, Victoria, BC (CD-ROM), 8 pp.

Rennie, C.D., Millar, R.G. and Church, M.A. (2002) Measurement of bedload velocity using an acoustic Doppler current profiler. *J. Hydraul. Eng.*, **128**, 473–483.

Ryan, S.E. and Porth, L.S. (1999) A field comparison of three pressure-difference bedload samplers. *Geomorphology*, **30**, 307–322

Sontek (1998). *Acoustic Doppler Profiler Technical Documentation*. Sontek, San Diego.

Soulsby, R.L. (1997). *Dynamics of Marine Sands*. Thomas Telford, London.

Sutton, D.W. and Jaffe, (1992) Acoustic bedload velocity estimates using a broadband pulse-pulse time correlation technique. *J. Acoust. Soc. Am.*, **92**(3), 1692–1698.

Thorne, P.D., Vincent, C.E., Hardcastle, P.J., Rehman, S. and Pearson, N. (1991) Measuring suspended sediment concentrations using acoustic backscatter devices. *Mar. Geol.*, **98**, 7–16.

Van Rijn, L.C. (1984) Sediment transport, part I: bed load transport. *J. Hydraul. Eng.*, **110**(10), 1431–1456.

Villard, P.V. (1995) Spatial and temporal behaviour of turbulent sand suspension over dunes: Fraser River, British Columbia. Unpublished MSc thesis, University of Guelph.

Villard, P.V. and Kostaschuk, R.A. (1998). The relation between shear velocity and suspended sediment concentration over dunes: Fraser Estuary, Canada. *Mar. Geol.*, **148**, 71–81.

Zedel, L. and Hay, A.E. (1999). A coherent Doppler profiler for high-resolution particle velocimetry in the ocean: Laboratory measurements for turbulence and particle flux. *J. Atmos. Ocean. Technol.*, **16**, 1102–1117.

Experimental and numerical modelling

The morphological and stratigraphical effects of base-level change: a review of experimental studies

FRANK G. ETHRIDGE*, DRU GERMANOSKI†, STANLEY A. SCHUMM* and LESLI J. WOOD‡

*Department of Earth Resources, Colorado State University, Fort Collins, CO, 80523, USA (Email: fredpet@cnr.colostate.edu);
†Department of Geology and Environmental Geosciences Lafayette College, Easton, PA 18042, USA; and
‡Bureau of Economic Geology, The University of Texas at Austin, University Station, Box X, Austin, TX 78713-8924, USA

ABSTRACT

Results obtained during 20 yr of experimental research at Colorado State University (CSU) provide partial answers to questions posed by the conveners of this symposium on the response of near-coastal fluvial systems to sea-level change. These questions do not have easy answers because response will be variable depending upon system size, magnitude of base-level change and initial condition. Most flume experiments were designed to observe the effects of base-level change on the evolution of channel and valley networks, and they provide general conclusions regarding the effects of base-level change. Base-level change at a river's terminus has a geographically limited influence on a drainage system and that influence is usually confined to the coastal plain. Small base-level change is commonly accommodated by changes in channel sinuosity, roughness and width. Larger base-level changes, such as lowering of base level below a major topographic break, results in down cutting and the formation of one or more upslope valleys or valley networks. Channels commonly adjust rapidly to base-level change especially at the point of change. Base-level fall usually has a more dramatic effect on channels and valleys than does base-level rise. Incision resulting from a base-level fall concentrates energy in a narrow, deep valley that extends rapidly up valley. Aggradation from base-level rise, however, is restricted spatially by the position of the static backwater profile and the effect is over the entire valley width.

Colorado State University experiments designed specifically to test the sequence stratigraphical paradigm reveal how variable and complex the morphological and sedimentological response on a continental shelf and slope can be depending on initial conditions and rates and magnitudes of base-level change. Shelf valleys formed only when base level dropped below a significant break in slope. Because the growth of these valleys occurred by headward erosion there was usually a lag time when no source-to-sink, cross-shelf bypass valley existed. During this time only fine-grained sediments were delivered to the deep-water offshore environment. This process may explain the muddy nature of early lowstand deposits in many deep-water basins around the world. Once a source-to-sink bypass valley developed, sediment from the hinterland drainage basin could be transported to form a coarse-grained lowstand fan. Experimental results suggest that late lowstand and early transgressive times are periods when the greatest volumes of coarse-grained sediment are transported to the downstream portions of the fluvial system. As base-level rise continued a lowstand wedge formed within the main bypass valley. When base-level rise accelerated and the shoreline moved onto the shelf, the transgressive systems tract within the bypass valley consisted of backstepping deltaic and marine deposits. Valley fills varied significantly between shelf bypass valleys and abandoned shelf break valleys. Valley cross-sections are modified during base-level rise by mass wasting and slumping of valley walls both in the flume experiments and in natural settings. Results of recent experiments at the University of Minnesota and Utrecht reveal many similarities and a few differences with CSU experiments. Furthermore, they add scaling factors that allow development of quantitative analogue experiments of real-world systems.

INTRODUCTION

The concept of base level has been basic to geological, geomorphological and engineering thought since it was enunciated by Powell (1875), Davis (1902) and Malott (1928). Changes of sea and lake levels (base level) are assumed to significantly influence river behaviour, sediment transport and deposition (Fisk, 1944; Born & Ritter, 1970; Blum & Törnqvist, 2000), as do dams and grade-control structures (Harvey, 1984; Harvey & Watson, 1986), which constitute artificial, local, base-level controls.

Base level is defined as 'The theoretical limit or lowest level toward which erosion of the Earth's surface constantly progresses but seldom, if ever, reaches; esp. the level below which a stream cannot erode its bed. The general or ultimate base level for the land surface is sea level, but temporary base levels may exist locally' (Bates & Jackson, 1980). This definition is in error in part because streams do erode below base level (Shepherd & Schumm, 1974; Schumm & Ethridge, 1994), but the general concept is reasonable, and there are considerable implications for river response and sediment delivery when base level is changed. An investigator attempting to elucidate system response to base-level change is hampered, however, by the slow rate of most base-level change and the large response time of fluvial systems. Nevertheless, the constraints of time in base-level investigation can be circumvented by experimental studies. The experimental approach also permits the evaluation of the influence of other significant variables upon base-level controls. For example, the effect of a base-level change can vary in magnitude depending on the type of river affected, valley morphology and rate of base-level change, among other factors (Koss, 1992; Wood, 1992; Wood et al., 1993, 1994; Koss et al., 1994; Van Heijst & Postma, 2001).

The concept of base level as a driving geological force experienced a rejuvenation in stratigraphical research with the publication of Posamentier & Vail (1988), who contended that depositional systems respond in a predictive manner to base-level change and that alluvial architecture preserved in the rock record is a reflection of that response. The proposed architecture that develops in lowstand, transgressive and highstand systems tracts is linked to relative base level or shoreline level and is used as a standard by the majority of stratigraphers today (for a review of these topics see Nystuen, 1998).

During the period 1970–1990, a series of experimental studies of drainage-network development, alluvial-fan evolution, channel response to sediment type, and base-level change, was performed in flumes and a rainfall–erosion facility (REF) at Colorado State University (Schumm et al., 1987). These studies were undertaken to address numerous and varied research objectives, all directly or indirectly associated with the evolution of drainage networks. A review of those experiments provides information on the effects of base-level change on the fluvial system. Subsequently, an additional series of experiments was begun with the more specific objective of addressing morphological, sedimentological and stratigraphical development in a single channel and on a continental shelf and slope under conditions of base-level change (Germanoski, 1990, 1991; Koss, 1992; Wood, 1992; Wood et al., 1993, 1994; Koss et al., 1994). These experiments were a first attempt to produce analogue models that might provide partial answers to real world problems. Second generation experiments by Milana (1998), Van Heijst & Postma (2001), Van Heijst et al. (2001), Heller et al. (2001) and Paola et al. (2001) have introduced scaling strategies and have attempted to quantify time-averaged erosion and deposition, factors not incorporated into the first generation experiments at Colorado State University. It is believed that a review of these investigations and an integration of the experimental results can provide at least partial answers to the questions that were posed to the participants of this symposium, as follows:

1 Do sea-level changes result in widespread aggradation and degradation throughout a drainage system?
2 Can sea-level changes be buffered by channel morphological change?
3 How rapidly can channels adjust to base-level change?
4 What are the differences in the impact of sea-level rise and fall on channels and drainage networks?

This review will deal with valleys, channels and sediment yield changes as a result of base-level change. The specific objective of this review is to respond to the four questions and to initiate a discussion of important sediment yield variations at the valley mouth that occur as a result of base-level change. It is assumed that answers will not be simple because responses will depend upon system size, magnitude of base-level change and initial conditions.

Before proceeding it is important to distinguish between valleys and channels. It is generally agreed among geomorphologists that the characteristics of alluvial adjustable channels are determined by discharges that occur at least once every 1.2 to 2 yr (Williams, 1978). Therefore, if the recurrence frequency of bankfull discharge is significantly greater than 2 yr (10–100 yr) the trough that contains the flow is actually a valley by hydraulic standards. Valleys, of the type discussed in this paper, originate when channels incise. The incision is the result of an increase in potential energy following base-level lowering or uplift, or an increase in kinetic energy following concentration of flow on an erosive surface (Schumm & Ethridge, 1994). Valleys and channels can be differentiated on the basis of the following criteria: (i) valleys are underlain by erosional surfaces having relief greater than the thickness of associated depositional elements (i.e. channel deposits); (ii) channels often have higher sinuosity than their host valleys; (iii) valleys are at least an order of magnitude greater in size than their formative channels; and (iv) valleys show a more dramatic juxtaposition of more landward facies over more basinward facies than do channels in the same relative spatial position (Van Wagoner et al., 1988). The incision of experimental channels into a simulated continental shelf forms a valley rather than a channel. However, at the bottom of the valley there will be a channel that is adjusting to changing hydraulic conditions (Schumm et al., 1984).

VALLEY NETWORKS

The initiation and evolution of drainage patterns have been of major interest to geomorphologists. Field studies of rapidly eroding badlands showed that dendritic drainage networks often grew by capture of adjacent channels (Horton, 1945; Morisawa, 1964). A logical extension of this work was experimental studies of drainage network development and evolution. McLane's (1978) experiments showed that surface irregularities, especially those that impounded or directed the flow of water, determined the eventual drainage pattern. Hence, the character of valley networks (pattern, depth) on an emerged shelf probably depends to a large extent on the initial micromorphology of the shelf. Phillips & Schumm (1987) determined that the slope of the surface upon which a pattern was developing had a major impact on the pattern (Fig. 1), and that the length of channels per unit area (drainage density) was determined by slope of the surface up to a maximum determined by the characteristics of the material composing the surface (Fig. 2). Results of

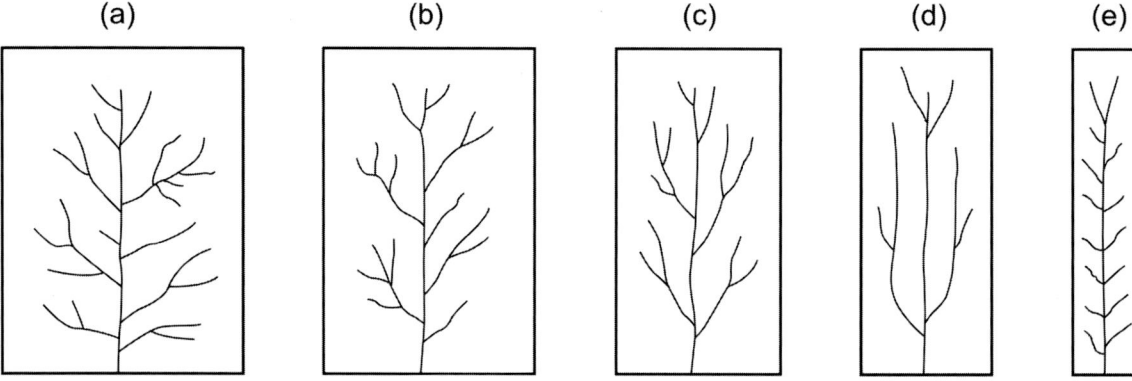

Fig. 1 Idealized drainage patterns that develop on a plane surface as slope increases from a to e: (a) dendritic pattern 1% slope; (b) subdendritic pattern 2% slope; (c) subparallel pattern 3% slope; (d) parallel pattern at 5%; (e) pinnate pattern > 5% slope. (From Phillips & Schumm, 1987.)

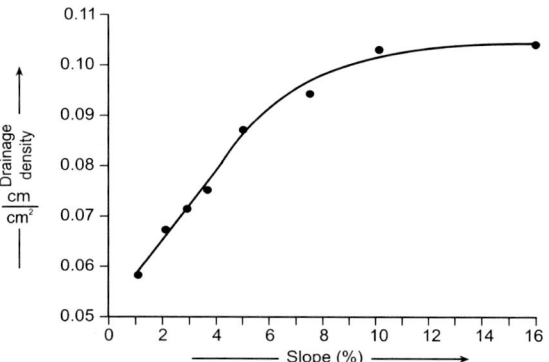

Fig. 2 Plot of drainage density (total length of channel/drainage area) at maximum extension of the drainage network as a function of slope. The lower part of the curve is linear and represents a direct relationship of drainage density to increasing slope. The upper part of the curve represents the maximum drainage density obtainable for the experimental material used. (From Phillips, 1986.)

Fig. 3 Photograph of experimental plateau–plain landform at end of experiment. Experiment performed in rainfall–erosion facility. (From Creager, 1991.)

the McLane and Phillips experiments did not involve base-level change, but their studies indicate the control that continental shelf and slope topography and composition can exert on developing drainage systems.

Creager (1991) simulated drainage network development at the foot of an escarpment (Fig. 3). Precipitation was applied to the scarp, plateau and piedmont, and drainage patterns developed on a 10% sloping plain. With the application of precipitation to the plateau and plain, sediment delivery to the plain increased as the drainage areas in the plateau enlarged and as the surface was undercut and collapsed. Initially, sediment was deposited as a sheet over the entire piedmont, but where streams exited the scarp, alluvial fans formed. As the fans developed, channels also formed on the plain, and sediment was transported and deposited further down the plain. The cross-sections of Figs 4a–e show deposition above the initial surface during the first 10 h. Between 10 and 70 h, the cross-sections show erosion, as sediment yields from the plateau decreased (Fig. 4). Creager's (1991) experiment produced some interesting observations concerning valley network

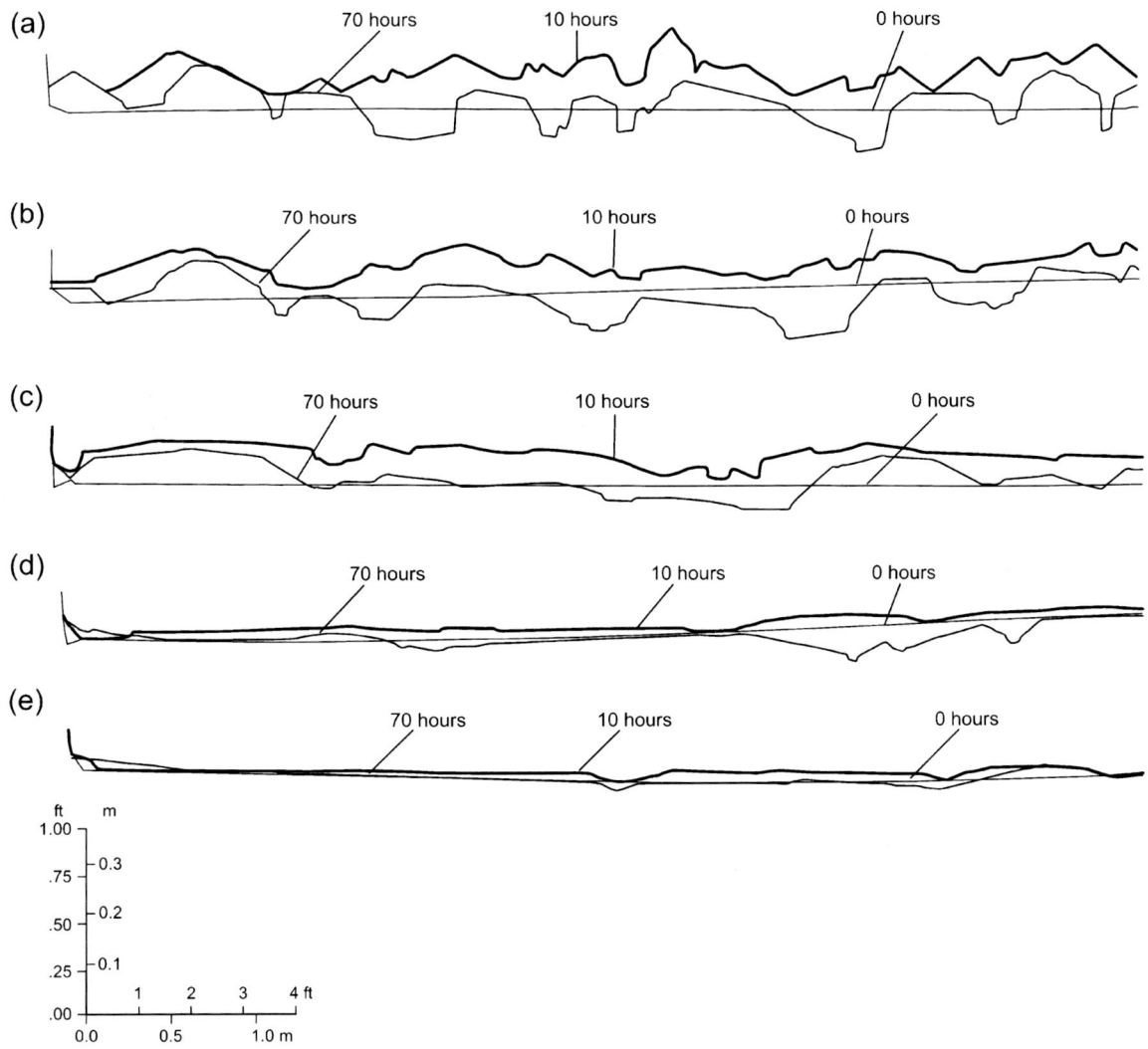

Fig. 4 Cross-sections showing change of piedmont surface from 0 to 70 h into the experiment. Numbers above each cross-section indicate the time in hours since the beginning of the experiment. (From Creager, 1991.) (a) Cross-sections 4.2 m from base level. Note formation of alluvial fans (10 h) and their subsequent dissection (70 h). Left side of cross-section is left side of plain looking upslope (Fig. 3). (b) Cross-sections 4 m from base level. Note deeper incision on right side as a result of drainage network integration by stream capture. (c) Cross-sections 3.4 m from base level. Dissection was less pronounced at this location, but it was greatest on the right side of the piedmont. (d) Cross-sections 2 m from base level. Incision was greatest on the right side of the piedmont. (e) Cross-sections 1 m from base level. Change at this location was minor, and the channels were less well defined.

evolution on a plain or piedmont. Initially stream channels were discontinuous on the upper and lower plain (Fig. 5a). When sediment yields from the plateau decreased, the alluvial fans were dissected, and channel incision extended farther down the piedmont. The two groups of channels merged, and better defined valleys formed on the lower piedmont (Fig. 5b, c, d). Finally, as a result of reduced sediment loads and the integration of piedmont channels by extension and capture, a dendritic valley pattern was incised into the piedmont surface.

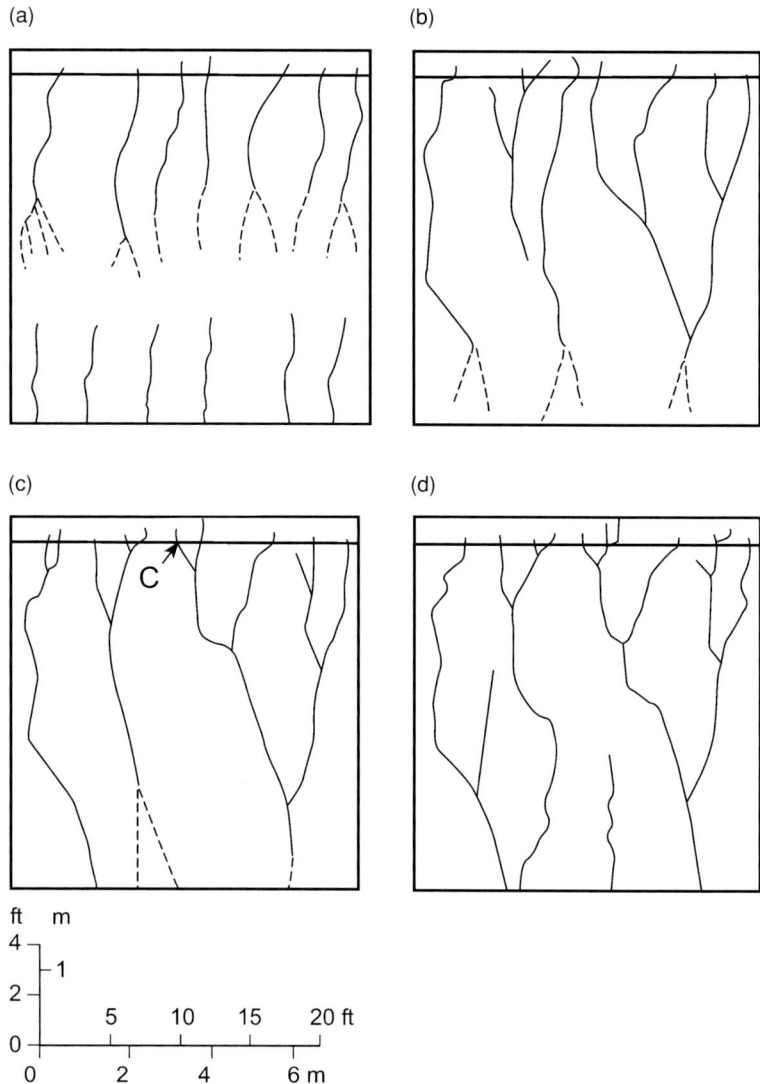

Fig. 5 Maps showing development of drainage pattern on plain during 70 h. (From Creager, 1991.) Rectangular area at top of each drawing is the escarpment at the head of the plain. (a) Initial development of channels on the plain. The upper group of channels cross alluvial fans or formed where flow has concentrated between fans. The lower group of channels formed at the base level. (b) Map of channels after 50 h. Note capture and integration of drainage at right side. Right side drainage network was well developed in contrast to that of the left side. (c) Map of channels after 60 h. Note capture at point C, which eliminated a major channel (Fig. 5b). (d) Map of channels after 70 h. All channels extended to the base-level control. The right side channels were well integrated (Fig. 1b) whereas the left side channels were not, which explains the different amount of piedmont dissection (Fig. 3).

Parker's (1977; Schumm et al., 1987) study of the development of drainage networks in the REF was designed to simulate the development of a drainage network, but some results have a bearing on the question of the propagation of base-level effects inland. Following base-level lowering, Parker showed that the rate of knickpoint migration was closely related to the drainage area above it and the discharge delivered to it. Therefore, the rate of knickpoint migration decreased, as it moved up channel.

The experiments reviewed above were not designed to provide information on fluvial, coastal plain and shelf response to base-level change. Nevertheless, they provide information that can be used to respond to questions 1 and 3 and an explanation of the variability of palaeochannel dimensions and location on an exposed continental shelf. During these experiments, response was rapid, and results were widespread. Even in these high-slope situations a base-level lowering that created knickpoints resulted in a decreased rate of knickpoint migration and a loss of energy in an upstream direction. A major lowering of base level can cause major incision for long distances, but moderate and small base-level changes probably will have only minor impacts because the increase of potential energy is low. As will be demonstrated later, however, the increase of sediment discharge can be significant, as the near base-level channels incise and valleys widen and deepen. The development of drainage patterns in response to irregularities of a surface can explain palaeo-drainage pattern anomalies and the variable depth of incision, as more aggressive channels capture drainage area (Figs 3 & 5).

VALLEYS

During a series of experiments involving base-level fluctuations on a simulated continental shelf and slope different types of valleys formed, as described by Koss (1992), Wood (1992), Wood et al. (1993, 1994) and Koss et al. (1994). Koss' experiments were performed in the REF. Wood's experiments were performed in a non-tilting, non-circulating outdoor flume. Characteristics of the experimental design and method are given in Tables 1–3.

Falling base level

The Koss experiment (Koss et al., 1994) commenced with a water-level highstand that covered the shelf (Table 1). The influx of sediments from the channels of the drainage basin onto the shelf resulted in a delta prograding across the shelf. Most deposition occurred near the shoreline as progradation forced the shoreline basinward (Fig. 6). During this time deltaic shelf deposition and delta-lobe switching occurred. Usually there were two or three lobes active at any one time, although one lobe was dominant. As the delta prograded, fluvial deposition occurred in the upper reaches of the deltaic plain.

When a delta had developed on the shelf and the fluvial channels in the upper drainage basin were no longer visibly growing (i.e. they had reached their maximum extension), base level was lowered to a point below the shelf–slope break to simulate a rapid sea-level fall. During the early part of this fall, the shoreline moved steadily basinward, although it remained landward of the

Table 1 Characteristics of the experimental method for experiments by Koss (1992) and Koss et al. (1994).

Experimental set-up	Properties
Dimensions	Rainfall–erosion facility: 9.2 m wide by 15.5 m long by 1.8 m deep
Sediment fill	Compacted silty sand with 13% silt and clay. Mean grain size was 2.9 ϕ and standard deviation was 0.7
Initial sediment fill lay-out	Sediment moulded to create a model fluvial drainage basin from 0 to 9 m; a coastal plain and continental shelf from 9 to 13 m; and a continental slope from 13 to 15.5 m
Gradients	Coastal-plain–continental-shelf and slope inclination remained constant for all runs at 0.01 and 0.22 respectively. Upper drainage basin inclinations were set at 0.02, 0.05 and 0.08 for successive runs
Coordinates	x, y and z coordinates were measured along eight transects with a point gauge mounted on a movable platform. Average standard deviation for the z coordinate was 0.0017 m
Water supply and yield	Water was supplied to the upper drainage basin using a sprinkler system. Water discharge was uneven over the drainage basin, but remained constant during and between runs. Rainfall values varied from 0.00 to 10.46 cm h^{-1}. Water discharge through the main channel of the drainage basin averaged 0.556 L s^{-1}
Froude number	Less than 1 in drainage basin and fluvial valleys during all experiments
Sediment supply and yield	Sediment supply derived from drainage basin. Sediment yield not measured
Base-level change	Initially water in the basin covered 75% of the coastal plain. After allowing a drainage basin and a delta to form and stabilize a sinusoidal base-level change was initiated. A rapid fall for 4 h (4.2 cm h^{-1}), followed by a stillstand for 4 h, and a rapid rise for 4 h (3.8 cm h^{-1}). This cycle was followed by a slow fall for 2 h of (1.0 cm h^{-1}) and a slow rise for 2 h (0.6 cm h^{-1}). During rapid fall base level fell below the shelf–slope edge. During slow base-level fall it did not. Each cycle ended with the shoreline 2 cm higher than when it started

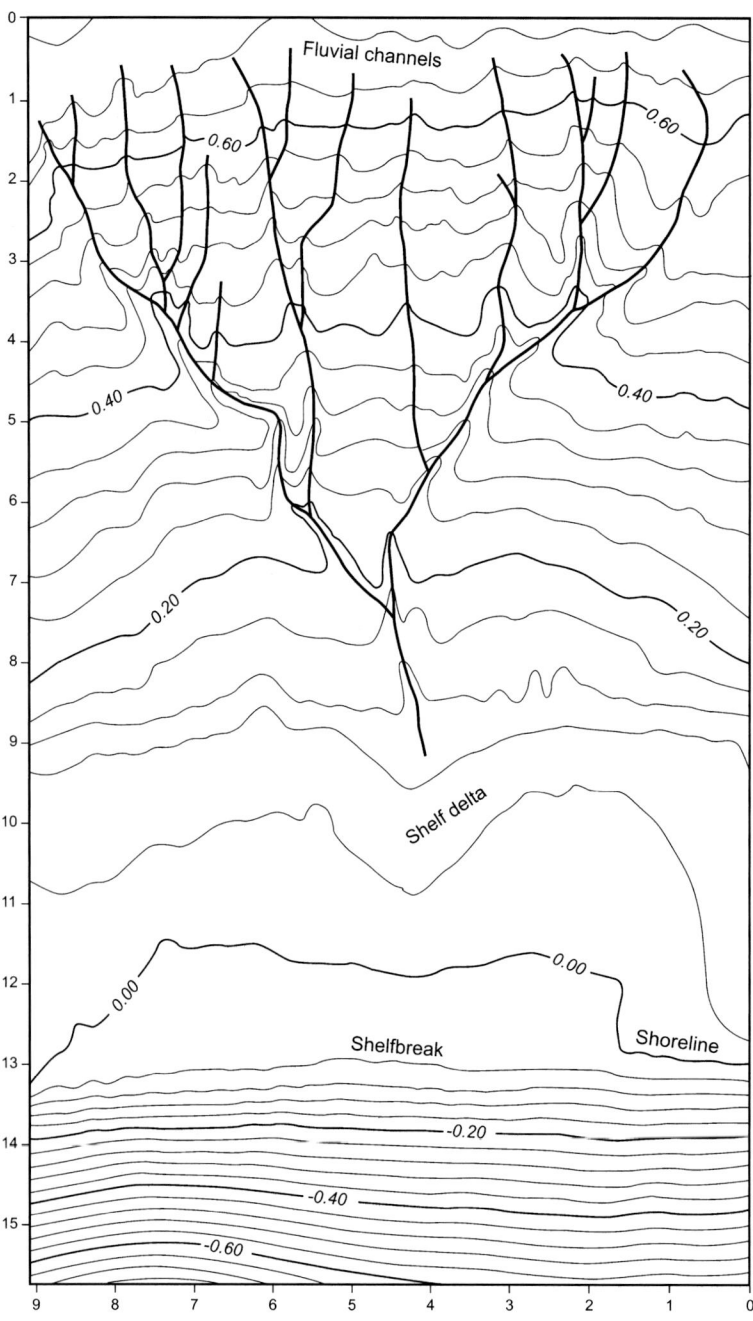

Fig. 6 Contour map of surface before water level fell below shelf break. All units are in metres. Fluvial channels occur from 0 to 9 m, feeding a delta at 9 to 10 m. Contour interval = 0.04 m. (From Koss, 1992.)

shelf–slope break. The delta prograded rapidly, with minor incision occurring on the upper portions of the delta plain. Streams deposited coarse-grained sediment on a shelf-bound braid plain, and they carried fine-grained sediment further basinward across the exposed shelf. Deposits on the exposed shelf consisted of widespread, but thin, fluvial braid-plain sediment and a shelf-phase delta. As water level dropped and the shoreline moved rapidly basinward, a single channel was not maintained and flooding was widespread over the coastal plain. This pattern of events is similar to the initial results of Creager's (1991) piedmont experiment, described in the previous section.

Once the shoreline reached the shelf–slope break and dropped below it, numerous shelf-break valleys began forming by headward migration of knickpoints (Fig. 7). It was commonplace for one valley to grow past another and to capture its flow

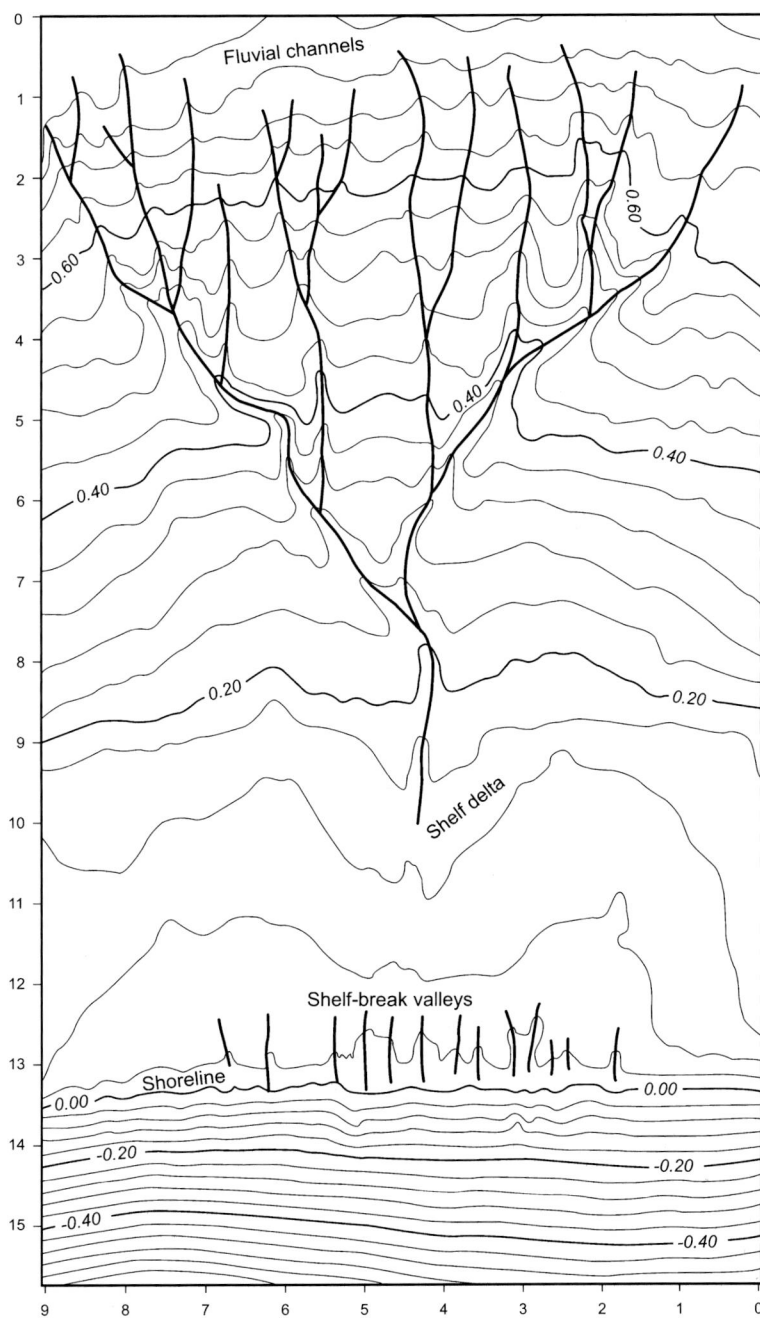

Fig. 7 Contour map of surface after water level fell below shelf break. All units are in metres. Delta front is at 12 m. Incised shelf-break valleys have reached the edge of the delta. Contour interval = 0.04 m. (From Koss, 1992.)

because the knickpoints migrated at different rates relative to water discharge. Slower growing shelf-break valleys were abandoned. These wide, flat-floored valleys were widened further as tributaries developed and as the channel shifted laterally. In addition it is likely that groundwater sapping of the valley sides exacerbated widening through sidewall failure. While the shelf-break valleys grew headward, the only sediments transported through them were cannibalized shelf sediment and suspended-sediment from the upland basin. No cross-shelf bypass channel existed during falling base level that would have transported coarse-grained sediment from the upland basin across the shelf.

During stable lowstand, shelf-break valleys continued to erode headward, depositing excavated shelf sediments at their mouths as slope-phase deltas with angle-of-repose delta-front clinoforms on the upper slope (Fig. 8). In a natural system these deposits probably would have much lower clinoform angles, be significantly reworked by marine processes, and slumping and turbidity currents would transport much of the sediment further down slope. Eventually one of these shelf-break valleys grew far enough upstream to capture the main fluvial channel near the apex of the highstand delta (Fig. 9). At this time, the other shelf-break valleys became inactive. The valley that captured the main fluvial channel then became a cross-shelf bypass valley transporting coarse-grained upper drainage-basin sediments into deep-water environments below the shelf break. As multiple shelf-edge valleys developed with shoreline fall over the shelf break, it was very difficult to predict which valley would eventually capture drainage basin flow. The final lowstand valley did not necessarily have the shortest or most direct path from hinterland to lowstand shoreline. In fact, in one case the valley looped around the highstand delta. There was always a significant lag between maximum lowstand and eventual capture and transport of coarse-grained drainage basin sediment to the shoreline.

Rising base level

A rapid base-level rise reflooded the shelf during the Koss *et al.* (1994) experiment. Abandoned shelf-break valleys were rapidly drowned and were filled with fluvial sediments, slumped wall material and fine-grained 'marine' sediments. The single incised valley that served as an active shelf-bypass conduit had a different fill history. As base level began rising, sediment ceased to be delivered to the slope delta and was instead deposited in deltaic lobes within the confines of the valley. As base-level rise continued and accelerated these deltaic lobes backstepped up the valley. The backstepping occurred as a delta lobe was abandoned in favour of a new lobe. The new lobe usually formed on the opposite side of the valley and further up valley than the abandoned lobe. Thus, the valley was filled with basal fluvial sediments overlain by backstepping deltaic deposits. Slumping of the valley walls occurred, but these deposits were less extensive than those in the abandoned valleys. Depending upon the interaction of sediment supply and rate of base-level rise, the valley could be either entirely filled by the deltaic deposits, or by a combination of deltaic and finer-grained estuary to 'marine' deposits. As rapid base-level rise slowed to a stable highstand, a new delta prograded over the shelf on top of the previous highstand delta.

Effects of rates of base-level change

Experiments designed by Wood *et al.* (1993, 1994) contained a coastal plain, shelf and slope. In the first series of experiments (Wood *et al.*, 1993), base

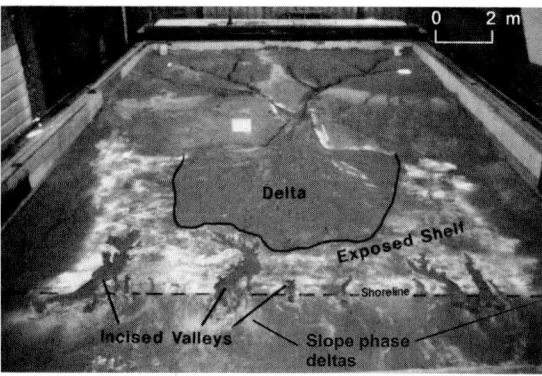

Fig. 8 Photograph showing shelf-break valleys and highstand delta. Large valley in the middle of the photograph has almost reached the front of the delta. (From Koss *et al.*, 1994.)

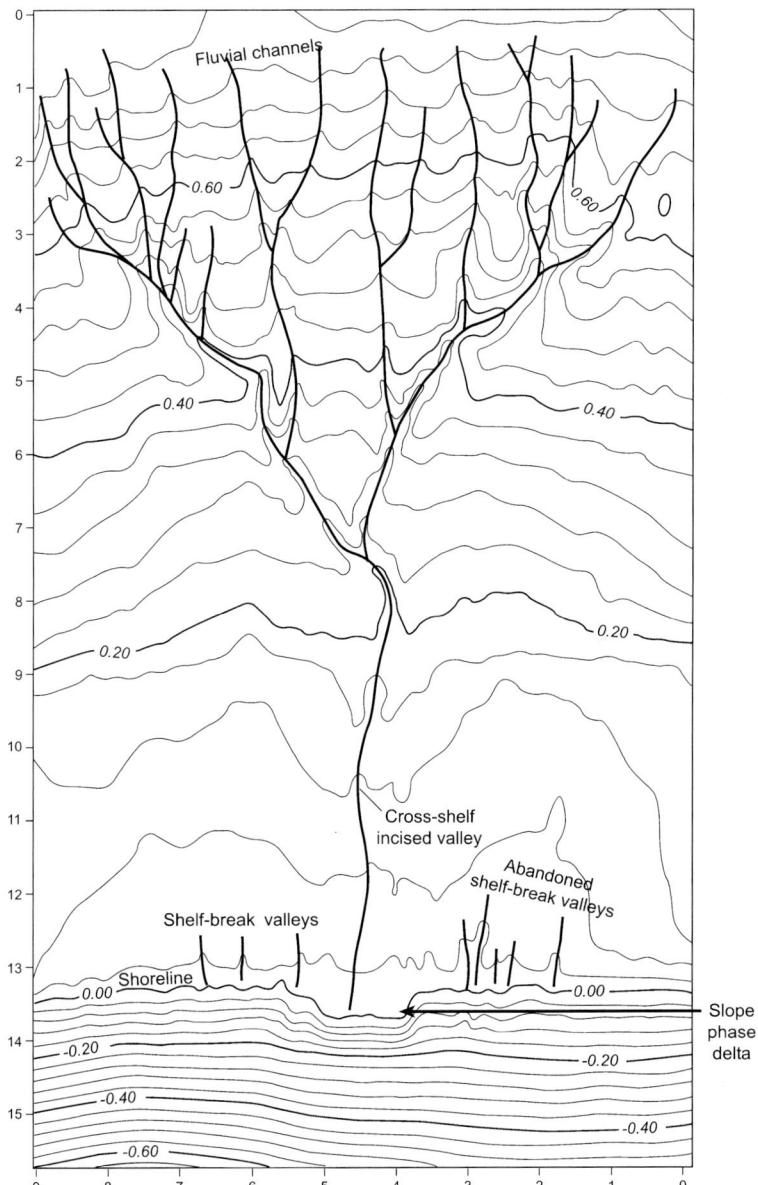

Fig. 9 Contour map of stable lowstand surface. All units are in metres. Cross-shelf valley connects drainage basin to shoreline by most direct route. Contour interval = 0.04 m. (From Koss, 1992.)

level was changed during three separate experiments to simulate three different rates of fluctuation, 26.5, 17.7, and 13.3 cm h^{-1} (Table 2).

Results of these experiments suggest that the rate at which base level fluctuates has a significant effect on erosional and depositional features that develop on the coastal plain, shelf and continental slope. The slow rate of fall allowed more time for deposition of thick shelf delta deposits, which created topographic breaks on the shelf, and for the channel to regrade its profile. These two factors combine to produce a single, continuous valley, which incised into the shelf. Each time the shoreline fell over these depositional breaks a knickpoint was initiated in response to the channel regrading its profile. This valley contained all

Table 2 Characteristics of the experimental method for base-level fluctuation experiments by Wood (1992) and Wood *et al.* (1993).

Experimental set-up	Properties
Dimensions	7 m long by 4.5 m wide by 0.9 m deep
Sediment fill	Compacted fine to very fine grained sand with 10% silt and clay. Mean grain size was 2.9 ϕ with a standard deviation of 0.7
Initial sediment fill lay-out	Sediment moulded to create a model fluvial drainage basin from 0 to 2 m; a coastal plain and continental shelf from 2 to 4.5 m and a continental slope from 4.5 to 5.5 m
Gradients	Coastal plain, continental shelf and slope inclination remained constant through all runs at 0.04, 0.14 and 0.36 respectively
Coordinates	x, y and z coordinates were measured with a point gauge mounted on a movable platform. Average standard deviation for the z coordinate was 0.0013 m
Water supply and yield	Water was supplied to the upper end of the flume through a V-shaped notch in a head box. Water discharge was held constant at 0.1 L s^{-1} for all runs
Froude number	Less than one in fluvial valleys during all experiments
Sediment supply and yield	An equal mixture of coarse sand, fine sand, and the material used to construct the experimental basin were added to the flow at the head of the fume using an electric sand feeder. Different colours of the coarse and fine sand were fed into the flow during base-level rise. Rate of sediment supply and yield not measured
Base-level change	Each experiment began with the shoreline held constant half-way up the shelf for 6 h to stabilize the system. Base level was then raised and lowered 106 cm during 4-h, 6-h and 8-h periods respectively. Average rate of base-level rise and fall for each period was 26.5 cm h^{-1}, 17.7 cm h^{-1} and 17.2 cm h^{-1}

of the discharge, which resulted in a high rate of deposition on the shelf and thick, shelf-margin delta deposits at the mouth of the valley. A series of diachronous deltas formed, which were successively younger toward the basin, as they stepped down the continental slope in response to falling base level. In contrast, during a rapid fall in base level the shoreline moved rapidly basinward and the associated coastal-plain channels lacked the time necessary to accomplish significant incision on the exposed shelf. Depocentres moved basinward at a faster rate and resulting shelf-delta deposits were significantly thinner than those deposited during a slow fall in base level. The initial highstand channel dispersed water and sediment onto the shelf as a broad shelf-delta apron. As base level fell below the shelf edge, several valleys formed at the shelf edge and grew landward by headward erosion (Fig. 10). Eventually one valley captured all of the discharge and funnelled sediment across the previous highstand shelf to the slope. Inactive valleys usually contained small amounts of coarse-grained sediments from reworked, previous-highstand deposits, but they were dominantly filled with fine-grained

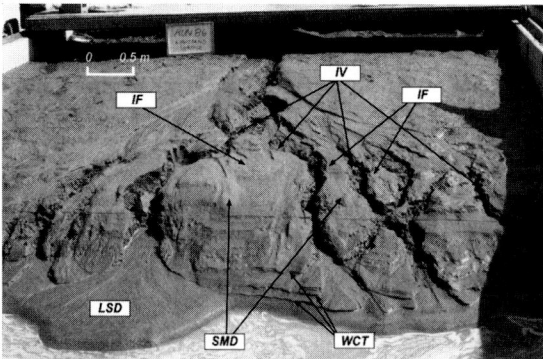

Fig. 10 Photograph showing multiple incised valleys (IV) that developed with a fast base-level fall, interfluvial areas (IF) separating incised valleys, multiple shelf-margin delta deposits (SMD), wave-cut terraces (WCT) and multiple, thin, lowstand deltas (LSD). (From Wood, 1992.)

muds and fine- to coarse-grained, mass-failure sediments. Submarine-fan deposits derived from these inactive valleys did not contain appreciable amounts of coarse-grained sediment derived from

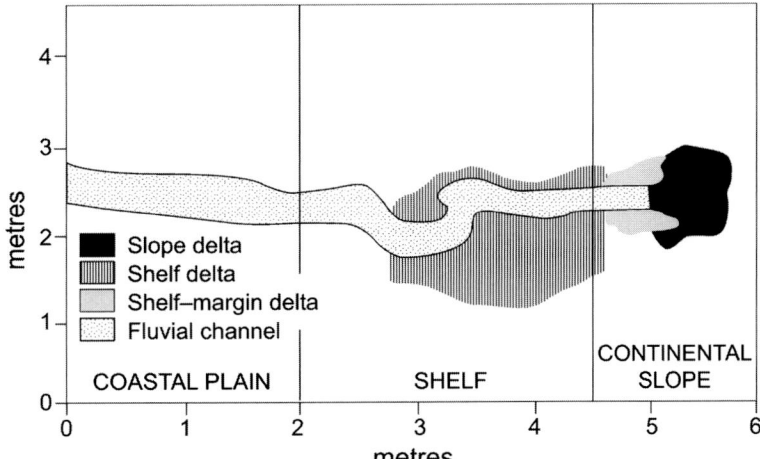

Fig. 11 Lithofacies map of the maximum lowstand surface. Map shows a single incised valley, one shelf-margin delta, and a narrow band of shelf-delta sediment formed during a slow base-level fall. (Modified from Wood et al., 1993.)

the hinterland. Any coarse-grained sediment contained within such submarine deposits were derived from a local, reworked coarse-grained sediment source.

Thus, a slow rate of base-level fall produced thick, areally restricted shelf-delta deposits, one well-defined, slightly sinuous, incised valley, and a few, thick, diachronous shelf-margin deltas (Fig. 11). In contrast a rapid base-level fall produced thin, widespread shelf-delta deposits, multiple, poorly defined, straight valleys and time-synchronous shelf-margin deltas (Fig. 12).

The rate at which the shelf and incised valleys were flooded also had a pronounced effect on the thickness and preservation of transgressive deposits. A slow transgression led to a longer period of accumulation, resulting in thicker transgressive deposits at any one depositional site. Slow flooding of valleys also extended the amount of time over which restricted marine deposition was occurring within these valleys. A slow transgression, however, exposed lowstand and transgressive deposits to a prolonged reworking by fluvial and shoreline processes, resulting in decreased preservation potential for these deposits. On the other hand, a rapid rise of water level resulted in less time to accumulate sediment at specific depositional sites, resulting in thinner transgressive deposits. Rapid flooding of incised valleys shortened the time interval over which

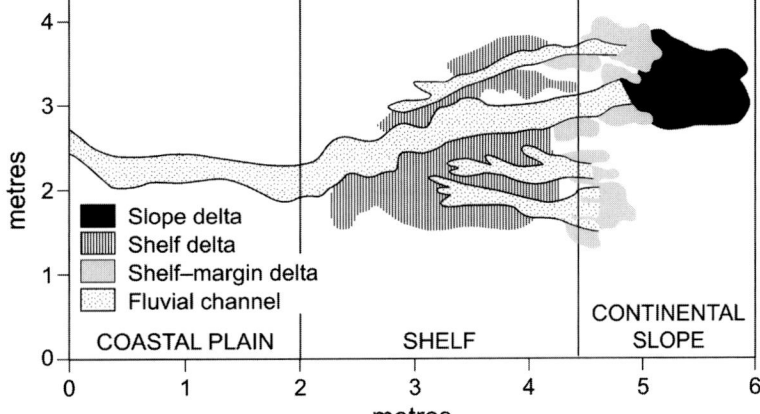

Fig. 12 Lithofacies map of the maximum lowstand surface. Map shows four incised valleys, multiple shelf-margin deltas, and a wide band of shelf-delta sediments formed during a rapid base-level fall. (Modified from Wood et al., 1993.)

Experimental set-up	Properties
Gradients	Coastal plain and slope inclination remained constant through all runs at 0.04 and 0.36 respectively. Shelf inclination varied as followed for three runs: 0.04, 0.07 and 0.14
Base-level change	Each experiment began with the shoreline held constant half-way up the shelf for 6 h to stabilize the system. Base-level change was kept relatively constant for each run by varying the amplitude and run time. Average rates of rise and fall were 12.00 cm h^{-1}, 10.0 cm h^{-1} and 13.25 cm h^{-1} for the low, intermediate and steep shelf inclinations

Table 3 Characteristics of the experimental method for influence of subaqueous shelf angle experiments by Wood (1992) and Wood et al. (1994). NOTE: data are the same as in Table 2 with the exception of gradients and base-level change noted below.

restricted marine facies were deposited within valleys. A rapid transgression also meant that lowstand and transgressive deposits were exposed to reworking by fluvial and shoreline processes for a shorter period, which resulted in increased preservation potential.

Effect of slope

Wood et al. (1994) completed several experiments in which base level fluctuated over shelves of different angles (Tables 2 & 3). These experiments were designed to assess the influence of the shelf angle on depositional sequence development (Fig. 13). Rates of base-level fluctuations were kept approximately constant between runs with an average rate of base-level rise and fall of 11.75 cm h^{-1}. As the shelf was exposed during a base-level fall it became the regional lowstand coastal plain. The angle of this surface influenced the development and character of depositional systems during lowstand and transgressive time.

The initial response of the main channel to a lowering of base level was vertical incision. In steeper shelf settings, vertical incision occurred rapidly, as streams attempted to adjust to the new lowstand base level. Deep, vertical incision confined the stream within the valley and made lateral erosion and meandering more difficult. The result was a deep valley with a low width/depth ratio and a low degree of sinuosity. In contrast, valleys that developed on shelves with a lower shelf angle incised less rapidly, because there was less difference between the coastal plain angle to which the stream was previously adjusted and the angle of the shelf over which it prograded. More of the stream's energy was expended in lateral erosion and widening of the valley, and in the development of meanders. The result was a shallow valley with a high width/depth ratio and a higher degree of sinuosity than those with steeper shelf settings (Fig. 14).

Steeper shelf angles increased the capacity of channels to erode and transport sediment. Rapid rates of vertical channel incision and straight, deep incised-valley geometries combined to produce higher flow velocities and very high sediment volumes. The result was thick, fine-grained, early-lowstand shelf and shelf-margin delta deposits, primarily composed of reworked shelf and shelf-margin sediments. In contrast, lower shelf angles result in lower sediment yields. Slower rates of vertical channel incision and

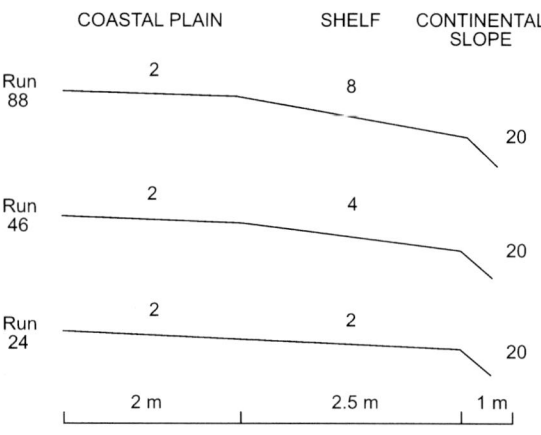

Fig. 13 Profile (in degrees) of the three initial sediment surfaces used prior to the start of each experiment. (From Wood et al., 1994.)

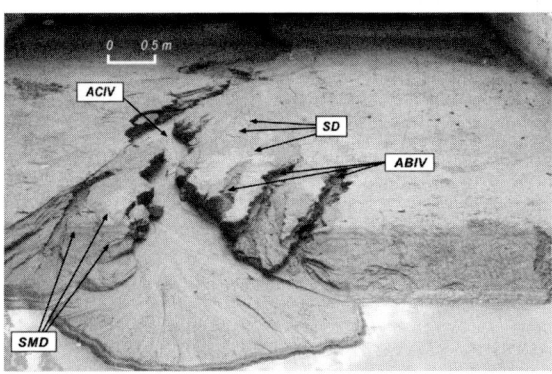

Fig. 14 Photograph showing an active valley (ACIV) that formed during base-level fall over a gently angled shelf. The valley is shallowly incised with a high width to depth ratio, and it has a slightly sinuous planform. On the shelf and at the shelf-margin are thin, early lowstand shelf and shelf-margin delta deposits (SD and SMD), which contain higher cumulative percentages of coarse-grained sediment than the same types of features sampled from runs on a more steeply angled shelf. Higher percentages indicate that these deposits are composed of a higher volume of hinterland sediment (sediment derived from the sediment feeder) than those occurring on lower angled shelves. Also noted are abandoned valleys located on the adjacent shelf (ABIV). (From Wood, 1992.)

sinuous, shallow valley geometries combined to yield lower sediment volumes, with less of the sediment being produced from the reworked shelf and shelf margin. The result was thin, early lowstand, shelf and shelf-margin delta deposits (Fig. 14) with a higher percentage of coarse-grained sediment.

A transgression of the shoreline over a steep-angled shelf resulted in rapid submergence of previous highstand and lowstand deposits. This rapid water deepening coupled with thick depositional sequences increases the probability of finding more complete stratigraphical sequences preserved along steep basin margins (Fig. 15a). In contrast, water depths increase more slowly during transgression over a low-angled shelf. Slow submergence and reworking resulted in incomplete preservation of previous thin highstand and lowstand deposits (Fig. 15b).

An additional series of experiments was completed in which base level fluctuated in a basin characterized by a ramp-style margin. The initial response of ramp shoreline depositional systems to a base-level fall was progradation. Incision by fluvial channels was initiated into previously deposited delta front or shoreface features, resulting in funnel-shaped valleys that regraded incrementally in response to shoreline movement over small topographic breaks created by shoreface and delta-front features. These valleys showed significantly higher width/depth ratios than those that developed in experiments carried out in a shelf-break setting under similar conditions. A delta apron formed at the mouth of each valley simultaneously with upstream incision. The

Fig. 15 Cartoon illustrating a steeply angled shelf (a) with thick, lowstand and highstand shelf and shelf-margin deltaic deposits, and a gently-angled shelf (b) with thin, lowstand and highstand shelf and shelf-margin deltaic deposits. Steeper shelves have higher volumes of accommodation space than gentle shelves. This factor increases the space available for thick deposits to develop. Wave base is less effective in reworking thick sediments deposited along steeply angled shelves where reworking elements of the shelf are more likely to come into contact with thin, shallowly deposited sediments. (Modified from Wood et al., 1994.)

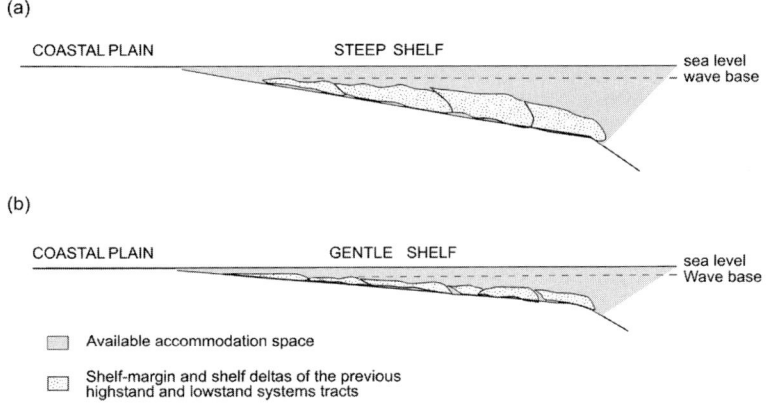

Table 4 Characteristics of the experimental method for single-channel experiments (Germanoski, 1990, 1991).

Experimental set-up	Properties
Dimensions	18.3 m long and 1.83 m wide
Sediment fill	Compacted medium to coarse grained sand. Mean grain size was 0.65 ϕ with a standard deviation of 1.41 ϕ
Initial sediment fill lay-out	Sediment moulded to create a model uniformly dipping surface
Gradients	Flume gradient of 0.02
Coordinates	x, y and z coordinates were measured for the equilibrium channel and at the conclusion of base-level fall and rise stillstands. Channel cross-sections were measured at 1 m intervals midway through the base-level change
Water supply and yield	Water was supplied to the upper end of the flume. Water discharge was held constant at 1.3 L s^{-1} for all runs
Froude number	Less than one during all experiments
Sediment supply and yield	Sediment was fed at the head of the channel using a vibrating sediment feeder. Sediment feed was established to maintain an equilibrium braided channel prior to base-level change and maintained at that rate throughout the entire run. Sediment delivery rates ranged from 2 to 5 g s^{-1}
Water to sediment delivery ratio	570 to 1430
Base-level change	Base level was raised and lowered a total of 17.2 cm in 1.9 cm increments. Base-level change rates were 7.2 cm h^{-1}, 3.6 cm h^{-1} and 1.8 cm h^{-1} for rapid, intermediate and slow rates of change

shallow nature of valleys on ramp-style basin margins encourages multiple episodes of abandonment and reoccupation throughout lowstand time. Eventually, a single valley captures all of the hinterland drainage through headward erosion and stream piracy, after which remaining ramp incisions were abandoned. Capture of the drainage by a single incised valley may not happen until well into late lowstand time.

Results of this series of experiments indicate that the angle of the continental shelf over which base level fluctuates has a significant effect on the development and geometry of incised valleys, sedimentation rates, sediment types and rates of knickpoint migration that occur in response to changing base level. The fact that a faster incision rate was noted when the initial change in slope between the fluvial and shelf systems was greatest is consistent with the diffusive fluvial morphodynamic theory used in numerical model studies by Paola et al. (1992) and Marr et al. (2000).

INCISED CHANNELS

In contrast to the experiments of Koss and Wood, which involved a simulated continental shelf and slope, experiments by Begin et al. (1980a,b, 1981) and Germanoski (1990, 1991) involved only a single channel that was affected by base-level change (Table 4). These studies were similar to those of Brush & Wolman (1960), Yoxall (1969) and Holland & Pickup (1976).

Falling base level

Begin et al. (1980a, 1981) lowered base level twice in a flume containing sediment with a high percentage of silt and clay. As knickpoints migrated through the flume, the cohesive sediment caused a narrow and deep incised channel to form. Base-level lowering generated a knickpoint that moved upstream, and secondary knickpoints formed and followed the primary knickpoint. The effect of the multiple knickpoints was to develop a channel that essentially paralleled the original channel gradient (Fig. 16). Initial stability of the cohesive banks prevented rapid widening of the valley after channel incision, but with time the channel impinged on the walls causing collapse and widening of the valley. Alternatively, in another series of experiments a small fraction of coarse sediment was added, which armoured the bed, limited channel incision and prevented full channel adjustment (Begin et al., 1980b; Fig. 17).

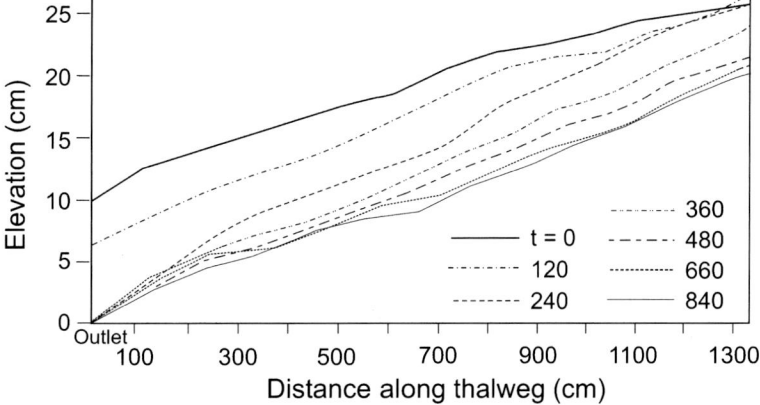

Fig. 16 Channel profile development following base-level lowering. Base level was lowered twice: at $t = 0$ and at $t + 120$ min. Numbers denote time in minutes. Profiles are based on mean bed elevations. (From Begin et al., 1981.)

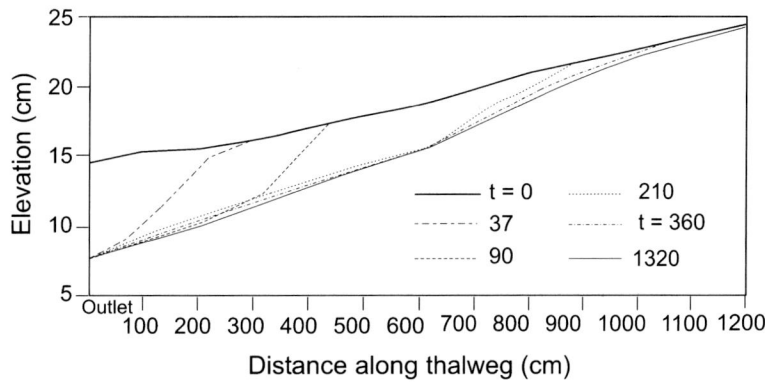

Fig. 17 Channel profile development following base-level lowering. Numbers denote time in minutes. Profiles are based on mean bed elevation (from Begin et al., 1981) and show effect of armouring.

Begin's experiments suggest that without the influence of armour or the effects of bedrock or resistant sediments base-level lowering would propagate far upstream. In contrast, experiments with non-cohesive sediment by Germanoski (1990, 1991) reveal a quite different result. Knickpoints quickly reclined and lost their identity as they formed a knickzone. As the knickzones migrated up channel, the magnitude of incision decreased progressively until the actively incising zone graded imperceptibly into the unaffected portion of the channel bed. The length of channel affected by successive base-level lowering increased through time as knickpoints migrated up channel, and the oversteepened reach extended further up system (Fig. 18). The extent of the influence of lowered base level decreased up channel, however, as illustrated by the longitudinal profiles. The decrease in the knickpoint migration rate and apparent

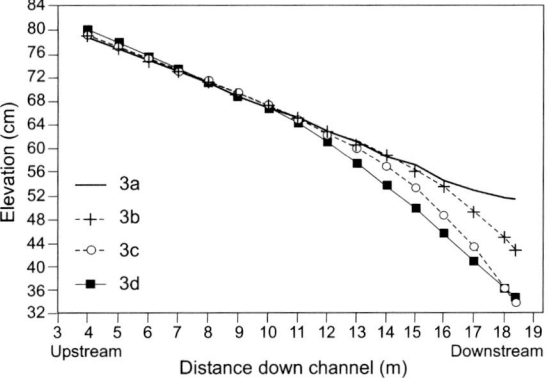

Fig. 18 Longitudinal profiles formed during rapid base-level lowering. Note oversteepened reach downstream and sharp inflexion point between unstable oversteepened reach and unaffected reach upstream (time progression from 3a to 3d).

limit of base-level influence results from a continued supply of sediment from upstream sources and the increase in sediment delivery due to base-level induced incision. Such a decrease in knickpoint migration has been observed and described by nearly all other workers in both laboratory (Holland & Pickup, 1976; Begin et al., 1980a,b, 1981) and field channels (Germanoski & Ritter, 1988).

Further evidence for this type of channel response is demonstrated by the experimental studies of Jeff Ware (personal communication, 1992). He lowered base level slowly to a maximum fall of 12 cm in a flume with a total length of 18.4 m. This change, in effect, doubled the surface gradient. In response the channel widened and roughness increased. The effect of base-level lowering extended only 4 m upstream, however, and the change in base level was accompanied by an increase in sinuosity from 1.2 to 1.5. Ware's experiments showed that a sinuosity increase, which resulted in a channel slope decrease, was only part of the adjustment, because width, depth and roughness adjusted to decrease velocity and stream power.

Another impact of base-level lowering and rapid incision was the development of terraces. Near the mouth of Germanoski's channel, at least one and sometimes two terraces formed in response to each base-level lowering. Terraces converged in the upstream direction, and terrace height was therefore greatest near the mouth of the channel. A complete terrace sequence was never preserved because of the high rate of bank erosion in the unstable degrading channels. Although new terraces formed as base level was lowered, the total number of terraces preserved was low because the new channels became progressively wider and existing terraces were destroyed (Fig. 19). Although as many as ten distinct terraces developed during base-level lowering, only two remained at the conclusion of the run. Thus, reconstruction of fluvial history on the basis of terrace sequences may be inherently flawed in many cases owing to incomplete terrace preservation.

Rising base level

Germanoski (1990, 1991) also documented the effect of raising base level on a single valley.

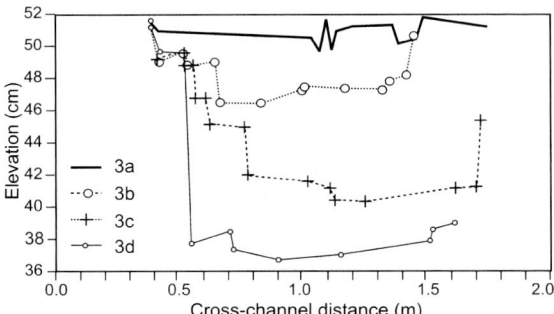

Fig. 19 Terraces formed during rapid base-level lowering 1.4 m upstream of base-level control. Terraces are poorly preserved owing to high rates of bank erosion. Terraces are destroyed primarily during the period of base-level stability (3c–3d), as the channel continues to adjust to the lowered base level (time progression from 3a to 3d).

Beginning with a degraded profile at the conclusion of a period of base-level stability, boards were inserted at the base of the flume. The base-level rise occurred in discrete increments, which is not typical of natural changes in sea level. The insertion of base-level boards caused a pool of water to form in the downstream portion of the channel, which simulated a sea-level rise. The reduction in flow velocity and shear stress associated with the ponded water resulted in sediment deposition and channel aggradation. Lobes of sediment prograded into the standing pool of water forming deltaic deposits.

Base-level rise outpaced the rate of deposition in the standing pool regardless of the rate of base-level rise. Thus, as evident in Fig. 20, the pre-degradation channel profile (Fig. 18, 3a) was never fully restored. Sediment discharge over the base-level control boards decreased to zero, and all sediment introduced into the channel was stored.

Although the greatest deposition occurred at the position of the ponded water, the locus of deposition shifted up channel, as base level rose and the edge of the ponded water moved up channel. Moreover, deposition reduced the channel gradient incrementally upstream, which led to a further upstream shift in the locus of deposition. Distinct lobes of sediment were deposited in the backwater pool marking the position of maximum profile recovery. Deposition of sediment at distinct points

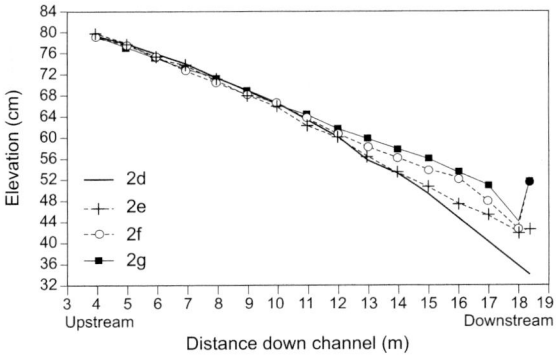

Fig. 20 Longitudinal profiles formed during slowly rising base level. Note upstream shift in locus of deposition through time and incomplete profile reconstruction relative to the position of base-level control boards (solid circle) (time progression from 2d to 2g).

in the channel resulted in the development of a distinctly stepped longitudinal profile in the downstream portion of the channel. This relationship demonstrates the importance of the backwater effect in controlling the locus of deposition in aggrading channels.

Comparison—rising and falling base level

Examination of longitudinal profile adjustment to falling base level (Fig. 18) suggests that there is a spatial limit to base-level influence. In fact, in experiments by Germanoski (1990, 1991), although base level was falling, the channel upstream of the 10-m mark was responding to sediment delivery from further upstream and was aggrading slightly (Fig. 18). Similar slight aggradation occurred in the upstream reaches of all channels, whereas downstream reaches were rapidly incising in response to falling base level. The lack of continuity between upstream and downstream reaches is further illustrated by Fig. 21, which shows cross-sections measured in both the upstream and downstream portions of a channel that was responding to a slowly falling base level. The upstream reach aggraded almost continuously while base level fell and the downstream reach was incising. These observations indicate clearly that upstream and downstream parts of the channel may be behaving differently

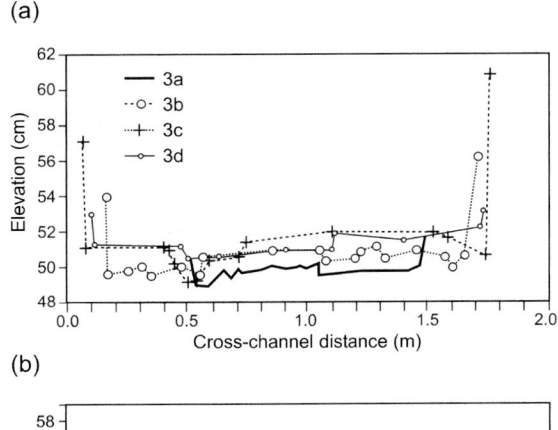

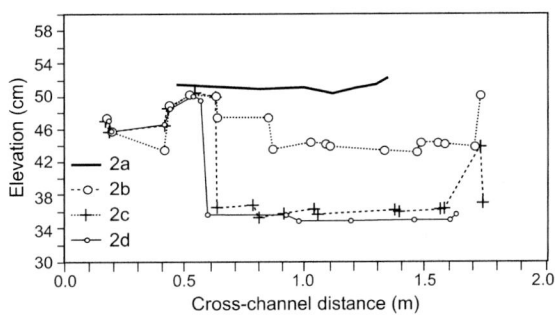

Fig. 21 (a) Cross-sections formed upstream during slowly falling base level. (b) Cross-sections formed downstream during slowly falling base level. Note aggradation occurring upstream whereas the downstream reach is incising (time progression from 2a to 2d).

at any given time and that a clear spatial limit to the effects of changing base level exists.

Once a channel incised in response to lowered base level, a subsequent commensurate rise in base level did not result in a complete restoration of the channel to its pre-incision position. Examination of channel cross-section profiles illustrates the disparity in channel response to changing base level (Fig. 22). It requires a much longer time for rising base level to return a degraded channel to its former position. The impact of falling base level was much more significant than the average impact of rising base level if impact is measured in terms of the average change in channel elevation in the affected length of channel. For these experiments the average depth of incision was 62.5% greater than the average increase in channel elevation resulting from aggradation throughout the affected reaches of

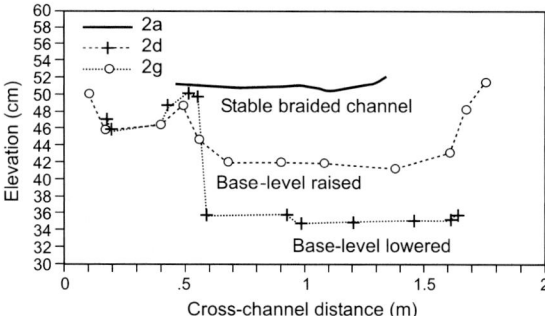

Fig. 22 Comparison of cross-profiles of initial channel and final profiles of degraded channel after lowered base level, and aggraded channel after rising base level (runs 2a, 2d, and 2g).

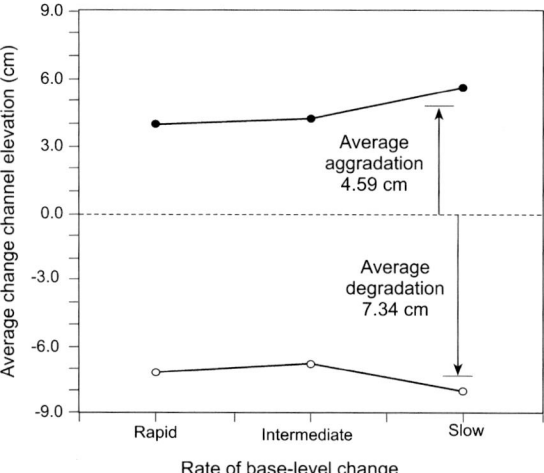

Fig. 23 Comparison of average change in channel elevation for degrading versus aggrading channels. Regardless of the rate of base-level change, the average depth of incision was much greater than the average height of aggradation.

the channels (Fig. 23). Viewed in these terms, the overall magnitude of geomorphological change and system disruption is far greater when base level is lowered than it is when base level rises.

In addition to having a greater impact at each individual cross-section, the effects of falling base level extended further up channel than the effects of rising base level. Falling base level affected an average of 61.7% of the channel upstream of base level, whereas the effects of rising base level influenced 51.5% of the channel. The differences in channel response to opposite directions of base-level change result from a fundamental difference in the way the downstream disturbances propagate upriver. The narrow and deep channel incision propagated rapidly upstream, whereas the compensating deposition not only fills the channel, but also spreads sediment across the valley floor.

The rate of base-level change did not have a clear effect on the changes in channel pattern. In each case, falling base level resulted in extreme channel instability, a reduction in braiding intensity, and a tendency for the channel to become more of a single-thread slightly sinuous channel. Subsequent increase in base level reversed these trends, and braiding intensity increased. The rate of base-level change also did not have a striking influence on the relative depth of degradation or aggradation, except that maximum degradation and aggradation did occur when the rate of base-level change occurred most slowly (Fig. 23). Likewise, the length of channel influenced by changing base level reached maxima when base-level change occurred most slowly. These results reflect the fact that, given the equivalent magnitude of total base-level change, the magnitude of the response increases, as the rate of base-level change decreases, because the total duration of the run increases. The channels simply have more time to adjust to the base-level changes.

SEDIMENT YIELD

Falling base level

An abrupt lowering of base level causes channel incision and a marked and rapid increase of sediment yield. After a rapid increase of sediment yield, there is a rapid decrease followed by additional pulses of sediment as temporarily stored sediment is flushed from the valley (Fig. 24; Schumm & Rea, 1995). Progressive lowering of base level at the mouth of an experimental channel resulted in rapid incision, channel instability and a major increase in sediment discharge (Germanoski, 1990, 1991). Sediment discharge increased tremendously when base level was lowered, and it increased further as more base-level boards were removed. Each decrease in base level was marked

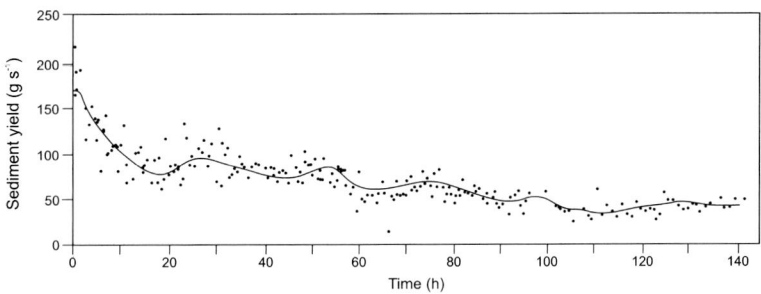

Fig. 24 Sediment yield variations following base-level lowering of 0.25 m at time 0, during 140 h of precipitation application to surface of rainfall–erosion facility. Total relief was 0.78 m; each point represents a sediment sample, and moving mean line shows secondary peaks at about 25, 55, 75 and 100 h, as stored sediment is flushed from the network. (From Parker, 1977; Schumm et al., 1987.)

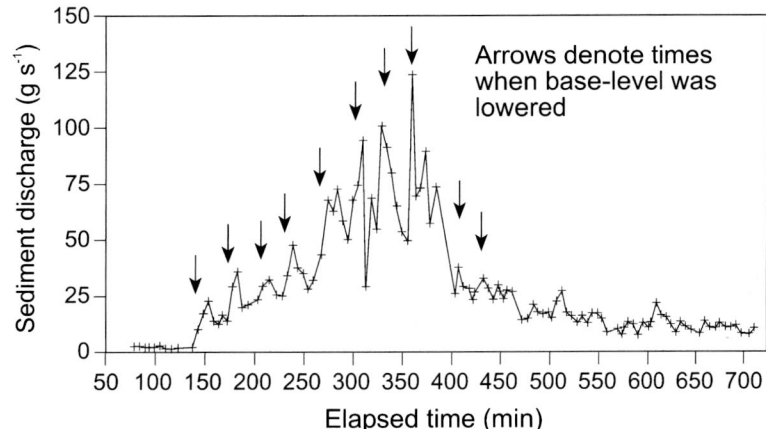

Fig. 25 Effect of base-level change on sediment yield. Run 1a-e, sediment discharge during intermediate rate of basel-level lowering.

by a distinct increase or spike in the sediment discharge curves (Fig. 25). Base-level lowering also caused the channels to become unstable laterally and bank erosion increased significantly. Therefore, sediment discharge was highly variable, and there were numerous peaks in sediment discharge curves, which were associated with periods of incision, bank erosion and bank failure.

SECOND GENERATION PHYSICAL EXPERIMENTS

Recently a series of physical experiments have been performed at the University of Leeds (Milana, 1998), University of Minnesota (Heller et al., 2001; Paola et al., 2001) and Utrecht University (Van Heijst & Postma, 2001; Van Heijst et al., 2001, 2002). Experiments by Milana (1998) were designed to investigate the effects of different rates of generation of accommodation space and discharge on the architecture of the resulting deposits. Although these experiments permitted the development of a model of alluvial sequences in proximal alluvial basins, the experimental set-up was so dissimilar to the experiments at CSU that comparisons are not readily apparent.

Experiments performed by Heller et al. (2001) involved an investigation of the effects of slow and fast base-level changes in an experimental subsiding basin. These experiments differed in a number of ways from the basin-scale experiments performed at CSU. Experimental topography (shelf-slope) was generated during the experiment rather than being constructed as in the CSU experiments. Sediment input was composed of a 50 : 50 mixture of quartz and coal sand and the experimental facility permitted creation of accommodation space by subsidence of the basin floor. In the channels the grain-size mixture resulted in steep slopes, and sheet flow and supercritical flows were common. However, base-level changes were similar to those generated in the experiments of Wood et al. (1993) and Koss et al. (1994). As a result of basin subsidence, growth faults were generated in the experimental depositional sequences. Valleys developed during both the slow and fast base-level cycles. The deepest and narrowest valley formed during rapid base-level fall and backfilling of this valley did not begin until rapid base-level rise started.

Experiments performed by Van Heijst et al. (2001) and Van Heijst & Postma (2001) were designed to quantify time-averaged erosion and deposition in fluvial and offshore environments as a result of sea-level change. Base-level change was the independent variable in the experiments by Wood et al. (1993) and Van Heijst & Postma (2001). Scaling of response time of the system was incorporated into the experiments by Van Heijst & Postma (2001) using the ratio between the period of change of a variable and the time at which the system reached equilibrium (Paola et al., 1992; Burns et al., 1997). The experimental set-up had similarities and a few significant differences to that of Wood et al. (1993). Experiments began with a coastal-plain–shelf-slope topography. The substrate and the sediment feed consisted of a uniform medium sand. Substrate sediment was slightly coarser grained, but much better sorted than that used by Wood et al. (1993). The coastal plain gradient was less steep than the shelf gradient in both sets of experiments, however, in Wood et al. (1993) the ratio of the two slopes was higher by an order of magnitude. In both sets of experiments the base-level change cycle was varied between experimental runs. In the Van Heijst & Postma (2001) experiments, however, the variation was formulated by changing the duration of the base-level fall portion of the curve relative to the base-level rise portion. The duration of base-level rise remained constant for most experiments. In the Wood et al. (1993) experiments the variation resulted from changing the total run time for the completion of one cycle. The duration of base-level fall always equalled the duration of base-level rise for a single experiment. Van Heijst & Postma (2001) introduced the term 'connection rate' to quantify the delay in response between the start of the fall of base level and the first features of erosion in the fluvial valley. The connection rate controlled the diachroneity of the lowstand sequence unconformity and, therefore, many depositional and erosional features on the shelf and in the fluvial valley. For fast base-level fall experiments the time transgressive nature of the lowstand unconformity is greater, the volume of lowstand deltas is smaller, cannibalized shelf material forms the major portion of the lowstand deltas, and the amount of transgressive deposits increases. In general shelf sequences and fluvial sequences are more out of phase in fast base-level fall experiments.

DISCUSSION AND STRATIGRAPHICAL SIGNIFICANCE

The experiments of Phillips & Schumm (1987) and Creager (1991) reveal how important initial conditions are with regard to the evolution of drainage patterns responding to a fall in base level. The concentration of water in depressions and general irregularities of a surface determine the direction of channel growth. Hence, the development of valleys on an exposed continental shelf will be heavily dependent upon slope and irregularities (i.e. faults, diapiric structures, etc.) that promote stream capture and integration of drainage networks.

Parker's (1977) and McLane's (1978) experiments show that channels lose energy in an upstream direction, and Germanoski showed that an increase in sinuosity, roughness and channel widening rather than vertical incision can compensate for the increased gradient, as a result of base-level lowering. Cohesive sediments that permit deep incision will allow base-level effects to be propagated further upstream than will low cohesive sediments in braided rivers and those with armour development.

The experiments of Koss et al. (1994) and Wood et al. (1993, 1994) reveal how variable and complex the morphology and sedimentology of the continental shelf and slope can be depending on initial conditions and rates of change. In general shelf-break valleys formed only when sea level dropped below a break in slope (gentle slope to steeper slope), as predicted by Posamentier & Vail (1988). Depending on sediment supply and the rate of sea-level fall, many shelf-break valleys or a single valley might form. As these valleys grew by headward erosion, there was a lag from the time when shoreline dropped below the shelf break until the time when one of these valleys grew across the shelf and captured the upper channel. During this time lag, no cross-shelf bypass channel existed, so the only sediment delivered to the deep-water environment (lowstand fan) was excavated from the eroding valley and from fine-grained suspended-load sediment from the upper drainage basin. Deposition of coarse-grained bed load from the upper drainage basin occurred at this time on a fluvial braid plain or shelf-phase delta. The fact that not all incised valleys contained shelf bypass channels led to important differences in the fill histories of adjacent, coeval valleys.

Once a shelf bypass valley developed, a lowstand fan was deposited with sediments from the upper drainage basin. The fans associated with the abandoned shelf-break valleys were smaller and finer-grained. As base level rose (but with shoreline remaining below the shelf break), a lowstand wedge formed, consisting of fluvial deposits within the shelf bypass valley, deltaic nearshore deposits and finer grained sediments on the slope overlying the lowstand fan. As base-level rise accelerated, and the shoreline moved onto the shelf, the transgressive systems tract, which was composed primarily of nearshore deposits reworked from older shelf sediments, formed. Backstepping deltaic deposits and marine sediments formed in the shelf bypass valley. The sequence of events observed in the CSU depositional basin experiments is almost identical to those noted in the second generation experiments of Van Heijst & Postma (2001).

Experiments by Wood et al. (1993, 1994) revealed how the rate of base-level change and inclination of the shelf control the variability in channel and valley development. Results of the CSU experiments involving the rate of base-level change are very similar to those of Van Heijst & Postma (2001) and contrast with those of Yoxall (1969) and Heller et al. (2001). In both the CSU experiments and those by Van Heijst & Postma (2001) a rapid fall in base level produced multiple fluvial valleys on the shelf and slope and widespread shelf-margin delta deposits. In contrast a slow fall produced one large valley and thicker areally restricted shelf-edge delta deposits. In rapid base-level fall experiments by Heller et al. (2001) a relatively narrow, deep valley developed quickly and lengthened by downcutting through deltaic deposits at the valley mouth as base level continued to fall. Differences between the CSU and the University of Minnesota experiments (Heller et al., 2001) were attributed to a combination of factors, including sediment type, initial set-up and subsidence. These differences emphasize that initial conditions influence results.

In spite of many similarities between the experiments of Wood et al. (1993) and Van Heijst & Postma (2001), there is one significant difference. Wood observed thicker transgressive deposits that were associated with a slower base-level rise. Van Heijst & Postma (2001) suggested that the thickness of the transgressive deposits was a function of the preceding base-level fall, with thicker deposits being associated with faster falls. They reasoned that fluvial erosion occurs earlier during a fast base-level fall, which results in higher sediment fluxes to the coastal plain and shelf. As most of their base-level rises were equal or varied only slightly relative to the large variations in base-level fall, a comparison of the effect of varying base-level rise in their experiments is not possible with available data.

The study of ramp-margin settings has recently increased as more foreland basins are recognized as having a ramp-style basin geometry (Posamentier & Vail, 1988; Van Wagoner et al., 1990; Van Wagoner & Bertram, 1995). Van Wagoner et al. (1990) and Posamentier & Vail (1988) discussed the cannibalization of late highstand beach and delta parasequences as a mechanism for generation of knickpoints. Discussion of processes responsible for the preserved rock record in the Cretaceous Western Interior, USA, was reviewed by Van Wagoner & Bertram (1995) following publication of the work of Schumm (1993), Wood et al. (1993) and Wood (1992). Quantitative analysis of valleys developed along ramp margins (Dolson, 1981; Jennette et al., 1991; Ethridge et al., 1994; Van Wagoner & Bertram, 1995) suggests that width/thickness ratios are significantly higher in ramp-margin settings relative to those incised valleys developed along shelf-break style basin margins (Thomas, 1991). Similar ratios were seen in the flume (Wood et al., 1993, 1994). The shallow, wide, funnel-shaped geometry of these ramp-margin valleys means that, in a real-world setting, incoming flood tides would be progressively compressed into a smaller cross-sectional area, resulting in gradual increase in the speed of the flood-tidal currents entering the estuary. These experimental results may help explain the abundance of tidally influenced fills documented in many valleys developed in foreland/ramp-margin basins such as the Western Interior Cretaceous Seaway of North America (Van Wagoner & Bertram, 1995) or the Pyrenean Foreland Basin (Nio & Yang 1991a,b; Crumeyrolle et al., 1993). Most settings characterized by tidal processes in valley fills occur where palaeogradients are interpreted to have been very low (i.e. Lower Triassic of Barles; Alpes-de-Haute-Provence, France (Richards, 1994); Eocene Ametlla Formation, northern Spain (Dreyer, 1994)) with no discernable shelf break, similar to the conditions simulated in the flume. Therefore, it is possible that the geometry of the valley, which is so influential in enhancing tidal influence, is a product of the basin geometry as much as regional oceanographic processes.

In addition to width/depth ratios, other differences were noted during Wood's experiments between stratigraphical sequences in ramp-margin basins and those in shelf-break margin basins. Valleys that develop along shelf-break margins become stable morphological features in early lowstand time. In contrast, valleys developed along ramp-style basins did not become stable morphological features until late lowstand time. Early transgressive deposition of fluvial sediments along shelf-break margins was confined to the last active lowstand incised valley. Early transgressive deposition of fluvial sediments along ramp-style margins was not confined to the last active lowstand valley.

Valleys incised deeply into older shelf sequences are common in stratigraphical sequences in a shelf-break basin (Fig. 26). Valley fill consists of thick, basal fluvial deposits, overlain by thick, transgressive or bayhead deltaic deposits. This lower fill is separated from the overlying units by a transgressive surface of erosion. This transgressive surface is overlain by restricted marine (estuarine) deposits, thick sidewall slump deposits and marine muds. The transgressive

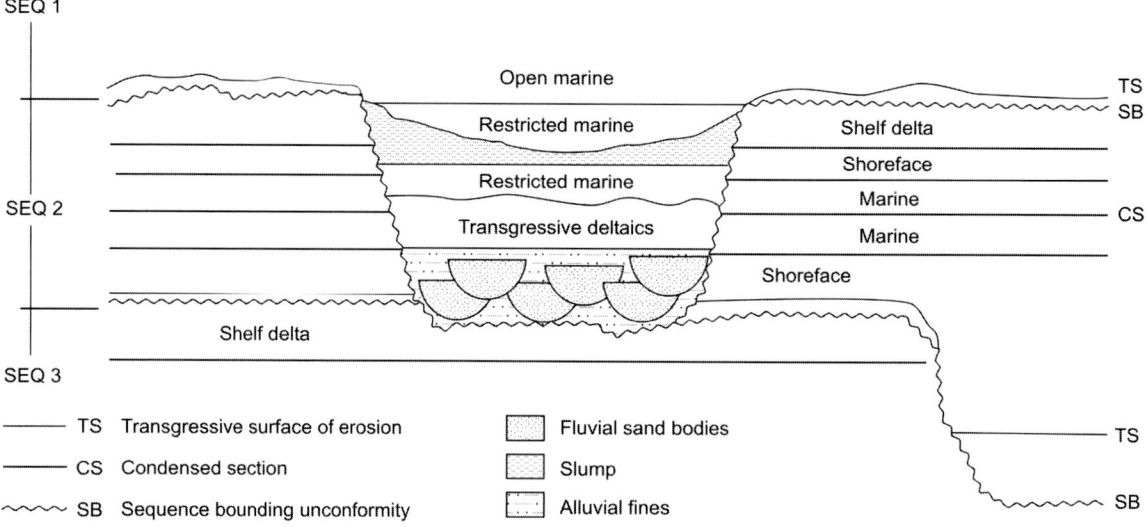

Fig. 26 Cartoon illustrating a typical shelf-break style stratigraphical sequence developed in the flume experiments. This sequence consists of shelf-delta deposits, and/or shoreface deposits of a much older sequence underlying the sequence bounding unconformity (SB). The valley will be filled from bottom to top with thick, fluvial deposits, thick transgressive deltaic deposits overlain by a transgressive surface of erosion (TS), and restricted marine (estuarine) facies. Thick, slump units deposited within the valley during rising base level will be extensive across the entire valley. Marine deposits will cap the incised valley. (From Wood, 1992.)

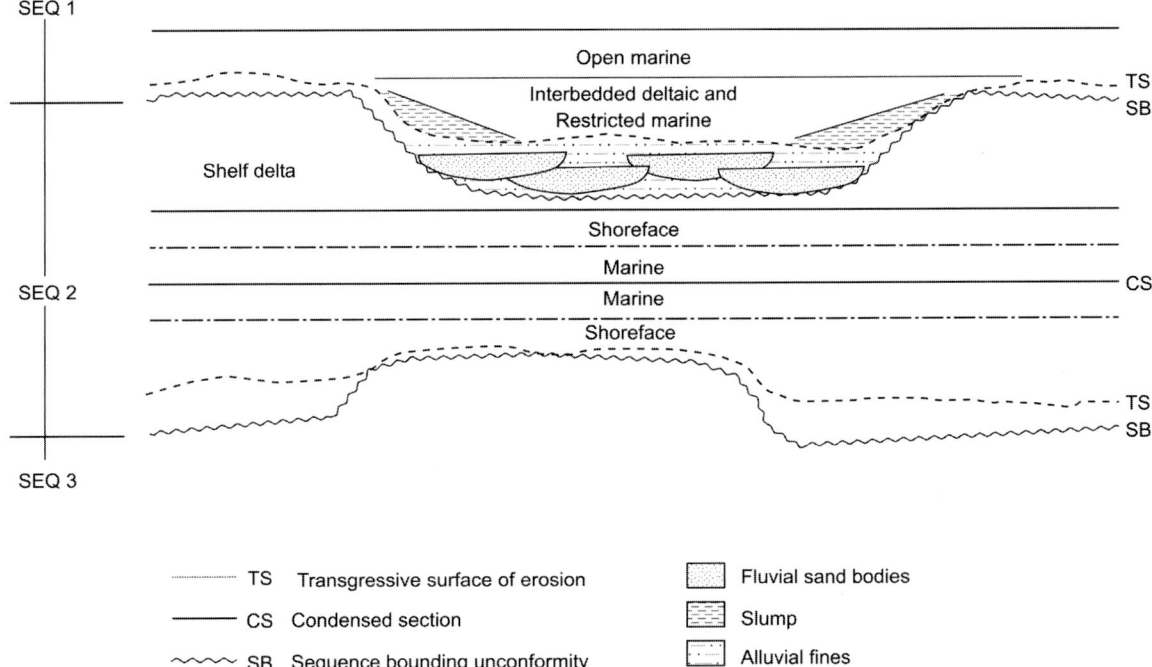

Fig. 27 Cartoon illustrating a typical ramp-style stratigraphical sequence developed in flume experiments. This sequence consists of shelf-delta deposits and/or shoreface deposits of the previous highstand underlying the sequence bounding unconformity (SB). The base of the valley contains thin, fluvial and deltaic deposits that may show abundant tidal indicators. These deposits may be overlain by tidally influenced deltaic and restricted marine facies and be separated from the overlying units by a transgressive surface of erosion. Thin, slump units deposited within the valley during rising base level will be confined to the side of the valley. Marine deposits will cap the incised valley. (From Wood, 1992.)

surface in this setting is characterized by significant stratigraphical separation between the interfluves and the valleys. In contrast, a common stratigraphical sequence in a ramp-style margin is characterized by shallow, valley incision into underlying shelf deltaic and shoreface sediments deposited in response to base-level fall (Fig. 27). Valley fills consist of thin, basal fluvial deposits overlain by thin, finer grained transgressive or bay-head deltaic deposits. These units are overlain by a transgressive surface of erosion. Valley fill above this transgressive surface consists of interbedded deltaic and restricted marine (estuarine) deposits. Thin sidewall slump deposits are confined to the sides of the valley, and marine muds cap the sequence.

Similarities, as well as differences, were noted between the experiments and published theoretical sequence stratigraphical models and subsurface data from the Gulf of Mexico, the Atlantic Shelf and offshore New Zealand. Many workers in nearshore depositional environments have noted a relationship between rates of base-level fluctuation and thickness of deposits associated with these environments. Fletcher et al. (1990) noted that migration rates of regional depocentres were determined by the topography of the pre-transgression surface and rates of sea-level rise. Thicker deposits are related to slower rates of base-level rise and thinner deposits with faster rates of rise. Using seismic and palaeontological data collected off the coast of New Zealand, Carter et al. (1991) noted that slow rates of sea-level rise produce thick, transgressive systems tracts. Results of the experimental flume work corroborate these observations (Wood, 1992). Other workers have noted the relationship between rapid rates of transgression and higher preservation potential for shelf and coastal-plain deposits (Belknap & Kraft, 1981; Suter & Berryhill, 1985; Sirigan et al., 1990). Experimental results suggest a higher preservation potential for deposits that are being rapidly transgressed. Suter & Berryhill (1985) and Suter et al. (1987) relate thin, widespread shelf deltaic deposits to rapid falls in base level, and the experimental results corroborate this observation.

Experiments completed in the flumes suggest that, although some minor scour of the shelf may occur as the systems prograde basinward, the major vertical component of valley incision occurs as the terminus of the fluvial valley reaches the topographic break separating the shelf and the continental slope. These valleys then continue to develop through incision headward across the shelf until the valleys either link with a fluvial channel issuing from the hinterland or are abandoned by processes of stream piracy. Many of these abandoned incised valleys have been documented in Quaternary-age deposits of the Atlantic Shelf (Knotts & Hoskins, 1968; Uchupi, 1970), the Gulf of Mexico Shelf (Thomas, 1991), and offshore Japan (Saito, 1991).

Differences between sequence stratigraphical models and the experimental results exist most obviously in the timing of transfer of coarse-grained hinterland sediment to the distal reaches of valleys. Posamentier & Vail (1988) suggested that the initiation of base-level fall marks the onset of a wave of coarse-grained fluvial sediment that progrades down the channel and is deposited at the channel mouth as coarse-grained, early lowstand deltaic deposits. In contrast, experimental results show that there is a significant lag between downstream impulses and upstream responses in a fluvial channel. Base-level lowering does not immediately result in a wave of upstream sediment prograding down the system. Much sediment may be temporarily stored within the incised valley. During the initial stages of base-level fall, any coarse-grained sediment contained in early lowstand deltaic deposits is derived from the reworking of previous highstand deposits.

Posamentier & Vail (1988) stated that the onset of base-level rise initiates a cessation of coarse-grained fluvial deposition at the mouth of the fluvial channel. Experimental results suggest that this is not true. Lag times documented in the CSU experiments suggest that the greatest volume of coarse-grained sediment derived from the upper drainage basin, is delivered to the depositional basin during late lowstand and transgressive time. This delay in the transfer of large amounts of drainage basin sediments ensures a continual supply of coarse sediment to the lower reaches of the incised valley into transgressive time and is in general agreement with experimental studies by Van Heijst & Postma (2001) and empirical data from Blum & Price (1998) and Blum & Törnqvist (2000).

Finally, sequence stratigraphical models suggest that valley cross-sections become modified owing to mass wasting and slumping of the valley walls, as these valleys are flooded by rising base level (Posamentier et al., 1991). Similar processes occurred in the flume experiments. Rising water tables increased pore-water pressure in the valley walls, already weakened by fluvial undercutting. These factors contributed to mass failure of the valley walls, which decreased valley-wall slopes, widened the valley and produced chaotic sedimentation in the valley itself.

The experimental studies described here provide useful insights into the impacts of base-level change and tentative answers to the four questions posed earlier as follows:

1 Base level has a limited upstream influence on a fluvial system. The extent of the impact is a function of the size of the system (Blum & Törnqvist, 2000; Van Heijst & Postma, 2001) and many other factors such as the gradient, grain size and bank stability of the system, to mention a few. The greatest impact of base-level change is near the site of the change.
2 Base-level change can be buffered by channel changes such as sinuosity, roughness and width.
3 Channels adjust rapidly to base-level change especially at the point of change.
4 Base-level fall has a more dramatic effect on channels and valleys than does base-level rise.

ACKNOWLEDGEMENTS

We thank the National Science Foundation, Army Research Office, ARCO Exploration and Production Technology, Amoco Production Company and Chevron Oil and Gas for financial support during various times over the 20 yr that the experiments were conducted. Figures were drafted by the Drafting Department, Bureau of Economic Geology, University of Texas. The co-authors are listed alphabetically. The paper has benefitted greatly from thoughtful reviews by John B. Swenson and George Postma and from organizational suggestions by co-editor Michael D. Blum.

REFERENCES

Bates, R.L. and Jackson, J.A. (1980) *Glossary of Geology*. American Geological Institute, Washington, DC, 749 pp.

Begin, Z.B., Meyer, D.F. and Schumm, S.A. (1980a) Knickpoint migration due to base level lowering. *Am. Soc. Civ. Eng. Proc. J., Water. Port Coast. Ocean Div.*, **106**, 369–387.

Begin, Z.B., Meyer, D.F. and Schumm, S.A. (1980b) Sediment production of alluvial channels in response to base level lowering. *Am. Soc. Agric. Eng. Trans.*, **23**, 1183–1188.

Begin, Z.B., Meyer, D.F. and Schumm, S.A. (1981) Development of longitudinal profiles of alluvial channels in response to base level lowering. *Earth Surf. Process. Landf.*, **6**, 49–68.

Belknap, D.F. and Kraft, J.C. (1981) Preservation potential of transgressive coastal lithosomes on the U.S. Atlantic shelf. In: *Sedimentary Dynamics of Continental Shelves* (Ed. C.A. Nittrouer), pp. 429–442. Elsevier, Amsterdam.

Blum, M.D. and Price, D.M. (1998) Quaternary alluvial plain construction in response to glacio-eustatic and climatic controls: a Late Quaternary example from the Colorado River, Gulf coastal Plain of Texas. In: *Siliciclastic Sequence Stratigraphy, Recent Developments and Applications* (Eds P. Weimer and H.W. Posamentier). *Am. Assoc. Petrol. Geol. Mem.*, **58**, 259–283.

Blum, M.D. and Törnqvist, T.E. (2000) Fluvial responses to climate and sea-level change: a review and look forward. *Sedimentology*, **47**, 2–48.

Born, S.M. and Ritter, D.F. (1970) Modern terrace development near Pyramid lake, Nevada, and its geologic implications. *Geol. Soc. Am. Bull.*, **81**, 1233–1242.

Brush, L.M., Jr. and Wolman, M.G. (1960) Knickpoint behavior in noncohesive material: a laboratory study. *Geol. Soc. Am. Bull.*, **71**, 59–74.

Burns, B.A., Heller, P.L., Marzo, M. and Paola, C. (1997) Fluvial response in a sequence stratigraphic framework: examples from the Montserrat fan delta, Spain. *J. Sediment. Res.*, **67**, 311–321.

Carter, R.M., Abbott, S.T., Fulthorpe, C.S., Haywick, D.W. and Henderson, R.A. (1991) Application of global sea-level and sequence–stratigraphic models in Southern Hemisphere Neogene strata from New Zealand. In: *Sedimentation, Tectonics and Eustacy: Sea-level Changes at Active Margins* (Ed. D.I.M. McDonald). *Spec. Publ. Int. Assoc. Sedimentol.*, **12**, 41–65.

Creager, M.G. (1991) *Experimental Study of Piedmont Landforms*. Unpublished Report, Colorado State University, Fort Collins, CO, 12 pp.

Crumeyrolle, P., Claude, D., Lesueur, J.-L. and Joseph, P. (1993) The Roda deltaic complex (Spain); reservoir stratigraphy of a low stand prograding wedge. *Am. Assoc. Petrol. Geol. Bull.*, **77**, 312.

Davis, W.M. (1902) Base level, grade and peneplain. *J. Geol.*, **10**, 77–111.

Dolson, J. (1981) *Depositional Environments and Petrography of the Dakota Group*. Unpublished M.S. thesis, Colorado State University, Fort Collins, Colorado, 342 pp.

Dreyer, T. (1994) Architecture of an unconformity-based tidal sandstone unit in the Ametlla Formation, Spanish Pyrenees. *Sediment. Geol.*, **94**, 21–48.

Ethridge, F.G., Kellison, L.B. and Jump-Ware, C.J. (1994) Unconformity related hydrocarbon production: Lower Cretaceous, Fall River Sandstone, Powell and Buck Draw Fields, Southern Powder River Basin, Wyoming. In: *Unconformity Related Hydrocarbons in Sedimentary Sequences* (Eds J.C. Dolson, W.L. Hendricks and W.A. Wescott), pp. 149–156. Rocky Mountain Association of Geologists, Denver, CO.

Fisk, H.N. (1944) *Geologic investigations of the alluvial valley of the lower Mississippi River*. U.S. Army Corps of Engineers, Mississippi River Commission, Vicksburg, Mississippi, 78 pp.

Fletcher, C.H.III, Knebel, H.J. and Kraft, J.C. (1990) Holocene evolution of an estuarine coast and tidal wetlands. *Geol. Soc. Am. Bull.*, **102**, 283–297.

Germanoski, D. (1990) Laboratory study of braided channel response to variable rates of baselevel change. *Trans. Am. Geophys. Union*, **71**, 513.

Germanoski, D. (1991) Flume study of the relative impact of falling baselevel in braided rivers. *Geol. Soc. Am., Abstr. Progr.*, **23**(5), A240.

Germanoski, D. and Ritter, D.F. (1988) Tributary response to local base level lowering below a dam. *Regulated Rivers*, **2**, 11–24.

Harvey, M.D. (1984) A geomorphic evaluation of a grade-control structure in a meandering channel. In: *River Meandering* (Ed. C.M. Elliott), pp. 284–294. American Society of Civil Engineers, New York.

Harvey, M.D. and Watson, C.C. (1986) Fluvial processes and morphologic thresholds in stream channel restoration. *Water Resour. Bull.*, **22**, 359–368.

Heller, P.L., Paola, C., Hwang, I.-G., John, B. and Steel, R. (2001) Geomorphology and sequence stratigraphy due to slow and rapid base-level changes in an experimental subsiding basin (XES96-1). *Am. Assoc. Petrol. Geol. Bull.*, **85**, 817–838.

Holland, W.N. and Pickup, G. (1976) Flume study of knickpoint development in stratified sediment. *Geol. Soc. Am. Bull.*, **87**, 76–82.

Horton, R.E. (1945) Erosional development of streams and their drainage basins. *Geol. Soc. Am. Bull.*, **50**, 275–370.

Jennette, D.C., Jones, C.R., Van Wagoner, J.C. and Larsen, J.E. (1991) High-resolution sequence stratigraphy of the upper Cretaceous Tocito sandstone: The relationship between incised valleys and hydrocarbon accumulation, San Juan Basin, New Mexico. In: *Sequence Stratigraphic Applications to Shelf Sandstone Reservoirs: Outcrop to Subsurface Examples* (Eds J.C. Van Wagoner, D. Nummedal, D.C. Jennette, C.R. Jones and G. Riley). Field Trip Guidebook, American Association of Petroleum Geologists, Tulsa, OK, 39 pp.

Knotts, S.T. and Hoskins, H. (1968) Evidence of Pleistocene events in the structure of the continental shelf off the northeastern United States. *Mar. Geol.*, **6**, 5–43.

Koss, J.E. (1992) Effects of sea-level change on fluvial and coastal plain systems. An experimental approach. Unpublished MSc thesis, Colorado State University, Fort Collins, CO, 157 pp.

Koss, J.E., Ethridge, F.G. and Schumm, S.A. (1994) An experimental study of the effects of base-level change on fluvial, coastal plain, and shelf systems. *J. Sediment. Res.*, **B64**, 90–98.

Marr, J.G., Swenson, J.B., Paola, C., Voller, V.R., Gupta, S. and Cowie, P.A. (2000) A two-diffusion model of fluvial stratigraphy in closed depositional basins. *Basin Res.*, **12**, 381–398.

Malott, C.A. (1928) Base-level and its varieties. *Indiana Univ. Studies*, **15**, 37–59.

McLane, C.F., III (1978) Channel network growth: an experimental study. Unpublished MSc thesis, Colorado State University, Fort Collins, Colorado, 100 pp.

Milana, J.P. (1998) Sequence stratigraphy in alluvial settings: a flume-based model with applications to outcrop and seismic data. *Am. Assoc. Petrol. Geol. Bull.*, **82**, 1736–1753.

Morisawa, M. (1964) Development of drainage systems on an upraised lake floor. *Am. J. Sci.*, **262**, 340–354.

Nio, S.D. and Yang, C.S. (1991a) Sea-level fluctuations and the geometric variability of tide-dominated sand bodies. *Sediment. Geol.*, **70**, 161–193.

Nio, S. D. and Yang, C. (1991b) Diagnostic attributes of clastic tidal deposits: a review. *Can. Soc. Petrol. Geol.*, **16**, 3–27.

Nystuen, J.P. (1998) History and development of sequence stratigraphy. In: *Sequence Stratigraphy—Concepts and Applications* (Eds F.M. Gradstein, K.O. Sandvik and N.J. Milton), pp. 31–116. Elsevier, Amsterdam.

Paola, C., Heller, P.I. and Angevine, C.I. (1992) The large-scale dynamics of grain-size variations in alluvial basins. 1: theory. *Basin Res.*, **4**, 73–90.

Paola, C., Mullin, J., Ellis, C., Mohrig, D.C., Swenson, J.B., Parker, G., Hickson, T., Heller, P.L., Pratson, L., Syvitski, J., Sheets, B. and Strong, N. (2001) Experimental stratigraphy. *GSA Today*, **11**, 4–9.

Parker, R.S. (1977) Experimental study of basin evolution and its hydrologic implications. Unpublished PhD dissertation, Colorado State University, Fort Collins, CO, 331 pp.

Phillips, L.F. (1986) The effect of slope on experimental drainage patterns: possible application to Mars. Unpublished MSc thesis, Colorado State University, Fort Collins, CO, 121 pp.

Phillips, L. and Schumm, S.A. (1987) Effect of regional slope on drainage networks. *Geology*, **15**, 813–816.

Posamentier, H.W. and Vail, P.R. (1988) Eustatic controls on clastic deposition II—sequence and systems tract models. In: *Sea-level Changes: an Integrated Approach* (Eds C.K. Wilgus, B.S. Hastings, C.G.St.C. Kendall, H.W. Posamentier, C.A. Ross and J.C. VanWagoner). *Soc. Econ. Paleontol. Mineral. Spec. Pub.*, **42**, 125–145.

Posamentier, H.W., Erskine, R.D. and Mitchum, R.W., Jr. (1991) Models for submarine fan deposition within a sequence stratigraphic framework. In: *Seismic Facies and Sedimentary Processes of Submarine Fans and Turbidite Systems* (Eds Paul Weimer and Martin H. Link), pp. 127–136. Springer-Verlag, New York.

Powell, J.W. (1875) *Exploration of the Colorado River of the West and its Tributaries.* U.S. Government Printing Office, Washington, DC, 291 pp.

Richards, M.T. (1994) Transgression of an estuarine channel and tidal flat complex; the Lower Triassic of Barles, Alpes de Haute Provence, France. *Sedimentology*, **41**, 55–82.

Saito, Y. (1991) Sequence stratigraphy on the shelf and upper slope in response to the latest Pleistocene–Holocene sea-level changes off Sendai, northeast Japan. In: *Sedimentation, tectonics and Eustasy; Sea-level Changes at Active Margins* (Ed. D.I.M. Macdonald). *Spec. Publ. Int. Assoc. Sedimentol.*, **12**, 133–150.

Schumm, S.A. (1993) River response to base level change: implications for sequence stratigraphy. *J. Geol.*, **101**, 279–294.

Schumm, S.A. and Ethridge, F.G. (1994) Origin, evolution and morphology of fluvial valleys. In: *Incised-Valley Systems: Origin and Sedimentary Sequences* (Eds R.W. Dalrymple, R. Boyd and B.A. Zaitlin). *Soc. Econ. Paleontol. Mineral. Spec. Pub.*, **51**, 11–27.

Schumm, S.A. and Rea, D.K. (1995) Sediment yield from disturbed earth systems. *Geology*, **23**, 391–394.

Schumm, S.A., Harvey, M.D. and Watson, C.C. (1984) *Incised Channels: Morphology, Dynamics and Control.* Water Resources Publications, Littleton, CO, 200 pp.

Schumm, S.A., Mosley, M.P. and Weaver, W.E. (1987) *Experimental Fluvial Geomorphology.* Wiley, New York, 413 pp.

Shepherd, R.G. and Schumm, S.A. (1974) Experimental study of river incision. *Geol. Soc. Am. Bull.*, **66**, 257–268.

Suter, J.R. and Berryhill, H.L., Jr. (1985) Late Quaternary shelf-margin deltas, northwest Gulf of Mexico. *Am. Assoc. Petrol. Geol. Bull.*, **69**, 77–91.

Suter, J.R., Berryhill, H.L., Jr. and Penland, S. (1987) Delineation of late Quaternary depositional sequences by high-resolution seismic stratigraphy, Louisiana continental shelf. *Am. Assoc. Petrol. Geol. Bull.*, **71**, 620.

Thomas, M.A. (1991) The impact of long-term and short-term sea level changes on the evolution of the Wisconsinan–Holocene Trinity/Sabine incised valley system, Texas continental shelf. Unpublished PhD dissertation; Rice University, Houston, TX, 314 pp.

Uchupi, E. (1970) Atlantic continental shelf and slope of the United States—shallow structure. *U.S. Geol. Survey Prof. Pap.*, **529-I**, 44 pp.

Van Heijst, M.W.I.M. and Postma, G. (2001) Fluvial response to sea-level changes: a quantitative analogue, experimental approach. *Basin Res.*, **13**, 269–292.

Van Heijst, M.W.I.M., Postma, G., Meijer, X.D., Snow, J.N. and Anderson, J.B. (2001) Quantitative analog flume-model of river-shelf systems: principles and verification exemplified by the Late Quaternary Colorado river-delta evolution. *Basin Res.*, **13**, 243–268.

Van Heijst, M.W.I.M., Postma, G., van Kesteren, W.P. and de Jongh, R.G. (2002) Control on syn-depositional faulting on systems tract evolution across growth-faulted shelf margins: an analogue experiment model of the Miocene Imo River Field Nigeria. *Am. Assoc. Petrol. Geol. Bull.*, **86**, 1335–1366.

Van Wagoner, J.C. and Bertram, G.T. (Eds) (1995) Sequence stratigraphy of foreland basin deposits: outcrop and subsurface examples from the Cretaceous of North America. *Am. Assoc. Petrol. Geol. Mem.*, **64**, 489 pp.

Van Wagoner, J.C., Posamentier, H.W., Mitchum, R.M., Vail, P.R., Sarg, J.F., Loutit, T.S. and Hardenbol, J. (1988) An overview of the fundamentals of sequence stratigraphy and key definitions. In: *Sea-Level Changes: an Integrated Approach* (Eds C.K. Wilkus, B.S. Hastings, H.W. Posamentier, C.A. Ross, J. Van Wagoner and C.G.St.C. Kendall). *Soc. Econ. Paleontol. Mineral. Spec. Publ.*, **42**, 39–45.

Van Wagoner, J.C., Mitchum, R.W., Campion, K.M. and Rahmanian, V.D. (1990) *Siliciclastic Sequence Stratigraphy in Well Logs, Cores and Outcrops: Concepts for High-Resolution Correlation of Time and Facies. Am. Assoc. Petrol. Geol. Methods Explor. Ser.*, **7**, 55 pp.

Williams, G. (1978) Bank full discharge of rivers. *Water Resour. Res.*, **14**, 1141–1154.

Wood, L.J. (1992) Influence of base-level change on coastal plain, shelf and slope depositional systems: Unpublished PhD dissertation, Colorado State University, Fort Collins, CO, 164 pp.

Wood, L.J., Ethridge, F.G. and Schumm, S.A. (1993) The effects of rate of base-level fluctuation on coastal plain, shelf and slope depositional systems: an experimental approach. In: *Sequence Stratigraphy and Facies Associations* (Eds H.W. Posamentier, C.P. Summerhayes, B.V. Haq and C.P. Allen). *Spec. Publ. Int. Assoc. Sediment.*, **18**, 43–53.

Wood, L.J., Koss, J.E. and Ethridge, F.G. (1994) Simulating unconformity development and unconformable stratigraphic relationships through physical experiments. In: *Unconformity-related Hydrocarbon Exploration and Exploitation in Sedimentary Sequences* (Eds J.C. Dolson, M.L. Hendricks and W.A. Wescott), pp. 23–34. Rocky Mountains Association of Geologists, Denver, CO.

Yoxall, W.H. (1969) The relationship between falling base level and lateral erosion in experimental streams. *Geol. Soc. Am. Bull.*, **80**, 1379–1384.

A mass-balance framework for quantifying downstream changes in fluvial architecture

NIKKI STRONG*, BEN SHEETS*, TOM HICKSON† and CHRIS PAOLA*

*Department of Geology and Geophysics and St Anthony Falls Laboratory, University of Minnesota, Minneapolis, MN 55455, USA (Email: stro0068@umn.edu); and
†Department of Geology, University of St Thomas, St Paul, MN, USA

ABSTRACT

It is still commonly believed that channel stacking density in alluvial cross-sections is controlled mainly by local subsidence rate, despite new models that emphasize the three-dimensionality of alluvial architecture. New data are presented from an experimental alluvial basin that show variation in the spatial distribution of deposition to be the main control on architecture, rather than subsidence *per se*. A simple coordinate transformation is proposed that maps downstream distance into the fraction of the sediment supply deposited to that point. Transforming measured sections into this 'mass balance' coordinate system removes much, although not all, of the observed variability in channel stacking density and grain size. Furthermore, removal of the dominant mass-balance effects via transformation to fraction deposited reveals more clearly those residual architectural effects that are not mass-balance controlled: for example, changes in channel size associated with fluctuations in water supply.

INTRODUCTION

Most workers think of physical stratigraphy as being controlled by three main allogenic drivers: climate, tectonics and eustasy. For analytical purposes, the first two of these must be distilled into specific governing variables, of which the main ones are rate and size distribution of sediment supply, space–time distribution of crustal deformation, and space–time distribution of water runoff. It follows that alluvial architecture should be controlled by these same variables.

One of the major effects of changes in any of these governing variables is to alter the distribution of deposition (referred to here as the depositional mass balance) within the fluvial system. Deposition preferentially extracts the coarser fraction of the sediment supply, so both the amount and the size distribution of sediment available to the fluvial system at a given point are strongly controlled by upstream sediment extraction. Thus a direct effect of changing the sediment mass balance is to cause longitudinal (streamwise) migration of facies belts in concert with changes in the distribution of deposition. In general one would expect the architecture at any point to reflect a superposition of these direct mass-balance effects and changes in other relevant variables, such as local sedimentation rate, avulsion frequency and pattern, and channel belt geometry. Note, however, that although sedimentation rate and depositional mass balance are related, they are not the same: sedimentation is a local variable, but depositional mass balance is meaningful only in the context of the whole fluvial system. The relationship between the two is made precise later in the paper. At this point, little is known about the relative importance for alluvial architecture of direct mass-balance effects versus the other variables listed above, although the latter have been the exclusive focus of architecture models to date (Allen, 1978, 1979; Bridge & Leeder, 1979;

Alexander & Leeder, 1987; Leeder & Gawthorpe, 1987; Bridge & Mackey, 1993a,b; Mackey & Bridge, 1995). In this paper, an experimental case study is presented in which changes in the depositional mass balance turn out to be the dominant control on alluvial architecture.

BACKGROUND AND GOALS

Alluvial architecture was one of the first problems in physical stratigraphy to be analysed quantitatively, thanks to the model proposed by Leeder (1978) and subsequently expanded upon by Allen (1978) and Bridge & Leeder (1979). These models and their many descendents will be referred to as the Leeder–Allen–Bridge (LAB) models. In Leeder's original model, the main control on architecture was sedimentation rate, assumed equal to subsidence rate, relative to avulsion frequency. The main result of this early work was that low sedimentation rates cause randomly avulsing channels to stack more densely upon one another, assuming constant avulsion frequency. The Leeder model led to a series of refinements and extensions (Allen, 1978, 1979; Bridge & Leeder, 1979; Alexander & Leeder, 1987; Leeder & Gawthorpe, 1987; Bridge & Mackey, 1993a,b; Mackey & Bridge, 1995) as well as a variety of tests (Behrensmeyer & Tauxe, 1982; Kraus & Middleton, 1987; Leeder et al., 1996, Guiseppe & Heller, 1998; Aslan & Blum, 1999; Ashworth et al., 1999; Kraus, 2002).

Two major shortcomings of the original LAB approach have been addressed in recent years. First, one would expect avulsion frequency to depend strongly on sedimentation rate, as avulsion is largely driven by sedimentation. This was shown experimentally by Bryant et al. (1995), who also pointed out that, even at a qualitative level, the relation between architecture and sedimentation depends on the relationship between avulsion frequency and sedimentation rate. Heller & Paola (1996) went on to examine the consequences of this in more detail. The second major shortcoming was that the original LAB models were two-dimensional, focusing on channel stacking in panels taken in the cross-stream direction, i.e. perpendicular to the mean flow direction (depositional strike). This was addressed by Mackey & Bridge (1995) and Heller & Paola (1996), who extended the domain to the third, streamwise, dimension. Two points about the LAB model series are particularly important to the work we describe here:

1 despite efforts to uproot it, the aspect of the LAB models that persists in common stratigraphical wisdom is that greater channel stacking density (i.e. a more sheet-like sandbody geometry) implies lower sedimentation (subsidence) rate and vice versa;

2 despite considerable effort, even the most basic predictions of the LAB models remain largely untested.

In this paper the importance of mass balance in the streamwise (depositional dip) direction is examined in alluvial architecture. Data from a recent experiment in a new subsiding-floor sedimentary basin (Paola, 2000; Paola et al., 2001; Sheets et al., 2002) are used to demonstrate that fluvial morphology and preserved alluvial architecture in the subsurface are strongly controlled by the spatial distribution of deposition in a basin. The importance of downstream (downdip) location in terms of deposition is studied as a control on channel stacking density and overall grain size, and it is suggested that vertical changes in channel stacking density seen in many field cases (e.g. Fig. 1) may simply reflect variation in depositional mass balance owing to any of a number of possible external causes. A simple coordinate transformation is proposed that allows for the comparison of cross-stream panels in a framework that accounts for changes in mass balance. Although applied here to alluvial architecture, it is believed that the approach can be applied to interpreting any mass-conserving stratigraphical system.

EXPERIMENTAL SET-UP

Experiments were conducted in the Experimental EarthScape (XES) Facility at St Anthony Falls Laboratory, University of Minnesota, USA. The XES facility is a large (6 m × 3 m × 1.3 m) experimental basin with a programmable subsiding floor. Water discharge and sediment discharge into the basin, as well as base level, are also fully controllable. This particular XES run modelled basin filling by a braided river system prograding

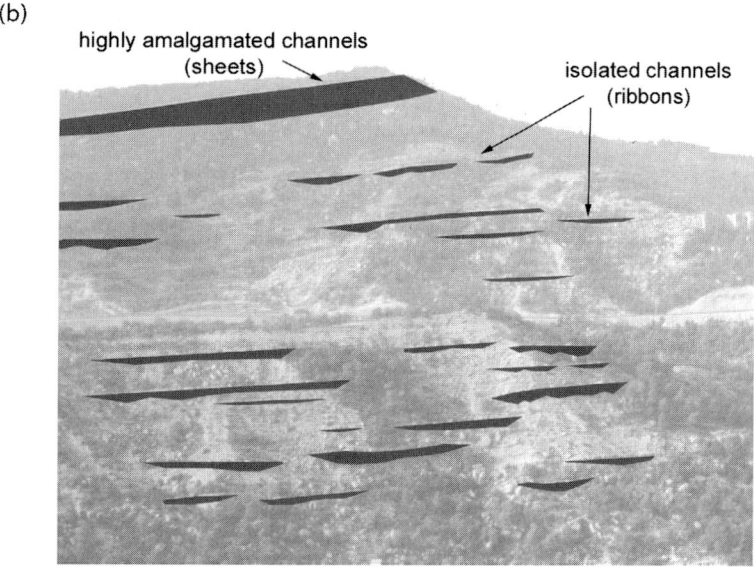

Fig. 1 Montsant alluvial fan (middle Eocene to late Oligocene: Anadón et al., 1986), Villanova de Prades, Spain.

into a standing body of water—a braided fluvial fan delta as defined by Nemec & Steele (1984). The main focus of this run was to test the effects of changes in subsidence and water supply on alluvial architecture. Base level was thus held constant throughout the experiment. Sediment composition was also held constant, the sediment mixture consisting of 40% by volume crushed anthracite coal and 60% by volume 120 μm white quartz sand. These materials served as experimental proxies for fine and coarse-grained sediment respectively, coal serving as the fine-grained proxy owing to its lower density (1.4 g cm^{-3}) thus greater mobility relative to the quartz sand (2.65 g cm^{-3}). The total volumetric sediment feed rate, i.e. integrated net deposition, was matched to the volumetric rate of accommodation produced by subsidence throughout the experiment. This choice was intended to minimize facies migration, but this was only partially

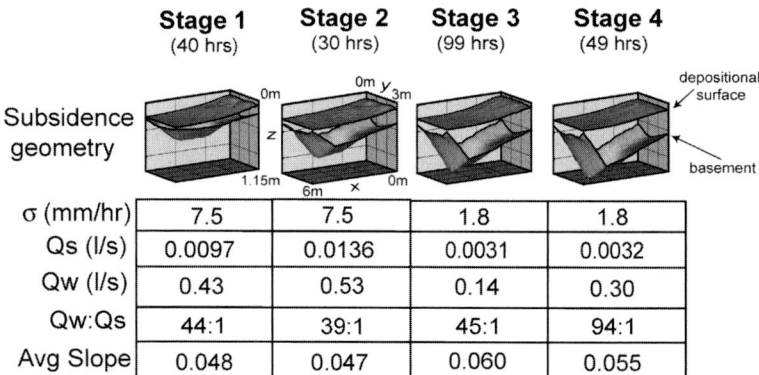

Fig. 2 Experimental parameters.

successful. This point will be discussed later in greater detail.

The experiment comprised four stages, analogous to four geological units in a sedimentary basin, each representing alluvial-basin filling under distinct climatic and tectonic conditions (Fig. 2; Sheets et al., 2002). Stages 1 and 2 were designed to study the sensitivity of the depositional system to changes in subsidence geometry. Stage 1 was characterized by a complex varying subsidence geometry, with both streamwise and cross-stream variability, whereas stage 2 used a simple linear-hinge subsidence pattern. Stages 2 and 3 were designed to study the sensitivity of the depositional system to changes in mean subsidence rate. As such, the rate of subsidence during stage 3 was four times slower than that of stage 2, and sediment and water supply were scaled back proportionally. Finally, stage 4 was designed to study the sensitivity of the depositional system to changes in water supply; the water discharge was approximately doubled relative to that of stage 3 whereas all other variables were held constant.

During the experiment, the surface flow pattern was recorded using video and still cameras. Surface topography was recorded with a laser scanning system, allowing the tracking of fluctuations in the spatial distribution of deposition and erosion within the basin. After the experiment was completed the resultant deposit was sectioned in the cross-stream (strike) direction to produce a series of parallel faces spaced 2 cm apart. These faces were imaged and these images compiled into a three-dimensional visualization of the basin stratigraphy.

EXPERIMENTAL OBSERVATIONS

This paper focuses on stages 2 and 3 of the experiment. As mentioned above, these two stages were designed to examine the effects of changes in mean subsidence rate on alluvial architecture. By holding constant the ratio of water supply to sediment supply, the ratio of sediment supply to accommodation, and base level, it was hoped to isolate the effects of reduction in subsidence on alluvial architecture.

Considering the 'synthetic outcrop' in Fig. 3, which consists of a cross-stream (strike) section located at a downstream distance of 2.64 m, it is evident that stage 3 has fewer channels than any of the other stages and is clearly richer in coal, the experimental proxy for fine-grained sediment. A similar scenario in a natural outcrop, an abrupt decrease in channel-stacking density, could be interpreted according to conventional wisdom as indicating an increase in subsidence rate. In the experiment, however, the subsidence rate in stage 3 was approximately four times lower than that of stage 2. Why then does stage 3 look less channelized? To answer this question the three-dimensional character of the experimental basin stratigraphy and experimental conditions prevalent during stage 3 must be compared with those in the other stages.

One advantage of experimental stratigraphy is that it can be sectioned at sufficiently close spacing so as to reconstruct a reasonably complete view of the three-dimensional structure of the deposit. For this experiment, the three-dimensional

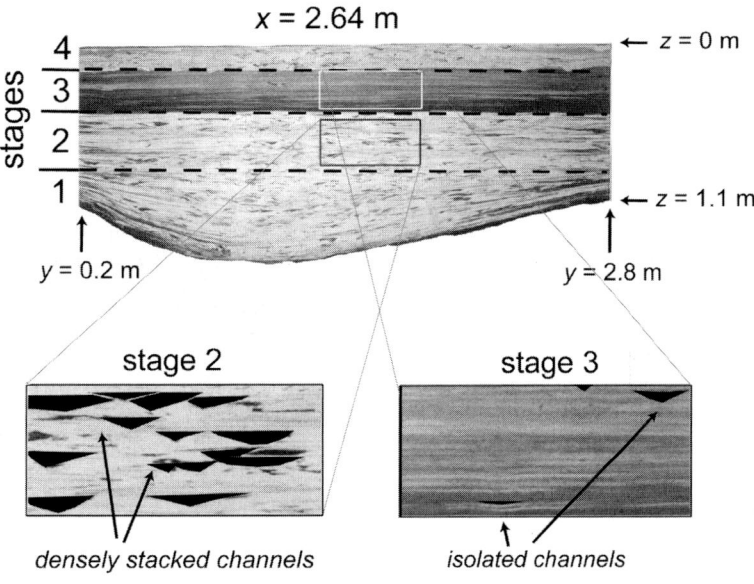

Fig. 3 Strike-section image of the experimental deposit at a downstream distance of $x = 2.64$ m. Notice that stage 3 has fewer channels and is clearly richer in coal than the other three stages.

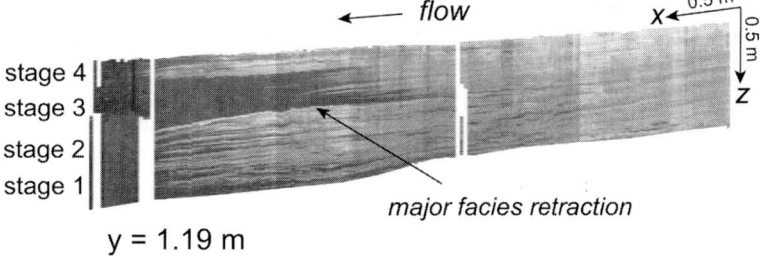

Fig. 4 Flow-parallel sectional image of the experimental deposit. Notice the upstream migration of the sand to coal transition in stage 3. Breaks in the image result from loss of data during the process of slicing the deposit.

reconstruction of the basin stratigraphy shows that the sand–coal transition in stage 3 is located significantly farther upstream than in any of the other stages (Fig. 4). The explanation for this facies shift lies in a small discrepancy between the behaviour of the experiment and theoretical expectations. Based on available theory (Paola et al., 1999) and experimental data (Whipple et al., 1998), the main control on river slope is the ratio of water discharge to sediment discharge. By holding this ratio constant, it was hoped to maintain a constant topographic slope during stages 2 and 3. Topographic measurements taken during the experiment, however, indicate that the slope of the upstream end of the fluvial system increased by about 28% in stage 3 relative to stage 2. Evidently, the reduced water discharge was less effective at transporting the proportionately reduced sediment supply of stage 3. As a result, relatively more sediment was deposited upstream in stage 3 than in the other stages, increasing the depositional slope and the ability of the system to transport sediment. This was most pronounced at the beginning of stage 3 when, relative to stage 2, most of the coarse (sand) component was being deposited rapidly in the upstream portion of the system. The relative amounts of coal and sand in the sediment mixture remained constant throughout the experiment, but sand was removed from the system more rapidly in stage 3 than in any of the other stages, causing the sand–coal transition (and, to a lesser extent, the shoreline) to retrograde abruptly at the beginning of stage 3. An abrupt facies retraction is shown clearly in the deposit (Fig. 4).

Thus, the spatial distribution of deposition in stage 3, i.e. the mass balance of deposition, was different than in the other stages. Hence facies changes observed between stages 2 and 3 reflect both the effects of facies migration induced by variation in the depositional pattern and any effects resulting purely from the reduced water and sediment discharges in stage 3. How can these two independent controls on the observed alluvial architecture be separated?

MASS-BALANCE TRANSFORMATION

Instinctively, 'proximal' and 'distal' are thought of in terms of distance from a sediment source. The terms are generally understood in a relative sense; for instance, 10 km from source might be proximal in a large basin and distal in a small one. An alternative conceptualization would be to think of proximity not in terms of physical distance, but rather in terms of the fraction of all supplied sediment deposited upstream of that point. The motivation for adopting this point of view is that one of the strongest drivers of changes in the morphodynamic character of sedimentary systems appears to be loss of sediment and the associated grain-size decrease induced by deposition. Of course, downstream sediment loss and associated fining are not the *only* causes of down-transport facies changes. The mass-balance signal is often sufficiently dominant, however, so as to mask other effects unless it is accounted for correctly. The goal here is to develop a transformation that facilitates this accounting. This effectively isolates that part of the depositional signal that is directly linked to changes in mass distribution in the depositional system. Removing the mass balance signal allows us to compare facies deposited under different depositional geometries on a consistent basis, and to distinguish those aspects of stratigraphy that are directly linked to changes in mass distribution from those that are not.

Consider a measure of downstream distance not in terms of an absolute length x, but in terms of a non-dimensional number $\chi(x)$, where $\chi(x)$ is equal to the fraction of all supplied sediment deposited between the sediment source at $x = 0$ and a distance x downstream. Formally $\chi(x)$ is defined by:

$$\chi(x) = \frac{\int_0^x r_{\Delta T}(x)dx}{\int_0^L r_{\Delta T}(x)dx} \quad (1)$$

where $r(x)$ is the rate of deposition in units of (L T^{-1}) at a given downstream distance x, measured over a time interval ΔT. Here ΔT is chosen to be long enough to average out intrinsic, flow-controlled (autogenic) fluctuations (Sheets et al., 2002) but shorter than the time-scales of changes in external variables. Variable L is the total length of the depositional system. The term in the numerator of equation (1) is equivalent to the width-averaged volume of sediment deposited between the sediment source and a distance x. The term in the denominator of equation (1) is equivalent to the width-averaged volume of sediment deposited over the entire length of the system and, for a closed system, is equal to the total volume of sediment supplied over the time interval in question. For example, if at some downstream distance x, $\chi(x) = 0.3$, then 30% of the sediment has been extracted from the system at that point, and 70% is still in transport. Likewise $\chi(x) = 0.5$ is the downstream distance at which half the sediment has been extracted from the system. The physical distance x_{50} corresponding to $\chi(x) = 0.5$ is in effect the 'depositional mid-point' of the system. It is important to note that $\chi(x)$ is a non-dimensional function that reflects the spatial distribution of the local rate of deposition $r_{\Delta t}(x)$ over the entire basin. It is not hard to imagine scenarios in which either $\chi(x)$ or $r_{\Delta t}(x)$ changes when other variables remain constant.

Figure 5 illustrates the transformation from x to $\chi(x)$. In the example basin in Fig. 5 sedimentation rate decreases exponentially downstream. At a downstream distance of one-quarter the total basin length, 63% of the sediment has been extracted from the system, and at a downstream distance of one-half the total basin length, 86% of the sediment has been extracted from the system. These downstream distances correspond to $\chi(x) = 0.63$ and 0.86 respectively. Figure 6 illustrates how the shape of a basin affects the downstream distance x corresponding to the depositional mid-point $\chi(x) = 0.5$. For example, in a tectonically foretilted basin (e.g. a passive margin) sediment deposition occurs farther downstream (Fig. 6a) than in a backtilted (e.g. foreland) basin (Fig. 6b) of the same size. Likewise $\chi(x) = 0.5$ occurs farther upstream in a short basin (Fig. 6c) than in a long basin of the same shape (Fig. 6d).

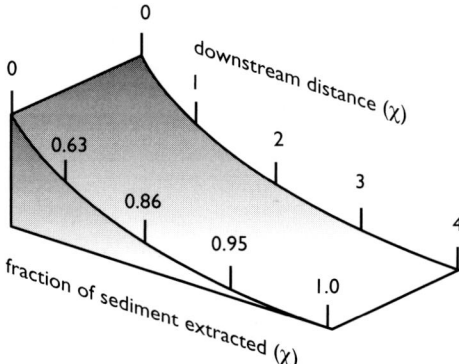

Fig. 5 A schematic representation of a sedimentary basin, where sediment thickness decreases exponentially downstream, illustrating the fundamental difference between measuring distance in terms of x (linear scale in back) and measuring distance in terms of $\chi(x)$ (non-linear scale in front).

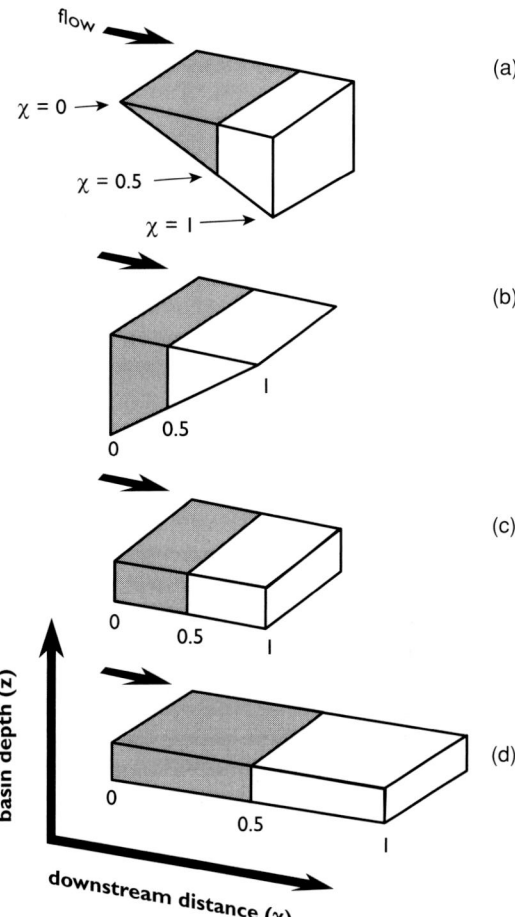

Fig. 6 Schematic representation of the effect of basin shape on the downstream location of $\chi(x) = 0.5$, corresponding to the downstream distance, x, where 50% of the sediment has been extracted from the system.

The most obvious difference between the two images of the experimental surface in Fig. 7(a, b) is that the sand–coal transition occurs farther upstream in stage 3 than it does in stage 2. This is important with respect to alluvial architecture, because at the sand to coal transition a few large fluvial channels bifurcate into numerous smaller channels, as evident in both the overhead images taken during the experiment (Fig. 7a, b) and the images of the topography of the experimental deposit (Fig. 7c & d). The power of the χ transformation can be seen by comparing the location of this transition in standard dimensional coordinates (left side of each image, Fig. 7a & b) with transformed χ coordinates (right side of each image, Fig. 7a & b). It can be seen that for similar values of χ the fluvial system seems to behave similarly. For instance in both images $\chi(x) = 0.25$ occurs upstream of the coal–sand transition, $\chi(x) = 0.5$ occurs near the transition, and $\chi(x) = 0.75$ occurs downstream of the transition.

This observation can be made quantitative by estimating the volume of coal versus that of quartz sand that has been deposited in a given stage for a given downstream location. To do so, mean luminosity, a measure of brightness that varies in value between 0 (black, i.e. pure coal) and 255 (white, i.e. pure quartz sand), may be used. For each stage total luminosity decreases downstream as the coal fraction increases, analogous to downstream fining in a natural system. This downstream decrease in luminosity occurs at different rates for stages 2 and 3 in x coordinates (Fig. 7e), whereas in $\chi(x)$ coordinates the rates of decrease in luminosity for stages 2 and 3 are nearly identical (Fig. 7f).

Transformation to $\chi(x)$ coordinates has a similar effect on downstream trends in channel stacking density. Although stage 3 shows generally fewer channel deposits than the other stages for a particular downstream location, x (Fig. 8a), much of that variability disappears upon transformation to

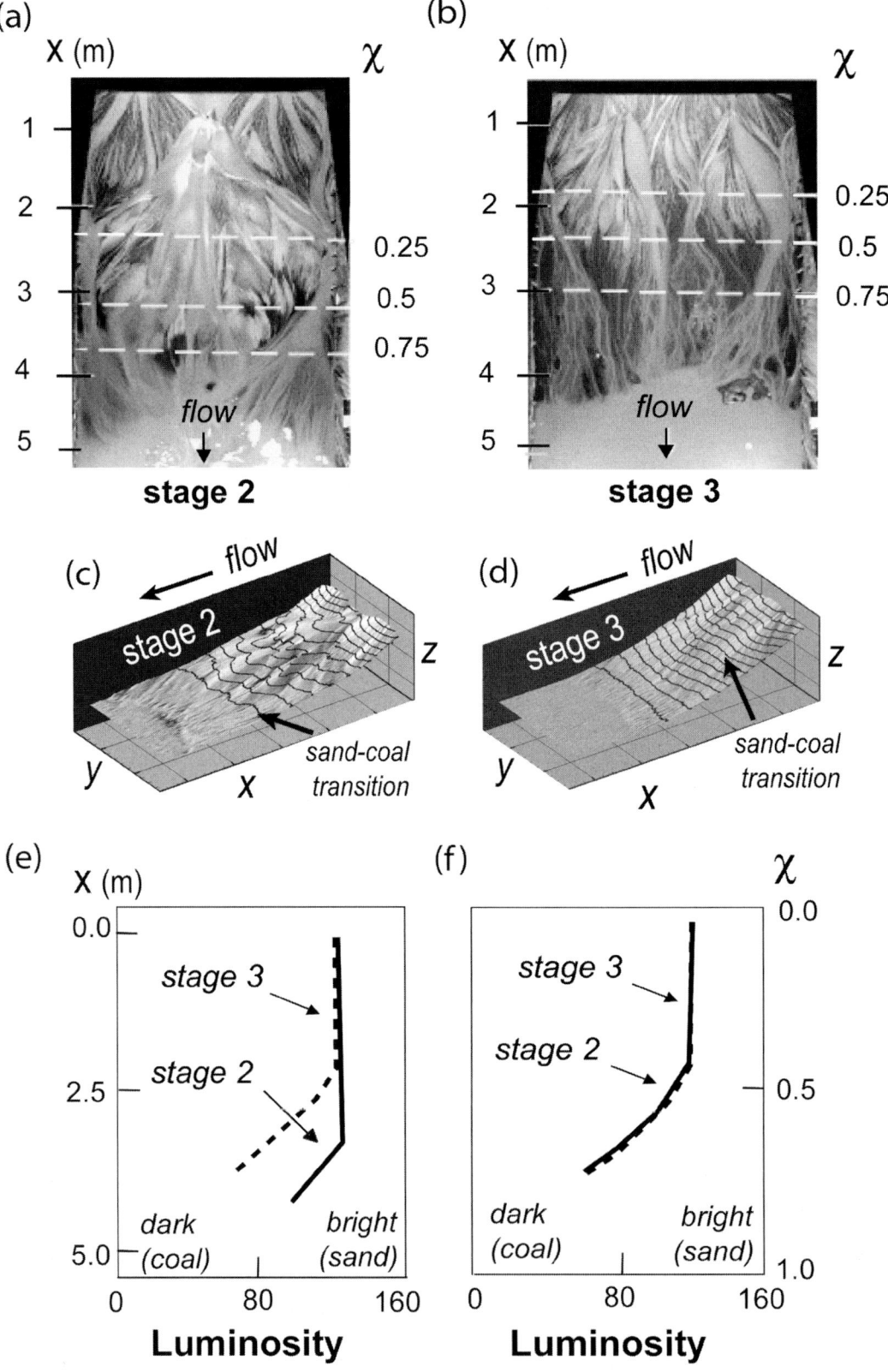

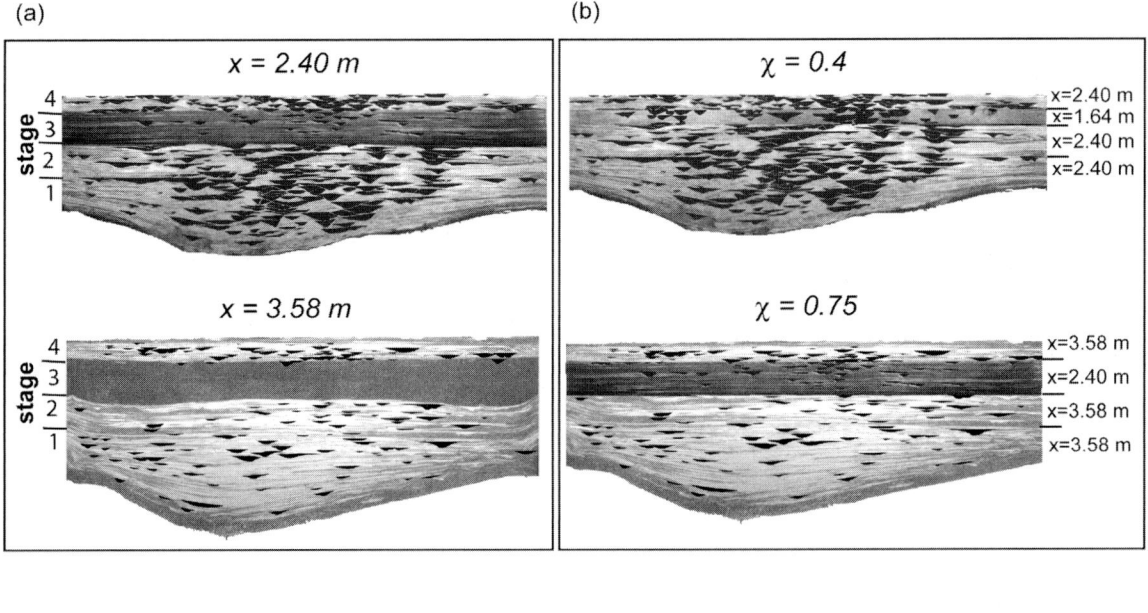

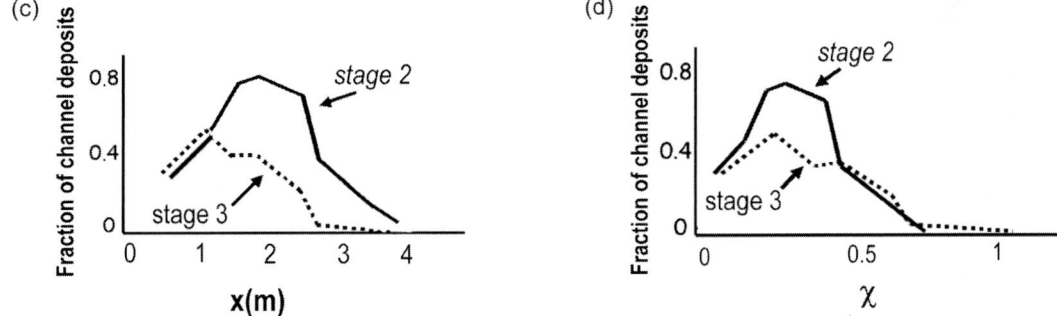

Fig. 8 (a, b) Channel mapping (channels are in black) for different downstream locations in terms of x and $\chi(x)$. (c) Channel stacking densities for stages 2 and 3 as a function of x and $\chi(x)$.

$\chi(x)$ coordinates (Fig. 8b). Further, general downstream trends in preserved channel fraction are more similar between stages in the $\chi(x)$ coordinate system than in the x coordinate system (Fig. 8c, d). Transformation into the $\chi(x)$ coordinate system does not, however, remove all of the observed variability: for instance stage 3 has a lower overall channel deposit density than stage 2 (Fig. 8 c & d) even for the same value of $\chi(x)$. This was due to the presence of more and larger fluvial channels during stage 2, accommodating sediment and water discharges four times those of stage 3. The important point is that transformation from x to $\chi(x)$ makes it possible to discriminate the effect of more and larger fluvial channels, owing to a relatively greater water discharge, from the stronger mass-balance-controlled variation in the preserved channel fraction.

Fig. 7 (*opposite*) Images of the experimental surface taken in plan view: (a) during stage 2 of the experiment (high subsidence rate, sediment supply and water supply); (b) during stage 3 of the experiment (low subsidence rate, sediment supply and water supply). (c & d) Topographic images of the depositional surface during stages 2 and 3. (e & f) Mean luminosities for stages 2 and 3 as a function of downstream distance, in terms of x and $\chi(x)$.

CONCLUSIONS

1 In an experiment with independently controlled subsidence, sediment supply and water supply, a four-fold reduction in rates of these three variables produced an abrupt reduction in channel stacking density in vertical section. This change was caused by an abrupt migration of facies towards the sediment source, which was in turn induced by upstream storage of sediment associated with a small increase in fluvial slope.

2 The effect of varying the distribution of basinal deposition can be quantified using a simple coordinate transformation, from downstream distance x to a non-dimensional mass-balance coordinate $\chi(x)$, where $\chi(x)$ is the fraction of all supplied sediment deposited upstream of distance x.

3 For the experimental data presented here, transformation into this mass-balance coordinate system leads to internally consistent downtransport decreases in sand content and channel stacking density. In particular, many of the vertical changes in architecture resulting from the imposed changes described in (1) above disappear upon transformation to mass-balance coordinates.

4 The most important control on alluvial architecture in the experiment reported here was the spatial distribution of sedimentation, i.e. the sediment mass balance. Other allogenic controls, such as sediment supply, water supply and subsidence, mainly exerted influence indirectly by changing this mass balance.

5 After removal of mass-balance effects by transformation into the $\chi(x)$ coordinate system, the main residual architectural effect of reducing water discharge in the experiment reported here was a decrease in channel stacking density associated with a decrease in channel size.

6 Transformation to a mass-balance coordinate system provides a unifying quantitative framework for analysing the role of external variables such as subsidence, sediment supply and water supply in controlling alluvial architecture. The dominant role played by changes in mass balance in controlling alluvial architecture also illustrates a basic limitation of the present generation of 'stand-alone' alluvial architecture models: architecture at any particular location has at least as much to do with relative net sediment extraction upstream as it does with avulsion dynamics or channel geometry. Future generations of architecture models should be tightly coupled with models for alluvial deposition and erosion on basin scales to account for these effects.

ACKNOWLEDGEMENTS

We are grateful for helpful comments by reviewers Mike Blum, Julio Freedman, and David Mohrig. We also thank Jim Mullin, Chris Ellis, and John Swenson for help with the enormous task of running 'Jurassic Tank'. Finally, we thank our sponsors; the National Science Foundation (grant EAR 97-25989), the University of Minnesota, the Office of Naval Research (grant N00014-99-1-0603), and our Industrial Consortium (Anadarko Petroleum Corporation, Conoco Inc., Exxon-Mobil Upstream Research Company, Japan National Oil Company, and Texaco Inc.).

REFERENCES

Alexander, J. and Leeder, M.R. (1987) Active tectonic control on alluvial architecture. In: *Recent Developments in Fluvial Sedimentology* (Eds F.G. Ethridge, R.M. Flores and M.D. Harvey). *Soc. Econ. Paleontol. Mineral. Spec. Publ.*, **39**, 243–252.

Allen, J.R.L. (1978) Studies in fluviatile sedimentation: an exploratory quantitative model for the architecture of avulsion-controlled alluvial suites. *Sediment. Geol.*, **21**, 129–147.

Allen, J.R.L. (1979) Studies in fluviatile sedimentation: an elementary geometrical model for the connectedness of avulsion-related channel sand bodies. *Sediment. Geol.*, **24**, 253–267.

Anadón, P., Cabrera, L.L., Colombo, F., Marzo, M. and Riba, O. (1986) Syntectonic intraformational unconformities in alluvial fan deposits, eastern Ebro basin margins (NE Spain). In: Foreland Basins (Eds P. Allen and P. Homewood). *Spec. Publ. Int. Assoc. Sedimentol.*, **8**, 259–271.

Ashworth, P.J., Best, J.L., Peakall, J. and Lorsong, J.A. (1999) The influence of aggradation rate on braided alluvial architecture: field study and physical scale-modelling of the Ashburton River gravels, Canterbury Plains, New Zealand. In: *Fluvial Sedimentology VI* (Eds N.D. Smith and J. Rogers). *Spec. Publ. Int. Assoc. Sedimentol.*, **28**, 333–346.

Aslan, A. and Blum, M.D. (1999) Contrasting styles of Holocene avulsion, Texas Gulf Coastal Plain, U.S.A.

In: *Fluvial Sedimentology VI* (Eds N.D. Smith and J. Rogers). *Spec. Publ. Int. Assoc. Sedimentol.*, **28**, 293–308.

Behrensmeyer, A.K. and Tauxe, L. (1982) Isochronous fluvial systems in Miocene deposits of northern Pakistan. *Sedimentology*, **29**, 331–352.

Bridge, J.S. and Leeder, M.R. (1979) A simulation model of alluvial stratigraphy. *Sedimentology*, **26**, 617–644.

Bridge, J.S. and Mackey, S. D. (1993a) A revised alluvial stratigraphy model. In: *Alluvial Sedimentation* (Eds M. Marzo and C. Puigdefábregas). *Spec. Publ. Int. Assoc. Sedimentol.*, **17**, 319–336.

Bridge, J.S. and Mackey, S.D. (1993b) A theoretical study of fluvial sandstone body dimensions. In: *Quantative Description and Modelling of Clastic Hydrocarbon Reservoirs and Outcrop Analogues* (Eds S. Flint and I.D. Bryant). *Spec. Publ. Int. Assoc. Sedimentol.*, **15**, 213–236.

Bryant, M., Falk, P. and Paola, C. (1995) Experimental study of avulsion frequency and rate of deposition. *Geology*, **23**, 365–368.

Guiseppe, A.C. and Heller, P.L. (1998) Long-term river response to regional doming in the Price River Formation, central Utah. *Geology*, **26**, 239–242.

Heller, P.L. and Paola, C. (1996) Downstream changes in alluvial architecture: an exploration of controls on channel-stacking patterns. *J. Sediment. Res.*, **66**, 297–306.

Kraus, M.J. (2002) Basin-scale changes in floodplain paleosols: implications for interpreting alluvial architecture. *J. Sediment. Res.*, **72**, 500–509.

Kraus, M.J. and Middleton, L.T. (1987) Contrasting architecture of two alluvial suites in different structural settings. In: *Recent Developments in Fluvial Sedimentology* (Eds F.G. Ethridge, R.M. Flores and M.D. Harvey). *Soc. Econ. Paleontol. Mineral. Spec. Publ.*, **39**, 253–262.

Leeder, M.R. (1978) A quantitative stratigraphic model for alluvium, with special reference to channel deposit density and interconnectedness. In: *Fluvial Sedimentology* (Ed. A.D. Miall). *Can. Soc. Petrol. Geol. Mem.*, **5**, 587–596.

Leeder, M.R. and Gawthorpe, R.L. (1987) Sedimentary models for extensional tilt block/half-graben basins. *Geol. Soc. Lond. Spec. Publ.*, **28**, 139–152.

Leeder, M.R., Mack, G.H., Peakall, J. and Salyards, S.L. (1996) First quantitative test of alluvial stratigraphic models: southern Rio Grande rift, New Mexico. *Geology*, **24**, 87–90.

Mackey, S.D. and Bridge, J.S. (1995) Three-dimensional model of alluvial stratigraphy: theory and application. *J. Sediment. Res.*, **B65**, 7–31.

Nemec, W. and Steel, R.J. (1984) Alluvial and coastal conglomerates: their significant features and some com-ments on gravelly mass-flow deposits. In: *Sedimentology of Gravels and Conglomerates* (Eds E.H. Koster and R.J. Steel). *Can. Soc. Petrol. Geol. Mem.*, **10**, 1–31.

Paola, C. (2000) Quantitative models of sedimentary basin filling. *Sedimentology*, **47** (suppl. 1), 121–178.

Paola, C., Parker, G., Mohrig, D.C. and Whipple, K.X. (1999) The influence of transport fluctuations on spatially averaged topography on a sandy, braided fluvial fan. In: *Numerical Experiments in Stratigraphy: Recent Advances in Stratigraphic and Sedimentologic Computer Simulations* (Eds J. Harbaugh, W.L. Watney, E.C. Rankey, R. Slingerland, R.H. Goldstein and E.K. Franseen). *Spec. Pub. Soc. Econ. Paleontol. Mineral.*, **62**, 211–218.

Paola, C., Mullin, J., Ellis, C., Mohrig, D.C., Swenson, J.B., Parker, G., Hickson, T., Heller, P.L., Pratson, L., Syvitski, J., Sheets, B. and Strong, N. (2001) Experimental stratigraphy. *GSA Today*, **11**, 4–9.

Sheets, B., Hickson, T. and Paola, C. (2002) Assembling the stratigraphic record: depositional patterns and time-scales in an experimental alluvial basin. *Basin Res.*, **14**, 287–301.

Whipple, K.X., Parker, G., Paola, C. and Mohrig, D.C. (1998) Channel dynamics, sediment transport, and the slope of alluvial fans: experimental study. *J. Geol.*, **106**, 677–693.

Quaternary fluvial systems

The linkage between alluvial and coeval nearshore marine successions: evidence from the Late Quaternary record of the Po River Plain, Italy

ALESSANDRO AMOROSI *and* MARIA LUISA COLALONGO

Università di Bologna, Dipartimento di Scienze della Terra e Geologico-Ambientali, Via Zamboni 67, 40127 Bologna, Italy (Email: amorosi@geomin.unibo.it)

ABSTRACT

Detailed facies analysis of cores, up to 200 m long, within the Late Quaternary fluvial to shallow-marine succession of the Po River Plain, reveals characteristic cyclic changes in lithofacies and channel stacking pattern. Transgressive surfaces are readily identifiable, showing a greater extent and correlation potential than sequence boundaries or maximum flooding surfaces. Stacked transgressive–regressive (T–R) sequences, 50–100 m thick and spanning time intervals of about 100 kyr, form the basic motif of the Late Quaternary Po Basin fill. At relatively seaward locations (coastal sections), the lower parts of these 4th-order cycles show coastal-plain aggradation and rapid shoreline transgression (retro-grading barrier–lagoon–estuary systems), forming thin transgressive systems tracts (TST). Transgressive deposits are overlain by characteristic shallowing-upward successions, related to delta and strandplain progradation (highstand systems tracts or HST). Subsequent long phases of sea-level fall are recorded by exceptionally thick (up to 60 m) successions of interbedded alluvial and coastal-plain deposits (falling-stage (FST) and lowstand (LST) systems tracts). At landward locations, within non-marine strata, bounding surfaces of T–R sequences are marked by abrupt facies changes from amalgamated fluvial-channel gravel and sand, formed mostly at lowstand conditions, to mud-dominated floodplain deposits, with isolated channel bodies and organic horizons (transgressive alluvial deposits or TST). This section grades upward into thick alluvial plain deposits, showing increased channel clustering and sheet-like geometries (regressive alluvial deposits, including HST, FST and LST). The sharp lower boundaries of T–R sequences identified within the alluvial sections can be traced physically into the transgressive surfaces recognized at seaward locations. The transgressive surfaces are characterized by well identifiable pollen assemblages. Correlation with the marine oxygen-isotope record documents strict relationships between T–R sequences and glacial–interglacial cycles, showing that transgressive surfaces correlate invariably with the onset of warm-temperate (interglacial) phases.

INTRODUCTION

The application of sequence stratigraphy concepts to non-marine successions is one of the most debated topics in current stratigraphical research. Relationships between fluvial and equivalent nearshore deposits are commonly unclear, so the widely established sequence-stratigraphy principles from marine packages of strata are generally applied very cautiously to alluvial successions.

Although a variety of models concerning the sequence stratigraphy of alluvial strata has been published after the initial Exxon sequence-stratigraphy model (Posamentier & Vail, 1988; Posamentier et al., 1988), the matter is still highly controversial (Miall, 1991; Schumm, 1993; Shanley & McCabe, 1993, 1994; Wright & Marriott, 1993; Helland-Hansen & Gjelberg, 1994; Helland-Hansen & Martinsen, 1996; Leeder & Stewart, 1996; Milana, 1998; Marriott, 1999; Posamentier & Allen, 1999). How far inland base-level changes may affect a fluvial system is one of the major points of debate. Shanley & McCabe (1993) stated that base-level changes can affect a valley up to 100–150 km updip from the shoreline, whereas Schumm (1993) extended this distance to perhaps 300 km. Blum & Törnqvist (2000) have recently defined the landward limit of sea-level influence as the upstream extent of coastal onlap resulting from sea-level rise. On the basis of data from fluvial systems of Europe and the USA, this landward limit of onlap has been observed to vary between 40 and 400 km, depending on the gradient of the onlapped floodplain surface.

Factors other than eustasy, however, may play a key role in controlling fluvial architecture. These include climate (Blum, 1993; Blum et al., 1994), tectonics (Crews & Ethridge, 1993; Leeder, 1993; Martinsen et al., 1993, 1999; Arche & López-Gómez, 1999) and changes in sedimentation rate and avulsion (Leeder, 1978; Törnqvist, 1994; Mackey & Bridge, 1995; Heller & Paola, 1996).

High-resolution stratigraphical correlations from shallow-marine to continental settings are often limited by several factors, including: (i) poor outcrop exposure; (ii) tectonic deformation, resulting in discontinuous stratigraphical surfaces; (iii) poorly known climatic and eustatic history; and (iv) difficulties in precise dating, especially of alluvial deposits. As recently observed by Blum & Törnqvist (2000), readily datable Late Quaternary sediments will more likely provide solutions to sequence-stratigraphy problems, and represent fertile ground for developing models of alluvial architecture.

The aim of this paper is to provide extensive documentation of the linkage between Late Quaternary fluvial and coastal depositional systems in the Po Plain (northern Italy), and suggest a model of fluvial interactions with marine deposits under a predominantly glacio-eustatic control. This model could be applied to the interpretation of older successions, where the glacio-eustatic signal cannot be estimated easily.

STUDY AREA AND GEOLOGICAL SETTING

The Late Quaternary deposits of the Po Plain, a foredeep system bounded by two mountain chains (the Alps to the north and the Apennines to the south), but in direct connection with the Adriatic Sea to the east (Fig. 1), represent an example where it is possible to investigate the relationships between fluvial–coastal sedimentation and sea-level changes over the past 400 kyr. Tectonically induced accommodation in this rapidly subsiding basin has allowed thick (up to 400 m) fluvial and shallow-marine successions to be preserved during this interval of time. The Po Basin thus includes an exceptionally detailed record of recent eustatic and climatic fluctuations, with only minor breaks in the stratigraphical succession.

As part of the geological mapping of Italy at 1 : 50 000 scale, a project undertaken by the Geological Survey of Italy, the Geological Survey of Regione Emilia-Romagna has drilled more than 100 continuous-core boreholes (50–200 m thick) and performed > 1000 cone penetration tests in the subsurface of the southern Po Plain. Stratigraphical analysis at the basin scale, based upon the interpretation of seismic profiles and well logs, has been carried out by Regione Emilia-Romagna & ENI-AGIP (1998). Facies analysis of the cores has led to the reconstruction of subsurface facies architecture, both in the coastal area (Amorosi et al., 1999a, 2003; Amorosi & Marchi, 1999) and at the basin margin (Amorosi & Farina,

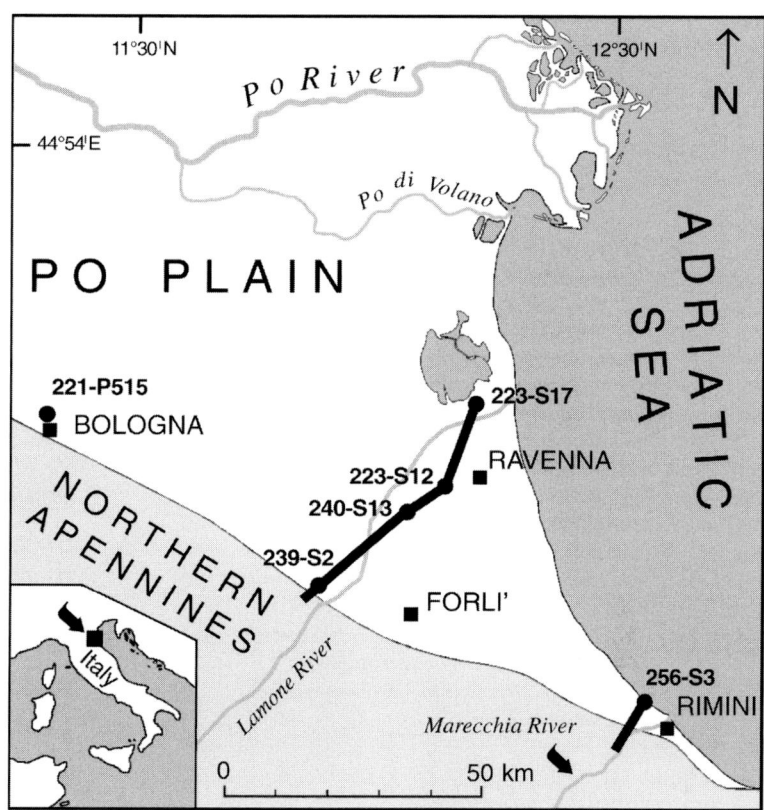

Fig. 1 Map of the study area with location of cores and section lines of Figs 3 & 5.

1995). Pollen analysis of cores (Amorosi et al., 1999b; 2004) has provided a chronological framework for the study units and helped correlate stratigraphical architecture with climatic fluctuations. Finally, sediment dispersal patterns in the Po Basin have been reconstructed from petrographic characteristics of sands (Marchesini et al., 2000), combined with geochemical and mineralogical analyses of fine-grained deposits (Amorosi et al., 2002). Integration of these studies has documented that cyclic facies patterns at Milankovitch-scales are the dominant feature of both coastal and fluvial deposits in the subsurface of the Po Plain. Correlations between the two domains, however, have not yet been performed.

METHODS

The data set of this study consists of several tens of continuously-cored boreholes, ranging in depth between 50 and 200 m, plus unpublished data, which were used for the construction of the cross-sections. For a detailed documentation of facies architecture, the reader is referred to pertinent papers by the authors (Amorosi et al., 1999a,b, 2003).

The importance of micropalaeontological analyses on benthic foraminifers and ostracods, is stressed as being a fundamental tool to refine facies interpretation of coastal deposits, providing detailed information on water depth and salinity. Mixed, benthic foraminifers and ostracod associations were identified and assembled into four groups (M, R, B and F), on the basis of their palaeoecological significance (Table 1). Attribution of microfossil associations to specific subenvironments (see Fig. 2) is based upon facies analysis and calibration with microfaunal distribution in modern environments. Lower case letters are used to define specific subenvironments, as shown in Table 1 & Fig. 2.

Table I Dominant benthic foraminiferal and ostracod taxa of microfossil associations in Late Quaternary deposits of the Po Basin. For depositional environment, see Fig. 2.

	Microfossil association	Depositional environment
F	*Candona* spp.	Coastal lake, swamp
Ba	*Trochammina inflata* (Montagu)	Salt marsh, head of estuary
Bb	*Cyprideis torosa* (Jones)	Inner lagoon, inner bay, inner estuary
Bc	*Ammonia tepida* (Cushman), *A. parkinsoniana* (d'Orbigny) and *Cyprideis torosa*	Central lagoon, central bay, central estuary
Bd	*Ammonia tepida* and *A. parkinsoniana* (dominant), *Cyprideis torosa*	Outer lagoon, outer bay, outer estuary
Rb	Mixed reworked marine and brackish-water microfossils	Washover
Rm	Reworked marine microfossils	Backshore, foreshore, upper shoreface
Ma	*Ammonia tepida* and *A. parkinsoniana*, *Loxoconcha stellifera* (G.W. Müller) and *Pontocythere turbida* (G.W. Müller)	Lower delta front, delta front–prodelta transition
Mb	*Ammonia tepida*, *A. parkinsoniana*, *Cribroelphidium* spp., *Semicitherura* spp. *Pontocythere turbida* and *Loxoconcha* spp.	Delta front–prodelta transition, prodelta
Mc	*Ammonia* spp., *Elphidium* spp. and *Pontocythere turbida*	Flood–tidal delta, spit
Md	*Miliolidae* spp., *Elphidium* spp. *Cribroelphidium* spp., *Pontocythere turbida* and *Callistocythere* spp.	Transgressive barrier, shoreface–offshore transition
Me	*Textularia* spp., *Miliolidae* spp., *Semicytherura* spp. and *Lepthocythere* spp.	Inner shelf

Associations in group M (Ma–Me) are indicative of normal marine salinity waters; associations in group R include reworked and transported microfauna, within marine (Rm) or brackish-water (Rb) environments; associations in group B (Ba–Bd) correspond to *in situ* fauna, diagnostic of brackish-water environments; group F is characteristic of freshwater settings.

LINKAGE BETWEEN FLUVIAL AND MARINE DEPOSITS IN THE PO BASIN

Given the difficulty in precise positioning of the key surfaces for sequence-stratigraphy interpretation within alluvial sections, stratigraphical descriptions will follow, from seaward to landward locations (Fig. 3).

Coastal section

The stratigraphical architecture of Holocene deposits of the Po coastal plain, between the Marecchia River and the Po River delta (Fig. 1), has been described in detail by Rizzini (1974), Bondesan et al. (1995) and Amorosi et al. (1999a, 2003). These authors identified a transgressive–regressive cycle, approximately 20–30 m thick, consisting of back-barrier, nearshore and shallow-marine strata, which overlie a monotonous alluvial plain succession of Late Pleistocene age (Fig. 3). On the basis of integrated sedimentological and micropalaeontological analyses of long (170 m), continuously-cored boreholes, Amorosi et al. (2004) have recently documented the presence, at an average depth of about 100 m, of an older but very similar transgressive–regressive cycle, characterized by a greater landward migration of the shoreline, and typically wedging out in a landward direction (Fig. 3). This cycle, which is separated from the Holocene sediments by approximately 60 m of alluvial deposits, shows the same arrangement of facies in a backward- and then forward-stepping stacking pattern, and has been confidently attributed to the last interglacial maximum (the Eemian = marine oxygen isotope substage 5e).

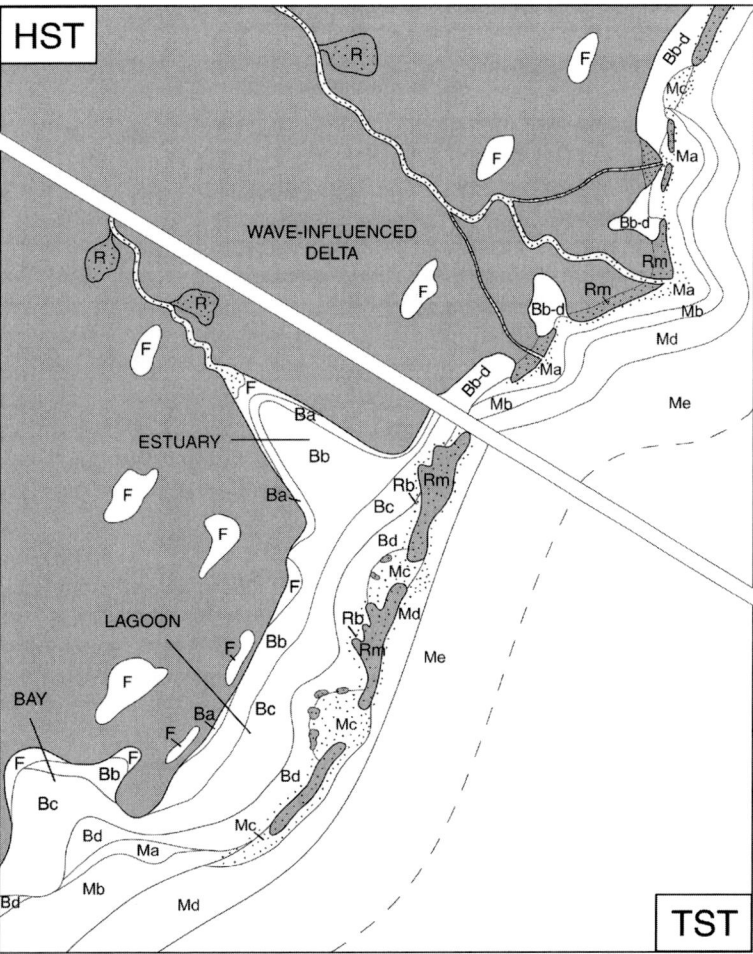

Fig. 2 Generalized distribution of microfossil associations (see Table 1), based upon data from cores of the modern coastal plain (after Amorosi et al., 2004): F, microfossil association indicative of freshwater environments; B, microfossil associations indicative of brackish-water environments, with decreasing salinity from Bd to Ba; M, microfossil associations indicative of marine environments (Ma is the shallowest and Me the deepest); R, reworked microfossils (Rm is dominated by large-sized marine species; Rb records the additional presence of brackish-water taxa; R includes microfossils reworked from older sequences); dots, sand; grey area, emerged lands; white area, sea. Transgressive palaeogeography (TST) is interpreted to reflect the landward migration of barrier–lagoon–estuary systems, whereas highstand palaeogeography (HST) records the progradation of a wave-influenced delta.

Investigation of deeper boreholes allows an obvious cyclic pattern of facies to be identified (Fig. 3). Particularly, the repeated alternation of coastal and alluvial deposits enables subdivision of the study succession into distinct transgressive–regressive sequences, with an average thickness of about 100 m, and a definite internal architecture. From a conceptual point of view, these correspond to 4th-order depositional sequences, and show strong similarities to the transgressive–regressive (T–R) sequences of Johnson et al. (1985) and Embry (1993, 1995).

Transgressive surfaces (TS), marking the abrupt facies change from stiff, pedogenically modified floodplain clays to overlying back-barrier (marsh, lagoonal and estuarine) deposits (Fig. 3), are readily identified in cores and form bounding surfaces to T–R sequences. Prominent ravinement surfaces (RS in Fig. 3—see Swift, 1968; Demarest & Kraft, 1987; Nummedal & Swift, 1987), separating back-barrier deposits from overlying nearshore sands, are generally observed just a few metres (0–4) above each TS. These laterally extensive erosional surfaces are typically overlain by a veneer of mollusc shells and grade upwards into shallow-marine clay–sand alternations. On the basis of three-dimensional studies from Holocene coastal deposits of the south-east Po Plain (Amorosi et al., 1999a, 2003) and the adjacent Adriatic Sea (Trincardi et al., 1994), and

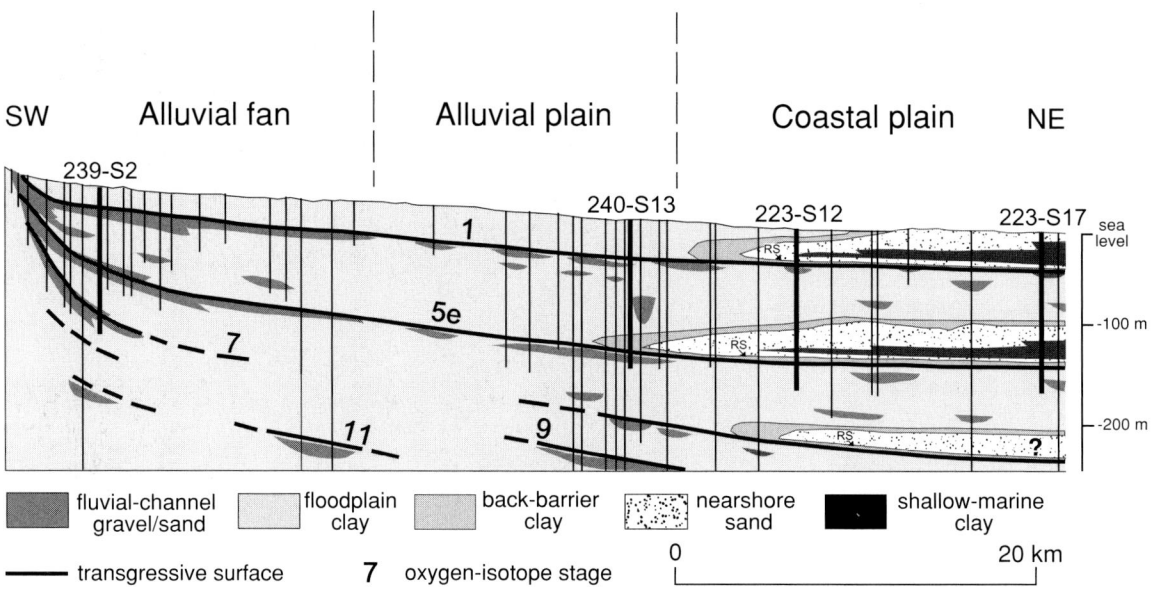

Fig. 3 Geological cross-section (see Fig. 1 for location), showing the linkage between alluvial fan, alluvial plain and nearshore deposits in the subsurface of the Po Plain. Note the use of transgressive surfaces for bracketing transgressive–regressive sequences. Continuous-core boreholes are in bold. RS, ravinement surface.

comparison with world-wide coeval successions (Amorosi & Milli, 2001), the retrogradational stacking pattern of facies in the lower parts of T–R sequences is interpreted to reflect the landward migration of barrier–lagoon–estuary systems, and defines the transgressive systems tracts (TST) (Fig. 2). The overlying progradational stacking pattern of facies, including, from base to top, shallow-marine, nearshore and back-barrier deposits, is interpreted to reflect progradation of wave-influenced deltas (prodelta, delta front and delta plain deposits, respectively—see Fig. 2) and adjacent strandplains, which took place when sea-level rise decreased, approaching relatively stable positions (highstand systems tracts—HST).

The upper parts of T–R sequences consist predominantly of floodplain muds, with subordinate fluvial-channel sands (Fig. 3). On the basis of pollen analyses (Amorosi et al., 1999b), these non-marine deposits are attributed to what has been termed the forced regressive (Hunt & Tucker, 1992) or falling-stage (Plint & Nummedal, 2000) systems tracts (FST) and to the overlying lowstand systems tract (LST). Lagoonal and paludal horizons (not shown in Fig. 3) present within the FST mark higher frequency sea-level cycles, which punctuate the prolonged phase of sea-level fall (Amorosi et al., 2004).

Amorosi et al. (1999b) have shown that this repetitive alternation of coastal and alluvial deposits is paralleled by a distinctive pollen signature. On the basis of correlation with long-core pollen series from southern Europe (Italy, France and Greece), the major culmination in arboreal pollen recorded in core 223-S17 at about −124 m (Fig. 4) has been attributed to the Eemian (cf. Woillard, 1978; de Beaulieu & Reille, 1992) and correlated to marine oxygen-isotope substage 5e (see Tzedakis et al., 1997). The correlation of coastal and shallow-marine deposits between 105 and 124 m with substage 5e, corresponding to the highest position of sea-level of the past 260 kyr (Aharon & Chappell, 1986; Bard et al., 1990; Pirazzoli, 1993), is fully consistent with the stratigraphical framework, showing no intervening marine deposits between this stratigraphical interval and the Holocene deposits.

The nearshore and shallow-marine deposits that characterize what are interpreted to represent TSTs and HSTs contain major peaks in pollen

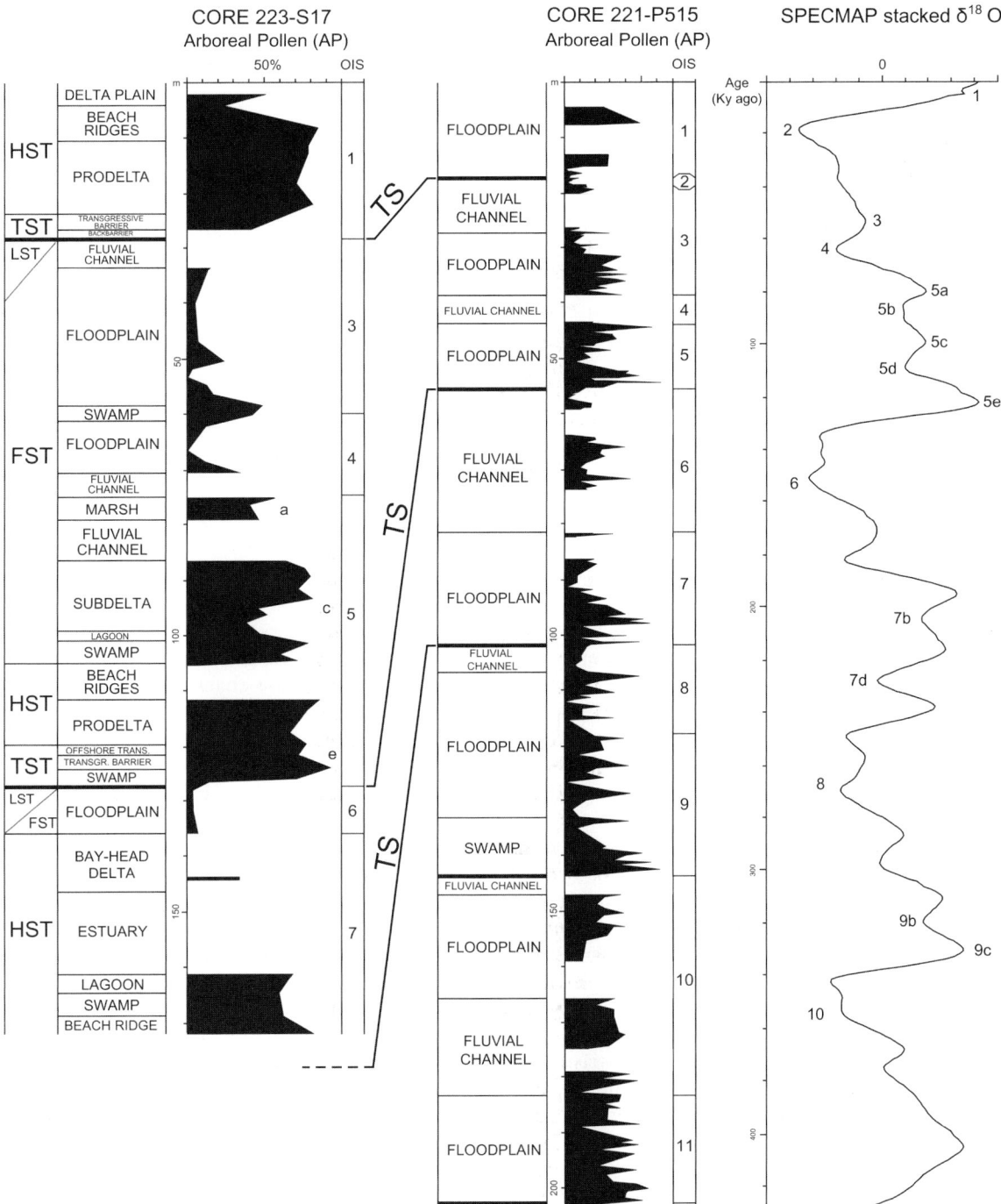

Fig. 4 Pollen spectra of transgressive–regressive sequences identified in the coastal area (core 223-S17) and close to the basin margin (core 221-P515), and their correlation to the SPECMAP time-scale (Imbrie et al., 1984). Note the sharp increase in arboreal pollen (excluding *Pinus*) above bounding surfaces (TS) of T–R sequences, corresponding to the onset of interglacial phases (marine oxygen isotope stages (OIS) 1, 5e, 7, 9 and 11). Palynostratigraphy of core 223-S17 after Amorosi et al. (1999b); palynostratigraphy of core 221-P515 after Amorosi et al. (2001).

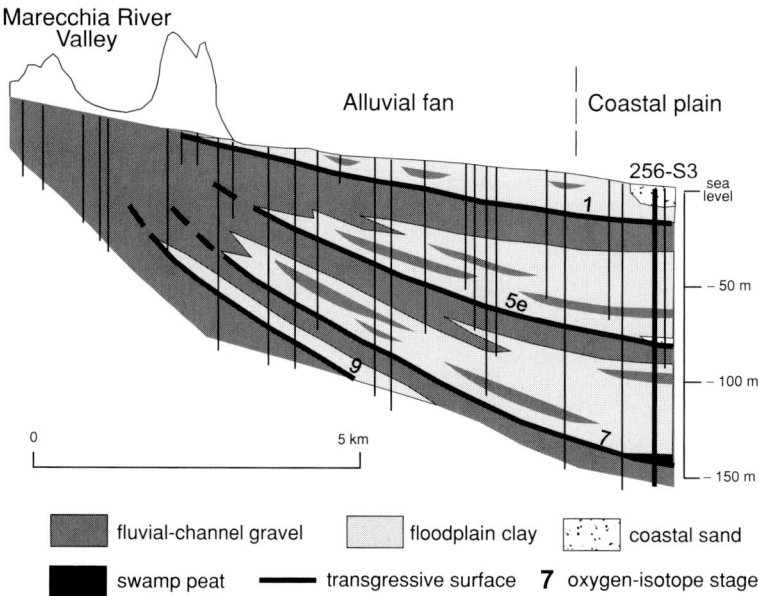

Fig. 5 Geological cross-section (see Fig. 1 for location), showing the linkage between alluvial fan and nearshore deposits in the subsurface of the Po Plain. Upper bounding surfaces of amalgamated fluvial-channel complexes (bases of T–R sequences) are much more easily identified than lower boundaries. After the Geological Map of Italy, Sheet 256—Rimini (in press), slightly modified (section constructed by U. Cibin and P. Severi—Geological Survey of Regione Emilia-Romagna).

concentration and the highest percentages of arboreal pollen (AP), suggesting wide development of mixed, deciduous broad-leaved forests, which are typical for interglacial conditions (stages 1, 5e and 7 in Fig. 4). By contrast, alluvial deposits interpreted to represent the FSTs and LSTs have very low pollen concentrations, indicating a strong reduction in vegetation cover, and show a vegetation dominated by *Pinus* and non-arboreal pollen (NAP), with AP scarce or absent (Fig. 4). These pollen spectra are characteristic of glacial intervals (stages 3, 4 and 6 in Fig. 4). It is remarkable that minor transgressive events within the FST, marked by the presence of lagoonal and marsh deposits sandwiched between alluvial plain strata (core depths 78 m and 100 m, core 223-S17 in Fig. 4), display a pollen signature characteristic of warm-temperate conditions, which can be related to substages 5c and 5a.

Alluvial section

The stratigraphical architecture described in the previous section is most pronounced adjacent to the present shoreline and thins away upstream over about 25 km (see Fig. 3), and is then replaced by a succession dominated by fluvial deposits.

The repeated alternation of coarse-grained (gravel–sand) and fine-grained (silt–clay) bodies, interpreted as fluvial-channel and floodplain deposits, respectively, is the basic motif of the entire alluvial succession of the Po Basin (Figs 3 & 5). Internal architecture of these cycles, which thicken in a seaward direction and generally are 50–70 m thick, varies as a function of the basin position. At landward locations, coinciding with the outlets of the intramontane valleys, the alluvial fan successions are gravel-dominated and consist of vertically stacked, generally amalgamated, gravel bodies (Figs 3 & 5), where identification of bounding surfaces of individual cycles is very difficult (see Miall & Arush, 2001). At more distal locations, fluvial-channel sand bodies occur at distinct stratigraphical levels, separated by comparatively thick sections dominated by floodplain deposits, with isolated channel bodies (Figs 3 & 5).

Stratigraphical architecture of Late Quaternary alluvial-fan and alluvial-plain complexes displays strong similarities across the entire study area (compare Figs 3 & 5). The similar stratigraphical patterns identified across the different fluvial systems along the Po Basin margin (see also Amorosi *et al.*, 2001 for the Bologna area) suggest a common sedimentary evolution, which developed under an external control on sediment flux.

Lower boundaries of sheet-like fluvial-channel complexes are transitional to the underlying mud-prone floodplain deposits, and form characteristic upward-coarsening successions, comprising increasingly amalgamated fluvial-channel gravels or sands (Figs 3 & 5). By contrast, the upper boundaries of the amalgamated channel bodies are very sharp and mark abrupt facies changes to organic-rich or paludal clays. This can be especially seen where the shoreline is close to the alluvial fan systems and channel abandonment owing to sea-level rise was followed by widespread swamp development (see swamp peat, 5 m thick, in core 256-S3—Fig. 5). Tidally influenced fluvial deposits (see Shanley & McCabe, 1993), showing the characteristic retrogradational stacking pattern of bay-head deltas depicted by Dalrymple et al. (1992) and Nichol et al. (1994), have been recorded in the Holocene of the Po Plain on top of the amalgamated fluvial-channel bodies, at the transition from alluvial to coastal sections (Amorosi & Milli, 2001; Centineo, 2001). The abrupt change from laterally amalgamated, sheet-like gravel bodies, with a high degree of interconnectedness, to predominantly muddy units, with mostly isolated, ribbon-shaped gravel and sand bodies, reflects a sudden lowering of the sediment-supply/accommodation ratio (Dreyer, 1993; Shanley & McCabe, 1994; Martinsen et al., 1999), which may provide the best record of sedimentary response to a base-level rise (cf. Olsen et al., 1995).

Repetitive or cyclic facies patterns in fluvial architecture have been widely described in the literature, and the lower boundaries of amalgamated fluvial-channel bodies have been taken in many instances as bounding surfaces of cycles (Aitken & Flint, 1995; Olsen et al., 1995; O'Byrne & Flint, 1996; Posamentier & Allen, 1999; Catuneanu & Elango, 2001). Early work on the alluvial fan successions of the Po Basin in the Bologna area (see Fig. 1) has followed the same tendency (Amorosi & Farina, 1995). Careful observation of stratigraphical architecture in the Po Plain, however, shows that well-expressed bounding unconformities, such as incised valleys or unincised sequence boundaries (see Holbrook, 1996; Posamentier, 2001), cannot be easily identified at the base of the amalgamated fluvial-channel gravels/sands (Figs 3 & 5). On the basis of core data alone, these unconformities might easily be confused with more localized channel scours (Aitken & Flint, 1995). By contrast, dramatic changes in facies and fluvial architecture are invariably recorded at the top of amalgamated fluvial-channel bodies. For this reason, alluvial successions have been subdivided into T–R sequences, similar to the coastal sections (see the transgressive–regressive facies tracts of Burns et al., 1997).

The cross-section of Fig. 3 shows correlations between the alluvial fan systems and the coastal area across a transect 50 km long. Lower boundaries of T–R sequences in the alluvial-fan and alluvial-plain sections appear to be stratigraphically equivalent to the transgressive surfaces overlain by coastal and shallow-marine deposits. This stratigraphical relationship between alluvial and coastal deposits is even more obvious in Fig. 5, where alluvial-plain deposits are lacking and transgressive swamp peats and nearshore sands at the base of T–R sequences are observed to overlie directly the amalgamated fluvial-channel bodies belonging to the alluvial-fan complexes.

Pollen data from core 221-P515 (Fig. 1), the stratigraphy of which has been recently correlated with that shown in Fig. 3 (Amorosi et al., 2001), are consistent with this stratigraphical framework, showing characteristic abundance peaks of AP (suggesting warm-temperate climatic phases) within floodplain clays and swamp peats, at the base of T–R sequences (Fig. 4). By contrast, AP percentages decrease significantly in the upper parts of T–R cycles, where amalgamated fluvial-channel bodies are the dominant feature. In this instance, pollen spectra are indicative of a cold-climate vegetation, dominated by NAP and *Pinus*.

KEY SURFACES FOR SEQUENCE STRATIGRAPHICAL INTERPRETATION

The sequence boundary (SB), the transgressive surface (TS) and the maximum flooding surface (MFS) allow identification of sedimentary packages of chronostratigraphical significance, and represent the three major key surfaces for sequence stratigraphical interpretation (Posamentier et al., 1988). It should be stressed, however, that these surfaces may be diachronous, because of changing accommodation and sediment supply in the different parts of a basin. In the subsurface of the Po Plain, SB, TS and MFS are identified by various criteria, most notably lithofacies changes that in the coastal sections correspond to important variations in the micropalaeontological associations.

Although none of the three surfaces can be traced confidently across the entire basin, one of them (the TS) is easier to correlate and can be traced over larger areas with a greater degree of confidence (Fig. 6). The next sections document the attributes and facies signatures of SB, TS and MFS in the late Quaternary deposits of the Po Basin, and explore the advantages and drawbacks in the use of these surfaces for bracketing stratigraphical packages of genetic significance.

Sequence boundary (SB)

The issue of where to place the sequence boundary of a depositional sequence has been a subject of debate in recent sequence-stratigraphy work. Two major alternatives exist for stratigraphical positioning of the SB (see recent discussion in Posamentier & Allen, 1999): (i) at the onset of relative sea-level fall (Posamentier et al., 1992; Kolla et al., 1995; Morton & Suter, 1996); or (ii) at its end (Hunt & Tucker, 1992, 1995), coincident with the maximum regressive surface of Helland-Hansen & Martinsen (1996) (Fig. 6). According to the Exxon approach, the SB should be placed at the base of the early lowstand systems tract. By contrast, placing the SB at the end of sea-level fall (Hunt & Tucker approach) implies its stratigraphical

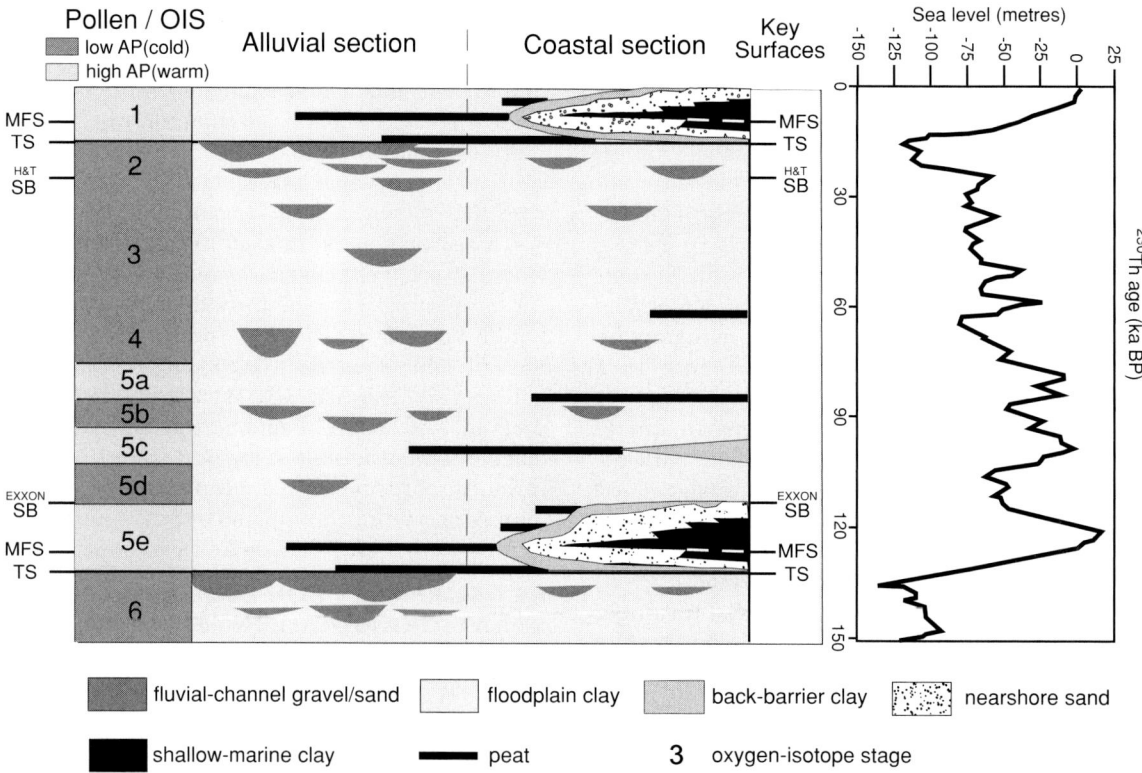

Fig. 6 Schematic representation of the post-Eemian (post-125 ka) stratigraphy (not to scale—approximate thickness 150 m) of the Po Plain, showing relationships between climate changes, stratigraphical architecture, sequence stratigraphy and sea-level fluctuations (sea-level curve after Bard et al., 1990): OIS, marine oxygen isotope stages; AP, arboreal pollen (excluding *Pinus*); SB, sequence boundaries of Exxon and Hunt & Tucker (1992); TS, transgressive surface; MFS, maximum flooding surface.

positioning at the boundary between the early lowstand systems tract (= the forced regressive systems tract of Hunt & Tucker (1992, 1995) or the falling-stage systems tract of Nummedal & Molenaar (1995) and Plint & Nummedal (2000)), and the overlying (late) lowstand systems tract.

When dealing with the sequence-stratigraphy interpretation of Late Quaternary depositional sequences formed during sea-level fluctuations with frequencies of 100 kyr, the choice between these two models becomes problematic, because of the strong asymmetry in the curves of sea-level variation, characterized by comparatively long periods (up to 90 kyr) of sea-level fall, punctuated by higher-frequency sea-level cycles (Fig. 6). A general assumption of sequence-stratigraphy models is that initiation of fluvial incision by relative sea level fall results in formation of the sequence-bounding unconformity, and that as relative sea level continues to fall the coastal plain is incised and sediment by-passes the highstand fluvial and coastal plains (Posamentier & Allen, 1999). As a consequence, when fluvial incision occurs, in alluvial plain and coastal plain areas outside of incised valleys an amalgamated unconformity develops. In this case the Exxon and Hunt & Tucker SBs would merge into a single surface. Blum & Törnqvist (2000) have argued recently against the theory of complete sediment by-pass in coastal plains during phases of fluvial incision related to glacio-eustatic sea-level fall. Examples of preserved forced regressive deposits in alluvial settings have been provided by Blum & Price (1998), Amorosi et al. (1999b) and Törnqvist et al. (2000).

The proposed interpretation of data from the last glacial–interglacial (post 125 ka) cycle (see Figs 3 & 4) differs from traditional sequence-stratigraphy models and is fully consistent with remarks by Blum & Törnqvist (2000). In the rapidly subsiding Po Basin, tectonically induced accommodation allowed accumulation of an extremely thick (60 m) succession of alluvial and coastal-plain deposits during the prolonged sea-level fall between 125 and 20 ka (i.e. FST). As recently shown by Amorosi et al. (2004), the FST in the Po Basin consists of vertically stacked transgressive and highstand deposits (5th-order depositional sequences), separated by closely spaced unconformities marking the progressive basinward shift of facies during the longer period, 4th-order sea-level fall. As a result, in the case of the Po Basin, the Exxon and Hunt & Tucker SBs are widely separated by 60 m of intervening falling stage deposits (Fig. 6).

Irrespective of which surface is chosen as a sequence boundary, both surfaces often are difficult to record, correlate and map in the Po Basin on the basis of the available data (Fig. 6). Exxon SBs are easily recognized within coastal sections, where they are defined locally by palaeosols associated with a remarkable basinward shift of facies at the HST–FST boundaries, and with non-marine deposits (bearing evidence of cold-climate vegetation) unconformably overlying shallow-marine strata, deposited during warm-temperate (interglacial) conditions (see substage 5e–5d transition in Fig. 6). In contrast, surfaces that Hunt & Tucker would define as an SB cannot be detected easily in core, owing to objective problems in determining the boundary between FST and LST within homogeneous alluvial plain deposits. Core analysis recognizes regionally significant surfaces of this type that coincide with weathered palaeosurfaces termed interfluve sequence boundaries (Van Wagoner et al., 1990; Aitken & Flint, 1996; McCarthy & Plint, 1998), where no lowstand deposition took place and this type of SB merges with a transgressive surface.

Recognition of sequence boundaries is increasingly difficult landwards, within non-marine successions consisting entirely of alluvial strata (Fig. 6). In these instances, amalgamated fluvial-channel bodies are invariably located below the transgressive surfaces (Figs 3 & 5), suggesting deposition during phases of increased sediment supply and slowed accommodation, immediately preceding the major transgressive phases, and leading to increased channel clustering (Leeder, 1978; Holbrook, 1996). Although most of these amalgamated fluvial-channel bodies can be broadly assigned to the LST, whether or not they are LST or partly late FST deposits is equivocal. This is probably the reason why some authors (see Legarreta & Uliana, 1998) have suggested grouping these highly interconnected fluvial sheets into a unique systems tract (forestepping systems tract). In landward positions, the Hunt & Tucker SBs have little objective expression in core and only an approximate position for this surface is possible. Concerning the Exxon SBs, apart from subtle features (local palaeosol develoment) in areas where mud-on-mud contacts dominate, their identification within alluvial sections is generally a very difficult task. The Exxon SBs easily can be confused with one of the closely spaced unconformities marking the subsequent steps of sea-level fall.

Owing to the high preservation of falling stage deposits, the general assumption that the Exxon SBs coincide with the lower boundaries to amalgamated fluvial-channel bodies (Shanley & McCabe, 1991, 1994; Posamentier & Allen, 1993, 1999; Van Wagoner, 1995) is invalid for the Po Basin, and should not be applied whenever subsidence (or high-frequency sea-level cycles) may create accommodation exceeding the effects of sea-level fall. Finally, it should be noted that pollen analyses may assist in the recognition of the Exxon SBs (Amorosi, 1999b), because these surfaces are commonly associated with significant changes in vegetation (lower boundaries of glacial–interglacial cycles—see 5e–5d boundary in Fig. 6). By contrast, the lower boundaries of amalgamated fluvial-channel bodies, corresponding broadly to the SBs of Hunt & Tucker, separate sedimentary units with the same cold-climate pollen signature.

Transgressive surface (TS)

The transgressive surface, corresponding to the 'initial transgressive surface' of Nummedal et al. (1993) and the 'maximum regressive surface' of Helland-Hansen & Martinsen (1996), has been proposed by Embry (1995) as a good alternative to the sequence boundary (Vail et al., 1977; Posamentier & Vail, 1988) for the subdivision of stratigraphical successions into T–R sequences, rather than depositional sequences. The example from the Po Plain shows that TSs are prominent features that mark the base of transgressive–regressive patterns, and are traceable across the entire study area (Figs 3 & 5). These surfaces are very useful for a pragmatic stratigraphical subdivision into mapping units (see also Weimer, 1984; Cross, 1988; Bhattacharya & Willis, 2001; Plint et al., 2001).

As observed from the cross-sections of Figs 3 & 5, transgressive surfaces are the most laterally persistent stratigraphical surfaces within the Late Quaternary sedimentary succession of the Po Basin. The TSs mark dramatic landward shifts of facies at almost any site in the basin. This is due to the very rapid sea-level rise (120–150 m in 10–15 kyr) that accompanied the 100-kyr Late Quaternary sea-level fluctuations, as opposed to the long phases of sea-level fall (Fig. 6). In coastal deposits, the TS generally corresponds to a sharp facies change from lowstand alluvial-plain deposits to the overlying back-barrier (marsh, lagoonal and estuarine) mud-dominated strata (Fig. 6). Close proximity between the TS and the ravinement surface, which represents another outstanding feature owing to its peculiar facies characteristics (Fig. 3), enables a ready identification of transgressive sedimentation, and facilitates delineation of T–R sequences. Where the LST is missing and the TS coincides with an interfluve sequence boundary (SB of Hunt & Tucker), the surface also has distinctive geotechnical characteristics; it is represented by an indurated, pedogenic horizon (Amorosi & Marchi, 1999).

The recognition of the TS in alluvial strata generally represents a major problem in sequence-stratigraphy models. In the Po Basin, on the basis of stratigraphical correlations (Figs 3 & 5) supported by radiometric dates and pollen data (Fig. 4), the TS is expressed as a sharply defined facies change. In alluvial-fan and alluvial-plain areas the TS is marked by an abrupt decrease in vertical amalgamation of coarse-grained (gravel/sand) bodies, which become increasingly isolated within fine-grained floodplain sediments (TST) (Fig. 6). The decrease in channel clustering and the change from sheet-like to ribbon-shaped bodies across the TS are consistent with increasing fluvial accommodation caused by relative sea-level rise. In the coastal plain, the TS is marked by the presence of laterally extensive peat horizons, which appear to correlate with initial marine flooding surfaces at coastal locations.

As shown in Fig. 6 (compare with Fig. 4), pollen analyses constitute an additional tool for identifying the TS. Particularly, in the Po Plain the TS is marked by a distinctive signature of: (i) maximum expansion of broad-leaved deciduous vegetation (AP); (ii) abrupt decrease in *Pinus* and NAP; and (iii) major peaks in pollen concentration (see Amorosi et al., 2004). These characteristics point to an obvious relationship between transgressive episodes and the onset of warm (interglacial) phases. The transgressive surface, however, may be difficult to identify locally in very proximal locations, where alluvial fan gravel bodies are amalgamated, with no intervening fine-grained units. In these settings, the TS is virtually unrecognizable (see Fig. 5). Stratigraphical correlations based upon core data may also fail in recognizing backstepping fluvial-channel bodies within incised valleys.

Maximum flooding surface (MFS)

In Galloway's (1989) model, the maximum flooding surface is chosen as the bounding surface for subdividing sedimentary successions into genetic stratigraphical sequences (see also the depositional complex of Frazier, 1974). Although the practical use of the MFS at the scale of 3rd-order cycles has been widely documented (Bhattacharya, 1993; Helland-Hansen & Martinsen, 1996; Cant, 1998), it must be stressed that some problems may arise when trying to locate the MFS within higher frequency (4th-order) cycles.

A variety of features characterizes the MFS in the Po Basin. Within coastal sections, the MFS is located a short distance above the TS and RS (Fig. 6). On the basis of facies analysis of cores alone, it would be logical to place the MFS coincident with the obvious deepening phase at the boundary between the transgressive nearshore sands and the overlying shallow-marine (offshore transition) clay–sand alternations. Laterally extensive, although not ubiquitous, shell concentrations within the homogeneous, shallow-marine unit may locally suggest a condensed section and the approximate position of the MFS (Amorosi & Marchi, 1999).

Microfaunal criteria allow refinement of this approximation, by locating the peak transgression within shallow-marine muds (Fig. 7). Palaeoecological data show that the MFS can be placed at the transition from benthic foraminifers associated with open-marine environments (microfossil association Me) to microfossils characteristic of shallower waters (association Md), and eventually to fossil associations Mb and Ma that contain an increasing abundance of species tolerant of salinity-stressed conditions and riverine influence (Fig. 2). In this respect, the MFS can be regarded as the surface reflecting the palaeoenvironmental change from retrograding barrier–lagoon–estuary systems (TST) to prograding deltas and strandplains (HST).

In coastal-plain sections, as predicted by recent models (Shanley et al., 1992; Hamilton & Tadros, 1994), the MFS is expressed as a tongue of lagoonal muds, bracketed by paludal, organic-rich clays (Fig. 6). Landward of this area (alluvial section in Fig. 6), the expression of this bounding surface is that of peat horizons (flooding surface equivalents of Cant, 1998) and the entire TST tends to contain abundant organic deposits (Cross, 1988). In these settings, the MFS can be easily mistaken for minor, but extensive, flooding surfaces forming other peat horizons, and the conceptual distinction between MFS and TS practically can be of minor significance. In addition, especially close to the TST–HST transition, subsidence and autocyclic mechanisms, such as delta-lobe abandonment and distributary avulsion, become increasingly important and may easily overwhelm the effects of sea-level change (Amorosi & Milli, 2001). This leads to significant diachroneity of the MFS, on the order of several thousands of years, with retrogradational and progradational stacking patterns developing synchronously in different parts of the basin (see Wehr, 1993; Martinsen & Helland-Hansen, 1994).

In very proximal alluvial sections, the MFS has no significant physical expression and lies presumably very close to the TS, within a section with a low degree of channel clustering.

CONCLUSIONS

Data presented and summarized herein point to a number of significant conclusions.
1 Transgressive–regressive (T–R) sequences formed during 4th-order (100 kyr) sea-level fluctuations are the dominant feature of Late Quaternary deposits of the Po Plain. The T–R sequences have a significant lateral persistence, and provide a powerful example of the linkage between coeval alluvial and nearshore strata. At relatively seaward locations, lower parts of T–R sequences show increased accommodation and shoreline transgression, interpreted to reflect the

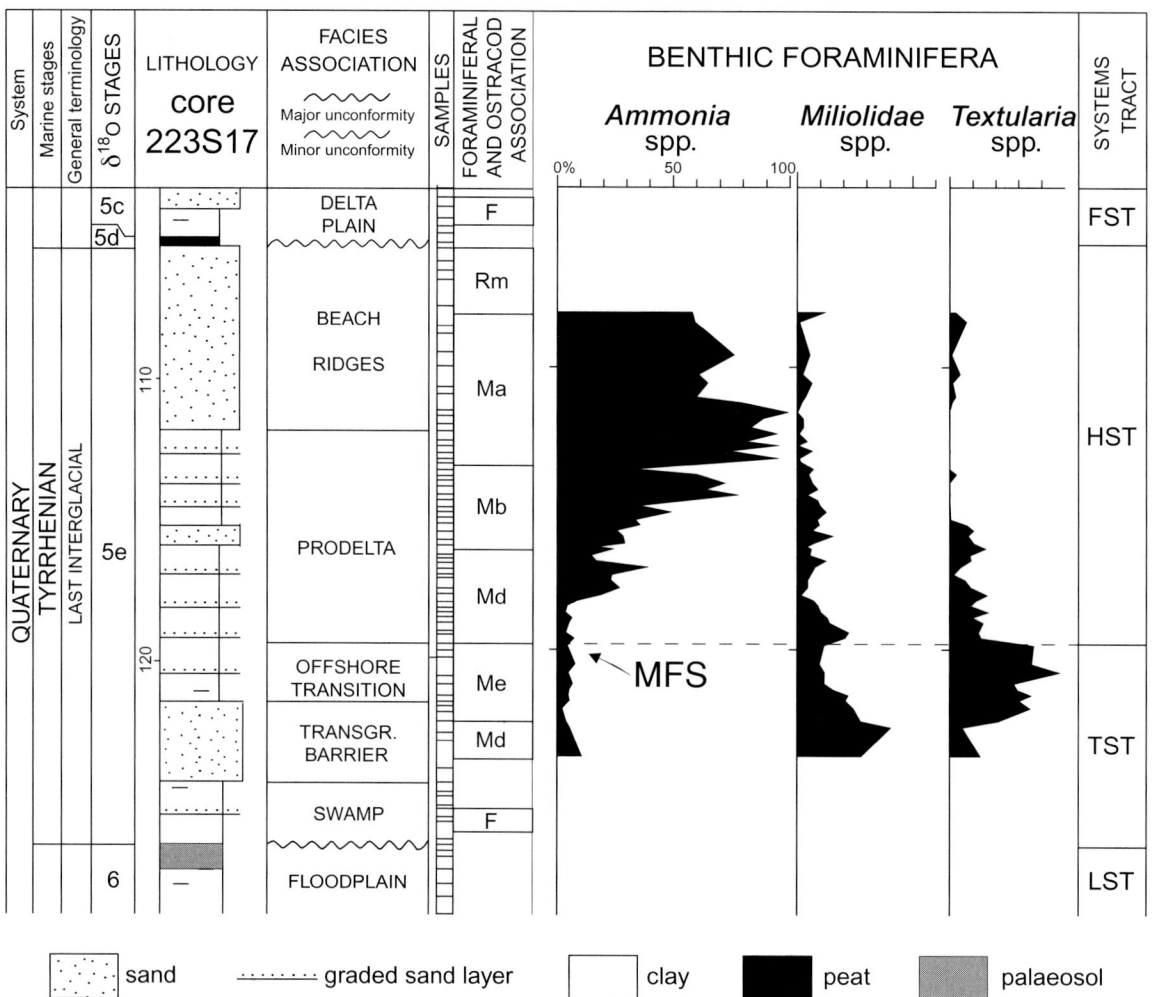

Fig. 7 Vertical distribution of selected benthic foraminiferal taxa, showing location of the maximum flooding surface (MFS) within homogeneous clay–sand alternations, based on major peaks in *Textularia* spp. and *Miliolidae* spp. (microfossil association Me—see Fig. 2). The progressive increase in *Ammonia* spp. within the overlying highstand systems tract (HST) indicates an increasing freshwater influence, suggesting delta progradation. TST, transgressive systems tract; FST, falling stage systems tract; LST, lowstand systems tract.

landward migration of barrier–lagoon–estuary systems (TST), followed by widespread delta and strandplain progradation (HST). This stratigraphical architecture is paralleled in the alluvial sections by an abrupt change in fluvial architecture, from amalgamated fluvial-channel bodies to predominantly organic-rich floodplain deposits with isolated channel bodies. The upper parts of T–R sequences, formed during periods of sea-level fall (FST) and subsequent sea-level lowstand (LST), are dominated by alluvial sedimentation, recording the transition from isolated ribbon-shaped channels to amalgamated, laterally extensive (sheet-like) sand and gravel fluvial bodies.

2 Pollen data show a close relationship between initial transgression and the development of mixed, broad-leaved vegetation, indicating the onset of warm (interglacial) periods. Similarly, a good correlation exists between middle and upper parts of T–R sequences and the development of

cold-climate vegetation. The amalgamated fluvial-channel deposits correspond to glacial (or stadial) conditions. The close match between stratigraphical architecture and pollen distribution suggests that sedimentation in the Po Basin was driven predominantly by combined eustatic sea-level changes and climatic variations.

3 This study has additional implications for application of sequence-stratigraphy concepts to non-marine strata, especially at the scale of 4th-order (100 kyr) cycles, where subsidence is expected to exert a significant degree of control on stratal architecture. Particularly, the ease of recognizing the transgressive surface in cores makes it very useful for basinwide correlations, and for subdividing stratigraphical sections dominated by fluvial and coastal successions into large-scale T–R sequences. The characteristic pollen signature of TS allows recognition of this surface even in cases where it is not readily apparent on the basis of sedimentological criteria alone.

ACKNOWLEDGEMENTS

We are indebted to Raffaele Pignone of the Geological Survey of Regione Emilia-Romagna for access to the cores and well data base. Some of the concepts and ideas developed in this paper were shared with U. Cibin and P. Severi (Geological Survey of Regione Emilia-Romagna). We thank A. Aslan and M. Guccione for their insightful and helpful reviews. Also, we thank F. Fiorini, F. Fusco G. Pasini and S.C. Vaiani for many informative discussions. This study was supported by MURST (40%) 2001 (Co-ordinator: F. Massari).

REFERENCES

Aharon, P. and Chappell, J. (1986) Oxygen isotopes, sea level changes and the temperature history of a coral reef environment in New Guinea over the last 10^5 years. *Palaeogeogr., Palaeoclimatol., Palaeoecol.*, **56**, 337–379.

Aitken, J.F. and Flint, S.S. (1995) The application of high-resolution sequence stratigraphy to fluvial systems: a case study from the Upper Carboniferous Breathitt Group, eastern Kentucky, USA. *Sedimentology*, **42**, 3–30.

Aitken, J.F. and Flint, S.S. (1996) Variable expressions of interfluvial sequence boundaries in the Breathitt Group (Pennsylvanian), eastern Kentucky, USA. In: *High Resolution Sequence Stratigraphy: Innovations and Applications* (Eds J.A. Howell and J.F. Aitken). *Spec. Publ. Geol. Soc. Lond.*, **104**, 193–206.

Amorosi, A. and Farina, M. (1995) Large-scale architecture of a thrust-related alluvial complex from subsurface data: the Quaternary succession of the Po Basin in the Bologna area (northern Italy). *Giorn. Geol.*, **57**, 3–16.

Amorosi, A. and Marchi, N. (1999) High-resolution sequence stratigraphy from piezocone tests: an example from the Late Quaternary deposits of the SE Po Plain. *Sediment. Geol.*, **128**, 69–83.

Amorosi, A. and Milli, S. (2001) Late Quaternary depositional architecture of Po and Tevere river deltas (Italy) and worldwide comparison with coeval deltaic successions. *Sediment. Geol.*, **144**, 357–375.

Amorosi, A., Colalongo, M.L., Pasini, G. and Preti, D. (1999a) Sedimentary response to Late Quaternary sea-level changes in the Romagna coastal plain (northern Italy). *Sedimentology*, **46**, 99–121.

Amorosi, A., Colalongo, M.L., Fusco, F., Pasini, G. and Fiorini, F. (1999b) Glacio-eustatic control of continental-shallow marine cyclicity from Late Quaternary deposits of the south-eastern Po Plain (Northern Italy). *Quat. Res.*, **52**, 1–13.

Amorosi, A., Forlani, M.L., Fusco, F. and Severi, P. (2001) Cyclic patterns of facies and pollen associations from Late Quaternary deposits in the subsurface of Bologna. *Geoacta*, **1**, 83–94.

Amorosi, A., Centineo, M.C., Dinelli, E., Lucchini, F. and Tateo, F. (2002) Geochemical and mineralogical variations as indicators of provenance changes in Late Quaternary deposits of SE Po Plain. *Sediment. Geol.*, **151**, 273–292.

Amorosi, A., Centineo, M.C., Colalongo, M.L., Pasini, G., Sarti, G. and Vaiani, S.C. (2003) Facies architecture and Latest Pleistocene–Holocene depositional history of the Po Delta (Comacchio area), Italy. *J. Geol.*, **111**, 39–56.

Amorosi, A., Colalongo, M.L., Fiorini, F., Fusco, F., Pasini, G., Vaiani, S.C. and Sarti, G. (2004) Palaeogeographic and palaeoclimatic evolution of the Po Plain from 150-ky core records. *Global Planet. Change*, **40**, 55–78.

Arche, A. and López-Gómez, J. (1999) Subsidence rates and fluvial architecture of rift-related Permian and Triassic alluvial sediments of the southeast Iberian Range, eastern Spain. In: *Fluvial Sedimentology VI* (Eds N.D. Smith and J. Rogers). *Spec. Publ. Int. Assoc. Sedimentol.*, **28**, 283–304.

Bard, E., Hamelin, B. and Fairbanks, R.G. (1990) U–Th ages obtained by mass spectrometry in corals from Barbados: sea level during the past 130 000 years. *Nature*, **346**, 456–458.

Bhattacharya, J.P. (1993) The expression and interpretation of marine flooding surfaces and erosional surfaces in core; examples from the Upper Cretaceous Dunvegan Formation, Alberta: core examples. In: *Sequence Stratigraphy and Facies Associations* (Eds H.W. Posamentier, C.P. Summerhayes, B.U. Haq and G.P. Allen). *Spec. Publ. Int. Assoc. Sedimentol.*, **18**, 125–160.

Bhattacharya, J.P. and Willis, B.J. (2001) Lowstand deltas in the Frontier Formation, Powder River basin, Wyoming: Implications for sequence stratigraphic models. *Am. Assoc. Petrol. Geol. Bull.*, **85**, 261–294.

Blum, M.D. (1993) Genesis and architecture of incised valley fill sequences: a Late Quaternary example from the Colorado River, Gulf Coastal Plain of Texas. In: *Siliciclastic Sequence Stratigraphy: Recent Developments and Applications* (Eds P. Weimer and H.W. Posamentier). *Am. Assoc. Petrol. Geol. Mem.*, **58**, 259–283.

Blum, M.D. and Price, D.M. (1998) Quaternary alluvial plain construction in response to glacio-eustatic and climatic controls, Texas Gulf Coastal Plain. In: *Relative Role of Eustasy, Climate, and Tectonism in Continental Rocks* (Eds K.W. Shanley and P.J. McCabe). *Soc. Econ. Paleontol. Mineral. Spec. Publ.*, **59**, 31–48.

Blum, M.D. and Törnqvist, T.E. (2000) Fluvial response to climate and sea-level change: a review and look forward. *Sedimentology*, **47** (Suppl. 1), 2–48.

Blum, M.D., Toomey, R.S. III and Valastro, S. (1994) Fluvial response to Late Quaternary climatic and environmental change, Edwards Plateau, Texas. *Palaeogeogr., Palaeoclimatol., Palaeoecol.*, **108**, 1–21.

Bondesan, M., Favero, V. and Viñals, M.J. (1995) New evidence on the evolution of the Po-delta coastal plain during the Holocene. *Quat. Int.*, **29/30**, 105–110.

Burns, B.A., Heller, P.L., Marzo, M. and Paola, C. (1997) Fluvial response in a sequence stratigraphic framework: example from the Montserrat fan delta, Spain. *J. Sediment. Res.*, **67**, 311–321.

Cant, D.J. (1998) Sequence stratigraphy, subsidence rates, and alluvial facies, Mannville Group, Alberta foreland basin. In: *Relative Role of Eustasy, Climate, and Tectonism in Continental Rocks* (Eds K.W. Shanley and P.J. McCabe). *Soc. Econ. Paleontol. Mineral. Spec. Publ.*, **59**, 49–63.

Catuneanu, O. and Elango, H.N. (2001) Tectonic control on fluvial styles: the Balfour Formation of the Karoo Basin, South Africa. *Sediment. Geol.*, **140**, 291 313.

Centineo, M.C. (2001) Stratigrafia ed evoluzione paleogeografica dei depositi tardoquaternari della pianura ferrarese. PhD thesis, University of Bologna.

Crews, S.G. and Ethridge, F.G. (1993) Laramide tectonics and humid alluvial fan sedimentation, NE Uinta Uplift, Utah and Wyoming. *J. Sediment. Petrol.*, **63**, 420–436.

Cross, T.A. (1988) Controls on coal distribution in transgressive–regressive cycles, Upper Cretaceous, Western Interior, U.S.A. In: *Sea Level Changes: an Integrated Approach* (Eds C.K. Wilgus, B.S. Hastings, C.G.St.C. Kendall, H.W. Posamentier, C.A. Ross and J.C. Van Wagoner). *Soc. Econ. Paleontol. Mineral. Spec. Publ.*, **42**, 371–380.

Dalrymple, R.W., Zaitlin, B.A. and Boyd, R. (1992) Estuarine facies models: conceptual basis and stratigraphic implications. *J. Sediment. Petrol.*, **62**, 1130–1146.

De Beaulieu, J.J. and Reille, M. (1992) The last climatic cycle at La Grand Pile (Vosges, France). A new pollen profile. *Quat. Sci. Rev.*, **11**, 431–438.

Demarest, II J.M. and Kraft, J.C. (1987) Stratigraphic record of Quaternary sea levels: implications for more ancient strata. In: *Sea-level Fluctuation and Coastal Evolution* (Eds D. Nummedal, O.H. Pilkey and J.D. Howard). *Soc. Econ. Paleontol. Mineral. Spec. Publ.*, **41**, 223–239.

Dreyer, T. (1993) Quantified fluvial architecture in ephemeral stream deposits of the Esplugafreda Formation (Palaeocene), Tremp-Graus Basin, northern Spain. In: *Alluvial Sedimentation* (Eds M. Marzo and C. Puigdefabregas). *Spec. Publ. Int. Assoc. Sedimentol.*, **17**, 337–362.

Embry, A.F. (1993) Transgressive–regressive (T–R) sequence analysis of the Jurassic succession of the Sverdrup Basin, Canadian Arctic Archipelago. *Can. J. Earth. Sci.*, **30**, 301–320.

Embry, A.F. (1995) Sequence boundaries and sequence hierarchies: problems and proposals. In: *Sequence Stratigraphy on the Northwest European Margin* (Eds R.J. Steel, V.L. Felt, E.P. Johannessen and C. Mathieu). *Spec. Publ. Nor. Petrol. Soc.*, **5**, 1–11.

Frazier, D.E. (1974) Depositional episodes: their relationship to the Quaternary stratigraphic framework in the northwestern portion of the Gulf Basin. *Univ. Texas Austin, Bur. Econ. Geol., Geol. Circ.*, **4**, 1–28.

Galloway, W.E. (1989) Genetic stratigraphic sequences in basin analysis I: architecture and genesis of flooding-surface bounded depositional units. *Am. Assoc. Petrol. Geol. Bull.*, **73**, 125–142.

Geological Survey of Italy (in press) *Geological Map of Italy, Sheet 256—Rimini*. Geological Survey of Italy—Regione Emilia-Romagna.

Hamilton, D.S. and Tadros, N.Z. (1994) Utility of coal seams as genetic stratigraphic sequence boundaries in non-marine basins: an example from the Gunnedah basin, Australia. *Am. Assoc. Petrol. Geol. Bull.*, **78**, 267–286.

Helland-Hansen, W. and Gjelberg, J.G. (1994) Conceptual basis and variability in sequence stratigraphy: a different perspective. *Sediment. Geol.*, **92**, 31–52.

Helland-Hansen, W. and Martinsen, O.J. (1996) Shoreline trajectories and sequences: description of variable depositional-dip scenarios. *J. Sediment. Res.*, **66**, 670–688.

Heller, P.L. and Paola, C. (1996) Downstream changes in alluvial architecture: an exploration of controls on channel-stacking patterns. *J. Sediment. Res.*, **66**, 297–306.

Holbrook, J.M. (1996) Complex fluvial response to low gradients at maximum regression: a genetic link between smooth sequence-boundary morphology and architecture of overlying sheet sandstone. *J. Sediment. Res.*, **66**, 713–722.

Hunt, D. and Tucker, M.E. (1992) Stranded parasequences and the forced regressive wedge systems tract: deposition during base-level fall. *Sediment. Geol.*, **81**, 1–9.

Hunt, D. and Tucker, M.E. (1995) Stranded parasequences and the forced regressive wedge systems tract: deposition during base-level fall–Reply. *Sediment. Geol.*, **95**, 147–160.

Imbrie, J., Hays, J.D., Martinson, D.G., McIntyre, A., Mix, A.C., Morley, J.J., Pisias, N.J., Prell, W.I. and Shackleton, N.J. (1984) The orbital theory of Pleistocene climate: support from a revised chronology of the marine ^{18}O record. In: *Milankovitch and Climate* (Eds A. Berger, J. Imbrie, J, Hays, G. Kukla and B. Saltsman), pp. 269–306. Reidel, Dordrecht.

Johnson, J.G., Klapper, G. and Sandberg, C.A. (1985) Devonian eustatic fluctuations in Euramerica. *Geol. Soc. Am. Bull.*, **96**, 567–587.

Kolla, V., Posamentier, H.W. and Eichenseer, H. (1995) Stranded parasequences and the forced regressive wedge systems tract: deposition during base-level fall–discussion. *Sediment. Geol.*, **95**, 139–145.

Leeder, M.R. (1978) A quantitative stratigraphic model for alluvium, with special reference to channel deposit density and interconnectedness. In: *Fluvial Sedimentology* (Ed. A.D. Miall). *Can. Soc. Petrol. Geol. Mem.*, **5**, 587–596.

Leeder, M.R. (1993) Tectonic controls upon drainage basin development, river channel migration and alluvial architecture: implications for hydrocarbon reservoir development and characterization. In: *Characterization of Fluvial and Aeolian Hydrocarbon Reservoirs* (Eds C.P. North and D.J. Prosser). *Spec. Publ. Geol. Soc. Lond.*, **73**, 7–22.

Leeder, M.R. and Stewart, M.D. (1996) Fluvial incision and sequence stratigraphy: alluvial responses to relative sea-level fall and their detection in the geological record. In: *Sequence Stratigraphy in British Geology* (Eds S.P. Hesselbo and D.N. Parkinson). *Spec. Publ. Geol. Soc. Lond.*, **103**, 25–39.

Legarreta, L. and Uliana, M.A. (1998) Anatomy of hinterland depositional sequences: Upper Cretaceous fluvial strata, Neuquen Basin, West-Central Argentina. In: *Relative Role of Eustasy, Climate, and Tectonism in Continental Rocks* (Eds K.W. Shanley and P.J. McCabe). *Soc. Econ. Paleontol. Mineral. Spec. Publ.*, **59**, 83–92.

Mackey, S.D. and Bridge, J.S. (1995) Three-dimensional model of alluvial stratigraphy: Theory and application. *J. Sediment. Res.*, **B65**, 7–31.

Marchesini, L., Amorosi, A., Cibin, U., Zuffa, G.G., Spadafora, E. and Preti, D. (2000) Sand composition and sedimentary evolution of a Late Quaternary depositional sequence, northwestern Adriatic Coast, Italy. *J. Sediment. Res.*, **70**, 829–838.

Marriott, S.B. (1999) The use of models in the interpretation of the effects of base-level change on alluvial architecture. In: *Fluvial Sedimentology VI* (Eds N.D. Smith and J. Rogers). *Spec. Publ. Int. Assoc. Sedimentol.*, **28**, 271–281.

Martinsen, O.J. and Helland-Hansen, W. (1994). Sequence stratigraphy and facies model of an incised valley fill: the Gironde Estuary, France–discussion. *J. Sediment. Res.*, **B64**, 78–80.

Martinsen, O.J., Martinsen, R.S. and Steidtman, J.R. (1993) Mesaverde Group (Upper Cretaceous). southeastern Wyoming: Allostratigraphy versus sequence stratigraphy in a tectonically active area. *Am. Assoc. Petrol. Geol. Bull.*, **77**, 1351–1373.

Martinsen, O.J., Ryseth, A., Helland-Hansen, W., Flesche, H., Torkildsen, G. and Idil, S. (1999) Stratigraphic base level and fluvial architecture: Ericson Sandstone (Campanian), Rock Sorings Uplift, SW Wyoming, USA. *Sedimentology*, **46**, 235–259.

McCarthy, P.J. and Plint, A.G. (1998) Recognition of interfluve sequence boundaries: integrating paleopedology and sequence stratigraphy. *Geology*, **26**, 387–390.

Miall, A.D. (1991) Stratigraphic sequences and their chronostratigraphic correlation. *J. Sediment. Petrol.*, **61**, 497–505.

Miall, A.D. and Arush, M. (2001). Cryptic sequence boundaries in braided fluvial successions. *Sedimentology*, **48**, 971–985.

Milana, J.P. (1998) Sequence stratigraphy in alluvial settings: a flume-based model with applications to outcrop and seismic data. *Am. Assoc. Petrol. Geol. Bull.*, **82**, 1736–1753.

Morton, R.A. and Suter, J.R. (1996) Sequence stratigraphy and composition of Late Quaternary shelf-margin deltas, Northern Gulf of Mexico. *Am. Assoc. Petrol. Geol. Bull.*, **80**, 505–530.

Nichol, S.L., Boyd, R. and Penland, S. (1994) Stratigraphic response of wave-dominated estuaries to different relative sea-level and sediment supply histories: Quaternary case studies from Nova Scotia, Louisiana and Eastern Australia. In: *Incised-Valley Systems: Origin and Sedimentary Sequences* (Eds W. Dalrymple, R. Boyd and B.A. Zaitlin). *Soc. Econ. Paleontol. Mineral. Spec. Publ.*, **51**, 265–283.

Nummedal, D. and Molenaar, C.M. (1995) Sequence stratigraphy of the Gallup Sandstone. In: *Sequence Stratigraphy of Foreland Basin Deposits* (Eds J.C. Van Wagoner and G.T. Bertram). *Am. Assoc. Petrol. Geol. Mem.*, **64**, 277–310.

Nummedal, D. and Swift, D.J.P. (1987) Transgressive stratigraphy at sequence-bounding unconformities: some principles derived from Holocene and

Cretaceous examples. In: *Sea-level Fluctuation and Coastal Evolution* (Eds D. Nummedal, O.H. Pilkey and J.D. Howard). *Soc. Econ. Paleontol. Mineral. Spec. Publ.*, **41**, 241–260.

Nummedal, D., Riley, G.W. and Templet, P.L. (1993) High-resolution sequence architecture: a chronostratigraphic model based on equilibrium profile studies. In: *Sequence Stratigraphy and Facies Associations* (Eds H.W. Posamentier, C.P. Summerhayes, B.U. Haq and G.P. Allen). *Spec. Publ. Int. Assoc. Sedimentol.*, **18**, 55–68.

O'Byrne, C.J. and Flint, S.S. (1996) Interfluve sequence boundaries in the Grassy Member, Book Cliffs, Utah: criteria for recognition and implications for subsurface correlation. In: *High Resolution Sequence Stratigraphy: Innovations and Applications* (Eds J.A. Howell and J.F. Aitken). *Spec. Publ. Geol. Soc. Lond.*, **104**, 208–220.

Olsen, T., Steel, R., Hogseth, K., Skar, T. and Roe, S.-L. (1995) Sequential architecture in a fluvial succession: sequence stratigraphy in the Upper Cretaceous Mesaverde Group, Price Canyon, Utah. *J. Sediment. Res.*, **B65**, 265–280.

Pirazzoli, P.A. (1993) Global sea-level changes and their measurement. *Global Planet. Change*, **8**, 135–148.

Plint, A.G. and Nummedal, D. (2000) The falling stage systems tract: recognition and importance in sequence stratigraphic analysis. In: *Sedimentary Responses to Forced Regressions* (Eds R.L. Gawthorpe and D. Hunt). *Spec. Publ. Geol. Soc. Lond.*, **172**, 1–17.

Plint, A.G., McCarthy, P.J. and Faccini, U.F. (2001) Nonmarine sequence stratigraphy: Updip expression of sequence boundaries and systems tracts in a high-resolution framework, Cenomanian Dunvegan Formation, Alberta foreland basin, Canada. *Am. Assoc. Petrol. Geol. Bull.*, **85**, 1967–2001.

Posamentier, H.W. (2001) Lowstand alluvial bypass systems: incised vs. unincised. *Am. Assoc. Petrol. Geol. Bull.*, **85**, 1771–1793.

Posamentier, H.W. and Allen, G.P. (1993) Siliciclastic sequence stratigraphic patterns in foreland ramp-type basins. *Geology*, **21**, 455–458.

Posamentier, H.W. and Allen, G.P. (Eds) (1999) Siliciclastic sequence stratigraphy—concepts and applications. *Soc. Econ. Paleontol. Mineral. Concepts Sedimentol. Paleontol.*, **7**.

Posamentier, H.W. and Vail, P.R. (1988) Eustatic controls on clastic deposition II—Sequence and systems tract models. In: *Sea Level Changes: an Integrated Approach* (Eds C.K. Wilgus, B.S. Hastings, C.G.St.C. Kendall, H.W. Posamentier, C.A. Ross and J.C. Van Wagoner). *Soc. Econ. Paleontol. Mineral. Spec. Publ.*, **42**, 125–154.

Posamentier, H.W., Jervey, M.T. and Vail, P.R. (1988) Eustatic controls on clastic deposition I: Conceptual framework. In: *Sea Level Changes: An Integrated Approach* (Eds C.K. Wilgus, B.S. Hastings, C.G.St.C. Kendall, H.W. Posamentier, C.A. Ross and J.C. Van Wagoner). *Soc. Econ. Paleontol. Mineral. Spec. Publ.*, **42**, 109–124.

Posamentier, H.W., Allen, G.P., James, D.P. and Tesson, M. (1992) Forced regressions in a sequence stratigraphic framework: concepts, examples and sequence stratigraphic significance. *Am. Assoc. Petrol. Geol. Bull.*, **76**, 1687–1709.

Regione Emilia-Romagna and ENI-AGIP (1998) *Riserve idriche sotterranee della Regione Emilia-Romagna*. S.EL.CA., Firenze.

Rizzini, A. (1974) Holocene sedimentary cycle and heavy mineral distribution, Romagna-Marche coastal plain, Italy. *Sediment. Geol.*, **11**, 17–37.

Schumm, S.A. (1993) River response to baselevel change: implications for sequence stratigraphy. *J. Geol.*, **101**, 279–294.

Shanley, K.W. and McCabe, P.J. (1991) Predicting facies architecture through sequence stratigraphy—an example from the Kaiparowits Plateau, Utah. *Geology*, **19**, 742–745.

Shanley, K.W. and McCabe, P.J. (1993) Alluvial architecture in a sequence stratigraphic framework: a case history from the Upper Cretaceous of southern Utah, USA. In: *The Geological Modelling of Hydrocarbon Reservoirs and Outcrop Analogues* (Eds S.S. Flint and I.D. Bryant). *Spec. Publ. Int. Assoc. Sedimentol.*, **15**, 21–56.

Shanley, K.W. and McCabe, P.J. (1994) Perspectives on the sequence stratigraphy of continental strata. *Am. Assoc. Petrol. Geol. Bull.*, **78**, 544–568.

Shanley, K.W., McCabe, P.J. and Hettinger, R.D. (1992) Significance of tidal influence in fluvial deposits for interpreting sequence stratigraphy. *Sedimentology*, **39**, 905–930.

Swift, D.J.P. (1968) Coastal erosion and transgressive stratigraphy. *J. Geol.*, **76**, 444–456.

Törnqvist, T.E. (1994) Middle and late Holocene avulsion history of the River Rhine (Rhine–Meuse Delta, Netherlands). *Geology*, **22**, 711–714.

Törnqvist, T.E., Wallinga, J., Murray, A.S., de Wolf, H., Cleveringa, P., de Gans, W. (2000) Response of the Rhine–Meuse system (west-central Netherlands) to the last Quaternary glacio-eustatic cycles: a first assessment. *Global Planet. Change*, **27**, 89–111.

Trincardi, F., Correggiari, A. and Roveri, M. (1994) Late Quaternary transgressive erosion and deposition in a modern epicontinental shelf: the Adriatic Semi-enclosed Basin. *Geo-Marine Letters*, **14**, 41–51.

Tzedakis, P.C., Andrieu, V., de Beaulieu, J.-L., Crowhurst, S., Follieri, M., Hooghiemstra, H., Magri, D., Reille, M., Sadori, L., Shackleton, N.J. and Wijmstra, T.A. (1997) Comparison of terrestrial and marine records of changing climate of the last 500 000 years. *Earth Planet. Sci. Lett.*, **150**, 171–176.

Vail, P.R., Mitchum, R.M., Jr. and Thompson, S., III (1977) Seismic stratigraphy and global changes of sea level, part 3: relative changes of sea level from coastal onlap. In: *Seismic Stratigraphy—Application to Hydrocarbon Exploration* (Ed. C.E. Payton). *Am. Assoc. Petrol. Geol. Mem.*, **26**, 63–81.

Van Wagoner, J.C. (1995) Sequence stratigraphy and marine to non-marine facies architecture of foreland basin strata, Book Cliffs, Utah, U.S.A. In: *Sequence Stratigraphy of Foreland Basin Deposits* (Eds J.C. Van Wagoner and G.T. Bertram). *Am. Assoc. Petrol. Geol. Mem.*, **64**, 137–223.

Van Wagoner, J.C., Mitchum, R.M., Campion, K.M. and Rahmanian, V.D. (Eds) (1990) Siliciclastic sequence stratigraphy in well logs, cores, and outcrops: concepts for high-resolution correlation of time and facies. *Am. Assoc. Petrol. Geol. Methods Explor. Ser.*, **7**.

Wehr, F.L. (1993) Effects of variations in subsidence and sediment supply on parasequence stacking patterns. In: *Siliciclastic Sequence Stratigraphy: Recent Developments and Applications* (Eds P. Weimer and H.W. Posamentier). *Am. Assoc. Petrol. Geol. Mem.*, **58**, 369–378.

Weimer, P. (1984) Relation of unconformities, tectonics, and sea-level changes, Cretaceous of Western Interior, U.S.A. In: *Interregional Unconformities and Hydrocarbon Accumulation* (Ed. J.S. Schlee). *Am. Assoc. Petrol. Geol. Mem.*, **36**, 7–35.

Woillard, G.M. (1978) Grand Pile peat bog: a continuous pollen record for the last 140 000 years. *Quat. Res.*, **9**, 1–21.

Wright, V.P. and Marriott, S.B. (1993) The sequence stratigraphy of fluvial depositional systems: the role of floodplain sediment storage. *Sediment. Geol.*, **86**, 203–210.

Depositional processes in latest Pleistocene and Holocene ephemeral streams of the Main Ethiopian Rift (Ethiopia)

M. BENVENUTI*, S. CARNICELLI†, G. FERRARI† and M. SAGRI*
*Dipartimento di Scienze della Terra, University of Florence, Via G. La Pira 4, 50121 Firenze, Italy
(Email: marcob@geo.unifi.it); and
†Dipartimento di Scienza del Suolo e Nutrizione della Pianta, University of Florence,
P.le delle Cascine 18, 50144 Firenze, Italy

ABSTRACT

Discontinuous ephemeral streams were described originally from the south-west USA as consisting of alternating reaches of incised rectangular channels and aggradational channel-fan systems. Modern ephemeral streams on the flanks of the Gademota Ridge, in the Lake Region of the Main Ethiopian Rift, show a similar pattern, and deposits exposed in the walls of modern stream valleys record the development of these systems since the latest Pleistocene. Streamflow, sheetflow and massflow deposits are recognized to be the basic sedimentary facies in palaeovalley fills, and within palaeochannel–fan systems. These deposits are stacked in a cyclic pattern within multiple and single valley fills, with basal streamflow deposits overlain by hyperconcentrated sheetflow and eventually by debris-flow deposits. Debris-fall and debris-flow deposits related to bank failure occur interbedded with streamflow and sheetflow facies.

A model for the cyclic incision and aggradation of the latest Pleistocene–Holocene ephemeral streams in this area is proposed based on incision during transition to a more moist climate, followed by aggradation during two main genetic stages that contrast in discharge volumes. During the early stage, relatively high and continuous seasonal discharge transports coarse-grained material, and covers the incised channel base with gravelly and sandy dune-scale bedforms. Bank undercutting and piping produce instability of the channel margins, with accumulation of relatively thick aprons of fine-grained material. The channel fan system is supplied with pebbly–sandy material flushed from the valleys. During the late stage, a decreased volume of flashy discharge redistributes sediment stored in the bank aprons within the valleys and to the next downstream channel–fan system through hyperconcentrated sheetflows and debris flows. Spreading of debris flows and sheetflows on interfluves marks the overfilling of the valleys.

INTRODUCTION

Erosion and aggradation in dryland fluvial and slope systems are highly dynamic processes over a variety of spatial and temporal scales. Particularly prone to rapid geomorphological change in drylands are the so-called discontinuous ephemeral streams (DES; Bull, 1997). First defined in the context of the *arroyos* in the south-west USA, and now considered to be a general category of

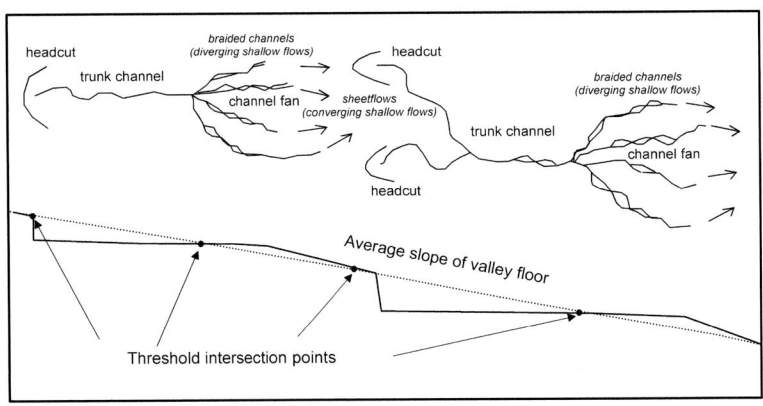

Fig. 1 Features of discontinuous ephemeral streams (DES) in plan view (top) and cross-section (bottom) (after Bull, 1997).

dryland rivers, DES are characterized by discharge that is intermittent during the year or, in particularly arid settings, over longer time periods. Discontinuous ephemeral streams' planforms are typified by the longitudinal recurrence of headcuts that define the upstream boundary to entrenched, rectangular trunk valley reaches. These headcuts are separated from each other by depositional zones where flushed sediment has accumulated as fan-shaped lobes (Fig. 1). The resulting stream has a markedly discontinuous and 'stepped' longitudinal profile, similar to those of gullies. Discontinuous ephemeral streams have, in fact, often been ascribed to gully erosion (Schumm & Hadley, 1957), although others have suggested DES are distinct because of valley geometry (broadly rectangular versus V- or U-shaped gullies; Graf, 1988) or because of their greater size and longer persistence in the landscape (Bull, 1997).

A peculiar feature of these streams is their fluctuation, both in space and time, between two end-member states of disequilibrium, incision and aggradation. In the early development of the *arroyo* concept in North America, cyclic incision and aggradation were considered to have started at the end of the nineteenth century when intensive land exploitation by European settlers began in the south-west USA. Detailed stratigraphical analyses and radiocarbon dating of valley floor alluvium in the same area, however, have subsequently demonstrated that multiple episodes of arroyo incision and aggradation have taken place since the late Pleistocene (Haynes, 1968; Waters & Haynes, 2001). Much attention has focused on the causes of alternating incision and aggradation, with explanations ranging from purely autocyclic mechanisms (e.g. the *semi-arid cycle of erosion* of Schumm & Hadley, 1957; Patton & Schumm, 1981), to interactions between climatic and anthropogenic forcing (as discussed in Graf, 1988; Bull, 1997; Waters & Haynes, 2001).

A common theme among DES valley fills of different ages is a fining upward succession of sediments, including boulder to pebble gravel at the base, overlain by sands and silts that normally form the bulk of fills that are from metres to tens of metres in thickness. Buried soils are commonly found in these successions as well. Beyond this level of generalization, and by comparison with the attention devoted to stratigraphical issues, few detailed sedimentological descriptions of valley-fill deposits have been produced. This paper aims to:

1 provide examples of modern DES in a dryland setting located in the Lake Region of the Main Ethiopian Rift (MER), with special emphasis on depositional processes within the trunk channel reaches;
2 describe the stratigraphical framework of, and sedimentary facies within, latest Pleistocene–Holocene valley fills;
3 suggest a depositional model for the Late Quaternary ephemeral streams of the study area.

GENERAL SETTING

Geological outline

Located in the northern portion of the East African Rift System (Fig. 2A), the Main Ethiopian Rift is a

Plate 1 Photographs of key depositional environments and facies described in text. (A) The bouldery channel bed in the rectilinear low-sinuosity reach of the stream in the Kedida area. Person in the upper left corner for scale. (B) Linear bouldery bedform (boulder berm), persons for scale, with indication of the stratigraphical units in the right bank of the stream. (C) Sinuous channel reach with a small point bar during the dry spell: thick arrow indicates the current direction; dotted arrow indicates a chute channel. (D) Detailed view of the point bar in (C). (E) Material accumulated at the base of the 9-m-high right bank; (F) The subrecent higher terraced valley fill deposits; the rod is 1 m long.

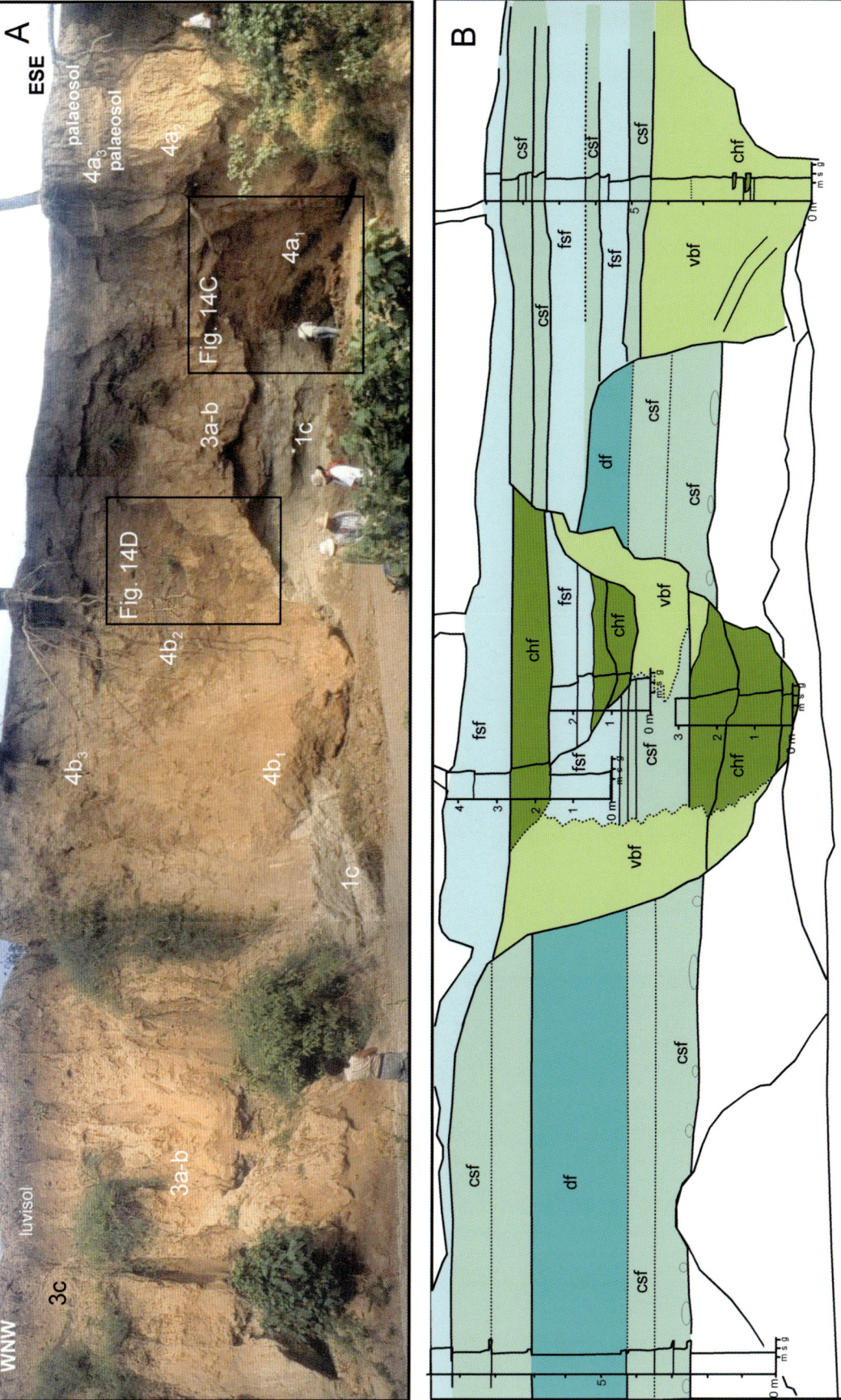

Plate 2 (A) Photomosaic and (B) line drawing of section G2b–3 (Fig. 4) to illustrate the stratigraphical architecture and distribution of facies associations. Unit 3 is interpreted to represent a channel-fan system. Note the different orientations of the various fluvial units with respect to their original flow directions: parallel to palaeoflow (to the right) for unit 3, transverse to palaeoflow (to the right) for subunit 4a (only the left valley side is exposed) and orthogonal to palaeoflow (out of the plane) for subunit 4b.

Plate 3 Photographs of key facies in stratigraphical context, as described in the text: (A) Angular pebbles and coarse sands at the base of subunit 4a in section G4 (Fig. 10): the whitish clasts (from subunit 1c) in the upper part indicate a debris fall deposit interbedded with the channel-fill gravels; hammer for scale. (B) Close-up view of the coarse-grained sheetflow deposits in subunit 4a (section G4); pencil is about 13 cm long. (C) Debris fall deposits at the base of subunit 4a in section G2b–3 (Plate 2); the rod is 1 m long. (D) Debris fall and channel-fill deposits in the mid-part of subunit 4b, section G2b–3: arrows indicate tension cracks on the right bank of the palaeovalley. (E) Matrix-supported gravels deposited from cohesive debris flows along the left bank of the palaeovalley filled by subunit 4a in section G4; pencil is about 13 cm long. Dotted lines indicate inclined beds. (F) Cohesive debris-flow deposits at the base of unit 3 resting on a pebble lag; pen is 12 cm long. (G) A large clast of subunit 1c at the base of a debris-flow deposit in unit 3, section G4; the pencil is 13 cm long.

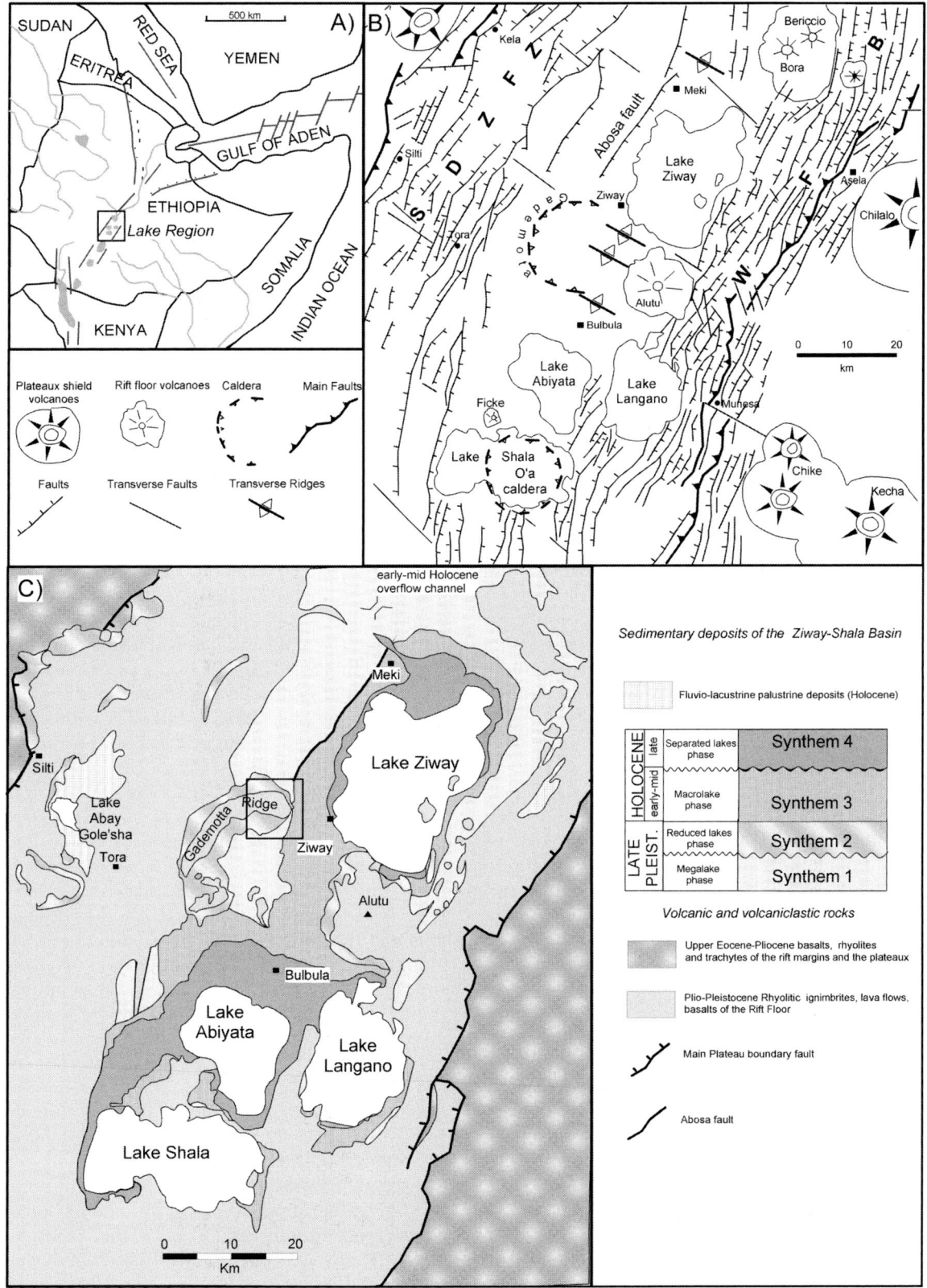

Fig. 2 Maps of study area. (A) Location of the Lake Region within the Main Ethiopian Rift (MER). (B) Tectonic and volcanic features in the central MER. (C) Schematic geological map of the Lake Region and location of the study area (rectangle) on the eastern end of Gademota Ridge.

NNE–SSW trending rift valley, 80 km wide and 700 km long (Mohr, 1962). The MER is bordered by the Ethiopian Plateau to the west and the Somali Plateau to the east. A dense network of NNE–SSW trending normal faults on the western and the eastern margins (Fig. 2B) accommodates total offsets of about 1500–2000 m between the rift floor and surrounding plateaux (Di Paola, 1972; Woldegabriel et al., 1990; Chorowicz et al., 1994): still active are the Silti-Debre Zeit Fault Zone (SDZFZ) to the west and the Wonji Fault Belt (WFB) to the east (Fig. 2B). In some places, northwest trending transverse faults truncate the Rift-orientated faults along the plateau margins (Di Paola, 1972; Woldegabriel et al., 1990).

From a lithological and stratigraphical point-of-view, the MER is dominated by thick volcanogenic successions that range in age from early Tertiary to recent (Fig. 2C). Rift walls, shoulders and surrounding plateaux are comprised of lower Eocene to Pliocene flood basalts (Trap Series) with interbedded ignimbrites, which are unconformably capped by massive rhyolites, tuffs and basalts (Di Paola, 1972; Merla et al., 1979; Woldegabriel et al., 1990; Di Paola et al., 1993). These rocks formed before rift initiation (Trap Series) and during subsidence associated with early-stage asymmetric rifting. Most of the rift floor consists of lower to middle Pliocene peralkaline rhyolitic ignimbrites, basalts and tuffs, erupted contemporaneously with development of a full symmetrical rift. Upper Pliocene to Middle Pleistocene alkaline and peralkaline rhyolites and domes, erupted from large calderas, also form the rift floor. The Gademota Ridge (Fig. 2B, C), an arc structure that rises up to 400 m above the plain west of Lake Ziway, is the remnant of an Early Pleistocene caldera 25–30 km in diameter (Laury & Albritton, 1975; Mohr et al., 1980; Woldegabriel et al., 1990). Middle Pleistocene to Holocene rhyolites and volcaniclastics originated from volcanoes and calderas that include the Bora-Bericcio complex, the Alutu volcano, and the Ficke, O'a and Corbetti calderas. Upper Pleistocene to Holocene basalts and scoria cones in the SDZFZ and the WFB record the latest volcanic activity in the MER (Di Paola, 1972; Kazmin et al., 1980).

Sedimentary rocks and deposits within the MER are subordinate to volcanics and volcaniclastics, and mainly consist of mid- to upper Quaternary fluvio-lacustrine and colluvial deposits that cover the central part of the rift floor (Street, 1979; Le Turdu et al., 1999; Benvenuti et al., 2002). These deposits rest on the Lower–Middle Pleistocene volcanic products of the Gademota caldera, and alluvial–colluvial and volcaniclastic materials accumulated on its outer and inner flanks after the caldera deactivation (Laury & Albritton, 1975). The stratigraphical and facies-scale characteristics of these units, especially around Gademota Ridge, are the primary focus of this paper and are described in a later section.

Hydrology and climate

The Lake Region in the central sector of the MER (Fig. 2) is characterized by the presence of four major lakes, which constitute the terminal part of a closed watershed (Chernet, 1982). From north to south, these are the shallow (7–40 m deep) Lakes Ziway, Langano and Abiyata, and the 250 m deep Lake Shala. With the exception of freshwater Lake Ziway, which is fed directly by the perennial Meki and the Katar Rivers flowing from the Ethiopian highlands, the other lakes are alkaline with salinity increasing to the south from Lakes Abiyata and Langano to Lake Shala.

The drainage network within the MER itself is dominated by short ephemeral streams, active only during the two rainy seasons, the so-called 'little rains' from March to May, and the 'big rains' from June to September. A marked rainfall–temperature gradient exists between the rift margins and the rift floor. The mean annual rainfall along the rift margins is about 1200 mm, whereas the rift floor is much drier, with average annual rainfall of c. 700 mm and mean annual temperatures of 18°C at Ziway. Overall, the study area belongs to the tropical wet and dry class of the Köppen climate classification, or a semi-arid dryland according to Thornthwaite's climatic index-based classification (Meigs, 1953; Adams et al., 1979; Graf, 1988).

Seasonal contrasts in precipitation correspond to the annual fluctuation of the intertropical convergence zone (ITCZ) over the Horn of Africa. Northward migration of the ITCZ during the boreal spring produces the little rainy season, whereas the big rains coincide with the northernmost position of the ITCZ during the boreal summer, and reflects inland penetration of humid airstreams from the Congo basin and the Indian Ocean. The dry period, by contrast, reflects the southernmost shift of the ITCZ during the boreal winter, which coincides with an enhanced subtropical high to the north of Ethiopia and an influx of dry north-easterly winds from the Arabian Peninsula.

MODERN FLUVIAL SYSTEMS AND DEPOSITS

Modern streams that drain the flanks of Gademota Ridge are quite short, less than 8 km (Chirgulo) in length, with catchment areas of 0.5–8.5 km^2 (Fig. 3). Analysis of 1972 aerial photographs reveals that some stream profiles, especially in the Kile Harsema and Chirgulo areas, are characterized by marked discontinuities similar to those described by Bull (1997), with unentrenched channel fans alternating with entrenched trunk stream channel segments.

Field observations in some localities demonstrate that the unentrenched channel fans from 1972 are still active. Surficial deposits here consist of pebbly coarse to fine sands distributed in fan-shaped bodies ranging from a few metres to some tens of metres in radius, and which accumulate during the rainy seasons when sediments are flushed from entrenched channels farther upstream, and flow expands over the fan surface. Unfortunately, intense human pressure on land, including grazing, use of streams as trails and subsistence agriculture, commonly inhibits preservation of detailed morphological characteristics and internal sedimentary structures.

Entrenched trunk channel segments can be a few metres to some tens of metres in width, and up to 30 m deep, with near-vertical banks and low-relief to flat channel floors. Entrenched channel segments are characterized by reaches with rectilinear and/or low-sinuosity planforms

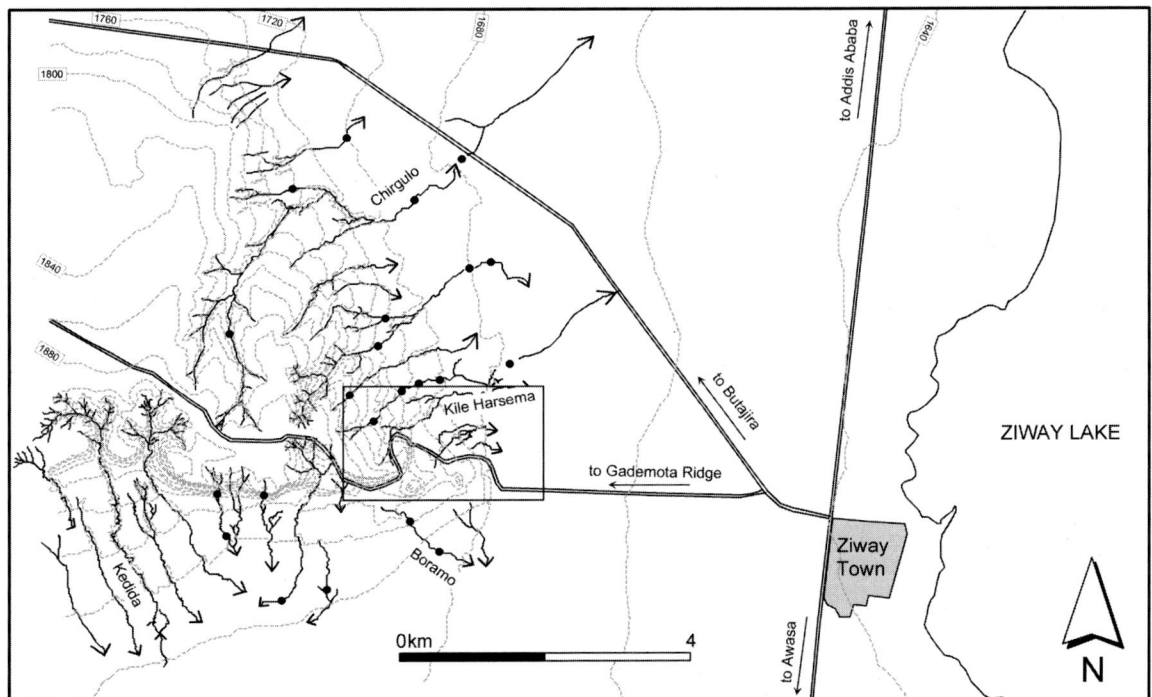

Fig. 3 Drainage network in the study area derived from aerial photographs taken during the 1972 dry spell. Large dots indicate the location of discontinuities between trunk channel reaches. Dashed lines are contours in metres above sea level. Rectangle indicates the location of Fig. 4.

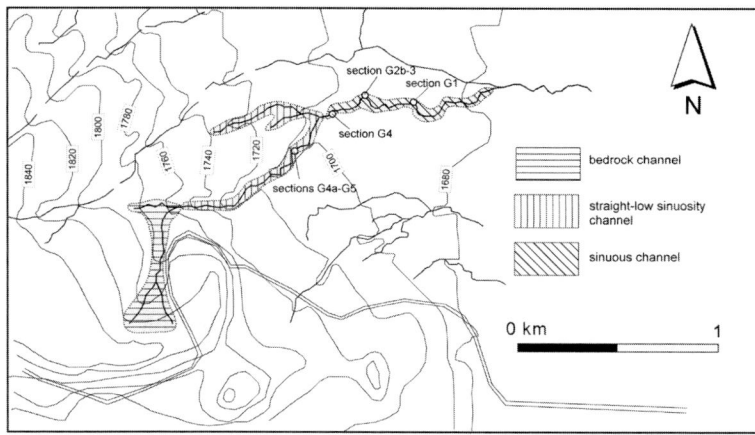

Fig. 4 Distribution of different channel reaches along an ephemeral stream valley in the Kile Harsema area, with location of stratigraphical sections discussed in the text.

that alternate with sinuous reaches, the latter frequently observed in the middle portion of the entrenched segment as a whole (Fig. 4). Characteristic bedforms and deposits vary with position within the entrenched reaches, as described below with examples from the Kile Harsema area.

Immediately downstream from the bedrock channel reach the bed of the low-sinuosity to straight entrenched channel is dominated by boulder- to cobble-size rhyolite clasts, with subordinate pebbles and coarse sand. Clasts are angular to poorly rounded, and are distributed both chaotically (Plate 1A) and in organized medium to small-scale transverse ribs and pebble clusters. Both of these small-scale bedforms are commonly recognized in gravel-bed rivers, not only in humid temperate settings but also in drylands (e.g. Koster, 1978; McDonald & Day, 1978; Allen, 1985; Billi, 1988; Reid & Frostick, 1994), and are thought to reflect low-sediment concentrations and high bed shear within the context of rapid fluctuations of flow regime during high-magnitude floods.

Zones of sudden channel expansion are characterized by bedforms up to 5 m long and 1 m high, orientated parallel-to-flow, and comprised of a framework of boulders, cobbles and plant debris (Plate 1B). These are similar to the boulder berms and related debris torrent deposits described by Scott & Gravlee (1968), Blair (1987) and Carling (1987), and are considered to result from sudden flow expansion during high-magnitude floods. The disorganized framework of these deposits is the result of rapid bedload and suspended load deposition, which Benvenuti & Martini (2001) argued occurs as an *en masse* frictional freezing, and produces sedimentological characteristics that are intermediate between streamflow and massflow deposits.

Further downstream, where the channel is more sinuous, small point bars dominate bed morphology (Plate 1C). At any location along the channel, point bars are characterized by a recurrent grain-size distribution at the surface (Plate 1D), with a boulder to pebble core, frequently vegetated by shrubs, that passes abruptly downstream to pebbly sands that in turn grade into medium sands toward the channel. Trenching of bars shows that sandy deposits are inclined transverse to the flow direction, indicating lateral accretion. Accretion of the coarse-grained bar cores most probably reflects the highest magnitude events, whereas lower magnitude events, or perhaps the waning stages of large floods, may winnow the bar core and accrete laterally and downstream through addition of sandy bedload. Chute channels, active only during flood events, frequently cross the bars as well. Although a precise age for the formation of the point bars is unknown, rainfall data recorded by the Ziway meteorological station from 1971 to 1997 (Fig. 5) indicate a rainfall peak from 1973 to 1976 (Sagri, 1998; Accetta, 1999) during the big rains period, which is more likely to trigger large floods, and a decreasing trend after that.

Terminal reaches of entrenched trunk channel segments are again rectangular in cross-section,

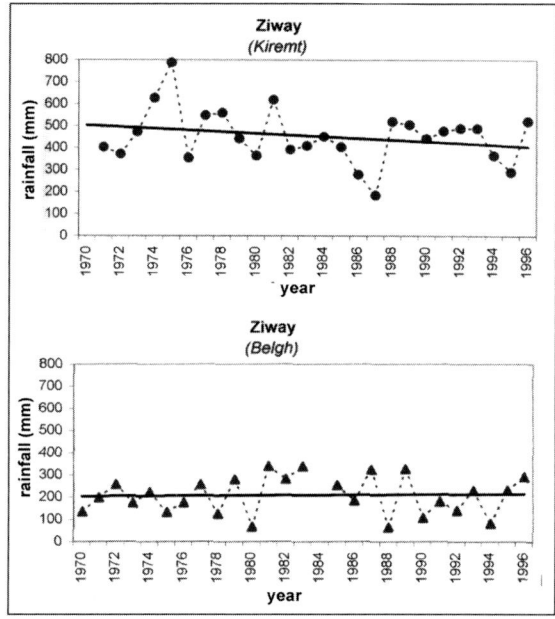

Fig. 5 Rainfall trends during the 'big rains' (top) and the 'little rains' (bottom) (respectively Kiremt and Belgh in amharic), as recorded by the Ziway meteorological station (after Sagri, 1998).

and low-sinuosity in plan-view. Beds are characterized by pebbly sands that, after rainy periods and when not disturbed by humans or cattle, occur as multiple lobes a few centimetres thick and up to tens of metres in length. This multilobe morphology suggests pulses, or kinematic waves, of sediment that have been flushed through the trunk channel.

UPPER QUATERNARY DEPOSITS AROUND THE GADEMOTA RIDGE

As noted above, Upper Quaternary fluvio-lacustrine and colluvial deposits cover the central part of the rift floor. The network of deeply entrenched ephemeral streams along the flanks of Gademota Ridge provides a series of exposures where these deposits can be fully examined (Fig. 6). A stratigraphical framework was developed through regional mapping in the area, as well as by measuring a series of sections located in the Kedida and Boramo areas to the south of the Gademota Ridge, and in the Chirgulo and Kile

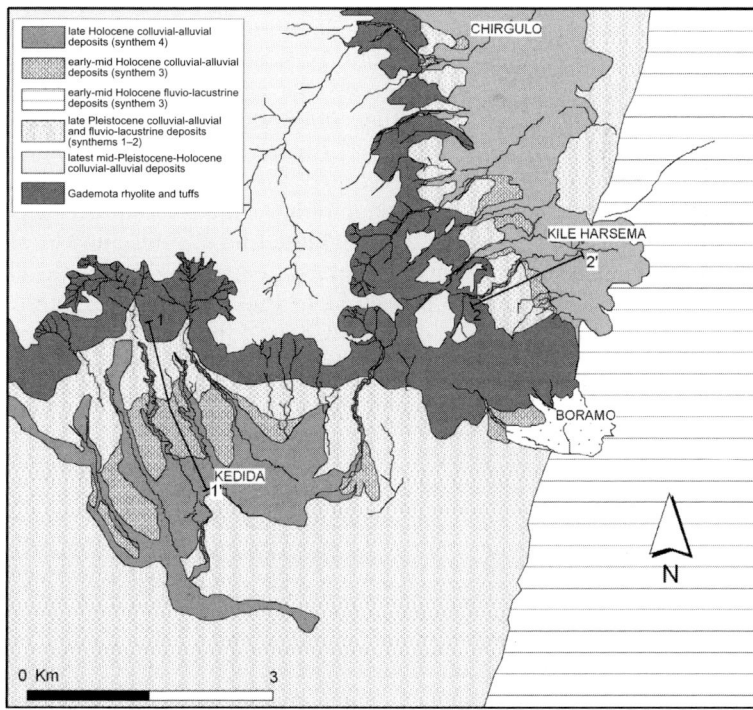

Fig. 6 Schematic geological map of the Gademota Ridge study area showing the location of sections 1-1' and 2-2' illustrated in Fig. 7.

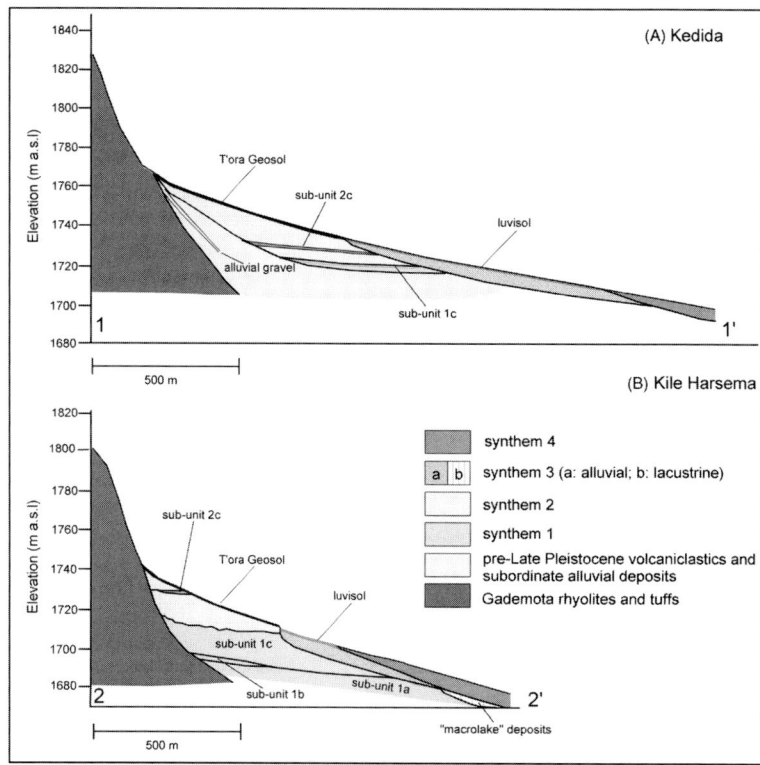

Fig. 7 Typical stratigraphical cross-sections on the (A) southern and (B) northern flanks of Gademota Ridge, illustrating relationships between synthems 1–4. For section locations see Fig. 6.

Harsema areas to the north. Upper Quaternary deposits have been subdivided into four major unconformity-bounded synthems (*sensu* Salvador, 1987; ISSC, 1994), with each synthem subdivided into distinct subunits. A chronological framework for these deposits has been developed from a series of ^{14}C ages, as well as stratigraphical and historical data (Benvenuti et al., 2002). This stratigraphical framework is summarized in Fig. 7 and Table 1. Lake-level data are presented in Fig. 8. The Upper Quaternary deposits are illustrated in Figs 9–13 and Plates 2 & 3, and discussed below.

Synthem 1

Synthem 1 consists of colluvial, fluvio-deltaic, and lacustrine gravel, sand and mud, as well as lacustrine diatomites and volcaniclastic material deposited during the last glacial period, between 100 and 22 ka. During this time, an extensive lacustrine system developed in the central part of the MER, which has been referred to as the Megalake phase (Benvenuti et al., 2002). This lake expanded during wetter interstadials, and contracted during the drier stadials.

Three subunits (1a–1c) bearing fluvial and lacustrine deposits have been differentiated at Boramo and Kile Harsema (Figs 6 & 7B), and bracketed to the period between 39 and 30 ka. At Kedida (Figs 6 & 7A) equivalent deposits are represented by alluvial–colluvial silts and lacustrine deposits at elevations of *c.* 1720 m a.s.l., which rest on a thick apron of pre-Last Glacial volcaniclastics. The lake incursion recorded at 1720 m a.s.l. is equated to the lake highstand recorded by subunit 1c in the Boramo and Kile Harsema areas.

Synthem 2

Synthem 2 is dominated by alluvial–colluvial and volcaniclastic deposits that accumulated in the basin during the last full glacial and late glacial, between 22 and 10 ka, when the lake system was, in general, contracted dramatically under very dry climatic conditions (Reduced Lake Phase, Benvenuti et al., 2002), as well as when intense

Table 1 Chronology of the Late Quaternary deposits on the flanks of the Gademota Ridge. Estimated age based on conventional (uncalibrated) ^{14}C ages, or from stratigraphical and historical data. Ziway–Shala lacustrine phases I–IV after Street (1979). Table does not include coarse-grained bedforms in modern valleys, which are thought to represent high rainfall during the 'big rains' of the early 1970s; terraced or streamflow deposits found in modern valleys, which are thought to represent the humid phase of the 1940s defined by high Ziway–Shala lake levels (Vatova, 1941).

Synthem	Subunit	Estimated age
4	b_3	Second half of 18th century to 1940
	b_2	Second half of 18th century to 1940
	b_1	250 ^{14}C yr BP‡
	a_3	520 ^{14}C yr BP†
	a_2	2400 ^{14}C yr BP†
	a_1	4600*–3560 ^{14}C yr BP†
3	c	5390 ^{14}C yr BP†
	b	10–5 ka
	a	10–5 ka
2	d	11–10 ka
	c	12–11 ka (Ziway–Shala IV)
	b	19–12 ka
	a	22 200–19 130 ^{14}C yr BP†
1	c	30 ka (Ziway–Shala II)
	b	39–30 ka (Ziway–Shala II)
	a	39 ka (Ziway–Shala I)

*Conventional ^{14}C age on clastic charcoal found in the streamflow facies.
†Conventional ^{14}C age for time of burial of soil on top of the subunits.
‡Conventional ^{14}C age of vertebrate bones within the streamflow facies.

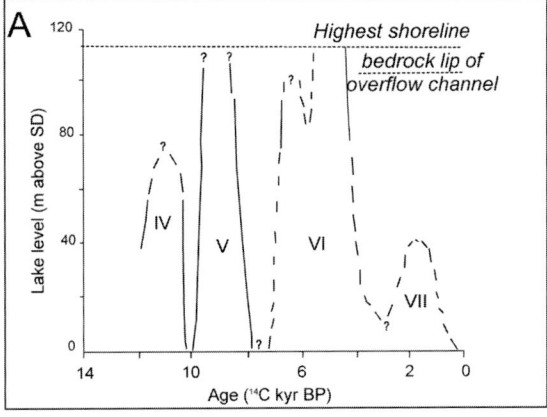

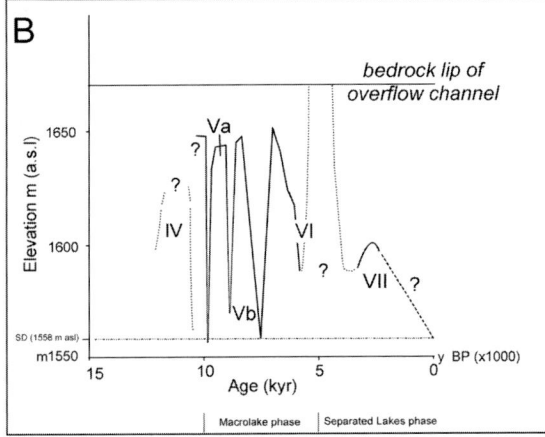

Fig. 8 Latest Pleistocene to Holocene fluctuations of the Ziway–Shala lakes. (A) Summary data for the whole basin, after Gillespie et al. (1983). (B) Fluctuations of Lake Shala, after Alessio et al. (1995). Roman numerals refer to the Ziway–Shala lacustrine phases of Street (1979).

volcanic activity from the Alutu and other silicic volcanoes affected the MER.

Four distinct vertically stacked subunits have been recognized in the Boramo, Kile Harsema (Fig. 7B) and Chirgulo areas. Basal subunit 2a is represented by alluvial sands and gravels deposited c. 20–19 ka, which are overlain by fluvially reworked fine-grained greyish tuffs (subunit 2b). Subunit 2c consists of thin (2 m) deltaic sands at an elevation of 1730 m a.s.l., which record one of the highest lake levels documented in the study area (see also Laury & Albritton, 1975), even though short-lived. These deposits are in turn erosively overlain by fluvial and colluvial sands and silts of subunit 2d, which record a contraction of the lake and the progradation of fluvial systems. At Kedida, the lower portion of Synthem 2 is represented by fluvial sands and silts arranged in thick tabular bedsets with intervening palaeosols, which correlate to subunit 2a at the other sites. Ash-bearing subunit 2b is apparently missing at Kedida (Fig. 7A), however, and subunit 2a is overlain by a thin bed of laminated silty sands of lacustrine origin at an elevation of ~1730 m a.s.l., and then by fluvial and colluvial sands and silts.

The T'ora Geosol developed on pre-Holocene volcanics, as well as on colluvial deposits of synthems 1 and 2. The T'ora Geosol is a compound palaeosol or pedocomplex in the terminology of Morrison (1977), and is a key stratigraphical marker in the study area (Benvenuti et al., 2002; Carnicelli et al., 2002).

Synthem 3

Synthem 3 consists of colluvial, fluvio-deltaic, and lacustrine gravel, sand, and mud, as well as lacustrine diatomites, and volcaniclastic materials that were deposited during the early to middle Holocene, between 10 and 5 ka. During this time, overall moist conditions established a wide lake system (the Macrolake phase of Benvenuti et al., 2002) that was smaller than the previous Megalake phase, and on which were superimposed fluctuations in level owing to short arid episodes punctuating the overall moist conditions (Gillespie et al., 1983; Alessio et al., 1996; Fig. 8). Maximum lake levels were controlled by an outlet to the north-west of Lake Ziway (Fig. 2C) at an elevation of 1670 m a.s.l., with excess water draining into the Awash River basin. This outlet could have been generated by a tectonically controlled narrowing and deepening of the basin between the end of the Pleistocene, when lake levels stood well above this elevation, and the beginning of the Holocene (Benvenuti et al., 2002).

In the study area, Synthem 3 has been subdivided into three subunits, with each representing incision into older deposits, then filling of palaeovalleys. Accordingly, each subunit is dominated by alluvial and colluvial sands and silts, which are discussed in the next section. Lacustrine deposits are represented in the Kile Harsema area by massive diatomites cropping out at about 1670 m a.s.l., thus marking a high level close to or at the Holocene overflow. Luvisols that lack both tephra cover and petrocalcic horizons, but have well-developed Bt horizons and thick, organic-rich A horizons, are often found on slope deposits of Synthem 3. These palaeosols were buried c. 5.3 ka.

Synthem 4

Synthem 4 consists of colluvial, fluvial, deltaic and lacustrine sediments that accumulated over the past 5 kyr when the early to middle Holocene Macrolake split into the present four lakes as a consequence of increasing aridity in the MER (the Separated Lakes Phase of Benvenuti et al., 2002).

In the study area, Synthem 4 has been subdivided into subunits 4a and 4b, with each representing incision into older deposits, then filling of palaeovalleys. At Boramo and Kile Harsema, deposits of synthems 3 and 4 mantle the eastern edge of Gademota Ridge, the base of which stands close to the maximum level reached by the early to middle Holocene Macrolake system. In these areas, lobes of Synthem 4 fluvial–deltaic deposits downlap lacustrine deposits of Synthem 3. At Chirgulo, all upper Quaternary deposits onlap the Gademota volcanics, although synthems 1–3 are widely eroded and/or buried by deposits of Synthem 4. At Kedida, Holocene synthems 3 and 4 occur within a depression that may be related to differential collapse of the Gademota Caldera during the late Pleistocene. Weakly developed soils characterized by Cambic or Mollic horizons and incipient Luvisol features are found on deposits of Synthem 4.

Facies analysis of latest Pleistocene–Holocene valley fills

Ancient valley fills are present within synthems 2 (subunit 2d), 3 and 4, indicating that ephemeral stream dynamics were active in the study area since the latest Pleistocene late-glacial period. In particular, individual fills range from a few metres to 15 m in thickness, and are comprised of a range of facies that recur within palaeovalley fills in a cyclic pattern. These facies are outlined below (see Figs 9–11 and Plates 2 & 3).

Streamflow facies

Coarse-grained deposits are typically found at the base of single (Fig. 11), as well as compound (Figs 9 & 10 and Plate 2), valley fills. This facies consists of clast-supported, angular to poorly rounded boulder to pebble gravels, mostly derived from the Gademota rhyolite, with poorly sorted interstitial sandy matrix and subordinate coarse sands (Plate 3A). Gravels can be imbricated, often in cobble–pebble clusters, massive and crudely cross-stratified in beds centimetres to decimetres thick. Both trough and planar cross-stratification

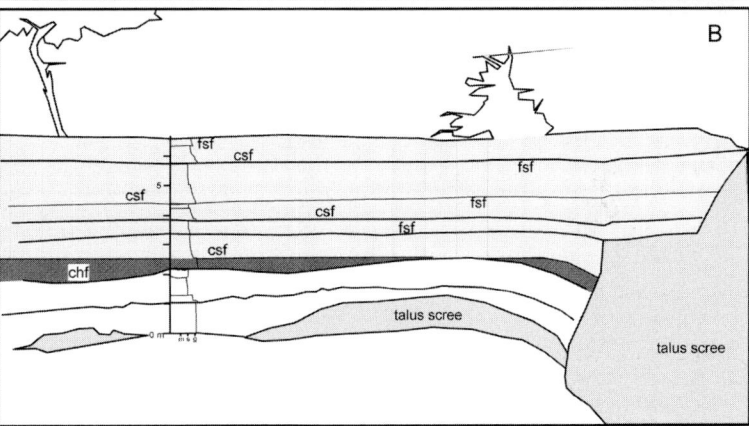

Fig. 9 (A) Photomosaic and (B) line drawing of section G1 (see location in Fig. 4) to illustrate stratigraphical architecture and distribution of facies associations. 3lac = Synthem 3 lacustrine deposits. Subunit 4b is interpreted to represent a channel-fan system. Palaeoflow to the right. Facies codes (also used for Figs 10 & 11 and Plate 2): chf, gravelly–sandy channel fill; csf, coarse-grained (sandy–pebbly) sheetflow deposits; fsf, fine-grained (silty–sandy) sheetflow deposits; df, massflow facies, horizontally bedded debris-flow deposits; vbf, massflow facies, debris fall deposits; vbd, massflow facies, debris-flow deposits.

were observed, the latter generally orientated at a high angle to the palaeoflow direction. In some cases gravelly planar cross-beds show an internal rhythmic alternation of openwork and sandy matrix-rich textures. This facies ranges from a lag a few clasts thick to one-third of the total sediment thickness.

Characteristics of these deposits point to bedload transport and deposition in flows similar to the flashy streamflows that deposited coarse-grained lags and gravelly bedforms in the present-day channel. Massive gravels could result from sudden inertial freezing of bedload and suspended load as a result of a hydraulic jump within the channel, whereas clusters of imbricated gravels represent preserved pebble cluster-like bedforms and similar to the massive gravels can result from bedload traction under low-concentration, high-discharge flood-flows. By analogy with the present-day bed morphology, cross-stratified gravels and sands are interpreted to result from migration and lateral accretion of bars in sinuous tracts of the ancient trunk valleys. The rhythmic grading observed in some cross-beds has been long described in gravel-bed river deposits (Smith, 1972, 1974; Steel & Thompson, 1983; Rust, 1984; Anketell & Rust, 1990). A general mechanism may be the differential sorting of bedload, perhaps through sweep–burst cycles with pulse-like arrival of different particle populations as sediment waves at the front crest of the bedform (Reid & Frostick, 1994; see also the experimental work of Iseya & Ikeda, 1987).

Sheetflow facies

Laterally extensive fine-grained deposits form the bulk of the Late Quaternary valley fills (see Figs 9–11 and Plate 2), and consist of horizontally bedded pebbly sands to sandy silts decimetres to

288 M. Benvenuti et al.

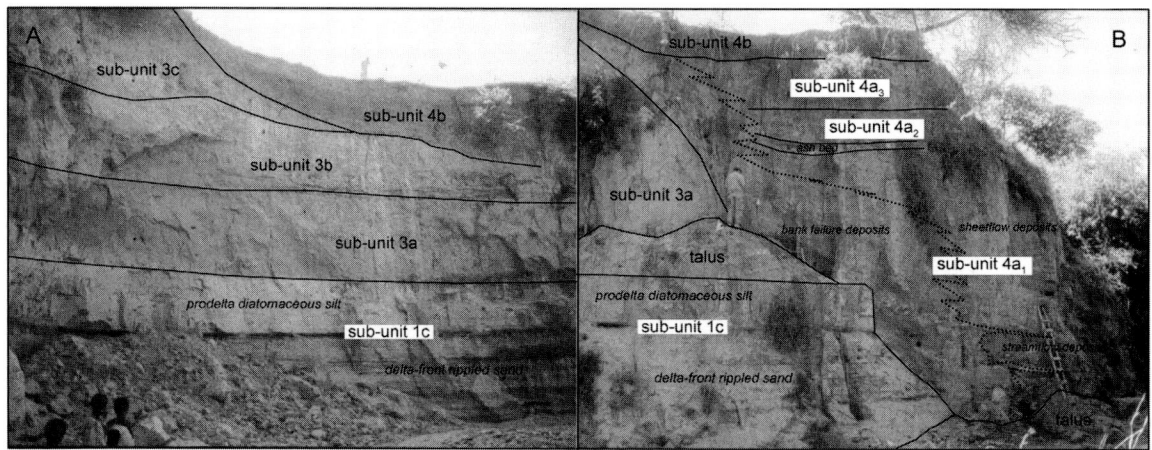

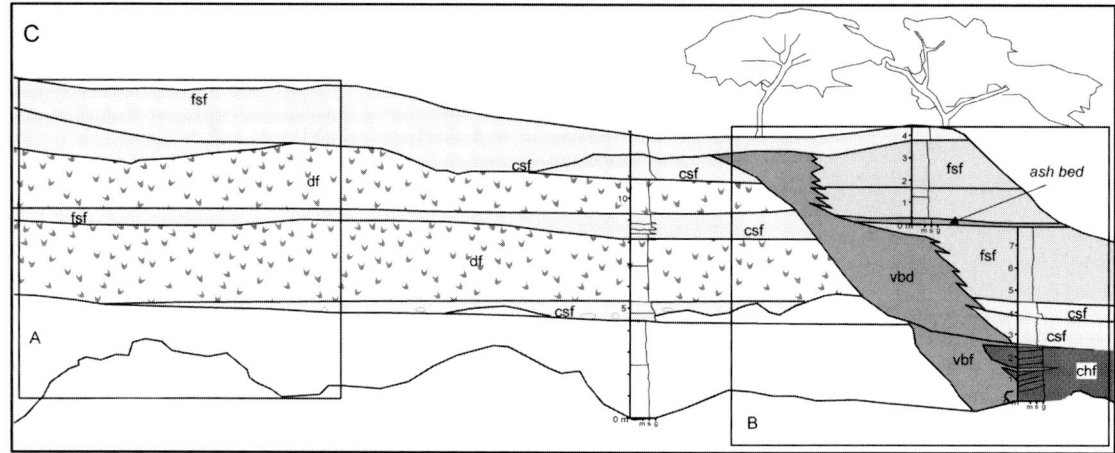

Fig. 10 (A, B) Photomosaics and (C) line drawing of section G4 to illustrate stratigraphical architecture and distribution of facies associations. Logs with heights in metres for scale. Subunits 3a–c are interpreted to represent a channel-fan system. Palaeoflow was to the left for subunit 3. Only the left side of the palaeovalley filled with subunit 4a is exposed, palaeoflow towards the viewer.

centimetres thick with overall sheet-like geometries. Individual beds fine upwards from coarse-grained lags comprised of a few pebble- to granule-sized clasts to poorly sorted sand, often with floating pebbles and granules, and then to faintly stratified sandy silts (Plate 3B).

Bed geometry and sedimentary structures point to an origin of these deposits from sheetflows that occupied the entire palaeovalley width, as zcompared with the streamflow facies described above that originated from flows confined to discrete channels. Sheetflow (or sheetflood) deposits have been described from dryland alluvial fans as sheets of gravels and sands deposited from supercritical flow conditions (Blissenbach, 1954; Hooke, 1967; Bull, 1972; Blair & McPherson, 1994). Other authors (Wasson, 1977, 1979; Nemec & Muszynski, 1982; Wells, 1984; Balance, 1984; Todd, 1989) have described sheet-like deposits displaying features intermediate between deposits of gravity and fluid flows, suggesting that multiple mechanisms can be important in these settings (Benvenuti & Martini, 2001). In the present case, the lack of traction-produced sedimentary structures, and the crude grading in coarse-grained beds, suggest that coarse sands and pebbles were segregated at the base of the flows, and fine-grained sediment formed a hyper-

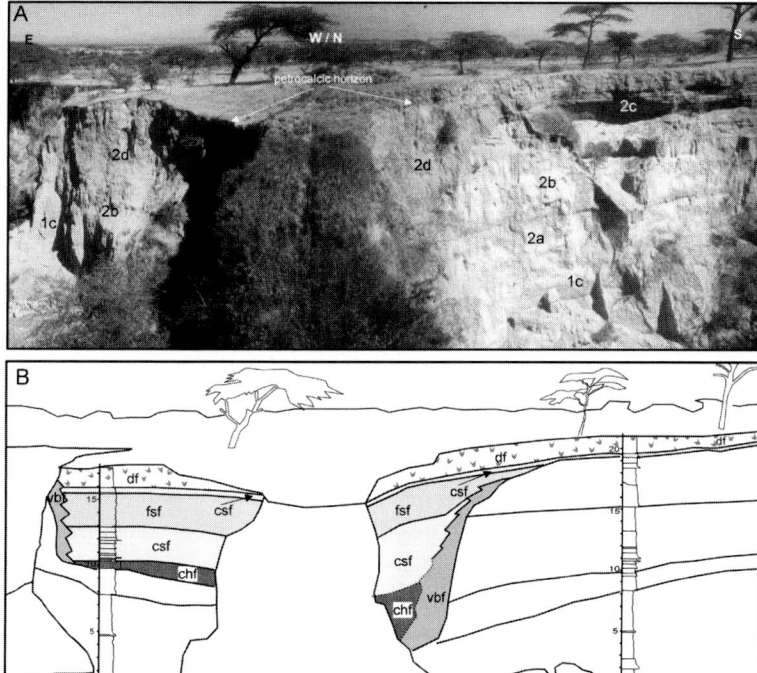

Fig. 11 (A) Photomosaic and (B) line drawing of section G4a–5 to illustrate stratigraphical architecture and distribution of facies associations within the palaeovalley of subunit 2d. Logs with heights in metres for scale, palaeoflow to the left.

concentrated suspension. Such flows may belong conceptually to the traction-carpet-bearing flow described by Sohn (1997), which was re-evaluated from the subaqueous traction carpet of Lowe (1982), as adapted to alluvial deposits (Todd, 1989). In Sohn's (1997) hypothesis a great variety of deposits characterized by different bedding and textural features can form from traction-carpet-bearing, density-stratified flows, depending on sediment concentration in, and rheological behaviour of, the different portions of the flow.

Massflow facies

Chaotic, matrix-rich deposits are found adjacent to the banks of the palaeovalleys, as well as interbedded within or on top of valley fills (see Figs 10 & 11 and Plate 2). Two distinct deposits are observed along the flanks of the palaeovalleys. The first is commonly interbedded with streamflow facies described above, and consists of a chaotic framework of mostly angular blocks, up to 1 m in diameter, comprised of the mostly fine-grained lithologies that constitute the bank material, or the clayey soils developed on interfluve areas (Plate 3C & D). These deposits are unstratified or crudely stratified in beds inclined toward the centre of the palaeovalleys. The block framework is supported by an abundant matrix derived from partial or total disruption of less cohesive sands and sandy silts. These deposits are interpreted to represent debris falls and bank wasting where blocks accumulate at the base of channel banks, and become interbedded with streamflow facies. By analogy to modern unstable banks, it seems likely that piping and undercutting would have been the primary mechanisms triggering ancient bank failures.

The second type is found in the mid-upper portion of the valley fills, and generally interbedded with sheetflow facies. It consists of cm–dm thick beds of silts with dispersed clasts, inclined toward the centre of the palaeovalleys (Plate 3E). Texture is massive with scattered clasts, ranging from a few millimetres to a few decimetres in size (Plate 3G) and made up of various lithologies (rhyolite and obsidian, lacustrine diatomaceous silts, petrocalcic horizon of the T'ora Geosol), chaotically distributed in or concentrated near the top of the bed. These massive silts, in both inclined and horizontal beds, are interpreted to represent cohesive debris flows in which clasts were sustained by buoyancy and matrix cohesion. In some cases clasts concentrated at the top of beds may indicate the rigid plug of cohesive debris flows. Similar deposits, characterized by horizontal beds centimetres to metres thick (Plate 3F), and found generally above the sheetflow facies, occur within the channel-fan systems (Figs 9 & 10 and Plate 2) or on top of valley fills (Fig. 11), and record the late overfilling of the valleys.

A DEPOSITIONAL MODEL FOR LATE QUATERNARY VALLEY FILLS

The late Pleistocene to Holocene valley fills of the study area are similar to DES alluvial sequences described from the south-west USA (Graf, 1988; Bull, 1997). Based largely on the work of Packard (1974) on late Holocene valley fills of southern Arizona, it has been argued that fining upward valley-fill successions, recording ancient phases of DES aggradation, represent backfilling of the trunk channel owing to base-level rise forced by fan-lobe retrogradation, fundamentally an autocyclic mechanism (complex response of Schumm, 1973). As an alternative, it is suggested here that cyclic incision and aggradation of ephemeral streams of the Gademota Ridge, as described above, was mostly driven by factors external to the fluvial systems (see Waters & Haynes, 2001 for a similar hypothesis). More specifically, the nature of sedimentary features observed in the stratigraphical framework outlined above points to variations of water discharge volume, driven by climatic and environmental change, as the most probable cause in cyclic valley incision and filling.

Studies of landscape sensitivity in East Africa during the Late Quaternary (e.g. Roberts & Barker, 1993) suggest that sediment yields increased during the Last Glacial Maximum (LGM) to Holocene transition, reaching a maximum at the beginning of the early Holocene *climatic optimum*. Vegetation cover during this time period was still depleted in response to the LGM aridity, but the runoff that drives erosion was increasing as a consequence of an increasingly moist climate. As a result, huge volumes of sediment were flushed through incised fluvial valleys to the lake systems. It is assumed here that similar responses occurred during the Holocene century scale arid–humid cycles as well, as documented by the Ziway–Shala lake-level fluctuations (Fig. 8). Such arid–humid cycles are envisaged to have forced repeated episodes of valley incision during the transition to a more moist climate, followed by aggradation with eventual valley filling as aridity increased again.

Episodes of valley aggradation are interpreted to have occurred through two main stages (Fig. 12). In an early stage, high-volume discharges characteristic of a more moist climate transport bedload within the trunk valley downstream to a channel fan. The coarser material, eroded from the upper catchments, is temporarily stored in medium- to small-scale bedforms such as pebble clusters, transverse ribs, boulder berms and point bars. At this time, bank wasting, mostly through debris fall, is an important process, with accumulation of debris aprons at the base of the banks. Examples of this early stage can be cited from both the present-day underfilled ephemeral stream system and ancient valley fills of the study area. Coarse-grained bedforms in the present valleys are, for example, not in equilibrium with the present hydrological regime, and are instead attributed to the last high discharge phase that characterized the study area. Within ancient valley fills, the early stage is best expressed by the streamflow facies interfingered with debris fall deposits that occur at the base of valley fills (Figs 10–12). Coeval channel fan systems are documented by coarse-grained sheetflow facies within the tabular deposits of subunits 3 (Figs 10 & 11) and 4a (Plate 2).

The main stage of valley filling is characterized by a progressive decrease of discharge, and a corresponding increase in the importance of sediment-laden sheetflows and debris flows. Sediment is mostly derived from the reworking of bedforms and the fine-grained sediments within the debris aprons, and is transported downstream to backstepping channel fan systems that eventually fill the previously cut valley. Eventually, the channel fan is deactivated, and sheetflow deposits and debris flow lobes spread over the filled valleys and onto the interfluves. The sandy–silty bedload of the present valleys may represent the beginnings of this stage in the study area. In older valley fills, this stage is represented by the widespread

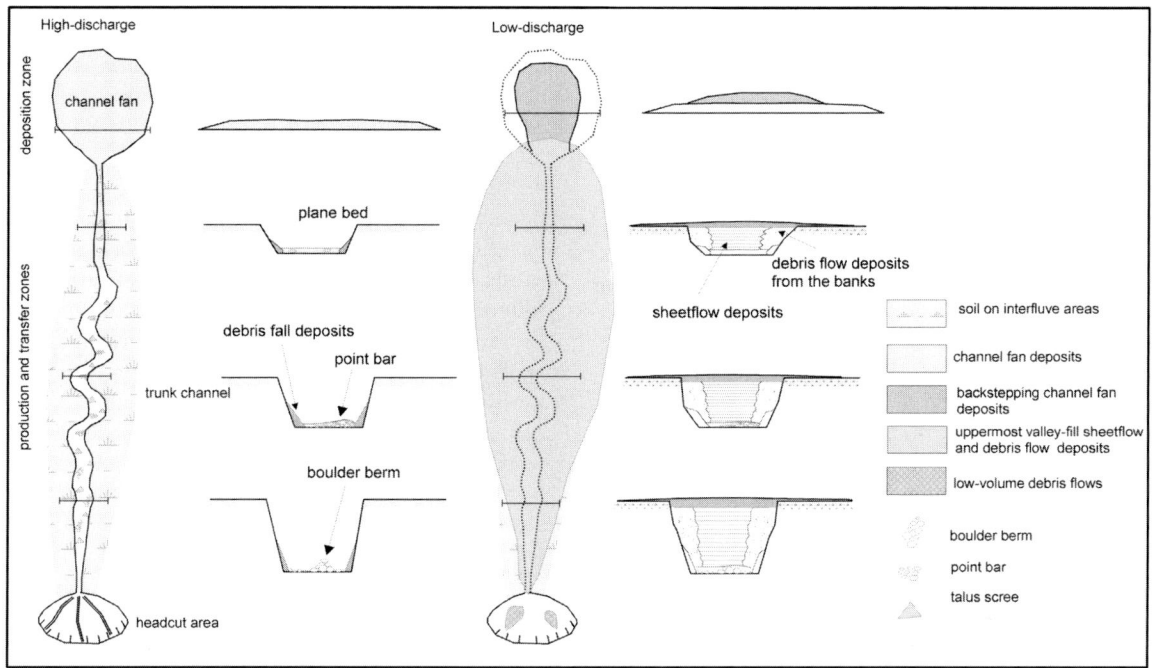

Fig. 12 Conceptual model for the filling of ephemeral stream valleys ('high-discharge' early stage and 'low-discharge' late stage, as discussed in the text) during the Late Quaternary on the sides of the Gademota Ridge.

coarse and fine-grained sheetflow facies interbedded with debris flow deposits of the banks (Figs 10–12). The channel fan systems of this stage are represented by horizontal debris flows and fine-grained sheetflow deposits in multiple channel-fans (subunit 3, Figs 10 & 11), as well as in the uppermost part of valley fills (subunit 2d, Fig. 12; subunit 4a, Plate 2).

The examples described point to variability in the type of valley fill and the duration of time represented by the different stages of Late Quaternary filling (Fig. 13). Subunit 2d, for example, is interpreted to represent a single valley fill that evolved without significant break, apparently over a very short period of time (century scale, see Table 1). By contrast, subunit 4a includes multiple valley fills that represent a millennial time-scale (Table 1), with episodes of valley filling punctuated by periods of stasis marked by soil development. Subunit 4b displays a similar pattern, but represents a century scale time period. The present-day valleys show a partial filling that records decadal-scale fluctuations between incision and aggradation. More data are needed to refer this configuration to specific conditions at the present time, such as a decrease of water and sediment discharge, rather than to the internal variability similar to those of subunits 4a and 4b.

CONCLUSIONS

This study has demonstrated that:

1 Modern ephemeral streams on the flanks of the Gademota Ridge in the Main Ethiopian Rift are characterized by alternation of straight to sinuous trunk valleys and unentrenched reaches, i.e. channel-fan systems, similar to that of dryland rivers described from the south-west USA. Morphological and sedimentary features are well preserved within the trunk valley reaches, and indicate that fluvial processes are responsible for different coarse-grained bedforms, but interact with mass wasting of banks, mainly through debris fall.

2 A relatively thick apron of Late Quaternary deposits has been subdivided into four unconformity-

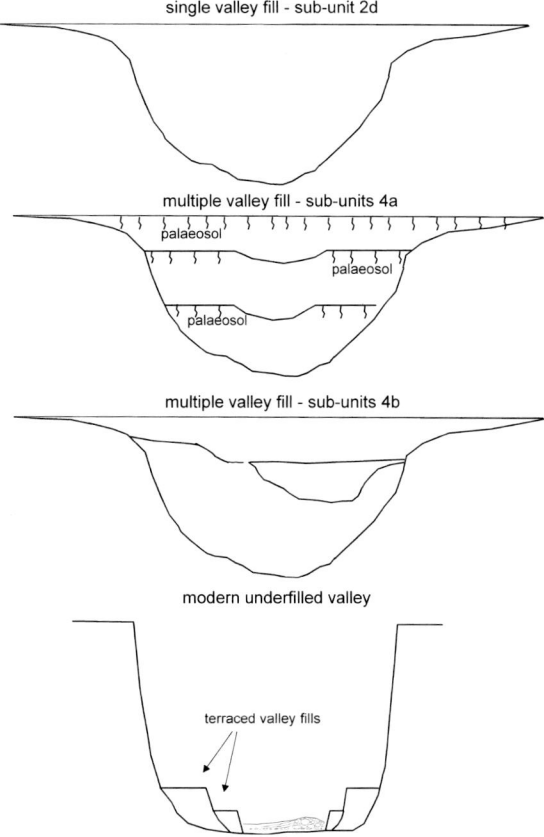

Fig. 13 Valley fill geometries recognized in this study (not to scale).

bounded synthems. Around the Gademota Ridge these deposits mostly accumulated in coalescing alluvial fans, with synthems representing successive episodes of entrenchment of older fan surfaces, but also include lacustrine deposits that record periods of lake level rise and fall. Palaeovalley fills are present within synthems 2 (subunit 2d), 3 and 4. These palaeovalley fills provide evidence of ephemeral stream dynamics that date back to the latest Pleistocene.

3 Streamflow, sheetflow and massflow facies are recognized in palaeovalley fills, as well as within palaeochannel-fan systems. Within multiple and single valley fills the three facies are stacked in a cyclic pattern with streamflow deposits, sedimentologically equivalent to modern channel lags, overlain by sheetflow, deposited from hyperconcentrated flows, and eventually by cohesive debris-flow deposits. Massflow facies also include inclined beds of debris-fall and debris-flow deposits that represent bank failures, and interfinger with streamflow and sheetflow facies.

4 A general model is proposed for aggradation of latest Pleistocene–Holocene ephemeral streams, based on a temporal variation of discharge rather than on internal autogenic variations, as hypothesized for periods of DES alluviation in the south-west USA. Periods of incision correspond to climatically controlled increases in moisture and discharge. An early stage of valley filling then coincided with relatively high and regular discharge that was competent enough to transport coarse-grained material and construct medium- to small-scale bedforms similar to those observed in the modern valleys. Relatively thick aprons of fine-grained material develop at the base of the banks, and represent an important sediment source for the stream. Further downstream, the channel-fan system is supplied with pebbly–sandy material flushed from the valleys. The later stage reflects a decrease of discharge, and redistribution of the debris-fall deposits through hyperconcentrated sheetflows and debris flows. Channel-fan systems, expected to retrograde during this stage, are supplied with sheetflow and debris flow deposits. Overfilling of the valleys finally occurs through spreading of debris flows and sheetflows on interfluves.

5 Palaeovalley fills reflect single or multiple valley-filling successions that occurred over decadal (the modern and sub-recent valley fills), century (subunits 2d and 4b) and millennial (subunit 4a) time-scales. Future research will focus on the significance of such a variable cyclicity in respect to Late Quaternary hydro-climatic fluctuations in the study area.

ACKNOWLEDGEMENTS

The study summarized in this paper was carried out under a MURST-PRIN project during the period 2000–2002. The authors are greatly indebted to Mike Blum and Peter Martini for their suggestions and critical reviews.

REFERENCES

Accetta, S. (1999) Fenomeni erosivi e produzione di sedimenti nella Regione dei Laghi della Rift Etiopica. Unpublished Laurea thesis, University of Firenze, 137 pp.

Adams, R., Adams, M., Willens, A. and Willens, A. (1979) *Drylands: Man and Plants*. St. Martin's, New York.

Alessio, M., Allegri, L., Belluomini, G., Benvenuti, M., Cerasoli, M., Improta, S., Manfra, L., Sagri, M. and Ventra, D. (1996) Le oscillazioni tardo-quaternarie del Lago Shala (Rift Etiopico): Analisi dell'evoluzione ambientale dall'integrazione di evidenze morfologiche, sedimentarie e cronologiche. *Il Quaternario*, **9**, 387–392.

Allen, J.R.L. (1985) A simplified cascade model for transverse stone-ribs in gravelly streams. *Proc. R. Soc. Lond.*, **A355**, 253–265.

Anketell, J.M. and Rust, B.R. (1990) Origin of cross-stratal layering in fluvial conglomerates, Devonian Malbaie Formation, Gaspé, Quebec. *Can. J. Earth. Sci.*, **27**, 1773–1782.

Ballance, P.F. (1984) Sheet-flow-dominated gravel fans of the non-marine Middle Cenozoic Simmler Formation, Central California. *Sediment. Geol.*, **38**, 337–359.

Benvenuti, M., Martini, I.P. (2001) Analysis of terrestrial hyperconcentrated flows and their deposits. In: *Floods and Megafloods processes and deposits* (Eds V. Baker, I.P. Martini and G. Garzon). *Spec. Publ. Int. Assoc. Sedimentol.*, **32**, 167–193.

Benvenuti, M., Carnicelli, S., Belluomini, G., Dainelli, N., Di Grazia, S., Ferrari, G.A., Iasio, C., Sagri, M., Ventra, D., Balemwald Atnafu and Seifu Kebede (2002) The Ziway-Shala Lake Basin (Main Ethiopian Rift, Ethiopia): a revision of basin evolution with special reference to the Late Quaternary. *J. Afr. Earth Sci.*, **35**, 247–269.

Billi, P. (1988) Forme di fondo grossolane. *Giorn. Geol.*, **50**, 15–26.

Blair, T.C. (1987) Sedimentary processes, vertical stratification sequences, and geomorphology of the Roaring River alluvial fan, Rocky Mountain National Park, Colorado. *J. Sediment. Petrol.*, **57**, 1–18.

Blair, T.C. and McPherson, J.G. (1994) Alluvial fans and their natural distiction from rivers based on morphology, hydraulic processes, sedimentary processes, and facies assemblages. *J. Sediment. Res.*, **A64**, 450–469.

Blissenbach, E. (1954) Geology of alluvial fan in Southern Nevada. *Geol. Soc. Am. Bull.*, **65**, 175–190.

Bull, W.B. (1972) Recognition of alluvial fan deposits in the stratigraphic record. In: *Recognition of Ancient Sedimentary Environments* (Eds W.K. Hamblin and J.K. Rigby). *Soc. Econ. Paleontol. Mineral. Spec. Pub.*, **16**, 63–83.

Bull, W.B. (1997) Discontinuous ephemeral streams. *Geomorphology*, **19**, 227–276.

Carling, P.A. (1987) Hydrodynamic interpretation of a boulder berm and associated debris-torrent deposits. *Geomorphology*, **1**, 53–67.

Carnicelli, S., Benvenuti, M., Bonaiuti, F., Iasio, C., Sagri, D., Mirabella, A., Sagri, M., Ferrari, G.A., Belluomini, G. and Wolf, U. (2002) The T'ora geosol(?) (Main Ethiopian Rift, Ethiopia): problems in defining palaeosols. *Il Quaternario*, **15**, 73–78.

Chernet, T. (1982) *Hydrogeology of the Lakes Region, Ethiopia*. Ministry of Mines and Energy, Addis Ababa.

Chorowicz, J., Collet, B., Bonavia, F.F. and Korme, T. (1994) Northwest to north-northwest extension direction in the Ethiopian Rift deduced from the orientation of extension structures and fault slip analysis. *Geol. Soc. Am. Bull.*, **105**, 1560–1570.

Di Paola, G.M. (1972) The Ethiopian Rift Valley (between 7°00′ and 8°40′ lat. North). *Bull. Volcanol.*, **36**, 517–560.

Di Paola, G.M., Seife, M.B. and Arno, V. (1993) The Kella horst: its origin and significance in crustal attenuation and magmatic processes in the Ethiopian Rift Valley. In: *Geology and Mineral Resources of Somalia and Surrounding Regions*. Ist. Agronom. Oltremare, Firenze, *Relaz. Monograf.*, **113**, 323–338.

Gillespie, R., Street-Perrot, A.F. and Switsur, R. (1983) Post-glacial arid episodes in Ethiopia have implications for climate prediction. *Nature*, **306**, 680–683.

Graf, W.L. (1988) *Fluvial Processes in Dryland Rivers*. Springer-Verlag, Berlin, 346 pp.

Haynes, C.V. (1968) Geochronology of late Quaternary alluvium. In: *Means of correlation of Quaternary successions; INQUA 7th Congress, Proceedings* (Eds R.B. Morrison and H.E. Wright, Jr), Vol. **8**, pp. 591–631. Salt Lake City, University of Utah Press.

Hooke, R.Leb. (1967) Processes on arid-region alluvial fans. *J. Geol.*, **75**, 438–460.

Kazmin, V., Seife, M.B., Nicoletti, N. and Petruccini, C. (1979) Evolution of the Northern Part of the Ethiopian Rift. In: *Geodynamic Evolution of the Afro-Arabic Rift System*. Accad. Naz. Lincei, Roma, *Atti Convegni*, **47**, 275–292.

Koster, E.H. (1978) Transverse ribs: their characteristics, origin and palaeohydraulic significance. In: *Fluvial Sedimentology* (Ed. A.D. Miall). *Can. Soc. Petrol. Geol. Mem.*, **5**, 161–186.

ISSC (1994) *A Guide to Stratigraphic Classification, Terminology, and Procedure*, 2nd edn. International Subcommission on Stratigraphic Classification, International Union of Geological Sciences, Boulder, 214 pp.

Iseya, F. and Ikeda, H. (1987) Pulsations in bedload transport rates induced by longitudinal sediment sorting: a flume study using sand and gravel mixtures. *Geog. Ann.*, **69A**, 15–27.

Laury, R.L. and Albritton, C.C. (1975) Geology of the Middle Stone Age archaeological sites in the Main Ethiopian Rift Valley. *Geol. Soc. Am. Bull.*, **86**, 999–1011.

Le Turdu, C., Tiercelin, J.J., Gibert, E., Travi, Y., Lezzar, K., Richert, J., Massault, M., Gasse, F., Bonnefille, R., Decobert, M., Gensous, B., Jeudy, V., Tamrat, E., Mohamed, M.U., Martens, K., Balemwal, A., Chernet, T., Williamson, D. and Taieb, M. (1999) The Ziway–Shala lake basin system, Main Ethiopian Rift: influence of volcanism, tectonics, and climatic forcing on basin formation and sedimentation. *Palaeogeog., Palaeclim. Palaeoecol.*, 150, 135–177.

Lowe, D.R. (1982) Sediment gravity flows: II. Depositional models with special reference to the deposits of high-density turbidity currents. *J. Sediment. Petrol.*, 52, 279–297.

McDonald, B.D. and Day, T.J. (1978) An experimental flume study on the formation of transverse ribs. *Geol. Surv. Can.*, 78-1A, 441–451.

Meigs, P. (1953) World distribution of arid and semi-arid homoclimates. In: *Reviews of Research in Arid Zone Hydrology*, pp. 203–209. UNESCO, Paris.

Merla, G., Abbate, E., Azzaroli, A., Bruni, P., Canuti, P., Fazzuoli, M., Sagri, M. and Tacconi, P. (1979) *A Geological Map of Ethiopia and Somalia, and Comment*. Consiglio Nazionale delle Ricerche, Firenze, 95 pp.

Mohr, P.A. (1962) The Ethiopian Rift System. *Bull. Geophys. Obs. Addis Ababa*, 5, 33–62.

Mohr, P.A., Mitchell, J.C. and Raynolds, R.G.H. (1980) Quaternary volcanism and faulting at O'A Caldera, Central Ethiopian Rift. *Bull. Volcan.*, 43, 173–189.

Morrison, R.B. (1977) Quaternary soil stratigraphy, concepts, methods and problems. In: *Quaternary Soils* (Ed. W.C. Mahaney), pp. 77–108. GeoAbstracts, Norwich.

Nemec, W. and Muszynski, A. (1982) Volcaniclastic alluvial aprons in the Tertiary of the Sofia district (Bulgaria). *Ann. Soc. Geol. Pol.*, 52, 239–303.

Packard, F.A. (1974) The hydraulic geometry of a discontinuous ephemeral stream on a Bajada near Tucson, Arizona. Unpublished PhD thesis, University of Arizona, 127 pp.

Patton, P.C. and Schumm, S.A. (1981) Ephemeral stream processes—implications for studies of Quaternary valley fills. *Quaternary Research*, 15, 24–43.

Reid, I. and Frostick, L.E. (**1994**) Fluvial sediment transport and deposition. In: *Sediment Transport and Depositional Processes* (Ed. K. Pye), pp. 89–155, Blackwell Science, Oxford.

Roberts, N. and Barker, P. (1993) Landscape stability and biogeomorphic response to past and future climatic shifts in intertropical Africa. In: *Landscape sensivity (British Geomorphological Research Group Symposia Series)*, (Eds D.S.G. Thomas and R.J. Allison), pp. 65–82. Wiley, Chichester.

Rust, B.R. (1984) Proximal braidplain deposits in the Middle Devonian Malbaie Formation of Eastern Gaspé, Quebec, Canada. *Sedimentology*, 31, 675–695.

Sagri, M. (ed.) and the staff of EU Project (1998) *Land Resources Inventory, Environmental Change Analysis and their Application to Agriculture in the Lakes Region (Ethiopia)*. Final Report, European Commission, DG XII, Bruxelles, 183 pp.

Salvador, A. (1987) Unconformity-bounded stratigraphic units. *Geol. Soc. Am. Bull.*, 98, 232–237.

Schumm, S.A. (1973) Geomorphic thresholds and complex response of drainage systems. In: *Fluvial Geomorphology* (Ed. M. Morisawa), Publ. in Geomorphology, pp. 299–310, Binghampton, State University New York.

Schumm, S.A. and Hadley, R.F. (1957) Arroyos and the semiarid cycle of erosion. *Am. J. Sci.*, 225, 161–164.

Scott, K.M. and Gravlee, G.C. (1968) Flood surge on the Rubicon River, California—hydrology, hydraulics and boulder transport. *U.S. Geol. Surv. Prof. Pap.*, 422-M, 40 pp.

Smith, N.D. (1972) Some sedimentological aspects of planar cross-stratification in a sandy braided river. *J. Sediment. Petrol.*, 42, 624–634.

Smith, N.D. (1974) Sedimentology and bar formation in the Upper Kicking Horse River, a braided outwash stream. *J. Geol.*, 82, 205–223.

Sohn, Y.K. (1997) On traction-carpet sedimentation. *J. Sediment. Res.*, 67, 502–509.

Steel, R.J. and Thompson, D.B. (1983) Structures and textures in Triassic braided stream conglomerates ('Bunter' Pebble Beds) in the Sherwood Sandstone Group, North Staffordshire, England. *Sedimentology*, 30, 341–367.

Street, F.A. (1979) Late Quaternary lakes in the Ziway-Shala Basin, southern Ethiopia. Unpublished PhD thesis, University Cambridge, 457 pp.

Todd, S.P. (1989) Stream-driven, high-density gravelly traction carpets: possible deposits in the Trabeg Conglomerate Formation, SW Ireland and some theoretical considerations of their origin. *Sedimentology*, 36, 513–530.

Vatova, A. (1941) Relazione sui risultati idrografici relativi ai laghi dell' Africa Orientale Italiana esplorati dalla Missione Ittiologica. In: *Esplorazione dei laghi della Fossa Galla*, Vol. 1, pp. 67–127. Ministero Africa Italiana, Roma.

Wasson, R.J. (1977) Last-glacial alluvial fan sedimentation in the Lower Derwent Valley, Tasmania. *Sedimentology*, 24, 781–799.

Wasson, R.J. (1979) Sedimontation history of the Mundi-Mundi alluvial fans, western New South Wales. *Sediment. Geol.*, 22, 21–51.

Waters, M.R. and Haynes, C.V. (2001) Late Quaternary arroyo formation and climate change in the American Southwest. *Geology*, 29, 399–402.

Wells, N.A. (1984) Sheet debris flow and sheetflood conglomerates in Cretaceous cool-maritime alluvial fans, south Orkney Islands, Antartica. In: *Sedimentology of gravels and conglomerates* (Eds E.H. Koster and R.J. Steel). *Mem. Can. Soc. Petrol. Geol.*, 10, 133–145.

Woldegabriel, G., Aronson, J. and Walter, R.C. (1990) Geology, geochronology and rift basin development in the central sector of the Main Ethiopian Rift. *Geol. Soc. Am. Bull.*, 102, 439–458.

Fluvio-deltaic floodbasin deposits recording differential subsidence within a coastal prism (central Rhine–Meuse delta, The Netherlands)

K.M. COHEN, M.J.P. GOUW and J.P. HOLTEN

Centre for Geo-ecological Research (ICG), Department of Physical Geography, Utrecht University, P.O. Box 80.115, 3508 TC Utrecht, The Netherlands (Email: k.cohen@geog.uu.nl)

ABSTRACT

In the central Netherlands, the Rhine follows a course imposed by Late Quaternary glaciation, forcing it to cross several tectonic blocks of the Roer Valley Graben system before entering the south-eastern North Sea Basin. Holocene sea-level rise resulted in the formation of a coastal prism (Holocene Rhine–Meuse delta). Across the Peel Boundary Fault-zone (PBF) in the central delta, differences in subsidence between the downstream Roer Valley Graben and its upstream shoulder influenced fluvial deposition. This study examines the sedimentary response upstream and downstream of the PBF and uses that to quantify differential subsidence rates. The local identification and quantification of the subsidence component within deltaic relative base-level rise may serve to determine to what extent coastal prism aggradation and resulting architecture are controlled by downstream sea-level rise, local tectonics and discharge coming from upstream.

A floodbasin-section shows both syn-depositional and post-depositional tectonic effects in the Late-glacial to Holocene (marine oxygen isotope stages 2 and 1) sedimentary record, and reveals an active fault of the Peel Boundary Fault-zone. Differential subsidence across the PBF is quantified for the Last Glacial Maximum subsurface (averaged rate c. 0.06 m kyr^{-1} over the past 15 kyr). The offset in Holocene basal peat yields a similar value (0.03–0.07 m kyr^{-1}, averaged over the past 7 kyr). Groundwater-level rise in the study area (as reconstructed from series of radiocarbon-dated basal peats) is used to quantify subsidence by comparison with relative sea-level rise at the river mouth. High rates of subsidence between 7000 and 5500 cal. yr BP together with syn-depositional sedimentary evidence suggest that the last major activity (palaeo-earthquakes) along this fault occurred in the middle Holocene. The timing of this and earlier fault activity may be related to the deglaciation and sea-level rise history. The results imply that early–middle Holocene deltaic back-filling should not be attributed solely to downstream relative sea-level rise dominating upstream controls, but that local differential subsidence is another independent control. Similar effects of differential subsidence in coastal prism sedimentary architecture can be expected in other coastal prisms.

296 K.M. Cohen, M.J.P. Gouw and J.P. Holten

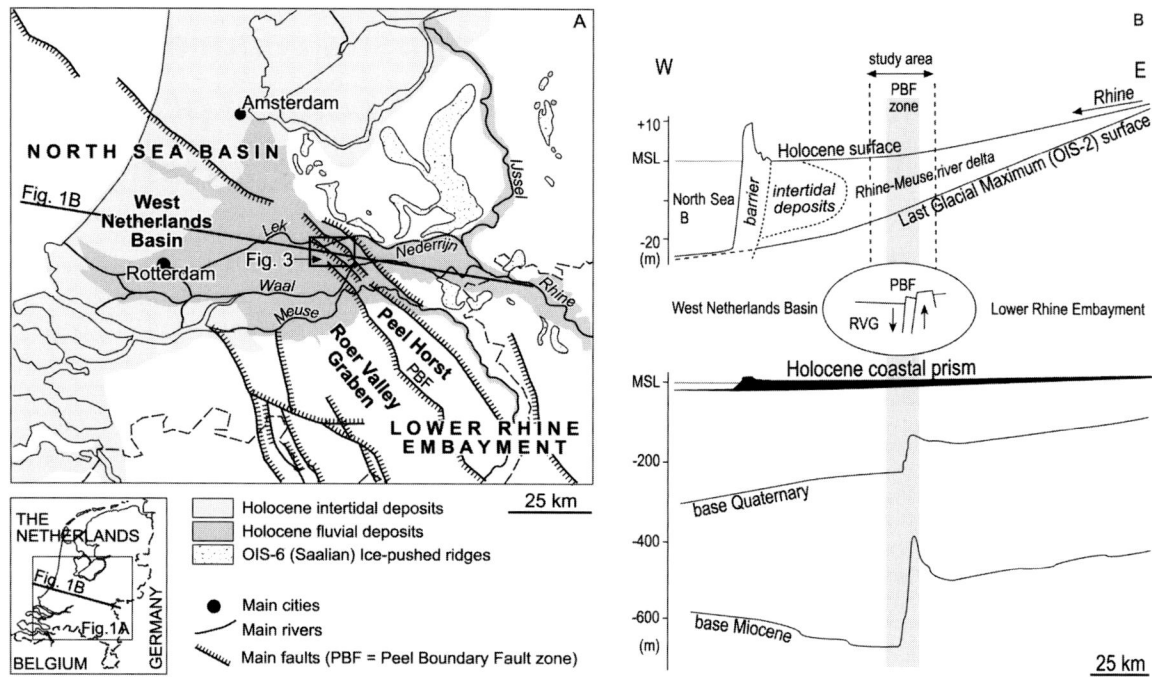

Fig. 1 Geological setting and Holocene coastal prism and subsurface tectonics. (A) Map of present delta and main neotectonic units. (B) Schematic illustration of subsurface neotectonic structure showing a much generalized cross-section. Base of Quaternary (base Maassluis Formation) after Zagwijn & Van Staalduinen (1975). Base of Miocene (base Breda Formation) after Zagwijn (1989).

INTRODUCTION

This study focuses on the response of the River Rhine to differential subsidence across fault zones in the central Netherlands. The study area (Fig. 1) is part of the coastal prism that formed during Holocene sea-level rise towards the present high stand, and covers the active Peel Boundary Fault-zone (PBF). The PBF is related to structural blocks of the Lower Rhine Embayment, an intracontinental rift system that grades into the southern North Sea Basin (West Netherlands Basin, Fig. 1; Geluk et al., 1994). It has a main normal fault that extends from the surface to a depth of ≥ 17 km (Camelbeeck & Meghraoui, 1998). Late Quaternary Rhine channels traversed the PBF as they entered the Roer Valley Graben (RVG, the main depo-centre of the Lower Rhine Embayment) from the north-east. South of the Rhine–Meuse delta, the PBF is narrow (less than 1 km wide) and separates the RVG from the Peel Block, which forms a topographic high. In the central Rhine–Meuse delta the fault-zone widens and is less well defined. The study area covers the north-western extension of the RVG, PBF and Peel Block (Fig. 1A).

The PBF is of Palaeozoic age, was reactivated in the Tertiary and has been active since (Fig. 1B). Studies showing Late Quaternary and Holocene activity include: seismic monitoring and earthquake hypocentre calculations (Ahorner, 1992), shallow seismographic surveying (Van den Berg et al., 1994), historic seismology (Alexandre, 1994), palaeoseismological studies in trenches and back-stripping of cores (Houtgast & Van Balen, 2000) and palaeogeographical studies (Van den Berg, 1994; Berendsen & Stouthamer, 2000; Cohen et al., 2002; Houtgast et al., 2002). Monitoring of seismicity and screening of historic sources has yielded estimates of earthquake activity (maximum local Richter magnitude observed: 5.8 ± 0.1; Roermond, 1992) for the past 100–1000 yr. Palaeoseismological data derived from exposures in trenches dug over topographic fault-scarps (Camelbeeck & Meghraoui, 1998; Vanneste et al., 1999; Lehmann et al., 2001; Van den Berg et al., 2002) cover a wider temporal range (along the RVG typically up to 200 kyr). Such studies generally have been conducted outside the Rhine–Meuse delta, in the RVG and upstream along the Meuse valley (Fig. 1). The delta itself is unsuitable for trenching because the groundwater table is near the surface. Moreover, the exact locations of faults with Holocene offsets are unknown. However, the Late-glacial to Holocene fluvial record in the study area is continuous, and covers the north-western extensions of the PBF (Cohen et al., 2002). This setting offers opportunities to quantify tectonic deformation and fill the gap between studies based on seismic monitoring and palaeoseismological interpretations of trenched Quaternary sediments, while exploring the fluvio-deltaic sedimentary response to differential subsidence.

Site selection: Holocene floodbasins

Although low-gradient channel belts preserved in the Rhine–Meuse delta indicate neotectonic activity (deformation in longitudinal profiles, asymmetric meander belts, distribution of avulsion sites; see Verbraeck, 1990; Berendsen & Stouthamer, 2000; Stouthamer & Berendsen, 2000; Stouthamer, 2001), it is hard to accurately quantify post-depositional vertical deformation from channel belt deposits. Quantification of aggradation rates by Törnqvist et al. (1998) and timing of the upstream migration of back-filling by Berendsen & Stouthamer (2000, 2001) demonstrated that differential subsidence is recorded in the floodbasins. In contrast to channel belts, fluvio-deltaic floodbasins are stable sedimentary environments: variation in surface elevation occurs over larger distances and the record is essentially continuous. Therefore, floodbasins are a more suitable environment to offer proxies for quantifying vertical neotectonics than channel belts, and hence this study focuses on the floodbasin sedimentary record.

Rising sea level is the primary control on aggradation rates within the floodbasins: the main body of the Rhine–Meuse coastal prism was deposited during middle Holocene back-filling in direct response to the rising sea level (Van Dijk et al., 1991; Törnqvist, 1993). During this sea-level rise, accommodation exceeded sedimentation: floodbasin peat could form extensively because of the relative absence of sediment. Curves of groundwater-level rise quantify aggradation at distal floodbasin locations. Accurately quantifying aggradation rates in floodbasins is possible only where compaction is near zero and hence can be neglected. Such sites (i.e. peat-covered flanks of buried aeolian dunes, Jelgersma, 1961; Van de Plassche, 1982; Van Dijk et al., 1991; Törnqvist et al., 1998) are available, so the compaction problem can be avoided. In the inland parts of the delta, upstream controls (discharge of water and sediment) interplay with the downstream controls, as is shown by the longitudinal gradient of isochrons within floodbasin peat (Van Dijk et al., 1991; Blum & Törnqvist, 2000, fig. 24). Towards the late Holocene, the rate of sea-level rise decreased. This is reflected by changes in fluvial style and increased clastic floodbasin sedimentation (Törnqvist, 1993; Berendsen & Stouthamer, 2001) and points to upstream controls becoming more significant and influencing sedimentation further downstream.

Neotectonic activity causing locally higher subsidence rates is a secondary control in this delta. Reach-to-reach variation in sedimentary response to *local* subsidence conditions can be identified separately from *general* response to upstream (discharge change) and downstream (relative base-level rise) controls, which have an impact over the whole delta. Local tectonic controls (Fig. 2) can lead to anomalies in subsurface elevation, sedimentary facies and aggradation rates in a longitudinal direction, which cannot be attributed to the primary controls. Post-depositional tectonic displacement can be quantified from deformed surfaces (Fig. 2A). Tectonic effects in the floodbasin sedimentary architecture, recorded, for example, by the changing distribution of organics and clastics within the sequence (Fig. 2B), are useful when tracing faults, but these are not suitable to quantify tectonic activity. Diverging curves of groundwater-level rise (Fig. 2C) quantify differential subsidence (i.e. relative movement between tectonic blocks). For two nearby sites absolute rise of groundwater levels is equal, whereas subsidence may differ significantly.

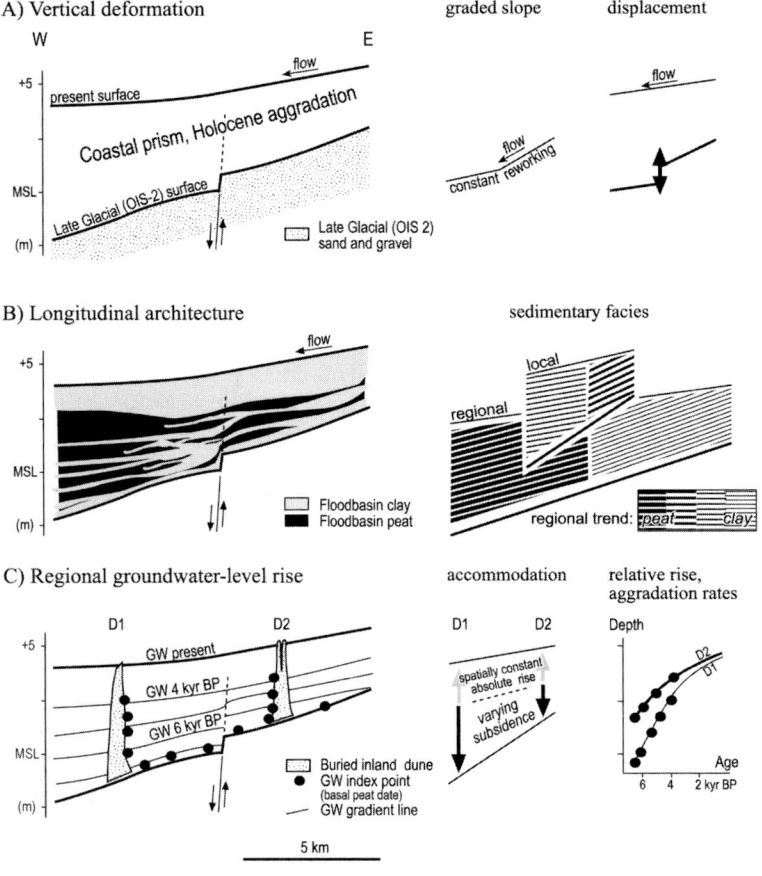

Fig. 2 Means by which post-depositional and syn-depositional controls can be identified. (A) Differential subsidence affects the graded slope of the river during deposition and causes local offsets at faults after deposition. (B) Differential subsidence affects the sedimentary architecture, causing local deviations from the general longitudinal trend in coastal-prism sedimentary architecture. (C) Differential subsidence affects accommodation. Buried dune sites allow the quantification of floodbasin aggradation rates, using age–depth relations of basal peat dates as a proxy for the rising groundwater (GW) level (Jelgersma, 1961; Van de Plassche, 1982; Van Dijk et al., 1991). Differential subsidence makes groundwater-rise curves diverge back in time.

Aims and approach

As fault-activity and fluvial deposition in the Rhine–Meuse delta occurred simultaneously during the Late-glacial and Holocene, a sedimentary response reflecting tectonic–alluvial interplay can be expected in the shallow subsurface. For the central Rhine–Meuse delta, indications for Holocene tectonics have been reported by Törnqvist et al. (1998), focusing on aggradation rates, and by Berendsen & Stouthamer (2000), Stouthamer & Berendsen (2000) and Stouthamer (2001), focusing on avulsion history. This study aims to identify the sedimentary response upstream and downstream of the PBF, and use it to quantify differences in rates of subsidence. Specific aims of this study are to:

1 trace active faults of the PBF within Holocene coastal prism and the Late-glacial subsurface (to recognize sedimentary features related to differential subsidence, a solid framework for the sedimentary architecture of the study area is needed and a database of over 200 000 borehole descriptions and 1250 ^{14}C ages is available);

2 quantify differential subsidence over time and space and constrain the timing of activity in relation to the fluvial response, and discuss if displacements are related to gradual tectonic creep or to episodic events (displacement related to earthquakes);

3 discuss interplay between local tectonic control and downstream and upstream controls, as well as its implications for coastal prism aggradation.

GEOLOGY OF THE STUDY AREA

Through most of the Quaternary, the Rhine valley in The Netherlands was oriented SE–NW, following the structural trend. Since the Saalian glaciation (marine oxygen isotope stage (OIS) 6), however, the Rhine has essentially followed an E–W course, bordered to the north by relict ice-pushed ridges (Fig. 1). Thus, the river followed a glacially imposed course, oblique to the structural trend, forcing it to cross relatively stable blocks before entering the Roer Valley Graben in the study area (Cohen et al., 2002). An alluvial valley formed during OIS 4–2 (Törnqvist et al., 2000; Wallinga, 2001) as sea level had dropped considerably, exposing the North Sea floor and extending the Rhine river several hundreds of kilometres (Gibbard, 1995). In the Late-glacial (OIS 2–1 transition) rivers incised in response to climatic amelioration. In the middle Holocene this was followed by back-filling and the formation of a coastal prism, creating well-preserved and datable deposits for each subsequent stage in the delta evolution (Figs 3 & 4).

In the study area, middle to late Holocene deposits fill the Late-glacial palaeovalley and reach a thickness of 4 m in the east and 8 m in the west. The present surface has a gentle westward slope ($c.$ 0.10 m km^{-1}) compared with the steeper Late-glacial (OIS 2) subsurface ($c.$ 0.30 m km^{-1}). Upstream of the study area, the coastal prism thins. Downstream of the study area, the valley-fill thickens and spreads over older deposits to the north and south, reaching a width of over 40 km (Fig. 1).

Sedimentary architecture of the coastal prism

The coastal prism in the study area (Fig. 3) consists of fine floodbasin deposits (intercalated beds of clay, silty clay, humic clay and peat) that are intersected by sandy channel belts and associated smaller channels (including some crevasse channels). Widespread peats and clayey peats occur in the central ('distal') parts of middle Holocene floodbasins, representing times when regional groundwater rise outran clastic deposition. Local intercalated clayey beds occur within these generally peaty floodbasins, which can be traced laterally to channel belt deposits (levee and crevasse-channel facies). At the base of the Holocene sequence a clayey peat is found. Its presence indicates the onset of aggradation, as the basal peat formed when groundwater levels rose

above the valley surface. Basal-peat dates from the study area range between 7200 and 6800 cal. yr BP. Isolated patches of older basal peat are also found (i.e. a strongly humic clay conventionally dated at 8500 cal. yr BP; Hofstede et al., 1989). These patches represent wet environments in small lows on the early Holocene floodplains, inherited from the underlying surface morphology. Floodbasin peat post-dating 4000 cal. yr BP is rare in this part of the Rhine–Meuse delta.

The Holocene record can be subdivided into two major units: a clay-dominated upper unit of late Holocene age; and a 2–4 m thick, clay–peat interbedded lower unit of middle Holocene age (Fig. 4, unit B). A stacked system of channel belts is embedded in the middle Holocene floodbasin deposits (Fig. 4, unit A), which by $c.$ 5.3 cal. yr BP was fully abandoned. Its deposits were covered by widespread peat, while the main Rhine discharge was routed north of the study area (see Berendsen & Stouthamer, 2001). Deposition of the upper 2 m of floodbasin clay (Fig. 4) started $c.$ 4.9 cal. yr BP, and coincided with late Holocene decreased aggradation rates. The abundant upper Holocene clay is attributed to changes in sediment discharge caused by both climatic change and human deforestation in the Rhine drainage basin and frequent nearby avulsions (Stouthamer & Berendsen, 2000), which brought a series of distributary channels into existence in the area (Fig. 4, unit C). The present 'Lek' and 'Linge' channels (Fig. 3) are the human-controlled end-members of these late Holocene channels.

Late-glacial (OIS 2, Late Weichselian) subsurface

Reconstruction of the Late Weichselian surface (OIS 2, Figs 3 & 4) is based primarily on archived lithological borehole descriptions. Various criteria were applied, discriminating a Last Glacial Maximum (LGM, OIS 2) terrace and incised Late-glacial and early Holocene (OIS 2–1 transition) channel belts. These include geometrical and lithostratigraphical relationships as well as sedimentary and pedogenetic features. The main criteria were: the relative elevation of the top of channel deposits; the lithology of these deposits; the presence, relative elevation, thickness, lithology and structure of the floodplain loam covering the channel deposits (Wijchen Member,

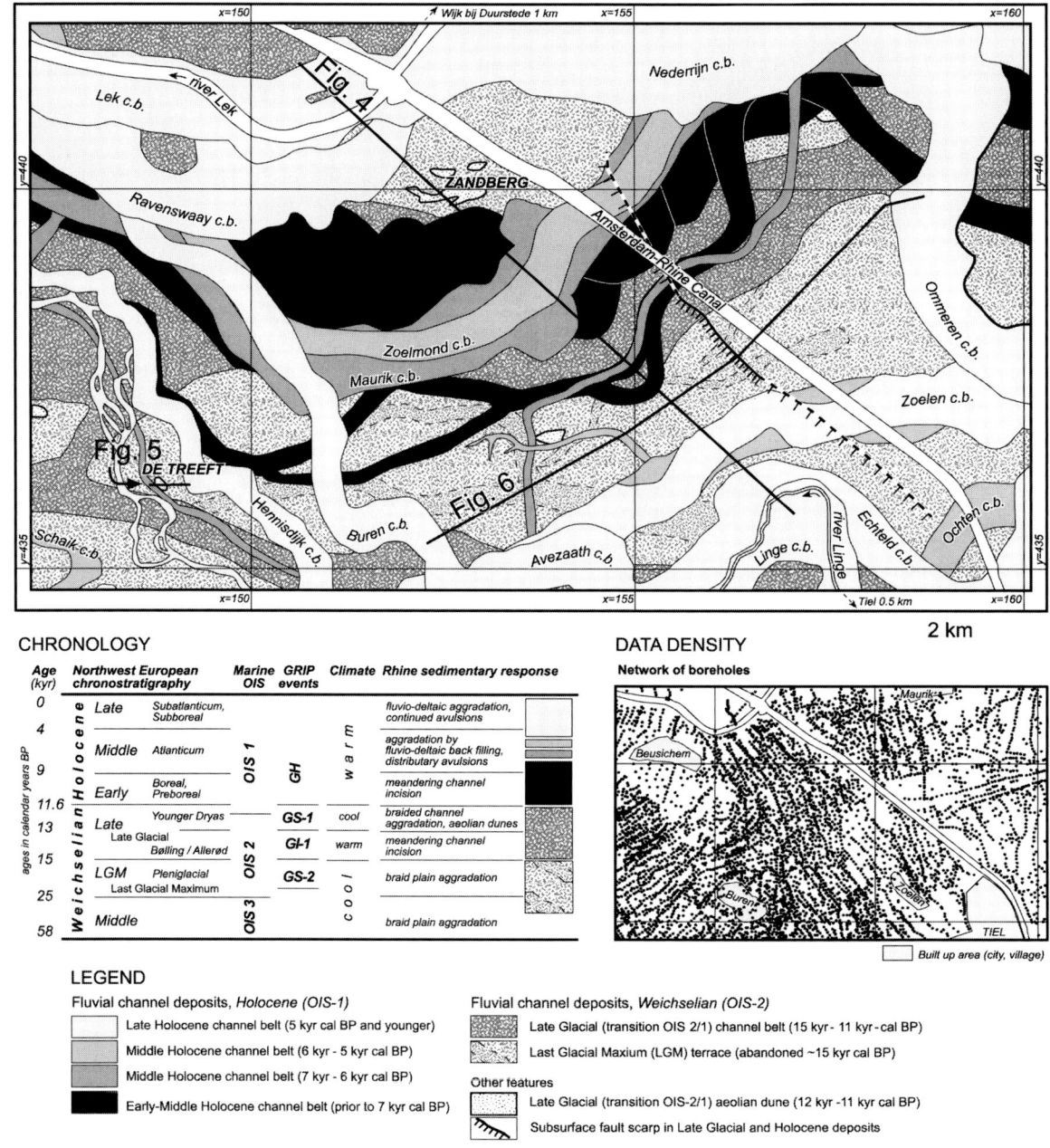

Fig. 3 Geological map and stratigraphy of the study area. The figure integrates new mapping of the Late-glacial, early and middle Holocene channel deposits at the base of the coastal prism with previous mapping of middle and late Holocene systems by Verbraeck (1984), Hofstede et al. (1989), Makaske (1998) and Stouthamer (2001). Names of channel belts are taken from Berendsen & Stouthamer (2001). A correlation of north-west European terrestrial chronostratigraphy with marine (SPECMAP, Martinson et al., 1987) and Greenland ice-core (following Walker et al., 1999) oxygen isotope stages is provided.

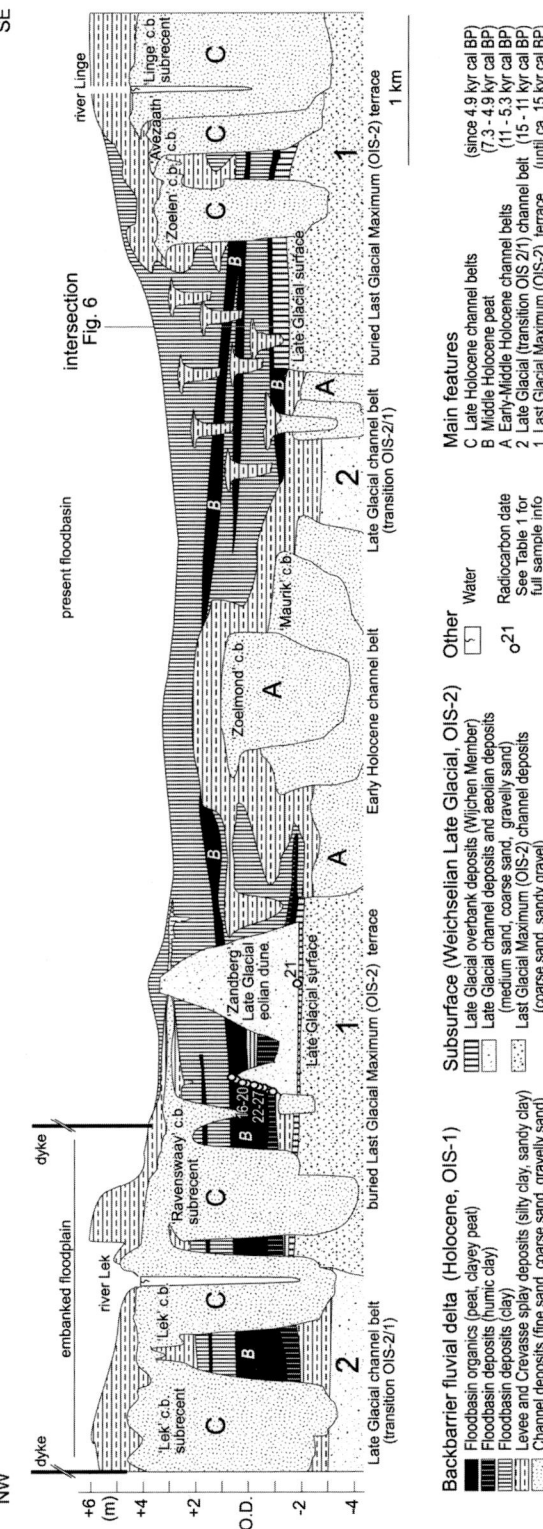

Fig. 4 North-west–south-east cross-section through the study area. The Late-glacial subsurface is characterized by two main levels: 1, a higher Late-glacial maximum level (abandoned braidplain terrace), overlain by Late-glacial overbank deposits (Wijchen Member) and locally Late-glacial dunes; 2, a lower Late-glacial level (incised in the LGM deposits). The local Holocene alluvial architecture is characterized by: A, an early–middle Holocene complex of aggrading channel belts; B, widespread floodbasin peat with interbedded clay (dated at c. 7300–4900 cal. yr BP), covering the Late-glacial surface and partly covering the early–middle Holocene channel-belt complex; C, two late Holocene channel-belt complexes of main delta distributaries in the north and south.

see below); the spatial distribution and thickness of the oldest floodbasin clay beds and longitudinal continuity of channel belts (see Cohen et al., 2002). The highest parts of the subsurface (Fig. 4, unit 1) form a terrace dating to the LGM (OIS 2). Incised channel belts date to the OIS 2–1 transition (Fig. 4, unit 2) and have surfaces at c. 1.5 m below the LGM terrace (Fig. 4). The Wijchen Member (Törnqvist et al., 1994), forming the top of the LGM terrace, is a clay-loam with admixed coarse sand c. 0.50 m thick. This deposit is an important stratigraphical marker in the Rhine–Meuse delta, and represents floodplain deposition during the OIS 2–1 transition, prior to the onset of Holocene aggradation. Scattered over the study area, isolated inland dunes of Late-glacial age (Kasse, 1995; Berendsen & Stouthamer, 2001) occur on top of the LGM terrace, along the younger incised channel belts. The Late-glacial surface is similar to upstream areas, where the OIS 2 terrace geomorphology is still preserved (Berendsen et al., 1995; Huisink, 1997; Tebbens et al., 1999, 2000). The reconstruction further suggests that early Holocene meandering channels reworked the preceding OIS 2–1 channel belt and at some locations laterally cut into the superelevated OIS 2 terrace, similar to early and middle Holocene meander scars found upstream along the Rhine near the Dutch–German border (Klostermann, 1992; Berendsen & Stouthamer, 2001).

METHODS

Field activities in the summer of 1999 and 2000 consisted of collecting 70 additional hand cores (auger holes). Using the method described by Berendsen & Stouthamer (2001), cores were described in the field at fixed intervals of 10 cm. Cores were numbered systematically and the geographical location (±5 m, Dutch coordinate system) and surface elevation (±0.1 m, relative to OD = NAP ≈ mean sea level) were registered. At locations where radiocarbon samples were taken, surface elevation and sample depth were obtained through levelling, with an accuracy of ±2 cm. Fieldwork concentrated on two sites: (i) at the 'Treeft' buried Late-glacial dune (Figs 3 & 5), where basal peat samples were collected to reconstruct Holocene aggradation rates; and (ii) along a longitudinal transect (Figs 3 & 6) in which the effects of neotectonic activity were studied.

Radiocarbon dating groundwater-level index points

Eleven basal peat samples were collected from cores 30–50 cm long spanning the transition of the Late-glacial subsurface to Holocene basal peat. The basal peat in the core was cut in slices 1 cm thick starting 2 cm above the interface with the underlying substrate, and organic macrofossils were selected using a 150 µm sieve. Specific macrofossils, mainly Alnus nuts (following Törnqvist et al., 1998), were radiocarbon dated at the R.J. van der Graaff Laboratory, Utrecht University, using accelerated mass spectrometry (AMS).

The age and depth of 21 basal peat samples (12 samples dated earlier and 9 samples collected for this study) were used as a proxy for groundwater level (Fig. 7). Each basal peat sample directly indicates a palaeogroundwater level. The largest source of uncertainty in this proxy is the tolerance of the dated plant species to variations of the groundwater level. Age and depth errors related to the sampling are of minor importance. Törnqvist et al. (1998) estimated the vertical accuracy of AMS-dated Alnus peat as a proxy for average groundwater levels in the floodbasin at ±10 cm, with a bias towards levels slightly above the sample, because of its preservation as peat. At the 'Treeft' dune site, the difference in elevation of some of the samples is < 30 cm, and the resulting dates do not overlap within a range of one standard deviation. This indicates that a vertical accuracy of 20 cm for this proxy is a fair estimate, at least for the period of relatively rapid groundwater rise in the middle Holocene. Table 1 presents all the radiocarbon dates used in this paper.

A series of dates from Holocene peat overlying the flanks of isolated inland dunes can be used to reconstruct the rise of groundwater. Age and depth of basal peat dates plot as groundwater-index points on groundwater-rise curves. This method was first applied to quantify relative sea-level rise from near-coastal sites (Jelgersma, 1961; Van de Plassche, 1982). Later it was used in inland fluvial floodbasins (Van Dijk et al., 1991). The location of the basal peat, directly overlying

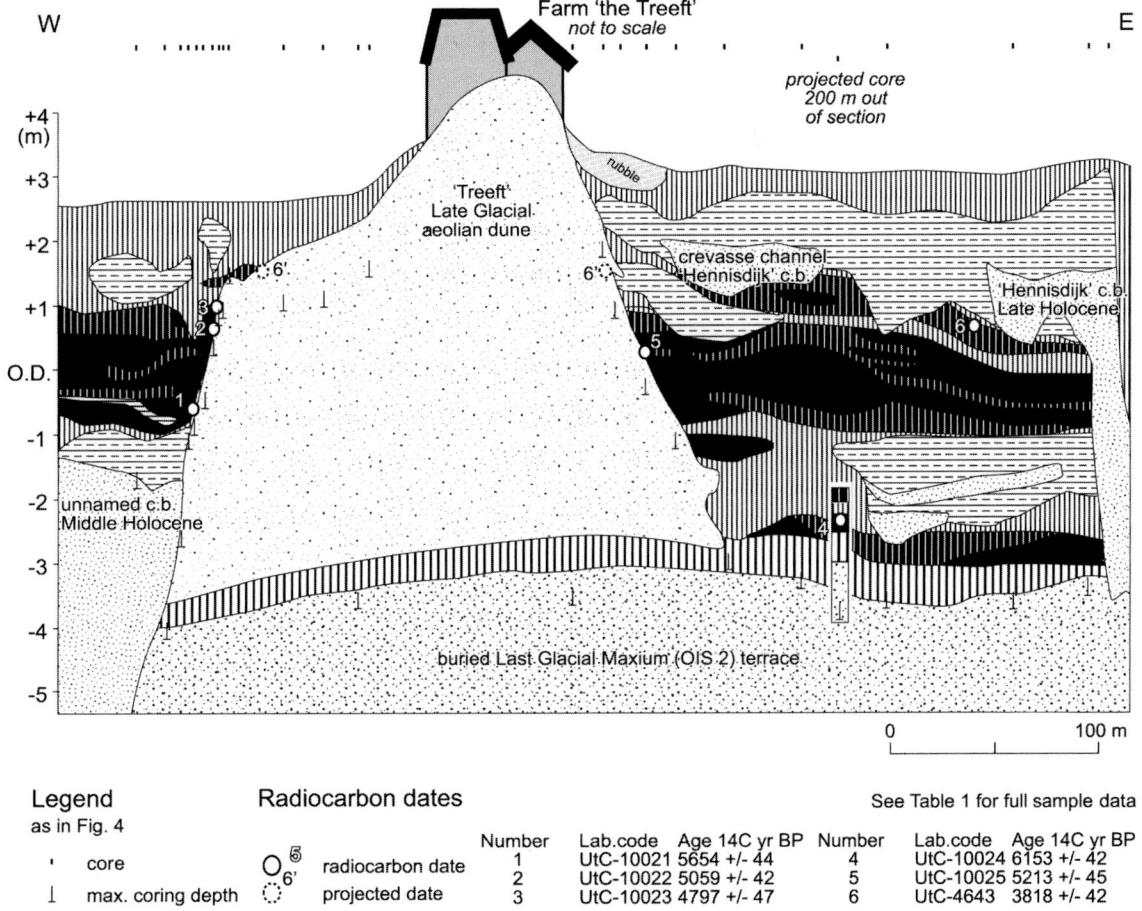

Fig. 5 Cross-section through the 'Treeft' aeolian dune. Holocene peat directly overlying a Late-glacial inland dune was dated at five locations to reconstruct groundwater-level rise. An earlier dated sample (sample 6; Makaske, 1998) was used as an additional (indirect, projected) groundwater-level index point. Ornamentation is the same as in Fig. 4.

dune sand, virtually excludes effects of post-depositional compaction, and the elevation of the basal peat samples directly represents the regional groundwater-table elevation at the time of peat formation. Alternatively, dates from basal peat overlying the LGM terrace may be used as groundwater index points. Prior to the formation of this peat, the terrace surface was superelevated above regional groundwater levels for a considerable period of time. The Wijchen Member loam on top of the LGM terrace represents a period of 3–6 kyr between abandonment (c. 15 000 cal. yr BP) and middle Holocene peat formation (c. 7000 cal. yr BP). The structure of the Late-glacial loam indicates pedogenesis (i.e. lessivage). The 0.5-m-thick loam is assumed to have compacted to this thickness mainly during stages of active pedogenesis and subaerial exposure, and did not compact after burial. The loam on average has the same thickness all over the Rhine–Meuse delta, whether it is buried by 15 m of Holocene clay and peat, by 5 m of Holocene clay, or hardly buried at all and remains partly subaerially exposed (Berendsen & Stouthamer, 2001). Hence, compaction of the loam after peat formation can be neglected, and basal peat samples from this type of location can be used as groundwater indicators, similar to basal peat at dune flanks.

Groundwater levels prior to the onset of aggradation cannot be reconstructed using basal peat

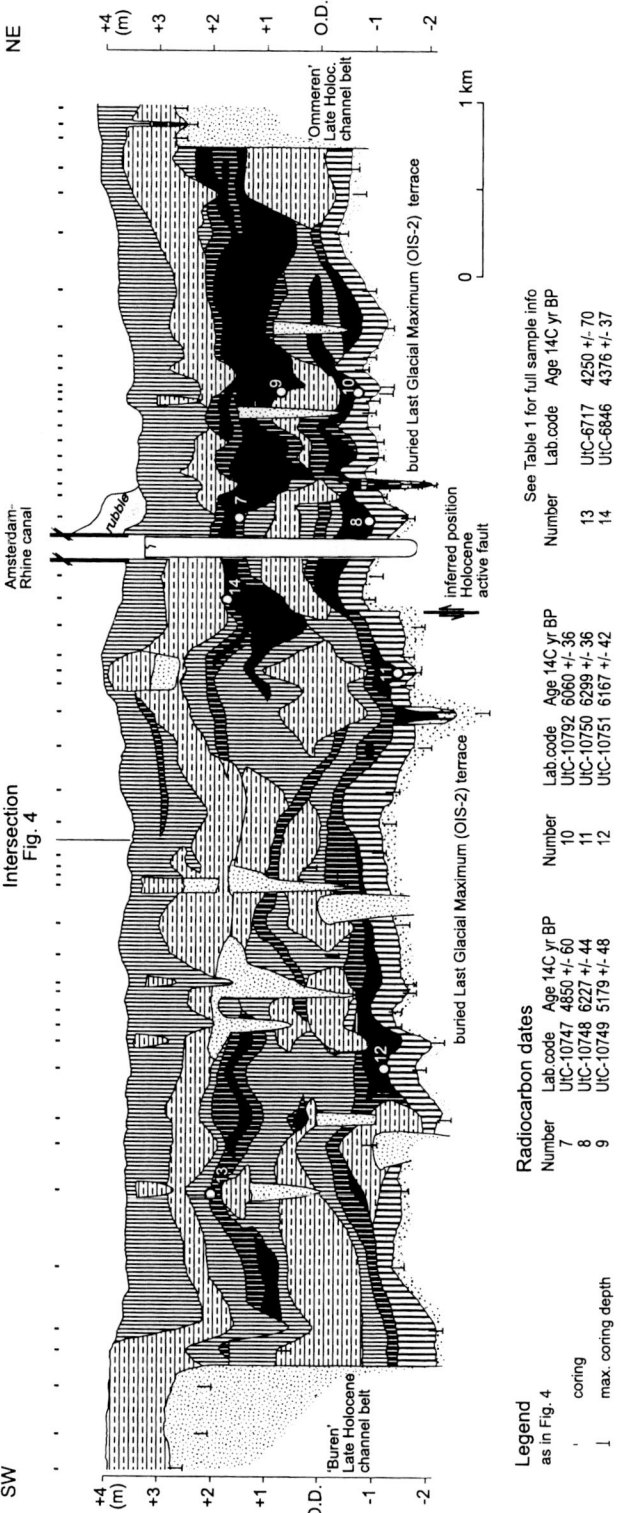

Fig. 6 South-west–north-east cross-section through the study area. The floodbasin sequence consists of a lower unit of interbedded clay and peat and a clay-dominated upper unit. The lower unit is clay-dominated in the downstream part, and peat-dominated in the upstream part. An abrupt change in facies is observed 200 m west of the Amsterdam–Rhine canal, and coincides with an irregularity in the LGM subsurface. At this position an active fault is inferred. Ornamentation is the same as in Fig. 4.

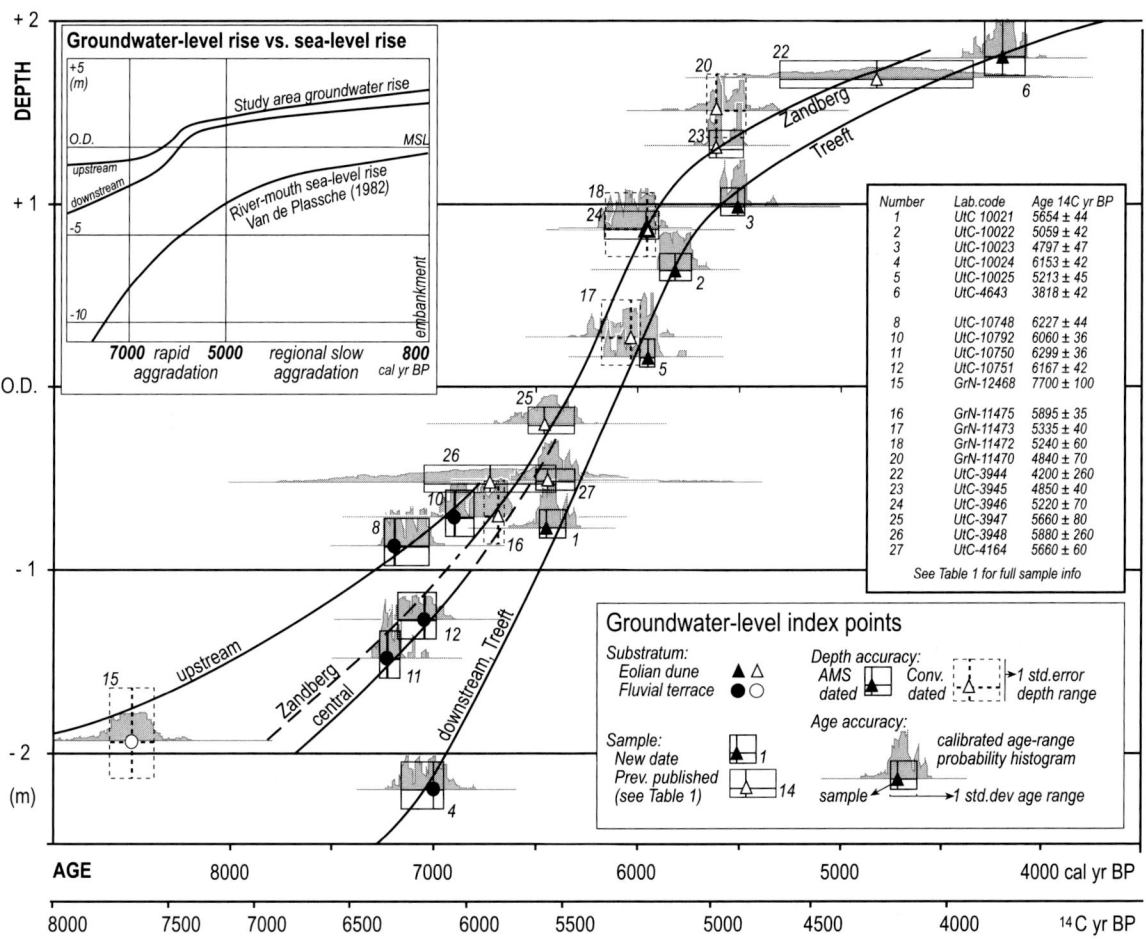

Fig. 7 Age–depth plot of radiocarbon dated basal peats. The age–depth relation for basal peat is a direct proxy for the yearly average palaeogroundwater level in the fluvial floodbasin. Basal peat samples are used as groundwater-index points, with standard deviation error boxes. The width of the error box represents the maximum and minimum of the calibrated calendar age range (Table 1), with the index point plotted at the mean of the largest calibrated-age subrange. Heights of error boxes (following Törnqvist et al., 1998): 0.15 m for AMS-dated samples in series, 0.25 m for isolated AMS-dated samples, 0.35 m for conventionally dated samples in series, 0.50 m for isolated conventionally dated samples. Of this height, two-thirds is above the index point and one-third is below it, incorporating bias towards palaeogroundwater levels slightly above the preservation depth of the sampled peat. Age-range probability histograms for individual samples were created using the OxCal program (Bronk Ramsey, 1995; version 3.5) and the latest calibration data set (Stuiver et al., 1998).

levels, but some general estimates can be made based on subaerial pedogenetic features in the Wijchen Member, and on the preservation of in-fills of Late-glacial residual channels. The Wijchen Member pedogenetic features point to groundwater levels (before the onset of aggradation) that must have been at least 0.5 m below the surface for some time. Residual channels filled with Late-glacial and early Holocene peat and gyttja in oxbow lakes have been preserved locally and indicate a minimum early Holocene groundwater level of 0.5–2.0 m below the LGM terrace surface (Cohen, 2003). Thus, unlike sea level, the groundwater-level was not tens of metres below the present level in the early Holocene, but at most 1–2 m below the OIS 2–1 valley surface, which is consistent with hydrological considerations of groundwater and fluvial discharge (De Vries, 1974).

Table 1 Radiocarbon-dated samples.

Site	Sample number*	Laboratory number	Date ± 1σ (^{14}C yr BP)	1σ calendar age range (cal. yr BP)	Sample name	x,y coordinates (km) and surface elevation (m)‡	Sample depth (m ± OD)	Source material and stratigraphical position†	Dated material: seeds, fruits, nuts, leaves, buds	References to full site descriptions
Buried dune 'Treeft' (Figs 5 & 7)	1	UtC 10021	5654 ± 44	6490–6400, 6365–6360	Treeft-1	148.611/436.265 2.47 m +NAP	0.77 m –NAP	Basal Alnus peaty clay on LG aeolian dune	6 Alnus glutinosa, nuts; 1 Solanum dulcamara, seed; 4 Urtica dioica, fruits; 3 leaf fragments; 98 bud scales; 4 buds	This paper, Fig. 5
	2	UtC 10022	5059 ± 42	5900–5740	Treeft-2	148.621/436.265 2.60 m +NAP	0.64 m +NAP	Basal Alnus clayey peat on LG aeolian dune	16 Alnus glutinosa, nuts; 1 Alisma plantago aquatica, fruit; 1 Typha, fruit; 1 Lycopus eurpaeus, mericarp; 2 buds; 91 bud scales	This paper, Fig. 5
	3	UtC 10023	4797 ± 47	5590–5575, 5540–5475	Treeft-3	148.627/436.264 2.34 m +NAP	0.99 m +NAP	Basal Alnus clayey peat on LG aeolian dune	16 Alnus glutinosa, nuts; 2 Alisma plantago aquatica, fruits; 1 Oenanthe aquatica, mericarp; 1 Typha, fruit; 1 Carex riparia, nut; 4 Urtica dioica, fruits; 4 leaf fragments; 1 bud; 11 bud scales	This paper, Fig. 5
	4	UtC 10024	6153 ± 42	7160–7120, 7090–6970, 6960–6950	Treeft-4	148.925/436.450 2.77 m +NAP	2.20 m –NAP	Strongly humic clay on LG terrace	2 Alnus glutinosa, cones; 11 Alnus glutinosa, nuts; 10 bud scales	This paper, Fig. 5
	5	UtC 10025	5213 ± 45	5990–5920	Treeft-5	148.830/436.270 3.05 m +NAP	0.16 m +NAP	Basal Alnus clayey peat on LG aeolian dune	2 Solanum dulcamara, seeds; 4 Alisma plantago aquatica, fruits; 1 Oenanthe aquatica, mericarp; 39 bud scales	This paper, Fig. 5
	6	UtC 4643	3818 ± 42	4290–4270, 4260–4140, 4110–4090	Schoon'wrd.-7	148.865/437.940 2.50 m +NAP	0.24 m +NAP	Humic clay underlying Holocene silty clay	4 Scirpus, seeds; 1 Solanum, seeds	Makaske (1998, appendix 4-[13])

	Lab code	Date	Calibrated range	Site	Coordinates / elevation	Depth	Description	Macrofossils	Reference
Longitudinal transect 'Maurik-Buren' (Figs 6 & 7)									
7	UtC 10747	4850 ± 60	5640–5630, 5610–5580, 5500–5490	Maurik-1	156.667/438.169 4.21 m +NAP	1.67 m +NAP	Alnus/Phragmites peat underlying Holocene clay	4 Alnus glutinosa, nuts; 50 Typha, fruits; 10 Carex rostrata, nuts; 1 Oenanthe aquatica, mericarp	This paper, Fig. 6
8	UtC 10748	6227 ± 44	7240–7155, 7130–7090, 7080–7025	Maurik-2	156.667/438.169 4.21 m +NAP	0.87 m −NAP	Basal Alnus clayey peat on LG terrace	5 Carex rostrata, nuts; 3 Solanum dulcamara, seeds; 19 budscales; 3 abcission layers	This paper, Fig. 6
9	UtC 10749	5179 ± 48	5990–5970, 5950–5910	Maurik-3	157.237/438.729 3.71 m +NAP	0.94 m −NAP	Alnus peaty clay underlying Holocene clay	6 Alnus glutinosa, nuts; 3 Solanum dulcamara, seeds; 1 Carex rostrata, nut; 7 Oenanthe aquatica, mericarps; 1 Urtica diocia, fruit; 4 leaf fragments	This paper, Fig. 6
10	UtC 10792	6060 ± 36	6950–6860, 6820–6800	Maurik-4	157.237/438.729 3.71 m +NAP	0.72 m −NAP	Basal Alnus clayey peat on LG terrace	24 Alnus glutinosa, nuts.	This paper, Fig. 6
11	UtC 10750	6299 ± 36	7270–7210, 7190–7180, 7170–7160	Maurik-5	156.160/437.527 4.03 m +NAP	1.49 m −NAP	Basal Alnus peat on LG terrace	2 Alnus glutinosa, nuts; 9 Urtica diocia, fruits; 2 Solanum dulcamara, seeds; 12 Oenanthe aquatica, mericarps; 3 Alisma plantago aquatica, fruits; 6 Carex sp., nuts; 1 bud scale.	This paper, Fig. 6
12	UtC 10751	6167 ± 42	7180–7170, 7160–7110, 7100–6990	Maurik-6	154.146/436.494 3.44 m +NAP	1.28 m −NAP	Basal Alnus clayey peat on LG terrace	5 Alnus glutinosa, nuts; 12 Polygonum, fruits; 1 Carex rostrata, nut; 100 Alisma plantago aquatica, fruits; 1 Ranunculus sceleratus, fruits; 14 Oenanthe aquatica, mericarps; 2 leaf fragments; 1 bud scale	This paper, Fig. 6
13	UtC 6717	4250 ± 70	4880–4800, 4770–4630	Buren	153.023/435.450 3.90 m +NAP	2.31 m +NAP	Peat underlying Holocene clay	90 Ranunculus sceleratus, fruits; 5 Alnus, nuts	This paper, Fig. 6

(cont'd)

Table 1 (cont'd)

Site	Sample number*	Laboratory number	Date ± 1σ (^{14}C yr BP)	1σ calendar age range (cal. yr BP)	Sample name	x,y coordinates (km) and surface elevation (m)‡	Sample depth (m ± OD)	Source material and stratigraphical position†	Dated material: seeds, fruits, nuts, leaves, buds	References to full site descriptions
	14	UtC 6846	4376 ± 37	4970–4860	Zoelen-244	156.230/437.870 3.50 m +NAP	1.74 m +NAP	Peat underlying Holocene clay	9 *Alnus glutinosa*, nuts; 2 *Oenanthe*, mericarps	This paper, Fig. 6
	15	GrN 12468	7700 ± 110	8600–8380	Broek-I	156.490/439.540 3.30 m +NAP	1.94 m –NAP	Stongly humic clay on LG terrace	Bulk sample	Hofstede et al. (1989, fig. 2)
Buried dune 'Zandberg' (Figs 4 & 7)	16	GrN 11475	5895 ± 35	6755–6715, 6710–6665	Zandberg-2	152.308/440.387 3.99 m +NAP	0.71 m –NAP	Basal *Alnus* peat on LG aeolian dune	Bulk sample	Van Dijk et al. (1991, fig. 6)
	17	GrN 11473	5335 ± 40	6180–6160, 6150–6090, 6080–5990	Zandberg-4	152.351/440.339 3.79 m +NAP	0.27 m +NAP	Basal *Alnus* peat on LG aeolian dune	Bulk sample	Van Dijk et al. (1991, fig. 6)
	18Ê	GrN 11472	5240 ± 60	6170–6140, 6120–6040, 6020–5920	Zandberg-5	152.378/440.307 3.82 m +NAP	0.86 m +NAP	Basal *Alnus* peat on LG aeolian dune	Bulk sample	Van Dijk et al. (1991, fig. 6)
	19§	GrN 11471	4575 ± 40 (rejected)	5450–5420, 5330–5280, 5170–5130, 5110–5070	Zandberg-6	152.385/440.290 3.85 m +NAP	1.08 m +NAP	Basal *Alnus* peat on LG aeolian dune	Bulk sample	Van Dijk et al. (1991, fig. 6)
	20	GrN 11470	4840 ± 70	5660–5570, 5550–5470	Zandberg-7	152.394/440.288 3.88 m +NAP	1.51 m +NAP	Basal *Alnus* peat on LG aeolian dune	Bulk sample	Van Dijk et al. (1991, fig. 6)

#	Lab code	14C age	Cal age	Site	Coordinates / m +NAP	Elevation	Context	Bulk sample	Reference
21	GrN 11469	11 700 ± 100	13 840–13 740, 13 710–13 470	Zandberg-10	152.621/440.044 3.99 m +NAP	2.20 m −NAP	Peat underlying LG aeolian dune		Van Dijk et al. (1991, fig. 6)
22	UtC 3944	4200 ± 260	5300–4350	Zandberg-I-1	152.387/440.281 3.63 m +NAP	1.68 m +NAP	Basal Alnus peat on LG aeolian dune	5 Alisma plantago, fruits	Törnqvist et al. (1998, fig. 4)
23	UtC 3945	4850 ± 40	5650–5580, 5510–5480	Zandberg-II-1	152.383/440.285 3.57 m +NAP	1.30 m +NAP	Basal Alnus peat on LG aeolian dune	2 Carex sp., nuts	Törnqvist et al. (1998, fig. 4)
24	UtC 3946	5220 ± 70	6170–6140, 6120–6070, 6060–6040, 6030–5900	Zandberg-III-1	152.37/440.300 3.57 m +NAP	0.86 m +NAP	Basal Alnus peat on LG aeolian dune	1 Carex rostrata, nut; 1 Carex sp. nut; 1 Solanum dulcamara, seed	Törnqvist et al. (1998, fig. 4)
25	UtC 3947	5660 ± 80	6540–6390, 6380–6310	Zandberg-V-1	152.326/440.354 3.62 m +NAP	0.21 m +NAP	Basal Alnus peat on LG aeolian dune	17 Ranunculus sceleratus, fruits	Törnqvist et al. (1998, fig. 4)
26	UtC 3948	5880 ± 260	7050–6400	Zandberg-VI-1	152.287/440.400 3.84 m +NAP	0.53 m −NAP	Basal Alnus peat on LG aeolian dune	14 Alnus glutinosa, nuts	Törnqvist et al. (1998, fig. 4)
27	UtC 4164	5660 ± 60	6500–6390, 6380–6340, 6330–6310	Zandberg-VI-2	152.287/440.400 3.84 m +NAP	0.52 m −NAP	Basal Alnus peat on LG aeolian dune	33 Alnus glutinosa, nuts	Törnqvist et al. (1998, fig. 4)

*Dates for samples 1–6 and 13–27 are also documented in Berendsen & Stouthamer (2001, appendices 1 & 2).
‡NAP = Dutch Ordnance Datum ≈ mean sea level.
†LG = Late-glacial (end of marine oxygen isotope stage 2).
§Sample 19 gives an anomalous date for its depth position, most probably owing to the sampled peat not being in situ but on a slumped part of the steep dune flank. Van Dijk et al. (1991) originally rejected samples 18 and 20 as being too old, but later AMS datings (samples 23 and 24; Törnqvist et al., 1998) reconfirmed those dates, making sample 19 the one to be rejected.

SITE DESCRIPTIONS

For the 'Zandberg' inland dune, located centrally in the study area, a series of samples was available (Table 1; Van Dijk et al., 1991; Törnqvist et al., 1998). More basal peat samples were collected in its vicinity, as well as from new sites to expand the age–depth range and to increase the spatial coverage and temporal resolution of the reconstructed aggradation rates.

'Treeft' buried dune site

Five basal peat samples were collected from the flanks of the 'Treeft' buried aeolian dune (Fig. 5) in the western part of the study area (Fig. 3). All samples were used as groundwater-index points. Samples 1, 2 and 3 (Table 1) from the western flank and sample 5 from the eastern flank were taken from a clayey peat layer on-lapping the dune between −0.5 and +1.0 m. The presence of a palaeosol directly beneath these samples suggests an undisturbed contact between peat and a palaeosurface. Disturbed basal-peat contacts (i.e. resulting from slumps along steep dune flanks) were avoided because macrofossils at such locations are not *in situ* and thus do not represent the groundwater level. Sampling from local depressions was avoided by drilling cores at distances of less than 5 m. Local depressions have to be avoided, because peat formation here may have started earlier (Van de Plassche, 1982; Törnqvist et al., 1998). Sample 4 (Table 1; Fig. 5; −2.20 m), at the lowest elevation, dates basal peat that formed at a relatively high elevation on the Late-glacial surface.

At the 'Treeft' site, the transition (at +1 m OD) between the peat–clay intercalated lower unit and the clayey upper unit marks the beginning of activity of the nearby 'Hennisdijk' channel belt (Fig. 3), which has been dated by various authors (Verbraeck, 1984; Makaske, 1998). Sample 6 (Table 1; Makaske, 1998) is from a humic clay layer that can be traced across the section. At the western flank it intersects the dune at +1.5 m OD, indicating a groundwater level at that stratigraphical position just before the 'Hennisdijk' channel formed. As this elevation is derived indirectly from laterally tracking a lithological transition, this index point is less accurate than data provided by directly dated basal peat samples.

Longitudinal section

To trace the PBF within the Holocene sediments and Late-glacial subsurface, a longitudinal section 7 km long was cored, at a location chosen using the data available (Fig. 3). The section aimed: (i) to traverse mainly distal floodbasin deposits, as far away as possible from Holocene channels; (ii) to cover a relatively old patch of the Late-glacial subsurface; (iii) to cover a large distance both upstream and downstream from the approximate position of the PBF. The section is located parallel to the present 'Lek' and 'Linge' channels (Fig. 3), and terminates at the subrecent 'Ommeren' and 'Buren' channel belts (Figs 3 & 6). It was constructed using new cores and additional selected cores from the database. In four cores, clayey basal peat was sampled to obtain radiocarbon dates for the onset of aggradation along the section (sample 8, 10, 11 and 12; Table 1). Four additional peat samples were used to further date the floodbasin sequence (samples 7, 9, 13 and 14; Table 1).

Within the section, several smaller channel belts and crevasse channels of middle Holocene age are recognized that have coexisted with the 'Maurik' and 'Zoelmond' (Fig. 3) channel belts, and some late Holocene crevasse channels were also recognized. The main part of the section consists of floodbasin deposits, with the previously described middle Holocene clay–peat intercalated lower unit (−1 to +2 m) and late Holocene clay-dominated upper unit (> 2 m). Following the radiocarbon dates (Table 1) the middle Holocene unit formed between 7200 cal. yr BP and 4900 cal. yr BP, covering the period of highest Holocene aggradation rates in this area.

RESULTS

Fault displacement in the longitudinal section

To find the Late-glacial and Holocene active faults of the PBF zone, longitudinal changes in sedimentary facies and vertical deformation of isochron surfaces were sought within the coastal prism (see Fig. 2). Three indications for neotectonic activity were found in the longitudinal section (Fig. 6).
1 East of the Amsterdam–Rhine canal, the lower floodbasin unit consists predominantly of peat,

whereas to the west clay beds are intercalated in the peat (Fig. 6). This is in contrast to the general trend in the Rhine–Meuse delta, which is a downstream decrease in clastics and an increase in peat (Törnqvist, 1993; Berendsen & Stouthamer, 2001). The floodbasin clays intercalated in the peat are the distal ends of the crevasse splays of crevasse channels that are also recognized in the lower floodbasin unit. Alternative ways of sediment delivery, such as discharge overtopping the levee instead of breaching it, will have contributed sediment too, but the number of crevasse channels and the lateral grading into distal clayey splays and floodbasin clayey peat point to splay formation as the most important contributor. The local, abrupt change of facies seems to be related to differences in aggradation rates in the upstream and downstream part of the section: clastic sediment entering the floodbasin from the surrounding channels was preferentially deposited in the most subsiding area (similar to Fig. 2B). This does not necessarily imply that fault activity caused breaching and created crevasse splays, but apparently splays developed preferentially in the more subsiding area.

2 Middle Holocene aggradation over the lower parts of the LGM (OIS 2) terrace surface started around 7200 cal. yr BP (samples 8 and 11, Table 1) and some 100 yr later covered the higher parts of the surface (samples 10 and 12). The four basal peat samples in the section thus all yielded similar ages (7200–6900 cal. yr BP, Table 1), but the downstream samples (11 and 12) and upstream samples (8 and 10) differ by c. 0.60 m in elevation (Fig. 6). This elevation difference occurs over a short distance (2 km between the mean coordinates of both sample pairs) and the apparent gradient in the 7100 cal. yr BP groundwater level is high (0.3 m km^{-1}) compared with later Holocene gradients (0.1–0.2 m km^{-1}, Van Dijk et al., 1991). The lateral continuity of the basal peat and its macrofossil composition suggest that the palaeogroundwater level was fluvially controlled and had a gradient that changed only very gradually over space. It is therefore concluded that the present c. 0.60 m basal peat elevation difference is anomalous: after the peat formed, downstream samples must have subsided relative to upstream samples (post-depositional deformation, as in Fig. 2A). Not all of the c. 0.60 m sample-offset can be attributed to vertical deformation; corrections need to be made (see below).

3 One of the major irregularities in the buried LGM (OIS 2) terrace coincides with the longitudinal facies change in the overlying Holocene sequence and the palaeogroundwater age–depth anomaly at the onset of aggradation. The top of the Late-glacial surface was mapped in a zone 4 km wide along the section. The offset could be traced for about 2 km perpendicular to the transect of Fig. 6 and was found to cross-cut the LGM (OIS 2) terrace as well as the Late-glacial (OIS 2–1) channel belt (Fig. 3). The upstream side is about 0.7–1.0 m higher than the downstream side. The NW–SE direction matches the strike of the faults known from the deeper subsurface (Van Montfrans, 1975). It is concluded that the irregularity marks one of the faults of the PBF that was active during the past 15 kyr, dividing the section into an upstream reach that was relatively elevated and a downstream reach that was relatively subsiding (similar to Fig. 2A).

The above three phenomena occur at the same location, which, when combined with the known existence of faults at depth, leads to the conclusion that differential subsidence in the study area not only caused post-depositional deformation along the inferred fault, but also was a syn-depositional control over the past 15 kyr.

Reconstructed groundwater-level rise

Holocene groundwater-level rise reflects increasing accommodation space (Van Dijk et al., 1991; Blum & Törnqvist, 2000). Curves representing regional (floodbasin) groundwater-level rise are presented in an age–depth graph (Fig. 7). All curves essentially have the same sigmoidal shape, with a rise that increased but remained relatively slow prior to the onset of extensive basal peat formation at 7200 cal. yr BP. The most rapid rise occurred after the onset of basal peat formation, and rates decreased after 5500 cal. yr BP. The difference in elevation between individual curves is primarily the effect of the river gradient: like the water-level gradient *in* the active channels, the groundwater level in the floodbasins *between* the channels is higher at sites further upstream (Van de Plassche, 1982; Van Dijk et al., 1991). To draw the curves, all dates from the 'Treeft' site (Fig. 5) and from the longitudinal section (Fig. 6) were accepted as correct. The previously published dates of the 'Zandberg' site (Van Dijk et al., 1991; Törnqvist et al., 1998) and surrounding area (Hofstede et al., 1989) were all accepted, except one sample that was rejected because it was clearly too young (date 19; Table 1).

The upstream curve covers the 8500–6500 cal. yr BP period, and represents the stage prior to extensive basal peat formation, when peat formation occurred only locally (date 15). The 'Treeft' and 'Zandberg' sites cover the period 7500–4000 cal. yr BP and show that by c. 7000 cal. yr BP (samples 4, 8 and 10–12) groundwater level had risen above the OIS 2 valley surface. The steep curve between 7000 and 5000 cal. yr BP represents the stage of most rapid groundwater-level rise, with widespread peat formation in the distal parts of the floodbasin, and crevasse-splay deposition (see above) downstream of the inferred fault (Fig. 6). After 5700 cal. yr BP rates decreased towards an approximately linear rate of 0.5 m kyr^{-1}. After 4000 cal. kyr BP peat is absent in the floodbasins, hence no basal peat index points are available from this period.

The 'Treeft' site is located 5 km downstream of the 'Zandberg' site. Groundwater levels at 7000 cal. yr BP differ by c. 1.2 m between the two sites (Fig. 7), yielding a floodbasin groundwater gradient of c. 0.2 m km^{-1}, approaching the valley gradient of the OIS 2–1 floodplain (0.25–0.30 m km^{-1}). At this time, aggradation had just set in and channel belts in the study area were adapting the gradient inherited from the Late-glacial valley to a lower, sea-level influenced, gradient (c. 0.1 m km^{-1}) of the coastal prism (Van Dijk et al., 1991). The groundwater curves show that by 6000 cal. yr BP the floodbasin gradient had decreased to c. 0.1 m km^{-1} and has remained constant ever since.

QUANTIFIED DIFFERENTIAL SUBSIDENCE

Vertical displacements along the inferred fault in the longitudinal section (Fig. 6) are quantified for two stratigraphical levels of known age, respectively the palaeogroundwater level in middle Holocene basal peat (Table 2) and the level of top-of-channel deposits in the LGM (OIS 2) buried terrace (see Table 3).

Table 2 Displacement rates from middle Holocene floodbasin deposits.

(a) Based on offset between two samples*, 0.8 km apart							(b) Based on offset between two means of sample pairs‡, 2.0 km apart						
Age difference samples (cal. yr)	Age difference offset† (m)	Displacement rate (m kyr^{-1}) assuming groundwater gradients (m km^{-1}) of:					Age difference pair-means (cal yr)	Age difference offset† (m)	Displacement rate (m kyr^{-1}) assuming groundwater gradients (m km^{-1}) of:				
		0.10	0.15§	0.20§	0.25§	0.30			0.10§	0.15§	0.20§	0.25	0.30
0	0.00	0.075	0.069	0.064	0.058	0.053	0§	0.00§	0.055	0.041	0.027	0.013	<0
50§	0.04§	0.069	0.064	0.058	0.053	0.047	50§	0.04§	0.049	0.035	0.021	0.007	<0
100§	0.08§	0.064	0.059	0.053	0.047	0.042	100§	0.08§	0.044	0.030	0.015	0.001	<0
150§	0.12§	0.059	0.053	0.048	0.042	0.036	150	0.12	0.038	0.024	0.010	<0	<0
200	0.16	0.053	0.048	0.042	0.037	0.031	200	0.16	0.032	0.018	0.004	<0	<0

*Samples 8 and 11; offset 0.62 m; mean age: 7220 cal. yr BP. Downstream sample 8: 7242–7026 cal. BP; 0.87 m OD. Upstream sample 11: 7267–7162 cal. BP; 1.49 m OD.
‡Sample pairs 8, 10 and 11, 12, offset 0.62 m; mean age: 7100 cal. yr BP. Downstream pair 8, 10: 7100 cal. BP; 0.80 m OD. Upstream pair 11, 12: 7100 cal. BP; 1.39 m OD.
†Calculated based on a mean rate of groundwater rise (for 7300–7000 cal. BP) of 0.8 m kyr^{-1} (cf. Fig. 7).
§Rows and columns with most realistic estimates.

Differential subsidence in fluvio-deltaic floodbasin deposits

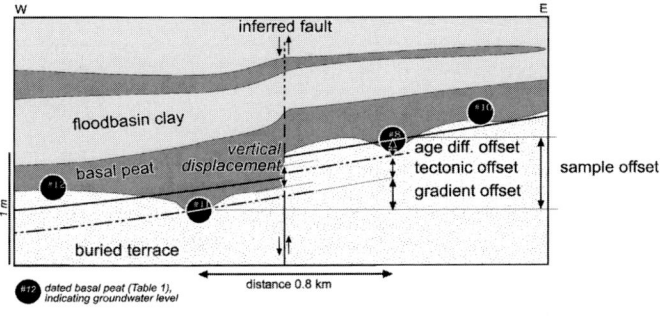

Fig. 8 Calculation of the vertical displacement (tectonic offset) in palaeogroundwater levels indicated by the basal peat. The gradient offset and age-difference offset between the groundwater levels were present at the time the peat was formed ('syn-depositional' situation). By subtracting these offsets from the present sample offset ('present' situation), the tectonic offset can be estimated.

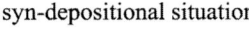

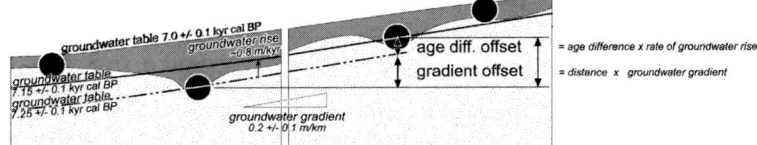

Vertical displacement rates, middle Holocene basal peat

The elevation difference in the 7300–7000 cal. yr BP basal peat (Figs 6 & 7: samples 8 and 10–12) is c. 0.60 m, but only part of this difference is a tectonic offset. To calculate fault displacement from the elevation difference between basal peat samples (sample-offset), a correction is made for: (1) the syn-depositional downstream groundwater gradient between the samples, i.e. the 'gradient' offset; and (2) for differences in age between the samples, i.e. the 'age difference' offset (Fig. 8).

1 The gradient offset (Fig. 8) is calculated by multiplying the known sample distance by a presumed gradient. This gradient ranges between the valley gradient (0.30 m km^{-1}) of the OIS 2–1 floodplain and the present (0.10 m km^{-1}) groundwater gradient (Van Dijk et al., 1991). The steeper the presumed palaeogroundwater gradient, the lower the displacement rate. Inferring the presence of a fault also introduces the possibility of additional local fault-related hydrological effects. In this case of longitudinal gradients in a fluvial floodbasin, however, with an average groundwater table at/above the surface and with highly permeable sand–gravel on either side of the fault in the first few metres below the surface, such effects could be neglected.

2 The calibrated calendar age ranges of the basal peat dates (samples 8 and 11 in particular) on either side of the fault partly overlap in age range. This may indicate a single palaeogroundwater level (zero age difference), but may also allow for an age difference of up to 200 yr. The back-filling setting that caused the peat to form means that it is probable that the downstream sample is older and the upstream sample is younger. For a range of presumed age differences, the 'age difference' offset is calculated by multiplying age-lag with the rate of groundwater rise in the study area (0.8 m kyr^{-1} between 7300 and 7000 cal. yr BP, based on Fig. 7). A larger presumed age difference yields a lower displacement rate estimate.

Estimates for vertical displacement rates of the basal peat were made for different combinations of values covering a realistic range for groundwater gradient and age difference (Table 2). A first estimate was based on the oldest dates on either side of the inferred fault, samples 8 and 11. Both are from relative lows in the buried LGM (OIS 2) terrace (Fig. 6) and are therefore directly comparable groundwater-index points. Alternatively, an estimate based on the means (age and depth) of sample-pairs on either side of the fault was made for the 7100 cal. yr BP groundwater level. This is a slightly younger level compared with the first approach and therefore its groundwater gradient probably was slightly lower. Furthermore, owing

Table 3 Displacement rates from the Last Glacial Maximum (OIS 2) subsurface.

Level of top-of-channel deposits	Displacement rate (m kyr^{-1}) assuming age of abandonment at (cal. kyr BP):		
	13	15*	22
(a) Near-fault subsurface offset: 0.87 m (Fig. 6) 0.80 m OD (two cores, within 200 m upstream of fault) 1.67 m OD (four cores, within 200 m downstream of fault)	0.067	0.058	0.040
(b) Gradient line offset: 0.55 m (Fig. 9) 0.45 m OD (62 cores, within 2.5 km upstream of fault) 1.00 m OD (135 cores, within 5 km downstream of fault)	0.042	0.037	0.025

*Most realistic value.

to the averaging of sample-pairs, the age difference also is less than in the first approach. Hence, in Table 2b, gradients exceeding 0.25 m km^{-1} and age differences exceeding 150 yr are not realistic assumptions, and yield negative values (technically implying reverse movement along the fault). Realistic values for displacement rates ranged between 0.03 and 0.07 m kyr^{-1} (Table 2), averaged over the past 7 kyr. They imply that 0.2–0.4 m of total relative displacement occurred, which is 5–10% of the 5 m of total aggradation just downstream of the inferred fault (Fig. 6).

Diverging groundwater-level curves (Fig. 7) may be an indication of differential subsidence (Fig. 2C). The resolution of basal-peat groundwater index points (depth error c. 0.20 m) only allows comparison of differential effects between older parts of the curves, which have been exposed to low post-depositional differential rates (c. 0.06 m kyr^{-1}) long enough (> 5 kyr) to be displaced by a significant amount (> 0.30 m). The 'upstream' and 'central' curves (Fig. 7) are based on the basal peat sample-pairs of the longitudinal section (Fig. 6). These curves do not span a wide enough temporal range to show clear divergence back in time. The older parts of the groundwater-level curves of the dune sites (7500–6000 cal. yr BP) do diverge back in time. As discussed above, this is mainly the result of the variable groundwater gradient (the older the samples, the steeper the floodbasin gradient and therefore the greater their difference in depth), but the divergence may also include some differential subsidence.

Vertical displacement rates, Last Glacial Maximum (OIS 2) subsurface

The offset in the top of the buried LGM terrace in Fig. 6 is c. 0.9 m, based on the descriptions of the boreholes within 200 m of the fault. An age of abandonment of 15 cal. yr BP is presumed for this terrace, which is consistent with available dating evidence (radiocarbon, OSL, palaeosols, pollen; Verbraeck, 1984; De Jong, 1995; Törnqvist et al., 2000; Wallinga, 2001; Berendsen & Stouthamer, 2001; Cohen et al., 2002). The maximum estimate for vertical displacement rates along the inferred fault is c. 0.06 m kyr^{-1} over the past 15 kyr (Table 3). Alternatively, the offset may be calculated from the analysis of longitudinal profiles of the top of Late-glacial sand, as recorded in 197 cores that penetrated the surface of the buried LGM terrace. The local offset (cf. Cohen et al., 2002) calculated from this gradient line is c. 0.55 m (Fig. 9), 33% less than the offset found in the longitudinal section (Fig. 6). The slope upstream of the inferred fault (0.33 m km^{-1}) is steeper than downstream (0.17 m km^{-1}). These values are respectively high and low, compared with the regional slope of the buried LGM terrace (c. 0.20–0.30 m km^{-1}; Berendsen & Stouthamer, 2001). Steep upstream slopes and gentle downstream slopes in the longitudinal profiles of the Late-glacial valley are also observed further south across the PBF (Stouthamer & Berendsen, 2001; Cohen et al., 2002). Syn-depositional tectonics might explain the lower gradient of the faster

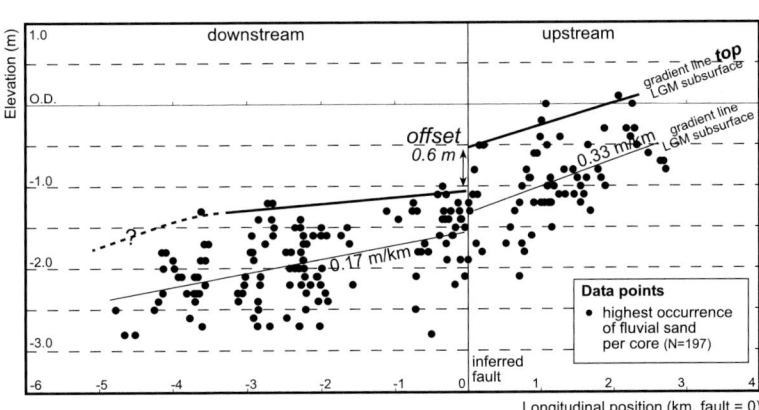

Fig. 9 Elevation and position of top-of-channel deposits in the Last Glacial Maximum (OIS 2) terrace relative to the inferred fault. Plotted is the highest occurrence of fluvial sand (top of channel deposits) in 197 boreholes that reached the LGM terrace and that are within 2 km lateral distance of the longitudinal section (Fig. 6), projected on a longitudinal axis perpendicular to the inferred fault. The standard deviation (relative to the downstream trend) in the elevation of the top of channel deposits is 0.42 m. A displaced gradient line for the top of the LGM surface is drawn based on local maxima in the data points, which represent bar tops. The fault offset estimated from this gradient line is c. 0.55 m.

subsiding downstream hanging wall (Fig. 2A), indicating that differential subsidence across the PBF was an active control in the LGM Rhine valley too. The present slope, however, is not necessarily identical to the river gradient during deposition (Cohen et al., 2002). Post-depositional tectonic deformation along unrecognized faults, tilting of blocks, and tectonic warping without faulting, may have changed the original slope. The gentle slope downstream of the fault may be caused by post-depositional back tilting, and may indicate that main subsidence occurred directly downstream of the fault whereas further downstream subsidence was less. This is not uncommon for fault zones, and would also explain why estimates of fault displacement based on basal peat and the buried terrace surface within a few hundred metres upstream and downstream of the fault yield higher values (Tables 2a & 3a) than estimates based on data selected over longer distances (Tables 2b & 3b).

Absolute subsidence rates

Subsidence rates were quantified from the relative rise of groundwater at the buried dune sites 'Zandberg' and 'Treeft' (Fig. 7). Rates of middle–late Holocene floodbasin aggradation at these sites approach rates of relative sea-level rise. Maximum rates of 2.0 m kyr^{-1} were reached 6500–6000 cal. yr BP. After 5500 cal. yr BP these decreased rapidly to c. 0.5 m kyr^{-1}. The latter rates are typical of areas of subsidence: global eustatic sea-level rise decreased to virtually zero, but subsidence continued to create accommodation space. Like relative sea-level rise, groundwater-level rise can be separated into a subsidence component (lowering the surface below the water table) and a component of absolute groundwater-table rise (controlled by either downstream eustatic sea-level rise or upstream 'climatic' changes in discharge). Groundwater-level gradients have been constant during the middle Holocene (Van Dijk et al., 1991), so the component of absolute groundwater level was spatially constant. Aggradation between 7000 and 5500 cal. yr BP at 'Treeft' dune site, however, *exceeds* that of relative sea-level rise (Van de Plassche, 1982; inset Fig. 7). Hence, during this period, local subsidence must have exceeded subsidence in the coastal area.

A significant effect of changes in upstream discharge on groundwater levels is not recorded in the basal peat record, but is suspected after 4000 cal. yr BP, when peat formation terminates in favour of clay deposition. Subrecent (after 4000 cal. yr BP) aggradation rates are linear and slightly greater at the coast than in the study area (respectively c. 0.7 and c. 0.5 m kyr^{-1}). These values suggest that subrecent subsidence was greater at the coast and that upstream controls did not add significantly to aggradation in the study area. Presuming zero eustatic sea-level rise since 4000 cal. yr BP, subrecent rates can be seen as maximum estimates of absolute subsidence. A more realistic estimate accounts for some eustatic rise ('melting tail', most probably ≤ 0.25 m kyr^{-1} sea-level rise; Peltier, 2002) and effects of upstream discharge change, leaving 0.2–0.4 m kyr^{-1} of the total 0.5 m kyr^{-1} to attribute to subsidence. Combining this rate with the differential subsidence (0.03–0.07 m kyr^{-1}) within the study area suggests that subsidence upstream of the inferred fault was 10–35% less than downstream.

DISCUSSION

The control of active tectonics on alluvial rivers has long been recognized (Schumm et al., 2000), and specific studies of syn-depositional fluvial response to tectonics are no longer rare. So far, syn-depositional response of active channels has been described in explorative studies, mapping changes of in-channel morphology of active rivers over different tectonic reaches of their alluvial valley. Fluvio-deltaic systems are usually excluded from these studies, because the sedimentary architecture there is primarily attributed to downstream controls (i.e. tides, storms, sea-level rise). Studies focusing on tectonics in overbank sediments of alluvial valley floodplains (such as the New Madrid Seismic Zone (NMSZ) in the Mississippi valley; Guccione et al., 2002) or deltaic floodbasins (such as this study) are still rare. Yet, deltaic systems have low-gradient reaches, and low-gradient channels are believed to be most sensitive to subtle neotectonic deformation (Holbrook & Schumm, 1999), and their Holocene records, particularly in the floodbasins, are relatively continuous and complete.

Local synsedimentary response and linearity of tectonic deformation

In the study area, both in the middle Holocene (rapid aggradation) and in the late Holocene (slower aggradation), crevasse splays preferentially developed downstream of the inferred fault, so the spatial distribution of the splays reflects the difference in subsidence rates. Linearly averaged over the middle–late Holocene, rates of differential subsidence are very low (< 0.07 m kyr^{-1}) and it is unlikely that, in reality, gradual deformation (tectonic creep) could have had such a persistent effect on crevasse-splay locations. Probably, most of the middle–late Holocene offset is essentially related to a shorter period of high fault activity. Given the observation that 7000–5500 cal. yr BP subsidence was greater in the study area than at the coast, whereas for the remaining Holocene it was not, the 0.20–0.40 m offset is probably related to displacement (tentatively related to PBF earthquakes driving fault offset) at this time. This activity may have caused significant lowering downstream of the fault, followed by the development of crevasse splays. As these splays cover basal peat, subsequent autocompaction (Allen, 1999) may have further amplified this response to tectonics in the floodbasin architecture. Effects of major middle Holocene displacement will not have been limited to the study area and may also have favoured middle Holocene avulsions at the delta's main nodal avulsion site (Wijk bij Duurstede, Fig. 3) directly to the north (Stouthamer & Berendsen, 2000).

Deglaciation-related tectonic activity

Although Holocene syn-depositional evidence points to non-linear displacement rates across the PBF, post-depositional offsets since 15 ka and 7000 cal. yr BP give roughly similar rates, pointing to linearity over the full time-scale of this study. In addition to normal extensional tectonics of the RVG (Houtgast & Van Balen, 2000), Lateglacial fault activity might have been increased owing to forebulge collapse (Lambeck et al., 1998; Van den Berg et al., 2002; Cohen et al., 2002). In that case, high Late-glacial rates, decreasing towards the Holocene, can be expected. Accepting a middle Holocene event causing most of the post 7 ka displacement would imply that the last major event occurred roughly 6000 yr ago. Apart from normal tectonics and forebulge collapse, this last activity also might be (partially) attributed to loading by the filling North Sea and aggrading coastal prism (hydro-isostasy; Lambeck et al., 1998; Peltier, 2002) At the present coast this *started* in the early–middle Holocene (Stanley & Warne, 1994) and thus lags Late-glacial forebulge collapse. Thus, although the syn-sedimentary effect of differential subsidence is independent of deglaciation-related controls, such as upstream climate change and sea-level rise, the timing of tectonic events causing post-depositional deformation might be related.

Coastal-prism differential subsidence

Coastal areas at the edge of sedimentary basins are areas of subsidence, and the build-up of coastal prisms by low-gradient rivers has continued over the second half of the Holocene (Posamentier et al., 1992; Stanley & Warne, 1994; Talling, 1998), despite the dramatic decrease in eustatic-driven sea-level rise. Subsidence rates can vary along the longitudinal profile of deltaic rivers, whereas there is only one rate of eustatic base-level rise for the entire coastal prism. This results in spatially variable sedimentary response to water-level rise. Many coastal prisms, in particular those of larger rivers, are aligned with the tectonic structure (e.g. Mississippi River, Po River, Orinoco River). In these cases, variation in subsidence along the delta axis is gradual, and differential subsidence will be hardly recognizable as an independent control. The glacially imposed valley of the Rhine, however, crosses the structural trend, which results in an abrupt downstream increase in subsidence. Hence significant syn-depositional effects in floodbasin architecture occurred, and tectonics may be identified as an independent local control on fluvio-deltaic deposition. This control locally affected aggradation rates and crevasse-splay development (this study) and regionally affected avulsion (Berendsen & Stouthamer, 2000).

Similar syn-depositional controls on floodbasin architecture have been identified in alluvial valley settings. For example, in the New Madrid Seismic Zone a crevasse splay has been documented that preferentially developed in the down-warped area directly north-east of the Lake County Uplift (Guccione et al., 2002), and within the Okavango Fan panhandle-region, the transition from meandering to avulsive anastomozing reaches has been related to the activity of the upstream fault-zone of a subsurface graben (Smith et al., 1997). The present study shows that across active normal faults, rivers in their deltaic reaches experience similar tectonic controls to those identified in alluvial valleys. Despite the strong sea-level control on the back-filling coastal prism reaches, the floodbasin faithfully records a tectonic component. Within the Lower Mississippi valley, given the distance to the river mouth, the South Louisiana Growth Fault region around Baton Rouge (Autin et al., 1991) might allow a direct comparison with this study. The tectonic control identified in the Rhine–Meuse delta may be of importance in many other high-stand coastal prisms.

CONCLUSIONS

The Holocene fluvial record in the study area reflects the tectonic control of differential subsidence on floodbasin aggradation. The location of an active fault belonging to the Peel Boundary Fault-zone (PBF) could be inferred from longitudinal changes in sedimentary facies, from deformation in sedimentary surfaces and from the analysis of dated basal peats as palaeogroundwater-level indicators. Differential subsidence was quantified from fault-related post-depositional vertical displacement of surfaces of known age (c. 0.9 m in the Last Glacial Maximum subsurface and 0.2–0.4 m in the middle Holocene basal peat, yielding 7 to 15 ka averaged rates of 0.03–0.07 m kyr^{-1}). High rates of subsidence between 7000 and 5500 cal. yr BP, together with sedimentological evidence of syn-depositional fault activity, suggest that the last major offset along this fault of the PBF occurred in the middle Holocene, and hence differential subsidence occurs at non-linear rates over time-scales < 7 kyr.

This study shows that active tectonics may cause syn-sedimentary response in coastal prism deltas, just like in alluvial valleys. Despite the strong sea-level control, aggrading deltaic floodbasins may record tectonics locally because subsidence rates across faults vary, whereas rates of eustatic/climatic base-level rise are uniform across the entire coastal prism. This implies that back-filling should not solely be attributed to downstream relative sea-level rise dominating upstream deposition, but that local differential subsidence must be considered as an additional independent control. The timing of tectonic events in coastal prisms, however, is not necessarily independent of sea-level rise, and models coupling coastal-prism fault activity to hydro-glacioisostasy should be considered.

ACKNOWLEDGEMENTS

We thank all the people who helped during our fieldwork in the study area, in particular M. Van Ree who prepared the basal peat samples and A.W. Hesselink who provided a cross-section through the 'River Lek' embanked floodplain to complete Fig. 4. The longitudinal section (Fig. 6) was presented at the session on 'Alluvial and tectonic system interactions' at the 7th International Congress of Fluvial Sedimentology (Nebraska, USA, 2001). This paper benefited from discussions at the conference and during the pre-conference fieldtrip along the Mississippi, including the New Madrid Seismic Zone, so all participants are thanked. Reviewers J. Alexander and A.E. Mather as well as editor M.D. Blum are thanked for their good comments. This research is part of a PhD project 'Neotectonics in the Rhine–Meuse delta', sponsored by the Netherlands Organisation for Scientific Research (NWO)—Earth and Life Sciences (ALW).

REFERENCES

Ahorner, L. (1992) *Das Erdbeben bei Roermond am 13. April 1992*. Bericht der Abteilung für Erdbebengeologie des Geologischen Instituts der Universität zu Köln, Bensberg, 11 pp.

Allen, J.R.L. (1999) Geological impacts on coastal wetland landscapes: some general effects on sediment autocompaction in the Holocene of northwest Europe. *The Holocene*, 9(1), 1–12.

Autin, W.J., Burns, S.F., Miller, B.J., Saucier, R.T. and Snead, J.I. (1991) Quaternary geology of the Lower Mississippi Valley. In: *Quaternary Nonglacial Geology, Conterminous United States* (Ed. R.B. Morrison). The Geology of North America, Vol. **K-2**, pp. 547–582. Geological Society of America, Boulder, CO.

Alexandre, P. (1994) Historical seismicity of the lower Rhine and Meuse valleys from 600 to 1525: a new critical review. *Geol. Mijnbouw*, **73**, 431–438.

Berendsen, H., Hoek, W. and Schorn, E. (1995) Late Weichselian and Holocene river channel changes of the rivers Rhine and Meuse in the Netherlands (Land van Maas en Waal). In: *European River Activity and Climate Change during the Lateglacial and Holocene* (Eds B. Frenzel, J. Vandenberghe, C. Kasse, S. Bohncke and B. Gläser). *Paläoklimaforschung/Paleoclim. Res.*, **14**, 151–172.

Berendsen, H.J.A. and Stouthamer, E. (2000) Late Weichselian and Holocene palaeogeography of the Rhine–Meuse delta, The Netherlands. *Palaeogeogr., Palaeoclimatol., Palaeoecol.*, **161**, 311–335.

Berendsen, H.J.A. and Stouthamer, E. (2001) *Palaeogeographical Development of the Rhine–Meuse Delta, The Netherlands*. Van Gorcum, Assen, 268 pp.

Blum, M.D. and Törnqvist, T.E. (2000) Fluvial response to climate and sea level change: a review and look forward. *Sedimentology*, **47**, 2–48.

Bronk Ramsey, C. (1995) Radiocarbon calibration and analysis of stratigraphy: the OxCal program. *Radiocarbon*, **37**, 425–430.

Camelbeeck, T. and Meghraoui, M. (1998) Geological and geophysical evidence for large palaeo-earthquakes with surface faulting in the Roer Graben (northwest Europe). *Geophys. J. Int.*, **132**, 347–362.

Cohen, K.M. (2003) Differential subsidence within a coastal prism. Published PhD thesis, Department of Physical Geography, Utrecht University. *Neth. Geogr. Stud.*, **316**, 176 pp.

Cohen, K.M., Stouthamer, E. and Berendsen, H.J.A. (2002) Fluvial deposits as a record for Late Quaternary neotectonic activity in the Rhine–Meuse delta, The Netherlands. *Neth. J. Geosci./Geol. Mijnbouw*, **81**(3/4), 389–405.

De Jong, J. (1995) Some palaeobotanical data on the fluviatile Rhine/Meuse Kreftenheye Formation in the Netherlands (Weichselian). In: *Neogene and Quaternary Geology of North-west Europe* (Eds G.F.W. Herngreen and L. Van der Valk). *Med. Rijks Geol. Dienst*, **52**, 369–385.

De Vries, J.J. (1974) Groundwater flow systems and stream nets in The Netherlands. PhD thesis, Free University Amsterdam, 226 pp.

Geluk, M.C., Duin, E.J.Th., Dusar, M., Rijkers, R.H.B., Van den Berg, M.W. and Van Rooijen, P. (1994) Stratigraphy and tectonics of the Roer Valley Graben. *Geol. Mijnbouw*, **73**, 129–141.

Gibbard, P.L. (1995) The formation of the Strait of Dover. In: *Island Britain: a Quaternary Perspective* (Ed. R.C. Preece). *Geol. Soc. Lond. Spec. Publ.*, **96**, 15–26.

Guccione, M.J., Mueller, K., Champion, J. Shepherd, S., Carlson, S.D., Odhiambo, B. and Tate, A. (2002) Stream response to repeated coseismic folding, Tiptonville dome, New Madrid seismic zone. *Geomorphology*, **43**, 313–349.

Hofstede, J.L.A., Berendsen, H.J.A. and Janssen, C.R. (1989) Holocene palaeogeography and palaeoecology of the fluvial area near Maurik (Neder-Betuwe, The Netherlands). *Geol. Mijnbouw*, **68**, 409–419.

Holbrook, J. and Schumm, S.A. (1999) Geomorphic and sedimentary response of rivers to tectonic deformation: a brief review and critique of a tool for recognizing subtle epeirogenic deformation in modern and ancient settings. *Tectonophysics*, **305**, 287–306.

Houtgast, R.F. and Van Balen, R.T. (2000) Neotectonics of the Roer Valley Rift System, The Netherlands. *Global Planet. Change*, **27**, 131–146.

Houtgast, R.F., Van Balen, R.T., Bouwer, L.M., Brand, G.B.M. and Brijker, J.M. (2002) Late Quaternary activity of the Feldbiss Fault zone, Roer Valley Rift System, The Netherlands, based on displaced fluvial terrace fragments. *Tectonophysics*, **352**, 295–315.

Huisink, M. (1997) Lateglacial sedimentological and morphological changes in a lowland river in response to climate change: the Maas, southern Netherlands. *J. Quat. Sci.*, **12**, 209–223.

Jelgersma, S. (1961) Holocene sea-level changes in The Netherlands. PhD thesis, University Leiden. *Med. Geol. Sticht.*, **CVI-1**.

Lehrman, K., Klostermann, J. and Pelzing, R. (2001) Paleoseismological investigations ate the Rurrand Fault, Lower Rhine Embayment. *Netherlands J. Geosci./Geol. Mijnbouw*, **80**, 139–154.

Kasse, C. (1995) Younger Dryas cooling and fluvial response (Maas River, The Netherlands) (extended abstract). *Geol. Mijnbouw*, **74**, 251–256.

Klostermann, J. (1992) *Das Quartär der Niederrheinische Bucht*. Geologisches Landesamt Nordrhein-Westfalen, Krefeld, 200 pp.

Lambeck, K., Smither, C. and Johnston, P. (1998) Sea-level change, glacial rebound and mantle viscosity for northern Europe. *Geophys. J. Int.*, **134**, 102–144.

Makaske, B. (1998) Anastomosing rivers. Forms, processes and sediments. PhD thesis. Utrecht University (Utrecht). *Ned. Geogr. Stud.*, **249**, 298 pp.

Martinson, D.G., Pisias, N.G., Hays, J.D., Imbrie, J., Moore, T.C. and Shackleton, N.J. (1987) Age dating and the orbital theory of the Ice Ages: development of a high resolution 0 to 300 000-year chronostratigraphy. *Quat. Res.*, **27**, 1–29.

Peltier, W.R. (2002) On eustatic sea level history: Last Glacial Maximum to Holocene. *Quat. Sci. Rev.*, **21**, 377–396.

Posamentier, H.W., Allen, H.W., James, D.P. and Tesson, M. (1992) Forced regressions in a sequence stratigraphic framework: Concepts, examples, and sequence stratigraphic significance. *Am. Assoc. Petrol. Geol. Bull.*, **76**, 1687–1709.

Schumm, S.A., Dumont, J.F. and Holbrook, J.M. (2000) *Active Tectonics and Alluvial Rivers*. Cambridge University Press, Cambridge, 276 pp.

Stanley, D.J. and Warne, A.G. (1994) Worldwide initiation of Holocene marine deltas by deceleration of sea-level rise. *Science*, **265**, 228–231.

Smith, N.D., McCarthy, T.S., Ellery, W.N., Merry, C.L. and Rüther, H. (1997) Avulsion and anastomosis in the panhandle region of the Okavango Fan, Botswana. *Geomorphology*, **20**, 49–65.

Stouthamer, E. (2001) Holocene avulsions in the Rhine–Meuse delta, The Netherlands. PhD thesis, Utrecht University. *Ned. Geogr. Stud.*, **283**, 212 pp.

Stouthamer, E. and Berendsen, H.J.A. (2000) Factors controlling the Holocene avulsion history of the Rhine–Meuse delta (The Netherlands). *J. Sediment. Res. Section A*, **70**(5), 1051–1064.

Stuiver, M., Reimer, P.J., Bard, E., Beck, J.W., Burr, G.S., Hughen, K.A., Kromer, B., McCormac, F.G., Van der Plicht, J. and Spurk, M. (1998) INTCAL98 Radiocarbon age calibration 24,000–0 cal BP. *Radiocarbon*, **40**, 1041–1083.

Talling, P.J. (1998) How and where do incised valleys form if sea level remains above the shelf edge? *Geology*, **26**, 87–90.

Tebbens, L.A., Veldkamp, A., Westerhoff, W. and Kroonenberg, S.B. (1999) Fluvial incision and channel downcutting as a response to Late-glacial and Early Holocene climate change: the lower reach of the River Meuse (Maas), the Netherlands. *J. Quat. Sci.*, **14**, 59–75.

Tebbens, L.A., Veldkamp, A., Westerhoff, W. and Kroonenberg, S.B. (2000) Reply: Fluvial incision and channel downcutting as a response to Late-glacial and Early Holocene climate change: the lower reach of the River Meuse (Maas), the Netherlands. Correspondence. *J. Quat. Sci.*, **15**, 95–100.

Törnqvist, T.E. (1993) Holocene alternation of meandering and anastomosing fluvial systems in the Rhine–Meuse delta (central Netherlands) controlled by sea-level rise and subsoil erodibility. *J. Sediment. Petrol.*, **63**, 683–693.

Törnqvist, T.E., Weerts, H.J.T. and Berendsen, H.J.A. (1994) Definition of two new members in the upper Kreftenheye and Twente formations (Quaternary, the Netherlands): a final solution to persistent confusion? *Geol. Mijnbouw*, **72**, 251–264.

Törnqvist, T.E., Van Ree, M.H.M., Van 't Veer, R. and Van Geel, B. (1998) Improving methodology for high-resolution reconstruction of sea-level rise and neotectonics by paleoecological analysis and AMS ^{14}C dating of basal peats. *Quat. Res.*, **49**, 72–85.

Törnqvist, T.E., Wallinga, J., Murray, A.S., De Wolf, H., Cleveringa, P. and De Gans, W. (2000) Response of the Rhine–Meuse system (West-central Netherlands) to the last Quaternary glacio-eustatic cycles: a first assessment. *Global Planet. Change*, **27**, 89–111.

Van Dijk, G.J., Berendsen, H.J.A. and Roeleveld, W. (1991) Holocene water level development in The Netherlands' river area; implications for sea-level reconstruction. *Geol. Mijnbouw*, **70**, 311–326.

Van de Plassche, O. (1982) *Sea-level change and water level movements in the Netherlands during the Holocene. Mededelingen Rijks Geologische Dienst*, **36**, 93 pp.

Van den Berg, M.W. (1994) Neotectonics of the Roer Valley rift system. Style and rate of crustal deformation inferred from syn-tectonic sedimentation. *Geol. Mijnbouw*, **73**, 143–156.

Van den Berg, M.W., Groenewoud, W., Lorenz, G.K., Brus, D.J. and Kroonenberg, S.B. (1994) Patterns and velocities of recent crustal movements in the Dutch part of the Roer Valley rift system. *Geol. Mijnbouw*, **73**, 157–168.

Van den Berg, M.W., Vanneste, K., Dost, B., Lokhorst, A., Van Eijk, M. and Verbeeck, K. (2002) Paleoseismic investigations along the Peel Boundary Fault: geologic setting, site selection and trenching results. *Neth. J. Geosci./Geol. Mijnbouw*, **81**, 39–60.

Van Montfrans, H.M. (1975) *Toelichting bij de ondiepe breuken kaart met de diepte van de formatie van Maassluis 1 : 600 000*. In: *Toelichtingen bij de geologische overzichtskaarten van Nederland* (Eds W.H. Zagwijn and C.J. Van Staalduijnen), pp. 103–109. Rijks Geologische Dienst, Haarlem.

Vanneste, K., Meghraoui, M. and Camelbeeck, T. (1999) Late Quaternary earthquake related soft-sediment deformation along the Belgian portion of the Feldbiss Fault, Lower Rhine Graben system. *Tectonophysics*, **309**, 57–79.

Verbraeck, A. (1984) *Toelichtingen bij de Geologische kaart van Nederland 1 : 50.000, Blad 39, Tiel West, Tiel Oost*. Rijks Geologische Dienst, Haarlem, 325 pp.

Verbraeck, A. (1990) De Rijn aan het einde van de laatste ijstijd: De vorming van de jongere afzettingen van de Formatie van Kreftenheye. *Geogr. Tijds. Nieuwe Reeks*, **XXIV**(4), 328–340.

Walker, M.J.C., Björck, S., Lowe, J.J., Cwynar, L.C., Johnsen, S., Knudsen, K.-L., Wohlfarth, B. and INTIMATE group (1999) Isotopic 'events' in the GRIP ice core: a stratotype for the Late Pleistocene. *Quat. Sci. Rev.*, **18**, 1143–1150.

Wallinga, J. (2001) The Rhine–Meuse system in a new light: optically stimulated luminescence dating and its application to fluvial deposits. PhD thesis. Utrecht University. *Ned. Geogr. Stud.*, **290**, 180 pp.

Zagwijn, W.H. (1989) The Netherlands during the Tertiary and the Quaternary: A case history of Coastal Lowland evolution. *Geol. Mijnbouw*, **68**, 107–120.

Zagwijn, W.H. and Van Staalduinen, C.J. (Eds) (1975) *Toelichtingen bij de geologische overzichtskaarten van Nederland*. Rijks Geologische Dienst, Haarlem.

Geomorphology and internal architecture of the ancestral Burdekin River across the Great Barrier Reef shelf, north-east Australia

C.R. FIELDING*,†, J.D. TRUEMAN*, G.R. DICKENS‡,§ and M. PAGE‡

*Department of Earth Sciences, University of Queensland, Qld 4072, Australia;
†Department of Geosciences, 214 Bessey Hall, University of Nebraska-Lincoln,
NE 68588-0340, USA (Email: cfielding2@unl.edu);
‡School of Earth Sciences, James Cook University, Townsville, Qld 4811, Australia; and
§Department of Earth Sciences, Rice University, Houston, TX 77005, USA

ABSTRACT

The ancestral channel of the Burdekin River has been traced across the Great Barrier Reef continental shelf of north-east Australia via a dedicated, high-resolution seismic reflection survey, and is continuous from the modern coastline to within c. 10 km of the shelf edge. The channel, entrenched into a relatively flat surface of interpreted compact Pleistocene alluvium and buried beneath up to a few metres of late Holocene marine sediment, is believed to record late Pleistocene drainage through the last glacial cycle. A plot of channel floor elevation against distance outboard from the coastline shows that the long profile of the palaeochannel is stepped, and can be divided into 14 segments alternating between gentle gradient (flats) and steeper gradient (ramps). The segments clearly delineate zones of similar channel width and depth, cross-sectional geometry, and sinuosity as interpreted from a constructed plan of the channel course. Channels in many segments are characterized by evidence for limited, systematic lateral accretion. The upper parts of channel fills have a concentric geometry and high-amplitude reflection character, suggesting final infilling during a different regime to that responsible for channel excavation and initial filling. The stepped profile is interpreted to be a consequence predominantly of falling sea level associated with the last glacial cycle. Channels were cut and partially filled during the fall and lowstand, and then backfilled during the Holocene sea level rise. The stepped long profile, the simple nature of channel cross-sections and the shallow entrenchment of the channel (as opposed to valley incision) indicate that the Burdekin was unable to reach equilibrium with environmental conditions during sea-level drawdown, but rather was constantly adjusting to repeated changes in base level and other variables. The contrasting character of the Burdekin palaeochannel relative to palaeochannels of similar age on other continental margins (many of which occupy pronounced incised valleys) can be attributed to the limited sediment supply and runoff to this margin, the physiography of the Great Barrier Reef, and the compact nature of the Pleistocene land surface into which the channel was entrenched.

INTRODUCTION

Lowstand river systems on subaerially exposed continental shelves play an important role in the evolution of continental margins. Generic depositional models (e.g. Van Wagoner et al., 1988; Posamentier et al., 1992) predict that rivers will incise into and across exposed shelves perpendicular to the coast during falling sea level because such a fall will produce convex longitudinal channel profiles. Numerous Quaternary examples of this phenomenon have been recognized around the World (e.g. Fisk, 1944; Colman & Mixon, 1988; Blum & Törnqvist, 2000; and see papers in Dalrymple et al., 1994). Shelf bathymetry and sediment supply, however, may complicate patterns of fluvial incision significantly (e.g. Thomas & Thorp, 1995; Talling, 1998). Most examples of lowstand incision have been termed 'incised valleys', large-scale geomorphological features (typically kilometres wide and 10s to 100s of m deep), within which significantly smaller river channels are nested. Such a geomorphological response to falling sea level is expected for rivers carrying substantial runoff throughout the year across relatively narrow and smoothly inclined shelves. However, what happens to rivers at lowstand when they discharge variable outputs onto broad shelves? Here, the possibility exists that rivers cannot incise valleys across the shelf. The situation can become even more complex on tropical mixed siliciclastic/carbonate margins where carbonate banks or reefs rim the outer shelf. In this case, and in contrast to most well-studied margins, cross-shelf profiles are not simple seaward-inclined gradients and river avulsion or incision parallel to the coastline might occur during lowstand (Woolfe et al., 1998; Ferro et al., 1999; Dunbar et al., 2000).

The north-east margin of Australia (Fig. 1), including the Great Barrier Reef (GBR) shelf, is the largest and perhaps best extant example of a tropical, mixed siliciclastic/carbonate margin (Maxwell, 1968; Davies et al., 1989). It is also an outstanding example of a margin where rivers discharge highly variable loads of water and sediment onto a broad (50–100 km) shelf (Neil et al., 2002). Regional and widely spaced seismic reflection data acquired on the GBR shelf during the late 1970s and 1980s show several buried channels on the shelf, presumably formed during lower sea level (Orme et al., 1978; Johnson et al., 1982; Searle, 1983; Johnson & Searle, 1984). Following emerging sequence-stratigraphy concepts, Johnson & Searle (1984, and unpublished maps held at School of Earth Sciences, James Cook University) connected some of these 'palaeochannels' to show rivers crossing the GBR shelf during lowstand, in a direction roughly perpendicular to the shoreline. The extensive reef network on the outer shelf, active today and during the penultimate interglacial highstand, was exposed and karstified during the last lowstand (e.g. Marshall & Davies, 1984; Davies et al., 1989; Davies & Peerdeman, 1998). Current reconstructions of the north-east Australian margin during lowstand thus have rivers incising across a broad shelf and bisecting a significant topographic barrier with minimal or no interaction.

The seismic lines on which such interpretations are based, however, are typically spaced greater than 10 km apart, and the resolution of the acquired data (typically mini-airgun or boomer data) is generally too poor to allow definition of internal channel architecture. As emphasized by Woolfe et al. (1998), no study to date has traced a palaeochannel across the shelf. Thus, Woolfe et al. (1998) suggested that rather than representing continuous rivers crossing the entire shelf, the imaged channels might only represent a loose patchwork of mainly small streams aligned in part parallel to the coast. Indeed, dated piston cores from the continental slope of the north-eastern Australian margin show that siliciclastic fluxes off the shelf were lowest during Pleistocene glacial lowstands, suggesting that riverine sediment loads may have been confined to the exposed shelf (Dunbar et al., 2000).

The Burdekin River (Fig. 1) dominates fluvial discharge on the north-eastern Australian margin, annually adding 9.8×10^9 m^3 water and $3-9 \times 10^6$ t of sediment to the GBR shelf (Burdekin Project Committee, 1977; Belperio, 1983; Neil et al., 2002). Much of the runoff, however, occurs in short-duration, high-magnitude events that have a recurrence interval of 5–10 yr and reach peak discharges of 20 000–40 000 m^3 s^{-1} (Alexander et al., 1999). The cross-sectional geometry of the river reflects this erratic, extremely variable discharge

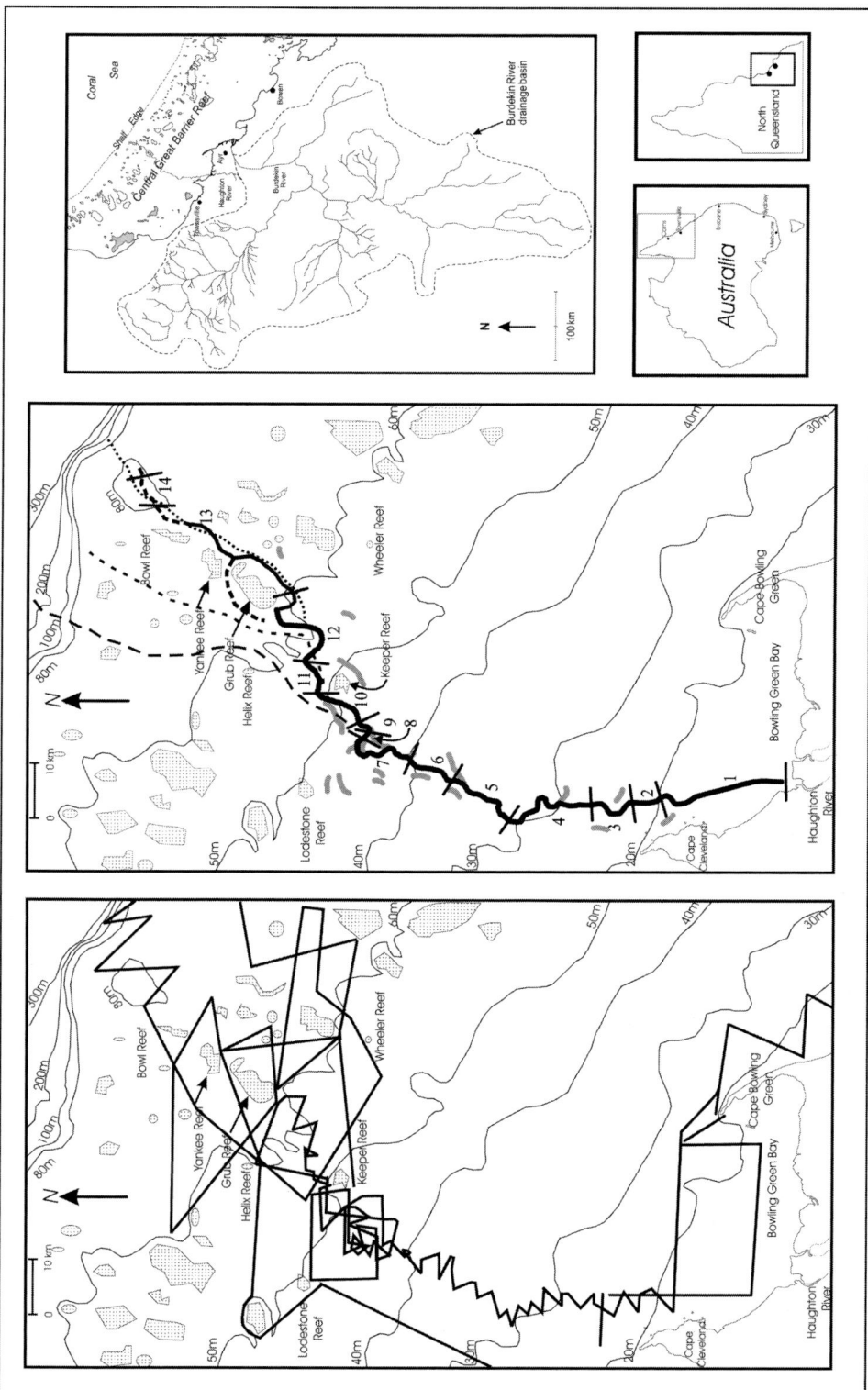

Fig. 1 Maps showing the location of the ancestral Burdekin palaeochannel in north-eastern Australia, the tracks of seismic surveys and the detailed course of the channel interpreted from those profiles (bold black line). The long-dashed line indicates the channel course as interpreted by Harris *et al.* (1990), the short-dashed line that of Johnson & Searle (1984), and the dotted line the unpublished interpretation of Carr & Johnson (data held at School of Earth Sciences, James Cook University of North Queensland). Numbers correspond to channel course segments (Table 1 and Fig. 3). Bathymetric contours are taken from published maps, and are considerably generalized.

regime, with a small, low-stage channel in the floor of a much larger channel scaled such as to contain major flow events (Plate 1). Connecting a few buried channel intersections seaward of the modern Burdekin Delta, Johnson & Searle (1984) and Harris et al. (1990) showed the 'palaeo-Burdekin' flowing north-east to the shelf edge. The two interpretations diverge on the outer shelf, however, with Johnson & Searle (1984) showing the channel passing west of Keeper and Grub Reefs, whereas Harris et al. (1990) plotted a somewhat different course due north from Helix Reef (Fig. 1). A further unpublished interpretation by Carr & Johnson (data and maps held by School of Earth Sciences, James Cook University) shows the palaeochannel passing east of Grub, Yankee and Bowl Reefs to the shelf edge (Fig. 1).

In this study, the lowstand channel of the ancestral Burdekin River is traced and characterized across the Great Barrier Reef shelf to examine the fate of a major river on an archetypal tropical, mixed siliciclastic/carbonate margin during lowstand. A companion paper (Fielding et al., 2003) provides a detailed analysis of the cross-shelf profile, and its interpretation, whereas the present paper emphasizes the geomorphology of the channel and internal character of its deposits, as far as can be surmised from seismic reflection data.

METHODS

Cruises KG-00/2 and KG-01/2 of the RV *James Kirby* were dedicated to mapping palaeochannels seaward of the Burdekin River. Seismic reflection data were acquired for a total of 12 days, using a side-mounted, Datasonics CAP6600 CHIRP II acoustic profiling system that generated a linear, FM 2–7 kHz pulse with a dominant frequency of 3.5 kHz. The positions of surveyed lines were determined accurately using a differential global positioning system (GPS). Working maps of channel location were compiled during cruises to facilitate navigation and efficiency.

Previous seismic work (e.g. Johnson & Searle, 1984; Orpin, 1999) indicated several palaeochannels immediately offshore of the Burdekin Delta, including a major channel north of the Haughton River (Fig. 1). Given the variable quality and wide spacing of earlier seismic lines, however, the number, size and course of channels remained uncertain. In the first survey (KG-00/2), 2 days were spent acquiring a comprehensive suite of data on the modern Burdekin Delta front to locate and characterise all palaeochannels extending from land. Four days were then spent following the most prominent palaeochannel by zig-zagging short, closely-spaced seismic lines. In this way, a trunk channel and some potential tributaries could be traced continuously across the shelf to the GBR. The direct tracing of the channel from near the shore outboard established that what was imaged is a single, continuous channel and not a patchwork of different, possibly discontinuous channels. In the second survey (KG-01/2), a pattern of lines was acquired to constrain possible channel courses over the outermost part of the GBR shelf. Following the cruises, data were downloaded, processed, and interpreted using a Seismic Micro-Technologies' Kingdom® software package. Although data quality was affected by sea-surface conditions, which varied from dead calm to 3 m swells, the new seismic profiles are significantly better quality than previous work.

Several critical channel characteristics were determined from the seismic data, including apparent width, depth (calculated using p-wave velocity in water: 1500 m s^{-1}), cross-sectional profile and fill reflection character. Apparent width was measured directly from one lip of the cross-section to the other, and depth taken as the vertical distance from the lip of the channel to the deepest point. Cross-sectional geometry and the orientation of accretionary bedsets in these mainly oblique intersections were then used to interpret the directional sense of curved reaches (cf. Willis, 1989), and to calculate true cross-sectional orientation. From these data, channel sinuosity was estimated by manually plotting the course of the channel over the length of the survey, and true width similarly estimated. All these properties have been plotted as a function of depth below present mean sea level (Australian Height Datum—AHD) and distance along the channel from the modern coastline (Fig. 2).

A long profile (Fig. 2) was constructed from channel intersections on seismic records. Three elevations are plotted as a function of along-channel distance downstream from the modern shoreline. Although it is customary to use floodplain

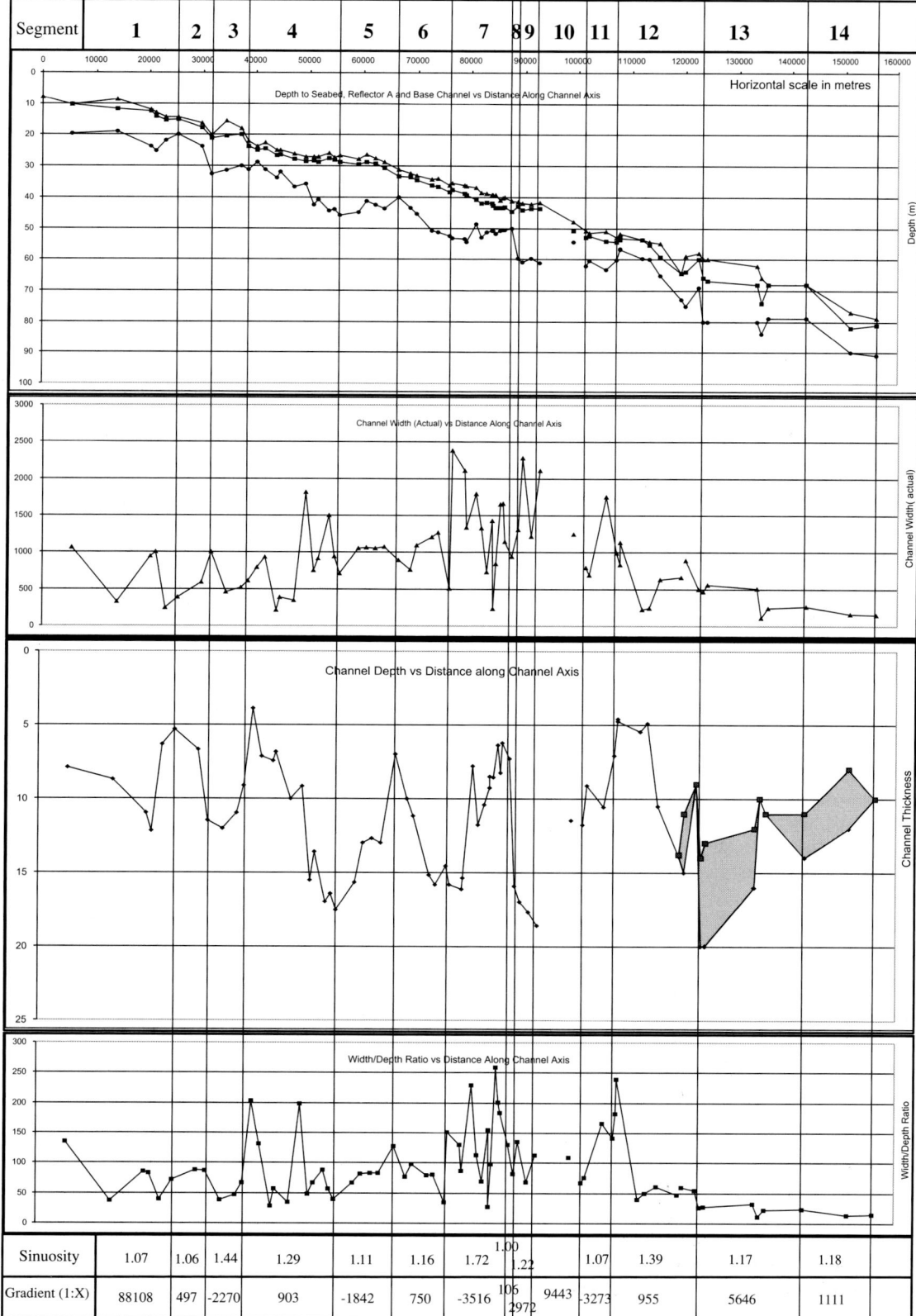

Fig. 2 Long profile along the ancestral Burdekin palaeochannel, showing variations in channel floor elevation (filled circles), top of channel fill (filled squares), sea-floor elevation (filled triangles), and channel dimensional parameters interpreted from seismic profiles. Gradient is expressed as $1/x$. Shaded area indicates regions where the palaeochannel is underfilled, line defined by filled squares denotes top of channel fill in these areas.

elevation to reconstruct channel and palaeochannel gradients, the base of the channel was used here because in many seismic records it was evident that the top of the channel fill had been truncated by erosion associated with the Holocene transgression. Furthermore, although it is accepted that the channel base elevation is subject to the irregularity of channel-floor scour, the profile as defined by base of channel in the present study shows a series of ramps and flats each < 20 km long that are clearly not artefacts. Such features would not be recognizable if the floodplain or the interfluve elevation was used as a datum. Nonetheless, the calculated gradient for individual segments is in some cases compromised by the apparently irregular channel floor surface. The overall gradient of the imaged channel, as calculated from the channel top, is 0.000474, and from the channel base it is 0.000465.

THE PALAEO-BURDEKIN CHANNEL

Surrounding the modern Burdekin Delta, one major buried channel (or channel belt) < 2.4 km across and several minor channels (< 200 m across) were observed between the 10 m and 30 m isobaths (Fig. 1). The largest palaeochannel observed occurs in the north-west corner of Bowling Green Bay, *c.* 55 km north-west of the modern river mouth (Fig. 1). This is entirely consistent with previous work.

All of the palaeochannels recognized around the Burdekin Delta are defined in part by a high-amplitude reflector that can be traced continuously from the relatively flat, more elevated surface to either side of the channel (the interfluve) then down into the base. On the interfluves, this reflector is overlain by a thin blanket of sediment. Vibrocores acquired by Orpin (1999) from Bowling Green Bay demonstrate that a widespread blanket of late Holocene mud overlies probable mid-Holocene transgressive sands and older palaeochannel deposits. The mud blanket unit can be traced across the shelf as an interval of typically 2–5 m thickness, acoustically quiet, having few coherent internal reflectors.

A well-defined, continuous, high-amplitude event colloquially termed 'Reflector A' has been identified across the GBR shelf by others (Orme *et al.*, 1978). This event has been interpreted ubiquitously as the Pleistocene–Holocene unconformity (e.g. Orme *et al.*, 1978; Johnson & Searle, 1984). The character of the reflector in the channel, however, suggests that the channels are not incised into Reflector A. Rather, the channels were formed during the period of time represented by the unconformity surface, and perhaps filled partly during that time and partly later, during the Holocene transgression.

The main palaeochannel extends offshore from the mouth of the present-day Haughton River (Fig. 1). This palaeochannel is at least two times wider and deeper than the modern Haughton River, however, being more closely comparable to the lower reaches of the modern Burdekin River (e.g. Plate 1). On the basis of geomorphological and limited drilling data, previous workers (e.g. Hopley, 1970) have suggested that the Haughton River represents one of the principal Pleistocene palaeochannels of the Burdekin River. Our research drilling activities on the coastal plain adjacent to the Haughton River found stiff, compacted Pleistocene clays near the surface, supporting this interpretation. It is suggested that the prominent buried channel is the trunk channel of the Burdekin system during the last glacial lowstand.

As well as being similar in width and depth to the modern Burdekin River, the palaeochannel is also similar in that it is entrenched within its own alluvium to a depth of 5–20 m below the floodplain level (Plate 1, and see Alexander *et al.*, 1999). 'Entrenchment' in this context is quite distinct from the sense used by some previous authors (e.g. Thorne, 1994), in that it does not imply a channel cut (inset) into the floor of a larger, incised alluvial valley, but more simply a channel on a relatively flat plain cut into earlier alluvial deposits of the same system. This alluvial cross-sectional style owes its character more to the extremely variable discharge regime of the region than to any change in base-level or other controlling variable.

CROSS-SHELF CHANNEL PROFILE

The palaeo-Burdekin channel has been traced for 160 km from the modern coastline in the

Plate 1 Oblique aerial photograph of the modern Burdekin River on the delta plain (Inkerman road bridge in background for scale, 800 m wide). In its lower reaches, the river is entrenched into its own alluvium on a low-lying plain, rather than occupying an incised valley. The interfluve is 5–10 m above the bed of the channel. Flow is towards the viewer.

south-west corner of Bowling Green Bay to the outermost part of the shelf (Fig. 1). The channel initially heads north before swinging north-east at about the −40 m isobath (Fig. 1). The north-east trend is roughly perpendicular to the modern coast, and to isobaths that probably approximate older shorelines. Although several small channels were noted close to the main channel (and therefore probably join it), no major tributaries have been recognized. The channel becomes difficult to trace on the outermost shelf but can be mapped to within 10 km of the shelf edge.

Properties of the palaeo-Burdekin channel change significantly across the shelf (Table 1). In particular, the channel floor gradient alternates between long, gently (or even negatively) sloping sections and shorter, steeper intervals. On this basis, the channel has been divided into 14 alternating 'ramp' and 'flat' segments (Fig. 2). Within most segments, channel widths and depths show a consistent trend (Table 1). Interpreted sinuosity within individual segments varies from 1 to 1.72, with no consistent relationship between gradient and sinuosity (Table 1 and Fig. 2). The outermost segment 14 of the profile (Fig. 2) drops into 80 m water, beyond which the sea floor shallows somewhat to the shelf edge and no channel was evident. Segment 10, *c.* 9 km long and located around Keeper Reef (Fig. 1), initially posed a problem. Only local evidence of channelling was found around the west and north-west side of Keeper Reef during KG-00/01, although a second channel was subsequently recorded on the southeast side of the reef during KG-01/02 (Fig. 1). The channel floor elevation in these channels is −60 m, similar to that in the immediately upstream segment 9 and the downstream segment 11, and on this basis segments 9–11 can be considered as one long flat reach. The channel can be interpreted to have avulsed or split into two coeval courses around the front of Keeper Reef, the first substantial reef structure encountered by the channel. A single, larger channel is re-established at the start of segment 11 and continues to the end of the survey (Fig. 1).

Channel geometry and fill vary considerably along the profile (Table 1). In many places, orthogonal cross-sections show a steeply incised (or entrenched), generally asymmetrical channel with a stepped channel floor and deepest point (thalweg) located close to the steeper bank. As noted by Johnson and Searle (1984), these channels often contain two units: a lower unit dominated by low reflectivity and reflectors that dip from the gently sloping bank to terminate against the steeper bank, and an upper unit with high reflectivity and reflectors that dip symmetrically toward the channel axis. Other orthogonal cross-sections show a steep-sided but symmetrical channel with a flat channel floor. As for asymmetrical channels, the lower parts of these channels typically have low reflectivity whereas the upper parts have higher reflectivity. In segments 5, 6 and 7, two or more steep-sided, mainly symmetrical channels were found.

For nearly all cross-sections, regardless of channel geometry, the top of channel fill is concave-up. This and the high reflectivity suggest that upper parts of the channel contain mud. The thickness of the late Holocene sediment blanket varies along the profile, and so Johnson et al.'s (1982) assertion that no significant relationship exists between modern bathymetry and palaeochannel location on the GBR shelf is upheld. An exception to this occurs on the outermost part of the channel course. Here, in segments 13 and 14, the channel has both a surface and subsurface expression (i.e. is underfilled), and in a small number of intersections has only a surface expression (i.e. no channel fill deposits could be recognized: Fig. 2). Elsewhere on the outer shelf, a scalloped sea floor truncates reflectors associated with the channel. Holocene erosion has therefore modified the palaeo-Burdekin channel in some parts of the shelf. A detailed description and interpretation of the cross-shelf channel profile is given by Fielding *et al.* (2003).

CHANNEL FORM AND FILL

Segment 1

Segment 1 is a very low gradient 'flat' that extends from the mouth of the present Haughton River to close to the tip of Cape Cleveland, and coincides approximately with the north–south extent of Bowling Green Bay (Fig. 1 and Table 1). Data are sparse over this reach, but the palaeochannel seems to have a relatively straight course that runs roughly parallel to, and 2–5 km offshore from, the

Table 1 Characteristics of the ancestral Burdekin palaeochannel in each of the segments recognised (Figure 2).

Segment	Morphology	Gradient (1/x)	Width (m)	Depth (m)	Width/depth	Cross-section	Sinuosity	Fill characteristics
1	Flat	88 108	250–1060	6.5–12	35–135	Variable (few data)	1.07 (minimum)	Possibly mud-dominated
2	Ramp	497	380–1000	5.5–11.5	70–90	Mostly narrow, symmetrical	1.06 (minimum)	Possibly mud-dominated
3	Flat	–2270	460–1000	4–11.5	40–205	Asymmetrical, lateral accretion with symmetrical upper part	1.44	Composite, ?coarse-grained lower part, muddy upper part
4	Ramp	903	210–1820	4–17.5	30–205	Slightly asymmetrical, minor lateral accretion, deep & steep-sided	1.29	Composite, as above
5	Flat	–1842	710–1070	7–17.5	40–130	Asymmetrical, lateral accretion, local anabranching	1.11 (minimum)	Composite, as above
6	Ramp	750	500–1270	7–16	35–130	Symmetrical to slightly asymmetric, anabranching channels rejoin to a single trunk channel	1.16	Composite, as above
7	Flat	–3516	230–2380	6–16	30–260	<4 symmetrical, anabranching channels	1.72	Composite, as above
8	Ramp	106	950–1310	7–16	80–130	Slightly asymmetrical, steep-sided	1.00	Composite, as above
9	Flat	2972	1210–2280	16–19	70–135	Slightly asymmetrical, possible anabranching	1.22	Composite, as above
10	?	9443	1250	11.5 (minimum)	110	Channel deposits eroded except in one profile, there anabranching	?	?
11	Flat	–3273	690–1750	7–12	70–165	Slightly asymmetrical, lateral accretion, steep-sided	1.07	Composite, as above
12	Ramp	955	220–1130	4.5–14	40–240	Asymmetrical, probably truncated by erosion	1.39	Composite, as above
13	Flat	5646	113–890	9–20	14–59	Symmetrical to slightly asymmetrical	1.17	Underfilled to unfilled
14	Ramp	1111	149–259	10–14	14–18	Nearly symmetrical	1.18	Underfilled

coastline. No particular reason for this course is evident from near-surface data, although the apparently straight planform and parallelism with the onshore Haughton River suggest a possible structural control. The channel appears to become slightly more sinuous towards the outer limit of segment 1 (Fig. 1). Available data indicate a mainly symmetrical, relatively steep-sided channel with concentric fill. Remnants of older, incised channels, severely truncated by the principal palaeochannel, are visible on some lines. The lower part (c. 50% of total stratigraphical thickness) of the fill is relatively reflection-free with a few, discontinuous, form-concordant, high-amplitude reflections. This character is particularly evident in the older, truncated palaeochannel sections. The upper part of the channel fill comprises more continuous, high-amplitude, downlapping and form-concordant reflections that flatten out progressively upward through the channel fill. In some channel cross-sections, the upper facies is volumetrically dominant.

Given the context, the lower seismic facies is interpreted as a coarse-grained lower channel fill, and the overlying, more strongly seismically-layered facies as a more mud-dominated, homogeneous upper channel fill. The minor high-amplitude reflectors observed within the otherwise acoustically quiet lower channel fill may record hiatuses in channel filling. A vibracore acquired close to the course of the palaeochannel in segment 1 by Orpin (1999) showed an interbedded sand–mud lithology with abundant plant and shell fossils above Reflector A (mid- to late Holocene bay facies), with a sand-dominated lithology immediately below Reflector A. The latter facies may be palaeochannel deposits, or (more likely) levee or splay deposits.

Segment 2

Segment 2 is a ramp, the beginning of which corresponds to the northward limit of waters protected by Cape Bowling Green and Cape Cleveland (i.e. the outer limit of Bowling Green Bay: Fig. 1). Interpreted channel sinuosity is similar to segment 1, virtually straight. Some intersections show several channel remnants truncated by the major (most recent) palaeochannel, indicating switching of the channel over time but no systematic lateral migration. Characteristics of the channel fill are similar to those described above for segment 1, but are better resolved by seismic data (e.g. Fig. 3). Line 20, located close to the outer limit of segment 2, shows a complex channel fill comprising at least three, nested, successively west-stepping channel remnants incised and truncated by a later channel form, the cross-section of which is slightly asymmetrical (Fig. 3). The fill of the later channel in an interpreted near-orthogonal cross-section shows downlap to the left (west), indicating limited lateral accretion. The lower part of the fill is acoustically quiet and characterized by low amplitude, discontinuous, dipping reflectors, whilst the upper part is characterized by higher amplitude, more continuous and gently dipping events. The stratigraphy is interpreted (in the absence of any corroborating lithological data) as a coarse-grained lower fill overlain by a more interbedded, heterolithic and fine-grained upper fill.

Segment 3

Segment 3 is a flat section with a calculated channel floor gradient that is slightly negative (possibly an artefact, related to seismic data quality and interpretation), extending across a zone of relatively flat sea-floor bathymetry (Fig. 2). Interpreted channel planform is somewhat more sinuous (Fig. 1 and Table 1), and several additional, smaller channel cross-sections were recorded in this sector, possible indications of tributary streams. Line 21 (Fig. 4) provides an illustration of channel geometry and internal fill in this segment. Three generations of channel can be seen in this cross-section, each of which is incised into and truncates the previous generation. The most recent of these channels is relatively narrow and steeply incised into the previous, shallower channel (Fig. 4). Dipping and downlapping, variable-amplitude reflectors within the most recent channel indicate a channel composed of principally coarse-grained material accumulated possibly by limited lateral accretion. The top of the channel, however, is filled concentrically and defined by high amplitude, continuous reflectors. The basal reflector of this upper facies continues across the interfluve as a

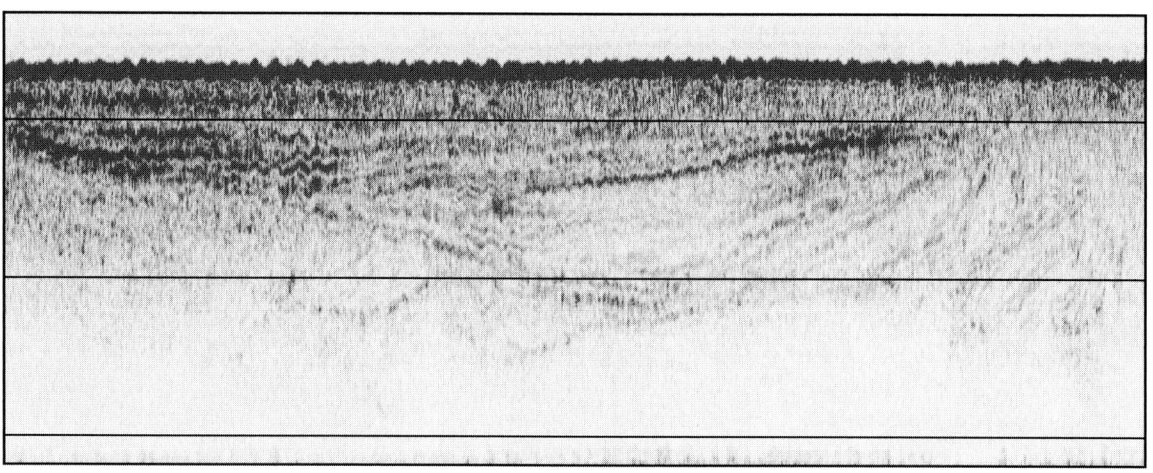

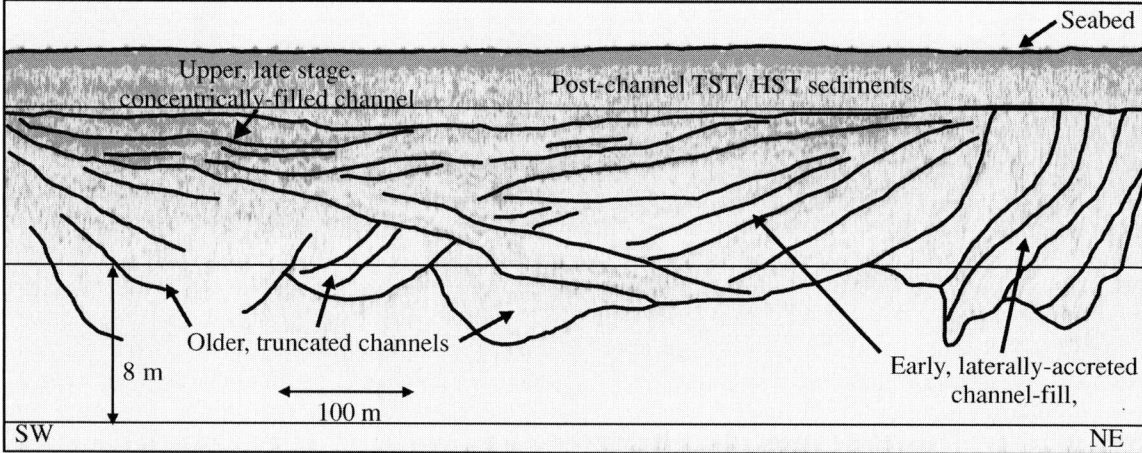

Fig. 3 Seismic reflection profile showing part of line 20 (boundary between segments 2 and 3). The main palaeochannel feature shows a clear distinction between an earlier fill composed of dipping and downlapping reflectors, and a later phase composed of higher amplitude, concentrically arranged reflectors. Note also the remnants of at least three older palaeochannels that have been erosionally truncated by the main (later) channel.

horizontal surface. These features may be indicative of channel abandonment, followed thereafter by passive filling during the Holocene transgression with fine-grained sediment.

Segment 4

Segment 4 is a ramp section in water depths of 25–30 m (Fig. 2 and Table 1). The channel over the upstream part of this segment is relatively straight, but becomes more sinuous downstream and deeper (Figs 1 & 2). Furthermore, segment 4 shows a considerable broadening of the channel belt constructed by the palaeochannel than in segments 1–3. In its upstream part, the channel cross-sectional profile is similar to segments 2 and 3 in being composed of at least two generations of channels, the most recent of which is relatively steeply incised and near-symmetrical, with only limited evidence of lateral accretion.

Downstream, however, the channel appears to become larger, and older generations of channels are either not developed or completely removed

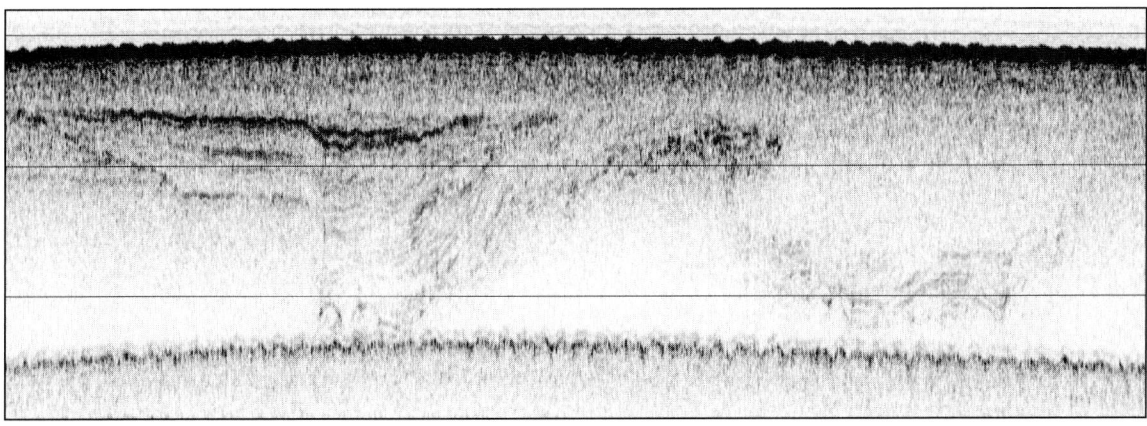
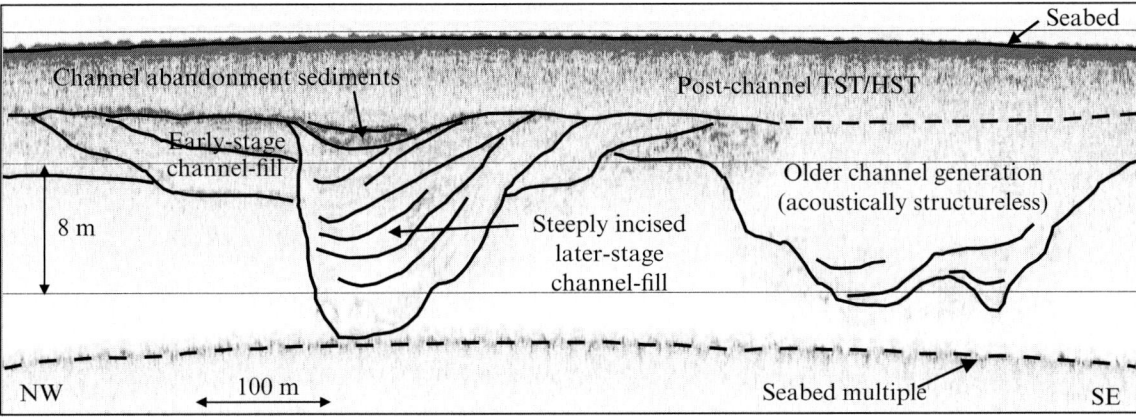

Fig. 4 Seismic reflection profile showing part of line 21 (segment 3). Here, an earlier shallow channel form has been eroded in the centre of the illustration by a later, steeply and deeply incised channel, which has incised to at least twice the depth of the earlier channel. Note also the presence of an interpreted residual channel at the level of the Pleistocene–Holocene interface (Reflector A—high amplitude, continuous event), which seems to have been left open until being covered by late Holocene marine sediment (top 3 m).

by erosion associated with the latest channel-forming event (some local indications of older channels are evident). The main channel feature in downstream data (Figs 1 & 5) is broad, deeply incised, but not as steeply, and slightly asymmetrical. Internally, the main channel is notable for the development of internally concordant, dipping and downlapping, variable amplitude reflectors in the lower part that may signify lateral accretion, and an upper fill that is more or less symmetrical, internally composed of higher amplitude reflectors some of which dip and downlap onto a basal surface (Fig. 5).

Segment 5

Segment 5 is a flat with a flat channel floor gradient, at the start of which the channel changes overall direction from north to north-east (Fig. 1 and Table 1). The size, cross-sectional geometry and internal character of the channel in segment 5 are entirely similar to those of segment 4 (Fig. 6), although in one line remnants of possibly two older channel generations are evident below the main channel cross-section, similar to segments 1–3. The principal palaeochannel evident in these lines displays well-developed dipping and

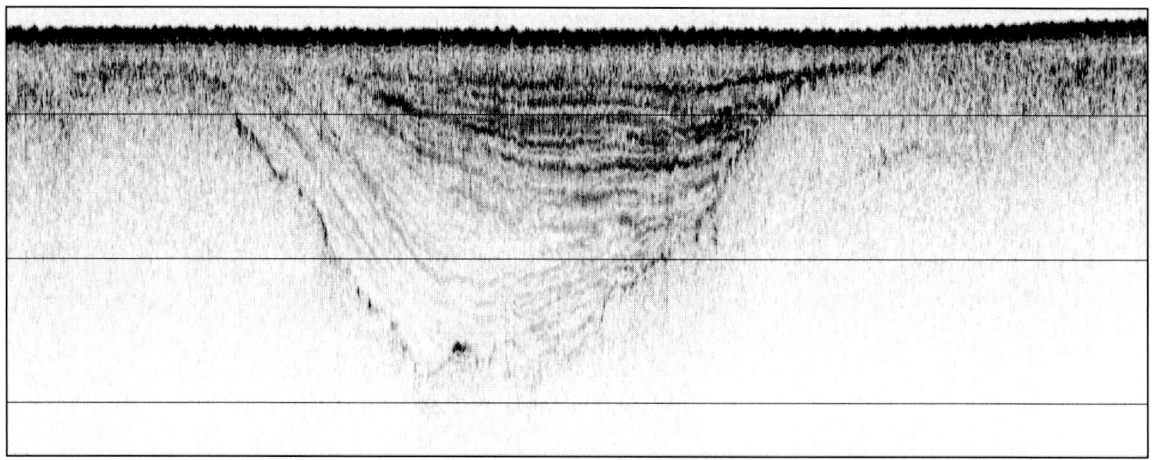

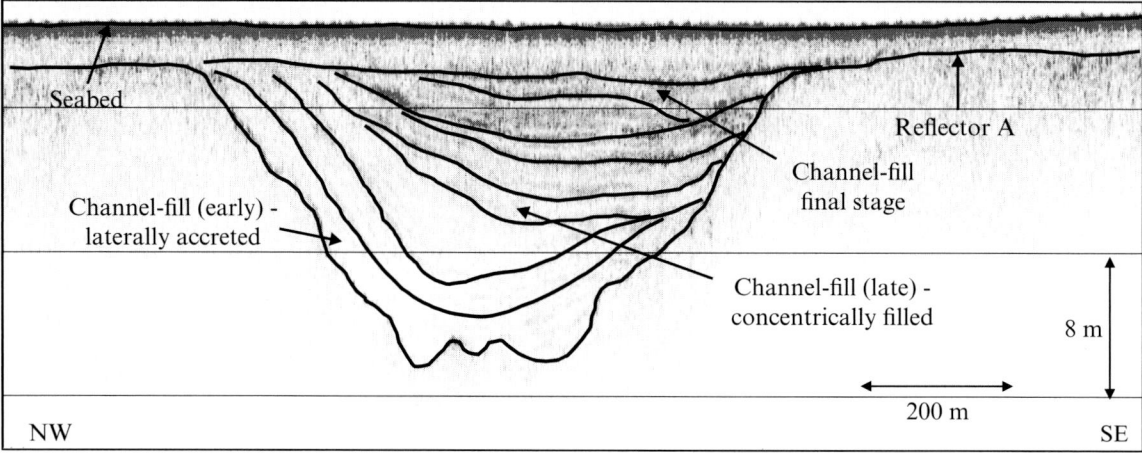

Fig. 5 Seismic reflection profile showing a portion of line 31 (segment 4). Here, a deeply incised channel shows evidence of limited, systematic lateral accretion in the lower part, and a concentrically filled upper part that is also characterized by high-amplitude reflectors.

downlapping reflectors that can be interpreted in terms of a single phase of channel incision, limited lateral migration and filling (Fig. 6). The cross-section shown in Fig. 6 also illustrates several steps and benches in the channel floor profile, features that are evident in other lines but particularly well-developed in this case. Towards the end of segment 5, the single channel diverges into up to five distinct, relatively shallow and mostly concentrically filled channels (Fig. 7). The similar depth of incision of the five channels and their simple cross-sectional geometry suggest that they are coeval, i.e. that the channel was anabranching at the end of segment 5.

Segment 6

Segment 6 is a relatively steep ramp section in 30–40 m water depth (Figs 1 & 2 and Table 1). The most upstream part of segment 6 shows a transition from the multichannel character noted above to a single channel. Some evidence of asymmetrical channel cross-section and lateral accretion geometry is evident. In the downstream portion, a single channel thread is re-established, with a large, deeply incised channel showing an asymmetrical fill geometry.

The upstream portion is suggestive of an anabranching channel planform, as above, the

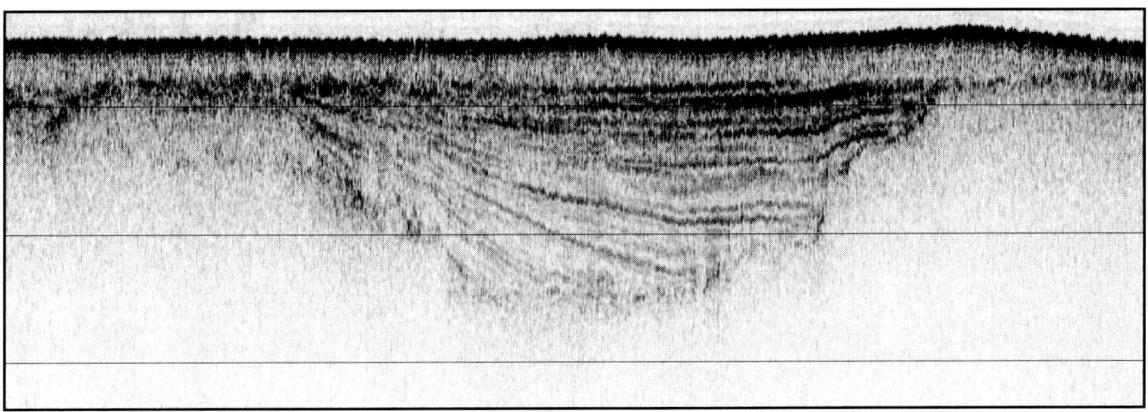
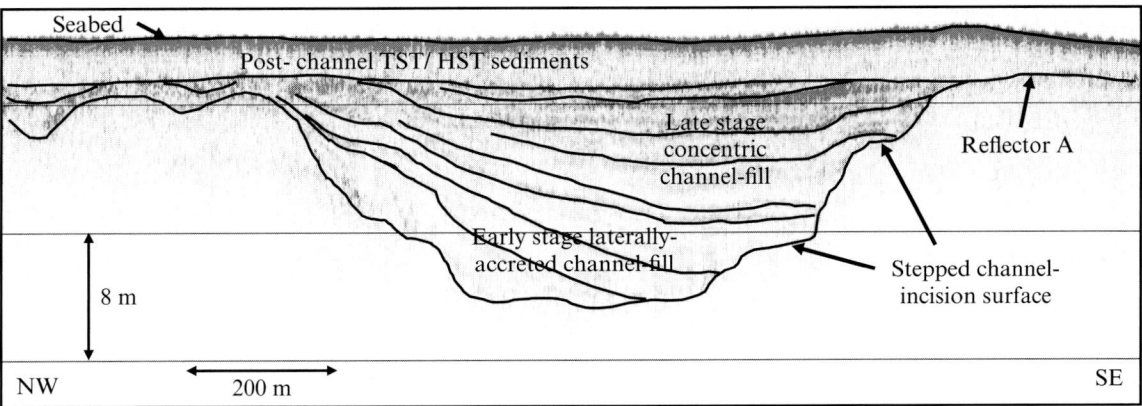

Fig. 6 Seismic reflection profile showing a portion of line 35 (segment 5). This view shows a spectacular example of lateral accretion and concentric upper fill, as in Fig. 5. Note also the stepped (or benched) nature of the basal incision surface.

channels coalescing downstream into a single channel thread. Further downstream in this segment, several distinct periods of channel-forming and -filling activity can be recognized from reflector geometry in some channel cross-sections, as for some upstream reaches (Fig. 8). The asymmetrical cross-sectional profile and dipping/downlapping reflection character in the central and downstream portions of this segment indicate a moderately sinuous channel filled in part by lateral accretion. A late-stage channel fill characterized by a dipping and downlapping, higher amplitude seismic facies is also evident in both of the channel sections shown in Fig. 8: the base of this channel clearly truncates the dipping and downlapping reflectors of the underlying, early stage channel fill, suggesting some repositioning of the channel late in its history. A channel abandonment facies bounded by very high-amplitude events is also present at the top of each channel cross-section (Fig. 8). It is not certain whether the two cross-sections in Fig. 8 record separate channels, or whether they are the same channel: both show a similar internal geometry and seismic facies distribution.

Segment 7

Segment 7 is a flat with a flat channel floor gradient, and encloses a large, right-hand bend in the course of the palaeochannel (Figs 1 & 2 and Table 1). Interpreted sinuosity over this segment is the highest of any segment in the survey (1.72). Channel (or channel belt) width increases abruptly at the start of the segment, then declines gradually downstream somewhat irregularly

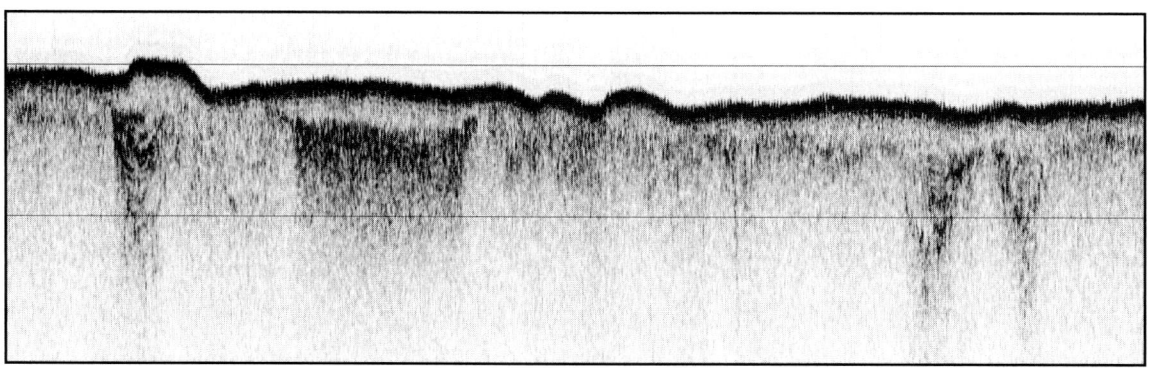
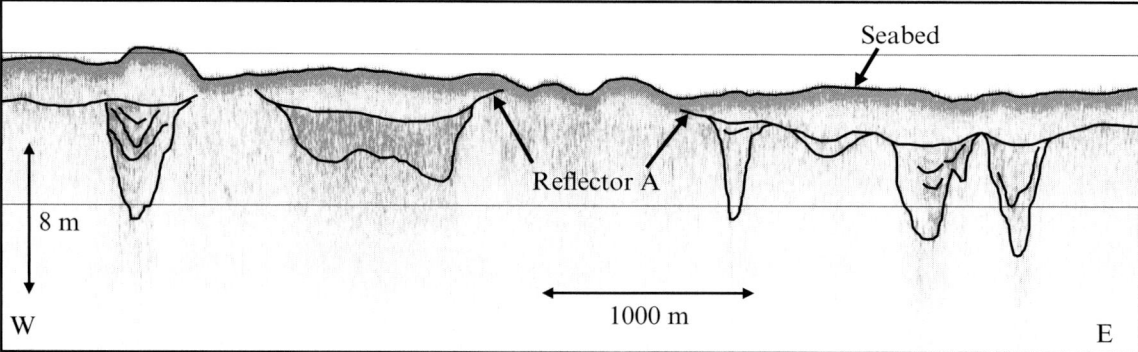

Fig. 7 Seismic reflection showing part of line 38 (end of segment 5: note compressed horizontal scale). In this view, five apparently coeval channels of reduced dimensions and relatively featureless fill are evident, and interpreted as evidence for anabranching channel planform towards the end of segment 5.

(Fig. 2). Channel depth also declines downstream across this segment. These changes may be related to the observations that many intersections indicate multiple channels at the same stratigraphical level, and that most channels intersected show a more or less symmetrical cross-sectional geometry. Thirdly, there is evidence of truncation of older, more deeply incised palaeochannels (e.g. Fig. 9).

The initial increase in channel belt width and the gradual decrease in channel depth can be interpreted as a response to the drop in gradient at the foot of segment 6, and the multichannel style is interpreted to represent anabranching of the channel across this low-gradient flat. The symmetrical cross-section of many channel intersections indicates a mainly vertical mode of channel filling, indicating channels that were positionally stable but perhaps prone to avulsion or anastomosis. Minor evidence of lateral accretion was noted in some channel intersections. The truncation of deeper channel-fill remnants again suggests more than one phase of channel activity, with reoccupation of earlier channel courses and excavation of older alluvium.

Segment 8

Segment 8 is a short ramp section, with the steepest gradient of any of the mapped channel reaches (Fig. 1 and Table 1). The calculated sinuosity over this segment is 1.0, although it is only based on two data points. Segment 8 records a return to a single, deeply incised channel, of apparently low sinuosity.

Segment 9

Segment 9 forms the upstream portion of a long flat interval (segments 9–11). Channel belt sinuosity within this reach is moderate (Fig. 1 and Table 1). The channel typically shows an asymmetrical

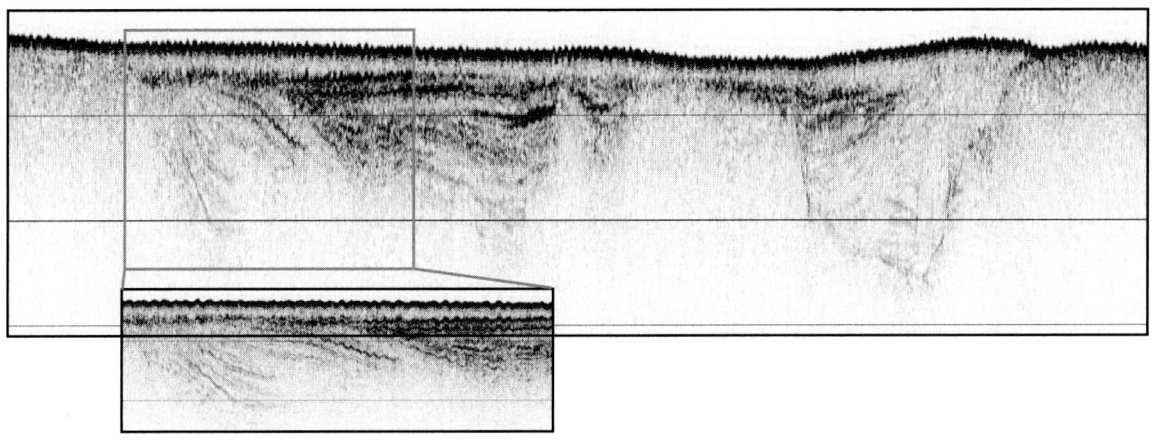
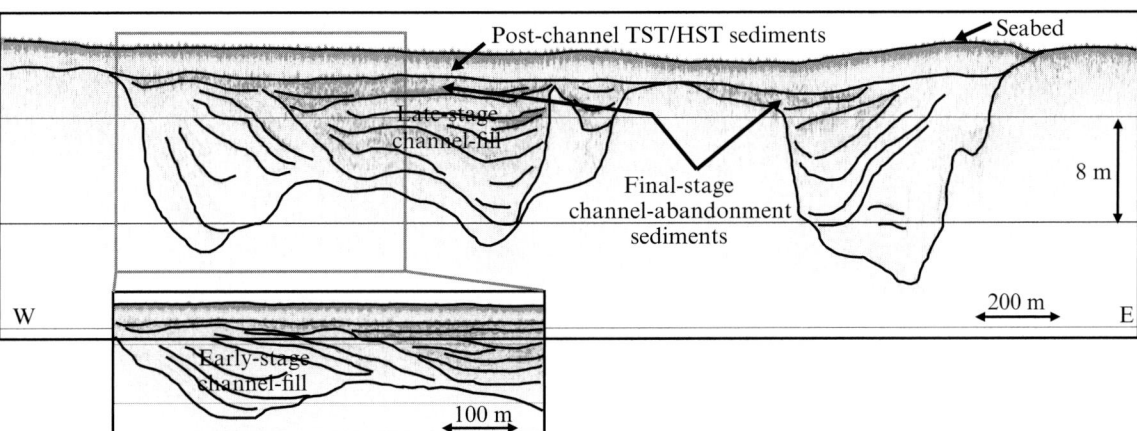

Fig. 8 Seismic reflection profile showing part of line 40 (segment 6). In this view, a complex, highly variable channel fill architecture is evident. Two channel intersections are shown, likely to be separate channel features in an area of anabranching.

cross-sectional profile and complex arrangement of dipping and downlapping reflectors (Fig. 10). This can be interpreted to indicate a multi-phase history of activity, involving several discrete periods of lateral migration and infilling. The resultant cross-sectional geometry is that of a classic meandering channel belt constructed from long-term lateral and vertical accretion. Although on a larger scale, the cross-sectional geometry of the lateral accretion sets is comparable to that imaged by the ground-penetrating radar method from modern meandering channels such as the upper Burdekin River (Fielding et al., 1999). One detail worthy of further comment is the packaging of distinct seismic 'sequences' of lateral accretion sets in Fig. 10, each sequence separated by a surface of discordance and a change in bed inclination. Bridge et al. (1995) noted similar features on a smaller scale in their ground-penetrating radar study of the meandering South Esk River in Scotland, and attributed them to periods of discrete point bar migration. Such an interpretation seems also applicable to the present case. Once again, hints of earlier channels are preserved below the main channel fill in some seismic lines, but were largely removed by erosion associated with excavation of the large channel belt.

Segment 10

Segment 10 is an enigmatic reach across which very little evidence of channels could be discerned

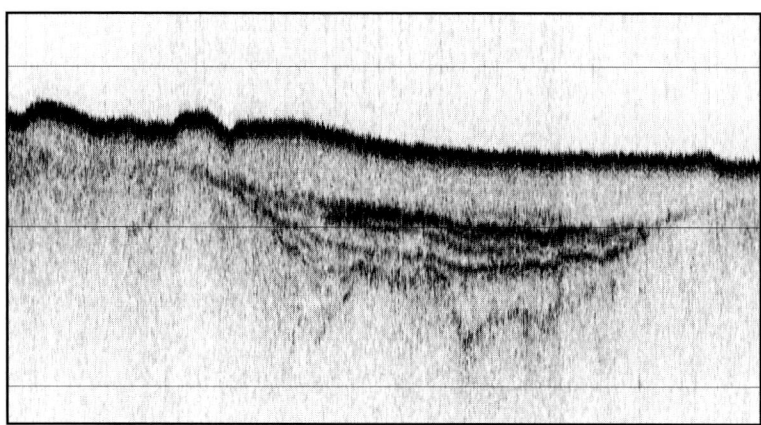

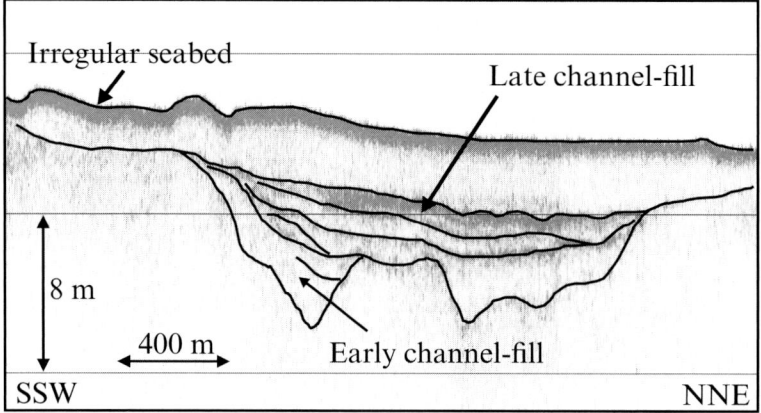

Fig. 9 Seismic reflection profile showing part of line 55 (segment 7). In this ramp section, a concentric channel fill truncates older channel cross-sections. Note the highly irregular sea-floor profile, indicative of an active erosional regime on the modern sea floor.

initially. Evidence of a channel was found only locally around the west and north-west of Keeper Reef. During the second survey, however, a further channel course was intersected to the south-east of Keeper Reef, suggesting that the channel bifurcated around that topographic feature. As the segments upstream and downstream of segment 10 are flats and as there is little difference in elevation between them, it seems likely that segment 10 is also a flat, forming the central portion of a long flat section in the channel (Figs 1 & 2 and Table 1). One line on the north-west side of Keeper Reef intersected multiple, small shallow channels. This may suggest an anabranching channel planform as in segments 5–7.

Two hypotheses may be raised to explain the abrupt disappearance of substantial channels at the start of segment 10. The first possible explanation is that channels were formed but subsequently eroded, perhaps during the Holocene transgression. Some support for this notion comes from the fact that the sea floor is strongly undulating in places through this area, and that the channel cross-sections seen in segment 10 appear to be truncated. Segment 10, furthermore, coincides with a region of greater than average sea-floor gradient down to the −50 m isobath (Fig. 2). Alternatively, channels may never have formed in this area, but rather fluvial runoff was dispersed into a network of shallow streams, wetlands and lakes as was suggested for the region as a whole by Woolfe et al. (1998) and Dunbar et al. (2000). The channel passes between and around substantial carbonate reef structures for the first time in segment 10, and could conceivably have been affected by these obstacles. Nonetheless, the former hypothesis is preferred, in view of the observations made above concerning sea-floor

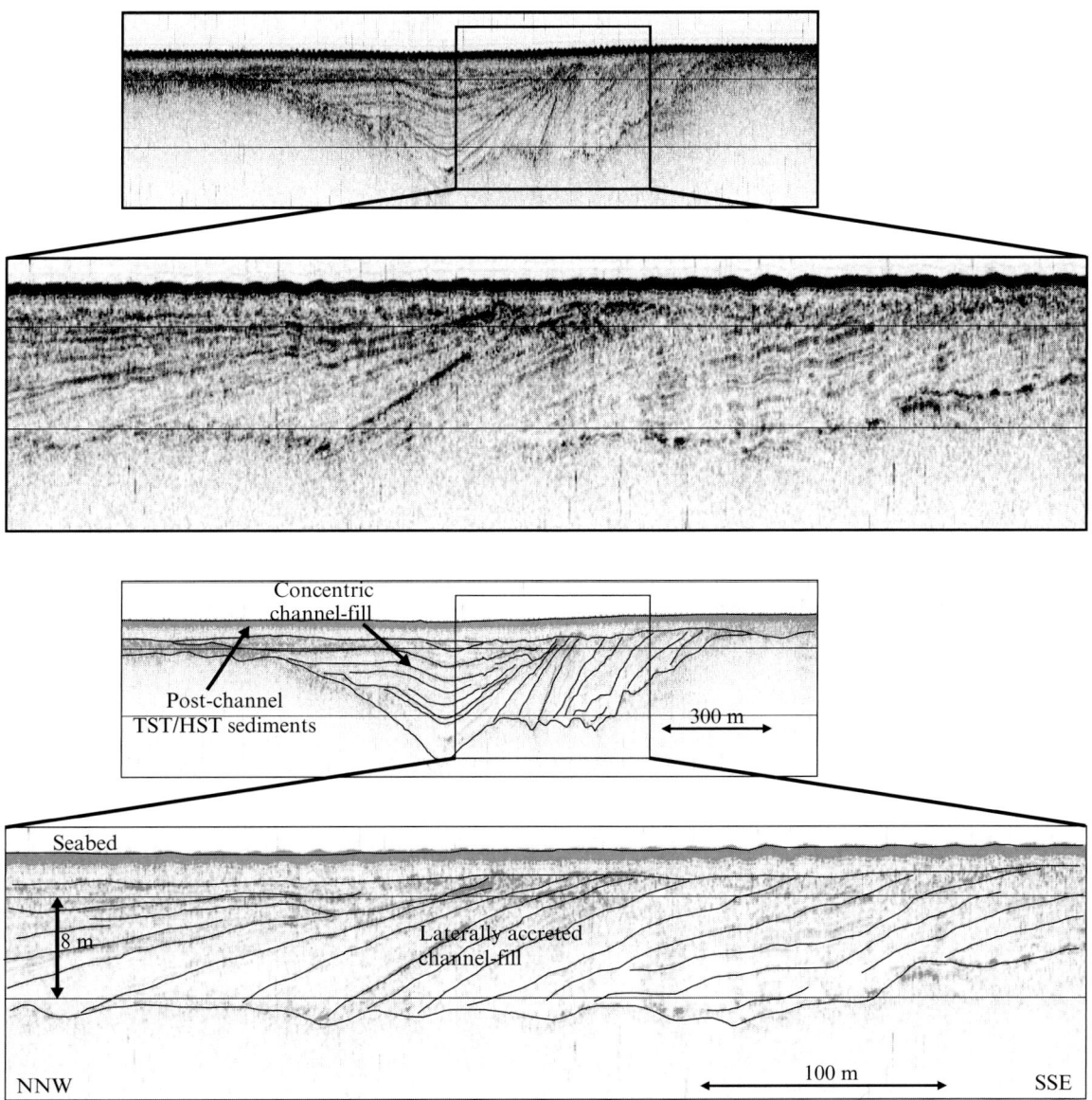

Fig. 10 Seismic reflection profile showing part of line 61 (segment 9). In this spectacular example of systematic lateral accretion, angular discordances between reflectors define packages indicating an episodic channel migration pattern.

topography, and the fact that the channel becomes re-established immediately downstream.

Segment 11

Segment 11 marks a return to clear records of channelization along the long profile, and is a flat that probably represents a continuation of segments 9 and 10 (Figs 1 & 2 and Table 1). In this region, the sea floor is strongly undulating, indicating a Holocene regime of predominantly erosion. A single channel is developed in line intersections, less deeply incised than upstream but nonetheless incised into the substrate. The channel has an asymmetrical cross-sectional profile and shows evidence of lateral accretion as before. The most striking characteristic, however, is the decrease in channel cross-sectional size relative to reaches upstream of segment 10.

Segment 12

Segment 12 is a ramp that straddles the −60 m isobath. Channels imaged over this reach vary in size but are generally smaller than those in segment 11. Overall, channel-belt sinuosity is increased, but individual channels show only limited evidence of lateral migration. Most channel cross-sections, particularly those further downstream, are strongly incised, narrow and have an asymmetrical cross-section with dipping and downlapping reflector character (Fig. 11). Many are located in topographic lows on the present sea floor, and a few are partly confined by reefal carbonate massifs. Many intersections again show evidence of a two-part fill, with an acoustically structureless lower part and an upper part defined by stronger, more continuous reflectors (Fig. 11). Figure 11 also shows a considerable thickness of interpreted post-channel sediments, perhaps suggesting incomplete filling or removal of the upper part of the channel fill by erosion.

Segment 13

Segment 13 is a relatively long, flat section, over which the sea floor is at about −70 m and is highly irregular. The palaeochannel over this segment is characterized by low width, width/depth ratio and interpreted channel sinuosity (Fig. 2), consistent with the immediately upstream segment 12. The palaeochannel in segment 13 is also notable for being underfilled (to locally unfilled) along

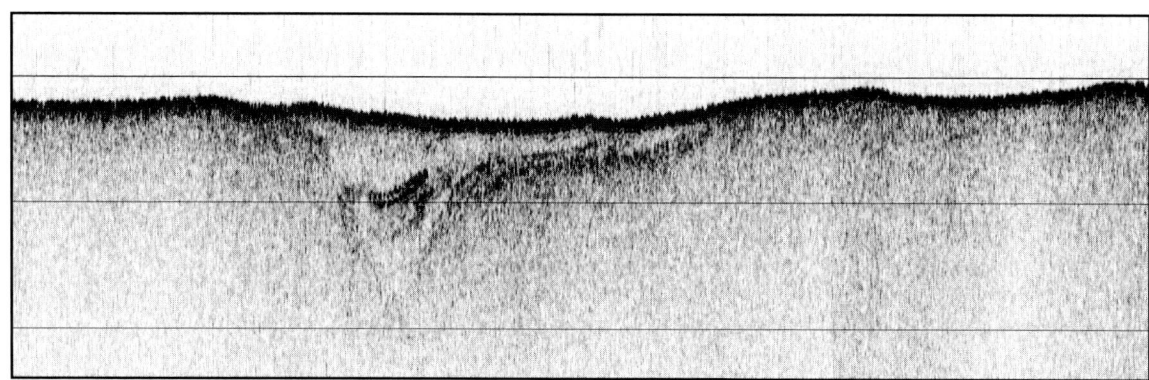

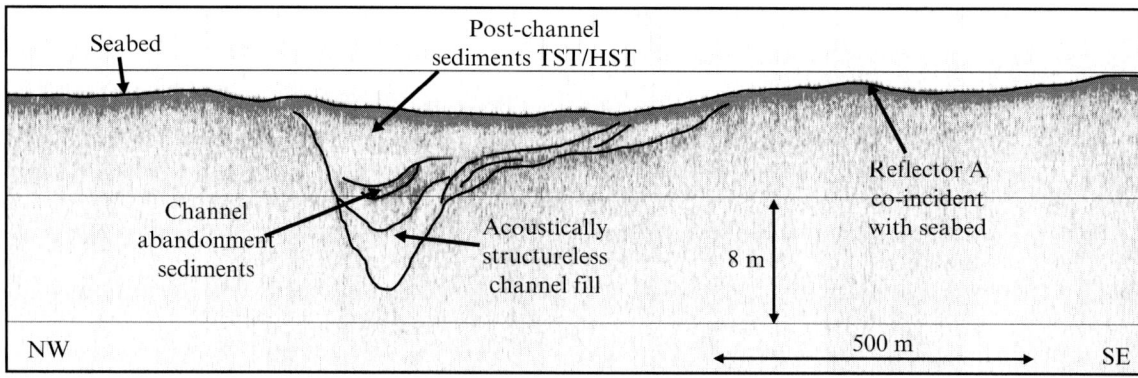

Fig. 11 Seismic reflection profile showing part of line 70 (segment 12). Note the reduced size of the channel in this outer reach, and relatively simple fill architecture. Note also the lack of late Holocene sediment overlying Reflector A, which here coincides with the sea floor.

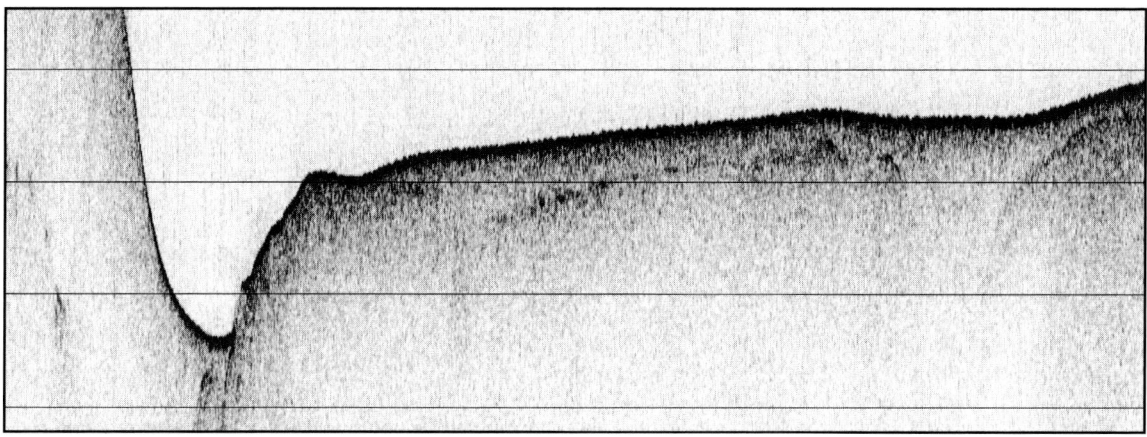

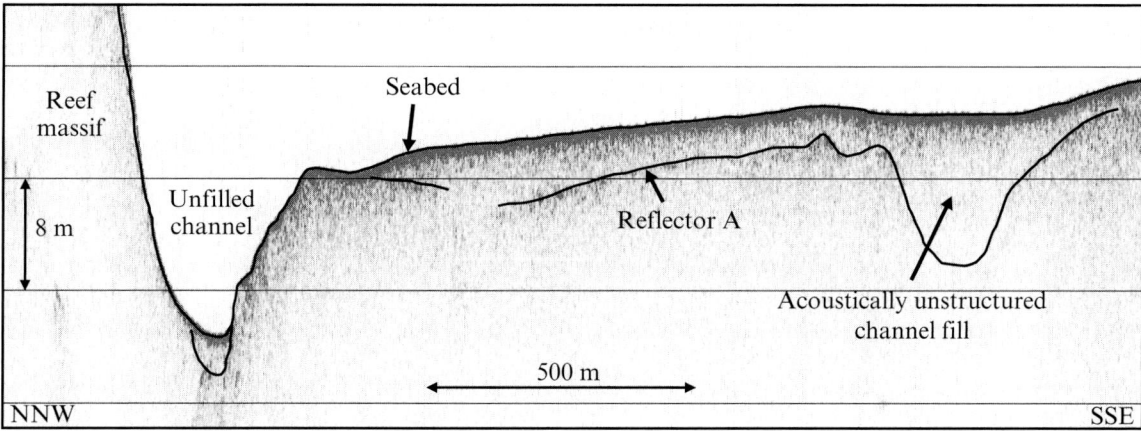

Fig. 12 Seismic reflection profile showing part of line 4 (second survey: segment 13). Note the virtually unfilled, relatively small channel located at the foot of a steep slope representing the edge of a large reefal limestone massif, an older palaeochannel fill truncated by the unfilled channel and the uneven sea-floor surface.

its entire length, having a surface as well as subsurface expression (Figs 2 & 12). In this segment, the channel also typically runs close to, or even at the foot of, steep slopes forming the edges of high-profile carbonate massifs (Fig. 12). It is apparent that the channel here is confined to a relatively narrow fairway between elevated reefal limestone highs.

Internally, the channel fill in segment 13 is typically acoustically bland and unstructured, with low amplitudes and few if any obvious reflectors (Fig. 12). In places, there is evidence of earlier channel fills (similar in dimensions and seismic fill character), truncated by the underfilled, latest channel (Fig. 12).

Segment 14

Segment 14 is another ramp section, at the base of which the sea floor is at about −80 m (Figs 1 & 2). This is the outermost part of the recognized palaeochannel course, with no channel cross-sections intersected seaward of the final data point on Fig. 2. Indeed, seaward of this point, the sea floor rises conspicuously to between −70 and −60 m, forming a shelf-edge high that extends to the shelf break. The sea-floor surface along the segment is once again highly irregular. The palaeochannel across segment 14 shows similar size and cross-sectional geometry (steep-sided, generally symmetrical) to segment 13, becoming

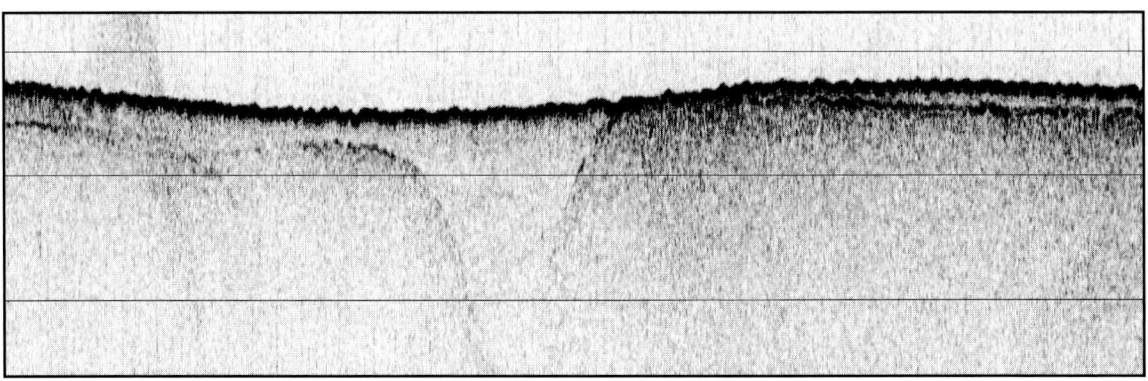

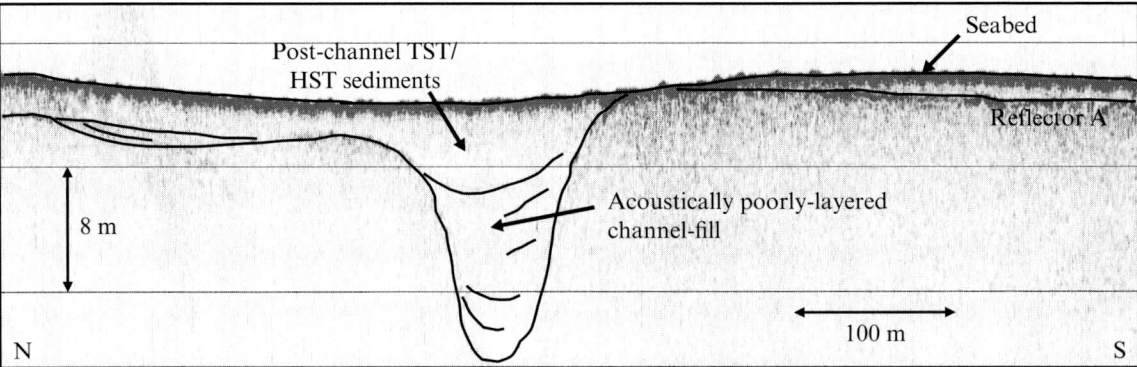

Fig. 13 Seismic reflection profile showing part of line 9 (second survey: segment 14). Note the small, steeply entrenched channel and irregular sea-floor surface.

gradually shallower and narrower downstream (Figs 2 & 13). The channel is also similar to segment 13 in being underfilled over much of the length of the segment. It differs from the latter, however, in being largely unconstrained by nearby reefal limestone massifs. Channel intersections show evidence of a generally symmetrical filling pattern, with generally only sparse, discontinuous, low to moderate amplitude reflectors visible on seismic records (e.g. Fig. 13).

INTERPRETATION

Channel fill

A two-part stratigraphy has been noted in channels from seismic data, the two units being defined by both reflection amplitude and (to a lesser extent) geometry:

1 A lower fill that can be either symmetrical or asymmetrical in cross-section, accounts for most of any given channel intersection, and is typically seismically well-stratified with low-amplitude reflections.

2 An upper fill that is typically symmetrical in cross-section, generally of limited vertical and lateral extent, and seismically displays high-amplitude reflections.

The mainly low-amplitude reflection character of asymmetrical fill units suggests that they were predominantly composed of coarse-grained sediments with perhaps thin mud partings. In symmetrical channel cross-sections the lower part of channel fills are also dominated by low-amplitude reflections, suggesting that similar coarse-grained sediments to those described above filled the lower parts of relatively straight channels. Channels interpreted as having an anabranching planform show a more variable reflection character between channels, which may indicate a variable fill lithology. No direct evidence of channel fill lithology is available outside Bowling Green Bay, owing to difficulties in acquiring cores from deep-water sites.

The asymmetrical channel cross-sections are interpreted as having formed by a period of systematic lateral accretion (and progressive infilling) of a meandering channel. The extent of these interpreted lateral accretion surfaces indicates that lateral migration and hence meander formation was limited, an impression corroborated by estimates of channel sinuosity, which in laterally accreted sections range from 1.2 to 1.7 (Fig. 2 and Table 1). This and the steep gradient of channel banks may in turn have been due to the compact nature of the substrate into which the channel was entrenched. The facts that remnants of older channels are extensively preserved beneath the main channel feature, and that no other major channels were intersected in the area, suggest that Pleistocene channels repeatedly reoccupied older courses rather than excavating new courses through stiff, compact interfluve materials.

The upper parts of channel fills were evidently accumulated in a stable channel of reduced size relative to the original dimensions. In some cases, this fill is symmetrical, indicating vertical accretion, and the high-amplitude reflection character in this context suggests a mud-rich lithology. Elsewhere, the upper part of channel fills displays a laterally accreted character as described above, but angular differences suggest a separate phase of channel migration and infilling. It is considered likely that the upper part of the channel fill was accumulated during the Holocene transgression, in a more stable channel with only local lateral migration. The more stable conditions and associated backstepping of the profile across the margin resulted in increased availability of mud during this time.

In nearly all cases, regardless of channel cross-sectional geometry, the top of the channel fill (where it is overlain by a late Holocene blanket) and/or the sea-floor surface are concave-up (Fig. 4). This further suggests that at least the upper parts of channel fills are composed largely of compactable mud. This relationship is also at variance with the observation by Johnson *et al.* (1982) that palaeochannel locations on the Great Barrier Reef shelf often coincide with sea-floor highs.

Channel planform and course

Previous seismic surveys and the current surveys have established that only one major channel crossed the lowstand shelf in this region. The channel extends from the mouth of a modern river (the Haughton) to the outermost shelf where it apparently dies out in front of a topographic barrier on the shelf edge (Fig. 1). Our surveys also indicate that although this channel was variable in width it did not widen downstream systematically in the manner typical of estuaries, and is therefore not interpreted as an estuary formed during Holocene rising sea level. Indeed, the channel gradually decreases in width over the lowermost segments, consistent with the finding that it dies out downstream. Rather, the channel is interpreted to represent an extension of the Burdekin system across the emergent shelf during sea-level fall and lowstand.

The sinuosity of the channel belt (as opposed to that of the channel itself) varies from virtually straight to 1.7 (Table 1). The sinuosity of the channel itself will be somewhat higher than the figure for the channel belt, but is impossible to determine accurately without three-dimensional seismic data. There is no consistent relationship between interpreted sinuosity and channel-floor gradient. In at least two ramp segments (4 and 6), the channel-belt planform seems to change from straight to more sinuous from the top to the base of the ramp. Segments in which anabranching planform has been tentatively identified (5, 7 and possibly 10) are all within relatively long flat reaches, which is consistent with the global relationship between anabranching/anastomosing planform and landscapes of very low gradient (Smith & Smith, 1980; Rust, 1981). Channel depth seems to show a consistent relationship with gradient, with downstream trends towards a shallower channel along flats, and towards a deeper channel down ramps (Fig. 2). Overall, there is no consistent trend in channel depth along the profile. Channel width, on the other hand, shows a progressive increase from the modern coast out to segments 7–11, and then a marked decrease from the end of segment 11 to the downstream end of the channel.

DISCUSSION

The channel floor profile (Fig. 2) is interpreted to provide a more or less faithful record of the original channel gradient, despite the shortcomings noted previously. The cited gradients for individual segments should be regarded only as approximations to the true gradient. Remnants of several discrete channel-forming events are preserved in some seismic lines, typically truncated by a later channel that overprints earlier records. Thus, the measured channel floor base may not, in all cases, refer to the same phase of channel activity, even though the traced channel belt is considered to be unique to the area and confined to a narrow belt. Furthermore, the channel floor will inevitably be an irregular surface, affected by local channel scouring patterns. Nonetheless, it is used herein for the reasons explained above. The channel floor gradient can be clearly segmented into relatively steep zones (ramps) and zones of negligible gradient (flats) that are not due to random variation. Downstream limits of flats, at the start of the succeeding ramp, are interpreted to define knick-points in the profile. The top of the channel fill shows a subdued version of the channel profile, presumably affected by compaction. The sea-floor elevation, on the other hand, is believed to be modified in places by wave/current erosion associated with the Holocene transgression and highstand, and therefore does not provide a reliable view of the original profile. The longitudinal profile reflects the initial incision/entrenchment of the channel, and the channel can be traced as far as the present shelf edge, indicating strongly that it is older than the recent transgression.

The stepped profile is interpreted to be a reflection predominantly of falling sea-level associated with the last glacial cycle (125–20 ka: Chappell *et al.*, 1996), in which a series of sea-level falls were punctuated by shorter periods of stasis or minor sea-level rise. This is considered more likely than the alternative scenario of formation during the post-glacial rise in sea level (< 20 ka), which, although also stepped (Beaman *et al.*, 1994; Larcombe *et al.*, 1995), would have produced a quite different longitudinal profile. No deltaic deposits can be identified, the channel shows no evidence of change in plan geometry, width/depth or internal fill character as would be expected in a coastal plain setting (other than in segment 7: see below), and the thickness of post-channel-fill sediments seems to show little relationship to the channel profile (Fig. 2). Accordingly, the incision of the channel profile is interpreted to reflect the progressive but punctuated fall in sea level associated with the onset of the last glacial period (Fig. 14A). Possible relationships between the elevations of flats in the channel long profile and the ages of hiatuses in the long-term sea-level drop during the last glacial cycle are explored by Fielding *et al.* (2003).

Given the punctuated nature of sea-level fall during the last glacial cycle, the complex pattern of channel cross-cutting relationships evident in segments 1–9 can be interpreted to reflect progressive lowering of base level. Many of the seismic lines that show older generations of channels cross-cut by the most recent channel also demonstrate that the most recent channel has incised to the greatest depth (e.g. Figs 3 & 4). This suggests that the principal channel illustrated and described herein was formed during the lowstand of the last glacial, and that the older, erosionally truncated channels were formed during intermediate stages in the long-term sea-level fall. It is interesting to note that the channel appears to decrease in cross-sectional area towards the present shelf edge, suggesting that discharge was decreasing downstream towards the Pleistocene lowstand shoreline. Indeed, this survey indicates that the channel terminates 10 km before the shelf edge. Thus, either (i) the channel never extended out as far as the lowstand shoreline, which would have been located on the upper part of the continental slope, or (ii) the channel aggraded to the point where it was able to overtop the shelf edge barrier (Fig. 14B). If the latter option is considered tenable, then the early stages of post-glacial transgression must have removed the accumulated sediments from the outermost 10 km of the shelf. Some support for this interpretation comes from the independent finding of Dunbar *et al.* (2000) that significant volumes of sediment were exported to the continental slope of northeast Australia during the early stages of the post-glacial transgression (Fig. 14C).

The lower, interpreted coarse-grained fill of the channel is considered to have accreted in part

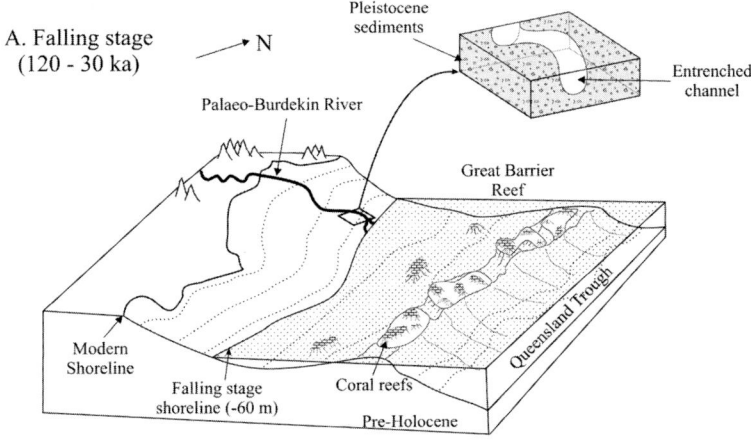

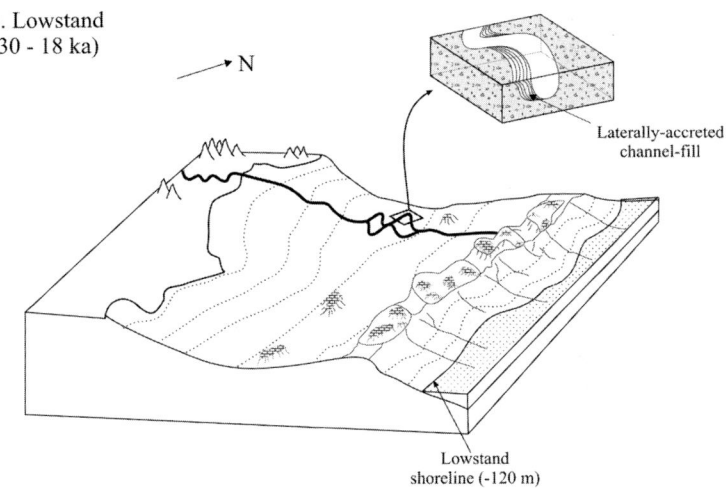

Fig. 14 Series of block diagrams summarizing the late Pleistocene to Holocene history of the Burdekin palaeochannel. (A) Late Pleistocene falling stage, in which the shoreline retreated across the shelf, causing the palaeo-Burdekin river channel to become entrenched into a low gradient land surface. (B) Late Pleistocene lowstand, with a small channel passing around exposed outer shelf reefal limestone massifs and extending to or close to the shoreline, which was located on the upper continental slope.

during the long-term decline in sea level from 120 to 20 ka, but given the elevation of channel fills must have also undergone significant vertical and lateral accretion during the Holocene sea-level rise. The younger phase of channel infilling with its interpreted finer grained deposits may be the product of the post-glacial transgression, which would have caused transformation of the channel into an estuary and progressive backfilling with mainly fine-grained sediment (Fig. 14C). The rate of sea-level rise is interpreted by Chappell et al. (1996) and others to have been much more rapid than the preceding drop, and there would have been insufficient time for the channel profile to re-equilibrate to the constantly and rapidly rising base level. Consequently, the lowstand channel was predominantly filled by vertical accretion, and by mainly fine-grained sediment, giving rise to the high-amplitude seismic facies noted in the data. This also explains why parts of the channel were left unfilled following inundation by the sea to form depressions on the mid-Holocene sea floor that were then filled by younger shelfal sediments (Fig. 14D).

The stepped channel floor profile has been interpreted as the result of repeated adjustment to a long-term (100 kyr) drawdown in sea level over the last glacial cycle (herein, and see Fielding

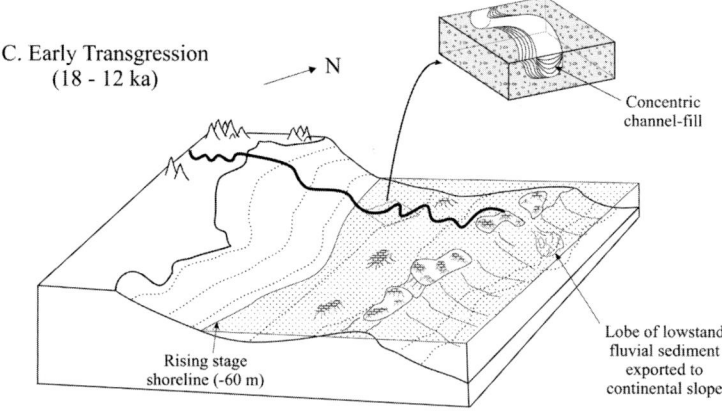

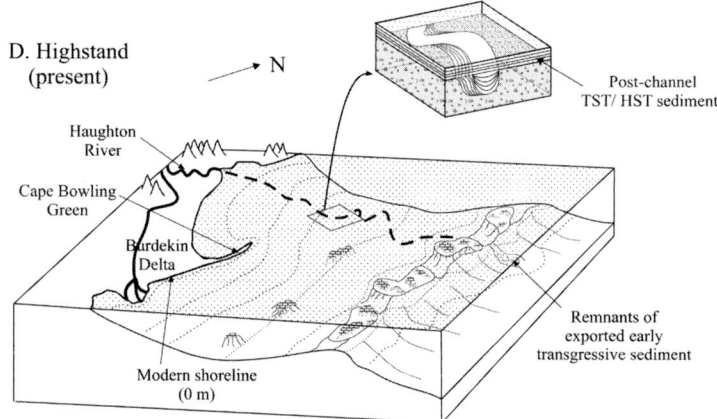

Fig. 14 (C) Latest Pleistocene transgression, in which sediment in the lower reaches of the channel was flushed out of the system and exported to the continental slope. (D) Holocene highstand, in which the channel was backfilled, sediment was partially reworked by marine processes and ultimately covered by marine sands and muds.

et al., 2003). The channel cross-sectional geometry illustrated in seismic reflection data indicates that the channel was not able to migrate freely to form a meander belt. Together, this suggests that the channel was in a state of disequilibrium, constantly adjusting to the drawdown in sea level. The decrease in channel size towards its downstream end also suggests that discharge during the sea-level drawdown was barely sufficient to sustain the channel as far as the shelf edge.

The late Pleistocene land surface, as recorded by 'Reflector A', further reinforces the impression of a channel that was unable to adjust sufficiently rapidly to keep pace with changing base level. The channel is clearly inset into a relatively smooth, even surface, rather than occupying part of a deeper valley incised during the drawdown of sea level. As such, it is better described as 'entrenched' rather than 'incised'. Indeed, the cross-sectional geometry resembles that summarized by Schumm & Ethridge (1994) as typical of the initial stages of valley formation, and also produced in the same way experimentally by Wood et al. (1993). Similar cross-sectional and plan geometry were also illustrated by Posamentier (2001) from the Neogene of Indonesia, although some examples described therein as 'incised' seem to be more consistent with the term 'entrenched' as used here.

The late Pleistocene Burdekin River was unable to incise a substantial alluvial valley as it crossed the exposed shelf during the last glacial sea-level drawdown. This is a quite different scenario to that described from Pleistocene records of other continental margins such as eastern and southern

USA (e.g. Colman & Mixon, 1988; Thomas & Anderson, 1994; Blum & Törnqvist, 2000), where Pleistocene palaeochannels occupied substantial (incised) valleys and extended as far seaboard as the lowstand shoreline.

CONCLUSIONS

A palaeochannel thought to represent a Pleistocene course of the Burdekin River has been traced continuously across the Great Barrier Reef shelf to within 10 km of the shelf edge. This data set demonstrates that at least one river was capable of extending across the northern Great Barrier Reef shelf, contrary to the proposal of Woolfe et al. (1998) that lowstand channels on the north-eastern Australian margin dissipated into a network of small channels and wetlands in front of exposed reef massifs. The channel-floor long profile, as established from over 100 seismic reflection record intersections, shows a series of flat and ramp segments that is interpreted as a record of long-term sea-level drawdown over the last glacial cycle (125–20 ka). This, together with the simple nature of the channel cross-section in most intersections, and the shallow entrenchment of the channel into the compact Pleistocene land surface, indicates that the Burdekin palaeochannel was unable to reach equilibrium with environmental conditions owing to constant adjustments to the declining base level. The channel was able only to entrench itself into the substrate, rather than to incise a substantial valley as rivers on other continental margins did at this time.

The contrasting behaviour of the Burdekin versus channels from temperate margins can be attributed to the nature of the substrate (compacted, partially lithified alluvium), the limited runoff and sediment supply to this continental margin, and the physiography of the Great Barrier Reef shelf, which is not a seaward-deepening gradient but punctuated by numerous reef massifs. Thus, even during a lowstand that exposed the shelf edge, the largest river on the north-eastern Australian margin was capable only of entrenching a channel into a low-relief land surface, a channel which was deflected around numerous exposed limestone hills and ultimately terminated against a shelf-edge barrier. This case history indicates the need for a broadening of genetic stratigraphical models for river behaviour during cycles of sea-level change, to include variants for situations characterized by limited sediment supply/runoff and irregular shelf topography.

ACKNOWLEDGEMENTS

This work was undertaken with financial support from the Australian Research Council (Grant A39937196 to CRF) and a JCU Earth Science Merit Grant (to GRD). Ship's master Don Battersby and Technical Officer Kevin Hooper are thanked for facilitating data acquisition on board the RV *James Kirby*. Seismic Micro-Technology are thanked for allowing the use of their Kingdom® seismic interpretation software. Reviews of the submitted manuscript by Henk Berendsen and Kathleen Farrell led to significant clarifications of our arguments. We dedicate this cruise and the results arising from it to our friend and colleague Kenneth J. Woolfe, who passed away in tragic circumstances in December 1999.

REFERENCES

Alexander, J. Fielding, C.R. and Pocock, G.D. (1999) Flood behaviour of the Burdekin River, tropical north Queensland, Australia. In: *Floodplains: Interdisciplinary Approaches* (Eds S.B. Marriott and J. Alexander). *Geol. Soc. Lond. Spec. Publ.*, **163**, 27–40.

Beaman, R., Larcombe, P. and Carter, R.M. (1994) New evidence for the Holocene sea level high from the inner shelf, central Great Barrier Reef, Australia. *J. Sediment. Res.*, **A64**, 881–885.

Belperio, A.P. (1983) Terrigenous sedimentation in the central Great Barrier Reef lagoon: a model from the Burdekin region. *Bur. Min. Res. J. Austr. Geol. Geophys.*, **8**, 179–190.

Blum, M.D. and Törnqvist, T.E. (2000) Fluvial responses to climate and sea-level change: a review and look forward. *Sedimentology*, **47**, (Suppl. 1), 2–48.

Bridge, J.S., Alexander, J., Collier, R.E.Ll., Gawthorpe, R.L. and Jarvis, J. (1995) Ground penetrating radar and coring used to study the large scale structure of point bar deposits in three dimensions. *Sedimentology*, **42**, 839–852.

Burdekin Project Committee (1977) *Resources and Potential of the Upper Burdekin River of North Queensland*. Australian Government Publication Service, Canberra, 195 pp.

Chappell, J., Omura, A., Ezat, T., McCulloch, M., Pandolfi, J., Ota, Y. and Pillans, B. (1996) Reconciliation of late Quaternary sea-levels derived from coral terraces at Huon Peninsula with deep sea oxygen isotope records. *Earth Planet. Sci. Lett.*, **141**, 227–236.

Colman, S.M. and Mixon, R.B. (1988) The record of major Quaternary sea-level changes in a large coastal plain estuary, Chesapeake Bay, eastern United States. *Palaeogeogr. Palaeoclimatol. Palaeoecol.*, **68**, 99–116.

Dalrymple, R.W., Boyd, R. and Zaitlin, B.A., Eds, (1994) *Incised-Valley Systems: Origin and Sedimentary Sequences*. Soc. Econ. Paleontol. Mineral. Spec. Publ., **51**, 39 pp.

Davies, P.J. and Peerdeman, F.M. (1998) The origin of the Great Barrier Reef—the impact of Leg 133 drilling. In: *Reefs and Carbonate Platforms in the Pacific and Indian Oceans* (Eds G.F. Camoin and P.J. Davies). Spec. Publ. Int. Assoc. Sedimentol., **25**, 23–38.

Davies, P.J., Symonds, P.A., Feary, D.A. and Pigram, C.J. (1989) The evolution of the carbonate platforms of northeast Australia. In: *Controls on Carbonate Platform to Basin Development* (Eds P. Crevello, J.F. Sarg, J.F. Read and J.L. Wilson). Soc. Econ. Paleontol. Mineral. Spec. Publ., **44**, 233–258.

Dunbar, G.B., Dickens, G.R. and Carter, R.M. (2000) Sediment flux across the Great Barrier Reef Shelf to the Queensland Trough over the last 300 ky. *Sediment. Geol.*, **133**, 49–92.

Ferro, C.E., Droxler, A., Anderson, J.B. and Mujcciarone, D. (1999) Late Quaternary shift of mixed siliciclastic environments induced by glacial eustatic sea-level fluctuations in Belize. In: *Advances in Carbonate Sequence Stratigraphy: Application to Reservoirs, Outcrops and Models* (Eds P.M. Harris, A.H. Saller and J.A. Simo). Soc. Econ. Paleontol. Mineral. Spec. Publ., **63**, 385–411.

Fielding, C.R., Alexander, J. and McDonald, R. (1999) Sedimentary facies from Ground-Penetrating Radar surveys of the modern, upper Burdekin River of north Queensland, Australia: consequences of extreme discharge fluctuations. In: *Fluvial Sedimentology VI* (Eds N.D. Smith and J. Rogers). Spec. Publ. Int. Assoc. Sedimentol., **28**, 347–362.

Fielding, C.R., Trueman, J.D., Dickens, G.R. and Page, M. (2003) Anatomy of the buried Burdekin River channel across the Great Barrier Reef shelf: how does a major river operate on a tropical mixed siliciclastic/carbonate margin during sea-level lowstand? *Sediment. Geol.*, **157**, 291–301.

Fisk, H.N. (1944) *Geological Investigation of the Alluvial Valley of the lower Mississippi River*. Mississippi River Commission, Vicksburg.

Harris, P.T., Davies, P.J. and Marshall, J.F. (1990) Late Quaternary sedimentation on the Great Barrier Reef continental shelf and slope east of Townsville, Australia. *Mar. Geol.*, **94**, 55–77.

Hopley, D. (1970) *The Geomorphology of the Burdekin Delta, North Queensland*. Monograph Series, **1**, Department of Geography, James Cook University, Townsville.

Johnson, D.P. and Searle, D.E. (1984) Post-glacial stratigraphy, central Great Barrier Reef, Australia. *Sedimentology*, **31**, 335–352.

Johnson, D.P., Searle, D.E. and Hopley, D. (1982) Positive relief over buried post-glacial channels, Great Barrier Reef province, Australia. *Mar. Geol.*, **46**, 149–159.

Larcombe, P., Carter, R.M., Dye, J., Gagan, M.K. and Johnson, D.P. (1995) New evidence for episodic post-glacial sea-level rise, central Great Barrier Reef, Australia. *Mar. Geol.*, **127**, 1–44.

Marshall, J. and Davies, P. (1984) Last interglacial reef growth beneath modern reefs in the southern Great Barrier Reef. *Nature*, **307**, 44–46.

Maxwell, W.G.H. (1968) *Atlas of the Great Barrier Reef*. Elsevier, Amsterdam, 258 pp.

Neil, D.T., Orpin, A.R., Ridd, P.V. and Yu, B. (2002) Sediment yield and impacts from river catchments to the Great Barrier Reef lagoon. *Mar. Freshwater Res.*, **53**, 1–20.

Orme, G.R., Webb, J.D., Kelland, N.C. and Sargent, G.E.G. (1978) Aspects of the geological history and structure of the northern Great Barrier Reef. *Phil. Trans. Roy. Soc. Lond., Ser. A*, **291**, 23–35.

Orpin, A.R. (1999) Sediment transport, partitioning, and unmixing relationships in the mixed terrigenous–carbonate system of the Great Barrier Reef, Burdekin shelf sector, Australia. Unpublished PhD thesis, James Cook University.

Posamentier, H.W. (2001) Lowstand alluvial bypass systems: incised vs. unincised. *Am. Assoc. Petrol. Geol. Bull.*, **85**, 1771–1793.

Posamentier, H.W., Allen, G.P., James, D.P. and Tesson, M. (1992) Forced regressions in a sequence stratigraphic framework: concepts, examples, and sequence stratigraphic significance. *Am. Assoc. Petrol. Geol. Bull.*, **76**, 1687–1709.

Rust, B.R. (1981) Sedimentation in arid-zone anastomosing fluvial system: Cooper's Creek, central Australia. *J. Sediment. Petrol.*, **51**, 745–755.

Schumm, S.A. and Ethridge, F.G. (1994) Origin, evolution and morphology of alluvial valleys. In: *Incised Valley Systems: Origin and Sedimentary Sequences* (Eds R.W. Dalrymple, R. Boyd and B.A. Zaitlin). Soc. Econ. Paleontol. Mineral. Spec. Publ., **51**, 11–27.

Searle, D.E. (1983) Late Quaternary regional controls on the development of the Great Barrier Reef—Geophysical evidence. *BMR J. Austr. Geol. Geophys.*, **8**, 267–276.

Smith, D.G. and Smith, N.D. (1980) Sedimentation in anastomosed river systems: examples from alluvial valleys near Banff, Alberta. *J. Sediment. Petrol.*, **50**, 157–164.

Talling, P.J. (1998) How and where do incised valleys form if sea level remains above the shelf edge? *Geology*, **26**, 87–90.

Thomas, M.A. and Anderson, J.B. (1994) Sea-level controls on the facies architecture of the Trinity/Sabine incised-valley system, Texas continental shelf. In: *Incised Valley Systems: Origin and Sedimentary Sequences* (Eds R.W. Dalrymple, R. Boyd and B.A. Zaitlin). *Soc. Econ. Paleontol. Mineral. Spec. Publ.*, **51**, 63–82.

Thomas, M.F. and Thorp, M.B. (1995) Geomorphic response to rapid climatic and hydrologic change during the late Pleistocene and early Holocene in the humid and sub-humid tropics. *Quat. Sci. Rev.*, **14**, 193–207.

Thorne, J. (1994) Constraints on riverine valley incision and the response to sea-level change based on fluid mechanics. In: *Incised Valley Systems: Origin and Sedimentary Sequences* (Eds R.W. Dalrymple, R. Boyd and B.A. Zaitlin). *Soc. Econ. Paleontol. Mineral. Spec. Publ.*, **51**, 29–43.

Van Wagoner, J.C., Posamentier, H.W., Mitchum, R.M., Vail, P.R., Sarg, J.F., Loutit, T.S. and Hardenbol, J. (1988) An overview of the fundamentals of sequence stratigraphy and key definitions. In: *Sea-level Changes, an Integrated Approach* (Eds C.K. Wilgus, C.G. Hastings, A.St.C. Kendall, H.W. Posamentier, C.A. Ross and J.C. Van Wagoner). *Soc. Econ. Paleontol. Mineral. Spec. Publ.*, **42**, 39–45.

Willis, B.J. (1989) Palaeochannel reconstructions from point bar deposits: a three-dimensional perspective. *Sedimentology*, **36**, 757–766.

Wood, L.J., Ethridge, F.G. and Schumm, S.A. (1993) The effects of rate of base-level fluctuation, on coastal plain, shelf and slope depositional systems: an experimental approach. In: *Sequence Stratigraphy and Facies Associations* (Eds H.W. Posamentier, C.P. Summerhayes, B.U. Haq and G.P. Allen). *Spec. Publ. Int. Assoc. Sedimentol.*, **18**, 43–53.

Woolfe, K.J., Larcombe, P., Naish, T. and Purdon, R.G. (1998) Lowstand rivers need not incise the shelf: an example from the Great Barrier Reef, Australia, with implications for sequence stratigraphic models. *Geology*, **26**, 75–78.

Quaternary alluvial stratigraphical development in a desert setting: a case study from the Luni River basin, Thar Desert of western India

M. JAIN*,1, S.K. TANDON*,
A.K. SINGHVI†, S. MISHRA‡ and S.C. BHATT*,2

*Department of Geology, Delhi University, Delhi-110058, India;
†Physical Research Laboratory, Navarangpura, Ahmedabad-380009, India; and
‡Department of Archaeology, Deccan College, Pune 411 006, India

ABSTRACT

This study describes the stratigraphical development of the Quaternary alluvial deposits in the Luni basin, Thar Desert of India. On the basis of mode of occurrence, lithofacies assemblages and overall alluvial motif, alluvial deposits have been differentiated into an older Type-1 and a younger Type-2 succession, with chronological control provided by optically stimulated luminescence dating. The Type-1 succession is composed of thick multistoreyed gravel–sand sheets and overbank heterolithic facies deposited in a subsiding basin by braided streams. Quartz in sediments of the Type-1 succession was in charge saturation, which implies a minimum age of 200 ka. It is argued that the Type-1 sequence was deposited during the Pliocene. The Type-2 succession consists of a diverse array of lithofacies assemblages that represents fluvial and aeolian depositional environments, which change in their relative importance through time. Optical ages from the Type-2 succession indicate deposition during the late Pleistocene and Holocene (oxygen isotope stages 5–1). The Type-2 succession indicates that high-amplitude climatic shifts during the last glacial cycle played a major role in determining the fluvial behaviour and resultant alluvial stratigraphy.

INTRODUCTION

Over the past two decades there has been an increased interest in understanding the fluvial response to past climate and environmental changes (see Blum & Törnqvist, 2000). Arid zone rivers, in particular, exist under marginal hydrological conditions such that even minor changes in climate can lead to significant changes in flow regime, sediment transport and associated channel style (Nanson & Tooth, 1999). Alluvial deposits have been used as significant records of environmental change in arid/semi-arid regions (e.g. Hereford, 1986; Bull, 1991; Nanson et al., 1995; Mass et al., 1998; Nanson & Tooth, 1999); however, the sedimentology and Quaternary history of arid-zone rivers remain somewhat poorly understood. Most recent studies have concentrated on modern systems (e.g. Thornes, 1994; Reid & Frostick, 1997; Tooth, 2000), and there exists a

[1]Present address: Institute of Geography and Earth Sciences, University of Wales, Aberystwyth, Ceredigion SY23 3DB, UK

[2]Present address: Department of Geology, Bundelkhand University, Jhansi, UP, India

need to understand the response of arid-zone rivers to climate changes over longer time-scales.

Previous studies in the semi-arid fluvial basins of western India indicate that climate change, sea-level change and active tectonics have influenced fluvial processes (e.g. Kale & Rajaguru, 1987; Merh and Chamyal, 1997; Tandon et al., 1997; Juyal et al., 2000; Mishra & Rajaguru, 2001). These studies have largely concentrated on the margins of the Thar Desert, and comprehensive attempts to understand the dominant controls on long-term fluvial behaviour within the desert itself have not yet been undertaken. This study examines the stratigraphy, sedimentology, and geochemistry of alluvial deposits of the Luni River basin. The overall purpose is to develop a more comprehensive understanding of climatic controls on alluvial stratigraphical development in this desert river system. To help accomplish this a geochronological framework for the alluvial deposits of the Thar Desert is developed using optically stimulated luminescence (OSL) dating (Jain et al., 1999; Tandon et al., 1999; Kar et al., 2000).

STUDY AREA

The River Luni and its tributaries constitute the major drainage system in the Thar Desert. The Luni originates in the Precambrian Aravalli range, then flows westward to Balotra, where it takes a southerly turn, and eventually discharges into the Arabian Sea (Fig. 1). Most of the Thar Desert landscape is covered by aeolian sands, but landforms reflecting surface water processes include hills, rocky/gravelly pediments, buried pediments, an older alluvial plain, a younger alluvial plain and the modern river bed.

Reaches of the River Luni examined during the course of this study receive precipitation of c. 250–300 mm yr^{-1}, which is largely concentrated during the summer monsoon (Fig. 1 inset). During the winter season, discharge is minimal to absent and aeolian sand accumulates within the main channel, only to be eroded by floods during the subsequent summer monsoon. Accordingly, the River Luni displays characteristics typical of many ephemeral streams, especially the downstream

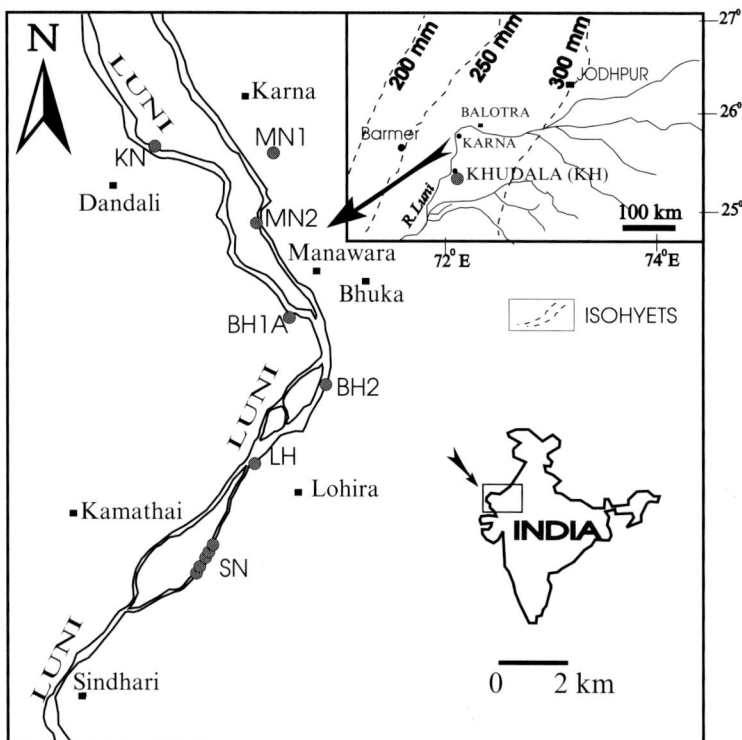

Fig. 1 Location map of the study area and key sections. Detailed facies and stratigraphical analyses were carried out in the stretch from Karna (KN) to Sindhari (SN), and near Khudala (KH; for Khudala see the inset).

Table 1 Mean grain size and sorting of different facies in the Type-1 and Type-2 successions.

Succession	Facies	Mean (ϕ)	Sorting (ϕ)
Type 1	Gravel	−0.3 to 1.3	1.03 to 2.5
	Coarse–medium sands	1.5 to 1.6	0.99 to 1.5
	Medium–fine sands	2.8 to 3.4	1.4 to 1.5
Type 2	Multistoried gravel sheets (MGS)	0.6 to −1.2; probability distribution shows 99% load in a single component between −2 and 4 ϕ or −4 and 4 ϕ range	1.3 to 3.4
	Gravel rich, red silty fine sands (RSFS)	2.8; 60% population in a single moderately well-sorted component between 3 and 5 ϕ	1.7
	Horizontally bedded fine to very fine sands (HBFS)	3.1 to 4.2; a dominant well-sorted component (c. 95%) between 2 and 4 ϕ	0.8 to 0.9
	Sand–silt alternations (SSA)	2.98 to 3.6; 90–99% population in a single well-sorted component between 4 and 5 ϕ	0.4 to 0.7
	Pedogenically modified silty fine sands (SFS)	3.4 to 3.9; 70–90% well-sorted component occurs in the fine to very fine sand range	1.4 to 1.6
	Pebbly coarse sand + medium to fine sand couplets (PCS + MFS)	PCS (0.8 to 0.6); MFS (1.06 to 1.44)	PCS (1.8 to 2.4); MFS (0.2 to 1.4)
	Well-sorted massive fine to very fine sands (MFS)	1.8 to 3.4 (a dominant c. 92% component between 3 and 4 ϕ)	1.4 to 3.4

concentration of sediment load as a result of transmission losses (Sharma et al., 1984), and the drainage network can appear locally disconnected owing to aeolian obstructions.

FIELD SUCCESSIONS

A stratigraphical framework for the alluvial deposits was developed by the examination and description of outcrops along the banks of the River Luni between Karna and Khudala (Fig. 1). Based on stratigraphical relations and lithofacies assemblages, these were categorized into Type-1 and Type-2 successions. The *Type-1 succession* is laterally extensive and, based on data from cores and well logs, extends for up to 300 m in the subsurface (Bajpai et al., 2001), whereas the *Type-2 succession* is less extensive, and inset within the Type-1 succession. The Type-1 succession typically lacks any fossil or archaeological materials, whereas these are common in the Type-2 succession. A geochronological framework for the alluvial deposits of the Type-2 succession has been developed using optically stimulated luminescence (OSL). The two major types of successions are described more fully below and summarized in Table 1.

Type-1 succession

The Type-1 succession is dominated by multistoreyed sheets of sandy gravel (Fig. 2a & b). Gravelly lithofacies contain abundant quartz, fresh feldspar and reworked calcrete nodules, and are dominantly planar and trough cross-stratified. Individual cross-strata can be up to 1.5 m thick, indicating deposition under normal streamflow conditions, and palaeocurrent directions are dominantly towards the south-southwest (Fig. 3A). Gravelly lithofacies also include disorganized conglomerates and massive to inversely graded gravels that are interpreted to represent sediment gravity flows (after Blair & McPherson, 1992). Towards the base of multistoreyed gravel sheets, isolated or cross-cutting ribbons (both the simple and complex ribbons of Friend et al., 1979) a few tens of metres in width may be present. Gravel sheets are, in some localities, overlain by medium to fine sands with calcified rhizoliths, which are in turn overlain by pedogenically modified interbedded fine sand and mud (heterolithic

facies) (Fig. 3A & B). Intraformational conglomerates are typically present at the contact between underlying heterolithic facies and the next gravel sheet, and consist of poorly sorted clasts of sandstone, siltstone and calcrete derived from the subjacent heterolithic facies. Some gravel bodies show very high concentrations of thick (5–12 mm diameter) discordant rhizocretions. Mean grain size and sorting for different facies are summarized in Table 1.

Fining upwards trends in the Type-1 succession can be observed at the bedform, macroform and sheet scales. At least three main sheet-scale depositional cycles consisting of gravel–sand sheets and overlying heterolithic facies can be observed in the surface exposures (Figs 2a & 3B). Channel bodies may be superimposed to form a thick, multistoreyed succession (Fig. 2b), or they may bifurcate into separate thinner sheets with interbedded muds (Fig. 3A). The final stage of sedimentation within a cycle occurs with a shift to an overbank regime and deposition of the heterolithic facies (Fig. 3B), which is followed in turn by subaerial exposure and pedogenic alteration, as indicated by reddening and the development of pedogenic calcrete and rhizocretions (Figs 2c & 3B). Successive depositional cycles begin with erosively based gravel above the heterolithic facies and palaeosols (Fig. 2c).

The thickness and lateral persistence of extensive multilateral and multistoreyed gravel-dominated bodies with low width-to-depth ratios, numerous internal scour surfaces and associated intraformational lags, unimodal palaeocurrent directions, and the near absence of lateral accretion elements are interpreted to indicate high energy floods and avulsive shifting of river channels (Turner, 1983; Godin, 1991; Eberth & Miall, 1991) within a gravel-dominated braided river system. The dominance of thick, cross-bedded deposits resulting from the migration of gravel dunes suggests a flashy discharge regime, with floods receding rapidly (Carling, 1996). Moreover, numerous internal scour surfaces with intraformational lags suggest that sheet floods were common (Rust, 1978). Abundant rhizoliths, dominantly reddish colours, colour mottling and dominantly fine nature of the heterolithic facies indicate a floodplain environment (Bown & Kraus, 1981) that may have remained dry for extended periods between flooding events. Moreover, the presence of immature calcrete profiles in floodplain facies indicates an overall semi-arid climate and high long-term net aggradation rates (Hill, 1989).

At the larger scale, the significant lateral extent of the Type-1 succession suggests deposition across an extensive alluvial plain, whereas the overall thickness in the subsurface supports the interpretation that the Type-1 succession was deposited during a period of subsidence (Bajpai et al., 2001). A significantly greater proportion of floodplain facies in the Type-1 succession, as compared with other gravelly braided systems described in the literature (e.g. Miall, 1996), might reflect the significant stage fluctuations typical of seasonally tropical rivers (Gupta, 1995) and the syndepositional accommodation provided by subsidence (Benthan et al., 1993). Finally, the compositional immaturity of the sediments, often including fresh angular feldspars plus micas and rock fragments, indicates a proximal source and the absence of prolonged weathering and reworking.

Fig. 2 (*opposite*) Photographs of key lithofacies within the Type-1 and Type-2 successions. The Type-1 succession: (a) the Sindhari section (SN) consisting of an alternation of multistoreyed gravel sheets (MGS) and heterolithic facies (HF); (b) thick MGS with discordant rhizocretions (R); (c) MGS overlying HF with an erosive contact—HF shows colour mottles, pedogenic calcrete nodules (N) and rhizocretions (R). The Type-2 succession: (d) Khudala section showing red silty fine sand (RSFS) overlain by sand–silt alternations (SSA) and finally capped by an aeolian accumulation (A); (e) surface gravel exposures above the Luni Gorge near Karna bear pottery fragments and gastropod shells—the inset shows MGS that laterally interfingers with vertic calcisols near Karna; (f) coexisting vertic calcisol with multistoreyed gravels of inset in (e)—calcrete-rich gravel sheet is sandwiched within the palaeosol; (g) sand–silt alternations in Khudala section showing lateral splitting and amalgamation of silt beds around sand lenses—CG indicates a calcrete-rich gravel lens; (h) the Bhuka section (BH1A) consisting dominantly of silty fine sands with thin intercalated gravels.

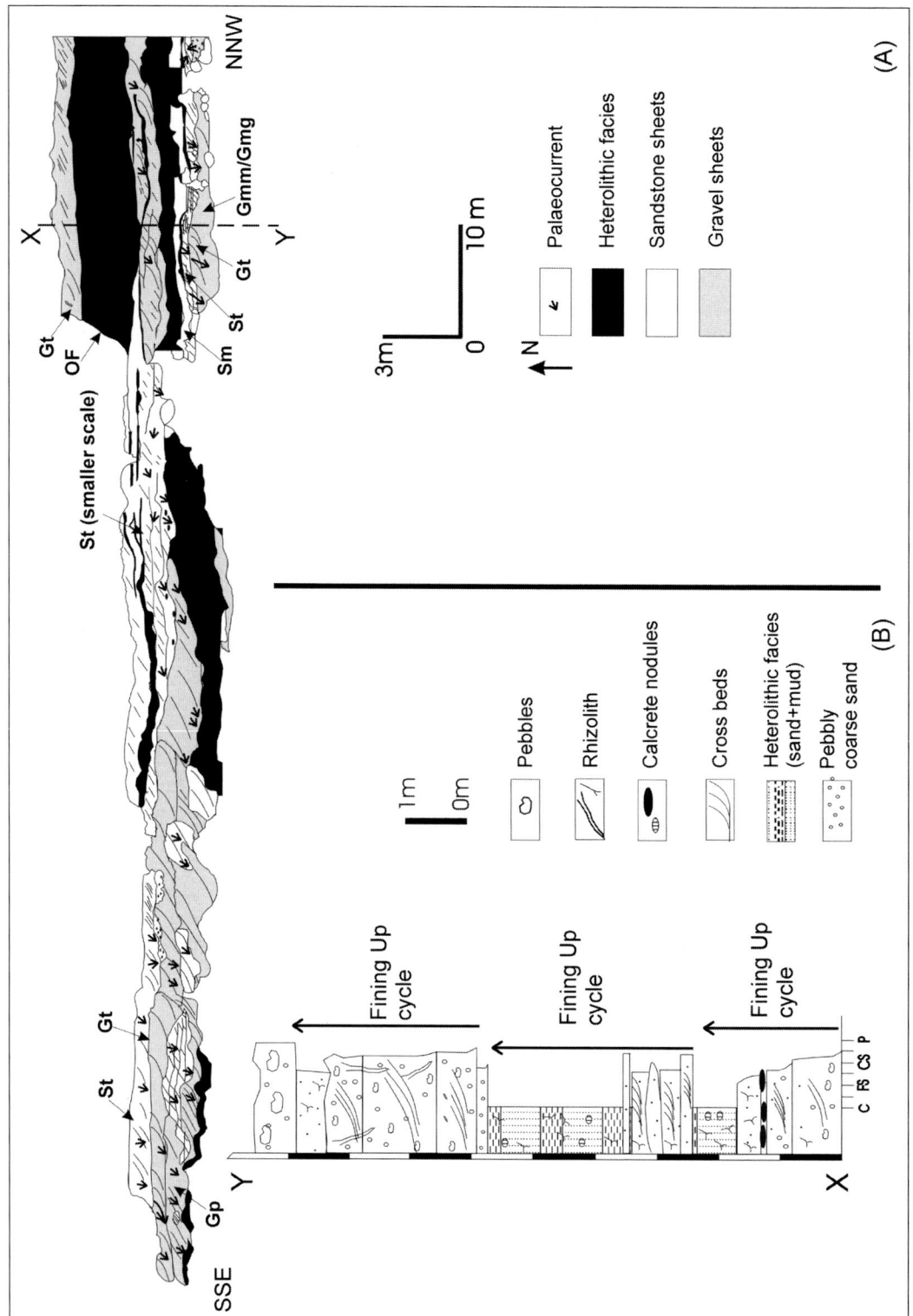

Fig. 3 Key stratigraphical section near Sindhari (SN) for Type-1 succession. (A) Lateral diagram showing an interfingering relationship between the overbank heterolithic facies and the multistoreyed gravel–sand bodies composed of different gravel and sand facies (B) Vertical log along XY in (A). Three sheet-scale fining upwards cycles (gravel–sand–heterolithic facies) can be observed. The sands and heterolithic facies show significant pedogenic alteration. Some gravel bodies show rhizocretion development.

Type-2 succession

The Type-2 succession includes valley-fills inset within the Type-1 succession, as well as unconsolidated surface gravels, and is best observed in the Karna (KN), Bhuka (BH1A), Khudala (KH) and Manawara (MN2) sections shown in Figs 2d–h & 4. The Type-2 succession consists of an array of distinct, laterally or vertically juxtaposed depositional environments (e.g. Fig. 2d), in sharp contrast to the monotonous Type-1 succession (see Figs 2a & 3). Grain-size distributions and facies associations in the Type-2 succession are summarized in Tables 1 & 2 and briefly described below.

Multistoreyed gravel sheets

Multistoreyed gravel sheets (MGS), up to 5 m in thickness, are generally present at the base of cliff sections (Figs 2e inset & 4A–C). Gravelly lithofacies are mostly low-angle to horizontally bedded (facies Gh of Miall, 1996), trough cross-stratified (facies Gt), or planar cross-stratified (facies Gp); low angle to horizontally bedded facies Gh generally grades laterally into the trough or planar cross-stratified gravels (facies Gt and Gp). The gravel fraction contains an average clast size of 1–3 cm (maximum *c.* 11 cm) and is comprised of 35–80% transported calcrete nodules. Cross-strata thicknesses in facies Gt and Gp generally range between 30 and 50 cm (rarely 1 m), with individual foresets 3–8 cm in thickness (Fig. 2e). There is a systematic intra- and interforeset clast-size variation in the Gp facies. Openwork textures dominate pebble-rich foreset layers, and flat tabular to equidimensional clasts may be crudely imbricated.

Volumetrically minor facies include matrix-supported gravels (facies Gmg) with gravel-sized clasts fining-upwards within a sandy matrix, and couplets of horizontally bedded gravel and horizontally bedded sand (facies Gh–Sh) occasionally topped by thin veneers of silt. Minor channel fills may be present in the upper parts of gravel sheets as well. Palaeocurrent directions vary from south-southeast to south-west. In places, gravel sheets interfinger with thick pedogenically modified mottled brown muds, sandy muds and calcic vertisols (Figs 2f & 4b & c: MGS + V, MGS + PM). X-ray diffraction (XRD) analysis shows smectite to be the dominant clay mineral (Jain, 2000). Carbonate rhizoliths, calcrete nodules, Fe/Mn stains and Fe/Mn concretions (*c.* 3 mm diameter) are common, and calcic vertisols display concavo-convex joints in the top 0.4 m of the profiles. Channel proximal settings show thin, interbedded calcrete gravel sheets (Fig. 2f).

Multistoreyed gravel sheets of the Type-2 succession are interpreted to indicate deposition by a braided stream with predominantly locally derived detritus. Different gravel facies indicate both traction dominated and sediment gravity flow processes. These gravels mainly consist of channel-bar assemblages accreting by avalanching, with intermittent high-density sheet-flood deposits. The macroform association includes transverse and linguoid bars with occasional incision of the bar tops filled by minor troughs. The presence and organization of the cross-beds and openwork texture indicate persistent flow conditions. Grain-size segregation in alternate foresets is perhaps due to discontinuous downstream accretion (growth increments of Smith, 1974). Thin veneers of silt and fining-up gravel–sand couplets resulted from waning of episodic sheet flows. The formation and preservation of cross-beds resulting from three-dimensional gravel dunes imply seasonal flashy floods with rapid recession (Carling, 1996). Interfingering of muds and calcic-vertisols, on the other hand, is interpreted to represent pedogenically modified overbank deposits. The thickness of the muds might reflect topographic differentiation owing to rapid aggradation of the channel belt, and a significant presence of fine-grained detritus in the streams.

Unconsolidated gravel sheets also occur on the topographically high surfaces above the present Luni River channel (Fig. 2e; here termed surface gravels). These surface gravels unconformably overlie the Type-1 succession, and contain unabraded gastropod shells, potsherds and animal bones (Fig. 5). Several cross-cutting trough fills (*c.* 30 cm thick and *c.* 1.2–5 m wide) are present. Palaeocurrent directions are similar to the Luni River flow in these stretches (Fig. 5). The spatial proximity and the palaeocurrent directions of the surface gravels indicate that they were palaeochannels of the River Luni (Fig. 5). Archaeological

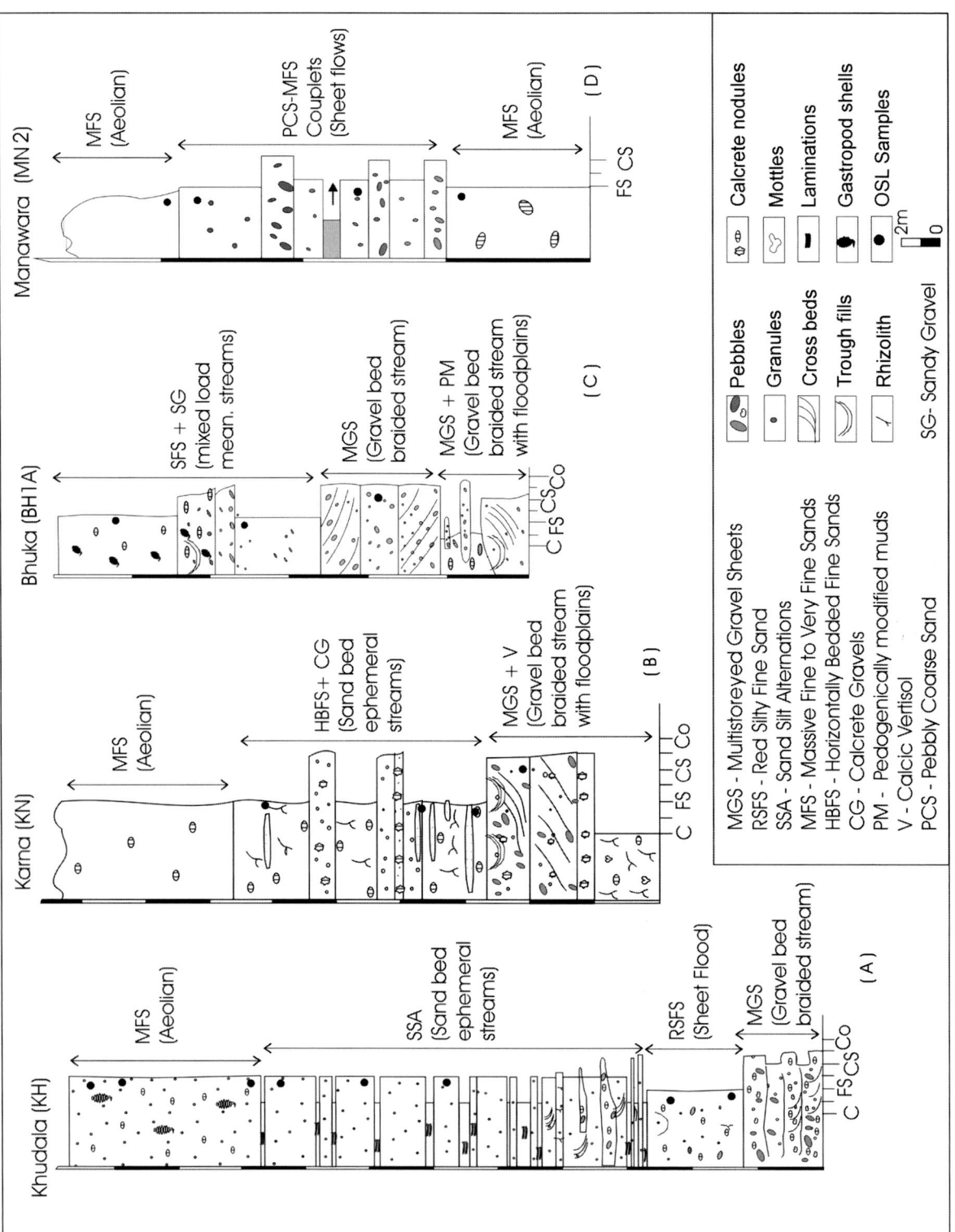

Fig. 4 Key measured sections within the Type-2 successions near (A) Khudala, (B) Karna, (C) Bhuka and (D) Manawara. Palaeoenvironmental interpretations are given against each facies assemblage. See the text for details. The OSL sample locations are shown as dots on the right of each section.

Table 2 Different lithofacies present in the Type-1 and Type-2 successions of the Luni basin. Environmental interpretations are given in parentheses.

Type 1 succession Sindhari (SN) to Bhuka (BH2) stretch	Type-2 succession Khudala (KH)	Karna (KN)	Bhuka (BH1A)	Manawara (MN1) and Lohida (LH)	Manawara (MN2)	Bhuka (BH2)
Multistoreyed gravel sheets: planar cross-stratified gravels (Gp); trough cross-bedded gravels (Gt); massive or graded gravels (Gmm, Gmg); horizontally stratified gravels (Gh); disorganized conglomerate (Gcm); intraformational conglomerate, trough cross-bedded coarse to medium sandstone (St); small scale trough cross-bedded medium to fine sandstone (St); massive medium-grained sandstone. Pedogenic alteration in the sandstones. Overbank Facies (OF): heterolithic facies (muds with intercalation of thin sand beds); muddy-silty fine sandstone; laminated silty clayey fine sands. Significant pedogenic alteration and calcrete development.	(1) Multistoreyed gravel sheet: Gp, Gt, Gh-Sh couplets (gravel-bed braided streams) (2) Red silty fine sands: massive and ill sorted (pedogenically modified sheet floods) (3) Sand–silt alternations (ephemeral sand-bed streams) (4) Massive very fine sands to coarse silt (aeolian)	(1) Multistoreyed gravel sheet (Gp, Gt, Gh) and overbank vertic calcisol association (gravel-bed braided streams with pedogenically modified floodplains) (2) Horizontally bedded sands + calcrete gravel association (ephemeral sand-bed streams) (3) Massive medium to very fine sands (Aeolian)	(1) Multistoreyed gravel sheet (Gp, Gt) and overbank mud association (gravel-bed braided streams with pedogenically modified floodplains) (2) Cross-stratified gravel sheet: Gp, Gt, Gmm (gravel-bed braided streams) (3) Pedogenically modified, gastropod-bearing silty fine sands + minor gravel lenses (mixed load meandering streams)	(1) Surface gravel exposures (Gt, Gh) associated with pottery, animal bones and gastropod shells (gravel-bed braided streams) (2) Massive medium to very fine sands (aeolian)	(1) Massive medium to very fine sands (aeolian) (2) Pebbly coarse sands + medium to fine sand couplets (sheet flows) (3) Massive medium to very fine sands (aeolian)	Fine sand–silts (slack-water deposits)

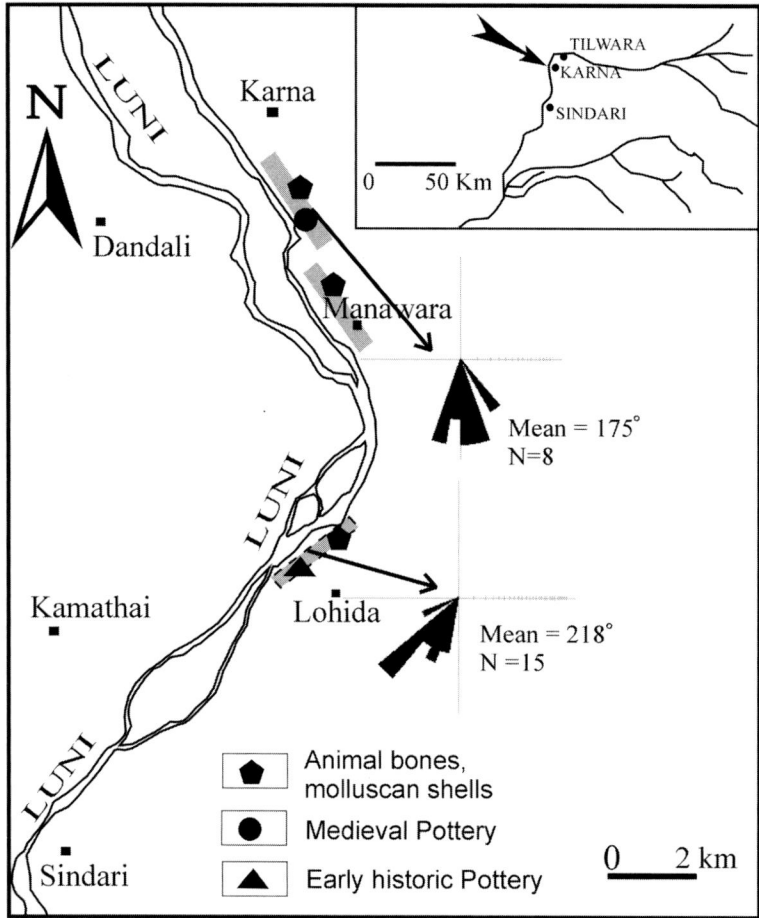

Fig. 5 Locations of exposures with palaeocurrent directions for surface gravels in the Type-2 succession.

evidence suggests that these surface gravels are of Holocene age (Mishra et al., 1999).

Red silty fine sands

Massive red silty fine sands (RSFS) with individual beds up to 2.5 to 3 m thick (Figs 2d basal unit & 4A) occur within the Type-2 succession. Generally, sands are poorly sorted and include disseminated coarse sand and pebbles, as well as a few mud intraclasts towards the base, with pebbles and mud clasts ranging in size from 1 to 6 cm. Rare isolated gravel troughs may be present (0.5 m high and c. 2.5 m wide), and merge laterally with the massive fine sands. Cross-beds within these abut against one of the trough walls. These massive sands are reddened, moderately cemented and show incipient calcrete nodule development.

The overall thick, massive and ill-sorted nature of the RSFS facies assemblage is interpreted to indicate hyperconcentrated flows, perhaps sediment gravity flows or sheet floods, where the sediment fallout rate was sufficiently high to prevent efficient sorting (Hjellbakk, 1997). Coarser particles within these high-density flows were perhaps maintained in suspension by turbulence, buoyant-support and dispersive pressures (Smith, 1986; Lowe, 1988). The troughs suggest either locally strong erosive traction as a result of discharge variations (Cant & Walker, 1978; Miall, 1978) or cut-and-fill aggradational processes during low discharge (Abdullatif, 1989). Post-depositional reddening and pedogenic calcrete

Horizontally bedded fine to very fine sands with interbedded calcrete gravel lenses

Moderate to well sorted horizontally bedded fine to very fine sands (HBFS) overlie gravel sheets or pedogenically modified muds in the Karna section (Fig. 4B). Sand beds are a few centimetres thick, with subtle variations in grain size and sorting. Sands are typically quartzo-feldspathic; however, some beds have high concentrations of opaque heavy minerals. Millimetre-scale parallel laminations are present. Biotite flakes typically show a preferred orientation along the laminae planes. Some non-cemented pockets in the upper parts of these sand beds are massive and very well sorted. Interbedded gravel lenses (CG) are 10–30 cm thick, massive, matrix-supported and pinch out within a few metres (Fig. 4B). Pebbles of calcrete, typically 'black pebble' (Ward, 1970), and rock fragments (*c.* 1 cm) are present within a matrix of grit and coarse sands. Some lenses are dominantly composed of calcrete nodules in both the coarse sand and gravel sizes, although mud intraclasts may be present. A well-developed calcrete profile (Stage 3 of Machette, 1985) and numerous rhizoliths occur in these deposits. Rhizoliths are mostly concentrated along bedding planes and give rise to concordant cementation. In places discordant rhizoliths (1–5 cm diameter) may be present.

Horizontally bedded fine to very fine sands are interpreted to represent shallow sheet flows deposited under upper flow regime conditions, as indicated by the orientated micas (Harms *et al.*, 1982). This facies is similar to that observed in sheet-flood and high-energy ephemeral stream deposits elsewhere (e.g. Williams, 1971; Tunbridge, 1981), where a major rainfall event within an unvegetated catchment causes rapid runoff with violent, short-lived sheets formed during the peak flow. Heavy mineral-rich beds represent a rapid fall of high-energy flood events characteristic of arid environments (Lucchitta & Suneson, 1981). Laboratory experiments by Bridge & Best (1988) and Paola *et al.* (1989) indicate that planar strata such as these result from the superposition of two processes: high-frequency erosion and deposition due to turbulence; and migration of low-amplitude bedforms. Pockets of massive very well-sorted fine sands perhaps represent remanant proximal aeolian dunes. Episodes of calcrete formation are interpreted to represent periods of non-deposition and stability. The grain-size range of dominant modes in Sh facies and the aeolian dune sands are nearly identical (Table 1), which together with their field association suggests that sediment load in the Sh facies was derived from the surrounding dune fields. Calcrete gravels indicate intermittent, relatively larger magnitude semi-stable flows (with dominant lateral inputs) characteristic of arid regions.

Sand–silt alternations

Interbedded well-sorted fine to very fine sands and silts (SSA) occur within the Type-2 succession (Figs 2d & 4A). The thickness of the silts ranges from 5 to 10 cm to more than 20 cm, and is laterally traceable for several tens of metres. Millimetre-scale laminations occur within the silts and are occasionally disrupted by fossilized root traces or cylindrical burrows. Sands have variable thickness and may show indistinct cross-stratification or horizontal laminations defined by micas. In the latter case, micas again reveal a primary current lineation. Locally, horizontal laminations in sands grade into small dunes draped with silt veneers. These silt beds show lateral splitting and amalgamation around the sand lenses (Fig. 2g). Minor gravel lenses (2–3 m in lateral extent), composed of large broken angular silt fragments and calcrete nodules in a matrix of medium to fine sands, are also present within this facies assemblage, and poorly sorted pebbly coarse sands may be present towards the base. Some dark horizons contain high concentrations of heavy minerals.

Horizontally laminated sands are interpreted to represent upper flow-regime high-stage flow in an ephemeral stream (*sensu* Picard & High, 1973; Frostick & Reid, 1977), whereas the indistinct cross-stratified units covered with silt caps (Fig. 2g) may represent dune-scale bedforms formed during waning flows and the low flow stage (Williams, 1971), or a reduction in flow energy behind obstacles, such as shrubs (Sneh, 1983). Lateral gradation and interfingering of the dune forms into horizontally bedded sands indicate rapid lateral variability in energy conditions, and also has been documented in ephemeral rivers (Abdullatif, 1989). The laminated silt facies is interpreted to represent deposition in pools during the waning stage (Stear, 1983). Thicker silts (> 20 cm thick) suggest deposition in abandoned secondary channels and pools. In general, flows were unable to carry coarse bedload, although small lenses of gravel indicate that flow concentration did occur in narrow localized zones. Heavy-mineral-rich black sand covered by quartzo-feldspathic sands represent a rapid fall of high-energy sediment-laden flood waters characteristic of arid environments (Frostick & Reid, 1977; Lucchitta & Suneson, 1981). Relatively weak pedogenic and/or erosional modification in this unit suggests high rates of aggradation, perhaps by flows with high sediment concentration. This could be as a result of rapid discharge variations in a semi-arid setting.

Pedogenically modified silty fine sands and sandy gravel association

Silty fine sands (SFS) occur intercalated with minor thin sandy gravel facies (SG), and display significant pedogenic modification (Figs 2h & 4C). The sands are texturally homogeneous (Table 1), as well as laterally extensive, as they overlie outcrops in the study area over distances of several kilometres. Disseminated coarse sand grains of quartz, rock fragments and micas, and numerous, unabraded gastropod shells are present. Nearly all silty sand horizons show a darker chroma and variable cementation. The upper layers show significant pedogenic modification with Stage-2 calcrete profiles (Fig. 4C; Machette, 1985). Sandy gravel lenses occur locally, generally pinching out in less than 10 m, and in some localities show multiple cross-cutting channel fills with a gravel lag deposited at the base of the channel. The channel widths and depths are about 4 m and 1 m, respectively. Occasionally, the pebbles may show a crude fining-up trend.

The overall fine-grained nature of the SFS and SG facies assemblage, coupled with its pedogenic modification, is interpreted to indicate a low-energy floodplain environment. The presence of *in situ* gastropod shells indicates shallow water puddles. Occasional high-magnitude floods incised into the floodplains and deposited small gravel ribbons/lenses. The overall depositional environment appears to be characteristic of mixed-load meandering streams as indicated by dominant, laterally extensive floodplain sediments. Sorting of the fine-sand component (Table 1) might have taken place in the trunk channel.

Pebbly coarse sand and medium to fine sand couplets

This lithofacies consists of thin (*c.* 30 cm) massive, matrix-supported sheets of pebbly coarse sands (PCS; Fig. 4D) that grade upwards into medium to fine sands (MFS) of comparable thickness, and which may, in turn, be overlain by a thin silt bed. The massive sheet-like nature and overall fining upward trend in each couplet is interpreted to reflect deposition by a single sheet-flow event, with the high-energy pulse represented by pebbly coarse sands, and waning flow stages represented by the transition to the medium to fine sands. The well-sorted nature of the medium-to-fine fraction (see Table 1) may be related to sorting of source materials by aeolian processes, and this sand was transported together with pebbles as a single sediment-laden sheet.

Well-sorted massive fine to very fine sands

Well-sorted massive fine to very fine sands (MFS) generally occur within, and cap, alluvial deposits of the Type-2 succession (Figs 2d top unit & 4 A,B & D). These massive caps are laterally continuous with aeolian dunes in the vicinity. There is generally a dominant well-sorted component of about 85% medium-to-fine sand (e.g. in section MN–2); however, in some cases the fine sand to coarse silt range comprises about 95% of the total (e.g. in section KH), and there is a conspicuous lack of any coarse sand and pebbles. Occasionally, these deposits show a reddish chroma and Stage-1 calcrete profile development (Machette, 1985) (Fig. 4A & D). Microliths and pottery fragments may be found on the surface.

These sands are interpreted to be of aeolian origin. The differences in the modal grain sizes in different deposits suggest that they were source-bordering dunes with most sediment derived from nearby abandoned channels. This might have occurred after the cessation of fluvial activity during a dry phase, as studies on modern wadi systems indicate that co-occurrence of these environments in the stratigraphical record is not expected and may require drastic environmental and climatic changes (Sneh, 1983).

Fine silts and fine sands

Successions of slack water deposits (SWD) some 2 m in thickness occur at the mouth of a tributary channel to River Luni near Bhuka. These deposits consist of alternating beds of medium to fine sand and silt. Previous workers have shown these deposits to be less than 1000 yr old (Kale et al., 2000). Additionally, some recent floodplain deposits are preserved along the modern Luni channel. These are composed of laminated silt and sandy silt couplets. Silt layers have yielded some recent pottery fragments. The predominantly fine-grained nature of these deposits indicates overbank sedimentation.

LUMINESCENCE GEOCHRONOLOGY

Considerable progress has been made in dating water-lain sediments using optically stimulated luminescence (OSL). Studies from modern settings indicate that the OSL signal in fluvial sediments is generally well zeroed (Olley et al., 1998; Murray & Olley, 1999; Jain et al., 2004). Moreover, in desert environments, the conditions for bleaching are even more favourable owing to the frequent emergence of bar surfaces and the stream bed during low-stage or dry conditions, and the local reworking of fluvial sediments by wind and vice versa. The results of luminescence studies of these successions are summarized below.

Type-1 succession

Jain et al. (1999) have shown that the Type-1 succession is too old to be dated by luminescence techniques, which implies a minimum age of 200 ka based on quartz (Fig. 6a). Both the OSL and TL palaeodoses calculated from feldspar were less than the saturation doses in the quartz and hence grossly underestimated (Fig. 6a). The age of the

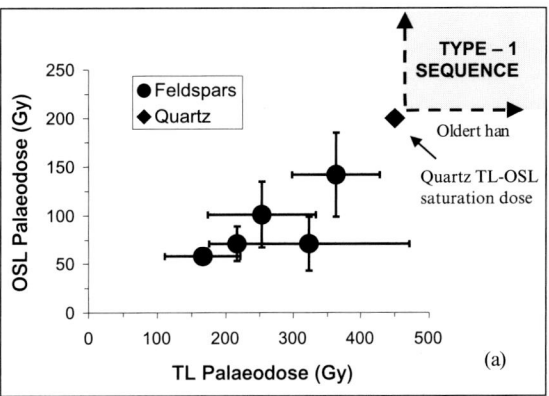

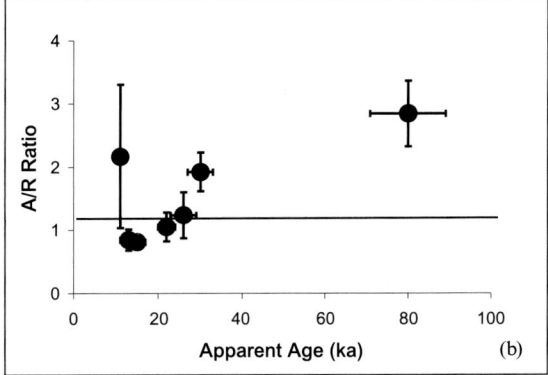

Fig. 6 (a) Equivalent doses (palaeodoses) obtained from quartz and feldspar of the Type-1 succession. The quartz TL and OSL signals are in the saturation region. Calculations from the regeneration growth curves gave a minimum palaeodose of 450 Gy (c. 200 ka) for the Type-1 succession. Feldspar palaeodoses were calculated using the additive-dose TL and OSL methods. The TL and OSL of feldspar (UV emission) show good dose-response and growth curves, however, calculated palaeodoses are lower than the saturation doses in quartz, and hence underestimated. (b) Ratio of palaeodoses obtained by using multiple-aliquot additive-dose (A) and regeneration (R) protocols versus the apparent additive-dose OSL ages of samples from the Type-2 succession.

Type-1 succession, therefore, remains uncertain, in the absence of any diagnostic fossils or archaeological material (discussed below).

Type-2 succession

For the Type-2 succession, the sample locations for OSL dating are shown in Figs 4 & 7, and

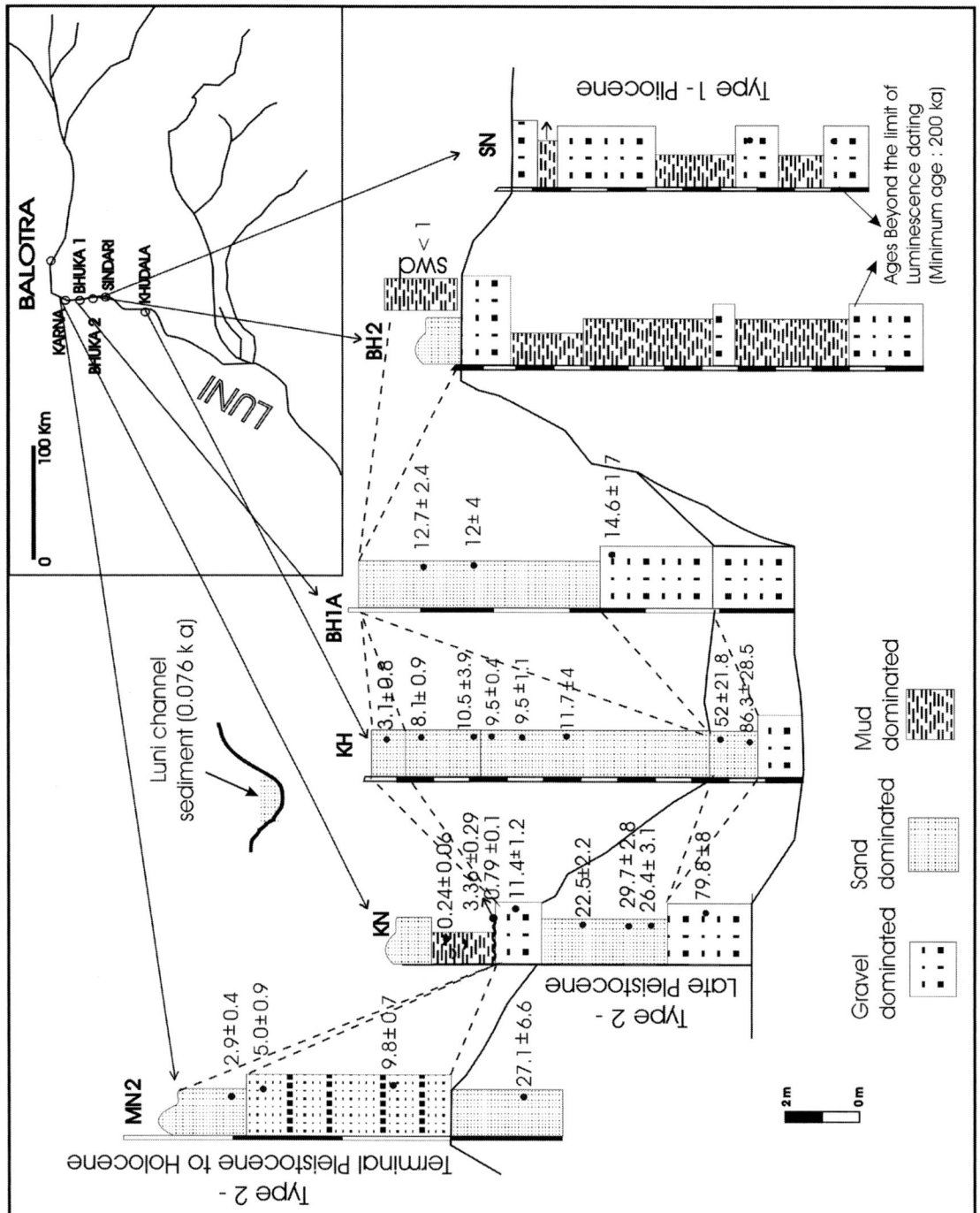

Fig. 7 Physical stratigraphy of Quaternary alluvial successions, Thar Desert. The OSL sample locations are shown. Dashed lines indicate various time-separated depositional environments. Chronology for the Khudala section (KH) is taken from Kar *et al.* (2000). Chronology for slack-water deposits (SWD) is from Kale *et al.* (2000). All ages reported as thousands of years ago (ka).

Table 3 Dose rate, equivalent dose (E_d) and calculated ages of sediments in the Type-2 succession. Ages for each locality are given in stratigraphical order: PT refers to pottery samples.

Location	Sample	K (%)	U (µg g^{-1})	Th (µg g^{-1})	'a' value	D_e (Gy)	Dose rate (Gy ka^{-1})	Age (ka)
Lohida (LH)	PT2	1.50	5.6 ± 1.3	10.0 ± 4.4	0.09	4.28 ± 0.53	5.40 ± 0.34	0.79 ± 0.098
Manawara (MN2)	SH1	1.40	2.5 ± 0.5	5.7 ± 1.6		6.5 ± 0.9	2.28 ± 0.16	2.9 ± 0.4
	SH2	1.65	2.5 ± 0.7	8.1 ± 2.3		13.24 ± 2.15	2.65 ± 0.22	5.0 ± 0.9
	SH3	1.70	2.3 ± 0.6	10.0 ± 2.1		27.4 ± 0.6	2.79 ± 0.15	9.8 ± 0.7
	SH4	1.54	2.4 ± 0.6	7.5 ± 2.2		67.8 ± 15.4	2.50 ± 0.21	27.1 ± 6.6
Manawara (MN1)	PT1	1.54	3.4 ± 1.0	12.6 ± 3.3	0.06	14.48 ± 1.24	4.3 ± 0.3	3.4 ± 0.3
	154	1.66	1.9 ± 0.5	6.8 ± 1.8		37.3 ± 2.7	3.26 ± 0.24	11.4 ± 1.2
Bhuka (BH1A)	161	1.45	2.3 ± 0.6	6.0 ± 1.9		27.7 ± 9.2	2.30 ± 0.19	12.1 ± 4.1
	160	1.50	2.4 ± 0.5	6.0 ± 1.8		33.9 ± 60	2.67 ± 0.19	12.7 ± 2.4
	159	1.70	2.2 ± 0.6	9.6 ± 2.0		40.04 ± 3.7	2.74 ± 0.20	14.6 ± 1.7
Karna (KN)	PT3	1.50	3.2 ± 0.9	11.6 ± 2.9	0.10	11.1 ± 0.3	4.60 ± 0.04	0.25 ± 0.07
	157	1.47	2.6 ± 0.1	11.4 ± 3.4		61.6 ± 3.0	2.74 ± 0.24	22.5 ± 2.3
	156	1.63	2.1 ± 0.7	8.7 ± 2.3		77.0 ± 3.0	2.59 ± 0.22	29.7 ± 2.8
	155	1.84	2.4 ± 0.9	11.6 ± 3.0		80.08 ± 60	3.03 ± 0.28	26.4 ± 3.1
	158	1.50	1.7 ± 0.3	4.0 ± 1.1		166.3 ± 15.4	2.08 ± 0.12	79.8 ± 8.7

presented in the same stratigraphical order in Table 3. In addition to samples from measured stratigraphical sections, a sample was collected from a depth of 1 m below the modern Luni River bed so as to test for partial bleaching.

The OSL dating of sediment samples focused on fine sand-sized quartz (105–150 µm). A RISØ TA-DA 15 reader was used for blue-green (420–550 nm) and infrared (IR; 880 ± 80 nm) stimulation, with detection optics consisting of BG39 + U340 filters. The OSL measurements were made at 125°C for 100 s in order to avoid OSL charge recycling via the 110°C peak (Wintle & Murray, 1997), and a preheat of 220°C for 300 s was used. Elevated-temperature IR cleaning was carried out if there was contamination from feldspar blue-green OSL (Jain & Singhvi, 2001). Component-specific dose normalization was used for greater accuracy and precision (Jain et al., 2003). Pottery samples were dated using thermoluminescence (TL) of fine grain polymineralic aliquots. The annual dose rates for OSL and TL samples were calculated using elemental concentrations of natural radionuclides U, Th (by thick source ZnS (Ag) alpha counting) and K (gamma counting), and by assuming a radioactive equilibrium in the decay chains of U and Th, average water contents of 10% and a cosmic dose rate of 150 ± 30 µGy yr^{-1}.

The additive-dose protocol with a saturating exponential fit was used for calculations of the equivalent dose (D_e). Regeneration doses were calculated to assess sensitivity changes; the apparent regeneration palaeodoses in quartz suggest progressively greater sensitivity change and age underestimation for the older samples (Fig. 6b). Further, differences in the functional form of the additive and regeneration dose growth curves caused errors in the Australian-slide method (Prescott et al., 1993).

The OSL results are presented in Table 3. To summarize, OSL ages on fluvial sediments from measured sections are stratigraphically consistent (Table 3), and indicate periods of deposition during the marine oxygen isotope substage (OIS) 5a interstadial (c. 80 ka), the OIS 3 interstadial (70–30 ka), the OIS 2 glacial (29–22 ka), the Late-glacial (14–10 ka), and the early to middle Holocene (10–3 ka). Analysis of the modern river bed sample produced an OSL age of 75 ± 32 yr and a TL age of 136 ± 32 yr (Jain et al., 1999). These very young ages attest to the likely resetting of both the OSL and TL signals during transport in this environment, and the overall reliability of the older OSL ages collected from similar facies. The TL ages on pottery range from 3.3 to 0.3 ka, which corresponds to the Prehistoric to Early Historic age.

PALAEOSOLS AND STABLE ISOTOPIC COMPOSITION OF PEDOGENIC CALCRETE

Mature palaeosols are absent in the floodplain deposits of Type-1 and Type-2 successions. Instead, pedogenic features include gleyed mottles and vertic features, but especially the development of Stage 1 to 3 calcrete profiles (*sensu* Machette, 1985). Micromorphological and geochemical investigations suggest that these calcretes are of pedogenic origin (Jain, 2000), and carbonate-enriched horizons do not cut across lithological surfaces, implying no post-burial pedogenic modification. The OSL results suggest that soil-forming intervals might have been as high as 1000–7000 yr or as low as a few hundred years to *c.* 2500 yr for the Stage 2 and 3 profiles, respectively (Jain, 2000). These estimates are generally consistent with previous results on time periods for calcrete formation in alluvial landscapes (Leeder, 1975; Wang *et al.*, 1996).

In general, these short soil-forming intervals suggest that it is possible to resolve climatic signals, at > 10 000 yr scale, using stable isotope signatures from pedogenic carbonates (see Jain, 2000). As shown in Fig. 8, carbonates in the Type-1 succession show closely clustered $\delta^{18}O$ values of −5.9‰ to −8‰, similar to that of the present summer monsoon-dominated regime. The $\delta^{13}C$ signatures of calcretes in both Type-1 and Type-2 successions record a mix of both C_3 and C_4 biomass, however, the Type-1 succession contains a greater concentration of C_3 flora ($\delta^{13}C$ = −4.8‰ to −7.7‰; average of −6.16‰) relative to the Type-2 succession ($\delta^{13}C$ = −0.42‰ to −4.3‰; average of −2.05‰). From these data it might be inferred that the Type-1 succession represents a persistently wetter interglacial climate with normal monsoon precipitation and a higher proportion of C_3 biomass.

Isotopic signatures in the Type-2 succession are more complex. The $\delta^{18}O$ signatures between units vary from −8.1‰ to −3‰, but by less than 1‰ within any individual unit. Similarly, $\delta^{13}C$ signatures between units vary from −0.42‰ to −4.3‰, but there is less than 0.5‰ variance within an individual unit. These data are interpreted to indicate fluctuations through time between

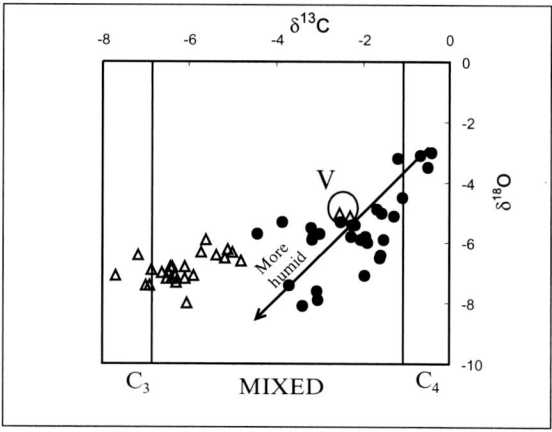

Fig. 8 Stable isotopic composition of pedogenic calcrete from the Type-1 (triangles) and the Type-2 (filled circles) successions. There are distinct $\delta^{13}C$ and $\delta^{18}O$ signatures within pedogenic calcrete of the Type-1 and Type-2 successions. The calcretes were sampled from below 30 cm depth. Micrite phase was separated for analysis after the cathodoluminescence and electron microprobe analysis of thick sections. Stable isotopic measurements were made at the Korean Basic Science Institute (KBSI), South Korea. V indicates the isotope signature of calcrete from vertisol. The composition of nodules from the laterally coexisting heterolithic facies (proximal floodplain situation) of the vertisol lies within the cluster defined by the Type-1 succession. This anomalous more-enriched composition in the vertisol nodules is probably the result of atmospheric CO_2 ingression through vertical fractures in the vertisols.

environments dominated by the normal summer monsoon and relatively increased C_3 biomass, and those dominated by isotopically enriched winter precipitation and a greater proportion of C_4 biomass. Both $\delta^{13}C$ and $\delta^{18}O$ values are, for example, depleted within calcrete profiles that formed in the OIS 3 interstadial and the Holocene interglacial ($\delta^{13}C$ = −2.58‰, $\delta^{18}O$ = −6.4‰), indicating summer monsoon precipitation and a dominantly C_3 biomass, but relatively enriched during the OIS 2 full glacial ($\delta^{13}C$ = −0.98‰, $\delta^{18}O$ = −3.24‰), indicating precipitation from winter rainfall (north-east monsoons) and an increase in C_4 biomass. These variations occur owing to a combination of increased C_4 biomass, amount effect, decreased rates of evaporation, and decreased soil respiration rate during glacial times and an opposite trend during interstadials.

Thus, within the Type-2 succession, stable isotope data suggest deposition during a series of humid-arid cycles, such that normal summer monsoons dominated relatively moist interstadials and the Holocene interglacial, but the summer monsoon was weakened during a relatively dry glacial period (Jain & Tandon, 2001). This interpretation is consistent with studies from the Indian Ocean and Arabian Sea (Sirocko et al., 1993), as well as previous work in the Thar Desert (Singh et al., 1974; see also Andrews et al., 1998).

DISCUSSION

From the above discussion, it is clear that Type-1 and Type-2 successions are significantly different, and represent a fundamental reorganization of fluvial systems in the study area. The physical stratigraphy of the Luni basin is shown in Fig. 7 and summarized in Table 4. Key differences between the Type-1 and Type-2 successions can be summarized as follows:

1 The Type-1 succession consists of extensive sheets of gravels deposited by a braided stream within a subsiding basin, and many details of the alluvial architecture most likely developed in response to autogenic channel avulsion processes. The Type-1 succession lacks aeolian facies, which might imply an absence of dunefields in the surrounding regions during deposition. By contrast, the Type-2 succession records a diverse range of fluvial and aeolian depositional environments that most likely changed through time in response to changing late Pleistocene and Holocene climates.

2 The Type-1 succession was deposited in an environment dominated by C_3 flora and persistent summer monsoon precipitation, as indicated by the stable isotopic composition of calcretes. Isotopic signatures in the Type-2 succession are more variable, and indicate fluctuations between environments dominated by the normal summer monsoon and relatively increased C_3 biomass, and those dominated by winter precipitation and a greater proportion of C_4 biomass.

Table 4 Summary of the physical stratigraphy of Quaternary alluvial successions, Luni basin, Thar in the context of global climate changes.

Succession	Time*	Fluvial pattern
Type 2	OIS 1 (present to 10 ka)	Floodplains (0.3 ka)
		Slack-water deposition (< 1 ka)
		Incision (3–1 ka)
		Aeolian activity (3 ka)
		Sheet flows (9–5 ka)
		Incision (c. 10 ka)
	OIS 1 (11–14 ka)	Aeolian activity (12–8 ka)
		Ephemeral sand bed (12–8 ka)
		Gravel braided (c. 11.5 ka)
		Mixed-load meandering (~12 ka)
		Gravel braided (c. 14 ka)
		Incision (14–22 ka)
	Terminal OIS 3 and OIS 2	Aeolian activity (27 ± 6 ka)
		Calcrete development
		Ephemeral sand bed (30–20 ka)
	OIS 4 and OIS 3	Pedogenesis (70–30 ka)
		Sheet floods
	OIS 5a	Gravel braided with floodplain development (80 ka)
	OIS 5e	Gravel braided (> 90 ka)?
Type 1	Late Miocene to Pliocene(?)	Gravel–sand braided stream within a subsiding basin

*OIS = (marine) oxygen isotope stage.

3 The Type-1 succession contains no archaeological material, whereas archaeological records are common in the Type-2 succession. Moreover, quartz TL and OSL dating indicate that the Type-1 succession has a minimum age of 200 ka. A number of OSL ages indicate that the Type-2 succession represents the late Pleistocene and Holocene (past 100 kyr).

It seems clear, therefore, that the Type-1 and Type-2 successions record a major shift in environment within what is now the Thar Desert. A compilation of global data indicates that the evolution of C_4 biomass occurred around the Miocene–Pliocene boundary as a response to CO_2 stress (Cerling et al., 1997), whereas general circulation models indicate that strong monsoons similar to the present day may have been initiated by the Late Miocene (Prell & Kutzbach, 1992). Based on available geochronology, the Type-1 succession would have been deposited between the onset of the strong monsoon in the Late Miocene, and the minimum age of 200 ka provided by TL dating. The completely different alluvial motif, tectonic setting and palaeoclimate, with an apparent absence of the wet–dry cycles characteristic of the Quaternary Period (inference based both on the stable isotopic record and the sedimentary archives), however, suggest that the Type-1 succession is of Pliocene age, deposited between the Late Miocene onset of the monsoon system and the onset of Neogene glacial cycles (Pillans et al., 1998). The character of the alluvial suite, plus the apparent absence of active fluvial–aeolian interaction, suggest that the Type-1 succession represents a distant ancestor of the present Luni River, one that existed in the region prior to development of the present Thar Desert system.

By contrast, the dominance of ephemeral stream and aeolian deposits, as well as the stable isotope signature of calcretes, suggest that the Type-2 succession was deposited within a desert environment, and under the influence of the high-frequency climate changes of the past 100 kyr. A more detailed sequence of events can be outlined as follows:

1 The prominent gravel–palaeosol associations at the base of the sections in the Type-2 succession (KN and KH, Fig. 4) were deposited by gravelly braided streams during the OIS 5 interstadials (Table 4). The presence of laterally extensive gravel bodies indicates highly competent river systems and a relatively wet climate. This was followed by sheet flood deposition (Khudala) and the formation of a regionally extensive reddish palaeosol (Tandon et al., 1997, 1999; Juyal et al., 2000; Kar et al., 2000). Available geochronological data constrain pedogenesis to the period between 70 and 30 ka, which suggests that it occurred during the warm-humid climate of OIS 3 (Andrews et al., 1998; Jain & Tandon, 2001).

2 The late OIS 3 to OIS 2 transition, around 30–22 ka, is dominated by the deposits of what is interpreted to have been ephemeral sand-bed streams. A variety of data suggests that this was dominantly an arid period with weak summer monsoon conditions (Sirocko et al., 1993; Andrews et al., 1998), and it is reasonable to assume that the hydrological regime also would have been very weak during this time. Fluvial deposits from the Last Glacial Maxima (LGM), a time period thought to represent peak aridity, have not been identified. Moreover, only one aeolian depositional phase has been identified, at 27.1 ± 6.6 ka, which corresponds to the OIS 3–2 transition.

3 Fluvial activity resumed following the LGM, perhaps with the re-establishment of summer monsoon precipitation during the Late-glacial, ca. 14 ka (OIS 1). Deposits from the Late-glacial period occur inset within deposits that represent OIS 5–2, hence there was an intervening period of incision that can be bracketed to the period 22–14.6 ka (Fig. 9). Owing to the general dormancy of processes during the LGM it is reasonable to infer that the incision occurred at the end of this period, which is consistent with the rejuvenation of monsoons in the region by c. 14 ka. The base of the OIS 1 succession (Fig. 9) then records gravel sheet deposition at 14.6 ± 1.7 ka, perhaps by several flood surges in a coarse gravelly braided river (Fig. 9).

4 A diverse range of lithofacies represents the period 14–11 ka, including deposits interpreted to represent a gravelly braided stream, a mixed-load meandering stream, a sand-bed ephemeral stream, and aeolian processes. This part of the Type-2 succession is interpreted to reflect the high-frequency and high-amplitude climatic changes associated with the Late-glacial (the Bølling, Allerød, Older Dryas and Younger Dryas periods, as initially defined in northern Europe) with, perhaps, gravelly braided streams during wet periods,

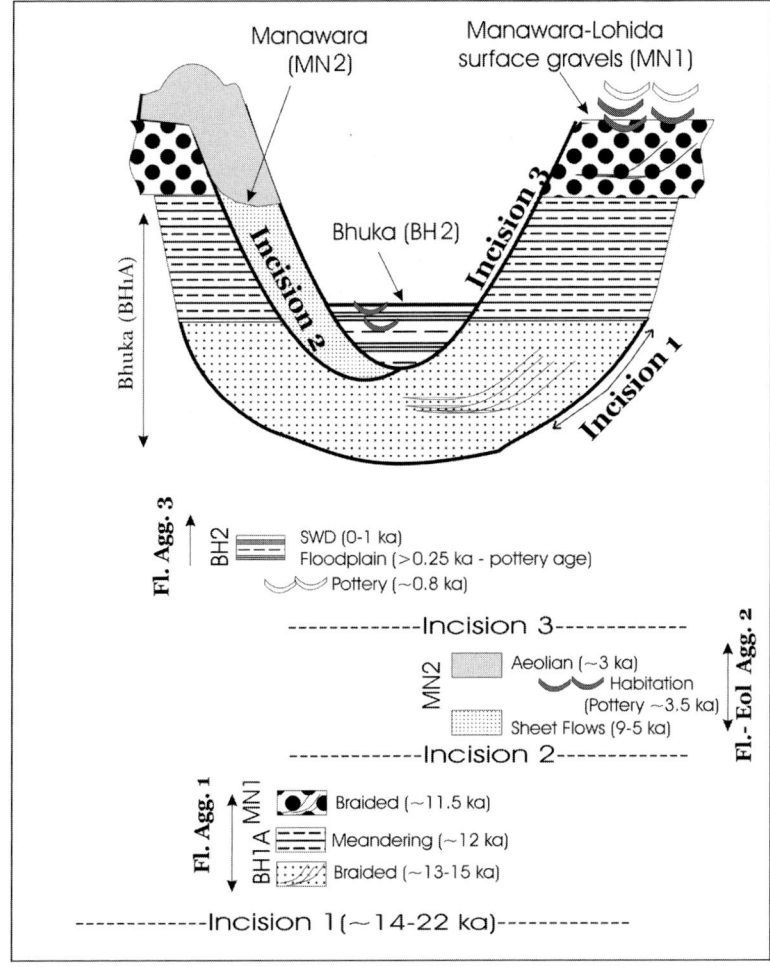

Fig. 9 Schematic representation of OIS 1 (Late-glacial and Holocene) deposits. There are three phases of aggradation and incision. The first period of incision occurred near the beginning of the Late-glacial followed by valley aggradation (BH1A). The deposition of surface gravels at MN-1 at around 11 ka was followed by a second phase of incision. Subsequently, there were sheet-flow deposits (9–5 ka, MN2) and a phase of human habitation on the gravels (c. 3.5 ka). An arid phase at c. 3 ka buried the alluvial deposits and the habitation site by aeolian dunes (MN2). A third phase of incision occurred during the post-dry-phase climatic amelioration. Present-day floodplain and slack-water deposits (< 1 ka) are developed along the Luni channel (BH2).

and sand-bed ephemeral streams and aeolian sediments during relatively dry phases (Table 4 and Fig. 9). The last depositional phase of the Late-glacial (OIS 1) is represented by the surface gravels (MN1: Figs 2e, 5 & 9), which at one time would have covered the valley floor (Fig. 4C & 9).

5 Incision through the surface gravels occurred during the early Holocene wet phase, around 11–9 ka (Fig. 9). This period of incision was followed by the accumulation of fluvial and aeolian deposits between 9 and 5 ka (Figs 4D & 9), perhaps resulting from a series of sheet-flow events with aeolian reworking. This interpretation is consistent with a relatively unstable climate indicated by the fluctuating lake levels and aeolian deposits in the Thar Desert (Enzel et al., 1999; Thomas et al., 1999). Aeolian deposits then blanketed alluvial sections at c. 3 ka during an arid phase (Fig. 9).

6 A third phase of incision occurred sometime during the phase of climatic amelioration, between 3 and 1 ka (Fig. 9). The fact that the slack-water deposits date to about 1 ka (Kale et al., 2000) perhaps indicates that incision/gullying was absent prior to this time. The presence of slack water deposits (< 1 ka) and floodplain silts (potsherds in the silts dated to < 0.3 ka) in the more confined reaches indicate that the River Luni in these parts has been stable for the past 1 kyr.

To summarize, the Type-2 succession of the lower Luni basin indicates that the fluvial regime was responding to climate change. Interglacial and interstadial moist periods were dominated by gravelly braided systems, whereas the LGM, and dry periods within the late Pleistocene and Holocene, were dominated by ephemeral sand-bed braided rivers and aeolian processes.

The stratigraphical record from the Late-glacial, between 14 and 11 ka, indicates that the Luni fluvial system was very sensitive to high-frequency climate changes, as channels oscillated from gravelly braided to ephemeral, to mostly inactive, during this short period of time. Three phases of incision occurred at the onset of wet phases during the Late-glacial and Holocene. Each of these distinct depositional and erosional environments left a distinct imprint, providing evidence for climatic forcing of the alluvial stratigraphical record.

One additional aspect of the Type-2 succession deserves mention. Large hiatuses in the stratigraphical records in any one location may be the result of enhanced sensitivity of the arid zone rivers to precipitation changes. Oscillations between periods of dormancy and activation, during which the local geomorphology changes in response to active aeolian processes, may lead to frequent changes in the river courses. This could cause stratigraphical records to be relatively randomly scattered and more chaotic than their humid counterparts. *Thus for dryland rivers, an absence of robust chronological control can cause gross stratigraphical misinterpretations.* A common element between some Australian (Central Australia) and the Thar rivers is the presence of peak aridity reflected as aeolian deposits during the OIS 2 full glacial period (e.g. Nanson *et al.*, 1995; Nanson & Tooth, 1999). The exact nature of the Thar fluvial record, however, appears to be unique and has no one-to-one correspondence with other studied rivers of the world (see Jain & Tandon (2003) for a detailed comparison).

CONCLUSIONS

The present study has identified two distinct alluvial successions in the lower Luni basin, Thar Desert. Sedimentological, stratigraphical, geochronological and geochemical analysis of these successions permit the following conclusions:

1 The Type-1 succession consists of an extensive suite of vertically stacked fining upward cycles deposited by gravelly braided streams within a subsiding basin context. Each cycle begins with gravel sheets that represent channel deposits, and terminates with floodplain deposition then pedogenesis. The Type-1 succession is thought to have been deposited during the Pliocene.

2 The Type-2 succession (OIS 1 to OIS 5) shows distinct depositional environments that suggest a strong climatic control on stratigraphical development. The record correlates with arid–humid cycles during the last glacial–interglacial cycle. Relatively humid phases are represented by gravelly braided streams (OIS 1 and OIS 5) and a reddening event (OIS 3). Relatively arid phases are represented by ephemeral sand-bed streams, sheet floods, sheet flows and aeolian sands (OIS 1, OIS 2 and late OIS 3). The rivers were particularly dynamic during the phase of climate instability associated with the Late-glacial period and transition to OIS 1 (between 11 and 14 ka).

3 Three incision events (around 14 ka, 9–11 ka and 3–1 ka) occurred as responses to increases in moisture (monsoon intensification).

4 There exists a large hiatus between the deposition of the Pliocene Type-1 and Late Pleistocene Type-2 successions, perhaps owing to overall low accommodation space during the Quaternary. It appears that the reworking of the sediment by each Quaternary glacial–interglacial cycle allows only the latest deposits to be best preserved.

5 A comparison between the Type-1 and Type-2 successions indicates that although climate is a dominant control on facies architecture, the overall control on preservation, and hence long-term stratigraphical development, may be tectonism (subsidence).

6 This study gives the first clear demonstration of the sensitivity of desert rivers to global climate change.

ACKNOWLEDGEMENTS

Mayank Jain acknowledges UGC, India for supporting his PhD work, and the Department of Geology, University of Delhi for providing infrastructural support. The Physical Research Laboratory in Ahmedabad is thanked for the laboratory facilities. Professor Yong IL Lee, KBSI, South Korea, is thanked for the stable isotope analysis. Fieldwork was supported by the Department of Science and Technology coordinated programme number ESS/CA/A3-08/92. Constructive reviews were provided by Darrel Maddy, David May and Mike Blum.

REFERENCES

Abdullatif, O.M. (1989) Channel-fill and sheet-flood facies sequences in the ephemeral terminal River Gash, Kassala, Sudan. *Sediment. Geol.*, **63**, 171–184.

Andrews, J.E., Singhvi, A.K., Kailath, J.A., Kuhn, R., Dennis, P.F., Tandon, S.K. and Dhir, R.P. (1998) Do stable isotope data from calcrete record Late Pleistocene monsoonal climate variation in the Thar Desert of India? *Quat. Res.*, **50**, 240–251.

Bajpai, V.N., Roy, T.K.S. and Tandon, S.K. (2001) Subsurface sediment accumulation patterns and their relationships with tectonic lineaments in the semi-arid Luni river basin, Rajasthan, Western India. *J. Arid Environ.*, **48**, 603–621.

Bentham, P.A., Talling, P.J. and Burbank, D.W. (1993) Braided stream and flood-plain deposition in a rapidly aggrading basin: the Escanilla formation, Spanish Pyrenees. In: *Braided rivers* (Eds Best, J.L. and Bristow, C.S.). *Geol. Soc. Lond. Spec Publ.*, **75**, 177–194.

Blair, T.C. and McPherson, J.G. (1992) The Trollheim alluvial fan and facies model revisited. *Geol. Soc. Am. Bull.*, **104**, 762–769.

Blum, M.D. and Törnqvist, T.E. (2000) Fluvial responses to climate and sea-level change: a review and look forward. *Sedimentology*, **47**, 2–48.

Bown, T.M. and Kraus, M.J. (1981) Lower Eocene alluvial paleosols (Wilwood Formation, northwest Wyoming, USA) and their significance for palaeoecology, palaeoclimatology, and basin analysis. *Palaeogeogr. Palaeoclimatol. Palaeoecol.*, **34**, 1–30.

Bridge, J.S. and Best, J.C. (1988) Flow, Sediment transport and bedform dynamics over the transition from dune to upper-stage plane beds: implications for the formation of planar laminae. *Sedimentology*, **35**, 753–763.

Bull, W.B. (1991) *Geomorphic Responses to Climatic Change*. Oxford University Press, New York.

Cant, D.J. and Walker, R.G. (1978) Fluvial processes and facies sequences in the sandy braided south Sasketchewan river, Canada. *Sedimentology*, **25**, 625–648.

Carling, P.A. (1996) Morphology, sedimentology and palaeohydraulic significance of large gravel dunes, Altai Mountains, Siberia. *Sedimentology*, **43**, 647–664.

Cerling, T.E., Harris, J.M., MacFadden, B.J., Leakey, M.G., Quade, J., Eisenmann, V. and Ehleringer, J.R. (1997) Global vegetation change through the Miocene/Pliocene boundary. *Nature*, **389**, 153–158.

Eberth, D.A. and Miall, A.D. (1991) Stratigraphy, sedimentology and evolution of a vertebrate-bearing, braided to anastomosed fluvial system, Cutler Formation (Permian–Pennsylvanian), north-central New Mexico. *Sediment. Geol.*, **72**, 225–252.

Enzel, Y., Ely, L.L., Mishra, S., Ramesh, R., Amit, R., Lazar, B., Rajguru, S.N., Baker and V.R., Sandler, A. (1999) High-resolution Holocene environmental changes in the Thar Desert, northwestern India. *Science*, **284**, 124–128.

Friend, P.F., Slater, M.J. and Williams, R.C. (1979) Vertical and lateral building of river sandstone bodies. Ebro Basin, Spain. *J. Geol. Soc. Lon.*, **136**, 39–46.

Frostick, L.E. and Reid, I. (1977) The origin of horizontal laminae in ephemeral stream channel fill. *Sedimentology*, **24**, 1–10.

Godin, P. (1991) Fining-upward cycles in the sandy-braided river deposits of the Westwater Canyon Member (Upper Jurassic), Morrison Formation, New Mexico. *Sediment. Geol.*, **70**, 61–82.

Gupta, A. (1995) Magnitude, frequency and special factors affecting channel form and processes in the seasonal tropics. *Geophys. Monogr.*, **89**, 125–136.

Harms, J.C, Southard, J.B. and Walker, R.G. (1982) Structures and sequences in clastic rocks. *Soc. Econ. Paleontol. Mineral. Short Course*, **9**, 279 pp.

Hereford, R. (1986) Modern alluvial history of the Paria River drainage basin, southern Utah. *Quat. Res.*, **25**, 293–311.

Hill, G. (1989) Distal alluvial fan sediments from the Upper Jurassic of Portugal: controls on their cyclicity and red beds, St. Mary's Bay, Nova Scotia. *Sedimentology*, **29**, 457–474.

Hjellbakk, A. (1997) Facies and Fluvial architecture of a high-energy braided river: the Upper Proterozoic Seglodden Member, Varanger Peninsula, northern Norway. *Sediment. Geol.*, **114**, 131–161.

Jain, M. (2000) Stratigraphic development of some exposed Quaternary alluvial sequences in the Thar and its margins: fluvial response to climate change, western India. Unpublished PhD thesis, University of Delhi.

Jain, M. and Singhvi, A.K. (2001) Limits to depletion of blue-green light stimulated luminescence in feldspars: implications for quartz dating. *Radiat. Meas.*, **33**, 883–892.

Jain, M. and Tandon, S.K. (2001) Palaeoclimatic reconstruction using $\delta^{13}C$ and $\delta^{18}O$ composition of pedogenic calcretes in some Quaternary alluvial palaeosols, Thar desert, Western India. Abstract volume, *7th International Conference on Fluvial Sedimentology*, 139.

Jain, M. and Tandon, S.K. (2003) Fluvial response to Late Quaternary climate changes, western India. *Quat. Sci. Rev.*, **22**, 2223–2235.

Jain, M., Tandon, S.K., Bhatt, S.C., Singhvi, A.K. and Mishra, S. (1999) Alluvial and aeolian sequences along the river Luni, Barmer district: physical stratigraphy and feasibility of luminescence chronology methods. In: *Vedic Saraswati* (Eds B.P. Radhakrishna and S.S. Merh). *Geol. Soc. Ind. Mem.*, **42**, 273–295.

Jain, M., Boetter-Jensen, L. and Singhvi, A.K. (2003) Dose evaluation using multiple-aliquot quartz OSL: test of methods and a new protocol for improved accuracy and precision. *Radiat. Meas.*, **37**, 67–80.

Jain, M., Murray, A.S. and Boetter-Jensen, L. (2004) Optically stimulated luminescence dating. How significant is incomplete light exposure in fluvial environments? *Quaternaire*, **15**, 143–157.

Juyal, N., Raj, R., Maurya, D.M., Chamyal, L.S. and Singhvi, A.K. (2000) Chronology of Late Pleistocene environmental changes in the lower Mahi basin, western India. *J. Quat. Sci.*, **15**, 501–508.

Kale, V.S. and Rajaguru, S.N. (1987) Late Quaternary alluvial history of the northwestern Deccan upland region. *Nature*, **325**, 612–614.

Kale, V.S., Singhvi, A.K., Mishra, P.K. and Banerjee, D. (2000) Sedimentary record and luminescence chronology of late Holocene paleofloods in the Luni valley, Thar Desert North west India. *Catena*, **49**, 337–358.

Kar, A., Singhvi, A.K., Rajguru, S.N., Juyal, N., Thomas, J.V., Banerjee, D. and Dhir, R.P. (2000) Reconstruction of the late Quaternary environments of the lower Luni plains, Thar Desert, India. *J. Quat. Sci.*, **15**, 1–8.

Leeder, M.R. (1975) Pedogenic carbonates and flood sediment accretion rates: a quantitative model for alluvial arid zone lithofacies. *Geol. Mag.*, **112**, 257–270.

Lowe, D.R. (1988) Suspended load fallout rate as an independent variable in the analysis of current structures. *Sedimentology*, **35**, 765–776.

Lucchitta, I. and Suneson, N. (1981) Flash flood in Arizona observations and their application to the identification of flash-flood deposits in the geologic record. *Geology*, **9**, 414–418.

Machette, M.N. (1985) Calcic soils of southwestern United States. In: *Soils and Quaternary Geology of the Southwest United States* (Ed. Weide DL). *Geol. Soc. Am. Spec. Pap.*, **203**, 1–21.

Mass, G.S., Macklin, M.G. and Kirkby, M.J. (1998) Late Pleistocene and Holocene development in Mediterranean steepland environments, Southwest Crete, Greece. In: *Palaeohydrology and environmental change* (Eds G. Benito, V.R. Baker and K.J. Gregory), pp. 153–165. Wiley, Chichester.

Merh, S.S. and Chamyal, L.S. (1997) The Quaternary Geology of Gujarat Alluvial Plains. *Proc. Ind. Nat. Sci. Acad.*, **63A**(1), 1–98.

Miall, A.D. (1978) Lithofacies types and vertical profile models in braided river deposits: a summary. In: *Fluvial Sedimentology* (Ed. A.D. Miall). *Can. Soc. Petrol. Geol. Mem.*, **5**, 597–604.

Miall, A.D. (1996) *The Geology of Fluvial Deposits*. Springer-Verlag: Berlin.

Mishra, S. and Rajaguru, S.N. (2001) Late Quaternary palaeoclimates of western India: A geoarchaeological approach. *Mausam*, **52**, 285–296.

Mishra, S., Jain, M., Tandon, S.K., Singhvi, A.K., Joglekar, P.P., Bhatt, S.C., Kshirsagar, A., Naik, S. and Mukerjee, A.D. (1999) Prehistoric Cultures and Late Quaternary environments in the Luni basin around Balotra. *Man Environ.*, **24**, 38–49.

Murray, A.S. and Olley, H.M. (1999) Determining sedimentation rates using luminescence dating. *GeoRes. Forum*, **5**, 121–144.

Nanson, G.C. and Tooth, S. (1999) Arid-zone rivers as indicators of climate. In: *Palaeoenvironmental reconstruction in Arid lands* (Eds A.K. Singhvi and E. Derbyshire), pp. 175–216. Balkema, Rotterdam.

Nanson, G.C., Chen, X.Y. and Price, D.M. (1995) Aeolian and fluvial evidence of changing climate and wind patterns during the past 100 k yrs in the western Simpson Desert, Australia. *Palaeogeogr. Palaeoclimatol. Palaeoecol.*, **113**, 87–102.

Olley, J., Caitcheon, G. and Murray, A.S. (1998) The distribution of apparent dose as determined by optically stimulated luminescence in small aliquots of fluvial quartz: Implications for dating young sediments. *Quat. Geochem.*, **17**, 1033–1040.

Paola, C., Wiele, S.M. and Reinhart, M.A. (1989) Upper regime parallel lamination as the result of turbulent sediment transport and low-amplitude bedforms. *Sedimentology*, **36**, 47–59.

Picard, M.D. and High, L.R. (1973) *Sedimentary Structures of Ephemeral Streams*. Developments in Sedimentology, **17**, Elsevier, Amsterdam.

Pillans, B., Chappell, J. and Naish, T.R. (1998) A review of the Milankovitch climatic beat: template for Plio-Pleistocene sea-level changes and sequence stratigrphy. *Sediment. Geol.*, **122**, 5–21.

Prell, W.L. and Kuzbach, J.E. (1992) Sensitivity of Indian monsoon to forcing parameters and implications for its evolution. *Nature*, **360**, 647–652.

Prescott, J.R, Huntley, D.J., Hutton, J.T. (1993) Estimation of equivalent dose in thermoluminescence dating—the Australian slide method. *Ancient TL*, **11**, 1–5.

Reid, I. and Frostick, L.E. (1997) Channel form, flows and sediments in deserts. In: *Arid Zone Geomorphology: Process, Form and Change in Drylands*, 2nd edn (Ed. D.S.G. Thomas), pp. 205–229. Wiley, Chichester.

Rust, B.R. (1978) Depositional models for braided alluvium. In: *Fluvial Sedimentology* (Ed. A.D. Miall). *Can. Soc. Petrol. Geol. Mem.*, **5**, 605–625.

Sharma, K.D., Vangani, N.S. and Choudhari, J.S. (1984) Sediment transport characteristics of the desert streams in India. *J. Hydrol.*, **67**, 261–272.

Singh, G., Joshi, R.D., Chopra, S.K. and Singh, A.B. (1974) Late Quaternary history of vegetation and climate of the Rajasthan desert, India. *Philos. Trans. Roy. Soc. Lond.*, **267B**, 467–501.

Sirocko, F., Sarnthein, M., Erlenkeuser, H., Lange, H., Arnold, M. and Duplessy, J.C. (1993) Century-scale events in monsoonal climate over the past 24,000 years. *Nature*, **364**, 322–324.

Smith, D.G. (1986) Anastomosing river deposits, sedimentation rates and basin subsidence, Magdalena River, northwestern Colombia, South America. *Sediment. Geol.*, **46**, 177–196.

Smith, N.D. (1974) Sedimentology and bar formation in the upper Kicking Horse River, a braided outwash stream. *J. Geol.*, **82**, 205–224.

Sneh, A. (1983) Desert stream sequences in the Sinai Peninsula. *J. Sediment. Petrol.*, **53**, 1271–1280.

Stear, W.M. (1983) Morphological characteristics of ephemeral stream channel and overbank splay sandstone bodies in the Permian Lower Beaufort Group, Karoo Basin, South Africa. In: *Modern and Ancient Fluvial Systems* (Eds J. Collinson and J. Lewin). *Spec. Publ. Int. Assoc. Sediment.*, **6**, 405–420.

Tandon, S.K., Sareen, B.K., Rao, M.S. and Singhvi, A.K. (1997) Aggradation history and luminescence chronology of Late Quaternary semi-arid sequences of the Sabarmati basin, Gujarat, western India. *Palaeogeogr. Palaeoclimatol. Palaeoecol.*, **128**, 339–357.

Tandon, S.K., Jain, M. and Singhvi, A.K. (1999) Comparative development of mid- to late Quaternary fluvial and fluvio-aeolian stratigraphy in the Luni, Sabarmati and Mahi river basins of western India. *Gond. Geol. Mag.*, **4**(Spec. Vol.), 1–16.

Thomas, J.V., Kar, A., Kailath, A.J., Juyal, N., Rajguru, S.N. and Singhvi, A.K. (1999) Late Pleistocene–Holocene history of aeolian accumulation in the Thar Desert, India. *Z. Geomorph. Suppl. Bd*, **116**, 181–194.

Thornes, J.B. (1994) Channel processes, evolution and history. In: *Geomorphology of Desert Environments* (Eds A.D. Abrahams and A.J. Parsons), pp. 288–317. Chapman and Hall, London.

Tooth, S. (2000) Process, form and change in dryland rivers: a review of recent research. *Earth Sci. Rev.*, **1**, 67–107.

Tunbridge, I.P. (1981) Sandy high-energy flood sedimentation—some criteria for recognition, with an example from the Devonian of SW England. *Sediment. Geol.*, **28**, 79–96.

Turner, B.R. (1983) Braidplain deposition of the Upper Triassic Molteno Formation in the main Karoo (Gondwana) Basin, South Africa. *Sedimentology*, **30**, 77–89.

Wang, Y., McDonald, E., Amundson, R., McFadden, L. and Chadwick, O. (1996) An isotopic study of soils in chronological sequences of alluvial deposits, Providence mountains, California. *Geol. Soc. Am. Bull.*, **108**(4), 379–391.

Ward, W.C. (1975) Petrology and diagenesis of carbonate eolianites of northeastern Yucatan, Mexico. In: *Belize Shelf: Carbonate Sediments, Clastic Sediments and Ecology* (Eds K.F. Wantland and W.C. Pussey III). *Am. Assoc. Petrol. Geol. Stud. Geol.*, **2**, 500–571.

Williams, G.E. (1971) Flood deposits of the sand bed ephemeral streams of central Australia. *Sedimentology*, **17**, 1–40.

Wintle, A.G. and Murray, A.S. (1997) The relationship between quartz thermoluminescence, photo-transferred luminescence and optically stimulated luminescence. *Radiat. Meas.*, **27**, 611–624.

The Middle Valley of the Tiber River, central Italy: Plio-Pleistocene fluvial and coastal sedimentation, extensional tectonics and volcanism

MARCO MANCINI* and GIAN PAOLO CAVINATO†

*Dipartimento di Scienze della Terra, Università di Roma 'La Sapienza',
P.le A. Moro, 5 box 11, 00185 Rome, Italy (Email: marco.mancini@uniroma1.it); and
†CNR- Istituto di Geologia Ambientale e Geoingegneria, Dipartimento Scienze della Terra,
Università di Roma 'La Sapienza', P.le A. Moro, 5 box. 11, 00185 Rome, Italy

ABSTRACT

The Middle Valley of the Tiber River, Italy (MVT), corresponds to a NNW–SSE trending extensional basin that has developed since the middle Pliocene along the western flank of the central Apennines. Stratigraphical, sedimentological, palaeontological and Sr isotope analyses have been conducted to detail the stratigraphy of the MVT and to reconstruct the history of Pliocene–Quaternary basin filling. Most of the basin-fill is composed of fluvial and deltaic deposits that are chronologically constrained by biostratigraphical data and Sr isotopes from marine deposits, and through relationships with volcanic and volcaniclastic units with K/Ar and Ar/Ar radiometric ages. This paper focuses on relationships between sedimentary phases, long-term (> 1 Myr) tectonic movements, shorter term (100 kyr) climatic and eustatic changes, and volcanism, within the overall extensional tectonic context of the MVT.

Two main tectonic phases are recognized in the MVT, each recorded by the responses of the mostly gravel-dominated fluvial and deltaic systems. The first phase encompasses the middle Pliocene to earliest Early Pleistocene, and was dominated by rapid subsidence. This phase was characterized by transverse rivers that fed cyclically prograding and retrograding fluvial–deltaic wedges, with interfingering marine deposits, and within an overall aggradational context. Superimposed on this overall aggradational trend is an inferred 4th order cyclicity that is interpreted to reflect eustatic and climatic changes. The second phase began in the late Early Pleistocene and extends to the present, and is linked to uplift of the Apennines. This phase was characterized by complete emergence of the MVT, the occurrence of volcanism, development of the Tiber River system, and the initiation of a long-term trend of uplift-driven net valley incision. Alternating aggradational and degradational phases are superimposed on this longer term trend, and are thought to be linked to Late Quaternary climatic forcing, with 100 kyr glacial–interglacial cycles important in the upper reaches of the fluvial system, and the related effects of glacio-eustasy important farther downstream.

INTRODUCTION

The Middle Valley of the Tiber River (MVT) is a hilly geographical area that straddles the Latium and Umbria provinces of Italy, approximately 30 km north of Rome (Fig. 1), and extends for 60 km in the NNW–SSE direction. The Tiber River is a N–S flowing axial drainage system (Fig. 2).

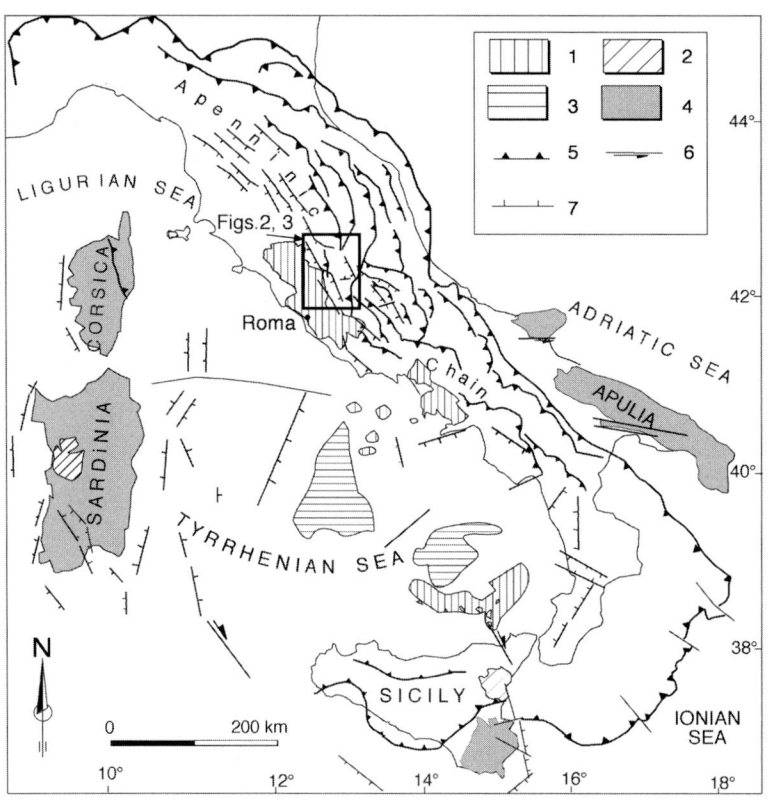

Fig. 1 Simplified tectonic map of peninsular Italy (from Cavinato & De Celles, 1999): 1, subduction-related and high-K volcanic rocks; 2, extension-related volcanic rocks; 3, oceanic crust; 4, foreland areas; 5, subsurface and surface thrust front; 6, strike-slip fault; 7, extensional fault.

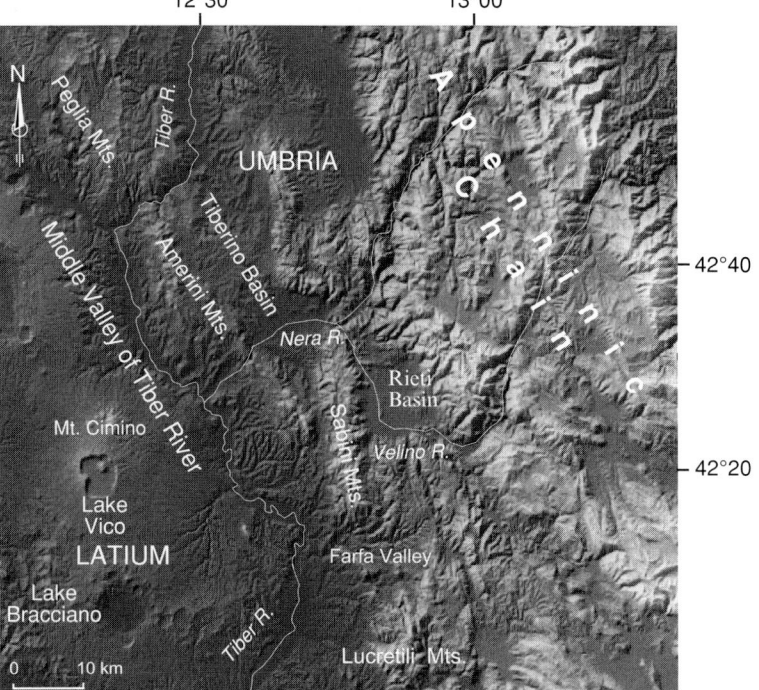

Fig. 2 Shaded relief image of the Middle Valley of the Tiber River.

Structurally the MVT corresponds in part to the 'Paglia-Tevere Graben' (Funiciello et al., 1981), one of the Neogene–Quaternary, NNW–SSE trending, extensional basins in west central Italy, and is filled with several kilometres of Plio-Quaternary, marine and continental siliciclastic and volcanic deposits (Martini & Sagri, 1993; Cavinato et al., 1994). The valley is bordered to the east by the Mount Peglia–Amerini–Narni–Sabini–Lucretili Mountains ridge of the central Apennines, and to the west by the Quaternary volcanic hills of the Vulsini Mountains, Vico, Mount Cimino, Sabatini Mountains Volcanic Districts (Fig. 2). Mesozoic–Cenozoic calcareous and siliciclastic successions lie underneath the Plio-Quaternary basin fill, and are laterally continuous with successions exposed on the Mount Peglia–Lucretili Mountains ridge (Barberi et al., 1994).

New geological mapping, structural observations, and stratigraphical, sedimentological, palaeontological and isotopic analyses have been conducted to better define the general Pliocene–Quaternary stratigraphy of the MVT (Mancini, 2000; Mancini et al., 2001) (Figs 3 & 4). In the MVT, as is the case in many extensional tectonic contexts (Miall, 1996, 2000; Gawthorpe & Leeder, 2000), fluvial and coastal systems play a prominent role in understanding the architecture and timing of the basin fill. This paper provides an overview of the fluvial and deltaic stratigraphy so as to help decipher the relationships between sedimentation, long-term (> 1 Myr) tectonic and volcanic controls, and the shorter-term climatic and eustatic signals of the Pliocene-Quaternary.

GENERAL TECTONIC SETTING

The MVT is located along the central western flank of the Apennines, an east-northeast-verging imbricate fold-thrust belt that has developed since the Early Miocene and progressively migrated eastward above the subducting Adriatic microplate (Patacca et al., 1990; Doglioni, 1991; Cavinato & De Celles, 1999) (Fig. 1). Thrust sheets consist of Triassic–Middle Miocene, pre-orogenic marine, calcareous and siliciclastic successions (Bally et al., 1986; Bigi et al., 1992), and of Miocene to Pleistocene synorogenic terrigenous clastics of shallow marine origin (Ricci Lucchi, 1986; Cipollari et al., 1998). The MVT area was affected by compressional tectonics during the Tortonian–Messinian (Cavinato & De Celles, 1999).

A WSW–ENE directed extensional regime, probably related to the opening of the Tyrrhenian Sea back-arc basin (Malinverno & Ryan, 1986) or to post-collisional orogenic collapse (Keller et al., 1994), has affected the western margin of the Apennines since the Miocene (Patacca et al., 1990). A progressive eastward migration of this extensional regime occurred just after formation of the fold-thrust belt, and caused the development of a series of grabens and half-grabens trending NW–SE and NE–SW (Cavinato & De Celles, 1999). An almost contemporary, widespread, eastward migrating volcanism also developed (Barberi et al., 1994), while thrusting continued toward the Adriatic margin (Fig. 1).

The MVT basin is a complex tectonic depression that formed by the conjunction of several NW–SE trending grabens and half-grabens, and has been affected by extensional tectonics since the earliest middle Pliocene (Cavinato et al., 1994; Cavinato & De Celles, 1999). It is filled with almost 1 km of Plio-Pleistocene continental and shallow marine siliciclastic deposits (De Rita et al., 1993; Barberi et al., 1994), beneath which Mesozoic–Cenozoic marine calcareous and terrigenous successions are arranged into a thrust pile trending NW–SE to N–S (Buonasorte et al., 1987). The Plio-Pleistocene sedimentary fill is overlain by a wide Early–Middle Pleistocene volcanic cover that is up to 1 km thick (Barberi et al., 1994).

Several extensional and transtensional fault systems affect the Plio-Quaternary sedimentary and volcanic successions. The NNW–SSE trending master faults of the MVT graben discountinuously border the western flank of the 'Mount Peglia–Lucretili Mountains ridge'. They are high-angle normal faults, dipping south-west, that in some cases have reactivated several sole thrusts in the reverse sense (Cavinato & De Celles, 1999). The MVT structural setting (Fig. 3) is also characterized by: (i) NNW–SSE antithetic faults; (ii) WSW–ESE and E–W striking normal faults, which generate local depressions, such as in the southern MVT, in the Farfa Valley and in the Rieti Basin; (iii) 'antiapenninic', NE–SW trending,

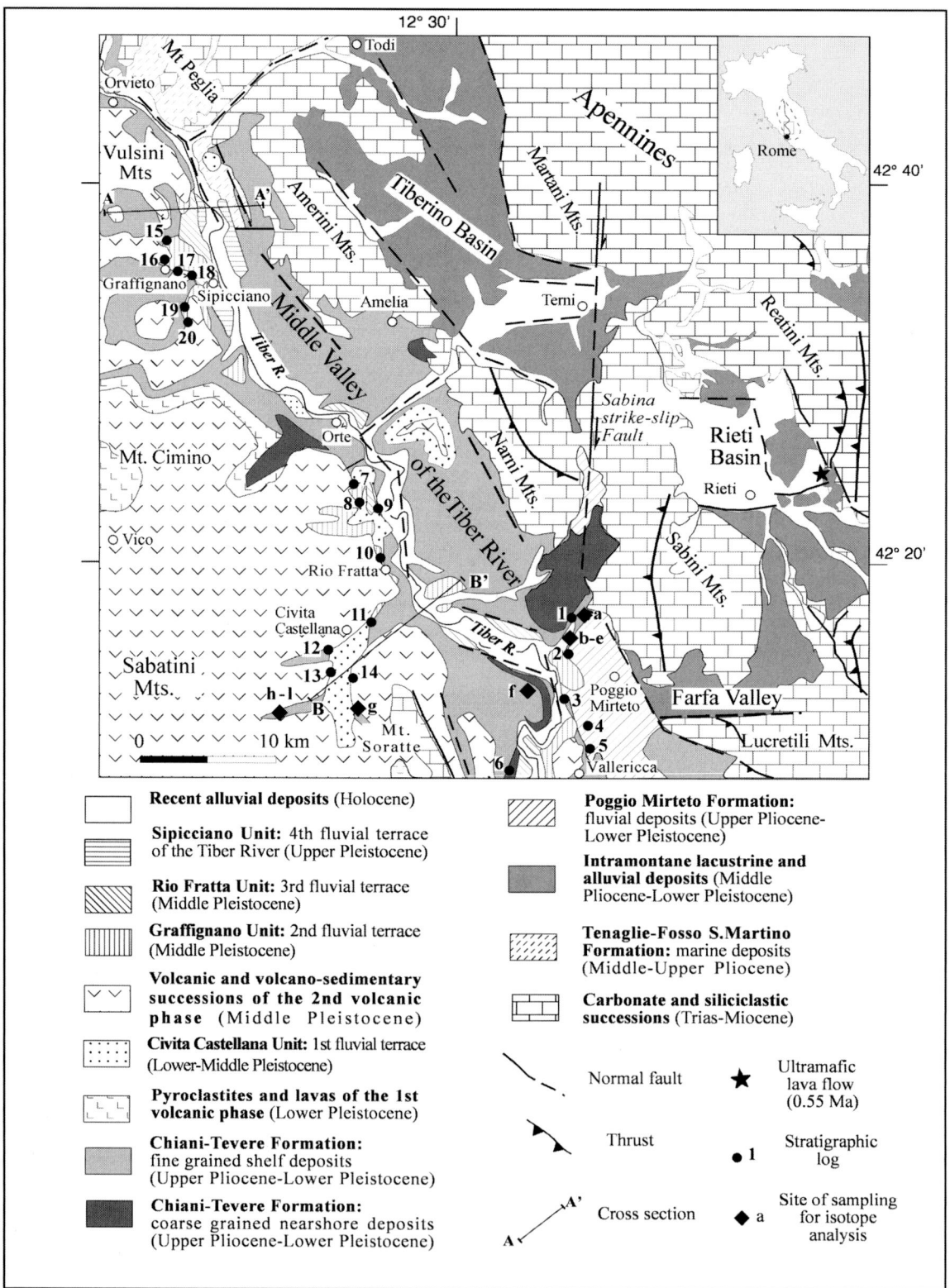

Fig. 3 Geological map of the Middle Valley of the Tiber River.

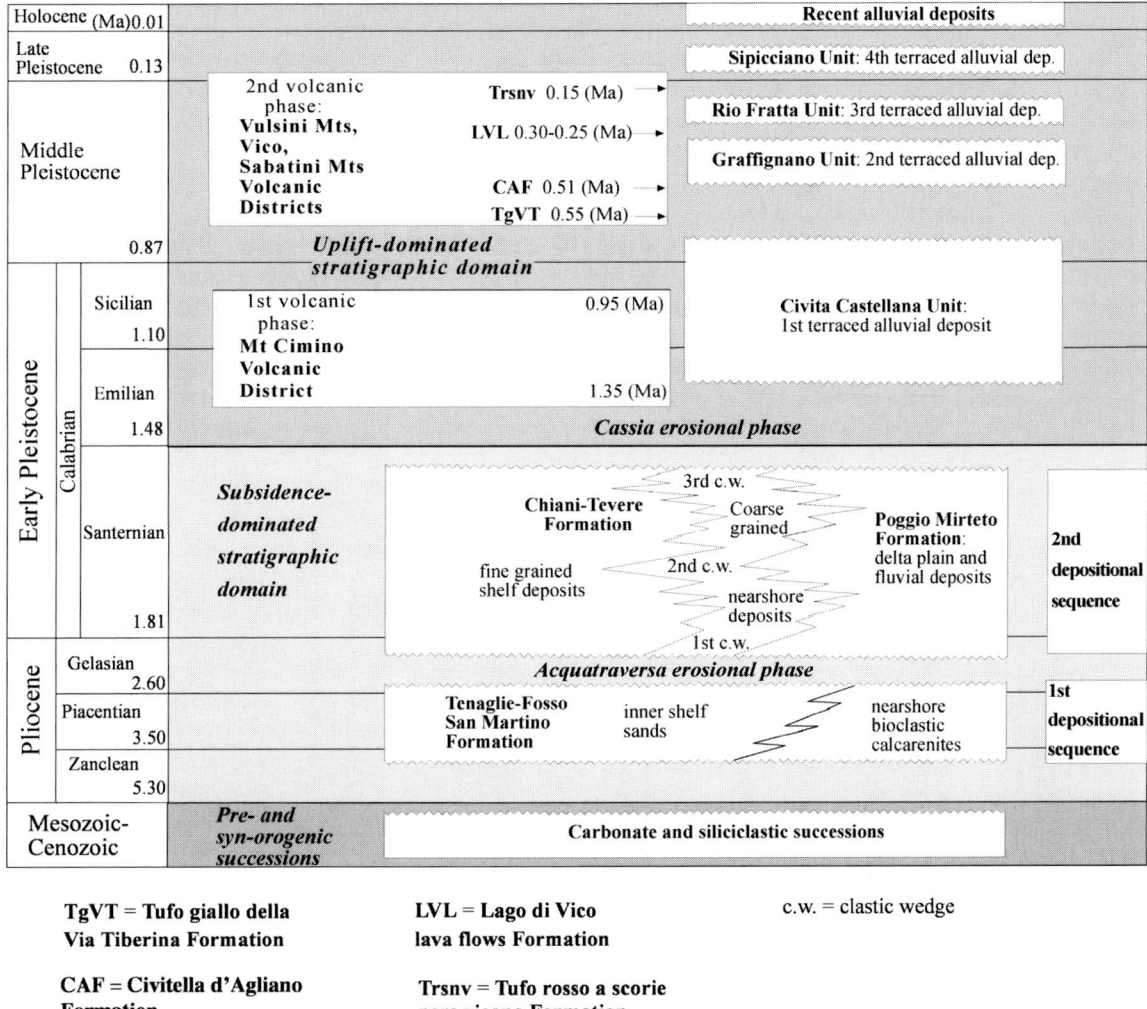

Fig. 4 Chronostratigraphy and lithostratigraphy of the MVT.

probably transfer, normal faults; (iv) the 'Sabina Fault' a N–S directed, dextral strike-slip fault (Alfonsi et al., 1991). Plio-Pleistocene strata generally have subhorizontal attitudes, with dips that very rarely exceed 10° towards the east or west (Fig. 5).

PALAEOCLIMATIC SETTING

The well-known Late Cenozoic global climate changes are expected to have had significant influence on MVT fluvial activity. However, relatively few data are available to precisely outline the MVT palaeoclimatic setting for earlier time periods of interest for this paper. Nevertheless, a general Pliocene through to Early–Middle Pleistocene palaeoclimatic record for the central Mediterranean region can be inferred from palynological and palaeoecological data on Foraminifera. The Middle to Late Pleistocene and Holocene are more well-known from local data, as palynological records from maar-lakes span the past 300 kyr (Follieri et al., 1988, 1998; Turner, 1998).

The Pliocene–Quaternary climate record for the central Mediterranean is noted for a significant change at 2.6–2.4 Ma (Channell et al., 1992;

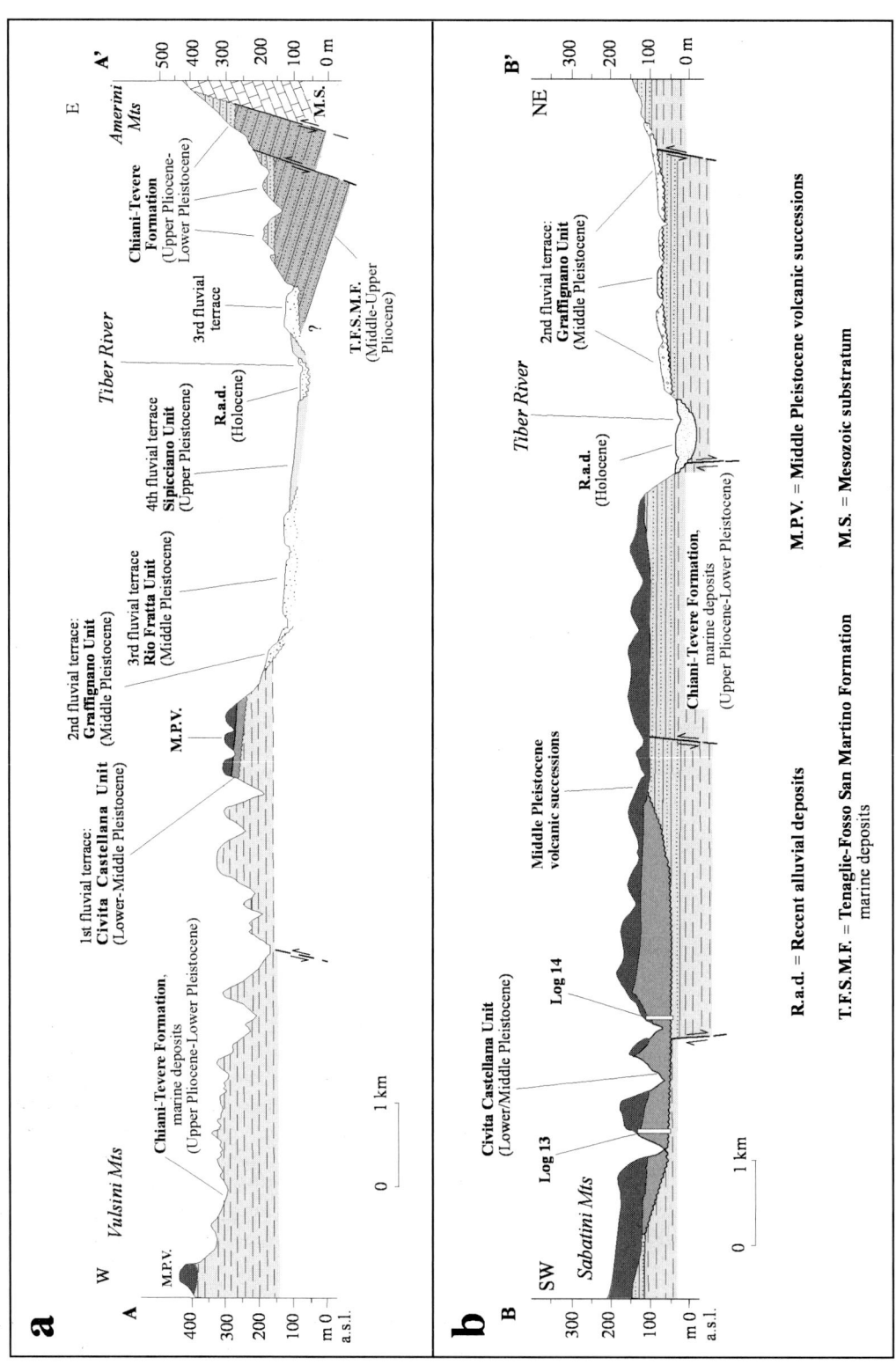

Fig. 5 Geological cross-sections showing the stratigraphical relations among Plio-Pleistocene successions (redrawn and simplified from Mancini *et al.*, 2001): (a) terraced fluvial deposits of the Tiber River unconformably overlying Plio-Pleistocene marine successions in the northern MVT; (b) deeply incised palaeovalley filled with the Civita Castellana Unit in the southern part of the MVT. Geographical location of the cross-sections is shown in Fig. 3.

Rio et al., 1994; Bossio et al., 1998), which is related to a major glacial event recorded in the Northern Hemisphere (Raymo et al., 1989; Bradley, 1999). Before that time the Mediterranean was humid subtropical, whereas after that the region was characterized by strong seasonality and periodic changes from cold and arid to warm and humid (Bertoldi et al., 1989; Combourie-Nebout, 1993; Bertini, 1995). Early–Middle Pliocene vegetation was mostly forest, whereas after that steppe vegetation dominated cold and arid periods, and forests dominated the warmer humid periods. Such changes are thought to be linked to glacial–interglacial alternations forced by 40 kyr obliquity-driven cyclicity (Imbrie, 1985; Raymo et al., 1989).

On a global scale, Middle–Late Quaternary climatic changes, after c. 850 ka, were dominated by the 100 kyr glacial–interglacial cyclicity (Bradley, 1999). Palaeobotanic data from maar-lakes indicate longer (> 50 kyr) arid steppe-dominated environments concomitant with glacial and partly with interglacial phases (Follieri et al., 1988, 1998; Turner, 1998), whereas shorter (10–20 kyr) forest-dominated periods are correlated to the interglacial peaks. In particular, forest environments were dominant during marine oxygen isotope substages 7c, 7a, 5e, 5c and 5a (Follieri et al., 1988; Amorosi et al., 1999).

METHODOLOGY AND DATA

The data base discussed here is derived from examination of 20 stratigraphical and sedimentological sections (Fig. 3). Each section has been measured in terms of lithology, grain size, and primary sedimentary and biogenic structures following the standard methods of field geology and facies analysis (Miall, 2000).

Micropalaeontological data were obtained by collecting 73 samples of pelite or sand for qualitative analyses of micromolluscs and Foraminifera (Mancini, 2000). Each sample had a volume of 200 cm^3, and was treated with a H_2O and H_2O_2 solution, then washed through a 125 μm sieve. The > 125 μm size fraction of the wash residue was used for the microfauna analysis; up to 300 specimens were counted for each sample. The biostratigraphical schemes used are those of Cita (1975) and Iaccarino (1985).

Measurements of $^{87}Sr/^{86}Sr$ ratios were carried out on 12 samples of marine molluscs and hydrozoa for chronostratigraphical purposes; Sr isotope ratios were measured on 15 mg of a calcitic or aragonitic shell fragment for each specimen. Each fragment was washed ultrasonically in ultrapure water, then gently crushed and rewashed in distilled water, so as to separate the organic and pelitic residues from the carbonate. Each sample was dissolved in 2.5 N ultrapure HCl, centrifuged and loaded onto a cation exchange resin. Sr was collected as $Sr(NO_3)_2$ and put on a tungsten double-filament of a multicollector mass-spectrometer. Strontium isotope analyses were performed by Professor Mario Barbieri, at the 'Istituto di Geologia Ambientale e Geoingegneria' (CNR-Rome). Spectrometric analyses were normalized to $^{88}Sr/^{86}Sr$ = 0.1194. More than 200 ratios were collected for each analysis, with errors expressed as two standard errors of the mean ($\times 10^{-6}$).

Radiometric (Nicoletti, 1969; Arias et al., 1980; Borghetti et al., 1981; Laurenzi & Villa, 1987; Cioni et al., 1993; Barberi et al., 1994; Nappi et al., 1995; Perini et al., 1997) and magnetostratigraphical (Florindo & Sagnotti, 1995; Borzi et al., 1998) data were used to directly date some volcanic and sedimentary units, and to provide an indirect age control on other formations. The geochronological and chronostratigraphical nomenclature used follows Sprovieri (1993) and Cita & Castradori (1995) for the Mediterranean region.

MVT BASIN STRATIGRAPHY

Several major lithostratigraphical and unconformity-bounded units were recognized; these are summarized in Fig. 4. They are organized into two main chronostratigraphically significant groups on the basis of sedimentological, chronostratigraphical and palaeoenvironmental characteristics, as well as previous work (Ambrosetti et al., 1987; Barberi et al., 1994; Bossio et al., 1998). For a more comprehensive discussion, the reader is referred to Mancini (2000).

Middle Pliocene–lowest Lower Pleistocene units

The MVT basin was filled by two cycles of marine and continental sedimentation during the early

Piacentian–Santernian (Ambrosetti et al., 1987; Bossio et al., 1998). The older cycle is middle–Late Pliocene in age (3.6–2.4 Ma), and corresponds to the Tenaglie-Fosso San Martino Formation (Fig. 3), where it crops out in the southernmost MVT, south of Mount Soratte, and to the Sabbie a Flabellipecten Formation near Orvieto to the north (Ambrosetti et al., 1987) (Fig. 3). It is a fully marine unit, more than 200 m thick, comprised of inner shelf, cross-bedded, fine to coarse-grained sand, with rare lenses of massive or cross-bedded pebbles. The sandy beds pass laterally into bioclastic calcarenite that includes *Amphistegina* spp. and onlaps the Mesozoic–Cenozoic substratum. The fauna consists of a rich macrofossil and foraminiferal assemblage with *Pecten flabelliformis*, *Chlamys latissima*, *Terebratula ampulla*, *Bulimina marginata* and *Globorotalia aemiliana*, which indicates a temperate-warm seawater environment. The Tenaglie-Fosso San Martino Formation is placed within the *G. punticulata* and *G. aemiliana* zones (MPl 4 and MPl 5 Mediterranean foraminiferal zones), and corresponds to the P2 and P3 units described in Bossio et al. (1998), spanning 1.2 Myr. The Tenaglie-Fosso San Martino Formation is commonly tilted to the east (Fig. 5a).

The younger cycle is widespread throughout the entire MVT, and unconformably overlies Triassic through to Miocene marine sedimentary rocks and the Tenaglie-Fosso San Martino Formation. This unconformity, known as the Acquatraversa erosional phase (Ambrosetti et al., 1987), is typically angular and considered to be of regional importance, as it has been recognized in almost all the Neogene–Quaternary basins of Latium (Buonasorte et al., 1991; Barberi et al., 1994). It partly corresponds to the Upper Pliocene *G. inflata* zone (MPl 6) and is thought to represent 0.2–0.3 Myr (Bossio et al., 1998).

Two laterally continuous units, the marine Chiani-Tevere Formation and the continental Poggio Mirteto Formation, constitute the second cycle (Fig. 4). A tephra layer exposed at Vallericca (Fig. 3), close to the base of the Chiani-Tevere Formation, has been dated at 2.1 ± 0.2 Ma by fission track analyses on volcanic glass shards (Arias et al., 1980). Magnetostratigraphical analysis of the same layer (Florindo & Sagnotti, 1995; Borzi et al., 1998) indicates that it was deposited during the C2r.1r polarity subchron of the Geomagnetic Polarity Time Scale, between 2.14 and 1.95 Ma (Cande & Kent, 1992). The occurrence of *Bulimina marginata* and *Globorotalia inflata* in the lowermost levels, and of *Bulimina etnea*, *Globigerina cariacoensis*, *Globigerina calabra* and *Globigerina* aff. *calida calida* in the upper part of the formation, indicates that these lithofacies were deposited during the late Gelasian and the Santernian before the Santernian–Emilian boundary, at 1.48 Ma (Pasini & Colalongo, 1997). Measurements of the $^{87}Sr/^{86}Sr$ ratio (Table 1) on molluscs and hydrozoa sampled in the upper part of the formation (Fig. 3) provided numerical values corresponding to 1.5–1.7 Ma

Table 1 Strontium isotope ($^{87}Sr/^{86}Sr$) ratios on molluscs and hydrozoa from the Chiani-Tevere Formation.

Sample	Specimen	Altitude (m a.s.l.)	Longitude E	Latitude N	Numerical value
a	Sinodia brocchii	102	12°38'36"	42°17'32"	0.709078 ± 6
b	Venus multilamella	80	12°27'13"	42°16'06"	0.709075 ± 3
c	Venus multilamella	80	12°27'13"	42°16'06"	0.709075 ± 5
d	Amyclina semistriata	80	12°27'13"	42°16'06"	0.709083 ± 2
e	Panopaea glyimeris	80	12°27'13"	42°16'06"	0.709082 ± 7
f	Callista italica	155	12°32'27"	42°15'21"	0.709073 ± 3
g	Cladocora coespitosa	188	12°27'38"	42°14'12"	0.709075 ± 2
h	Natica tigrina	90	12°25'12"	42°14'39"	0.709068 ± 7
i	Archimediella spirata	90	12°25'12"	42°14'39"	0.709080 ± 4
j	Natica tigrina	90	12°25'12"	42°14'39"	0.709073 ± 3
k	Amyclina semistriata	90	12°25'12"	42°14'39"	0.709075 ± 8
l	Archimediella spirata	90	12°25'12"	42°14'39"	0.709072 ± 5

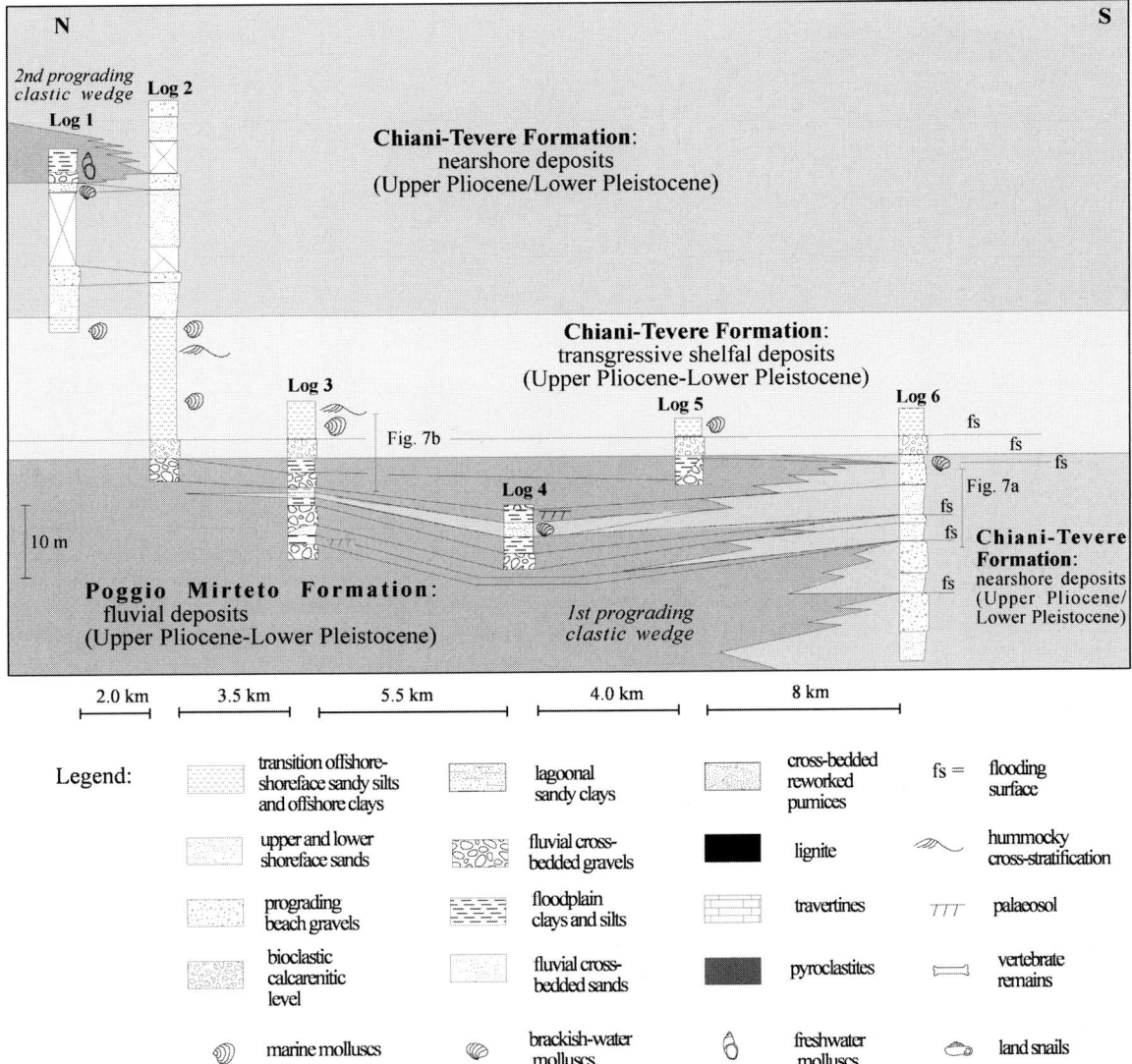

Fig. 6 Detailed stratigraphical and sedimentological correlations between the Chiani-Tevere and Poggio Mirteto formations, showing progradational fluvial–deltaic deposits alternating with transgressive marine shelfal deposits. The legend is common to Figs 6, 8 & 10. Location of logs is shown in Fig. 3.

(Mancini, 2000; see also Amorosi et al., 1998; Barbieri et al., 1998), in good agreement with the biostratigraphical data. Hence the second cycle represents a period of about 0.65 Myr.

Chiani-Tevere Formation

The Chiani-Tevere Formation is more than 350 m thick, and corresponds to the Q1 unit described in Bossio et al. (1998). The Chiani-Tevere Formation is composed of two main lithofacies associations (Figs 4 & 6).

Fine-grained shelf lithofacies. These lithofacies are the most widespread in the entire basin, occurring along both the western and eastern bank of the Tiber River. They are composed of outer to inner-shelf, massive or plane-bedded sandy clays

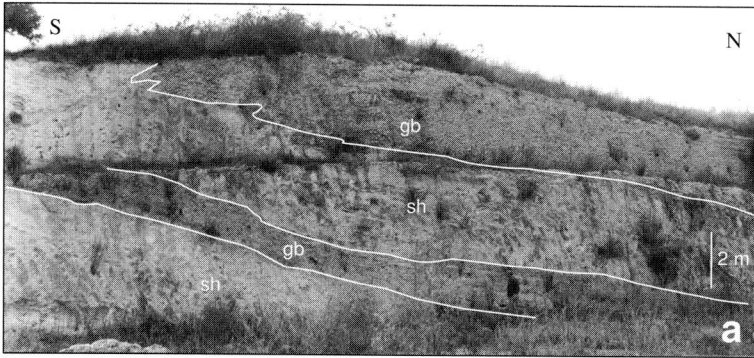

Fig. 7 (a) Two facies successions in nearshore deposits of the Chiani-Tevere Formation (Upper Pliocene–Lower Pleistocene), composed of shoreface sands (sh) alternating with beach gravels (gb). The outcrop corresponds to log 6 of Fig. 6. (b) Fluvial–deltaic deposits of the Poggio Mirteto Formation (PMF) (Upper Pliocene–Lower Pleistocene) overlain by transgressive marine deposits of the Chiani-Tevere Formation (CTF): (ts) transition shoreface–offshore sandy silts, (cl) calcarenitic bioclastic level, (fp) floodplain muds, (cg) cross-bedded fluvial gravels, (la) lagoonal sandy muds. The outcrop corresponds to log 3.

that pass laterally and vertically to plane- or cross-bedded silty sands, very rich in quartz and muscovite. Structures such as hummocky cross-stratification are frequent. These deposits contain a marine faunal assemblage that indicates temperate conditions, including *Bulimina elegans marginata*, *Bulimina etnea*, *Cladocora caespitosa*, *Sinodia brocchii* and *Panopaea glycimeris*. These fine-grained shelf deposits are laterally continuous southward to the 3rd order 'Monte Mario Sequence' (Milli, 1997), partly exposed in Rome.

Coarse-grained nearshore lithofacies. Coarse-grained nearshore deposits cover a 30 km² area and are located to the east of Mount Soratte and in the Orte-Amelia area. These deposits are composed of coarsening upward facies successions, *sensu* Walker (1990), with the thickness of each succession ranging from 3 to 10 m. The base is a planar surface that may correspond to a flooding surface (Fig. 6), and is covered by bioturbated, massive, lower shoreface silty sand, overlain by well-sorted, cross-bedded, upper shoreface coarse sand. The sands are interfingered with westward and south-westward clinostratified gravels (Fig. 7a), vertically passing to tabular beds of seaward imbricated, cobbles and pebbles. The clasts are calcareous and cherty, indicating a provenance from the Mesozoic–Cenozoic successions bordering the eastern margin of the basin. Such facies successions have been interpreted as progradational gravel beach sequences, *sensu* Massari & Parea (1988), and were deposited along wave-dominated palaeocoasts at the eastern edge of the basin in a microtidal regime. Gravel imbrication and clinoform orientations indicate progradation toward the centre of the basin, to the west-southwest. At the top of some gravel beach deposits, rare tabular beds of thinly laminated sandy clay have been found, commonly with an abundant oligohaline fauna indicative of a lagoonal environment (Fig. 6): *Cerastoderma glaucum*, *Potamides tricinctus*, *Bittium deshayesi*, Hydrobiidae and *Ammonia tepida*.

Most of the coarsening upward facies successions are stacked into three main sets, with thicknesses ranging between 60 and 100 m. Each set represents a clastic wedge that interfingers with shelf deposits to the west and the south (Figs 3 & 6). It is inferred that the oldest clastic wedge may span a time interval comprising the Gelasian–Santernian boundary, as it is overlain by shelf deposits bearing Santernian markers (*B. etnea*, *G. calabra*, *G.* aff. *calida calida*).

Poggio Mirteto Formation

The Poggio Mirteto Formation crops out along the south-eastern edge of the basin and is more than 300 m thick (Figs 4 & 6). It is composed of cross-stratified coarse gravels of fluvial origin, and arranged into multistorey–multilateral bodies several tens of metres thick. Lenses and planar sheets of trough-stratified silty sands, interspersed with gravels, are less frequent. Locally, the gravelly–sandy channel-related lithofacies are vertically and laterally interbedded with tabular or lenticular bodies of thinly laminated sandy silt and clay, rarely pedogenically-modified, but interpreted to represent floodplain deposits (Fig. 7b). The abundant fossil content is characterized by an assemblage that contains *Melanopsis affinis*, *Theodoxus groyanus*, *Belgrandia* spp. and *Prososthenia* spp., and by rare vertebrate remains such as *Equus stenonis*, typical of the earliest Early Pleistocene (Gliozzi *et al.*, 1997). The Poggio Mirteto Formation is laterally continuous to the east with coeval fluvial deposits filling the intermontane Tiberino and Rieti basins (Cavinato *et al.*, 1994; Barberi *et al.*, 1995), whereas seaward it is laterally continuous with the nearshore deposits of the Chiani-Tevere Formation.

Summary

Fluvial deposits and the nearshore clastic wedges of the second cycle within the MVT are interpreted to represent coarse-grained fluvial–deltaic systems prograding into a marine basin. The Poggio Mirteto Formation is considered to represent upper delta and alluvial plain environments, whereas nearshore deposits of the Chiani-Tevere Formation may represent the lower delta plain and delta front, and inner shelf sediments of the Chiani-Tevere Formation record periodic transgressions (Fig. 6). The early stage of each transgression is in some cases evidenced by bioclastic calcarenitic levels, up to 4 m thick, interpreted as transgressive lag deposits related to the shore retreat (Figs 6 & 7b). No evidence of significant incision has been observed at the boundaries between clastic wedges and transgressive deposits. The overall depositional architecture of the Chiani-Tevere and Poggio Mirteto formations is one of an aggradational fill, punctuated by major progradational and retrogradational episodes (Fig. 6).

Uppermost Lower Pleistocene–Holocene units

Marine sedimentation in the MVT ended at the late Santernian and was replaced by alternating periods of volcanic activity and continental sedimentation within an overall period of net erosion (Ambrosetti *et al.*, 1987; Cavinato *et al.*, 1994). The unconformity that separates Gelasian–Santernian units from younger strata is referred to as the Cassia erosional phase (Ambrosetti *et al.*, 1987), and is interpreted to indicate retreat of the sea followed by a period of non-deposition and erosion that spans at least 130 kyr, between the Santernian–Emilian boundary at *c.* 1.48 Ma, and the beginning of volcanic activity at *c.* 1.35 Ma (Barberi *et al.*, 1994).

Explosive volcanism developed in two distinct phases along the western margin of the MVT (Figs 3 & 4) (Barberi *et al.*, 1994). The first phase was limited to the Mount Cimino stratovolcano (Nicoletti, 1969; Borghetti *et al.*, 1981; Sollevanti, 1983), and produced mostly rhyodacitic ignimbrites that crop out along the central western bank of the Tiber. These ignimbrites exceed 100 m thick, form a wide plateau that covers the Chiani-Tevere Formation, and were emplaced at around 1.30 Ma (Barberi *et al.*, 1994). The second phase, which was derived from the Vulsini, Vico and Sabatini Mountains Volcanic Districts, started at 600 ka, peaked between 400 and 300 ka, and terminated *c.* 100 ka (Barberi *et al.*, 1994; Cavinato *et al.*, 1994). Key formations within this second phase include:

1 the Tufo giallo della Via Tiberina Formation (Mattias & Ventriglia, 1970), a phonolitic ignimbrite derived from the Sabatini Mountains Volcanic District with a radiometric age of 550 ± 10 ka (Cioni *et al.*, 1993);

2 the Civitella d'Agliano Formation (Aurisicchio *et al.*, 1992), another phonolitic ignimbrite that crops out in the north-western MVT, with a radiometric age of 505 ± 6 ka (Nappi *et al.*, 1995).

3 the Lago di Vico lava flow Formation (Perini *et al.*, 1997), a leucite-bearing phonolite that crops out in the central western MVT, and has a radiometric age range of 305 ± 9 to 258 ± 2 ka;

4 the Tufo rosso a scorie nere vicano Formation (Mattias & Ventriglia, 1970), a tephritic-phonolitic ignimbrite derived from the Vico strato-volcano with a radiometric age of 151 ± 3 ka (Laurenzi and Villa, 1987).

Latest Early Pleistocene–Holocene continental sedimentation was characterized by phases of alluvial aggradation alternating with periods of incision by an axial, as opposed to a transverse, fluvial system that is interpreted to represent the ancestral Tiber River. Chronological constraints on volcanic products, as described above, are critical in developing the stratigraphical framework for these deposits. Four gravel-dominated, unconformity-bounded units (Salvador, 1994) of fluvial origin can be differentiated flanking the present Tiber course (Figs 3 & 4), as described more fully below.

The Civita Castellana Unit

The Civita Castellana Unit unconformably overlies the Upper Pliocene–Lower Pleistocene Chiani-Tevere Formation (Fig. 5), and forms the first terraced deposit of the ancestral Tiber River. The altitude of the subplanar terrace tread decreases from 280 to 160 m a.s.l. from Orvieto to Civita Castellana, where the unit is buried below a thick Middle Pleistocene volcanic cover. The deeply incised palaeovalley, almost completely filled with fluvial sediments (Figs 5b & 8), borders the modern course of the Tiber up to Civita Castellana, beyond which it continues along the N–S direction, west of Mount Soratte (Fig. 3). The Civita Castellana Unit is 120 m thick, corresponds in part to the Palaeotiber gravels of Alvarez (1972), and was deposited during the Early to earliest Middle Pleistocene based on stratigraphical relations with the older Chiani-Tevere Formation and younger ignimbrites.

The Civita Castellana Unit is mainly composed of cross-bedded, clast-supported coarse gravels, organized into amalgamated, multistorey–multilateral channel form and tabular bodies (Fig. 9 a & b). Gh, Gt and Gp lithofacies, and CH and GB architectural elements (Miall, 1996) have been recognized. The largest channelized macroforms (CH elements) are up to 3 m deep and 10 m wide, and commonly filled with massive or cross-stratified gravels. Rare tabular bodies of trough cross-stratified silty sand, interpreted as SB elements (Miall, 1996), are interbedded with the gravels. Coarse clasts are mainly calcareous and cherty, with rare arenaceous clasts coming from the Miocene terrigenous successions exposed at Mount Peglia. The sandy matrix is abundant and rich in biotite and sanidine, as well as a few ignimbrite clasts (Brandi *et al.*, 1970) derived from the Mount Cimino Volcano. Gravels are commonly imbricated and show a prevalent N–S palaeocurrent direction. The overall lithofacies and the architectural style suggest that the Civita Castellana Unit was deposited by a gravel-bed braided river.

Fine-grained lithofacies occur interbedded with the channel-related deposits described above. These consist of tabular bodies of massive or thinly laminated pelite (the Fm and Fl facies of Miall, 1996), in some cases lignite-bearing or alternating with travertines and rare palaeosols, which are a few metres thick (Fig. 8). These fine-grained, organic and carbonate-rich lithofacies are interpreted to indicate local floodplain environments, perhaps small lakes or swamps (FF element). In the lower portion of this unit, these lithofacies contain a typical late Early Pleistocene freshwater fossil assemblage, with *Melanopsis affinis* and *Theodoxus groyanus*, as well as vertebrate remains such as *Bison* cf. *B. degiulii* (Gliozzi *et al.*, 1997), and land snails such as *Vallonia* spp. and Helicidae spp. associated with palaeosol horizons.

The basal unconformity and overlying deposits of the Civita Castellana Unit record a major cut-and-fill cycle, with deep incision followed by aggradation of coarse-grained sediments. Fine-grained lithofacies, travertines and palaeosols locally interrupt the vertical continuity of these coarser deposits, which indicate local changes in channel-belt position or orientation. After

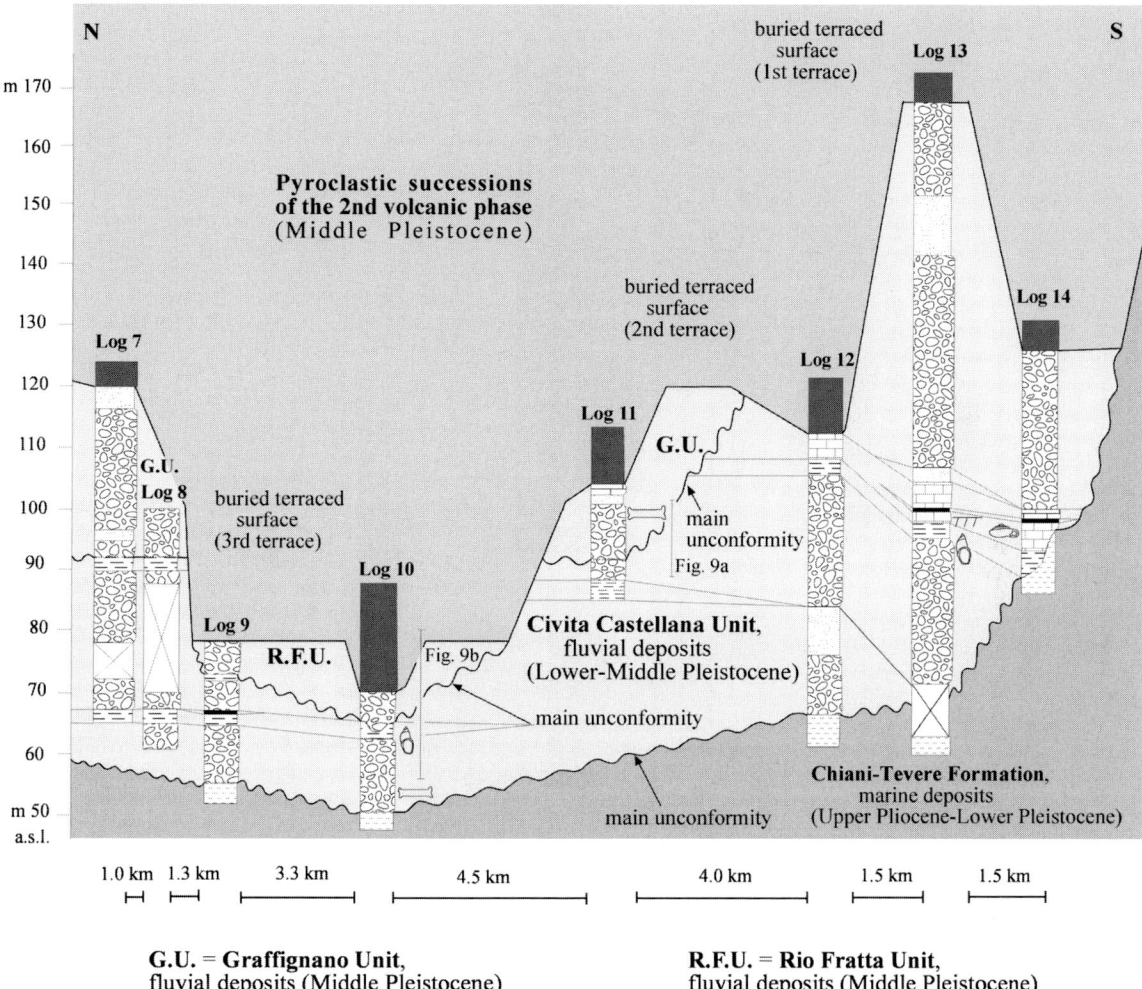

Fig. 8 Detailed stratigraphical and sedimentological correlations in the Civita Castellana area. The marine Chiani-Tevere Formation is deeply incised and unconformably overlain by gravelly fluvial terraced successions and volcanites, which represent the palaeovalley fill. Location of logs is shown in Fig. 3. Symbols as in Fig. 6.

deposition of floodplain-related facies, a new phase of gravel and sand aggradation occurred, without any evident deep incision of the underlying fine-grained deposits.

Although no direct stratigraphical relations between the Civita Castellana Unit and the Mount Cimino volcanites are evident, such units are considered contemporary in part because of radiometric and biochronological data. The top of the fluvial unit probably extends into the lower part of the Middle Pleistocene.

Graffignano Unit

The Graffignano Unit represents the second terraced fluvial deposit of the MVT (Figs 4, 8 & 10). Terrace tread altitudes decrease from 210 to 65 m a.s.l., along the tract between Orvieto and Poggio Mirteto, and underlying fluvial deposits are up to 50 m thick. This unit in part corresponds to the Obere Terrasse described by Vinken (1963), and is Middle Pleistocene in age because it is younger than the Tufo giallo della Via Tiberina and

Fig. 9 (a) Fluvial deposits of the Lower–Middle Pleistocene Civita Castellana Unit (CCU) overlain by the Middle Pleistocene Graffignano Unit (GU). The vertical passage between units is well evidenced by the abundance of dark ferromagnesian minerals in the upper formation: (cg) cross-bedded fluvial gravels, (cs) cross-bedded fluvial sands, (tr) travertines. The outcrop corresponds to log 11. (b) Civita Castellana Unit (CCU) unconformably overlain by the latest Middle Pleistocene Rio Fratta Unit (RFU) and the 'Tufo rosso a scorie nere vicano' Formation (TRF): (cg) cross-bedded fluvial gravels, (fp) floodplain muds, (ig) tephritic–phonolitic ignimbrite. The outcrop corresponds to log 10.

Civitella d'Agliano ignimbrites, but overlain by Lago di Vico lava flows.

The Graffignano Unit is composed of cross-stratified, calcareous, cherty and arenaceous gravels interbedded with tabular cross-laminated, silty sands rich in magnetite and clinopyroxene (Fig. 9a); lenses of reworked pumice are fairly frequent (Fig. 10). Facies Gh, Gt and St have been observed, and GB, SB, CH and, more rarely, LA elements are the main macroforms, which indicate an ancient braidplain environment with the local occurrence of laterally accreting bars.

At the top of the succession, in the northern MVT, channel-related, coarse-grained lithofacies are replaced by silty clay and fine volcanic sand deposited in a floodplain environment, and locally interbedded with travertines and palaeosols (Fig. 10). This vertical lithological change corresponds to two distinct fossil assemblages:

1 molluscs such as *Margaritifera auricularia*, *Corbicula fluminalis* and *Theodoxus isselii*, as well as reworked remains of *Elephas antiquus*, have been found in the coarse-grained deposits;

2 land snails such as *Chondrula tridens*, *Monacha cantiana* and *Granaria illirica*, as well as palustrine molluscs such as *Planorbis planorbis*, *Valvata piscinalis* and *Bithynia* spp., are abundant in the uppermost levels.

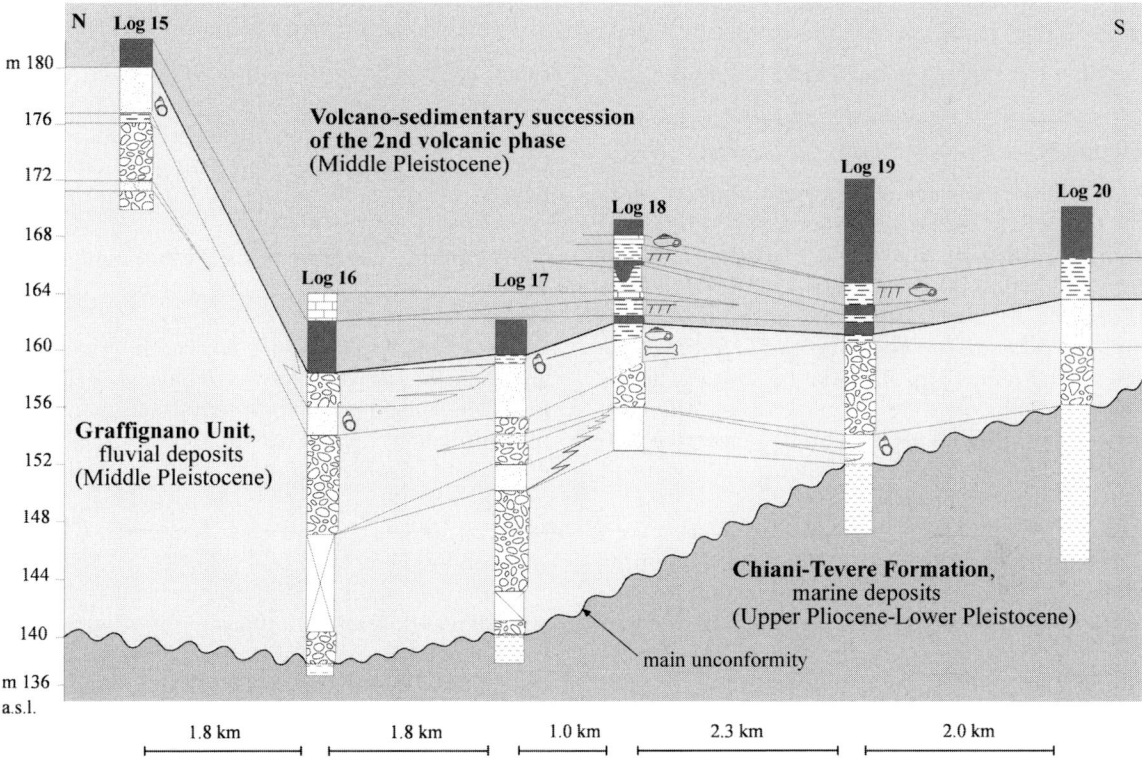

Fig. 10 Detailed stratigraphical and sedimentological correlations in the Graffignano Unit and Middle Pleistocene volcanic and volcano-sedimentary successions of the 2nd phase (see text for description). Location of logs is shown in Fig. 3. Symbols as in Fig. 6.

The Rio Fratta Unit

The Rio Fratta Unit represents the third terraced fluvial deposit of the MVT (Figs 3 & 5). Terrace tread altitudes decrease from 150 to 40 m a.s.l. along the N–S direction, and underlying fluvial deposits are up to 30 m thick. This unit in part corresponds to the Mittlere Terrasse of Vinken (1963), and is also Middle Pleistocene in age, as it is younger than the Lago di Vico lava flow Formation but older than the Tufo rosso a scorie nere vicano Formation (Fig. 9b).

The Rio Fratta Unit is composed of cross-stratified calcareous, cherty and arenaceous coarse gravels (Gt and Gp lithofacies), with abundant coarse sandy matrix almost entirely composed of ferromagnesian minerals. Lenses of cross-stratified coarse sands are more rare. Sedimentological data are interpreted to indicate a braidplain environment.

The Sipicciano Unit

The Sipicciano Unit represents the lowest and youngest terraced fluvial deposit of the MVT (Figs 3 & 5). Terrace tread altitudes decrease from 100 to 35 m a.s.l. in the N–S direction, and underlying fluvial deposits are up to 15 m thick. This unit corresponds to the Untere Terrasse of Vinken (1963), and is Late Pleistocene in age, as it is younger than the Tufo rosso a scorie nere vicano Formation, but older than the Holocene deposits described below (Fig. 9b). The Sipicciano Unit consists of cross-stratified coarse gravels and sands of braided-river origin, with many reworked fragments of the Tufo rosso a scorie nere vicano Formation.

Recent alluvial deposits

Recent deposits consist of fluvial sands and gravels, which constitute the modern floodplain, were probably deposited just after the last deglaciation, and are Holocene in age. The altitude of the floodplain decreases from 70 to 25 m a.s.l.

Summary

The Middle Pleistocene-Holocene alluvial units of the MVT represent four major cycles of significant degradation and aggradation that rest unconformably on the Chiani-Tevere Formation (Figs 5a & 8). Each cycle starts with fluvial incision into Chiani-Tevere strata, and is followed by lateral migration and aggradation, which leaves behind straths overlain by thick valley fills with a complex internal architecture. Successive episodes of incision have cut deeper into Pliocene to Early Pleistocene strata, dissected older straths, and eroded previously deposited valley fills, with resultant formation of a downward-stepping staircase of dissected terraces. Older terrace surfaces are commonly buried by volcanic material or travertines, and only the top surface of the Sipicciano Unit is always well preserved and uncovered.

TECTONICS, SEDIMENTATION AND PALAEOGEOGRAPHICAL EVOLUTION OF THE MVT

The basin fill of the MVT and the Plio-Pleistocene palaeogeographical evolution are closely linked to the long-term (> 1 Myr) trends of tectonically driven subsidence and uplift that affected the Latial region. Two main tectono-sedimentary phases have been recognized, as discussed more fully below.

Fluvial–deltaic sedimentation during the subsidence-dominated phase

The MVT basin was initiated with earliest Middle Pliocene activation of NW–SE and NE–SW trending fault systems, as a result of regional NE–SW extension. Just to the east, this same pattern of tectonic activity resulted in the formation of the intramontane Tiberino and Rieti basins (see Cavinato et al., 1994; Cavinato & De Celles, 1999). Subsidence controlled the early stage of the MVT basin-fill, and permitted marine transgression from the ancestral Tyrrhenian Sea to the western margin of the Mount Peglia-Lucretili Mountains ridge. At this time, the entire MVT basin would have been submerged, with the small horsts of Ferento, Mount Razzano, Cesano, Mount Soratte, Cornicolani Mountains remaining as emergent islands (Fig. 11). Pliocene to Early Pleistocene marine units of the MVT can be correlated with coeval marine deposits of the western and south-western bordering areas in northern Latium (Ambrosetti et al., 1987; Barberi et al., 1994; Bossio et al., 1998). By contrast, intermontane basins farther east are known to have experienced multiple episodes of continental sedimentation interrupted by erosional phases (see Abbazzi et al., 1997; Ambrosetti et al., 1987, 1995; Barberi et al., 1995; Basilici, 1997; Cavinato et al., 2000; Mancini, 2000), but no clear evidence of transgression has been found.

The two main sedimentary cycles of the subsidence-dominated phase, represented by the Tenaglie-Fosso San Martino Formation, and the Chiani-Tevere and Poggio Mirteto Formations, may be considered 3rd order depositional sequences on the basis of their time spans, thickness and lateral relationships with strata in the south-west bordering areas (Milli, 1997). Vertical continuity between the two sequences has been detected only beneath the Sabatini Mountains Volcanic District, from the 'Bracciano drilled hole' (Carboni & Palagi, 1998), and in the lower Tiber Valley, at Vallericca (Arias et al., 1990). Elsewhere in northern Latium, however, the two sequences are separated by the Acquatraversa erosional phase (Barberi et al., 1994). This significant and widespread unconformity is thought to reflect the well-known major glacio-eustatic fall at 2.6–2.4 Ma (Haq et al., 1987; Bossio et al., 1998). The angular nature of this unconformity may correlate to magmatic activity in the Ceriti-Manziate Volcanic District (De Rita et al., 1997) of north-western Latium (Fig. 11), which interrupted the overall long-term subsiding trend, and resulted in tilting of the Tenaglie-Fosso San Martino Formation.

Reactivation of NW–SE and NE–SW trending normal fault systems in the late Gelasian is inferred to have reinitiated the overall trend of subsidence, and provided accommodation for the Chiani-Tevere and Poggio Mirteto formations.

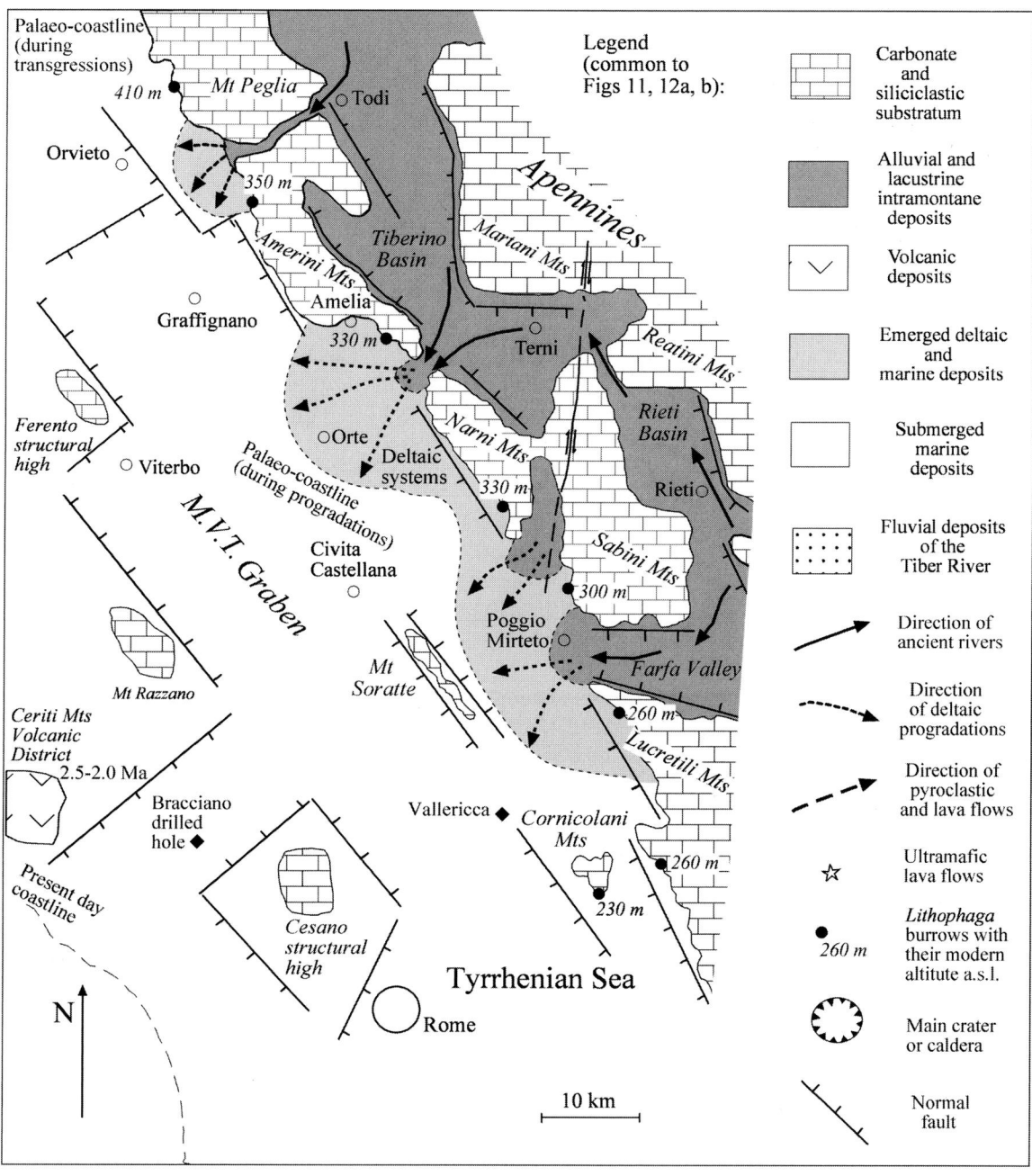

Fig. 11 Palaeogeographical reconstruction of the Latium Region during the Late Pliocene to earliest Early Pleistocene (2.2–1.5 Ma). Legend is common to Figs 11 & 12.

This major 3rd order sequence is interpreted to reflect transverse rivers that delivered sediments from the Apenninic chain and bordering intramontane basins to the east, and fed coarse-grained prograding fluvial–deltaic systems in the MVT basin. Internally, the Chiani-Tevere and Poggio Mirteto formations consist of alternating prograding clastic wedges and transgressive shelf deposits, which are interpreted to represent a 4th order cyclicity. Episodes of progradation proceeded as far as 15 km seaward, almost reaching the axis of the basin (Fig. 11), whereas transgressions extended eastward to the Mount Peglia-Lucretili Mountains ridge, based on the alignment of *Lithophaga* borings that indicate the position of the Gelasian–Santernian coastline (Girotti & Piccardi, 1994).

It is possible to envisage several combinations of causal mechanisms for the 4th order cyclicity evident in the Chiani-Tevere and Poggio Mirteto formations. One combination might hold eustatic sea-level and climate constant, and suggest that episodes of fluvial–deltaic progradation were related to ongoing uplift of the Apennines, with respect to coeval subsidence of the marine basin. Episodes of uplift may have resulted in increased production of coarse debris that was first delivered into the intramontane basins, and then into the eastern MVT. Transgressions, with deposition of shelf strata, would then be linked to time periods when tectonically forced sediment delivery was low, and subsidence resulted in continued relative sea-level rise.

A second scenario might hold tectonically forced sediment delivery and climate constant, and vary relative sea level. The inferred palaeotopographic setting would have been characterized by steeper upland gradients relative to a lower gradient shelf, and in this scenario would have promoted progradation and aggradation without any significant incision (*sensu* Emery & Myers, 1996; Blum & Törnqvist, 2000) during periods of relative sea-level fall (eustatic rates of fall > rates of subsidence). By contrast, relative sea-level rise (eustatic sea-level rise, or rates of subsidence > rates of eustatic fall) may have generated accommodation along the western margin of the Mount Peglia-Lucretili Mountains ridge.

A third combination might again hold tectonically forced sediment delivery constant, but considers how climatic and eustatic changes may have influenced progradational and transgressive phases. In the Mediterranean region, steppe-dominated glacial periods have been linked to enhanced sediment delivery from the uplands to the marine basin, at time periods that correspond to sea-level lowstands (Leeder *et al.*, 1998; Gawthorpe & Leeder, 2000). On the other hand, forest-dominated interglacial phases may have been characterized by decreased sediment supply at time periods that correspond to sea-level highstands (Leeder *et al.*, 1998). Thus the progradational episodes in the MVT may have recorded cold phases, whereas transgressions may be correlated to warm periods. Without precise time contraints, however, it remains impossible to correlate 4th order cycles of the MVT with the 40 kyr Pliocene and Early–Middle Pleistocene cycles of climate change. It is only possible to correlate the first prograding clastic wedge to the cooling event recorded at the Plio-Pleistocene boundary (Aguirre & Pasini, 1985; Bertoldi *et al.*, 1989; Bertini, 1995).

Finally, local activity along several faults inside the MVT graben, such as the Mount Soratte eastern bordering fault, may have forced progradation by adding accommodation along downthrown hangingwalls.

Fluvial sedimentation during the uplift-dominated phase

Uplift in the MVT that began at around 1.48 Ma, at the Santernian–Emilian boundary, is still ongoing (Karner *et al.*, 2001), and has affected the whole of northern Latium. The main palaeogeographical and sedimentary consequences are shown in Fig. 12A, and included emergence of the MVT, a rapid south-west and downward shift of the coastline and related facies, and development of the ancestral Tiber River axial drainage. Uplift was more pronounced in the northern region, as suggested by the N–S decrease in altitudes of the Santernian coastline (Fig. 11) and the southward decreasing altitudes of the Tiber fluvial terraces. Similarly, shift of the coastline some 30 km to the south-west and downward suggests greater uplift in the inner, axial, mountainous areas when compared with the western piedmont region. It is inferred that regional uplift was particularly relevant during the late Early Pleistocene. For example, the youngest Santernian coastline closest to Rome is at 230 m a.s.l. (Fig. 11) and formed at around 1.50 Ma. By contrast, the terrace surface for the oldest and highest post-Santernian coastal unit, the Monte Ciocci Formation as exposed in Rome (see Karner *et al.*, 2001), occurs at 70 m a.s.l. This surface marked the sea level during OIS 21, at

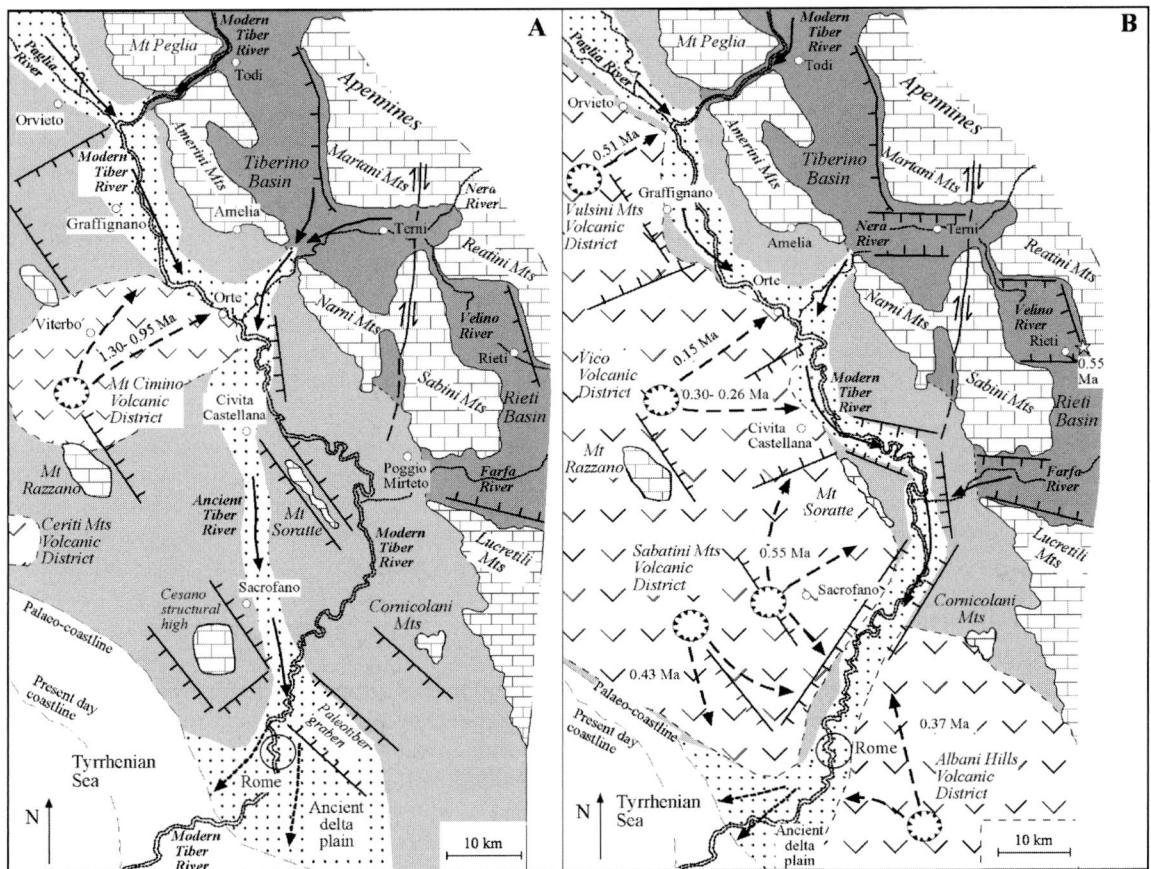

Fig. 12 Palaeogeographical reconstruction of the Latium Region: (A) during the latest Early Pleistocene to earliest Middle Pleistocene (1.5–0.6 Ma); (B) during the Middle–Late Pleistocene (0.6–0.1 Ma). Symbols as in Fig. 11.

0.85 Ma, close to the Early–Middle Pleistocene boundary (Karner et al., 2001), and 160 m altitudinal difference between these two units suggests rapid uplift over a period of 650 kyr.

Rapid uplift in the region represents a tectonically forced relative base-level fall of significant magnitude, and is interpreted to have initiated a long-term trend of fluvial incision into bedrock, as well as integration of the Tiber drainage. It seems probable that the juvenile Tiber River incised the Chiani-Tevere Formation and extended headward from its mouth in the area of Rome along the longitudinal axis of the MVT Graben west of Mount Soratte, being confined to the west by the uprising Mount Cimino Volcano and the Mount Razzano and Cesano horsts. The river filled this new palaeovalley with the Civita Castellana Unit, as well as correlative deltaic sands and gravels in Rome (Marra et al., 1998). Most of the coarse-grained sediments were derived from the Apennines and also from the 'cannibalized' Gelasian–Santernian fluvial–deltaic deposits exposed in the Orte-Amelia area.

A significant increase in extensional tectonism occurred around 600 ka, and is closely related to the beginning of the second volcanic phase (Fig. 12B). Localized areas of enhanced uplift were associated with upwelling of magma, and emission centres such as the eastern Vulsini Mountains, the Vico and the Sacrofano Centres, developed along or proximal to the western bordering faults of the graben. Large volumes of pyroclastic debris reached the western bank of the Tiber, causing a gradual eastward shift in its course between the towns of Orvieto and Civita Castellana. The NNE–SSW and WNW–ESE trending fault systems were activated for the first time, generating local tectonic depressions with axes that were transverse to the MVT Graben. In particular the tectonics acted with a contemporary twofold movement around the Mount Soratte: west of the ridge, the Sabatini Mountains Volcani District rose, whereas the northern and eastern bordering areas collapsed, causing deviation of the Tiber River from the ancient course to the modern. The same tectonic phase also affected the Rieti Basin, with the collapse of its central-northern part and the extrusion of ultramaphic lava flows at 550 ka (Cavinato et al., 1994). As volcanism ended in the early Late Pleistocene, the Umbrian-Latial region became tectonically less active but its pattern of uplift continued.

Inferences on the MVT terrace formation

Traditionally, aggradational phases have been associated with glacial periods, whereas phases of incision were assigned to interglacials, in the general context of uplift (Vinken, 1963). Moreover the MVT was considered too far inland to be influenced by sea-level change. Here, we suggest the combined effects of uplift, long-term (100 kyr) climatic cyclicity, and perhaps glacio-eustasy drove development of the staircase of terraces in the MVT.

The chronostratigraphical setting of the MVT suggests a 100 kyr recurrence for the last three alluvial units, which may be related to the late Middle Pleistocene–Holocene glacial–interglacial cycles. More specifically, the recent alluvial deposits correspond to the present interglacial, the Sipicciano Unit may be correlated to the penultimate interglacial and the last glacial, corresponding to marine oxygen isotope stages (OIS) 5 and 4–2, respectively, and the Rio Fratta Unit may have recorded the previous interglacial–glacial cycle (OIS 7–6). The Graffignano and Civita Castellana Units span longer intervals than a single glacial–interglacial cycle, most likely recording multiple stacked cut-and-fill cycles that progressively filled the palaeovalley. As very precise time constraints are lacking for the latest Early Pleistocene to earliest Middle Pleistocene, however, no singular glacial–interglacial cycle has been detected within these units.

As far as climatic influences are concerned, it is inferred that events of intense runoff and consequent sediment discharge may have occurred during each long-term (50 kyr) glacial and shorter-term (10–20 kyr) stadial, in a steppe-dominated palaeolandscape. Moreover it is probable that an increasing sediment supply was provided from uplands during deglacial periods, which led to aggradation downstream in the middle valley. Volcanism also supplied sediments to the river system, although the ashy pyroclastic deposits were better preserved at the top of every terrace. In fact most of the volcanic supply was presumably taken away by strong streams, during the main fluvial, gravel-dominated, aggradational phases.

Incision and terrace formation may be a result of the contemporary effect of high water-discharge versus low sediment-supply superimposed on constant uplift. The potential effect of glacio-eustasy should not be disregarded, however, as incision by knickpoint recession during sea-level lowstands may have affected the southernmost MVT. On the other hand, no sedimentological or palaeontological data have been found that indicate direct sea-level influence on the MVT.

By contrast, the lower valley of the Tiber, from Rome to the coast, is characterized by 100 kyr spaced, terraced, fluvial–deltaic deposits, interpreted as 4th order glacio-eustatically controlled depositional sequences (Milli, 1997). They progressively recorded several aggradational phases, developed during glacial–interglacial intervals. Furthermore, they are bounded by unconformities related to sea-level falls, and concomitant with interglacial–glacial transitions (Milli, 1997; Karner et al., 2001). As in the middle reaches, coastal terrace formation was concomitant with the regional uplift; however, the direct correlation between coastal and inner terraces is made difficult by the modern, wide, urban cover of Rome. The lowermost deposits underlying each coastal terrace correlate to late lowstand phases, which correspond to the period of deglaciation much farther upstream. Therefore the late lowstand coastal deposits may be correlated hypothetically to the coarse-grained deposits left in the MVT during the deglaciations.

On the whole, it is likely that the MVT terraces record a complex sedimentary and erosive response to competing external forcing mechanisms, such as climate, eustasy and uplift (*sensu* Vandemberge, 1995). The cut-and-fill cyclicity and the morphology of each terrace may not be solely the result of interglacial–glacial alternation. Although the climatic factor should be always considered as one of the dominant causes, uplift exerted a major role in driving long-term incision and formation of the succession of terraced deposits and landforms (see also Maddy *et al.*, 2001). Furthermore, the 100 kyr recurrent glacio-eustasy may have influenced the MVT terrace formation. More detailed chronological data, as well as detailed correlations throughout the entire catchment, are needed to fully understand linkages between landforms, deposits and interactions between the various forcing mechanisms.

CONCLUSIONS

The MVT provides a long, chronologically well-constrained and mostly continuous Late Pliocene–Quaternary record of fluvial–deltaic and fluvial sedimentation. Geochronological control is based on biostratigraphy and the Sr-isotope stratigraphy of interfingering marine deposits, and on radiometrically dated volcanics and volcaniclastics.

The overall stratigraphical record in the MVT extensional basin was affected by two main, long-term (> 1 Myr) tectono-sedimentary phases:
1 the earliest Middle Pliocene to the earliest Early Pleistocene was subsidence-dominated, which favoured net aggradation;
2 the late Early Pleistocene to the Holocene was uplift-dominated, and favoured formation of a flight of terraces within an overall period of net valley incision.

The MVT stratigraphical record places the major tectonic change at about 1.48 Ma (Santernian–Emilian boundary), which also helps constrain the timing of uplift and emergence of the Apennines.

The responses of fluvial and deltaic deposits to such a major tectonic change, and the related palaeogeographical settings, were very different. During the first tectono-sedimentary phase, the basin was filled with interfingering marine and continental deposits. The eastern margin of the basin, however, was fed by transverse rivers flowing east-northeast to west-southwest, and witnessed episodes of progradation and retrogradation by coarse-grained fluvial and deltaic systems. It is inferred that this progradational–retrogradational cyclicity is of 4th order, and may be related to eustatic and climatic forcing, within the context of active basin subsidence and uplift of the source terrain to the east-northeast. During this time, uplifting source regions supplied large volumes of coarse sediments, in particular during cold phases that were dominated by reduced vegetation cover. On the other hand, accommodation was being produced in front of the chain, farther to the west, also in concomitance with eustatic sea-level falls.

During the second tectono-sedimentary phase, the gravel-dominated fluvial deposits recorded the development of the ancestral to modern Tiber River. Several cut-and-fill cycles were syn-uplift and synvolcanic, and fluvial aggradation and incision led to the formation of a flight of terraces and underlying fills that border the ancient and the modern course of the Tiber. As most of these cycles appear to be 100 kyr apart, it seems reasonable to suggest that they may correlate to Late Quaternary glacial–interglacial cycles. In this scenario, aggradational phases may represent glacial periods, in particular periods of deglaciation, whereas periods of incision and terrace formation may be linked to the warmest interglacial maxima. It is not improbable, however, that dissection of MVT terraces may be linked in part to glacio-eustatic effects, especially in the southernmost MVT.

ACKNOWLEDGEMENTS

We would like to thank Professors Michael D. Blum, Frank J. Pazzaglia and Tammy Rittenour, whose useful comments and suggestions have greatly improved the manuscript. Special thanks to Professor Mario Barbieri for the isotope analyses, and in particular to Professor Odoardo Girotti for all the useful discussions and for having given the senior author the chance to join the 7th ICFS.

REFERENCES

Abbazzi, L., Albianelli, A., Ambrosetti, P., Argenti, P., Basilici, G., Bertini, A., Gentili, S., Masini, F.,

Napoleone, G. and Pontini, M.R. (1997) Paleontological and sedimentological record in Pliocene distal alluvial fan at Cava Toppetti (Todi, Central Italy). *Boll. Soc. Paleont. It.*, **36**, 5–22.

Aguirre, E. and Pasini, G. (1985) The Pliocene–Pleistocene Boundary. *Episodes*, **8**, 87–92.

Alfonsi, L., Funiciello, R., Girotti, O., Maiorani, A., Martinez, M.P., Mattei, M., Trudu, C. and Turi, B. (1991) Structural and geochemical features of the Sabina strike-slip fault. *Boll. Soc. Geol. It.*, **110**, 217–230.

Alvarez, W. (1972) The Treia Valley north of Rome: volcanic stratigraphy, topographic evolution and geological influences on human settlement. *Geol. Romana*, **11**, 153–176.

Ambrosetti, P., Basilici, G., Capasso Barbato, L., Carboni, M.G., Di Stefano, G., Esu, D., Gliozzi, E., Petronio, C., Sardella, R. and Squazzini, E. (1995) Il Pleistocene inferiore nel ramo sud-occidentale del Bacino Tiberino (Umbria): aspetti litostratigrafici e biostratigrafici. *Il Quaternario*, **8**, 19–36.

Ambrosetti, P., Carboni, M.G., Conti, M.A., Esu, D., Girotti, O., La Monica, G.B., Landini, B. and Parisi, G. (1987) Il Pliocene ed il Pleistocene inferiore del Bacino del Fiume Tevere nell'Umbria meridionale. *Geogr. Fis. Dinam. Quat.*, **10**, 10–33.

Amorosi, A., Barbieri, M., Castorina, F., Colalongo, M.L., Pasini, G. and Vaiani, S.C. (1998) Sedimentology, micropaleontology, and strontium-isotope stratigraphy of a lower–middle Pleistocene marine succession ('Argille azzurre') in the Romagna Apennines, northern Italy. *Boll. Soc. Geol. It.*, **117**, 789–809.

Amorosi, A., Colalongo, M.L., Fusco, F., Pasini, G. and Fiorini, F. (1999) Glacio-eustatic control of continental–shallow marine cyclicity from Late Quaternary deposits of the southeastern Po Plain, northern Italy. *Quat. Res.*, **52**, 1–13.

Arias, C., Azzaroli, A., Bigazzi, G. and Bonadonna, F.P. (1980) Magnetostratigraphy and Pliocene–Pleistocene boundary in Italy. *Quat. Res.*, **13**, 65–74.

Arias, C., Bigazzi, G., Bonadonna, F.P., Iaccarino, S., Urban, B., Dal Molin, M., Dal Monte, L. and Martolini, M. (1990) Vallericca Late Neogene stratigraphy (Lazio Region, Central Italy). *Paleobiol. Cont.*, **17**, 61–80.

Aurisicchio, C., Nappi, G., Renzulli, A. and Santi, P. (1992) Mineral chemistry, glass composition and magma fractionation of the welded ignimbrite in the 'Civitella d'Agliano Formation', Vulsini Volcanic District (Italy). *Mineral. Petrogr. Acta*, **35**, 157–182.

Bally, A.W., Burbi, C., Cooper, C. and Ghelardoni, R. (1986) Balanced cross sections and seismic reflection profiles across the Central Apennines. *Mem. Soc. Geol. It.*, **35**, 257–310.

Barberi, F., Buonasorte, G., Cioni, R., Fiordelisi, A., Foresi, L., Iaccarino, S., Laurenzi, M.A., Sbrana, A., Vernia, L. and Villa, I.M. (1994) Plio-Pleistocene geological evolution of the geothermal area of Tuscany and Latium. *Mem. Descr. Carta Geol. It.*, **49**, 77–134.

Barberi, R., Cavinato, G.P., Gliozzi, E. and Mazzini, I. (1995) Late Pliocene–Early Pleistocene palaeoenvironmental evolution of the Rieti Basin (Central Apennines). *Il Quaternario*, **8**, 515–534.

Barbieri, M., Castorina, F., Colalongo, M.L., Pasini, G. and Vaiani, S.C. (1998) Worldwide correlation of the Pliocene–Pleistocene GSSP at Vrica (Southern Italy) confirmed by strontium isotope stratigraphy. *Newsl. Stratigr.*, **36**, 177–187.

Basilici, G. (1997) Sedimentary facies in an extensional depositional system: the Pliocene Tiberino Basin, Central Italy. *Sediment. Geol.*, **109**, 73–94.

Bertini, A. (1995) La vegetation et le climat en Italie centro-septentrionale entre 7 et 1.6 Millions d'annees (du Messinien au Pleistocene Inferieur). In: *La Mediterranee: variabilites climatiques, environnement et biodiversitè, Actes du Colloque Scientifique*, 6–9 April, Maison de l'Environment de Montpellier, 91–97.

Bertoldi, R., Rio, D. and Thunell, R. (1989) Pliocene–Pleistocene vegetational and climatic evolution of the south-central Mediterranean. *Palaeogeogr., Palaeoclimatol., Palaeoecol.*, **72**, 263–275.

Bigi, G., Cosentino, D., Parotto, M., Sartori, R. and Scandone, P. (1992) *Structural Model of Italy*, scale 1 : 500 000, 6 sheets. *CNR, Quaderni de La Ricerca Scientifica*, **114**(3).

Blum, M.D. and Törnqvist, T.E. (2000) Fluvial responses to climate and sea-level change: a review and look forward. *Sedimentology*, **47**, 2–48.

Borghetti, G., Sbrana, A. and Sollevanti, F. (1981) Vulcano-tettonica dell'area dei Monti Cimini e rapporti cronologici tra vulcanismo Cimino e Vicano. *Rend. Soc. Geol. It.*, **4**, 253–254.

Borzi, M., Carboni, M.G., Cilento, G., Di Bella, L., Florindo, F., Girotti, O., Piccardi, E. and Sagnotti, L. (1998) Bio- and Magneto-stratigraphy in the Tiber Valley revised. *Quat. Int.*, **47/48**, 65–72.

Bossio, A., Costantini, A., Foresi, L.M., Lazzarotto, A., Mazzanti, R., Mazzei, R., Pascucci, V., Salvatorini, G., Sandrelli, F. and Terzuoli, A. (1998) Neogene–Quaternary sedimentary evolution in the western side of the northern Appennines (Italy). *Mem. Soc. Geol. It.*, **52**, 513–525.

Bradley, R.S. (1999) *Paleoclimatology*, 2nd edn. Academic Press, San Diego, 613 pp.

Brandi, G.P., Cerrina Feroni, A., Decandia, F.A., Giannelli, L., Monteforti, B. and Salvatorini, G. (1970) Il Pliocene del Bacino del Tevere fra Celleno (Terni) e Civita Castellana (Viterbo). Stratigrafia ed evoluzione tettonica. *Atti Soc. Toscana Sci. Nat.*, **77**, 308–326.

Buonasorte, G., Carboni, M.G. and Conti, M.A. (1991) Il substrato plio-pleistocenico delle vulcaniti sabatine: considerazioni stratigrafiche e paleoambientali. *Boll. Soc. Geol. It.*, **110**, 35–40.

Buonasorte, G., Fiordelisi, A., Pandeli, E., Rossi, U. and Sollevanti, F. (1987) Stratigraphic correlations and structural setting of the pre-neoautochtonous sedimentary sequences of Northern Latium. *Per. Mineral.*, **56**, 111–122.

Cande, S.C. and Kent, D.V. (1992) A new geomagnetic polarity time scale for the Late Cretaceous and Cenozoic. *J. Geophys. Res.*, **97**, 13917–13951.

Carboni, M.G. and Palagi, I. (1998) Biostratigrafia del sottosuolo plio-pleistocenico a Sud del Lago di Bracciano: il sondaggio Sabatini 9. *Il Quaternario*, **11**, 107–114.

Cavinato, G.P., Cosentino, D., De Rita, D., Funiciello, R. and Parotto, M. (1994) Tectono-sedimentary evolution of intrapenninic basins and correlation with the volcano-tectonic activity in Central Italy. *Mem. Descr. Carta Geol. It.*, **49**, 63–76.

Cavinato, G.P. and De Celles, P.G. (1999) Extensional basins in the tectonically bimodal central apennines fold-thrust belt, Italy: response to corner flow above a subducting slab in retrograde motion. *Geology*, **27**, 955–958.

Cavinato, G.P., Gliozzi, E. and Mazzini, I. (2000) Two lacustrine episodes during the late Pliocene–Holocene evolution of the Rieti Basin (Central Apennines, Italy). In: *Lake basins through space and time* (Eds E.H. Gierlowski-Kordesch and K.R. Kelts). *Am. Assoc. Petrol. Geol. Stud. Geol.*, **46**, 527–534.

Channell, J.E.T., Di Stefano, E. and Sproveri, R. (1992) Calcareous plankton biostratigraphy, magnetostratigraphy and paleoclimatic history of the Plio-Pleistocene Monte S. Nicola section (Southern Sicily). *Boll. Soc. Paleont. Ital.*, **31**, 351–382.

Cioni, R., Laurenzi, M., Sbrana, A. and Villa, I.M. (1993) $^{40}Ar/^{39}Ar$ chronostratigraphy of the initial activity in the Sabatini Volcanic Complex. *Boll. Soc. Geol. It.*, **112**, 251–263.

Cipollari, P., Cosentino, D. and Gliozzi, E. (1998) Extension- and compression related basins in central Italy during the Messinian Lago-Mare event. *Tectonophysics*, **315**, 163–185.

Cita, M.B. (1975) Planktonic foraminiferal biozonation of the Mediterranean Pliocene deep sea record. A revision. *Riv. Ital. Paleontol. Stratigr.*, **81**, 527–544.

Cita, M.B. and Castradori, D. (1995) Rapporto sul workshop 'Marine sections from the Gulf of Taranto (southern Italy) usable as potential stratotypes for the GSSP of the Lower, Middle and Upper Pleistocene' (29 Settembre–4 Ottobre 1994). *Boll. Soc. Geol. It.*, **114**, 319–336.

Combourieu-Nebout, N. (1993) Vegetational response to Upper Pliocene glacial/interglacial cyclicity in the central Mediterranean. *Quat. Res.*, **40**, 228–236.

De Rita, D., Di Filippo, M. and Sposato, A. (1993) Carta Geologica del Complesso Vulcanico sabatino. In: *Sabatini Volcanic Complex* (Ed. M. Di Filippo), CNR Quaderni de La Ricerca Scientifica 114, Progetto Finalizzato: Geodinamica, monografie finali, **11**, Rome.

De Rita, D., Bertagnini, A., Faccenna, C., Landi, P., Rosa, C., Zarlenga, F., Di Filippo, M. and Carboni, M.G. (1997) Evoluzione geopetrografica-strutturale dell'area tolfetana. *Boll. Soc. Geol. It.*, **116**, 143–175.

Doglioni, C. (1991) A proposal for the kinematic modelling of W-dipping subduction—possible applications to the Tyrrhenian–Apennines system. *Terra Nova*, **3**, 423–434.

Emery, D. and Myers, K.J. (1996) *Sequence Stratigraphy*. Blackwell Science, Oxford, 295 pp.

Florindo, F. and Sagnotti, L. (1995) Paleomagnetism and rock magnetism in the upper Pliocene Valle Ricca (Rome, Italy) section. *J. Geophys. Int.*, **123**, 340–354.

Follieri, M., Giardini, M., Magri, D. and Sadori, L. (1998) Palynostratigraphy of the last glacial period in the volcanic region of Central Italy. *Quat. Int.*, **47–48**, 3–20.

Follieri, M., Magri, D. and Sadori, L. (1988) 250,000-year pollen record from Valle di Castiglione (Roma). *Pollen Spores*, **30**, 329–356.

Funiciello, R., Parotto, M. and Praturlon, A. (1981) Carta Tettonica d'Italia. *CNR Progetto Finalizzato: Geodinamica*, publ. **269**, Rome, Italy.

Gawthorpe, R.L. and Leeder, M.R. (2000) Tectono-sedimentary evolution of active extensional basins. *Basin Res.*, **12**, 195–218.

Girotti, O. and Piccardi, E. (1994) Linee di riva del Pleistocene inferiore sul versante sinistro della media Valle del Fiume Tevere. *Il Quaternario*, **7**, 525–536.

Gliozzi, E., Abbazzi, L., Argenti, P., Azzaroli, A., Caloi, L., Capasso Barbato, L., Di Stefano, G., Esu, D., Ficcarelli, G., Girotti, O., Kotsakis, T., Masini, F., Mazza, P., Mezzabotta, C., Palombo, M.R., Petronio, C., Rook, L., Sala, B., Sardella, R., Zanalda, E. and Torre, D. (1997) Biochronology of selected mammals, molluscs and ostracods from Middle Pliocene to the Late Pleistocene in Italy. The state of the art. *Riv. It. Paleont. Strat.*, **103**, 369–388.

Haq, B.U., Hardenbol, J. and Vail, P.R. (1987) Chronology of the fluctuating sea levels since the Triassic. *Science*, **235**, 1156–1167.

Iaccarino, S. (1985) Mediterranean Miocene and Pliocene planktic foraminifera. In: *Plankton Stratigraphy* (Ed. H.M. Bolli), pp. 283–314. Cambridge University Press, Cambridge.

Imbrie, J. (1985) A theoretical framework for the Pleistocene ice ages. *J. Geol. Soc. Lond.*, **142**, 417–432.

Karner, D.B., Marra, F., Florindo, F. and Boschi, E. (2001) Pulsed uplift estimated from terrace elevations in the coast of Rome: evidence for a new phase of volcanic activity? *Earth Planet. Sci. Lett.*, **188**, 135–148.

Keller, J.V.A., Minelli, G. and Pialli, G. (1994) Anatomy of late orogenic extension: the Northern Apenninic case. *Tectonophysics*, **238**, 275–294.

Laurenzi, M.A. and Villa, I.M. (1987) $^{40}Ar/^{39}Ar$ chronostratigraphy of Vico ignimbrites. *Per. Mineral.*, **56**, 285–293.

Leeder, M.R., Tracey, H. and Kirkby, M.J. (1998) Sediment supply and climate change: implications for basin stratigraphy. *Basin Res.*, **10**, 7–18.

Maddy, D., Bridgland, D. and Westaway, R. (2001) Uplift-driven valley incision and climate-controlled river terrace development in the Thames Valley, UK. *Quat. Int.*, **79**, 23–36.

Malinverno, A. and Ryan, W.B. (1986) Extension in the Tyrrhenian Sea and shortening in the Apennines as results of arc migration driven by sinking of the lithosphere. *Tectonics*, **18**, 108–118.

Mancini, M. (2000) Stratigrafia dei depositi fluviali e costieri pleistocenici nella Media Valle del Tevere. Unpublished PhD thesis, XII cycle, Università di Roma 'La Sapienza', Dipartimento di Scienze della Terra, Rome, 209 pp.

Mancini, M., Girotti, O. and Cavinato, G.P. (2001) *Carta Geologica della Media Valle del Tevere*. Università di Roma 'La Sapienza', Dipartimento di Scienze della Terra, CNR CSQEA, Autorità di Bacino del Fiume Tevere, Rome.

Marra, F., Rosa, C., De Rita, D. and Funiciello, R. (1998) Stratigraphic and tectonic features of the Middle Pleistocene sedimentary and volcanic deposits in the area of Rome (Italy). *Quat. Int.*, **47–48**, 51–63.

Martini, I.P. and Sagri, M. (1993) Tectono-sedimentary characteristics of Late Miocene–Quaternary extensional basin of the northern Apennines. *Earth Sci. Rev.*, **34**, 197–233.

Massari, F. and Parea, G.C. (1988) Progradational gravel beach sequences in a moderate- to high-energy, microtidal environment. *Sedimentology*, **35**, 881–913.

Mattias, P.P. and Ventriglia, U. (1970) La regione vulcanica dei monti Sabatini e Cimini. *Mem. Soc. Geol. It.*, **9**, 331–384.

Miall, A.D. (1996) *The Geology of Fluvial Deposits: Sedimentary Facies, Basin Analysis and Petroleum Geology*. Springer-Verlag, Berlin, 582 pp.

Miall, A.D. (2000) *Principles of Sedimentary Basin Analysis*, 3rd edn (updated and enlarged). Springer-Verlag, Berlin, 616 pp.

Milli, S. (1997) Depositional setting and high-frequency sequence stratigraphy of the Middle–Upper Pleistocene to Holocene deposits of the Roman Basin. *Geol. Romana*, 33, 99–136.

Nappi, G., Renzulli, A., Santi, P. and Gillot, P.Y. (1995) Geological evolution and geochronology of the Vulsini Volcanic District (Central Italy). *Boll. Soc. Geol. It.*, **114**, 599–613.

Nicoletti, M. (1969) Datazioni argon-potassio di alcune vulcaniti delle regioni vulcaniche Cimina e Vicana. *Per. Mineral.*, **38**, 1–20.

Pasini, G. and Colalongo, M.L. (1997) The Pleistocene boundary stratotype at Vrica, Italy. In: *The Pleistocene Boundary and Beginning of the Quaternary* (Ed. J.A. Van Couvering), pp. 15–45. Cambridge University Press, Cambridge.

Patacca, E., Sartori, R. and Scandone, P. (1990) Tyrrhenian Basin and Apenninic Arcs: kinematics relations since Late Tortonian times. *Mem. Soc. Geol. It.*, **45**, 425–451.

Perini, G., Conticelli, S. and Francalanci, L. (1997) Inferences of the volcanic history of the Vico volcano, Roman Magmatic Province, Central Italy: stratigraphical, petrographic and geochemical data. *Mineral. Petrogr. Acta*, **40**, 67–93.

Raymo, M.E., Ruddiman, W.F., Backman, J., Clement, B.M. and Martinson, D.G. (1989) Late Pliocene variation in northern hemisphere ice sheets and North Atlantic deep water circulation. *Paleoceanography*, **4**, 413–446.

Ricci Lucchi, F. (1986) The Oligocene to recent foreland basins of the northern Apennines. In: *Foreland Basins* (Eds P.A. Allen and P. Homewood). *Spec. Publ. Int. Assoc. Sedimentol.*, **8**, 105–139.

Rio, D., Sprovieri, R. and Di Stefano, E. (1994) The Gelasian Stage: a proposal of a new chronostratigraphic unit of the Pliocene Series. *Riv. It. Paleont. Strat.*, **100**, 103–124.

Salvador, A. (1994) *International Stratigraphic Guide*, 2nd edn. International Union of Geological Sciences, Trondheim, and Geological Society of America, Boulder, 214 pp.

Sollevanti, F. (1983) Geologic, volcanologic and tectonic setting of the Vico-Cimino area, Italy. *J. Volcanol. Geotherm. Res.*, **17**, 203–217.

Sprovieri, R. (1993) Pliocene–Early Pleistocene astronomically forced planktonic foraminifera abundance fluctuations and chronology of Mediterranean calcareous plankton bio-events. *Riv. It. Paleont. Strat.*, **99**, 371–414.

Turner, C. (1998) Volcanic maars, long Quaternary sequences and the work of the INQUA Subcommission on European Quaternary Stratigraphy. *Quat. Int.*, **47–48**, 41–49.

Vandemberghe, J. (1995) Timescales, climate and river development. *Quat. Sci. Rev.*, **14**, 631–638.

Vinken, R. (1963) Uber das Altpleistozan und die Flussterrassen im Gebiet des mittleren Tibers. *Eiszeitalter Ggw.*, **14**, 35–52.

Walker, R.G. (1990) Facies modeling and sequence stratigraphy. *J. Sediment. Petrol.*, **60**, 777–786.

Pre-Quaternary fluvial systems

Transport modes and grain-size patterns in fluvial basins

PETER F. FRIEND* and W. BRIAN DADE†

*Department of Earth Sciences, University of Cambridge, Cambridge, CB2 3EQ, UK
(Email: pff1000@esc.cam.ac.uk); and
†Department of Earth Sciences, Fairchild Hall, Dartmouth College, Hanover, NH 03755, USA

ABSTRACT

Fluvial transport modes have been defined using arbitrary quantitative limits to separate bedload, mixed-load and suspended-load dominant modes. If estimates of bed grain-size, flood depth and flood slope are available for particular reaches of present-day rivers, this makes it possible to allocate a dominant transport mode to each reach. These fluvial transport modes represent equilibrium flow-transport states, separated by relatively abrupt transitions. Examination of sediment transport relationships using a data set from present-day rivers shows that mixed-load rivers are less efficient than bedload rivers, where efficiency is defined in terms of the flow strength needed to transport sediment load of a given weight. This may be a result of the more important presence in mixed-load rivers of dissipative features, particularly sinuous channel bends and sand bedforms (ripples and dunes). Some further aspects of the transport mode approach are examined in the context of two very different kinds of data sets from fluvial basins.

The present-day, intensively studied, Allt Dubhaig river, Highland Scotland, exhibits an abrupt downstream fining from gravel, via a few hundred metres of bimodal gravel and sand, to sand with bedforms. This downstream change corresponds to the transition from bedload to mixed-load dominant transport mode. Within the bedload dominant gravel reaches, Shields parameter and dimensionless width vary around constant values, and the downstream profile decreases in elevation, as predicted. The abrupt downstream transition in mode appears to reflect exhaustion of gravel supply, relative to subsidence and sand accumulation, owing to the raising of downstream base level, in this case produced partly by construction of a local dam.

A margin of the Old Red Sandstone (Devonian) basin in East Greenland has been examined using patterns in the fluvial sedimentary structures to identify and characterise the fluvial systems. For these ancient rocks, direct flow measurements are not available, but the transport mode approach still provides important insights. Abrupt downstream transitions from gravel to sand appear to represent changes in fluvial transport modes and made the toes of the basin-margin fans. Exhaustion of gravel downstream appears to have determined the build-out distance of the gravels, as with the Allt Dubhaig, and the decrease of slope produced by general filling of the basin centre by axial rivers may also have been important. Decrease of stream power owing to divergence of channels on the marginal fans, as well as soak-in and evaporation of river water, may also have limited the downstream extent to which the bedload dominant mode existed. In both cases, the transport mode approach provides valuable understanding of the basin-filling systems.

INTRODUCTION

Careful studies of the bed sediments of active rivers tend to show considerable variation in grain-size and morphological features. Changing flood patterns also add to this complexity. Nevertheless, many workers have generalized the grain-size in particular river reaches, and these generalizations provide an essential basis for attempts to reconstruct the properties of ancient rivers that are now largely represented by their deposits. In this paper, recent work based on a generalization of sediment transport theory is reviewed, along with published general data from present-day rivers. The model emerging provides valuable insights into the way that river flows transport and deposit sediment of different grain-sizes. One remarkable insight is that, in terms of the weight of sediment being carried, flows that are transporting sand are less efficient than those that are transporting gravel. Possible reasons for this are discussed below.

Pioneering studies by Schumm (1963, 1968a,b) established by direct measurement that the cross-sectional form of many river channels is correlated with the grain-size of their beds and banks. He then interpreted this discovery in terms of three river transport modes dominated by bed-load, mixed-load or suspended-load sediment flux. It has been difficult to apply this classification of transport modes objectively to present-day rivers. This is particularly because of the practical problems of measuring the proportion of bedload under the flood conditions that largely determine the morphology and evolution of channels. Recent reconsideration of the physics of alluvial sediment transport (Dade & Friend, 1998; Dade, 2000), however, has provided a framework, using arbitrarily defined limits, that allows classification by sediment transport mode, provided that flood hydraulic and bed grain-size measurements are available. To be more specific, it is now possible, provided that values of flood depth, slope and bed grain-size are generalized, to determine whether a reach of a present-day river is likely to have been characterized by flood transport dominated by bedload, mixed load or suspended load. This paper examines some further implications of recognizing these three modes of transport, and outlines how this approach can be applied to the stratigraphy of a fluvial basin.

TRANSITIONS BETWEEN DISTINCTIVE BEDLOAD, MIXED-LOAD AND SUSPENDED-LOAD DOMINATED MODES

The data set

Various published data sets on present-day rivers have provided a basis for testing the above approach (Dade & Friend, 1998; Dade, 2000). Some idea of the range and distribution of data used in this work is provided as a scatter diagram of flood depth and grain size in Fig. 1; sources of these data are listed in the caption. Reaches of pebble to cobble grade predominate, sand bed river reaches provide a relatively less numerous second group of reach data, and there are only a small number of reaches with coarse sand to granule bed material or with mud or silt-grade bed material.

In alluvial sediments world-wide, the lack of grain-size modes in the range from very coarse sand to granule grades (1–4 mm, or 0 to −2 φ) has been discussed by many authors, for example, Pettijohn (1957, p. 47) and Russell (1968). Pettijohn believed that part of the reason for this lack is that this grain-size range tends to be under-represented in the products of river erosion and weathering. Russell, in contrast, felt that it is due to 'hydraulic sorting' reflecting the ease with which particles of this size range are entrained and transported out of the river systems. Whatever the reason, the sand–gravel 'gap' is clear in this data set.

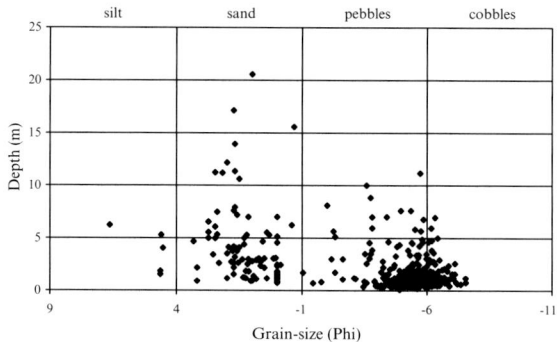

Fig. 1 Grain-size and depth of more than 400 river reaches analysed by Dade (2000), using data on alluvial rivers compiled by Schumm (1968), Chitale (1970), Church & Rood (1983) and Van der Berg (1995).

Arbitrary limits

Arbitrary limits for the definition of the different transport modes are based on the use of a dimensionless, Rouse-like parameter, as described by Dade & Friend (1998). This parameter is defined as w_s/u_*, where w_s is the settling velocity of bed particles of average size, and u_* is the friction velocity of the flood flow, calculated as $(\tau_0/\rho)^{0.5}$. This parameter (w_s/u_*) is described as the 'settling ratio'. The arbitrary limits of the settling ratio used are as follows (Fig. 2):

1 $w_s/u_* < 0.3$ defines the suspended-load dominant mode;
2 $0.3 < w_s/u_* < 3$ defines the mixed-load dominant mode;
3 $w_s/u_* < 3$ defines the bedload dominant mode.

The limits were chosen so that, in suspended-load dominant transport, < 10–20% of the total load is likely to be bedload. In bedload dominant transport more than 80–90% of the total load is likely to be bedload. Further details of the simplified sediment transport and concentration relationships used are provided by Dade & Friend (1998).

In the data set, all of the bedload dominant river reaches have average grain-sizes of pebble or cobble grade. The mixed-load dominant reaches range from cobble, through pebble to sand-grade, and the few suspended-load dominant reaches have either sand or silt-grade beds. It is clear that the data set includes too few examples of suspended-load dominated river reaches to allow a useful examination of this class.

TRANSPORT EFFICIENCY DISTINCTION BETWEEN MODES

A conventional way of assessing the ability of a river flow to move its bed sediment is the Shields plot (Dade & Friend, 1998, Fig. 2). In such a plot (Fig. 3), the Shields parameter ($\theta = sh/Rd$) is plotted against particle Reynolds number ($Re_* =$

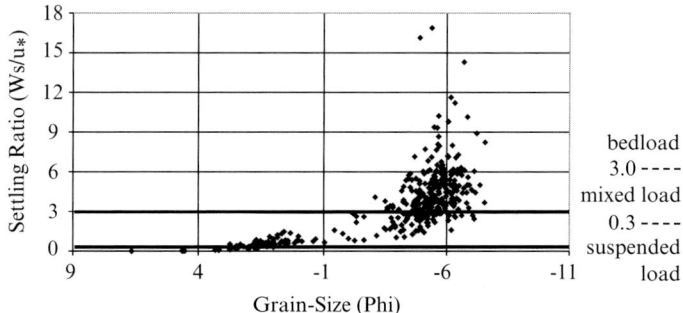

Fig. 2 Grain-size and a settling ratio (w_s/u_*) for the same data set as used in Fig. 1.

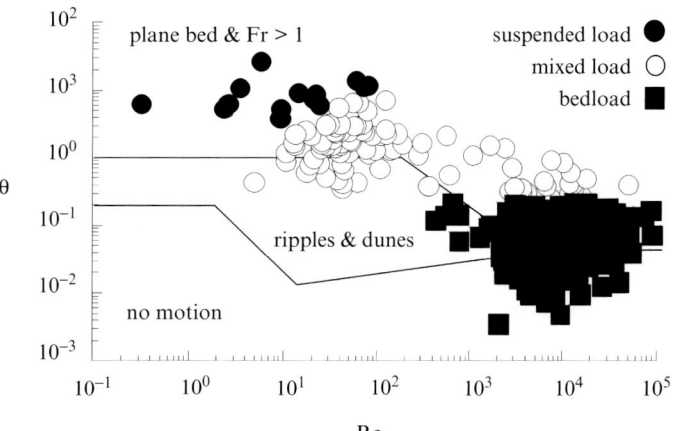

Fig. 3 Shields plot for the same data set as Fig. 1. The Shields parameter ($\theta = sh/Rd$) for each reach is plotted against the grain Reynolds number ($Re_* = u_*d/\upsilon$).

($u_*d)/\upsilon$) for the present-day alluvial reach data set. The Shields parameter provides a dimensionless measure of the strength of river flow relative to the size and density of a typical bed particle. In Fig. 3 transitional boundaries between various bed states are also plotted with respect to the same parameters, using data assembled by Allen (1982): (i) no sediment motion; (ii) ripples and dunes; (iii) upper stage plane bed. Although there is scatter in the data points, one of the striking features of this plot is the clustering of the Shields parameter values around constant values of 0.04, 1.9 and 12 for bedload, mixed-load and suspended-load dominant reaches, respectively, and this is confirmed by one standard error values of ±0.003, ±0.2 and ±2, respectively (Dade & Friend, 1998, p. 667).

It is concluded, therefore, that even though the data do not provide uniform cover of bed grain-sizes, the transport modes are relatively distinct, and can be roughly characterized by constant values of Shields parameter. It seems that there are stable transport modes separated by transitions in physical behaviour. Modes of distinctive flow and transport behaviour, separated by transitions, are recognized at other length scales in the complex world of sediment transport. Examples are the settling of individual grains, changing from Stokes to Impact 'Laws' of behaviour, and the change in grain support system from bedload to suspended load.

It is remarkable that the bedload value of the Shields parameter ($\theta = 0.04 \pm 0.003$, quoting 1 standard error; Dade & Friend, 1998, p. 667) coincides so closely with the boundary marking the onset of sediment motion (Fig. 3). This implies that bedload dominated rivers can transport their bed sediment with a flow strength that is just sufficient to produce bed-transport, i.e. the sediment is being transported by rivers that are just competent to carry the cobble or pebble grain-size present. In contrast, mixed- and suspended-load dominated rivers produce flood strengths that are distinctly greater than the minimum values needed to move material of their bed grain-size. The amount of sediment they transport depends on their capacity to transport this generally finer sediment, and the availability of the sediment.

Further insight on this contrast in behaviour comes from consideration of the relationship between bed-shear stress, or resistance to flow

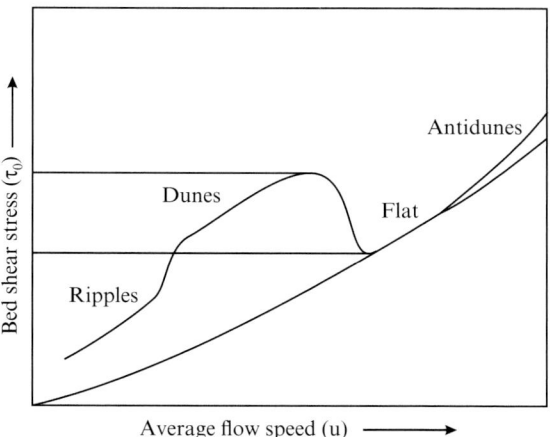

Fig. 4 Generalized and diagrammatic representation (Middleton and Southard, 1984) of the relationship of shear stress ($\tau_0 = \rho ghs$) against average flow speed (u) for a sandy bed.

(roughness), and mean flow velocity. Not only is it very much to be expected that this relationship will show an association between roughness and the grain-size of the bed, but it may also show that roughness is associated with bedforms and with channel form in two dimensions (width/depth ratio) and three dimensions (sinuosity). The effect of the presence of bedforms can be seen in flume work in which the same value of shear-stress may be measured at two, or even three, different values of mean flow velocity depending on the bedform state (Fig. 4; Middleton & Southard, 1984).

It seems that the higher levels of Shields parameter that characterize mixed and suspended-load modes, may represent new states of stability in which some of the high flow energy is dissipated by the presence of bedforms, and channel geometry. With the coarser grain-sizes of most bedload dominant rivers, local particle-scale friction and turbulence effects are likely to influence the bedform scaling turbulence, so that bedforms on a ripple-scale, which are typical of sand-grade beds, do not occur.

Bedload dominated rivers have transported their sediment at flow strengths that were only slightly above their competence, and the predominance of deposits of this relatively coarse grade implies that any finer grained sediment being transported was carried downstream and not

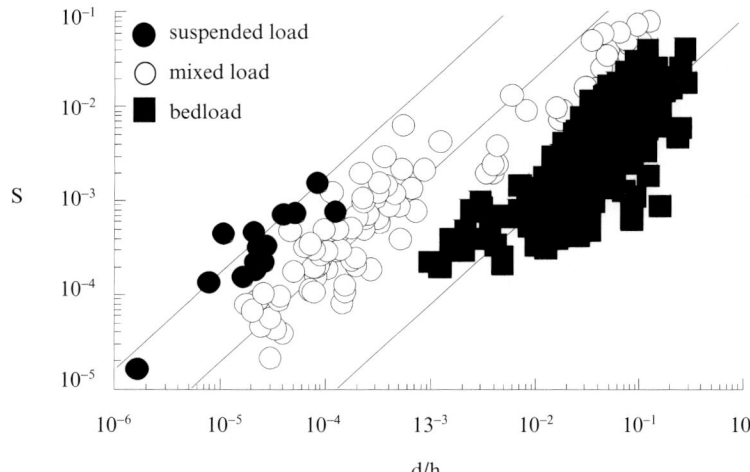

Fig. 5 Relationship between slope and relative grain-size (d/h), for the data set of Figs 1–3.

deposited in the reach studied. This bedload behaviour can be regarded as being highly efficient, because the low values of Shields parameter indicate that relatively low values of flow strength will move the bed material.

In contrast, mixed- and suspended-load rivers are more than simply competent in relation to the grain-size of their deposits. This implies an absence of available coarser sediment, as well as the availability of the finer material. The finer sediment may have been transported in sand-grade bedforms, with values of flow shear stress higher than those expected in a sediment-free (perhaps non-alluvial, bedrock) channel that has the same slope and depth. Transport in this mode can be described as having relatively low efficiency, and this may be attributed to dissipation of some of the power available, owing to the presence of bedforms and also possibly the sinuous channel features typical of this mode (Dade, 2000).

Availability of sediment of the different grain-sizes can clearly influence the transport mode dominance in many local situations (Friend, 1993), but changes in modes, with their characteristic Shields parameters, will also change patterns of sediment mass flux and the downstream distribution of deposits of different grain-size.

Figure 5 presents data for the river reaches plotted on axes of relative grain-size (grain-size/depth) against slope (s). These three simple parameters (grain-size, flood-depth and slope) are probably the most readily visualized measures of the controls that may influence changes in transport modes.

Downstream transitions from one transport mode to another are likely to be fruitful situations for examination, and in the remainder of this paper attention will be focused on the transition from bedload to mixed-load domination. Moving from the bedload to the mixed load cluster of points on a plot such as Fig. 5 can clearly involve many combinations of change, in one or more of the three variables. Independent information will be necessary to further develop the analysis in any specific situation. Decrease in grain-diameter owing to exhaustion of gravel supply, as a result of gravel burial in an otherwise sandy subsiding bed, is likely to be a particularly important element of this process.

EXAMPLE OF A DOWNSTREAM CHANGE IN TRANSPORT MODE, THE ALLT DUBHAIG RIVER, SCOTLAND

Sambrook Smith & Ferguson (1995) analysed a number of rivers in which an abrupt downstream fining in bed material grain-size (over several phi) exists over a downstream distance of 10^2 to 10^4 m, with no significant change of water or sediment availability. There is a significant change of river channel slope in most (but not all) of the cases they considered. The river primarily measured by Sambrook Smith & Ferguson is the small

($10 \text{ m}^3 \text{ s}^{-1}$ bankfull discharge) Allt Dubhaig, of Highland Scotland, which, although flowing in a valley floor between bedrock mountains, is alluvial in character in the reaches studied. Figure 6 shows, in general terms, the decrease in slope and in relative grain-size associated with a downstream change from gravel via bimodal sediment to sand bed in the Allt Dubhaig, which occurs over a distance of a few hundred metres, and marks the downstream change in transport mode.

The detailed study of the Allt Dubhaig has provided much information about the gravel-grade, bedload dominant part of the downstream profile (Fig. 6), and T. Hoey provided further data to augment that published by Ferguson & Ashworth (1991), Hoey & Ferguson (1994) and Sambrook Smith & Ferguson (1995). In outline, these data confirm the constant value of the Shields parameter (0.04) over the 2.5 km of gravel-grade reach that are assigned to the bedload mode. Downstream plotting of the data also confirms the downstream decrease in slope and relative grain-size over this distance. This is predicted assuming that some subsidence has acted to trap and retain the coarser, gravel-grade sediment that is being supplied to the upstream end of the river.

Downstream from the transition to mixed-load domination, sandbars are characterized by bedforms, so that the increased Shields parameter in these reaches is likely to have involved some energy dissipation on bedform movement, and the sinuosity of the channel may also have had a similar dissipative effect. The downstream transition is also associated with a decrease in river slope (Fig. 6), and this partly may be a result of the raising of downstream base level produced by construction of a local dam. Changes of river depth seem to have had little significance in creating the downstream threshold. It is concluded that exhaustion of the supply of gravel, owing to its concealment within the increasingly sand-grade bed, is likely to have been the major factor in causing the transition.

OLD RED SANDSTONE OF EAST GREENLAND AS AN EXAMPLE OF THE USE OF TRANSPORT MODES IN THE ANALYSIS OF AN ANCIENT, TECTONICALLY ACTIVE BASIN

An example of an ancient fluvial basin-fill is now used to explore the application of the transport mode approach to ancient basin stratigraphy. The example comes from the Old Red Sandstone (Devonian) of East Greenland, which has been shown to be the preserved relic of a fluvial basin, with a variable history that resulted in an aggregate thickness of 10.5 km of sediment, a width of about 100 km and a length parallel to its N–S axis of at least 300 km. A systematic study of this basin-fill (Friend et al., 1983) divided it into ten different stratigraphical units, each of which is broadly chronological. Tectonic events, involving the uplift, subsidence, compression and extension, occurred before, during and after the deposition of the sediment. The use of local sedimentary structures (mainly sand-grade bedforms) as keys to the geographical patterns of river flow directions, and to the local environments of deposition in each unit, was a fundamental part of the field investigation.

The patterns to be discussed here are features of units 5 and 6 (middle and upper Kap Kolthoff Supergroup), which were deposited with an aggregate thickness of about 3 km, over a period of about 5 Myr. This great thickness is clear evidence of the importance of continuing tectonic subsidence during this episode in the basin's history. Figure 7c is a diagrammatic representation of a vertical-plane reconstruction of the sedimentary geometry present along the western margin of the N–S trending basin. Between about 20 km and 60 km from the northern end of the reconstruction, a basement ridge of pre-Old Red Sandstone

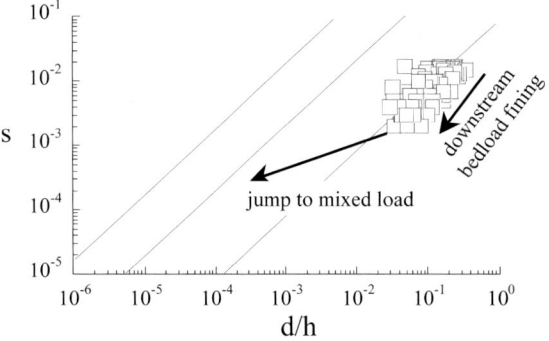

Fig. 6 Relationship between slope and relative grain-size (d/h) for data from the Allt Dubhaig river, Scotland.

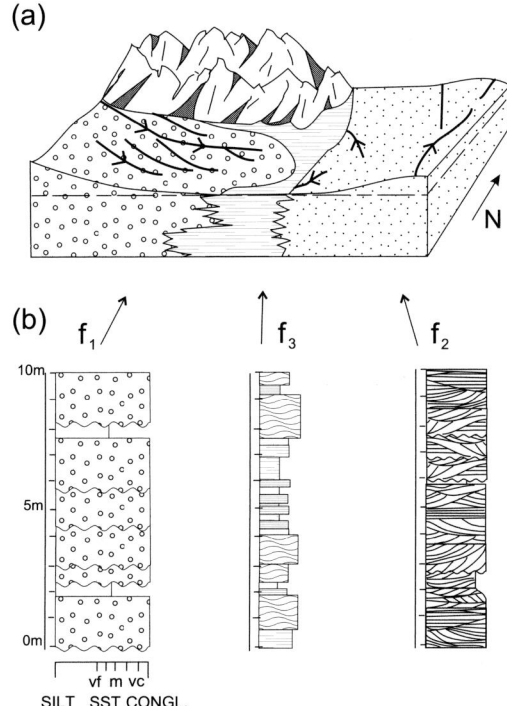

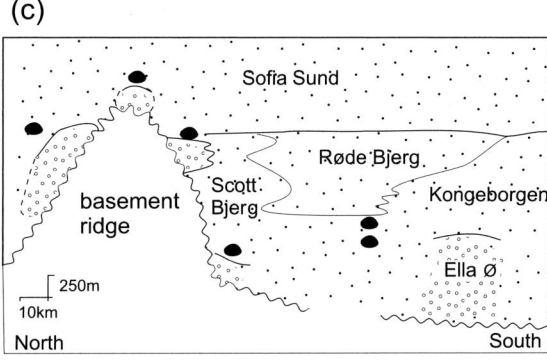

Fig. 7 (c) Diagrammatic vertical profile along the western margin of the basement, showing the geometrical relationship between marginal alluvial fan deposits formed by bedload dominant streams (f_1), deposits of the axial mixed-load dominant river deposits (f_2) and relatively minor deposits of the suspended-load lacustrine facies (f_3), representing suspended-load dominant rivers.

Fig. 7 Basin-margin sediment patterns in the Old Red Sandstone of East Greenland (Friend et al., 1983), illustrating the distinctive facies (f_1–f_3) generated by different modes of sediment transport. (a) Block diagram illustrating the basin-margin relationship between the outcropping basement of the margin and the three main facies (f_1–f_3). Typical sediment transfer directions are shown. (b) Comparison of typical 10-m logs in sediments of the three facies.

age is present, presumably reflecting a large cross-cutting tectonic structure.

Three distinct types of sediment accumulation are recognized.

f_1 Discrete conglomerate bodies, five in number on this transect (Fig. 7). These represent the deposition of localized alluvial fans that formed in the basin where river systems entered from the source mountains to the west. The conglomerates are typically of framework type with average clast size of 50 mm, and an average conglomerate bed thickness of 2 m (Fig. 7b). Using this bed-thickness as a statistical measure of channel depth, bar height or flood depth, a relative grain size (d/h) of 3×10^{-2} is suggested.

f_2 A very large sandstone unit, occupying the axis of the basin, and formed by braided rivers flowing southwards along the axis of the basin, and parallel to the margin being considered. The sandstones are typically of medium-sand grade (0.5 mm grain-size), and have an average sand-bed thickness of 1 m. Using the same logic as in the preceding paragraph, a relative grain size (d/h) of 5×10^{-4} is suggested.

f_3 Thin and isolated siltstone units, generally red but also containing dark grey calcareous lacustrine sediments. These siltstones appear to have been deposited in and around isolated lakes that formed between the marginal alluvial fans and the major axial sand system. Generalising on the average grain-size (0.004 mm) and taking bed-thickness of the coarse siltstones as an indicator of flood depth (0.3 m), a relative grain-size (d/h) of 1×10^{-5} is obtained.

The interpretation offered in terms of river transport modes is that the conglomerate bodies were formed by bedload dominant rivers, and that the transition from gravel to sand grade radially down the marginal alluvial fans represents the change to mixed-load dominant deposition. There is no clear evidence from the data on sedimentary structures or alluvial architecture that there was a significant change in channel water-depth at this transition, nor is there evidence of the new availability of sediment of a different grain-size. The change in mode is therefore taken to be the result of a change of slope, probably in response to relative rise of local base level as alluviation of the basin centre proceeded. It may be that the exhaustion of the gravel-grade material, owing to its incorporation in the subsiding basin-fill succession, reduced the mean grain size. The changes in slope and grain-size were sufficient to bring about a transition to mixed-load Shields parameter along with a decrease in efficiency and increased power dissipation by bedforms and increasingly sinuous channel forms.

The transition from sandstones accumulated within the marginal fan systems and sandstones formed from the axial system involved no change of transport mode but an important change in the direction of local sediment flux. The discrete siltstone bodies between the conglomerates and the marginal unconformity clearly represent an unusually sharp transition from bedload dominant to suspended-load dominant rivers. It seems certain that this transition must have involved a major decrease in slope, but it also seems likely that channel depth also decreased, and that coarser sand-grade sediment became exhausted, so that the siltstones formed in hollows between the bedrock topography of the basin margin and the actively growing alluvial fans. These hollows seem to have been sheltered from the powerful flows that were active on the surface of the gravel fans and also below their surface.

In this example of the examination of an ancient stratigraphical basin-fill, the inevitable lack of independent direct measurements of flood regimes or of channel slopes or even the build-out distances of the fans imposes severe limits on the analysis. The new use of the transport mode approach, however, makes an important contribution by providing more powerful insights on the range of different variables that may have been in control.

CONCLUSION

Rivers and their sediments are highly variable in space and time, and attempts to summarise this variability are likely to appear not only difficult, at the present state of knowledge, but even misleading. Further development of the distinctions between bedload, mixed-load and suspended-load domination, however, provides the promise of a generally improved understanding of the controls of river behaviour, particularly the importance of the availability of sediment of different grain-sizes. Further studies of carefully collected data will place speculations such as these on a firmer foundation.

ACKNOWLEDGEMENT

We are grateful to the two reviewers of our first submission for their careful and constructive comments.

NOMENCLATURE

R relative excess density of particles, $(\rho_p - \rho)/\rho$
Re_* grain Reynolds number, $u_* d/\upsilon$
d median grain-size of bed material
h channel depth
s channel slope
u average flow speed
u_* friction velocity
υ kinematic viscosity
ρ density
θ Shields parameter, sh/Rd
τ_0 bed shear stress, $\rho g h s$

REFERENCES

Allen, J.R.L. (1982) *Sedimentary Structures*. Developments in Sedimentology, **30**, Elsevier, Amsterdam, 1256 pp.

Dade, W.B. (2000) Grain size, sediment transport and alluvial channel pattern. *Geomorphology*, **35**, 119–126.

Dade, W.B. and Friend, P.F. (1998) Grain-size, sediment transport regime, and channel slope in alluvial rivers. *J. Geol.*, **106**, 661–675.

Chitale, S.V. (1970) River channel patterns. *Proc. Am. Soc. Civil Eng., Hydrol. Div.*, **96**(HY1), 201–221.

Church, M. and Rood, K. (1983) *Catalogue of Alluvial River Channel Regime Data*. Department of Geography, University of British Columbia, Vancouver, 99 pp.

Ferguson, R.I. and Ashworth, P. (1991) Slope-induced changes in channel character along a gravel bed stream: the Allt Dubhaig, Scotland. *Earth Surf. Proc. Landf.*, **16**, 65–82.

Friend, P.F. (1993) Control of river morphology by the grain-size of sediment supplied. *Sediment. Geol.*, **85**, 171–177.

Friend, P.F., Alexander-Marrack, P.D., Allen, K.C., Nicholson, J. and Yeats, A.K. (1983) Devonian sediments of East Greenland, VI, review of results. *Medd. Grønl.*, **206**(6), 96 pp.

Hoey, T.B. and Ferguson, R.I. (1994) Numerical simulation of downstream fining by selective transport in gravel bed rivers. *Water Resources Res.*, **30**, 2251–2260.

Middleton, G.V. and Southard, J.B. (1984) *Mechanics of Sediment Movement. Soc. Econ. Paleontol. Mineral. Short Course*, **3**, 401 pp.

Pettijohn, F.J. (1957) *Sedimentary Rocks*, 2nd edn. Harper Brothers, New York, 718 pp.

Russell, R.J. (1968) Where most grains of very coarse sand and fine gravel are deposited. *Sedimentology*, **11**, 31–38.

Sambrook Smith, G.H. and Ferguson, R.I. (1995) The gravel–sand transition along river channels. *J. Sediment. Res.*, **A65**, 423–430.

Schumm, S.A. (1963) A tentative classification of alluvial river channels. *U.S. Geol. Surv. Circ.*, **477**, 10 pp.

Schumm, S.A. (1968a) Speculations concerning paleohydrologic controls of terrestrial sedimentation. *Geol. Soc. Am. Bull.*, **79**, 1573–1588.

Schumm, S.A. (1968b) River adjustment to altered hydrologic regime—Murrumbidgee River and paleochannels, Australia. *U.S. Geol. Surv. Prof. Pap.*, **598**, 65 pp.

Van der Berg, J. (1995) Prediction of alluvial channel pattern of perennial rivers. *Geomorphology*, **12**, 259–279.

Gulf of Mexico Basin depositional record of Cenozoic North American drainage basin evolution

WILLIAM E. GALLOWAY

Institute for Geophysics, The University of Texas at Austin, Austin, TX, USA
(Email: galloway@mail.utexas.edu)

ABSTRACT

A comprehensive synthesis of the Cenozoic depositional history of the Gulf of Mexico basin has integrated data from the coastal plain, shelf, slope and deep basin. Twenty widely recognized Cenozoic genetic-stratigraphy sequences record major and minor depositional episodes of the northern Gulf basin. Results of this synthesis provide a picture of the location and relative importance of principal fluvial systems that drained the interior of North America. Major observations include:

1 Five principal and three secondary, long-lived, extrabasinal fluvial–deltaic axes provided the bulk of the sediment that infilled the northern Gulf basin.
2 Paleocene through to middle Eocene pulses of Laramide uplift along the Central and Southern Rockies and Sierra Madre Oriental supported the early Cenozoic depositional episodes. Late Eocene through to early Oligocene crustal heating, volcanism and consequent uplift and erosion of much of central Mexico and the south-western USA nourished major Oligocene through to early Miocene depositional episodes.
3 Initiation of erosion during the early to middle Miocene of the Cumberland Plateau and Appalachians invigorated supply to the east-central Gulf basin. At the same time, the high-standing Rocky Mountain uplands experienced continued regional exhumation. Whether climate change or uplift triggered this Miocene phase of erosion remains controversial; current literature favours climatic causes.
4 Pliocene uplift of the western High Plains further rejuvenated north-western sources and created a broad eastward slope that converged with the west-sloping alluvial apron of the eastern interior. Converging streams were variously combined and directed southward, forming the distinct Red, Mississippi and Tennessee fluvial axes that dominated middle Pliocene through to Holocene sedimentation.
5 High rates of Pleistocene sediment accumulation reflect rapid Quaternary climate cycling, and glacial erosion and runoff directly into the principal sediment transport systems.

INTRODUCTION

A comprehensive synthesis of the Cenozoic depositional history of the Gulf of Mexico basin has integrated literature, well, and seismic data from the North American coastal plain, shelf, slope, and deep basin (Galloway *et al.*, 2000). Using widely recognized marine shale units and palaeontological markers, 18 Cenozoic genetic stratigraphical sequences, recording major and minor depositional episodes of the northern and north-western Gulf basin, were mapped and interpreted regionally (the minor Midway episode and the latest Pleistocene–Holocene episode have not been mapped in this study). Results of this synthesis provide, in addition to the history of Gulf deposition, a picture of the location and relative importance of principal extrabasinal fluvial systems, the drainage basins of which extended far beyond the geological and topographic boundaries of the Gulf basin into the interior of North America. It is these extrabasinal rivers that provided the bulk of the sediment that has prograded the northern gulf margin hundreds of kilometres into the ancestral Mesozoic basin (Galloway, 1981). This summary builds on and updates previous syntheses of Winker (1982) and Galloway et al. (1991).

Methodology and database

The synthesis compiled and rationalized information from more than 240 published cross-section folios, maps, papers, monographs and theses, and more than 800 well logs, in order to create a geographical information system (GIS) database for the northern Gulf of Mexico Cenozoic basin. Using these data, a suite of lithofacies, thickness and palaeogeographical maps was compiled for each of the pre-modern depositional episodes. Twenty-three depositional system categories, including fluvial, deltaic and shore-zone system types, were differentiated on palaeogeographical maps. Together, the map suite defines location, areal extent, and total sediment volume associated with the major sediment dispersal systems within the Gulf basin during each depositional episode.

CENOZOIC DEPOSITIONAL EPISODES

Recognition and correlation of genetic stratigraphical units used for mapping and quantification of sediment accumulation and supply rates to the Gulf utilized the concepts of depositional episodes (Frazier, 1974) and genetic-stratigraphy sequences (Galloway, 1989). Genetic sequences are bounded by transgressive marine beds and their contained flooding surfaces and condensed beds. A genetic sequence records a regional episode of high sediment supply and accompanying shoreline and continental margin out-building. Commonly, substantial changes in patterns of sediment dispersal and accumulation accompany the transgressions that bound depositional episodes; these changes typically reflect changes in the location and/or significance of principal fluvial input axes (Galloway, 1989).

Twenty regional depositional episodes can be recognized throughout the northern Cenozoic basin fill (Fig. 1). The lowermost depositional episode, the Midway, is a regional marine shale recording minimal influx of sediment throughout the northern Gulf. The uppermost depositional episode includes the latest Pleistocene–Holocene (Wisconsin) glacial cycle, and was not mapped in this work. Each of the remaining 18 depositional episodes was characterized by substantial sediment input and coastal plain progradation. Neogene depositional episodes were dated by faunal content, primarily Foraminifera. Depositional episodes were named using common outcrop or subsurface nomenclature (Lower Wilcox, Middle Wilcox, Upper Wilcox, Queen City, Sparta, Yegua/Cockfield, Jackson, Frio-Vicksburg, Lower Miocene 1 and 2, Middle Miocene, Upper Miocene) or the upper bounding faunal datum (*Buliminella* 1, *Globoquadrina altispira*, *Lenticulina* 1, *Angulogerina* B, *Trimosina* A, Sangamon fauna) (Fig. 1).

In the Paleogene section, time intervals between depositional episodes are recorded by marine shale tongues that accumulated slowly over periods of as much as 2 Myr. Depositional episodes ranged from 2 to 8 Myr in duration. By contrast, Neogene flooding pulses were brief, and transgressive strata rarely extend up-dip to current outcrops; they are commonly identified and correlated by use of specific microfossil datums (Fig. 1). Miocene depositional episodes were long lived, ranging from 2 to 6 Myr. Increasingly frequent, eustatically forced transgressions of the late Neogene, however, created high-frequency depositional episodes. Stratigraphical resolution for Plio-Pleistocene sequences is 1 Myr or less.

Each depositional episode created a genetic-stratigraphy sequence containing deposits of a unique array of depositional systems, including fluvial, delta, shore-zone, shelf, slope and basinal

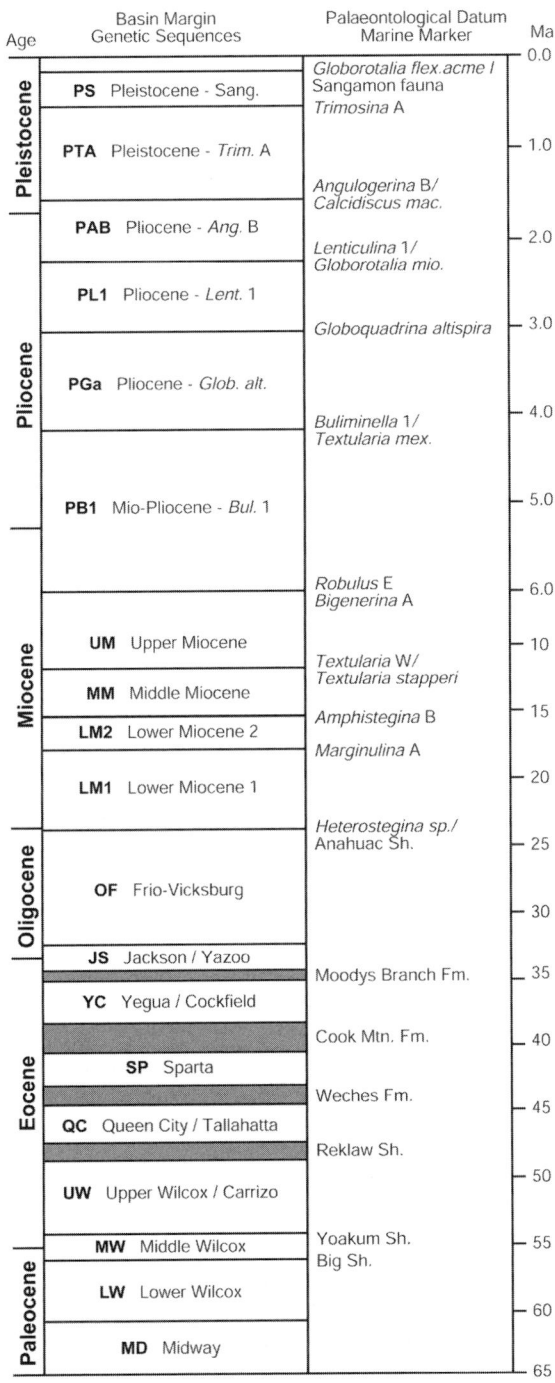

Fig. 1 Genetic sequences of the northern Gulf of Mexico basin margin that record principal episodes of sediment influx and deposition. Bounding transgressive marine markers are named for prominent outcropping shales or palaeontological marker fossil-zone tops (modified from Galloway et al., 2000).

systems (Galloway et al., 2000). Principal sediment storage for most episodes occurred on the outer depositional platform and upper slope. Deposits, however, were typically thicker and sandier beneath principal delta systems and their subjacent slope aprons (Winker, 1982).

PATTERNS OF CENOZOIC SEDIMENT SUPPLY

Sediment has been supplied to the northern Gulf of Mexico basin primarily via major extrabasinal rivers that had large drainage basins located within south-central and south-western North America (Galloway, 1981; Winker, 1982). These rivers constructed large fluvial- or wave-dominated delta systems. In turn, sediments were reworked by waves and marine currents from deltaic headlands, forming shore-zone and shelf systems, and by gravitational mass transport processes to form the thick continental slope aprons, fans and basinal systems of the deep Gulf.

The Cenozoic history of sediment supply to the Gulf of Mexico provides a record of sediment yield from the adjacent continent. Overall sediment supply has been largely modulated by continental tectonism and climate. More specifically, key variables probably included changing drainage basin area, average and maximum drainage basin elevation, tectonism and climate, as manifested in overall temperature, rainfall and weather patterns (Milliman & Syvitski, 1992; Hovius, 1998).

Continental source areas

The modern generalized elevation of North America (Fig. 2) provides a framework for interpreting probable Cenozoic sediment sources. The principal uplands include the remanant Palaeozoic core of the Appalachian Mountains and adjacent Cumberland Plateau, which have modern elevations of 0.5 to 1 km. To the west, the Central and Southern Rocky Mountains and Sierra Madre Oriental of northern Mexico form relatively young and high-standing (> 2 km) headwaters for east-flowing streams. These mountain belts had their origin in the late Cretaceous through to Paleocene Laramide compressional tectonics. Whereas the

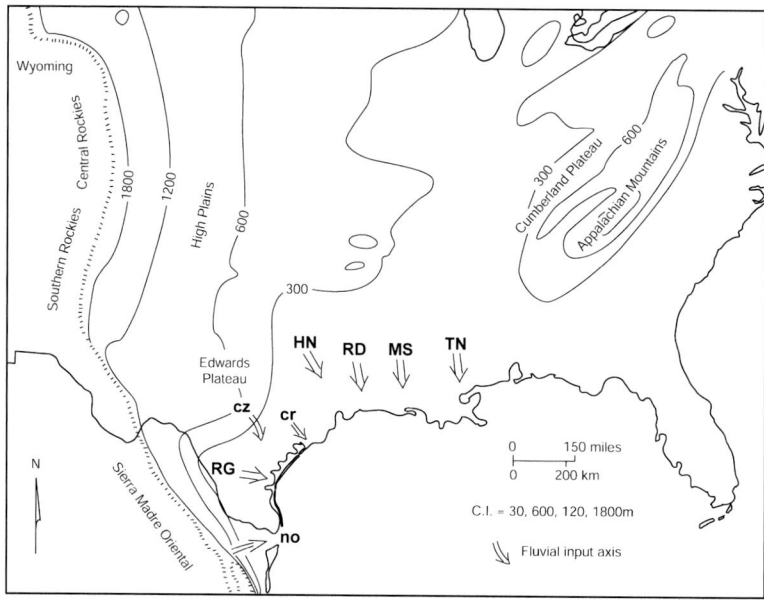

Fig. 2 Location of principal fluvial sediment axes for Cenozoic sediments of the Gulf of Mexico (modified from Galloway *et al.*, 2000) and of North American uplands that provided sediment sources. Contours show generalized modern elevations of the uplands (modified from Doering, 1956). Fluvial axes: no, Norias; RG, Rio Grande; cz, Carrizo; cr, Corsair; HN, Houston; RD, Red; MS, Mississippi; TN, Tennessee.

Rockies stand far removed from the Gulf basin periphery, the Sierra Madre impinges directly onto the Gulf margin. Standing between the Rockies and the Gulf are the east-sloping High Plains, with average elevation of 0.5 to 1.5 km, and the Edwards Plateau of Texas, which stands at about 0.5 km above present sea level. The Appalachian–Cumberland upland, Rockies, High Plains and northern Sierra Madre provide potential elevated source terrains for large continental rivers draining into the Gulf. The Edwards Plateau and southern Sierra Madre would contain headwaters for basin-fringe rivers with headwaters along the margin of the topographic basin.

Geography of Gulf sediment input

Mapping of Gulf Cenozoic genetic sequences has revealed consistent patterns in the location of major extrabasinal and basin-fringe fluvial axes (Galloway, 1981; Winker, 1982; Galloway *et al.*, 2000). Six principal extrabasinal axes, named after their modern manifestations where such exist, are now differentiated. From south-west to east, these include the Rio Grande, Carrizo, Houston, Red, Mississippi and Tennessee (Fig. 2). Two basin-fringe axes, the Norma and Corsair, put in relatively brief appearances and are further distinguished by unusual sand compositions that tie them to specific basin-margin sources. Geographical positions suggest that the headwaters of the Norma and Rio Grande systems lay in the adjacent Sierra Madre Oriental. The Carrizo and Houston axes are logical outlets for the Southern Rockies, the Houston and Red axes for the Central Rockies, the Mississippi axis for the Central Rockies, Wyoming and Appalachian–Cumberland uplands, and the Tennessee for the Appalachian–Cumberland upland.

Rates of sediment supply

Rates of sediment influx and accumulation to the Gulf of Mexico basin have varied substantially during the Cenozoic (Galloway & Williams, 1991; Galloway, 2002). Although capturing only a subset of total sediment accumulated within a sequence, calculated total volume of sediment deposited and stored in delta and shore-zone systems per million years for 18 of the Cenozoic depositional episodes (Fig. 3) quantifies the variability. The volumetric rates were calculated from generalized isochore maps constructed for the fluvial, deltaic and shore-zone deposits of each of the 18 genetic sequences. Details are described in Galloway (2002).

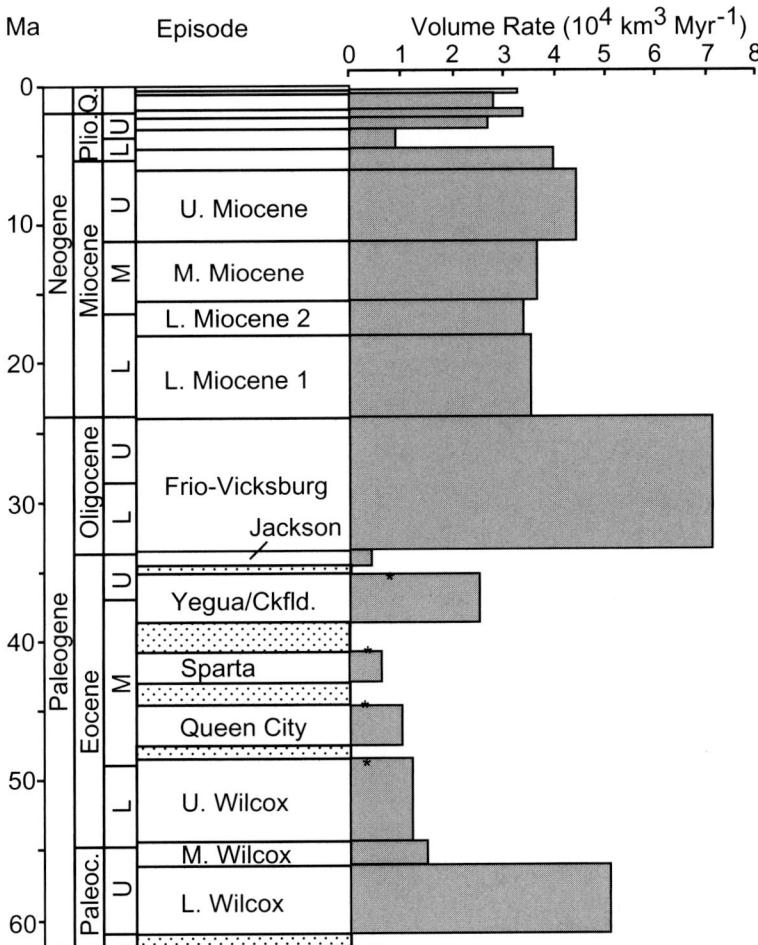

Fig. 3 Volume rate of Cenozoic sediment accumulation in deltaic and shore-zone systems along the northern Gulf of Mexico margin (from Galloway, 2002). Rates are calculated for 18 principal depositional episodes. Asterisks indicate time spans associated with the development of widespread transgressive marine shale tongues that extend to outcrop.

Some general patterns are noteworthy.
1 Upper Paleocene (Lower Wilcox) rates are high, then decrease progressively through lower and middle Eocene depositional episodes. Concomitantly there is an increase in the time span recorded by marine transgressive shale tongues that bound depositional episodes.
2 Upper Eocene supply was rejuvenated, but quickly waned again.
3 A massive surge of sediment influx deposited the Oligocene Frio depositional episode. More detailed estimation of Oligocene accumulation rates in the north-west Gulf suggests that supply increased abruptly in the early Oligocene and waned progressively during the later Oligocene (Galloway & Williams, 1991).

4 Influx remained high, with little change, through the Miocene depositional episodes and into the earliest Pliocene.
5 The Lower to Middle Pliocene *Globoquadrina altispira* depositional episode was characterized by low rates of input, which then rapidly returned to high values through the remainder of the Pliocene and Pleistocene depositional episodes.

DRAINAGE BASIN EVOLUTION AND SUPPLY HISTORY

The spatial and temporal patterns of Cenozoic sediment influx into the Gulf of Mexico basin provide insight into the history of drainage basin evolution and sediment yield of the adjacent North American continent. Ideally, the Gulf 'tape recorder' should, when merged with the history of continental tectonics and climate change, create a logical story of continental drainage basin evolution. The following discussion attempts such a synthesis. Cenozoic drainage history is broken down into five phases that reflect principal steps in drainage evolution as recorded in the receiving basin. The spatial and temporal distribution of fluvial axes are summarized in Figs 4 & 5.

Paleocene and early Eocene

The Lower Wilcox depositional episode records the first major Cenozoic influx of clastic sediment into the western and central Gulf basin. Progradation of the central Gulf margin by as much as 40 miles from its Cretaceous precursor clearly records this major pulse of sediment supply. Lithofacies maps outline two prominent fluvial-dominated delta systems in the northern Gulf rim (Fig. 4). The Houston delta, the largest and sandiest depocentre, was fed by an extrabasinal, bedload-rich fluvial system centred on the Houston axis. It is separated by a broad shore-zone system from a secondary delta located in central Louisiana, the first of a long succession of Cenozoic deltas to align along the Mississippi fluvial–deltaic axis (Fig. 5). In the north-west Gulf, several small deltas merged into and nourished an extensive wave-dominated shore-zone system that extended from northern Mexico into South Texas.

Sediment supply was dominated by syntectonic Laramide uplands, which extended from the Central and Southern Rockies into northern Mexico. Tectonic uplift had created uplands with elevations of 3–4 km, in the Southern Rockies, to > 2.2 km in the Central Rockies (Chase et al., 1998). These syn- to early post-tectonic uplands were probably sites of high sediment yield that was collected by southeast- and east-flowing rivers (Crabaugh, 2001) that were tributary to the Houston fluvial axis and, to a lesser degree, the palaeo-Mississippi fluvial system. Sediment from the southernmost Rockies and Sierra Madre uplifts were flushed into the Gulf through smaller, basin-fringing rivers, creating a thick sediment prism of coastal and slope strata, but no single large delta system (Fig. 4).

The uppermost Paleocene to lowermost Eocene Middle Wilcox depositional episode records a progressive evolution of sediment dispersal patterns. A single wave-dominated delta prograded into the North-west Gulf basin along the Rio Grande axis, indicating integration of southernmost Rocky Mountain and Sierra Madre upland drainage. In the central Gulf, fluvial–deltaic systems built from the Houston and Mississippi axes, forming a composite deltaic coastal plain. Together these fluvial–deltaic systems, which drained most of the Southern and Central Rocky Mountain uplands, provided sufficient volumes of sediment to prograde the Gulf shelf break as much as 32 km beyond the previous Lower Wilcox margin. Nonetheless, the surge of syntectonic supply that created extremely high rates of Lower Wilcox sediment influx had waned.

The early Eocene Upper Wilcox depositional episode records pulsed rejuvenation and/or reorganization of Southern Rocky Mountain sources (Crabaugh, 2001). The dominant fluvial system, which constitutes the Carrizo Sand of outcrop and shallow subsurface (Hamlin, 1988), shifted to the central Texas coast. This sand-rich Carrizo axis supplied the Rosita delta system of the down-dip Upper Wilcox; together the Carrizo and Rosita fluvial–deltaic axes existed as significant dispersal systems only during the 5 Myr of the Upper Wilcox depositional episode (Galloway et al., 2000) (Fig. 5). The Houston and Mississippi systems persisted, but continued to shrink in relative importance during Upper Wilcox deposition.

In summary, Upper Paleocene through to Lower Eocene Wilcox depositional episodes record the

Fig. 4 (*opposite*) Simplified outlines of deltaic and shore-zone systems formed during each of the 18 principal Cenozoic depositional episodes (modified from Galloway, 2002). Principal fluvial axes operating during each episode are labelled: no, Norias; RG, Rio Grande; cz, Carrizo; cr, Corsair; HN, Houston; RD, Red; MS, Mississippi; TN, Tennessee.

MS
HN
RG
Oligocene Frio-Vicksburg
HN
RG
Upper Eocene Jackson
MS
HN
RG
Middle-Upper Eocene Yegua/Cockfield
HN
MS
Middle Eocene Sparta
HN
RG
Middle Eocene Queen City
MS
HN
CZ
RG
Lower Eocene Upper Wilcox
HN
MS
RG
Eocene-Paleocene Middle Wilcox
MS
HN
Upper Paleocene Lower Wilcox
LEGEND
Delta system area
Shore-zone system area
3 Thickness contour (x10²m)
C.I.: 150, 300, 600, 1200, 1800, 2400, 3000, 3600 m.

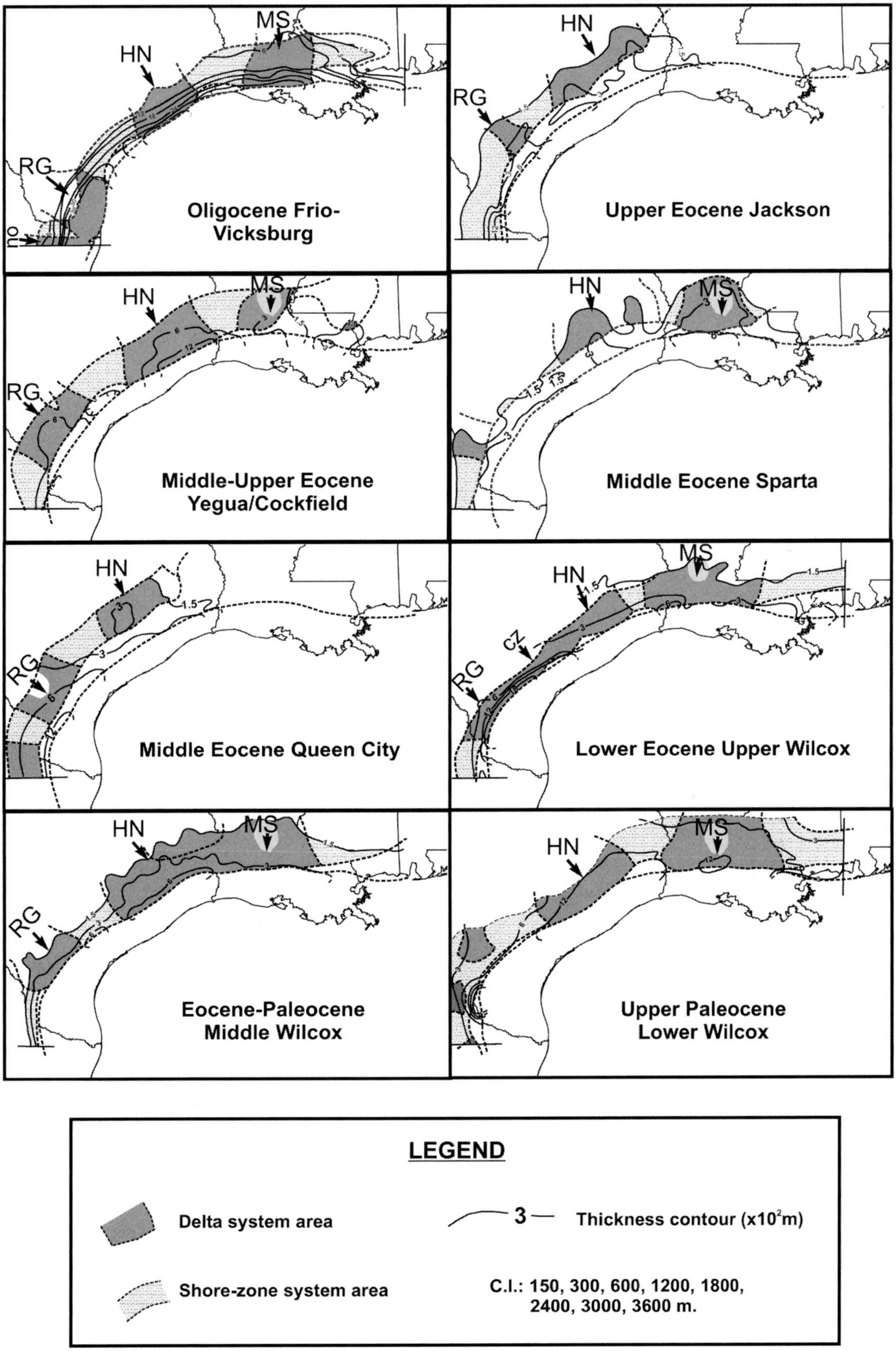

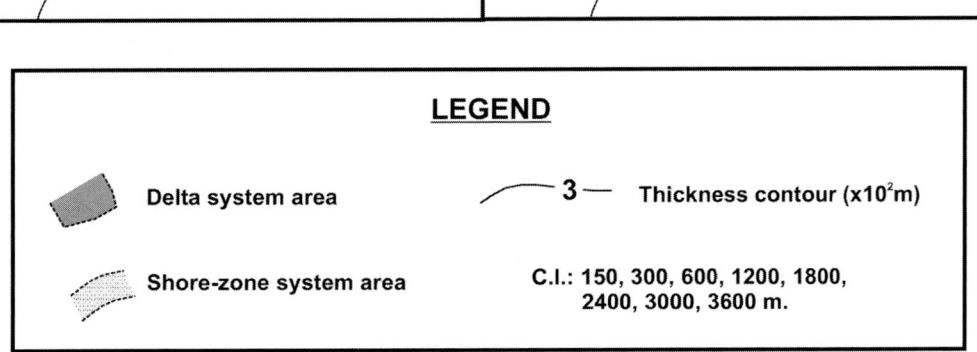

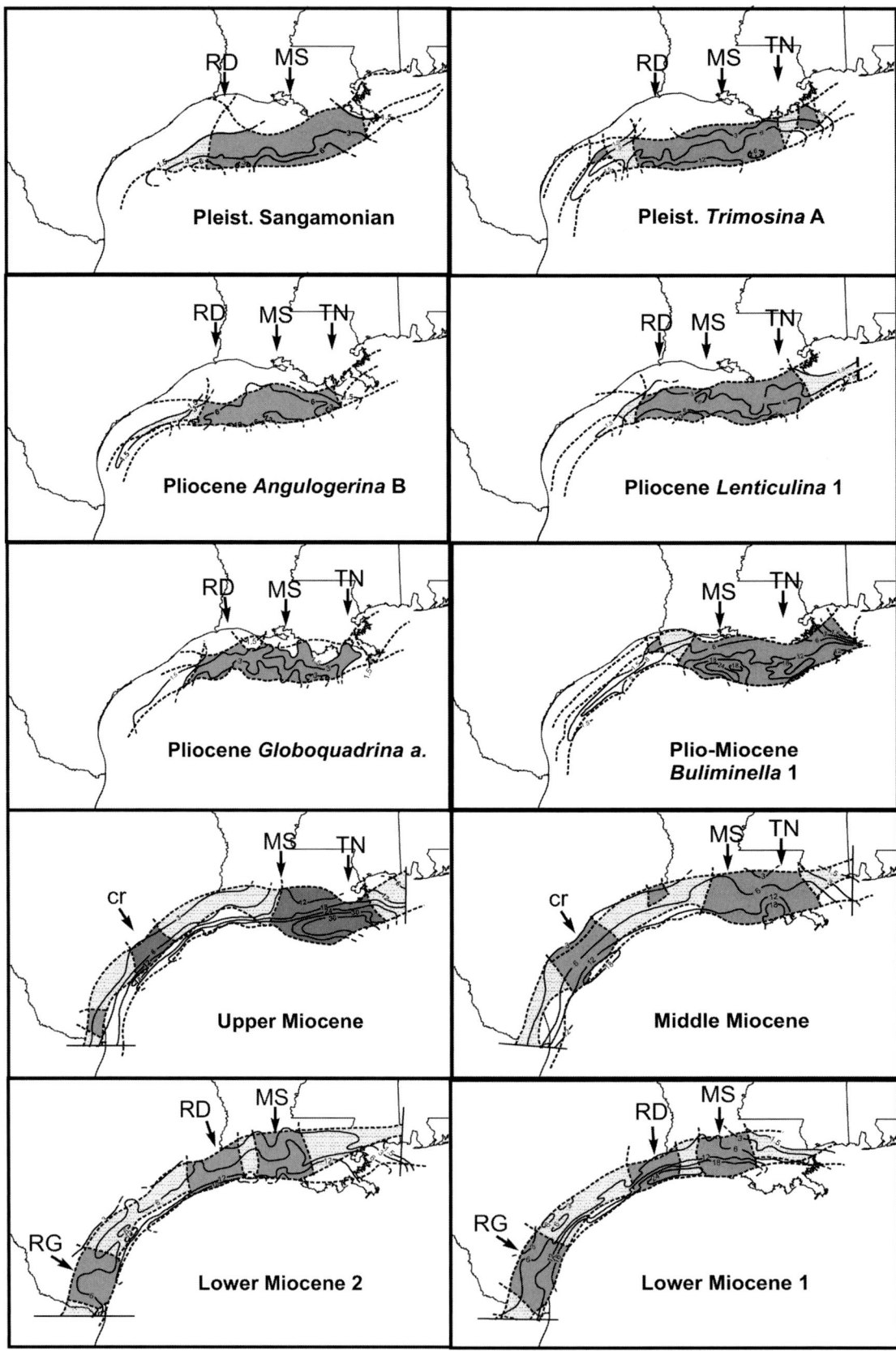

Fig. 4 (*continued*)

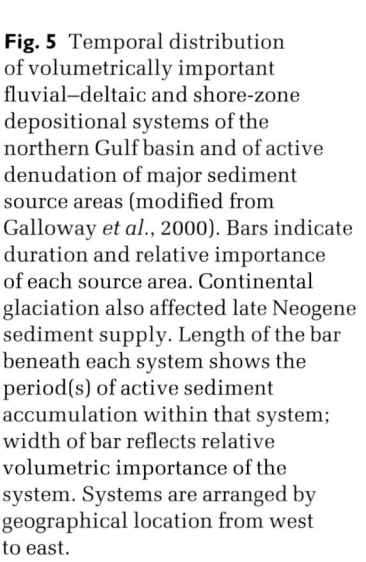

Fig. 5 Temporal distribution of volumetrically important fluvial–deltaic and shore-zone depositional systems of the northern Gulf basin and of active denudation of major sediment source areas (modified from Galloway *et al.*, 2000). Bars indicate duration and relative importance of each source area. Continental glaciation also affected late Neogene sediment supply. Length of the bar beneath each system shows the period(s) of active sediment accumulation within that system; width of bar reflects relative volumetric importance of the system. Systems are arranged by geographical location from west to east.

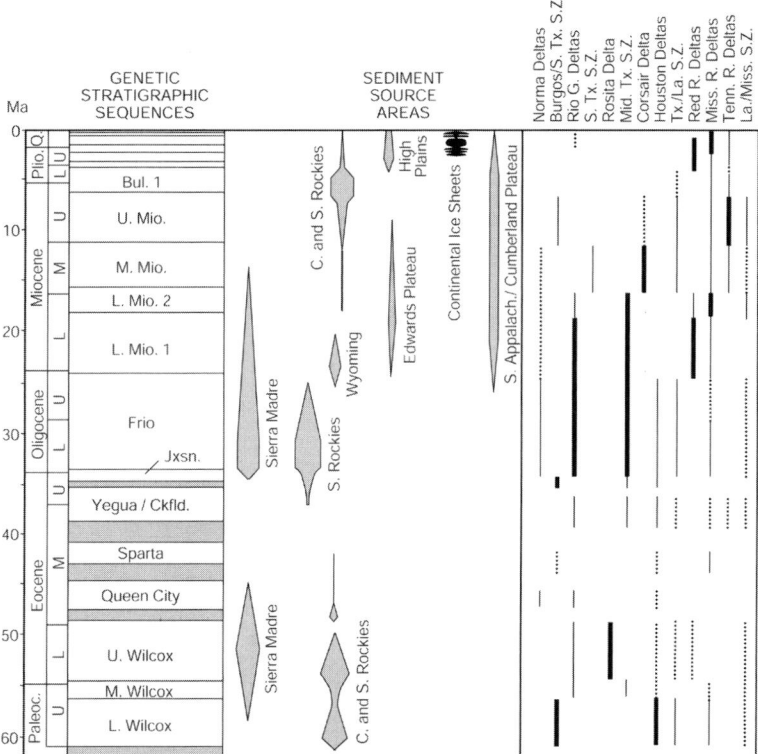

direct impact of Laramide tectonism and consequent uplift on the Gulf. Principal fluvial axes and sediment supply were directed to the north-western basin margin. Generally, one large extrabasinal fluvial system, the Houston and later the Carrizo, collected discharge primarily from the Southern Rockies (Fig. 6). Along the north-west Gulf margin, numerous small streams flowed from southernmost Rocky Mountain and Sierra Madre uplifts directly into the Gulf. Through time, these drainage systems became integrated and focused along the Rio Grande fluvial axis. The Central Rockies, also high standing due to Laramide uplift, supplied tributaries to continental rivers draining both north, into Canada, and south, into the Gulf of Mexico (Fig. 6). Tectonically elevated uplands resulted in high sediment yields; Gregory & Chase (1994) estimated that 2–4 km of material was removed from the Front Range of the central Rockies during this time. Rejuvenated, moderate to large drainage systems exported much of this sediment to the Gulf basin, resulting in high volume-rates of accumulation in deltas and their marginal shore zones (Fig. 3).

Middle–late Eocene

Four depositional episodes punctuate middle–upper Eocene stratigraphy of the Northern Gulf. Three were minor in terms of sediment supply (Fig. 3). Progressive integration and grading of early Eocene drainage basins coupled with stable, warm, humid climate throughout the Central and Southern Rockies created a relatively low-relief surface consisting of in-filled intermontane basins and smoothed uplands (Fig. 6) (Gregory & Chase, 1994; Chapin and Kelley, 1997). Despite the low relief and low sediment yield, much of this broad Rocky Mountain surface remained well above 2 km in elevation (Gregory & Chase, 1994; Chase *et al.*, 1998; Wolfe *et al.*, 1998). Drainage basins and axes were re-established along the older

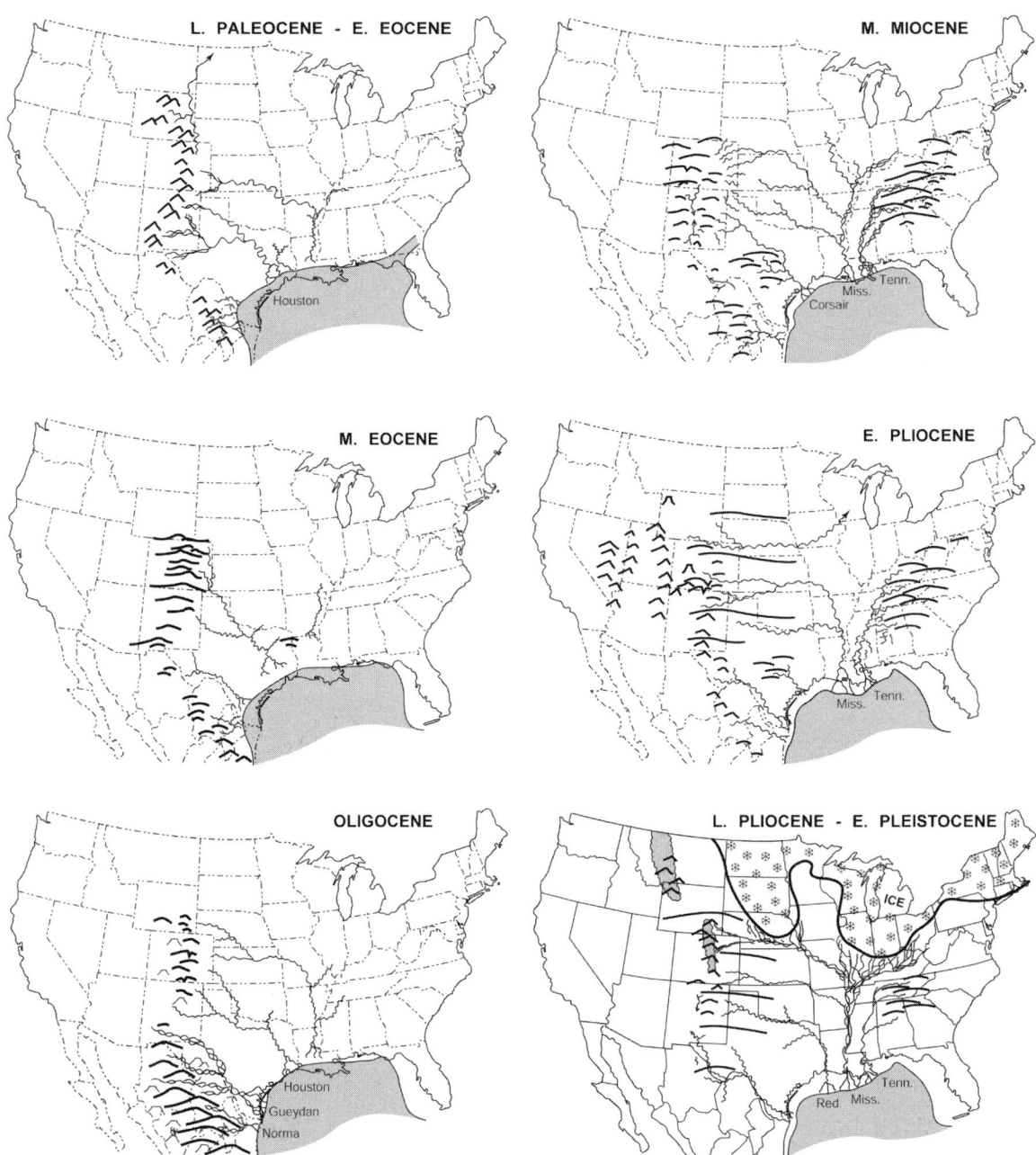

Fig. 6 Sketch maps schematically showing principal fluvial–deltaic systems of the northern Gulf of Mexico and their inferred drainage basins at six representative time steps.

Paleocene template (Fig. 6), with the Rio Grande, Houston and Mississippi axes dominating input to the Gulf (Fig. 4). High mud content of fluvial systems reflects the preponderance of chemical weathering across the western uplands as inferred by Chapin & Kelley (1997).

The Yegua/Cockfield depositional episode recorded briefly rejuvenated supply rates, perhaps reflecting the thermal uplift and volcanism and/or climate change that would soon dominate the Oligocene record. A coeval phase of repeated canyon excavation and filling, interspersed with effusive volcanic events, punctuates the late Eocene stratigraphical record in the Colorado and Wyoming Rockies (Evanoff, 1990; Chapin & Kelley, 1997). However, supply reverted to normally low values during the latest Eocene Jackson depositional episode (Fig. 3).

Oligocene

The Oligocene Frio depositional episode records the second great tectonically induced flood of sediment into the Gulf of Mexico (Fig. 3). The epoch began with extensive crustal heating, explosive volcanism, and continental-scale uplift of source areas in northern Mexico and the south-western U.S. (Galloway, 1977; Yurewicz et al., 1997). Volcanism extended northward along the Rockies as far as Colorado (Fig. 6). Pre-existing high elevations, combined with additional uplift, created an elevated source terrain ranging from 2.2 to 3.2 km along the Southern Rockies (Wolfe et al., 1998; Chase et al., 1998). Tectonic inversion of 2–3 km of the western margin of the Gulf in Mexico elevated new upland source areas along the Sierra Madre in Northern Mexico (Yurewicz et al., 1997). In addition to the regional tectonic uplift, vast amounts of air-fall volcanic ash spread across the inland drainage basins of the Gulf rivers. Most of this readily erodible ash found its way into the Gulf as reworked vitric debris or clay pedogenically derived from the ash temporarily stored in floodplains or interfluves (Galloway, 1977). At the same time, along the north-west Gulf margin, the subtropical late Eocene palaeoclimate was replaced by a subarid Oligocene climate regime (Galloway, 1977). In summary, numerous variables combined to create a multi-million-year surge of sediment input to the entire northern Gulf of Mexico not matched before or since (Fig. 3).

Four major fluvial axes dominated Oligocene Frio deposition (Galloway et al., 2000) (Figs 4 & 5). The large, sand-rich Norias delta system of the Rio Grande embayment was fed by a palaeo-Rio Grande fluvial system, the tributaries of which probably extended into the volcanic uplands of the south-western USA and northern Mexico (Fig. 6). Proximal uplift of Mesozoic strata along the western Gulf margin also initiated influx along the gravelly, sand-rich Norma axis, which drained directly from these newly elevated sources (Fig. 6). Although more distant, the Southern and Central Rocky Mountain highlands, rejuvenated by local volcanic heating and perhaps by climate change, supplied tributaries of the large, mixed-load dominated Houston axis and suspended-load dominated Mississippi axis. Note that the textural composition characteristic of each fluvial axis fines west-to-east across the Gulf rim from gravelly and sand-rich in Mexico to mud-dominated in Louisiana, reflecting the increasing distance between successive fluvial systems and their source uplands. As erosion reduced and graded these uplands, and as volcanism waned in the later Oligocene, the rate of Frio depositional episode accumulation in the Gulf progressively decreased (Galloway & Williams, 1991).

Miocene

Miocene sediment input rates decreased by half from the extreme highs of the syn- to early post-orogenic Oligocene depositional episode (Fig. 3); nonetheless, rates remained substantial throughout the Miocene, indicating ongoing active erosion in large continental drainage basins. The lower Miocene depositional episodes (LM1 and LM2) record the evolution from a Paleogene Gulf of Mexico dominated by sediment supply through north-western fluvial axes to the Neogene Gulf, which was and remains dominated by northerly fluvial input axes (Galloway et al., 2000). During the early Miocene the Rio Grande axis continued as a prominent locus of fluvial–deltaic deposition in the north-west Gulf (Fig. 4). The Houston axis, however, was abandoned in favour of the more easterly Red River axis, which is focused along the present Texas–Louisiana border. At the same time, the Mississippi axis assumed its role as a major depocentre. Together, the shift to the Red River axis and growth in relative importance of the Mississippi axis at the expense of the Rio Grande axis record the declining importance of the Sierra Madre uplands and the rejuvenation of sediment yield from the mature uplands of the Wyoming and Central Rockies and Appalachians.

By the middle Miocene, a new North American drainage regime was fully established (Fig. 6). Principal source areas included the Central and Southern Rockies along with the southern Appalachian–Cumberland Plateau uplands. Broad alluvial aprons built out from these uplands. To the west, during middle to late Miocene times, an aggrading, thin, but widespread apron of coarse-grained alluvial deposits spread eastward from the Central and Southern Rockies across the High Plains, forming the Ogallala Formation (Gustavson, 1996; Pazzaglia & Kelley, 1998). To the east, a broad belt of coarse, gravelly sands derived from the southern Appalachians now forms a broad, dissected and poorly dated alluvial apron that may be, in part, as old as Miocene (Saucier, 1994). The east and west flowing tributaries forming these aprons converged in the midcontinent before flowing southward along the Mississippi and Tennessee axes (Figs 4 & 5). A secondary basin-fringe fluvial system had integrated drainage from the Edwards Plateau by the middle Miocene, and entered the Gulf basin along the Corsair axis, located on the central Texas coastal plain (Figs 4 & 6).

Drainage basin evolution records a complex history of Miocene tectonics and climate change. Tectonic factors include the onset of Basin and Range deformation, uplift and tilting along the east flank of the Rocky Mountains and Edwards Plateau of Texas, and uplift of the southern Appalachian–Cumberland Plateau uplands. Basin and Range tectonics initiated formation of the Rio Grande rift, fragmenting the integrated drainage basins that extended across the south-western USA to form the Oligocene palaeo-Rio Grande (Galloway, 1989) (Fig. 6). Uplift of mountain blocks and High Plains east of the rift led to an estimated local exhumation of 2 km of early Cenozoic strata (Pazzaglia & Kelley, 1998) and rejuvenated Miocene drainage basins along the Colorado Front Range north into southern Wyoming (Bolyard, 1997; Steven et al., 1997). These sources became tributaries to the Corsair and Mississippi systems (Fig. 6). Reorientation of Central Rocky Mountain tributary axes from their south-eastern Oligocene trend to east and north-east trends (Steven et al., 1997) fits well with the observed abandonment of the Oligocene Houston axis and progressive shift to the lower Miocene Red River and then to the middle to upper Miocene Mississippi axis (Fig. 4). Concomitantly, the Edwards Plateau was uplifted, leading first to an initial flood of reworked Cretaceous sediment in Lower Miocene strata (Galloway et al., 1982) and then to the well-defined Middle–Upper Miocene Corsair fluvial axis as the many small basin-margin streams matured into an integrated system (Galloway et al., 2000). Miocene erosional unroofing of the Appalachian–Cumberland Plateau upland is well documented by influx of sediment onto the Atlantic margin (Poag, 1992) and fission track analysis (Boettcher & Milliken, 1994). The cause of the onset of aggressive erosion of the uplands after tens of million years of Paleogene passivity remains enigmatic. Both climate change and consequent acceleration of mechanical erosion accompanied by passive flexural and isostatic uplift (Boettcher & Milliken, 1994; Pazzaglia & Gardner, 1994) or crustal uplift (Poag, 1992; Pazzaglia & Brandon, 1996) are suggested. Regardless of origin, this newly activated source of sediment both initiated the palaeo-Tennessee River as a major extrabasinal sediment transport system and provided significant new tributary systems to the palaeo-Mississippi system (Fig. 6).

Pliocene and Pleistocene

Sediment influx into the Gulf records the combined effect of further intracratonic tectonic adjustment and pronounced climate change, culminating in glaciation in the headwaters of several drainage basins. Sediment influx to the Gulf remained high, with the singular exception of the mid-Pliocene *Globoquadrina altispira* (PGa) depositional episode (Fig. 3). Note that illustrated late Neogene supply rates are probably underestimated as increasing frequency and amplitude of late Neogene glacio-eustatic sea-level change led to increased proportional bypass to slope and basin systems where it is not reflected in calculations of delta and shore-zone system volumes (Galloway, 2002).

Evolution of fluvial supply to the northern Gulf of Mexico during the past 6 Myr records three basic drainage patterns (Galloway et al., 2000). Latest Miocene to earliest Pliocene drainage (PB1 depositional episode) reflects consolidation of principal supply through the Mississippi and Tennessee axes (Fig. 4). A south-flowing palaeo-Tennessee River was first postulated by Brown (1967). Together with local basin-fringe streams that drained the southern Appalachians, the palaeo-Tennessee deposited a gravelly sand veneer, the Citronelle Formation, which rests on a regional unconformity across the east-central Gulf coastal plain (Doering, 1956; Rosen, 1969; Autin et al., 1991; Autin, 1996). Concomitant with the diminished input of the PGa depositional episode, the Red River axis was revitalized; it then remained a substantial extrabasinal fluvial system until the latest Pleistocene (Fig. 4). Latest Pleistocene–Holocene depositional episodes record the increasing consolidation of the Red and Tennessee into the Mississippi.

Source uplands for fluvial supply to the Gulf were established in their present form by early Pliocene time (Fig. 6). A final phase of tectonic tilting probably further elevated the Rockies and adjacent western High Plains by up to several hundred metres and increased the eastward gradient established by the Miocene Ogallala apron (Bolyard, 1997; Steven et al., 1997; McMillan et al., 2002). In addition, the latest Miocene to early Pliocene continental interior was characterized by increased precipitation and storminess (Chapin & Kelley, 1997). Together, uplift and increased runoff initiated excavation of intermontane basin fills and the Ogallala apron (Bolyard, 1997; Chapin & Kelley, 1997; Steven et al., 1997). Runoff from these sources first entered northeast flowing tributaries inherited from the late Miocene (Fig. 6), then was reorientated by increasing eastward tilt to become tributary to the nascent Red River, which assumed prominence during the Pliocene PGa depositional episode. To the east, sandy, braided channels emerging from the Appalachian–Cumberland upland were tributary to the south-flowing palaeo-Tennessee (Fig. 6).

Appearance of alpine and then continental glaciation in the late Pliocene significantly impacted sediment supply and drainage patterns (Fig. 6). The PAB through to PS depositional episodes record glacially influenced fluvial discharge through the palaeo-Mississippi River into the Gulf (Fig. 5) (Galloway et al., 2000). The North American ice-sheet permanently diverted northern mid-continent rivers southward into the Mississippi drainage basin, significantly increasing discharge and sediment flux (Saucier, 1994). Cyclic incision of the Mississippi Valley during glacial stages resulted in periodic capture of the Red River and Tennessee, but these rivers resumed separate courses as the valley refilled with outwash. Only in the late Pleistocene (around 800 ka) did the Mississippi Valley become sufficiently incised that capture of these separate systems was complete and the modern Mississippi 'Father of Waters' was born, permanently integrating drainage from the Rockies to the Appalachians (Saucier, 1994).

Climatically enhanced runoff and erosion in Southern Rocky Mountain uplands (Formento-Trigilio & Pazzaglia, 1998; Dethier, 2001) rejuvenated supply through an again integrated Rio Grande drainage system, which once more delivered notable sediment into the Gulf. The modern rugged topographic relief in these areas is largely a product of this late Neogene erosion (Dethier, 2001).

CONCLUSIONS

This paper presents a preliminary synthesis of the Gulf of Mexico Cenozoic depositional record with the geological and climate record of potential North American sources for those sediments. The results of this synthesis confirm and expand upon observations of Winker (1982) and Galloway et al. (1991).

1 Principal extrabasinal fluvial systems traverse the northern Gulf coastal plain along a few, geologically long-lived axes. Five fluvial–deltaic axes provided the bulk of the sediment that infilled the northern Gulf basin (Fig. 2).
2 Two major Paleogene phases of sediment influx into the Gulf are clearly related to tectonic uplift of principal source uplands (Fig. 5). Paleocene through to middle Eocene pulses of Laramide

uplift along the Central and Southern Rockies and Sierra Madre Oriental supported the early Cenozoic depositional episodes. Crustal heating, volcanism and consequent uplift and erosion extending from central Mexico through to the south-west USA supplied the Oligocene Frio through to early Miocene depositional episodes.

3 Climate change is now accorded an important role in sustaining high rates of sediment supply to the Gulf, primarily through distinct Red, Mississippi and Tennessee transport axes, during the Neogene. Miocene incision of the Cumberland Plateau and Appalachians (Fig. 5) invigorated supply to the east-central Gulf basin. At the same time, the high-standing Rocky Mountain upland experienced continued regional exhumation. Pliocene uplift and tilting of the western High Plains further rejuvenated Rocky Mountain sources. High rates of Pleistocene sediment accumulation reflect rapid Quaternary climate cycling, and consequent pulses of glacial erosion and runoff directly into the principal sediment transport systems.

In summary, the strata of the northern Gulf of Mexico margin record the long-term tectonic and climatic history of the interior North American continent. At the multimillion year time-scale, tectonism and consequent uplift primarily controlled the time–space distribution of Paleogene through to early Neogene extrabasinal rivers that infilled the Gulf. Some major shifts in sediment supply axes and the inherent multimillion year pulse of late Paleogene depositional episodes, however, remain poorly explained. Climate, expressed as continental and montane glaciation, clearly dominated patterns of sediment supply in the Pleistocene. Debate continues about the relative roles of climate versus epeirogenic tectonism in the middle Miocene–Pliocene depositional episodes.

REFERENCES

Autin, W.J. (1996) Pleistocene stratigraphy in the southern Lower Mississippi Valley. *Eng. Geol.*, **45**, 87–112.

Autin, W.J., Burns, S.F., Miller, B.J., Saucier, R.T. and Snead, J.I. (1991) Quaternary geology of the Lower Mississippi Valley. In: *The Geology of North America*, Vol. K-2, *Quaternary Nonglacial Geology: Conterminous U.S.* (Ed. R.B. Morrison), pp. 547–582. Geological Society of America, Boulder, CO.

Boettcher, S.S. and Milliken, K.L. (1994) Mesozoic–Cainozoic unroofing of the southern Appalachian Basin: apatite fission track evidence from Middle Pennsylvanian sandstones. *J. Geol.*, **102**, 655–663.

Bolyard, D.W. (1997) Late Cainozoic history of the northern Colorado Front Range. In: *Geologic History of the Colorado Front Range* (Eds D.W. Bolyard and S.A. Sonnenberg), pp 125–134. 1997 Rocky Mountain Section–American Association of Petroleum Geologists Field Trip No. 7, Rocky Mountain Association of Geologists, Denver, CO.

Brown, B.W. (1967) A Pliocene Tennessee River hypothesis for Mississippi. *Southeast. Geol.*, **8**, 81–84.

Chapin, C.E. and Kelley, S.A. (1997) The Rocky Mountain erosion surface in the Front Range of Colorado. In: *Geologic History of the Colorado Front Range* (Eds D.W. Bolyard and S.A. Sonnenberg), pp. 101–113. 1997 Rocky Mountain Section–American Association of Petroleum Geologists Field Trip No. 7, Rocky Mountain Association of Geologists, Denver, CO.

Chase, C.G., Gregory-Wodzicki, K.M., Parrish, J.T. and DeCelles, P.G. (1998) Topographic history of the Western Cordillera of North America and controls on climate. In: *Tectonic Boundary Conditions for Climate Reconstructions* (Eds T.J. Crowley and K.C. Burke), pp. 73–99. Oxford University Press, New York.

Crabaugh, J.P. (2001) Nature and growth of nonmarine-to-marine clastic wedges: examples from the upper Cretaceous Iles Formation, western interior (Colorado) and the lower Paleogene Wilcox Group of the Gulf of Mexico Basin (Texas). Unpublished PhD thesis, University of Wyoming, Laramie, 201 pp.

Dethier, D.P. (2001) Pleistocene incision rates in the western United States calibrated using Lava Creek B tephra. *Geology*, **29**, 783–786.

Doering, J. (1956) Review of Quaternary surface formations of Gulf Coast region. *Am. Assoc. Petrol. Geol. Bull.*, **40**, 1816–1862.

Evanoff, E. (1990) Early Oligocene paleovalleys in southern and central Wyoming: evidence of high local relief on the late Eocene unconformity. *Geology*, **18**, 443–446.

Formento-Trigilio, M.L. and Pazzaglia, F.J. (1998) Tectonic geomorphology of the Sierra Nacimiento: traditional and new techniques in assessing long-term landscape evolution in the Southern Rocky Mountains. *J. Geol.*, **106**, 433–453.

Frazier, D.E. (1974) Depositional episodes: their relationship to the Quaternary stratigraphic framework in the northwestern portion of the Gulf basin. *Univ. Texas Austin, Bur. Econ. Geol. Geol. Circ.*, **71–1**, 28 pp.

Galloway, W.E. (1977) Catahoula Formation of the Texas coastal plain: depositional systems, composition, structural development, ground-water flow history and uranium mineralization. *Univ. Texas Austin, Bur. Econ. Geol. Rep. Investigations*, **87**, 59 pp.

Galloway, W.E. (1981) Depositional architecture of Cainozoic Gulf Coastal Plain fluvial systems. In: *Recent and Ancient Nonmarine Depositional Environments: Models for Exploration* (Eds F.G. Ethridge and R.M. Flores). *Soc. Econ. Paleontol. Mineral. Spec. Publ.*, **31**, 127–155.

Galloway, W.E. (1989) Genetic sequences in basin analysis II: application to northwest Gulf of Mexico Cainozoic basin. *Am. Assoc. Petrol. Geol. Bull.*, **73**, 143–154.

Galloway, W.E. (2002) Cenozoic evolution of sediment accumulation in deltaic and shore-zone depositional systems, Northern Gulf of Mexico Basin. *J. Mar. Petrol. Geol.*, **18**, 1031–1040.

Galloway, W.E. and Williams, T.A. (1991) Sediment accumulation rates in time and space: Paleogene genetic stratigraphic sequences of the northwestern Gulf of Mexico basin. *Geology*, **19**, 986–989.

Galloway, W.E., Bebout, D.B., Fisher, W.L., Cabrera-Castro, R., Lugo-Rivera, J.E. and Scott, T.M. (1991) Cainozoic. In: *The Geology of North America: the Gulf of Mexico Basin*, Vol. J, (Ed. A. Salvador), pp. 245–324. Geological Society of America, Boulder, CO.

Galloway, W.E., Ganey-Curry, P.E., Li, X. and Buffler, R.T. (2000) Cainozoic depositional history of the Gulf of Mexico basin. *Am. Assoc. Petrol. Geol. Bull.*, **84**, 1743–1774.

Galloway, W.E., Henry, C.D. and Smith, G.E. (1982) Depositional framework, hydrostratigraphy, and uranium mineralization of the Oakville Sandstone (Miocene), Texas coastal plain. *Univ. Texas Austin, Bur. Econ. Geol. Rep. Investigations*, **113**, 51 pp.

Gregory, K.M. and Chase, C.G. (1992) Tectonic significance of paleobotanically estimated climate and altitude of the late Eocene erosion surface, Colorado. *Geology*, **20**, 581–585.

Gustavson, T.C. (1996) Fluvial and eolian depositional systems, paleosols, and paleoclimate of the upper Cainozoic Ogallala and Blackwater Draw Formations, southern High Plains, Texas and New Mexico. *Univ. Texas Austin, Bur. Econ. Geol. Rep. Investigations*, **239**, 62 pp.

Hamlin, H.S. (1988) Depositional and ground-water flow systems of the Carrizo-Upper Wilcox, south Texas. *Univ. Texas Austin, Bur. Econ. Geol. Rep. Investigations*, **175**, 61 pp.

Hovius, N. (1998) Controls on sediment supply by large rivers. In: *Relative Role of Eustasy, Climate, and Tectonism in Continental Rocks* (Eds K.W. Shanley and P.J. McCabe). *Soc. Econ. Paleontol. Mineral. Spec. Publ.*, **59**, 1–16.

McMillan, M.E., Angevine, C.L. and Heller, P.L. (2002) Postdepositional tilt of the Miocene–Pliocene Ogallala Group on the western Great Plains: evidence for late Cainozoic uplift of the Rocky Mountains. *Geology*, **30**, 63–66.

Milliman, J.D. and Syvitski, J.P.M. (1992) Geomorphic/tectonic control of sediment discharge to the ocean: the importance of small mountainous rivers. *J. Geol.*, **100**, 525–544.

Pazzaglia, F.J. and Brandon, M.T. (1996) Macrogeomorphic evolution of the post-Triassic Appalachian Mountains determined by deconvolution of the offshore basin sedimentary record. *Basin Res.*, **8**, 255–278.

Pazzaglia, F.J. and Gardner, T.W. (1994) Late Cainozoic flexural deformation of the middle U.S. Atlantic passive margin. *J. Geophys. Res.*, **99**(B6), 12,143–12,157.

Pazzaglia, F.J. and Kelley, S.A. (1998) Large-scale geomorphology and fission-track thermochronology in topographic and exhumation reconstructions of the Southern Rocky Mountains. *Rocky Mount. Geol.*, **33**, 229–257.

Poag, C.W. (1992) U.S. Middle Atlantic continental rise: provenance, dispersal, and deposition of Jurassic to Quaternary sediments. In: *Geologic Evolution of Atlantic Continental Rises* (Eds C.W. Poag and P.C. de Graciansky), pp. 100–154. Van Norstad Reinhold, New York.

Rosen, N.C. (1969) Heavy minerals and size analysis of the Citronelle Formation of the Gulf Coastal Plain. *J. Sediment. Petrol.*, **39**, 1552–1565.

Saucier, R.T. (1994) *Geomorphology and Quaternary Geologic History of the Lower Mississippi Valley.* Mississippi River Commission, Vicksburg, 205 pp.

Steven, T.A., Evanoff, E. and Yuhas, R.H. (1997) Middle and late Cainozoic tectonic and geomorphic development of the Front Range of Colorado. In: *Geologic History of the Colorado Front Range* (Eds D.W. Bolyard and S.A. Sonnenberg), pp. 115–124. 1997 RMS–American Association of Petroleum Geologists Field Trip No. 7, Rocky Mountain Association of Geologists, Denver, CO.

Winker, C.D. (1982) Cainozoic shelf margins, northwestern Gulf of Mexico. *Gulf Coast Assoc. Geo. Soc. Trans.*, **32**, 427–448.

Wolfe, J.A., Forest, C.E. and Molnar, P. (1998) Paleobotanical evidence of Eocene and Oligocene paleoaltitudes in midlatitude western North America. *Geol. Soc. Am. Bull.*, **110**, 664–678.

Yurewicz, D.A., Chuchla, R.J., Richardson, M., Pottorf, R.J., Gray, G.G., Kozar, M.G. and Fitchen, W.M. (1997) Hydrocarbon generation and migration in the Tampico-Misantla basin and Sierra Madre Oriental, east central Mexico: evidence from an exhumed oil field in the Sierra de el Abra. In: *Sedimentation and Diagenesis of Middle Cretaceous Platform Margins, East Central Mexico*, American Association of Petroleum Geologists Annual Meeting Field Trip Guidebook, Dallas Geological Society and Society of Economic Paleontologists and Mineralogists, 24 pp.

Fluvial–estuarine transitions in fluvial-dominated successions: examples from the Lower Pennsylvanian of the Central Appalachian Basin

STEPHEN F. GREB* and RONALD L. MARTINO†
*Kentucky Geological Survey, University of Kentucky, Lexington, KY 40506, USA
(Email: greb@uky.edu); and
†Department of Geology, Marshall University, Huntington, WV 25755, USA

ABSTRACT

Early Pennsylvanian sedimentation in the Central Appalachian Basin was dominated by the successive development of south- to southwest-flowing, low-sinuosity streams in broad, longitudinal braidplains, which deposited a series of quartzarenites. Successive quartzarenite belts are locally separated by grey shales with brachiopods and other body fossils interpreted to represent marine- to brackish-water facies. Local features indicative of tidal sedimentation occur between fluvial facies and the marine- to brackish-water shales. Tidal features occur in transgressive successions between fluvial and overlying marine to brackish-water shales, and significant wave-generated features are absent, indicating that tide-dominated estuaries developed during transgressions. The boundary between fluvial facies and recognizable estuarine tidal facies represents a fluvial–estuarine transition. Tidal sedimentary features in the fluvial–estuarine transition, however, can be subtle, because upper estuarine channels may record only the most headward tidal effects in an otherwise fluvially dominated system.

Some of the possible tidal indicators noted in upper estuarine channel facies include local occurrences of opposing palaeoflow indicators, noncyclic rhythmites, lenticular bedding, small reversing ripples on the crests of underlying current ripples, sigmoidal cross-strata, cross-strata with rising troughs, thick–thin laminae pairs and bundled laminae in ripple cross-lamination. None of these features is diagnostic for tidal sedimentation. Where multiple tidal indicators are found within otherwise fluvial facies, within a probable transgressive succession, interpretation of an upper estuarine channel facies becomes more tenable.

Recognition of fluvial–estuarine transitions is important in fluvial-dominated basins, especially in the upper reaches of longitudinal basins, because the transitions may be the only evidence of correlative down-dip marine flooding surfaces. Identification of the transition zone facilitates the distinction between lowstand and transgressive systems tracts. In turn, such sequence analyses can increase the potential for predicating lateral changes in fluvial channel continuity, and vertical changes in porosity and permeability characteristic of lithological changes from fluvial to estuarine facies, both of which are important in exploration for hydrocarbons.

INTRODUCTION

The location and character of the transition between fluvial and estuarine strata are somewhat dependent on the definition of estuary used. Estuaries can be defined based on geomorphological, hydrological, biological, and chemical attributes (summarized in Perillo, 1995). Estuaries can be defined as being partly enclosed by land, having an open connection to the sea and showing significant salinity dilution landward (e.g. Pritchard, 1967). A more geological definition describes estuaries as drowned river valleys that receive sediment from marine and fluvial sources, which are influenced by tide, wave and fluvial processes (Zaitlin & Shultz, 1990; Dalrymple et al., 1992). Using the latter definition, the landward limit of the inner, river-dominated part of the estuary will extend farther inland than a definition based purely on salinity (Fig. 1) because tidal influences tend to occur farther headward than salinity influences (Dionne, 1963; Nichols & Briggs, 1985; Dalrymple et al., 1992; Perillo, 1995).

Many tide-dominated estuaries can be divided into three parts based on salinity and the hydrological effects of wave, tidal and current processes, although form varies owing to differences in tidal range, and the relative effects of tides, waves, and rivers (Dionne, 1963; Dalrymple et al., 1991, 1992; Allen, 1991). Those that have three parts consist of:

1 a marine-influenced lower estuary where waves and tides interact at the outer limit of estuarine sands, dominated by tidal sand bars and upper-flow-regime sand flats;

2 a zone of salt- and freshwater mixing called the middle estuary, which is dominated by tidal processes and is often muddy;

3 an upper estuary or fluvial-dominant zone marked by fresh water and the landward limit of tidally influenced sediments (Fig. 1; Dionne, 1963; Dalrymple et al., 1991, 1992; Allen, 1991; Perillo, 1995).

The upper estuary encompasses the fluvial–estuarine transition, which is the focus of this study.

In the tide-dominant estuary model of Dalrymple et al. (1992), the fluvial–estuarine transition occurs within the upper estuarine channel, the boundary of which is marked by the

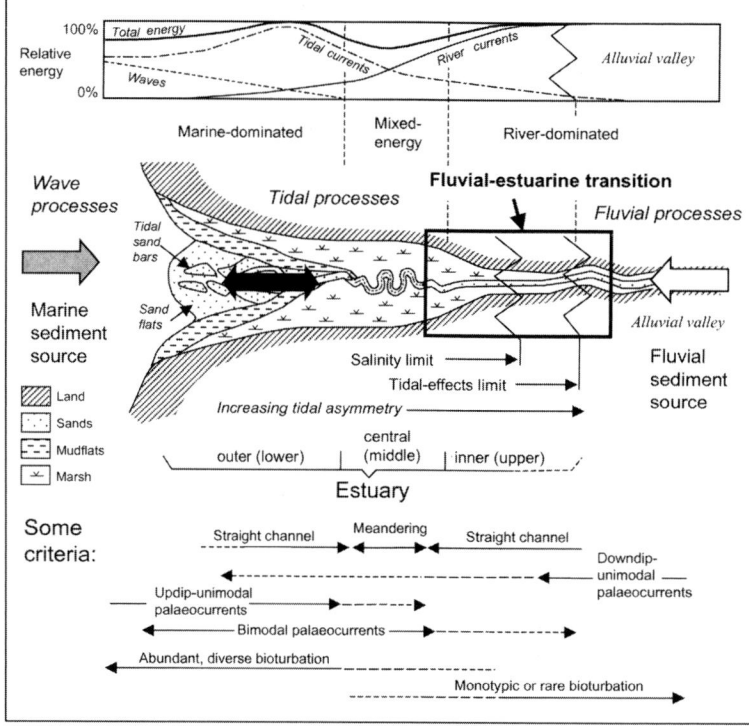

Fig. 1 Diagram showing typical features of a tide-dominated estuary (after Dalrymple et al., 1992; Perillo, 1995). The fluvial–estuarine transition is highlighted in the black rectangle. No scale implied.

headward limit of tidal effects on sedimentation. Limited studies of modern fluvial–estuarine transitions in tide-dominated estuaries indicate relatively narrow transition zones between fluvial and tidal facies (e.g. Smith, 1988; Lanier & Tessier, 1998). A certain amount of temporal variability in the relative influence of tides, river discharge, waves, climates, rates of transgression, or net sediment supply could cause lateral migration of the transition based on the relative contribution of influencing factors. In addition, there are several modern tide-dominated estuaries that are associated with rivers with large discharge (Wells, 1995). In large, tidally influenced rivers tidal influences extend well up-river. For example, macrotidal conditions exist in the St Lawrence River 500 km inland from the Gulf of St Lawrence (Archer, 1995). In the Amazon River, flood tides reach 800 km inland and may influence sedimentation at least 200 km inland (Archer, this volume, pp. 17–39). If these rivers were transgressed, and estuaries developed within the drowned river system, the potential length of the upper estuary, if defined by the most landward occurrence of tidal sedimentation, could extend much farther inland than is typical for what is commonly considered an estuary.

Ancient estuarine strata have been interpreted in several basins, generally based on identification of a transgressive succession, and bimodal current indicators and/or trace-fossil suites suggestive of mixed or changing salinities (Rahmani, 1980; Diemer & Bridge, 1988; Smith, 1988; Shanley et al., 1992; Archer et al., 1994; Kvale & Barnhill, 1994; Greb & Chesnut, 1996; Els & Mayer, 1998; Lanier & Tessier, 1998). In ancient fluvial-dominated successions, sequence-stratigraphy analysis of fluvial successions is difficult where marine- and brackish-water indicators are absent because transgressive and maximum flooding surfaces cannot be defined (Emery & Myers, 1996). The purpose of this paper is to describe fluvial–estuarine transitions and their variability in fluvial-dominated Lower Pennsylvanian strata of the Central Appalachian Basin within a facies framework based on a geological definition of tide-dominated estuaries (sensu Dalrymple et al., 1992). Recognition of the variability in this transitional facies may aid in the recognition of similar facies in fluvial-dominated successions in other basins.

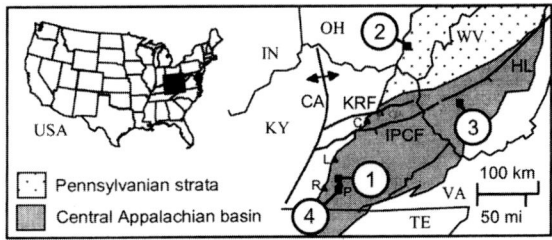

Fig. 2 Location of the Central Appalachian Basin showing major structures and study locations (numbered 1–4): CA, Cincinnati Arch; KRF, Kentucky River Fault Zone; IPCF, Irvine-Paint Creek Fault Zone; HL, hingeline; triangles, locations discussed in Greb & Chesnut (1996); C, Corbin Sandstone; L, Livingston palaeovalley; P, Pine Creek Sandstone; R, Rockcastle Sandstone; IN, Indiana; KY, Kentucky; OH, Ohio; TN, Tennessee; WV, West Virginia; VA, Virginia. *Pine Creek Sandstone is the informal name for the upper Bee Rock Sandstone in the study area.

STRATIGRAPHY AND PALAEOGEOGRAPHY

The Central Appalachian Basin is a foreland basin, which covers parts of Tennessee, Kentucky, West Virginia and Ohio. The basin is bordered on the west by the Cincinnati Arch (CA in Fig. 2) and is elongated in a NE–SW orientation (Fig. 2). The Kentucky River Fault System (KRF in Fig. 2) and Irvine–Paint Creek Fault System (IPCF in Fig. 2) in Kentucky, and a structural hingeline (HL in Fig. 2) in West Virginia, represent a structural hingeline on the northern margin of the basin. Strata thicken significantly south of this hingeline and continue to thicken to the south-east toward the preserved axis of the foreland basin in south-eastern Kentucky and western Virginia (Fig. 3A & B; Chesnut, 1992; Greb et al., 2002).

The stratigraphical nomenclature of Lower Pennsylvanian strata is shown in Fig. 4.

Lower Pennsylvanian strata are dominated by thick quartzarenite tongues (Figs 3B & 4). In plan view, the quartzarenite tongues occur as elongate belts, which successively overlap the basin margin to the west and north (Fig. 3B). Successive quartzarenite belts both truncate and interfinger with coal-bearing strata to the south-east (Figs 3B & 4). Lower Pennsylvanian quartzarenites were previously interpreted as beach barriers (e.g. Ferm et al., 1971), tidal strait deposits (e.g. Cecil & Englund, 1989) and fluvial facies (e.g. Potter &

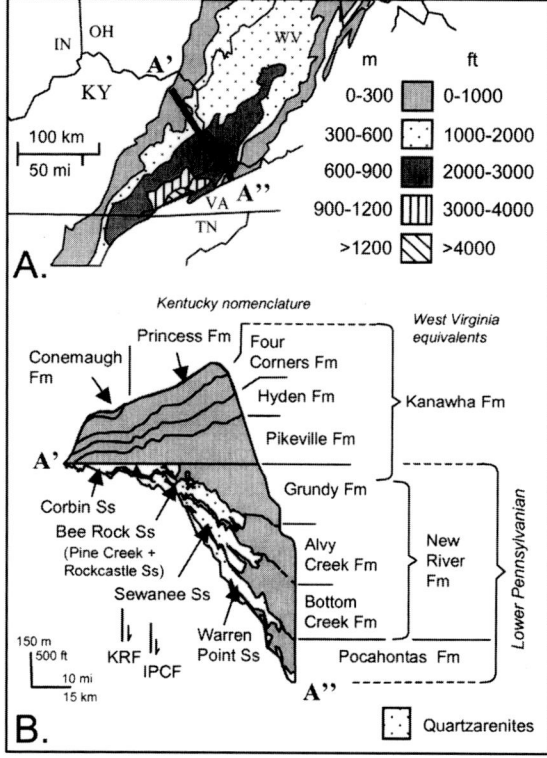

Fig. 3 Pennsylvanian strata in the Central Appalachian Basin. (A) Isopach showing south-eastward thickening and position of cross-section A′–A″. (B) Cross-section of Pennsylvanian strata showing the westward overlapping of quartzarenite belts on the western margin of the basin (modified from Chesnut, 1992). Datum is the Betsie Shale at the Lower–Middle Pennsylvanian boundary. KRF, Kentucky River Fault Zone; IPCF, Irvine–Paint Creek Fault Zone; IN, Indiana; KY, Kentucky; OH, Ohio; TN, Tennessee; WV, West Virginia; VA, Virginia.

Siever, 1956). Recent investigations have favoured fluvial deposition with most investigators interpreting the thick quartzarenites as braidplain deposits (Bement, 1976; Rice, 1984, 1985; Chesnut, 1992; Rice & Schwietering, 1988; Wizevich, 1991; Barnhill, 1994). Greb & Chesnut (1996) showed that although many of the quartzarenite belts are dominated by fluvial facies, tidal facies occur locally toward the top of each unit.

Figure 5 is a generalized view of Early Pennsylvanian palaeogeography in the Central Appalachian Basin and surrounding areas. Longitudinal, braided-stream trunk systems were developed on the western side of the basin. Quartz-pebble-bearing quartzarenites deposited within the braidplains had a north-eastern source (Donaldson et al., 1985; Chesnut, 1992, 1994; Archer & Greb, 1995; Greb & Chesnut, 1996). Lithic arenites and sublitharenites were deposited on the eastern side of the basin within transverse drainages with a source to the south-east (Ferm, 1974; Englund, 1974; Houseknecht, 1980). These streams flowed toward the north-west, presumably draining into the longitudinal braidplains (Fig. 5). Greb & Chesnut (1996) noted that sedimentary structures typical of tidal sedimentation and local bioturbation were developed at the top of four different quartzarenites on the western side of the basin, which in turn were overlain by dark grey shales with marine fauna. These features were inferred to indicate Early Pennsylvanian transgressive successions. The identification of tidal sedimentary structures in these fluvial-dominated strata is important for the identification of transgressive successions as

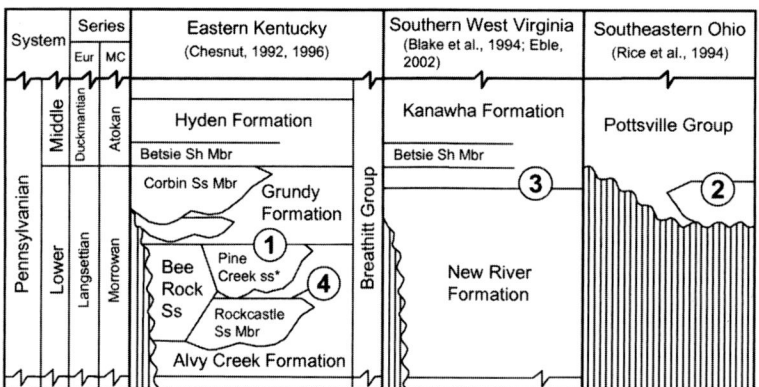

Fig. 4 Stratigraphical correlations of Lower and lower Middle Pennsylvanian strata in the Central Appalachian Basin. Numbers 1 to 4 refer to the stratigraphical positions of locations used in this study.

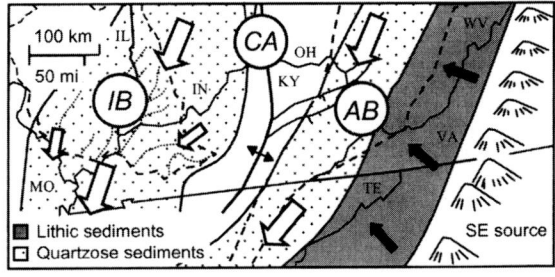

Fig. 5 Palaeogeographical map during the Early Pennsylvanian showing development of longitudinal and transverse drainages in the Central Appalachian Basin (AB). The Central Appalachian Basin was separated from the Illinois Basin (IB) by the Cincinnati Arch (CA). Quartzarenites occur in both basins and were sourced from the north-east (white arrows), whereas the source of lithic sediments in the Central Appalachian Basin was mostly to the south-east (black arrows) (modified from Chesnut, 1992; Greb & Chesnut, 1996). IN, Indiana; KY, Kentucky; OH, Ohio; TN, Tennessee; WV, West Virginia; VA, Virginia.

marine fauna are rare in Lower Pennsylvanian strata of the basin (e.g. Chesnut, 1991).

Early Pennsylvanian seas were interpreted to have transgressed from the south-west up the longitudinal drainage belts (Greb & Chesnut, 1996). In one example, the tidal facies succeeded fluvial facies within the Livingston Palaeovalley (L in Fig. 2), a bedrock-confined palaeovalley, indicating estuarine conditions during transgression. Wave-formed structures are rare to absent in these successions, so that if estuaries were developed as part of the succession, they were tide-dominated. In other parts of the basin, burrowed, compound cross-bedded sandstones in the Kanawha Formation of West Virginia below the Betsie Shale have been interpreted as tidal sand-bar facies (Martino, 1996; Hamrick, 1996). Such sand bar facies are typical of lower estuarine settings in modern macrotidal estuaries (Dalrymple et al., 1992), which lends support to the application of tide-dominated estuary models.

Herein, facies are interpreted for Lower Pennsylvanian transgressive successions and examples are illustrated from different parts of the basin. Examples of the variety of vertical successions are shown in order to illustrate differences in fluvial–estuarine transitions in these fluvial-dominated sequences. Well-defined transitions are compared with more subtle transitions in order to determine the varied appearance of fluvial–estuarine transitions in individual vertical sections, which may aid in interpreting similar transitions in other basins, or in less well-exposed areas. Documentation of the variability in fluvial–estuarine transitions is important in fluvial-dominated successions because these may be the only facies that can be used to infer the up-dip equivalents of marine flooding surfaces, which are important for regional sequence analyses.

COMMON LITHOFACIES

Typical facies at the top of Lower Pennsylvanian quartzarenite belts in the Central Appalachian Basin are shown in Table 1 and described below.

Fluvial cross-bedded sandstone facies

Each of the Lower Pennsylvanian quartzarenite belts is dominated by coarse- to fine-grained, cross-bedded sandstones, arranged in a variety of downstream accreting-bar and channel macroforms (Wizevich, 1991, 1992, 1993; Barnhill, 1994). Scour-based, cross-bedded sandstones may fine upward into ripple-bedded sandstones. Palaeocurrent modes are unimodal to the south and south-west. In general, the cross-bedded sandstones have mostly been interpreted as fluvial deposits (Table 1). For the purpose of analysing fluvial–estuarine transitions, herein, cross-bedded sandstones with unimodal current indicators orientated in a down-dip direction within the quartzarenite belts are interpreted as fluvial cross-bedded sandstone facies. An attempt is not made to interpret individual macroforms or lithofacies within the all-inclusive fluvial sandstone facies for the purpose of defining fluvial–estuarine transitions herein.

Peat-mire coal facies

Fluvial cross-bedded sandstone facies may be immediately overlain or may fine upwards into thin rooted horizons overlain by coals. Coal beds occur above each of the major quartzarenite formations (Chesnut, 1992), and are widely accepted as the accumulations of topogenous to ombrogenous peat mires (Table 1).

Table 1 Facies used in this study.

Facies	Description	Physical structures	Palaeoflow	Interpretation
Fluvial cross-bedded sandstone channel	Scour-based, cross-bedded, fine to coarse-grained quartzose sandstones, locally conglomeratic with quartz pebbles. Generally multistorey deposits	Planar and trough cross-bedding common, local current-ripple lamination. Channel and down-stream-accreting, compound macroforms common. Occasional fossil plant debris	Unimodal down-dip	Fluvial bedload (braided) channels (Bement, 1976; Rice, 1984; Wizevitch, 1991, 1992, 1993; Greb & Chesnut, 1996)
Peat-mire coal	Thin (< 1.5 m) coal beds. Underlain by rooted palaeosols or seat earths	Layers of different texture and brightness. Underlain by root structures, seat earths/palaeosols	No flow	Different kinds of peat mires (Donaldson et al., 1985; Eble et al., 1991; Eble, 1996)
Marine- to brackish-water shale	Dark grey to medium grey laminated, silty shales. Sharp-based and coarsening upward	Planar lamination. Lenticular, flaser, and wavy bedding where coarsening upward. Marine- to brackish-water body and trace fossils toward base	Few flow indicators	Marine- to brackish-water seaways (Chesnut, 1991, 1992; Martino, 1994, 1996)
Heterolithic tidal flat	Interlaminated silty shales, siltstones, and very fine- to fine sandstones. May fine upward. Only defined where bioturbated or tidal-bedding features are noted	Flaser to lenticular lamination and rhythmic lamination. Local climbing-ripple lamination. Biogenic structures and soft-sediment deformation rare to uncommon. Locally rooted at top	Unimodal down-dip to variable or bimodal	Intertidal flats (Greb & Chesnut, 1996), similar to modern flats (Klein, 1977; Terwindt, 1988; Nio & Yang, 1991; Dalrymple et al., 1991)
Upper estuarine cross-bedded sandstone channel	Scour-based, cross-bedded, fine to medium-grained quartzose sandstones, locally interbedded with ripple-bedded sandstone. Only defined where tidal bedding features are noted	Bedding similar to fluvial cross-bedded facies, but cross-beds exhibit local sigmoidal bedding, bundled foresets, shale-draped foresets, thick–thin foreset couplets or rhythmic lamination. Generally more shale inter-lamination than fluvial channel facies. Bioturbation absent to rare	Unimodal down-dip to variable with rare to uncommon up-dip modes preserved locally	Fluvial-dominant, straight channels that become more tidal seaward characterize the upper estuary (Allen, 1991; Dalrymple et al., 1992). Shale-draped bedding more common than in fluvial channel (Rahmani, 1988; Smith, 1988)
Middle estuarine heterolithic channel	Scour-based, interbedded silty shales, siltstones, and very fine- to fine sandstones as inclined stratification with lateral accretion. May be interstratified with cross-bedded sandstones	Parallel laminated to rhythmically laminated, local flaser to lenticular bedding, soft-sediment deformation, and cross-bedding as in the upper estuarine channel facies. Bioturbation rare to abundant. Top may be rooted	Unimodal down-dip, unimodal up-dip, bimodal, or variable	Heterolithic meandering channels common in the middle estuary of tide-dominant estuaries (Allen, 1991; Dalrymple et al., 1992) and tidal channels (Thomas et al., 1987)

Marine- to brackish-water grey-shale facies

Fluvial facies at the top of the quartzarenites, or the coals atop the quartzarenites, are locally succeeded by dark grey shales. These shales may contain local marine- or brackish-water fauna towards their base. Shales coarsen upward into laminated siltstones and sandstones. Similar Middle Pennsylvanian shales are widely interpreted as marine zones deposited in seaways during transgressions (Table 1). The bases of Middle Pennsylvanian marine zones have been interpreted as marine flooding surfaces and have been used to define third-order sequences (Aitken & Flint, 1995). At least one Lower Pennsylvanian shale (above the Pine Creek Sandstone, Fig. 4) has a fossiliferous siderite layer at its base, also interpreted as a marine flooding surface (P in Fig. 2, Greb & Chesnut, 1996). The fact that marine fauna are much lower in diversity and abundance in Lower Pennsylvanian transgressive grey shales than lithologically similar Middle Pennsylvanian transgressive grey shales does not preclude the Lower Pennsylvanian grey shales from representing transgressive deposits, but indicates that Early Pennsylvanian transgressions were more diluted or less extensive than their Middle Pennsylvanian counterparts. The coarsening upward part of the facies is interpreted to represent progradation of coastal–deltaic facies during highstands, similar to interpretations for the Middle Pennsylvanian.

Middle estuarine heterolithic channel facies

Greb & Chesnut (1996) identified scour-based, bioturbated, inclined heterolithic strata (IHS of Thomas et al., 1987) at the top of the Rockcastle Sandstone (R in Fig. 2), one of the Lower Pennsylvanian quartzarenites (Figs 3B & 4). Inclined heterolithic strata form in meandering mixed-load channels in fluvial and tidal environments (Thomas et al., 1987). Where non-bioturbated IHS occurs in vertical succession above the fluvial cross-bedded sandstone facies, the IHS could represent a change in channel morphology from bedload fluvial-braided streams to mixed-load meandering streams. Such a change could occur as rising base level decreased stream gradient. In cases where the IHS are bioturbated with brackish- to marine-trace fossils, a tidal channel origin is probable (Table 1). Likewise, shaly IHS, with shales draped across the macroform, may be more common in tidal IHS than fluvial IHS. Shale layers extending from the upper to lower margins of IHS have been used to infer tidal influences in other deposits (Rahmani, 1988). Deposition of mud on fluvial point bars occurs where suspended loads are high and system energy is relatively low, but generally is restricted to the upper portion of the point bar (Collinson, 1996). Such mud layers become more common and more extensive where even minor tidal influence occurs (Smith, 1988; Allen, 1991).

In the depositional model developed by Dalrymple et al. (1992) for tide-dominated estuaries, the alluvial channel enters the inner estuary as a tidally influenced, but fluvial-dominated straight channel. There is a net seaward transport in the channel owing to the prevalence of fluvial currents. Mud-rich meandering channels are more likely to develop in the central, low-energy zone of tide-dominated estuaries (Smith, 1988; Allen, 1991; Dalrymple et al., 1991, 1992). Such facies are a distinctive component of the tide-dominated estuarine model, and are not found in wave-dominated estuaries.

Estuarine tidal-flat facies

Greb & Chesnut (1996) also noted local occurrences of bioturbated, parallel-laminated to ripple-laminated heterolithic strata at the top of one of the Lower Pennsylvanian quartzarenites (L in Fig. 2). Some of these facies contain rhythmic lamination typical of tidal facies, although non-cyclic rhythmites (*as defined by* Greb & Archer, 1998) dominate. Laminated heterolithic strata may grade upward into rooted strata capped by coals. In a transgressive succession, this type of bedding could be interpreted as tidal-flat deposits (Table 1). In modern, tide-dominated estuaries, tidal flats occur from the lower to upper estuary, and are often bounded landward by marshes (e.g. Dalrymple et al., 1992). In general, tidal flats become sandier, exhibit increasing sedimentation rates, and become less bioturbated headward in the estuary (Dalrymple et al., 1991, 1992). Rhythmites recording daily to neap–spring cyclicity are perhaps best recorded on tidal flats in the inner or upper estuary where there is little bioturbation and increased tidal amplification owing to funnelling effects of the estuary (Dalrymple et al., 1991, 1992; Tessier, 1993, 1998). Decreasing preservation potential of sedimentation within rhythmic lamination is noted higher on flats, which can lead to the deposition of non-cyclic, rather than cyclic tidal rhythmites (Archer, 1998; Tessier, 1998).

Upper estuarine sandstone channel facies

Upper estuarine channels near the transition zone with fluvial channels would be expected to be river-dominated straight channels, exhibiting some evidence of tidal sedimentation (Fig. 1 and Table 1). In Lower Pennsylvanian transgressive successions, down-dip-orientated planar cross-beds typical of the fluvial cross-bedded sandstone facies are locally interbedded with flaser- to lenticular-bedded sandstones that contain rare brackish-water trace fossils, non-cyclic rhythmites, and/or sedimentary structures with up-dip-orientated current indicators. Cross-beds may show increased foreset concavity, crude thick–thin foreset alternations, shale-draped foresets, sigmoidal-shaped foresets or rising trough levels. If bioturbation, rhythmic lamination or bedding indicates tidal effects, in an otherwise fluvial-dominated channel, the deposits are re-interpreted as upper estuarine sandstone channel facies. The headward limit of the upper estuary is the headward limit of tidal influence (Dalrymple et al., 1992), so the headward limits of upper estuarine channels might exhibit only rare tidal signatures and be difficult to interpret.

TRANSGRESSIVE SUCCESSIONS

Examples of Lower Pennsylvanian transgressive successions at the tops of the quartzarenites are documented around the basin to show their varied appearance. In some cases, there is abundant evidence of tidal sedimentary features, which can be used to interpret an upper estuarine sandstone channel facies. In other cases, delineation of the facies is based upon more speculative evidence, and the facies is difficult to differentiate from underlying fluvial deposits.

Example 1

Greb & Chesnut (1996) previously used a vertical section at the top of the Pine Creek sandstone along Kentucky Highway 80 in Pulaski County to illustrate a gradual transition from fluvial to tidal bedding within previously inferred fluvial facies (location 1, Figs 2 & 6). Herein, strata within the transgressive succession are used as an example of a gradual change from fluvial to tidal deposition within a tidally dominated estuary framework.

Bedding within the lower two-thirds of the sandstone is dominated by unimodal, southwest-orientated cross-bedding (Fig. 7A) of the fluvial cross-bedded sandstone facies (Table 1). The upper part of the sandstone is shown in Fig. 6. At this location, there is a vertical shift from unimodal, down-dip-orientated palaeocurrents to variable and then up-dip-orientated palaeocurrents (Fig. 6), assigned to upper estuarine sandstone channel facies, middle estuarine heterolithic channel facies, and estuarine tidal-flat facies.

Upper estuarine channel facies at this location occur above two scours at the top of the Pine Creek sandstone. The lower scour fill is sheet form, fines upward (Fig. 7B), and exhibits a low angle of dip to the south-west. Sheets contain isolated trough cross-beds down-dip and are dominated by flaser bedding up-dip. Some ripples exhibit crude bundling of foreset laminae (alternating thicker and thinner sets) separated by shale drapes (Fig. 7C). Some ripples in the upper channel exhibit possible smaller scale ripples orientated in opposing directions on their crests or in troughs between ripples (Fig. 7D). Rare vertical burrows also occur but are not very distinct (Fig. 7D). Some down-dip-orientated foresets in isolated cross-beds (Fig. 7E) exhibit possible thick–thin alternations (Fig. 7F). A single up-dip-orientated cross-bed occurs at the top of the sandstone (Fig. 6).

The upper channel is incised sharply into the underlying sheet-form sandstones (Fig. 6). The upper channel is also dominated by ripple bedding and cross-bedding, but contains more parallel lamination and shale clasts than the underlying sheet sandstone. Asymmetric cross-beds have mostly up-dip orientations (Figs 6 & 8A–C). Some cross-beds exhibit crude bundling of

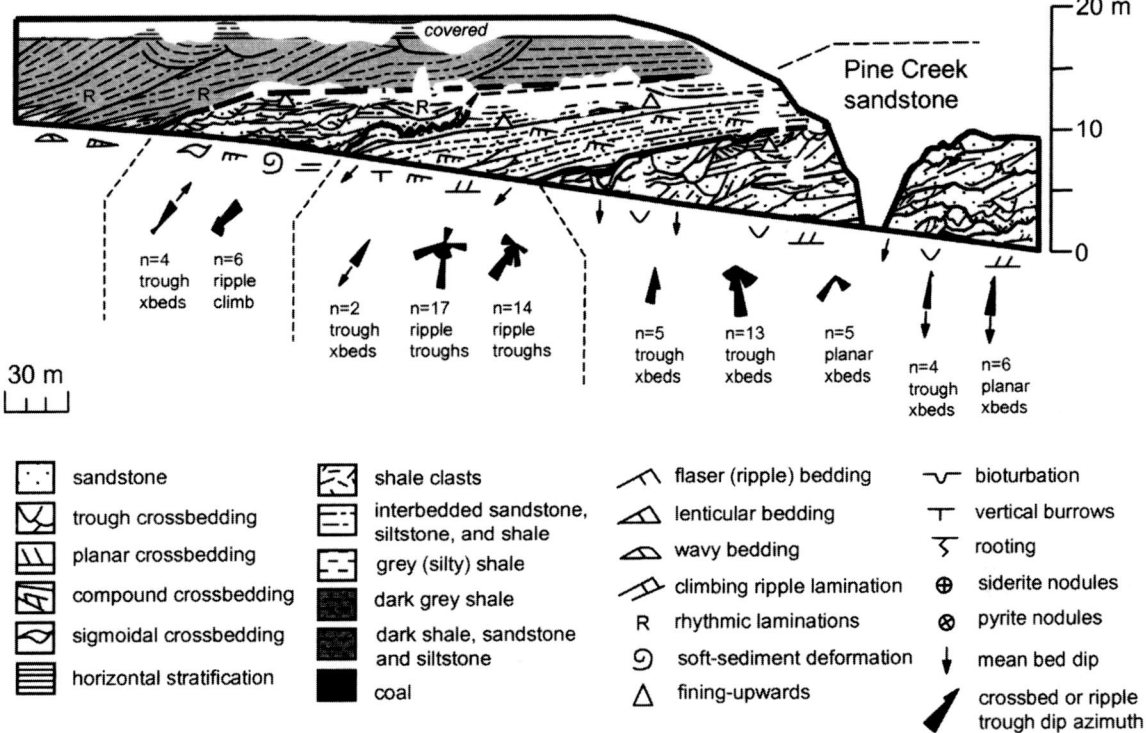

Fig. 6 Line drawing of upper Pine Creek sandstone at location 1, Kentucky Highway 80, east of the Rockcastle River, Laurel County, Kentucky.

thicker, sandy foresets, and thinner, shale-draped foresets (black arrows in Fig. 8A), thick–thin foreset alternations (Fig. 8B), laterally rising trough levels and sigmoidal foresets (Fig. 8C). Non-cyclic, rhythmic lamination occurs above a small scour at the top of the sandstone (Figs 6 & 8D). Crude thick–thin bundling occurs within climbing ripples within this scour (arrows in Fig. 8E).

The sandstone is sharply overlain by dark grey, shaly heterolithic strata correlated to the Dave Branch Shale (Figs 6 & 8F). At other locations marine fauna have been found in this shale (Chesnut, 1991; Greb & Chesnut, 1996). The lower part of the heterolithic facies here contains bundles of sandier and shalier strata (Fig. 8F & G). Sandier bundles consist of rhythmic lamination and lenticular bedding. Individual ripple laminae are separated by packets of 5–8-mm-scale rhythmic laminations (Fig. 8G). Some bundles show thickening of ripples toward the centre of bundles, and possible reversing crest ripples (black arrow in Fig. 8H).

Interpretation

The base of the upper estuarine sandstone channel facies is placed along the scour at the base of the sheet-form sandstones because this marks a major bedding change. Dalrymple *et al.* (1992) noted that facies contacts within estuarine successions were likely to coincide with erosional channel bases. Ripple cross-laminae azimuths are more varied than in the underlying fluvial facies, but reversing current indicators are rare (Fig. 6). Likewise, cross-beds are still mostly orientated down-dip, but show thick–thin foreset alternations. Thick–thin foreset alternation is a common feature of diurnal inequality in tides formed in semi-diurnal settings (Rahmani, 1988; Smith, 1988; DeBoer *et al.*, 1989; Kvale *et al.*, 1989; Nio & Yang, 1991). Visser (1980) considered persistent

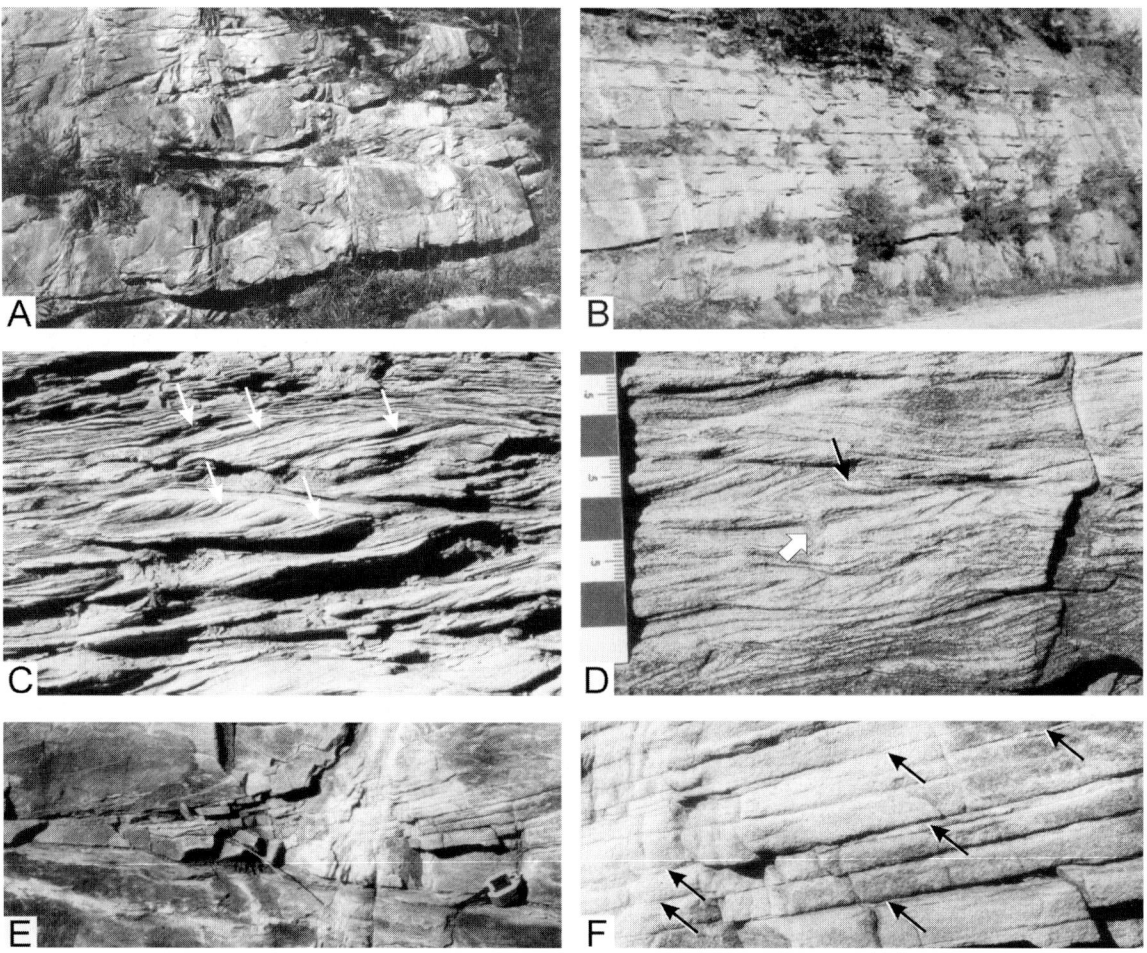

Fig. 7 Pine Creek exposures, location 1. (A) Planar and trough cross-beds in fluvial facies. Hammer scale = 30 cm. (B) Sheet sandstones at the top of the upper main sandstone. Yard stick scale = 0.9 m. (C) Ripples in sheet sandstones showing alternation between thicker, sandier bundles of foresets, and thinner, shalier bundles of foresets (white arrows). Large ripples are *c.* 2 cm thick. (D) Vertical burrow (white arrow). Note possible reversing crest ripple on overlying ripple (black arrow). Scale in millimetres. (E) Isolated down-dip-orientated cross-bed in ripple-bedded sandstone sheets. Brunton scale. (F) Alternating thick–thin laminae foresets (arrows point to thin parts of each pair) in cross-bed. Thick laminae are *c.* 5 to 6 mm thick.

thick–thin laminae alternations in cross-beds diagnostic of estuarine facies. The thick–thin laminations here are not persistent, but may be the first headward indicator of tidal influence in this succession. In the headward part of upper estuarine channels, where fluvial conditions dominate, persistent tidal features would not be expected.

Up-dip-orientated cross-beds in the upper channel are more easily inferred to be tidally influenced. This channel is still dominated by ripple-bedding and cross-bedding so is still inferred to represent an upper estuarine channel, although it contains some features more typical of the inner estuary. The increase in up-dip palaeocurrent indicators (Fig. 6), parallel lamination, shale clasts and cross-beds with tidal indicators suggest a position more seaward than the underlying sheet sandstones. Bundled foresets are common in tidal-estuarine cross-beds (Visser, 1980; Boersma & Terwindt, 1981; Nio & Yang, 1991). Thinner,

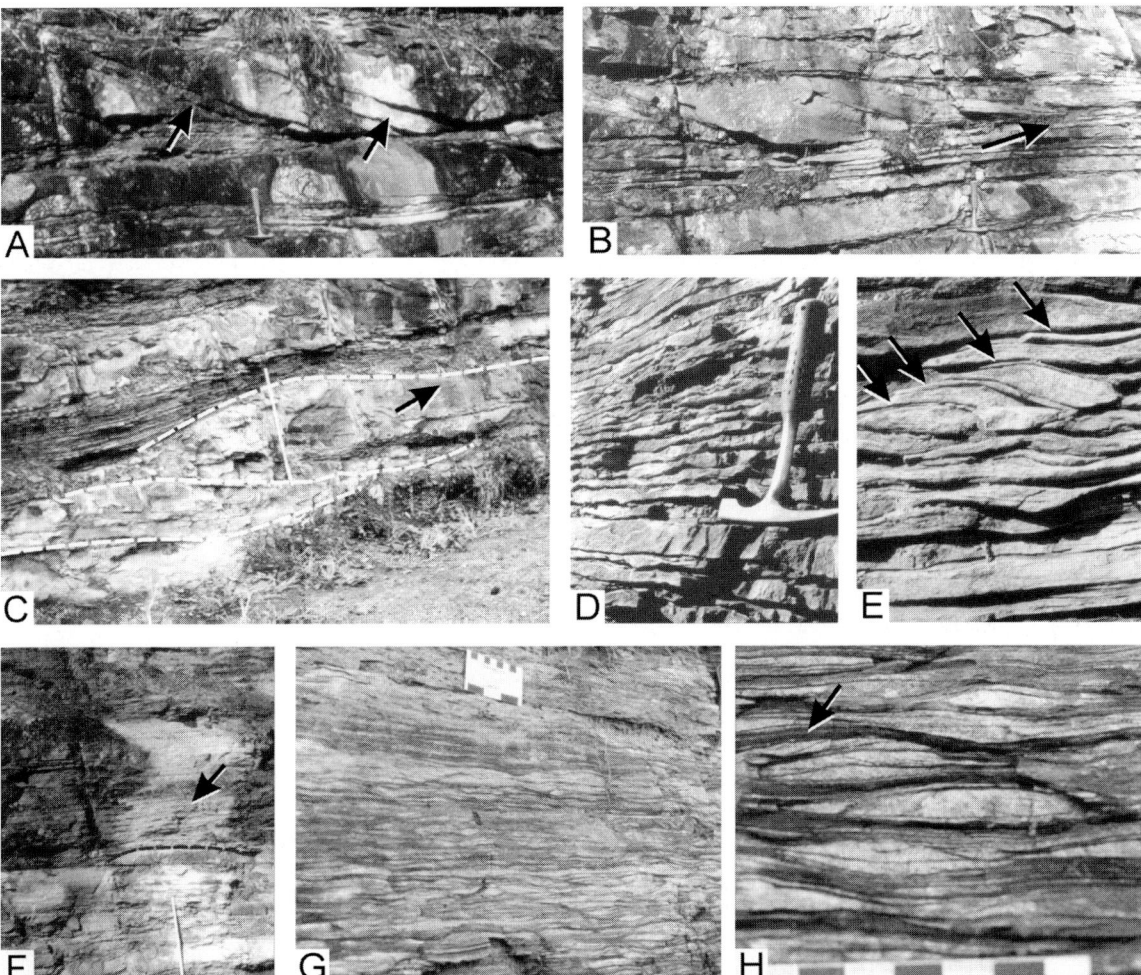

Fig. 8 Pine Creek exposures, location 1. (A) Up-dip-orientated asymmetric cross-bed showing crude bundling of thinner, shale-draped foresets (black arrow) with thicker bundles of sandier foresets. Hammer scale = 30 cm. (B) Small, up-dip-orientated asymmetric cross-bed (black arrow) showing crude thick–thin foreset alternation and soft-sediment deformation (white arrow). (C) Sigmodial foresets (dashed lines) in up-dip-orientated asymmetric cross-bed. Yard stick scale = 0.9 m. (D) Rhythmic lamination in small scour fill. Hammer scale = 30 cm. (E) Climbing ripples in same scour fill showing thick–thin foreset alternations along sigmoidal reactivation surfaces (black arrows). (F) Base of shaly, inclined heterolithic strata (dashed line) with bundles of rhythmically laminated strata (black arrow). Yard stick scale = 0.9 m. (G) Bundles of laminae at the base of the shaly heterolithic strata. Note alternating bundles of sandier, ripple-bedded strata with shalier flaser- to lenticular-bedded sandstone. (H) Detail of bundle showing ripple cross-lamination alternating with millimetre-scale rhythmic lamination. Some ripples have possible reversing crests (black arrow). Scale in centimetres.

shalier foresets bundles represent neap-tide migration, and thicker, sandier foreset bundles represent spring-tide migration. Likewise, rising trough cross-beds with increasing foreset concavity can be formed by rapid upgrowth of troughs during neap periods (Boersma & Terwindt, 1981).

Sigmodial bedding (Kreisa & Moiola, 1986; Nio & Yang, 1991) and thick–thin foreset lamination are also common in tidal cross-bedding.

The unusual occurrence of bundled climbing ripples in the small scour at the top of the upper channel suggests periodic alternations in deposition across a relatively short time span, which also would be more typical of tidal influences than fluvial influences. Climbing ripples with neap–spring–neap changes in shale concentration and climb angle related to neap–spring–neap changes in energy have been documented in the Mont-St-Michel estuary (Lanier & Tessier, 1998). Ebb-dominant climbing ripples are only common in chute channels and point bars of the meandering middle estuary in that analogue. The example herein is ebb-orientated (Fig. 6) and occurs in a small scour at the top of the sandstone, which could represent a chute channel at the inner–middle estuary transition. Overlying shaly heterolithic strata above the sandstone are inclined and represent middle estuarine heterolithic channel facies, although better examples of this facies occur in examples 2 and 3.

Within the middle estuarine heterolithic channel facies, bundles of five to eight laminae couplets overlain or truncated by ripple cross-lamination (as in Fig. 8H) could be interpreted as individual neap–spring deposits. Bundles of neap–spring deposits are arranged in alternating bundles of shalier, rhythmic-lamination-dominated and sandier, ripple cross-bed-dominated bundles, interpreted as annual deposits, similar to bundling described in tidal channels and tidal flats elsewhere (Greb & Archer, 1998; Tessier, 1998).

Example 2

In some transgressive successions, the succession is not as gradational but a middle estuarine heterolithic channel facies can be readily identified and used to re-examine previously interpreted fluvial facies as possibly parts of upper estuarine channels. The top of a Pottsville Sandstone, a Lower Pennsylvanian quartzarenite (Location 2 in Fig. 2), is exposed along US Highway 35 near Jackson, Ohio (Fig. 9A), on the northern margin of the basin (Fig. 2). The fluvial cross-bedded sandstone facies is exposed at outcrops D1 and D2 (Figs 9A & 10). It exhibits mean palaeoflow azimuths to the west, which are typical of fluvial cross-bedded sandstone facies in this area (e.g. Dominic, 1992). Palaeocurrent measurements for vertically adjacent storeys varied by 45° or less, but flow divergence between trough and planar cross-beds within storeys was as much as 90°. A

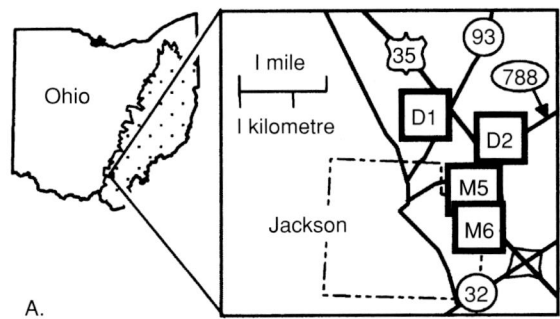

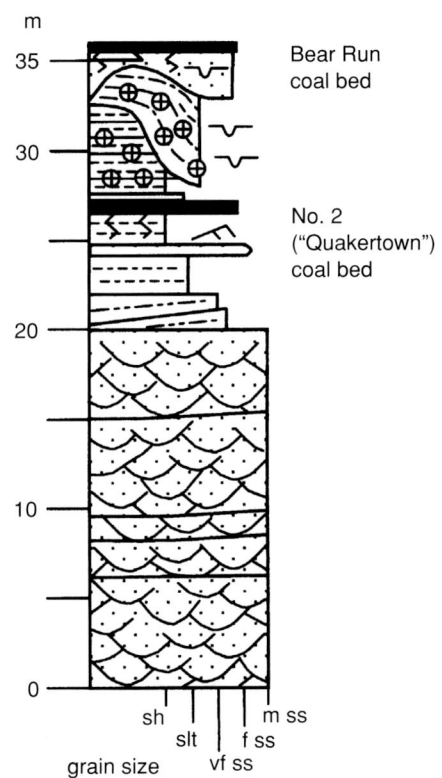

Fig. 9 A lower Pottsville Sandstone at location 2, near Jackson, Ohio. (A) Map of sample outcrops discussed in text (from Martino et al., 1992, Fig. 2.7, p. 13). (B) Stratigraphical section through lower Pottsville strata at exposures shown on the map (after Dominic, 1992, p. 22). See Fig. 6 for legend.

single cross-bed climbs reactivation surfaces in the opposite direction of underlying planar cross-beds (Dominic, 1992; Fig. 11).

A short distance away, at outcrops M5 and M6 (Fig. 9A), the middle estuarine heterolithic

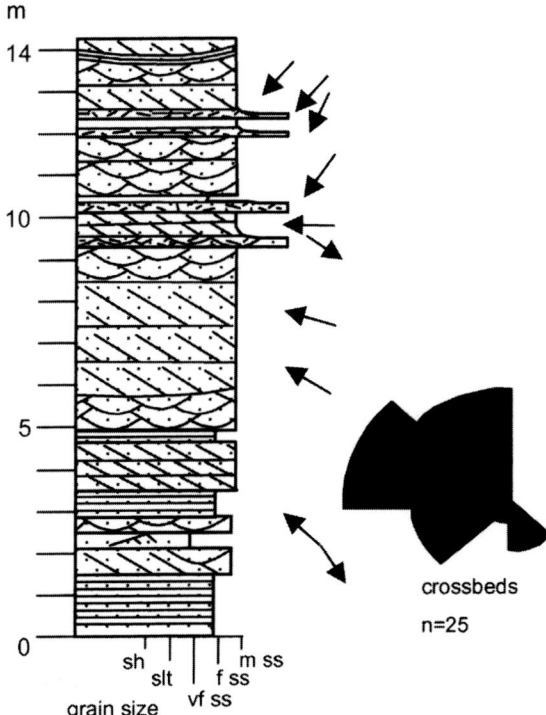

Fig. 10 Measured section of a lower Pottsville Sandstone at location 2, outcrop D2 in Fig. 9A (after Dominic, 1992, p. 16). See Fig. 6 for legend.

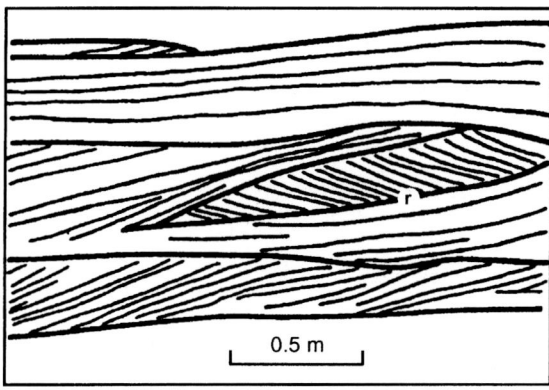

Fig. 11 Line drawing of planar cross-beds with opposing cross-beds above reactivation surface (r), location 2, outcrops D1 and D2 in Fig. 9A (from Dominic, 1992, Fig. 2.7, p. 19).

channel facies occurs between the Quakertown coal and the fluvial cross-bedded sandstone facies. The channel containing the heterolithic facies is 10.5 m deep and contains 7.25-m-thick,

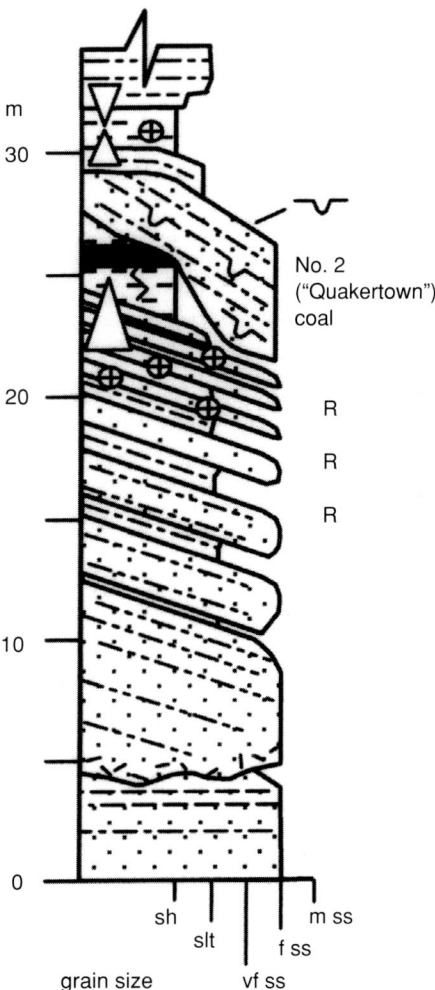

Fig. 12 Measured section of inclined heterolithic strata at location 2, outcrop M6 in Fig. 9A (after Martino et al., 1992). See Fig. 6 for legend.

large-scale accretion surfaces dipping toward the north-west (320°) at angles of as much as 13° (Figs 12 & 13A). Ripple cross-lamination within inclined sandstone-dominated sets indicates flow toward the south-west (230°). Accretion surfaces are defined by very fine to fine sandstone-dominated intervals, as much as 30 cm thick, which alternate with dark grey, interlaminated very fine sandstone to shale. A crude cyclicity is evident, with sand-dominated intervals alternating with widely spaced shale layers, and more heterolithic intervals alternating with thinner sand layers

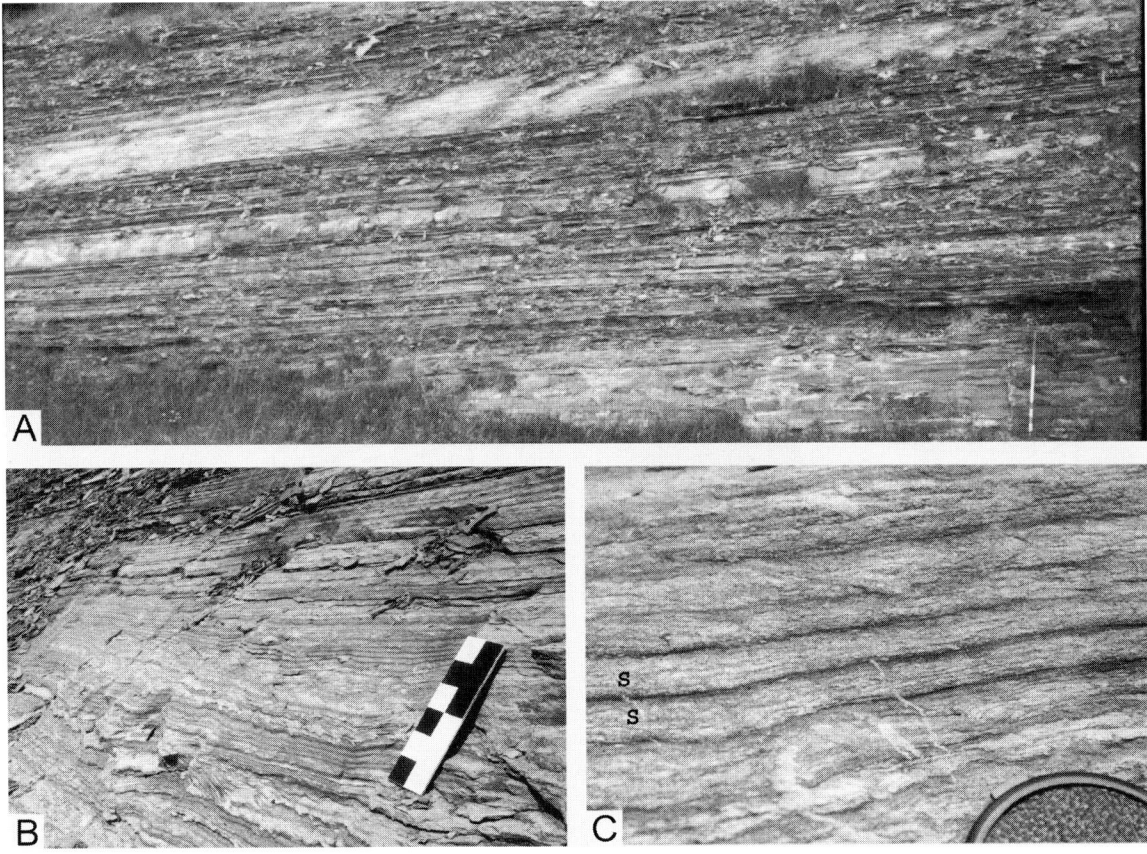

Fig. 13 Inclined heterolithic strata (HIS) at location 2, outcrop M6 in Fig. 9A. (A) Low-angle dipping heterolithic strata. Jacobs staff at lower right 1.5 m. (B) Rhythmic laminations within IHS at M6. Note bundled laminations and alternating larger scale bundles of sandier and shalier strata. Bar scale = 30 cm. (C) Sand–shale couplets in sandstone bedding within IHS. Note sandier parts of bundles (S) are composed of millimetre-scale laminae couplets. These are interpreted as spring tidal deposits that alternate with dark shale intervals formed during neap tides when currents were too weak to entrain sand. Lens cap for scale.

(Fig. 13B). Sand/silt–shale laminae couplets occur in bundles within the heterolithic intervals, which are separated by a persistent shale laminae. Five to eight laminae couplets occur in each laminae bundle (Fig. 13C). Thicker sandstone laminae occur toward the middle of many bundles, although there is local truncation of laminae within bundles. Some of the thicker sandstone units truncate underlying strata. Bioturbation is sparse to abundant, although specific ichnogenera could not be identified. The inclined sandstone–shale interval grades vertically and laterally into thin-bedded dark grey shale and sideritic siltstone. The channel-fill is capped by root-traced seatrock of the overlying Quakertown coal, which is overlain by burrowed sandstones, and shales and limestones with marine invertebrates (Fig. 12).

Interpretation

In this series of outcrops, typical fluvial cross-bedded sandstone facies and middle estuarine heterolithic channel facies can be documented. Both are capped by the peat-mire coal facies and the marine- to brackish-water grey-shale facies, so that a transgressive succession is identified. The middle estuarine heterolithic channel facies here contains decimetre-scale mud couplets. Smith (1988) inferred that there was a transition between fluvial and tidal creek point bars, largely reflected in increased development and continuity of decimetre-scale sand–mud couplets and frequency in tidal creeks and from micro- to mesotidal conditions within an estuary. The increased frequency of mud drapes in tidally influenced point bars is related to influences of turbidity maximum in the middle estuary (e.g. Thomas et al., 1987). Within the decimetre-scale couplets, shale-draped bundles of laminae on accretion surfaces are similar to non-cyclic rhythmites interpreted as tidal rhythmites in other parts of the basin (Martino & Sanderson, 1993; Martino, 1996; Adkins & Eriksson, 1998; Greb & Archer, 1998). The preservation of five to eight laminae couplets within each bundle does not match a complete tidal cycle duration, but could represent incomplete preservation of neap–spring cycles.

At this particular series of outcrops, the middle estuary heterolithic channel facies succeeds the fluvial cross-bedded sandstone facies, and may actually be laterally equivalent to that facies, because both are overlain by the Quakertown coal. Therefore, in vertical section, a fluvial–estuarine transition can at least be defined across the scour at the base of the middle estuary heterolithic channel facies. A lateral transition may also occur between the two outcrops. Interestingly, the upper part of the fluvial cross-bedded sandstone facies contains a cross-bed orientated opposite to the dominant fluvial mode. Rare opposing palaeocurrent modes can occur in fluvial channels from reversing eddies in unidirectional flow (Allen, 1982a). This example, however, does not appear to occur on the slope or in front of a larger bedform. Additionally, it occurs in an interval of increased overall current variability, relative to more uniform, unimodal palaeoflow indicators lower in the fluvial facies. Hence, it is possible that the feature represents preservation of tidal flood currents in an otherwise fluvial-dominant channel. In that case, the sandstone is part of an upper estuarine sandstone channel facies, and the fluvial–estuarine transition would occur across that facies, at least to the base of the channel scour in which the reversing cross-bed is noted.

Example 3

In some cases, heterolithic strata indicative of tidal flat origins supercede fluvial facies and can be used to interpret the fluvial–estuarine transition. The Kanawha Formation at Low Gap in Boone County, West Virginia (location 3, Fig. 2), is exposed in a series of channels 40 m below the base of the Betsie Shale Member, which defines the Lower–Middle Pennsylvanian boundary (Blake et al., 1994). The lower channel fill is 10 m thick and contains three different types of bedding (Fig. 14). The lower part of the channel fill is typical of the fluvial cross-bedded sandstone facies in the area. It consists of 1.75 m of cross-stratified, medium-grained sandstone, with compound cross-bedding orientated to the south-west. The

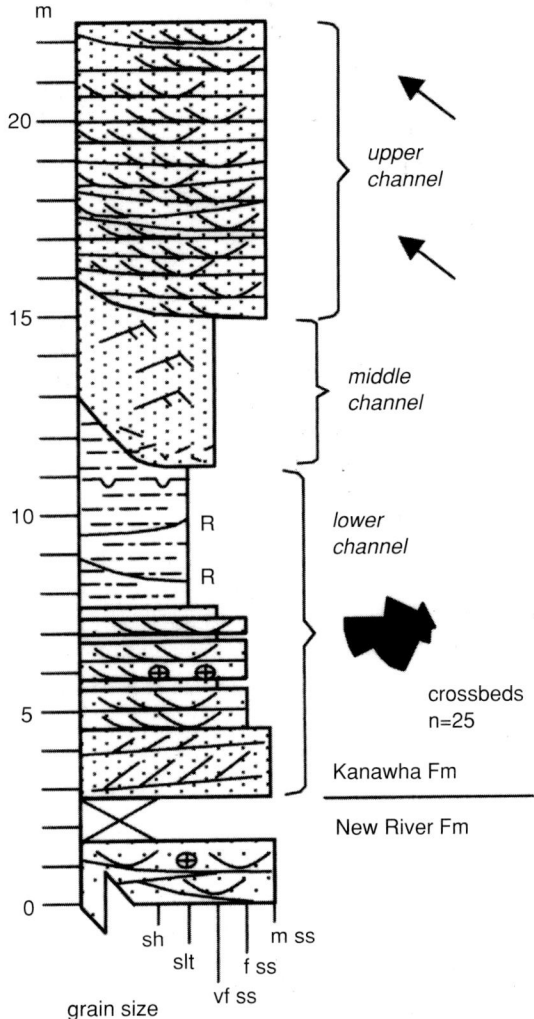

Fig. 14 Measured section through lower Kanawha Formation at location 3, Low Gap, West Virginia. See Fig. 6 for legend.

middle part of the lower channel fill consists of 3.35 m of trough cross-stratified medium- to very-fine sandstone with unimodal palaeocurrents to the south-west similar to the base of the channel fill. Shale-bounded isolated trough sets occur upwards within the middle part of the channel.

The upper part of the lower channel fill consists of 5 m of interlaminated very fine sandstone, siltstone, and shale with well-developed rhythmic lamination (R, Fig. 14). Horizontal, sand-filled burrows, ripple cross-lamination, and broad scour surfaces occur locally. Analysis of sand lamination thickness shows abundant sand/silt–shale laminae couplets arranged in a hierarchy of bundled thickening and thinning cycles (Fig. 15A; Martino, 1996). Minor cycles contain five to eight layers; intermediate-scale cycles contain 15 to 18 layers; Major cycles contain 50 to 62 layers (Fig. 16; modified from Martino, 1996). In addition, thick–thin couplets are also common (Figs 15B & 16). The rhythmically laminated part of the channel fill is interpreted as the estuarine tidal-flat facies.

Interpretation

Rhythmites in intertidal flats are most commonly preserved in the upper middle to inner (upper) estuary (Dalrymple *et al.*, 1991; Tessier, 1993, 1998). Furthermore, thick–thin alternation of laminae couplets would be uncommon in purely fluvial environments, and suggest diurnal inequality in tides formed in semi-diurnal settings (e.g. Kvale *et al.*, 1989). The rhythmites at Low Gap preserve distinct orders or scales of cyclicity but do not preserve complete tidal sedimentation records. In an intertidal setting, tidal currents may be too weak to entrain sediments during the neap portion of spring–neap cycles (Fig. 17), in the higher portions of intertidal flats, or headward within the fluvial–estuarine transition, resulting in increasingly non-cyclic (incomplete) rhythmites (Dalrymple *et al.*, 1991, 1992; Tessier, 1993, 1998; Archer, 1998). The partial preservation or amalgamation of tidal laminae in rhythmites at Low Gap could have resulted in spring–neap cycles that contain only five to eight mud-draped sand layers (minor cycles in Fig. 16). Likewise, lunar perigee–apogee cycles might contain less than 30 layers owing to the weakness of neap tides at this posi-

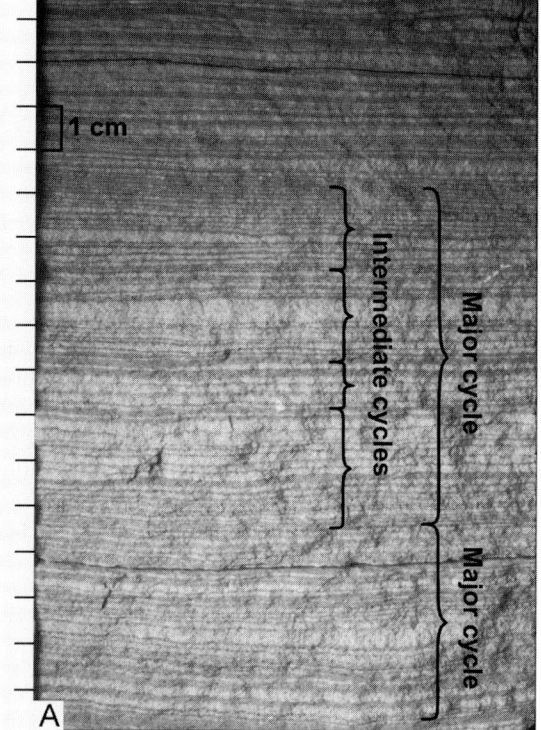

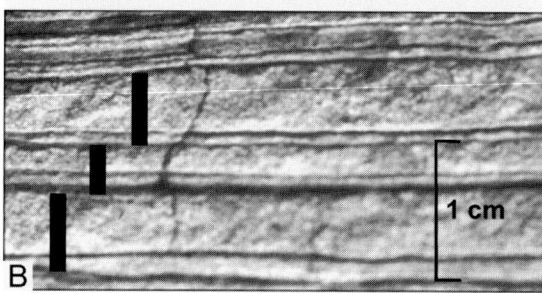

Fig. 15 Rhythmic laminations at Low Gap. (A) Sandstone–shale laminae couplets arranged in rhythmic successions from a block that was collected and analysed; see Fig. 16 for analysis. Vertical thickness = 16 cm. (B) Thick–thin laminae pairs and alternating thick–thin pairs in couplets (black bars).

tion in the palaeo-estuary and be represented by the intermediate cycles of 15 to 18 layers (Figs 15A & 16). The major cycles of 50 to 62 layers (Figs 15A & 16) may represent attenuated 6-month (solstice–equinox) cycles. Seasonal tidal cycles with maximum tidal ranges occurring during June and December have been interpreted for

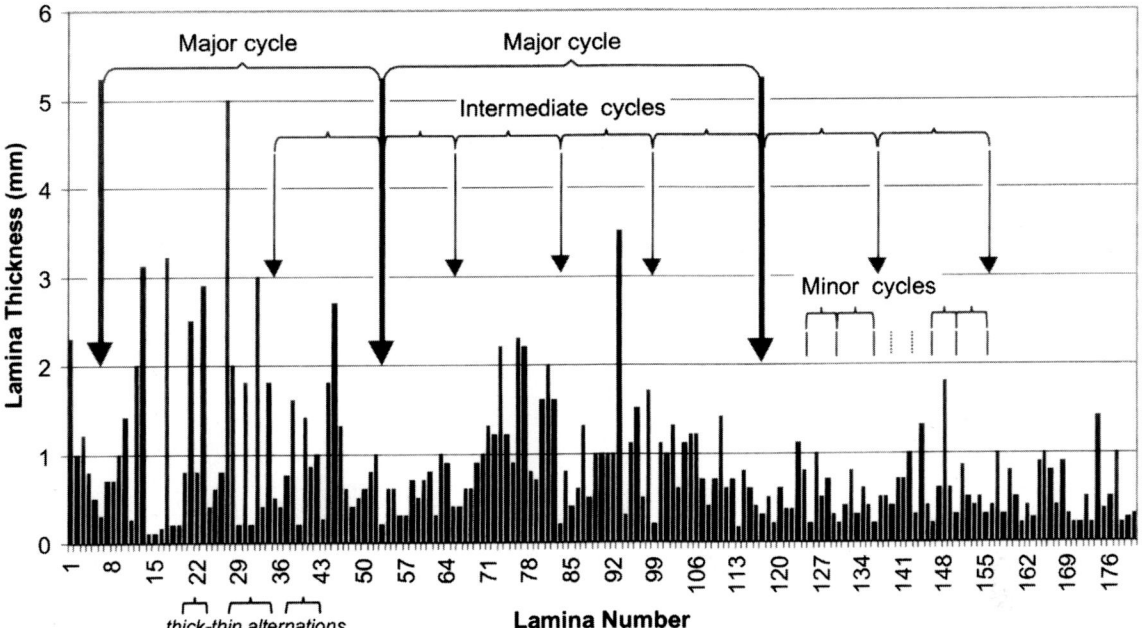

Fig. 16 Plot of laminae thickness measured from collected block, a portion of which is illustrated in Fig. 15A. Coarse layers (very fine sandstone, siltstone) in each couplet are plotted.

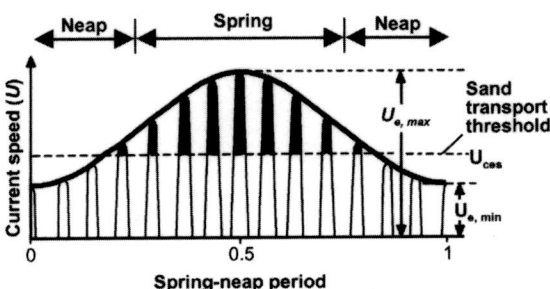

Fig. 17 Plot of current speed variation during a hypothetical spring–neap cycle. Sandstone laminae may only be preserved above the sand transport threshold (after Allen, 1982b).

other Carboniferous rhythmites (Kvale *et al.*, 1989; Archer, 1991; Martino & Sanderson, 1993).

The tidal rhythmites occur above a channel with otherwise fluviatile characteristics. The lower part of the channel fill contains cross-bedding orientated in the inferred direction of fluvial flow. Hence, the fluvial–estuarine transition occurs at least across the scour in the upper sandstone overlain by the estuarine tidal-flat facies. The trough cross-beds within the middle interval of the channel fill are draped by continuous shale laminae. These could represent seasonal fluctuation in river discharge within the fluvial facies, but in a transgressive succession they could indicate rapid sedimentation from suspension caused by the interaction of flood tidal currents in the fluvial–estuarine transition. If so, the fluvial–estuarine transition would occur across this part of the channel fill, rather than at the scour at the top of the trough cross-bedded interval. There is no supporting evidence within the trough cross-bedded interval to substantiate tidal influences, but such evidence might not be preserved in the most headward part of an upper estuarine channel.

Example 4

In some cases, diagnostic tidal features are not found, but possible tidal features are noted within transgressive successions at stratigraphical horizons where estuarine facies are noted elsewhere. Greb & Chesnut (1996) identified a transgressive succession at the top of the Rockcastle Sandstone on the western margin of the basin. Bioturbated IHS (herein referred to as the middle estuarine

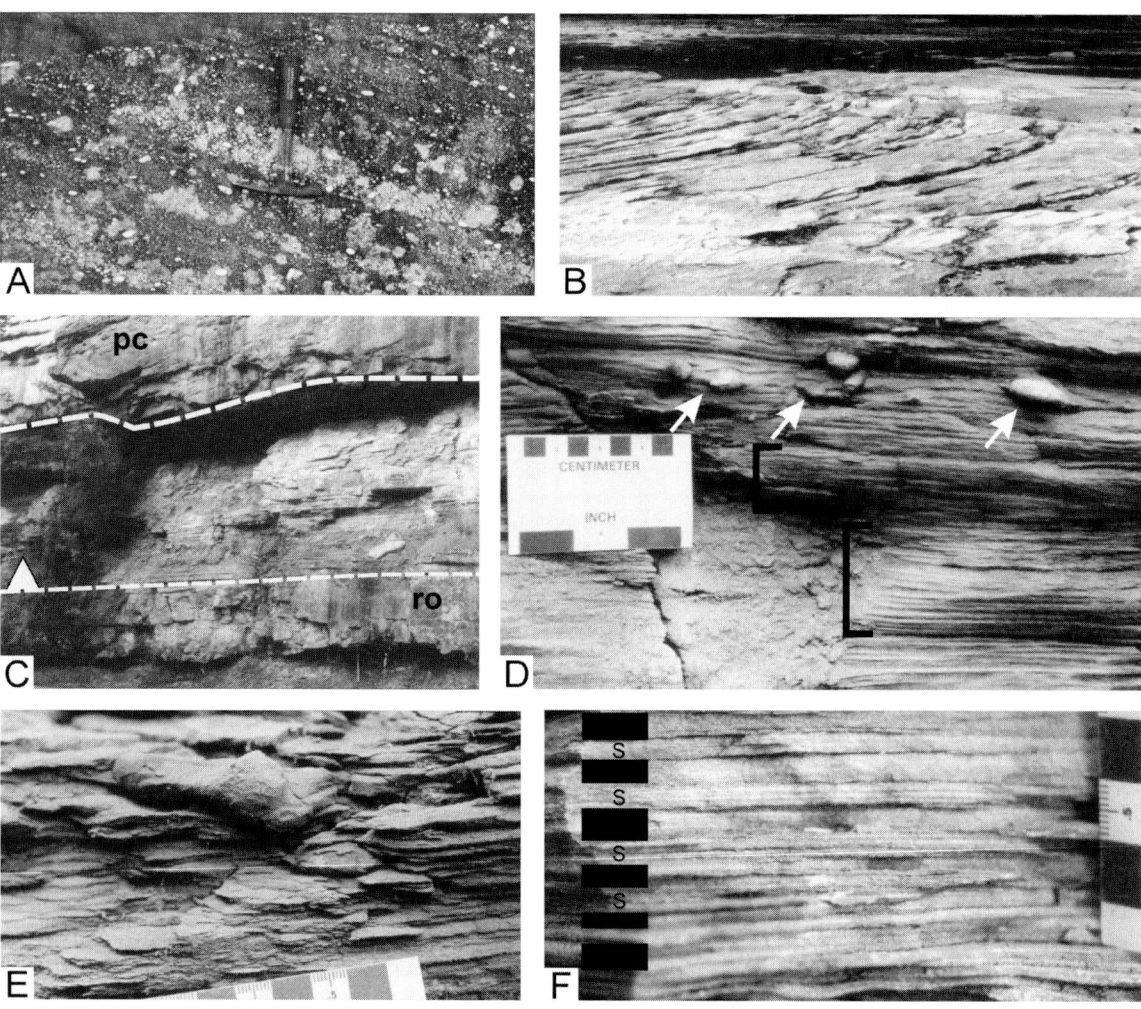

Fig. 18 Rockcastle Sandstone at Cumberland Falls, location 4. (A) Conglomeratic cross-bed facies typical of fluvial facies in the Rockcastle Sandstone. Hammer scale = 30 cm. (B) Bedding-plane exposures of south-easterly orientated, broadly arcuate, planar-tabular cross-beds in fluvial facies. (C) Channel exposed at the top of the Rockcastle Sandstone (ro) north of the entrance to Cumberland Falls State Park: bf, Barren Fork coal; pc, Pine Creek Sandstone. (D) Siderite nodules (white arrows) in non-cyclic rhythmites. Rhythmites consist of bundled laminae couplets (brackets). (E) Siderite nodule with *Rosselia* shape. Scale in centimetres. (F) Detail of sandy part of rhythmite bundle showing alternating groups (s) of sandier couplets and shalier (black boxes) couplets. Scale in centimetres.

heterolithic channel facies) was noted in a transgressive succession. Examination of other localities within 15 km of the IHS locality shows that there is considerable variability in facies beneath the coal and grey shale. An example is described here (location 4, Fig. 2).

The upper 3 m of the Rockcastle Sandstone is dominated by the fluvial cross-bedded sandstone facies at most outcrops (Fig. 18A & B; Wizevitch, 1991; Greb & Chesnut, 1996). Locally, however, a fining-upward, heterolithic interval occurs between the fluvial facies and Barren Fork coal bed (Figs 18C & 19A & B). Also, at many localities, the marine- to brackish-water grey shale facies that overlies the coal in other areas (Greb & Chesnut, 1996) is replaced by scour-based fluvial cross-bedded sandstone facies of the overlying sandstone (Figs 18C & 19A & B). The heterolithic

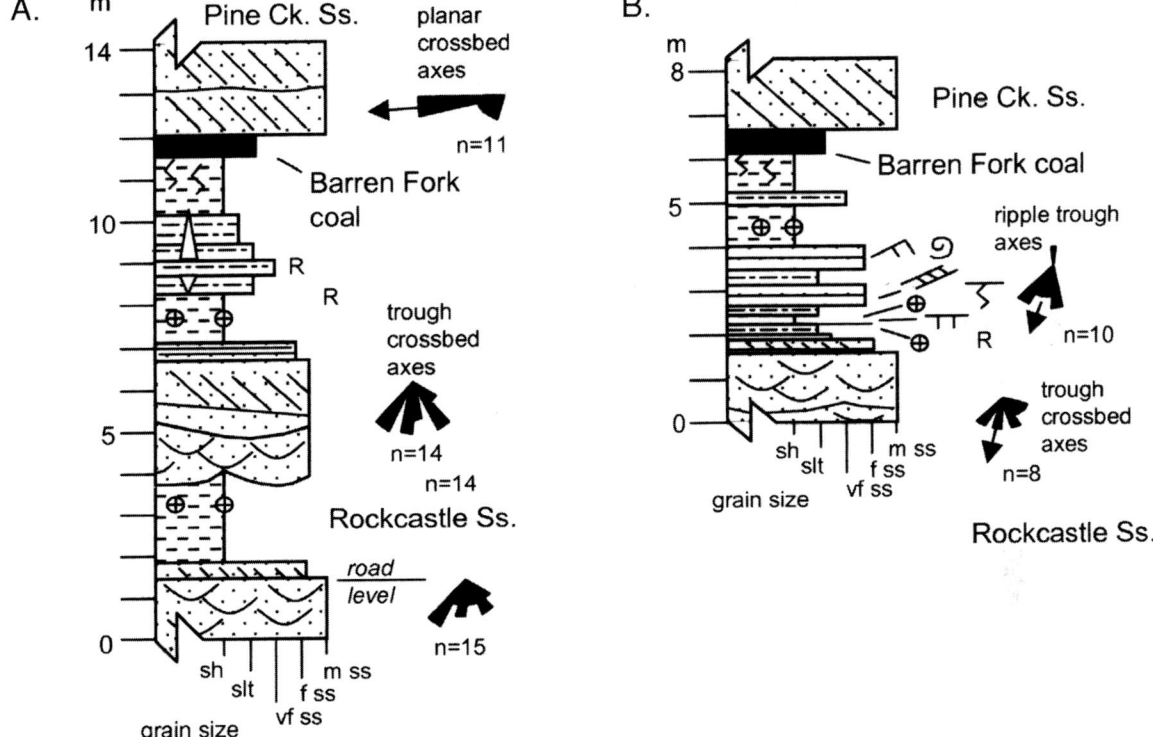

Fig. 19 Vertical sections from the upper Rockcastle Sandstone, location 4. (A) North side of Cumberland River near the entrance to Cumberland Falls car park at Cumberland Falls State Park. (C) South side of the Cumberland River near the entrance to Cumberland Falls car park at Cumberland Falls State Park. See Fig. 6 for legend.

strata at the top of the Rockcastle Sandstone exhibit ripple-bedded sandstones with climbing ripples, crude rhythmic laminations and contorted bedding. Large *Stigmaria* roots penetrate the sandstone and heterolithic layers (Fig. 19A & B). Within the heterolithic layers, siderite nodules occur along distinct bedding planes. In one layer, sideritized vertical, cylindrical structures occur above an exposed *Stigmaria* root. Some nodules are tubular and parallel to bedding, such that they are similar to *Planolites*. Some are more conical or irregular in shape (Fig. 18D), with one having the shape of *Rosselia* (Fig. 18E). Rhythmic laminations are ordered in bundles of laminae 3 to 4 cm thick, consisting of a sandier and a shalier half (brackets in Fig. 18D). These bundles are composed of smaller scale bundles, each 0.2 to 0.7 cm thick (Fig. 18F). Sandy bundles (s in Fig. 18F) alternate with shalier bundles (black boxes in Fig. 18F). The small-scale sandy bundles consist of three to five laminae couplets, each couplet consisting of a very fine sandstone or siltstone and shale laminae. Some bundles show vertical thickening and thinning of laminae couplets, although discontinuities are common.

Interpretation

Local, rooted, fining upward, heterolithic channel fills could be interpreted as floodplain deposits of the typical fluvial facies reported for Lower Pennsylvanian sandstones. In fluvial floodplain deposits, however, fining upward laminations from suspension deposition or coarsening upward laminations from bedform progradation might be more typical than the sand–shale laminae couplets arranged in bundles preserved here. As a transgressive succession was interpreted at this stratigraphical horizon nearby (Greb & Chesnut, 1996), the non-cyclic rhythmites in this example are interpreted as incomplete records of tidal sedimentation, similar to the non-cyclic rhythmites at the other locations previously discussed. If the laminae couplets represent daily deposition, then the alternating sandy and shaly small-scale bundles might represent incomplete neap–spring cycles. The occurrence of five to six neap–spring cycles in the sandy part of each larger scale bundle represents a higher energy five- to six-month seasonal influence. Larger scale bundles with sandy and shaly halves could represent annual deposits consisting of a high- and low-energy seasonal deposit, similar to annual rhythmites described from Carboniferous tidal channels at other locations (Greb & Archer, 1998). The non-cyclic rhythmites are crudely ordered, indicating that the fining upward heterolithic interval should be placed within the estuarine tidal-flat facies. Rooting structures in these rhythmites may indicate a relatively high position on tidal flats, which would preclude preservation of the high tidal signature, and lead to incomplete tidal records of tidal deposition (Dalrymple et al., 1991; Tessier, 1993, 1998).

Where climbing ripples have been found in tidal deposits, they are restricted to the fluvial–estuarine transition (Lanier & Tessier, 1998). Interestingly, in the Mont-St-Michel estuary, flood-dominant climbing ripples (as occur here) are restricted to areas along the margins of inner straight channels of the fluvial–estuarine transition, where they are also associated with rhythmites, soft-sediment deformation and rooting (Lanier & Tessier, 1998). Rooting is common in fluvial–estuarine transition zones because of the common development of marshes in the inner estuary (Lanier & Tessier, 1998; Tessier, 1998). Most of the siderite nodules at this location are rootlets, but some could represent invertebrate bioturbation. Sideritization would have destroyed the internal structure of any structures that might have been burrows. At least one of the siderite nodules is similar in shape to *Rosselia*, which is common in Carboniferous marine and tidal facies in the basin (Martino, 1989, 1992, 1994; Greb & Chesnut, 1992, 1994). In this case, the nodule may represent a rootlet, but the point of similarity is brought up because differentiating infaunal invertebrate burrows from nodular root traces can be difficult in the fluvial–estuarine transition.

DISCUSSION

Controls on fluvial–estuarine transitions

There are numerous controls on the development of palaeofluvial–estuarine transitions in tide-dominated estuaries. As estuaries are coastal features, the relative position of the palaeocoastline and orientation of valleys and lowlands along that coast are critical to the development of estuaries. If river valleys are transgressed, the overall shape and slope of the valleys will influence the development of tidal prisms and resultant tidal sedimentation. Estuaries become longer as coastal gradient decreases and/or tidal range increases (Dalrymple et al., 1992). Upstream changes in valley shape could cause changes in the character of tidal sedimentation as the seas transgressed across those changes. Likewise, the amount of sediment flux, seasonality and other variables controlling fluvial discharge will influence whether a delta or estuary forms, and in tide-dominated estuaries, the preservation potential of tidal features in the most landward areas of tidal influences (Dalrymple et al., 1992; Wells, 1995). Moreover, the position and extent of fluvial–estuarine transitions in the rock record will be dependent on the rate and amount of relative sea-level rise, which will be a function of eustasy, sediment flux and tectonic accommodation.

During transgression and into highstand, the overall size of drainage basins decreases so that fluvial input into up-dip-advancing estuaries is decreased if climate is unchanged. Overall river grade also decreases. As base level rises, increasing numbers of tributaries can be converted to estuaries, because of the upstream-branching in pre-existing fluvial systems (Archer & Greb, 1995). The increase in the number of transgressed tributaries causes an increase in the number of locales in which there can be fluvial–estuarine transitions and the number of sediment catchments in which fluvial–estuarine transitions can be preserved. The preservation potential of tidal-estuarine facies in these catchments will be dependent on the geometry of the valley as it is drowned, and the amount of fluvial input during transgression. The relative amount of fluvial and tidal energy will determine the length of the fluvial- and tidal-dominant parts of the estuary (Dalrymple et al.,

1992), as well as the nature of the fluvial–estuarine transition. Some tributaries (secondary and tertiary drainages) will have little fluvial input so that tidal influences might leave a greater imprint on the sedimentology of the preserved fill, as long as the tributaries are large enough or shaped in such a way that they can generate a tidal prism. First-order drainages with active discharge will be more likely to be fluvially dominated, even if influenced by flood-tidal currents. Studies of modern estuaries with fluvial inflow have shown that during floods:

1 more sediment may be supplied to the estuary in a matter of days than is supplied by months or years of average sedimentation;
2 there is lateral translation of salt/fresh-water interface;
3 there is translation of zones of suspended sediment accumulation (Nichols and Briggs, 1985).

All of these factors could lead to the net dominance of fluvial sedimentation at the expense of tidal sedimentation across the fluvial–estuarine transition, even if fluvial dominance was temporally subordinate. Additionally, as individual estuaries translate landward, the upper portion of the transgressive succession can be removed by tidal channel erosion (Davis & Clifton, 1987; Dalrymple et al., 1992), further complicating identification of transgressive successions and fluvial–estuarine transition facies. Considering all of these factors, variability in fluvial–estuarine transitions might be expected both spatially and temporally if transgressions flooded broad alluvial braidplains, such as those envisioned for the Early Pennsylvanian of the Central Appalachian Basin.

Early Pennsylvanian estuaries

During the Early Pennsylvanian a series of longitudinal braidplains was developed in the Central Appalachian Basin (Chesnut, 1992, 1994; Archer & Greb, 1995; Greb & Chesnut, 1996). Each of the braidplains was succeeded by marine- to brackish-water grey-shale facies, indicating that the braidplains were transgressed. The development of estuarine facies between the typical fluvial facies of the braidplains and the overlying transgressive shale indicates local estuarine development during transgression (Greb & Chesnut, 1996). Some inferences can be made about the development of Early Pennsylvanian estuaries, based upon comparison to Pleistocene analogues. During Pleistocene trangressions, the slope of the flooded coastline dramatically affected the development of estuaries. In general, estuaries developed within valleys where there was an increase in the slope of the transgressed surface, whereas bays and lagoons developed on flat, or low, continuous slopes (Emery, 1967).

Most vertical sections through the Lower Pennsylvanian quartzarenite belts of the Central Appalachian Basin show a succession from the fluvial cross-bedded sandstone facies to peat-mire coal facies to marine- to brackish-water grey-shale facies, suggesting that broad floodplains and abandoned braidplains within the quartzarenite belts were paludified and capped by peats (especially in topographic depressions and coastal lowlands), which accumulated as water tables rose. Peats were transgressed across broad areas where underlain by floodplains and abandoned parts of the braidplain. Facies containing tidal sedimentary structures occur only locally, mostly in channel facies. This indicates that Early Pennsylvanian transgressions initially followed the paths of abandoned or active channels within valleys. Fluvial channels within valleys were locally converted to upper estuarine channels.

Upper estuarine channels, by definition, are fluvially dominant with some tidal influences (Allen, 1991; Dalrymple et al., 1992). The upper estuarine sandstone channel facies defined herein is difficult to interpret at any outcrop if it contains only a few, possible, tidal indicators. In cases of more limited outcrop, isolated evidence for opposing currents might only represent reversing eddies in a fluvial environment. Where a transgressive succession can be defined at a specific stratigraphical interval and there are at least middle estuarine heterolithic channel facies or estuarine tidal-flat facies at the same horizon, then interpreting fluvially dominant channels with subtle tidal indicators as upper estuarine sandstone channel facies is tenable.

Recognition of tidal-estuarine indicators

In analyses of Cretaceous fluvial–estuarine transitions, Shanley et al. (1992) noted that the preponderance of evidence was sometimes needed to infer tidal influences in the fluvial–estuarine transition; such is the case in the Pennsylvanian examples studied herein. In the examples studied, tidal indicators were sometimes subtle, and often individually inconclusive. The greatest diversity and abundance of possible tidal structures not surprisingly occurred in the upper part of the preserved succession, either in middle estuarine heterolithic channel facies or estuarine tidal-flat facies. With decreasing frequency and variety of possible tidal features down-section, the possible lower (headward) limit of the fluvial–estuarine transition within the underlying fluvial deposit is difficult to delineate precisely.

In a complete transgressive succession, a vertical section through a palaeoestuary might preserve upper, middle and lower estuary facies. In the headward regions of estuaries, however, the succession might not contain lower estuarine facies. Dalrymple et al. (1992) illustrated hypothetical vertical sections through estuarine fills produced by transgression followed by progradation. In the headward examples of these fills, the maximum flooding surface occurred within the estuary. A vertical section through a similar succession in which the middle estuarine heterolithic channel facies was the point of maximum transgression is shown in Fig. 20A. This is similar to the succession noted at location 1 (Fig. 2), where the Dave Branch Shale may contain (in part) the middle estuary heterolithic channel facies (Fig. 6). In most cases, however, the marine- to brackish-water grey shale facies caps estuarine facies and a wide variety of vertical successions is noted.

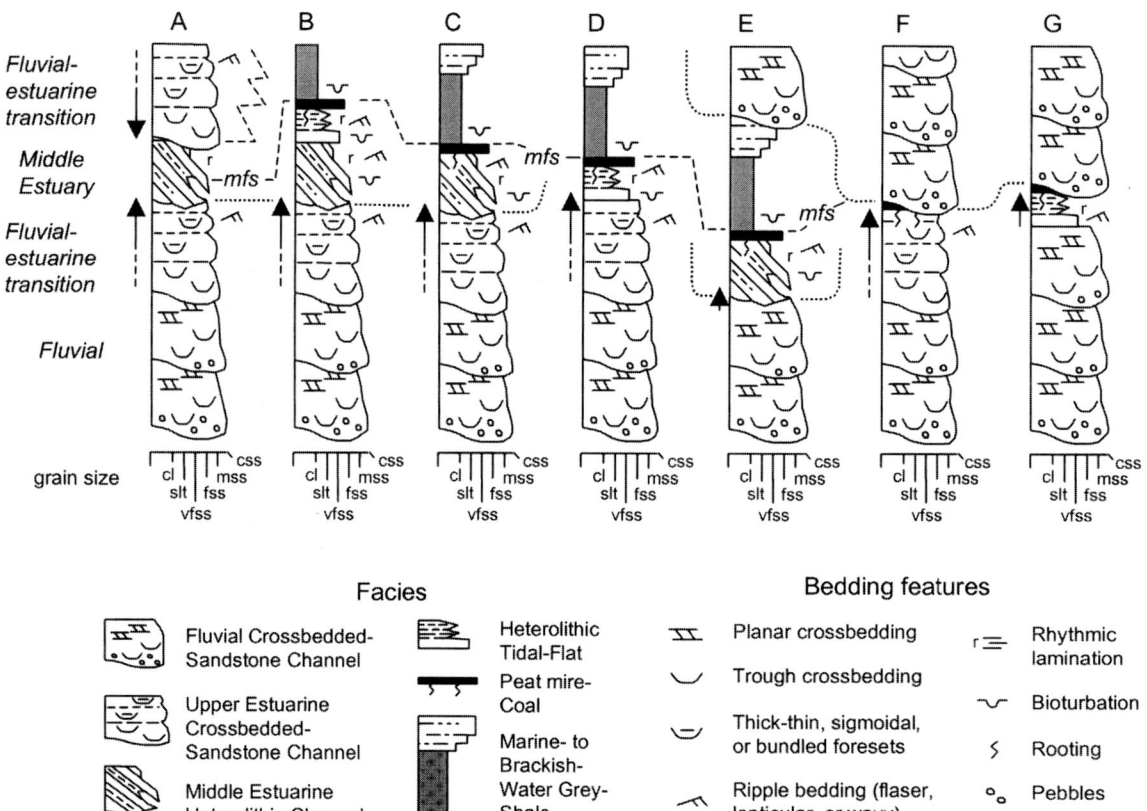

Fig. 20 Schematic vertical sections (A–G) showing variability in transgressive successions in fluvial-dominated strata. The fluvial–estuarine transition is easier to infer when middle estuarine heterolithic channel facies can be defined in a transgressive succession (A, B, C & E). Where that facies is missing (D, F & G) or the overlying marine facies are missing (F & G), the fluvial–estuarine transition is more difficult to interpret.

If each of the middle and upper estuarine facies described among the various outcrops studied (Table 1) were stacked in vertical succession, they might appear in the succession illustrated in Fig. 20B. In some cases, middle estuarine heterolithic channel facies overlie upper estuarine channel facies (Fig. 20C), as in Example 2, or estuarine tidal-flat facies overlie upper estuarine channel facies (Fig. 20D), as in Example 3. In some cases, the middle estuarine heterolithic channel facies might directly overlie fluvial facies without an intervening upper estuarine channel transition (Fig. 20E). This is what could be interpreted for Example 2 if the up-dip cross-bed in underlying 'fluvial' facies was not noted, or not considered a tidal indicator. Where middle estuarine heterolithic channel facies or estuarine tidal-flat facies can be identified (Fig. 20B, C & E), detailed examination of underlying strata can be undertaken to determine if there is any evidence for tidal sedimentation that could lead to reinterpretation of fluvial cross-bedded sandstone facies as upper estuarine cross-bedded sandstone channel facies. It is more complicated to interpret estuarine facies where the transgressive shale or middle estuarine facies is missing (Fig. 20F & G). If marine facies are truncated by overlying sequences, then the transgressive succession may not be apparent, and subtle tidal indicators of the upper estuarine cross-bedded sandstone channel facies might be overlooked.

Bioturbation, one of the most commonly used indicators of salinity change, is not pervasive in the fluvial–estuarine transition of modern environments (Howard & Frey, 1973, 1975; Allen, 1991; Dalrymple et al., 1991, 1992; Tessier, 1993, 1998), and is uncommon in the upper estuarine facies interpreted herein. Unless the study area is at a location where the transgression continued past the fluvial–estuarine transition into the middle or lower estuary, bioturbation may not be common. Additionally, in quartzarenites, there may not be sufficient variations in grain colour or texture to distinguish burrows in outcrop. Weathering accentuated the burrow shown in Fig. 7D. If bioturbation did occur lower in the section, where interlaminated shales are less common, it might not be noticed. Another complication to recognition of invertebrate bioturbation in the fluvial–estuarine transition is the common occurrence of rooting, as occurs in Example 4. Many Lower Pennsylvanian fluvial facies are capped by coal beds such that paludification and rooting of immediately underlying facies might destroy important tidal evidence. Root disturbance of tidal flats in the fluvial–estuarine transition is common in modern estuaries (Dalrymple et al., 1992; Tessier, 1993, 1998). Geochemical indicators of salinity changes, such as carbon isotope ratios in siderite (Weber et al., 1979), or pyritization and carbon/sulphur ratios of shales (Berner & Raiswell, 1984), may be needed to help support a fluvial–estuarine transition where bioturbation is lacking or inconclusive.

Bipolar cross-bedding is probably the most commonly cited criterion used to interpret tidal influences in fluvial–estuarine transitions (e.g. Thomas et al., 1987; Rahmani, 1988). Features formed by opposing tidal modes in sediments within the fluvial–estuarine transition may be very subtle. Rather than truly bipolar palaeocurrent distributions, preserved flow indicators of fluvial–estuarine transitions are generally unimodal with only a few opposing orientations. It is difficult to conclusively interpret these as tidally modulated without supporting tidal evidence; however, their limited occurrence in a transgressive succession might indicate the most headward limits of tidal influences in the palaeoestuary.

Several studies of Cretaceous fluvial–estuarine transitions have indicated that sigmoidal reactivation surfaces within cross-strata where they bound thickening and thinning foreset bundles are a tidal indicator (Kreisa & Moiola, 1986; Shanley et al., 1992). In several of the Pennsylvanian outcrops studied here, sigmoidal bedding and asymmetrically-filled cross-beds with rising troughs were noted stratigraphically below the lowermost occurrence of bidirectional dips or bioturbation. Bundled cross-beds in these facies, however, were uncommon. If these features record tidal influences, then the fluvial cross-bedded sandstone facies at the top of quartzarenite belts in some areas may actually belong to the upper estuarine cross-bedded sandstone channel facies.

Another type of feature reported for other fluvial–estuarine transitions is so-called tidal lamination, which can include bundled laminae bounded by shale laminae, thick–thin laminae alternations and rhythmic lamination (e.g. Rahmani, 1988). Part of the evidence for showing cyclicity in tidal rhythmites is counting continuous repetitive thickening and thinning laminae couplets related to known tidal cyclicities as was done for the rhythmites at location 3 (Fig. 16). The inner or upper estuary is a zone in which tidal amplification can be favourable for the deposition of cyclic rhythmites on intertidal flats. Yet even in these areas, the preservation potential of the rhythmites is dependent upon local accommodation, sediment flux, tidal asymmetry and position of deposition relative to the level of the highest tides. In general, cyclic rhythmites will become increasingly non-cyclic (incomplete) higher on tidal flats or channel margins where weaker tides do not reach (Dalrymple *et al.*, 1991; Archer, 1998; Greb & Archer, 1998; Tessier, 1998). Most of the rhythmites noted in the Lower Pennsylvanian transgressive successions were non-cyclic. By themselves, non-cyclic rhythmites are not diagnostic tidal indicators (Greb & Archer, 1995). Likewise, thick–thin alternations in foreset laminae were noted in many of the examples studied, but none were persistent throughout several trough cross-beds, such that they also are not diagnostic tidal indicators.

Subtle, relatively isolated tidal features may be the only indication of tidal influences in many fluvial–estuarine transitions. If marine flooding surfaces had not been defined previously for each of the study intervals, subtle tidal features might not have been looked for in underlying strata and facies would have been interpreted as wholly fluvial. This is particularly true in areas where incision of the following fluvial sandstone truncates the marine- to brackish-water grey-shale facies or peat-mire coal facies, as happens at location 4. In areas of low accommodation, or in the up-dip parts of basins where fluvial sequences dominate, subtle tidal features in upper estuarine facies may be the only record of transgressions and down-dip marine flooding surfaces for sequence-stratigraphy implications. Recognition of these transitions can enable more accurate delineation of lowstand and transgressive systems tracts within fluvial-dominated strata, and will improve the ability to anticipate vertical and lateral changes in fluvial style, and channel-fill geometry, lithology and continuity. Recognition of fluvial–estuarine transitions may also help in interpreting or predicting lateral and vertical changes in porosity and permeability owing to lithological changes from fluvial to estuarine facies in hydrocarbon reservoirs.

ACKNOWLEDGEMENTS

The authors would like to thank R. Dalrymple and an anonymous reviewer for their thoughtful review. Thanks also to M. Blum for his review and helpful suggestions.

REFERENCES

Adkins, R.M. and Eriksson, K.A. (1998) Rhythmic sedimentation in a mid-Pennsylvanian delta-front succession, Magoffin Member (Four Corners Formation, Breathitt Group), eastern Kentucky: a near-complete record of daily, semi-monthly, and monthly tidal periodicities. In: *Tidalites—Processes and Products* (Eds C.R. Alexander, R.A. Davis and V.J. Henry). *Soc. Econ. Paleontol. Mineral. Spec. Publ.*, **61**, 85–94.

Aitken, J.F. and Flint, S.S. (1995) The application of high-resolution sequence stratigraphy to fluvial systems: A case study from the Upper Carboniferous Breathitt Group, eastern Kentucky, U.S.A. *Sedimentology*, **42**, 3–30.

Allen, G.P. (1991) Sedimentary processes and facies in the Gironde estuary: a recent model for macrotial estuarine systems. In: *Clastic Tidal Sedimentology* (Eds D.G. Smith, G.E. Reinson, B.A. Zaitlin and R.A. Rahmani). *Can. Soc. Petrol. Geol. Mem.*, **16**, 29–39.

Allen, J.R.L. (1982a) *Sedimentary Structures: their Character and Physical Basis*. Developments in Sedimentology **30A**, **B**, Elsevier, New York, 593 pp. (A), 663 pp. (B).

Allen, J.R.L. (1982b) Mud-drapes in sand wave deposits: a physical model with application to the Folkstone Beds (early Cretaceous) southwest England. *Proc. Royal Soc. Lond. Ser. A*, **306**, 291–345.

Archer, A.W. (1991) Modeling of tidal rhythmites using modern tidal periodicities and implications for short-term sedimentation rates. In: *Sedimentary Modeling: Computer Simulations for Improved Parameter Definition* (Eds E.K. Franseen, W.L. Watney, C.G. St. C. Kendall and W.C. Ross). *Kansas Geol. Surv. Bull.*, **233**, 185–194.

Archer, A.W. (1995) Modeling of tidal rhythmites based on a range of diurnal to semidiurnal tidal-station data: *Mar. Geol.*, **123**, 1–10.

Archer, A.W. (1998) Hierarchy of controls on cyclic rhythmite deposition: Carboniferous basins of eastern and mid-continental U.S.A. In: *Tidalites—Processes and Products* (Eds C.R. Alexander, R.A. Davis and V.J. Henry). *Soc. Econ. Paleontol. Mineral. Spec. Publ.*, **61**, 59–68.

Archer, A.W. (2005) Review of Amazonian depositional systems. In: *Fluvial Sedimentology VII* (Eds M.D. Blum, S.B. Marriott and S.F. Leclair). *Spec. Publ. Int. Assoc. Sediment.*, **35**, 17–39.

Archer, A.W. and Greb, S.F. (1995) An Amazon-scale drainage system in the early Pennsylvanian of central North America. *J. Geol.*, **103**, 611–627.

Archer, A.W., Lanier, W.P. and Feldman, H.R. (1994) Stratigraphic and depositional history within incised-paleovalley fills and related facies, Douglas Group (Missourian/Virgillian; Upper Carboniferous) of Kansas, U.S.A. In: *Incised-valley systems: origin and sedimentary sequences* (Eds R.W. Dalrymple, Boyd, R. and B.A. Zaitlin). *Soc. Econ. Paleontol. Mineral. Spec. Publ.*, **51**, 175–190.

Barnhill, M.L. (1994) The sedimentology of the Corbin Sandstone Member, Lee Formation, eastern Kentucky, and a comparison to the age-equivalent rocks of the Illinois Basin, southwestern Indiana. Unpublished PhD dissertation, University of Cincinnati, Cincinnati, OH, 271 pp.

Bement, W.O. (1976) Sedimentological aspects of Middle Carboniferous sandstones on the Cumberland overthrust sheet. Unpublished PhD dissertation, University of Cincinnati, Cincinnati, OH, 182 pp.

Berner, R.A. and Raiswell, R. (1984) C/S method for distinguishing freshwater from marine sedimentary rocks. *Geology*, **12**, 365–368.

Blake, B.M., Keiser, A.F. and Rice, C.L. (1994) Revised stratigraphy and nomenclature for the Middle Pennsylvanian Kanawha Formation in southwestern Virginia. In: *Elements of Pennsylvanian Stratigraphy, Central Appalachian Basin* (Ed. C.R. Rice). *Geol. Soc. Am. Spec. Pap.*, **294**, 41–53.

Boersma, J.R. and Terwindt, J.H.J. (1981) Neap–spring tidal sequences of intertidal shoal deposits in a mesotidal estuary. *Sedimentology*, **28**, 151–170.

Cecil, C.B. and Englund, K.J. (1989) Origin of coal deposits and associated rocks in Carboniferous of the Appalachian Basin. In: *Carboniferous Geology of the Eastern United States; 28th International Geologic Congress Field Trip* (Eds C.B. Cecil, J.C. Cobb, D.R. Chesnut, Jr., H. Damberger and K.J. Englund). *Am. Geophys. Union, Guidebook.*, **T143**, 84–88.

Chesnut, D.R., Jr. (1991) Paleontological survey of the Pennsylvanian rocks of the Eastern Kentucky Coal Field. *Kentucky Geol. Sur. Ser. 11, Info. Circ.*, **36**, 71 pp.

Chesnut, D.R., Jr. (1992) Stratigraphic and structural framework of the Carboniferous rocks of the Central Appalachian Basin in Kentucky. *Kentucky Geol. Sur. Ser. 11, Bull.*, **3**, 42 pp.

Chesnut, D.R., Jr. (1994) Eustatic and tectonic control of deposition of the Lower and Middle Pennsylvanian strata of the central Appalachian basin. In: *Tectonic and Eustatic Controls on Sedimentary Cycles* (Eds J.M. Dennison and F.R. Ettensohn). *Soc. Sediment. Geol. Concepts Sedimentol. Paleontol.*, **4**, 51–64.

Collinson, J.D. (1996) Alluvial sediments. In: *Sedimentary Environments: Processes, Facies and Stratigraphy* (Ed. H.G. Reading), pp. 37–82. Blackwell Science, Oxford.

Dalrymple, R.W., Makino, Y. and Zaitlin, B.A. (1991) Temporal and spatial patterns of rhythmite deposition on mudflats of the macrotidal Cobequid Bay–Salmon River estuary, Bay of Fundy, Canada. In: *Clastic Tidal Sedimentology* (Eds D.G. Smith, G.E. Reinson, B.A. Zaitlin and R.A. Rahmani). *Can. Soc. Petrol. Geol. Mem.*, **16**, 137–160.

Dalrymple, R.W., Knight, R.J., Zaitlin, B.A. and Middleton, G.V. (1992a) Dynamics and facies model of a macrotidal sand bar complex, Cobequid Bay–Salmon River estuary (Bay of Fundy). *Sedimentology*, **37**, 577–617.

Dalrymple, R.W., Zaitlin, B.A. and Boyd, R. (1992b) Estuarine facies models: conceptual basis and stratigraphic implications. *J. Sediment. Petrol.*, **62**, 1130–1146.

Davis, R.A., Jr. and Clifton, H.E. (1987) Sea-level change and the preservation potential of wave-dominated and tide-dominated coastal sequences. In: *Sea-level Fluctuation and Coastal Evolution* (Eds D. Nummedal, O.H. Pilkey and J.D. Howard). *Soc. Econ. Paleont. Mineral. Spec. Publ.*, **41**, 167–178.

De Boer, P.L., Ooost, A.P. and Visser, M.J. (1989) The diurnal inequality of the tide as a parameter for recognizing tidal influence. *J. Sediment. Petrol.*, **59**, 912–921.

Diemer, J.A. and Bridge, J.S. (1988) Transition from alluvial plain to tide-dominated coastal deposits associated with the Tournaisian marine transgression in southwest Ireland. In: *Tide-influenced Sedimentary Environments and Facies* (Eds P.L. de Boer, A. van Gelder and S.D. Nio), pp. 359–388. D. Reidel, Dordrecht.

Dionne, J.C. (1963) Towards a more adequate definition of the St. Lawrence estuary: *Z. Geomorph.*, **7**, 36–44.

Dominic, D.F. (1992) Facies architecture of sandstones below the No. 2 Coal bed between Stops 2 and 3, Jackson County Ohio. In: *Regional Aspects of Pottsville and Allegheny Stratigraphy and Depositional Environments, Ohio and Kentucky* (Eds C.L. Rice, R.L. Martino and E.R. Slucher). *U.S. Geol. Surv. Open File Rep.*, 17–20.

Donaldson, A.C., Renton, J.J. and Presley, M.W. (1985) Pennsylvanian deposystems and paleoclimates of the Appalachians. *Int. J. Coal Geol.*, **5**, 167–193.

Eble, C.F. (1996) Lower and lower Middle Pennsylvanian palynofloras, southwestern Virginia. *Int. J. Coal Geol.*, **31**, 67–114.

Eble, C.F., Greb, S.F. and Chesnut, D.R., Jr. (1991) *Coal-bearing rocks along the western margin of the Eastern Kentucky Coal Field.* Field Trip Guidebook, The Society of Organic Petrologists 34 pp.

Els, B. and Mayer, J. (1998) Coarse clastic tidal and fluvial sedimentation during a large Late Archaen sea-level rise: the Turffontein Subgroup in the Vredefort Structure, South Africa. In: *Tidalites—Processes and Products* (Eds C.R. Alexander, R.A. Davis and V.J. Henry). *Soc. Econ. Paleontol. Mineral. Spec. Publ.*, **61**, 155–166.

Emery, D. and Myers, K. (1996) *Sequence Stratigraphy.* Blackwell Science, London, 297 pp.

Emery, K.O. (1967) Estuaries and lagoons in relation to continental shelves. In: *Estuaries* (Ed. G.H. Lauff), pp. 9–11. American Association for the Advancement of Science, Washington, DC.

Englund, K.J. (1974) Sandstone distribution patterns in the Pocahontas Formation of Southwest Virginia and southern West Virginia. In: *Carboniferous of the southeastern United States* (Ed. G. Briggs). *Geol. Soc. Am. Spec. Pap.*, **148**, 31–45.

Ferm, J.C. (1974) Stratigraphic evidence for the position of Pocahontas Basin source terranes. In: *Carboniferous Depositional Environments in the Appalachian Region* (Eds J.C. Ferm and J.C. Horne), pp. 280–282. Carolina Coal Group, Department of Geology, University of South Carolina, Columbia, SC.

Ferm, J.C., Horne, J.C., Swinchatt, J.P. and Whaley, P.W. (1971) Carboniferous depositional environments in northeastern Kentucky (Roadlog for Geological Society of Kentucky (1971 Field Excursion). *Kentucky Geol. Surv. Ser.*, **10**, 30 pp.

Greb, S.F. and Archer, A.W. (1995) Rhythmic sedimentation in a mixed tide and wave deposit, eastern Kentucky, U.S.A. *J. Sediment. Res.*, **B65**, 96–106.

Greb, S.F. and Archer, A.W. (1998) Annual sedimentation cycles in rhythmites of Carboniferous tidal channels. In: *Tidalites—Processes and Products* (Eds C.R. Alexander, R.A. Davis and V.J. Henry). *Soc. Econ. Paleontol. Mineral. Spec. Publ.*, **61**, 75–83.

Greb, S.F. and Chesnut, D.R., Jr. (1992) Transgressive channel filling in the Breathitt Formation (Upper Carboniferous), Eastern Kentucky Coal Field, U.S.A. *Sediment. Geol.*, **75**, 209–221.

Greb, S.F. and Chesnut, D.R., Jr. (1994) Paleoecology of an estuarine sequence in the Breathitt Formation (Pennsylvanian) central Appalachian basin. *Palaios*, **9**, 388–402.

Greb, S.F. and Chesnut, D.R., Jr. (1996) Lower and lower Middle Pennsylvanian fluvial to estuarine deposition, Central Appalachian basin: effects of eustasy, tectonics, and climate. *Geol. Soc. Am. Bull.*, **108**, 303–317.

Greb, S.F., Eble, C.F. and Chesnut, D.R., Jr. (2002) Comparison of the Eastern and Western Kentucky Coal Fields, U.S.A—why are coal distribution patterns and sulfur contents so different in these coal fields?. *Int. J. Coal Geol.*, **50**, 89–118.

Hamrick, T. (1996) Paleoecology and sedimentology of a tidally influenced, marginal marine facies of the Lower Kanawha Formation (Middle Pennsylvanian) near Danville, West Virginia, U.S.A. *Abstracts for Tidalites '96, International Conference on Tidal Sedimentology*, Savannah, GA, 46–47.

Houseknecht, D.W. (1980) Comparative anatomy of a Pottsville lithic arenite and quartz arenite of the Pocahontas Basin, southern West Virginia; petrogenetic, depositional, and stratigraphic implications. *J. Sediment. Petrol.*, **50**, 3–20.

Howard, J.D. and Frey, R.W. (1973) Characteristic physical and biogenic sedimentary structures in Georgia estuaries. *Am. Assoc. Petrol. Geol. Bull.*, **57**, 1169–1184.

Howard, J.D. and Frey, R.W. (1975) Regional animal–sediment characteristics of Georgia estuaries. *Senckenb. Marit.*, **7**, 33–103.

Kreisa, R.D. and Moiola, R.J. (1986) Sigmoidal tidal bundles and other tide-generated structures of the Curtis Formation, Utah. *Bull. Geol. Soc. Am.*, **97**, 381–387.

Kvale, E.P. and Barnhill, M.L. (1994) Evolution of Lower Pennsylvanian estuarine facies within two adjacent paleovalleys, Illinois Basin, Indiana. In: *Incised-valley Systems: Origin and Sedimentary Sequences* (Eds R.W. Dalrymple, R. Boyd and B.A. Zaitlin). *Soc. Econ. Paleontol. Mineral. Spec. Publ.*, **51**, 191–207.

Kvale, E.P., Archer, A.W. and Johnson, H.R. (1989) Daily, monthly, and yearly tidal cycles within laminated siltstones (Mansfield Formation: Pennsylvanian) of the Illinois Basin. *Geology*, **17**, 365–368.

Lanier, W.P. and Tessier, B. (1998) Climbing-ripple bedding in the fluvio-estuarine transition: a common feature associated with tidal dynamics (modern and ancient analogues). In: *Tidalites—Processes and Products* (Eds C.R. Alexander, R.A. Davis and V.J. Henry). *Soc. Econ. Paleontol. Mineral. Spec. Publ.*, **61**, 110–117.

Martino, R.L. (1989) Trace fossils from marginal marine facies of the Kanawha Formation (Middle Pennsylvanian), West Virginia. *J. Paleontology*, **63**, 389–403.

Martino, R.L. (1992) Marine rocks: Their recognition and role in coal correlation and quality in the Kanawha Formation (Middle Pennsylvanian) of West Virginia. In: *1.2: New Perspectives on Central Appalachian Low-sulfur Coal Supplies* (Eds J. Platt, J. Price, M. Miller and S. Suboleski), pp. 125–142. Techbooks, Coal Decisions Forum Publication, Fairfax, VA.

Martino, R.L. (1994) Facies analysis of Middle Pennsylvanian marine units, southern West Virginia. In: *Elements of Pennsylvanian Stratigraphy, Central Appalachian Basin* (Ed. C.R. Rice). *Geol. Soc. Am. Spec. Pap.*, **294**, 69–86.

Martino, R.L. (1996) Stratigraphy and depositional environments of the Kanawha Formation (Middle

Pennsylvanian), southern West Virginia, U.S.A. *Int. J. Coal Geol.*, **31**, 217–248.

Martino, R.L. and Sanderson, D.D. (1993) Fourier and autocorrelation analysis of estuarine tidal rhythmsites, lower Breathitt Formation (Pennsylvanian), eastern Kentucky, USA. *J. Sediment. Petrol.*, **63**, 105–119.

Martino, R.L., Rice, C.L., Slucher, E.R. (1992) Stop 2, basal Pennsylvanian strata, near Jackson, Ohio. In: *Regional Aspects of Pottsville and Allegheny Stratigraphy and Depositional Environments, Ohio and Kentucky* (Ed., C.L. Rice, R.L. Martino and E.R. Slucher). *U.S. Geol. Surv. Open File Rep.*, 13–16.

Nichols, M.M. and Briggs, R.B. (1985) Estuaries. In: *Coastal Sedimentary Environments* (Ed. R.A. Davis, Jr.), pp. 77–186, Springer-Verlag, New York.

Nio, S.D. and Yang, C.S. (1991) Diagnostic attributes of clastic tidal deposits—a review. In: *Clastic Tidal Sedimentology* (Eds D.G. Smith, G.E. Reinson, B.A. Zaitllin and R.A. Rahmani). *Can. Soc. Petrol. Geol. Mem.*, **16**, 3–28.

Perillo, G.M.E. (1995) Definitions and geomorphological classifications of estuaries. In: *Geomorphology and Sedimentology of Estuaries* (Ed. G.M.E. Perillo), pp. 17–47. Developments in Sedimentology **53**, Elsevier, Amsterdam.

Potter, P.E. and Siever, R. (1956) Sources of basal Pennsylvanian sediments in the Eastern Interior Basin–Part 1. Cross-bedding. *J. Geol.*, **64**, 225–244.

Pritchard, D.W. (1967) What is an estuary? Physical viewpoint. In: *Estuaries* (Ed. G.H. Lauff), pp. 3–5. American Association for the Advancement Science, Washington, DC.

Rahmani, R.A. (1988) Estuarine tidal channel and nearshore sedimentation of a Late Cretaceous epicontinental sea, Drumheller, Alberta, Canada. In: *Tide-influenced Sedimentary Environments and Facies* (Eds P.L. de Boer, A. van Gelder and S.D. Nio), pp. 433–471. D. Reidel, Dordrecht.

Rice, C.L. (1984) Sandstone units of the Lee Formation and related strata in eastern Kentucky. *U.S. Geol. Surv. Prof. Pap.*, **1151-G**, 53 pp.

Rice, C.L. and Schwietering, J.F. (1988) Fluvial deposits in the central Appalachians during the Early Pennsylvanian. *U.S. Geol. Surv. Bull.*, **1839**, 1–10.

Shanley, K.W., McCabe, P.J. and Hettinger, R.D. (1992) Tidal influence in Cretaceous fluvial strata from Utah, USA: a key to sequence stratigraphic interpretation. *Sedimentology*, **39**, 905–930.

Smith, D.G. (1988) Tidal bundles and mud couplets in the McMurray Formation, northeastern Alberta, Canada. *Bull. Can. Petrol. Geol.*, **36**, 216–219.

Tessier, B. (1993) Upper intertidal rhythmites in the Mont-Saint-Michel Bay (NW France): perspectives for paleoreconstruction. *Mar. Geol.*, **110**, 355–367.

Tessier, B. (1998) Tidal cycles: Annual versus semilunar records. In: *Tidalites—Processes and Products* (Eds C.R. Alexander, R.A. Davis and V.J. Henry). *Soc. Econ. Paleontol. Mineral. Spec. Publ.*, **61**, 69–74.

Thomas, R.G., Smith, D.G., Wood, J.M., Visser, J., Calverley-Range, E.A. and Koster, E.H. (1987) Inclined heterolithic stratification–Terminology, description, interpretation, and significance. *Sediment. Geol.*, **53**, 123–179.

Visser, M.J. (1980) Neap–spring cycles reflected in Holocene subtidal large-scale bedform deposits: a preliminary note. *Geology*, **8**, 543–546.

Weber, J.N., Williams, E.G. and Keith, M.L. (1979) Paleoenvironmental significance of carbon isotopic composition of siderite nodules in some shales of Pennsylvanian age. In: *Carboniferous Depositional Environments in the Appalachian Region* (Eds J.C. Ferm, J.C. and J.C. Horne), pp. 110–113. Carolina Coal group, Department of Geology, University of South Carolina, Columbia, SC.

Wells, J.T. (1995) Tide-dominated estuaries and tidal rivers. In: *Geomorphology and Sedimentology of Estuaries* (Ed. G.M.E. Perillo), pp. 179–205. Developments in Sedimentology **53**, Elsevier, Amsterdam.

Wizevich, M.C. (1991) Photomosaics of outcrops; useful photographic techniques. In: *The Three-dimensional Facies Architecture of Terrigenous Clastic Sediments and its Implications for Hydrocarbon Discovery and Recovery* (Eds A.D. Miall and T. Noel). *Soc. Econ. Paleontol. Mineral. Concepts Sedimentol. Paleontol.*, **3**, 22–24.

Wizevich, M.C. (1992) Sedimentology of Pennsylvanian quartzose sandstones of the Lee Formation, central Appalachian Basin—fluvial interpretation based on lateral profile analysis. *Sediment. Geol.*, **78**, 1–45.

Wizevich, M.C. (1993) Depositional controls in a bedload-dominated fluvial system-internal architecture of the Lee Formation, Kentucky. *Sediment. Geol.*, **85**, 537–556.

Zaitlin, B.A. and Shultz, B.C. (1990) Wave-influenced estuarine sand body, Senlac heavy oil pool, Saskatchewan, Canada. In: *Sandstone Petroleum Reservoirs* (Eds J.H. Barwis, J.G. McPherson and J.R.J. Studlick), pp. 363–387. Springer-Verlag, New York.

Palaeogeography and fluvial to estuarine architecture of the Dakota Formation (Cretaceous, Albian), eastern Nebraska, USA

R.M. JOECKEL*, G.A. LUDVIGSON†, B.J. WITZKE†,
E.P. KVALE‡, P.L. PHILLIPS§, R.L. BRENNER§, S.G. THOMAS¶
and L.M. HOWARD**

*Conservation and Survey Division, School of Natural Resources, University of Nebraska-Lincoln, 113 Nebraska Hall, Lincoln, Nebraska 68588-0517, USA (Email: rjoeckel3@unl.edu);
†Iowa Department of Natural Resources, Geological Survey Bureau, 109 Trowbridge Hall, Iowa City, Iowa 52242-1319, USA;
‡Indiana Geological Survey, Indiana University, 611 North Walnut Grove, Bloomington, IN 47405, USA;
§Geology and Geography Program, University of North Carolina-Pembroke, Pembroke, North Carolina 28372-1510, USA;
¶Department of Geosciences, University of Nebraska-Lincoln, Lincoln, NE 68588-0340, USA; and
**Conservation and Survey Division, School of Natural Resources, University of Nebraska-Lincoln, 113 Nebraska Hall, Lincoln, Nebraska 68588-0517, USA

ABSTRACT

Regional mapping in easternmost Nebraska indicates that the Dakota Formation (Cretaceous) fills broad palaeovalleys incised into a Pennsylvanian bedrock surface with at least 115 m of regional relief. These palaeovalleys contain successions of cross-stratified sandstones, conglomerates and mudrocks, with evidence for multiple episodes of cut-and-fill. Palaeocurrent measurements from large sandstone bodies indicate dominant westward to northward flow paralleling the orientations of the mapped sub-Dakota palaeovalleys. Coarse clasts from both local and distant sources appear in coarse sandstones and conglomerates at the base of the Dakota at multiple sites. These observations are compatible with the interpretation of a dominant fluvial component in Dakota deposition. There are also features, however, that are compatible with marine tidal influence, such as laminites, flaser and linsen bedding, mud drapes, reactivation surfaces in cross-stratified sandstones and marine to brackish trace fossils. Well-logs through complete Dakota sections west of the study area show smaller scale coarsening upward and fining upward trends, as well as medium-scale stratigraphical trends. In these wells, the Dakota typically shows 7–10 discrete, medium-scale packages, each a few to several tens of metres thick. Thick valley-filling channel sandstones, meandering fluvial deposits and possible deltaic or shoreface deposits are interpreted from electronic logs of these wells. At least one well, represented by a thorough lithological description, electronic logs and partial core, also appears to contain tidal facies in the upper Dakota. Thus, the deposition of the Dakota in eastern Nebraska, particularly the filling of palaeovalleys in the outcrop belt, involves a complex history of fluvial and estuarine deposition.

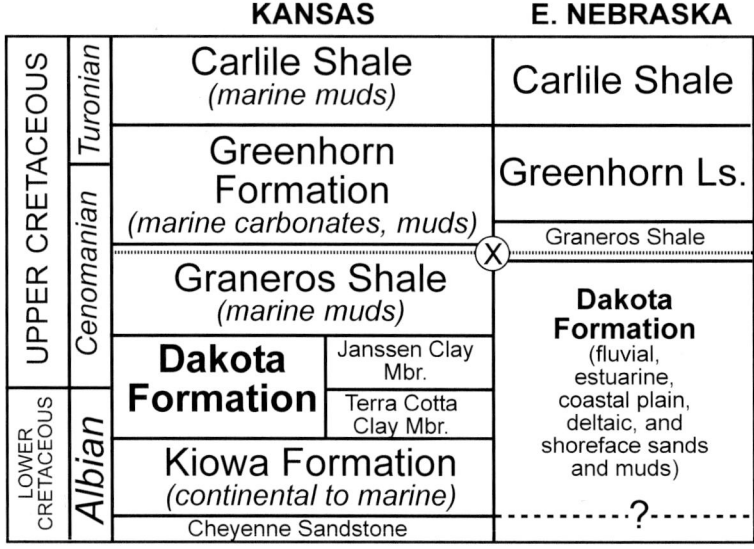

Fig. 1 Revised stratigraphy of mid-Cretaceous sediments in study area and adjacent areas (modified after Brenner et al., 2000). Possible age-equivalence of lowermost Dakota Formation in eastern Nebraska and Cheyenne Sandstone in Kansas is based on lower palaeotopographic position of lowermost Dakota sediments near Douglas–Sarpy county line, Nebraska, relative to Albian sediments at Ash Grove Quarry (AGQ). 'Graneros Shale' in traditional Nebraska usage probably refers mostly to the lower, shaly part of Greenhorn Formation elsewhere (e.g. Colorado). Interpretations of depositional environments emerge from this paper and summary in Hattin et al. (1978).

INTRODUCTION

The Dakota Formation (officially the 'Dakota Group' in Nebraska (Condra & Reed, 1959) but retaining formation status in adjacent states) is the basal Cretaceous deposit at the eastern edge of the Western Interior Seaway (WIS), extending from central Kansas northwards and eastwards (Fig. 1). The interpretation of Dakota history in eastern Nebraska forms an essential linkage between the relatively poorly known cratonic eastern margin of the WIS and the thick, well-studied Cretaceous sections of the western High Plains and Rocky Mountains. The Dakota's significance is underscored by its increasing value as a source of groundwater, its yield of industrial clays and long-term oil and gas production.

The first facies-based interpretations of the Dakota Formation in Nebraska (e.g. Karl, 1976) envisaged an upward intraformational transition from aggrading braided to meandering streams in a purely 'continental' setting. Recently, a resurgence of interest in the Dakota on the eastern margin of the WIS (e.g. Brenner et al., 2000) has initiated a reinterpretation that centres on the filling of incised fluvial valleys and episodic marine influence on fluvial sedimentation.

The purposes of this paper are to:

1 present a detailed geographical information system (GIS)-based reconstruction of sub-Dakota incised bedrock drainage in a part of the study area where wells and exploratory boreholes into the Dakota are common;
2 describe the architecture of the Dakota Formation from outcrops and well logs;
3 interpret the regional palaeogeography and depositional history of the unit.

Work in the study area, however, is made challenging by a lack of continuous exposures, extensive post-Dakota erosion and the local absence of correlative marker horizons, such as the X-bentonite (Fig. 1). Nonetheless, over 200 shallow-well logs, a few excellent and comparatively large exposures and scattered deep petroleum-exploration wells provide sufficient data for expanding interpretations of mid-Cretaceous sedimentation at the eastern margin of the WIS.

PALAEOGEOGRAPHY OF THE SUB-DAKOTA BEDROCK SURFACE

In the easternmost Platte River Valley (EPRV), the Dakota overlies a > 200 Myr unconformity on Upper Pennsylvanian cyclothems with more than 115 m of relief (Fig. 2 and Plate 1). This figure is comparable to relief on the modern land surface in

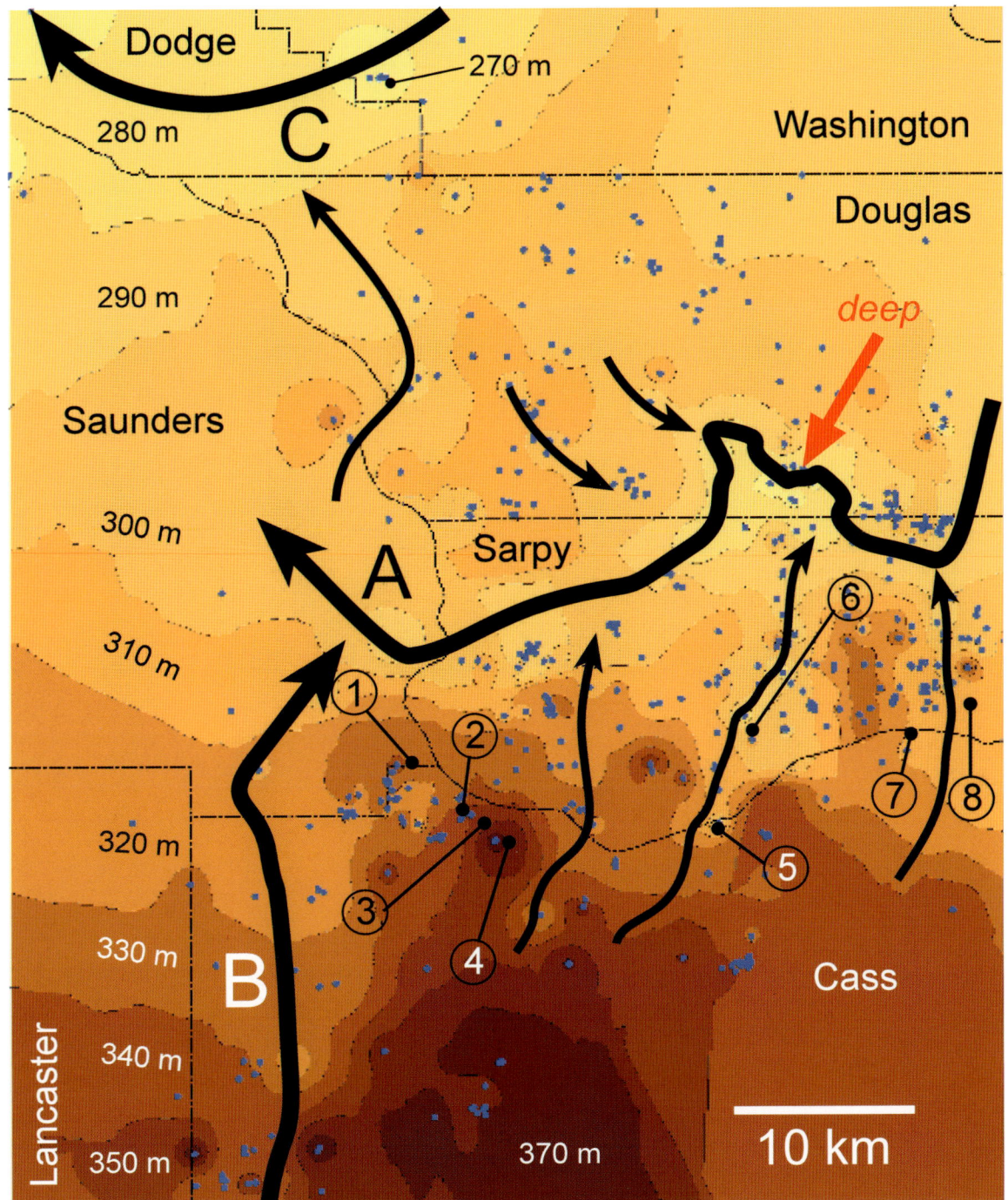

Plate I Sub-Dakota palaeotopographic surface reconstruction based on extrapolation of the Dakota–Pennsylvanian contact from existing well and outcrop data points (blue dots) across Plio-Pleistocene erosional topography. Present-day north is to the top of the map. Contour interval is 10 m; lighter tones represent lower elevations and darker tones represent higher elevations (highest contour is 370 m); total relief across area exceeds 115 m. Major interpreted palaeovalleys (A, B & C) and palaeoflow directions are indicated by broadest black arrows, and smaller interpreted palaeovalleys by finer black arrows; deepest part of trunk palaeovalley A (approximately 260 m) indicated by red arrow. Note apparent Pennsylvanian bedrock uplands to the south, in Cass County. Compare with positions of structural features in Fig. 2. Measured sections numbered as in Fig. 2.

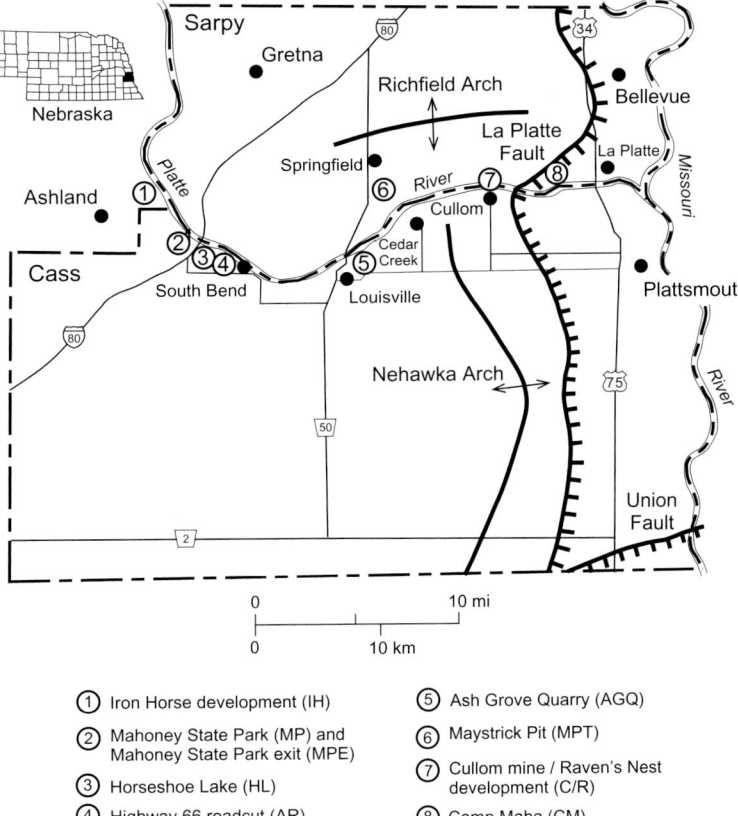

Fig. 2 Location map of measured sections and selected structural features (after Condra & Scherer, 1939; Harden, 1959; Nelson, 1958; Burchett, 1970) in the easternmost Platte River Valley (EPRV), defined herein as the bedrock gorge or trench in which the river flows between Ashland and Plattsmouth, Nebraska. Additional geological structures have been identified very recently in the La Platte, Nebraska, area.

the same area. The lower contact of the Dakota in the EPRV either cuts across strata affected by faults and gentle flexures (Condra & Scherer, 1939; Nelson, 1958; Harden, 1959; Burchett, 1970, 1971), or shows response to other geological structures in underlying strata (Fig. 2 and Plate 1). Thus, the Dakota is useful in interpreting Early Cretaceous palaeogeography, and as an indicator of the relative timing of deformation events.

The sub-Dakota bedrock palaeosurface in the EPRV area was reconstructed by contouring continuously from data points (over 200 logs of registered water wells, test holes and construction borings, in addition to outcrops) on the Dakota–Pennsylvanian contact, using a GIS database. Late Cenozoic erosion features such as the modern Platte River gorge or trench were disregarded, thus extrapolating from intact remnants of the originally extensive contact across interspersed eroded areas. The resulting map (Plate 1) reveals at least two scales of sub-Dakota palaeovalleys and adjacent bedrock uplands composed of Pennsylvanian bedrock. A broad (2–10 km wide) sub-Dakota palaeovalley (A in Plate 1) lies near the Sarpy–Douglas county line, trending from the southern suburbs of Omaha west-southwest between Gretna and Springfield, then turning west-northwest and crossing under the modern Platte about 10 km north-northeast of Ashland, Nebraska (Fig. 2 and Plate 1). This large palaeovalley is infilled with basal Dakota sediments and is interpreted as a late Early Cretaceous trunk drainage. Its orientation is compatible with the western Iowa palaeovalley pattern proposed in Witzke *et al.* (1996) and Brenner *et al.* (2000). Its position parallels the trend of the 'Richfield Arch' (Fig. 2), a gentle structure in underlying Pennsylvanian rocks. Smaller palaeovalleys appear on Pennsylvanian bedrock uplands to the south of the trunk palaeovalley in southern Sarpy and northern Cass counties (Plate 1). Bedrock uplands of lesser relief appear north of the trunk palaeovalley in southern Douglas County. Along the western flank of the southern bedrock uplands of Cass County, parallel and slightly east of the Lancaster County line, there is a second major palaeovalley trending S–N (B in Plate 1). There appears to be a third major palaeovalley, with a strong westward component to flow, in Washington and Dodge counties (C in Plate 1).

OUTCROP-SCALE ARCHITECTURE

Dakota sedimentation in western Iowa and adjacent areas has been interpreted as the filling of incised bedrock valleys as early as the Albian Kiowa–Skull Creek transgression, and possibly even earlier (Witzke & Ludvigson, 1994; Witzke et al., 1996; Brenner et al., 2000). Palaeovalley fills in the Dakota show evidence for multiple episodes of deposition and therefore record complex local histories of sedimentation.

Although sandstones and conglomerates typically appear at the base of valley fills, directly on top of or just above the Dakota–Pennsylvanian, mudrocks also appear in several places directly on top of the eroded Pennsylvanian surface. In the EPRV, coarse-grained sediments dominate the Dakota as a whole: 75–80% of the Dakota sediments logged in EPRV wells and outcrops are sandstones and pebbly sandstones to conglomerates. Subangular to subrounded cobbles and boulders of characteristic Pennsylvanian chert, as much as 750 mm in maximum diameter, have been found at or just above the Dakota–Pennsylvanian contact at different elevations in the EPRV from Cedar Creek, near the eastern edge of the Dakota in Nebraska, to Ashland, Nebraska, nearly 20 km westward. Pebbly sands and conglomerates have also been found at the base of the Dakota in wells up to 100 km westward in Saunders and Platte counties (Figs 2 & 3). A small percentage (< 2%) of the coarsest clasts in

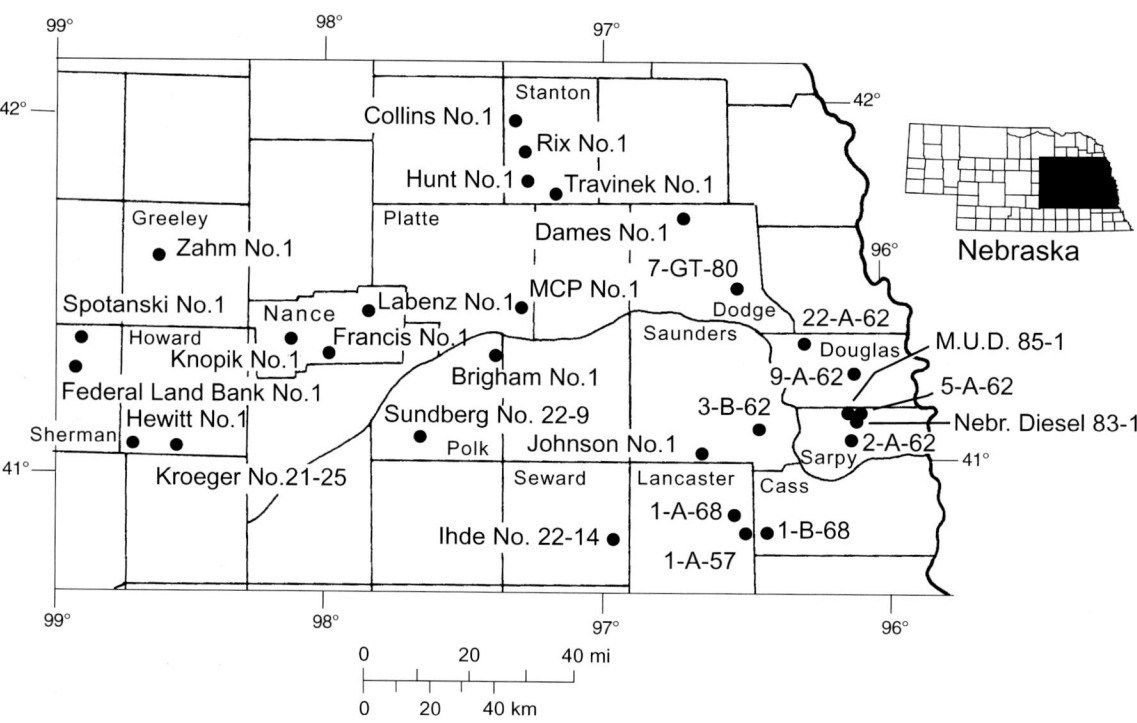

Fig. 3 Key Dakota penetrating rotary test-holes and deep petroleum exploration wells in Nebraska east of 99°W longitude. In general, these are pre-1970 holes having only resistivity and spontaneous potential logs and lack corresponding cores.

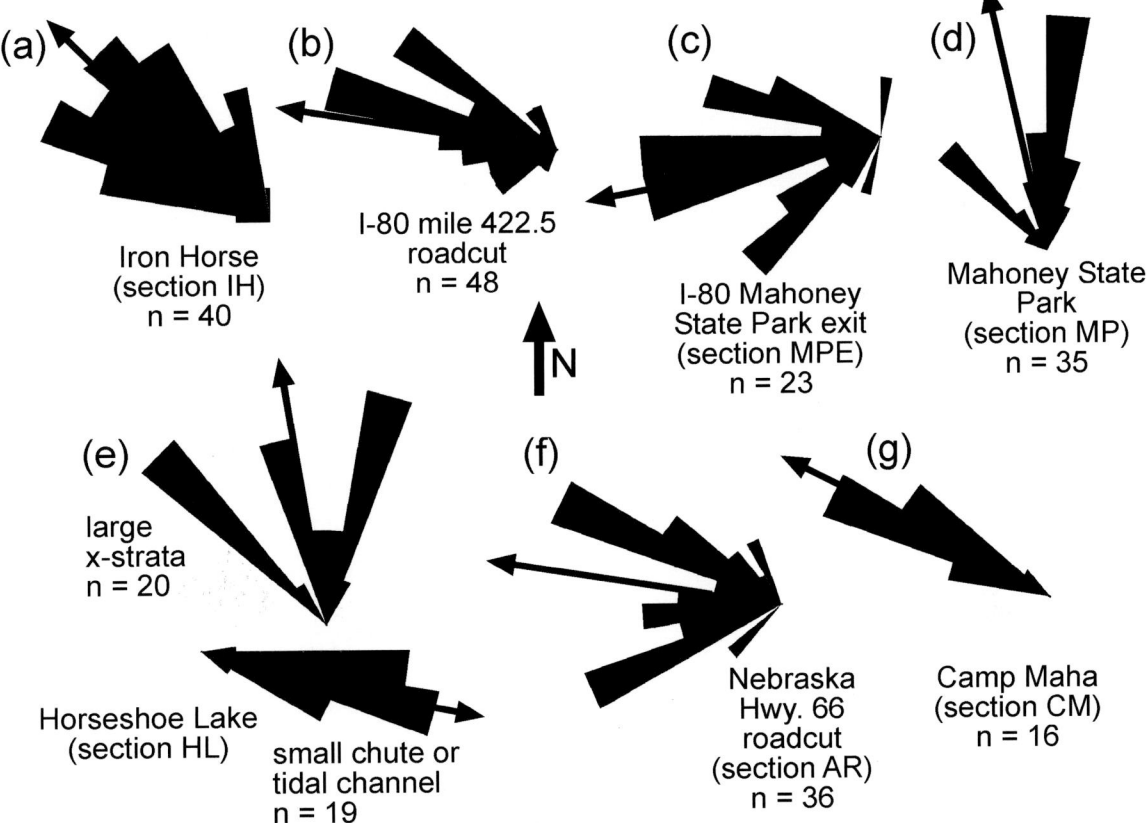

Fig. 4 Palaeocurrent measurements (foreset dip azimuths) from Dakota Formation exposures in the EPRV; means are represented by arrows. Note dominance of west to northward flow. Interstate 80 mile 422.5 exposure (b) is small Dakota sandstone exposure about 5 km west-southwest of section MPE (c).

lower Dakota strata consist of exotic rock types (banded iron formation, jasper, non-local chert, quartzite, vein quartz), which could only be derived from the Canadian Shield and environs, 1000 km or more distant.

Matching interpretive cross-sections derived from borehole logs with the reconstructed sub-Dakota palaeosurface is largely unproductive. In multiple instances, thin Dakota (deeply eroded during the late Cenozoic) overlies the Cretaceous–Pennsylvanian in a palaeovalley. Conversely, it is not unusual to find a much thicker Dakota section in the adjacent modern uplands. These conditions make the construction of informative cross-sections very difficult, if not impossible, in most places.

Western end of EPRV

At the western end of the EPRV, sandstones constitute the majority of Dakota outcrops. These sandstones are dominated by planar tabular cross-stratification in sets 0.1 to 0.5 m thick, although thicker (up to 1.6 m) sets also appear. Palaeocurrent measurements show a relatively narrow westerly to northerly range (Fig. 4), which matches the general orientation of sub-Dakota palaeovalleys (Plate 1 and Fig. 4). The lateral extent of sandstones (≤ 200 m) typically corresponds to the lengths of individual outcrops.

Exposures near Mahoney State Park (Figs 2 & 5a–d: sections MP, MPE, HL, AR; Fig. 6a) are in close proximity to each other, similar in elevation

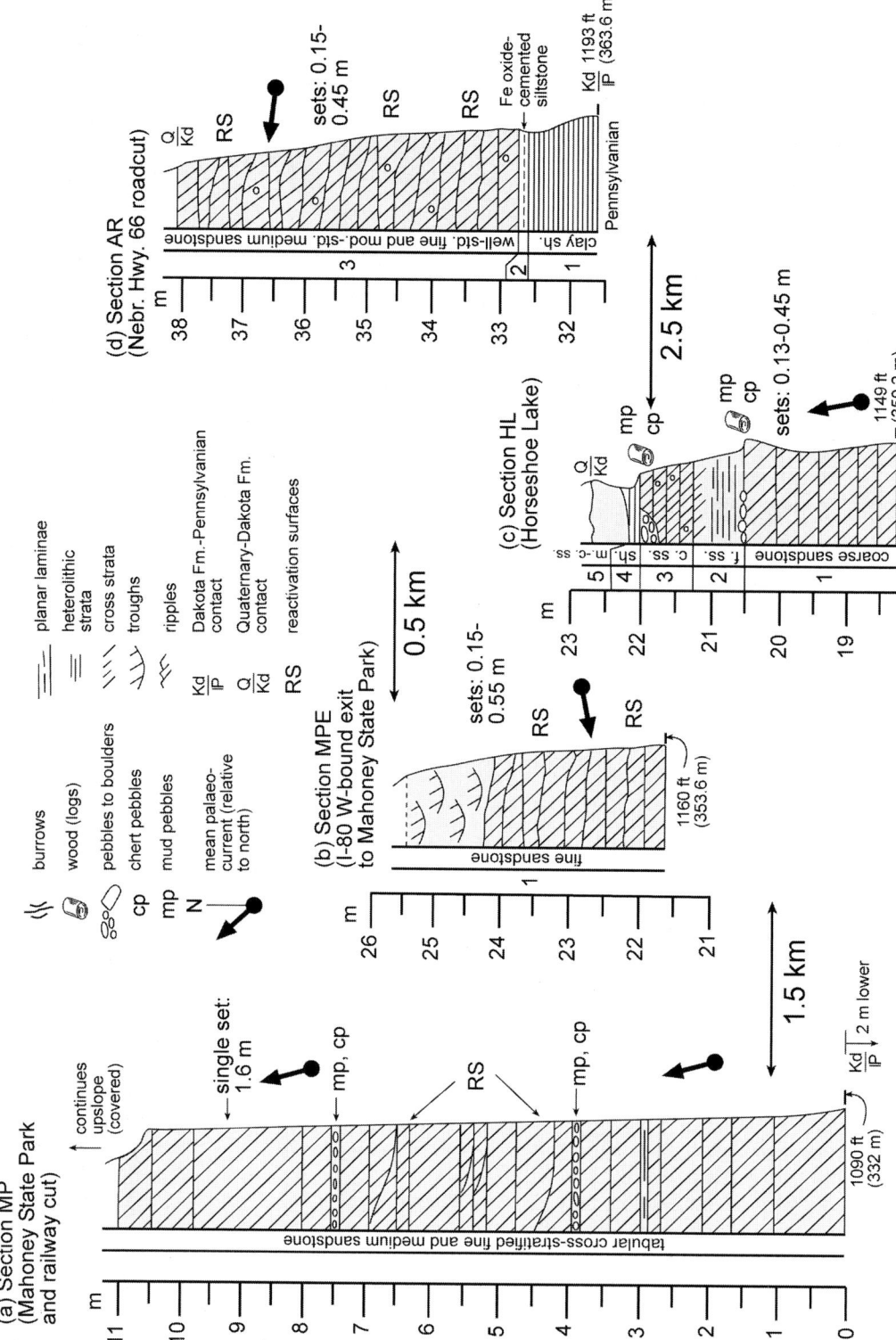

Fig. 5 Measured sections at the western end of the EPRV, in the vicinity of Mahoney State Park (MP) section (a). Section scales in metres are numbered continuously from base of section MP. Note dominance of planar cross-stratified sandstones; tabular sets and reactivation surfaces are common. Representative cross-strata set thicknesses are provided. See Fig. 2 (numbers 2–4) for locations. Abbreviations: coarse (c), fine (f), medium (m), moderately (mod.), sandstone (ss), shale (sh), sorted (std.).

Fig. 6 Features of Dakota exposures at western end of EPRV. (a) Tabular cross-stratified sands with reactivation surfaces at section MP; note cross-strata at arrow. Section contains laterally continuous horizon of mud pebbles (mp). Rod Root (lower foreground) is 1.7 m tall. (b) Imbricated rip-up clasts of shale–sand laminite at section IH (see Fig. 7, Unit 2b); bar scale is 100 mm long.

and have the same facies composition. Thus, they are probably part of the same, large, palaeovalley fill at the margin of a large sub-Dakota bedrock palaeovalley (B in Plate 1), which would be at least 38 m thick. These exposures are dominated by relatively uniform tabular cross-stratified (sets reach 1.6 m at MP), fine to medium-grained sandstone. Individual cross-strata at all four sections are characteristically tangential, low-angle reactivation surfaces, with heights exceeding 1 m in some cases, and produce wedge-shaped sets of cross-strata which could easily be mistaken for troughs. Thin (< 1 m) lenses or horizons of chert and quartz gravels, or stringers of clay pebbles (Fig. 6a), are rare within the sandstones, but are nonetheless prominent when present. Two such horizons of mud pebbles at MP (Fig. 5a) extend laterally for at least 200 m and therefore are likely to mark genetically significant erosion surfaces. Two small channel fills at the top of section HL

(Fig. 5c) show palaeocurrents that differ significantly from those in the remainder of the putative sand-dominated valley fill (Fig. 4b–e), and indeed from palaeocurrent measurements made elsewhere in the study area. Orientated limonitized logs also appear in the upper part of section HL.

An exposure 110+ m long at a former limestone quarry (IH: Figs 6b & 7) shows at least two, and very likely three, storeys of fluvial deposits filling a palaeovalley eroded into Pennsylvanian cyclothems (Fig. 7). Section IH is also at the eastern side of a major N–S palaeovalley on the reconstructed pre-Dakota bedrock map (B in Plate 1). Tabular cross-stratified sands dominate, as at MP, MPE, and HL. Very large-scale, inclined mud shale drapes (or lateral-accretion units) and subtle surfaces within sandstones delineate major scour surfaces between successive fills. Deposits at the margins of a large, sheet-like sand body (Fig. 7, Unit 2a) at this locality are matrix-supported

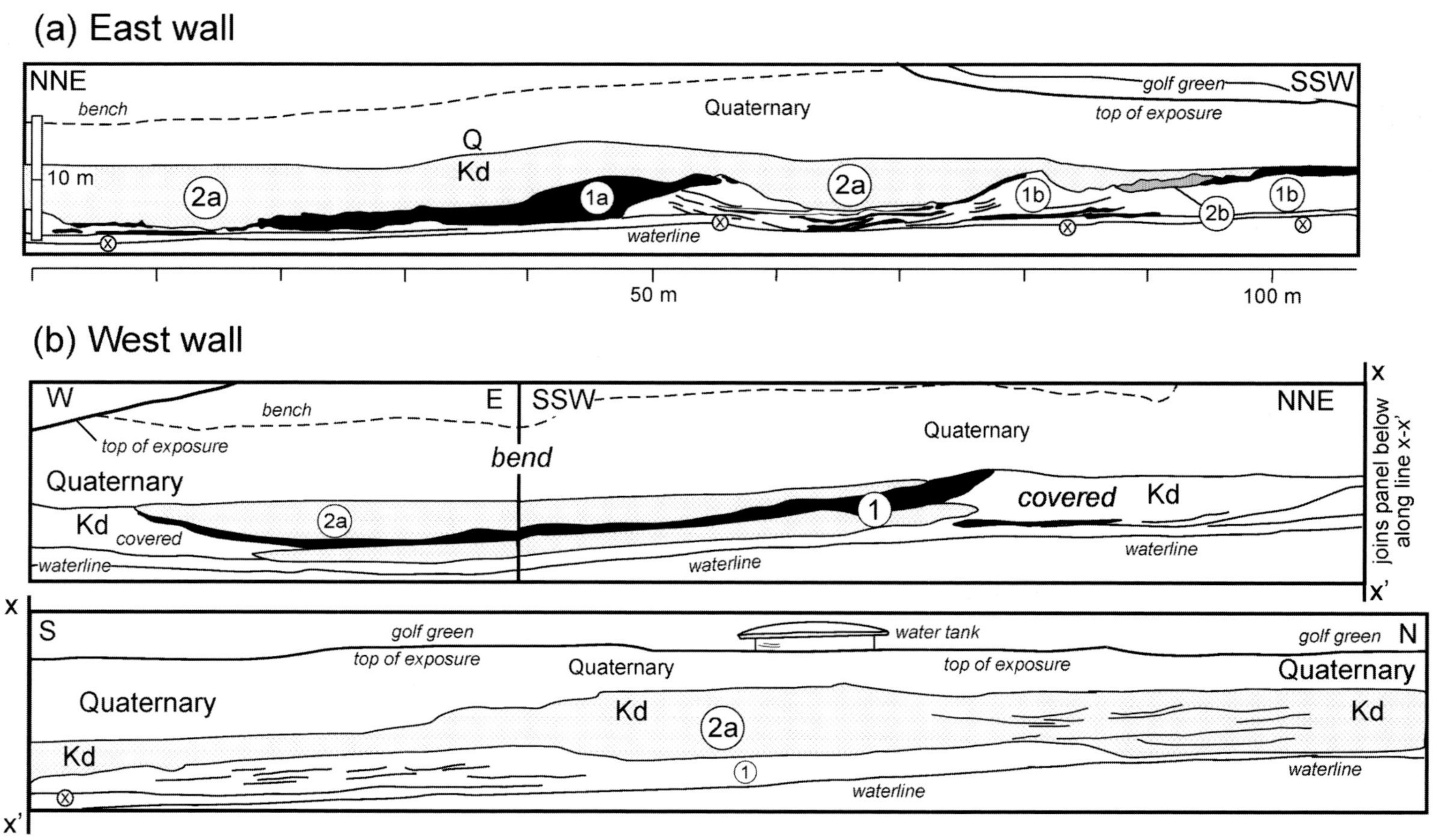

Fig. 7 Panoramas of east (a) and west (b) walls of partially reclaimed quarry at IH (Fig. 2, number 1), which lies near the northern end of a roughly N–S bedrock palaeovalley (see Plate 1, B). Unit 2a is a sheet-like sandstone dominated by tabular cross-stratification, and with an irregular lower contact. Unit 2b is a sand-matrix-supported laminite–pebble conglomerate. Unit 1, differentiated into subunits by lower case letters where possible, is an earlier fill, which contains mud units (black) that are frequently orientated at low angles to the horizontal (presumably due to lateral accretion). The circled 'x' represents the top of what is very likely a third (and even older) fill of tabular cross-stratified sandstone (directly overlying Pennsylvanian cyclothems), now covered by waters of a recreational lake. Formerly, lenses of chert pebble to boulder conglomerate were visible in the lowermost Dakota here.

conglomerates of 40–60% imbricated, oblate, grey shale pebbles to cobbles 10–250 mm in maximum diameter (Fig. 6b). Current-orientated limonitized logs were visible in the Dakota at IH in the past, before a recreational lake covered the lower part of the exposure.

Central to eastern EPRV

Exposures in the central to eastern EPRV are more varied in facies composition and stratal geometry, although sandstones still dominate.

Ash Grove Quarry (AGQ)

An extensive (> 1 km) linear (and locally three-dimensional) exposure at the Ash Grove Cement Company's quarry (AGQ) at Louisville, Nebraska (Figs 8 & 9) was used as the basis of recent biostratigraphical and sequence stratigraphical reinterpretation of regional Dakota deposition (see Brenner *et al.*, 2000). At least 18 m of Dakota sediments infill a broad palaeovalley (exposed obliquely to flow) cut into slightly deformed Pennsylvanian cyclothems (Fig. 9). The sub-Dakota erosion surface locally shows at least 5 m of relief, but total relief across the entire quarry area is about 10 m. Underlying Pennsylvanian limestones (Raytown and Argentine limestone members) show as much as 3° of apparent dip, and this structure, however subdued, appears to have influenced the position of the palaeo-valley (Fig. 9). Sub-Dakota strata also show very gentle flexures, small-offset normal faults and stratabound calcite-megaspar-infilled joints, all of which reflect post-Late Pennsylvanian and pre-Cretaceous deformation.

The stratigraphical succession at AGQ consists of basal sands (Fig. 8, Unit 1; Fig. 10) overlain by prominent inclined heterolithic strata (IHS) (Fig. 8, Unit 2; Figs 11 & 12a) and then a second sand (Fig. 8, Unit 3a) that also contains basal lenses to lags of matrix-supported chert-pebble conglomerate with some clay pebbles (Fig. 8, Unit 3b). Laterally adjacent to and above this sand were laminites (Fig. 8, Unit 4; Fig. 12b) that have been interpreted as tidal rhythmites (Brenner *et al.*, 2000), and a third sand (Fig. 8, Unit 5). Units 4 and 5 have been completely mined out since they were discovered and cored in 1994–1995.

Post-Cretaceous oxidation fronts (yellowish brown colours resulting from iron oxides and hydroxides) are visible in Units 2, 3 and 5, whereas the other units retain grey colours and contain unweathered pyrite and siderite.

Unit 1 at AGQ is light grey, fine to medium, cross-stratified, very friable sandstone with common low-angle scours several metres long, some of which are draped with thin shales as much as 50 mm thick. Other scours are filled by prominent concentrations of pebbles to small boulders (Fig. 10a). These prominent 'lags' contain both locally derived clasts (weathered, tripoli-like and unweathered Pennsylvanian cherts) and a few exotic clasts (long-transported Lower Palaeozoic or Proterozoic quartzites from terranes to the north-east). Smaller-scale cross-stratification (20–70 mm sets) is common in the unit, but large cross-strata locally infill entire scours/troughs and range between 0.8 and 1.2 m in individual thickness. Some groups of cross-strata near the top of Unit 1 contain 250–400 mm coarse–fine couplets in which the much thicker (240–380 mm) coarse layers have an elongate, lenticular form. Fine (1–1.5 mm) laminae of organic matter drape these cross-strata, and chevron folding (soft-sediment deformation) of such strata is visible at the very top of the unit. Large siderite-cemented oblate mud pebbles mark a few foreset laminae. Groups of closely spaced, nearly horizontal to moderately dipping (30°) carbonized logs appear in many scour fills. Typically, these logs are compressed parallel to bedding, but a few early cemented (pyritized and/or sideritized) examples approach 1 m in diameter. *Skolithos*-like burrows are locally common in the upper part of Unit 1 (Fig. 10b).

Unit 2 is the most prominent unit currently exposed. In most of the operating quarry, it rests directly on top of Pennsylvanian limestone, indicating that some of the basal sands of Unit 1 were eroded prior to the deposition of Unit 2. The latter is an interval of IHS up to 8 m thick with complex internal stratification and soft-sediment deformation; lateral accretion surfaces (LAS) bounding IHS units dip as much as 60°. Multiple generations of IHS and bounding LAS, representing point bars of different ages, can be seen in one wall of the quarry (Fig. 9). *Teichichnus*, *Chondrites* and *Arenicolites* (considered to be brackish or marine

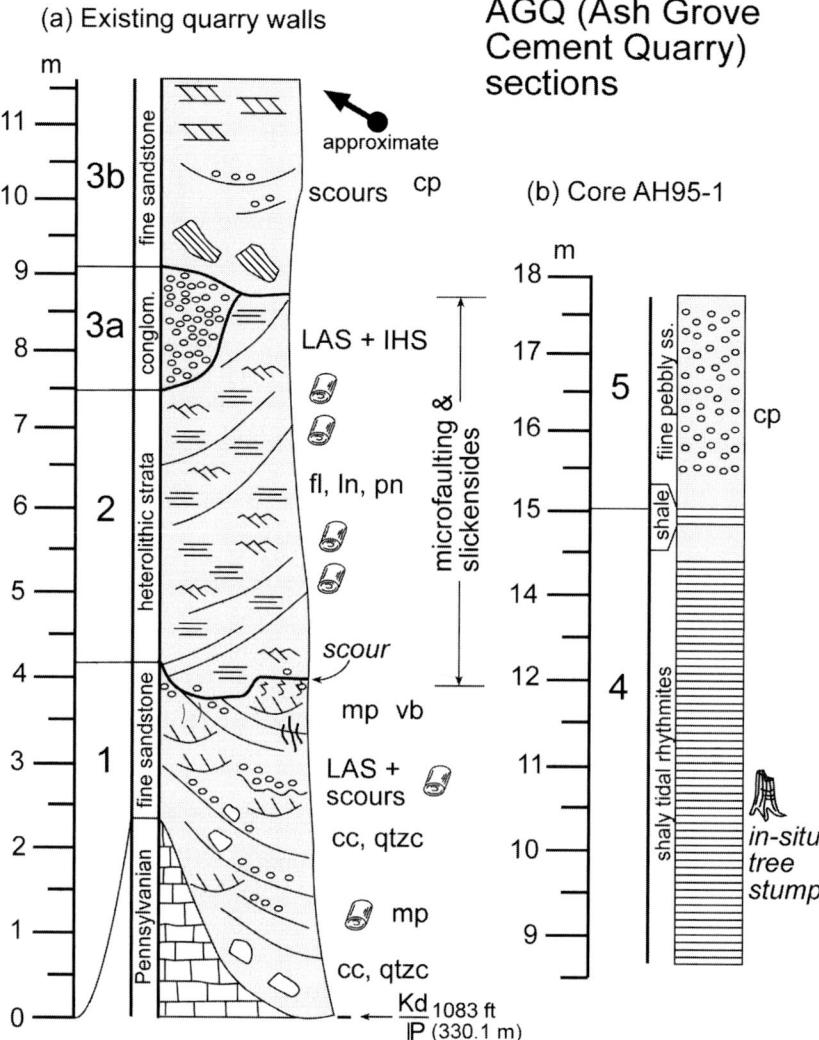

Fig. 8 Measured sections at Ash Grove Cement Company Quarry (AGQ), Louisville, Nebraska (Fig. 2, number 5). (a) Existing highwall section (Land Survey Reference $S^{1/2}$ $SE^{1/4}$ section 13, T.12 N., R.11E., Cass County, Nebraska). (b) Interval cored at same quarry 1995 (in now mined-out area), which overlaps in metre scale with part of currently exposed section (a). Note contrast between fluvial-dominated facies (sands) and tidally dominated facies (Units 2 and 4). Unit 1 has many metre-scale scour surfaces and larger lateral accretion units; chert cobbles (cc) and quartzite cobbles (qtzc) are common along such surfaces. Many chert pebbles and cobbles in Unit 1 are heavily weathered to a tripolitic consistency. Exotic, far-travelled clasts are also found in Unit 1. Drab-coloured mud pebbles (mp) in the unit are frequently surrounded by siderite-cemented rinds. Trough cross-strata immediately below the upper contact of Unit 1 show chevron folding, which is interpreted as soft sediment deformation. Vertical burrows (vb) also appear near upper contact. Unit 2 contains prominent lateral accretion surfaces (LAS) with common siderite cementation, and inclined heterolithic strata (IHS); internal stratification includes flaser (fl), linsen (ln) and pinstripe (pn) stratification. Accumulations of carbonized logs along LAS are frequently observed in Unit 2. Scour surface at base of Unit 2 is cemented with siderite. Upper sandstone and underlying conglomerate (Units 3a and 3b) pinch out laterally and were absent at site of core AH95-1, where shaly tidal rhythmites directly overlaid heterolithic strata of Unit 2. Unit 3b contains large scours, possibly channel forms, and large clasts of laminite similar in characteristics to Unit 4. Tidal rhythmites (Unit 4) and upper pebbly sandstone (Unit 5) have been completely mined out since their coring in 1995. See Fig. 5 for symbol key and text for discussion.

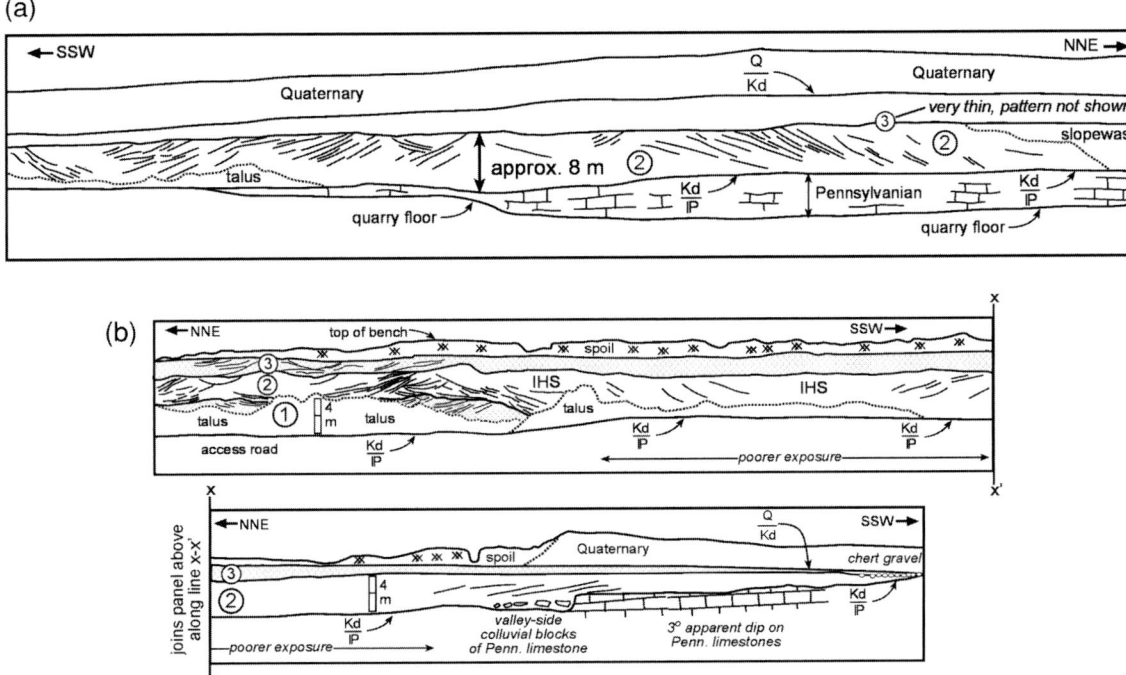

Fig. 9 Simplified panoramas of adjacent walls exposing Dakota Formation strata filling a bedrock palaeovalley incised into Pennsylvanian strata at Ash Grove Quarry. Circled numbers identify stratigraphical units discussed in text and shown in Fig. 8. (a) Unit 2, with prominent lateral accretion surfaces (LAS) and inclined heterolithic strata (IHS), directly overlying Pennsylvanian limestones (P); LAS are dipping in more than one direction, indicating multiple generations of point-bar deposition by cut-and-fill. (b) Opposite wall showing stratigraphy (Units 1–3) from left-hand column in Fig. 8; panels are continuous and join at line x–x′; cross-section of palaeovalley is more apparent because of rising Dakota Formation–Pennsylvanian contact to the south-southwest, visible pinchout of Dakota Formation, and presence of siderite-impregnated colluvial blocks of Pennsylvanian limestone near edge of palaeovalley.

traces; Bromley, 1990), are locally common in the IHS. Millimetre- to decimetre-scale microfaulting (Figs 12a & 13), large slickensides, deformed logs and deformed sedimentary strata are pervasive (Fig. 13), the latter appearing even at the microscopic scale. Carbonized logs are concentrated in groups along LAS.

At the southern margin of the AGQ palaeovalley fill, IHS units interfinger with thicker (> 1 m thick) sands, which are themselves truncated within a short distance by Pleistocene erosion (Fig. 9). Numbers of chert and exotic pebbles to small boulders (including Canadian Shield-derived vein quartz) lie directly above a single LAS within Unit 2 (Fig. 11a), and there is even a large pothole filled with these coarse clasts (Fig. 11b). Siderite cementation is common in Unit 2: the basal contact in one face of the quarry (Fig. 10b) is cemented by siderite, and there are irregular bands and large (up to 0.5 m in diameter) nodules of siderite throughout the unit. The superjacent relationship of the siderite-cemented lower contact of Unit 2 with the deformed cross-strata at the top of Unit 1 indicates very early (syndepositional) cementation, as does the presence of siderite-cemented rip-up clasts within Unit 1 itself. Siderite from the cemented zone at the contact between Units 1 and 2 yielded relatively depleted $\delta^{18}O$ values clustering around −5 to −6‰ (VPDB) and $\delta^{13}C$ values between −1 and −6‰ (VPDB).

Unit 3 (Fig. 8) is a laterally extensive well-sorted, fine sandstone with < 2% chert and clay

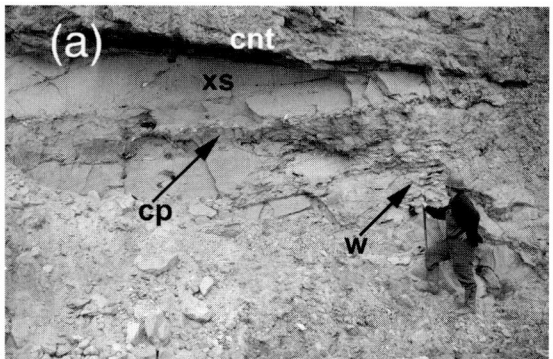

Fig. 10 Features of lowermost Dakota sediments (Unit 1) at Ash Grove Quarry. (a) Unit 1, showing scour surface lined with chert pebbles to cobbles (cp), and carbonized wood (w), probably entangled logs; Greg Ludvigson is 1.6 m tall. Top part of unit has large-scale cross-stratification (xs) with organic matter drapes and soft-sediment deformation (chevron folding, see Fig. 9) underneath siderite-cemented contact (cnt) with Unit 2. (b) *Skolithos* in upper part of Unit 1 (see Fig. 9).

Fig. 11 Features of Unit 2 (IHS unit, Fig. 9) at Ash Grove Quarry. (a) Cobbles and small boulders of chert and far-travelled clasts lining a LAS (arrows). Lee Phillips (1.7 m tall) is standing at large pothole eroded into underlying IHS unit. (b) Close-up of large pothole infilled with the same types of clasts.

pebbles, which are typically concentrated on large scour surfaces. A few, very large (up to 200 × 1450 mm in outcrop face) clasts of light grey, interlaminated mud and fine silt–sand, identical to the sediments in Unit 4 (see discussion below), are concentrated near the bases of large scour surfaces. These clasts are orientated obliquely to the horizontal (10–45°) and may have been imbricated relative to palaeocurrents. Stratification in Unit 3 is strongly obscured by Liesegang banding, but its uppermost part shows faint, decimetre-scale tabular cross-stratification. Lenses to thin, discontinuous sheets of matrix-supported conglomerate (dominated by weathered chert pebbles) mark the base of Unit 3.

Unit 4 is thinly and rhythmically laminated shale and fine sand or silt, which contains at least one *in situ* tree stump (Fig. 12b). Approximately 200 different Cretaceous palynomorph taxa (including unassigned or indeterminate species) have been identified from this unit, making it one of the richest known Albian palynomorph assemblages in the world (Witzke *et al.*, 1996). It includes forms assigned to the bryophytes, lycophytes, pterophytes (fern), gymnosperms, angiosperms, freshwater algal spores, and marine acritarchs and dinoflagellates (Witzke *et al.*, 1996).

Maystrick Pit (MPT)

Maystrick Pit (Figs 14 & 15) shows at least two storeys of fluvial fill: (i) a lower, coarse, gravelly

Fig. 12 (a) Soft-sediment deformation in Unit 2 (see Fig. 9) at Ash Grove Quarry. Note fault (arrows) truncating what appear to be linsen-bedded strata. Overlying concentration of carbonized logs under overhang is crushed and deformed, and slickensides are common in surrounding sediments. Log at right is compressed and sheared along near-vertical plane. Scale at bottom is 150 mm. (b) *In situ* tree stump in tidal rhythmites (Unit 4) at Ash Grove Quarry; scale is 80 mm long.

succession and (ii) a distinctly different upper succession of mixed sandstone and mudrock, interfingering laterally with mottled mudrocks (hydromorphic palaeosols), which were formerly mined as brick and tile clays. The upper succession incorporates shallowly dipping surfaces that must be depositional in origin, because underlying strata are horizontal and no evidence for structural dip exists in the Dakota Formation in the study area.

The upper succession at MPT is now mostly obscured by vegetation and slumping. An archival photograph, however, shows large-scale, low-angle surfaces bounded by sandstone- and mudstone-dominated intervals (Fig. 14a). Laterally, thin sands, stratified mudrocks, and thin, lenticular gravels in the upper succession appear to interfinger with thick, mottled mudstones (palaeosols). A dark-coloured shale unit in this upper sequence yielded the marine palynomorph *Micrhystridium* sp.

The lower, coarse succession at MPT has exceedingly complex stratification (Fig. 14b). The lowermost exposed unit is a large-scale (> 1 m set), tabular cross-stratified pebbly sand with pebble lenses on foresets dipping roughly west-northwestward (Figs 14b, 1 & Fig. 15b; no palaeocurrent measurements could be made). Thin (5–50 mm) mud shale drapes (locally burrowed) appear above this unit and between other units up-section in the coarse fill (Figs 14 & 15a, md). Above the pebbly sand is a large-scale tabular cross-stratified, friable, matrix-supported pebble conglomerate (Fig. 14b: 2); pebbles are dominantly

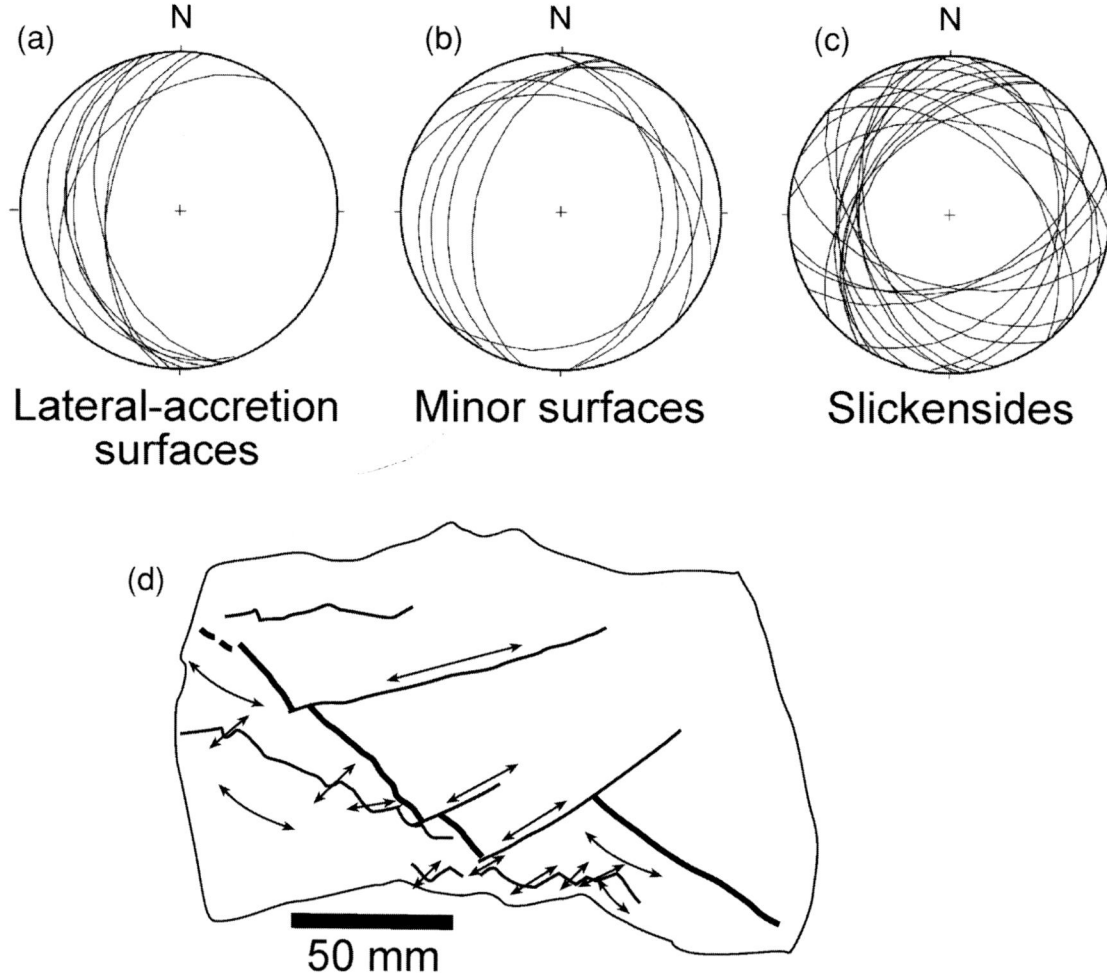

Fig. 13 Deformation features in Unit 2 at Ash Grove Quarry (see Fig. 10). Stereographic plots (a–c) are projected onto the upper hemisphere. (a) Sterogram showing orientation of selected bundle of LAS; note strong preferred orientation (depositional dip with strong eastward component). (b) Orientation of minor surfaces, probably secondary fracture planes, which are smaller in scale than LAS. (c) Orientation pattern of large slickensides, which shows some weak preference parallel to LAS. (d) Diagram (from photograph) of pervasive small-scale faulting in vertically orientated slab cut from hand specimen from Unit 2. Arrows indicate apparent planes of movement.

Palaeozoic cherts and silicified corals (pre-Pennsylvanian, and, therefore, derived from terranes eastward of the study area), as well as yellowish and pinkish quartz. This unit contains two or more distinct bands of closely packed (fitted in places), coarser, white-rinded chert pebbles and cobbles (Fig. 14b, bold arrows). Thin lenses of poorly sorted coarse to very coarse, cross-stratified sand (Fig. 14b, s) partially bounded by mud shale drapes appear within the unit. Foresets in some of these sand lenses are covered by thin (3–5 mm) mud drapes (Fig. 15a). The pebble conglomerate is overlain by moderately to poorly sorted, medium and coarse sand (Fig. 14b, 3), and the lower succession is capped by a thinly and irregularly bedded ('flaggy'), fine, silty sandstone with abundant load casts (Fig. 14b, 4). These features are lobate in vertical section and may be dinosaur tracks (S. Hasiotis, University of Kansas, personal communication, 2001).

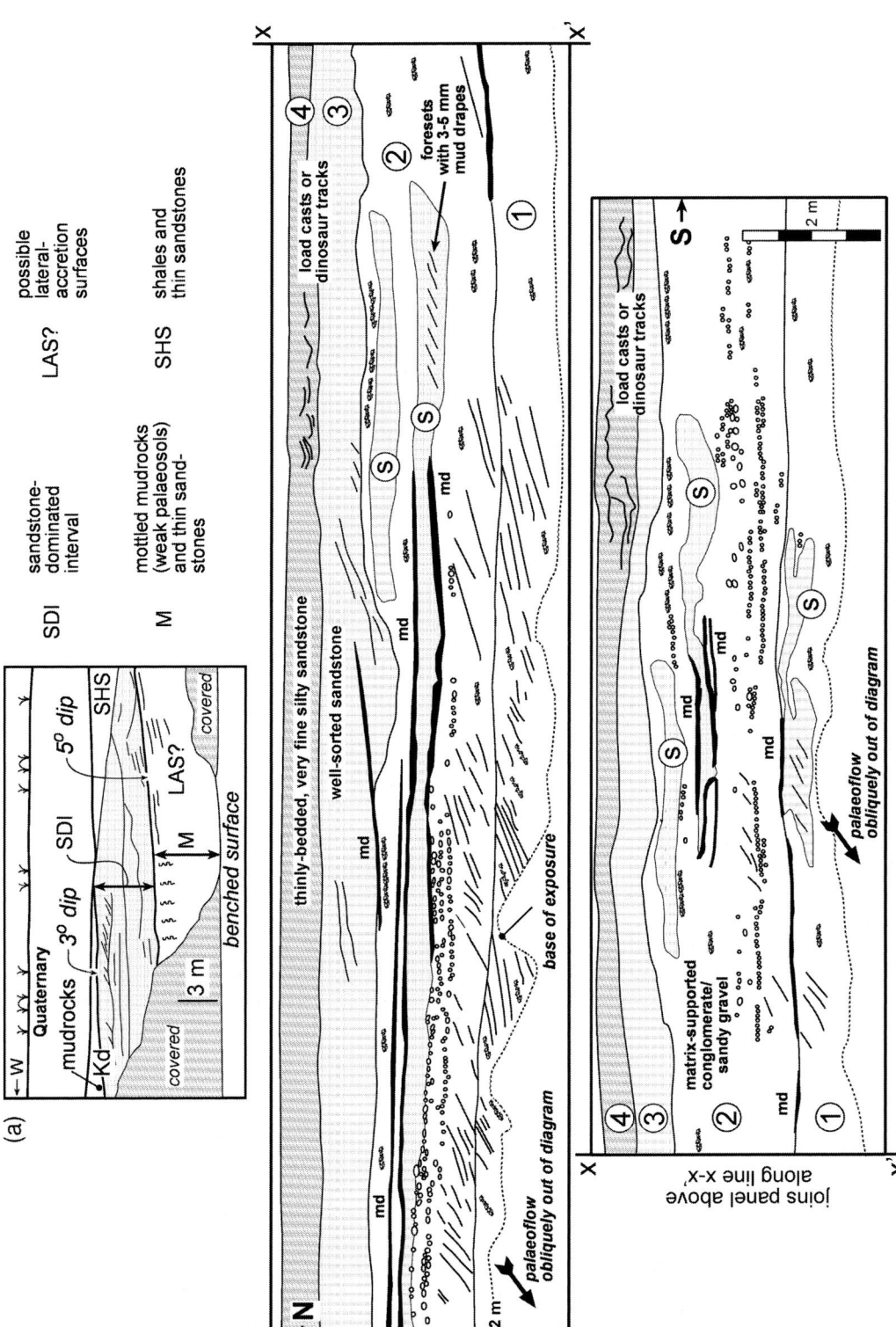

Fig. 14 (a) Upper sequence of strata from Maystrick Pit (Fig. 2, number 6), drawn from archival photograph, showing gently dipping strata, probably lateral-accretion units in large, sinuous stream. Smaller scale LAS appear in mudstone-dominated unit at base of section. (b) Lower, gravelly sequence directly below bench surface shown in (a). Complex internal stratification and abundant mud (shale) drapes (md) demonstrate a mixture of fluvial and tidal influence, although relatively unidirectional flow to west-northwest is indicated. Bold arrows mark prominent bands of coarse chert pebbles to cobbles, which are probably erosional lags on bar tops.

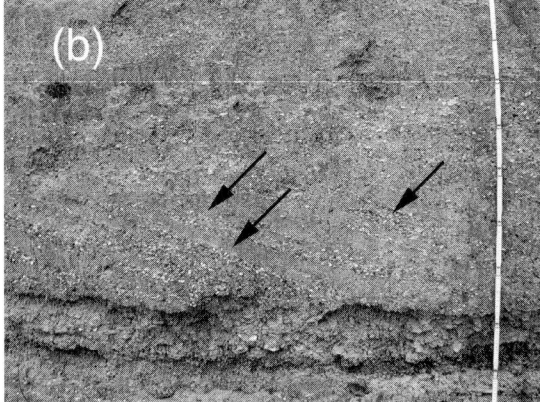

Fig. 15 Features at Maystrick Pit. (a) Small sand lens with horizontal mud (shale) drapes (md) and diagonal drapes on foresets (arrow). Scale in centimetres. First segment on folding rule at right is 150 mm long. (b) Lenticular gravels on large-scale foresets at base of section at Maystrick Pit.

Other localities

Other exposures at the middle to eastern end of the EPRV prove helpful in completing a reconstruction of regional Dakota sedimentation. At Camp Maha (CM; Fig. 16a), at the eastern edge of Dakota exposure in Nebraska, a highly oblique cross-section through a Dakota valley fill is visible along the north bank of the Platte River. The Dakota–Pennsylvanian contact rises at least 4 m over less than 50 m (actual palaeovalley relief is undoubtedly greater), and the Dakota itself exceeds 15 m in total thickness in the immediate area. The well-exposed part of the Dakota at CM is distinguished by:

1 a thick (2.9 m) laterally continuous medium to coarse sandstone consisting of a single tabular set of cross-strata, which is nearly at the base of the exposed section;
2 overlying sets of much smaller scale tabular cross-strata, including a package of 50-mm-scale tabular cross-strata (Fig. 16a).

The lowermost Dakota strata at the site are poorly exposed. They consist mostly of pebbly sands to conglomerates consisting almost entirely of quartz and chert. The CM section has an overall fining upward trend.

The Cullom Mine/Raven's Nest (C/R) exposures (Fig. 16b), where underground hydraulic gravel mining was once carried out (Condra, 1908), show a vertical transition from thick, basal, gravelly deposits into well-sorted, cross-stratified sandstones (complete with reactivation surfaces and burrowing) like those seen at the western end of the EPRV. Subrounded boulders, up to 300 mm in diameter, of locally derived Pennsylvanian chert nodules (with characteristic white rinds) have been removed from the lowermost Dakota exposures in the area. The overall trend at the site is fining upward.

LARGER SCALE ARCHITECTURE

Unfortunately, a sizeable gulf exists between detailed observations and interpretations made at the relatively small scale of outcrops in the EPRV and a result approaching a comprehensive image of Dakota sedimentation. Outcrop-scale observations, however, greatly facilitate the interpretation of less-detailed subsurface data (sparse lithological logs and electric logs). In this manner, a larger scale architecture of the Dakota can be postulated.

Cross-sections in the EPRV

Constructing meaningful cross-sections of the Dakota in the study area is difficult because of Plio-Pleistocene erosion. Nonetheless, two interpretive cross-sections were assembled from miscellaneous well logs having highly variable degrees of detail in their lithological descriptions (Fig. 17a & b). These cross-sections suggest multiple storeys of channel-form to sheet sands and associated mudrocks within deeply incised

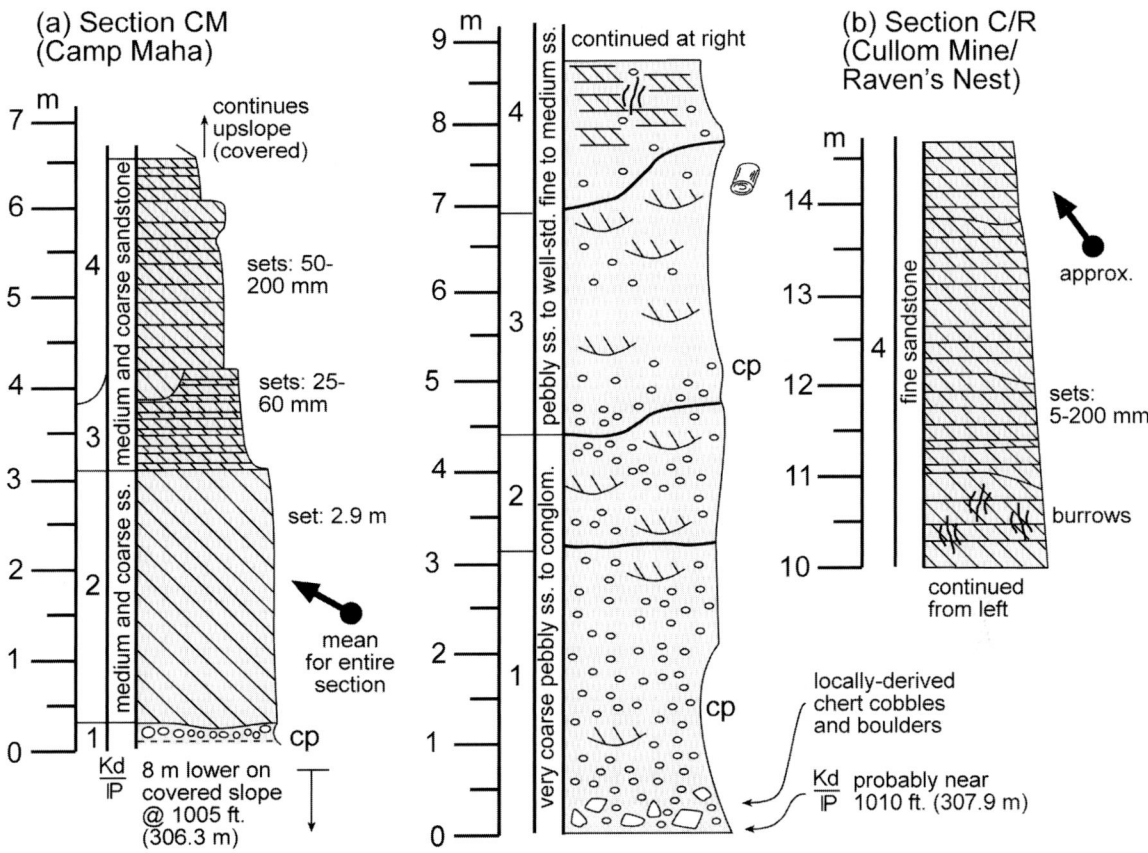

Fig. 16 Measured sections at eastern end of EPRV showing overall fining upward trends. (a) Section at Camp Maha (CM; Fig. 2, number 8); Unit 2 is a single set of tabular cross-strata 2.9 m thick. (b) Section at Cullom Mine (CM; Fig. 2, number 7), now Raven's Nest housing development. Site is an early 1900s underground hydraulic gravel mine with associated excavations. Largest clasts of locally derived, unweathered Pennsylvanian chert at the base of this section are several tens of millimetres in diameter. Representative cross-strata set thicknesses are provided. See Fig. 5 for symbol key.

bedrock palaeovalleys (Fig. 17a & b). Basal strata in these palaeovalley fills are not uniformly gravelly; mudrocks fill the lowest part of a deeply incised palaeovalley in one cross-section (Fig. 17).

Well logs in the EPRV

There are a few, uncorrected electric logs (mostly resistivity) of rotary boreholes through the Dakota in the EPRV (Figs 4 & 18), but no corresponding cores exist. Close comparison with lithological descriptions (e.g. Burchett & Smith, 1992, 1999a,b) indicates that the resistivity logs record coarsening upward (CU) and fining upward (FU) trends (Fig. 19). Exposures are limited, so these logs assume vital importance in interpreting medium-scale stratigraphical trends.

Several boreholes show thick (up to 20 m) sandstones dominating incised valley fills in the immediate vicinity of the EPRV. Within these sandstones, interpretive CU and FU packages range from 2 to 6 m thick (most commonly 2 to 4 m). These packages are similar in scale to the stratigraphical units visible at several outcrops (e.g. AGQ, IH and C/R). The log from test hole 5-A-62 (Fig. 19) seems to show a succession of CU packages alone, but the remainder show a combination of interpreted CU and FU packages. The FU trends, which might be strictly fluvial in origin, do not dominate any single log.

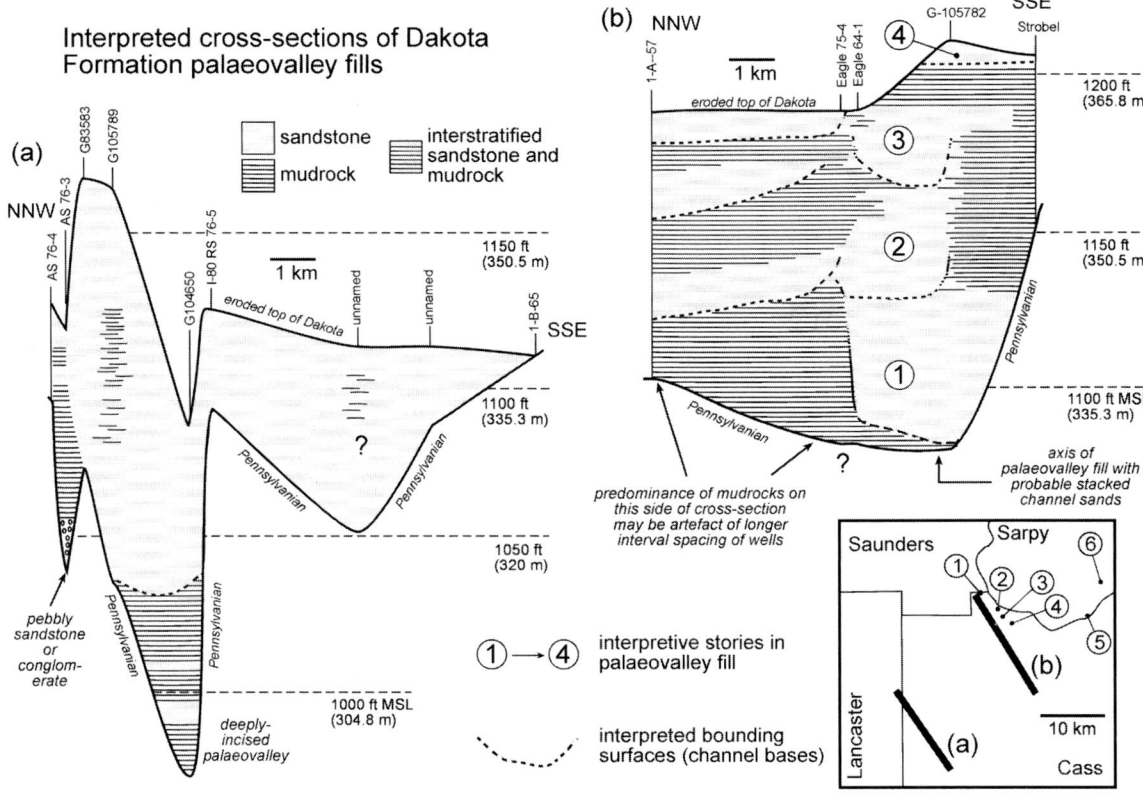

Fig. 17 Interpretive cross-sections of Dakota Formation in south-western part of palaeogeographical map shown in Plate 1; cross-sections cut obliquely across parts of major palaeovalley (Plate 1, B); map (inset) shows approximate locations of cross-sections and selected measured sections, as numbered in Fig. 2.

Well logs from adjacent parts of Nebraska

Westward in Nebraska there are untruncated Dakota sections, both exposed and in the subsurface, which are overlain by the laterally consistent and easily recognized Graneros–Greenhorn marine interval. The resistivity logs from deeper petroleum exploration wells through the Dakota east of 99°W longitude (Figs 3, 19–22), but outside of the outcrop belt, show a distinctive double excursion in the Graneros–Greenhorn interval (e.g. Fig. 20) that can be used as an effective marker. There is only one incomplete core from this area: the Ben Franklin Zahm No. 1, from Greeley County (Fig. 19). Corresponding lithological descriptions from wells are lacking in most cases, although both the Zahm No. 1 and MCP wells (Fig. 20) have reliable lithological description. Nonetheless, gross stratigraphical patterns are decipherable from the electric logs, and the untruncated sections represented by these logs afford the best opportunities in the foreseeable future to reconstruct a more complete history of regional Dakota sedimentation. Three closely spaced wells in Stanton

Fig. 18 (*opposite*) Resistivity logs of rotary test holes in the immediate vicinity of the EPRV. See Fig. 4 for locations. Fining upward and coarsening upward trends are interpreted from patterns in resistivity logs and, in most cases, accompanying lithological logs. Abbreviations in this and subsequent figures are: fining upward (FU) and coarsening upward (CU) trends, sandstone-dominated intervals (ss), mudrock-dominated intervals (mr), mixed sandstone–mudrock dominated by sandstone (ss/mr), mixed sandstone–mudrock dominated by mudrock (mr/ss), transgressive surface or up-dip equivalent (tr), subaerial unconformity (su), interbedded (itbd), bayhead (bayhd).

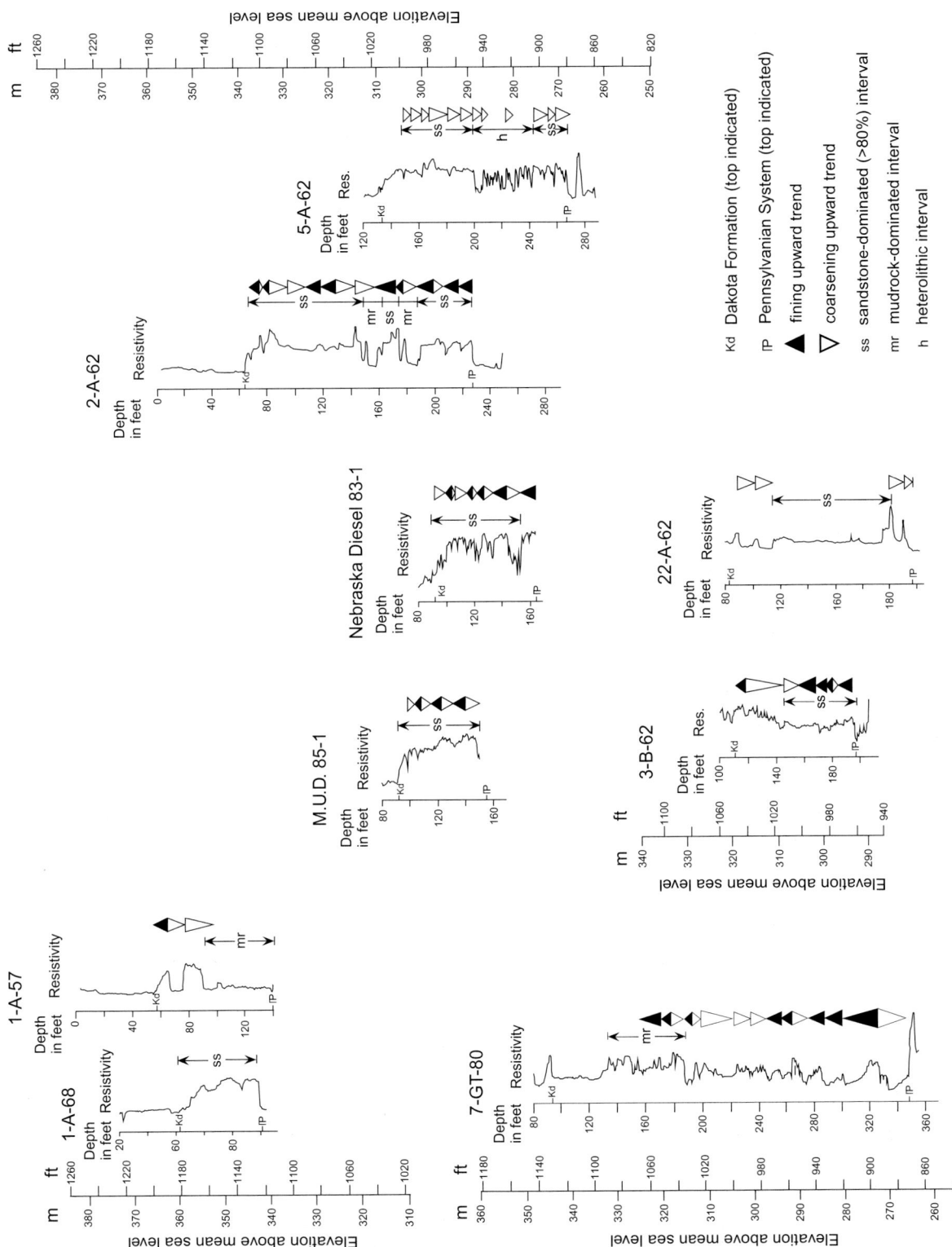

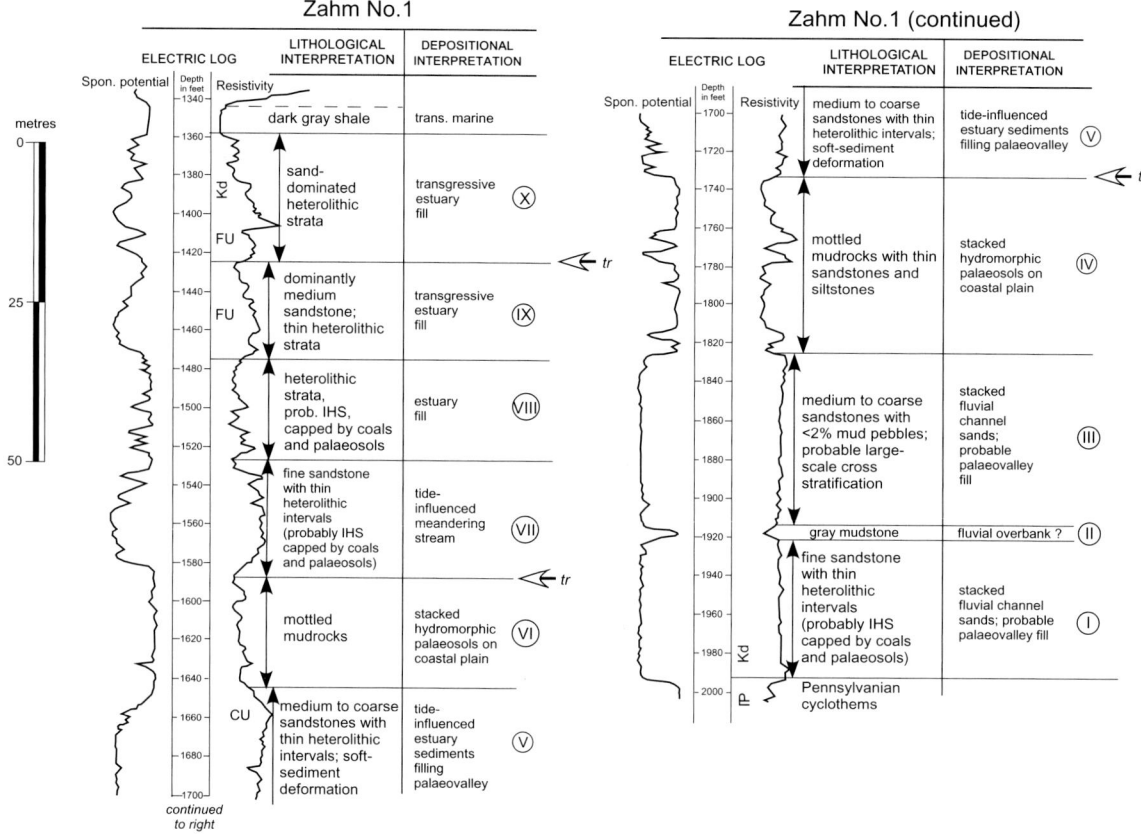

Fig. 19 Log of Dakota Formation in Ben Franklin Zahm No. 1 well, Greeley County, Nebraska (Fig. 3) and interpretation. Partial core from this well, coupled with a detailed original lithological log, permits a more accurate interpretation of depositional environments and sequence stratigraphy from electric logs than in other cases. Intervals with distinctive characteristics are indicated by Roman numerals. See Fig. 18 for abbreviations.

County (Figs 3 & 21), which form a line approximately along strike, allow a 'cross-referenced' interpretation of laterally equivalent depositional environments impossible in the analysis of single, isolated wells.

Five general types of intervals within the Dakota can be interpreted from electric logs (Figs 19–22):

1 mixed sandstone–mudrock intervals characterized by stacked, dominantly FU packages, each 1.5 to 7 m thick;

2 intervals dominated by one or two prominent, thick (10–25 m), sand-dominated CU units;

3 intervals containing one or more thinner CU sands, each 1.5 to 7 m thick;

4 thick (20–40 m, but generally > 30 m) intervals dominated by sand, which only rarely show distinct internal packaging (CU and/or FU), or which may be represented as relatively straight-sided excursions on resistivity and spontaneous potential logs;

5 mudrock-dominated intervals (generally 3–20 m) containing either greyish shales or red-mottled mudstones (hydromorphic palaeosols), probably similar to mudrock-bearing strata visible in outcrop at MPT.

The Zahm No. 1 well (Fig. 19) stands out because it contains a sixth type of log pattern: large-scale (c. 20 m thick) fining upward trends. Inspection of incomplete cores recovered from this well indicates that these intervals contain sands, heterolithic strata (some of which appear to be rhythmic), thin low-chroma palaeosols, and thin lignites. Sandstone-dominated facies are

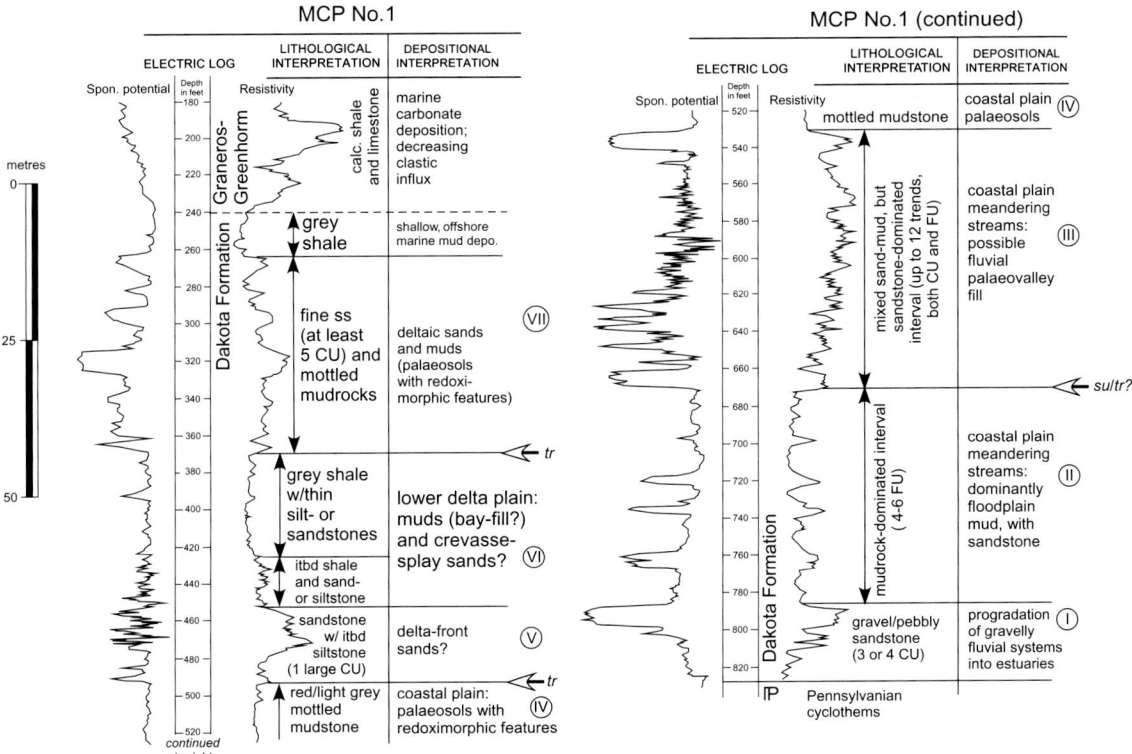

Fig. 20 Interpretation of Dakota Formation in Maystrick Pit No. 1 well log, Platte County, Nebraska, which is supported by original, but cursory, lithological descriptions. Intervals with distinctive characteristics are indicated by Roman numerals. Contact between the Dakota Formation and Graneros–Greenhorn formations is uncertain, and may be at the base of the grey shale shown at the top of the Dakota Formation in this diagram. See Fig. 3 for location and Fig. 18 for abbreviations.

at the base of these intervals, whereas mudrocks with palaeosols and lignites cap them. Electric logs from wells penetrating the Dakota east of 99° W in Nebraska show a relatively consistent pattern of 7–10 distinct packages of sediment (Figs 19–21: Roman numerals I–X).

DISCUSSION

Palaeogeography

The palaeodrainage map represents the surface upon which the first Dakota sediments were deposited. Broad, incised valleys of at least two general size ranges are discernable in what can be interpreted as a palaeodrainage network. The temporal relationship between the three main palaeodrainages (A, B & C in Plate 1) cannot be determined, although a first-order assumption, for which there is no evidence to the contrary, is that they were all present at the onset of Dakota deposition, rather than being incised at different stages. It is noteworthy that the modern Platte River (the Sarpy–Cass county line in Plate 1) skirts the northern edge of the putative Early Cretaceous Pennsylvanian bedrock uplands (perhaps even being deflected by them at two points (3 & 5 in Plate 1).

Depositional history

The incised Dakota palaeovalleys reconstructed by GIS analysis were filled with clastic sediments, initially in response to Albian sea-level rise (Brenner *et al.*, 2000). A predominantly fluvial

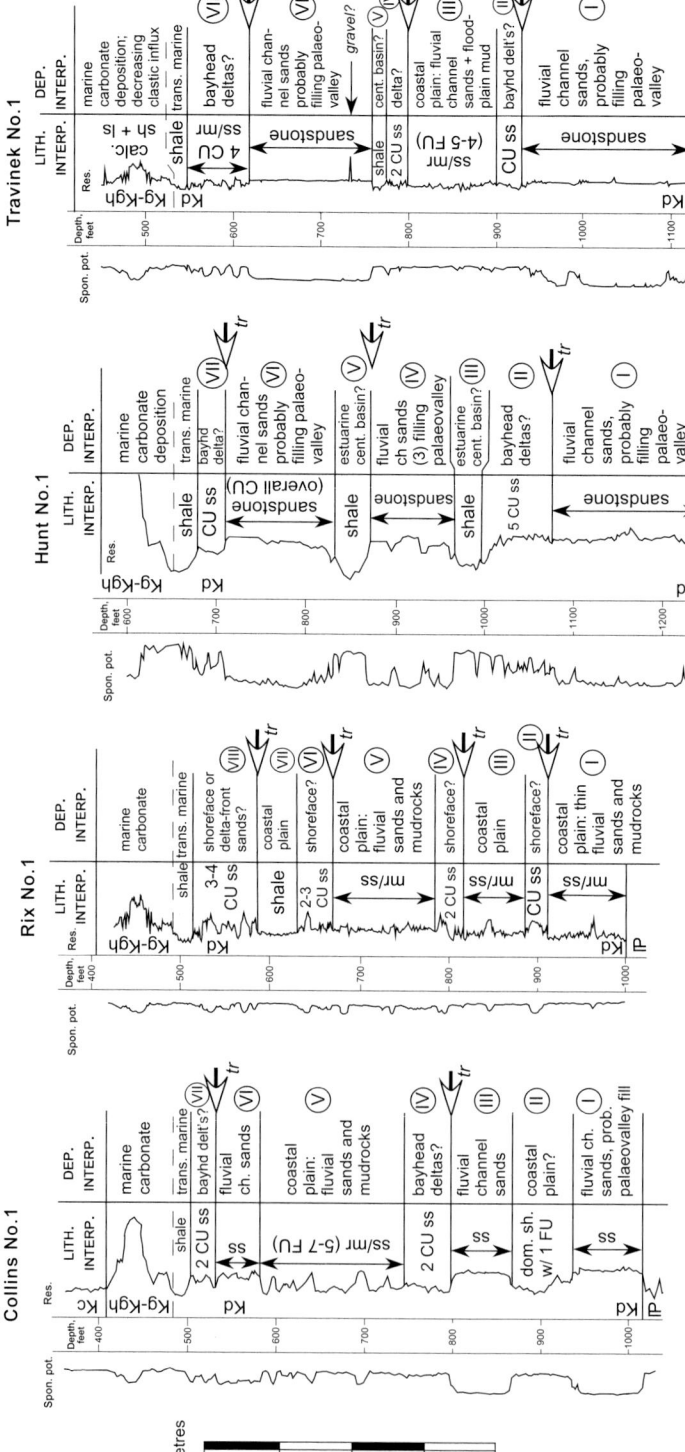

Fig. 21 Stanton County, Nebraska petroleum exploration wells arranged NNW–SSE and roughly along strike. Intervals with distinctive characteristics are indicated by Roman numerals, but numbering does not imply correlation between wells. Graneros Shale–Greenhorn Limestone interval (Kg–Kgh) includes part or all of shale underneath prominent double excursion on resistivity log, which is correlated in all wells by dashed line. Outcrops of uppermost Dakota Formation in Jefferson County, Nebraska show contact between characteristic Dakota shales and characteristic Graneros shales, and therefore some of the shale interval below the double excursion may be part of the Dakota. See Fig. 3 for locations and Fig. 18 for abbreviations.

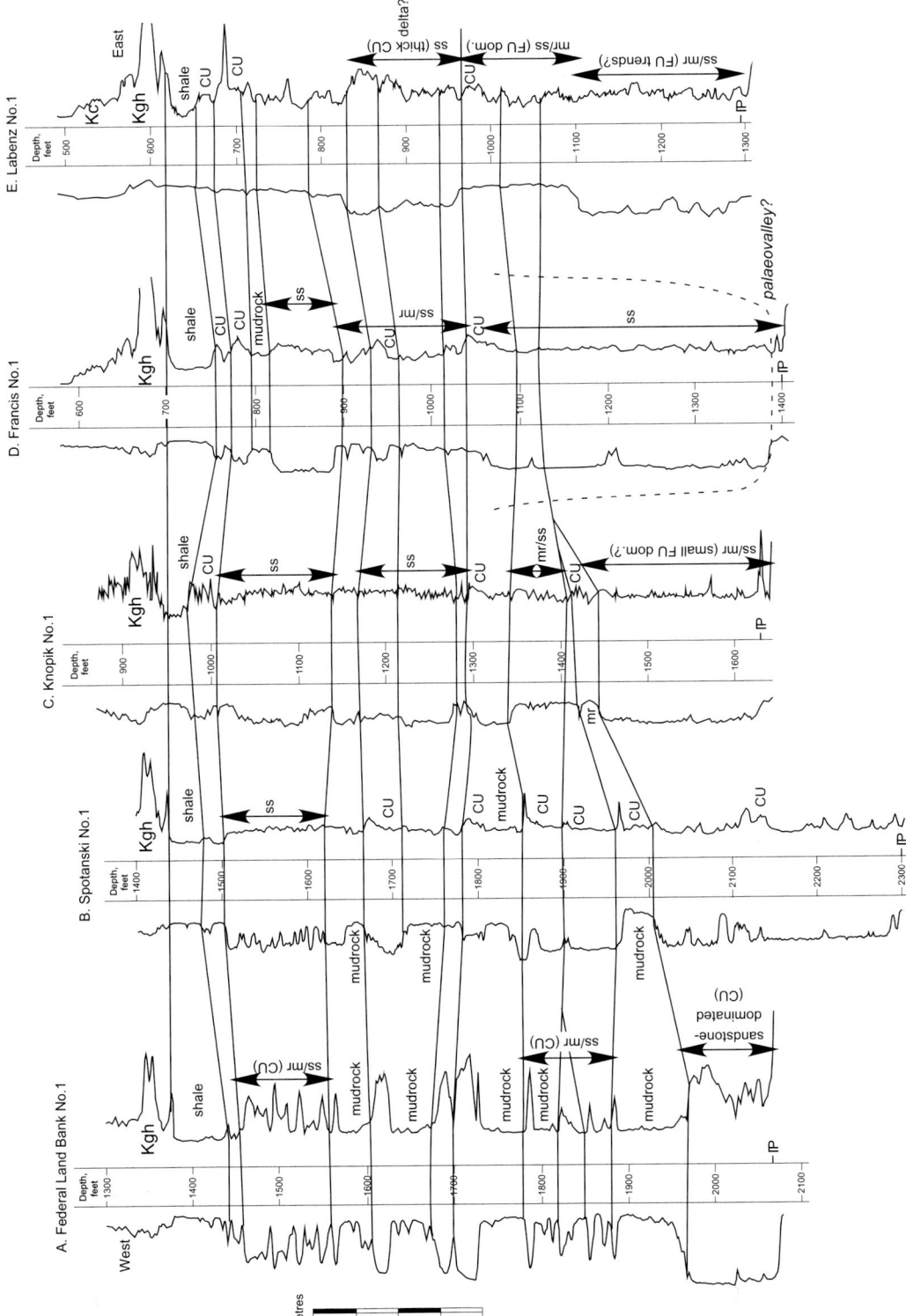

Fig. 22 Tentative west–east correlation and interpretation of possible genetic surfaces (sequences and/or parasequence boundaries) in the Dakota Formation west of the eastern Nebraska outcrop belt. See Fig. 3 for locations and Fig. 18 for abbreviations.

origin for the palaeovalley fills and superjacent Dakota strata is strongly supported by:

1 the essentially consistent and relatively narrow range of palaeocurrents determined from sandstones in the EPR;

2 the presence of coarse clasts that originated from the north-east and up-dip from the shoreline of the WIS, rather than seaward (cf. Siemers, 1976; Hattin et al., 1978, fig. 3; Brenner et al., 2000).

Outcrop-scale details also lend support to this interpretation. Low-angle surfaces in the upper part of the section at MPT are depositional in origin, and therefore are interpreted as lateral accretion surfaces in a large (> 50 m wide) sinuous stream. Lateral accretion surfaces at AGQ, although occurring within sediments that have flaser and linsen bedding, as well as typically non-freshwater traces, also imply an underlying fluvial mechanism of deposition. Thick, cross-stratified gravels at MP and other undescribed localities in the EPRV have narrow apparent palaeocurrents, and appear to record the migration of metre-scale, coarse-grained, straight-crested bars in a large fluvial channel. Mud drapes within these gravels, and concentrations of thin, horizontal lags of coarse gravel and cobbles, indicate major fluctuations in flow regime: both slackwater periods and high-velocity flow. The large scale of cross-strata at CM is unmatched in any known Dakota exposure in the EPRV, and probably represents downstream migration of a large fluvial unit bar, over which smaller bedforms simultaneously migrated.

Despite the evidence for a predominantly fluvial origin of Dakota deposits in the study area, multiple observations indicate that marine and/or tidal influence was widespread. Evidence for tidal influence on deposition in these incised fluvial systems is both direct and indirect. Mathematical analyses have determined that the Unit 3 laminates from the Dakota valley fill at AGQ are tidal rhythmites (Brenner et al., 2000). Both the number of laminae per cycle and the lack of alternating thick and thin laminae suggest a diurnal preservation of tidal deposits. Alternatively, this pattern could represent deposits from a high intertidal position in a semidiurnal tidal system, with minimal semidiurnal inequality (e.g. Tessier et al., 1995), where only the highest semimonthly tides would have deposited sediment (Archer, 1996). The occurrence of an *in situ* tree stump (Fig. 12b) within the deposits may support this hypothesis.

Compared with the AGQ laminites, laminated sediments in the Zahm No. 1 well have similarly regular stratification, although there is insufficient core material for a mathematical analysis. Rip-up pebbles of finely interstratified mud and silt to very fine sand laminae, such as those at IH, are also similar in appearance to the tidal rhythmites at AGQ, although at IH there are no corresponding *in situ* strata.

The IHS packages at AGQ contain abundant features compatible with tidal stratification (wavy, flaser and linsen bedding; brackish to marine trace fossils). Gastaldo et al. (1995) and others have described similar heterolithic sediments in point bars on modern tidally influenced meandering streams; such settings are also frequently characterized by slumping. Much of the deformation in the IHS appears to have occurred simultaneously with deposition, as saturated, mud- and organic-matter-rich sediments began to dewater and slump down steep point-bar surfaces. Flow-stage fluctuations may have triggered slumping. Nonetheless, occurrences of cobbles and small boulders (including far-travelled exotics) on LAS and in potholes in the same strata are interpreted to represent major increases in fluvial flow: epsiodes of fluvial dominance in a river system typically influenced by tides.

Isotopic data from the siderite-cemented bed at the Unit 1–2 contact suggest a purely meteoric origin for the siderite cements. This bed, however, is overlain by marine tidally influenced deposits, and there are *Skolithos* burrows below it. Thus, there is evidence for fluctuations in marine-influenced and non-marine conditions (cf. Taylor et al., 2000).

Reactivation surfaces, such as those that are common in the tabular cross-stratified sandstones of the Dakota in the western EPRV, result from changes in flow conditions entirely within a fluvial regime, or from tidal influence in the lower stretches of large rivers (cf. Bristow, 1995; Johnson & Levell, 1995; Willis et al., 1999). Marine palynomorphs and trace fossils typical of brackish to marine settings further indicate some form of marine influence on Dakota sedimentation. Small filled channels at HL, which have palaeocurrents that differ markedly from those in thick sands below (Figs 5b–e & 6c), could represent fluvial chutes or small tidal channels, perhaps infilled by flood deltas.

Interpretation of well logs

Numerous authors (e.g. Bhattacharya, 1993; Plint et al., 2001) have used, or proposed using (Bridge & Tye, 2000), electric logs either as a primary or secondary data set in the interpretation of similar stratigraphical successions. Following such practice, a generalized interpretation of characteristic interval types can be made in the context of existing knowledge about Dakota depositional environments. Interval type 1 is interpreted as point bar and overbank deposits of mixed-load, meandering fluvial systems on low coastal plains. Type 4 intervals probably represent stacked fluvial channel sands, almost certainly filling palaeovalleys, considering their tens of metres of thickness. These deposits would be similar to the thick sandstones visible in outcrop at MP, MPE, HL and AR. Incised valley fills have been reported from well-exposed Dakota strata several hundred kilometres to the south-west in central Kansas (Siemers, 1976).

Interval types 2 and 3 may represent a variety of environments: large constructional deltas, bayhead deltas and even sandy shorefaces. Larger CU intervals represent larger deltas at the mouths of river systems, whereas the smaller CU intervals, particularly when lying on top of thick sands (interval type 4), are bayhead deltas deposited during marine transgressions. Evidence for these types of features to the east in the outcrop belt is sparse or lacking, so these interpretations of successions further basinward are speculative. Mudrock of interval type 5, which are either true fissile shales or mottled mudrocks, as indicated in some lithological logs, also have a wide range of potential interpretation. The characteristic outcrop appearance of mottled massive Dakota mudstones with palaeosols instantly defines the 'red-white' or 'red-grey' mottled references in lithological descriptions as hydromorphic palaeosols, whether in a low coastal-plain setting or on a subaerially exposed, prograded delta. Other intervals described simply as 'shale' may indeed be just that: stratified muds deposited in bay fills, estuarine basins, or even in very shallow, shoreline-proximal marine prodeltaic settings.

The lack of domination of Dakota successions by FU packages probably indicates that fluvial influence, although certainly important, did not predominate in the resultant sedimentary succession anywhere in the study area. Similarly, Scott et al. (1998) noted a dominance of CU cycles, with subordinate FU cycles, in the Dakota in the Bounds No. 1 core from westernmost Kansas. The comparatively unique log pattern of interval type 6, which shows large-scale FU in the Zahm No. 1 well (Fig. 19), could be a fluvial fining upward package, but in view of the presence of possible tidal rhythmites and the superposition of the marine Graneros Shale it appears more likely to represent the transgressive filling of tidal estuarine valleys (cf. Bhattacharya, 1993; Ainsworth & Walker, 1994).

In the absence of cores (and, sometimes, lithological logs), however, making more specific interpretations of these intervals is impossible. It is also almost certain that no additional subsurface data will be forthcoming, because all of the wells were unsuccessful 'wildcat' ventures, generally completed at least 30 yr ago. A large body of literature (e.g. Ainsworth & Walker, 1994; Allen & Posamentier, 1994; MacEachern & Pemberton, 1994; Zaitlin et al., 1994; Borrego et al., 1995; Walker, 1995; Nichol et al., 1997; Uličný, 1999), however, interprets similar but better-documented successions and sets compatible interpretative precedents.

At a larger scale and using a combination of well and outcrop data, Brenner et al. (2000) defined three depositional sequences within the Dakota. The more numerous, relatively discrete packages of sediment interpreted from Dakota well logs are thinner than these sequences but thicker than the CU and FU packages previously described. Thus, particularly if genetically significant surfaces are interpreted, these intermediate-scale packages can be identified tentatively as parasequences (Fig. 22).

SUMMARY AND CONCLUSIONS

The Dakota Formation in the easternmost Platte River Valley (EPRV) exhibits characteristics of modern and ancient incised fluvial valleys. Direct, large-scale evidence for incised palaeovalleys is provided by the reconstruction of the sub-Dakota palaeosurface: in large measure, palaeocurrents seem to correspond with the sub-Dakota palaeodrainage pattern. There is also abundant evidence for complex cut-and-fill within the Dakota itself. The overwhelming dominance of westward to northward palaeocurrent indicators and the occurrence of rare, far-travelled clasts suggest fluvial sediment transport down-dip from the east to north-east. Very coarse clasts (up to boulder size) at the base of the Dakota, although locally derived, are compatible with episodic fluvial flows of notable competence, as are the large clasts covering lateral accretion surfaces at Ash Grove Quarry.

Small- to mid-scale evidence for tidal influence consists of inclined heterolithic strata (IHS) with tidal stratification, tidal rhythmites, rip-up clasts of rhythmite-like shales, common reactivation surfaces and brackish to marine burrows. Thus, brackish to marine influence on Dakota fluvial systems was common, and the history of Dakota deposition was relatively complicated. Small-scale coarsening upward trends in resistivity logs from the EPRV can be interpreted as coarse (possibly braided) fluvial facies prograding into estuarine environments. This interpretation is compatible in scale and stratigraphical succession with the pattern of coarse-sands–IHS–sands–laminites–sands (a basal coarse interval with two coarsening upward packages) visible in outcrop at Ash Grove Quarry. In light of the demonstration of considerable sub-Dakota relief, and the direct observation of a bedrock palaeovalley at Ash Grove Quarry, it is all but certain that coarsening upward trends elsewhere in the EPRV area also represent the progradation of bedload-rich fluvial systems into drowned river valleys (estuaries) during small-scale sea-level falls. Fining upward trends, although they do not dominate logs, are interpreted as representations of near-complete fluvial domination of estuarine systems.

These outcrop observations can be extrapolated to some degree westward into the subsurface. Outside of the EPRV, well logs suggest a variety of depositional environments, although no outcrop or core data are available to support these interpretations. Logs from deep wells fully penetrating the Dakota also suggest that there is a relatively consistent pattern of large-scale sediment packaging, and that there are genetically significant surfaces (sequence or parasequence boundaries). Dakota sequence stratigraphy is therefore very likely to be more complicated than the three sequences proposed by Brenner *et al.* (2000).

The standing Nebraska Geological Survey concept of a Dakota Group consisting of three distinct and laterally continuous formations (Omadi Sandstone, Fuson Shale and Lakota Sandstone), as defined by Condra & Reed (1959) is not tenable, and neither is the twofold division of the Dakota more recently favoured (e.g. Brenner *et al.*, 2000). Only additional, detailed subsurface data (unlikely to be generated anytime soon) would allow a reliable, comprehensive sequence-stratigraphy interpretation of the Dakota in eastern Nebraska. The Kiowa–Skull Creek transgressive deposits identified by Brenner *et al.* (2000) are several tens of metres above the topographically lowest Dakota in the EPRV and environs. The most parsimonious interpretation of reconstructed sub-Dakota palaeogeography, then, is that an even earlier record of Cretaceous sedimentation (equivalent to the Albian Cheyenne Sandstone of Kansas) exists within the Dakota of eastern Nebraska. Alternatively, but probably less likely, there may have been major base-level changes (several tens of metres) and episodes of aggradation and degradation during Dakota times, producing a complex of valley fills of different ages and in different topographic positions, which may not be laterally correlative.

Given the local to regional scale of relief on the Dakota–Pennsylvanian contact, it is likely that upland areas continued to be sites of erosion during the earliest phase of Dakota sedimentation, while valleys were filling with fluvial and estuarine sediments. Eventually, these uplands were stripped of any pre-Dakota regolith and buried by fluvial facies as additional accommodation space was created. Although weathered chert clasts are common in basal Dakota sediments in the EPRV, *in situ* upland weathering profiles on the sub-Dakota unconformity, if present at all, are likely to be preserved only up-dip into Iowa.

ACKNOWLEDGEMENTS

Messrs A. Fedde, K. Maystrick and D. Wolkins, and the Girl Scouts Great Plains Council have been eminently cooperative and gracious in allowing access to their properties. Mr Mark Dietz of the Great Plains Council coordinated our visits to Camp Maha and was instrumental in assuring our continued access to the area. We are unreservedly thankful to the Ash Grove Cement Company for their long-standing support of our research on the Dakota; in particular, we wish to commend Mr Howard Stubbendieck for his cordiality, his many kindnesses and his earnest assistance. Mike Ramsey, also of Ash Grove, has been extremely helpful.

RMJ wishes to express his deepest appreciation to Ann Mack (CSD) for incredibly patient and protracted efforts in drafting. RMJ's efforts have benefited greatly from conversations with Marvin Carlson, Vince Dreeszen and Dave Gosselin of the CSD. Moreover, ongoing research on the Dakota would have been considerably more difficult without the benefit of the meticulous stratigraphical data storage and compilation of RMJ's predecessor at CSD, Ray Burchett. Rod Root and Erik Waiss contributed much-appreciated assistance in fieldwork. Two anonymous reviewers provided suggestions that greatly improved the original manuscript.

REFERENCES

Ainsworth, R.B. and Walker, R.G. (1994) Control of estuarine valley-fill deposition by fluctuations of relative sea-level, Cretaceous Bearpaw-Horseshoe Canyon transition, Drumheller, Alberta, Canada. In: *Incised-valley Systems: Origin and Sedimentary Sequences* (Eds R.W. Dalrymple, R. Boyd and B.A. Zaitlin). *Soc. Econ. Paleontol. Mineral. Spec. Publ.*, **51**, 159–173.

Allen, G.P. and Posamentier, H.W. (1994) Transgressive facies and sequence architecture in mixed tide- and wave-dominated incised valleys: example from the Gironde Estuary. In: *Incised-valley Systems: Origin and Sedimentary Sequences* (Eds R.W. Dalrymple, R. Boyd and B.A. Zaitlin). *Soc. Econ. Paleontol. Mineral. Spec. Publ.*, **51**, 225–240.

Archer, A.W. (1996) Reliability of lunar orbital periods extracted from ancient cyclic tidal rhythmites. *Earth Planet. Sci. Lett.*, **141**, 1–10.

Bhattacharya, J.P. (1993) The expression and interpretation of marine flooding surfaces and erosional surfaces in core: examples from the Upper Cretaceous Dunvegan Formation, Alberta foreland basin, Canada. In: *Sequence Stratigraphy and Facies Associations* (Eds B.U. Haq and G.P. Allen). *Spec. Publ. Int. Assoc. Sedimentol.*, **18**, 125–160.

Borrego, J., Morales, J.A. and Pendon, J.G. (1995) Holocene estuarine facies along the mesotidal coast of Huelva, south-western Spain. In: *Tidal Signatures in Modern and Ancient Sediments* (Eds B.W. Flemming and A. Bartholomä). *Spec. Publ. Int. Assoc. Sedimentol.*, **24**, 151–170.

Brenner, R.L., Ludvigson, G.A., Witzke, B.J., Zawistoski, A.N., Kvale, E.P., Ravn, R.L. and Joeckel, R.M. (2000) Late Albian Kiowa–Skull Creek marine transgression, lower Dakota Formation, eastern margin of Western Interior Seaway, U.S.A. *J. Sediment. Res.*, **70**, 868–878.

Bridge, J.S. and Tye, R.S. (2000) Interpreting the dimensions of ancient fluvial channel bars, channels, and channel belts from wireline-logs and cores. *Am. Assoc. Petrol. Geol. Bull.*, **84**, 1205–1228.

Bristow, C.S. (1995) Internal geometry of ancient tidal bedforms revealed using ground penetrating radar. In: *Tidal Signatures in Modern and Ancient Sediments* (Eds B.W. Flemming and A. Bartholomä). *Spec. Publ. Int. Assoc. Sedimentol.*, **24**, 313–328.

Bromley, R.G. (1990) *Trace Fossils: Biology and Taphonomy.* Unwin Hyman, London, 280 pp.

Burchett, R.R. (1970) *Guidebook to the Geology along the Missouri River Bluffs of Southeastern Nebraska and Adjacent Areas.* University of Nebraska–Lincoln Conservation and Survey Division, Nebraska Geological Survey, 23 pp.

Burchett, R.R. (1971) *Guidebook to the Geology along Portions of the Lower Platte River Valley and Weeping Water Valley of Eastern Nebraska.* University of Nebraska–Lincoln Conservation and Survey Division, Nebraska Geological Survey, 38 pp.

Burchett, R.R. and Smith, F.A. (1992) *Lancaster County Test-hole Logs.* Nebraska Water Survey Test-Hole Report No. 55, Conservation and Survey Division, Institute of Agriculture and Natural Resources, University of Nebraska-Lincoln, 138 pp.

Burchett, R.R. and Smith, F.A. (1999a) *Cass County Test-hole Logs.* Nebraska Water Survey Test-Hole Report No. 13, Conservation and Survey Division, Institute of Agriculture and Natural Resources, University of Nebraska-Lincoln, 180 pp.

Burchett, R.R. and Smith, F.A. (1999b) *Sarpy County Test-hole Logs.* Nebraska Water Survey Test-Hole Report No. 77, Conservation and Survey Division, Institute of Agriculture and Natural Resources, University of Nebraska-Lincoln, 75 pp.

Condra, G.E. (1908) *The Sand and Gravel Resources of Nebraska*, Vol. 3, Part 1. Nebraska Geological Survey, Woodruff-Collins, Lincoln, Nebraska, 206 pp.

Condra, G.E. and Reed, E.C. (1959) *The Geological Section of Nebraska.* Nebraska Geological Survey Bulletin 14A, Conservation and Survey Division, University of Nebraska, Lincoln, NB, 82 pp.

Condra, G.E. and Scherer, O.J. (1939) *Upper Carboniferous Formations in the Lower Platte Valley.* Nebraska Geological Survey Paper 16, Conservation and Survey Division, University of Nebraska, Lincoln, NB, 18 pp.

Gastaldo, R.A., Allen, G.P. and Huc, A.Y. (1995) The tidal character of fluvial sediments of the modern Mahakam River delta, Kalimantan, Indonesia. In: *Tidal Signatures in Modern and Ancient Sediments*

(Eds B.W. Flemming and A. Bartholomä). *Spec. Publ. Int. Assoc. Sedimentol.*, **24**, 172–181.

Harden, R.W. (1959) A structural investigation in Sarpy County, Nebraska, and adjacent areas. Unpublished MSc thesis, University of Nebraska, Lincoln, 37 pp.

Hattin, D.E., Siemers, C.T. and Stewart, G.F. (1978) *Guidebook: Upper Cretaceous Stratigraphy and Depositional Environments of Western Kansas.* Guidebook Series 3, Kansas Geological Survey, Lawrence, KS, 55 pp.

Johnson, H.D. and Levell, B.K. (1995) Sedimentology of a transgressive estuarine sand complex: the Lower Cretaceous Woburn Sands (Lower Greensand), southern England. In: *Sedimentary Facies Analysis: a Tribute to the Research and Teaching of Harold G. Reading* (Ed. A.G. Plint). *Spec. Publ. Int. Assoc. Sedimentol.*, **22**, 17–46.

Karl, H.A. (1976) Depositional history of the Dakota Formation (Cretaceous) sandstones, southeastern Nebraska. *J. Sediment. Petrol.*, **46**, 124–131.

MacEachern, J.A. and Pemberton, G.S. (1994) Ichnological aspects of incised-valley fill systems from the Viking Formation of the western Canada sedimentary basin, Alberta, Canada. In: *Incised-valley Systems: Origin and Sedimentary Sequences* (Eds R.W. Dalrymple, R. Boyd and B.A. Zaitlin). *Soc. Econ. Paleontol. Mineral. Spec. Publ.*, **51**, 129–157.

Nelson, H.R. (1958) A structural investigation in eastern Cass County and northeastern Otoe County, Nebraska. Unpublished MSc thesis, University of Nebraska, Lincoln, 55 pp.

Nichol, S.L., Zaitlin, B.A. and Thom, B.G. (1997) The upper Hawkesbury River, New South Wales, Australia: a Holocene example of an estuarine bayhead delta. *Sedimentology*, **44**, 263–286.

Plint, A.G., McCarthy, P.J. and Faccini, U.F. (2001) Nonmarine sequence stratigraphy: up-dip expression of sequence boundaries and systems tracts in a high-resolution framework, Cenomanian Dunvegan Formation, Alberta foreland basin, Canada. *Am. Assoc. Petrol. Geol. Bull.*, **85**, 1967–2001.

Scott, R.W., Franks, P.C., Evetts, M.J., Bergen, J.A. and Stein, J.A. (1998) Timing of mid-Cretaceous relative sea-level changes in the Western Interior: Amoco No. 1 Bounds core. In: *Stratigraphy and Palaeoenvironments of the Cretaceous Western Interior Seaway, USA* (Eds W.E. Dean and M.A. Arthur). *Soc. Econ. Paleontol. Mineral. Concepts Sedimentol. Palaeontol.*, **6**, 11–34.

Siemers, C.T. (1976) Sedimentology of the Rocktown channel sandstone, upper part of the Dakota Formation (Cretaceous), central Kansas. *J. Sediment. Petrol.*, **46**, 97–123.

Taylor, K.G., Gawthorpe, R.L., Curtis, C.D., Marshall, J.D. and Awwiller, D.N. (2000) Carbonate cementation in a sequence-stratigraphic framework: Upper Cretaceous sandstones, Book Cliffs, Utah-Colorado. *J. Sediment. Res.*, **70**, 360–372.

Tessier, B., Archer, A.W., Lanier, W.P. and Feldman, H.R. (1995) Comparison of ancient tidal rhythmites (Carboniferous of Kansas and Indiana, U.S.A.) with modern analogues (Mont-Saint-Michel Bay, France). In: *Tidal Signatures in Modern and Ancient Sediments* (Eds B.W. Flemming and A. Barholomä). *Spec. Publ. Int. Assoc. Sedimentol.*, **24**, 259–271.

Uličný, D. (1999) Sequence stratigraphy of the Dakota Formation (Cenomanian), southern Utah: interplay of eustasy and tectonics in a foreland basin. *Sedimentology*, **46**, 807–836.

Walker, R.G. (1995) An incised valley in the Cardium Formation at Ricinus, Alberta: reinterpretation as a valley fill. In: *Sedimentary Facies Analysis: a Tribute to the Research and Teaching of Harold G. Reading* (Ed. A.G. Plint). *Spec. Publ. Int. Assoc. Sedimentol.*, **22**, 47–74.

Willis, B.J., Bhattacharya, J.P., Gabel, S.L. and White, C.D. (1999) Architecture of a tide-influenced river delta in the Frontier Formation of central Wyoming, USA. *Sedimentology*, **46**, 667–688.

Witzke, B.J. and Ludvigson, G.A. (1994) The Dakota Formation in Iowa and the type area. In: *Perspectives on the Eastern Margin of the Cretaceous Western Interior Basin* (Eds G.W. Shurr, G.A. Ludvigson and R.H. Hammond). *Geol. Soc. Am. Spec. Pap.*, **287**, 43–78.

Witzke, B.J., Ludvigson, G.A, Brenner, R.L. and Joeckel, R.M. (1996) Regional Dakota sedimentation. In: *Mid-Cretaceous Fluvial Deposits of the Eastern Margin, Western Interior Basin: a Field Guide to the Cretaceous of Guthrie County* (Eds B.J. Witzke and G.A. Ludvigson), pp. 15–19. Iowa Geological Survey Bureau Guidebook Series No. 17, Iowa Department of Natural Resources, Geological Survey Bureau, Iowa City, IA.

Witzke, B.J., Ravn, R.L., Ludvigson, G.A., Joeckel, R.M. and Brenner, R.L. (1996) Age and correlation of the Nishnabotna Member. In: *Mid-Cretaceous Fluvial Deposits of the Eastern Margin, Western Interior Basin: A Field Guide to the Cretaceous of Guthrie County* (Eds B.J. Witzke and G.A. Ludvigson), pp. 13–18. Iowa Geological Survey Bureau Guidebook Series No. 17, Iowa Department of Natural Resources, Geological Survey Bureau, Iowa City, IA.

Zaitlin, B.A., Dalrymple, R.W. and Boyd, R. (1994) The stratigraphic organization of incised-valley systems associated with relative sea-level change. In: *Incised-valley Systems: Origin and Sedimentary Sequences* (Eds R.W. Dalrymple, R. Boyd and B.A. Zaitlin). *Soc. Econ. Paleontol. Mineral. Spec. Publ.*, **51**, 45–60

Improved understanding of fluvial architecture using three-dimensional geological models: a case study of the Westphalian A Silkstone Rock, Pennine Basin, UK

KEVIN J. KEOGH*, JOHN H. RIPPON†, DAVID HODGETTS, JOHN A. HOWELL‡ and STEPHEN S. FLINT

STRAT Group, Department of Earth Sciences, University of Liverpool, Liverpool L69 3GP, UK

ABSTRACT

The Westphalian A and B of the Pennine Basin is one of the most intensely studied sedimentary successions in the UK owing to its economic importance in the coal and hydrocarbon industries. Previous studies of this interval have utilized the wealth of mining data available to accurately correlate sand bodies and deterministically map their regional extents in a two-dimensional framework. This paper expands on one particular study by taking the same data used by previous workers and digitally processing them into a format that is readable by an oil industry reservoir modelling programme (IRAP-RMS). A three-dimensional deterministic model has then been constructed to allow visualization and analysis of the resulting facies geometries in three dimensions, with the aim of gaining an improved understanding of the geometry and architecture of a major multistorey fluvial sandbody. Application of these techniques has allowed for enhanced visualization and interrogation of the Pennine Basin interval, previously not possible using traditional two-dimensional techniques. Understanding of the Silkstone Rock and its relationship to other facies has improved, because erosion at the base of the unit can be demonstrated to truncate coals and single-storey channel-fills that represent an earlier depositional phase. The visualization of this relationship is not possible from a single two-dimensional section, and the three-dimensional model therefore clarifies stratal relationships.

INTRODUCTION

In fluvial successions accurate correlation and mapping of major multistorey sandbodies is critical to the correct interpretation of the geometry, aerial extent and stratigraphical context of these bodies. These interpretations are necessary for understanding the evolution of the depositional systems and their external controls. The correlation of one-dimensional borehole logs and two-dimensional outcrops into a three-dimensional visualization is commonly difficult on paper. Current computing power and software tools allow relatively complex geological relationships to be represented realistically in a three-dimensional

*Present address: Statoil ASA, TEK F&T UTV TOS, Forushagen, 4035 Stavanger, Norway (Email: keke@statoil.com).

†Present address: IMC Ltd, Common Road, Huthwaite, Nottinghamshire, NG17 2NS, UK.

‡Present address: Geologisk Institutt, Universitetet i Bergen, 5007 Bergen, Norway.

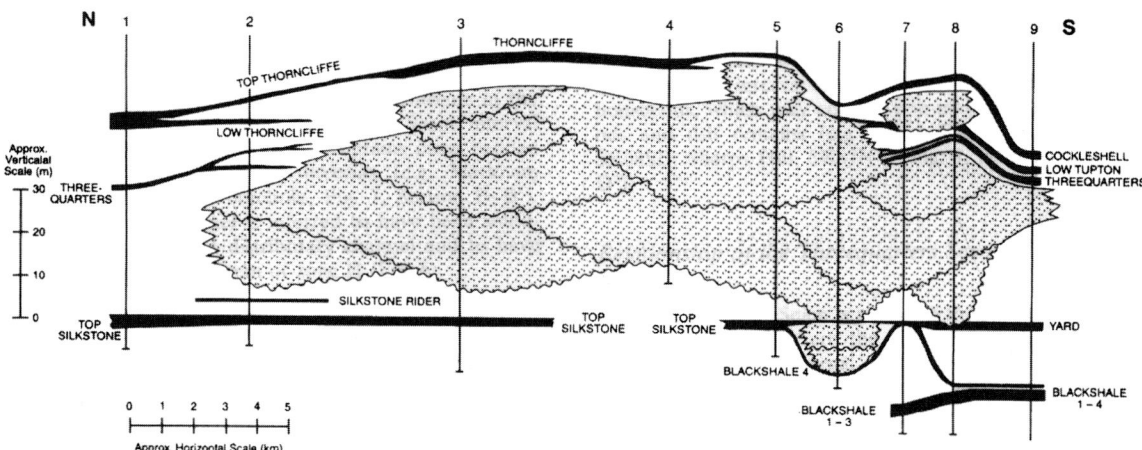

Fig. 1 A typical diagram from Guion *et al.* (1995a) showing the interpreted borehole sections in a two-dimensional view. By using digital three-dimensional visualization software, the Silkstone Rock can be better understood in terms of its spatial and temporal relationships with the adjacent facies.

framework, such that a 'close to deterministic' (i.e. very little uncertainty or interpolation) model can be built that allows visualization and interrogation of spatial and temporal relationships between identified facies. The three-dimensionality of the models and the functionality of the software allows for quick, user-defined 'field of view' and filtering on specific facies to be generated without the need to redraw sections or maps. With this functionality, three-dimensional modelling techniques can be a powerful tool to reconstruct and interpret ancient depositional systems (e.g. MacDonald *et al.*, 1992, 1998; Dreyer *et al.*, 1993; Mackey & Bridge, 1995; Ritchie *et al.*, 1998).

Many studies over the past few decades have been undertaken on the Westphalian in the Pennine Basin (e.g. Bailey *et al.*, 2002; Guion, 1987a,b; Guion *et al.*, 1995b; Rippon, 1996). Data used in these studies were generated from the underground extraction of coal seams and the same database is used in this study, which concentrates on one major, multistorey sandbody from the mid-Westphalian A, the Silkstone Rock. The purpose of the paper is to demonstrate the usefulness of three-dimensional digital techniques in assisting sedimentologists to understand the complex geometries and nature of the fill of multistorey fluvial units. The Silkstone Rock (Fig. 1) has previously been studied in detail (Guion *et al.*, 1995a) but this is the first attempt to develop a digital three-dimensional model for the unit.

DATA SET

The data set for this study is derived from a 20 km × 20 km × 600 m thick data volume within the East Pennine Coalfield, England (Fig. 2). The total interval studied extends from mid-Westphalian A (Black Shale seam) to top Westphalian B (base *Aergiranum* Marine Band) (Fig. 3). The data comprise 1100 cored boreholes, many of which cover the whole stratigraphical section and are spaced at a few hundred metres. Core from these boreholes, described and recorded graphically for lithological, coal and faunal characteristics by British Coal geologists, provides the basis for facies correlation across the study area. Borehole data are supplemented by mine plans and geological records, and plans that systematically record the occurrences of channel washouts (channel erosion into coal seams), channel belts, channel abandonment deposits ('swilleys') and other sedimentological details. These data alone can provide absolute correlations over vertical intervals of up to 150 m along the length of the underground roadways. Natural outcrops generally only expose the major sandbodies, which provided useful data on architecture, whereas opencast mines reveal fresh exposures including the fine-grained lithologies. Table 1 details the facies associations identified within the study area and the recognition criteria used to classify them. These facies associations are used to define the modelling 'objects' and

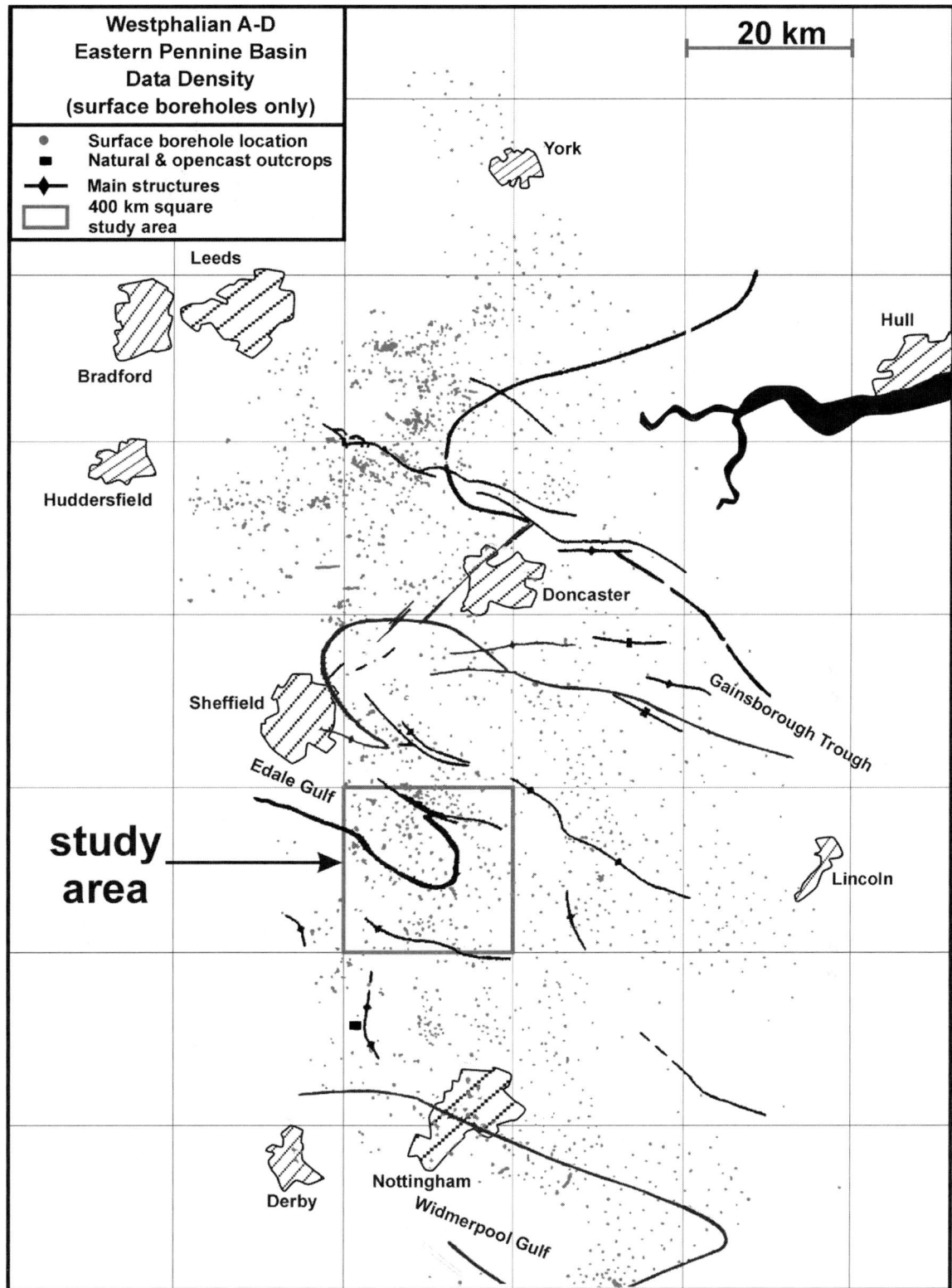

Fig. 2 The extent of borehole coverage (dots) throughout the East Pennine coalfield, showing the high density of borehole data captured within the study area, delimited by the arrowed box.

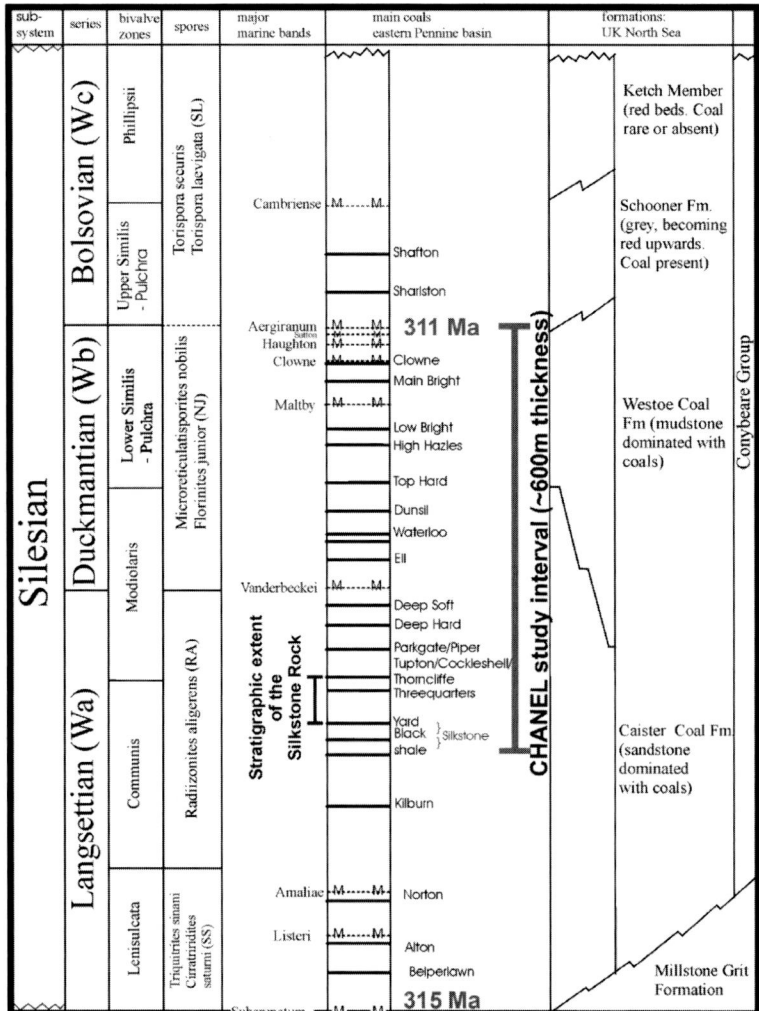

Fig. 3 The stratigraphy covered by the CHANEL study interval. The 28 economic (named) seams are represented in the section, as are the marine bands. These deterministically mapped horizons are very useful markers both stratigraphically and for zone boundary definition within the 3D model. The base Westphalian A and top Westphalian B ages are taken from O'Mara & Turner (1999).

as such form the building blocks of the three-dimensional stratigraphical model.

Figure 4 is taken from Guion et al. (1995a) and shows the areal extent of the Silkstone Rock within the study area. Although the entire extent of the Silkstone Rock is not captured within the study area, the model details changes in lateral and vertical facies associations that are important in understanding the evolution of the unit.

MODELLING METHODOLOGY

Correlation considerations, data density requirements and the conceptual application of such a high quality data set have already been addressed by Rippon (2000). The scope of this paper is insufficient to fully explain the modelling methods used in the study, however, the modelling workflow used in this study for integrating these data is highlighted below and in Plate 1.

Despite the high density of data available, all data types were in a paper format and therefore unsuitable for immediate input into conventional three-dimensional model building software as these applications require data to be in a digital format. A key challenge was therefore to develop a method by which all the different data types could be integrated and digitized and then output in a format suitable for reading into a three-dimensional visualization application (Keogh, 2002).

Table 1 Table of facies associations identified within the study area using the descriptions from borehole data combined with detailed outcrop studies.

Description	Scale†, geometry and overall percentage model volume	Model properties (input code, porosity and permeability values)			Facies association/ model object name	Other comments
		Model code	*Porosity (%)	*Permeability (mD)		
Strongly erosive bases, commonly eroding through fine-grained floodplain material and, less commonly, coals and condensed sections. Concentrations of mudstone rip-up clasts at channel stacking surfaces. Multilateral and multistorey. No apparent grain-size increase. Fill types include mappable units of sandy, heterolithic and muddy lithofacies	Width (km): –:–: 10.5‡ Thickness (m): 21:8:42 (40% > 20) Length (km): always > study area Sheet-like (high lateral amalgamation) to steep-sided bodies. Predominantly very straight mappable edges, occasionally curvilinear over large distances Volume percentage: 8.14	0	0.27	950	Multistorey channel—sandy	The thickness and aerial extent of these units provide the potential for excellent hydrocarbon reservoir units within the study area. However, the muddy facies association and rip-up clast concentrations would potentially act as baffles or barriers to flow both internally and out from the units. Published examples include: Guion et al. (1995b) and Ritchie et al. (1998)—time equivalent to Priest Rock
		1	0.18	250	Multistorey channel—heterolithic	
		2	0	0	Multistorey channel—muddy	
Channels vary from highly sinuous, meandering to almost straight. Lateral accretion point bars evident in subsurface and outcrop. Bifurcations common on minor distributaries. Channel depth commonly only single-storey. Lateral amalgamation of channels can occur, resulting in thin, wide-bodied units. Fills can vary from clean sand, through heterolithic (common) to mud and coal-fills. Occurrence of channels persistent throughout stratigraphical interval	Width (km): 0.44:0.1:4 (75% < 0.45) Thickness (m): 9.5:1:18 (95% < 15) Length (km): –:0.375: > study area (75% < 9.7)§ Channel cross-section modelled as straight-sided objects in the stratigraphical three-dimensional model. Mapped planform geometry exactly replicated in the stratigraphical three-dimensional model. Mapped thickness also exactly represented (see Chapter 3 for explanation of gridding techniques) Volume percentage: 11.21	3	0.22	400	Single-storey channel—sandy	The single-storey channels would potentially provide a good hydrocarbon reservoir unit within the study area. However, more care is needed in assigning reservoir quality to these units as the overall higher volume fraction of heterolithic facies association would mean classifying some of these units as only marginal reservoir units. The presence of a muddy facies association would produce potential flow baffles and barriers in these units. Published examples: Guion (1987a,b), Guion et al. (1995a) and Rippon (1996)
		4	0.17	125	Single-storey channel—heterolithic	
		5	0	0	Single-storey channel—muddy	

Table 1 (*continued*)

Description	Scale†, geometry and overall percentage model volume	Model properties (input code, porosity and permeability values)			Facies association/ model object name	Other comments
		Model code	*Porosity (%)	*Permeability (mD)		
Highly mature (heavily rooted and oxidized) examples represent greater time spans of floodplain inactivity and are known from other studies to represent candidate interfluves to low accommodation fluvial complexes	Width (km): generally > 8 Thickness (m): c. 1–2 Length (km): generally > 8 Thin but laterally extensive sheet-like geometry Volume percentage: 0.29	6	0.05	20	Overbank—very well drained	Published examples of interfluve palaeosol development: Besley & Fielding (1989), Aitken & Flint (1996) and McCarthy & Plint (1998)
Sheet sandstones, commonly 0.5–2 m thick, associated with floodplain fines. Their extensive lateral correlation is interpreted as the amalgamation of adjacent relatively confined crevasse-style sheets. Erosion of these sheets can occur from downcutting of younger single-storey channels	Width: tens of kilometres Thickness (m): 0.5–2 Length: tens of kilometres Thin but laterally extensive sheet-like geometry Volume percentage: 0.59	7	0.20	180	Overbank—sandy	The sheet-like sandstones and sand-dominant heterolithic units are potentially important reservoir units, especially in intervals where channels are isolated/poorly connected. A recent study has shown that the presence of these extensive non-channel sands can be important for better connecting poorly channelized reservoir intervals (Bailey et al., 2002)
Grey, finely laminated shales, silty shales and siltstones with siderite nodules and occasional thin fine-grained sandstones. Basal shales may contain rare, non-marine bivalves. Bioturbation rare. Fine-grained sandstones occur as laminae or thin beds with current and wave ripples. Palaeosols and root structures typically cap successions. In the mining industry, the rooted facies types are described as seatearths in the core descriptions	Width: tens of kilometres Thickness (m): 0.5–10 Length: tens of kilometres Laterally extensive sheet-like geometry. Has the largest volume percentage and as such is regarded as 'background' material in the stratigraphical three-dimensional model Volume percentage: 70.46	8	0.02	5	Overbank—heterolithic	Predominantly, these units are non-reservoir. Their volumetric abundance and sheet-like geometry can completely encase single-storey channels and restrict communication between them. Published example: Besley & Fielding (1989)
		9	0	0	Overbank—muddy	
		10	0	0	Overbank—muddy and rooted	

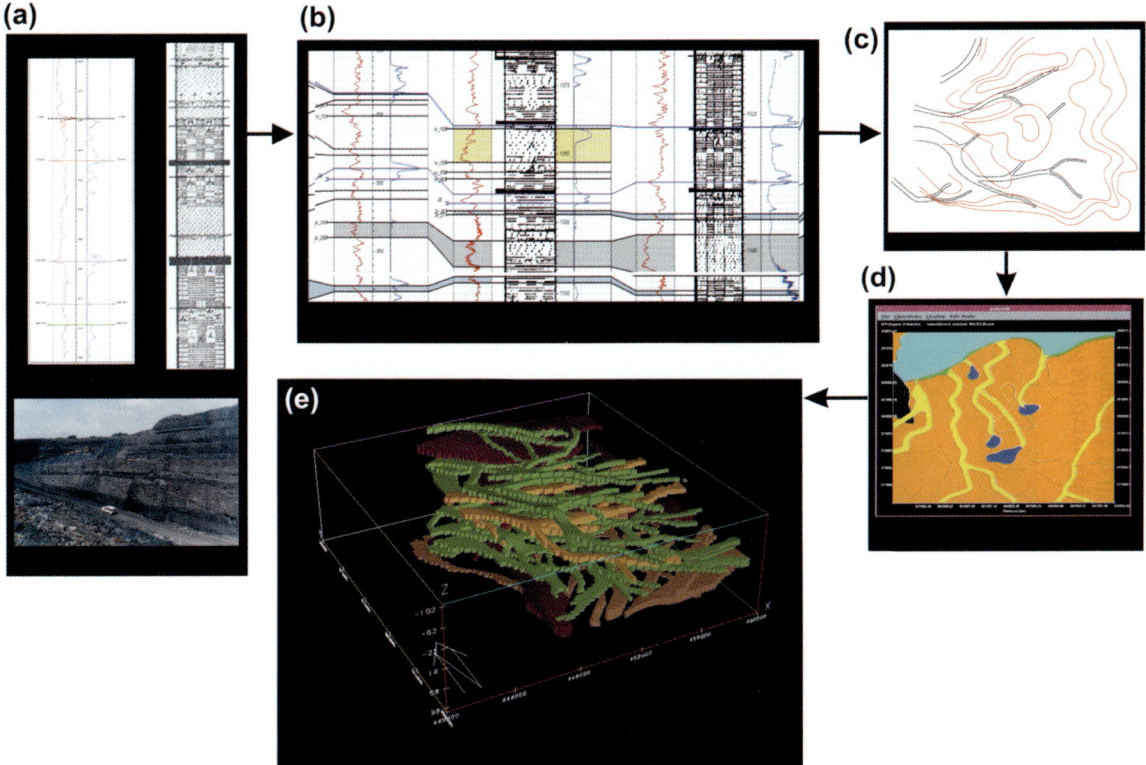

Plate 1 Workflow showing the various stages of data processing involved in building the three-dimensional model. The various data types are used to correlate the identified facies (a & b), which are then deterministically mapped at an interseam interval scale (c). These maps are then digitized along with the corresponding isopach data for the interval and read into the in-house software (d). The in-house software will then combine the two-dimensional map data with the vertical facies variability descriptions and output a three-dimensional volume in a format that can be imported into three-dimensional visualization software (e).

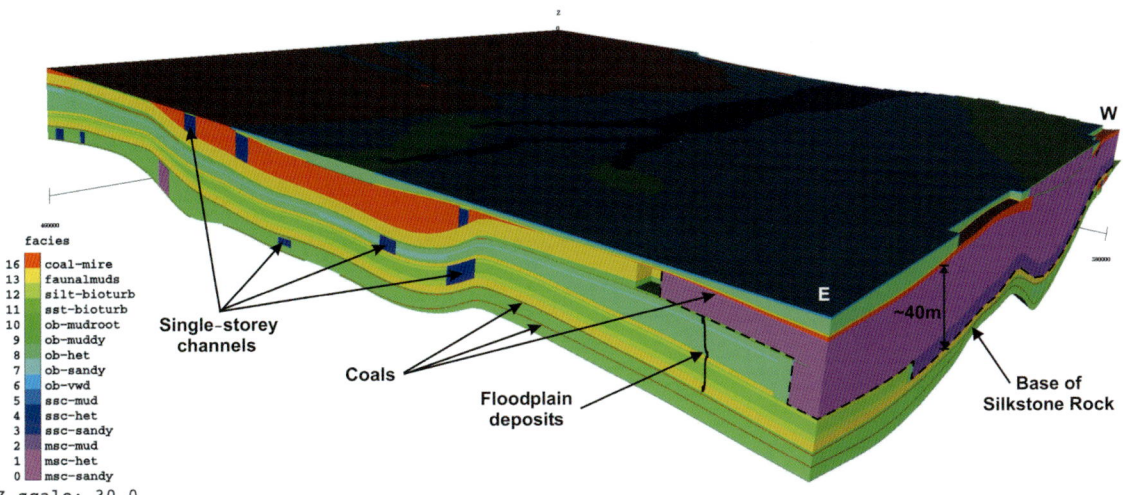

Plate 2 Three-dimensional view looking approximately south-west of the complete Silkstone Rock interval. The three-dimensional model enables visualization of the vertical and lateral relationship of the Silkstone Rock with the adjacent facies. The multistorey Silkstone Rock unit can be seen to be spatially associated with single-storey channels, coals and overbank deposits. Also demonstrated is the significant amount of basal erosion of the unit both down depositional dip and along depositional strike.

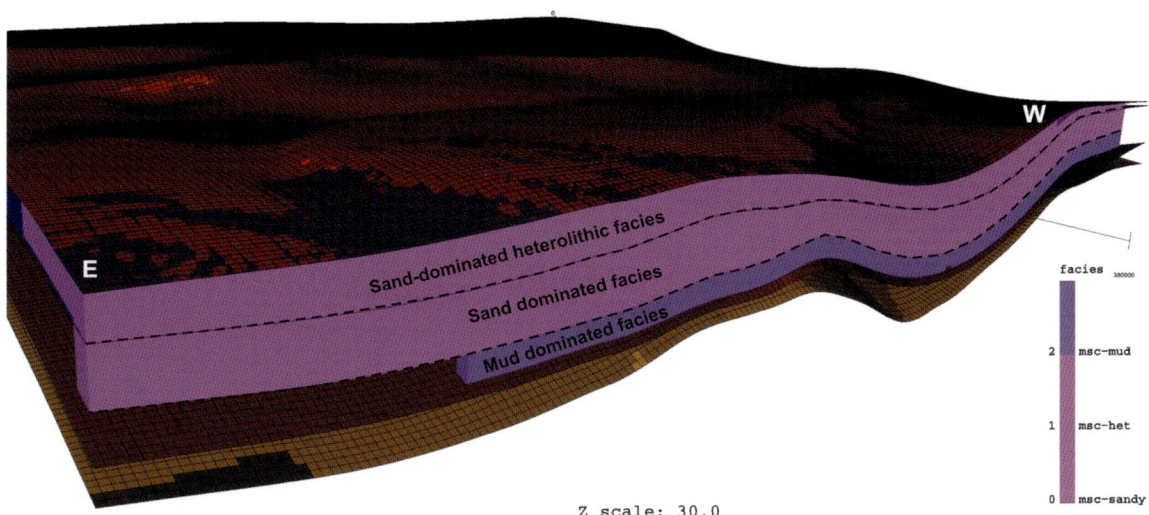

Plate 3 This three-dimensional view, looking south, shows the internal fill of the Silkstone Rock (internal divisions separated by dashed lines). The basal portion is dominantly a muddy facies, with a sand-rich middle portion and a sand-dominant heterolithic facies for the upper portion of the fill. The model shows that the muddy facies at the base of the Silkstone Rock is not present along its entire length. The model also shows that there is a slight thickening of the middle and upper portions of the fill from the west to the east (right to left respectively).

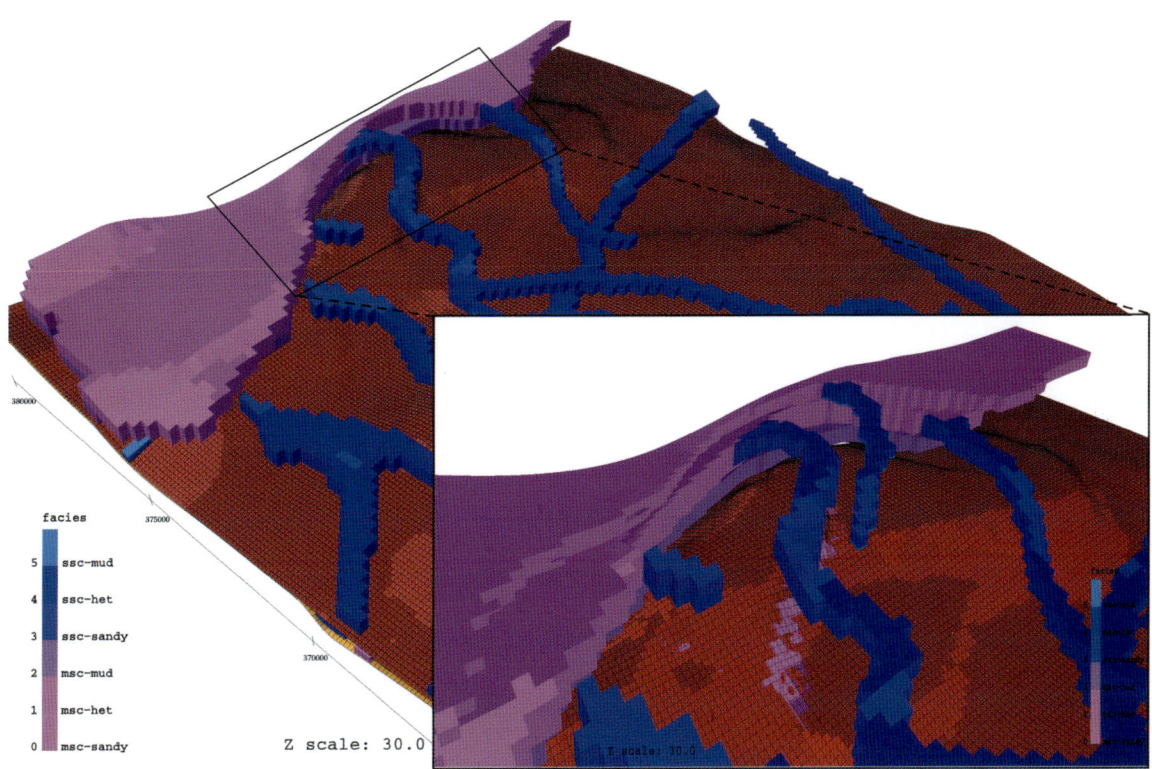

Plate 4 Three-dimensional view showing the regional relationship between the Silkstone Rock and a series of single-storey channels. The figure shows the orthogonal nature of the channels compared with the easterly palaeoflow of the major, multistorey Silkstone Rock. This view of the model interval also shows the highly sinuous and bifurcating nature of the channels, not previously represented by the work of Guion et al. (1995a). This close-up three-dimensional view of the relationship between the two channel types highlights the non-organized stacking relationship between the adjacent single-storey channels rather than the offset stacking relationship as proposed by Guion et al. (1995a).

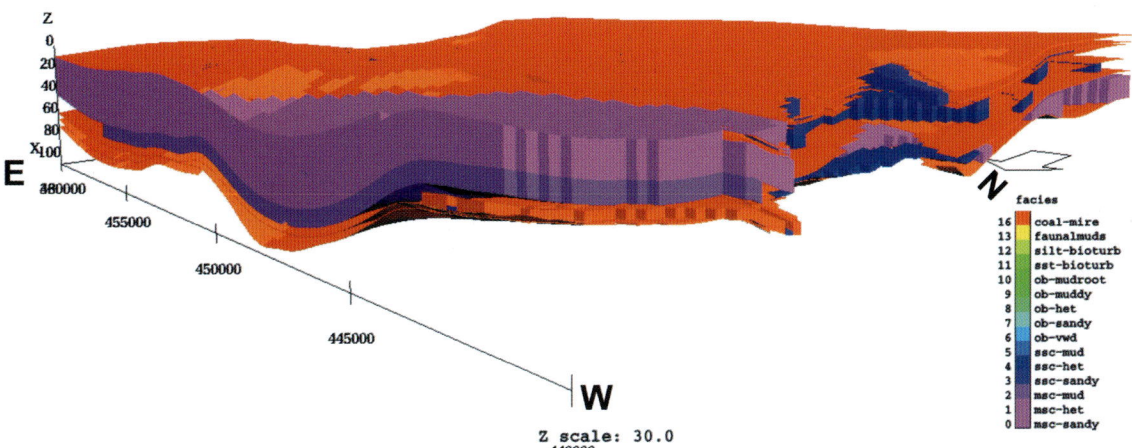

Plate 5 Three-dimensional view of the Silkstone Rock interval, looking approximately south-east, with the volume filtered to show only the coal seams and channels. This view demonstrates the complexity of the coal seam geometries and their thickness variations both in a vertical and lateral context. The base of the model in the north-west corner shows the vertical amalgamation of three seams that can be mapped as individual seams elsewhere in the study area. The youngest coal seam can be seen to thin uniformly away from the margin of the Silkstone Rock, but thicken locally again towards the southern margin of the study area. This view also demonstrates that the coal seams are not uniformly flat over the model area, but exhibit a variable topographic expression over the whole of the study area. The visualization of these aspects is not readily achievable through the use of simple two-dimensional sections.

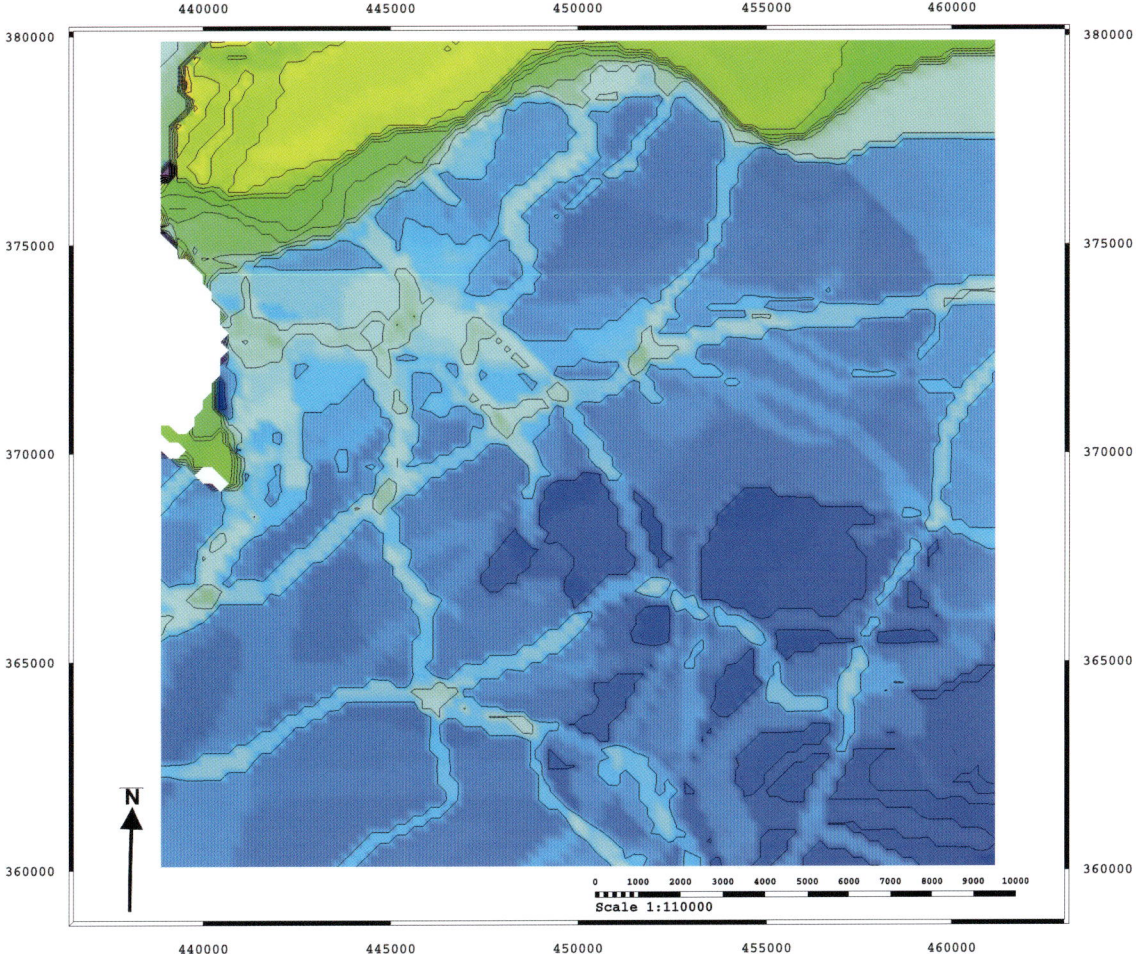

Plate 6 Net sandstone map generated from the three-dimensional volume using the three-dimensional software application. Bright coloured areas represent regions with high net sandstone values. The map shows that the highest net sandstone deposition has occurred within the Silkstone Rock channel belt. This figure also highlights the ease with which multiple visualizations of the same data can be achieved within a three-dimensional environment by using the built-in functionality of the software applications.

Description	Dimensions	#			Facies	Comments
Quasi-marine shales and occasional sandstones containing marine assemblages, but not always exclusively marine. Useful as local correlation markers. Examples include the 'Estheria' bands	Width: tens of kilometres; Thickness (m): 0.5–1; Length: tens of kilometres; Thin, extensive sheet-like geometry; Volume percentage: 4.3	11	0.15	100	Sandstone—heavily bioturbated	The sheet-like sandstones are potentially important reservoir units, especially where channels are isolated. 'Estheria' inhabit environments of varying salinity, but are found most commonly in close proximity to marine bands, indicating prevalence of brackish conditions
		12	0	0	Siltstone—heavily bioturbated	
		13	0	0	Faunal mudstone	
Coarsening upward successions from claystone to fine sandstone, 2–10 m thick with sheet to lobate geometries		14	0	0	Lagoonal	The high-resolution stratigraphical interpretation precludes the identification of this facies association as an individual modelling object. Lagoonal facies are present, but are split into the individual overbank/floodplain facies associations
Characterized by a distinctive faunal assemblage that allows confident regional correlation to other coal basins in the UK and beyond. Westphalian B records more frequent occurrence of marine incursions than Westphalian A strata	Width: hundreds of kilometres; Thickness (m): 1–10; Length: hundreds of kilometres; Laterally extensive, sheet-like geometry; Volume percentage: 2.04	15	0	0	Marine	Correlatable on intercontinental scale. Published examples: Calver (1968), Riley & Turner (1995) and Davies et al. (1999)
The regional (named) coal seams form the main zone boundaries in the reservoir model. Their distribution and geochemistry are fully deterministic. Cannel channels present but rare	Width: tens to hundreds of kilometres; Thickness (m): 0.10–2; Length: kilometres to tens of kilometres; Volume percentage: 2.93	16	0	0	Coal—mire	
		17	0	0	Coal—cannel	

*Values taken from analogous facies associations of the Ness Formation (Livera, 1989).
†Where available scale measurements for width, thickness and length are given as: average:minimum:maximum.
‡Minimum width not recordable within study area, therefore average value cannot be calculated.
§Average length not calculable due to maximum length being greater than study area extent.

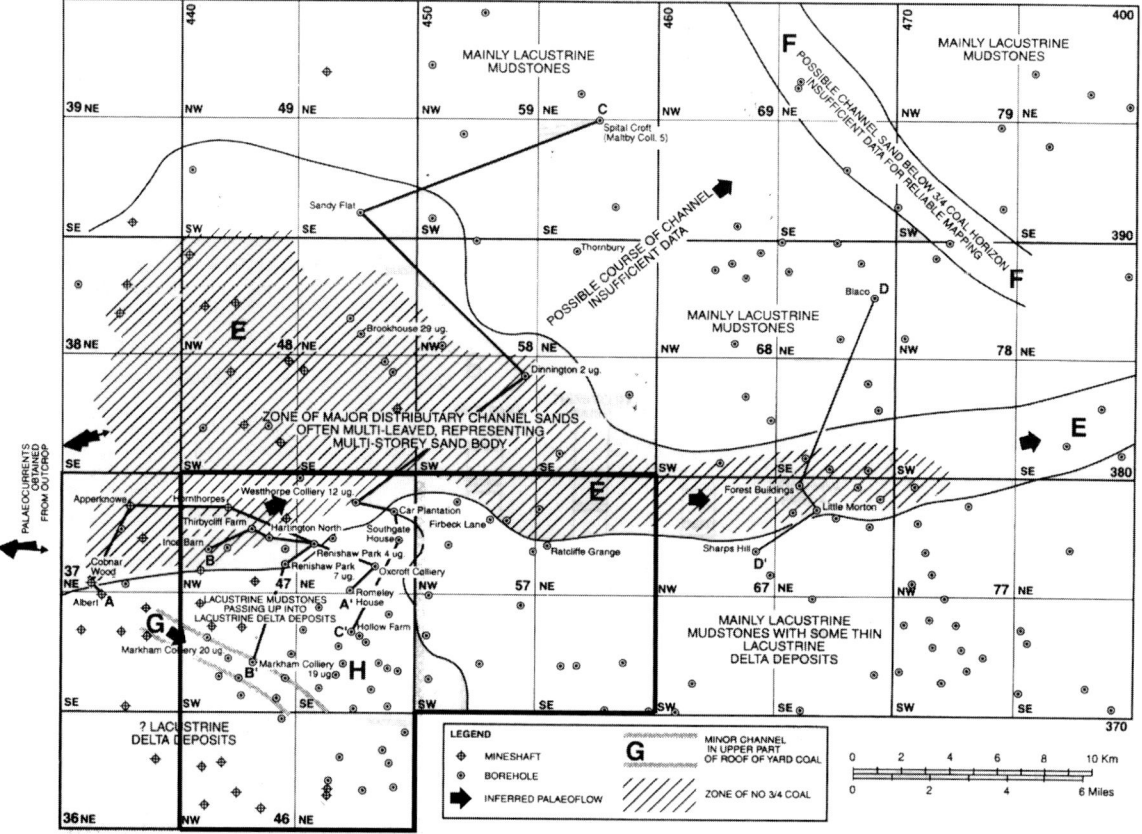

Fig. 4 Diagram taken from Guion *et al.* (1995a) showing the regional extent of the Silkstone Rock, mapped from correlation of borehole data. The area of the present study is highlighted in bold.

The data interpretations were presented in a map format that accurately captured the spatial arrangement of the facies associations in two dimensions (Keogh, 2002). Thickness variations of the mapped intervals were represented by isopach lines on the maps and the vertical variability of the mapped facies associations were represented by written descriptions on a per facies basis. In-house software was purpose-written to import the digitized maps of facies and thickness variations along with functionality to incorporate the written descriptions of vertical variability.

THE THREE-DIMENSIONAL MODEL

The view of the model in Plate 2 shows the whole stratigraphical interval covered by the Silkstone Rock. Within the study area the Silkstone Rock is up to 40 m thick, but Guion *et al.* (1995a) measured a maximum thickness of 50 m further north. This three-dimensional view demonstrates the facies relationships both in a lateral and vertical sense. The Silkstone Rock is spatially associated with single-storey channels, coals and overbank deposits, and demonstrates significant basal erosion both down depositional dip and along depositional strike. Plate 3 is a three-dimensional view to the south-southeast and highlights the multiple phases of fill in the Silkstone Rock. The basal portion of the fill is predominantly a muddy facies that does not extend the whole length of the unit in the study area. The upper portion of the fill is dominated by a sand-rich heterolithic facies, and this portion of the fill extends south along strike the furthest. The lateral extent of this unit defines the mappable southern edge of the Silkstone Rock. The main thickness of the Silkstone Rock is a sand-rich package and is seen in the model to be laterally more restricted than the upper heterolithic unit. Guion et al. (1995a) described the oldest mappable component of the Silkstone Rock as the first major sandstone above the Yard coal seam, but the three-dimensional model reveals a basal, discontinuous mud-rich facies.

The southern margin of the Silkstone Rock shows a complex relationship with a series of single-storey channels that are orientated orthogonal to the regional palaeoflow of the unit (Plate 4). Guion et al. (1995a) interpreted these single-storey channels as crevasse channels, fed from and therefore coeval with the Silkstone Rock. Investigation of the three-dimensional model suggests instead that these single-storey channels were truncated by the basal Silkstone Rock incision surface and thus represent a slightly older depositional cycle characterized by minor channels within a predominantly muddy to heterolithic floodplain (Plate 4a). There are no multistorey channel complexes at this older stratigraphical level in the 20 km × 20 km study area (Plate 2).

Guion et al. (1995a) described the minor, heterolithic channel-fills as having an offset stacking relationship owing to differential compactional subsidence, but viewing the same data in the three-dimensional model indicates a less well-organized stacking relationship; the minor channels occur at various stratigraphical levels along the 20 km length of the Silkstone Rock in the study area (Plate 4b). The three-dimensional model also shows the complex geometry of the channels, which could not be as well represented by two-dimensional maps (Guion et al., 1995a). These channel-fills are highly sinuous and bifurcate to the south. The basal portions of the channel-fills are sand with a sand-rich heterolithic upper portion of the fill.

From the deterministic mapping of the coal seams, it has been known for some years that their vertical and lateral geometries and their splitting relationships across the study area are complex, and thus their representation using only two-dimensional sections is inadequate to describe such features (Guion et al., 1995a; Rippon, 1996).

Plate 5 is an example of how these relationships can be viewed and examined better in a three-dimensional environment. The model identifies a region in the north-west corner of the model where there is vertical amalgamation of three seams that, elsewhere in the study area, can be mapped as individual seams. Thickness variations of individual coal seams can also be identified in the study area. The youngest coal seam in the study interval, the Top Thorncliffe/Cockleshell seam, can be seen to thin uniformly southwards away from the Silkstone Rock, but then locally thicken again substantially over a short distance on the westerly edge of the study area. The variable topographic expression of the coal seams over the entire study area is also easily visualized in the three-dimensional environment. None of these relationships can be well represented by using two-dimensional techniques such as cross-sectional diagrams (Fig. 1).

A net sand map (a two-dimensional map calculated from the three-dimensional volume) shows that the high net sandstone regions of the interval are contained entirely within the Silkstone Rock (Plate 6). The map shows that the mappable southern margin of the Silkstone Rock is not defined by the areas of high net sandstone, but by the sand-rich heterolithic facies. The map also shows that the single-storey channels in the southern part of the area are encased in a predominantly muddy floodplain facies. Multiple visualizations of the same data can be easily generated within the three-dimensional environment by using the built-in functionality of the software application. As well as two-dimensional maps, fence diagrams, cross-sections and volume slices can also be created very quickly and efficiently without any further manipulation of the data (Roxar, 2000).

SUMMARY AND CONCLUSIONS

Different data types can be easily integrated and manipulated within the three-dimensional software environment, allowing for well-constrained models to be built with minimal uncertainty. The spatial relationship between the data is retained within the three-dimensional environment. Commonly with two-dimensional sections, the data spacing is shown equally even though that is not often the case. The ability within the three-dimensional environment to zoom, rotate and filter on specific facies and regions of a model is a very powerful tool for testing spatial and temporal facies relationships.

Application of these techniques has allowed for enhanced visualization and interrogation of the Pennine Basin interval, previously not possible using traditional two-dimensional techniques. Understanding of the Silkstone Rock and its relationship to other facies have improved because:

1 erosion at the base of the unit can be demonstrated to truncate coals and single-storey channel-fills that were deposited earlier—the visualization of this relationship is not possible from a single two-dimensional section;

2 the three-dimensional model demonstrates that the older, southerly flowing single-storey channels do not show lateral offset stacking relationships to the Silkstone Rock as reported in earlier work;

3 the model allows the visualization of the three-dimensional complexity of coal seam geometries and their splitting/reuniting relationships, which previously has not been possible through the use of two-dimensional techniques (Fig. 1).

ACKNOWLEDGEMENTS

This paper derives from the 2-year CHANEL project at Liverpool University with the Fault Analysis Group (FAG), now relocated to University College Dublin. We thank FAG staff, in particular Wayne Bailey and Philip Nell, for their support and discussions. CHANEL was funded by Exxon, JNOC, Statoil, Texaco and Total, whose interest in the project and permission to publish is gratefully acknowledged. We are indebted to Roxar for their generous provision of academic licenses for the software used in this project.

REFERENCES

Aitken, J.F. and Flint, S.S. (1996) Variable expression of interfluvial sequence boundaries in the Breathitt Group (Pennsylvanian), eastern Kentucky, USA. In: *High Resolution Sequence Stratigraphy: Innovations and Applications* (Eds J.A. Howell and J.F. Aitken). *Geol. Soc. Lond. Spec. Publ.*, **104**, 193–206.

Bailey, W.R., Manzocchi, T., Walsh, J.J., Keogh, K.J., Hodgetts, D., Rippon, J.H., Nell, P.A.R. and Flint, S.S. (2002) The effect of faults on the 3-D connectivity of reservoir bodies: a case study from the East Pennines Coalfield, U.K. *Petrol. Geosci.*, **8**, 263–277.

Besly, B.M. and Fielding, C.R. (1989) Paleosols in Westphalian coal-bearing and red-bed sequences, central and northern England. *Palaeogeogr. Palaeoclimatol. Palaeoecol.*, **70**, 303–330.

Calver, M. (1968) Distribution of Westphalian marine faunas in northern England and adjoining areas. *Proc. Yorks. Geol. Soc.*, **37**, 1–72.

Davies, S.J., Hampson, G.J., Flint, S.S. and Elliott, T.E. (1999) Continental-scale sequence stratigraphy of the Namurian, Upper Carboniferous and its applications to reservoir predication. In: *Petroleum Geology of Northwest Europe: Proceedings of the 5th Conference* (Eds A.J. Fleet and S.A.R. Boldy), pp. 757–770. Geological Society Publishing House, Bath.

Dreyer, T., Falt, L.-M., Hoy, T., Knarud, R., Steel, R. and Cuevas, J-L. (1993) Sedimentary architecture of field analogues for reservoir information (SAFARI): a case study of the fluvial Escanilla Formation, Spanish Pyrenees. In: *The Geological Modelling of Hydrocarbon Reservoirs and Outcrop Analogues* (Eds S. Flint and I. Bryant). *Spec. Publ. Int. Assoc. Sedimentol.*, **15**, 57–80.

Guion, P.D. (1987a) Palaeochannels in mine workings in the High Hazles Coal (Westphalian-B), Nottinghamshire Coalfield, England. *J. Geol. Soc. Lond.*, **144**, 471–488.

Guion, P.D. (1987b) The influence of a palaeochannel on seam thickness in the coal measures of Derbyshire, England. *Int. J. Coal Geol.*, **7**, 269–299.

Guion, P.D., Banks, N.L. and Rippon, J.H. (1995a) The Silkstone Rock (Westphalian A) from the East Pennines, England—implications for sand body genesis. *J. Geol. Soc. Lond.*, **152**, 819–832.

Guion, P.D., Fulton, I.M. and Jones, N.S. (1995b) Sedimentary facies of the coal-bearing Westphalian A and B north of the Wales-Brabant High. *Eur. Coal Geol.*, **82**, 45–78.

Guion, P.D., Jones, N.S., Fulton, I.M. and Ashton, A.J. (1995c) Effects of a Westphalian channel on coal-seam geometry: a re-appraisal of the 'Dumb Fault' of north Derbyshire. *Proc. Yorks. Geol. Soc.*, **50**, 317–332.

Keogh, K.J. (2002) Sequence stratigraphy and 3-D Modelling of the East Pennine coalfield, U.K: a deterministic and stochastic approach. Unpublished PhD thesis, University of Liverpool, 281 pp.

Livera, S. (1989) Facies associations and sand-body geometries in the Ness Formation of the Brent Group, Brent Field. In: *Deltas: Sites and Traps for Fossil Fuels* (Eds M.K.G. Whateley and K.T. Pickering). *Geol. Soc. Lond. Spec. Publ.*, **41**, 269–288.

MacDonald, A.C., Falt, L.-M. and Hektoen, A.-L. (1998) Stochastic modelling of incised valley geometries. *Am. Assoc. Petrol. Geol. Bull.*, **82**, 1156–1172.

MacDonald, A.C., Hoye, T.H., Lowry, P., Jacobsen, T., Aasen, J.O. and Gringheim, A.O. (1992) Stochastic flow unit modelling of a North Sea coastal-deltaic reservoir. *First Break*, **10**.

Mackey, S. and Bridge, J. (1995. A three-dimensional model of alluvial stratigraphy: theory and application. *J. Sediment. Res.*, **B65**: 7–31.

McCarthy, P.J. and Plint, A.G. (1998) Recognition of interfluve sequence boundaries: integrating paleopedology and sequence stratigraphy. *Geology*, **26**, 387–390.

O'Mara, P.T. and Turner, B.R. (1999) Sequence stratigraphy of coastal alluvial plain Westphalian B coal measures in Northumberland and the Southern North Sea. *Int. J. Coal Geol.*, **42**, 33–62.

Riley, N. and Turner, N. (1995) The correlation of mid-Westphalian marine bands between the central Appalachian basin (USA) and the United Kingdom. In: *The XIII International Congress on the Carboniferous and Permian*, Krakow, Poland, 122 pp.

Rippon, J.H. (1996) Sand body orientation, palaeoslope analysis and basin-fill implications in the Westphalian A–C of Great Britain. *J. Geol. Soc. Lond.*, **153**, 881–900.

Rippon, J.H. (2000) The Westphalian A to mid-C of Great Britain: new conceptual modelling for hydrocarbon reservoir analogues. *Zbl. Geol. Paläontol. Teil I*, **3**, 217–231.

Ritchie, J.S., Pilling, D. and Hayes, S. (1998) Reservoir development, sequence stratigraphy and geological modelling of Westphalian fluvial reservoirs of the Caister C Field, U.K. Southern North Sea. *Petrol. Geosci.*, **4**, 203–211.

Roxar (2000) *RMSbase*. RMS User Manuals, Roxar AS.

Changing alluvial style in response to changing accommodation rate in a proximal foreland basin setting: Upper Cretaceous Dunvegan Formation, north-east British Columbia, Canada

MATTHEW P. LUMSDON-WEST* and A. GUY PLINT

Department of Earth Sciences, University of Western Ontario, London, Ontario N6A 5B7, Canada

ABSTRACT

The early–mid-Cenomanian Dunvegan Formation represents a major deltaic complex that prograded > 400 km from the north-west to the south-east, parallel to the rising Rocky Mountain Cordillera, over about 2 Myr. The formation comprises at least ten regressive–transgressive allomembers, separated by regional marine transgressive surfaces and, further up-dip, by lacustrine flooding surfaces. The lower five allomembers (J–F) have a sigmoidal geometry, thinning up-dip towards the north-west. Allomember E differs in showing in the north-west a pronounced isopach 'moat', into which strata thicken from 35 to 80 m over 120 km, suggestive of renewed subsidence along the basin margin. Collectively, allomembers D–A show an even more pronounced thickening towards the active basin margin, from 30 to 100 m over 250 km, and this geometry persists in the lower units of the overlying Kaskapau Formation. Outcrop sections of the Dunvegan Formation in the Chetwynd–Pine River area are located within the thickened isopach 'moats' of allomembers E–A, and are dominated by alluvial deposits about 160 m thick. Non-channelized alluvial facies represent lake, mouth-bar, levee and crevasse splay environments, and organic histosol (coal), poorly drained, intermediate redox and better drained palaeosols. Siderite, organic debris and dinosaur tracks are ubiquitous elements of these facies. Channelized facies are dominated by non-migrating (probably anastomosed) sandstone channel-fills. Non-channelized facies form two associations: A, which is dominated by crevasse splays, levees, intermediate redox and better drained palaeosols; B, which is dominated by lakes, mouth-bars, organic and poorly drained palaeosols. Facies association A, indicating relatively well-drained conditions, dominates allomembers G to the lower part of E. From near the middle to near the top, allomember E is progressively dominated by facies association B, but the uppermost 5 m shows an abrupt return to better drained palaeosols. Allomembers D to A show alternations between associations A and B, although overall, association B predominates upward. The basal units of the overlying Kaskapau Formation comprise a transitional succession 40–50 m thick dominated by facies of association B, but including brackish-water lagoonal deposits, overlain by marine deposits.

*Present address: 54 Belorun Court, London, Ontario N6K 3K8, Canada (Email: redtail@sympatico.ca).

The thinning of allomembers H–F towards the north-west suggests minimal accommodation in that direction, and this is reflected in their constituent alluvial facies (association A), dominated by relatively well-drained environments. The upward change within allomember E, from association A to relatively poorly drained association B, is interpreted to record an upward increase in accommodation rate. This change is independently suggested by the isopach 'moat' in allomember E, attributed to renewed flexural subsidence. A pedocomplex 5 m thick at the top of allomember E suggests a final phase of very low accommodation rate. Allomembers D–A are again dominated by poorly drained freshwater environments, which pass upward into rocks of the basal Kaskapau Formation which show progressively increasing marine influence. This facies succession reflects an increasing accommodation rate, also suggested by regional isopach patterns.

INTRODUCTION

Controls on alluvial deposition

Alluvial style is controlled by numerous factors, of which the most important are climate, rate of tectonic movement and base level (Ethridge et al., 1998). The rate of sediment supply, although important (Hovius, 1998), is nevertheless a function of the foregoing controls (Ethridge et al., 1998). Much attention has been devoted to interpreting vertical and lateral changes in ancient alluvial successions. Changes in lithofacies and alluvial style have, in relatively few instances, been attributed to climatic variation (e.g. Olsen, 1990, 1994; Legarreta & Uliana, 1998; Plint, 2002). More commonly, changes in tectonically controlled accommodation rate have been interpreted as the principal control on alluvial style (Kraus & Middleton, 1987; Blakey & Gubitosa, 1984; Martinsen et al., 1999). Vertical changes in alluvial style in coastal plain deposits have also been attributed to relative, or possibly eustatic, sea-level changes, which may have caused incision or aggradation, or changes in channel pattern. A clear relationship with eustatic sea-level change has been established in only a few, well-dated Holocene cases (e.g. Törnquist, 1993; Törnqvist et al., 2000; Aslan & Autin, 1999). Interpretations of older coastal plain successions have been hampered by an inability to establish precise age relationships between marine and non-marine strata. Nevertheless, correlation of stratigraphical surfaces has permitted marine facies to be traced up-dip into coeval alluvial facies in several cases (e.g. Shanley & McCabe, 1991, 1993; Burns et al., 1997; Cant, 1998; Rogers, 1998; Hampson et al., 1999; McCarthy et al., 1999; McLaurin & Steel, 2000; Plint, 2000; Plint et al., 2001).

Purpose and key features of the study

The study presented here describes a succession of coastal plain deposits from the deltaic Upper Cretaceous Dunvegan Formation, which were deposited in the orogen-proximal part of the Western Canada Foreland Basin. The key features of this study are as follows.

1 The coastal plain strata can be placed in a well-established allostratigraphical framework that approximates a time framework (Plint, 2000; Plint et al., 2001). It is possible to map a series of transgressive surfaces and subaerial unconformities that approximate time lines, and which define ten successive deltaic packages. The lateral distribution of alluvial delta plain, delta-front and prodelta facies can be mapped between each set of bounding surfaces.

2 The high-resolution stratigraphical control makes it possible to draw a series of detailed palaeogeographical maps (Plint, 2000) that show how the delta complex evolved over about 2 Myr. It is also possible to construct isopach maps for each successive allostratigraphical unit, which not only show how successive deltas filled the marine accommodation, but also reveal dramatic changes in accommodation rate in the orogen-proximal part of the basin. In essence, the lower five allomembers of the Dunvegan Formation *thin* up-dip, which we interpret to indicate very low rates of up-dip subsidence adjacent to the orogen. In contrast, the upper five allomembers *thicken* towards the orogen in the north-west and west. Although this thickening pattern *could* be interpreted to reflect the construction of an elevated alluvial wedge adjacent to the orogen, there is no facies evidence to support such a change in alluvial slope. The alternative, and favoured, interpretation is that the isopach thickening reflects increased rates of subsidence, probably related to increased rates of thrusting and resultant flexural subsidence.

3 Excellent outcrop sections are available in the orogen-proximal part of the basin. These expose the entire up-dip portion of the Dunvegan Formation which is known, from regional isopach mapping, to lie near the centre of a flexural depocentre.

The available stratigraphical, palaeogeographical and outcrop control allows us to address the question: 'Were temporal changes in subsidence rate accompanied by changes in alluvial style?'

GEOLOGICAL SETTING

Palaeogeography and subsidence

The lower to middle Cenomanian Dunvegan Formation was deposited over about 2 Myr on the western margin of the Cretaceous Western Interior Seaway of North America. The western margin of this basin is a classic example of a retro-arc foreland basin, which experienced about 3–3.5 km of subsidence during the Late Jurassic and Cretaceous (Beaumont, 1981). Figure 1 shows the generalized palaeogeography of the Dunvegan delta complex, and a simplified dip cross-section (located in Fig. 1) is shown in Fig. 2. Erosion of Palaeozoic and Mesozoic sedimentary rocks exposed on intermittently active thrust sheets along the south-west margin of the basin provided abundant detrital sediment to the adjacent foreland basin (Monger, 1993).

Sandy delta-front strata of the Dunvegan Formation overlie, and are intergradational with,

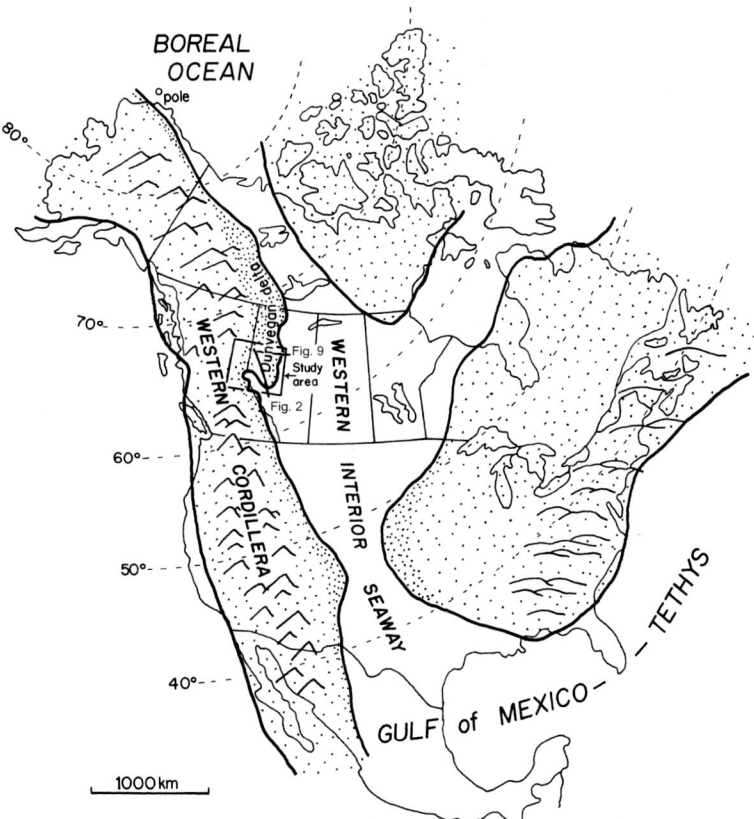

Fig. 1 Summary palaeogeography of the Western Interior Seaway during the middle Cenomanian, including the generalized distribution of the Dunvegan delta complex in Alberta and British Columbia (after Plint, 2000). The north-west–south-east line indicates the position of the dip cross-section shown in Fig. 2, and the E–W line indicates the position of the cross-section shown in Fig. 9.

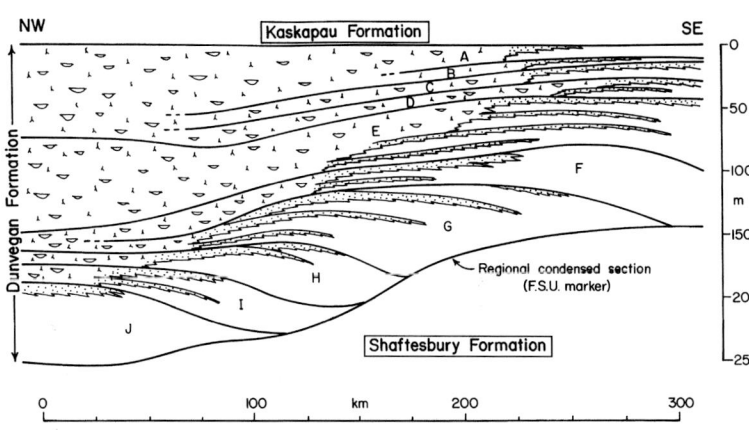

Fig. 2 Summary north-west–south-east trending dip cross-section (drawn to scale, located in Fig. 1), illustrating the broad facies distribution and stratal geometry of the Dunvegan Formation. The datum for the section is the regional marine flooding surface at the top of Dunvegan allomember A. The Dunvegan deltaic units downlap onto the FSU marker, which comprises a highly phosphatic fine sandstone representing a condensed section. Note the pronounced up-dip (north-west) thickening of Dunvegan allomembers E–A, relative to older units.

offshore marine mudstones of the upper Albian to lower Cenomanian Shaftesbury Formation. The Shaftesbury Formation forms a south-west-thickening wedge adjacent to the Cordillera, and is interpreted to record rapid syndepositional flexural subsidence of the basin. In contrast, the overlying Dunvegan Formation as a whole initially records very limited subsidence, although successively younger allomembers show progressively more widespread subsidence along the orogenic margin of the basin. This trend is further developed in the overlying marine mudstones of the Kaskapau Formation, which were deposited during the Late Cretaceous (late Cenomanian to mid-Turonian) eustatic peak (Kauffman & Caldwell, 1993). Additional biostratigraphical details are provided in Plint (2000).

Study area

The regional subsurface study area, on which the isopach mapping is based, is located in north-west Alberta and north-east British Columbia (Fig. 3). The smaller, outcrop study area discussed here lies within the proximal, up-dip portion of the Dunvegan coastal plain, and at least 60 km to the east of the contemporaneous deformation front. The palaeogeographical setting of these sites (Fig. 4a–d) shows that alluvial strata of Dunvegan allomembers G to A, discussed herein, were deposited on a coastal plain, about 70 to 140 km landward of the transgressive limits of the contemporaneous marine shoreline. In the vicinity of Chetwynd, north-east British Columbia (Fig. 5), 37 measured sections, with a cumulative thickness of about 1400 m, were measured along Dickebusch Creek (13 sections), Pine River (8 sections) and Coldstream Creek (16 sections; Fig. 6). Exposure along the streams was not continuous, but composite sections were constructed based on correlation of local marker beds and facies associations (Lumsdon, 2000).

A composite section from Pine River and Coldstream Creek represents the entire thickness of the Dunvegan Formation, and includes the basal units of the Kaskapau Formation (Fig. 7; Plint, 2000). Dickebusch Creek, 60 km to the south-west, also exposes a near-complete section, spanning allomembers H to A (Fig. 8; Plint, 2000).

STRATIGRAPHICAL AND PALAEOGEOGRAPHICAL FRAMEWORK

Allomembers

In deltaic strata of the Dunvegan Formation, Bhattacharya & Walker (1991a) and Plint (2000) defined 10 allomembers, J through to A in ascending order, that were defined by regionally mappable marine transgressive surfaces that pass up-dip into approximately correlative subaerial unconformities (McCarthy et al., 1999; Plint et al., 2001). Stratigraphical control is provided by > 2400 well logs and about 100 outcrop sections (Fig. 3). The Dunvegan Formation as a whole represents about 2 Myr (Plint, 2000), and it is

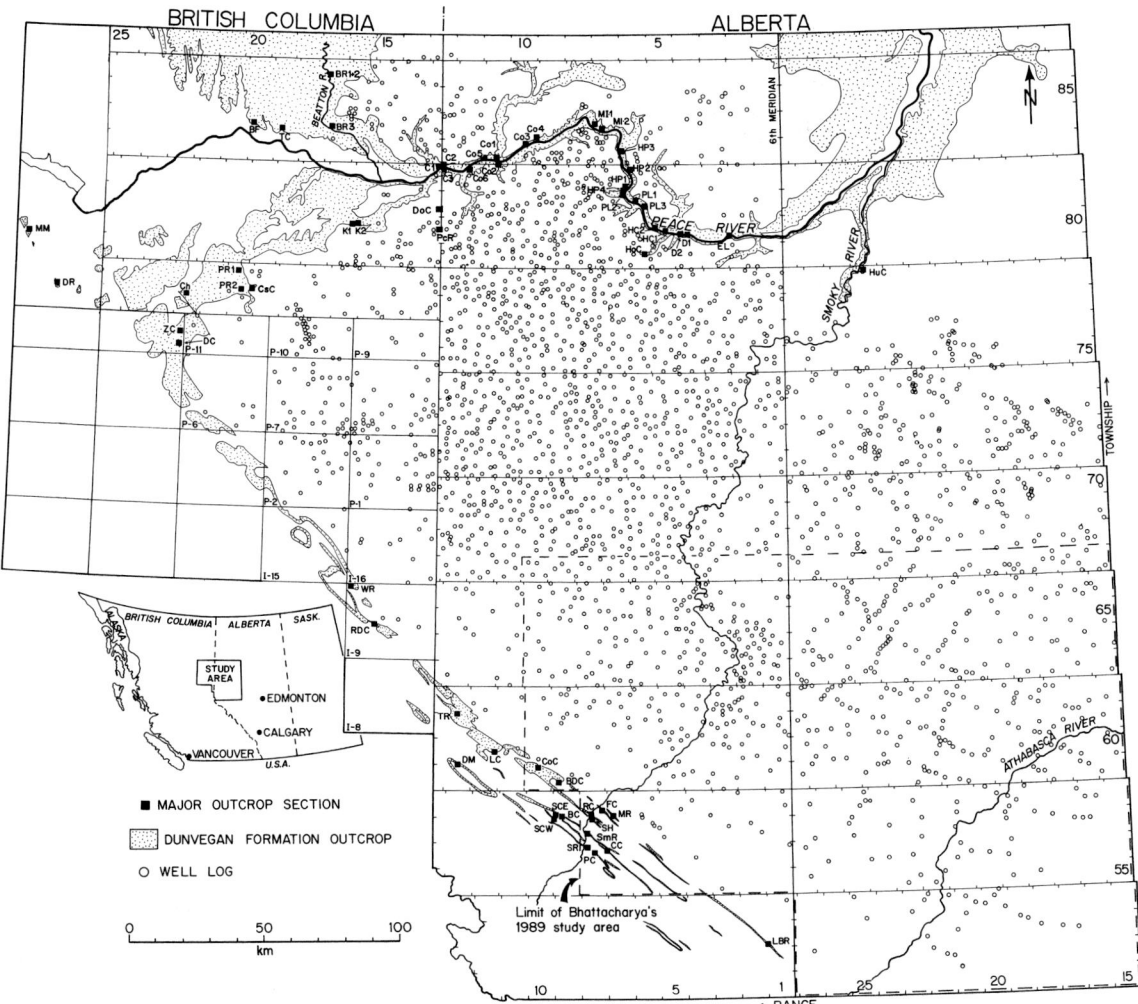

Fig. 3 Map of north-west Alberta and adjacent British Columbia, showing the distribution of well logs (2340) and principal outcrop sections used to define the stratigraphy and palaeogeography of the Dunvegan Formation. Locality abbreviations are explained in Plint (2000). These data were used to construct the isopach and palaeogeographical maps shown in Fig. 4A–D. The outcrop study area discussed in this paper is shown in the north-west corner of the map, where localities are indicated as follows: CsC, Coldstream Creek; PR, Pine River; DC, Dickebusch Creek.

assumed that each of the ten allomembers represents an average of about 200 kyr.

The stratigraphical geometry of the Dunvegan Formation is summarized in Fig. 2. The top of the Dunvegan Formation is defined by a regionally mappable marine transgressive surface that marks the base of the mainly marine Kaskapau Formation, whereas the base is defined by the regional downlap surface below the Dunvegan deltaic units (Fig. 2). For most allomembers, marine transgressive surfaces can be correlated up-dip into coastal plain strata on the basis of distinctive well-log deflections, which correspond to lacustrine deposits that commonly overlie interfluve palaeosols. It proved impossible to trace far onto the coastal plain the transgressive surfaces that define the tops of allomembers B, C and D. As a result, allomembers D to A have been

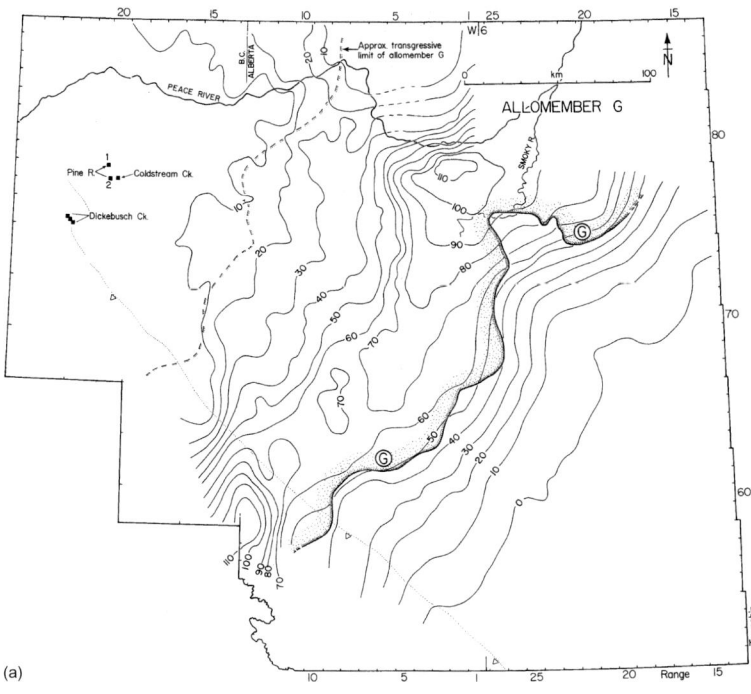

(a)

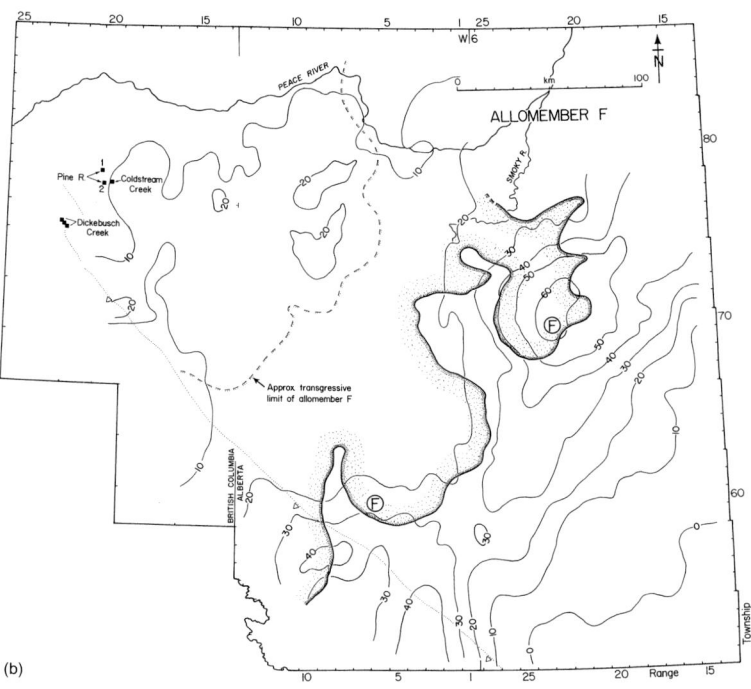

(b)

Fig. 4 Combined palaeogeographical and isopach maps for allomembers G–A. Maximum regressive limits of delta-front sandstones and maximum transgressive limits of marine mudstones are based primarily on subsurface data in Plint (2000, 2002). (a) Palaeogeography of allomember G, where the maximum progradational limit of delta-front sandstone is shown by the stippled zone, and the broken line indicates the maximum transgressive limit of marine mudstone. Isopachs show steady up-dip thinning onto the coastal plain. The allomember G–F boundary cannot be distinguished in well logs in the far north-west part of the map area. (b) Palaeogeography of allomember F, showing two major delta lobes in the south-east. Note the very consistent thickness of the unit, between 10 and 20 m, over much of the delta plain. Allomembers G and F cannot be differentiated with confidence in the far north-west (Dickebusch Creek area).

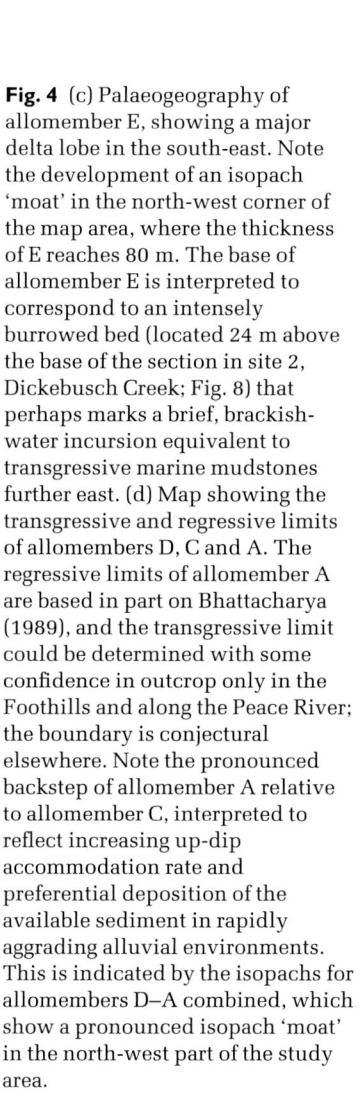

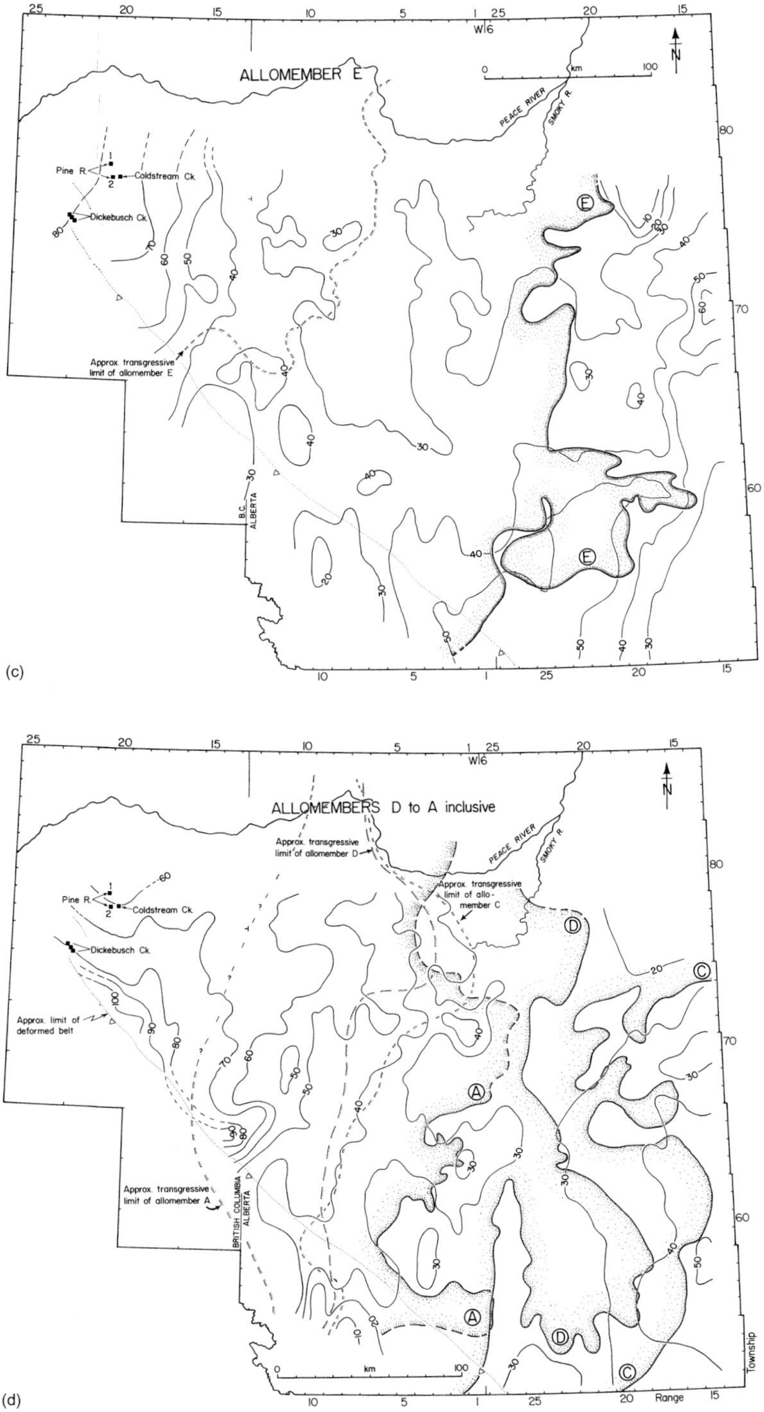

Fig. 4 (c) Palaeogeography of allomember E, showing a major delta lobe in the south-east. Note the development of an isopach 'moat' in the north-west corner of the map area, where the thickness of E reaches 80 m. The base of allomember E is interpreted to correspond to an intensely burrowed bed (located 24 m above the base of the section in site 2, Dickebusch Creek; Fig. 8) that perhaps marks a brief, brackish-water incursion equivalent to transgressive marine mudstones further east. (d) Map showing the transgressive and regressive limits of allomembers D, C and A. The regressive limits of allomember A are based in part on Bhattacharya (1989), and the transgressive limit could be determined with some confidence in outcrop only in the Foothills and along the Peace River; the boundary is conjectural elsewhere. Note the pronounced backstep of allomember A relative to allomember C, interpreted to reflect increasing up-dip accommodation rate and preferential deposition of the available sediment in rapidly aggrading alluvial environments. This is indicated by the isopachs for allomembers D–A combined, which show a pronounced isopach 'moat' in the north-west part of the study area.

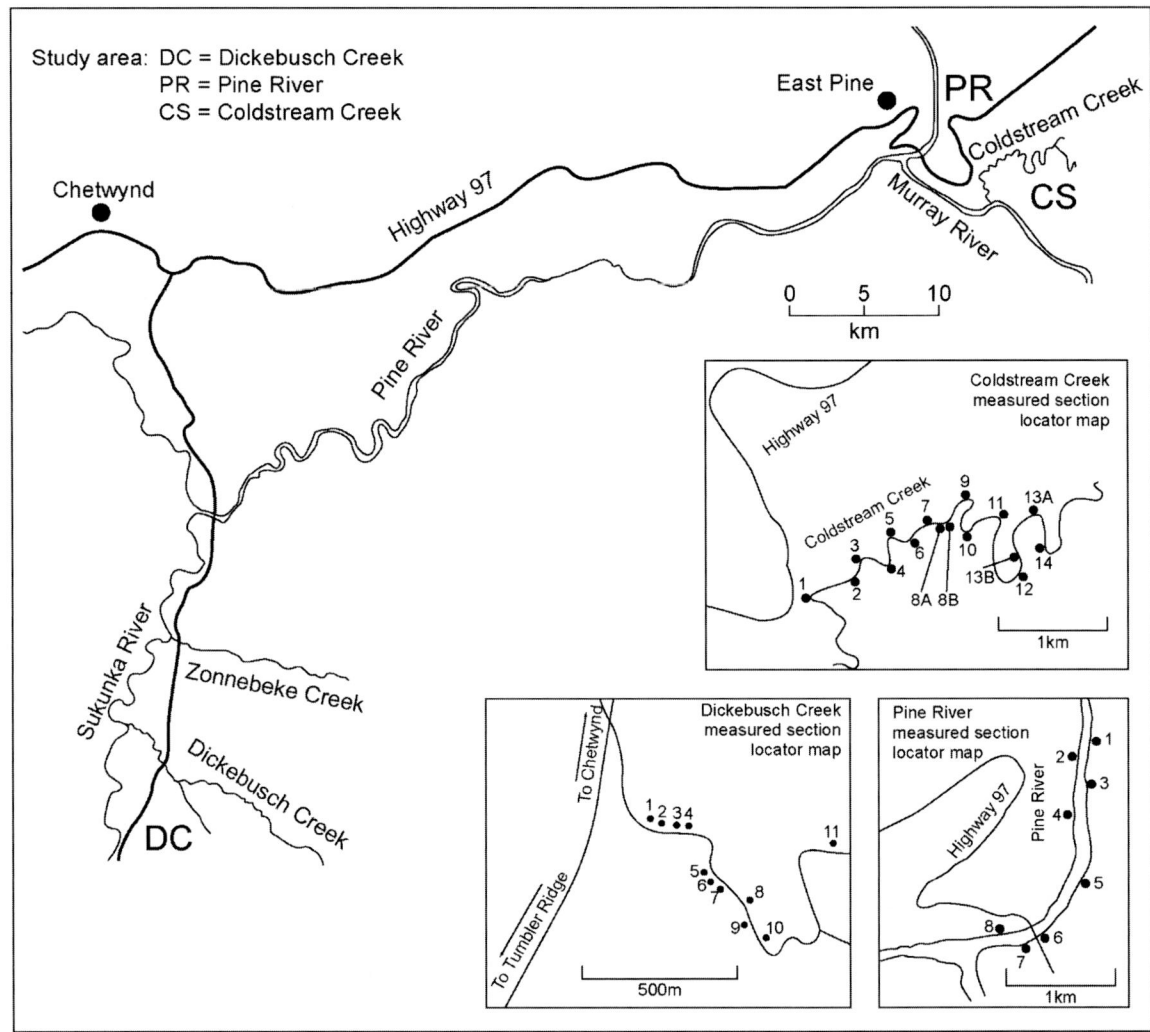

Fig. 5 Map of the Chetwynd–Pine River area of north-east British Columbia, showing location of the outcrop sections described in the text. Note that the outcrop section numbers given here are those defined in Lumsdon (2000), and do not correspond to the site numbering scheme used in Figs 7 & 8.

mapped as a single unit in areas where they consist exclusively of non-marine facies (i.e. mainly to the west of the Alberta–British Columbia border).

The Dunvegan Formation has been divided broadly into prodelta, delta front and non-marine facies associations (Bhattacharya & Walker, 1991b; Plint, 1996, 2000; Plint et al., 2001). This study focuses on the coastal plain deposits of allomembers G through to A, which consist entirely of non-marine deposits (Figs 4a–d, 7 & 8).

SEDIMENTARY FACIES

Previous work

Bhattacharya (1989) and Bhattacharya & Walker (1991b) focused primarily on the marine delta-front, distributary and prodelta facies that characterize the more seaward portion of the Dunvegan delta complex, some 200–300 km south-east of the present outcrop study area (Fig. 3). They recognized a variety of non-marine facies deposited

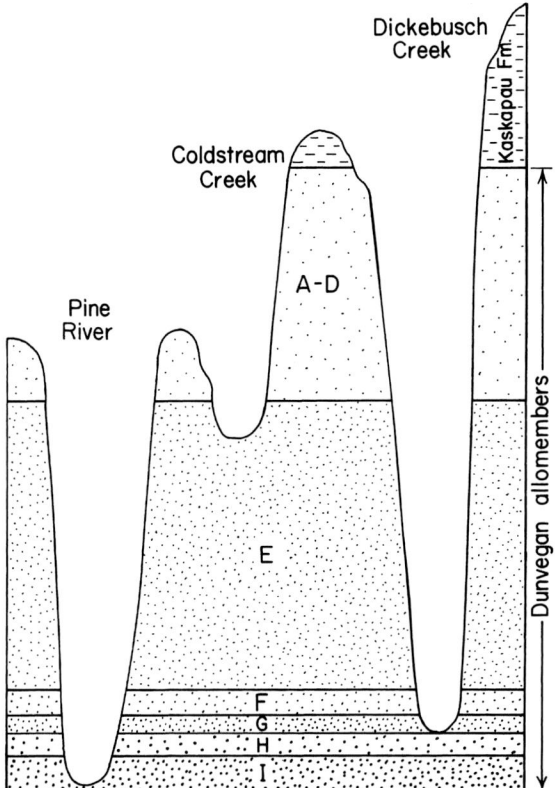

Fig. 6 Schematic cross-section to show the stratigraphical range of the various Dunvegan allomembers as they are exposed in Pine River, Coldstream Creek and Dickebusch Creek.

in lacustrine, lagoonal, floodplain and distributary channel settings. Plint (1996) described the Dunvegan Formation in outcrop sections some 200 km to the north of Bhattacharya's (1989) study area. There, a variety of coastal plain environments, including fluvial channels, lakes, floodplains, lagoons and valley-fills, were briefly documented. Micromorphological descriptions of Dunvegan floodplain palaeosols from sites in the Peace River Plains were provided by McCarthy & Plint (1999), who showed that they developed under generally poorly drained conditions, under a cool-temperate climate. McCarthy & Plint (1998, 2003), McCarthy et al. (1999) and Plint et al. (2001) examined outcrop sections at Kiskatinaw River, Beatton River and Clayhurst, located 50–100 km east of the localities described here (Fig. 3). In these more orogen-distal locations, a variety of floodplain facies, including channels, levees, crevasse deposits, valley-fills and, in particular, mature interfluve palaeosols, was documented. Lumsdon (2000) and this study focus on alluvial facies deposited in the most orogen-proximal localities available within the study area. Lumsdon (2000) recognized 11 facies, based on lithology, physical and biological structures that are distinguishable in the field. Brief descriptions and interpretations of these facies are given below. More comprehensive documentation of geometrical relationships will be presented elsewhere.

Alluvial plain and fluvio-lacustrine facies

1. Platy mudstone (lacustrine silt and mud)

This facies comprises laminated or platy mudstone, silty mudstone, or muddy siltstone that commonly forms units from a few decimetres to about 3 m thick, extending laterally for tens of metres to > 1 km. Units are commonly dark grey or brown, and may show black, brown and orange colour banding on a centimetre scale. Abundant fine-grained organic debris is ubiquitous. Rare freshwater molluscs, including ?corbulids, unionids and *Viviparus*, are present. In some places, the top few centimetres of units have symmetrical ripple cross-lamination in centimetre-scale beds of silty sandstone. Siderite spherules and millimetre- to centimetre-scale siderite nodules are occasionally observed. Units of this facies commonly incorporate poorly preserved, sand-filled dinosaur tracks at or near the top. The upper surface is occasionally blocky grey siltstone with root traces (Plate 1).

Interpretation. The symmetrical ripple cross-lamination, platy siltstone and mudstone, lack of bioturbation, abundance of terrestrial organic debris and occasional presence of roots and freshwater molluscs all suggest deposition in a shallow lake. The thickness and lateral extent of platy mudstone units, as well as their association with various palaeosol, crevasse splay, crevasse delta and mouth-bar facies (see below), are consistent with a lacustrine environment. Coleman (1966) described similar deposits from recent shallow delta-plain lakes ranging from < 20 km to < 1 km

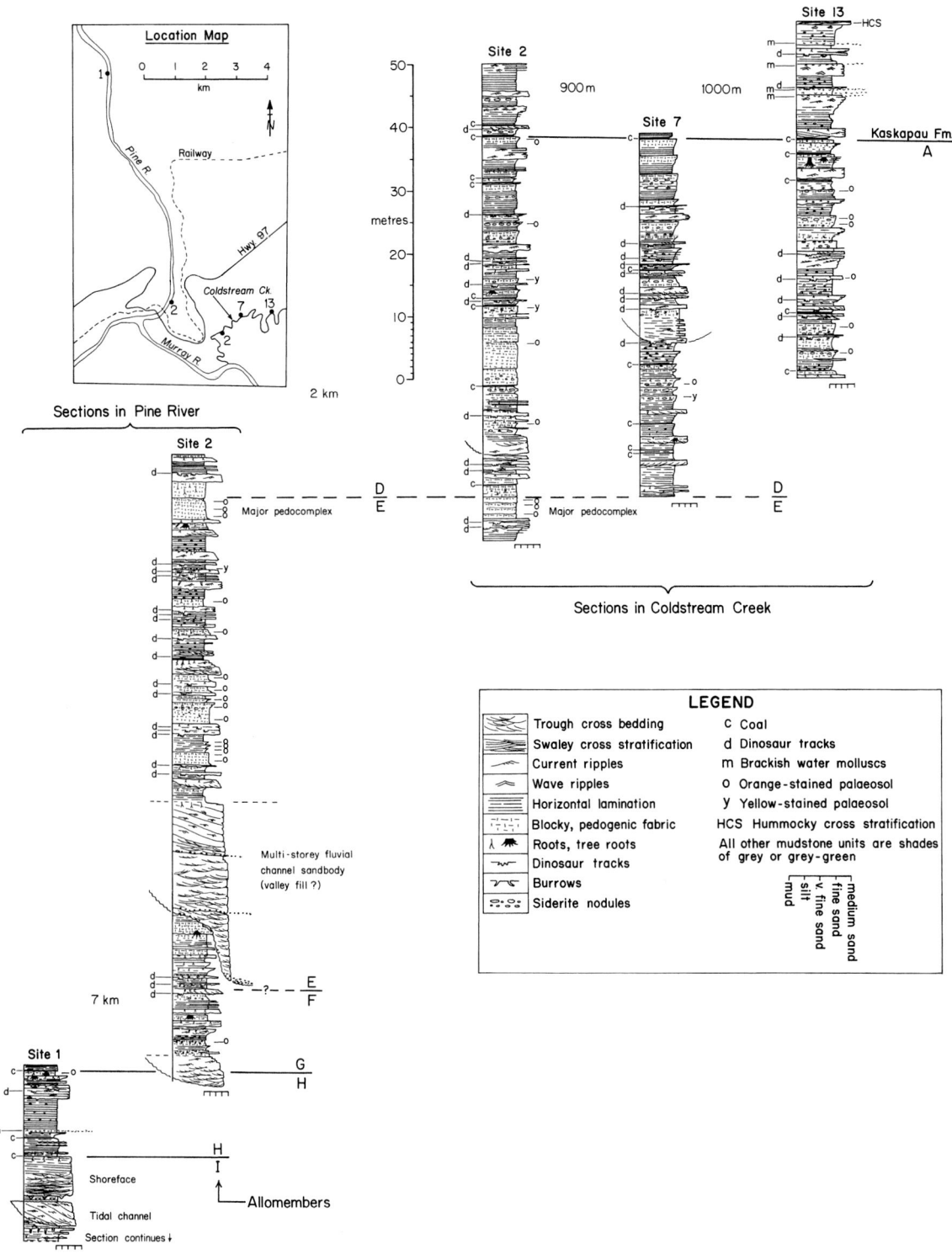

Plate 1 Facies 1, lacustrine mud. Laminated and thinly bedded, weakly sideritic grey mudstone sharply overlying blocky, grey-green rooted sandy siltstone, upon which the shotgun is resting. The base of the overlying sandstone is irregular as a result of dinoturbation.

Plate 2 Facies 1, lacustrine mudstone, grading up into laminated siltstone and sandstone (facies 2) of a crevasse delta/mouth bar. Assistant is holding a natural cast of a three-toed hadrosaur track from the sandstone beds at the top of the succession.

Plate 3 Overview of about 30 m of the upper part of Dickebusch Creek site 2 (Fig. 8), showing sharp-based, tabular crevasse-splay sandstones of facies 3. Note shallow, sandstone-filled crevasse channel at middle right. Blocky, pedogenic mudstones between sandstones show alternating grey and orange colours, interpreted to reflect subtle variations in palaeodrainage conditions.

Plate 4 Facies 7, intermediate-redox palaeosol, dominated by grey-green colour and weak blocky fabric. A large tree root is present immediately to the right of the ruler, with major marks at 10 cm intervals.

Plate 5 Facies 8, better drained palaeosol showing strong orange colour and well-developed blocky fabric. Top of palaeosol is just above assistant's hand, above which is a grey, poorly drained palaeosol containing large siderite nodules. The deformed layers at the bottom of photograph are the result of dinoturbation.

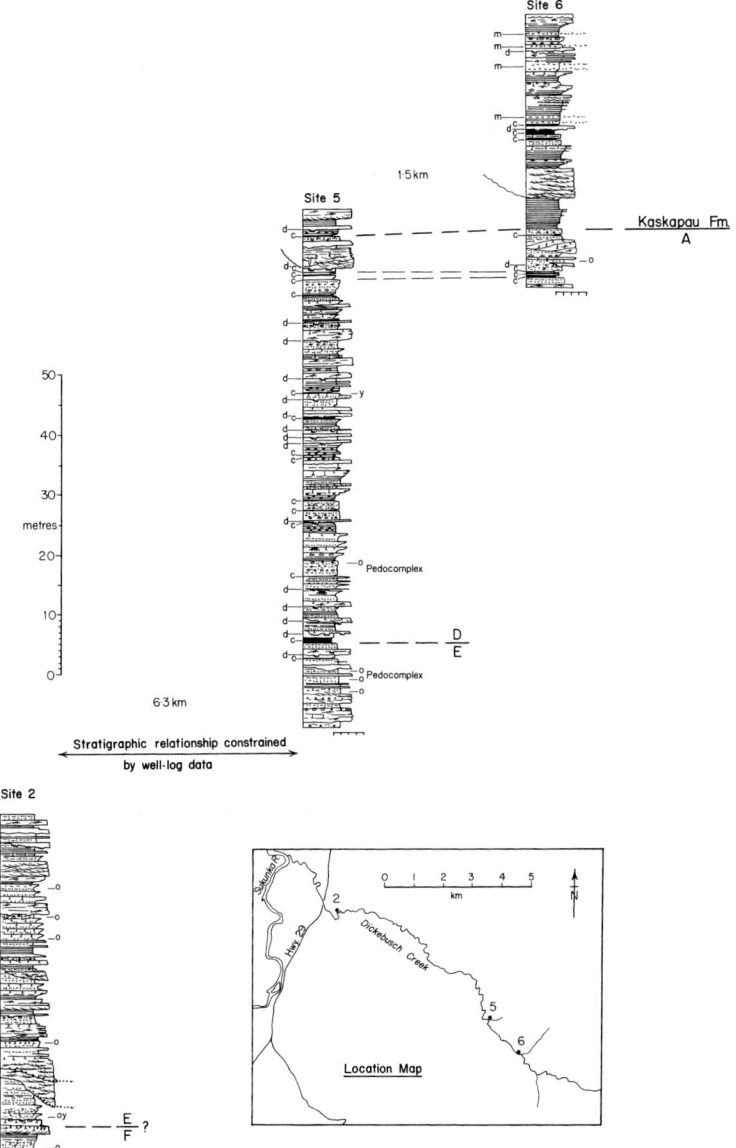

Fig. 8 Composite stratigraphical section from Dickebusch Creek, where inset map shows relative position of measured sections. Like the Pine River–Coldstream Creek section, allomembers G–E are dominated by facies association A, with lacustrine deposits more common in the upper half of allomember E. A major orange-stained pedocomplex marks the top of allomember E, which is overlain by almost entirely grey and grey-green rocks representing progressively more poorly drained environments. As in Coldstream Creek, coals are present only above the allomember E–D boundary. Legend as in Fig. 7.

Fig. 7 (*opposite*) Composite stratigraphical section from Pine River and Coldstream Creek. Inset location map shows relative position of the sections. Pine River site 1 exposes marine sediments of allomember I, overlain by lagoonal and lacustrine deposits of allomember H deposited only 10–20 km landward of the maximum marine transgressive shoreline (Plint, 2000). Overlying deposits of allomembers G–E and basal D are exposed in the vicinity of Pine River site 2. Note that coals (facies 5) are absent from this section, but become common above the top of allomember E in sections exposed along Coldstream Creek. The position of the Kaskapau Formation–Dunvegan allomember A boundary, and the boundaries of other Dunvegan allomembers are constrained by nearby well-log data, coupled with facies changes evident in the field. These sections provide the basis for the vertical facies variations illustrated in Fig. 9.

in diameter. Comparable facies, interpreted to represent lacustrine deposits, were also described by Smith & Smith (1980), Smith (1986), Smith & Perez-Arlucea (1994), McCarthy et al. (1999) and Perez-Arlucea & Smith (1999). These lakes probably formed when local subsidence and compaction lowered an area below the water table.

2. Sandier upward successions (crevasse delta or mouth-bar)

This facies comprises a sandier upward succession of very fine to fine, wave rippled sandstone beds with centimetre- to decimetre-scale mudstone and siltstone interbeds, and it always overlies the platy mudstone facies. Sandstone beds are 1–3 cm thick, thicken upward and contain planar and current-ripple cross-lamination. Siltstone interbeds are wave rippled, laminated and commonly contain leaf and wood impressions, and dinosaur tracks (Plate 2). Units range from 1 to 3 m thick, and extend laterally from < 100 m to about 1 km. Sandier upward successions locally contain sharp-based lenticular sandstones 0.5 to 1 m thick and about 2 m wide. The tops of the units sometimes contain root traces and are massive and/or dinoturbated (Lockley, 1991).

Interpretation. The sandier-up succession, vertical relationship with lacustrine silt and mud, and the interbedding of ripple cross-laminated and wave-rippled sandstone and siltstone indicates a gradual upward shallowing, probably owing to progradation of a lacustrine delta mouth-bar. The small sandstone-filled channels may represent components of a delta distributary system. Coleman (1966) and Perez-Arlucea & Smith (1997) described similar facies where rivers build small deltas into lakes.

3. Tabular or lenticular, sharp-based sandstone (crevasse splay)

This facies is highly variable in form, scale and features. Units occur as tabular or lenticular, very fine to fine-grained, yellow, orange or grey sandstones that lack vertical grain-size trends. Individual beds contain ripple and trough cross-stratification, occasional wave ripples and planar lamination. Upper surfaces are commonly irregular on centimetre to decimetre scales. Facies units range from < 10 cm to > 2 m thick, averaging 50 cm, and they extend laterally from tens to hundreds of metres (Plate 3). Beds commonly contain organic fragments and the tops are commonly rooted, with scattered *in situ* coalified tree stumps up to 30 cm wide. The base of each unit is usually sharp and locally scoured, commonly preserving dinosaur undertracks in the underlying floodplain mudstone. Units are commonly incised by decimetre to metre scale, non-migrating, sand-filled channels (Plate 3).

Interpretation. The sheet-like form and sharp base of these units, combined with the presence of a variety of current-formed structures are suggestive of non-channelized flow across a floodplain (Bridge, 1984; Ten Brinke et al., 1998). The presence of small channel-fills within the tabular body is suggestive of a crevasse-splay distributary system similar to that described by Smith & Perez-Arlucea (1994). Comparable crevasse-splay deposits have been described by Smith & Smith (1980), Bridge (1984), Flores & Hanley (1984), Smith (1986), Smith et al. (1989) and Nadon (1993).

4. Interbedded rooted silty sandstone and sandstone (levee)

This facies comprises centimetre- to decimetre-scale planar-bedded, current- and climbing-ripple cross-laminated fine sandstone with rare wave-ripple lamination. Sandstone beds are interstratified with siltstone and/or < 1 cm thick carbonaceous mudstone, and units generally become sandier upwards. Units are 0.5 to 3 m thick and extend laterally for tens to hundreds of metres. Roots are common throughout, and locally *in situ* coalified tree stumps up to 30 cm across are encountered. Locally the upper surfaces of units are indistinctly dinoturbated, or preserve discrete dinosaur tracks, and may be overlain by a thin carbonaceous layer. In places, units form 'wings' attached to large non-migrating sand-filled channels (described below).

Interpretation. The sheet-like geometry of individual beds within these units, combined with the presence of a variety of current-formed structures, especially climbing ripples, is suggestive of rapidly decelerating, non-channelized flow. The ubiquitous presence of roots suggests that units aggraded episodically, permitting repeated colonization by vegetation. Thin carbonaceous interbeds and carbonaceous mudstone on the upper surface of some units suggest the presence of forest or marsh litter that was subsequently buried by flooding. These features are all consistent with deposition on levees (cf. Smith & Smith, 1980; Flores & Hanley, 1984).

5. Coal (organic soil)

This facies is black with a texture varying between fissile (papery) to friable (blocky) or vitreous and ranges from 3 to 50 cm thick, averaging about 20 cm. Laterally coals extend for tens of metres to over 3 km, with thicker units tending to be more extensive. Units may contain detrital fragments of leaves, twigs, bark and fine-grained organic matter, together with a significant proportion of fine-grained clastic material, resulting in low-grade coal or coaly shale.

Interpretation. The high organic content is indicative of an organic soil/histosol that developed in a waterlogged area of relatively low clastic influx (Smith & Smith, 1980; Flores & Hanley, 1984; Leckie *et al.*, 1989; Ugolini & Spaltenstein, 1992; Retallack, 1997). The low grade of the coal nevertheless indicates that the swamp or marsh received a regular influx of clastic sediment from river floods.

6. Pale grey to grey sideritic mudstone (poorly drained soil)

This facies comprises massive to weakly blocky, pale grey to grey silty mudstone and muddy siltstone with minor reddish brown mottling. It contains centimetre-scale carbonaceous roots, plant debris, centimetre-scale 'smooth' siderite concretions and spherulitic siderite. Units are typically 30 to 50 cm thick and extend tens to hundreds of metres laterally.

Interpretation. The grey colour, spherulitic siderite, roots and plant debris are indicative of a reducing (poorly drained) or gleyed, vegetated environment (Retallack, 1990, 1997; McCarthy *et al.*, 1999). The weak blocky structure and mottling are indicative of incipient pedogenesis, consistent with a poorly drained or hydromorphic soil (Retallack, 1997; McCarthy *et al.*, 1999; Lumsdon, 2000). Lateral associations with intermediate-redox soil, coal and lacustrine silt and mud indicate that drainage conditions were variable over tens to hundreds of metres. The lateral variability is probably related to subtle variations in topography and water-table elevation (Aslan *et al.*, 1995).

7. Dark grey to greenish grey mottled blocky mudstone (intermediate redox soil)

This facies is the most common mudstone facies in the study area. It consists of weakly to strongly blocky-weathering, medium to dark grey to greenish-grey silty mudstone and muddy siltstone with orange, reddish-brown or light tan mottling. Some units are pale in the lower half. Facies units are typically 40 to 60 cm thick and extend laterally for tens to hundreds of metres (Plate 4). Carbonaceous roots on a millimetre- to centimetre-scale, and nodular 'smooth' and spherulitic siderite are widespread. Siderite spherules occur most commonly along root traces and are usually oxidized.

Interpretation. The blocky texture, presence of roots and oxidized siderite spherules are indicative of pedogenic processes above the water table (Besley & Fielding, 1989; Retallack, 1990; Mack *et al.*, 1993; McCarthy *et al.*, 1997). The abundant mottling, however, suggests iron migration under variable to poor drainage conditions (Mack *et al.*, 1993; Retallack 1990, 1997), and the presence of spherulitic siderite and the dark grey colour indicate reducing conditions below the water table. This combination of features suggests that the water table fluctuated. The weak development of peds and possibly seasonal changes in drainage conditions are consistent with an intermediate redox soil, showing features of both oxidation and reduction (Besley & Fielding, 1989; Retallack, 1990). McCarthy *et al.* (1999) attributed the formation of mottles to segregation of iron oxides as a result of wetting and drying over numerous redox cycles, and also to oxidation of earlier-formed siderite.

8. Orange to greenish grey blocky mudstone (better drained soil)

This facies comprises blocky-weathering, clay-rich muddy siltstone to silty mudstone with well-developed ped structure. Units are typically 30 to 60 cm thick, and extend tens to hundreds of metres laterally (Plate 5). Clay coatings, some with slickensides, surround the peds. The colour is typically brown or yellowish-brown to grey to greenish-grey with weakly developed colour horizons and a downward increase in clay and iron content (indicated by a gradual increase in reddening or mottling). Orange and reddish-brown mottles, varying from a few millimetres to a few centimetres diameter, and spherulitic siderite are present, but not common. Carbonaceous roots 2–20 mm in diameter and up to a decimetre in length are present locally, and insect (bee or beetle?) burrows were noted in one unit at Dickebusch Creek.

Interpretation. The blocky texture and well-developed peds are diagnostic of pedogenesis (Retallack, 1990). The presence of colour horizons indicates translocation of minerals, and the presence of clay coatings and overall clay enrichment relative to the other palaeosol facies indicates significant clay illuviation, and, perhaps, *in situ* weathering of the parent material. The warm colours (yellow, orange, brown) and deep rooting are indicative of a relatively well-drained, aerobic environment (Mack *et al.*, 1993; Retallack, 1990, 1997). These palaeosols, however, also contain features indicative of alternate wetting and drying, and of water logging, including slickensides, mottling and, in some units, spherulitic siderite. These features suggest that, overall, only moderately well-drained conditions prevailed, due in part to local topographic position (Kraus, 1987; McCarthy *et al.*, 1999).

Channel fills

9. Lenticular cross-bedded sandstone (non-migrating channel)

Lenticular, sharp-based, predominantly trough cross-stratified, very fine to medium-grained sandstone bodies range from 4.5 to 9 m thick and 10 to 200 m wide. Width/thickness ratios of fully exposed channel-fills are typically < 30, and occasionally < 10. The bases of sandbodies are commonly marked by intraclasts of mudstone, sideritic mudstone and carbonized plant fragments, and unit tops are commonly sharp. Subordinate sedimentary structures include ripple cross-lamination, planar lamination and wave ripples. Units are partitioned by subhorizontal internal erosion surfaces, sometimes mud-draped, separating decimetre scale packages of trough cross-sets. There are no vertical trends in either grain size or type and scale of sedimentary structure. Some units contain *in situ* coalified tree stumps up to 30 cm across. Sandstone bodies are typically spaced hundreds of metres to a few kilometres apart, and are encased in floodplain fines.

Interpretation. These isolated lenticular bodies of sandstone incised into floodplain facies are interpreted as non-migrating fluvial channels, and meet the sedimentary criteria of Nanson & Knighton (1996) for a type 1b anabranching river. The presence of mud drapes and internal discontinuities indicates that the competence of the flow was variable. The features of this facies are similar to other interpreted anastomosed fluvial facies described by Miall (1996), Smith (1983, 1986) and Perez-Arlucea & Smith (1999).

10. Inclined heterolithic stratification (migrating channel)

This facies comprises sharp-based, upward fining medium and fine sandstone to very fine sandstone and siltstone with mudstone drapes. Inclined heterolithic stratification (Thomas *et al.*, 1987) is a characteristic component of the facies. The length of inclined strata varies from about 30–50 m. Units are 2 to 7 m thick and tens to a few hundred metres in lateral extent. This facies was observed only above the allomember E–D boundary.

Interpretation. The sharp base and inclined heterolithic strata are suggestive of a laterally accreting point bar in a meandering channel (Smith, 1987; Thomas *et al.*, 1987). The heterolithic lithology indicates variable discharge, although a significant tidal influence seems unlikely, given the palaeogeographical setting, 60 to > 170 km landward of the most transgressive marine shoreline (Fig. 4d). The lengths of inclined accretion strata suggest channel widths of about 50–75 m, based on empirical relationships (Ethridge & Schumm, 1975).

11. Lenticular mudstone (non-migrating mud-filled channel)

This facies comprises sharp-based, symmetrical, lenticular mudstone bodies composed of laminated to blocky dark grey mudstone and siltstone. Units occasionally contain centimetre- to decimetre-scale beds of siltstone to very fine sandstone. The bodies are < 10 m wide and < 2 m thick. The upper contact is usually gradational. They are incised into, and enclosed by, floodplain palaeosols and crevasse splays of facies 2 and 3.

Interpretation. The lenticular morphology and mudstone lithology suggest that this facies represents the fill of small, suspended load streams on the floodplain. Channels may have been sufficiently distant from active trunk streams that they did not receive sand during floods.

ALLUVIAL RESPONSES TO ACCOMMODATION CHANGE

Vertical facies trends

Closely spaced vertical sections at the various study sites (Fig. 5) show that facies change laterally over horizontal distances of a few tens to hundreds of metres, probably reflecting subtle topographic variation related to autogenic factors such as channel proximity, local deposition rates and compaction. Vertical trends in facies are less obvious, except that there is an upward increase in the frequency and thickness of lacustrine lithologies, coals and poorly drained soils. In order to quantify this observation, a moving 3 m averaging window was passed over representative sections of each allomember in the Pine River–Coldstream Creek area. Sections that included channelized facies (9–11 above) were excluded from this analysis. The proportion of each of the non-channelized alluvial plain facies (1–8 above) was calculated for the window, then the window was moved 1 m up-section and the relative proportions were determined again.

This analysis revealed that the succession falls naturally into two non-channelized facies associations. Facies association A is dominated by crevasse splays, levees and better drained types of palaeosol (facies 3, 4, 7 and 8 above), whereas facies association B is dominated by lakes and crevasse deltas (facies 1 and 2 above), with subordinate poorly drained palaeosols (facies 5 and 6 above). The relative proportions of facies associations A and B were plotted against one another in a vertical profile, where the horizontal axis represents the relative proportions of associations A and B, and the vertical axis represents the thickness in metres (Fig. 9).

Palaeodrainage changes

Facies associations A and B reflect differences in the average position of the water table, relative to the sediment surface. Most deposits of association A were subaerially emergent for sufficient time to inhibit coal preservation, and for palaeosols to develop, although none of the palaeosols can be considered well-developed (Lumsdon, 2000; McCarthy & Plint, 1999). In contrast, facies of association B are dominantly subaqueous.

On an alluvial plain, the top of the accommodation envelope is approximated by the position of the flood-stage water table. Facies association A indicates, in broad terms, that accommodation on the alluvial plain was filled 'immediately' (on a geological time-scale), maintaining the sediment surface at or just above the average water table. Facies association B represents times when the accommodation rate was higher, when lakes represented accommodation that remained unfilled, perhaps for hundreds of years.

Figure 9 shows that the relative proportions of facies associations A and B vary dramatically through the approximately 1.5 Myr represented by the sections studied (Fig. 7). Overall, Fig. 9 shows that the accommodation rate, as reflected in the relative proportions of facies associations A and B, increased through the interval studied. Allomembers G, F and lower E are dominated by sandy crevasse splays, levees and palaeosols of facies association A. Palaeosols are dominantly intermediate redox or better drained, and

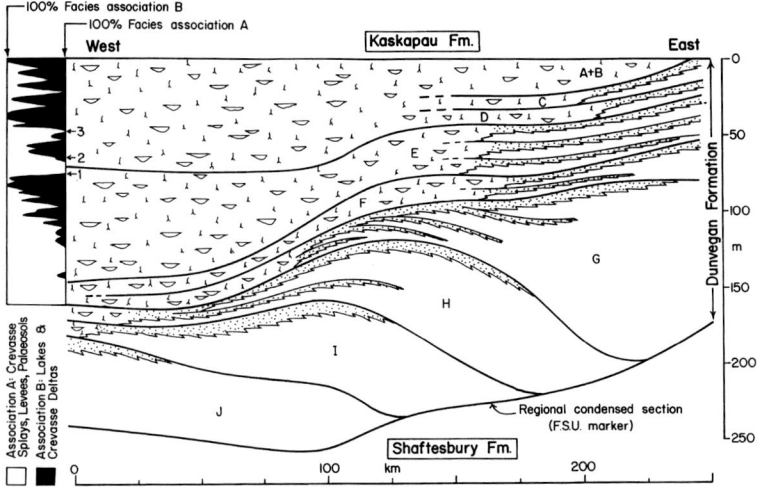

Fig. 9 Summary cross-section (drawn to scale, located in Fig. 1), illustrating the broad facies distribution and stratal geometry of the Dunvegan Formation in an E–W transect across the northern part of the study area. The datum for the section is the regional marine flooding surface at the top of Dunvegan allomember A. The western end of the cross-section lies close to the outcrop sections in Pine River and Coldstream Creek. The vertical changes in the relative proportions of well-drained facies association A, and poorly drained facies association B, as observed in those localities, is shown on the left side of the diagram. Note the steady upward increase in the proportion of the poorly drained to subaqueous facies association B (shown in black) through allomember E, up to arrow 1. This change is interpreted to record an increasing accommodation/supply ratio, controlled by accelerating subsidence. Between arrows 1 and 2, a thick pedocomplex suggests that the accommodation rate was relatively low. Above arrow 3, poorly drained floodplain and lake facies dominate, up to the top of allomember A, above which both fresh and brackish-water deposits are present (Fig. 7). This facies succession suggests a progressively increasing accommodation/supply ratio, which led eventually to marine transgression.

commonly amalgamate to form pedocomplexes. Collectively, these facies suggest a predominantly low accommodation rate. It is particularly noteworthy that coals are *absent* from this interval. Most channel facies are sandy and represent non-migrating streams, although some smaller, probably minor crevasse-splay distributary channels show limited lateral accretion. Well-preserved dinosaur tracks are common on the bases of most sandstone beds, and indistinct 'dinoturbation' is ubiquitous.

A dramatic facies change is observed in the upper half of allomember E (Fig. 9), in which sandy crevasse splays, levee deposits and better drained palaeosols give way to mouth-bar and lacustrine facies, the latter tending to form thicker units up-section. Channel sandbodies, where present, represent non-migrating river channels. Palaeosols represent better drained to intermediate redox varieties, although palaeosols become thinner and less well-developed up-section; coals are absent. Dinosaur tracks become increasingly better preserved up-section. In Pine River and Coldstream Creek, this wetter upward trend is reversed in the uppermost 5 m of allomember E by the reappearance of a strongly orange-stained pedocomplex, representing relatively well-drained palaeosols (Fig. 7). The top of the pedocomplex is sharp, and marked by a return to grey, poorly drained palaeosols. The *field* aspects of this pedocomplex (Fig. 7) closely resemble those of interfluve pedocomplexes described from allomember E in localities 50–100 km east of the present study site (McCarthy & Plint, 1998, 2003; McCarthy *et al*., 1999). The pedocomplex capping allomember E in Pine River has not yet been subject to micromorphological analysis, however, and hence the top surface cannot be interpreted confidently as an interfluve.

Allomembers D–A (Fig. 9) are dominated by the lacustrine mud and mouth-bar facies of association B, which alternates with thinner packages of facies association A. Palaeosols are dominantly poorly drained to intermediate redox, and in this interval coals were first deposited and are thicker and more closely spaced up-section. Well-preserved dinosaur tracks, mostly on the bottom of sandstone beds, are common throughout the D–A interval. Channel sandbodies represent both migrating and non-migrating varieties, with the latter type dominant.

In Dickebusch Creek and Coldstream Creek, the basal facies of the Kaskapau Formation (Figs 7 & 8) resemble those of the underlying allomember A of the Dunvegan Formation. They differ, however, in that they include laminated mudstones with more abundant wave-rippled sandstones, and also contain abundant brackish-water molluscs, including *Brachydontes*, *Ostrea* and *Corbula*, and serpulid worms. These basal units of the Kaskapau Formation form a succession 45 m thick that is transitional to fully marine mudstone and sandstone above (Plint, 2000, figs 4 & 5). These observations further corroborate the interpretation that accommodation rate progressively increased from allomember E time into the lower part of the Kaskapau Formation, by which time, brackish lagoons and finally open-water marine conditions prevailed across the study area. The distribution of coal in the Dunvegan sections studied appears to support the contention of Bohacs & Suter (1997) that conditions for optimum coal accumulation develop at some accommodation rate intermediate between 'low', where oxidation prevents organic accumulation, and 'high' when mires are inundated by water and/or siliciclastics. In the example considered here, the former condition pertained below the top of allomember E, whereas the latter condition tended to prevail in the basal part of the Kaskapau Formation. Only in allomembers D–A was the accommodation rate conducive to widespread coal accumulation.

Accommodation rate and regional isopach patterns

Plint (2000, figs 4 & 5) established a correlation between the regional subsurface well-log stratigraphy and the sections seen in outcrop on Dickebusch Creek, Pine River and Coldstream Creek. These outcrop sections are shown in more detail in Figs 7 & 8, and are crucial to the interpretations presented because they allow essentially one-dimensional, vertical facies changes observed in outcrop to be related to three-dimensional changes in the regional geometry of successive allomembers. These geometric changes are interpreted to reflect changes in tectonically controlled accommodation patterns.

Figure 4a–d illustrates the regional isopach patterns for allomembers G, F, E and A–D inclusive. Allomembers G and F thin up-dip towards the north-west, and isopach lines trend north-east–south-west, approximately perpendicular to the Rocky Mountain deformation front (Fig. 4a & b). These isopach patterns are also observed in the underlying allomembers J, I and H (Plint, unpublished data). The fact that isopach lines for allomembers J to F trend *perpendicular* to the orogenic margin of the basin suggests that, for these allomembers, accommodation for successive deltaic units was provided primarily by water-filled space in the pro-delta area of the underlying unit. Flexural subsidence, which would have resulted in isopach trends *parallel* to the orogenic margin, does not appear to have been significant during this interval. Plint *et al.* (2001) showed that the up-dip portions of allomembers H–F comprise thin, aggradational shelf deltas and coeval coastal plain deposits that correspond to transgressive and highstand systems tracts. During the falling stage and lowstand systems tracts, progressive valley incision resulted in up-dip sediment bypass and the development of well-developed interfluve palaeosols on those coastal plain deposits that formed interfluves between valleys (McCarthy *et al.*, 1999; Plint *et al.*, 2001; Plint, 2002).

The isopach map for allomember E (Fig. 4c), shows a dramatic change relative to older allomembers. A semi-circular zone of thickening about 120 km in radius is evident in the north-west corner of the study area, across which allomember E thickens from about 35 to 80 m. This semi-circular sedimentary 'moat', first recognized in allomember E, expands in allomembers D–A to a radius of about 250 km, across which allomembers D–A thicken from about 30 to 100 m (Fig. 4d). It is pertinent to note that, relative to allomembers J–E, the marine deltas in allomembers D–A (located in the south-east corner of the study area; Fig. 4d) are thin and sand-starved. This is interpreted to reflect sediment partitioning between alluvial and marine depocentres, with the balance between the two progressively shifting in favour of the alluvial realm, as the orogen-proximal subsidence rate increased through allomembers E to A, as discussed below.

Figure 9 shows that in Pine River, the facies that constitute allomembers G, F and the lower part of E can be interpreted in terms of a relatively low accommodation rate, in which sedimentation kept pace with accommodation. A low accommodation rate is also suggested by the up-dip thinning of the regional isopach patterns (Fig. 4a & b). The upward change in the proportion of subaerial to subaqueous facies through allomember E (Fig. 9) is interpreted, *on sedimentological grounds*, to indicate an upward increase in accommodation rate. This interpretation is supported by the thickening of allomember E in the north-west part of the study area, where our outcrop sections are located (Fig. 4c). The vertical facies changes in allomember E (Fig. 9) indicate that the increase in subsidence rate that produced the isopach 'moat' in the north-west corner of Fig. 4c took place relatively late in allomember E time. A return to low accommodation rate late in allomember E time (arrow 1 in Fig. 9) is indicated by the thick, relatively well-drained pedocomplex at the top of the unit (Fig. 7). In Dickebusch Creek (Fig. 8), allomembers H to E are also dominated by relatively well-drained facies of association A, although lacustrine-dominated facies are more common in the upper part of E at site 2. Like allomember E in Pine River, a thick, strongly orange-stained pedocomplex is present at the top of allomember E in site 5 (Fig. 8), suggesting a period of low accommodation rate.

In overall terms, allomembers D–A suggest continued high accommodation rates, interspersed with intervals of lower accommodation rate. Figure 9, based on exposures of allomembers D–A in Coldstream Creek (Fig. 7), shows relatively abrupt alternations between dominant facies associations on a vertical scale of 5–10 m. Unlike allomember E below, there is no clear, long-term trend in the relative proportions of the two facies associations, although six points can be made.
1 Coals appear in the D–A interval, and become thicker and more laterally extensive upwards.
2 The allomember E–D boundary is marked by an abrupt change from orange, relatively well-drained, to grey, relatively poorly drained palaeosols. Arrow 2 in Fig. 9 indicates an abrupt increase in the proportion of lacustrine-dominated, facies association B, suggesting an increase in accommodation rate, followed by a decline to arrow 3.
3 Between arrow 3 and the top of the Dunvegan Formation the overall trend is towards complete dominance by lacustrine facies, and ultimately, in the basal Kaskapau Formation, by brackish-water, lagoonal facies (Fig. 7).
4 Within the interval between arrow 3 and the top of Fig. 9, it is possible to discern six horizons at which facies association B is abruptly superimposed on facies association A. Each of these 'lacustrine flooding surfaces' might represent a rapid increase in accommodation rate (tectonic subsidence), followed by partial filling of the space created. Alternatively, they might reflect autogenic effects. Regardless of mechanism, the palaeosols remained poorly drained, rarely achieving the extent or maturity observed at the top of allomember E.
5 The appearance of brackish-water fauna a few metres above the Dunvegan–Kaskapau boundary (Figs 7 & 8) signifies a continued relative sea-level rise, when marine waters gained access to formerly freshwater environments. This suggests a continued increase in accommodation rate, to the point where sedimentation was no longer able to maintain the surface above sea level.
6 Despite the deteriorating drainage conditions, dinosaur tracks are widely observed on the bases of crevasse splay sandstones, on the top of mouth-bar successions, and even in association with brackish-water deposits. This observation indicates that these animals continued to inhabit the swampy coastal lowlands, despite a gradual transition to brackish-water conditions.

The points noted above for Coldstream Creek also apply to the D–A interval exposed in Dickebusch Creek (Fig. 8), although a quantitative analysis of the proportions of facies associations A and B was not undertaken for those sections.

Sequence-bounding unconformities

In the Kiskatinaw River, Beatton River and Clayhurst areas, 50–100 km east of the present study area, McCarthy & Plint (1998, 2003) and McCarthy *et al.* (1999) documented well-developed palaeosols on interfluves between valley systems (Plint, 2002). The micromorphological and textural characteristics of these palaeosols suggested that pedogenesis took place during a depositional hiatus of several tens of thousands of years (McCarthy et al., 1999). In the study area described here, Dunvegan allomembers H, G and F thin up-dip, and the top surface of each allomember is extensively incised by valleys (Plint, 2002). Valleys were interpreted by Plint (2002) to have been incised in response to both relative (possibly eustatic) sea-level changes of about 20 m, coupled with climatically driven changes in the ratio of sediment load to discharge. The existence of valleys implies: (i) that up-dip subsidence rates were low during deposition of allomembers H, G and F; and (ii) that interfluve surfaces should exist between valleys. On the basis of field evidence (but not micromorphological investigation), an interfluve palaeosol is tentatively identified at the top of allomember H at Pine River (Fig. 7). To date, it has not been possible to identify, in Pine River or Dickebusch Creek, interfluve palaeosols corresponding to the tops of allomembers G and F. This is partly because of poor exposure along critical parts of the Pine River, and the presence of major fluvial sandbodies that have cut out much of the section.

The remainder of the section in Pine River, however, encompassing allomembers E–A, is well-exposed. As noted above, the top 5 m of allomember E comprises a pedocomplex of better drained palaeosols which has a very sharp upper contact with more poorly drained palaeosols. Regional mapping (Plint, 2002) reveals extensive valley systems at the top of allomember E, and it is *possible* that the top of the pedocomplex represents an interfluve surface, with significant sedimentary hiatus. Lack of micromorphological information (such as documented by McCarthy et al., 1999), however, precludes definitive identification at present. Despite good exposure of allomembers D–A, no palaeosol or pedocomplex that can be demonstrated to represent an interfluve has been identified, despite the fact that contemporaneous valley-fills are known to exist about 100 km to the east of the study sites (Plint, unpublished data). Instead, the sedimentary succession indicates more-or-less continuous aggradation, albeit at varying rates.

This study demonstrates that interfluve surfaces (sequence-bounding unconformities) can develop, even in the relatively orogen-proximal part of a foreland basin, provided that the subsidence rate is sufficiently low. Interfluve surfaces, however, are not recognized in the upper part (allomembers D–A) of the Dunvegan Formation in sections close to the basin margin (Pine River–Chetwynd area). Posamentier & Allen's (1993) conceptual model of sequence-boundary development in a foreland basin is appealing in its simplicity, predicting the disappearance of subaerial unconformities with proximity to the foredeep. The vertical changes in the proportions of better drained to poorly drained alluvial facies, and of spatial and stratigraphical variation in the distribution of interfluve surfaces reported here, suggest that the development of a sequence-bounding unconformity at any point within the basin depends strongly upon relatively short-term (few hundred thousand years or less), spatial and temporal variations in subsidence rate, which here are postulated to have been linked to short-term variations in rates of thrust-sheet emplacement along the basin margin. The observations reported here suggest that it is unrealistic to assume temporally or spatially constant rates of subsidence in a foreland basin.

As the study area lies 60 to > 170 km landward of the nearest marine palaeoshoreline, it is unlikely that eustatic effects were an important influence on accommodation rate. A climatic influence on sedimentation is, however, difficult to dismiss. The length (> 330 km) of some Dunvegan valley systems led Plint (2002) to conclude that valley incision was in part controlled by changes in fluvial discharge, perhaps related to Milankovitch-band climatic cycles. It is therefore possible that climatic cycles also effected changes in the rate of sediment supply. Changes in the accommodation/supply ratio would be reflected in the resulting sedimentary facies. Nevertheless, it is concluded that the tectonically controlled accommodation rate was the major influence on the observed facies succession.

SUMMARY AND CONCLUSIONS

1 The Dunvegan Formation represents a delta complex that prograded from north-west to south-east, along the foredeep of the Western Canada foreland basin, parallel to the ancestral Rocky Mountains, over about 2 Myr. Deltaic and coeval coastal plain strata are up to about 240 m thick, and can be traced at least 400 km parallel to, and 300 km perpendicular to, the orogen.

2 Ten allomembers, J–A in ascending order, are defined by regional marine transgressive surfaces. In well-logs, marine transgressive intervals can be traced landward onto the coastal plain on the basis of a high gamma-ray and high resistivity response that represents lacustrine mudstone. In most areas, these lacustrine mudstones overlie interfluve palaeosols that separate valleys. In areas closest to the orogenic margin of the basin, discrete interfluve surfaces are not recognized in the upper four Dunvegan allomembers.

3 Isopach maps can be constructed for each of allomembers J–E, and for allomembers D–A collectively. Isopach maps show that the subsidence rate along the active margin of the basin was very low during deposition of allomembers J–F. In contrast, allomember E shows a semicircular isopach 'moat' in the north-west, with a radius of about 120 km, into which alluvial strata thicken from 35 to 80 m. Collectively, allomembers D–A also show pronounced thickening adjacent to the orogen, with a 'moat' about 250 km in radius, across which alluvial strata thicken from 30 to 100 m.

4 Near-complete outcrop sections in the north-west corner of the study area are dominated by alluvial plain facies, which were deposited within the isopach 'moats' described above. Non-channelized alluvial deposits include lake, mouth-bar, levee and crevasse splay facies, and organic histosol (coal), poorly drained, intermediate redox and better drained palaeosols. Channelized facies are dominated by non-migrating (probably anastomosed) sand-filled channels; laterally migrating meandering channels and non-migrating mud-filled channels are relatively rare. These sites of deposition lay about 60 to 170 km landward of the up-dip limits of contemporaneous transgressive marine shorelines.

5 Non-channelized facies fall into two associations. Facies association A is dominated by crevasse splays, levees, intermediate-redox and better drained palaeosols. Facies association B is dominated by lakes, mouth-bars, organic and poorly drained palaeosols. Facies association A dominates the alluvial succession in allomembers H–F, and the lower half and the uppermost 5–10 m of allomember E. Most of the upper half of allomember E is progressively dominated by facies association B. Allomembers D–A show alternations between facies associations A and B, although overall, facies association B predominates upward. The basal units of the overlying Kaskapau Formation comprise a transitional succession 40–50 m thick dominated by facies of association B, and also includes brackish-water lagoonal deposits. These marginal marine deposits are eventually overlain by open-marine deposits.

6 The up-dip (north-west) thinning of allomembers H–F suggests minimal accommodation in that direction. These units are dominated by alluvial facies (association A) deposited in relatively well-drained environments, in which accommodation space was filled as fast as it was created. The upward change within allomember E, from relatively well-drained facies association A to relatively poorly drained facies association B, is interpreted to record an upward increase in subsidence rate. This is independently suggested by the development of an isopach 'moat' in allomember E, attributed to renewed flexural sub-sidence of the basin margin. The presence of a relatively well-drained pedocomplex at the top of allomember E suggests a major decrease in accommodation rate, tentatively attributed to a period of local tectonic quiescence. Allomembers D–A are dominated by poorly drained freshwater environments, which grade upward, in the basal Kaskapau Formation, into brackish, marginal-marine and finally shallow-marine deposits. This succession of facies is developed within a broad isopach 'moat' that is interpreted to record progressively accelerating flexural subsidence through allomembers D–A, and continuing in the basal units of the Kaskapau Formation. The upward change to more subaqueous deposits, with lakes giving way to lagoons and finally marine deposits, reflects the changing ratio of accommodation to sedimentation over a period of about 1 Myr.

7 Valleys and interfluve palaeosols marking subaerial unconformities are well-developed 50 to > 100 km east of the study area. Within the study area, the top of allomember H appears to be marked by an interfluve palaeosol; allomembers G and F are poorly exposed and interfluves have not been recognized to date, although isopach maps predict their occurrence. The top of allomember E is marked by a thick pedocomplex, the top of which might represent an interfluve. Although allomembers D–A are well-exposed within the study area, candidates for interfluve palaeosols have not been recognized within this interval. This apparent lack of depositional hiatuses is attributed to higher rates of subsidence during deposition of these allomembers, an inference supported by their isopach patterns, and by widespread coal and lake deposits. The Dunvegan Formation shows that in a foreland basin the spatial and temporal distribution of subaerial sequence-bounding unconformities can change over relatively short time-scales (c. 100–200 kyr). This is likely to reflect corresponding spatial and temporal variations in tectonic activity along the active margin of the basin.

ACKNOWLEDGEMENTS

This research was funded by grants to AGP from the Natural Sciences and Engineering Research Council of Canada; additional logistical, technical and financial support was provided by British Petroleum (Canada), Canadian Hunter, Home Oil, Husky Energy, Pan Canadian Petroleum, Petro-Canada, Union Pacific Resources and Wascana Energy. We are very grateful to all these companies and institutions for their support. MPL acknowledges receipt of an NSERC postgraduate studentship, a Postgraduate Award from the Canadian Society of Petroleum Geologists, and support from Dr Gordon Heuser. Joyia Chakungal, Ubiratan Faccini, Michael Kreitner, Paul McCarthy, Chantale McIntosh, Jennifer McKay, Joanna Moore and Jennifer Wadsworth provided cheerful and energetic field assistance at various stages of this project. We are particularly grateful to Paul McCarthy for his help in characterizing and identifying palaeosols, and Martin Lockley and Richard McCrae for their identification of dinosaur and other vertebrate tracks. We thank Robert Brenner and Jeff Crabaugh for their reviews which significantly improved the manuscript.

REFERENCES

Aslan, A. and Autin, W.J. (1999) Evolution of the Holocene Mississippi River floodplain, Ferriday, Louisiana: Insights on the origin of fine-grained floodplains. *J. Sediment. Res.*, **69**, 800–815.

Aslan, A., Autin, W.J. and Törnqvist, T.E. (1995) *Holocene to Wisconsinan Sedimentation, Soil Formation, and Evolution of the Mississippi River Flood Plain, Southern Lower Mississippi Valley (LMV)*, pp. 59–93. Field Trip Guide Book No. 3, Geological Society of America Annual Meeting.

Beaumont, C. (1981) Foreland basins. *Geophys. J. Roy. Astronom. Soc.*, **65**, 291–329.

Besley, B.M. and Fielding, C.R. (1989) Palaeosols in Westphalian coal-bearing and red-bed sequences, central and northern England. *Palaeogeogr. Palaeoclimatol. Palaeoecol.*, **70**, 303–330.

Bhattacharya, J.P. (1989) Allostratigraphy and river- and wave-dominated deltaic sediments of the Upper Cretaceous (Cenomanian) Dunvegan Formation, Alberta. Unpublished PhD thesis, McMaster University, Hamilton, Ontario, 588 pp.

Bhattacharya, J.P. and Walker, R.G. (1991a) Allostratigraphic subdivision of the Upper Cretaceous Dunvegan, Shaftesbury and Kaskapau formations in the north-western Alberta subsurface. *Bull. Can. Petrol. Geol.*, **39**, 145–164.

Bhattacharya, J.P. and Walker, R.G. (1991b) River- and wave-dominated depositional systems of the Upper Cretaceous Dunvegan Formation, north-western Alberta. *Bull. Can. Petrol. Geol.*, **39**, 165–191.

Blakey, R.C. and Gubitosa, R. (1984) Controls of sandstone body geometry and architecture in the Chinle Formation (Upper Triassic), Colorado Plateau. *Sediment. Geol.*, **38**, 51–86.

Bohacs, K. and Suter, J. (1997) Sequence stratigraphic distribution of coaly rocks: fundamental controls and paralic examples. *Am. Assoc. Petrol. Geol. Bull.*, **81**, 1612–1639.

Bridge, J.S. (1984) Large scale facies sequences in alluvial overbank environments. *J. Sediment. Petrol.*, **54**, 583–588.

Burns, B.A., Heller, P.L., Marzo, M. and Paola, C. (1997). Fluvial response in a sequence stratigraphic framework: Example from the Montserrat fan delta, Spain. *J. Sediment. Res.*, **B67**, 311–321.

Cant, D.J. (1998) Sequence stratigraphy, subsidence rates, and alluvial facies, Mannville Group, Alberta foreland basin. In: *Relative Role of Eustasy, Climate and Tectonism in Continental Rocks* (Eds K.W. Shanley and P.J. McCabe). *Soc. Econ. Paleontol. Mineral. Spec. Publ.*, **59**, 49–63.

Coleman, J.M. (1966) Ecological changes in a massive freshwater clay sequence. *Trans. Gulf Coast Assoc. Geol. Soc.*, **16**, 159–174.

Ethridge, F.G. and Schumm, S.A. (1975) Reconstructing paleochannel morphologic and flow characteristics: methodology, limitations and assessment. In: *Fluvial Sedimentology* (Ed. A.D. Miall). *Can. Soc. Petrol. Geol. Mem.*, **5**, 703–721.

Ethridge, F.G., Wood, L.J. and Schumm, S.A. (1998) Cyclic variables controlling fluvial sequence development: problems and perspectives. In: *Relative Role of Eustasy, Climate and Tectonism in Continental Rocks* (Eds K.W. Shanley and P.J. McCabe). *Soc. Econ. Paleontol. Mineral. Spec. Publ.*, **59**, 17–29.

Flores, R.M. and Hanley, J.H. (1984) Anastomosed and associated coal-bearing fluvial deposits: Upper Tongue River Member, Paleocene Fort Union Formation, northern Powder River Basin, Wyoming, U.S.A. In: *Sedimentology of Coal and Coal-bearing Sequences* (Eds R.A. Rahmani and R.M. Flores). *Spec. Publ. Int. Assoc. Sedimentol.*, **7**, 85–103.

Hampson, G., Stollhofen, H. and Flint, S.S. (1999) A sequence stratigraphic model for the Lower Coal Measures (Upper Carboniferous) of the Rhur district, north-west Germany. *Sedimentology*, **46**, 1199–1231.

Hovius, N. (1998) Controls on sediment supply by large rivers. In: *Relative Role of Eustasy, Climate and Tectonism in Continental Rocks* (Eds K.W. Shanley and P.J. McCabe). *Soc. Econ. Paleontol. Mineral. Spec. Publ.*, **59**, 3–16.

Kauffman, E.G. and Caldwell, W.G.E. (1993) The Western Interior Basin in space and time. In: *Evolution of the Western Interior Basin* (Eds W.G.E. Caldwell and E.G. Kauffman). *Geol. Assoc. Can. Spec. Pap.*, **39**, 1–30.

Kraus, M.J. (1987) Integration of channel and floodplain suites, II. Vertical relations of alluvial paleosols. *J. Sediment. Petrol.*, **57**, 602–612.

Kraus, M.J. and Middleton, L.T. (1987) Contrasting architecture of two alluvial suites in different structural settings. In: *Recent Developments in Fluvial Sedimentology* (Eds F.G. Ethridge, R.M. Flores and M.D. Harvey). *Soc. Econ. Paleontol. Mineral. Spec. Publ.*, **39**, 253–262.

Leckie, D., Fox, C. and Tarnocai, C. (1989) Multiple palaeosols of the late Albian Boulder Creek Formation, British Columbia, Canada. *Sedimentology*, **36**, 307–323.

Legarreta, L. and Uliana, M.A. (1998) Anatomy of hinterland depositional sequences: Upper Cretaceous fluvial strata, Nequen Basin, West-Central Argentina. In: *Relative Role of Eustasy, Climate and Tectonism in Continental Rocks* (Eds K.W. Shanley and P.J. McCabe). *Soc. Econ. Paleontol. Mineral. Spec. Publ.*, **59**, 83–92.

Lockley, M. (1991) *Tracking dinosaurs: a New Look at an Ancient World*. Cambridge University Press, Cambridge, 238 pp.

Lumsdon, M.P. (2000) Alluvial architecture and paleoenvironments in a proximal foreland basin setting: Upper Cretaceous (Cenomanian) Dunvegan Formation, B.C. MSc thesis, University Western Ontario, London, 299 pp.

Mack, G.H., James, W.C. and Monger, H.C. (1993) Classification of paleosols. *Geol. Soc. Am. Bull.*, **105**, 129–136.

Martinsen, O.J., Ryseth, A., Helland-Hansen, W., Flesche, H., Torkildsen, G. and Idil, S. (1999) Stratigraphic base level and fluvial architecture: Ericson Sandstone (Campanian), Rock Springs Uplift, SW Wyoming, USA. *Sedimentology*, **47**, 235–259.

McCarthy, P.J. and Plint, A.G. (1998) Recognition of interfluve sequence boundaries: Integrating paleopedology and sequence stratigraphy. *Geology*, **26**, 387–390.

McCarthy, P.J. and Plint, A.G. (1999) Floodplain palaeosols of the Cenomanian Dunvegan Formation, Alberta and British Columbia, Canada: Micromorphology, pedogenic processes and palaeoenvironmental implications. In: *Floodplains: Interdisciplinary Approaches* (Eds S. Marriott and J. Alexander). *Geol. Soc. Lond. Spec. Publ.*, **163**, 289–310.

McCarthy, P.J. and Plint, A.G. (2003) Spatial variability of palaeosols across Cretaceous interfluves in the Dunvegan Formation, NE British Columbia, Canada; palaeohydrological, palaeomorphological and stratigraphic implications. *Sedimentology*, **50**, 1187–1220.

McCarthy, P.J., Martini, I.P. and Leckie, D.A. (1997) Anatomy and evolution of a Lower Cretaceous alluvial plain: sedimentology and palaeosols in the Upper Blairmore Group, south-western Alberta, Canada. *Sedimentology*, **44**, 197–220.

McCarthy, P.J., Faccini, U.F. and Plint, A.G. (1999) Evolution of an ancient coastal plain: palaeosols, interfluves and alluvial architecture in a sequence stratigraphic framework, Cenomanian Dunvegan Formation, NE British Columbia, Canada. *Sedimentology*, **46**, 861–891.

McLaurin, B.T. and Steel, R.J. (2000) Fourth-order nonmarine to marine sequences, middle Castlegate Formation, Book Cliffs, Utah. *Geology*, **28**, 359–362.

Miall, A.D. (1996) *The Geology of Fluvial Deposits: Sedimentary Facies, Basin Analysis, and Petroleum Geology*. Springer-Verlag, Berlin, 582 pp.

Monger, J.W.H. (1993) Cretaceous tectonics of the North American Cordillera. In: *Evolution of the Western Interior Basin* (Eds W.G.E. Caldwell and E.G. Kauffman). *Geol. Assoc. Can. Spec. Pap.*, **39**, 31–47.

Nadon, G.C. (1993) The association of anastomosed fluvial deposits and dinosaur tracks, eggs, and nests: implications for the interpretation of floodplain environments and a possible survival strategy for ornithopods. *Palaios*, **8**, 31–44.

Nanson, G.C. and Knighton, A.D. (1996) Anabranching Rivers: their cause, character, and classification. *Earth Surf. Process. Landf.*, **21**, 217–239.

Olsen, H. (1990) Astronomical forcing of meandering river behaviour: Milankovitch cycles in the Devonian of East Greenland. *Palaeogeogr. Palaeoclimatol. Palaeoecol.*, **79**, 99–115.

Olsen, H. (1994) Orbital forcing on continental depositional systems—lacustrine and fluvial cyclicity in the Devonian of East Greenland. In: *Orbital Forcing and Cyclic Sequences* (Eds P.L. de Boer and D.G. Smith). *Spec. Publ. Int. Assoc. Sediment.*, **19**, 429–438.

Pérez-Arlucea, M. and Smith, N.D. (1997) Crevasse-splay and levee development following the avulsion of the Saskatchewan River (Cumberland Marshes, Canada). *18th IAS Regional Meeting of Sedimentolgy*, Heidelberg, 2–4 September, Abstract Vol., 269–270.

Pérez-Arlucea, M. and Smith, N.D. (1999) Depositional patterns following the 1870's avulsion of the Saskatchewan River (Cumberland Marshes, Saskatchewan, Canada). *J. Sediment. Res.*, **69**, 62–73.

Plint, A.G. (1996) Marine and non-marine systems tracts and fourth-order sequences in the Early–Middle Cenomanian Dunvegan Alloformation, northeastern British Columbia, Canada. In: *High-resolution Sequence Stratigraphy: Innovations and Applications* (Eds J.A. Howell and J.F. Aitken). *Geol. Soc. Lond. Spec. Publ.*, **104**, 159–191.

Plint, A.G. (2000) Sequence stratigraphy and paleogeography of a Cenomanian delta complex: the Dunvegan and Lower Kaskapau formations in subsurface and outcrop, Alberta and British Columbia, Canada. *Bull. Can. Petrol. Geol.*, **48**, 43–79.

Plint, A.G. (2002) Paleo-valley systems in the Upper Cretaceous Dunvegan Formation, Alberta and British Columbia. *Bull. Can. Petrol. Geol.*, **50**, 277–296.

Plint, A.G., McCarthy, P.J. and Faccini, U.F. (2001) Nonmarine sequence stratigraphy: Up-dip expression of sequence boundaries and systems tracts in a high-resolution framework, Cenomanian Dunvegan Formation, Alberta Foreland Basin, Canada. *Am. Assoc. Petrol. Geol. Bull.*, **85**, 1967–2001.

Posamentier, H.W. and Allen, G.P. (1993) Variability of the sequence stratigraphic model: effects of local basin factors. *Sediment. Geol.*, **86**, 91–109.

Retallack, G.J. (1990) *Soils of the Past: an Introduction to Paleopedology.* Unwin Hyman, Winchester, 520 pp.

Retallack, G.J. (1997) *A Colour Guide to Palaeosols.* Wiley, Chichester, 175 pp.

Rogers, R.R. (1998) Sequence analysis of the Upper Cretaceous Two Medicine and Judith River formations, Montana: nonmarine response to the Claggett and Bearpaw marine cycles. *J. Sediment. Res.*, **68**, 615–631.

Shanley, K.W. and McCabe, P.J. (1991) Predicting facies architecture through sequence stratigraphy—an example from the Kaiparowits Plateau, Uath. *Geology*, **19**, 742–745.

Shanley, K.W. and McCabe, P.J. (1993) Alluvial architecture in a sequence stratigraphic framework: a case history from the Upper Cretaceous of southern Utah, U.S.A. In: *The Geological Modelling of Hydrocarbon Reservoirs and Outcrop Analogues* (Eds S.S. Flint, and I.D. Bryant). *Spec. Publ. Int. Assoc. Sedimentol.*, **15**, 21–56.

Smith, D.G. (1983) Anastomosed fluvial deposits: modern examples from Western Canada. In: *Modern and Ancient Fluvial Systems* (Eds J. Collinson and J. Lewin). *Spec. Publ. Int. Assoc. Sediment.*, **6**, 155–168.

Smith, D.G. (1986) Anastomosing river deposits, sedimentation rates and basin subsidence, Magdalena River, northwestern Colombia, South America. *Sediment. Geol.*, **46**, 177–196.

Smith, D.G. (1987) Meandering river point bar lithofacies models: Modern and ancient examples compared. In: *Fluvial Sedimentology* (Eds F.G. Ethridge, R.M. Flores and M.D. Harvey). *Soc. Econ. Paleont. Mineral. Spec. Publ.*, **39**, 83–91.

Smith, D.G. and Smith, N.D. (1980) Sedimentation in anastomosed river systems: Examples from alluvial valleys near Banff, Alberta. *J. Sediment. Petrol.*, **50**, 157–164.

Smith, N.D. and Pérez-Arlucea, M. (1994) Fine-grained splay deposition in the avulsion belt of the lower Saskatchewan River, Canada, *J. Sediment. Res.*, **B64**, 159–168.

Smith, N.D., Cross, T.A., Dufficy, J.P. and Clough, S.R. (1989) Anatomy of an avulsion. *Sedimentology*, **36**, 1–23.

Ten Brinke, W.B.M., Schoor, M.R., Sorber, A.M. and Berendsen, H.J.A. (1998) Overbank sand deposition in relation to transport volumes during large-magnitude floods in the Dutch sand-bed Rhine River system. *Earth Surf. Process. Landf.*, **23**, 809–824.

Thomas, R.G., Smith, D.G., Wood, J.M., Visser, J., Calverley-Range, E.A. and Koster, E.H. (1987) Inclined heterolithic stratification—terminology, description, interpretation and significance. *Sediment. Geol.*, **53**, 123–179.

Törnqvist, T.E. (1993) Holocene alternation of meandering and anastomosing fluvial systems in the Rhine–Meuse Delta (central Netherlands) controlled by sea-level rise and subsoil erodibility. *J. Sediment. Petrol.*, **63**, 683–693.

Törnqvist, T.E., Wallinga, J., Murray, A.S., de Wolf, H., Cleveringa, P. and de Gans, W. (2000) Response of the Rhine–Meuse system (west-central Netherlands) to the last Quaternary glacio-eustatic cycles: a first assessment. *Global Planet. Change*, **27**, 89–111.

Ugolini, F.C. and Spaltenstein, H. (1992) Pedosphere. In: *Global Biogeochemical Cycles* (Ed. S.S. Butcher), pp. 123–148. Academic Press.

A new evaluation of fining upward sequences in a mud-rock dominated succession of the Lower Old Red Sandstone of South Wales, UK

SUSAN B. MARRIOTT*, V. PAUL WRIGHT†
and BRIAN P.J. WILLIAMS‡

*School of Geography and Environmental Management, University of the West of England,
Bristol BS16 1QY, UK (Email: susan.marriott@uwe.ac.uk);
†Department of Earth Sciences, Cardiff University, Cardiff, CF10 3YE and BG Group, 100 Thames Valley
Park Drive, Reading RG6 1PT, UK; and
‡Department of Geology and Petroleum Geology, University of Aberdeen, King's College, Aberdeen
AB24 3UE, UK

ABSTRACT

As much as 80% of the 4.5-km-thick succession of alluvial sediments of late Silurian to early Devonian age in South Wales is composed of mudstone, generally regarded as overbank suspension deposits. Studies of selected formations reveal a variety of lithofacies associations in these mud-dominated intervals, which suggest diverse mechanisms for mud emplacement. For example, there are distinct fining upward units, 2.5–5 m thick, that represent ephemeral channel-zone deposits in which extensive reworking of palaeo-Vertisols took place as clay pellet bedload material. The overall depositional system has many similarities with that described from the Channel Country of central Australia, where anastomosing fluvial systems are associated with distal mud sheetflood deposits. The exact channel planforms for the Old Red Sandstone systems, however, cannot be readily determined, although where larger channel sandbodies exist they have a sheet-like occurrence with a high width/depth ratio, and possibly may be the result of large-scale, low-frequency sheet-flood events. The development of palaeosols with vertic features, calcretes, alternations of desiccation and burrowing in the channel deposits suggests marked seasonality with a flashy regime. The Channel Country represents a fairly stable low-relief cratonic interior, whereas the Old Red Sandstone in South Wales was deposited on a coastal-plain in a subsiding basin setting. There are, however, similarities in terms of depositional products despite the disparity in morpho-tectonic setting, which may reflect extremely low gradients and climatic influences such as frequency and magnitude of flood events.

INTRODUCTION

Regionally, the Lower Old Red Sandstone (ORS) of the Anglo-Welsh Basin is a coarsening upward, mainly continental basin-fill sequence, representing the Upper Silurian (Pridolian) to Lower Devonian (Emsian) (Friend et al., 2000). The stratigraphy (Fig. 1) was described originally by Dixon (1921), with later detailed work on the sedimentology by Allen (e.g. 1963, 1974), Allen & Williams (e.g. 1978, 1979) and Williams et al. (1982). The Lower ORS sequences of the

Pembroke Peninsula, South Wales, were assigned to three formations of the Milford Haven Group (Williams et al., 1982) (Fig. 1). The Moor Cliffs and Freshwater West Formations have been the subject of detailed study more recently. For example, Love & Williams (2000) described vertical and lateral variations in alluvial architecture in the Moor Cliffs Formation and Marriott & Wright (1993) used features in the two upper formations of the Milford Haven Group to demonstrate the dynamic behaviour of geomorphological surfaces during deposition.

Locally, in south-west Pembrokeshire, the Lower ORS is a sequence of red beds up to 4.5 km thick comprising mainly alluvial sediments, with occasional tuff horizons (Figs 1 & 2) that act as regional stratigraphical markers (e.g. Townsend Tuff Bed, Allen & Williams, 1981, 1982). The Milford Haven Group is dominated by fine sediments, mainly mudstones, that display varying stages of palaeosol development (Allen, 1974; Wright & Marriott, 1996). Much previous work, however, has concentrated on the features in the sandstones that form a much smaller proportion of the succession (< 25%). In this paper, the results of some preliminary work that aims eventually to interpret and characterize the mudrock facies in the Lower ORS are presented as an example of an ancient, mud-dominated dryland deposystem.

The sedimentary succession discussed here is the Rat Island Mudstone that forms the upper Member of the Freshwater West Formation and is mid- to late Lochkovian in age (Figs 1 & 2). The Rat Island Mudstone is dominated by brownish red to purple mudstone–siltstone, with relatively thin

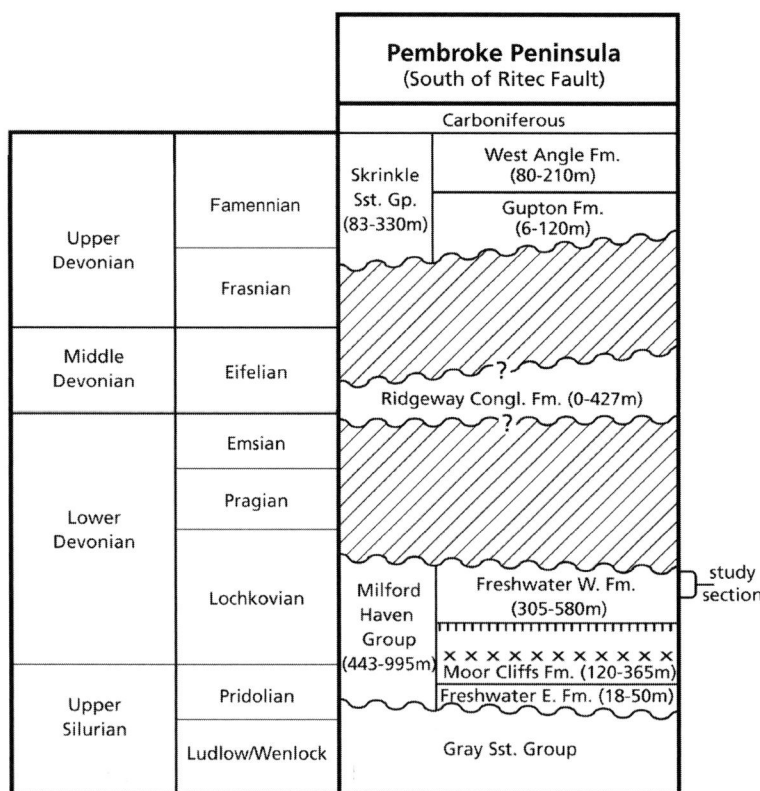

Fig. 1 The stratigraphy of the Pembroke Peninsula. (After Williams et al., 1982.)

Plate 1 The Rat Island Mudstone at Manorbier. The 25-m-thick sequence illustrated is represented by Fig. 3 and youngs to the left.

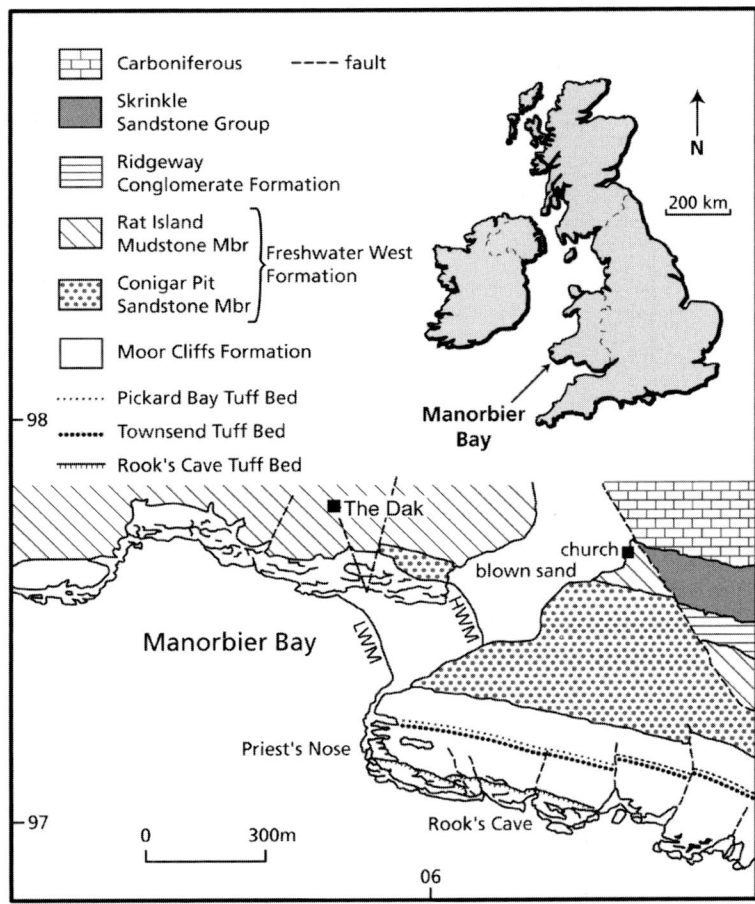

Fig. 2 Geology of the study area, Manorbier Bay, South Wales.

(< 2 m), sheet-like sandstones. The mudstones have been extensively bio- and pedoturbated with carbonate concretions locally abundant (Plate 1). Earlier work identified a strong cyclicity of fining upward units in the Rat Island Mudstone and interpreted the sequence as being deposited by ephemeral stream floods in a more or less continuously aggrading environment (Marriott & Wright, 1993). Soil development in this succession is mainly in simple or composite palaeosol profiles. Wright & Marriott (1996) compared the stage of development found here with that in other ORS sequences to erect a model that linked palaeosol maturity to frequency and magnitude of flood events.

Extensive exposure (c. 1 km) of the Rat Island Mudstone was examined over a lateral distance of c. 0.5 km and vertical thickness of 50–70 m on the western side of Manorbier Bay, Pembrokeshire (Fig. 2). Figure 3 is a representative section of part of the Rat Island Mudstone taken from a longer section measured on the foreshore below The Dak (National Grid Reference SS 0580 9775) (see also Plate 1). The section can be divided into fining upward cycles containing four distinctive facies that are characteristic of a wide range, geographically and stratigraphically, of mud-dominated Lower ORS successions in South Wales. Eight such cycles (I–VIII) are shown on Fig. 3.

FACIES DESCRIPTIONS

Facies 1—conglomerate/sandstone

These are intraformational conglomerates and sandstones. The conglomerates predominantly

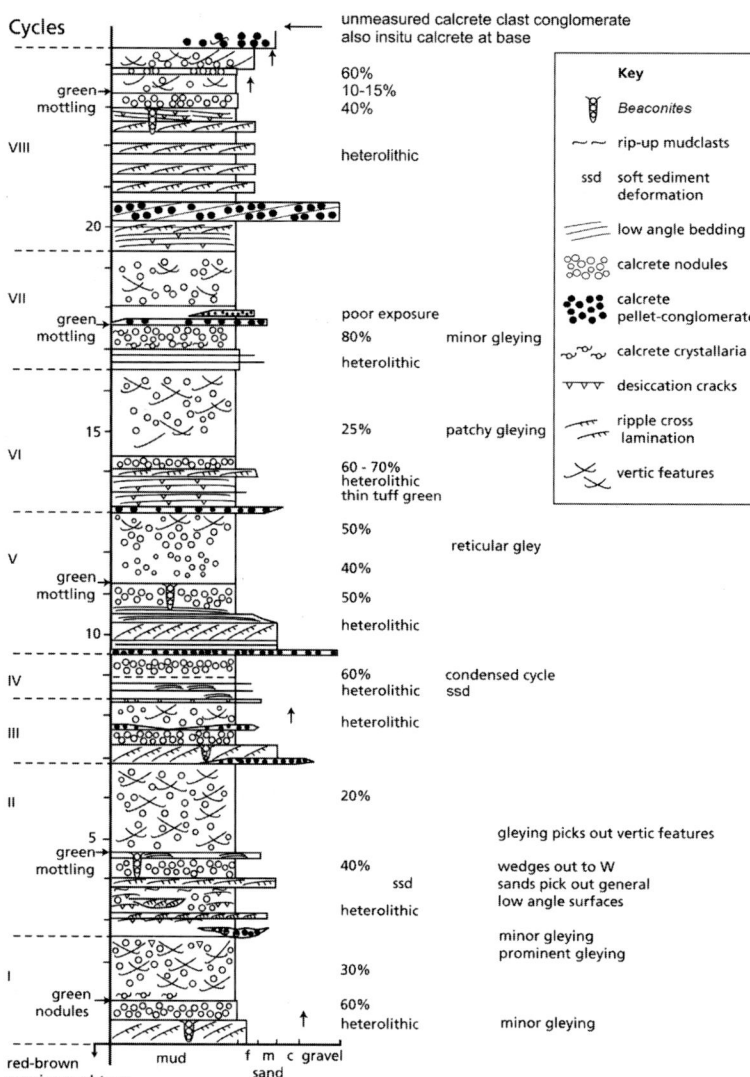

Fig. 3 Representative section of the Rat Island Mudstone, selected from a larger section measured on the foreshore below The Dak (National Grid Reference SS 0580 9775). Percentages refer to amount of carbonate nodules present. Vertical scale is in metres.

contain reworked calcrete (pedogenic carbonate) and mudstone clasts generally less than 10 mm in diameter. They are mainly clast-supported, with a red mudstone matrix and range from 0.05 to 0.50 m thick. They form thin lenses at the base of, or within, the sandstones, or are found as more continuous sheets where thin pulses of conglomerate pick out low-angled accretion surfaces. The sandstones are also sheet-like and generally thin, ranging from only 400 to 600 mm thick. They are red and are composed of fine- to medium-grained, micaceous sand with planar lamination and ripple cross-lamination. The sandstones and conglomerates form only a small proportion of the sequence examined (< 25%) and in many cases they pinch out laterally over distances of less than 100 m, particularly the conglomerates. Thicker units, e.g. cycle V on Fig. 3, tend to show a fining upward sequence. This facies has an erosive base with distinct lower boundaries but, despite the occurrence of rip-up mud clasts in the basal sediment, there is no evidence of channelized scours.

Facies 2—heterolithic

This facies is represented by heterolithic, interbedded red sandstones and mudstones, and occurrences range from 0.05 to 3.5 m thick. The sandstones are thin, mainly composed of very fine-grained sand and planar laminated or current ripple cross-laminated. These heterolithic units have flat bases and sometimes overlie a thicker, coarser sandstone or conglomerate of facies 1. They tend to form sequences with a stratification surface at a low angle (c. 3°) to the horizontal and some of the sandier beds within these units wedge out laterally over distances of 10–20 m. Some reveal rip-up clasts of mudstone at their bases, or contain thin lenses of intraformational conglomerate as described above. In some horizons loading structures press down into the mudstone beneath.

The interbedded units are laminated mudstones that commonly contain desiccated horizons, picked out by sand-filled desiccation polygon cracks on exposed surfaces or infilled cracks that can be viewed in vertical section. Carbonate nodules are also present in some units. The nodules tend to be horizontally elongated and slightly distort the laminar structure around them. No diagnostic pedogenic features occur at these levels and preservation of lamination in the mudstones suggests passive cementation by carbonate.

Both sandstones and mudstones within these heterolithic deposits are bioturbated. There are small vertical and horizontal burrows, ranging in size from 4 to 10 mm in diameter, which can be picked out on exposed surfaces by an apparent accumulation of mica. There are also frequent large-scale burrows (up to 0.15 m diameter) that occasionally have a meniscus fill, picked out by tiny clasts of reworked calcrete. These extend vertically through the heterolithic beds for distances up to 1.5 m and occasionally are found with horizontal extensions at the base, running along unit boundaries. Similar trace fossils of Devonian age in Antarctica were assigned to the ichnogenus *Beaconites* by Bradshaw (1981). Exposed surfaces occasionally reveal arthropod trackways of various scales.

The heterolithic units generally overlie deposits of facies 1, although this is sometimes represented only by thin lenses of intraformational conglomerate at the base of the heterolithics (e.g. cycle II on Fig. 3).

Facies 3—pedified mudstone

Facies 3 is a purplish-red mudstone, 0.25–0.50 m thick, that has been extensively pedified, containing calcareous nodules up to 30 mm in diameter and having a discrete-columnar or coalescent-columnar form (stage II or III of Machette, 1985). Some of these rod-like nodules are up to 60 mm in length. This facies always has a sharp top that is often picked out by green mottling and/or staining of the nodules. Thin blue/purplish-grey, vein-like reticulate patterns are found in some of these units, similar to the 'drab haloes' or local pseudogleying around burrows or root traces described by Retallack (2001).

Facies 4—mudstone

This facies is characterized by brick-red mudstone that has a pelleted appearance in a fresh hand-specimen. The mud 'pellets' are fine-sand sized and appear to be well-sorted. These units are 1.3 to 2.5 m thick, and have been weakly pedified. Drab haloes and vertic, pseudo-anticlinal features are present, although generally they are only poorly developed. Small (5–10 mm diameter), discrete carbonate nodules occur within the vertic horizons representing only stage I–II on Machette's (1985) scale of development. The nodules are not confined to a single B_k or C_k horizon below a structural B horizon within these units. Instead the calcrete nodules overprint the weakly developed pseudo-anticlinal structures, making these composite soil profiles. The bases of some of these units contain very thin lenses of intraformational conglomerate, or occasional small (2–5 mm) rip-up clasts. There is no evidence of primary lamination.

CYCLICITY OF FACIES

The representative section of the Rat Island Mudstone presented in Fig. 3 has been divided into eight fining upward cycles. Each cycle begins with an intraformational conglomerate or sandstone bed (facies 1) of varying thickness, which

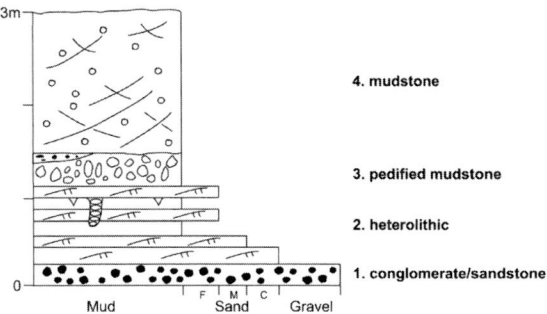

Fig. 4 Summary log, showing facies 1–4 in sequential order.

then passes into facies 2, the heterolithic unit. In some cycles (e.g. II and VI) the conglomerate lens is very thin or lensoid, and in others (e.g. cycle VIII) a thick heterolithic unit forms the base with a sheet of conglomerate making up part of the unit. Above the heterolithics in every cycle lie the truncated well-developed palaeosols of facies 3, which in turn are usually topped by thick, weakly pedified mudstones (facies 4). Cycle IV is a condensed cycle with thin heterolithics at the base, overlain by a siltstone containing a stage II calcrete and topped by a few centimetres of weakly pedified mudstone. This whole cycle is only 1 m thick. The sequence is summarized in Fig. 4.

INTERPRETATION AND FACIES ASSOCIATIONS

Facies 1—conglomerate/sandstone

These units, which occasionally fine upward, probably represent the deposits from channel systems that may have been slightly sinuous, in view of the presence of low-angled surfaces that could imply lateral accretion. The sandstones and conglomerates are thin, however, so it is likely that the rivers that deposited them were shallow and that flow was short-lived. The sand in this facies forms only a minor constituent and was probably brought in by the stream from elsewhere in the catchment. The calcrete and mudstone clasts present in the intraformational conglomerates could have been derived from two possible sources. The clasts might represent reworking of exposed B_k or C_k soil horizons that entered the channel during sheet-flood erosive events when up to 2 m of overlying soil may have been removed from nearby floodplains (see Retallack, 2001, p. 115). An alternative explanation is that the clasts were incorporated into bedload sediment during bank caving as ephemeral streams widened their channels to cope with flood discharge (Smith, 1990; Bendix, 1992). Ephemeral streams tend to have a high width/depth ratio because their annual discharge tends to be very variable (Patton & Schumm, 1981).

Facies 2—heterolithic

These units were interpreted previously as possible levee or crevasse splay deposits, as the thin sand beds contained within them occasionally occur as isolated lenses and some coarsen up, although no crevasse channel deposits were observed (Marriott & Wright, 1993). Additional work on the outcrop has revealed subtle low-angled (c. 3°) inclination of the beds contained in these units that assists with further interpretation.

Facies 1 and 2 together form fining upward sequences that could be interpreted as point bars. The 'classic' meandering river point-bar sequences (Allen, 1965) are produced by lateral accretion as meanders migrate. They typically have a coarse gravel lag at the base, topped by fining upward, cross-bedded and ripple-laminated sands, with silts and clays deposited by overbank flows at the top of the sequence. Later investigations discovered some complex variations in this sequence that related to changes in flow regime and flow pattern around the meanders (e.g. Bridge & Jarvis, 1976; Jackson, 1978; Nanson & Page, 1983). Although there are some similarities in the Rat Island Mudstone sequences, there are also significant differences, mainly in relation to facies 2, that depart from the 'classic' model and also do not fit with the later documented variations. The earlier models tended to be derived from studies of perennial rivers, whereas the ORS environments are thought to have been semi-arid with ephemeral stream flow.

Similar, gently inclined, sequences of interbedded ripple cross-laminated sands and muds with desiccation cracks have, however, been described from the Barwon River of New South Wales (Taylor & Woodyer, 1978) and from the arid to semi-arid Channel Country of central Australia by Gibling et al. (1998). These were interpreted as accretionary bench deposits, formed along the banks of both sinuous and straight streams, that have dimensions of up to 100 m long and 30 m wide. The benches are built up from laterally accreted layers of sand and mud during ephemeral channel flow, with several millimetres of clay being deposited on the surface by floods. This is rapidly cracked into desiccation polygons as the flood retreats (Gibling et al., 1998).

The sediments on the Australian benches are bioturbated by deposit-feeding gastropods and deeper burrows (down to the water table, to a depth of about 3.5 m) are made by crayfish (*Euastacus serratus*) when the benches are exposed (Horwitz & Richardson, 1986; Hasiotis et al., 1993). There is ample evidence for exposure in the heterolithic deposits of facies 2, with infilled cracks evident in section, desiccation polygons of various scales and arthropod trackways on surfaces. *Beaconites* burrows resemble those of the crayfish and may indicate similar behaviour of the Lower Devonian animal during periods of drying out. As with the heterolithic deposits in the Rat Island Mudstone, the accretionary benches form the bulk of the channel sediments in the Channel Country.

Facies 3—pedified mudstone

These units represent the lower B_k or C_k horizons of palaeo-Vertisols (described in detail in Allen, 1986; Marriott & Wright, 1993) that developed in fine sediments deposited on the surface of point bars, accretionary benches or floodplains. Where calcrete development has reached stage II or III, it indicates a relatively lengthy period of exposure with minor new increments of sediment (Machette, 1985; Wright & Marriott, 1996). In these cases (e.g. cycles IV and VII in Fig. 3), it is most likely that the calcretes developed in distal regions of the floodplain, where only infrequent, major floods had any impact. The sharp top indicates that a large part (possibly up to 2 m) of the original soil profile has been removed to expose the surface of the lower horizon. These lower horizons, because of the armouring effect of the calcrete, act as a local erosional base-level, so the number of events that caused the stripping of upper horizons cannot be ascertained. Possible agents of erosion are discussed below.

Facies 4—mudstone

The pelleted appearance of these mudstones suggests that they were deposited as fine sand-sized soil aggregates in bedload rather than as muds from suspension. Evidence has been found in the ORS in other localities in South Wales that this mode of transport and deposition could have prevailed (Êkes, 1993; Marriott & Wright, 1996). The presence of Vertisols in these sequences is similar to the Channel Country of central Australia, where shrinking and swelling of the clays in Vertisols give rise to an ample supply of soil aggregates that are readily entrained and transported as fine sand-size particles (Rust & Nanson, 1989). Some of the units have thin lenses of intraformational conglomerate, or small rip-up clasts at their bases, again testifying to the scouring action of the flow. Additionally, the bases of these units sharply overlie facies 3 and there is frequently a change in colour intensity, with facies 3 being purplish-red and facies 4 brick red. The green mottling and staining of the upper surface of facies 3 may be related to temporary waterlogging during deposition of facies 4. Subsequent pedogenesis has removed any lamination or bedforms that may have been present.

These mudstones also represent palaeo-Vertisols, but the stage of development of their carbonate horizons is not as advanced as facies 3 and many distinguishing features, for example, pseudo-anticlines, are only poorly developed. The presence of carbonate nodules within the vertic horizon makes these composite profiles where soil development has occurred progressively during aggradation (Marriott & Wright, 1993; Wright & Marriott, 1996). This being the case, these sediments are not likely to have been deposited in a single episode but it is apparent that they were exposed between flows long enough for some pedogenic processes to act. One assumption could be that they aggraded during shallow, sheet-flood events, which deposit a 'sheet' of sediment across the floodplain as the flood progresses down the valley, similar to alluvial fan processes. These events are different from 'overbank' floods where rising water transfers material from the channel onto the floodplain laterally. Apart from the lenses of conglomerate at the base of this facies, no other evidence of erosion surfaces or desiccation occurs within the unit at this locality, which tends to suggest that subsequent flows may have come through in gentler pulses or slurries. This type of deposition has been described by Bendix (1992) for ephemeral streams of Utah, where bank caving results in an increase in the sediment concentration within the flow, and this would tie in with the sheet-flood scenario. Flow transmission loss by infiltration into dry ground and evaporation in the semi-arid environment (Renard & Keppel, 1966) results in a slurry of alluvial mud, soil aggregates and reworked calcrete clasts.

An alternative interpretation is that the upper parts of facies 4 could have resulted from aeolian processes. Dust storms are common in arid and semi-arid areas of the present day (Goudie, 1978, 1983) and it is not unlikely that wind would have been an important agent of erosion on the extensive flat floodplains of the ORS environment. Dust clouds are granulometrically stratified with coarser material moving mainly at the bottom (Goossens, 1985). Soil aggregates could be transported as grains of various sizes in this manner, or if broken down, particles of < 20 μm can travel at heights of up to 100 m and can be transported considerable distances (Tsoar & Pye, 1987). Deposition occurs when wind velocity and turbulence decreases, or dust can be brought down in rain (e.g. Collyer et al., 1984). Deposition rates are highest near the source of the dust (McTainsh, 1989), however, and the coarser grains travelling at the bottom of the cloud can be entrapped by surface obstacles, vegetation or moist ground (Tsoar & Pye, 1987).

Aeolian dust deposition produces a thin blanket over the landscape that can easily be remobilized or incorporated into the solum. Once deposited, bio- or pedoturbation will obliterate any identifiable sedimentary structures (McTainsh, 1989). Hence dust deposition is difficult to identify or measure from the sedimentary record. While we have no direct evidence for aeolian dust transport on the ORS floodplains, the environment as interpreted would be conducive to entrainment, transport and deposition of soil aggregates as particles in a dust cloud. McTainsh (1989) considered that alluvial floodplains are likely sites for dust entrainment, and mentions particularly the Channel Country of central Australia where there is a plentiful supply of fine sediment. Alluvial sediment is more at risk of entrainment just after flooding, when freshly deposited material (here in the form of mud aggregates) starts to desiccate, but biological and pedological processes are not active. The dust storms are more frequent in the Channel Country than flood events (McTainsh & Pitblado, 1987), although it was observed that dust storms tend to be more frequent when dry periods follow flooding (Marshall, 1937). This area is strongly subject to the influence of El Niño, which results in a prolonged drought and prominent dust entrainment (M. Gibling, personal communication, 2002)

A tentative suggestion therefore can be made that facies 4 actually represents redistribution of alluvial sediment by dust transport processes. Given that the environment of deposition of the Rat Island Mudstone was analogous to that of the Channel Country, dust storms that occurred more frequently than flooding could give rise to thicker deposits than sheet flood or overbank flow over a similar time frame. The frequency could also account for the relative weakness of pedogenic development and the composite profiles found.

The facies associations described above and illustrated in Fig. 3 would appear to represent a multistage deposition process, where facies 1–3 were deposited within an ephemeral channel system probably during a major pluvial event and then exposed for a sufficiently long period with minimum increments of sediment to enable soil development (see model in Wright & Marriott, 1996). There then followed a period of stripping, where the surface was lowered, either by fluvial erosion or dust entrainment, eventually down to the calcretized horizon. Subsequent deposition of alluvial sediment took place over this surface, which has resulted in reduction of iron and green mottling at the top of facies 3 attributable to waterlogging. This distinct marker that occurs in most of the cycles is a clear indication that facies 3 is not the B_k or C_k horizon of a palaeosol in which the upper horizons are contained in facies 4. Additionally, the nature of pedogenesis within facies 4 is different and separate from that in facies 3, and indicates a higher aggradation rate.

A FACIES MODEL FOR THE RAT ISLAND MUDSTONE

Traditionally, fining upward cycles were interpreted as representing point-bar deposits from moderately sinuous river systems (e.g. Allen, 1965). The fining upward conglomerate/sandstone sequences in the Rat Island Mudstone have already been discussed by Marriott & Wright (1993) who proposed that, as lateral accretion is suggested by the low-angled sets, the depositional setting could have been in meandering rivers. They made the observation that, as the sandbodies are not vertically or laterally extensive and there is substantial evidence of periods of exposure within the finer units (i.e. desiccation polygons and soil formation), the rivers were likely to have been shallow and probably ephemeral. As has been noted above, however, other work has suggested that some mudstone units in the ORS could have been deposited from bedload transport of sand-sized, pedogenic aggregates (Êkes, 1993; Marriott & Wright, 1996). These studies, with field observations on texture and sedimentary structures in the conglomerates and thick mudstones, suggest a bedload, soil aggregate origin for at least some of the finer grained units in the Rat Island Mudstone.

The transport of sand-sized pedogenic aggregates as bedload has been studied, amongst others, by Rust & Nanson (1989) and Gibling et al. (1998). They based their work on studies of the mud-dominated rivers of the Channel Country of central Australia. They described a fixed channel system with accretionary benches and sandy channel-bed deposits that becomes extensively desiccated and bioturbated during dry periods. Distal mud-sheets composed mainly of mud aggregates are laid down on the floodplain during flooding when braid bars made of sand-sized mud aggregates are reactivated. During dry periods the floodplain muds are subject to pedogenesis, with formation of gilgai, desiccation cracks and concretions of carbonate and gypsum. The facies described in their work bear several similarities to those described here from the Rat Island Mudstone and suggest that a different interpretation could be placed on this sequence.

Figure 5 shows the depositional environment envisaged. The system has a channel 'zone' and

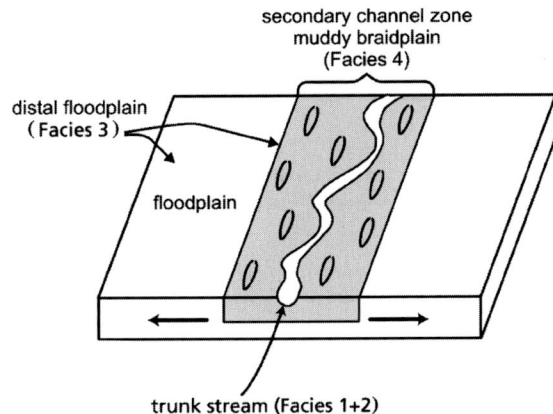

Fig. 5 Schematic diagram of depositional setting for the Rat Island Mudstone. Facies numbers as described in text.

wide floodplain instead of a single ephemeral channel. At low-stage, ephemeral flow is concentrated in a slightly sinuous trunk stream channel. The channel zone takes the form of a muddy braidplain, where high-stage flow spreads and reactivates anastomosing secondary channels separated by bars built up from sand-sized mud aggregates. Infrequent severe flooding will reach distal parts of the floodplain.

Ephemeral streams tend to develop wider, shallower channels because of the regime of flash floods, and the peak flows tend to be high (Wolman & Gerson, 1978; Osterkamp, 1980). The additional characteristic interpreted for the Rat Island Mudstone streams is that they carried a high bedload, and these streams also tend to have wider, shallower channels (Parker, 1979). It is also likely that high-stage flows could result in avulsion because of channel shape and the scarcity of stabilizing vegetation during Lochkovian times. The flashy regime and tendency to high peak flows mean that streams in semi-arid to arid environments are unlikely to develop the dynamic equilibrium state that rivers in humid environments tend toward (Howard, 1988). The changes in ephemeral streams tend to be event-based, with the channel responding to its variable flow regime by adjusting morphology and pattern (Bourke & Pickup, 1999). Some of these adjustments can be dramatic, as shown by Graf (1988) for the Salt River in Phoenix, which has a low-flow-stage meandering channel and at flood-stage becomes

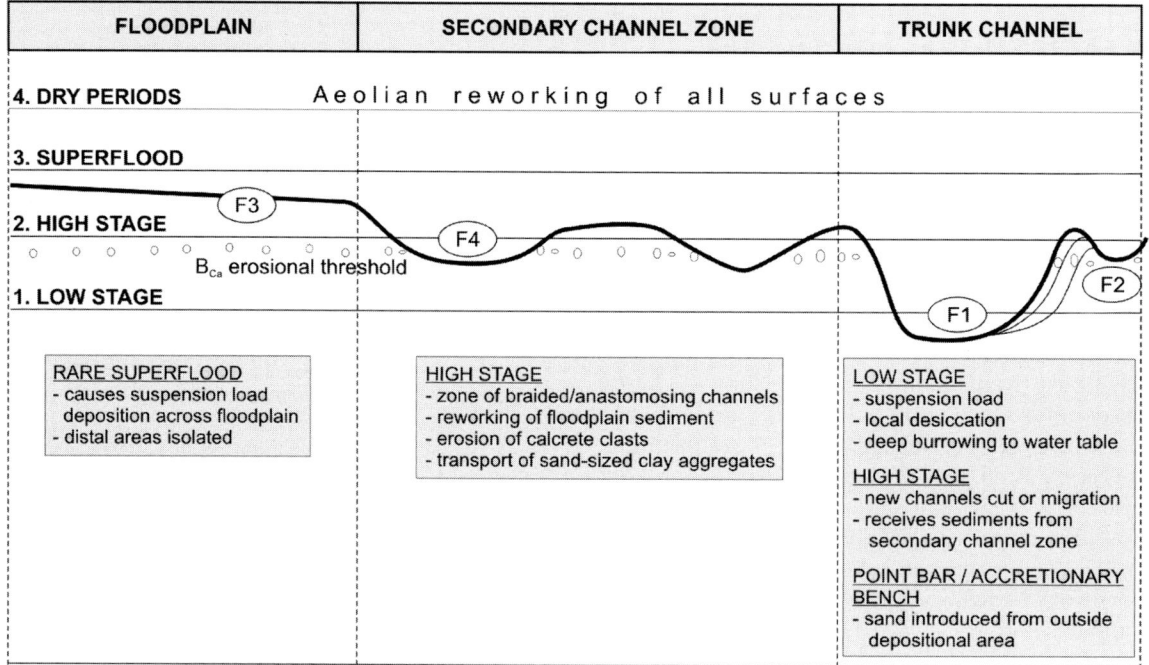

Fig. 6 Summary diagram of flow stages (1–4) and sedimentology. Facies numbers (F1–F4) as described in text. Compare with Taylor & Eggleton (2001, fig. 12.9).

braided. Rust & Nanson (1986) have described similar variations for Cooper Creek, central Australia, which has both braided and anastomosing networks. Superfloods in arid or semi-arid systems make major changes, both by the creation of large-scale landforms and by causing dramatic avulsions (Bourke & Pickup, 1999). The landforms created may then remain inactive for long periods of time (Pickup, 1991).

As shown on Fig. 5, facies 1 and 2 are associated with the main channel, where they represent channel-base gravel or sand sheets and accretionary benches, respectively. Facies 3 forms as pedogenically altered fine sediments on top of the benches and/or floodplain. Distal floodplain areas are where a higher stage of calcrete development is likely to occur because of the infrequency of superfloods that would affect them, providing increments of sediment on a very irregular basis. Facies 4 is most likely to occur in the secondary channel zone or proximal floodplain, where more frequent flooding deposits more regular increments of sediment leading to composite soil profiles. Both facies 3 and 4 could be affected by frequent dust storms causing stripping and redeposition of surface sediment. These aspects are summarized in Fig. 6.

The cycles identified in the Rat Island Mudstone relate to various stages of flow (Fig. 6) with facies 1–3 identified as related to 'normal' ephemeral channel flow. High stage events would lead to increased stripping down to the calcrete armoured B_k horizon thus producing calcrete clasts for incorporation in bedload. Facies 4, which makes up the greatest part of the succession (see Fig. 3), relates to deposition within the secondary channel zone (Figs 5 & 6) from reworking of floodplain mud aggregates in a muddy braidplain during higher stage flow. Higher stage flow events would occur infrequently, however, and further reworking during dust storms, which have a higher frequency of occurrence, could account for the thicknesses found in this facies. Marked differences in cycle thickness probably relate to climatic influences that would govern the return period of superfloods.

A further influence on stream gradients and the nature of sedimentation is the tectonic setting.

The Channel Country is part of the Lake Eyre Basin, which lies on the stable Australian craton. The gradients of rivers in the basin are very low, in the region of < 0.0002 (Bonython, 1963). It is envisaged that, as for the Channel Country, ORS floodplains had low gradients. This would account for low sinuosity and the tendency for anastomosing reaches (Knighton & Nanson, 1993). The ORS probably represents a coastal plain setting in a flexurally subsiding foreland basin (James, 1987; King, 1994; Friend et al., 2000) rather than the stable cratonic interior of the Lake Eyre Basin.

CONCLUSIONS

The Rat Island Mudstone shows a sequence of fining upward cycles that have been interpreted as relating to ephemeral channel flow. An analogy has been drawn with the ephemeral streams of the Channel Country of central Australia where planform changes between slightly sinuous anastomosing channels interchange with braided channels. The channels carry a high bedload made up of sand-sized mud aggregates reworked from the floodplain. Relatively thick heterolithic deposits found in the Rat Island Mudstone are very similar in lithology and internal structure to the accretionary benches of the Channel Country.

A model that builds on this analogy has been presented that relates the development of the four facies identified to changes in flow regime in a dual-channel–secondary-channel system. This leads to the isolation of distal areas that are reactivated during occasional superflood extreme events, which also give rise to channel avulsion and extensive stripping of floodplain sediment. More frequent reworking of floodplain, and within-channel sediment when dry, occurred during dust storms. Any lamination or sedimentary bedforms related to this reworking and redeposition have been overprinted by subsequent pedogenesis.

Climate and tectonic setting that influence channel style and gradient have given rise to a range of depositional forms, allowing a modification of the conventional channel–floodplain model for ephemeral streams to accommodate the more dynamic, event-driven deposystem.

ACKNOWLEDGEMENTS

We are indebted to Martin Gibling and Gerald Nanson for helpful discussions on their field areas. David Catherine and Alun Rogers provided the neat graphic reproduction of our sketchy diagrams.

REFERENCES

Allen, J.R.L. (1963) Depositional features of Dittonian rocks: Pembrokeshire compared with the Welsh Borderland. *Geol. Mag.*, **100**, 385–400.

Allen, J.R.L. (1965) A review of the origin and characteristics of recent alluvial sediments. *Sedimentology*, **5**, 89–191.

Allen, J.R.L. (1974) Studies in fluviatile sedimentation: implications of pedogenic carbonate units, Lower Old Red Sandstone, Anglo-Welsh outcrop. *Geol. J.*, **9**, 181–207.

Allen, J.R.L. (1986) Pedogenic calcretes in the Old Red Sandstone facies (Late Silurian–Early Carboniferous) of the Anglo-Welsh area, southern Britain. In: *Paleosols: their Recognition and Interpretation* (Ed. V.P. Wright), pp. 58–86. Blackwell Scientific, Oxford.

Allen, J.R.L. and Williams, B.P.J. (1978) The sequence of the earlier Lower Old Red Sandstone (Siluro-Devonian), north of Milford Haven, southwest Dyfed (Wales). *Geol. J.*, **13**, 113–136.

Allen, J.R.L. and Williams, B.P.J. (1979) Interfluvial drainage on Siluro-Devonian alluvial plains in Wales and the Welsh Borders. *J. Geol. Soc. Lond.*, **136**, 361–269.

Allen, J.R.L. and Williams, B.P.J. (1981) Sedimentology and stratigraphy of the Townsend Tuff Bed (Lower Old Red Sandstone) in South Wales and the Welsh Borders. *J. Geol. Soc. Lond.*, **138**, 15–29.

Allen, J.R.L. and Williams, B.P.J. (1982) The architecture of an alluvial suite: rocks between the Townsend Tuff and Pickard bay Tuff Beds (early Devonian) Southwest Wales. *Phil. Trans. Roy. Soc. Lond.*, Series B, **297**, 51–89.

Bendix, J. (1992) Fluvial adjustments on varied timescales in Bear Creek Arroyo, Utah, U. S. A. *Z. Geomorph.*, **36**, 141–163.

Bonython, C.W. (1963) Further light on river floods reaching Lake Eyre. *Proc. S. Aust. Branch Roy. Geogr. Soc. Aust.*, **64**, 9–22.

Bourke, M.C. and Pickup, G. (1999) Fluvial form and variability in arid central Australia. In: *Varieties of Fluvial Form*. (Eds A.J. Miller and A. Gupta), pp. 249–271. Wiley, Chichester.

Bradshaw, M.A. (1981) Palaeoenvironmental interpretations and systematics of Devonian trace fossils from the Taylor Group (lower Beacon Supergroup), Antarctica. *N. Z. J. Geol. Geophys.*, **24**, 615–652.

Bridge, J.S. and Jarvis, R.S. (1976) Flow and sedimentary processes in the meandering River South Esk, Glen Clova, Scotland. *Earth Surf. Process.*, **1**. 303–306.

Collyer, F.X., Barnes, G.B., Churchman, G.J., Clarkson, T.S. and Steiner, T.I. (1984) A trans-Tasman dust transport event. *Weather Climate*, **4**, 42–46.

Dixon, E.E.L. (1921) *Geology of the South Wales Coalfield. Part XIII. The Country around Pembroke and Tenby.* Memoirs of the Geological Survey of England and Wales, HMSO, London.

Êkes, C. (1993) Bedload transported pedogenic mud aggregates in the Lower Old Red Sandstone in southwest Wales. *J. Geol. Soc Lond.*, **150**, 469–472.

Friend, P.F., Williams, B.P.J., Ford, M. and Williams, E.A. (2000) Kinematics and dynamics of Old Red Sandstone basins. In: *New Perspectives on the Old Red Sandstone.* (Eds P.F. Friend and B.P.J. Williams). *Geol. Soc. Lond. Spec. Publ.*, **180**, 29–60.

Gibling, M.R., Nanson, G.C. and Maroulis, J.C. (1998) Anastomosing river sedimentation in the Channel Country of central Australia. *Sedimentology*, **45**, 595–619.

Goossens, D. (1985) The granulometric characteristics of a slowly moving dust cloud. *Earth Surf. Process. Landf.*, **10**. 353–362.

Goudie, A.S. (1978) Dust storms and their geomorphic implications. *J. Arid Environ.*, **1**, 291–310.

Goudie, A.S. (1983). Dust storms in space and time. *Prog. Phys. Geogr.*, **7**. 502–530.

Graf, W.L. (1988) Definition of flood plains along arid-region rivers. In: *Flood Geomorphology* (Eds V.R. Baker, R.C. Kochel and P.C. Patton), pp. 231–242. Wiley, New York.

Hasiotis, S.T., Mitchell, C.E. and Dubiel, R.E. (1993) Application of burrow morphologic interpretations to discern continental burrow architects: lungfish or crayfish? *Ichnos*, **2**, 1–19.

Horwitz, P.H.J. and Richardson, A.M.M. (1986) An ecological classification of the burrows of Australian freshwater crayfish. *Aust. J. Mar. Freshwat. Res.*, **37**, 237–242.

Howard, A.D. (1988) Equilibrium models in geomorphology. In: *Modelling Geomorphological Systems* (Ed. M.G. Anderson), pp. 49–72. Wiley, Chichester.

Jackson, R.G. (1978) Preliminary evaluation of lithofacies models for meandering alluvial streams. In: *Fluvial Sedimentology* (Ed. A.D. Miall). *Can. Soc. Petrol. Geol. Mem.*, **5**, 543–576.

James, D.M.D. (1987) Tectonics and sedimentation in the Lower Palaeozoic back-arc basin of South Wales, UK: some quantitative aspects of basin development. *Norsk. Geol. Tidskr.*, **67**, 419–426.

King, L.M. (1994) Subsidence analysis of Eastern Avalonian sequences: implications for Iapetus closure. *J. Geol. Soc. Lond.*, **151**, 647–657.

Knighton, A.D. and Nanson, G.C. (1993) Anastomosis and the continuum of channel pattern. *Earth Surf. Process. Landf.*, **18**, 613–625.

Love, S.E. and Williams, B.P.J. (2000) Sedimentology, cyclicity and floodplain architecture in the Lower Old Red Sandstone of SW Wales. In: *New Perspectives on the Old Red Sandstone.* (Eds P.F. Friend and B.P.J. Williams). *Geol. Soc. Lond. Spec. Publ.*, **180**, 371–388.

Machette, M.N. (1985) Calcic soils of the southwestern United States. In: *Soils and Quaternary Geology of the South West United States.* (Ed. D.L. Weide). *Geol. Soc. Am. Spec. Pap.*, **203**, 1–21.

Marriott, S.B. and Wright, V.P. (1993) Palaeosols as indicators of geomorphic stability in two Old Red Sandstone alluvial suites, South Wales. *J. Geol. Soc. Lond.*, **150**, 1109–1120.

Marriott, S.B. and Wright, V.P. (1996) Sediment recycling on Siluro-Devonian floodplains. *J. Geol. Soc. Lond.*, **153**, 661–664.

Marshall, A. (1937) Where the Diamantina flows. *Walkabout*, **4**, 57–61.

McTainsh, G.H. (1989) Quaternary aeolian dust processes and sediments in the Australian region. *Quat. Sci. Rev.*, **8**, 235–253.

McTainsh, G.H. and Pitblado, J.R. (1987) Dust storms and related phenomena measured from meteorological records in Australia. *Earth Surf. Process. Landf.*, **12**, 414–424.

Nanson, G.C. and Page, K. (1983) Lateral accretion of fine grained concave benches on meandering rivers. In: *Modern and Ancient Fluvial Systems* (Eds J. Collinson and J. Lewin). *Spec. Publ. Int. Assoc. Sedimentol.*, **6**, 133–143.

Osterkamp, W.R. (1980). Sediment-morphology relations of alluvial channels. *Proceedings of the Symposium on Watershed Management*, American Society of Civil Engineers, Boise, ID, 21–23 July, 201–204.

Parker, G. (1979) Hydraulic geometry of active gravel rivers. *J. Hydraul. Div. Am. Soc. Civ. Eng.*, **105**(HY9), 1185–1201.

Patton, P.C. and Schumm, S.A. (1981) Ephemeral stream processes: implications for studies of Quaternary valley fills. *Quat. Res.*, **15**, 24–43.

Pickup, G. (1991) Event frequency and landscape stability on the floodplain systems of arid central Australia. *Quat. Sci. Rev.*, **10**, 463–473.

Renard, K.G. and Keppel, R.V. (1966) Hydrographs of ephemeral streams in the southwest. *J. Hydraul. Div. Am. Soc. Civ. Eng.*, **92**(HY2), 33–52.

Retallack, G.J. (2001) *Soils of the Past, an Introduction to Paleopedology*, 2nd edn. Blackwell Science, Oxford, 404 pp.

Rust, B.R. and Nanson, G.C. (1986) Contemporary and palaeochannel patterns and late Quaternary stratigraphy of Cooper Creek, southwest Queensland, Australia. *Earth Surf. Process. Landf.*, **11**, 581–590.

Rust, B.R. and Nanson, G.C. (1989) Bedload transport of mud as pedogenic aggregates in modern and ancient rivers. *Sedimentology*, **36**, 291–306.

Smith, R.M.H. (1990) Alluvial palaeosols and pedofacies sequences in the Permian Lower Beaufort of the southwestern Karoo Basin, South Africa. *J. Sediment. Petrol.*, **60**, 258–276.

Taylor, G. and Eggleton, R.A. (2001) *Regolith Geology and Geomorphology*. Wiley, Chichester, 375 pp.

Taylor, G. and Woodyer, K.D. (1978) Bank deposition in suspended-load streams. In: *Fluvial Sedimentology* (Ed. by A.D. Miall). *Can. Soc. Petrol. Geol. Mem.*, **5**, 257–275.

Tsoar, H. and Pye, K. (1987). Dust transport and the question of desert loess formation. *Sedimentology*, **34**, 139–153.

Williams, B.P.J., Allen, J.R.L. and Marshall, J.D. (1982) Old Red Sandstone facies of the Pembroke peninsula, south of the Ritec fault. In: *Geological Excursions in Dyfed, South-west Wales* (Ed. M.G. Bassett), pp. 151–174. National Museum of Wales, Cardiff.

Wolman, M.G. and Gerson, R. (1978) Relative scales of time and effectiveness of climate in watershed geomorphology. *Earth Surf. Process.*, **3**, 189–208.

Wright, V.P. and Marriott, S.B. (1996) A quantitative approach to soil occurrence in alluvial deposits and its application to the Old Red Sandstone of Britain. *J. Geol. Soc. Lond.*, **153**, 907–913.

Reservoir scale sequence stratigraphy for hydrocarbon production and development: Tarbat–Ipundu Field, south-west Queensland, Australia

ROBERT S. ROOT*, SIMON C. LANG* and DENNIS HARRISON†
*Australian School of Petroleum and the Australian Petroleum Cooperative Research Centre, University of Adelaide, SA, 5005, Australia (Email: slang@ncpgg.adelaide.edu.au); and
†Santos House, Level 14, 60 Edward Street, Brisbane, Qld, 4000, Australia

ABSTRACT

The value of sequence stratigraphy to petroleum exploration stems from understanding spatial stratigraphical relationships and developing new play concepts. In contrast, the value of sequence stratigraphy to reservoir development stems from providing stratigraphical divisions that facilitate geologically meaningful models of reservoir heterogeneity. Understanding the relationship between reservoir properties and depositional facies allows modern analogues to be used as tools for selecting meaningful geological models that guide reservoir development.

This study uses an oil field data set from Tarbat–Ipundu Field in South-west Queensland, Australia to demonstrate how sequence stratigraphy can be applied at the reservoir scale. The Wyandra Sandstone Member of the Cadna-owie Formation is a thin (c. 20 m), fluvial to marginal-marine volcaniclastic sandstone reservoir. Reservoir heterogeneity results almost exclusively from diagenetic processes, but the spatial distribution of medium- to very coarse-grained sandstone within the reservoir is a significant influence on the occurrence and intensity of diagenesis.

Sequence-stratigraphy surfaces mapped across the field define a lower fluvio-lacustrine highstand systems tract, a middle sheet-like, fluvial lowstand systems tract, and an upper estuarine to marginal-marine transgressive systems tract. The majority of reservoir sandstone (horizontal permeability > 10 mD, porosity > 20%) is contained within the lowstand systems tract. To understand the internal architecture of the reservoir, however, modern fluvial depositional analogues are used to explain the distribution of medium- to very coarse-grained sandstone. The implication for reservoir development is that a genetically meaningful sequence-stratigraphy framework is essential to understand reservoir heterogeneity.

INTRODUCTION

Numerous studies document the success of the sequence-stratigraphy approach for understanding large-scale, spatial stratigraphical relationships and for developing new stratigraphical play concepts in a variety of depositional settings (e.g. Van Wagoner *et al.*, 1990; Bowden *et al.*, 1993; Posamentier & Allen, 1993; Shanley & McCabe, 1994; Allen *et al.*, 1996; Legarreta & Uliana,

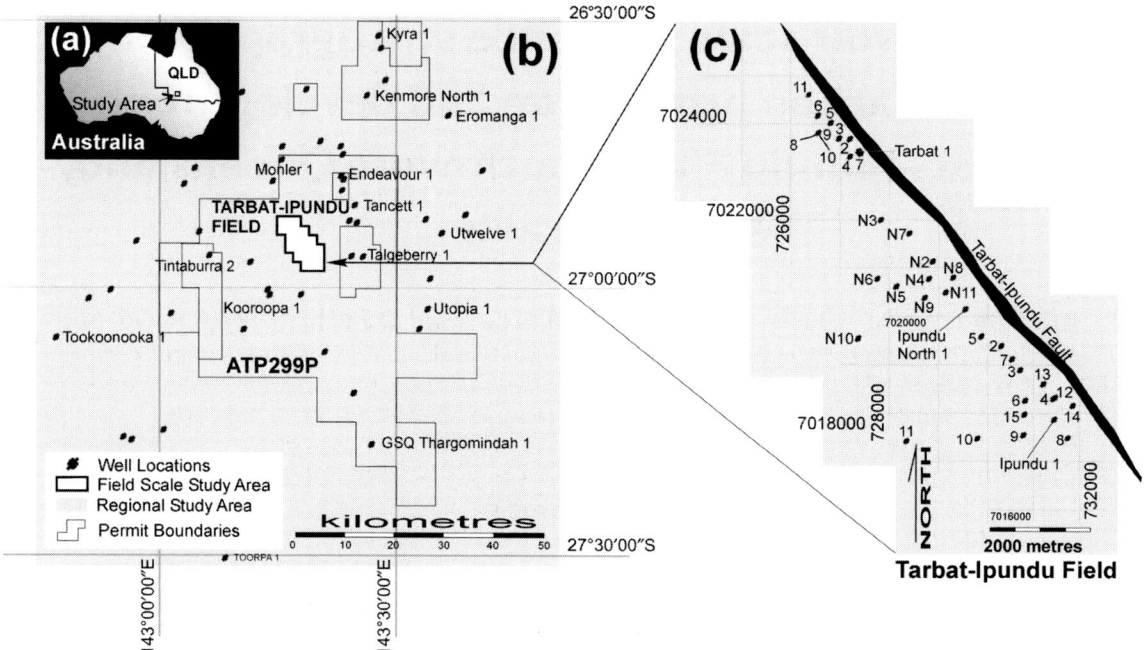

Fig. 1 The study area in (a) south-west Queensland (QLD), Australia; (b) regional and (c) field scale maps showing well locations. The Tarbat–Ipundu is a 37-well oil-field that exists within permit ATP299P.

1998; Kennard et al., 1999). The versatility and success of the sequence-stratigraphy approach has led to optimism in the literature regarding the application of sequence stratigraphy to reservoir development (Posamentier & Allen, 1999; Kennard et al., 1999; Posamentier & Weimer, 1993).

The value of first generation sequence-stratigraphy models stems largely from constraining the stratigraphical position of reservoir intervals by using seismic reflection data. In reservoir development the stratigraphical position of the reservoir interval is generally already established, however, and at typical field-scales (kilometres laterally and a few tens to a few hundreds of metres vertically), sequence-stratigraphy surfaces may be below seismic resolution. Furthermore, at typical field-scales sequence-stratigraphy surfaces are commonly identified with the same data set that is used to indicate reservoir properties (e.g. full-hole cores and wireline suites). The identification of sequence-stratigraphy surfaces, therefore, does not precede the identification of reservoir rock in a predictive or commercially meaningful way. Consequently, the application of sequence stratigraphy for improving recovery efficiency and reservoir management, which are the primary objectives of reservoir development, differs from the application of sequence stratigraphy to petroleum exploration. The aim of this paper is to document the application of sequence stratigraphy to improving the geological model for a small-scale volcaniclastic reservoir—the Wyandra Sandstone Member of the Eromanga Basin at Tarbat–Ipundu Field in south-west Queensland, Australia (Fig. 1). In addition, some general conclusions are drawn regarding the utility of the sequence-stratigraphy approach for reservoir development.

Background

The Wyandra Sandstone Member of the Cadna-owie Formation is a Lower Cretaceous, volcaniclastic reservoir that is interpreted to overlie a regional sequence boundary (Musakti, 1997; Lonergan et al., 1998; Lang et al., 2001). The Lower Cretaceous Cadna-owie Formation records a change from fluvio-lacustrine conditions characterized by internal drainage and infill of large interior cratonic lakes (e.g. 'Lake Murta') to open-marine conditions represented by the Bulldog Shale overlying the Cadna-owie Formation. The Early Cretaceous in the Eromanga Basin was characterized by warm temperate climatic conditions (Frakes, 1979) with large-scale basin architecture controlled by regional crustal downwarping (Zhou, 1989; Moussavi-Harami, 1996). Regional sediment transport directions during the deposition of the Wyandra Sandstone Member are interpreted to trend from the north and north-east to the south and south-west (Green et al., 1996; Musakti, 1997).

The Wyandra Sandstone is the upper member of the Cadna-owie Formation and is the uppermost oil reservoir of the Eromanga Basin succession in south-west Queensland (Fig. 2). The upper flooding surface that separates the Wyandra Sandstone Member from the Bulldog Shale is regionally resolvable on seismic profiles. The basal sequence boundary of the Wyandra Sandstone Member and the internal architecture of the Wyandra Sandstone Member, however, are below seismic resolution.

The structural play at Tarbat–Ipundu Field is a fault-related anticline with three-way dip closure on the western, southern and northern flanks of the structure, and fault closure on the eastern to north-eastern flank of the structure. Three associated anticlinal structures with four-way dip closure define three subaccumulations: Tarbat, Ipundu North and Ipundu with 1939 ha of total closure (Lonergan et al., 1998). Although the Wyandra Sandstone Member is a high sandstone to mudstone ratio interval across much of the Tarbat–Ipundu Field, the effectiveness of development drilling has been limited by significant reservoir heterogeneity, with permeability within the reservoir interval ranging from < 0.001 mD to > 1000 mD.

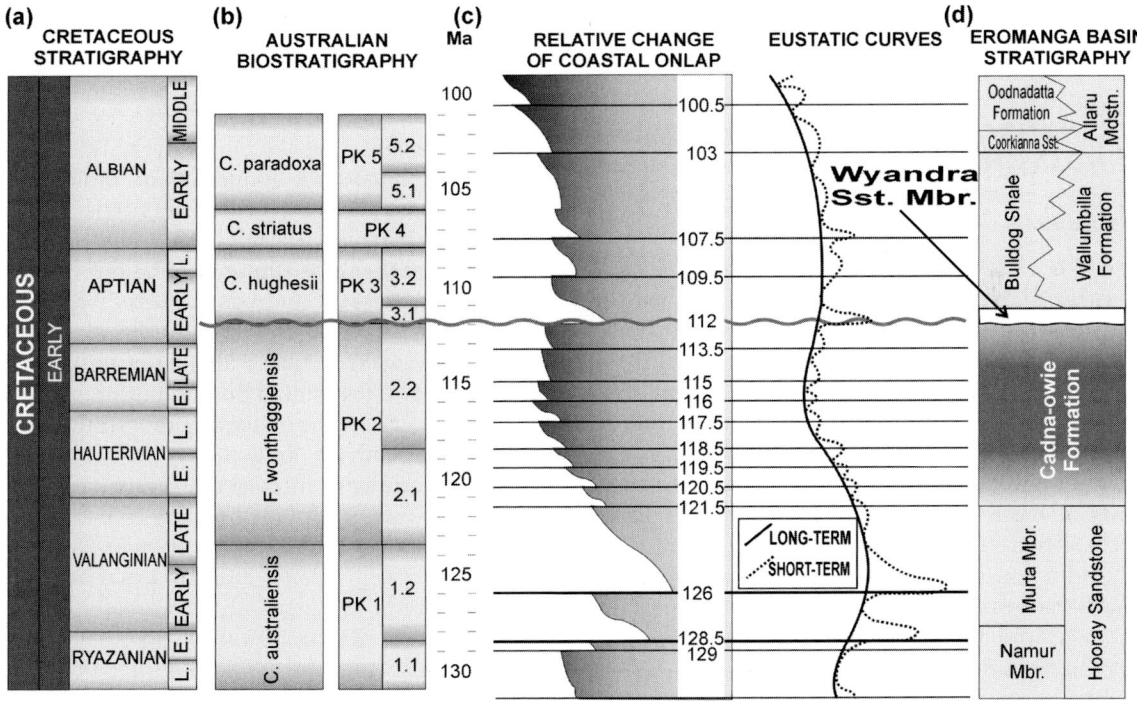

Fig. 2 Stratigraphical charts for the Early Cretaceous. (a) Cretaceous stratigraphical divisions. (b) Australian biozonation from Helby et al. (1987) and Price et al. (1985). (c) Coastal onlap and eustatic curves from Haq et al. (1988). (d) The stratigraphy of the Eromanga Basin. The Wyandra Sandstone Member of the Cadna-owie Formation overlies the unnamed, lower Cadna-owie member and is overlain by the Bulldog Shale. (Modified from Gorter, 1994.)

METHODOLOGY

Sequence-stratigraphy surfaces in the Wyandra Sandstone Member were identified and mapped, both regionally and at field scales, using 13 full-hole cores, wireline suites from 100 wells (37 within Tarbat–Ipundu Field) and palynological data. Formation microscanner (FMS) and fullbore formation micro-imager (FMI) image logs were used for palaeocurrent interpretation and some facies interpretation. The FMS and FMI image logs provide orientated, high-resolution resistivity images of the borehole and can be used for sedimentological analysis in a similar way as orientated, full-hole core. Over 1000 palaeocurrent determinations were made in the Wyandra Sandstone Member from the image logs. The integration of petrographic studies (71 thin-sections), core analysis data and production data served to establish two relationships: (i) the link between primary rock composition and texture and reservoir properties; and (ii) the link between sequence-stratigraphy units and the subsurface spatial distribution of reservoir rock.

SEDIMENTOLOGY OF THE WYANDRA SANDSTONE MEMBER

Description

The lower unit of the Wyandra Sandstone Member has a high ratio of sandstone to mudstone and overlies a regional unconformity. The unit is characterized by the amalgamation of fining upward sandstone units that average 1.5 to 2 m thick and are represented by blocky, fining upwards gamma-ray log motifs in north-east parts of the Tarbat–Ipundu Field. In south-western parts of the field, with increased distance from Tarbat–Ipundu Fault, the lower unit of the Wyandra Sandstone Member is thinner and log motifs have a higher gamma value and may be characterized by coarsening or fining upward log motifs of less than 5 m thick.

The lower sections of fining upward sandstone units in the eastern and north-eastern parts of Tarbat–Ipundu Field consist of fine- to very coarse-grained sandstone that is characterized by multiple erosive conglomeratic lags, unidirectional FMI and FMS palaeocurrents, parallel stratification, trough cross-bedding and planar-tabular cross-bedding (Figs 3a–c & 4). The upper sections of fining upward sandstone units are generally very fine- and fine-grained sandstone with ripple cross-lamination, flaser bedding and multiple layers of carbonaceous detritus. Graded beds, climbing ripple cross-lamination and water escape structures (e.g. flame structures) are also associated with the upper section of fining upward sandstone units (Fig. 3d). Fining upward sandstone units are commonly capped by thin mudstone units.

Fining upward, low-value log motif sandstone units of the lower unit of the Wyandra Sandstone Member are commonly associated with interbedded mudstone and sandstone deposits with individual beds of several centimetres to several decimetres thick, and thin (< 3 cm) coal beds. Sandstone beds are characterized by parallel stratification, climbing ripple cross-lamination, ripple cross-lamination and load casts at sandstone–mudstone bed boundaries. Mudstone units may have thin (< 1 cm) interbedded lenticular bedded or starved ripple bedded siltstone and very fine-grained sandstone beds and may also be associated with root traces (Fig. 5a & b).

Core coverage is poor in the western and south-western parts of Tarbat–Ipundu Field, but core is available from regional wells (Kooroopa 1 and GSQ Thargomindah 1A; see Fig. 1) with similar log motifs for the same stratigraphical level as the wells in the western and south-western parts of Tarbat–Ipundu Field. Based on full-hole cores from nearby wells, higher value, coarsening and fining upward log motifs in the lower unit of the Wyandra Sandstone Member represent intervals of interbedded mudstone and sandstone commonly arranged in coarsening upward vertical successions. Sandstone beds are 1–2 dm thick and are parallel stratified, lenticular bedded and trough and planar-tabular cross-bedded. Mudstone beds are generally bioturbated. Thin coal beds and root traces are absent.

The middle unit of the Wyandra Sandstone Member has a variable sandstone to mudstone ratio ranging from 10 to 80% within 2 km. Individual sandstone beds in the middle unit of the Wyandra Sandstone Member are thinner than fining upward sandstone units of the lower sec-

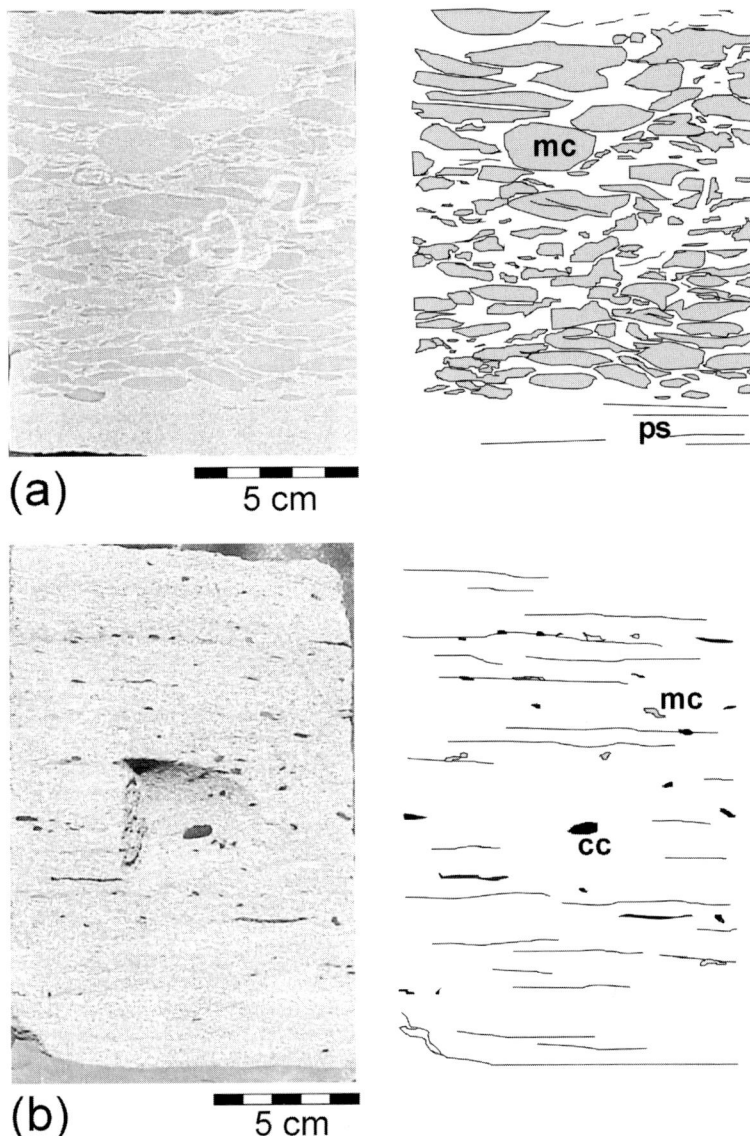

Fig. 3 Core photographs and interpretive diagrams for selected lithofacies. (a) Conglomerate consisting of imbricated mudstone clasts (*mc*) overlying parallel stratified sandstone (*ps*). (b) Parallel stratified sandstone associated with coarse-grained to pebble-sized coal chips (*cc*) and mudstone clasts (*mc*).

tion of the Wyandra Sandstone Member, with the typical thicknesses between 0.3 and 0.8 m. Sandstone facies are commonly associated with planar-tabular cross-bedding that has mudstone drapes at foreset boundaries (Fig. 5c), bi-directional FMI and FMS palaeocurrents (Fig. 6) and a greater proportion of glauconitic accretionary and peloidal particles than the lower unit of the Wyandra Sandstone Member. Mudstone-dominated facies in the middle unit of the Wyandra Sandstone Member are characterized by starved ripple bedding and lenticular bedding, rarely herringbone cross-lamination, and bioturbation.

The upper unit of the Wyandra Sandstone Member is an intensely bioturbated, lithological mixture of sandstone and mudstone. Consequently, primary sedimentary structure is not commonly resolvable. The basal boundary of the upper unit may be represented by a heavily cemented, very laterally continuous pebbly sandstone unit a few

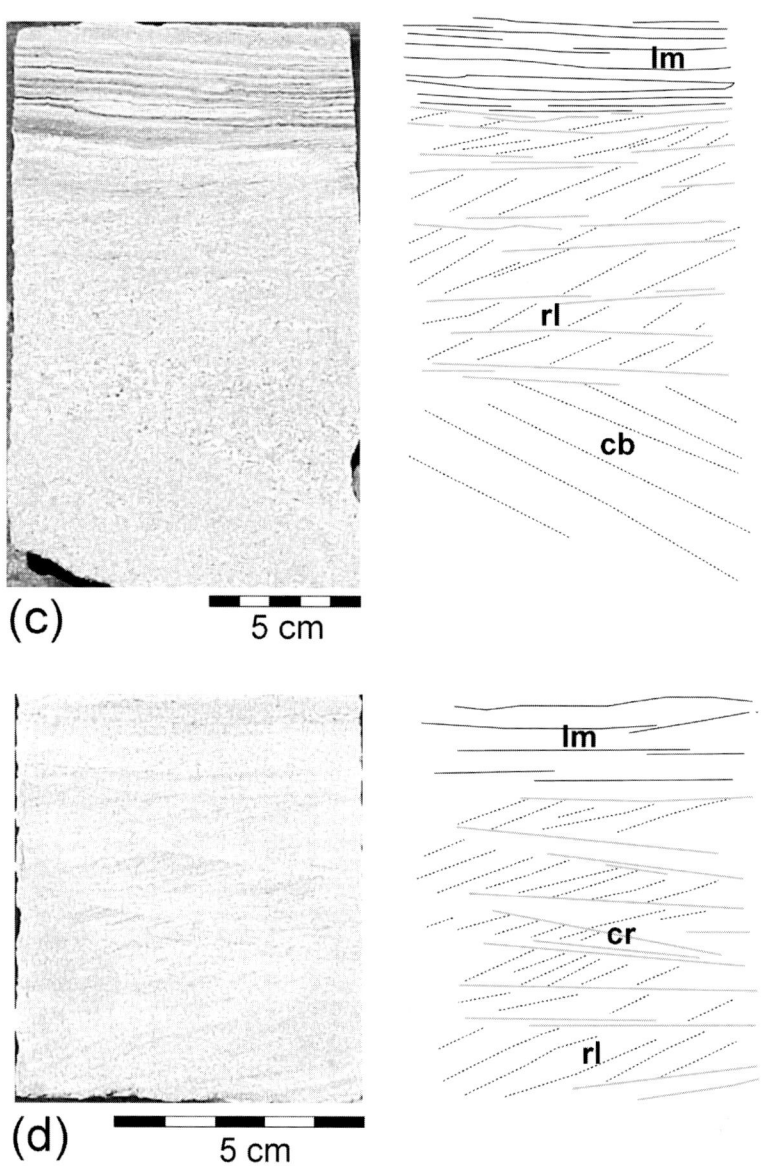

Fig. 3 (c) Vertical succession of planar-tabular cross-bedded sandstone (*cb*), ripple cross-laminated sandstone (*rl*), horizontally laminated interbedded sandstone and siltstone (*lm*). (d) Ripple cross-laminated sandstone (*rl*) and climbing ripple cross-laminated sandstone (*cr*) overlain by horizontally laminated sandstone (*lm*).

decimetres thick. Where primary structures can be identified, mudstone intervals are finely interbedded with ripple cross-laminated and wave-rippled sandstone beds (Fig. 5d). Siderite nodules with reduction rings are common in the upper unit of the Wyandra Sandstone Member.

Trace fossils in the lower, middle and upper units of the Wyandra Sandstone Member are dominantly cylindrical or elliptical, unlined burrows, several millimetres in diameter, orientated horizontally or obliquely to bedding. The burrows are filled with siltstone or very fine- to fine-grained sandstone in a mudstone substrate. These burrows are interpreted as *Planolites* (Fig. 7a). In the middle unit of the Wyandra Sandstone Member two other kinds of trace fossils are found along with *Planolites*. The first are closely spaced burrows less than 2 mm in diameter that are obliquely orientated to bedding and reflect a downward branching geometry; the second are thin, tube-like,

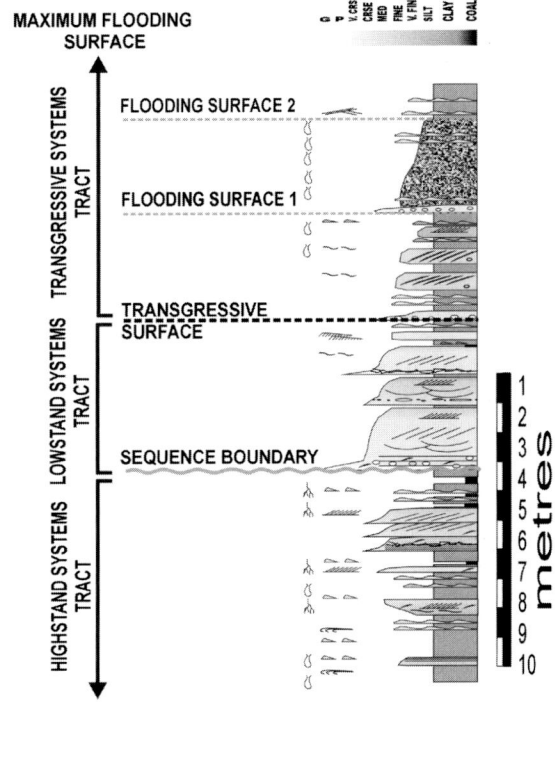

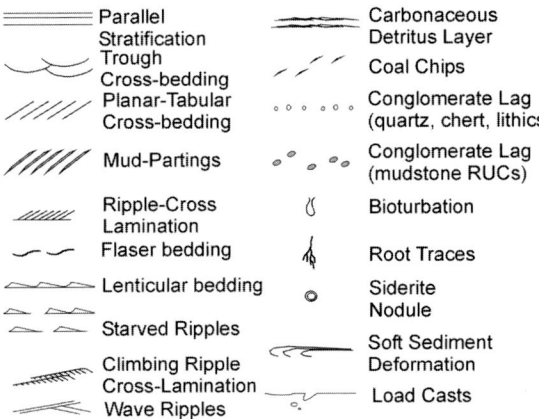

Fig. 4 Composite stratigraphical log of the Wyandra Sandstone Member based on 13 full-hole cores through the Wyandra Sandstone Member. The sequence-stratigraphy surfaces identified in the Wyandra Sandstone Member define a lower highstand systems tract, a middle sheet-like, fluvial lowstand systems tract, and an upper estuarine to marginal marine transgressive systems tract.

vertical burrows a few millimetres in diameter and a few centimetres in length. These trace fossils are interpreted as *Chondrites* and *Skolithos* (Fig. 7b), respectively. In the upper unit of the Wyandra Sandstone Member larger burrows up to 20 mm in diameter have a graded fill consisting of very fine- to coarse-grained sandstone and glauconitic mudstone, with burrows that are both unlined and lined with finer grained material than the substrate. The larger burrows may represent ?*Thalassinoides* (Fig. 7c) and *Ophiomorpha* (Fig. 7d).

Interpretation

Multiple erosive lags, parallel lamination and trough cross-bedding in the lower unit of the Wyandra Sandstone Member suggests a flow velocity and flow turbulence that is consistent with channelized deposition. Single fining upward units of 1.5–2 m thick are interpreted to represent single fluvial channel deposits. The vertical associations of sedimentary structures within these fining upward units indicate a decrease in flow velocity with time that suggests lateral channel accretion or channel abandonment. Evidence for rapid flow deceleration and rapid deposition such as graded beds, climbing ripple cross-lamination and water-escape structures suggests that many of the preserved fluvial channel fills were deposited as a result of fluvial flooding events.

Thin parallel stratified, climbing ripple cross-laminated sandstone units with load casts interbedded with mudstone units characterized by root traces and thin coals suggest sudden bed-load injection into relatively low-energy areas that were dominated by deposition from suspension and subaerial exposure. These deposits are interpreted as overbank crevasse-splay deposits where bedload sediment was probably derived from nearby channels during overbank flooding events. In contrast, interbedded sandstone and mudstone deposits associated with relatively high-value log motifs in the western and south-western parts of Tarbat–Ipundu Field lack signs of subaerial exposure and very shallow-water deposition such as root traces and coals. Instead, coarsening upward successions of interbedded sandstone and mudstone in the western and south-western parts of Tarbat–Ipundu Field are characterized

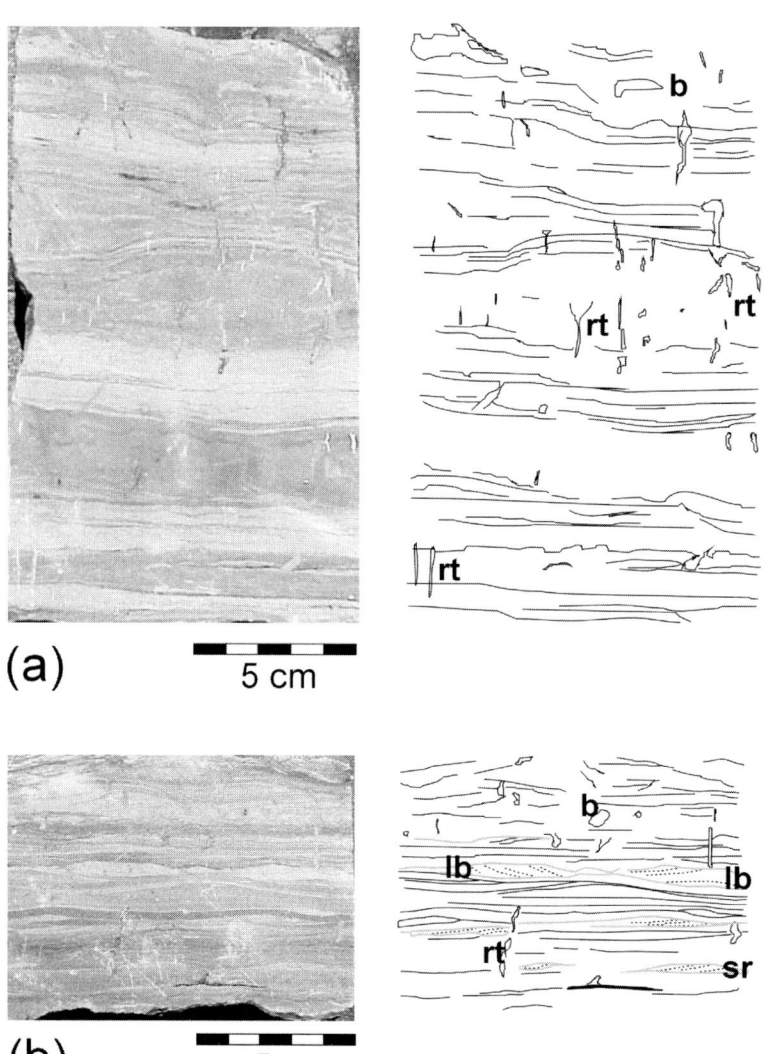

Fig. 5 Core photographs and interpretive diagrams for selected lithofacies. (a) Interbedded horizontally laminated sandstone and mudstone with burrows (*b*) and root traces (*rt*). (b) Interbedded lenticular bedded (*lb*) and starved ripple bedded (*sr*) sandstone and mudstone with burrows (*b*) and root traces (*rt*).

by bioturbation and indications of confined or semi-confined flow (trough and planar tabular cross-bedding). The deposits in the western and south-western parts of Tarbat–Ipundu Field were deposited under lower energy depositional conditions than the stacked fluvial channel deposits that are proximal to the fault, in the northern and north-eastern parts of Tarbat–Ipundu Field. The lower energy deposits in the western and south-western parts of Tarbat–Ipundu Field are interpreted as mouth-bar deposits of the lower deltaic plain and thin, distal distributary channel deposits near the terminus of the channel.

Indications of systematic variations in flow velocity and flow reversals in the middle unit of the Wyandra Sandstone Member include lenticular and flaser bedding, mud drapes on cross-bed foresets and herringbone ripple cross-lamination. Additionally, a significant increase in the proportion of glauconitic accretionary particles relative to the lower unit of the Wyandra Sandstone Member and the presence of *Chondrites* trace fossils may suggest an increased marine influence during the deposition of the middle unit of the Wyandra Sandstone Member. Accordingly, the FMI and FMS palaeocurrent

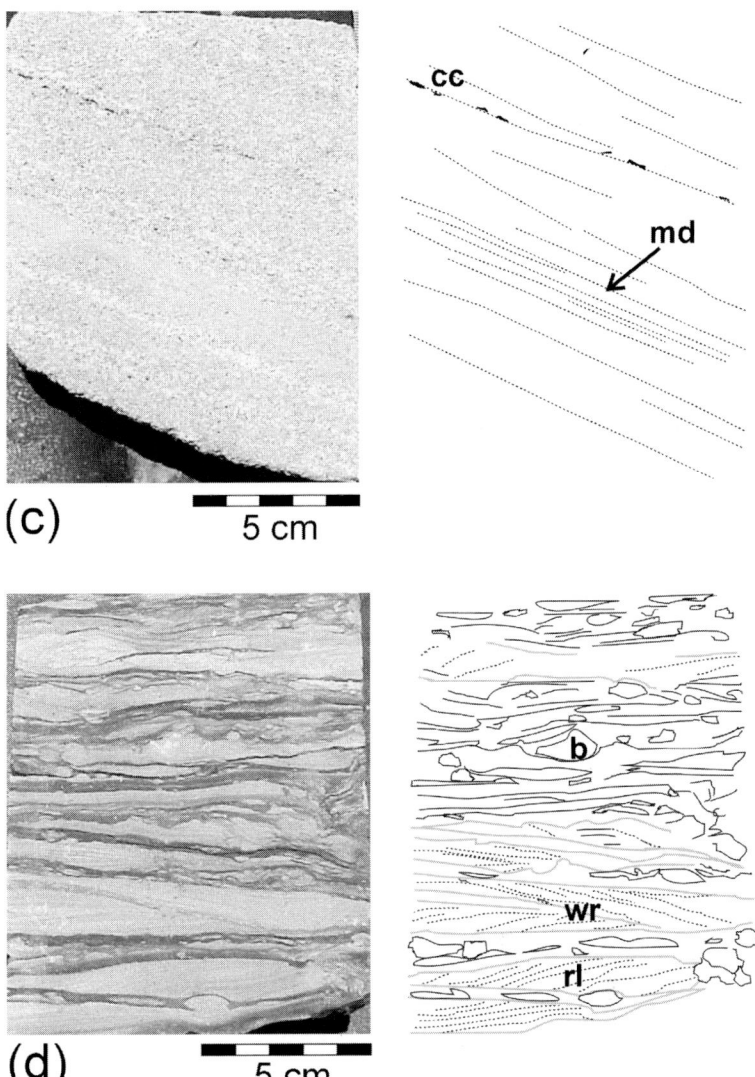

Fig. 5 (c) Planar-tabular cross-bedded sandstone with mudstone drapes (*md*) and coal chips (*cc*) on cross-bed foresets. (d) Wave-ripple bedded (*wr*), ripple cross-laminated (*rl*) and bioturbated (*b*) sandstone interbedded with thin mudstone layers.

interpretation shows bi-directional palaeocurrents and a greater variation in palaeocurrent dip azimuth directions in the middle unit of the Wyandra Sandstone Member. The sedimentary structures, the palaeoichnology, and the change in palaeocurrent patterns between the lower and middle units of the Wyandra Sandstone Member are consistent with the introduction of tidal currents to the depositional system.

The upper unit of the Wyandra Sandstone Member is characterized by *Ophiomorpha* and ?*Thalassinoides* trace fossils, evidence of reducing conditions and indications of a wave influence on depositional processes in the uppermost sections of the unit. The upper unit of the Wyandra Sandstone Member is directly overlain by open-marine shale of the Wallumbilla Formation. In concert, the sedimentological evidence suggests that the upper unit of the Wyandra Sandstone Member is comprised of more distal, deeper water facies that the lower and middle units of the Wyandra Sandstone Member.

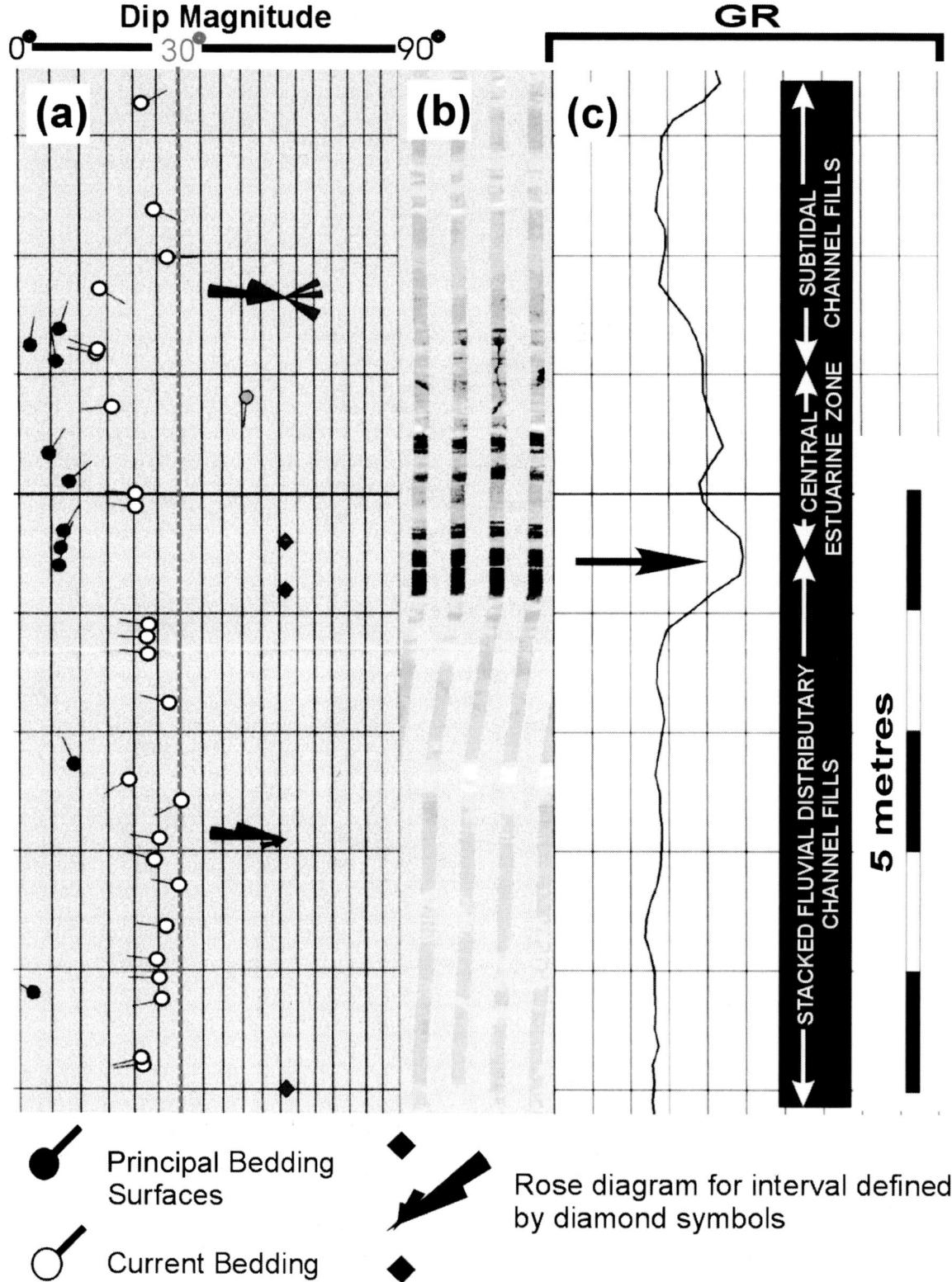

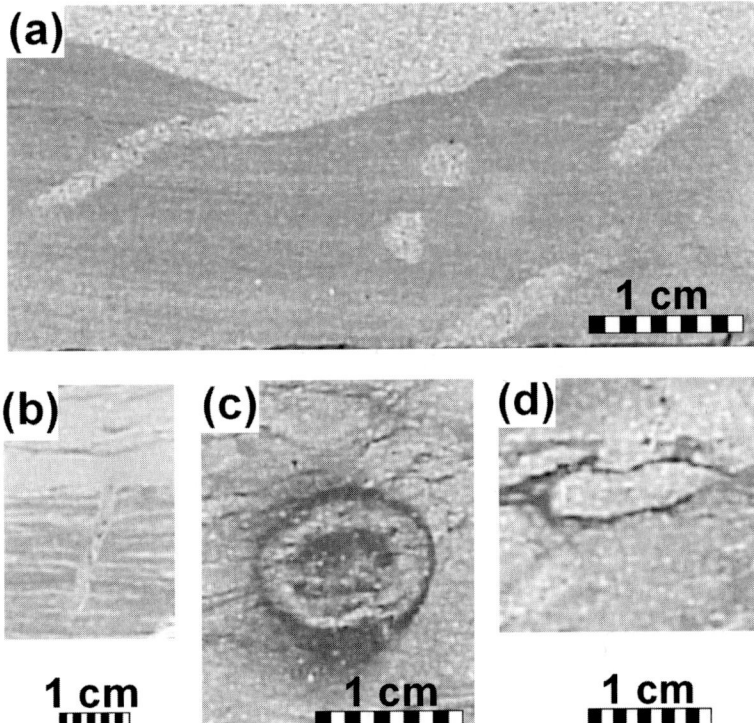

Fig. 7 Trace fossils identified in full-hole core. (a) *Planolites*, (b) *Skolithos*, (c) *?Thalassinoides* and (d) *Ophiomorpha*.

SEQUENCE STRATIGRAPHY OF THE WYANDRA SANDSTONE MEMBER

Key sequence-stratigraphy surfaces identified within the Wyandra Sandstone Member are a basal sequence boundary, a transgressive surface and two higher-order flooding surfaces (Fig. 4). Sequence-stratigraphy surfaces define linkages of contemporaneously deposited depositional systems called systems tracts (Brown & Fisher, 1977). In the Wyandra Sandstone Member a lowstand systems tract (LST) is recognized bounded by a lower sequence boundary and an upper transgressive surface. The strata overlying the transgressive surface, which is formed at the first landward shift of the shoreline (Posamentier & Allen, 1999), is part of the transgressive systems tract (TST). The TST strata of the Wyandra Sandstone Member may be divided into two subunits defined by two higher-order flooding surfaces, both of which represent sudden increases in palaeobathymetry. The key sequence-stratigraphy surfaces identified in the Wyandra Sandstone Member are described below.

The Wyandra Sandstone Member and the underlying strata of the unnamed lower Cadna-owie Member are separated by a regional unconformity interpreted as a sequence boundary. Characteristics of the surface are a palynologically resolvable depositional hiatus, a basinward shift in facies tracts, an increase in grain size above the

Fig. 6 (*opposite*) Sample FMS interpretation for the Wyandra Sandstone Member. (a) An image log interpretation. (b) An FMS image log (darker colours indicate more conductive rock). (c) A gamma-ray log with facies interpretation. The transgressive surface (arrow) is represented by a gamma-ray spike, a conductive unit on the image log (representing a thin mudstone deposit), and a change up-section from unidirectional, westward trending palaeocurrents to bi-directional, westward and eastward trending palaeocurrents.

surface, and a change in palaeocurrents above the surface. The strata directly overlying the sequence boundary, previously referred to in this paper as 'the lower unit of the Wyandra Sandstone Member', are interpreted to represent deposits of a LST.

The LST of the Wyandra Sandstone Member is dominantly comprised of amalgamated fluvial distributary channel-fill deposits. During the formation of the basal Wyandra Sandstone Member sequence boundary, stable base-level conditions forced the lateral amalgamation and the preferential preservation of channelized deposits during deposition of the LST of the Wyandra Sandstone Member. Vertical space for the development of an aggradational stacking geometry was restricted, resulting in a fairly lithologically homogeneous LST, however, there is significant variability of fluid-flow properties within the LST. Lowstand systems tract reservoirs composed of amalgamated fluvial reservoirs typically have one of two large-scale geometries: a ribbon-like geometry where LST fluvial deposits are confined to incised valleys; or a sheet-like geometry where unconfined fluvial channel amalgamation results in a laterally continuous sandstone body (Posamentier, 2001).

Strata overlying the LST (the middle and upper units of the Wyandra Sandstone Member) show indications of increasing tidal and wave influence and increasing palaeobathymetry up-section. Both factors are interpreted to result from transgression and the consequent landward shift in facies tracts. The stratigraphical implication of the change in depositional style brought about by transgression is that the rate of accommodation creation has exceeded the rate at which sediment is supplied to the depositional system. Therefore, the strata deposited during this period may be viewed as part of a TST. The middle unit of the Wyandra Sandstone Member was deposited concurrently with the Aptian eustatic sea-level rise and a sudden increase in the rate of basin subsidence (Morgan, 1980; Zhou, 1989; Moussavi-Harami, 1996). The retrogradation stacking pattern represented in the middle and upper units (TST) of the Wyandra Sandstone Member is probably a result of the combination of eustatic sea-level rise and an increased rate of tectonic subsidence. Characteristically, the effect of rising relative sea/lake level on tidally dominated coastlines is to segment the fluvial depositional system into areas of dominant fluvial influence, areas of dominant tidal influence and areas of mixed fluvial and tidal influence. The segmentation of the depositional system commonly occurs along both depositional dip and depositional strike, which is probably the cause of highly variable sandstone to mudstone ratios in the middle unit of the Wyandra Sandstone Member.

The introduction of basinal influences (tides and waves) on deposition necessitates the formation of the transgressive surface that represents the first landward shift of the shoreline. In the Wyandra Sandstone Member the transgressive surface is represented by either an erosive lag comprised of quartz and chert pebbles or, more commonly, a thin, carbonaceous mudstone unit. Regardless of its character, the transgressive surface separates fluvial influenced strata of the LST with unidirectional palaeocurrents below, from tidally and wave influenced strata of the TST with both unidirectional palaeocurrents and bi-directional palaeocurrents above. The pebble lag is interpreted to represent the reworking and winnowing of underlying LST strata by tidal currents. The thin mudstone unit is interpreted to have been deposited in the central estuarine zone where the destructive interference of tidal and fluvial processes and increased flocculation owing to saline conditions promote deposition from suspension. The transgressive surface is difficult to identify without full-hole core, but is commonly characterized by a high gamma spike and a sudden deviation of dip azimuth directions on image logs (Fig. 6).

The presence of wave ripples, deeper water trace fossils and the stratigraphical position of the upper unit of the Wyandra Sandstone Member are indicative of continued increase in palaeobathymetry and the continued landward migration of facies tracts. Both the upper and lower bounding surfaces of the upper unit of the Wyandra Sandstone Member are higher order flooding surfaces that represent sudden increases in palaeobathymetry. Neither of the flooding surfaces represents the maximum landward position of the shoreline and therefore neither flooding surface is identified as the maximum flooding surface. Instead, the maximum flooding surface, which represents the upper surface of the TST, is

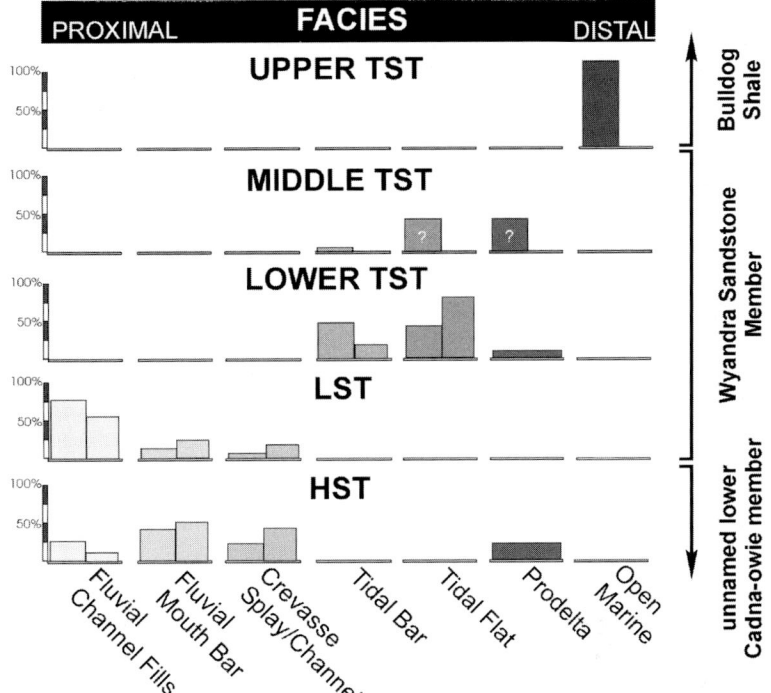

Fig. 8 Depositional facies for each systems tract of the Wyandra Sandstone Member shown as a percentage of total rock volume (left bar graph), and as a percentage of total facies occurrences (right bar graph). Up-section from the lowstand systems tract (LST), the proportion of more distal facies increases because of transgression and the increasingly landward position of the shoreline with time. The LST consists dominantly of fluvial channel deposits whereas the upper transgressive systems tract (TST) consists entirely of open marine deposits. HST, highstand systems tract.

interpreted to exist within the Bulldog Shale. Therefore, both the middle and upper unit of the Wyandra Sandstone Member as well as the lower section of the Bulldog Shale are part of the TST.

In summary, the Wyandra Sandstone Member represents a regressive–transgressive succession that dominantly reflects the superposition of increasingly distal facies tracts with the landward migration of the shoreline. Four sequence-stratigraphy surfaces are identified: a sequence boundary, a transgressive surface and two higher order flooding surfaces. The four sequence-stratigraphy surfaces define three units of genetically related strata, each with distinct depositional facies distributions and reservoir properties (Figs 8 & 9).

RESERVOIR HETEROGENEITY

Reservoir heterogeneity in the Wyandra Sandstone Member stems from the occlusion of primary porosity and permeability by kaolinite, illite, smectite, chlorite, vermiculite and calcite cements. The clays are the products of the diagenetic degradation of chemically unstable volcaniclastic grains and account for the majority of the porosity and permeability occlusion. Petrographic study suggests that good reservoir properties in the Wyandra Sandstone Member result from the secondary dissolution of porosity- and permeability-occluding clays and cements. Evidence for secondary dissolution includes anomalous grain-packing arrangements, dissolution rims, honeycomb structures and insoluble remnants in porosity, including empty, lath-shaped clay rims (Fig. 10). The occurrence and intensity of secondary dissolution is the main control on reservoir heterogeneity in the Wyandra Sandstone Member.

The spatial variation in the occurrence and intensity of secondary dissolution in the Wyandra Sandstone Member is of considerable commercial importance because of its relationship with reservoir properties. A natural permeability divide at c. 10 mD separates sandstone samples with petro-

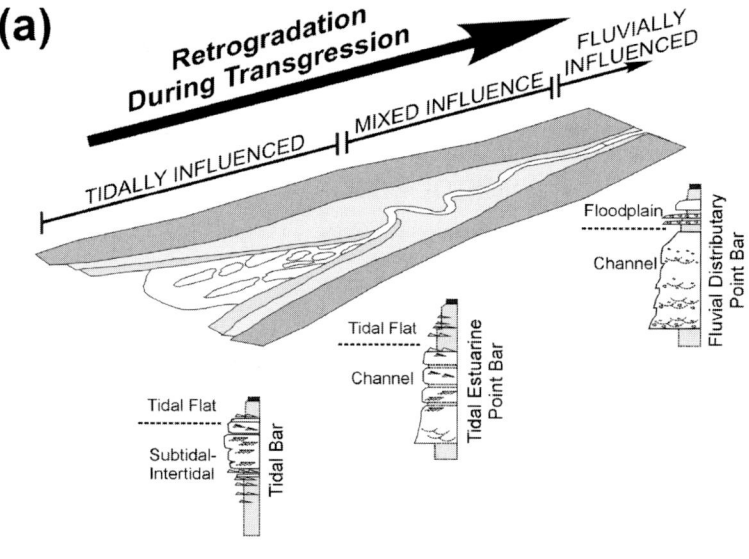

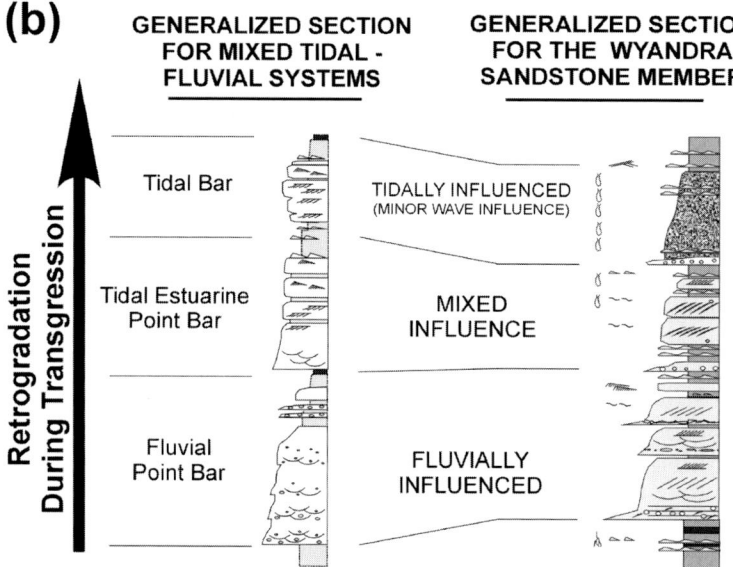

Fig. 9 Comparison of stratigraphical logs from a modern mixed, fluvial–tidal system (modified from Dalrymple et al., 1994) and the generalized stratigraphical log for the Wyandra Sandstone Member. The sedimentological characteristics and the stratigraphical architecture of the Wyandra Sandstone Member are interpreted to result from the superposition of increasingly distal facies tracts in a mixed fluvial–tidal system. Medium- to very coarse-grained sandstone is largely confined to the lowstand systems tract, which represents the fluvially influenced, basal part of the succession.

graphic indications of secondary dissolution from sandstone samples without petrographic indications of secondary dissolution (Fig. 11). Image analyser measurements coupled with petrographic studies indicate that secondary dissolution in the Wyandra Sandstone Member occurred preferentially in medium- to very coarse-grained sandstones (Fig. 12). Very fine- and fine-grained sandstones are likely to have hindered the extensive movement of diagenetic fluids and, therefore, the removal of porosity- and permeability-occluding clays and cements was restricted. Consequently, grain-size partitioning during the deposition of the Wyandra Sandstone Member is a significant primary depositional control on the subsurface distribution of reservoir properties (Fig. 13).

Understanding the relationship between the primary sedimentological characteristics (sandstone composition and texture) and reservoir properties is important for evaluating the utility of the sequence-stratigraphy approach. In the case

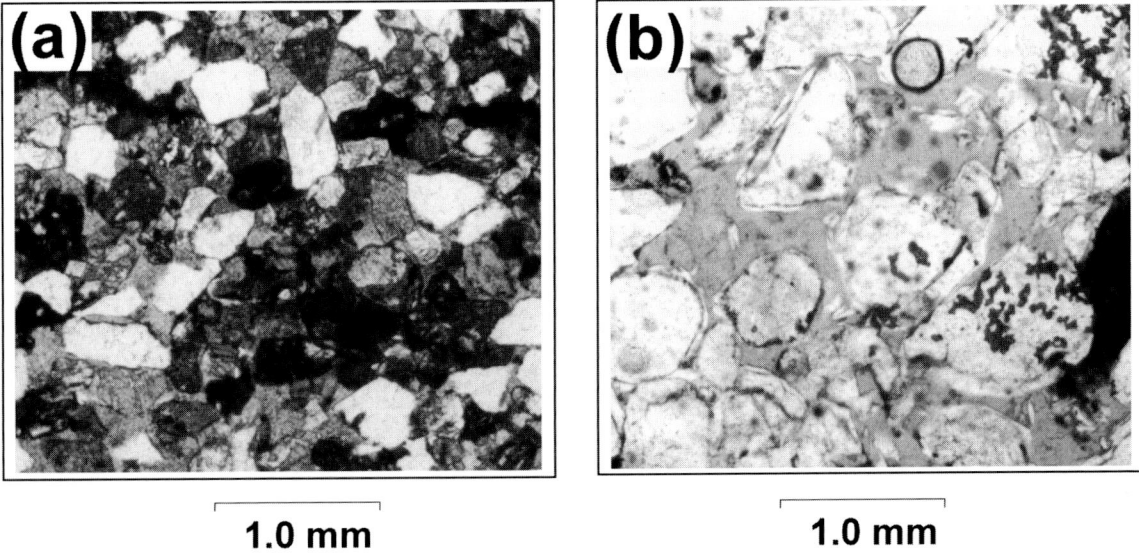

Fig. 10 Thin-section photomicrographs (plain polarized light) of (a) non-reservoir and (b) reservoir rock in the Wyandra Sandstone Member. Both samples are from the basal sections of fluvial channel deposits. The non-reservoir sample (a), however, is characterized by low compositional maturity and authigenic-clay filling porosity, which results in poor reservoir properties (porosity = 16.5%, permeability = 0.39 mD). In contrast, the reservoir sample (b) is compositionally mature and has less authigenic-clay filling poor space resulting in good reservoir properties (porosity = 20.1%, permeability = 1863 mD). Feldspathic and volcanogenic framework grains and clays are interpreted to have been removed from the reservoir sample by secondary dissolution.

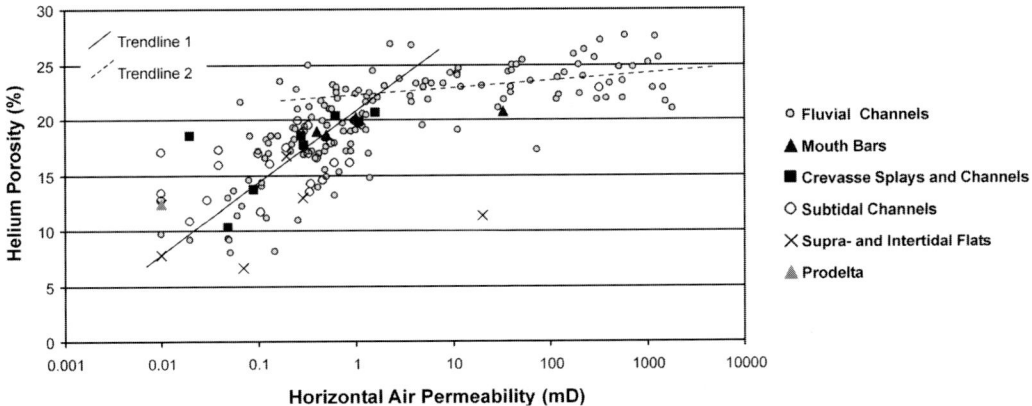

Fig. 11 Core plug porosity and permeability (measured from full-hole core) of the Cadna-owie Formation is plotted from full-hole cores taken at Tarbat–Ipundu and other nearby oil fields. Facies associations are also represented on the graph. Fluvial channel deposits are more likely to display good reservoir properties than other facies associations. The difference in slope between trendlines 1 and 2 is the result of enhanced permeability through the secondary dissolution of authigenic clay and cement in the more permeable samples. Secondary dissolution affects permeability more than porosity because diagenetic processes typically have a greater affect on pore-throat geometry than on the whole intergranular volume.

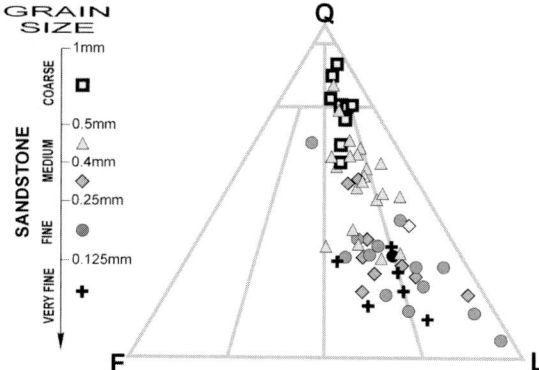

Fig. 12 Quartz (Q), feldspar (F) and (L) lithic ternary plot of petrographic samples from the Wyandra Sandstone Member with grain size indicated. In the Wyandra Sandstone Member, medium- to coarse-grained samples are the more compositionally mature than finer grained samples. This is interpreted to result from preferential dissolution of chemically unstable volcanolithic and feldspathic framework grains in medium- to coarse-grained sandstone.

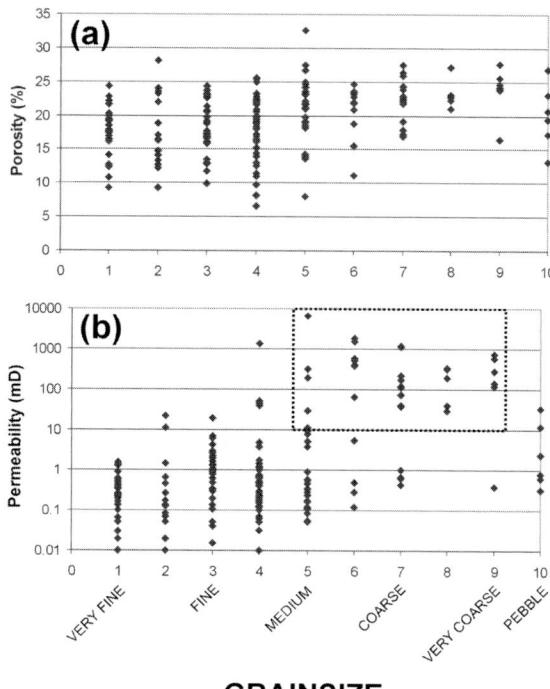

Fig. 13 Relationship between grain size and reservoir properties in the Wyandra Sandstone Member. Grain size was determined from visual estimates during analysis of full-hole cores. (a) Grain size plotted against helium injection porosity. (b) Grain size plotted against air permeability. Permeability > 10 mD is more likely to occur in medium- to very coarse-grained sandstones.

of the Wyandra Sandstone Member, medium- to very coarse-grained sandstones exist almost exclusively in the LST, which is defined only by mapping key sequence-stratigraphy surfaces (Fig. 14). Partitioning of medium- to very coarse-grained sandstones to the LST is interpreted to result from two processes. First and most significantly, stable relative sea/lake level conditions during sequence boundary formation caused high-energy, channelized facies to shift basinward through progradation, resulting in the preservation of coarse-grained deposits in the LST. Second, sequence boundary formation triggered the delivery of a greater proportion of compositionally mature, coarse-grained sediment sourced dominantly from a metamorphic terrain. Most of the high permeability rock (> 10 mD) in the Wyandra Sandstone Member (73% of the total rock volume) exists in the LST. Internal heterogeneity within the LST is significant, however, despite the fact that the LST is laterally extensive across the study area—reservoir rock is confined to specific facies and specific locations within the LST. Consequently, identifying and mapping the LST is only the first step in using sequence stratigraphy to aid reservoir development.

The internal, lateral reservoir heterogeneity within the LST stems from grain-size partitioning during deposition. Significant secondary dissolution occurs only in thick (> 5 m) successions of stacked, high-energy fluvial distributary channel deposits with palaeocurrents trending to the west and south-west. Stacked fluvial distributary channel deposits occur preferentially in areas of the LST that are characterized by thick sediment accumulations relative to areas dominated by finer grained, non-reservoir deposits.

Incised valley deposits typically have a similar pattern of reservoir rock distribution and isopach thickness to that which is seen in the Tarbat–Ipundu Field. Additionally, the Wyandra Sandstone Member vertical succession is similar to many incised valley-fill successions described in the literature (Dalrymple et al., 1992; Allen & Posamentier, 1993; Archer et al., 1994; Zaitlin

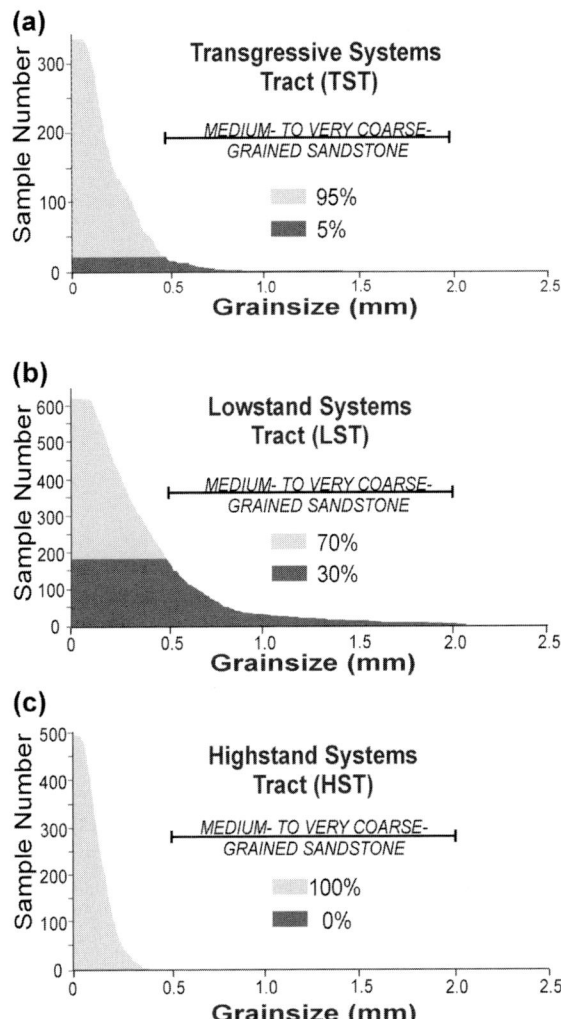

Fig. 14 Image analyser grain-size measurements for (a) transgressive, (b) lowstand and (c) highstand systems tracts of the Wyandra Sandstone Member and the unnamed lower Cadna-owie Member. The number of image analyser measurements are plotted on the vertical axis and grain size is plotted on the horizontal axis. Medium- to very coarse-grained sandstones (dark grey) exist dominantly in the LST.

et al., 1994; Feldman et al., 1995; Fig. 15). In a classic incised valley scenario, lowstand fluvial channel deposits are confined within the valley, whereas areas outside the incised valley are characterized by exposure and non-deposition (the interfluve). In the Tarbat–Ipundu Field a case can be made for either several incised valleys trending NE–SW across the field, or a case could be made for a single larger incised valley trending SE–NW across the field, parallel with the main structural grain of the basin. Although there is a palaeocurrent component to the north-west, it is not associated with thick LST fluvial deposits. For thick fluvial LST deposits with permeabilities of > 10 mD, a strong west to south-west palaeocurrent trend exists transverse to the main structural grain of the basin (Fig. 16).

Despite some similarities to classic incised-valley models, the patterns of isopach thickness variation between the LST and the underlying, non-reservoir units of the highstand systems tract (HST) suggests that the degree of fluvial incision does not control the distribution of depositional facies in the Wyandra Sandstone Member (Fig. 17). Where stacked, relatively high-energy channelized deposits are well developed, detailed correlations suggest that the depth of fluvial incision at the sequence boundary does not exceed 5 m within 1 km laterally. Detailed sequence-stratigraphy correlations suggest that the thickness of the LST is controlled by a combination of incision and subsidence (Fig. 18). Additionally, all image logs and full-hole cores that intersect the sequence boundary show that the surface is everywhere erosive, associated with pebble lags and mudstone rip-up clasts. There is no evidence for an incised valley interfluve that is characterized by non-deposition and signs of subaerial exposure rather than fluvial incision. Instead, LST strata in the western and south-western parts of Tarbat–Ipundu Field have log motifs that are indicative of more distal mouth-bar facies, which were deposited contemporaneously with the stacked fluvial distributary channels in the eastern and north-eastern parts of the field.

Although some fluvial incision does occur at Tarbat–Ipundu Field in positions proximal to Tarbat–Ipundu Fault, LST fluvial channels are not confined within incised valleys, and therefore the areal distribution of reservoir rock is not exclusively controlled by the position, shape and orientation of incised valleys. Instead, reservoir heterogeneity is controlled by internal heterogeneity within a 'sheet-like' reservoir of amalgamated fluvial channels. Outcrop and subsurface examples of sheet-like 'mosaics' of LST channelized deposits that may be similar to the LST of the

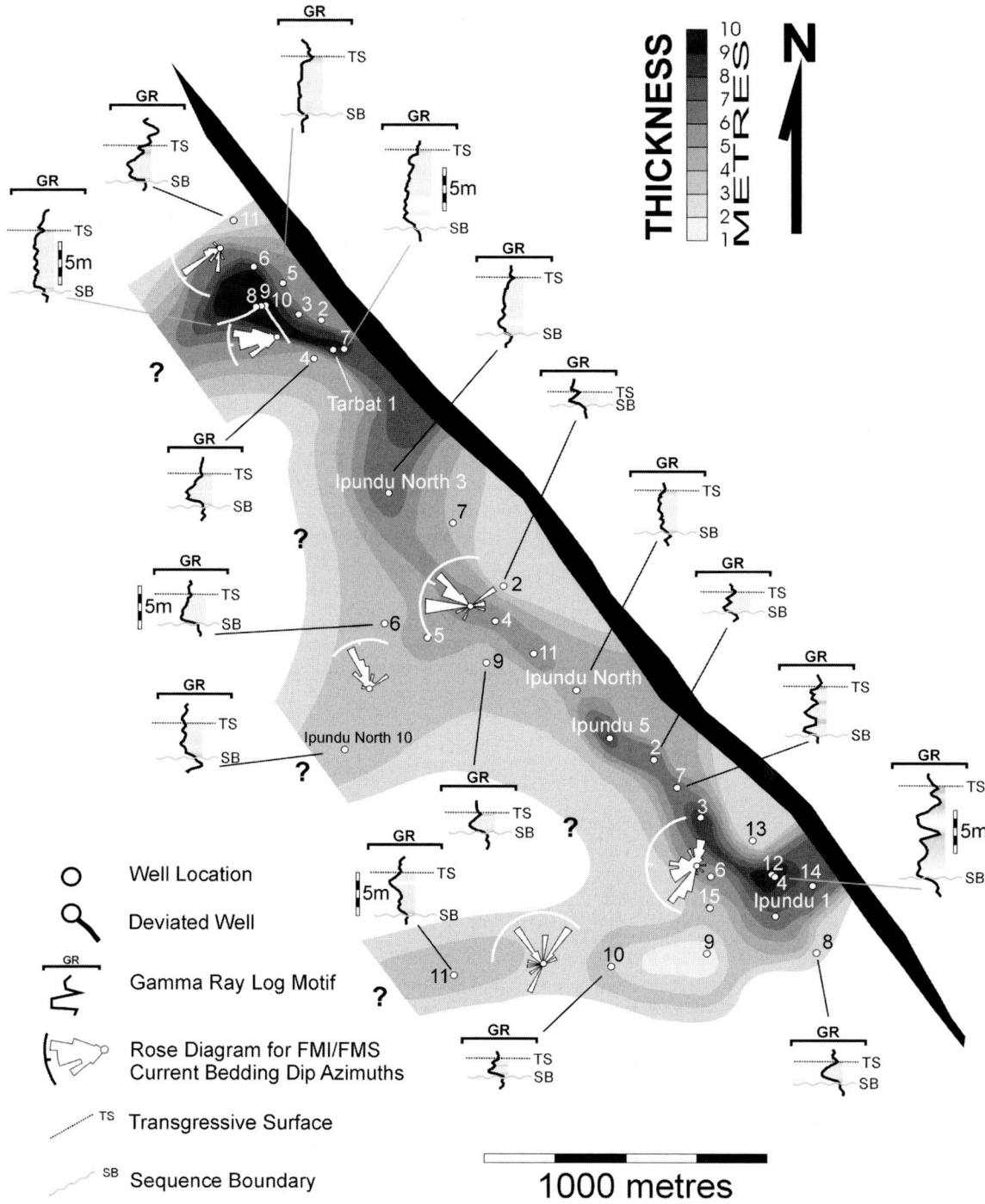

Fig. 15 Isopach map of the lowstand systems tract with FMI and FMS palaeocurrent diagrams and gamma-ray log motifs. Medium- to very coarse-grained reservoir sandstones are largely confined to areas of thick accumulations of high-energy stacked fluvial channel deposits with low variability palaeocurrents trending to the west (e.g. Tarbat 1, 5, 8 and Ipundu 4). Stacked fluvial channel deposits are characterized by relatively thick, low value, blocky or fining upward log motifs (e.g. Ipundu 4, Ipundu North 3 and Tarbat 1 and 5), whereas interdistributary, lower deltaic plain and prodelta deposits are characterized by higher value, thinner, serrated or coarsening upward log motifs (e.g. Ipundu 7 and 2, and Ipundu North 2 and 10).

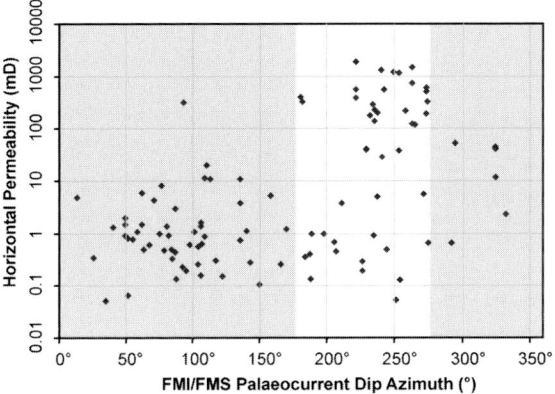

Fig. 16 Palaeocurrent determinations by FMI and FMS plotted against horizontal permeability (measured from core plugs). The highlighted area (white) shows the mean dip azimuth ±1 SD for palaeocurrents associated with permeability > 10 mD. High permeability rock is associated with south to south-west trending palaeoflow.

Wyandra Sandstone Member have been described in the literature (Van Wagoner, 1995; Holbrook, 1996; Hampson et al., 1999; Martinsen et al., 1999).

Modern environments that are deemed similar to the LST depositional system of the Wyandra Sandstone Member lend insight into the internal distribution of medium- to very coarse-grained sandstones, which controls reservoir heterogeneity. The LST of the Wyandra Sandstone Member was deposited in a relatively tectonically quiescent interior cratonic setting probably with low depositional slopes and with negligible marine influence. Climatic conditions are thought to have been warm and temperate, with flooding events that resulted in the deposition of much of the preserved LST sediment. Lastly, the LST Wyandra Sandstone Member depositional systems associated with reservoir-quality rock appear to have run transverse to the main structural grain of the basin. Modern depositional systems of central Australia are therefore similar to the interpreted depositional environment of the LST of the Wyandra Sandstone Member in many respects.

Relatively small, flood-prone, fluvial fan depositional systems associated with localized structurally controlled highs are characterized by low depositional slopes, sandy sediment supply and no marine influence in modern basins of central Australia (Hicks, 1998). The most important architectural element of the fluvial fans is the fluvial fan apex, which is characterized by the amalgamation of high-energy fluvial channels and local incision in sections of the depositional system proximal to the structural high (Fig. 19). Like the lowstand reservoir rock of the Wyandra Sandstone Member, deposits of the fluvial fan apex are characterized by coarse grain sizes relative to other fan environments, and low variability in palaeocurrent dip azimuths. Away from the fan apex, reservoir deposits typically decrease in grain size, have a greater palaeocurrent variance, have fewer indications of channelization, and interfinger with finer-grained sandstone and mudstone deposited in interdistributary, lower deltaic plain and prodelta environments.

Thick accumulations of fluvial distributary channel deposits in the LST of the Tarbat–Ipundu Field are interpreted to be analogous to fluvial fan apices of many modern depositional systems running transverse to the main structural grain of the basin. During deposition, the areas of reservoir rock at Tarbat–Ipundu Field were characterized by a slightly higher rate of subsidence than surrounding areas that may have provided more efficient depositional slopes for fluvial distributary channels, but were not sufficiently high to result in a standing body of water. Additionally, initial incision during sequence boundary formation may have confined early LST fluvial deposits to specific positions along depositional strike, but the resulting reservoir geometry is significantly different to the reservoir geometry that results from a classic incised-valley scenario (Fig. 20). The depositional model proposed for the Wyandra Sandstone Member is shown in Fig. 21.

IMPLICATIONS FOR RESERVOIR PRODUCTION AND DEVELOPMENT

Stratigraphical units representing single depositional systems are the most compatible kind of stratigraphical division for the application of modern analogues because our knowledge of depositional processes stems directly from the observation of modern systems. Clearly, mapping single depositional systems is unrealistic, however, sequence-stratigraphy units (i.e. systems tracts) represent linkages of contemporaneous

Hypothetical, Incised Valley Data

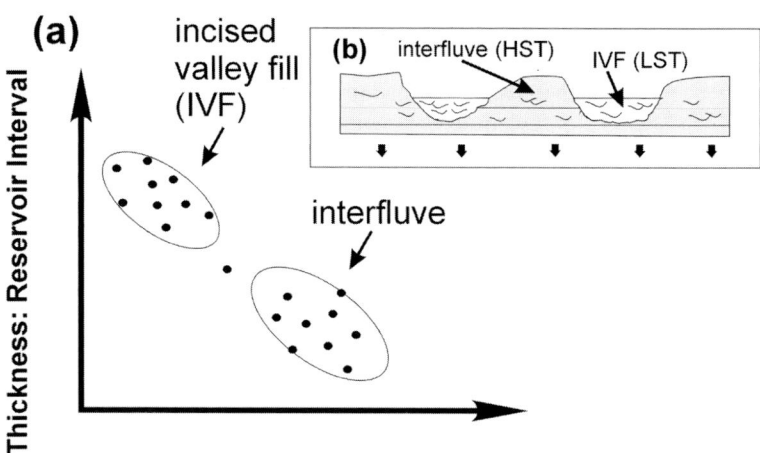

Actual Data

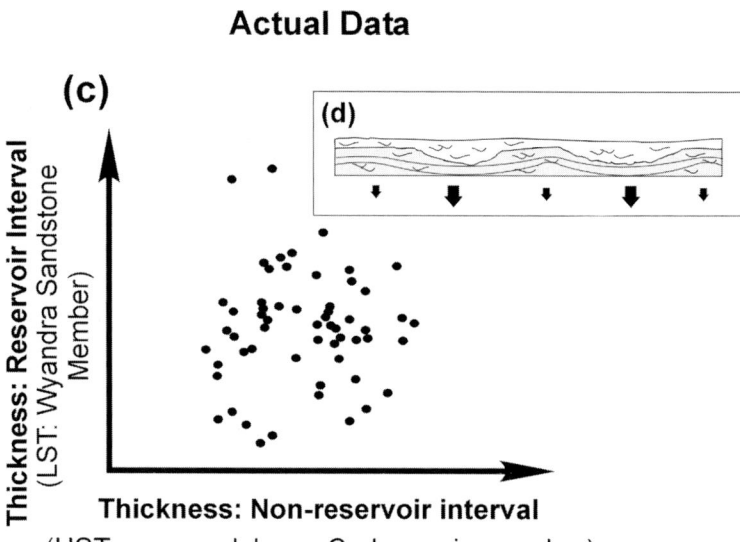

Fig. 17 Graphs of isopach thickness for lowstand systems tract (LST) reservoir intervals versus underlying highstand systems tract (HST) non-reservoir intervals are plotted. (a) A hypothetical incised valley scenario. (c) Wyandra Sandstone Member at Tarbat–Ipundu Field. For the hypothetical incised valley model in (a), the degree of incision strongly controls the distribution of the overlying reservoir unit resulting in a negative trend on the graph. (a & b) Two distinct suites of data may be evident that represent the incised valley fill (IVF) and the interfluve. (c & d) In contrast with the incised valley model, the actual data at Tarbat–Ipundu Field show no trend, suggesting that the degree of incision did not exclusively control the distribution of reservoir rock. The interpreted geometry of the reservoir rock at Tarbat–Ipundu Field is provided (d).

depositional systems (Brown & Fisher, 1977). Therefore, sequence-stratigraphy units are geologically meaningful stratigraphical divisions that are compatible with modern analogues. By comparison, lithostratigraphical units will generally group non-contemporaneous depositional systems that may have been deposited under a variety of depositional influences. Consequently, a number of different kinds of depositional environments will be represented in a lithostratigraphical unit and a single modern analogue will not accurately reflect depositional processes and the resulting facies distribution. A lithostratigraphical correlation of the Cadna-owie Formation treats the Wyandra Sandstone Member and some of the underlying strata as a single stratigraphical unit. At Tarbat–Ipundu Field, a lithostratigraphical correlation combines HST, LST and TST

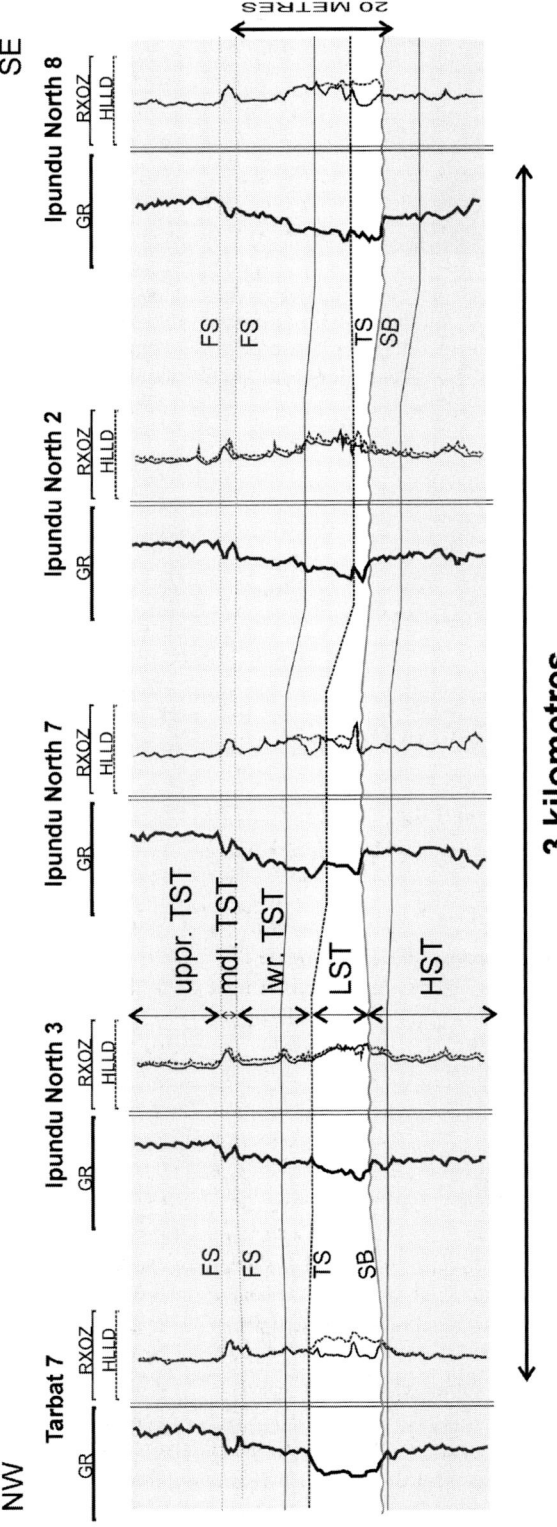

Fig. 18 North-west to south-east wireline correlation of sequence-stratigraphy surfaces in the Tarbat–Ipundu Field. The sequence boundary (SB), transgressive surface (TS), two flooding surfaces (FS) and two unspecified marker horizons are shown. The lowstand systems tract (LST) is highlighted. Detailed sequence-stratigraphy correlation suggests that thickness variation of the LST results from a combination of differential subsidence and fluvial incision.

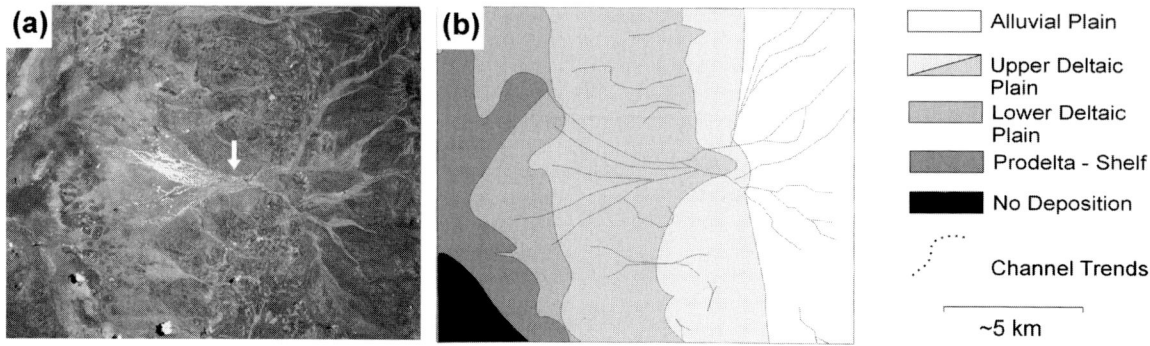

Fig. 19 (a) Aerial photograph and (b) facies distribution in the Grey Ranges of central Australia. One of many fluvial fan systems that drain into shallow lakes and floodplains in central Australia is shown. Thick sandstone accumulations are most likely to occur in the upper deltaic plain at the fluvial fan apex (arrow) where fluvial distributary channels amalgamate. The fluvial fan is an analogue for the lowstand systems tract of the Wyandra Sandstone Member at Tarbat–Ipundu Field.

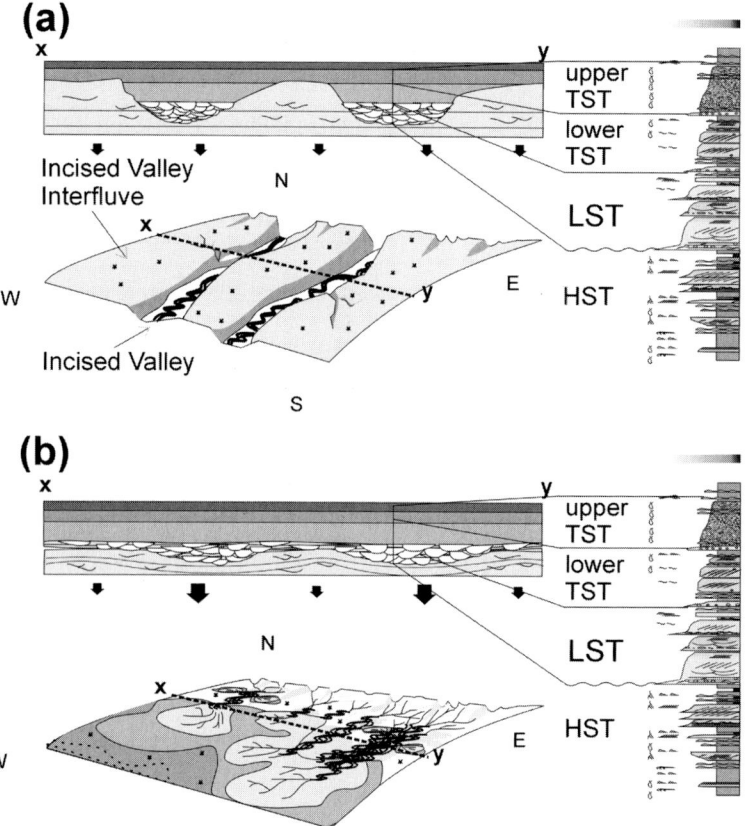

Fig. 20 (a) Conceptual incised valley model and (b) a depositional model based on modern systems for the lowstand systems tract of the Wyandra Sandstone. In the incised valley case (a), all lowstand fluvial deposits are confined to palaeovalleys and form a ribbon-like reservoir geometry. In the modern analogue case (b), fluvial deposits are unconfined by palaeovalleys and form a sheet-like reservoir geometry.

Fig. 21 (*opposite*) Conceptual block diagrams showing the depositional history of the Wyandra Sandstone Member and the underlying, unnamed lower Cadna-owie Member. The accompanying bar graphs are explained in Fig. 3. A eustatic sea-level fall is interpreted to have triggered sequence boundary formation and the deposition of the lowstand systems tract. By comparison with the highstand systems tract (a), the lowstand systems tract (b) is characterized by a basinward shift in facies tracts and a higher density of fluvial channels. The deposition of transgressive systems tract sediment was triggered by a rapid rise in sea level (c) that is interpreted to have converted fluvial channel systems to mixed fluvial and tidal estuarine systems. Continued rise in sea level resulted in the drowning of the depositional system (d) and the deposition of open marine shale.

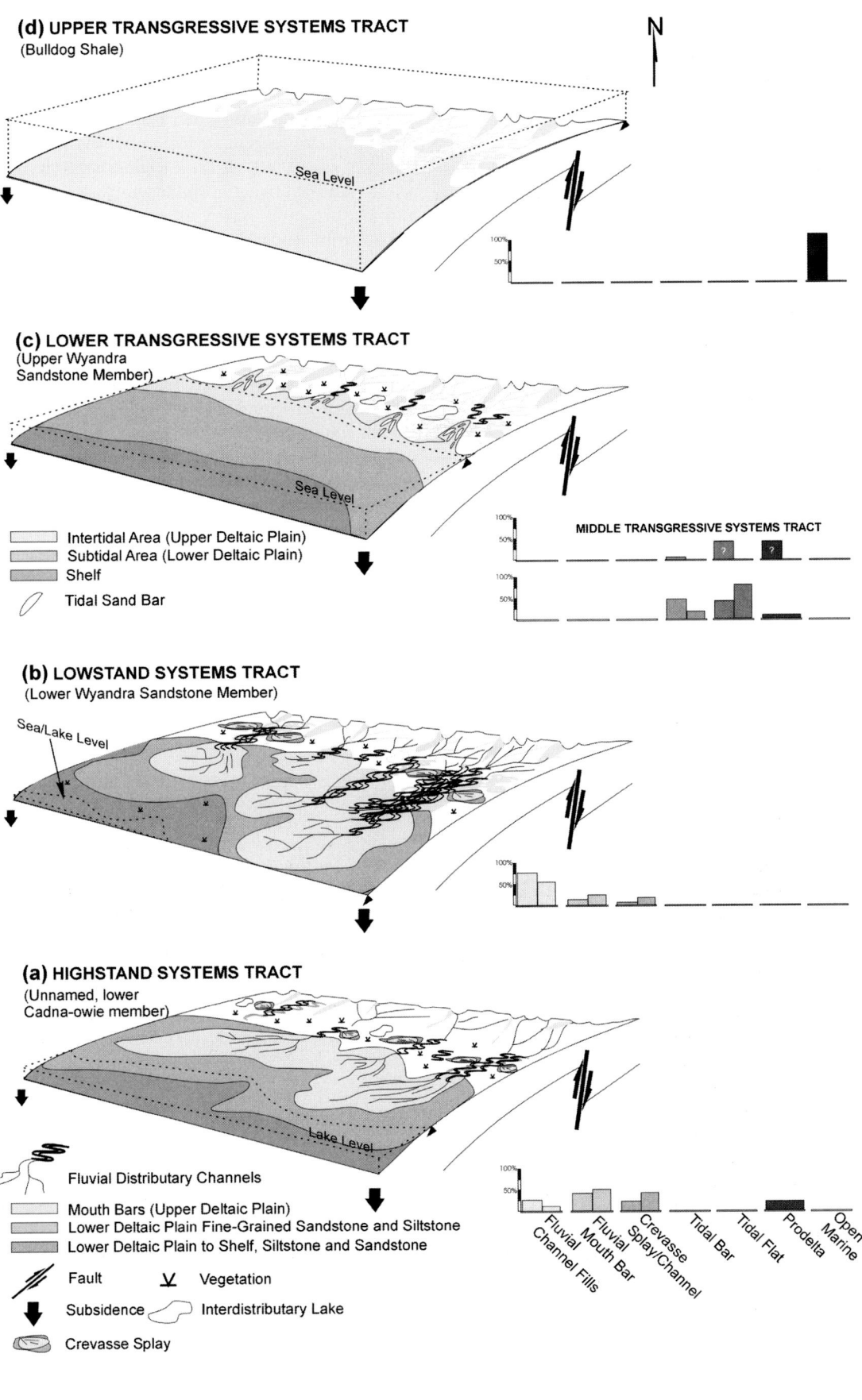

strata, each represented by different kinds of depositional environments and different facies distributions. However, the LST, which hosts the majority of reservoir strata, is defined only through the identification and mapping of sequence-stratigraphy surfaces.

Although mapping sequence-stratigraphy surfaces at Tarbat–Ipundu Field results in a more accurate subsurface map of the reservoir unit, the most pressing problem with the development of the Tarbat–Ipundu Field is the lateral reservoir heterogeneity within the LST. Although an increasingly accurate model of the subsurface is produced through an iterative process of drilling development-wells that both test and refine the geological model, the risk of drilling uneconomic development wells is not significantly reduced with each successive well. The sequence-stratigraphy approach facilitates a more efficient route to reducing development risk by incorporating modern analogues in the construction of geological models (Lang et al., 2000).

In the case of the Wyandra Sandstone Member, the comparison of modern depositional systems with the LST suggests that sandstone with good reservoir properties will be restricted to proximal positions in the palaeodepositional system, equivalent to the position of the fluvial fan apex in similar modern depositional systems of central Australia. Consequently, finer grained, more distal facies in the eastern and south-eastern portions of the Tarbat–Ipundu Field are likely to be non-prospective in terms of reservoir development. The occurrence of sandstones showing good reservoir properties (palimpsest fluvial fan apices) along depositional strike is likely to be dependent on the localized variations in the rate of subsidence and related patterns of incision. The pattern of local subsidence is probably controlled by the style of basement tectonics during deposition of the LST and differential compaction in cover units underlying the LST.

The use of modern analogues for a given sequence-stratigraphy unit (i.e. systems tract) is valuable for the construction of geological models to guide reservoir development. Information regarding the scale, interconnectivity, geometry and lateral distribution of various depositional elements is critical for the development of a geological model, and often can be obtained only through comparison with modern analogues (Lang et al., 2000). A clear understanding of the relationship between primary rock composition and texture, and reservoir properties is critical in usefully comparing hydrocarbon reservoirs with modern depositional systems.

ACKNOWLEDGEMENTS

The authors would like to thank Santos Ltd, Drillsearch Energy Limited, and CPC Energy Pty Ltd for financial assistance, provision of data and permission to publish. Thorough and constructive reviews by John Howell and Howard Feldman greatly improved this manuscript. We are grateful to staff and students from the Australian School of Petroleum and from Queensland University of Technology who provided input and support.

REFERENCES

Allen, G.P. and Posamentier, H.W. (1993) Sequence stratigraphic and facies model of an incised valley fill: the Gironde Estuary, France. *J. Sediment. Petrol.*, **63**(3), 378–391.

Allen, G.P., Lang, S., Musakti, O. and Chirinos, A. (1996) Application of sequence stratigraphy to continental successions: implications for the Mesozoic cratonic interior basins of Australia. *Mesozoic Geology of Eastern Australia Plate Conference,* Brisbane, 23–26 September. *Geological Society of Australia, Extended Abstracts,* **43**, 22–26.

Archer, A.W., Lanier, W.P. and Feldman, H.R. (1994) Stratigraphy and depositional history within incised palaeovalley fills and related facies, Douglas Group (Missourian/Virginian; Upper Carboniferous) of Kansas, USA. *Soc. Econ. Paleontol. Mineral. Spec. Publ.*, **51**, 175–190.

Bowden, D.W., Weimer, P. and Scott, A.J. (1993) The relative success of siliciclastic sequence stratigraphic concepts in exploration: examples from incised valley fill and turbidite system reservoirs. In: *Siliciclastic Sequence Stratigraphy: Recent Developments and Applications* (Eds Weimer, P. and H.W. Posamentier). *Am. Assoc. Petrol. Geol. Mem.*, **58**, 259–282.

Brown, L.F. and Fisher, W.F. (1977) Seismic stratigraphic interpretations of depositional systems: examples from Brazil rift and pull-apart basins. In: *Seismic Stratigraphy—Applications to Hydrocarbon Research* (Ed. C.E. Payton). *Am. Assoc. Petrol. Geol. Mem.*, **26**, 213–248.

Dalrymple, R.W., Zaitlin, B.A. and Boyd, R. (1992) Estuarine facies models: conceptual basins and stratigraphic implications. *J. Sediment. Petrol.*, **62**, 1030–1146.

Feldman, H.R., Gibling, M.R., Archer, A.E., Wightman, W.G. and Lanier, W.P. (1995) Stratigraphic architecture of the Tonganoxie paleovalley fill (Lower Virginian) in northeastern Kansas. *Am. Assoc. Petrol. Geol. Bull.*, **79**(7), 1019–1043.

Frakes, L.A. (1979) *Climates Throughout Geologic Time*. Elsevier, Amsterdam.

Gorter, J.D. (1994) Sequence stratigraphy and the depositional history of the Murta Member (upper Hooray Sandstone), South-eastern Eromanga Basin, Australia: implications for the development of source and reservoir facies. *Austral. Petrol. Prod. Explor. J.*, **32**, 644–673.

Green, P.M. and McKellar, J.L. (1996) Stratigraphic relationship between latest Triassic–Early Cretaceous basins of South Australia. *Mesozoic Geology of Eastern Australia Plate Conference, Geological Society of Australia, Extended Abstracts*, **43**, pp. 218–223. Queensland.

Hampson, G., Elliott, T., Flint, S. and Stollhofen, H. (1999) Incised valley fill sandbodies in Upper Carboniferous fluvio-deltaic strata: recognition and reservoir characterisation of Southern North Sea analogues. In: *Petroleum Geology of Northwest Europe* (Eds A.J. Fleet and S.A.R. Boldy). *Proceedings of the 5th Conference*, Vol. **2**, pp. 771–788. Geological Society Publishing House, Bath.

Haq, B.U., Hardenbol, J. and Vail, P.R. (1988) Mesozoic and Cenozoic chronostratigraphy and eustatic cycles. In: *Sea Level Changes: an Integrated Approach* (Eds C.K. Wilgus, B.S. Hastings, C.G.St.C. Kendal, H.W. Posamentier, C.A. Ross and J.C. van Wagoner). *Soc. Econ. Paleontol. Mineral. Spec. Publ.*, **42**, 39–45.

Helby, R., Morgan, R. and Partridge, A.D. (1987) Palynological zonation of the Australian Mesozoic. *Assoc. Australas. Palaeontol. Mem.*, **4**, 1–94.

Hicks, T.R. (1998) Stratigraphy, reservoir geometry and facies architecture of a modern arid playa basin lacustrine delta, Neales River Delta, Lake Eyre, Australia. Honours thesis, Queensland University of Technology, Brisbane.

Holbrook, J.M. (1996) Complex fluvial response to low gradients at maximum regression: a genetic link between smooth sequence-boundary morphology and architecture of overlying sheet sandstone. *J. Sediment. Petrol.*, **66**(4), 713–722.

Kennard, J.M., Allen, G.P. and Kirk, R.B. (1999) Sequence stratigraphy: a review of fundamental concepts and their application to petroleum exploration and development in Australia. *Austral. Geol. Surv. Org. J. Geol. Geophys.*, **17**(5/6), 77–104.

Lang, S.C., Kassan, J., Benson, J.M., Grasso, C.A. and Avenell, L.C. (2000) Applications of modern and ancient geological analogues in characterisation of fluvial and fluvial–lacustrine deltaic reservoirs in the Cooper Basin. *Austral. Petrol. Prod. Explor. J.*, **40**, 393–416.

Lang, S.C., Grech, P., Root, R., Hill, A. and Harrison, D. (2001) The application of sequence stratigraphy to exploration and reservoir development in the Cooper–Eromanga–Bowen–Surat basin system. *Austral. Petrol. Prod. Explor. J.*, **41**, 223–249.

Legarreta, L. and Uliana, M.A. (1998) Anatomy of hinterland depositional sequences: upper Cretaceous fluvial strata, Neuquen Basin, West-central Argentina. In: *Relative Role of Eustasy, Climate and Tectonism in Continental Rocks* (Eds K.W. Shanley and P.J. McCabe). *Soc. Econ. Paleontol. Mineral. Spec. Publ.*, **59**, 83–93.

Lonergan, T.P., Ryles, P.G., McClure, S.T. and McMillan, D.W. (1998) The Tarbat–Ipundu Field, a case study in identifying bypassed oil. *Austral. Petrol. Prod. Explor. J.*, **36**, 36–50.

Martinsen, O.J., Ryseth, A., Helland-Hansen, W., Flesche, H. Torkildsen, G. and Idil, S. (1999) Stratigraphic baselevel and fluvial architecture: Ericson Sandstone (Campanian), Rock Springs Uplift, SW Wyoming, USA. *Sedimentology*, **46**, 235–259.

Morgan, R. (1980) *Eustacy in the Australian Early and Middle Cretaceous*. Bulletin No. 27, Department of Mineral Resources, Brisbane.

Moussavi-Harami, R. (1996) Burial history of the Eromanga Basin in northeast South Australia. *Mesozoic Geology of Eastern Australia Plate Conference, Geological Society of Australia, Extended Abstracts*, **43**, pp. 400–403. Queensland.

Musakti, O.T. (1997) Regional sequence stratigraphy of a non-marine intracratonic succession; the Hooray Sandstone and the Cadna-owie Formation, Eromanga Basin, Queensland. Unpublished Masters thesis, Queensland University of Technology, Brisbane.

Posamentier, H.W. (2001) Lowstand alluvial bypass systems: incised vs. unincised. *Am. Assoc. Petrol. Geol. Bull.*, **85**(10), 1771–1793.

Posamentier, H.W. and Allen, G.P. (1993) Variability of the sequence stratigraphic model: effects of local basin factors. *Sediment. Geol.*, **86**, 91–109.

Posamentier, H.W. and Allen, G.P. (1999) Siliciclastic sequence stratigraphy—concepts and application. *Soc. Econ. Paleontol. Mineral. Concepts Sedimentol. Paleontol.*, **7**, 216 pp.

Posamentier, H.W. and Weimer, P. (1993) Siliciclastic sequence stratigraphy and petroleum geology—where to from here?. *Am. Assoc. Petrol. Geol. Bull.*, **77**(5), 731–742.

Price, P.L., Filatoff, J., Williams, A.J., Pickering, S.A. and Wood, G.R. (1985) *Late Palaeozoic and Mesozoic Palynostratigraphical Units*. Oil and Gas Division Report 274/25, Colonial Sugar Refining Company, Sydney.

Shanley, K.W. and McCabe, P.J. (1994) Perspectives on the sequence stratigraphy of continental strata. *Am. Assoc. Petrol. Geol. Bull.*, **78**(4), 544–568.

Van Wagoner, J.C. (1995) Sequence stratigraphy and marine to non-marine facies architecture of foreland basin strata, Book Cliffs, Utah, U.S.A. In: *Sequence Stratigraphy of Foreland Basin Deposits: Outcrop and Sub-Surface Examples from the Cretaceous of North America* (Eds J.C. Van Wagoner and G.T. Bertram). *Am. Assoc. Petrol. Geol. Mem.*, **64**, 137–223.

Van Wagoner, J.C., Mitchum, Jr., R.M., Campion, K.M. and Rahmanian, V.D. (1990) Siliciclastic sequence stratigraphy in well logs, cores and outcrops: concepts for high-resolution correlation of time facies. *Am. Assoc. Petrol. Geol. Methods Explor. Ser.*, **7**, 55.

Zaitlin, B.A., Dalrymple, R.W. and Boyd, R. (1994) The stratigraphic organisation of incised-valley systems associated with relative sea level change. *Soc. Econ. Paleontol. Mineral. Spec. Publ.*, **51**, 175–190.

Zhou, S. (1989) Subsidence history of the Eromanga Basin, South Australia. In: *The Cooper and Eromanga Basins Australia* (Ed. B.J. O'Neil), *Proceedings of the Petroleum Exploration Society of Australia, Society of Petroleum Engineers, Australian Society of Exploration Geophysicists (SA Branches), Adelaide*, 251–264.

Recognition of a floodplain within braid delta deposits of the Oligocene Minato Formation, north-east Japan: fine deposits correlated with transgression

KOJI YAGISHITA* and OSAMU TAKANO†

*Department of Geology, Iwate University, Morioka-City, Iwate, Japan 020-8550
(Email: yagi@iwate-u.ac.jp); and
†Japan National Oil Corporation, Mihama-ku, Chiba-City, Chiba, Japan 261-0025

ABSTRACT

The fluvial Oligocene Minato Formation, north-east Japan, consists mostly of clast-supported conglomerate beds interpreted to represent deposition by a braided river system. Tuffaceous sandstone beds, a facies interpreted to indicate a nearshore sedimentary environment, however, are interbedded with conglomerates at the southern basin margin, suggesting that the distal Minato Formation represents a coalescing braid delta. Moreover, laterally extensive fine-grained sediments that include *in situ* lignite beds were recently exposed at a large quarry in the northern part of the basin, and are interpreted to be of floodplain origin. The recognition of fine-grained sediments within thick conglomerates enables the correlation of alluvial plain and marine strata, and provides an excellent chronostratigraphical basis for interpreting the two different depositional facies in terms of sequence stratigraphy.

The Minato Formation is interpreted to represent part of a transgressive systems tract, and correlations suggest that base-level rise can have a profound effect on inland alluvial sedimentation in a coarse-grained braided system. As a result of sea-level rise, widespread but thin shallow-marine sediments may extend a few tens of kilometres inland over fluvial deposits. The generally high gradient of braided rivers, however, means that shallow-marine sediments are unable to spread over gravelly braided systems. Instead, in the case of the Minato Formation, a slight change of stream gradient produced by the sea-level rise caused fining of the river gravels, resulting in floodplain formation.

INTRODUCTION

Facies changes in alluvial sediments inland, including braided channel deposits, may correlate with sea-level changes along basin margins. Such correlations, however, have not been fully documented and discussed to date, except for a few case studies (Shanley *et al.*, 1992; McLaurin & Steel, 2000). Indeed, because of discontinuous outcrops, or poor preservation of isochronous deposits, chronostratigraphical correlation between marine and coeval alluvial strata within a basin is generally difficult. Concepts of base-level control and sequence stratigraphy, however, provide the means to understand and correlate between these different facies successions.

The Oligocene Minato Formation, north-east Japan, consists mostly of clast-supported conglomerates typical of a braided river system. In this paper tuffaceous sandstones that occur interbedded with conglomeratic facies in the south-eastern portion of the outcrop belt are described, as well as sandy and silty sediments with lignite beds that are newly exposed in a quarry farther up-dip in the northern part of the basin. We argue that the Minato Formation was formed as part of a braided river and coalescing braid-delta system, and that these facies occur within a transgressive systems tract that records a period of relative sea-level rise. By correlating the two successions of fine-grained facies, it can be shown how the effects of sea-level rise were transmitted up-dip into the fluvial system.

GEOLOGICAL SETTING, LITHOFACIES AND PALAEOCURRENT ANALYSIS

The Oligocene Noda Group is distributed along the Pacific coast of north-east Japan (Fig. 1), and consists of the lower Minato and the overlying Kuki formations. The Minato Formation is about 180 m thick, with the Kuki Formation about 170 m thick, and both units are dominantly conglomerate. The contact between these two units is conformable.

Using the facies nomenclature of Miall (1996), the Minato Formation is dominated by clast-supported conglomerate (facies Gm, Table 1) of fluvial origin, in which most of the cobble-sized clasts are acidic volcanic rocks (Shimazu & Teraoka, 1962). Facies Gm has been interpreted to represent superimposed broad longitudinal bar deposits that occasionally interfinger with facies Gt (Yagishita, 1997). Facies Gt displays concave-up bedding that indicates small-scale channels dissecting obliquely into the main longitudinal bars (Table 1; Teisseyre, 1975; Yagishita, 1997). The absence of debris-flow deposits (facies Gms) is noticeable in the Minato Formation.

Imbrication of gravel clasts within facies Gm from the Minato Formation suggests an eastward-orientated palaeocurrent direction (Fig. 2; Yagishita, 1997). Although some palaeocurrent plots show a deflection to the south (points 11, 12 and 13 in Fig. 2), most data suggest that sediments were derived from topographical highs to the west of the basin. The highs most likely constituted a N–S trending range dominated by acidic volcanic rocks (Yagishita, 1997). Multiple sources for the gravelly sediments of the formation indicate a broad system of braided rivers. Both the northernmost (points 1, 2 and 3) and southernmost proximal source areas (17, 18 and 24 in Fig. 2) show an eastward or north-eastward palaeoflow direction. The distance between the two areas, however, is too far to suggest a linked drainage system.

The Kuki Formation also consists of clast-supported conglomerates, but clast compositions are dominated by chert and hornfels, indicating that they were derived from Lower Cretaceous basement rocks (Yagishita, 1997). In this case, imbrication measurements of facies Gm from this formation record sediments derived from a topographic high located to the north or north-east (points 19, 20 and 21 in Fig. 2).

Downslope decreases in maximum particle size (MPS), based on an average of 20 clasts, occur over a relatively long distance in the Minato Formation

Table 1 Facies classification in this study.

Facies code	Facies	Interpretation
Gm	Clast-supported gravel, crudely stratified	Longitudinal bar deposits in braided rivers
Gms	Matrix-supported gravel	Debris flow deposits
Gt	Clast-supported gravel with concave-up structure	Small-scale channel-fills dissecting longitudinal bar
Th	Tuffaceous sandstone bed with parallel laminations and inverse-grading in individual sand laminae	Swash-zone lamination
Tt	Tuffaceous sandstone bed with trough cross-bedding	Longshore current deposits
Sp	Medium-grained sandstone with planar cross-bedding	Laterally accreted channel bars
Sh	Fine-grained sandstone with parallel lamination	Overbank deposits
Fl	Siltstone bed with parallel lamination	Overbank deposits and crevasse splay

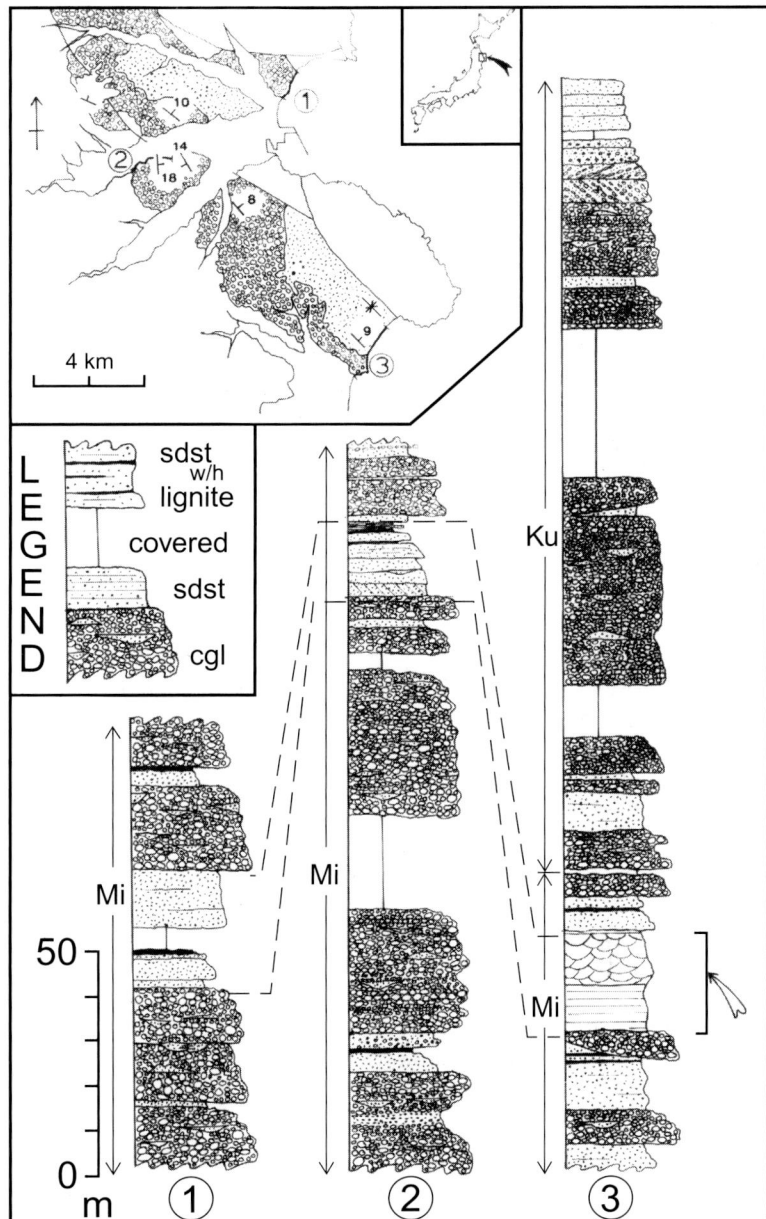

Fig. 1 Graphic logs of the Minato and Kuki formations of the Oligocene Noda Group. Arrow in log 3 indicates tuffaceous beds of shallow-marine origin, which are thought to be coeval with inland floodplain sediments in logs 1 and 2. The interval of fine-grained sediments in log 2 is observed in the newly excavated quarry. Mi, Minato Formation; Ku, Kuki Formation.

(Fig. 3). The MPS decreases downcurrent in lines a and b of Fig. 3 at rates of about 1 cm km^{-1}. This is much lower than the rate of decrease of MPS on other alluvial fans or fan-deltas. For example, according to calculations from data presented in Rust & Koster (1984), the rate of decrease for such fans ranges from 9 to 45 cm km^{-1}.

MARINE SEDIMENTS AT THE SOUTHERN BASIN MARGIN

Tuffaceous sandstone and tuff beds occur within the Minato Formation along the southernmost margin of the basin. These beds are bounded by

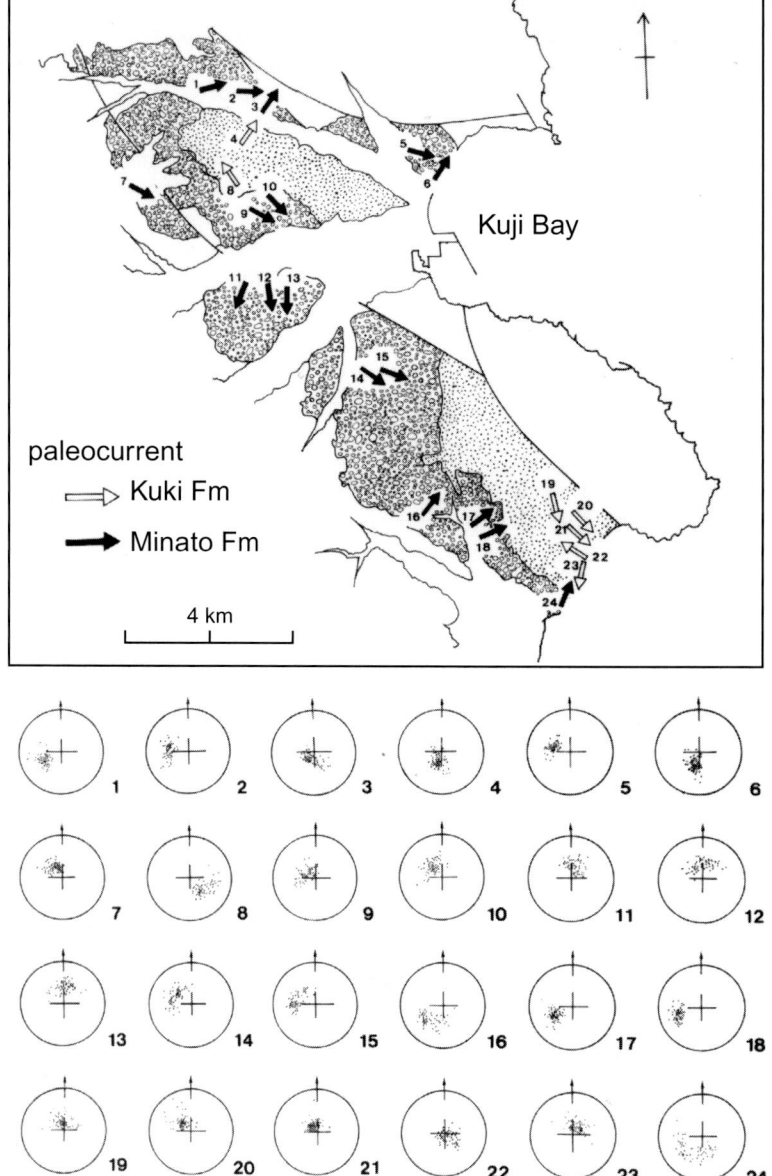

Fig. 2 Palaeocurrent directions deduced from plots of imbricated clasts in conglomerates, facies Gm. Each plot comprises 50 imbricated clasts selected along the bedding plane. All the data obtained in the field were corrected for structural dip, and all plots are shown in upper-hemisphere projection (after Yagishita, 1997).

fluvial conglomerates (facies Gm) as well as coarse-grained sandstones and lignite-bearing beds (Figs 1 & 4A). The absence of microfossils means that interpretation of the environment of deposition for the tuffaceous sandstone beds remains uncertain. A variety of data suggest, however, that these tuffaceous sandstone beds are of marine origin, as described below.

Facies analyses

The tuffaceous sandstone and tuff beds are about 20 m thick (Fig. 1), and consist of a lower parallel-laminated unit with significant lateral continuity (facies Th), and an upper trough cross-stratified unit (facies Tt, Fig. 4B & C). The parallel laminated coarse- to medium-grained sandstone and

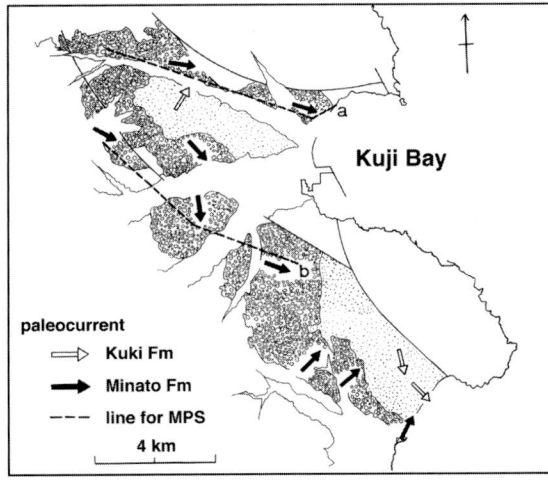

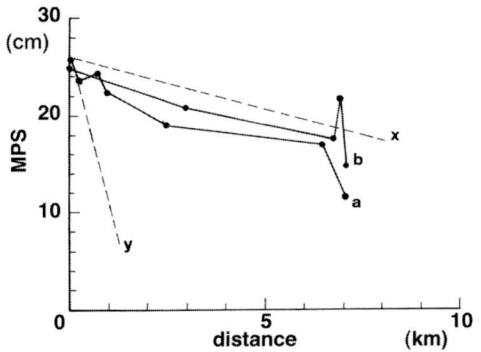

Fig. 3 Representative palaeoflow directions from Fig. 2 and the decreasing rate of MPS (maximum particle size) (Yagishita, 2001). The MPS of gravels was measured along NW–SE traverses a and b in the Minato Formation. Average rates of decreasing MPS in braided rivers (x) and in alluvial fans (y) are calculated from fig. 4 of Rust & Koster (1984).

tion is visible on many beaches, and inverse-grading within individual laminae is characteristic of swash lamination (Clifton, 1969). By contrast, currents flowing through longshore troughs generally tend to winnow finer materials and show coarse sand/or gravel including longshore-dipping medium-scale trough cross-bedding (Hunter et al., 1979). The trough cross-stratified tuffaceous beds in this study are therefore interpreted to represent dunes formed by longshore currents (Clifton et al., 1971; Davidson-Arnott & Greenwood, 1976).

Although the tuffaceous facies Th and Tt are bounded by fluvial conglomerate and sandstone beds along the southernmost margin of the basin (Fig. 1), the tuff beds are unknown farther inland to the north. This is one of the critical pieces of evidence that suggests the tuffaceous beds are of subaqueous origin. Subaerial airfall tuffs on the northern land areas might have been mostly eroded away (Yagishita, 1997) but, owing to rapid regression and burial in the shallow marine environment, voluminous airfalls mixed with nearshore marine sands probably survived. The lateral continuity (> 400 m) of the parallel-laminated tuffaceous sandstone beds of facies Th is another significant criterion that supports a subaqueous origin. The bedding produced in subaqueous environments, particularly in shallow-marine settings, tends to be more laterally uniform and less lenticular than that in fluvial environments (Clifton, 1973). Moreover, the average orientation of trough cross-bed axes in facies Tt indicates sediment transport from the south-east, which is at a high angle to palaeocurrent directions deduced from fluvial gravel imbrication (point 24 in Fig. 2). Sediment transport by longshore currents therefore provides an explanation for facies Tt.

Coastal onlap and downlap

The tuffaceous beds described above onlap the underlying fluvial conglomerate beds (Fig. 4A). These beds are then overlain by more fluvial conglomerate beds that downlap onto the tuffaceous beds. Judging from the outcrop, then, these tuffaceous beds represent a short-term transgression of the eastern open sea.

tuffaceous sandstone beds (facies Th) display a remarkable continuity (> 400 m), and generally are inverse-graded within laminae. By contrast, sediments of facies Tt include abundant pumices with sands, and they are generally coarser than those of facies Th.

Horizontally laminated tuffaceous sandstones (facies Th) are interpreted to represent sedimentation in a foreshore environment (Fig. 4B), particularly the inner planar facies formed in the swash zone (sensu Clifton et al., 1971). Parallel lamina-

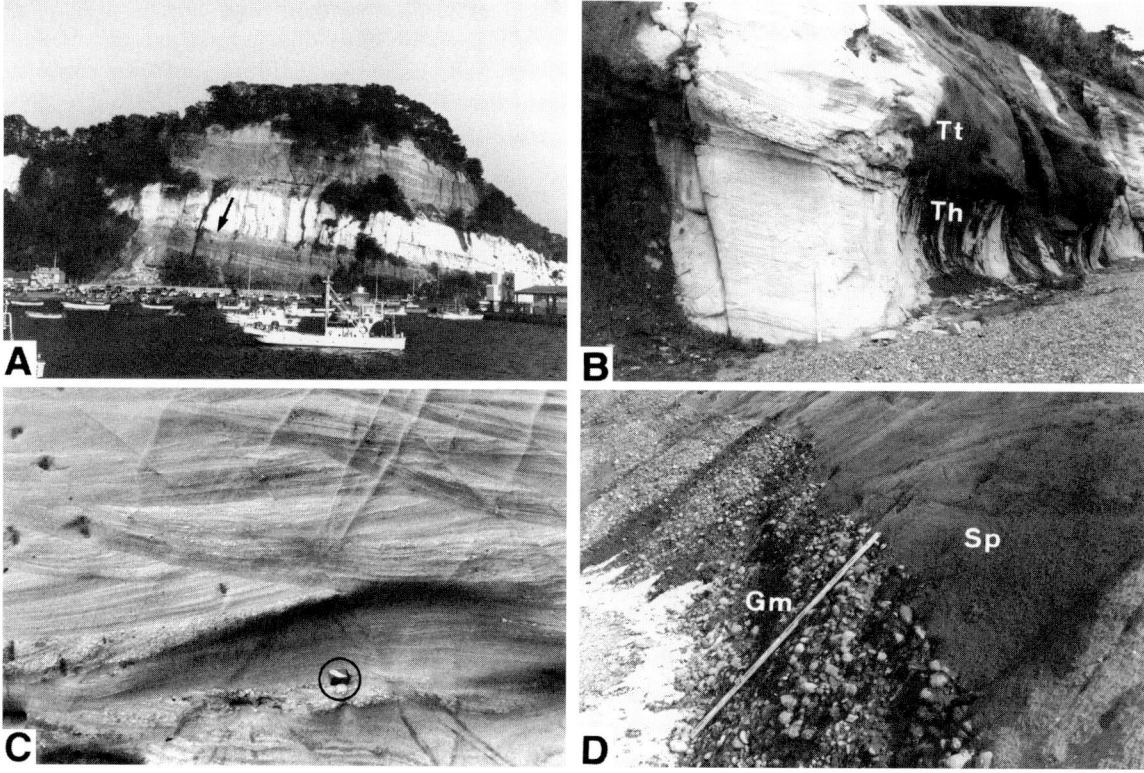

Fig. 4 Minato Formation lithofacies. (A) Tuffaceous sandstone beds of nearshore marine sediments. The bed displays coastal onlap (arrow) onto the underlying fluvial conglomerate and sandstone beds. The outcrop orientation is almost east (right)–west (left). (B) Tuffaceous sandstone beds showing parallel lamination (facies Th), probably formed in a swash zone, and the overlying trough cross-bedded (facies Tt). The scale bar is 1.5 m long. (C) Upper part of the tuffaceous beds (facies Tt), probably produced as longshore current deposits. The 35 mm film box (encircled) provides scale. (D) Abrupt facies change from clast-supported conglomerate bed (facies Gm) to planar cross-stratified sandstone bed (facies Sp) in the newly excavated quarry (see also Fig. 5A). The Jacob's staff is 1.5 m long.

Geochemical evidence

Sulphur contents in lignite and lignite-bearing siltstone beds occurring a few metres below the tuffaceous beds on the southern basin margin are much greater than those of the beds below the conglomerate beds of the quarry in the northern part of the basin (Fig. 5 and Table 2). Coal or lignite-bearing fine sediments immediately overlain by marine rocks typically display relatively high sulphur contents, whereas those overlain by non-marine sediments have low sulphur contents (e.g. Williams & Keith, 1963; Koma, 1992). Previous work has shown that much of the sulphur in high sulphur coals (lignites) can be produced by sulphate reduction after a peat swamp was flooded by sea water (McCabe, 1984). The high sulphur content in lignite and lignite-bearing siltstone beds below the tuffaceous sandstone beds of the Minato Formation is therefore interpreted to represent invasion of a peat swamp by marine waters and the resultant increased availability of the sulphate ion.

A variety of data, therefore, suggests that the tuffaceous sandstone beds are of shallow-marine origin, and the distal Minato Formation can be interpreted to represent a coalescing braid delta system.

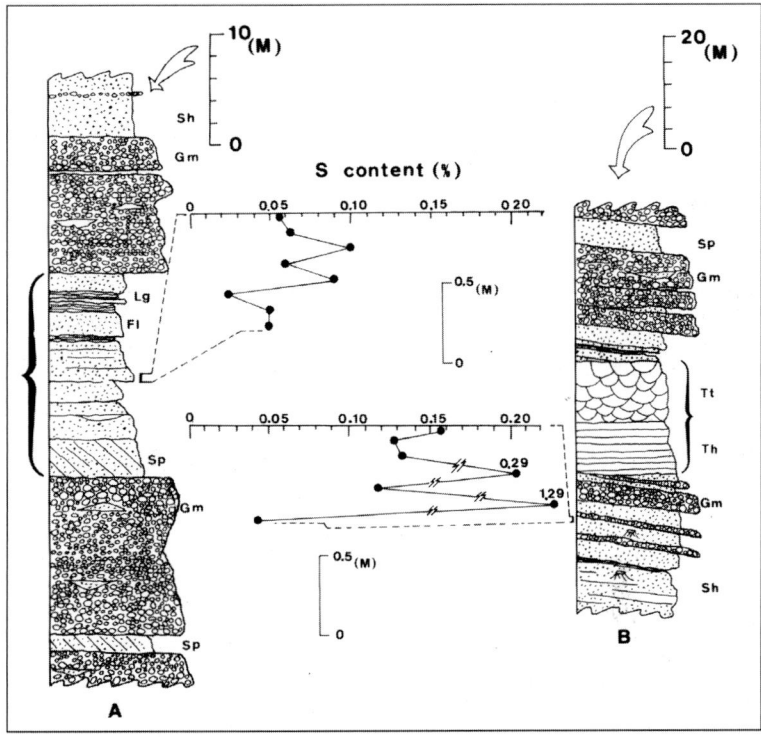

Fig. 5 Total sulphur contents from (A) the fine-grained sediments interval of log 2 in Fig. 1, and from (B) the sediment interval below the tuffaceous beds of log 3 in Fig. 1. Note that the high sulphur contents from sediments below the marine tuffaceous beds of Th and Tt (Table 2). Samples were taken from intervals 80 cm thick, with 10 cm spacing (see also Table 2).

Table 2 Sulphur contents* (wt %) in fine-grained sediments of the inland quarry and basin margin sections. See Figs 1 & 5 for details of logs 2 and 3.

Locality	Sample number	1	2	Average
Quarry (log 2)	NHC-1	0.052	0.053	0.053
	NHC-2	0.06	0.064	0.062
	NHC-3	0.106	0.098	0.102
	NHC-4	0.057	0.063	0.060
	NHC-5	0.086	0.094	0.090
	NHC-6	0.026	0.02	0.023
	NHC-7	0.051	0.044	0.048
	NHC-8	0.048	0.05	0.049
Basin margin (log 3)	NPORT-1	0.151	0.162	0.157
	NPORT-2	0.129	0.129	0.129
	NPORT-3	0.132	0.133	0.133
	NPORT-4	0.282	0.298	0.290
	NPORT-5	0.115	0.122	0.119
	NPORT-6	1.30	1.28	1.29
	NPORT-7	0.044	0.046	0.045

*Determined by X-ray fluorescence spectrometry (Phillips PW1480).

INLAND FLOODPLAIN ENVIRONMENT

A large quarry has been excavated recently (Figs 1 & 6A) about 9 km away from the southern outcrop of the marine beds. The rocks in the quarry contain very extensive, thick (about 19 m), medium- to fine-grained sandstone and siltstone beds with several *in situ* lignite seams. This unit is bounded above and below by clast-supported conglomerates (facies Gm, Fig. 6A), and is a fining and thinning upward succession that includes sharp-based medium-grained planar cross-strata (facies Sp, Fig. 4D), overlain by parallel-laminated sandstone (facies Sh), and parallel-laminated siltstone and mudstone (facies Fl, Fig. 6A), including a scoured channel-form structure. The top of this unit also includes several *in situ* lignite seams (Figs 1 & 5A).

There is no field evidence to suggest tidal influence, such as reactivation surfaces, sigmoidal bedding or bidirectional cross-beds. There is also no evidence for a prograding delta origin, such as a coarsening upward bay to mouth-bar succession

Fig. 6 Newly excavated quarry of the Minato Formation (log 2, Fig. 1). (A) General view of the section. The white arrows (c) and (l) display crevasse-splay channel and lignite seams, respectively. (B) Interlayered sand–mud bedding just below the lignite seams in the quarry. The hammer is 0.3 m long.

(Horne *et al.*, 1978), and the low sulphur content is characteristic of a non-marine environment of deposition. Hence, the sharp-based planar cross-stratified sandstone (facies Sp) is interpreted to represent dunes within a laterally accreting channel (Fig. 4D). Moreover, the sheet-like geometry of laminated fine-grained sediments (facies Sh and Fl) are interpreted to have been deposited in an overbank environment by sheet flow and settling from suspension (Fig. 6B). The scoured channel form structure in facies Fl is interpreted to represent a crevasse channel in the floodplain, and the lignite beds in the uppermost part of the section are interpreted to represent backswamp deposits (Fig. 6A). The overall fining upward succession is therefore thought to have been formed in the 'upper delta plain–fluvial environment' of Horne *et al.* (1978).

DISCUSSION

The palaeocurrent pattern of the Minato Formation differs significantly from that of a typical alluvial fan or a fan-delta, in which the palaeoflow directions generally show a radial pattern from a point source (e.g. Collinson, 1996; Miall, 1996). The gently downcurrent decrease of MPS (Fig. 3) and the absence of debris-flow sediments are also consistent with interpretation of a lower gradient braided river system, rather than steep alluvial fans dominated by deep flows (McPherson *et al.*, 1987). Interpreted nearshore marine sediments at the southern margin of the basin, however, suggest that the depositional environment was that of a coalescing braid delta system.

Many workers have argued that the formation of coal beds in coastal areas suggests sea-level rise (e.g. Cross, 1988; van Wagoner et al., 1990; Wright & Marriott 1993; Gibling & Bird, 1994; Aitken & Flint, 1995; Breyer 1997), hence fine-grained sediments with lignites are commonly placed in either the transgressive (TST) or highstand systems tracts (HST). Wright & Marriott (1993) claimed that accommodation space will be created during base-level rise, allowing floodplains to store sediment. In their model, aggradation but not lateral accretion will dominate during the transgressive phase.

Shanley et al. (1992) showed that Upper Cretaceous shallow-marine sediments (tidal facies) within an overall ramp setting are interbedded with thick fluvial deposits, with facies changes some 65 km inland correlated with coeval transgressive shoreline deposits. Shanley et al. (1992) claimed that these facies changes are the result of increasing accommodation space as a result of the transition from the late transgressive to early highstand phase. The increasing accommodation space is thought to reflect a landward horizontal shift in the equilibrium point (Posamentier & Vail, 1988). McLaurin & Steel (2000) have also suggested that transgressive and maximum flooding surfaces extend about 80 km inland in the middle Castlegate Formation (Campanian). Yoshida et al. (1996, 2001), however, recognized that the shallow-marine sediments of the formation cannot be followed far landward. Owing to uplift in the source areas, sediment slope and sediment yield could not provide enough accommodation space for nearshore sediments far inland (Yoshida et al., 1996, 2001).

Generally, relative sea-level rise and fall can affect sedimentation significantly on a gently sloping ramp setting margin (Fig. 7). For example, a relative sea-level rise and fall of only 10 m will broadly expose or cover a gently sloping offshore shelf profile (0.02°), resulting in a seaward or landward shift of the shoreline of about 29 km (Posamentier et al., 1992).

Owing to the generally high gradient in a braided river system, however, nearshore marine sediments are not expected to occur interbedded with fluvial deposits very far inland. Posamentier et al. (1992) have estimated only 1 km shift of the shoreline for 10 m sea-level fall on a relatively steep (0.5°) gradient profile. In this study the lack of continuous outcrop inhibits the determination of how far the interpreted marine tuffaceous sandstone beds (Fig. 4A) extend over fluvial conglomerate beds. Fluvial plains for the Minato Formation were obviously of gentler slope than those of a fan-delta system. Nevertheless the transgressive tuffaceous beds probably pinch out over a very short distance. In this respect this study differs from that of Shanley et al. (1992), in which nearshore (tidally influenced) marine sediments were shown to extend a considerable distance inland.

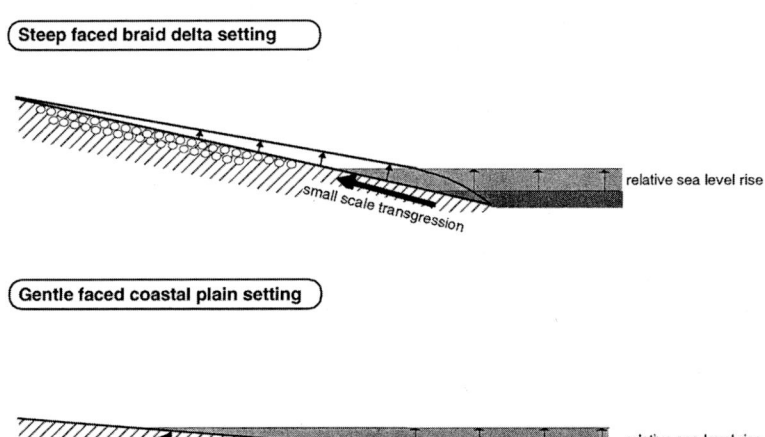

Fig. 7 Two cross-sections of fluvial response to sea-level rise: (A) shallow-marine transgression occurs on a relatively steep gradient fluvial profile; (B) a transgression occurring on a gently inclined fluvial profile.

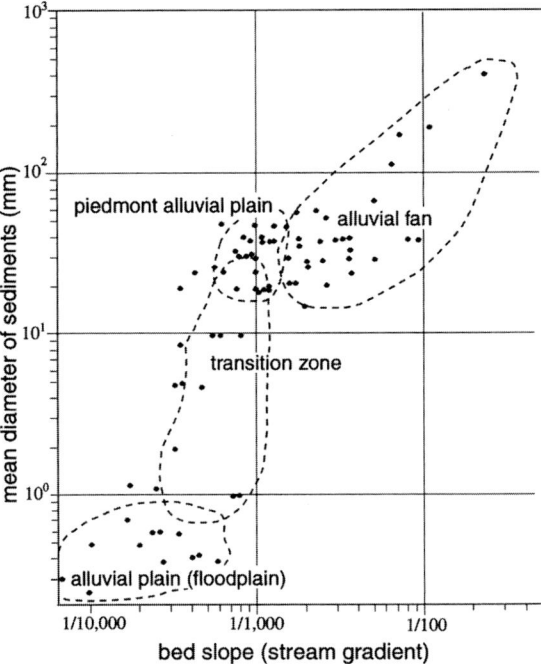

Fig. 8 Clast size of sediments on river beds versus stream gradient. Data were obtained from a number of modern rivers running across the Japanese Archipelago (Yamamoto, 1994). Note that a slight change of stream gradient (from 1 : 1000 to 1 : 8000) may cause fining of sediments on river beds.

Although the relatively high gradient of braided rivers does not provide substantial, widespread accommodation space for marine sediments during rising sea level, a slight change of stream gradient produced by sea-level rise may be very effective for causing fining of the braided river gravels. Based on numerous data, Yamamoto (1994) compiled a diagram of clast-size in river beds versus stream gradient (bed slope) in the Japanese Archipelago (Fig. 8). According to these data, the critical value for stream gradient to produce fining river gravels lies between 1 : 1000 and 1 : 8000. Sea-level rise, even if only a small-scale rise, can therefore change stream gradient, and it may also cause a fining of river gravels inland. Schumm (1993) suggested that a typical fluvial response to sea-level rise would be backfilling of the valley, where backfilling sediments thicken downstream, and the sediments 'would fine upward as the coarser sediments are progressively deposited farther upstream' (see fig. 12 of Schumm, 1993). In this case the stream gradient decreases as the result of sea-level rise, and gravelly materials would be trapped farther upstream. Thus the occurrence of floodplain sediments in the inland quarry is thought to have been a response to transgression away from the quarry.

In each of the lithological logs shown in Fig. 1, fine-grained floodplain sediments or nearshore marine tuffaceous beds occur near the top of the stratigraphical level of the Minato Formation. Although coarse gravels of facies Gm (about 25 m thick) overlie fine sediments that include an *in situ* lignite seam in one case (also see Yagishita, 1997), it is suggested that the fine sediments and the tuffaceous interval in these sections are the product of a short-term transgression. An alternative interpretation would be that this part of the succession represents avulsion of the main channel in the braided system. No such thick unit (more than 19 m thick) of fine sediments, however, can be found elsewhere at other stratigraphical levels. The isolated occurrence of fine sediments in the upper part of each measured section, together with an abrupt facies change between these sediments, suggests that during deposition of the Minato Formation there was no significant avulsion that could exist for a long period and produce such thick fine-grained sediments.

CONCLUSION

This paper reports on a study of the Oligocene Minato Formation in north-east Japan. Within this overall conglomeratic unit, tuffaceous marine beds along the basin margin can be correlated with laterally persistent, fine-grained floodplain deposits farther inland that include several lignite seams. These strata are interpreted to represent a transgressive systems tract that formed during a short-term rise in relative sea level. Transgressive shallow-marine sediments might not be expected to correlate with inland floodplain sediments in the case of a high-gradient braided fluvial system. These deposits, however, are interpreted to represent the overall fining of a braided river system and a slight decrease of stream gradient owing to sea-level rise, with resultant floodplain formation.

ACKNOWLEDGEMENTS

We wish to thank Professor A.D. Miall for his careful review and instructive comments on our revised draft of this paper. The original manuscript was greatly improved by critical comments of two reviewers, Drs J. Isbell and S. Davies-Vollum. Professor H. Miwa gave stimulating suggestions on clast-size variations of river beds, and in an attempt to find microfossils, Dr A. Obuse kindly analysed several rocks of facies Th and Tt. Financial support was in part defrayed by JNOC (Japan National Oil Corporation).

REFERENCES

Aitken, J.F. and Flint, S.S. (1995) The application of high-resolution sequence stratigraphy to fluvial systems: a case study from the upper carboniferous Breathitt Group, eastern Kentucky, U.S.A. *Sedimentology*, **42**, 3–30.

Breyer, J.A. (1997) Sequence stratigraphy of Gulf Coast lignite, Wilcox Group (Palaeogene), south Texas. *J. Sediment. Res.*, **67**, 1018–1029.

Clifton, H.E. (1969) Beach lamination-nature and origin. *Mar. Geol.*, **7**, 553–559.

Clifton, H.E. (1973) Pebble segregation and bed lenticularity in wave-worked versus alluvial gravel. *Sedimentology*, **20**, 173–187.

Clifton, H.E., Hunter, R.E. and Phillips, R.L. (1971) Depositional structures and processes in the non-barred high-energy nearshore. *J. Sediment. Petrol.*, **41**, 651–670.

Collinson, J.D. (1996) Alluvial sediments. In: *Sedimentary Environments*, 3rd edn (Ed. H.G. Reading), pp. 37–82. Blackwell Science, Oxford.

Cross, T.A. (1988) Controls on coal distribution in transgressive–regressive cycles, Upper Carboniferous, Western Interior, U.S.A. In: *Sea-level Changes: an Integrated Approach* (Eds C.K. Wilgus, B.S. Hastings, H. Posamentier, J. Van Wagoner, C.A. Ross and C.G.St.C. Kendall). *Soc. Econ. Paleontol. Mineral. Spec. Publ.*, **42**, 371–380.

Davidson-Arnott, R.G.D. and Greenwood, B. (1976) Facies relationships on a barred coast, Kouchibouguac Bay, New Brunswick, Canada. In: *Beach and nearshore sedimentation* (Eds R.A. Davies and R.L. Ethington) *Soc. Econ. Paleontol. Mineral. Spec. Publ.*, **24**, 149–168.

Gibling, M.R. and Bird, D.J. (1994) Late Carboniferous cyclothems and alluvial palaeovalleys in the Sydney Basin, Nova Scotia. *Geol. Soc. Am. Bull.*, **106**, 105–117.

Horne, J.C., Ferm, J.C., Caruccio, F.T. and Baganz, B.P. (1978) Depositional models in coal exploration and mine planning in Appalachian region. *Am. Assoc. Petrol. Geol. Bull.*, **62**, 2379–2411.

Hunter, R.E., Clifton, H.E. and Phillips, R.L. (1979) Depositional processes, sedimentary structures, and predicted vertical sequences in barred nearshore systems, southern Oregon coast. *J. Sediment. Petrol.*, **49**, 711–726.

Koma, T. (1992) Studies on depositional environments from chemical components of sedimentary rocks—with special reference to sulfur abundance. *Bull. Geol. Surv. Jpn*, **43**, 473–548. (In Japanese with English abstract.)

McCabe, P.J. (1984) Depositional environments of coal and coal-bearing strata. In: *Sedimentology of Coal and Coal-bearing Sequences* (Eds R.A. Rahamani and R.M. Flores). *Spec. Publ. Int. Assoc. Sediment.*, **7**, 3–42.

McLaurin, B.T. and Steel, R.J. (2000) Fourth-order nonmarine to marine sequences, middle Castlegate Formation, Book Cliffs, Utah. *Geology*, **28**, 359–362.

McPherson, J.G., Shanmugam, G. and Moiola, R.J. (1987) Fan-deltas and braid deltas: varieties of coarse-grained deltas. *Geol. Soc. Am. Bull.*, **99**, 331–340.

Miall, A.D. (1996) *The Geology of Fluvial Deposits*. Springer-Verlag, New York, 582 pp.

Posamentier, H.W. and Vail, P.R. (1988) Eustatic controls on clastic deposition II—sequence and system tract models. In: *Sea-level Changes: an Integrated Approach* (Eds C.K. Wilgus, B.S. Hastings, H. Posamentier, J. Van Wagoner, C.A. Ross and C.G.St.C. Kendall). *Soc. Econ. Paleontol. Mineral. Spec. Publ.*, **42**, 125–154.

Posamentier, H.W., Allen, G.P., James, D.P. and Tesson, M. (1992) Forced regressions in a sequence stratigraphic framework: concepts, examples, and exploration significance. *Am. Assoc. Petrol. Geol. Bull.*, **76**, 1687–1709.

Rust, B.R. and Koster, E.H. (1984) Coarse alluvial deposits. In: *Facies Models* (Ed. R.G. Walker), pp. 53–69. Geoscience Canada, St John's, Newfoundland.

Schumm, S.A. (1993) River response to baselevel change: implications for sequence stratigraphy. *J. Geol.*, **101**, 279–294.

Shanley, K.W., McCabe, P.J. and Hettinger, R. (1992) Tidal influence in Cretaceous fluvial strata from Utah, U.S.A: a key to sequence stratigraphic interpretation. *Sediementology*, **39**, 905–930.

Shimazu, M. and Teraoka, Y. (1962) *Geology of Rikuchu-Noda*, scale 1 : 50,000. Geological Survey of Japan, Tsukuba, 53 pp. (In Japanese with English abstract.)

Teisseyre, A.K. (1975) Pebble fabric in braided stream deposits with examples from Recent and 'frozen' Carboniferous channels (Intrasudetic Basin, central Sudetes). *Geol. Sudetica*, **10**, 7–46.

Van Wagoner, J.C., Mitchum, R.M., Campion, K.M. and Rahmanian, V.D. (1990) Siliciclastic sequence stratigraphy in well logs, cores, and outcrops. *Am. Assoc. Petrol. Geol. Methods Explor. Ser.*, **7**, 55 pp.

Williams, E.G. and Keith, M.L. (1963) Relationship between sulfur in coals and the occurrence of marine roof beds. *Econ. Geol.*, **58**, 720–729.

Wright, V.P. and Marriott, S.B. (1993) The sequence stratigraphy of fluvial depositional systems: the role of floodplain sediment storage. *Sediment. Geol.*, **86**, 203–210.

Yagishita, K. (1997) Palaeocurrent and fabric analyses of fluvial conglomerates of the Palaeogene Noda Group, northeast Japan. *Sediment. Geol.*, **109**, 53–71.

Yagishita, K. (2001) *Facies Analyses and Sedimentary Structures*. Kokon-Shoin, Tokyo, 222 pp. (In Japanese.)

Yamamoto, K. (1994) *Alluvial Rivers*. Sankaido, Tokyo, 480 pp. (In Japanese.)

Yoshida, S., Willis, A. and Miall, A.D. (1996) Tectonic control of nested sequence architecture in the Castlegate Sandstone (Upper Cretaceous), Book Cliffs, Utah. *J. Sediment. Res.*, **66**, 737–748.

Yoshida, S., Willis, A. and Miall, A.D. (2001) Fourth-order nonmarine to marine sequences, middle Castlegate Formation, Book Cliffs, Utah: comment and reply. *Geology*, **29**, 187–188.

Index

Page numbers in *italic* refer to figures, those in **bold** refer to tables

Abiyata, Lake (Ethiopia), 280
abundance peaks, 265
accelerated mass spectrometry (AMS), 302, *305*
accommodation change
 and alluvial architecture, 507–11
 rates, 509–10
acoustic Doppler profilers (ADPs)
 applications, bedload transport estimation, 197–209
 bottom tracking, 197, 200–1
 noise figures, 200
Acquatraversa erosional phase, 380, 388
Acre (Brazil), 162, 167, 176
acritarchs, 464
active layer models, 75
Adriatic Sea, 258, 261, 375
aeolian bedforms, 103
aeolian deposits, 367, 368
aeolian dunes, *367*
Aergiranum isp. (trace fossil), 482
aerial photography, 9, *12*, *149*, *552*
Africa, tectonics, 102
agate, 103
aggradation, 141, 277, 311
 floodplains, 3
Alberta (Canada), *495*, *496*, *497*, 500
 sediments, 160, 161, 173
Albian, 160, 177–8, *454*, 456, 464, 473
 mudstones, 496
 sandstones, 478
algae, 464
Allt Dubhaig (river) (Scotland), 399
 fluvial transport modes, 403–4

alluvial architecture
 and accommodation change, 507–11
 Dunvegan Formation, 493–515
 mass-balance effects, 243–53
 scale invariance, 148–51
alluvial channels, classification, 155
alluvial fans, 8, *262*, *264*, 406
 and anastomosis, 4, 6–7
 classification, 115–16
alluvial plains, 501–6
alluvial sediments, 351–61, 518, 524
 grain size, **351**
 recent, 387–8
 successions
  Type-1, 349, 351–3, *354*, 361, 364–8
  Type-2, 349, 355, *356*, 361–3, 364–8
alluvial stratigraphy, development, 349–71
alluviation rates, 14
alluvium, Pleistocene, 321
Alnus spp. (alder), 302
Alpes-de-Haute-Provence (France), 236
alpha counting, 363
Alps (Italy), 258
Alps (Switzerland), rivers, 135–6
Alutu Volcano (Ethiopia), 280, 285
Alvarez (Italy), 384
Amazon Cone, 36
 use of term, 19
Amazon Plume, 35
Amazon River system
 bathymetry, *19*
 bedforms, 32
 channel sediments, 175–7, *178*
 coastal environments, 35–6
 coastal estuarine environments, 33–5

 and coastal sediment supply, 17
 deep-sea fans, 36
 depositional systems, 17–39
  estuarine, 29–35
  fluvial, 22–6
  ria, 26–9
 discharge, 30, 35
  fluctuations, 21–2, 177
  total, 18
 drainage basin, 18
  geological evolution, 19
 erosional headlands, 34–5
 estuaries, 34–5
 flood tides, 427
 flooded valleys, 17
 freshwater outflow, 30
 freshwater systems, 31–3
 glacial sediments, 18–19
 inner estuary, 31–2
 laminae, 36
 meandering channels, 22–6
 mouth area, 29–30, 32–3, 160
  geological setting, 161
  grain-size parameters, 171, **172**, 176–7
 mudcapes, 33–4
 North Channel, 29, *31*, 32, 33
 rivers
  blackwater, 20–1
  clearwater, 21
  types of, 17, 20
  whitewater, 20
 sediment transport, 36
 sedimentation rates, 36
 sequence stratigraphy, 18–19
 shelf environments, 35–6
 South Channel, 29, *31*, 32, 33
 statistics, 18
 straight channels, 26
 subaqueous deltas, 35–6
 tidal bores, 22, *23*, 29, 32
 tidal fluctuations, 21, 22
 tidal ranges, 17

Amazon River system (cont'd)
  topography, *19*
  water-level fluctuations, 17, 21
Amazonas, Rio, 20, 21, *25*, 27
  channels, 26
  hydrology, 32
  mouth, 29–30, 31
  sedimentation rates, 23
  tidal range, 29
  see also Solimões, Rio (Amazonia)
Amazonia, 159
  study area, *162*
Ambrosiusberg Fault (Namibia), 102
Ametlla Formation (Spain), 236
Ammonia spp. (foraminifers), *270*
Ammonia tepida (foraminifer), 383
Amphistegina spp. (foraminifers), 380
AMS (accelerated mass spectrometry), 302, *305*
Amsterdam–Rhine Canal, 310–11
anabranching, 4
anastomosis, 517
  definitions, 3, 4
  degree of, 7
  generic model, 10–13, *14*
  longitudinal variations, 6–10
  mechanisms, 4
  origins, 3–15
    hypotheses, 4
Andean Foreland Basin, 194
Andean thrust belt, 161
Andes Mountains, 17, 18, 19, 36, 176
  drainage, 20
angiosperms, 464
Anglo-Welsh Basin (UK), 517
Angulogerina spp. (foraminifers), 410, *416*
Antarctica, 521
Apennines (Italy), 258, 375
  sediments, 389, 391
  uplift, 390, 393
  western flank, 373
Appalachian–Cumberland Plateau (USA), 412, 420
Appalachian Mountains (USA), 409, 411–12, 420, 421, 422
Aptian, 160, 177–8, 542
aquicludes, 187–8
aquifers, 187–8
Arabian Peninsula, 281

Arabian Sea, 350, 365
Aravalli Range (India), 350
Arc/Info GIS, 64–5, 68
Arc/View, 68
Archipelago de Anavilhanas (Amazonia), 27, *29*
archipelagos
  fluvial, 26–9
  tidally influenced, 29
Arctic, fans, 116
Arenicolites isp. (trace fossils), 461
Argentine Limestone (USA), 461
Aripuanã, Rio (Amazonia), *29*
Arizona (USA), 290
arkosic sands, 19
arroyos, 277, 278
arthropods, 523
Ash Grove Quarry (AGQ) (USA), *454*, *465*, *466*, *476*, *478*
  fluvial-estuarine architecture, 461–4
Ashland (NE, USA), 455, 456
Atacama Desert (Chile), 101
Athabasca River (Canada), 173
Atlantic Ocean, 30, 99, 103, 115
  coastal sediment supply, 17, 420
  fans, 115
  river discharges, 161
  tectonics, 102
Atlantic Shelf, 237, 238
Attenberg Cylinders, 163
auger holes, 302
Australia, 517
  mudstones, 522–3
  oil fields, sequence stratigraphy, 531–56
  river geomorphology, 321–47
  rivers, 368
Australian Height Datum (AHD), 324
Australian-slide method, 363
Austria, rewidening schemes, 137
Avicennia spp. (mangroves), 34
avulsions, 9
  and anastomosis, 4, 13–14
  Baghmati River, 192–3, 194
  and tectonism, 192
Awash River (Ethiopia), 286

Baghmati River (Himalayas)
  abandoned-channel fills, 187
  anabranching system, 181
    long-term history, 193–4

  avulsion dynamics, 192–3, 194
  borehole logs, 187–92
  channel sediments, 184
  early studies, 181–2
  floodplains, near-surface sediments, 184–7
  geological setting, 182–4
  satellite images, 186–7
  sedimentation model, *195*
  sedimentology, 181–96
  study area, *182*
Bagnold sediment transport formula, 12–13, *14*
Bahadurabad (Bangladesh), sand dunes, 45
Baie d'Oyapack, 34
Balotra (India), 350
Bangladesh
  river morphology, 149–50
  sand dunes, 41, 45
bankfull channel width, 146
bar length, 146
bar platform shape, 145
bar shape
  data, **147**
  scale invariance, 147, 154–5
bar width, 146
Baraila Tal (India), 187
Barcellos (Brazil), *21*
Barles (France), 236
Barren Fork (USA), 442
barrier–lagoon–estuary systems
  migration, *261*, 262, 270
  retrograding, 257, 269
bars
  elevation, 153
  longitudinal, 106–7
  scroll, 23
  slipfaces, 153–4
  vegetation on, 154, 155
  see also braid bars; point bars
Barwon River (Australia), 522–3
basalts, 107, 280
  see also flood basalts
base levels, concept of, 214
base-level change
  early studies, 214–15
  experimental criteria, **224**, **228**
  morphological effects, 213–41
  second-generation experiments, 233–4
  and sediment transport, 214
  and sediment yield, 232–3
  slope effects, 226–8
  stratigraphical effects, 213–41

stratigraphical significance, 234–8
and valleys, 219–28
see also sea-level change
basins
Mesozoic, 9, 19, 410
proximal foreland, 493–515
see also drainage basins; floodbasins; fluvial basins
Bathy 1500 (echo sounder), 199
bathymetry, 329
shelf, 322
Baton Rouge (USA), 317
Battery Point Formation (Canada), 146
Beaconites isp. (trace fossils), 521
burrows, 523
Beatton River (Canada), 501, 510–11
beaver dams, 5
bed shear stress, 124–6
bed shear velocity, 201–2
bedforms
Amazon River system, 32
morphology, in dune-related macroturbulence studies, **44**
bedload, definition, 198
bedload capacity, limited-width braided rivers, 137–8
bedload transport, 400–1
estimation, 197–209
determination, 204–6
via acoustic Doppler profilers, 199–201
formulae, 137–8, 197, 198, 201–2, 205, 207
high vs. low, 141, *142*
homogeneity, 199
maximum, 138
measurement, 198
sand-bed channels, 197–209
see also sediment transport
bedload velocity, 204
bed-surface elevations
computations, 124–6
experimental methods, 124
probability distribution, 121–34
and dune deposits, 128–31
and dune geometry, 126–8
recent developments, 123–4
variations, *122*
Belém (Brazil), 20, 30, 33, 34
Belgh (Ethiopia), *283*
Belgrandia spp. (molluscs), 383

Belwa (India), 183, 184
Ben Franklin Zahm No. 1 (USA), 470, 472, 477
Benguela Current, 101, 103
Benibad (India), 184
Bergwinds, 102
Berne (Switzerland), *136*
Betsie Shale (USA), 429, 439
Bhuka (India), 353, 355, *356*, 361
Bihar (India), 181, 182
Binghamton University (USA), 124
biostratigraphy, 393
biotite, 384
bioturbation, 447
Bison spp. (bison), 384
Bison cf. *B. degiulii* (bison), 384
Bithynia spp. (molluscs), 386
Bittium deshayesi (gastropod), 383
Black Shale (UK), 482
blackwater rivers, 37
characteristics, 20–1
mixing zones, 22
origins, 20
sediments, 26
boils (water surface), 41, 42, 45
mechanisms, 50–2, *53*, *54*
Bolivia, sedimentary rocks, 161
Bologna (Italy), 265
Boone County (WV, USA), 439
Bora-Bericcio complex (Ethiopia), 280
Boramo (Ethiopia), 282–4, 285, 286
Boreal Sea, 160
borehole logs, 181, 187–92, 259
boreholes, 454, 469, 482, *483*
Bovenrijn, River (Netherlands), 80, 84, 85
deposits, *86*
dunes, *81*
sediment transport, **83**
Bowl Reef (Australia), 324
Bowling Green Bay (Australia), 326, 327, 341
brachiopods, 425
Brachydontes spp. (molluscs), 509
Brahmaputra River (Bangladesh), 146, 149
see also Jamuna River (Bangladesh)
braid bars, 184
formation, 155
migration, 155
sedimentology, 146

braid delta deposits, 557–68
braided channels, 61
morphology, 137
braided rivers, 349, 367, 565
depositional model, 155
and fan morphology, 99–120
maximum erosion depth, 148
morphology, 145
planforms, fractal analysis, 146
scale invariance, 147–8
scours, 137
surface morphology, 147–8
width–depth ratios, 154
see also limited-width braided rivers; sandy braided rivers
braidplains, 425, 428, 445
Branco, Rio (Amazonia), 27
Brazil, 162, 167
mudcapes, 34
sedimentary rocks, 160, 161
Brazilian Shield, 18, 19, 21, 26
Breda Formation (Netherlands), 296
Brenno (river) (Switzerland), 139
brick manufacture, 185
brick pits, 185, 187
British Coal, 482
British Columbia (Canada), 197
alluvial architecture, 493–515
anastomosis origins, 3–15
sediment transport studies, 63
see also Dunvegan Formation (Canada)
bryophytes, 464
Bulimina spp. (foraminifers), 410, *416*
Bulimina elegans marginata (foraminifer), 382
Bulimina etnea (foraminifer), 380, 382, 383
Bulimina marginata (foraminifer), 380
Bulldog Shale (Australia), 533, 543
Burdekin Delta (Australia), 324, 326
Burdekin River (Australia)
geological setting, 322–4
geomorphology, 321–47
palaeochannel, 321–47
characteristics, 326, **328**
course, 341
cross-shelf profiles, 326–7
evolution, 342–5

Burdekin River (Australia) (cont'd)
  forms and fills, 327–40
  planform, 341
  seismic reflection surveys, 324–6
Buren channel belt (Netherlands), 310
Burhi Gandak River (India), 183, 187, 192
burrows, *538*
  fossil, 523, 536–7
  insect, 506
  see also trace fossils

Cabo Cassipore (Amazonia), 34
cabos, 33
Cadna-owie Formation (Australia), 531, 532–3, 541–2, *545*, *547*, *553*
Calamus River (USA), 128, 146
  aerial photograph, *149*
  bars, 154, 155
  cross-stratification
    high-angle planar, 150–1, 152
    trough, 150
  discharge rates, 153
  facies, 151, *152*
  low-angle stratification, 151
  morphology, 148–9, 150
  sedimentology, 150–1
  width–depth ratio, 154
calcite cements, 543
calcretes, 517, 523, 524
calderas, 280
cambic horizons, 286
Camp Maha (USA), 468, *469*
Campanian, 565
Canada, 176, 417
  alluvial architecture, 493–515
  anastomosis origins, 3–15
  river morphology, 149
  sediment transport studies, 63
  sedimentary rocks, 160, 161
Canadian Shield, 160, 457, 463
Canal do Norte, 32, 33
Canal do Sul, 32, 33
Canal Perigoso, 33
Canoe Pass (Canada), 198, 207
canyons, 102
  shelfal submarine, 19
  submarine, 36
Cape Bowling Green (Australia), 329

Cape Cleveland (Australia), 327, 329
carbonate reefs, 322
carbonates, 160, 166, 188, 525
  nodules, 521
Carboniferous
  rhythmites, 441
  tidal channels, 444
Carrizo, River (USA), 412
Carrizo Sand, 414, 417
Cass County (NE, USA), 455–6, *462*, 473
Cassia erosional phase, 383
Cassiar–Columbia Mountain (Canada), 4
Castledale (Canada), 8
Castlegate Formation (USA), 565
Cedar Creek (USA), 456
Cenomanian, 495–6
Cenozoic, 18, 19
  depositional history, 409–23
  episodes, 410–11
  late, 377
  sediment supply, 411–13
  sequence stratigraphy, 410
Central Appalachian Basin (USA)
  estuaries, 445
  fluvial–estuarine transitions, 425–45
  geological setting, 427–9
  palaeogeography, 427–9
  stratigraphy, 427–9
  see also Lower Pennsylvanian (USA)
Central Namib Desert (Namibia), 102
Central Water Commission (CWC) (India), 183
Cerastoderma glaucum (lagoon cockle), 382–3
Ceriti-Manziate Volcanic District, 388
Cesano (Italy), 388, 391
CHANEL study, *484*
channel belts, 310
  definition, 7
Channel Country (Australia), 517, 523, 524, 525, 526–7
channel facies
  middle estuarine heterolithic, 431
  transgressive, 432–6
  upper estuarine sandstone, 432
channel fills, 506–7
  palaeochannels, 340–1

channel gradient, and anastomosis, 6–7
channel planforms, 341, 517
  surveys, 61, 62–3, 64–5
channel sediments, 175–6, 178
  Baghmati River, 184
  Miocene, 176–7
channels
  alluvial, 155
  ancestral, 321–47
  main, 7
  meandering, 22–6, 302
  migration, 183, 506–7
  morphology, 5
    limited-width braided rivers, 137
    and sediment transport, 62–3
  Pleistocene, 341
  straight, 26
  and valleys compared, 215
  see also braided channels; incised channels; palaeochannels; river channels; sand-bed channels
charcoal, 186
cherts, 558
  Palaeozoic, 466
Chetwynd (Canada), 493, 496, *500*, 511
Cheviot Hills (UK), 64
Cheyenne Sandstone (USA), *454*, 478
Chézy constant, 12
Chézy formula, 12
Chiani-Tevere Formation (Italy), 380, 381–3, 384, 388–90
Chile, deserts, 101
Chirgulo (Ethiopia), 281, 282–4, 285, 286
Chlamys latissima (lamellibranch), 380
chlorite, 103, 159, 175, 178, 543
Chondrites isp. (trace fossils), 461, 537, 538
Chondrula tridens (land snail), 386
chronostratigraphy, Tiber River, *377*, 379
chute channels, 282
Cimino, Mount (Italy), 375, 383, 384, 385, 391
Cincinnati Arch (USA), 427
Citronelle Formation (USA), 421
Civita Castellana (Italy), 392

Civita Castellana Unit (Italy), *378*, 384–5, *386*, 391, 392
Civitella d'Agliano Formation (Italy), 384, 386
Cladocora caespitosa (coral), 382
clasts, 282, 461, 464, 521, 547
   calcrete, 524
   gravel, 558
   size, *566*
clay minerals, 35, 185–6, 506, 543
   analyses, 163, 172, *173*, 178
   classification, 188
   distribution, 175
   floodbasin, 311
   floodplain, 265
   grain-size distribution analysis, 159–80
   inclined heterolithic stratification, 172
   paludal, 265
Clayhurst (Canada), 501, 510–11
Clearwater Formation (Canada), 161
clearwater rivers, 21
   mixing zones, 22
   sediments, 26
climate
   classification schemes, 280
   fans, 116–18
   Main Ethiopian Rift, 281–2
   Namibia, 101–2
climate change, 349
   India, 350
   Late Cenozoic, 377
   Pliocene–Quaternary, 377–9
   Quaternary, 379, 409
climatic indices, Thornthwaite's, 280
clinopyroxene, 103
coal, 504, 505, 512, 562
   accumulation, 509
   experimental stratigraphy, 246–7
   formation, 565
   peat-mire facies, 429
   seams, 482, 489
coalfields, boreholes, 482, *483*
coastal environments, 35–6
coastal estuarine environments, 33–5
coastal prisms
   differential subsidence, 295, 317
   sedimentary architecture, 299
coasts, sediment supply, 17

Cockleshell Coalseam (UK), 489
Coldstream Creek (Canada), 496, 507, 508, 509, 510
   cross-sections, *501, 502–3*
   geological setting, *497*
Colombia, sedimentary rocks, 161
Colorado (USA), 419, *454*
Colorado Front Range (USA), 420
Colorado State University (CSU), 233, 235, 238
   base-level change studies, 213–14, 215
   rainfall–erosion facility, 214
Columbia Lake (Canada), 4
Columbia River (Canada)
   anastomosis, 3, 4, 7
   bar shape, 147
   discharge data, *6*
   morphology, *8*
   sediment load, 5
   structure, 5
   see also Upper Columbia River (Canada)
combing depth, 148
conglomerates, 102, 353, 405, 453, *535*
   clast-supported, 557–68
   Old Red Sandstone, 519–20, 521, 522
Congo Basin, 281
Congo River, total discharge, 18
continental margins, evolution, 322
Cooper Creek (Australia), 526
coppice dunes, 103
Coquet, River
   sediment transfers, 61–74
   study area, 64
corals, 466
Corbetti caldera (Ethiopia), 280
Corbicula fluminalis (mollusc), 386
Corbula spp. (molluscs), 509
corbulids, 501
cores, 302, 453
   see also vibracores
Cornicolani Mountains (Italy), 388
Corsair (USA), 412, 420
costa de ria, 34–5
Coulter LS-200, 166
crayfish, 523
Cretaceous, 149, 177, 235, 236, 454
   fluvial–estuarine transitions, 447

   palaeovalleys, 457
   palynomorphs, 464
   sands, 176
   sedimentary rocks, 159, 160, 478
   subsidence, 495
   volcanic rocks, 102
   see also Early Cretaceous; Late Cretaceous; Lower Cretaceous; Upper Cretaceous
crevasse deltas, 504
crevasse splays, 181, 183, 311, 493, 504, 512
   Columbia River, 3, 5
   deposits, 186–7, 188
   development, 316
   distribution, 6–7, 9, 192
   formation, 14
cross-bedded deposits, 76–8, 87
   formation, 76
   and sediment sorting, 94
cross-set formation model, 122–3
cross-sets, 121, 129–31
   formation, 122
   mean thickness, 126
   preservation, 128
   and vertical sorting, 128–9
cross-stratification
   high-angle planar, 145, 150–1, 152–4, 155
   trough, 145, 150
crustaceans, 187
Cruzeiro do Sul (Brazil), *21*
Cullom Mine/Raven's Nest (USA), 468, *469*
Cumberland Falls State Park (USA), *442, 443*
Cumberland Marshes (Canada), 192, 193
Cumberland Plateau (USA), 409, 411–12, 422
Cumberland River (USA), *443*

Dakota Formation (USA)
   economic significance, 454
   fluvial–estuarine architecture, 453–80
   large-scale, 468–73
   outcrop-scale, 456–68, 476
   palaeogeography, 453–80
   sub-Dakota bedrock, 454–68
Dakota–Pennsylvanian contact (USA), 455, 456, 468, 478
Damara rocks (Namibia), 103

Damaran basement ridge (Namibia), 107, 114
dams, 27
DANTEC systems, *43*, 44–5
Datasonics CAP6600 CHIRP II (acoustic profiling system), 324
Dave Branch Shale (USA), 433, 446
debris flows, 99, 116, 277
deglaciation, and tectonics, 316
deltas, 414
    bayhead, 21
    bird's-foot, 4, 27
    crevasse, 504
    highstand, 222
    progradation, 219–21, 262, 269, 270, 511–12
    shelf-margin, 225
    subaqueous, 35–6
DEMs *see* digital elevation models
depositional cycles, and sedimentation, 17
depositional episodes, 413
    Neogene, 410
    sediment supply, 413–21
depositional models, generic, 322
depositional systems, Amazon River, 17–39
deserts
    alluvial stratigraphical development, 349–71
    climate, 101
    Devonian, 117, 146, 517
    carbonates, 160, 166
    Lower, 517
    sandstones, 399, 404–6
    trace fossils, 521
DGPS *see* differential global positioning system
Dhengbridge (India), 182–3
diagenesis, 531
Dickebusch Creek (Canada), 496, *498–9*, 509, 510, 511
    cross-sections, *501, 503*
    geological setting, *497*
    insect burrows, 506
differential global positioning system (DGPS), 197, 200–1, 202–4
differential subsidence
    mechanisms, *298*
    quantified, 312–16
    rates, **312**
    Rhine–Meuse delta, 295–320
    vertical displacement rates, 313–15
digital elevation models (DEMs), 61–2
    accuracy, 67–70
    advantages, 71
    differencing, 69–70, 71
    error analysis, **68**
    generation, 65–7
    reaches, 63
    residual analysis, 69
    vs. profile budgeting, 70–1
dinoflagellates, 464
dinosaur tracks, 501, 508
discharge rates
    and dune height, *93*
    effects on facies, 153
    and sediment transport, *93*
discharge waves, 91, *93*
    and sediment sorting, 77–8
    and sediment transport, 77–8
discontinuous ephemeral streams (DESs)
    alluvial sequences, 290
    alluviation, 292
    characteristics, *278*
    definitions, 277–8
    depositional processes, 277–94
    early studies, 277
distributed terrain-sensitive surveys, 63
Dodge County (NE, USA), 456
double integrals, 69
Douglas County (NE, USA), *454*, 455–6
downstream velocity profiles, 47
drainage basins
    evolution, 409–23
    sediment supply, 413–21
drainage density, *216*
drainage patterns
    evolution, 215
    idealized, *216*
    modelling, 215–19
dune crest
    fluid motion, *46*, 47, 49–50, *51*
    uv vectors, 49–50
dune deposits, and bed-surface elevation probability distribution, 128–31
dune dimensions, average, 78
dune formation, and macroturbulence, 42–3
dune geometry, and bed-surface elevation probability distribution, 126–8
dune height, 121, *127*, 146
    and discharge rates, *93*
    effects on bed-surface elevation, 127–8
    and flow depth, 129–30
    mean, *130*
    measurement, 80–4, *92*
    modelling, 91
    prediction, 88–9
    variability effects, 87
dune lee
    fluid motion, 46–7, 48–9
    sediment sorting, 75, 77
dune length, 146
dune troughs, 77–8, 79, 123
    scour paths, 130–1
dune-phase sediment transport, 75–97
    flume models, 78–80
    sorting processes, 75
dune-related macroturbulence, 41–60
    bedform morphology, **44**
    dune crest, *46*, 47, 49–50, *51*
    dune lee, 46–7, 48–9
    field studies, 43, 45, 50–2
    kinematics, 43
    laboratory studies, 43–5, 46–50
    mechanisms, 41–2
    topology, 43, 52–6
    and water surface structure, 50–2
dunes, 302, 399, 402
    aeolian, *367*
    buried, 310, 315
    coppice, 103
    evolution, 42
    fibreglass, 43–4
    flow dynamics, 42
    temporal, 42–3
    and flow resistance, 42
    migration, 75, 78, 122, 123, 145, 149
    modelling, 43–5, 85–6
    morphology, 42
    horseshoe, 53
    occurrence, 42
    scour depth, *86*, 87
    sediment transport, 42, 56, 121–34
    shape factor, 130, 132
    three-dimensional, 87

vorticity, 53–6
see also sand dunes
Dunvegan Formation (Canada)
  alluvial architecture, 493–515
  facies, 500–7
  geological setting, 495–6
  palaeogeography, 495–500
  stratigraphy, 496–500
  study area, 496, *497*
Dunvegan–Kaskapau boundary (Canada), 510
dust plumes, 102
dust storms, 524

Early Cretaceous, 19, 533
  palaeogeography, 455
  tectonics, 102
Early Eocene, drainage basins, sediment supply, 414–17
Early Miocene, 161, 375
Early Palaeozoic, 99
  fans, 103
Early Pleistocene, 280, 373
earthquakes, 183, 296
East Africa, 290
East African Rift System, 278–80
East Greenland, 399, 404–6
East Pennine Coalfield (UK), 482, *483*
easternmost Platte River Valley (EPRV), 454–7, 476, 477–8
  central–eastern, 461–8
  cross-sections, 468–9
  well logs, 469
  western end, 457–61
Edwards Plateau (USA), 412, 420
Eemian, 260, 262, *266*
efficiency coefficients, 137–8
ejection-inrush sequences, 56
El Niño, 524
Elephas antiquus (elephant), 386
Ells River (Canada), 161
  cross-section, *168*
  facies, 167
  grain-size distribution, 167, *168*
  grain-size parameters, **169**
  samples, 162–3
  study area, *162*
embayment–estuary systems, 160
Emilia-Romagna (Italy), 258
Emme, River (Switzerland), *136*
Encontro das Águas, 22
Energy and Utilities Board (EUB) (Alberta, Canada), Core Laboratory, 161

Environment Canada, 7
Eocene
  Early, 414–17
  flood basalts, 280
  Middle–Late, 417–19
  uplifts, 409
  Upper, 413
ephemeral streams, 350–1, 524, 525
  see also discontinuous ephemeral streams (DESs)
equilibrium channel theory, 10
Equus stenonis (horse), 383
Eromanga Basin (Australia), 532, 533
erosion, 277, 409
  Himalayas, 183
  Holocene, 327, 337
  in hydraulic modelling, 141
  Palaeozoic, 495
  Pleistocene, 463
  Plio-Pleistocene, 468
erosional headlands, 34–5
estreitios (narrows), 32
estuaries, 17, 34–5
  brackish-water, 33
  definitions, 30–1, 426, 427
  depositional systems, 29–35
  freshwater, 31–3
  lengthening, 444
  Lower Pennsylvanian, 445
  sedimentation, 160
  see also fluvial–estuarine architecture; fluvial–estuarine transitions; tide-dominated estuaries
estuarine sediments
  grain-size distribution analysis, 159–80
  Miocene, 169–71, 176
Etendeka Plateau (Namibia), 99, *101*, 102, 103
  fans, 114, 116
  faults, 105
  volcanic rocks, 113
ETH (Swiss Federal Institute of Technology), 135
Ethiopia
  discontinuous ephemeral streams, depositional processes, 277–94
  volcanism, 280, 285–6
Ethiopian Plateau, 280
ethylene glycol, 163
Euastacus serratus (crayfish), 523

Exner equation, 121, 130, 132
Experimental EarthScape (XES) Facility (USA), 244
Exxon sequence-stratigraphy model, 258, 266, 267, 268
Eyre, Lake (Australia), 526–7

facies, *152–3*
  cyclicity, 521–2
  Dunvegan Formation, 500–7
  estuarine tidal-flat, 431–2
  fluvio-lacustrine, 501–6
  heterolithic, 521, 522–3
  levees, 299
  Lower Old Red Sandstone, 519–21
  McMurray Formation, 163–7
  massflow, 289–90, 292
  minor, 151
  models, 525–7
    scaling issues, 146
  scale invariance, 151–5
  seismic, 329
  sheetflow, 287–9, 292
  streamflow, 286–7, 292
  valley fills, 286–90
  vertical trends, 507
  Westphalian A, **485–7**
  see also channel facies; lithofacies
faecal pellets, 36
falling-stage systems tracts (FSTs), 257, 264, 270
  structure, 267
  use of term, 262
fan-deltas, 115
fans
  classification, 115–18
  climate, 116–18
  deep-sea, 36
  interactions, with marine environments, 113–14
  morphology, 99–120
  terminal, 115
  vegetation, 116–18
  see also alluvial fans
Farfa Valley (Italy), 375
faults, 375–7
fauna, marine, 428–9
Federal Office for Water and Geology (Switzerland), 135
feldspar, 103, *546*
  grains, 36
Ferento (Italy), 388

Feshie, River (Scotland), 67
FESWMS (Finite Element Surface Water Modeling System), 13
fibreglass dunes, in macroturbulence modelling, 43–4
Ficke caldera (Ethiopia), 280
fine sands, 360–1, 504
  horizontally bedded, 359
  red silty, 358–9
Finite Element Surface Water Modeling System (FESWMS), 13
fish, fossils, 438, 521
flats, 342
flood basalts, 99, 280
  tholeiitic, 103
flood depth, 399
flood prisms, 115
flood slope, 399
flood tides, 427
floodbasin deposits, fluvio-deltaic, and differential subsidence, 295–320
floodbasins, 3
  Holocene, 297, 299
flooding, 101, 113, 524, 542–3, 549
  annual cycles, 22, 23, 350
  and rainfall, 21
  rainforests, 17
  valleys, 225–6
  see also maximum flooding surface
flooding surfaces, 551
floodplain deposits
  distal, 185–6
  proximal, 184–5, 188
floodplain lakes, 17, 25, 183
  deposits, 187, 194
floodplains, 445, 522, 523, 527
  aggradation, 3
  environment, 563–4
  forested, 22–3
  Minato Formation, 557–68
  near-surface sediments, 184–7
  sediments, 566
  whitewater rivers, 20
floods, 9–10
  effects, 193
flow depth, 146
  and dune height, 129–30
  effects on bed-surface elevation, 127–8

flow fields
  approach, 41
  holistic mapping, 48
  low/high-speed wedges, 56
flow visualization, 44, 56
flume models, 75, 76, 90, 214, 238
  dune-phase sediment transport, 78–80
  in limited-width braided-river studies, 138–9
  in macroturbulence studies, 43–4
fluvial architecture
  downstream changes, mass-balance transformations, 243–53
  models, three-dimensional, 481–91
  Pennine Basin, 481–91
fluvial basins
  fluvial transport modes, 399–407
  grain-size patterns, 399–407
fluvial deposits
  grain-size distribution, 172–3
  Po Basin, 260–5
fluvial–estuarine architecture
  Dakota Formation, 453–80
  large-scale, 468–73
  outcrop-scale, 456–68, 476
fluvial-estuarine transitions
  Central Appalachian Basin, 425–45
  controls, 444–5
  occurrence, 426–7
  recognition, 425
fluvial sedimentology, scale invariance in, 145, 146
fluvial systems, near-coastal, and sea-level change, 213
fluvial transport modes
  applications, 404–6
  arbitrary limits, 401
  definition, 399
  fluvial basins, 399–407
  transitions, 400–1, 403–4
  transport efficiency, 401–3
  see also bedload transport; sediment transport
Fonte Boa, 25
Foraminifera, 377, 379, 380
  benthic, 259, **260**, 269, 270
  Neogene, 410
foresets, sigmoidal, 186

forests, 379
  flooded, 20–1
  rainforests, 17
formation micro-imagers (FMIs), 534, 535
  palaeocurrent diagrams, 548, 549
formation microscanners (FMSs), 534, 535
  image logs, 540
  palaeocurrent diagrams, 548, 549
fossils, 329, 383, 425
  fish, 438, 521
  freshwater, 384
  ichnofossils, 160
  macrofossils, 302, 311, 380
  microfossils, 259–60, 261, 269
  see also trace fossils
Foz do Amazonas Basin, 19
fractal analysis
  braided river planforms, 146
  meanders, 146
Fraser River (Canada), 197–209
  discharge rates, 198–9
  South Arm, 198, 199
  study area, 198
French Guiana, 34
freshwater environments, 17, 30–1
freshwater systems, 31–3
Freshwater West Formation (UK), 518–19
Frio-Vicksburg (depositional episode), 410, 413, 415, 419, 422
Front Range, 417
Froude number, 124, 129
furos (canals), 30, 32
Furos de Breves (Brazil), 31, 32, 33

Gademota Caldera (Ethiopia), 286
Gademota Ridge (Ethiopia), 277, 279, 280, 292
  ephemeral streams, 281, 290, 291
  facies analyses, 286–90
  geological setting, 283
  Upper Quaternary deposits, 283–90
gamma counting, 363
gamma-ray logs, 540, 542, 548
Gandak Fan (Himalayas), 181
Ganga River (India), 181, 183, 187
  floodplains, 184

Ganges, River (Asia), 193
garnets, 99, 103
gastropods, 186, 187, 383, 384, 386, 523
Gelasian, 388
Gelasian–Santernian boundary, 383, 390, 391
geochemistry, marine deposits, 562
geochronology, 379
  luminescence, 361–3
geographical information systems (GIS), 454, 473
  database construction, 410
  in sediment transfer surveys, 64–5
Geological Survey of Italy, 258
Geological Survey of Regione Emilia-Romagna (Italy), 258
Geomagnetic Polarity Time Scale, 380
geomorphology
  Burdekin River, 321–47
  Upper Columbia River, 4–6
Georgia, Strait of (Canada), 198
Gérine (river) (Switzerland), 139
gilgai, 525
Gironde, River (France), estuary, 35
glacial sediments, 18–19
glaciation, 18
  Late Quaternary, 295
  Pleistocene, 322, 410
  Pliocene, 421
  Saalian, 299
Glauconitic Member (Canada), 173
global positioning system (GPS)
  applications, 324
  equipment, 8
Globigerina aff. *calida calida* (foraminifer), 380, 383
Globigerina calabra (foraminifer), 380, 383
Globigerina cariacoensis (foraminifer), 380
Globoquadrina altispira (foraminifer), 410, 413, *416*, 420
Globorotalia aemiliana (foraminifer), 380
Globorotalia inflata (foraminifer), 380
Globorotalia punticulata (foraminifer), 380

gneisses, 103
Golden (Canada), 3, 4, 5, 8
Gondwanaland, fragmentation, 19
grabens, 375, 392
Graffignano Unit (Italy), 385–6, 392
grain flow dynamics, 76
grain shear stress, 89, 91, *92*
grain size
  alluvial sediments, **351**
  analysers, 160
  and anastomosis, 6–7, 9
  and slope, 403
  variations, *112*
grain-size distribution analysis
  applications, 159–60
  field studies, 161–2
  laboratory studies, 162–3
  tidally influenced systems, 159–80
grain-size patterns, fluvial basins, 399–407
Granaria illirica (land snail), 386
Graneros Shale (USA), *454*, 477
Graneros Shale–Greenhorn Limestone interval (USA), 470, *474*
gravel lag layers, 88, 91, 95
  factors affecting, 87
gravel layers, 85–7
gravel lenses, 359
gravelly sand, dune-phase fluvial transport and deposition model, 75–97
Great Barrier Reef (GBR) (Australia), 321–47
  geological setting, 322
  palaeochannels, evidence, 322
Great Escarpment (Namibia), 102
Greeley County (NE, USA), 470, *472*
Greenhorn Formation (USA), *454*, 470
Grey Ranges (Australia), aerial photography, *552*
ground-penetrating radar (GPR)
  applications, 145
  braided-river studies, 146, 148, 150–2, 155
groundwater, level rise, 311–12
Grub Reef (Australia), 324
Guama, Rio (Amazonia), 33
Guiana Shield, 18, 19, 26, 161
Gulf of Mexico, 237

Gulf of Mexico Basin
  Cenozoic depositional history, 409–23
  drainage basin evolution, 413–21
  sediment supply, 411–13
    continental sources, 411–12
    rates, 412–13
  sequence stratigraphy, *411*
Gulf of Mexico Shelf, 238
Gulf of St Lawrence, 427
gullies, erosion, 278
Gurupá Arch, 19
gymnosperms, 464
gypsum, 103, 525

Hanaupah Fan (USA), *117*
Haughton River (Australia), 324, 326, 327, 329
Hayaghat (India), 182–3
Helicidae (snails), 384
Helix Reef (Australia), 324
Helley–Smith samplers, 80–1, *92*, 197, 199
  applications, 198, 205, 207
Hennisdijk channel belt (Netherlands), 310
High Plains (USA), 409, 412, 420, 421, 422, 454
Highland (Scotland), 399, 403–4
highstand systems tracts (HSTs), *261*, 267, 269, 270, 509, 565
  definition, 257
  fluvio-lacustrine, 531
  pollen analysis, 262–4
  reservoirs, *543*, *547*, 550–4
Himalayan Foreland Basin, 181–96
  study area, *182*
Himalayan Front, 182
Himalayas, erosion, 183
histosols, 505
Hoanib River (Namibia), 102, 107
Hoarusib River (Namibia), 102
Holland *see* Netherlands, The
Holocene, 36, 105, 113, 265, 296
  alluvial sediments, 349
  depositional processes, 277–94
  deposits, 260, 261
  erosion, 327, 337
  floodbasins, 297, 299
  interglacial, 364–5
  Late, 181, 193, 194
  meandering channels, 302
  Middle, 311

Holocene (*cont'd*)
  muds, 326
  peats, 313–14, 317
  sea floor, 343
  sea-level changes, 22–3, 35, 494
  sediments, 184, 321
  stratigraphy, 383–4
  transgressions, 326, 330, 336, 341
  valley fills, 290–1
  vegetation, 290
  wet phase, 367
  see also Pleistocene–Holocene
Holystone (UK), 64, *65, 67*, 70, 71–2
Hope (Canada), 199
Horingbaai Fan (Namibia), 114–15
Horn of Africa, 280
hornfels, 558
Houston Delta (USA), 414, 417
Houston River (USA), 412, 419, 420
Humboldt Current, 101
humic acids, 20, 27
hydraulic modelling
  limited-width braided rivers, 135–44
  principles, 138
hydraulic sorting, 400
Hydrobiidae (molluscs), 383
hydrocarbons
  exploration, 425
  production and development, 531–56
  see also coal
hydro-electric power plants, 135
hydrogen peroxide, in sample preparation, 163
hydrology
  Amazonas, 32
  Main Ethiopian Rift, 281–2
hydrozoa, 380

ichnofossils, 160
igapó, 20–1
ignimbrites, 280, 383, 384
Ilha de Marajó, 29, 31
illite, 159, 175, 178, 543
  abundance, 172
incised channels
  base-level change studies, 228–32
    falling, 228–30, 231–2
    rising, 230–2

incised valleys, *224*, 225–6, 236, 238
  model, *552*
  use of term, 321
inclined heterolithic stratification (IHS), 161, *162, 438*, 461, 463, 478
  analyses, 163–71
  applications, 431
  clay minerals, 172
  deformation, 476
  deposits, 173–5
  sands, 178
  tidal point-bars, 160
India
  alluvial stratigraphical development, 349–71
  climate change, 350
  Government of, 183
  sedimentology, 181–96
Indian Ocean, 102, 281, 365
Indonesia, 344
infrared cleaning, 363
insect burrows, 506
interglacial transitions, 392, 393
  Holocene, 364–5
intertropical convergence zone (ITCZ), 280–1
invertebrates, marine, 438
Iowa (USA), 455, 456, 478
Ipundu (Australia), 533, *548*
Ipundu North (Australia), 533, *548*
Ipururo Formation (Amazonia), 162, **164**
  clay minerals, 172, *173*, 175
  estuarine sediments, 176
  sands, 173
Iquitos (Peru), 18, 20, 162, 167–71, 176
IRAP-RMS (software), 481
iron oxides, 461, 505
Irvine–Paint Creek Fault (IPCF) System, 427
isobaths, 327
Istituto di Geologia Ambientale e Geoingegneria (Italy), 379
Itacoatiara (Brazil), 25
Italy
  geological mapping, 258
  marine successions, 257–75
  Tiber River, 373–96
ITCZ (intertropical convergence zone), 280

Jackson (USA), 436
Jackson (depositional episode), 410, *415*, 419
Jamaica, fans, 99
James Cook University (Australia), School of Earth Sciences, 322, 324
James Kirby (ship), 324
Jamuna River (Bangladesh), 41, 130, 146
  bars, 153–4, 155
  cross-stratification
    high-angle planar, 150–1, 152
    trough, 150
  discharge rates, 154
  dune-related macroturbulance studies, 45, 50–2, 56
  dunes, 150
  facies, 151, *153*
  low-angle stratification, 151
  morphology, 149–50
  sedimentology, 150–1
  width–depth ratio, 154
  see also Brahmaputra River (Bangladesh)
Janar (India), 185
Japan, floodplains, 557–68
Japanese Archipelago, 566
Japanese Earth Resources Satellite-1 (JERS-1), synthetic aperture radar (SAR) images, *24–6, 28, 30–1, 33, 35*
Japurá, Rio, 20
Jefferson County (NE, USA), *474*
Jurassic
  Late, 19, 495
  tectonics, 102
Juruá, Rio (Amazonia), *21, 23, 24, 25*
Jutaí Arch (Amazonia), *25*

Kanawha Formation (USA), 429, 439
kankar, 188
Kansas (USA), 454, 477, 478
kaolinite, 159, 175, 543
  abundance, 172
Kap Kolthoff Supergroup (Greenland), 404
Karna (India), *350*, 351, *352*, 355, *356*

Kaskapau Formation (Canada), 493, 494, 496, 509, 512
  cross-sections, *503*
  geological setting, 497
Katar, River (Ethiopia), 280
Kedida (Ethiopia), 282–4, 285, 286
Keeper Reef (Australia), 324, 327, 336
Kentucky (USA), 427
  Highway 80, 432, *433*
Kentucky River Fault (KRF) System, 427
Khorixas (Namibia), 103
Khudala (India), 350, 351, 355, 366
  cross-sections, *356*, *362*
  lithofacies, *352*
Kicking Horse River (Canada), 4, 5, 7, 8
Kiel University (Germany), 162
Kile Harsema (Ethiopia), 281, 282, 283–4, 285, 286
Kingdom software, 324
Kiowa–Skull Creek (USA), 456, 478
Kiremt (Ethiopia), *283*
Kiskatinaw River (Canada), 501, 510–11
knickpoints, 342, 392
  formation, 223, 225
  migration, 218–19, 221–2, 228–30
knickzones, formation, 229
Koigab Canyon (Namibia), 103, 105
Koigab Fan (Namibia), 99–120
  active channels, 105–7, 109, 113
  ancient fluvial sequences, 107–9
  channel characteristics, 105
  classification, 115–18
  climate, 116–18
  comparisons, 114–15
  drainage network, *104*
  geodynamic framework, 102
  geology map, *101*
  interactions with marine environment, 113–14
  lagoons, 113–14
  morphology, 103–9
  satellite images, *101*, *106*
  sedimentological processes, 109–14
  setting, 103
  shoreline barriers, 113–14
  shrub coppices, *109*
  significance, 103
  study area, 100
  surface characteristics, 105–6
    factors affecting, 109–13
  vegetation, 116–18
Koigab River (Namibia), 102, *104*, *105*
  abandoned channels, 107
  catchment, 99, 103
  floods, 107
  migration, 113
Koigabmond (Namibia), 103, 111
Kola Bridge (India), 187, 192
Kooroopa 1 (Australia), 534
Kootenay Range (Canada), *5*
Köppen climate classification, 280
Korean Basic Science Institute (KBSI), *364*
Kosi Fan (Himalayas), 116, *117*, 181, 187, 188, 194
Kosi River (Himalayas), 182
KRF (Kentucky River Fault) System, 427
kriging, applications, 67
Kuiseb Canyon (Namibia), 102
Kuiseb schists, 113
Kuki Formation (Japan), 558, *559*
Kunene River, 101, 102

Laboratory of Hydraulics, Hydrology and Glaciology (VAW) (Switzerland), 135
  braided-river studies, 138
  river-widening studies, 136–7
lag deposits, 76–8, 87
  formation, 77
  modelling, 91–4
Lago di Vico lava flow Formation (Italy), 384, 387
Lago Grandé de Curuai, *25*
lagoons, 113–14, 509
Lake County Uplift (USA), 317
Lake Eyre Basin (Australia), 526–7
lakes, 29
  oxbow, 23, *25*
  várzea, 26
  see also floodplain lakes
Lalbakeya River (Himalayas), 183
Lancaster County (NE, USA), 456
LANDSAT images, *106*
Langano, Lake (Ethiopia), 280
Laramide (North America), 409, 411, 414, 421–2
  tectonics, 417
laser diffraction, in grain-size distribution analysis, 159–80
laser Doppler anemometers (LDAs), applications, 47–8
Last Glacial Maximum (LGM), 290, 295, 299, 317, 366–7
  displacement rates, 312, **314**
  terraces, 302, 303, 311, *315*
  vertical displacement rates, 314–15
Late Cenozoic, climate change, 377
Late Cretaceous, 496
  tectonics, 102
Late Holocene, 181, 193, 194
Late Jurassic, 19, 495
Late Miocene, 161, 178, 366
Late Ordovician, 117
Late Pleistocene, 105
  alluvial sediments, 368
  braided rivers, 367
Late Quaternary, 296
  chronology, *285*
  glaciation, 295
  marine successions, 257–75
  valley fills, 290–1
Late Weichselian, 299–302
lateral accretion surfaces (LAS), 461, 463, *466*, *467*, 476
Latium (Italy), 373, 380, 388, *391*, 392
Laurel County (KY, USA), *433*
lava flows, 384, 387
leaves, 505
Leeder–Allen–Bridge (LAB) models
  development, 244
  limitations, 244
Leeder model, 244
Leica 500 (GPS), 8
Lek channel belt (Netherlands), 310
Lenticulina spp. (foraminifers), 410, *416*
levees, 135, 183, 184, 193–4
  Amazonian river system, 17, 23–6
  Columbia River, 5
  Dunvegan Formation, 493, 504–5, 512
  facies, 299

LGM *see* Last Glacial Maximum (LGM)
Liesegang banding, 464
lignite, 161, 562, 563, 565
lignite beds, 558
limestones, 438, 459
　Pennsylvanian, 461
limited-width braided rivers
　approaches, 137–8
　bedload capacity, 137–8
　channel morphology, 137
　design criteria, 136
　future research, 143
　hydraulic modelling, 135–44
Linge channel belt (Netherlands), 310
lithofacies
　coarse-grained nearshore, 382–3
　fine-grained shelf, 381–2
　Lower Pennsylvanian, 429–32
　Luni River Basin, **357**
　Minato Formation, 558–9
lithologs, 187–92
Lithophaga spp., 390
lithostratigraphy, Tiber River, *377*
Livingston Palaeovalley (USA), 429
Lochkovian, 525
Louisiana (USA), 414, 419, 420
Louisville (NE, USA), 461, *462*
Low Gap (USA), 439, 440
low-angle stratification, 151, 152–4, 155
Lower Cretaceous, 558
　mountains, 103
　sandstones, 532–3
Lower Devonian, 517
Lower Miocene, 410
Lower Mississippi Valley (LMV) (USA), 317
Lower Old Red Sandstone (UK)
　facies, 519–21
　　cyclicity, 521–2
　　interpretation, 522–4
　geological setting, 517–19
　sequence stratigraphy, 517–29
　study area, *519*
Lower Pennsylvanian (USA)
　estuaries, 445
　fluvial–estuarine transitions, 425–45
　lithofacies, 429–32
　stratigraphy, 427–9
　transgressive successions, 432–44

interpretation, 433–6, 438–9, 440–1, 443–4
Lower Pleistocene, stratigraphy, 379–84
Lower Rhine Embayment, 296
Lower Triassic, 236
Lower Wilcox (depositional episode), 410, *415*
　sediment supply, 414
　rates, 413
lowstand systems tracts (LSTs), 262, 264, 270
　boundaries, 267
　definition, 257
　reservoirs, 541–2, *543*, 546–9, 550–4
luminescence geochronology, 361–3
Luni Gorge (India), *352*
Luni River Basin (India)
　alluvial sediments, 351–61
　　Type-1, 349, 351–3, *354*, 361, 364–8
　　Type-2, 349, 355, *356*, 361–3, 364–8
　alluvial stratigraphical development, 349–71
　lithofacies, **357**
　study area, 350–1
luvisols, 286
lycophytes, 464

Maassluis Formation (Netherlands), 296
Macapá (Brazil), 31, 32
McMurray Formation (Canada)
　clay minerals, *173*
　cross-sections, *163*, *166*
　early studies, 160
　facies, 163–7
　fluvial deposits, 172–3
　geological setting, 160–1
　grain-size distribution analysis, 159–80
　grain-size parameters, **164, 165, 174**
　inclined heterolithic stratification, 173–5, 177
　study area, *162*
macrofauna, 36
macrofossils, 302, 311, 380
Macrolake phase, 286
macroturbulence
　and dune formation, 42–3
　evolution, 41

　and sand dunes, 41
　use of term, 42
　see also dune-related macroturbulence
Madeira, Rio (Amazonia), 20, 23, *25*, 29
Madre de Dios (Peru), 162, 167–71, 172, 173
　estuarine sediments, 176
　sands, 177
Magdalena River (Andes), 194
magma, 391
magnetic inductive discharge (MID) meters, 138
magnetites, 99, 103, 109
Mahoney State Park (USA), 457–9
Main Ethiopian Rift (MER)
　climate, 281–2
　discontinuous ephemeral streams, depositional processes, 277–94
　geological setting, 278–80
　hydrology, 281–2
　Lake Region, 277, 278, *279*, 280
　modern fluvial systems, 281–3
　study area, *279*
　tectonics, *279*
Manaus (Brazil), 18, 19, 21, 22, 23, 27
　satellite images, *24–5*
Manawara (India), 355, *356*
mangroves, 34
Manning constant, 10, 12
Manorbier Bay (UK), 519
Marajó Basin, 19
Maranhão (Brazil), 34
Marañón, Rio (Amazonia), 20, *24*
Marecchia River (Italy), 260
Margaritifera auricularia (bivalve), 386
marine deposits, 222
　geochemistry, 562
　Minato Formation, 559–62, 565
　Po Basin, 260–5
marine environments, 160–1
　interactions with fans, 113–14
marine successions, alluvial vs. coeval, 257–75
markers, 454
mass-balance effects
　on alluvial architecture, 243–53
　and sedimentation rates, 243–4
　studies, 244–8
mass-balance transformations, 248–54

massflows, 277
  facies, 289–90, 292
Maurik channel belt (Netherlands), 310
maximum flooding surface (MFS), 265–6
  characteristics, 269
maximum particle size (MPS), 558–9, *560*, 564
Maystrick Pit (MPT) (USA), *467, 468, 473*
  fluvial–estuarine architecture, 464–6
mean particle diameter, 9
meandering–braiding transition criteria, 137
meanders, 17
  channels, 22–6, 302
  fractal analysis, 146
  large-scale, 26
Mediterranean, 377, 379, 390
Megalake phase, 284, 286
Meki, River (Ethiopia), 280
Melanopsis affinis (gastropod), 383, 384
Melezza River (Switzerland), 139
Mesozoic, 17, 18
  basins, 19, 410
  erosion, 495
Mesozoic–Cenozoic, successions, 375, 380, 382
Messum Fan (Namibia), 114–15
metamorphic rocks, 103, 109–11
Mexico, 414, 419, 422
Meyer-Peter–Mueller formula, 138
Micrhystridium spp. (palynomorphs), 465
microfossils, subenvironmental associations, 259–60, *261*, 269
micromolluscs, 379
Middle Holocene, aggradation, 311
Middle–Late Eocene, drainage basins, sediment supply, 417–19
Middle Miocene, 410
Middle Pennsylvanian (USA), 431
Middle Pleistocene, 280, 384
Middle Pliocene, stratigraphy, 379–83
Middle Valley of the Tiber River (MVT) (Italy), 373–96
  chronostratigraphy, *377*, 379
  geological setting, 373–5, *376*

lithostratigraphy, *377*
palaeoclimatic setting, 377–9
palaeogeographical evolution, 388–93
sedimentation, 388–93
  fluvial, 390–2
  fluvial–deltaic, 388–90
  stratigraphy, 379–88
  tectonic setting, 375–7
  tectonics, 388–93
  terrace formation, 392–3
  uplift, 390–2
Middle Wilcox (depositional episode), 410, *415*
  sediment supply, 414
Midway (depositional episode), 410
migmatite, 103
Milankovitch scales, 259
Milankovitch-band climatic cycles, 511
Milford Haven Group (UK), 518
Miliolidae spp. (foraminifera), *270*
Minapur (India), 193
Minato Formation (Japan)
  floodplains, 557–68
  environment, 563–4
  geological setting, 558–9
  lithofacies, 558–9
    classification, **558**
  marine deposits, 559–62, 565
  geochemistry, 562
Minneapolis (USA), 85, 87
Miocene, 384
  basins, 19
  channel sediments, 176–7
  drainage basins, sediment supply, 419–20
  Early, 161, 375
  erosion, 409
  estuarine sediments, 169–71, 176
  grain-size analyses, 177–8
  Late, 161, 178, 366
  Lower, 410
  marine incursions, 161, *162*, 168
  Middle, 410
  sands, 175
  sedimentary rocks, 159, 160
  Upper, 410
Miocene–Pliocene boundary, 366
Mississippi River (USA), 409, 412, 419, 420, 421, 422
  avulsions, 192
  bar shape, 147

coastal prisms, 317
deltas, 414
total discharge, 18
Mississippi Valley (USA), 316, 421
Missouri River (USA), bar shape, 147
Mittlere Terrasse (Italy), 387
mixed-load transport, 399, 400–1
mollic horizons, 286
molluscs, 380, 386, 501, 509
  shells, 261
Monacha cantiana (land snail), 386
Mont St. Michel (France), 436, 444
Monte Alegre (Brazil), *25*, 26
Monte Carlo Sequence, 382
Montsant alluvial fan (Spain), *245*
Moor Cliffs Formation (UK), 518
morphodynamic models, 75
morphology
  base-level change effects, 213–41
  surface, 147–8
  see also geomorphology
Mount Ciocci Formation (Italy), 390
Mount Peglia–Amerini–Narni–Sabini–Lucretili Mountains (Italy), 375, 388, 390
Möwe Bay (Namibia), 102
mudcapes, 33–4, 37
mudrocks, 453, 465, 472, 473, 477
  sequence stratigraphy, 517–29
muds, 172, 175, 185, 193
  accumulation, 181
  floodplain, 262
  fluid, 35
  Holocene, 326
  lacustrine, 501–4
mudstones, 181, 505–6, 512, 534, 542, 547
  Albian, 496
  lenticular, 507
  pedified, 521, 523
  platy, 501–4
  sequence stratigraphy, 517–29
multistoreyed gravel sheets (MGS), *352*, 355–8
  interpretations, 355
Murta, Lake, 533
muscovite, 99
Muzaffarpur (India), 184, 185

Namib Desert (Namibia), environment, 100–2

Namib Sand Sea, 107
Namibia
  climate, 101–2
  coastline, 113, 114–15
  drainage evolution, 102
  fans, 99–120
  river catchments, 102
  tectonics, 102
Nanson–Huang hypothesis, 12, 13, 14
NAP (non-arboreal pollen), 264, 265, 268
Napo, Rio (Amazonia), 20
Nawada (India), 186, 192
Nebraska (USA), 148
  fluvial–estuarine transitions, 453–80
  well logs, 470–3
Nebraska Geological Survey (USA), 478
Negro, Rio (Amazonia), 20
  discharge fluctuations, 21–2
  rias, 26, 27, 29
  sediment transport, 22
Neogene, 344, 366, 419
  depositional episodes, 410
  flooding pulses, 410
  Foraminifera, 410
  sediment supply, 420–1
Neogene–Quaternary, 375, 380
Neoproterozoic, 117
neotectonics, 297
Nepal, 182
Netherlands, The
  differential subsidence studies, 295–320
  dune-phase sediment transport studies, 75–97
New Madrid Seismic Zone (NMSZ), 316, 317
New South Wales (Australia), 522–3
New Zealand, 237
Nicholson (Canada), 4, 5, 6, 7, 9
Nikuradse grain roughness length, 89
Noda Group (Japan), 558, *559*
non-arboreal pollen (NAP), 264, 265, 268
Norias Delta (USA), 419
Norma (USA), 412
North America, 160
  drainage basin evolution, 409–23
  see also Canada; United States (USA)

North Fork Toutle River (USA), 128
North Sea Basin, 296
Northumberland (UK), 61–74

O'a caldera (Ethiopia), 280
Obere Terrasse (Italy), 385
Óbidos (Brazil), *21, 22, 25*, 26, 31
obsidian, 289
Ogallala Formation (USA), 420, 421
Ohau River (New Zealand), 137
Ohio (USA), 427
  Highway 35, 436
Oiapoque, Rio (Amazonia), 34
oil fields, sequence stratigraphy, 531–56
OIS (oxygen isotope stage), 299–302, 311
Okavango Fan (Botswana), 117, 317
Okavango River (Botswana), 194
Old Red Sandstone, 399, 404–6, 519–29
Oligocene
  conglomerates, 557
  drainage basins, sediment supply, 419
  floodplains, 557–68
  sandstones, 557
  sediment supply rates, 413
  volcanism, 409
Omaha (USA), 455
Omaruru Canyon (Namibia), 102
Ommeren channel belt (Netherlands), 310
Ophiomorpha isp. (trace fossils), 537, 539, *541*
optically stimulated luminescence (OSL), 350, 351, 364, 365–8
  applications, 349
  geochronology, 361–3
Orange River, 102
Orinoco Delta (Venezuela), 33
Orinoco River (Venezuela), 317
Orte-Amelia area (Italy), 382, 391
orthopyroxenes, 103
Orvieto (Italy), 380, 384, 385, 392
OSL (optically stimulated luminescence), 350, 351, 364, 366–8
ostracods, 259, **260**
Ostrea spp. (molluscs), 509
overbank flooding, 183–4
oxbow lakes, 23, *25*

oxygen isotope stage (OIS), 299–302, 311
oxygen isotopes, 262
Oyapack, Rio, 34

Pacific Ocean, 161
Paglia-Tevere Graben (Italy), 375
Palaeocene
  drainage basins, sediment supply, 414–17
  uplifts, 409
palaeochannels, 192, 321
  Burdekin River, 321–47
  channel fills, 340–1
  entrenched, 326, 344
  in Great Barrier Reef, 322
palaeocurrent diagrams, *548, 549*
palaeocurrents, 355, 542, *560, 561*
  analysis, 558–9
palaeodrainage, changes, 507–9
Palaeogene, 419
palaeogeography
  Central Appalachian Basin, 427–9
  Dakota Formation, 453–80
  Dunvegan Formation, 495–500
  Western Interior Seaway, 495
Palaeoproterozoic, 117
palaeosols, 473, 493, 501, 512, 522
  clay-rich, 186
  ephemeral streams, 285, 286
  floodplain, 507
  formation, 511
  hydromorphic, 465
  Luni River Basin, 364–5
  redox, 505, 507–9
  Tiber River, 384, 386
Palaeotiber gravels, 384
palaeovalleys, 453, 456
  Cretaceous, 457
Palaeozoic, 18, 296
  basins, 19
  cherts, 466
  erosion, 495
  fans, 118
palynomorphs, 464, 476
pampas, 23
Pannerden Channel (Netherlands), 80
Pannerdensche Kop (Netherlands), 80
Panopaea glycimeris (bivalve), 382
Paola–Borgman model, 122
  modified, 122–3

Pará, Rio (Brazil), 29–30, 31, 32, 33, *35*, 36
Paranan Sea, *162*
paranás, 23, *25*
Parson (Canada), 4
particle imaging velocimetry (PIV), 41, 52
  applications, 43
  data collection, 44
    schematic diagram, *43*
  in dune-related macroturbulence studies, 44–5
  future research, 56
  images, *45*
  precautions, 45
Paru, Rio (Amazonia), *26*
Peace River Plains (Canada), 501
peats, 265, 299, 311, 312, 445
  Holocene, 313–14, 317
  radiocarbon dating, 302–5, **306–9**
  swamps, 562
Pebas Formation (Amazonia)
  channel sediments, 176–7, *178*
  estuarine sediments, 176
  facies, 167–71
  geological setting, 161
  grain-size distribution analysis, 159–80
  grain-size parameters, **164**, **174**
  inclined heterolithic stratification, 173–5, 177
  study area, *162*
pebbles, 188
Pecten flabelliformis (mollusc), 380
pedogenesis, 303, 511
pedogenic calcrete, isotopic composition, 364–5
Peel Boundary Fault-zone (PBF), 296, 297, 310, 317
  differential subsidence, 295
Peglia, Mount, 384
pegmatites, 103
Pembroke Peninsula (UK), 518
Pennine Basin (UK), fluvial architecture, 481–91
Pennsylvanian (USA), 453, 455
  limestones, 461
  see also Lower Pennsylvanian (USA)
Permo-Carboniferous, 117

Peru, 18, 162, 167
  pampas, 23
  sedimentary rocks, 160, 161
petroleum, stratigraphy, 531
pH, 27
Phoenix (USA), 525–6
Piacentian–Santernian (Italy), 380
piedmonts, surface change, 216–18
Pine Creek Sandstone (USA), 431, 432, *433*, *434*, *435*, *442*
Pine River (Canada), 493, 496, *500*, 509, 510
  allomembers, 507, 508, 511
  cross-sections, *501*, *502–3*
  geological setting, *497*
Pinus spp. (trees), 264, 265, 268
plagioclase, 103
planforms
  surface, 145, 154–5
  see also channel planforms
Planolites isp. (trace fossils), 443, 536–7, *541*
Planorbis planorbis (ramshorn snail), 386
plants
  fossil, 329
  see also vegetation
Platte County (NE, USA), 456, *473*
Platte River (USA), 152, 455, 473
  see also easternmost Platte River Valley (EPRV)
platy mudstones, 501–4
Pleistocene, 34, 102, 113
  alluvial sediments, 349
  alluvium, 321
  channels, 341
  depositional processes, 277–94
  drainage basins, sediment supply, 420–1
  Early, 373, 380
  erosion, 463
  glaciation, 322, 410
  land surface, 344
  Lower, 379–84
  Middle, 280, 384
  palaeochannels, 192, 344–5
  sediments, 184, 409
  shorelines, 342
  Tiber River, 373–96
  transgressions, 445
  valley fills, 290–1
  volcanism, 280, 375
  see also Late Pleistocene; Plio-Pleistocene

Pleistocene–Holocene
  episode, 410
  sedimentation, 384
  unconformity, 326, *331*
Pliocene
  alluvial sediments, 349, 368
  drainage basins, sediment supply, 420–1
  flood basalts, 280
  glaciation, 421
  Middle, 379–83
  Tiber River, 373–96
  uplifts, 409
  Upper, 280
  vegetation, 379
Pliocene–Quaternary, 375
  basin filling, 373
  climate change, 377–9
Plio-Pleistocene
  erosion, 468
  sequence stratigraphy, 410
  Tiber River, 373–96
Po Basin (Italy)
  alluvial section, 264–5
  coastal section, 260–4
  fluvial deposits, 260–5
  marine deposits, 260–5
Po River (Italy), 317
  delta, 260
Po River Plain (Italy)
  geological setting, 258–9
  marine successions, 257–75
  sequence stratigraphical interpretation, 265–9
  stratigraphical architecture, 265
  study area, 258–9
podzolic soils, 20
Poggio Mirteto Formation (Italy), 380, *382*, 383, 385, 388–90
  cross-sections, *381*
point bars, 183
  inclined heterolithic stratification, 160
pollen analysis, 259, 262–4, 268–9, 270–1
pool–riffle units, 64
pororoca, 32, 33
Potamides tricinctus (mollusc), 383
pottery fragments, 360, **363**
Pottsville Sandstone (USA), 436, *437*
Precambrian, 99, 350
  fans, 103, 116–18
  shields, 19

Pridolian, 517
probability density function
 (PDF), 122, 132
profile budgeting, vs. digital
 elevation models, 70–1
Prososthenia spp. (gastropods),
 383
pterophytes, 464
Pulaski County (KY, USA), 432
Purcell Mountains, 5
Purús Arch (Brazil), 19, *25*
Purús, Rio (Amazonia), *24*
Putumayo, Rio (Amazonia), 20
p-waves, velocity, 324
Pyrenean Foreland Basin, 236
pyrites, 461
pyroxenes, 99, 109

Quakertown Coal, 437, 439
quartz, 35, 103, 363, 542, *546*
 occurrence, 349
quartz grains, 36, 103
quartz latites, 103, 107, 109, 113
quartzarenites, 425, 427–8, 431,
 436, 445
quartzites, 461
quartzose, 20
Quaternary, 102, 149, 238
 alluvial stratigraphical
  development, 349–71
 channels, 322
 climate change, 379, 409
 Upper, 283–90
 see also Holocene; Late
  Quaternary; Pliocene–
  Quaternary
Queen City (depositional
 episode), 410, *415*
Queensland (Australia), sequence
 stratigraphy, 531–56

radiocarbon dating, 10, 32, 193
 peats, 302–5, **306–9**
radiometry, 379
radionuclides, 363
Radium (Canada), 4
rainfall, 183
 and flooding, 21
 seasonal variations, 17, 21,
  280–1
 trends, *283*
rainfall–erosion facility (REF),
 214, *216*, 218
 valley formation studies, 219
rainforests, flooding, 17

ramps, 342
Rat Island Mudstone (UK),
 518–19, *520*, 521–2, 523,
 524
 facies model, 525–7
Raytown Limestone (USA), 461
Razzano, Mount (Italy), 388, 391
reaches
 anabranching, 181
 digital elevation models, 63
Red River (USA), 409, 412,
 419–20, 421, 422
reefs
 carbonate, 322
 palaeochannels, 321–47
Reflector A (event), 326, 329, *331*,
 338, 344
repeated flood deposits, 193
reservoirs
 heterogeneity, 543–9
 modelling programmes, 481
 production and development,
  549–54
 sequence stratigraphy, 531–56
resistivity logs, *471*
Reynolds number, *401*, 401–2
Reynolds stresses, *Plate 1, Plate 2*,
 41, 48, 50, 52
 instantaneous, 56
rheoignimbrites, quartzlatitic, 103
Rhine–Meuse delta (Netherlands)
 cross-sections, *301, 303, 304*,
  310–11
 differential subsidence,
  295–320
 geological setting, 298–302
 study area, 298–302
Rhine–Meuse system, avulsions,
 192
Rhine, River (Netherlands)
 course, 295
 differential subsidence,
  295–320
 dune-phase sediment transport
  studies, 75–97
 data, 80–7
 dunes, 75
 gravel layers, 85–7
 sediment sorting, 88, 124
 sediment sorting model, 91–4,
  95
 sediment transport rate
  hysteresis, 84–5
 study area, 80
 vibracores, 75, 80, 87

Rhine Valley (Netherlands),
 orientation, 298–9
rhizocretions, *352*, 353
rhizoliths, 351, 353
Rhizophora spp. (mangroves), 34
Rhône, River, 139
rhyolites, 280, 282, 286, 289
rhythmites, 431–2, 440–1, 444,
 447–8, 476
rias, 21
 deposition, 26–9
Richfield Arch (USA), 455
Rieti Basin (Italy), 375, 383, 388,
 392
Río de la Plata, 161
Rio Fratta Unit (Italy), *386*, 387,
 392
Rio Grande (North America), 412,
 414, 417, 420, 421
ripple–dune transition, 42, 56
ripples, 399, 402
river channels
 morphology, 61
 modelling, 137
river systems, lowstand, 322
riverbanks, stable, 20
rivers
 anastomosed, 3, 194, 517
 ancestral, 321–47
 management schemes
  ideal, 135–6
  limitations, 135
 rewidening schemes, 135, 136
 types of, 17
 wandering, 64
 see also blackwater rivers;
  braided rivers; clearwater
  rivers; whitewater rivers
Roberts Bank (Canada), 198
Rockcastle River (USA), *433*
Rockcastle Sandstone (USA), 431,
 441, 442, 443
Rocky Mountain Cordillera, 483,
 496, 509
Rocky Mountain Trench, 4
Rocky Mountains, 5, 420, 422,
 454, 512
 drainage, 417, 419
 geological setting, 411–12
 sediment supply, 414
 uplifts, 409
Roer Valley Graben (RVG)
 (Netherlands), 295, 296,
 297
 geological setting, 299

Rome (Italy), 373, 382, 390, 391, 392
roots, 501, 504, 505, *538*
Rosita delta system (USA), 414
Rosselia isp. (trace fossil), *442*, 443, 444
Rouse sediment transport formula, 12–13, 14
Rouse-like parameters, 401

Saalian, glaciation, 299
Sabatini Mountains Volcanic District (Italy), 375, 383, 384, 388, 392
Sabbie a Flabellipecten Formation, 380
Sabina Fault (Italy), 376
sabkhas, 103–5, 113–14
Sacrofano Centre (Italy), 391
St Lawrence River (North America), 427
Saint-Anthony Falls Laboratory (USA), 85, 87, 94, 124
  Experimental Earthscape Facility, 244
salinity, 35
Salinopolis (Brazil), 34–5
Salt Fan (Namibia), 114, *115*
Salt River (USA), 525–6
salts, 103
samples, preparation, 162–3
sand–coal transitions, 246–7, 249, *250*
sand dunes
  and macroturbulence, 41
  separation zones, 41
  shear layers, 41
sand fluxes *see* sand transport
sand–silt alternations, 359–60
sand transport, types of, 197–8
sandbanks, 33
sand-bed channels
  bedload estimation, 197–209
  sediment transport, 197–209
  velocity, 197, 201, 202–3
sands
  arkosic, 19
  coarse, 360
  see also fine sands
sandstones, 453, 457–9, 465
  Albian, 478
  crevasse-splay, 510
  cross-bedded, 506
  Devonian, 399, 404–6
  fluvial cross-bedded, 429
  Lower Cretaceous, 532–3
  Oligocene, 557
  reservoirs, 531
  sedimentology, 534–9
  sequence stratigraphy, 517–29, 541–3
  sharp-based, 504
  silty, 504–5
  tuffaceous, 559–61, *562*, *563*
  wave-rippled, 509
sandwaves, 32
  composition, 33
sandy braided rivers
  facies, 145–58
  morphology, 145–58
Sangamonian, *416*
Santarém (Brazil), *25*, 26, 27, 29
Santernian
  coastline, 390
  markers, 383
Santernian–Emilian boundary, 380, 383, 390, 393
Sarpy County (NE, USA), *454*, 455–6, 473
Saskatchewan River (Canada)
  avulsions, 192
  see also Cumberland Marshes (Canada); South Saskatchewan River (Canada)
satellite imaging
  Baghmati River, 183
  JERS-1, *24*–6, *28*, *30*–1, *33*, *35*
  LANDSAT, *106*
Saunders County (NE, USA), 456
scale invariance
  alluvial architecture, 148–51
  bar shape, 147, 154–5
  facies, 151–5
  in fluvial sedimentology, 145, 146
  sandy braided rivers, 145–58
  scour depths, 148, 155
  sediment deposition, 151–5
schists, 103, 113
Scirpus lacustris (bulrush), 10
Scotland, 67
  rivers, 335, 399, 403–4
scour depths, 137, 145, 146
  scale invariance, 148, 155
scour-and-fill structures, measurement, 62
scours
  braided rivers, 137
  channel-floor, 326
scroll bars, 23
sea-level change, 22–3, 26, 35, 542
  effects, 215, 350
  Holocene, 22–3, 35, 494
  and near-coastal fluvial systems, 213
sediment budgets
  Locking's, 5
  morphological approach, 63
  techniques, 61, 63
sediment continuity models, 123
sediment deposition
  dune-phase, 75–97
  scale invariance, 151–5
sediment fluxes *see* sediment transport
sediment mixture, effects on bed-surface elevation, 126–7
sediment sorting
  and cross-bedded deposits, 94
  and discharge waves, 77–8
  dune lee, 75, 77
  mechanisms, 77
  modelling, 89–91
    boundary conditions, 91
    future development, 94–5
    and Rhine, 91–4
    sensitivity analysis, 94
  processes, 75
  relevance, 75–6
  river bed model, 77–8
  see also vertical sorting
sediment supply
  Cenozoic, 411–13
  coasts, 17
  drainage basins, 413–21
sediment transfer surveys
  cross-profiles, 64–5
  digital elevation models, 65–72
  planforms, 64–5
  volumetric change calculations, 65, *66*, 69–70
sediment transfers, annual reach-scale, derivation, 61–74
sediment transport, 36
  and anastomosis, 13–14
  and base-level change, 214
  and channel morphology, 62–3
  and discharge rates, *93*
  and discharge waves, 77–8
  dunes, 42, 56, 121–34
  formulae, 12–13, 14
  measurement, 80–4
  measurement issues, 62
  mechanisms, 243

sediment transport (*cont'd*)
  modelling, 10–13, 91
  predictive models, 75, 121
  rates, hysteresis, 84–7
  in sand-bed channels, 197–209
  theoretical magnitudes, **13**
  theory, 400
  and valley formation, 222
  Waal, **82**, **83**
  see also bedload transport; dune-phase sediment transport; mass-balance effects; sand transport
sediment yield, and base-level change, 232–3
sedimentary facies *see* facies
sedimentation, and depositional cycles, 17
sedimentation rates, 23
  and mass-balance effects, 243–4
sedimentology
  Baghmati River, 181–96
  Wyandra Sandstone Member, 534–9
sediments
  classification, 163
  near-surface, 184–7
  see also alluvial sediments; channel sediments; cross-bedded deposits; estuarine sediments; floodplain deposits; fluvial deposits; lag deposits; marine deposits
Seismic Micro-Technologies, Kingdom software, 324
seismic reflection surveys, 324–6
  high-resolution, 321
seismographic surveys, 296
Senckenberg Institut (Germany), 163
Separated Lakes Phase, 286
sequence boundary (SB), 265–6, *551*
  characteristics, 266–8
sequence stratigraphy
  Amazon River system, 18–19
  boundary recognition, 267–8
  Cenozoic, 410
  classification, 146
  interpretation, 265–9
  models, assumptions, 267
  non-marine successions, 258
  Tarbat–Ipundu Field, 531–56

Wyandra Sandstone Member, 541–3
Serravallian, 161
settling ratio, 401
Severn, River (UK), estuary, 35
Shaftesbury Formation (Canada), 496
Shala, Lake (Ethiopia), 280, *285*
  fluctuations, 290
shales, 425, 431
sheetflows, 277, 517
  facies, 287–9, 292
shelf bathymetry, 322
shelf environments, 35–6
shelf–slope breaks, 219–22
shelf valleys, 213
shelfal submarine canyons, 19
shelf-margin sediments, 226–7
shells, 186, 379
  fossil, 329
Sheopuri Range (Nepal), 182
Shields criterion, 88, 130, 402, 403, 406
Shields curve, 202, 401
shipping, 29–30
shoreface deposits, 175
shoreline barriers, 113–14
siderites, 444, 461, 463, 501
  nodules, 536
  spherulitic, 505, 506
Sierra Madre Oriental (USA), 409, 420, 422
  drainage, 417
  geological setting, 411–12
  sediment supply, 414
  uplift, 419
siliciclastic deposits, 375
Silkstone Rock (UK), fluvial architecture, 481–91
Silti-Debre Zeit Fault Zone (SDZFZ) (Ethiopia), 280
  volcanism, 280
silts, 35, 161, 285
  fine, 361
  lacustrine, 501–4
siltstones, 405, 504, 506, 518, 522, 562
Silurian, 117, 517
Simpson's rule, 69
Sindhari (India), 350, *352*, *354*
Sinodia brocchii (clam), 382
Sipicciano Unit (Italy), 387, 392
Skeleton Coast Erg (Namibia), 107, 109

Skeleton Coast National Park (Namibia), 99, 101, 102, 103, 114
  fans, 116
Skolithos isp. (trace fossils), 461, *464*, 476, 537, *541*
slack water deposits, 361
slope
  base-level change effects, 226–8
  flood, 399
  and grain size, 403
smectite, 159, 161, 175, 178, 355, 543
  abundance, 172
snails, land, 386
soils
  aggregates, 524
  drained, 506
  organic, 505
  podzolic, 20
  see also palaeosols
Sokkia Set 5F Total Station, 64
Soles Basin (Canada), *5*, 9
Solimões, Rio (Amazonia), 22, *24*, *25*, 27
  sediments, 26
Solimões basin, 19
Solimões Formation (Amazonia), 161
  see also Pebas Formation (Amazonia)
Somali Plateau, 280
Sontek Acoustic Doppler Profiler, 198, 199
Soratte, Mount (Italy), 380, 382, 384, 388, 390, 391
sorting (sediment) *see* sediment sorting
Sout Fan (Namibia), 114–15, *115*
South America, 160, 161
South Atlantic Ocean, 103, 113
  climate, 101
South Esk River (Scotland), 335
South Korea, *364*
South Louisiana Growth Fault (USA), 317
South Saskatchewan River (Canada), 146
  aerial photographs, *149*, *154*
  bars, 154, 155
  cross-stratification
    high-angle planar, 150–1, *152*
    trough, 150
  discharge rates, 154

facies, 151, *152*
low-angle stratification, 151
morphology, 149, 150
sedimentology, 150–1
width–depth ratio, 154
South Wales (UK), 517–29
Sparta (depositional episode), 410, *415*
SPECMAP time-scale, *263*
spectrometry, 379
Spillimacheen (Canada), 3, 4, 8, 9
Spillimacheen River (Canada), 5, 6
discharge data, *6*
grain size, 9
sediments, 8, 13–14
splays, *see also* crevasse splays
Squamish River (Canada), 50
Stanton County (NE, USA), 470–2, *474*
staurolite, 99
Stigmaria spp. (root casts), 443
stratigraphy
allogenic drivers, 243
base-level change effects, 213–41
Central Appalachian Basin, 427–9
Dunvegan Formation, 496–500
Tiber River, 379–88
see also sequence stratigraphy
streamflows, 277
facies, 286–7, 292
strike sections, modelling, 246–7
strontium isotopes, 373, 379
Student's *t*, 205
subsidence, 388, 542
absolute rates, 315–16
Cretaceous, 495
see also differential subsidence
sulphur, 562, **563**
surface morphology, braided rivers, 147–8
SURFER GIS, 66, 69–70
suspended load
definition, 197–8
transport, 399, 400–1
suspended sediment formula, Rouse's, 12–13
Swiss Federal Institute of Technology (ETH), 135
Switzerland, 135
rewidening schemes, 137
synthems, 284–6

tals *see* floodplain lakes
Tapajós, Rio (Amazonia), 21, *25*
rias, 26, 27–9
Tarbat–Ipundu Field (Australia)
geological setting, 531–3
sequence stratigraphy, 531–56
study area, *532*
tectonic deformation, 316
tectonics, 297
control, 316
deglaciation-related, 316–17
effects, 350
Gulf of Mexico Basin, 417, 420
Main Ethiopian Rift, *279*
Namibia, 102
Tiber River, 375–7, 388–93
tectonism, and avulsions, 192
Teichichnus isp. (trace fossils), 461
Tenaglie-Fosso San Martino Formation (Italy), 380, 388
Tennessee (USA), 427
Tennessee River (USA), 409, 412, 420, 421, 422
Terebratula ampulla (brachiopod), 380
terraces, 302, 303, 311, *315*
formation, 230, 392–3
terras caidas, 20
Tertiary, 18, 280
Texas (USA), 412, 414, 419, 420
Texas Gulf Coastal Plain (USA), avulsions, 192
Textularia spp. (foraminifera), *270*
Thalassinoides isp. (trace fossils), 537, 539, *541*
thalwegs, 3, 152
Thar Desert (India), 349–71
Thargomindah 1A (Australia), 534
theodolite–EDM surveys, 61
Theodoxus groyanus (gastropod), 383, 384
Theodoxus isselii (gastropod), 386
thermoluminescence (TL), 363
Tiber River (Italy), 391
geological setting, 373–5
see also Middle Valley of the Tiber River (MVT) (Italy)
Tiber Valley (Italy), 388
Tiberino Basin (Italy), 383, 388
tidal bores, 22, *23*, 29, 32, 33
tidal erosion, 32
tidal laminations, 447–8
tidal ranges, 17, 34

tidally influenced systems, grain-size distribution analysis, 159–80
tide-dominated estuaries
models, 426–7
recognition, 445–8
structure, 426
Tocantins, Rio (Amazonia), 21, 36
rias, 26, 29
toluene, in sample preparation, 162–3
Top Thorncliffe Coalseam (UK), 489
T'ora Geosol (Ethiopia), 285–6, 289–90
Torra Bay Granite (Namibia), 103
Tortonian–Messinian (Italy), 375
Total Station, 63, 64, 152
trace fossils, 427, 453, 536–7, 538–9, *541*
Devonian, 521
tracer data, 62, 123
transgressions
Holocene, 326, 330, 336, 341
Lower Pennsylvanian, 432–44
Pleistocene, 445
post-glacial, 342
transgressive–regressive (T–R) sequences, 257, 260, 261–2, 265, 269–71
pollen analysis, 262
transgressive surfaces (TSs), 265–6, 269, 493, *551*
characteristics, 268–9
identification, 261
transgressive systems tracts (TSTs), *261*, 262, 269, 509, 565
formation, 257
pollen analysis, 262–4
reservoirs, 542–3, 550–4
trapezoidal rule, 69
tree stumps, 504
Treeft dune, 302, *303*, 310, 311, 312, 315
triangulation with linear interpolation (TIN), 67
Triassic–Middle Miocene, successions, 375
Trimble AgGPS Model 122, 199, 201
Trimosina spp. (foraminifers), 410, *416*
Trollheim Fan (USA), *117*

tuffs, 280, 285, 559–60
Tufo giallo della Via Tiberina Formation, 383–4, 385–6
Tufo rosso a scorie nere vicano Formation (Italy), 384, *386*, 387
turbidity currents, 36
turbidity maximum zones, 35
Turonian, 496
Tyrrhenian Sea, 375, 388

Ucayali, Rio (Amazonia), 20, *24*
Umbria (Italy), 373, 392
unconformities
    Pleistocene–Holocene, 326, *331*
    sequence-bounding, 510–11
unionids, 501
United Kingdom (UK)
    sediment transfer studies, 61–74
    see also Scotland
United States (USA)
    alluviation, 292
    discontinuous ephemeral streams, 277–8
    drylands rivers, 291
    settlers, 278
University of Leeds (UK), 233
University of Minnesota (USA), 124, 214, 233, 235, 244
Untere Terrasse (Italy), 387
uplifts, 409
Upper Columbia River (Canada)
    anastomosed reaches
        geomorphology, 4, 5
        hydraulic properties, 4
        hydrology, 6
        longitudinal profile, 8
        sedimentology, 5
        steady-state flow field, 13
        upper vs. lower, **10**
    anastomosis origins, 3–15
    geomorphology, 4–6
    sediment transport modelling, 10–13
    study area, 4–6
Upper Cretaceous, 565
    alluvial architecture, 493–515
Upper Eocene, sediment supply rates, 413
Upper Miocene, 410
Upper Palaeocene, sediment supply rates, 413
Upper Pliocene, 280
Upper Proterozoic, 103

Upper Quaternary, deposits, 283–90
Upper Silurian, 517
Upper Wilcox (depositional episode), 410, *415*
    sediment supply, 414
Utah (USA), 524
Utrecht University (Netherlands), 214, 233
    R.J. van der Graaff Laboratory, 302
Uttar Pradesh (India), 193
Utzensdorf (Switzerland), *136*

Vallericca (Italy), 380, 388
valley fills
    depositional models, 290–1
    facies analysis, 286–90
valley networks, 219
valley width, and anastomosis, 6–7, 9
valleys, 511
    active, *227*
    and base-level change, 219–28
        falling, 219–22
        rates, 222–6
        rising, 222
        slope effects, 226–8
    and channels compared, 215
    flooded, 17, 26
    flooding, 225–6
    formation, 219
    inactive, 224–5
    shelf, 213
    shelf-break, 222, 234–5, 236
    see also incised valleys; palaeovalleys; rias
Vallonia spp. (land snails), 384
Valvata piscinalis (gastropod), 386
Van Rijn bedload transport formula, 197, 198, 201–2, 205, 207
Van Rijn functions, 13, 14
várzea, 20, 22–3, *24*
    forests, 23
    lakes, 26
Vedder River (Canada), 61–3
vegetation
    on bars, 154, 155
    cold-climate, 265
    and fans, 99, 116–18
    Holocene, 290
    Pliocene, 379

vertebrates, 383, 384
vertical sorting, 75, 79, *80*, 84, 121
    and cross-sets, 128–9
    effects on bed-surface elevation, 126
    process model, 76, 88–95
vertisols, 355, *364*, 517, 523
vibracores, 75, 80, 85, 87, 92–3, 326, 329
Vico, Monte di (Italy), 375, 383, 384, 391
Villanova de Prades (Spain), *245*
Virginia (USA), 427
Viviparus spp. (gastropod), 501
volcanic hills, 375
volcanic rocks, 99, 102, 103, 109, 113
volcaniclastics, 284–5
volcanism, 419
    Ethiopia, 280, 285–6
    Oligocene, 409
    Pleistocene, 280, 375
    Tiber River, 373, 393
vorticity, dunes, 53–6
vulcanism *see* volcanism
Vulsini Mountains (Italy), 375, 383, 391

Waal, River (Netherlands), 80, 84
    deposits, *86*
    discharge waves, 91
    dunes, 81, 87
        height prediction, 88–9
    sediment transport, **82**, **83**
    vibracores, 85, 92–3
Wabash River (USA), 50
Wabiskaw Formation (Canada), 161
Wallumbilla Formation (Australia), 539
Washington County (NE, USA), 456
water surface structure
    and dune-related macroturbulence, 50–2
    see also boils (water surface)
Water Survey of Canada, 7
waves
    velocity, 34
    wind-generated, 34
    see also discharge waves; sandwaves

well logs, 453, 454, 469–73, *497*
    interpretation, 477
West Netherlands Basin, 296
West Virginia (USA), 427, 429, 439
Western Canada Foreland Basin, 494, 512
Western Canadian Sedimentary Basin, 159, 178
Western Interior (USA), 235, 236
Western Interior Seaway (WIS), 454, 476
    palaeogeography, 495
Westphalian A (UK)
    facies, **485–7**
    fluvial architecture, 481–91
Westphalian B (UK), 481, *484*
White–Colebrook equation, 89
whitewater rivers, 27, 37
    characterization, 20
    floodplains, 20
    meandering, 23, *24*
    mixing zones, 22
    sediments, 20, 26
width/depth (W/D) ratios, 10–12, 13
    reduction, 14

Wijchen Member (Netherlands), 299–302, 303, 305
Wilhelmshaven (Germany), 163
Willapa River (USA), 173
wind, as geological factor, 102
Wisconsin (depositional episode) (USA), 410
WL/Delft Hydraulics (Netherlands), 124
Wonji Fault Belt (WFB) (Ethiopia), 280
    volcanism, 280
worms, 509
Wyandra Sandstone Member (Australia), 531
    geological setting, 532–3
    sedimentology, 534–9
    sequence stratigraphy, 541–3
Wyoming (USA), 412, 419, 420

X-bentonite, 454
Xingu, Rio (Amazonia), 19, 21, *26*, 31
    mouth, 32
    rias, 26, 29

X-ray diffraction (XRD)
    applications, 355
    clay mineral analyses, 163, 172, *173*, 178
    in grain-size distribution analysis, 159–80

Yalin's theory, 130
Yallahs Fan-delta (Jamaica), 99, 117, 118
Yana Fan (Alaska), 116, *117*
Yankee Reef (Australia), 324
Yard Coalseam (UK), 489
Yegua/Cockfield (depositional episode), 410, *415*, 419
Yellow River (USA), avulsions, 192
yttrium aluminum garnet (YAG) lasers, applications, 44

Zandberg dune, 310, 312, 315
Ziway, Lake (Ethiopia), 280, 282, *283*, *285*, 286
    fluctuations, 290
Zoelmond channel belt (Netherlands), 310
Zurich (Switzerland), 135